For reference

Not To Be Taken
From the Room

Telecommunications
Transmission
Handbook

WILEY SERIES IN TELECOMMUNICATIONS AND SIGNAL PROCESSING

John G. Proakis, Editor
Northeastern University

Telecommunications Transmission Handbook

Fourth Edition

Roger L. Freeman

A Wiley-Interscience Publication
JOHN WILEY & SONS, INC.
New York • Chichester • Weinheim • Brisbane • Toronto • Singapore

This book is printed on acid-free paper. ♾

Copyright ©1998 by Roger L. Freeman. All rights reserved.

Published by John Wiley & Sons, Inc.

Published simultaneously in Canada.

Library of Congress Cataloging in Publication Data:

Freeman, Roger L.
    Telecommunications transmission handbook / Roger L. Freeman. —4th
ed.
      p. cm.
    "A Wiley-Interscience publication."
    Includes index.
    ISBN 0-471-24018-4 (cloth : alk. paper)
    1. Telecommunication.   I. Title.
TK5101.F66 1998
621.382--dc21   97-20305
                                                            CIP

Printed in the United States of America

10 9 8 7 6 5 4 3 2 1

For my daughter,
Cristina

CONTENTS

PREFACE

Traditionally, *telecommunications* has been broken down into two basic disciplines: *transmission* and *switching*. A hard and fast dividing line between the two is rapidly disappearing. This book deals with transmission: however, certain elements of switching must be included such as network synchronization and timing. These certainly fall under transmission, but derive from network switches. Thus a system approach is emphasized throughout the text.

This book, over its four editions, has been prepared primarily as a reference text and tutorial. It assumes the reader has an elementary background of the physics of electricity, modulation, and mathematics through first-year college algebra including logarithms. Its primary thrust is toward the practitioner. Somebody has to design and build the network; somebody has to make it operate up to expectations and set reasonable performance objectives. Interfaces based on recognized world standards are stressed.

A major standard-setting body is the International Telecommunication Union (ITU) based in Geneva, Switzerland. It is an international treaty organization. As such, it prepares and publishes the Radio Regulations. The ITU underwent a major reorganization on January 1, 1993. Prior to that date, its two standard-setting organizations were called the CCITT (International Consultive Committee on Telephone and Telegraph) and the CCIR (International Consultive Committee on Radio). After that date, the CCITT became the ITU Telecommunication Standardization Sector and the CCIR became the ITU Radiocommunication Sector. In this text we have denoted them the ITU-T Organization and the ITU-R Organization. ITU references dating from before January 1, 1993 are designated CCITT and CCIR; for those issued after that date, we use ITU-T and ITU-R, reflecting the organizational changes.

Since previous editions, the discussion has kept clear of what one might call the entertainment industry such as radio, television broadcast and community antenna television (CATV). I have broken with that regime and have added a chapter on CATV. For several years, CATV has been poised to offer, and in some cases, is offering, supplemental services. These include conventional telephone service and data connectivity including downstream service (i.e., toward the subscriber) in the kilobit to megabit range for the internet.

Each chapter, from 2 through 16, covers a separate transmission discipline. The chapter order is purposeful. The intent of Chapter 1 is leveling. Here the reader is introduced to transmission measurement units such as the dB and related units, and basic impairments, such as noise and distortion.

Chapter 2 deals with basic telephone transmission. Multiplexing is covered in Chapter 3. Here the primary thrust is time division multiplexing and the digital network. Chapters 4 through 10 cover radio system design, from wideband systems such as line-of-sight microwave through narrowband systems including high-frequency radio and meteor burst. Fiber optics is discussed in Chapter 11. This chapter has been completely revised, Chapter 12 is new and provides a general overview of transmission factors in cellular radio and personal communication services (PCS). Chapter 13 addresses data transmission; ISDN and the asynchronous transfer mode (ATM) are included in this chapter. Chapters 14 through 16 deal essentially with image transmission and include CATV, video/TV including conference television, and still image such as facsimile. Appendix A provides a glossary of acronyms and abbreviations.

A book of this size can only treat a discipline to a limited depth. Therefore, references and bibliographies are provided at the end of each chapter to assist readers who want more information as well as to refer them to sources used in the preparation of the text.

ACKNOWLEDGMENTS

I am deeply indebted to a large group of friends and colleagues who have graciously provided input, reviewed, and suggested changes to the fourth edition of this book. Among this group are Marshall Cross, president of Megawave Corporation, Boylston, MA, John Lawlor, Principal, John Lawlor Associates, Sharon, MA and Dr. Ronald Brown, independent consultant, Melrose, MA. In the preparation of the new chapter on cable television, I am grateful for the help provided by the Society of Cable Telecommunications Engineers (SCTE), in particular Ted Woo, director of Standards and Marv Nelson, vice president. Robert Russell, chairman of the IEEE 802.14 committee, supplied the basic input for the text regarding that key standard. He also suggested a work-around for an understanding of the MPEG-2 video compression standard. Joe Golden who is product manager of Telco Systems, Inc., Norwood, MA. and Dan Danbeck of the University of Wisconsin both graciously provided input for the section on digital subscriber line.

I am obliged to Dr. Len Wagner of Naval Research Laboratories, to my dear friend John Ballard, president of TCI International, Sunnyvale, CA, for updated ionograms and Tom Adcock of Lukas, Nace . . . of Washington, DC, for providing information on microwave databases. Dr. Enric Vilar of Portsmouth University (UK) was always there when I needed help. Enric provided data on the characteristics of rainfall and scintillation fading. I am appreciative of the encouragement of Susan Hoyler of EIA/TIA to use current standards; and to my wife, Paquita, for being patient with me during the book production process, and to our son, Bob, of Axis Communications, for his help on the data transmission chapter.

ROGER L. FREEMAN

Scottsdale, Arizona
March 1998

1

INTRODUCTORY CONCEPTS

1.1 TRANSMISSION—DEFINITIONS

Telecommunication simply means "communications at a distance," from the Greek *tele*, at a distance. With reference to this text, by communication we mean electrical communication by radio, copper conductor, or light.

Telecommunications traditionally has been broken down into two distinct* disciplines: switching and transmission. Switching allows us to select and direct our communication to a specific user or family of users. Transmission deals with the delivery of that signal in some fashion where the signal quality is acceptable to the far-end recipient. The two key words here are *delivery* and *quality*. This is what this entire text is about.

As transmission engineers, we are responsible for delivery from source to destination of three media—voice, data, and image. If you wish, call the source a transmitter and the destination, the receiver.

To be effective, transmission engineers should have some knowledge of signaling and data communications. They should also have a familiarity with traffic engineering, electrical power, and civil engineering such as topographic maps and surveying.

Thus transmission engineering deals with the production, transport, and delivery of quality electrical signals from source to destination. The following chapters describe methods of carrying out this objective.

1.2 A SIMPLIFIED TRANSMISSION SYSTEM

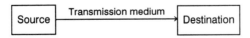

The simple drawing illustrates a transmission system. The source may be a telephone mouthpiece (transmitter) and the destination may be the telephone

---

*That distinction seems to be disappearing rapidly. The two disciplines are becoming more and more intertwined and inseparable.

earpiece (receiver). The source converts the human intelligence, such as voice, data information, or video, into an electrical equivalent or electrical signal. The destination accepts the electrical signal and reconverts it to an approximation of the original human intelligence. The source and destination are electrical transducers. In the case of printer telegraphy, the source may be a keyboard, where each key, when depressed, transmits to the destination distinct electrical impulses. The destination in this case may be a teleprinter, which converts each impulse grouping back to the intended character keyed or depressed at the source.

The transmission media can be represented as a network,

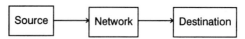

or as a series of electrical networks,

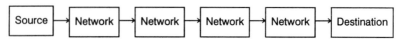

Some networks show a gain in level, others a loss. We must be prepared to discuss these gains and losses as well as electrical signal levels and the level of disturbing effects such as noise and/or distortion. To do this we must have a firm and solid knowledge of the decibel and related measurement units.

1.3 THE DECIBEL

The decibel (dB) is a measurement unit widely used in transmission systems. It is logarithmic with the number base 10. As we proceed through this book, we will see that transmission, whether copper conductor, radio, or light, deals with very large and very small quantities. Using decibels keeps these unit sizes to more manageable proportions. There are things we can do with decibels that we cannot do with conventional linear units. For example, we can add gains and losses. In our context, a plus sign in front of a unit in dB means a gain, while a minus sign indicates a loss.

If decibels confuse you, try to use powers of 10. For example, in the power domain, 10 dB is 10^1, 20 dB is 10^2, 30 dB is 10^3, 40 dB is 10^4, and so on; -10 dB is 10^{-1}, -20 dB is 10^{-2}, -30 dB is 10^{-3}, and so on. This is shown in tabular form in Table 1.1.

TABLE 1.1

Power Ratio		dB	Power Ratio	dB
10^1	(10)	+10	10^{-1}(1/10)	−10
10^2	(100)	+20	10^{-2}(1/100)	−20
10^3	(1,000)	+30	10^{-3}(1/1,000)	−30
10^4	(10,000)	+40	10^{-4}(1/10,000)	−40
10^5	(100,000)	+50	10^{-5}(1/100,000)	−50
10^6	(1,000,000)	+60	10^{-6}(1/1,000,000)	−60

With decibels, the reader must use care. Much of the time in this text we will be in the power domain and talk about power gains and losses, or we will be using derived decibel units, which express power levels in watts (W), milliwatts (mW), and microwatts (μW). We can also be in the voltage or ampere domain when we compare values in volts and amperes. Recall the power formulas: $P = EI = E^2/R = I^2R$. For the first expression we will use $10 \log_{10} X$ and for the next two expressions, $20 \log_{10} X$, and X is some value we wish to express in dB (note that we use $20 \log X$ because of the squared value of E or I).*

The decibel is a unit that describes a ratio. It is a logarithm with a base of 10. Consider first a power ratio. The number of decibels (dB) = $10 \log_{10}$ (the power ratio).

Let us look at the following network:

The input is 1 W and its output 2 W, in the power domain. Therefore we can say the network has approximately a 3-dB gain. In this case

$$\text{Gain (dB)} = 10 \log \frac{\text{output}}{\text{input}} = 10 \log \frac{2}{1} = 10(0.30103) = 3.0103 \text{ dB}$$

or, approximately, a 3-dB gain.

Now let us look at another network:

In this case there is a loss of 30 dB:

$$\text{Loss (dB)} = 10 \log \frac{\text{input}}{\text{output}} = 10 \log \frac{1}{1000} = -30 \text{ dB}$$

(Note that the negative sign shows that it is a loss.) Or in general we can state

$$\text{Power (dB)} = 10 \log \frac{P_2}{P_1} \tag{1.1}$$

where P_1 = input level and P_2 = output level.

A network with an input of 5 W and an output of 10 W is said to have a 3-dB gain:

---

*E is the traditional notation for voltage; I is the traditional notation for current (amperes, etc.).

$$\text{Gain (dB)} = 10\log\frac{10}{5} = 10\log\frac{2}{1} = 10(0.30103)$$

$$= 3.0103 \text{ dB} \simeq 3 \text{ dB}$$

This is a good figure to remember. Doubling the power means a 3-dB gain; likewise, halving the power means a 3-dB loss.

Consider another example, a network with a 13-dB gain:

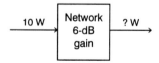

$$\text{Gain (dB)} = 10\log\frac{P_2}{P_1} = 10\log\frac{P_2}{0.1} = 13 \text{ dB}$$

Then

$$P_2 \simeq 2 \text{ W}$$

Table 1.1 may be helpful. All values in the power ratio column are $X/1$, or compared to 1.

It is useful to be able to work with decibels without pencil and paper or a calculator. Relationships of 10 and 3 have been reviewed. Now consider the following:

What is the power output of this network? To do this without a calculator, we would proceed as follows. Suppose that the network attenuated the signal 30 dB. Then the output would be 1/1000 of the input, or 1 mW. Also, 27 dB is 3 dB less than 30 dB. Thus the output would be twice 1 mW, or 2 mW. It really is quite simple. If we have multiples of 10, as in Table 1.1, or 3 up or 3 down from these multiples, we can work it out in our heads, without a calculator.

Look at this next example:

Working it out with a calculator, we see that the output is approximately 40 W. Here we have a multiple of 4. A 6-dB gain represents approximately a fourfold power gain. Likewise, a 6-dB loss would represent approximately one-fourth the power output. Now we should be able to work out many combinations without resorting to a calculator.

Consider a network with a 33-dB gain having an input level of 0.15 W. What would the output be? We see that 30 dB represents multiplying the input power by 1000, and 3 additional decibels doubles it. In this case the input power is multiplied by 2000. Thus the answer is $0.15 \times 2000 = 300$ W.

The following table may further assist the reader regarding the use of decibels as power ratios.

	Approximate Power Ratio	
Decibels	**Losses**	**Gains**
1	0.8	1.25
2	0.63	1.6
3	0.5	2.0
4	0.4	2.5
5	0.32	3.2
6	0.25	4.0
7	0.2	5.0
8	0.16	6.3
9	0.125	8.0
10	0.1	10.0

The decibel is a useful tool for calculating gains, losses, and power levels of networks in series. Consider this example. There are three networks in series. The first is an attenuator with a 12-dB loss (−12 dB), the second network is an amplifier with a 35-dB gain, and the third has an insertion loss of 10 dB. The input to the first network is 4 mW. What is the output of the third network in watts?

There are several ways to calculate the output. One of the simplest is to algebraically add the decibel values (here, −12, +35, and −10), which in this case equals +13 dB. Note that the plus sign indicates gain and the minus sign, loss. Thus we have

The +13 dB can be represented by a multiplier of 10 and a multiplier of 2 if we learn the 10- and 3-dB rules. We now have 10×4 mW $= 40$ mW and $40 \times 2 = 80$ mW or 0.08 W.

When applying decibels in the current and voltage regimes, we handle the problem somewhat differently. Turn to the power law formula in electricity where

$$\text{Power (watts)} = I^2 R = \frac{E^2}{R} \tag{1.2}$$

where I is the current in amperes, R is the resistance in ohms, and E is the voltage in volts.

Thus

$$\text{dB (voltage)} = 20 \log \frac{E_2}{E_1} \tag{1.3}$$

$$\text{dB (current)} = 20 \log \frac{I_2}{I_1} \tag{1.4}$$

The relationships between E_2 and E_1 and between I_2 and I_1 are as follows:

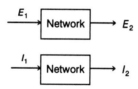

The network is any device that has a gain or a loss such as an amplifier or an attenuator.

When using the current and voltage relationships shown above, keep in mind that they must be compared against like impedances. For instance, E_2 may not be taken at a point of 600-ohm (Ω) impedance and E_1 at a point of 900 Ω.

Example 1. How many decibels correspond to a voltage ratio of 100?

$$\text{dB} = 20 \log \frac{E_2}{E_1}$$

When $E_2/E_1 = 100$,

$$\text{dB} = 20 \log 100 = 40 \, \text{dB}$$

(Same impedances are assumed.)

Example 2. If an amplifier has a 30-dB gain, what voltage ratio does the gain represent? Assume equal impedances at input and output of the amplifier.

$$30 = 20 \log \frac{E_2}{E_1}$$

$$\frac{E_2}{E_1} = 31.6$$

Thus the ratio is $31.6:1$. (Again the same impedances are assumed.)

1.4 BASIC DERIVED DECIBEL UNITS

1.4.1 The dBm

Up to now all reference to decibels has been made in terms of ratios or relative units. We *cannot* say the output of an amplifier is 33 dB. We *can* say that an amplifier has a gain of 33 dB or that a certain attenuator has a 6-dB loss. These figures or units give no idea whatsoever of the absolute level. Several derived decibel units do.

Perhaps dBm is the most common of these. By definition dBm is a power level related to 1 mW. A most important relationship to remember is 0 dBm = 1 mW. The formula may then be written:

$$\text{Power (dBm)} = 10 \log \frac{\text{power (mW)}}{1 \text{ mW}} \qquad (1.5)$$

Example 1. An amplifier has an output of 20 W. What is its output in dBm?

$$\text{Power (dBm)} = 10 \log \frac{20 \text{ W}}{1 \text{ mW}}$$

$$= 10 \log \frac{20 \times 10^3 \text{ mW}}{1 \text{ mW}} \approx +43 \text{ dBm}$$

(The plus sign indicates that the quantity is above the level of reference, 0 dBm.)

Example 2. The input to a network is 0.004 W. What is the input in dBm?

$$\text{Power (dBm)} = 10 \log \frac{0.0004 \text{ W}}{1 \text{ mW}}$$

$$= 10 \log 4 \times 10^{-1} \text{ mW} \approx -4 \text{ dBm}$$

(The minus sign in this case tells us that the level is below reference, 0 dBm or 1 mW.)

1.4.2 The dBW

The dBW is used extensively in microwave applications. It is an absolute decibel unit and may be defined as decibels referred to 1 W:

$$\text{Power level (dBW)} = 10\log\frac{\text{power (W)}}{1\text{ W}} \tag{1.6}$$

Remember the following relationships:

$$+30\text{ dBm} = 0\text{ dBW} \tag{1.7}$$

$$-30\text{ dBW} = 0\text{ dBm} \tag{1.8}$$

Consider this network:

Its output level in dBW is +20 dBW. Remember that the gain of the network is 20 dB or 100. This output is 100 W or +20 dBW.

Another way to look at the network is to convert the input to dBW and add dB values to get the output. One watt equals 0 dBW. Then 0 dBW + 20 dB = +20 dBW. Table 1.2, a table of equivalents, may be helpful.

TABLE 1.2

dBm	dBW	Watts	dBm	dBW	Milliwatts
+66	+36	4000	+30	0	1000
+63	+33	2000	+27	-3	500
+60	+30	1000	+23	-7	200
+57	+27	500	+20	-10	100
+50	+20	100	+17	-13	50
+47	+17	50	+13	-17	20
+43	+13	20	+10	-20	10
+40	+10	10	+7	-23	5
+37	+7	5	+6	-24	4
+33	+3	2	+3	-27	2
+30	0	1	0	-30	1
			-3	-33	0.5
			-6	-36	0.25
			-7	-37	0.20
			-10	-40	0.1

1.4.3 The dBmV

The absolute decibel unit dBmV is used widely in video transmission. A voltage level may be expressed in decibels above or below 1 mV across 75 Ω, which is said to be the level in decibel-millivolts or dBmV. In other words,

$$\text{Voltage level (dBmV)} = 20 \log_{10} \frac{\text{voltage (mV)}}{1 \text{ mV}} \qquad (1.9)$$

when the voltage is measured at the 75-Ω *impedance level.* Simplified,

$$\text{dBmV} = 20 \log_{10} \text{ (voltage in mV at 75-}\Omega \text{ impedance)} \qquad (1.10)$$

Table 1.3 may prove helpful.

1.4.4 The Use of Decibels to Express Antenna Gain or Loss

The dBi, "decibels with reference to an isotropic antenna," is the decibel unit used to express antenna gain (or loss). An isotropic antenna is a reference antenna with zero gain/zero loss. It is completely and uniformly omnidirectional.

The dBd, "decibels with reference to a dipole antenna," also expresses antenna gain or loss. A dipole antenna has a 2.15-dB gain over an isotropic antenna, or 0 dBd = +2.15 dBi.

1.4.5 Absolute Level of an Electromagnetic Field

An electromagnetic field set up by a transmitter is a concern under certain circumstances. At considerable distances from the transmitter's antenna, this

TABLE 1.3

Root-Mean-Square Voltage Across 75 Ω	dBmV
10 V	+80
2 V	+66
1 V	+60
10 mV	+20
2 mV	+6
1 mV	0
500 μV	−6
316 μV	−10
200 μV	−14
100 μV	−20
10 μV	−40
1 μV	−60

field is generally defined by its electric component, E, for which we often use the logarithmic scale.

For a nonguided wave propagated in a vacuum, or in practice in the atmosphere, there is a defined relationship between the electric field, E, and the power flux density, p:

$$E^2 = Z_0 P \qquad (1.11)$$

where Z_0 is the intrinsic impedance of a vacuum, having a fixed numerical value of 120π ohms. For reference, keep in mind that 1 microvolt per meter equals a power flux density of -145.8 dB (W/m^2).

The absolute value of an electric field can then be defined as

$$N = 20 \log (E/E_0) \qquad (1.12)$$

where E_0 is the reference field, usually taken as 1 μV per meter. In this case, N represents "decibels with respect to 1 μV per meter," the symbol for which is dB (μV/m).

1.5 THE NEPER

A transmission unit used in a number of northern European countries as an alternative to the decibel is the neper (Np). To convert decibels to nepers, multiply the number of decibels by 0.1151. To convert nepers to decibels, multiply the number of nepers by 8.686. Mathematically,

$$\text{Np} = \frac{1}{2} \log_e \frac{P_2}{P_1} \qquad (1.13)$$

where P_2, P_1 = higher and lower powers, respectively, and e = 2.718, the base of the natural or Naperian logarithm. A common derived unit is the decineper (dNp). A decineper is one-tenth of a neper.

1.6 THE ADDITION OF POWER LEVELS IN dB (dBm/dBW) OR SIMILAR ABSOLUTE LOGARITHMIC UNITS

Adding decibels corresponds to the multiplying of power ratios. Care must be taken when adding or subtracting absolute decibel units such as dBm, dBW, and some noise units. Consider the combining network below, which is theoretically lossless:

What is the resultant output? Answer: $+33$ dBm.

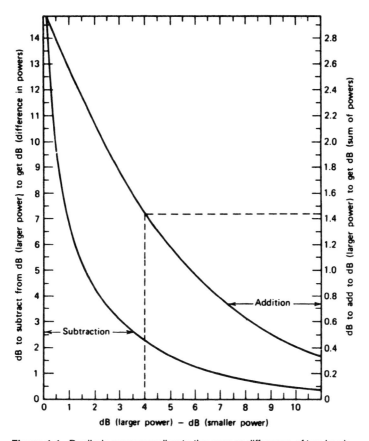

Figure 1.1. Decibels corresponding to the sum or difference of two levels.

Figure 1.1 is a curve for directly determining the level in absolute decibel units corresponding to the sum or difference of two levels, the values of which are known in terms of decibels with respect to some reference.

As an example, let us add two power levels, 10 dBm and 6.0 dBm. Take the difference between them, 4 dB. Spot this value on the horizontal scale (the abscissa) on the curve. Project the point upward to where it intersects the "addition" curve (the upper curve). Take the corresponding number to the right and add it to the larger level. Thus

$$10 \text{ dBm} + 1.45 \text{ dB(m)} = 11.45 \text{ dBm}$$

Suppose we subtract the 6.0-dBm signal from the 10-dBm signal. Again the difference is 4 dB. Spot this value on the horizontal scale as before. Project the point upward to where it meets the "subtraction" curve (the lower curve).

Take the corresponding number and subtract it from the larger level. Thus

$$10 \text{ dBm} - 2.3 \text{ dB(m)} = 7.7 \text{ dBm}$$

When it is necessary to add equal absolute levels expressed in decibels, add 10 log (the number of equal powers) to the level value. For example, add four signals of +10 dBm each. Thus

$$10 \text{ dBm} + 10 \log 4 = 10 \text{ dB(m)} + 6 \text{ dB} = +16 \text{ dBm}$$

When there are more than two levels to be added and they are not of equal value, proceed as follows. Pair them and sum the pairs, using Figure 1.1. Sum the resultants of the pairs in the same manner until one single resultant is obtained.

Another, more exact method of summing power levels expressed in dB is to convert the dB value to the equivalent numeric value, add, and then convert back to the equivalent decibel value. Consider the example below:

Convert the input dBm values to their numeric equivalent.

$\log^{-1}(7/10)$	=	5.01 mW
$\log^{-1}(11/10)$	=	12.59 mW
Sum	=	17.60 mW

Convert the sum to the equivalent dBm value or $10 \log 17.60 \text{ mW} = +12.45$ dBm. If the network in question had an insertion loss of 3 dB, the output would be 3 dB lower or +9.45 dBm.

If we wish to convert this value back to milliwatts, we would use our calculator. Divide the 9.45 mW by 10 and then get its antilog. This is represented by the equation

$$\text{Power (mW)} = \log^{-1}\left(\frac{9.45}{10}\right) = 8.81 \text{ mW}$$

1.7 NORMAL DISTRIBUTION—STANDARD DEVIATION

A normal or Gaussian distribution is a binomial distribution where n, the number of points plotted (number of events), approaches infinity (Ref. 3, pp. 121, 122, and 143). A distribution is an arrangement of data. A frequency distribution is an arrangement of numerical data according to size and magnitude. The normal distribution curve (Figure 1.2) is a symmetrical

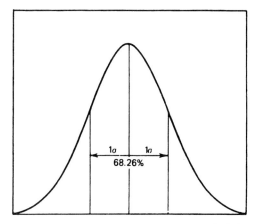

Figure 1.2. A normal distribution curve showing one standard deviation measured off either side of the arithmetic mean.

distribution. A nonsymmetrical frequency distribution curve is one in which the distributions extend further in one direction than in the other. This type of distortion is called skew. The peak of the normal distribution curve is called the point of central tendency, and its measure is its average. This is the point where the group of values tends to cluster.

The dispersion is the variation, scatteration of data, or the lack of tendency to congregate (Ref. 4). The range is the simplest measure of dispersion and is the difference between maximum and minimum values of a series. The mean deviation is another measure of dispersion. In a frequency distribution, ignoring signs, it is the average distance of items from a measure of the central tendency.

The standard deviation is the root mean square (rms) of the deviations from the arithmetic mean and is expressed by the lowercase Greek letter sigma:

$$\sigma = \sqrt{\frac{\Sigma(X'^2)}{N}} \qquad (1.14)$$

where X' = deviations from the arithmetic mean $(X - \bar{X})$, and N = total number of items.

The following expressions are useful when working with standard deviations; they refer to a "normal" distribution:

- The mean deviation $= 0.7979\sigma$.
- Measure off on both sides of the arithmetic mean one standard deviation; then 68.26% of all samples will be included within the limits.

- For two standard deviations measured off, 95.46% of all values will be included.
- For three standard deviations, 99.73% will be included.

These last three items relate to exact normal distributions. In cases where the distribution has moderate skew, approximate values are used, such as 68% for 1σ, 95% for 2σ, and so on.

1.8 THE SIMPLE TELEPHONE CONNECTION

Two people may speak to one another over a distance by connecting two telephone subsets together with a pair of wires and a common microphone battery supply. As the wires are extended (i.e., the distance between the talkers is increased), the speech power level decreases until at some point, depending on the distance, the diameter of the wire, and the mutual capacitance between each wire in the pair, communication becomes unacceptable. For example, in the early days of telephony in the United States it was noted that a telephone connection including as much as 30 mi (48 km) of 19-gauge nonloaded cable was at about the limit of useful transmission.

Suppose that several people want to join the network. We could add them in parallel (bridge them together). As each is added, however, the efficiency decreases because we have added subsets in parallel and, as a result, the impedance match between subset and line deteriorates. Besides, each party can overhear what is being said between any two others. Lack of privacy may be a distinct disadvantage at times.

This lack of privacy can be solved by using a switch so that the distant telephone may be selected. Now a signaling system must be developed so that the switch can connect the caller to the distant telephone. A system of monitoring or supervision will also be required so that on-hook (idle) and off-hook (busy) conditions may be known by the switch as well as to permit line seizure by a subscriber.

Now extend the system again, allowing several switches to be used interconnected by trunks (junctions). Because of the extension of the two-wire system without amplifiers, a reduced signal level at many of the subscribers' telephone subsets may be experienced. Now we start to reach into the transmission problem. A satisfactory signal is not being delivered to some subscribers owing to line losses because of excessive wire line lengths. Remember that line loss increases with length.

Before delving into methods of improving subscriber signal level and satisfactory signal-to-noise ratio, we must deal with basic voice channel criteria. In other words, just what are we up against? Consider also that we may want to use these telephone facilities for other types of communication such as telegraph, data, facsimile, and video transmission. The voice channel

Figure 1.3. A simplified telephone transmission system.

(telephone channel) criteria covered below are aimed essentially at speech transmission. However, many parameters affecting speech most certainly have bearing on the transmission of other types of signals, and other specialized criteria are peculiar to these other types of transmission. These are treated in depth in later chapters, where they become more meaningful. Where possible, cross reference is made.

Before going on, refer to the simplified sketch (Figure 1.3) of a basic telephone connection. The sketch contains all the basic elements that will deteriorate the signal from source to destination. The medium may be wire, coaxial cable, optical fiber, radio, or combinations of the four. Other transmission equipment may be used to enhance the medium by extending or expanding it. This equipment might consist of amplifiers, multiplex devices, and other signal processors such as compandors, voice terminals, and so forth.

1.9 THE PRACTICAL TRANSMISSION OF SPEECH

The telephone channel, hereafter called the voice channel, may be described technically using the following parameters:

- Nominal bandwidth
- Attenuation distortion (frequency response)
- Phase distortion
- Noise and signal-to-noise ratio
- Level

Return loss, singing, stability, echo, reference equivalent, and some other parameters deal more with the voice channel in a network and are discussed at length when we look at a transmission network later.

1.9.1 Bandwidth

The range between the lowest and highest frequencies used for a particular purpose may be defined as bandwidth. For our purposes we should consider bandwidth as those frequencies within which a performance characteristic of a device is above certain specified limits. For filters, attenuators, and amplifiers, these limits are generally taken where a signal will fall 3 dB below the average level in the passband or below the level at a reference frequency. The voice channel is a notable exception. In North America it is specifically defined at the 10-dB points about the reference frequency of 1000 hertz (Hz), approximately the band 200–3300 Hz. The International Consultive Committee for Telephone and Telegraph (CCITT) traditionally uses the reference frequency of 800 Hz and the CCITT voice channel occupies the band 300–3400 Hz at implied 10-dB points. In either case, we often refer to the nominal 4-kilohertz (kHz) voice channel.

1.9.2   Speech Transmission—The Human Factor

Frequency components of speech may be found between 20 Hz and 20 kHz. The frequency response of the ear (i.e., how it reacts to different frequencies) is a nonlinear function between 30 Hz and 30 kHz; however, the major intelligence and energy content exists in a much narrower band. For energy distribution see Figure 1.4. The emotional content, which transfers intelligence, is carried in a band that lies above the main energy portion. Tests have shown that low frequencies up to 600–700 Hz add very little to the intelligibility of a signal to the human ear, but in this very band much of the voice energy is transferred (solid line in Figure 1.4). The dashed line in Figure 1.4 shows the portion of the frequency band that carries emotion. From this it can be seen that for economical transfer of speech intelligence, a band much narrower than 20 Hz to 20 kHz is necessary. In fact the standard bandwidth of a voice channel is 300–3400 Hz (CCITT Recs. G.132 and G.151). However, the generally accepted voice channel frequency band in North America is 200–3200 Hz (Ref. 5, p. 32). As is shown later, this bandwidth is a compromise between what telephone subscribers demand

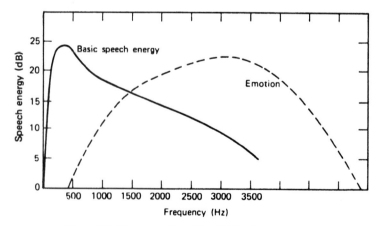

Figure 1.4. Energy and emotion distribution in speech.

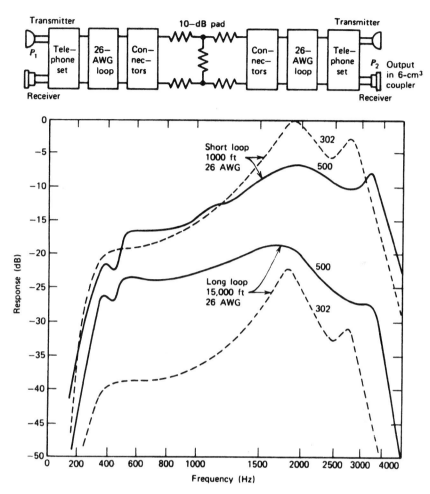

Figure 1.5. Comparison of overall response. (From Ref. 6; copyright © 1951 by Bell Telephone Laboratories.)

(Figure 1.4) and what can be provided to them economically. However, many telephone subsets have a response range no greater than approximately 500–3000 Hz. This is shown in Figure 1.5, where the response of the more modern Bell System 500 telephone set is compared to the older 302 set (Ref. 6).

1.9.3 Attenuation Distortion

A signal transmitted over a voice channel suffers various forms of distortion. That is, the output signal from the channel is distorted in some manner such that it is not an exact replica of the input. One form of distortion is called attenuation distortion and this is the result of less than perfect amplitude–frequency response. If attenuation distortion is to be avoided, all frequencies within the passband should be subjected to the same loss (or gain). On typical wire systems higher frequencies in the passband are attenuated more than lower ones. In carrier equipment the filters used tend to attenuate frequencies around band center the least, and attenuation increases as the band edges are approached. Figure 1.6 is a good example of this. The cross-hatched areas in the figure express the specified limits of attenuation distortion, and the solid line shows measured distortion on typical carrier (multiplex) equipment for the channel band (see Chapter 3). It should be remembered that any practical communication channel will suffer some form of attenuation distortion.

Attenuation distortion across the voice channel is measured compared to a reference frequency. CCITT specifies the reference frequency at 800 Hz. However, 1000 Hz is used more commonly in North America (see Figure 1.6).

For example, one requirement may state that between 600 and 2800 Hz the level will vary by not more than -1, $+2$ dB, where the plus sign means more loss and the minus sign means less loss. Thus if a signal at -10 dBm is placed at the input of the channel, we would expect -10 dBm at the output at 800 Hz (if there was no overall loss or gain), but at other frequencies we could expect a variation between -1 and $+2$ dB. For instance, we might measure the level at the output at 2500 Hz at -11.9 dBm and at 1000 Hz at -9 dBm.

CCITT recommendations for attenuation distortion may be found in volume III, Recs. G.132, G. 151, and G. 232. Figure 2.19 is taken from Rec. G.132 and shows permissible variation of attenuation between 300 and 3400 Hz. Often the requirement is stated as a slope in decibels. The slope is the maximum excursion that levels may vary in a band of interest about a certain

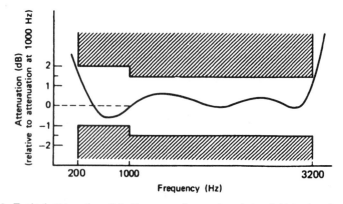

Figure 1.6. Typical attenuation distortion curve for a voice channel. Note that the reference frequency in this case is 1000 Hz.

frequency. A slope of 5 dB may be a curve with an excursion from −0.5 to +4.5 dB, −3 to +2 dB, and so forth. As links in a system are added in tandem, to maintain a fixed attenuation distortion across the system, the slope requirement for each link becomes more severe.

1.9.4 Phase Distortion and Envelope Delay Distortion

One may look at a voice channel or any bandpass as a bandpass filter. A signal takes a finite time to pass through the filter. This time is a function of the velocity of propagation. The velocity of propagation tends to vary with frequency, increasing toward band center and decreasing toward band edge, usually in the form of a parabola (see Figure 1.7).

The finite time it takes a signal to pass through the total extension of a voice channel or any other network is called delay. Absolute delay is the delay a signal experiences passing through the channel at a reference frequency. But we see that the propagation time is different for different frequencies. This is equivalent to phase shift. If the phase shift changes uniformly with frequency, the output signal will be a perfect replica of the input and there will be no distortion, whereas if the phase shift is nonlinear with respect to frequency, the output signal is distorted (i.e., it is not a perfect replica of the input). Delay distortion, or phase distortion as it is

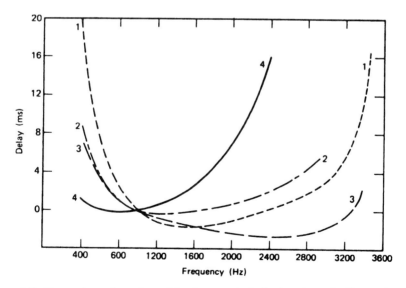

Figure 1.7. Comparison of envelope delay in some typical voice channels. Curves 1 and 3 represent the delay in several thousand miles of a toll-quality carrier system. Curve 2 shows the delay produced by 100 mi of loaded cable. Curve 4 shows the delay in 200 mi of heavily loaded cable. (From Ref. 7. Courtesy of Siemens, Inc.)

often called, is usually expressed in milliseconds (ms) or microseconds (μs) about a reference frequency.

We can relate phase distortion to phase delay. If the phase shift characteristic is known, the phase delay T_p at any frequency β_1 can be calculated as follows:

$$T_p = \frac{\beta_1 \,(\text{rad})}{\omega_1 \,(\text{rad/s})} \qquad (1.15)$$

The difference between phase delays at two frequencies in a band of interest is called delay distortion (T_d) and can be expressed as

$$T_d = \frac{\beta_2}{\omega_2} - \frac{\beta_1}{\omega_1} \qquad (1.16)$$

where β_2/ω_2 and β_1/ω_1 are the phase delays at ω_2 and ω_1, with frequency being expressed as angular frequency ($2\pi f$ rad).

In essence, therefore, we are dealing with the phase linearity of a circuit. The resulting phase distortion is best measured by a parameter called envelope delay distortion (EDD). Mathematically, envelope delay is the derivative of the phase shift with respect to frequency and expresses the instantaneous slope of the phase shift characteristic. Envelope delay T_{ed} can be stated as follows:

$$T_{ed} = \frac{d\beta}{d\omega} \qquad (1.17)$$

and the expression is valid for very small bandwidths, often referred to as apertures. For instance, the U.S. Bell System uses $166\frac{2}{3}$ Hz and CCITT, $83\frac{1}{3}$ Hz as standard apertures in measurement equipment.

The measurement of envelope delay is useful in television and facsimile transmission systems and is used in data transmission as some measure of intersymbol interference. It is a major limitation to maximum bit rate over a transmission channel. EDD is discussed in greater detail in Chapter 13.

In commercial telephony the high-frequency (HF) harmonic components, produced by the discontinuous nature of speech sounds, arrive later than the fundamental components and produce sounds that may be annoying but do not appreciably reduce intelligibility. With present handset characteristics the evidence is that the human ear is not very sensitive to phase distortions that develop in the circuit. Although a phase delay of 12 ms between the band limits is noticeable, the transmission in commercial telephone systems often contains distortions greatly in excess of this minimum.

Owing to the large amount of delay distortion in a telephone channel, as measured in its band and relative to a point of minimum delay, the usefulness of the entire telephone channel between its 3-dB cutoff points is severely restricted for the transmission of other than voice signals (e.g., data —see Chapter 13).

For the transmission of information that is sensitive to delay distortion, such as medium-speed digital signals, it is necessary to restrict occupancy to that part of the telephone channel in which the delay distortion can be tolerated or equalized at reasonable cost.

The applicable CCITT recommendation is Rec. G.133.

1.9.5   Level

1.9.5.1   General. In most systems when we refer to level, we refer to a power level that may well be in dBm, dBW, or other power units. One notable exception is video, which uses voltage, usually measured in dBmV.

Level is an important system parameter. If levels are maintained at too high a point, amplifiers become overloaded, with resulting increases in intermodulation products or crosstalk. If levels are too low, customer satisfaction may suffer.

1.9.5.2   Reference Level Points. System levels usually are taken from a level chart or reference system drawing made by a planning group or as part of an engineered job. On the chart a 0 TLP (test level point) is established. A TLP is the location in a circuit or system at which a specified test tone level is expected during alignment. A 0 TLP is the point at which the test tone level should be 0 dBm.

From the 0 TLP other points may be shown using the unit dBr (dB reference). A minus sign shows that the level is so many decibels below reference, and a positive sign that the level is so many decibels above reference. The unit dBm0 is an absolute unit of power in dBm referred to the 0 TLP. dBm can be related to dBr and dBm0 by the following formula:

$$dBm = dBm0 + dBr \qquad (1.18)$$

For instance, a value of -32 dBm at a -22-dBr point corresponds to a referenced level of -10 dBm0. A -10-dBm0 signal introduced at the 0-dBr point (0 TLP) has an absolute value of signal level of -10 dBm.

In North American practice the 0 TLP was originally defined at the transmission jack of a toll (long-distance) switchboard. Many technical changes, of course, have occurred since the days of manual switchboards. Nevertheless it was deemed desirable to maintain the 0 TLP concept. As a result, the outgoing side of a switch to which an intertoll trunk is connected (see Figure 2.11) is designated a -2-dB TLP, and the outgoing side of the switch at which a local area trunk is terminated is defined as 0 TLP (Ref. 8).

To quote from Ref. 8 in part:

> In the layout of four-wire trunks, a patch bay, called the four-wire patch bay, is usually provided to facilitate test, maintenance, and circuit rearrangements between trunks and the switching machine terminations. TLPs at these four-wire patch bays have been standardized for all four-wire trunks. On the transmitting side the TLP is −16 dB, and on the receiving side the TLP is +7 dB. Thus a four-wire trunk, whether derived from voice frequency or from carrier facilities, must be designed to have 23-dB gain between four-wire patch bays. These standard TLPs are necessary to permit flexible telephone plant administration.
>
> In four-wire circuits, the TLP concept is easily understood and applied because each transmission path has only one direction of transmission. In two-wire circuits, however, confusion or ambiguity may be introduced by the fact that a single point may be properly designated as two different TLPs, each depending on the assumed direction of transmission.

Refer to Section 2.6.3 for a discussion of two-wire and four-wire transmission, to CCITT Rec. G.141 for transmission reference point, and to Figure 2.13 for a definition of virtual switching points $a–t–b$.

1.9.5.3 *The Volume Unit (VU).* One measure of level is the volume unit (VU). Such a unit is used to measure the power level (volume) of program channels (broadcast) and certain other types of speech or music. VU meters are usually kept on line to measure volume levels of program or speech material being transmitted. If a simple dB meter or voltmeter is bridged across the circuit to monitor the program volume level, the indicating needle tries to follow every fluctuation of speech or program power and is difficult to read; besides, the reading will have no real meaning. To further complicate matters, different meters made by different manufacturers will probably read differently because of differences in their damping and ballistic characteristics.

The indicating instrument used in VU meters is a dc millimeter having a slow response time and damping slightly less than critical. If a steady sine wave is suddenly impressed on the UV meter, the pointer or needle will move to within 90% of the steady-state value in 0.3 s and overswing the steady-state value by no more than 1.5%.

The standard volume indicator (U.S.), which includes the meter and an associated attenuator, is calibrated to read 0 VU when connected across a 600-Ω circuit (voice pair) carrying a 1-mW sine wave power at any frequency between 35 and 10,000 Hz. For complex waves such as music and speech, a VU meter will read some value between average and peak of the complex wave. The reader must remember that there is no simple relationship between the volume measured in VUs and the power of a complex wave. It can be said, however, that for a continuous sine wave signal across 600 Ω, 0 dBm = 0 VU by definition, or that the readings in dBm and VU are the same for continuous simple sine waves in the voice-frequency (VF) range. For a complex signal subtract 1.4 from the VU reading, and the result will be approximate talker power in dBm.

Talker volumes, or levels of a talker at the telephone subset, vary over wide limits for both long-term average power and peak power. Based on comprehensive tests by Holbrook and Dixon (Ref. 2), "mean talker" average power varies between -10 and -15 VU, with a mean of -13 VU.

1.9.6 Noise

1.9.6.1 General. "Noise, in its broadest definition, consists of any undesired signal in a communication circuit" (Ref. 9). The subject of noise and its reduction is probably the most important problem that a transmission engineer must face. It is noise that is the major limiting factor in telecommunication system performance. Noise may be divided into four categories:

1. Thermal noise
2. Intermodulation noise
3. Crosstalk
4. Impulse noise

1.9.6.2 Thermal Noise. Thermal noise is the noise occurring in all transmission media and in all communication equipment arising from random electron motion. It is characterized by a uniform distribution of energy over the frequency spectrum and a normal (Gaussian) distribution of levels.

Every equipment element and the transmission medium contribute thermal noise to a communication system, provided the temperature of that element of medium is above absolute zero. Thermal noise is the factor that sets the lower limit for the sensitivity of a receiving system. Often this noise is expressed as a temperature referred to absolute zero (kelvin, K).

Thermal noise is a general expression referring to noise based on thermal agitations. The term *white noise* refers to the average uniform spectral distribution of energy with respect to frequency. Thermal noise is directly proportional to bandwidth and temperature. The amount of thermal noise to be found in 1 Hz of bandwidth in an actual device is

$$P_n = kT \quad (\text{W/Hz}) \tag{1.19}$$

where k = Boltzmann's constant = $1.3803 \, (10^{-23}) \, \text{J/K}$
　　　　T = absolute temperature (K) of thermal noise

At room temperature, $T = 17°C$ or 290 K,

$$P_n = 4.00(10^{-21}) \text{ W/Hz of bandwidth}$$
$$= -204 \text{ dBW/Hz of bandwidth}$$
$$= -174 \text{ dBm/Hz of bandwidth}$$

For a band-limited system (i.e., a system with a specific bandwidth),

$$P_n = kTB \quad \text{(W)} \tag{1.20}$$

where B is the bandwidth (Hz). At 0 K,

$$P_n = -228.6 \text{ dBW/Hz of bandwidth} \tag{1.21}$$

and for a band-limited system,

$$P_n = -228.6 \text{ dBW} + 10 \log T + 10 \log B \tag{1.22}$$

The thermal noise threshold of a receiver operating at room temperature (17°C or 290 K) by an equation derived from equation 1.22:

$$P_n = -228.6 \text{ dBW} + 10 \log 290 + NF_{dB} + 10 \log B_{Hz} \tag{1.22a}$$

where NF is the receiver noise figure in dB (see Section 1.12). The term "$10 \log 290$" brings the receiver noise temperature from absolute zero to standard room temperature, 290 K.

$$P_n = -204 \text{ dBW} + NF_{dB} + 10 \log B_{Hz} \tag{1.22b}$$

Example 1. Given a receiver with an effective noise temperature of 100 K and a 10-MHz bandwidth, what thermal noise level may we expect at its output?

$$P_n = -228.6 \text{ dBW} + 10 \log 1 \times 10^2 + 10 \log 1 \times 10^7$$
$$= -228.6 + 20 + 70$$
$$= -138.6 \text{ dBW}$$

Example 2. Given an amplifier with an effective noise temperature of 10,000 K and a 10-MHz bandwidth, what thermal noise level may we expect at its output?

$$P_n = -228.6 \text{ dBW} + 10 \log 1 \times 10^4 + 10 \log 1 \times 10^7$$
$$= -228.6 + 40 + 70$$
$$= -118.6 \text{ dBW}$$

Example 3. A certain receiver has a noise figure of 4 dB and operates at room temperature. Its bandwidth is 20 MHz. What is its thermal noise threshold? In this case use equation 1.22b:

$$P_n \text{ (dBW)} = -204 \text{ dBW} + 4 \text{ dB} + 10 \log 20 \times 10^6$$

$$= -204 \text{ dBW} + 4 \text{ dB} + 73 \text{ dB}$$

$$= -127 \text{ dBW}$$

From the examples it can be seen that there is little direct relationship between physical temperature and effective noise temperature (Ref. 10).

1.9.6.3 Intermodulation Noise.

Intermodulation noise is the result of the presence of intermodulation (IM) products. Let us pass two signals with frequencies F_1 and F_2 through the nonlinear device or medium. Intermodulation products will result, which are spurious frequencies. These frequencies may be present either inside or outside the band of interest for the device. Intermodulation products may be produced from harmonics of the signals in question, either as products between harmonics or as one or the other or both signals themselves.

The products result when the two (or more) signals beat together or "mix." Look at the "mixing" possibilities when passing F_1 and F_2 through a nonlinear device. The coefficients indicate first, second, and third harmonics:

- Second-order products $2F_1, 2F_2, F_1 \pm F_2$
- Third-order products $2F_1 \pm F_2; 2F_2 \pm F_1$
- Fourth-order products $2F_1 \pm 2F_2; 3F_1 \pm F_2$

Devices passing multiple signals, such as multichannel radio equipment, develop intermodulation products that are so varied that they resemble white noise.

Intermodulation noise may result from a number of causes:

- Improper level setting; too high a level input to a device drives the device into its nonlinear operating region (overdrive)
- Improper alignment causing a device to function nonlinearly
- Nonlinear envelope delay

To sum up, intermodulation noise results from either a nonlinearity or a malfunction having the effect of nonlinearity. The cause of intermodulation noise is different from that of thermal noise; however, its detrimental effects and physical nature are identical to those of thermal noise, particularly in multichannel systems carrying complex signals.

1.9.6.4 Crosstalk. Crosstalk refers to unwanted coupling between signal paths. Essentially there are three causes of crosstalk. The first is the electrical coupling between transmission media, for example, between wire pairs on a VF cable system. The second is poor control of frequency response (i.e., defective filters or poor filter design), and the third is the nonlinearity performance in analog (FDM) multiplex systems. Crosstalk has been categorized into two types:

1. *Intelligible Crosstalk.* At least four words are intelligible to the listener from extraneous conversation(s) in a 7-s period (Ref. 5, 3rd ed., p. 46).
2. *Unintelligible Crosstalk.* Any other form of disturbing effects of one channel upon another. Babble is one form of unintelligible crosstalk.

Intelligible crosstalk presents the greatest impairment because of its distraction to the listener. One point of view is that the distraction is caused by fear of loss of privacy. Another is that the annoyance is caused primarily by the user of the primary line consciously or unconsciously trying to understand what is being said on the secondary or interfering circuits; this would be true for any interference that is syllabic in nature.

Received crosstalk varies with the volume of the disturbing talker, the loss from the disturbing talker to the point of crosstalk, the coupling loss between the two circuits under consideration, and the loss from the point of crosstalk to the listener.

As far as this discussion is concerned, the controlling element is the coupling loss between the two circuits under consideration. Talker volume or level is covered in Section 1.9.5. The effects of crosstalk are subjective, and other factors also have to be considered when the crosstalk impairment is to be measured. Among these factors are the type of people who use the channel, the acuity of listeners, traffic patterns, and operating practices.

Crosstalk coupling loss can be measured quantitatively with precision between a given sending point on a disturbing circuit and a given receiving point on a disturbed circuit. Essentially, then, it is the simple measurement of transmission loss in decibels between the two points. Between carrier circuits, crosstalk coupling is most usually flat. In other words, the amount of coupling experienced at one frequency will be nearly the same for every other frequency capacitive, and coupling loss usually has an average slope of 6 dB/octave. See CCITT Rec. G.227 and the annex to CCITT Rec. G.134, which treat crosstalk measurements.

Two other units are also commonly used to measure crosstalk. One expresses the coupling in decibels above "reference coupling" and uses dBx for the unit of measure. The dBx was invented to allow crosstalk coupling to be expressed in positive units. The reference coupling is taken as a coupling loss of 90 dB between disturbing and disturbed circuits. Thus crosstalk coupling in dBx is equal to 90 minus the coupling loss in dB. If crosstalk coupling loss is 60 dB, we have 30 dBx crosstalk coupling. Thus, by definition, 0 dBx = -90 dBm at 1000 Hz.

The second unit is the crosstalk unit (CU). When the impedances of the disturbed and disturbing circuits are the same, the number of CUs is one million times the ratio of the induced crosstalk voltage or current to the disturbing voltage or current. When the impedances are not equal,

$$CU = 10^6 \sqrt{\frac{\text{crosstalk signal power}}{\text{disturbing signal power}}} \tag{1.23}$$

Figure 1.8 relates all three units used in measuring crosstalk. CCITT recommends crosstalk criteria in Rec. G. 151 (p. 4).

The percentage change of intelligible crosstalk on a circuit is defined by the crosstalk index. North American practice is to allow arbitrarily that on no more than 1% of calls will a customer hear a foreign conversation, which we have defined as intelligible crosstalk. The design objective is 0.5%. The graph in Figure 1.9 may be used for guidance. It relates customer reaction to crosstalk, and crosstalk index to crosstalk coupling.

All forms of unintelligible crosstalk are covered above; its nature is very similar to intermodulation noise.

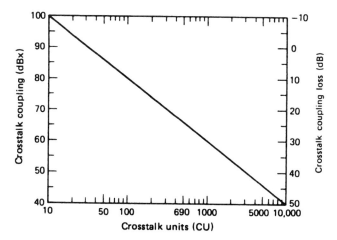

Figure 1.8. Relations between crosstalk measuring units. (From Ref. 11; copyright © 1961 by American Telephone & Telegraph Company.)

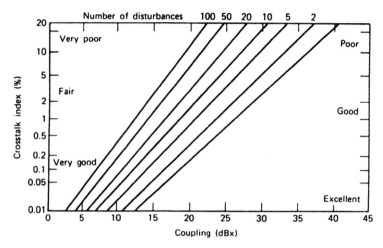

Figure 1.9. Crosstalk judgment curves. (From Ref. 5, 3rd ed.; copyright © 1964 by Bell Telephone Laboratories.)

1.9.6.5 Impulse Noise. This type of noise is noncontinuous, consisting of irregular pulses or noise spikes of short duration and of relatively high amplitude. Often these spikes are called "hits." Impulse noise degrades voice telephony only marginally, if at all. However, it may seriously degrade the error rate on a data transmission circuit, and the subject is covered in more depth in Chapter 13.

1.9.6.6 Noise Measurement Units. The interfering effect of noise on speech telephony is a function of the response of the human ear to specific frequencies in the voice channel as well as of the type of subset used.

When noise measurement units were first defined, it was decided that it would be convenient to measure the relative interfering effect of noise on the listener as a positive number. The level of a 1000-Hz tone at -90 dBm or 10^{-12} W (1 pW) was chosen by the U.S. Bell System because a tone whose level is less than -90 dBm is not ordinarily audible. Such a negative threshold meant that all noise measurements used in telephony would be greater than this number, or positive. The telephone subset than in early universal use in North America was the Western Electric 144 handset. The noise measurement unit was the dB-rn or dBrn, the rn standing for reference noise; 0 dBrn = -90 dBm at 1000 Hz.

With the 144-type handset as a test receiver and with a wide distribution of "average" listeners, it was found that a 500-Hz sinusoidal signal had to have its level increased by 15 dB to have the same interfering effect on the "average" listener over the 1000-Hz reference. A 3000-Hz signal required an 18-dB increase to have the same interfering effect, 6 dB at 800 Hz, and so on.

A curve showing the relative interfering effects of sinusoidal tones compared to a reference frequency is called a weighting curve. Artificial filters are made with a response resembling the weighting curve. These filters, normally used on noise measurement sets, are called weighting networks.

Subsequent to the 144 handset, the Western Electric Company developed the F1A handset, which had a considerably broader response than the older handset but was 5 dB less sensitive at 1000 Hz. The reference level for this type of handset was −85 dBm. The new weighting curve and its noise measurement weighting network were denoted by F1A (i.e., an F1A line weighting curve and F1A weighting network). The noise measurement unit was the dB adjusted (dBa).

A third, more sensitive handset (500 type) is now in use in North America, giving rise to the C-message line weighting curve and its companion noise measurement unit, the dBrnC. It is 3.5 dB more sensitive at the 1000-Hz reference frequency than the F1A, and 1.5 dB less sensitive than the 144-type weighting. Rather than choosing a new reference power level (−88.5 dBm), the reference power level of −90 dBm was maintained.

Figure 1.10 compares the various noise weighting curves now in use. Table 1.4 compares weighted noise units.

One important weighting curve and noise measurement unit has yet to be mentioned. The curve is the CCIR (CCITT) psophometric weighting curve. The noise measurement units associated with this curve are dBmp and pWp

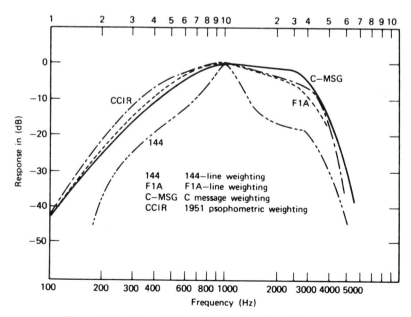

Figure 1.10   Line weightings for telephone (voice) channel noise.

TABLE 1.4   Conversion Chart, Psophometric, F1A, and C-Message Noise Units

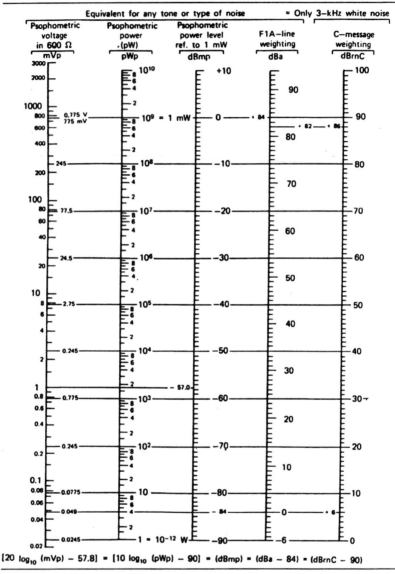

$$[20 \log_{10}(\text{mVp}) - 57.8] = [10 \log_{10}(\text{pWp}) - 90] = (\text{dBmp}) = (\text{dBa} - 84) = (\text{dBrnC} - 90)$$

Chart basis:

dBmp = dBa − 84

1 mW unweighted 3-kHz white noise reads 82 dBa = 88.5 dBrnC (C-message) rounded off to 88.0 dBrnC. 1 mW into 600 Ω = 775 mV = 0 dBm = 10^9 pW.

Readings of noise measuring sets when calibrated on 1-mW test tone:

F1A at 1000 Hz reads 85 dBa.

C-message at 1000 Hz reads 90 dBrn.

Psophometer at 800 Hz reads 0 dBm.

F1A and DBa shown   for historical purposes only.

(dBm psophometrically weighted and picowatts psophometrically weighted, respectively). The reference frequency in this case is 800 Hz rather than 1000 Hz.

Consider now a 3-kHz band of white noise (flat, i.e., not weighted). Such a band is attenuated 8 dB when measured by a noise measurement set using a 144 weighting network, 3 dB using F1A weighting, 2.5 dB for CCIR/CCITT weighting, and 1.5 dB rounded off to 2.0 dB for C-message weighting. Table 1.4 may be used to convert from one noise measurement unit to another.

CCITT states in Rec. G.223 that

> If uniform-spectrum random noise is measured in a 3.1-kHz band with a flat attenuation frequency characteristic, the noise level must be reduced 2.5 dB to obtain a psophometric power level. For another bandwidth B, the weighting factor will be equal to

$$2.5 + 10 \log B/3.1 \text{ dB} \qquad (1.24)$$

> When $B = 4$ kHz, for example, this formula gives a weighting factor of 3.6 dB.

1.10   SIGNAL-TO-NOISE RATIO

The transmission system engineer deals with signal-to-noise ratio probably more frequently than with any other criterion when engineering a telecommunication system.

The signal-to-noise ratio expresses in decibels the amount by which a signal level exceeds its corresponding noise.

As we review the several types of material to be transmitted, each will require a minimum signal-to-noise ratio to satisfy the customer or to make the receiving-end instrument function within certain specified criteria. We might require the following signal-to-noise ratios with corresponding end instruments:

Voice 30 dB ⎫
Video 45 dB ⎬ based on customer satisfaction
Data 15 dB   based on a specified error rate and type of modulation

In Figure 1.11 the 1000-Hz signal has a signal-to-noise ratio (S/N) of 10 dB. The level of the noise is 5 dBm and the signal, 15 dBm. Thus

$$(S/N)_{dB} = \text{level}_{\text{signal (dBm)}} - \text{level}_{\text{noise (dBm)}} \qquad (1.25)$$

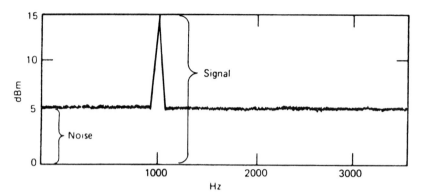

Figure 1.11. Signal-to-noise ratio S/N.

1.11   THE EXPRESSION E_b/N_0

For digital transmission systems the expression E_b/N_0 is a more convenient term to qualify a received digital signal than the signal-to-noise ratio under many circumstances. E_b/N_0 expresses the received signal energy per bit per hertz of thermal noise. Thus

$$\frac{E_b}{N_0} = \frac{C}{kT \text{ (bit rate)}} \qquad (1.26)$$

where C = the receive signal level (RSL). Expressed in decibel notation,

$$\frac{E_b}{N_0} = C_{dBW} - 10 \log(\text{bit rate}) - (-228.6 \text{ dBW}) - 10 \log T_e \quad (1.27)$$

where T_e = effective noise temperature of the receiving system (see Section 1.9.6).

Example. If the RSL for a particular digital system is -151 dBW and the receiver system effective noise temperature is 1500 K, what is E_b/N_0 for a link transmitting 2400 bps?

$$\frac{E_b}{N_0} = -151 \text{ dBW} - 10 \log 2400 - 10 \log 1500 + 228.6 \text{ dBW}$$

$$= 12 \text{ dB}$$

Depending on the modulation scheme, the type of detector used, and the coding of the transmitted signal, E_b/N_0 required for a given bit error rate (BER) may vary from 4 to 25 dB.

1.12   NOISE FIGURE

It has been established that all networks, whether passive or active, and all other forms of transmission media contribute noise to a transmission system. The noise figure (NF) is a measure of the noise produced by a practical network compared to an ideal network (i.e., one that is noiseless). For a linear system, the NF is expressed by

$$\text{NF} = \frac{(S/N)_{\text{in}}}{(S/N)_{\text{out}}} \tag{1.28}$$

It simply relates the signal-to-noise ratio of the output signal from the network to the signal-to-noise ratio of the input signal. From equation 1.20 the thermal noise may be expressed by the basic formula kTB, where $T = 290$ K (room temperature). As we can see, NF can be interpreted as the degradation of the signal-to-noise ratio by the network.

By letting the gain of the network G equal $S_{\text{out}}/S_{\text{in}}$,

$$\text{NF} = \frac{N_{\text{out}}}{kTBG} \tag{1.29}$$

It should be noted that we defined the network as fully linear, so NF has not been degraded by intermodulation noise. More commonly, NF is expressed in decibels, where

$$\text{NF}_{\text{dB}} = 10 \log_{10} \text{NF} \tag{1.30}$$

Example. Consider a receiver with a NF of 10 dB. Its output signal-to-noise ratio is 50 dB. What is its input equivalent signal-to-noise ratio?

$$\text{NF}_{\text{dB}} = (S/N)_{\text{dB input}} - (S/N)_{\text{dB output}}$$

$$10 \text{ dB} = (S/N)_{\text{input}} - 50 \text{ dB}$$

$$(S/N)_{\text{input}} = 60 \text{ dB}$$

1.13   RELATING NOISE FIGURE TO NOISE TEMPERATURE

The noise temperature of a two-port device, a receiver, for instance, is the thermal noise that device adds to a system. If the device is connected to a noise-free source, its equivalent noise temperature is given by

$$T_{\text{e}} = \frac{P_{\text{ne}}}{Gk\,df} \tag{1.31}$$

where G = gain and df = specified small band of frequencies. T_e is referred to as the effective input noise temperature of the network and is a measure of the internal noise sources of the network, and P_{ne} is the available noise power of the device (Ref. 5).

The noise temperature of a device and its NF are analytically related. Thus

$$NF = 1 + \frac{T_e}{T_0} \tag{1.32}$$

where T_0 = equivalent room temperature or 290 K.

$$T_e = T_0 (NF - 1) \tag{1.33}$$

To convert NF in decibels to equivalent noise temperature (T_e) in kelvins, use the following formula:

$$NF_{dB} = 10 \log_{10}\left(1 - \frac{T_e}{290}\right) \tag{1.34}$$

Example 1. Consider a receiver with an equivalent noise temperature of 290 K. What is its NF?

$$NF_{dB} = 10 \log\left(1 + \frac{290}{290}\right)$$

$$NF = 10 \log 2 = 3 \text{ dB}$$

Example 2. A receiver has a NF of 10 dB. What is its equivalent noise temperature in kelvins?

$$10 \text{ dB} = 10 \log\left(1 + \frac{T_e}{290}\right)$$

$$10 = 10 \log X$$

where $X = 1 + (T_e/290)$. Thus $\log X = 1$, $X = 10$, and

$$T_e = 2900 - 290 = 2610 \text{ K}$$

Several NFs are given with their corresponding equivalent noise temperatures in Table 1.5.

TABLE 1.5   **Noise Figure–Noise Temperature Conversion**

NF_{dB}	T (K) (approx.)	NF_{dB}	T (K) (approx.)
15	8950	6	865
14	7000	5	627
13	5500	4	439
12	4300	3	289
11	3350	2.5	226
10	2610	2.0	170
9	2015	1.5	120
8	1540	1.0	75
7	1165	0.5	35.4

1.14   EFFECTIVE ISOTROPICALLY RADIATED POWER (EIRP)

EIRP is a tool we use to describe the performance of a radio transmitting system. There are three basic elements in the system: a transmitter with a certain output power, an antenna with a gain (or perhaps a loss), and a transmission line that connects the transmitter to the antenna. The transmission line has a loss. It is convenient to express all values using a decibel notation. Thus

$$\text{EIRP}_{dBW} = P_t + G_{ant} - L_L \qquad (1.35)$$

where P_t = the radiofrequency (RF) output power of the transmitter in dB units

G_{ant} = the gain of the antenna (or loss) in dB

L_L = the transmission line loss in dB

Equation 1.35 is written using the dB power notation in dBW. It also can be written in dBm. Consistency is urged. If dBW is used, then dBW must be used throughout any related calculation procedure; if dBm is used, we have to be equally consistent.

Consider the following worked examples.

Example 1. A high-frequency (HF) transmitter has an output of 20 kW; its associated rhombic antenna has a gain of 12 dB and the balanced transmission line has a loss of 1.1 dB. What is the EIRP of this transmitting installation? 20 kW = +43 dBW.

$$\text{EIRP}_{dBW} = +43 \text{ dBW} + 12 \text{ dB} - 1.1 \text{ dB}$$

$$= +53.9 \text{ dBW}$$

Example 2. A microwave transmitter has an output of 1 W; the waveguide connecting this transmitter to its antenna has a loss of 4.6 dB and the antenna has a 36-dB gain. What is the EIRP in dBm? 1 W = −30 dBm.

$$\text{EIRP}_{dBm} = +30 \text{ dBm} + 36 \text{ dB} - 4.6 \text{ dB}$$

$$= +61.4 \text{ dBm}$$

1.14.1 The Concept of the Isotropic Antenna

An isotropic antenna is an antenna with a gain of 1 (0 dB) that radiates uniformly in all three dimensions. It is a fictitious antenna because we cannot build an antenna with such characteristics. It is, however, a very useful tool to describe the performance of a real antenna when compared to an isotropic. Gain of an antenna is most often given in dBi. This means the number of dB above or below the isotropic reference. The plus or minus sign in front of the dB value indicates whether the gain is greater or less than an isotropic.

Some texts and some of our peers will specify or give antenna gain in dB. We must remember that decibels express a ratio. We must then ask, dB relative to what? The isotropic is most convenient, but in some cases a dipole is inferred. Sometimes we will find ERP (effective radiated power) used rather than EIRP. That power may be relative to an isotropic or a dipole, or even some other antenna. To relate a half-wave dipole (in free space) to an isotropic in free space, we should keep in mind that the dipole has a gain of 2.15 dBi (decibels over an isotropic).

1.15 WAVELENGTH–FREQUENCY RELATIONSHIP

If we are given the frequency of a RF wave, we can calculate its wavelength; and conversely, given the wavelength, we can calculate the frequency by the following equation:

$$F\lambda = 3 \times 10^8 \text{ m/s} \tag{1.36}$$

where F = the frequency in hertz and λ = the wavelength in meters.

Example 1. If the wavelength of a RF emission is 40 m, what is the equivalent frequency?

$$40F = 3 \times 10^8 \text{ m/s}$$
$$F = 3 \times 10^8/40$$
$$= 7,500,000 \text{ Hz or } 7.5 \text{ MHz}$$

Example 2. If the wavelength of the international calling and distress frequency is 600 m, what is its equivalent frequency?

$$600F = 3 \times 10^8 \text{ m/s}$$
$$F = 500 \text{ kHz}$$

1.16 LONGITUDINAL BALANCE

Longitudinal balance is an important transmission parameter when dealing with transmission lines carrying baseband signals such as subscriber loops, metallic trunks, and metallic switching circuits. Longitudinal balance is a measure of each leg of a balanced circuit's symmetry to ground. With good balance, induced noise can be reduced or nearly eliminated entirely.

We define a balanced circuit as one in which two branches are electrically alike and symmetrical with respect to a common reference point, usually ground. A metallic circuit is defined as one where ground forms no part of that circuit. Longitudinal balance is measured in dB and is expressed by the following formula:

$$\text{Longitudinal balance (dB)} = 20\log\left(\frac{\text{open circuit longitudinal voltage}}{\text{metallic voltage}}\right)$$

$$(1.37)$$

The IEEE (Ref. 12) defines the degree of longitudinal balance measured in dB as the ratio of the disturbing longitudinal voltage and the resulting metallic voltage of the network under test.

These concepts are shown graphically in Figure 1.12, where we distinguish between longitudinal and metallic current. The figure shows a telephone switch, which is the terminating circuit (network) connected to a signal generator (E) by means of a balanced transmission line. Examples of such transmission lines are subscriber loops, metallic pair trunks, and talk paths terminating in switch signal ports in the case of a digital switch and through-switch ports in the case of analog space division switches.

The figure shows that currents flowing in the same direction on the two conductors of the transmission line are called longitudinal currents (Figure 1.12a), while currents flowing in opposite directions in the two conductors are called metallic currents (Figure 1.12b).

In Figure 1.12, the signal generator (voltage source) is coupled to the balanced transmission line by a coupling mechanism Z_c. The resulting currents are transmitted to the switch through a load impedance Z_L through common battery feed circuits through impedance Z_s. Z_s may represent switch components such as relays, transformers, and common battery leads.

Figure 1.12a shows a perfectly balanced condition, meaning that the balanced leads are electrically alike and symmetrical with respect to ground. In this case interference voltages cause no interference current through Z_L. However, if the Z_L and Z_s networks are not balanced, unequal currents flow on the two sides of the circuit. The difference between them is the metallic current that appears in load Z_L as interference. This metallic component of the current can be represented as originating in the equivalent circuit, as shown in Figure 1.12b, where E may represent the source of the unbalanced current, some system-generated interference, or a wanted signal source.

Some examples of good longitudinal balance follow. For a subscriber loop, over 50 dB (Ref. 8, vol. 3) and four-wire switches (space division) should exceed 53 dB (Ref. 13).

When one reviews vendor longitudinal balance values for a certain equipment, care must be taken to assure the test method used. Our experience

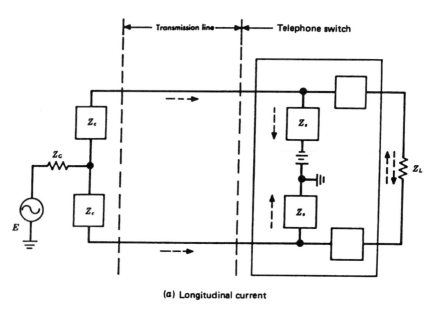

(a) Longitudinal current

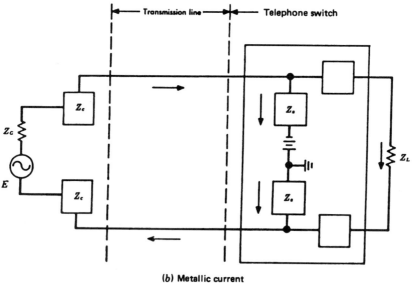

(b) Metallic current

Figure 1.12. Longitudinal and metallic current distinguished.

shows that the most widely accepted longitudinal balance test method is presented in ANSI/IEEE Std. 455-1985 (Ref. 12).

1.17 SOME COMMON CONVERSION FACTORS

To Convert	Into	Multiply by	Conversely, Multiply by
acres	hectares	0.4047	2.471
Btu	kilogram-calories	0.2520	3.969
°Celsius	°Fahrenheit	$9°C/5 = °F - 32$ $9(°C + 40)/5 = (°F + 40)$	
circular mils	square centimeters	5.067×10^{-6}	1.973×10^{5}
circular mils	square mils	0.7854	1.273
degrees (angle)	radians	1.745×10^{-2}	57.30
kilometers	feet	3281	3.048×10^{-4}
kilowatt-hours	Btu	3413	2.930×10^{-4}
liters	gallons (liq. U.S.)	0.2642	3.785
$\log_e$ or ln	$\log_{10}$	0.4343	2.303
meters	feet	3.281	0.3048
miles (nautical)	meters	1852	5.400×10^{-4}
miles (nautical)	miles (statute)	1.1508	0.8690
miles (statute)	feet	5280	1.890×10^{-4}
miles (statute)	kilometers	1.609	0.6214
nepers	decibels	8.686	0.1151
square inches	circular mils	1.273×10^{6}	7.854×10^{-7}
square millimeters	circular mils	1973	5.067×10^{-4}

Boltzmann's constant $(1.38044 \pm 0.00007) \times 10^{-16}$ erg/deg

velocity of light in free space 2.998×10^{8} m/s

186,280 mi/s
984×10^{6} ft/s

1 degree of longitude at the equator 68.703 statute mi or 59.661 nautical mi

1 rad $180°/\pi = 57.2958°$

1 m 39.3701 in. = 3.28084 ft

1° 17.4533 mrad

e 2.71828

REVIEW EXERCISES

1. Distinguish *transmission* from *switching*. List four subdisciplines of each.

2. A network has an input of 12 mW and a gain of 26 dB. What is the output of the network in dBm?

3. There are four networks in series. The first network has a gain of 15 dB, the second a loss of 4 dB, the third a gain of 35 dB, and the fourth a loss of 5 dB. The input of the first network is +3 dBm. What is the output to the last network in mW?

4. There are three networks in series. The first network has a gain of 19 dB, the second a loss of 23 dB, and the third a gain of 11 dB. The output of the third network is +23 dBm. What is the input to the first network in mW?

5. A combining network has two inputs: +20 dBm and +6 dBm. It has an insertion loss of 3 dB. What is the combined output in dBm?

6. What is the equivalent mW value of +23.65 dBm?

7. A microwave transmitter has an output of 500 mW. What is its output in dBW?

8. Determine the following equivalents: +27 dBm = ? dBW; +36 dBW = ? watts; +34 dBm = ? dBW.

9. What is the standard impedance in the CATV plant? *Note*: The dBmV is based on that impedance.

10. Express the gain in dBi of the standard dipole antenna. A cellular radio base station antenna has a +10 dBd gain. What is its gain in dBi?

11. In a population of 3500 subscribers, 2σ of them are satisfied with the transmission level. How many subscribers are unsatisfied with transmission level? The 3σ of subscriber lines have a return loss of 11 dB or better. How many subscriber lines in this population have a return loss of less than 11 dB?

12. As a subscriber loop is extended in length, what are the two electrical constraints on loop length and how do these constraints affect the subscriber?

13. What are the three basic *impairments* (not echo or singing) we have to deal with regarding the voice channel?

14. Define the CCITT voice channel and give its bandwidth.

15. Explain the cause of phase distortion.

16. Phase distortion, in general, does not impair speech transmission. It does impair one important type of information transmitted across the voice channel. What type of information is it?

17. Define 0 TLP.

18. What is the more common unit of measurement of speech level? Relate the unit to dBm. Include certain restrictions placed on that relationship.

19. Give the four basic types of noise that we must deal with in transmission as provided in the text.

20. What is the thermal noise threshold of a *perfect* receiver operating at absolute zero? At room temperature? Use the unit dBW in the answers.

21. What is the thermal noise threshold of a receiver with a 2-MHz bandwidth and an effective noise temperature of 2000 K?

22. Two signals, A and B, mix in a nonlinear device. Give the two most probable values of third-order products. Use the notation given in the text.

23. Of the four types of noise discussed, one type is not generally an impairment to speech transmission but can seriously affect data bit error rate. What type of noise is this?

24. Why are dB-related noise units referenced to -90 dBm and not some other value?

25. Give the dB difference between psophometric noise weighting and flat weighting when dealing with the standard voice channel.

26. It is often useful to give S/N using a formula based on dB-derived units. Write the formula for S/N using dBm.

27. Determine the following equivalents: $+13$ dBW = ? dBm; -3 dBm = ? dBW.

28. Define noise figure in an equation with decimal values (in some texts this is called noise factor). Now define noise figure using dB units.

29. (a) A certain receiver has a 3-dB noise figure. Give its equivalent noise temperature in K. (b) The effective noise temperature of a receiver is 2000 K. What is its equivalent noise figure in dB?

30. Calculate N_0 for a receiving system with a noise figure of 7 dB.

31. What is the value of E_b/N_0 at a receiver where the bit rate is 20 Mbps and the noise figure is 13 dB.

32. The noise temperature of a receiver is 290 K. What is its equivalent effective noise temperature?

33. An antenna has a 15-dB gain and is fed by a transmitter with 2-kW output. Transmission line losses are 0.6 dB. What is the EIRP (dBW) of the peak main beam of the antenna?

34. A transmitter has an output of 500 mW. Transmission line losses are 3.7 dB and the gain of the antenna is 36 dB. What is the EIRP of the peak main beam?

35. Discuss the importance of longitudinal balance and its measurement. Name at least three places in the telecommunications plant where longitudinal balance would be important to the transmission engineer.

36. (a) A transmitter has an output at 3 MHz. What is the equivalent wavelength of the signal? (b) A certain signal has a wavelength of 6 cm. What is its equivalent frequency?

REFERENCES

1. Ken Simons, *Technical Handbook for CATV Systems*, 3rd ed., Jerrold Electronics Corp., Hatboro, PA, 1969.

2. B. D. Holbrook and J. T. Dixon, "Load Rating Theory for Multichannel Amplifiers," *Bell Syst. Tech. J.*, vol. 18, 624–644, Oct. 1939.

3. H. Arkin and R. R. Colton, *Statistical Methods*, 5th ed., Barnes and Noble College Outline Series, New York, 1972.

4. R. C. James, *Mathematics Dictionary*, 3rd ed., Van Nostrand, Princeton, NJ, 1966.

5. *Transmission Systems for Communications*, 5th ed., Bell Telephone Laboratories, Holmdel, NJ, 1982.

6. W. F. Tuffnell, "500-Type Telephone Set," *Bell Lab. Rec.*, vol. 29, 414–418, Sept. 1951.

7. *Lenkurt Demodulator*, Lenkurt Electric Corp., San Carlos, CA, Dec. 1964, June 1965, and Sept. 1965.

8. *Telecommunication Transmission Engineering*, vols. 1–3, 2nd ed., American Telephone & Telegraph Co., New York, 1977.

9. *Reference Data for Radio Engineers*, 6th ed., Howard W. Sams, Indianapolis, IN, 1977.

10. Roger L. Freeman, *Reference Manual for Telecommunications Engineering*, 2nd ed., Wiley, New York, 1994.

11. *Principles of Electricity Applied to Telephone and Telegraph Work*, American Telephone & Telegraph Co., New York, 1961.

12. *IEEE Standard Test Procedure for Measuring Longitudinal Balance of Telephone Equipment Operating in the Voice Band*, ANSI/IEEE Std. 455-1985, IEEE, New York, 1985.

13. USTA Symposium, Apr. 1970, Open Questions 18–37, USTA, Washington, DC, 1970.

2

TELEPHONE TRANSMISSION

2.1 GENERAL

Section 1.8 introduced the simple telephone connection. An actual telephone call may route through a number of switches in tandem. For instance, a typical call in the United States, coast-to-coast, may traverse four switches after the first serving exchange, as shown in Figure 2.1*a*. The call may encounter congestion at one of the switches and require alternative routing, as shown in Figure 2.1*b*. We will find that we can roughly equate the deterioration in transmission quality to the number of links in tandem. A link connects one switch to another. We call it a link in this instance, not a trunk.

In Figure 2.1, a local serving switch (LSS) carries out the local switching function and provides subscriber interface with the public switch telecommunication network (PSTN). A tandem exchange is a higher-layer in the network hierarchy and provides switching for traffic relations that have low traffic densities; a transit exchange is similar to a tandem switch but operates in the long-distance network. See Ref. 3 for a more detailed discussion of switching nodes.

This chapter delves into problems of telephone transmission. It exclusively treats speech transmission over wire systems. However, implications of a larger network are also included, which involve transmission media other than wire pair. That necessary subscriber loop is also covered.

2.2 TRANSMISSION QUALITY

In Section 2.1 we mentioned *deterioration in transmission quality*. Let us qualify quality in basic terms for each of the media that we may wish to transmit. The point of measurement will be at the far-end receiver.

Speech Telephony. Quality is basically measured on how well we hear the reproduced voice. The unit of measure is loudness rating and optimum values range from 8 to 12 dB. (subjective). (See Section 2.5.2.)

43

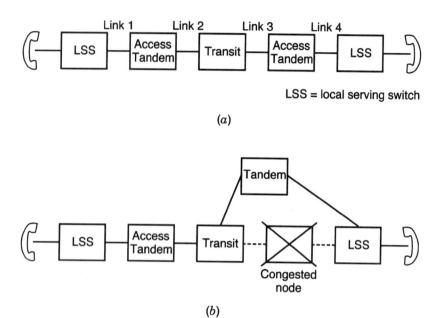

Figure 2.1. (a) A trans–United States telephone connection showing four links in tandem. (b) A trans–United States telephone connection with congestion at one node. Alternative routing to bypass the congested node is illustrated.

Video/TV. Weighted signal-to-noise (S/N) ratio at home receiver of ≥ 45 dB (subjective). (See Chapters 12 and 13.)

Data. Bit error rate (or ratio). (See Chapter 13.) One error in a million bits expressed as 1×10^{-6}. This value is elaborated on in Chapter 13.

2.3 THE TELEPHONE INSTRUMENT

The input/output (I/O) device that provides the human interface with the telephone network is the telephone instrument or subset. It converts sound energy into electrical energy and vice versa. The degree of efficiency and fidelity with which it performs these functions has a vital effect on the quality of telephone service provided. The modern telephone subset consists of a transmitter (mouthpiece), a receiver (earpiece), and an electrical network for equalization, sidetone circuitry, and devices for signaling and supervision.

In North America modern telephone subsets are based on the standard 500-type subset (circa 1953). Some of the problems that confronted the designers of the 500-type set were as follows:

1. The set is connected to a two-wire loop, yet it must separate the transmitter signal from that of the receiver, directing each of them to a different path.

2. Some of the transmitter signal power has to be diverted to the receiver to create sidetone.
3. The transmitter has to be fed direct current from the loop; the receiver has to be isolated from this current.
4. The set has to present low dc resistance to the loop in order to activate the supervisory circuit at the local switch when the handset is removed from the switchhook (taken out of its cradle). On the other hand, the ac impedance of the set has to match the large loop impedance to increase the efficiency of the coupling between the loop and the set.
5. The set has to interface with a wide variety of loops and yet perform properly without any need for customized adjustment.

These conflicting requirements had to be satisfied by the least-expensive available means. The result of that design is shown in Figure 2.2. The three-winding transformer serves as a four-wire to two-wire hybrid to satisfy the first requirement. Sidetone is generated as an echo by purposely mismatching this hybrid. The fourth requirement is met by using the transformer as a matching network. The capacitors isolate the receiver from the direct current, though that is not their only function. The varistors (V_1 and V_2 in the figure) help meet the last requirement.

Varistors are nonlinear devices whose dc resistance and ac impedance (which is resistive in the voice-frequency range), though different from each other, are decreasing functions of the current through them. In this application the signal currents are much smaller than the direct current, justifying

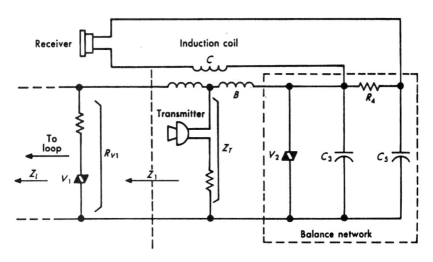

Figure 2.2. Transmission circuit of the 500-type telephone set. (From Ref. 1, Figure 11.3; copyright © 1982 by Bell Telephone Laboratories. Reprinted with permission.)

the use of the small-signal linear models for the ac behavior of the varistors. The direct current acts as a bias, setting the value of the ac impedance (Ref. 1).

2.3.1 Transmitters

The transmitter converts acoustic energy into electric energy by means of a carbon granule transmitter. The transmitter requires a dc potential, usually on the order of 3–5 V, across its electrodes. We call this the talk battery, and in modern systems it is supplied over the line (central battery) from the switch (see Section 1.8). Current from the battery flows through the carbon granules or grains when the telephone is lifted off its cradle (off-hook). When sound impinges on the diaphragm of the transmitter, variations of air pressure are transferred to the carbon, and the resistance of the electrical path through the carbon changes in proportion to the pressure. A pulsating direct current results. The frequency response of carbon transmitter peaks between 800 and 1000 Hz.

2.3.2 Receivers

A typical receiver consists of a diaphragm of magnetic material, often soft iron alloy, placed in a steady magnetic field supplied by a permanent magnet, and a varying magnetic field, caused by the voice currents flowing through the voice coils. Such voice currents are alternating (ac) in nature and originate at the far-end telephone transmitter. These currents cause the magnetic field of the receiver to alternately increase and decrease, making the diaphragm move and respond to the variations. As a result an acoustic pressure wave is set up, reproducing, more or less exactly, the original sound wave from the distant telephone transmitter. The telephone receiver, as a converter of electrical energy to acoustic energy, has a comparatively low efficiency, on the order of 2–3%.

Sidetone is the sound of the talker's voice heard in his/her own receiver. The sidetone level must be controlled. When the level is high, the natural human reaction is for the talker to lower his/her voice. Thus by regulating the sidetone, talker levels can be regulated. If too much sidetone is fed back to the receiver, the output level of the transmitter is reduced owing to the talker lowering his/her voice, thereby reducing the level (voice volume) at the distant receiver, deteriorating performance.

2.3.3 The Speech Signal

The speech signal arriving on the wire pair at the local serving switch has most of its energy concentrated in the band 100–5000 Hz. This is the combined result of both the characteristics of the human voice and the band-limiting introduced by the typical telephone set and subscriber loop.

The bandwidth is much wider than is needed for intelligibility and it is further constrained by filters in multiplex equipment to 300–3400 Hz for the standard CCITT voice channel; Bell Telephone Laboratories (Ref. 1) states a range from 200 to 3300 Hz.

Speech energy is very hard to measure for several reasons. The basic speech signal is amplitude modulated at a syllabic rate (several times per second). There are speaker pauses between phrases and sentences. These result in speech energy being concentrated in spurts of about 1-s average duration separated by gaps of a second or so. Thus a speech signal consists of randomly spaced bursts of energy of random duration.

The VU (volume unit) meter measures the magnitudes of speech peaks and obtains estimates of long-term average speech power. It averages the talker power over a period of time corresponding to approximately one syllable. When a speech signal is measured in VU, the person making the measurement visually averages five or six of the highest meter readings over a 3- to 10-second interval. The average of ten such measurements defines the signal magnitude in VU. The VU measurement thus roughly corresponds to the average of the peak-volume syllables.

Empirical relationships have been established between VU measurements and the peak and long-term average power of talkers using a telephone system. Peak powers have been found to be approximately 15 dB greater than the power-equivalent of the VU measurement. Approximate long-term average powers are obtained from VU measurements by first converting to the average power for a continuous talker. The following empirically derived relationship is used:

$$\text{Average power of a continuous talker} \approx \text{VU} - 1.4 \text{ dBm} \qquad (2.1)$$

Studies of speech signal magnitudes in the telephone system use a measure called *equivalent peak level* (EPL). The EPL can be described as approximately the 95% point on the cumulative probability distribution of instantaneous talker power. Periods of silence are excluded from the distribution. The EPL of a talker is measured automatically by equipment consisting of a power meter, an analog-to-digital (A/D) converter, and a personal computer (PC). The EPL is thus obtained without subjective interpretation.

In a survey of speech signal power in the telephone network, both EPLs and long-term average conversational powers were measured (*Bell Syst. Tech. J.*, Sept. 1978). The survey results indicate that EPLs of long-distance calls are characterized by a mean of -11.1 dBm0 and a standard deviation of 4.7 dB. Similarly, the average power for long-distance calls are characterized by a mean level of -26.5 dBm0 and a standard deviation of 5.3 dB. The corresponding levels for local calls are a mean EPL of -11.8 dBm0 with a standard deviation of 4.7 dB and a mean average power of -26.5 dBm0 with a standard deviation of 5.4 dB. The distribution of EPL and average power are approximately log-normal (i.e., normal when the independent variable is expressed in dB).

Signal-to-noise ratio has limited application in speech telephony. As we mentioned, individual talker signal power fluctuates widely so that the S/N ratio is far from constant from one call to another. In addition, subjective tests have shown that noise or any disturbance on a telephone channel is most annoying during quiet intervals when there is no one talking. Thus it is much more important to specify idle channel noise than S/N for speech telephony. Data circuits are another matter where signal-to-noise ratios are very significant. The reason for this is that most data circuits employ a tone (usually at 1800 Hz) that has a constant amplitude and 100% duty cycle (Ref. 1).

2.4 INTRODUCTION TO THE SUBSCRIBER LOOP OR TELEPHONE LOOP

We speak of the telephone subscriber as the user of the subset. As mentioned in Section 1.8, subscribers' telephone sets are interconnected via a switch or network of switches. Present commercial telephone service provides for transmission and reception on the same pair of wires that connect the subscriber to the local switch. Let us now define some terms.

The pair of wires connecting the subscriber to the local switch that serves him/her is the *subscriber loop*. It is a dc loop in that it is a wire pair typically supplying a metallic path for the following:

1. Talk battery for the telephone transmitter.
2. An ac ringing voltage for the bell on the telephone instrument supplied from a special ringing source voltage.
3. Current to flow through the loop when the telephone instrument is taken out of its cradle, telling the switch that it requires "access" and causing line seizure at the switching center.
4. The telephone dial that, when operated, makes and breaks the dc current on the closed loop, which indicates to the switching equipment the number of the distant telephone with which communication is desired.

The typical subscriber loop is supplied battery by means of a battery feed circuit at the serving switch. Such a circuit is shown in Figure 2.3. One important aspect of battery feed is lie balance. The telephone battery voltage has been fairly well standardized around the world at −48 Vdc. It is a negative voltage to minimize cathodic reaction.

Figure 2.4 shows the functional elements of the subscriber loop, local switch termination, and the subscriber subset.

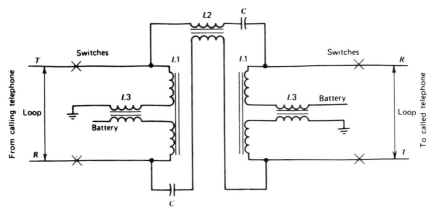

Figure 2.3. Battery feed circuit. Battery and ground are fed through inductors *L*3 and *L*1 through switch to loops. (From Ref. 4; copyright © 1961 by Bell Telephone Laboratories. Reprinted with permission.)

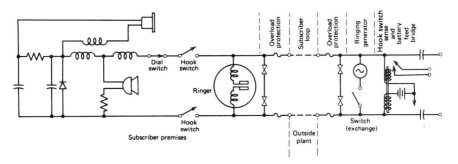

Figure 2.4. Signaling with a conventional telephone subset. Note functions of hook switch, dial, and ringer.

2.4.1   Subscriber Loop Length Limits

It is desirable from an economic viewpoint to permit subscriber loop lengths to be as long as possible. Thus the subscriber area served by a single switching center may be much larger. As a consequence, the total number of switches or telephone central offices may be reduced to a minimum. For instance, if loops were limited to 4 km in length, a switching center could serve all subscribers within a radius of something less than 4 km. If 10 km were the maximum loop length, the radius of an equivalent area that one office could cover would be extended an additional 6 km, out to a total of nearly 10 km. It is evident that to serve a large area, fewer switches (switching centers) are required for the 10-km situation than for the 4-km. The result is fewer buildings, less land to buy, fewer locations where maintenance is required, and all the benefits accruing from greater centralization, which become even more evident as subscriber density decreases, such as in rural areas.

The two basic criteria that must be considered when designing subscriber loops, and that limit their length, are the following:

- Attenuation limits
- Signaling limits

Attenuation in this case refers to loop loss in decibels (or nepers) at

- 1000 Hz in North America
- 800 Hz in Europe and many other parts of the world

As a loop is extended in length, its loss at reference frequency increases. It follows that at some point as the loop is extended, the level will be attenuated such that the subscriber cannot hear sufficiently well.

Likewise, as a loop is extended in length, some point is reached where signaling (supervision) is no longer effective. This limit is a function of the IR drop of the line. We know that R increases as length increases. With today's modern telephone sets, the first to suffer is usually the "supervision." This is a signal sent to the switching equipment requesting "seizure" of a switch circuit and, at the same time, indicating the line is busy. *Off-hook* is a term more commonly used to describe this signal condition. When a telephone is taken "off-hook" (i.e., out of its cradle), the telephone loop is closed and current flows, closing a relay at the switch. If current flow is insufficient, the relay will not close or it will close and open intermittently (chatter) such that line seizure cannot be effected.

Signaling (supervision) limits are a function of the conductivity of the cable conductor and its diameter or gauge. For this introductory discussion we can consider that the loss limits are controlled by the same parameters.

Consider a copper conductor. The larger the conductor, the higher the conductivity, and thus the longer the loop may be for signaling purposes. Copper is expensive, so we cannot make the conductor as large as we would wish and extend subscriber loops long distances. These economic limits of loop length are discussed in detail below. First we must describe what a subscriber considers as hearing sufficiently well, which is embodied in "transmission loss design" (regarding subscriber loop).

2.5 TRANSMISSION FACTORS IN SPEECH TELEPHONY

2.5.1 Loudness

Hearing "sufficiently well" on a telephone connection is a subjective matter under the blanket heading of customer satisfaction (regarding transmission).

The overriding factor here is level, which can be rated to "volume" or how loud one hears the distant subscriber. A secondary but important factor is idle-channel noise. This is the noise one hears between speech spurts or while listening. (See Table 2.3.)

2.5.2 Methods of Rating a Telephone Connection for Loudness

2.5.2.1 Reference Equivalent. Historically, *reference equivalent* was an artifice used to rate a telephone connection. The unit of measure was the decibel. The lower the decibel value, the louder the connection. The reference equivalent (RE) value consisted of three factors: the transmit reference equivalent (TRE), receive reference equivalent (RRE), and the intervening connection losses, all in dB. Thus the overall reference equivalent (ORE) equals the TRE + RRE + the intervening losses in a connection. CCITT recommended no more than 33 dB ORE for better than 97% of international connections. The median ORE for a typical U.S. long-distance connection had a median of about 26 dB, and a great percentage of listeners rated a 6-dB ORE (or smaller) too loud. To aid in ORE calculations, a country's standard telephone set TRE and RRE were rated using a CCITT-approved laboratory standard called a "NOSFER."

2.5.2.2 Corrected Reference Equivalent. In 1984 CCITT introduced the *corrected reference equivalent* (CRE). It had been found that if a local system were to be connected with a circuit having a loss of x and without distortion, the reference equivalent of the system increases by a value smaller than x.

CCITT, in Rec. G.111 (Ref. 5), stated that the introduction of a new subjective test method was not justified, since reference equivalent values (q) are available for many local systems and the corresponding CREs may be calculated by the formula below. The CRE of a local system or complete system is termed y. Then

$$y = 0.0082q^2 + 1.148q - 0.48 \quad \text{(dB)} \tag{2.2}$$

where, again, q is the conventional value in dB for reference equivalent discussed in Section 2.5.2.1.

2.5.2.3 Loudness Rating. CCITT in its IXth Plenary Assembly (1988) revised its standard for telephone speech quality and now recommends the use of "loudness rating" (LR). It is conceptually similar to reference equivalent. OLR (overall loudness rating) has become the international standard for measuring customer satisfaction of a speech telephone connection.

The ITU-T Organization discusses loudness loss (and loudness rating) regarding objective and subjective models. They describe two basic objectives (Ref. 15):

1. "To give a meaningful and reliable measure of the speech transmission quality for the acoustic loss between the talking subscriber's mouth and the listening subscriber's ear."
2. "To characterize the parts of a connection by simple electroacoustical and electrical quantities in such a way that the measure for each part can be easily combined by simple algebraic addition to give the measure for the total connection."

To introduce loudness rating (LR) and sidetone masking rating (STMR), ITU-T provides the following definitions:

Overall Loudness Rating (OLR). The loudness loss between the speaking subscriber's mouth and the listening subscriber's ear via a connection.

Send Loudness Rating (SLR). The loudness loss between the speaking subscriber's mouth and an electric interface in the network. The loudness loss here is defined as the weighted (dB) average of driving sound pressure to measured voltage.

Receive Loudness Rating (RLR). The loudness loss between an electric interface in the network and the listening subscriber's ear. The loudness loss is here defined as the weighted (dB) average of the driving electromotive force (emf) to measured sound pressure.

Circuit Loudness Rating (CLR). The loudness loss between two electrical interfaces in the network, each interface terminated by its nominal impedance, which may be complex. The loudness loss here is equivalent to the weighted (dB) average of the composite electrical loss.*

The following concerns the electric sidetone path, which is determined by two parameters.

Sidetone Masking Rating (STMR). The loudness of a telephone sidetone path compared with the loudness of the intermediate reference system (IRS), in which the comparison is made incorporating the speech signal heard via the human sidetone path (L_{MEHS}) as a masking threshold.

Listener Sidetone Rating (LSTR). The loudness of a diffuse room noise source as heard at the subscriber's (earphone) ear via the electric sidetone path in the telephone instrument, compared with the loudness

---

*This is the dB attenuation (e.g., attenuation or loss of the intervening telecommunications network along the connectivity).

of the intermediate reference system (IRS), in which the comparison is made incorporating a speech signal heard via the human sidetone path (L_{MEHS}) as a masking threshold (Ref. 5).

2.5.2.3.1 *Loudness Ratings of Telephone Sets.* Conventional analog telephone sets are laboratory rated objectively by an instrument setup described in ITU-T Recs. P.64, P.65, and P.79 for the physical implementation and computational algorithm. The measurement setup provides a representative current feeding bridge and may or may not include different lengths of (artificial) unloaded subscriber lines. The parameters measured are SLR, RLR, and STMR.

2.5.2.3.2 *LRs for Actual Connections.* LRs are evaluated by their individual parts of a total connection. Figure 2.5 shows LRs in a normal speech connection, mouth to ear.

The send and receive loudness ratings of the telephone sets themselves are designated SLR(Set) and RLR(Set), respectively, and the circuit loudness ratings are designated CLR_n. Then at interface $1 = n$ in the direction S to R we express

$$\text{SLR} = \text{SLR(Set)} + \sum_{i=1}^{n} \text{CLR}_i \qquad (2.3)$$

$$\text{RLR} = \text{RLR(Set)} + \sum_{i=n+1}^{N} \text{CLR}_i \qquad (2.4)$$

$$\text{OLR} = \text{SLR} + \text{RLR} \qquad (2.5)$$

The CLRs are equal to the insertion loss of the circuits at the reference frequency of 1020 Hz, using the nominal impedance appropriate to the particular interface. Unless there is a highly skewed attenuation distortion, the CLR is then the average decibel loss over the frequency band 300–3400 Hz.

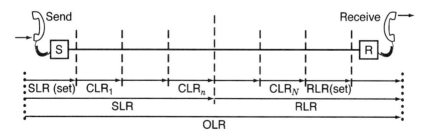

Figure 2.5. LRs in a normal speech connection, mouth to ear. (From Ref. 2, Figure 1/3.1.)

The optimum loudness of a connection is about OLR = 9 dB loss. For this value the average insertion acoustic loss from a speaker's mouth to the listener's ear is about 0 dB, taken over a logarithmic frequency scale.

For unloaded subscriber cable sections the CLRs do not change (with sufficient accuracy for planning purposes) the composite loss at the frequency 1020 Hz. This value can also be estimated from the cable parameters by the following rule-of-thumb relationship:

$$\text{CLR} = L(K\sqrt{RC}) \tag{2.6}$$

where R = cable resistance in ohm/km
   C = cable capacitance in nF/km
   K = a constant that is dependent on the cable termination:
      $K = 0.014$ for $Z_0 = 900 \ \Omega$
      $K = 0.015$ for $Z_0 = 600 \ \Omega$
      $K = 0.016$ for complex impedance
   Z_0 = termination impedance of the cable
   L = length in kilometers

2.5.2.3.3 Recommended Values for LRs

On International Connections and on National Systems Seen to/from Virtual Interconnecting Points (VICPs). The VICP (see Figure 2.6 for definition of VICP) has a 0 dBr relative level in the send direction and 0 dBr (-0.5 dBr for analog and mixed analog/digital circuits) in the receive direction. The relationship between the VICP and the previously used VASP (virtual analog switching point) is discussed in ITU-T Rec. G.111. A summary of recommended LR values on international connections is given in Table 2.1.

LRs on National Systems. The appropriate subdivision of the overall loudness requirement is obtained by the following long-term objectives referred to the VICP (see Figure 2.6).

SLR 7–9 dB

RLR 1–3 dB

For the STMR, the preferred range is 7–12 dB (two-wire sets) or 15 $\pm$5 dB (for digital and four-wire sets), while listener sidetone rating values above 13 dB are recommended. It is noted that for connections totally free of echo and sidetone problems, investigations have shown the optimum OLR to be somewhat lower, about 5 dB, but the optimum is rather flat so that moderate deviations from the given value have little subjective effect. Maximum and mean LR values for national systems are given in Table 2.2.

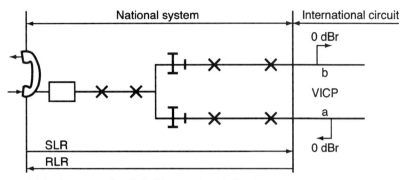

(*a*) In the case of a digital International circuit

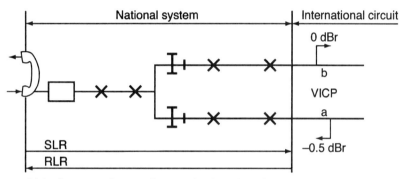

(*b*) In the case of an analog or mixed analog digital circuit

Figure 2.6. Reference points for defining the SLR and RLR for a national system. (From Ref. 5, Figure 1/G.111, ITU-T Rec. G.111.)

TABLE 2.1   Values (dB) of SLR, RLR, and OLR Cited in ITU-T Recs. G.111 and G.121

	Recommended in 1984			Recommended in 1988					
	SLR	RLR	OLR	SLR		RLR		CLR	OLR
	VASP	VASP		0 dBr	VASP	0 dBr	VASP		
Optimum value			≈ 5						≈ 10
Traffic-weighted mean values:									
Long-term objective (minimum)	6.5	−2.5	8	7	10.5	1	−3	(Note 1)	8
(maximum)	8	−1	11	9	12.5	3	−1	(Note 1)	12
Short-term objective (maximum)	14	2.5	20.5	15	18.5	6	2	(Note 1)	21
Maximum values for an average-sized country	20	9		16.5	20	13	9	$n \times 0.5$ (Note 2)	
Minimum for sending	2			−1.5[a]	2				

Note 1. CLR = 0 for a digital international circuit, 0.5 dB for an analog one. The average number of international circuits is about 1.
Note 2. n is the number of analog international circuits.
Note 3. The VASPs are defined in Rec. G.101. (Ref. 12)
[a]The 1.5-dB value is substituted by +2 dB (1992).
Source: Reference 2, Table 1/3.1.

TABLE 2.2 Maximum and Mean LRs for National Systems

Case	Loss (800 Hz)		Sending		Receiving		Means over Subscribers	
	Subscriber Line	Local Circuit	Local System SLRL	National System SLRN	Local System RLRL	National System RLRN	SLRN	RLRN
a	9.5	0	13.5	15	−0.5	8	8.5	2
b	8	2	12	16	−2	9		
c	7	3	11	16	−3	9		
CCITT limits				16.5		13		
CCITT maxima for the traffic-weighted means: long-term objective							7–9	1–3
short-term objective							7–15	1–6

Source: Reference 2, Table 2/3.1.

TABLE 2.3 Opinion Results, Circuit Noise

Circuit Noise at Point 0 dB RLR (dBmp)	Representative Opinion Results[a]	
	Percent "Good + Excellent"	Percent "Poor + Bad"
− 65	> 90	< 1
− 60	85	2
− 55	75	5
− 50	65	10
− 45	45	20

[a]Based on opinion relationship derived from the transmission quality index (see Ref. 6, Annex A).
Source: Reference 6, Table 3/P.11, ITU-T Rec. P.11.

2.5.2.4 Circuit Noise. As discussed earlier, circuit noise is another impairment to speech transmission. Such noise can be made up of components of thermal noise, intermodulation noise, impulse noise, and single-frequency tones.

The subjective effect of circuit noise measured at a particular point in a telephone connection depends on the electrical-to-acoustical loss or gain from the point of measurement to the output of the telephone receiver. As a convenience in assessing the contributions from different sources, circuit noise is frequently referred to the input of a receiving system with a specified loudness rating. A common reference point is the input of a receiving system having a RLR of 0 dB. When circuit noise is referred to this point, circuit noise values less than − 65 dBmp* have little effect on transmission quality in typical room noise environments. Transmission quality decreases with higher values of circuit noise. Table 2.3 provides opinion results for circuit noise referenced to the 0-dB RLR point.

*dBm psophometrically weighted.

2.6 TELEPHONE NETWORKS

2.6.1 General

The next logical step in our discussion of telephone transmission is to consider the large-scale interconnection of telephones. As we have seen, subscribers within a reasonable distance of one another can be interconnected by wire lines and we can still expect satisfactory communication. A switch is used so that a subscriber can speak with some other discrete subscriber as he/she chooses. As we extend the network to include more subscribers and circumscribe a wider area, two technical/economic factors must be taken into account:

1. More than one switch must be used.
2. Wire-pair transmission losses on longer circuits must be offset by amplifiers, or the pairs must be replaced by other, more efficient means.

Let us accept item 1. The remainder of this section concentrates on item 2. The reason for the second statement becomes obvious when the salient point of Section 2.5 regarding loudness loss is reviewed.

2.6.2 Basic Considerations

Considering item 2 in Section 2.6.1, what are some of the more common approaches that may be used to extend the network? We may use:

1. Coarser gauge cable (larger diameter conductors).
2. Amplifiers in the present wire-pair system; also inductive loading.
3. Carrier transmission techniques (Chapter 3).
4. Radio transmission techniques (Chapters 4–7).
5. Fiber optic transmission (Chapter 10).

Items 1 and 2 may be applied to extend a subscriber loop but are not economically feasible beyond the subscriber loop application. Items 3, 4, and 5 become attractive for multichannel transmission over longer distances, particularly so after the concentration that takes place in a local serving switch.* Discussion of the formation of a multichannel signal is left for Chapter 3. Particularly in North America, such a multichannel transmission technique is referred to as *carrier* transmission. Multichannel or multiplex signal formats are used over wire pairs, radio systems, and fiber optics.

---

*Local serving switches concentrate for outgoing trunk calls and expand for incoming trunk calls. This is aptly described in Ref. 3, Chap. 4.

These multichannel or multiplex signals require four-wire transmission techniques. In other words, the transmit portion of the signal uses one wire pair and the receive portion, a second wire pair. Two-wire/four-wire transmission is described in Section 2.6.3.

2.6.3 Two-Wire/Four-Wire Transmission

2.6.3.1 Two-Wire Transmission. By its basic nature a telephone conversation requires transmission in both directions. When both directions are carried on the same wire pair, we call it two-wire transmission. The telephones in our homes and offices are connected to a local switching center by means of two-wire circuits. A more proper definition for transmitting and switching purposes is that when oppositely directed portions of a single telephone conversation occur over the same electrical transmission channel or path, we call this two-wire operation.

2.6.3.2 Four-Wire Transmission. Carrier and radio systems require that oppositely directed portions of a single conversation occur over separate transmission channels or paths (or using mutually exclusive time periods). Thus we have two wires for the transmit path and two wires for the receive path, or a total of four wires for a full-duplex (two-way) telephone conversation. For almost all operational telephone systems, the end instrument (i.e., the telephone subset) is connected to its intervening network on a two-wire basis.*

Nearly all long-distance telephone connections traverse four-wire links. From the near-end user the connection to the long-distance network is two wire. Likewise, the far-end user is also connected to the long-distance network via a two-wire link. Such a long-distance connection is shown in Figure 2.7. Schematically the four-wire interconnection is shown as if it were wire line, single channel with amplifiers. More likely, it would be multichannel carrier on cable and/or multiplex on radio. However, the amplifiers in the figure serve to convey the ideas that this chapter considers.

As shown in Figure 2.7, conversion from two-wire to four-wire operation is carried out by a terminating set, more commonly referred to in the industry as a *term set*. A term set is a hybrid circuit usually consisting of two transformers connected as shown in Figure 2.8. In this figure transformers T_1 and T_2 each consist of at least three tightly coupled windings. If $Z_1 = Z_2$ and $Z_3 = Z_4$, a proper choice of turns ratios will make port 1 conjugate to port 2 and port 3 conjugate to port 4. This means that if Z_1 is a source delivering power to port 1, a negligible part of this power will be received by impedance Z_2 and vice versa. Power flowing into the circuit at either port 1 or port 2 will be delivered to impedances Z_3 and Z_4 equally.

---

*A notable exception is the U.S. military telephone network AutoVon, where end users are connected on a four-wire basis.

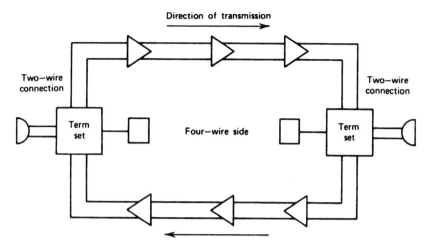

Figure 2.7. Typical long-distance telephone connection.

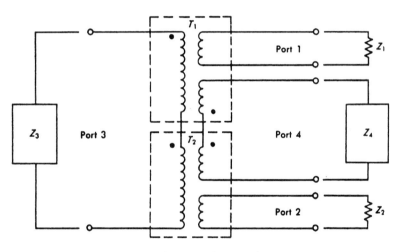

Figure 2.8. Hybrid circuit consisting of two transformers. (From Ref. 1, Figure 2-5; copyright © 1982 by Bell Telephone Laboratories. Reprinted with permission.)

In our practical application, Z_3 is a two-wire port, often extending through a switch to any one of a large family of subscriber lines.* Z_4 is a fixed network (often called a compromise network) whose only function is to match Z_3 and provide the necessary conjugacy (i.e., port to opposite port). Impedances Z_1 and Z_2 represent a four-wire line using separate pairs for the two directions of transmission.

*The family of subscriber lines could be 10,000 or more possible two-wire connectivities.

The isolation between ports 1 and 2, called the *transhybrid loss* or *transhybrid balance*, is a measure of the effectiveness of this circuit and may be as great as 50 dB when Z_3 is well matched by Z_4. Unfortunately, for a common application of interest here, where the two-wire side of the hybrid connects through a switch to any one of a large family of users, the effectiveness drops to 11 dB as a median with a 2- or 3-dB standard deviation. However, if the local serving switch is digital, a hybrid is inserted on each subscriber line prior to entering the appropriate digital channel bank and can be optimized, if required. Here we mean that Z_4 can be "adjusted" to more nearly match Z_3 (Ref. 1).

2.6.3.3 *Operation of a Hybrid—A Plain Discussion.* As we described in the paragraphs above, a hybrid, for telephone work (at VF), is a transformer. For a simplified description, a hybrid may be viewed as a power splitter with four sets of wire-pair connections. A functional block diagram of a hybrid is shown in Figure 2.9. Two of these wire-pair connections belong to the four-wire path, which consists of a transmit pair and a receive pair. The third pair is the connection to the two-wire line (L) to the subscriber subset. The last wire pair connects the hybrid to a resistance–capacitance balancing network (N), which electrically balances the hybrid with the two-wire connection (i.e., its conjugate) to eventually connect with a subscriber's subset over the frequency range of the balancing network. An artificial line may also be used for this purpose.

The hybrid function permits signals to pass from any pair through the transformer to adjacent pairs (conjugate pairs) but blocks signals to the opposite pairs as shown in Figure 2.9. Signal energy entering from the four-wire side divides equally, half dissipating into the balancing network (N) and half going to the desired two-wire connection. *Ideally*, no signal energy in this path crosses over to the four-wire transmit side. This is an important point, which we take up later.

Signal energy entering from the two-wire subset connection divides equally, half of it dissipating in the impedance of the four-wire side receive path, and half going to the four-wire side transmit path. Here the *ideal* situation is that no energy is to be dissipated by the balancing network (i.e., there is a perfect

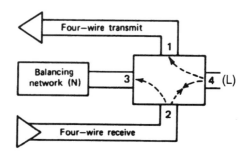

Figure 2.9. Operation of a hybrid transformer.

balance). The balancing network is supposed to display the characteristic impedance of the two-wire line (subscriber connection) to the hybrid.

The reader notes that in the description of the hybrid, in every case, ideally half of the signal energy entering the hybrid is used to advantage and half is dissipated, or wasted. Also keep in mind that any passive device inserted in a circuit such as a hybrid has an insertion loss. As a rule of thumb we say that the insertion loss of a hybrid is 0.5 dB. Hence there are two losses here of which the reader must not lose sight:

Hybrid insertion loss   0.5 dB
Hybrid dissipation loss   3.0 dB (half-power)
_____
3.5 dB total

As far as this chapter is concerned, any signal passing through a hybrid suffers a 3.5-dB loss. Resistive hybrids have higher losses and electronic hybrids have lower losses.

2.6.4 Echo, Singing, and Design Loss

2.6.4.1 General. The operation of the hybrid with its two-wire connection on one end and four-wire connection on the other leads us to the discussion of two phenomena that, if not properly designed for, may lead to major impairments in communication. These impairments are echo and singing.

Echo. As the name implies, echo in telephone systems is the return of a talker's voice. The returned voice, to be an impairment, must suffer some noticeable delay.

Thus we can say that echo is a reflection of the voice. Analogously, it may be considered as that part of the voice energy that bounces off obstacles in a telephone connection. These obstacles are impedance irregularities, more properly called impedance mismatches.

Echo is a major annoyance to the telephone user. It affects the talker more than the listener. Two factors determine the degree of annoyance of echo: its loudness and how long it is delayed.

Singing. Singing is the result of sustained oscillations due to positive feedback in telephone amplifiers or amplifying circuits. Circuits that sing are unusable and promptly overload analog multichannel carrier equipment (FDM, see Chapter 3).

Singing may be thought of as echo that is completely out of control. This can occur at the frequency at which the circuit is resonant. Under such conditions the circuit losses at the singing frequency are so low that oscillation will continue even after the impulse that started it ceases to exist.

The primary cause of echo and singing generally can be attributed to the mismatch at a hybrid between the balancing network and its two-wire connection associated with the subscriber loop. It is at this point that the major impedance mismatch usually occurs and an echo path exists. To understand the cause of the mismatch, remember that we often have at least one two-wire switch between the hybrid and the subscriber. Ideally the hybrid balancing network must match each and every subscriber line to which it may be switched. Obviously the impedances of the four-wire trunks (lines) may be kept fairly uniform. However, the two-wire subscriber lines may vary over a wide range. The subscriber loop may be long or short, may or may not have inductive loading (see Section 2.7.4), and may or may not be carrier-derived (see Chapter 3). The hybrid imbalance causes signal reflection or signal "return." The better the match, the more the return signal is attenuated. The amount that the return signal (or reflected signal) is attenuated is called the *return loss* and is expressed in decibels. The reader should remember that any four-wire circuit may be switched to hundreds or even thousands of different subscribers. If not, it would be a simple matter to match the four-wire circuit to its single subscriber through the hybrid. This is why the hybrid to which we refer has a compromise balancing network rather than a precision network. A compromise network is usually adjusted for a compromise in the range of impedance that is expected to be encountered on the two-wire side.

Let us consider now the problem of match. For our case the impedance match is between the balancing network N and the two-wire line L (see Figure 2.9). With this in mind,

$$\text{Return loss}_{dB} = 20 \log_{10} \frac{Z_N + Z_L}{Z_N - Z_L} \qquad (2.7)$$

If the network perfectly balances the line, $Z_N = Z_L$, and the return loss would be infinite.

The return loss may also be expressed in terms of the reflection coefficient:

$$\text{Return loss}_{dB} = 20 \log_{10} \frac{1}{\text{reflection coefficient}} \qquad (2.8)$$

where the reflection coefficient is the ratio of reflected signal to incident signal.

The CCITT uses the term *balance return loss* (see CCITT Rec. G.122, Ref. 7) and classifies it as two types:

1. Balance return loss from the point of view of echo.* This is the return loss across the band of frequencies from 300 to 2500 Hz.
2. Balance return loss from the point of view of stability. This is the return loss between 0 and 4000 Hz.

The band of frequencies that is most important from the standpoint of echo for the voice channel is that between 300 and 2500 Hz. A good value for the

---

*Called echo return loss (ERL) in via net loss (VNL; North American practice, Section 2.6.5.1), both use a weighted distribution of level.

echo return loss (ERL) for a toll telephone plant is 11 dB, with values on some connections dropping to as low as 6 dB. For the local telephone network, CCITT recommends better than 6 dB, with a standard deviation of 2.5 dB (Ref. 17).

For frequencies outside the 300–2500-Hz band, return loss values often are below the desired 11 dB. For these frequencies we are dealing with return loss from the point of view of stability. CCITT recommends that balance return loss from the point of view of stability (singing) should have a value of not less than 2 dB for all terminal conditions encountered during normal operation (CCITT Rec. G.122, p. 2.2). For further information the reader should consult Appendix A of CCITT Recs. G.122 and G.131.

Echo and singing may be controlled by

- Improved return loss at the term set (hybrid)
- Adding loss on the four-wire side (or on the two-wire side)
- Reducing the gain of the individual four-wire amplifiers

The annoyance of echo to a subscriber is also a function of its delay. Delay is a function of the velocity of propagation of the intervening transmission facility. A telephone signal requires considerably more time to traverse 100 km of a voice-pair cable facility, particularly if it has inductive loading, than 100 km of radio facility.

Delay is expressed in one-way or round-trip propagation time measured in milliseconds. CCITT recommends that if the mean round-trip propagation time exceeds 50 ms for a particular circuit, an echo canceler should be used. Bell System practices in North America use 45 ms as a dividing line. In other words, where the echo delay is less than that stated above, the echo will be controlled by adding loss.

An echo canceler utilizes an adaptive filter to model the near-end echo path and predict the echo that will result from the reflection of the far-end signal (usually at the far-end hybrid). A replica of the predicted signal is subtracted from the actual echo signal and, through an interactive process, the canceler converges toward full cancellation. Because only the echo signal replica is subtracted in the sending direction, signals originating at the near-end pass through unimpeded.

Figure 2.10 shows the echo path on a four-wire circuit.

Echo Return Loss, North American Definition. Return loss, as we discussed earlier, is a measure of impedance match or mismatch. Return losses on the order of 20–33 dB are indicative of a good match, whereas match deteriorates as the dB value gets smaller.

Our main area of interest here is the mismatch between a hybrid (term set) and a telephone channel. Return loss is defined rigidly in terms of the ratio of sum and difference of the complex impedances where we would expect an impedance discontinuity (equation 2.7). Because of the complexity

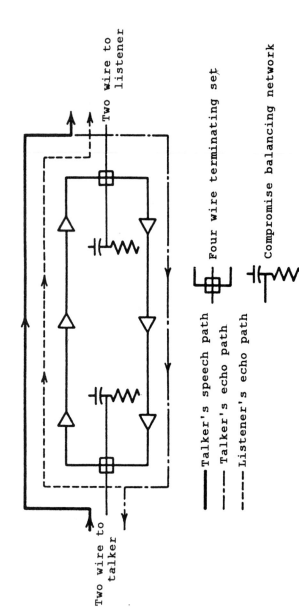

Two wire to listener

Two wire to talker

■■■ Talker's speech path
—·— Talker's echo path
----- Listener's echo path

⊞ Four wire terminating set

⊣⊢〰 Compromise balancing network

Figure 2.10. Echo paths in a four-wire circuit.

64

of phase relationships in the incident and reflected voltage waves, it makes it impractical to express return loss over a band of frequencies except by averaging the performance over the band of interest. Echo return loss (ERL), then, is weighted *power* averaged over the band 500–2500 Hz.

One method of determining return loss in this situation is the measurement of single-frequency return loss in 500-Hz increments from 500 to 2500 Hz. Other methods sweep each 500-Hz subband or use random noise measurements. One weighting method halves the value of each frequency extreme (e.g., 500 and 2500 Hz). If we look at the four bands of interest, calling then $B(1)$, $B(2)$, $B(3)$, and $B(4)$, and the five frequencies $f(1), \ldots, f(5)$, we can then define ERL as

$$\mathrm{ERL} = 10 \log \frac{B}{4B} \left(\frac{\mathrm{RL}_{f1} + \mathrm{RL}_{f5}}{2} \mathrm{RL}_{f2} + \mathrm{RL}_{f3} + \mathrm{RL}_{f4} \right) \mathrm{dB} \quad (2.9)$$

where $B = 500$ Hz and RL_{fn} is the return loss for each of the five frequencies expressed as a *power* ratio (Ref. 13).

In the North American switched network, ERL measurements are made at various switching centers (exchanges) and impedances are corrected to improve performance where required. The measurements are made against standard impedances. Such measurements are called *through balance* and *terminal balance*.

The reader should note the subtle differences between CCITT and North American practice.

Singing Return Loss. Singing return loss measurements are made to give assurance of the necessary stability and are made at all sample frequencies at which a circuit may display instability. Instability usually occurs in the bands from 200 to 500 Hz and 2500 to 3200 Hz (North American practice)—for the CCITT voice channel, the lower frequency is 300 Hz and the higher is 3400 Hz. It has been found that frequencies in the band 500–2500 Hz that meet ERL requirements are usually satisfactory for singing return loss.

Transmission Design to Control Echo and Singing. As stated previously, echo is an annoyance to the subscriber. Figure 2.11 relates echo path delay to echo path loss. The curve in Figure 2.11 is a group of points at which the average subscriber will tolerate echo as a function of its delay. Remember that the greater the return signal delay, the more annoying it is to the telephone talker (i.e., the more the echo signal must be attenuated). For instance, if the echo path delay on a particular circuit is 20 ms, an 11-dB loss must be inserted to make echo tolerable to the talker. The careful reader will note that the 11 dB designed into the circuit will increase the end-to-end reference equivalent by that amount, something quite undesirable. The effect of loss design on reference equivalents and the trade-offs available are discussed later.

To control singing, all four-wire paths must have some loss. Once they go into a gain condition, and we refer here to overall circuit gain, we will have

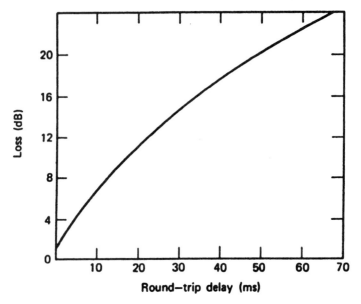

Figure 2.11. Talker echo tolerance for average telephone users.

positive feedback and the amplifiers will begin to oscillate or "sing." North American practice calls for a 4-dB loss on all four-wire circuits to guard against singing.

Almost all four-wire circuits have some form of amplifier and level control. Often such amplifiers are embodied in the channel banks of the frequency division multiplex (FDM) carrier equipment. For a discussion of FDM carrier equipment, see Chapter 3.

An Introduction to Transmission Loss Planning. One major aspect to transmission system design for a telephone network is to establish a transmission loss plan on a national basis. Such a plan, when implemented, is formulated to accomplish three goals:

• Control singing
• Keep echo levels within limits tolerable to the subscriber
• Provide an acceptable overall loudness rating to the subscriber

For North America the via net loss (VNL) concept embodies the transmission loss plan idea. VNL is covered in Section 2.6.5.1.

From our earlier discussions, we have much of the basic background necessary to develop a transmission loss plan. We know the following:

1. A certain minimum loss must be maintained in four-wire circuits to prevent singing.
2. Up to a certain limit of round-trip delay, echo is controlled by loss.

3. It is desirable to limit these losses as much as possible to improve the reference equivalent.

National transmission plans vary considerably. Obviously the length of circuit, as well as the velocity of propagation of the transmission media, is important. Two approaches are available in the preparation of a loss plan:

• Variable loss plan (i.e., VNL)
• Fixed plan (i.e., as used in Europe and recommended in CCITT Rec. G.122 and now being imported for North America's all-digital network).

A national transmission loss plan for a small country (i.e., small in extension) such as Belgium could be quite simple. Assume that a 4-dB loss is inserted in all four-wire circuits to prevent singing (see Figure 2.11). Here 4 dB allows for 5 ms of round-trip delay. If we assume carrier transmission for the entire length of the connection and use 105,000 mi/s for the velocity of propagation, we can then satisfy Belgium's echo problem. The velocity of propagation used comes out to 105 mi (168 km)/ms. By simple arithmetic we see that a 4-dB loss on all four-wire circuits will make echo tolerable for all circuits extending 262 mi (420 km). This is an application of the fixed-loss type of transmission plan. In the case of small countries or telephone companies operating over a small geographical extension, the minimum loss inserted to control singing controls echo as well for the entire country.

Let us try another example. Assume that all four-wire connections have a 7-dB loss. Figure 2.11 indicates that 7 dB permits an 11-ms round-trip delay. Assume that the velocity of propagation is 105,000 mi/s. Remember that we deal with round-trip delay. The talker's voice goes out to the far-end hybrid and is then reflected back. This means that the signal traverses the system twice, as shown:

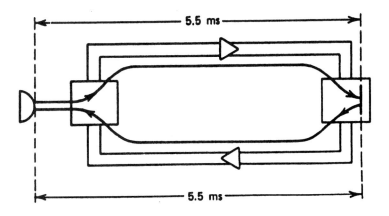

In this example the round-trip delay is 5.5 + 5.5 = 11 ms. Thus 7-dB loss for the velocity of propagation specified allows about 578 mi of extension (i.e., 5.5 × 105) or, for all intents and purposes, the distance between subscribers.

Two major goals of the transmission loss plan are to improve overall loudness ratings (OLRs) and to apportion more loss to the subscriber plant so subscriber loops can be longer, or to allow the use of less copper (i.e., smaller diameter conductors). What measures can be taken to reduce losses and still keep echo within tolerable limits? One obvious target is to improve return losses at the hybrids. If all hybrid return losses are improved, then the echo tolerance curve gets shifted. This is so because improved return losses reduce the intensity of the echo returned to the talker. Thus the subscriber is less annoyed by the echo effect.

One way of improving return loss is to make all possible two-wire connectivity combinations out of a hybrid look alike, that is, have the same impedance. The switch at the other end of the hybrid (i.e., on the two-wire side) connects two-wire loops of varying lengths, causing the resulting impedances to vary greatly from one connectivity to another. One approach is to extend four-wire transmission to the subscriber loop side of the local serving exchange. As each analog local serving exchange is replaced by a digital counterpart, each subscriber loop must terminate in a hybrid, usually mounted as part of the requisite channel bank equipment. Because now each hybrid is dedicated to just one subscriber loop, the compromise balancing network can be optimized. This should notably increase the balance return loss at each hybrid to values over 20 dB. Another approach is to extend four-wire transmission right to the customer subset. Theoretically this is done with the U.S. Department of Defense AutoVon network.

Reflecting these improved conditions where the PSTN is entirely digital, a simplified loss plan is being invoked with much smaller losses used to control echo than with North America's VNL plan, for example. Loss plans are discussed in Section 2.6.5.

As we know, the four-wire loop gives rise to echo. The signal is reflected twice (see Figure 2.10) and reaches the listener sometime later than the original signal. The phenomenon is called *listener echo* and will be perceived as *hollowness* in a telephone conversation and may cause bit errors in data transmission. The degradation of the transmission is dependent on the loss around the four-wire loop (*open-loop loss*, OLL).

2.6.4.2 *Stability, Hollowness, Singing, and Open-Loop Loss.* We define instability in the PSTN as a condition where singing or near-singing occurs. Singing derives from the same cause as echo. This can be seen in Figure 2.7, where if excessive receive signal from the four-wire receive side of a hybrid is fed through the hybrid to the four-wire transmit side, a possibility exists that a singing condition will be set up. The receive signal passes down the amplifiers on the four-wire transmit side causing positive feedback. With sufficient positive feedback, singing will occur. For a FDM-based analog network, the results could be catastrophic.

The *stability*, S, is defined as the amount of extra gain that must be introduced in each of the amplifiers (or four-wire circuits) to reach the limit of instability, that is, when

$$A_1 + A_2 - (G_1 + S) - (G_2 + S) = 0 \qquad (2.10)$$

(See Figure 2.12 to illustrate the notation used.)

$$S = \tfrac{1}{2}(A_1 + A_2 - G_1 - G_2) \qquad (2.11)$$

Usually the following approximations are valid:

$$A_1 = A_{B1} + 2A \qquad (2.12)$$

$$A_2 = A_{B2} + 2A \qquad (2.13)$$

A_{B1} and A_{B2} are the balance return losses of the hybrids, and A is the hybrid loss. This gives

$$S = \tfrac{1}{2}(A_{B1} + 2A - G_1 + A_{B2} + 2A - G_2) \qquad (2.14)$$

$$S = \tfrac{1}{2}(A_{B1} + A_{B2} + A_{01} + A_{02}) \qquad (2.15)$$

The *open-loop loss* (around the four-wire loop) will be

$$M = 2S = A_{B1} + A_{B2} + A_{01} + A_{02} \qquad (2.16)$$

Here A_{01} and A_{02} are the overall losses between the two-wire points of the hybrids, in the two directions of transmission. This means that the stability of the circuit is governed by the overall loss of the circuit and the balance return loss of the hybrids. If S is less than 0, instability results (i.e., singing). It

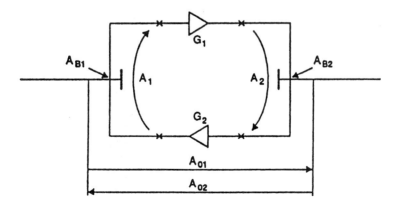

Figure 2.12. An aid to the illustration of the concept of stability. (From Ref. 2, Figure 3/3.2.)

should be noted that both the overall loss and balance return loss will vary with frequency; that is, the circuit may be stable for one frequency and unstable for another.

Although the circuit is stable if S is larger than 0, the four-wire loop will have a delay (different for different frequencies), and the transmission will be impaired because of listener echo. It is therefore necessary to require a stability value larger than 0, especially when the circuit is used for data transmission.

In a real telephone connection, the attenuation in the four-wire part of the connection will vary with time and be different for different circuits, and the value of the balance return loss will vary from case to case.* To calculate the stability in the PSTN, it is therefore necessary to use a statistical approach (Ref. 2).

The stability of a telephone connection depends on (Ref. 2):

- The transmission loss plan adopted
- The variation of transmission level with time
- The attenuation-frequency characteristics of the links in tandem
- The distribution of balance return loss

2.6.4.3 Talker Echo in Digital Networks. With the introduction of digital networks, the overall loss on a connection is reduced compared to conventional analog equipment. One disadvantage of this is that the loss of the talker echo path is reduced. Typically an overall loss reduction in one circumstance might be 6 dB by the deployment of a digital network; then the echo path loss is decreased by 12 dB.[†] In addition, the signal propagation delay within a digital connection is unlikely to be less than the delay observed on a similar analog connection.

The one-way transmission delays through a digital local exchange and a digital tandem/transit exchange are typically 1.2 and 0.45 ms, respectively. Delays here are significantly longer than equivalent analog network delays.

2.6.4.4 Talker Echo Loudness Rating. The loudness of an echo path may be expressed by means of the "talker echo loudness rating" (TELR). TELR = SLR + RLR + loss within the echo loop. It is normal practice to use minimum SLR and RLR values to achieve a worst-case echo value for a connection.

ITU-T Rec. G.111 (Ref. 5) recommends

$$SLR + RLR \geq 8 \, dB \tag{2.17}$$

The long-term minimum objective is 8 dB for equation 2.17. ITU-T Rec. G.

---

*Of course, this only refers to the traditional location of a hybrid; that is, outward from the serving switch looking toward the toll network, not on the subscriber side of the switch.
[†]Remember that the echo signal traverses the network twice, once going and once coming.

122 (Ref. 7) recommends an echo loss mean value of $(15 + n)$, where n is the number of four-wire circuits in the national extension.

Suppose there are two four-wire circuits in a national extension (i.e., $n = 2$). Then a mean echo loss $(a–b)$ of 17 dB will be required at the local exchange. (*Note:* $(a–b)$ is defined in Figure 2.13.) Therefore the TELR for this circuit will be

$$\text{TELR} = 8 + 17 = 25 \text{ dB}$$

For this hypothetical example connection, the total one-way transmission delay may well be 12 ms, when the one-way delay of, say, an 1800-km line plant and the 3-ms one-way delay introduced by the digital exchanges are taken into account. The customer perception of the echo performance using the echo tolerance curves of CCITT Rec. G.131 indicates that more than 10% of subscribers would express dissatisfaction. Another 8-dB loss would have to be added within the echo path to reduce echo to the point where less than 1% of subscribers would express dissatisfaction with the connection.

2.6.5 Transmission Loss Plans

Each national telecommunications administration sets forth a transmission plan.* A major thrust of a transmission plan is the loss plan. There are variable-loss, quasi-fixed-loss, and fixed-loss plans. We will give an example of each.

2.6.5.1 *Via Net Loss (VNL).* In North America via net loss (VNL) has been in force since 1953 (see Ref. 11) with the No. 4A crossbar switch. VNL is a variable-loss plan. It is now being phased out in favor of a fixed-loss plan due to the implementation of an all-digital network.

VNL takes into consideration that echo tolerance of a subscriber is most affected by delay.† Delay is a function of distance and the velocity of propagation. For example, the velocity of propagation can vary from 186,000 mi/s for propagation in free space to as low as 20,000 mi/s for propagation across heavily loaded wire-pair cable. VNL took this into account with its VNLF (via net loss factor). It took into account the number of trunks in a connection, the expected random deviations in trunk losses from design values, and the expected ERL (echo return loss, Section 2.6.4) at the distant-end serving switch.

The VNL plan assigned loss to a trunk based on the amount of round-trip delay the trunk was expected to contribute. Again, since delay is a function of the type of facility as well as trunk length, both of these characteristics were taken into account. To ensure a minimum value of loss on all connections, each tandem (toll) connecting trunk was assigned a loss of 2.5 dB in addition

---

*For the United States, the transmission plan may be found in Ref. 8.
†The other annoyance factor is echo intensity or level.

TABLE 2.4   VNL Values for Trunks on Carrier Transmission Facilities

Trunk Length in Miles	Via Net Loss (VNL)
0–165	0.5 dB
166–365	0.8 dB
366–565	1.1 dB

Note 1. Inserted connection loss (ICL) = VNL.
Note 2. When echo cancelers are used, VNL = 0 dB and valid for any length.
Note 3. Table is based on VNL formula for carrier facilities or VNL = (0.0015 × average length + 0.4) dB.
Source: Reference 8.

to the applicable VNL. A maximum connection loss of approximately 11 dB was permitted for round-trip delays approaching 45 ms. For greater delays, echo cancelers are deployed.

Reference 8 points out that the VNL plan is rarely used as more recent loss plans have been applied to modern networks. Table 2.4 gives VNL values for trunks employing carrier transmission used on intraLATA applications.

2.6.5.2 ITU-T Loss Planning. Figure 2.13 defines points *a–t–b* and just points *a–b*. ITU-T states that the transmission loss introduced between *a* and *b* by the national system, referred to as the loss (*a–b*), is important from three viewpoints:

1. It contributes to the margin that the international connection has against oscillation during the setup and clear-down of a connection. A minimum loss over the band of 0–4 kHz is the characteristic value.

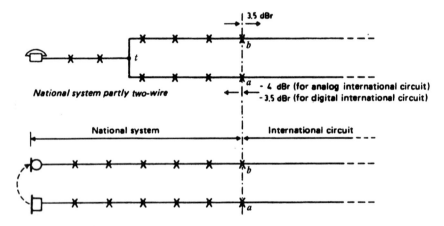

National system partly two-wire

National system wholly four-wire

Note: a, b are the virtual analog switching points of the international circuit.

Figure 2.13. Definition of points (*a–b*) and points (a-t-b) and virtual analog switching point (VASP). (From Ref. 7, Figure 1/G.122, ITU-T Rec. G.122.)

2. It contributes to the margin of stability during a communication. Again, a minimum loss in the band 0–4 kHz is the characteristic value, but in this case the subscribers' apparatuses (e.g., telephone, modem) are assumed to be connected and in an operating condition.

3. It contributes to the control of echo and, with respect to the subjective effect of talker echo, a weighted sum of the loss $(a-b)$ over the band 300–3400 Hz is the characteristic value.

ITU-T Rec. G.122 (Ref. 7) sets up three scenarios for echo/stability control:

1. Loss to avoid instability during setup, clear-down, and changes in a connection.
2. Unweighted loss on established connections.
3. Echo loss on established connections.

All references are made to loss $a-b$; however, with networks using a hybrid, the loss would be $a-t-b$.

2.6.5.2.1 Loss (a–b) to Avoid Instability During Setup, Clear-Down, and Changes in a Connection (Ref. 7). To ensure adequate stability on international connections during setup, clear-down, and connection changes, the distribution of loss $(a-b)$ during the worst situation should be such that the risk of loss $(a-b)$ of 0 dB or less does not exceed 6 in 1000 calls when using the calculation method described below. This requirement should be met over the frequency band 0–4 kHz.

The limit described above can be achieved if the following simultaneous conditions on the network are met:

1. The sum of the nominal transmission losses in both directions of transmission $a-b$ and $t-b$ measured between the two-wire input of the terminating set (hybrid) t, and one or the other of the virtual switching points on the international circuit, a or b, should not be less than $(4 + n)$ dB, where n is the number of analog or mixed analog/digital four-wire circuits in the national chain.
2. The stability balance return loss at the terminating set t should have a value of not less than 2 dB for the terminal conditions encountered during normal operation.
3. The standard deviation of variations of transmission loss of a circuit should not exceed 1 dB.

In a calculation to verify these values, it may be assumed that:

- There is no significant difference between nominal and mean value of the transmission losses of circuits.

- Variations of losses for both directions of transmission of the same circuit are fully correlated.
- Distributions are Gaussian.

For the loss $(a-b)$, we then have

$$\text{Mean value: } 2 + 4 + n = 6 + n \text{ dB}$$

$$\text{Standard deviation: } \sqrt{4n} \text{ dB}$$

With $n = 4$, the mean value becomes 10 dB and the standard deviation is 4 dB, resulting in a probability for values lower than 0 dB of 6×10^{-3}.

2.6.5.2.2   Unweighted Loss (a-b) on Established Connections. Here the objective is that the risk of loss $(a-b)$ reaches low values at any frequency in the range 0–4 kHz and should be as small as practicable. The objective is obtained by a national system sharing a mean value of at least $(10 + n)$ dB together with a standard deviation not larger than $\sqrt{6.25 + 4n}$ dB in the band 0–4 kHz, where n is the number of analog or mixed analog/digital four-wire circuits in the national chain.

It should be noted that wholly digital circuits may be assumed to have a transmission loss with mean value and standard deviation equal to zero. CODECS (see Chapter 3) in circuits or exchanges are expected to offer smaller variations in transmission loss than FDM carrier circuits. For the variations in transmission loss of a coder–decoder (codec) combination, standard deviations of 0.4 dB have been reported to the ITU-T Organization.

The distribution of stability loss $(a-b)$ recommended above, for example, could be achieved if, in addition to meeting the conditions in Section 2.6.5.2.1, the mean value of the stability balance return loss at the terminating set (hybrid) is not less than 6 dB and the standard deviation is not larger than 2.5 dB.

2.6.5.2.3   Echo Loss (a-b) on Established Connections. The mean value of echo loss $(a-b)$ for international calls established over the national system should have a mean value of not less than $(15 + n)$ dB with a standard deviation not exceeding $\sqrt{9 + 4n}$ dB, where n is the number of analog or mixed analog digital four-wire circuits in the national chain.

An example of how the requirements in the paragraph above may be achieved would be for the mean value of the sum of the transmission losses $a-t$ and $t-b$ not to be less than $(4 + n)$ dB with a standard deviation from the mean not exceeding $2\sqrt{n}$ dB, accompanied by an echo balance return loss at the terminating set (hybrid) t of not less than 11 dB with a standard deviation not exceeding 3 dB.

2.6.5.3   North American Loss Plan for the Digital Network. The loss plan now being implemented on the North American PSTN is based on

ANSI Standard T1.508-1992 (Ref. 9). The loss plan takes into account the increase in round-trip delay caused by inherent processing delay of digital elements such as digital switches, facility multiplexers, and digital cross-connects.

As we will discuss in Chapter 3, digital signal processing, such as digital-to-analog conversion or the insertion of digital loss, precludes the transmission of unmodified bit streams across the network. The maintenance of bit integrity* is necessary so that customers who use the network to transmit digital data can expect a signal to arrive at the destination without modification. To accomplish this, it is recommended that loss insertion, where required, be achieved as near to the end-user as possible. A goal, when connection is all-digital from an end-user terminal to another end-user terminal, is to migrate the control of loss insertion to the end-user terminal. Until that time, loss values that are dependent on the type of connection are usually administered at the point of switching nearest the customer. Loss values that are not dependent on the type of connection can be inserted at the final D/A conversion point such as a digital loop carrier (DLC) remote terminal or at the last point of switching. For digital connections terminated in analog access lines, loss values are dependent on the connection architecture:

- For interLATA or interconnecting network connections, the requirement is 6 dB.
- For intraLATA connections involving different LECs (local exchange carriers), 6 dB is the preferred value, although 3 dB may apply to connections not involving a tandem switch.
- For intraLATA connections involving the same LEC, the guidelines are 0–6 dB (typically 0 dB, 3 dB, or 6 dB).

The ANSI standard allows a choice of network loss values depending on performance considerations, administrative simplicity, and current network design.

2.7 SUBSCRIBER LOOP DESIGN

2.7.1 Introduction

Subscriber loop design was introduced in Section 2.4. Detailed design was delayed until this point so we could lay the groundwork of how subscriber loop loss affects loudness ratings (LRs) and the impact on subscriber satisfaction.

---

*Bit integrity means we cannot change the bit sequence in 8-bit PCM timeslots after A/D conversion or before D/A conversion, that is, mid-network somewhere.

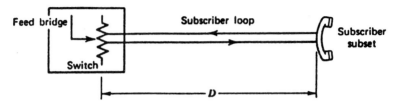

Figure 2.14. Subscriber loop model.

The subscriber loop connects a subscriber telephone subset with a local serving switch. A subscriber loop, in nearly all cases,* is two wire with simultaneous transmission in both directions. Figure 2.14, a simplified drawing, will help illustrate the problem. Distance D, the loop length, is most important. As we discussed in Section 2.4, D must be limited in length owing to (1) attenuation of the voice signal and (2) dc resistance for signaling.

The value for maximum attenuation is taken from the national transmission plan covered in Section 2.6.5. For example, in the United States it is 8 dB (objective); in many European countries the value is 7 dB. The maximum resistance is a function of the type of serving switch installed. For some switches this value is as low as 1500 Ω. Other switches on special order (e.g., Nortel DMS-100) can be as high as 2400 Ω. From this value we must subtract a budgetary value for the telephone subset resistance. Let that be 400 Ω. For the next maximum resistance allowed on a particular loop, we must subtract the 400 Ω from the switch maximum resistance value (e.g., 1500 Ω − 400 Ω = 1100 Ω). In the paragraphs below we will use the following values:

8 dB for the loop attenuation limit

1100 Ω for the resistance limit

2.7.2 Calculation of the Resistance Limit

To calculate the dc loop resistance for copper conductors, the following formula is applicable:

$$R_{dc} = 0.1095/d^2 \tag{2.18}$$

where R_{dc} = loop resistance (Ω/mi) and d = diameter of the conductor (inches).

*An exception is the U.S. Department of Defense AutoVon mentioned earlier in this chapter.

TABLE 2.5   Loop Resistance of Several Copper Conductors

Gauge of Conductor	Ω/1000 ft of Loop	Ω/mi of Loop	Ω/km of Loop
26	83.5	440	268
24	51.9	274	168.5
22	32.4	171	106
19	16.1	85	53

If we want a 10-mi loop and allow 110 Ω/mi of loop* (for 1100-Ω limit), what diameter copper wire would we need?

$$110 = 0.1095/d^2$$

$$d^2 = 0.1095/110 = 0.0009545$$

$$d = 0.0309 \text{ in. or } 1.216 \text{ mm}$$

If we wish to calculate maximum loop lengths for the 1100-Ω signaling resistance, use Table 2.5. As an example, for a 26-gauge loop,

$$1100/83.5 = 13.17 \text{ kft or } 13,700 \text{ ft}$$

This, then, is the supervisory signaling limit and not the loss (attenuation) limit, or what some call the "transmission limit." To assist in relating American wire gauge (AWG) to cable diameter in millimeters, Table 2.6 is presented.

Another guideline in the design of subscriber loops is the minimum loop current off-hook for effective subset operation. For instance, the traditional North American 500-type subset requires at least 23 mA for efficient operation.

2.7.3   Calculation of the Loss (Attenuation) Limit

The second design consideration mentioned earlier was attenuation or loss. The attenuation of a wire pair used on a subscriber loop varies with frequency, resistance, inductance, capacitance, and leakage conductance. The resistance of the line will depend on temperature. For example, with open-wire lines, attenuation of the line may vary ±12% between winter and summer conditions. For buried cable, with which we are more concerned, loss variations due to temperature are much less.

Table 2.7 gives losses of some common subscriber cable per 1000 ft. In our model, the loss limit was 8 dB. Then by simple division we can derive

---

*Be careful here of terminology. Mile of *loop* would be 2 mi of a single copper-wire lead. A loop consists of two wires that "loop" around.

TABLE 2.6 American Wire Gauge (B & S) Versus Wire Diameter and Resistance

American Wire Gauge	Diameter (mm)	Resistance (Ω/km)[a] at 20°C
11	2.305	4.134
12	2.053	5.210
13	1.828	6.571
14	1.628	8.284
15	1.450	10.45
16	1.291	13.18
17	1.150	16.61
18	1.024	20.95
19	0.9116	26.39
20	0.8118	33.30
21	0.7229	41.99
22	0.6439	52.95
23	0.5733	66.80
24	0.5105	84.22
25	0.4547	106.20
26	0.4049	133.9
27	0.3607	168.9
28	0.3211	212.9
29	0.2859	268.6
30	0.2547	338.6
31	0.2268	426.8
32	0.2019	538.4

[a]These figures must be doubled for loop/km. Remember it has a "go" and "return" path.

TABLE 2.7 Loss per Unit Length of Subscriber Cable

Cable Gauge[a]	Loss/1000 ft (dB)	dB/km	dB/mi
26	0.51	1.61	2.69
24	0.41	1.27	2.16
22	0.32	1.01	1.69
19	0.21	0.71	1.11
16	0.14	0.46	0.74

[a]Cable is low-capacitance type (i.e., under 0.075 nF/mi).

maximum loop length permissible for transmission design considerations for the wire gauges shown:

Gauge	Loop Length
26	8/0.51 = 15.68 kft
24	8/0.41 = 19.51 kft
22	8/0.32 = 25.0 kft
19	8/0.21 = 38.09 kft
16	8/0.14 = 57.14 kft

2.7.4 Loading

In many situations it is desirable to extend subscriber loop lengths beyond the limits described in Section 2.7.3. Common methods to attain longer loops without exceeding loss limits include:

1. Increasing conductor diameter
2. Using amplifiers and/or loop extenders*
3. Using inductive loading
4. Using carrier equipment (Chapter 3)
5. Using digital loop carrier (Chapter 3)

Loading tends to reduce the transmission loss on subscriber loops and other types of voice pairs at the expense of good attenuation–frequency response beyond 3000–3400 Hz. Loading a particular voice-pair loop consists of inserting series inductances (loading coils) into the loop at fixed intervals. Adding load coils tends to

• Decrease the velocity of propagation
• Increase impedance

Loaded cables are coded according to the spacing of the load coils. The standard code for load coils regarding spacing is shown in Table 2.8. Loaded cables typically are designated 19H44, 24B88, and so forth. The first number indicates the wire gauge, the letter is taken from Table 2.8 and is indicative of the spacing, and the third item is the inductance of the coil in millihenrys (mH). 19H66 is a cable commonly used for long-distance operation in Europe. Thus the cable has 19-gauge voice pairs loaded at 1830-m intervals with coils of 66-mH inductance. The most commonly used spacings are B, D, and H.

Table 2.9 will be useful in calculating the attenuation of loaded loops for a given length. For example, for 19H88 (last entry in table) cable, the attenuation per kilometer is 0.26 dB (0.42 dB/statute mile). Thus for our 8-dB loop loss limit, we have 8/0.26, limiting the loop to 30.77 km in length (19.23 statute miles).

Load coils are constructed of bifilar windings on a toroidal core made from highly compressed permalloy. The load coil is placed in series with each conductor. The first load coil outward from the serving switch is placed at one-half distance (one-half section), typically 2550 ft with D loading and 3000 ft with H loading. This allows loaded loops to be joined at the local serving switch and to be cascaded.

---

*A loop extender is a device that increases the battery voltage on a loop, extending its signaling range. It may also contain an amplifier, thereby extending the transmission loss limits.

TABLE 2.8   Code for Load Coil Spacing (United States)

Code Letter	Spacing (ft)	Spacing (m)
A	700	213.5
B	3000	915
C	929	283.3
D	4500	1372.5
E	5575	1700.4
F	2787	850
H	6000	1830
X	680	207.4
Y	2130	649.6

2.7.4.1   Line Build-Out. Often it is not possible to place a load coil within the right spacing tolerance. This is particularly true for the half-distance sections described earlier. When this occurs, *line build-out* is employed. Line build-out may consist of just capacitors or build-out lattice networks to simulate the missing length of line. They add capacitance resistance or just capacitance in the required amounts to add length *electrically.*

2.7.5   Subscriber Loop Design Plans—North American Practice

2.7.5.1   Introduction. Between 30% and 50% of a telephone company's (administration's) investment is tied up in what is generally referred to as *outside plant.* Outside plant, for this discussion, is defined as that part of the telephone plant that takes the signal from the local serving switch and delivers it to the subscriber.* Much of the expense can be attributed to the cost of copper in the subscriber cable. Another large expense is cable installation, such as that incurred in tearing up city streets to augment present installation or install a new plant. Much work in the recent past has been done to reduce these costs.

In the following paragraphs we present several "cookbook" methods for the design of the subscriber loop portion of the outside plant. It can be shown that almost universally, if a subscriber loop meets its signaling requirements, it will also meet its loss requirements. All the design plans described below are based on this concept. Before we launch into short-cut methods of subscriber loop design, we define a *bridged tap* (BT).

2.7.5.2   Bridged Taps. Whitham Reeves (Ref. 12) defines a *bridged tap* as any unused, open-circuited, parallel-connected cable pair on a subscriber loop. Conceptually, a bridged tap is illustrated in Figure 2.15.

---

*The IEEE defines outside plant as that part of the plant extending from the line side of the main distribution frame to the line side of the station (i.e., subset) or PABX protector or connecting block or to the line side of the main distribution frame in another central office (switch) building.

TABLE 2.9   Some Properties of Cable Conductors

Diameter (mm)	AWG No.	Mutual Capacitance (nF/km)	Type of Loading	Loop Resistance (Ω/km)	Attenuation at 1000 Hz (dB/km)
0.32	28	40	None	433	2.03
		50	None		2.27
0.40		40	None	277	1.62
		50	H66		1.42
		50	H88		1.24
0.405	26	40	None	270	1.61
		50	None		1.79
		40	H66	273	1.25
		50	H66		1.39
		40	H88	274	1.09
		50	H88		1.21
0.50		40	None	177	1.30
		50	H66	180	0.92
		50	H88	181	0.80
0.511	24	40	None	170	1.27
		50	None		1.42
		40	H66	173	0.79
		50	H66		0.88
		40	H88	174	0.69
		50	H88		0.77
0.60		40	None	123	1.08
		50	None		1.21
		40	H66	126	0.58
		50	H88	127	0.56
0.644	22	40	None	107	1.01
		50	None		1.12
		40	H66	110	0.50
		50	H66		0.56
		40	H88	111	0.44
0.70		40	None	90	0.92
		50	H66		0.48
		40	H88	94	0.37
0.80		40	None	69	0.81
		50	H66	72	0.38
		40	H88	73	0.29
0.90		40	None	55	0.72
0.91	19	40	None	53	0.71
		50	None		0.79
		40	H44	55	0.31
		50	H66	56	0.29
		50	H88	57	0.26

Source: Reference 10. Courtesy of ITT.

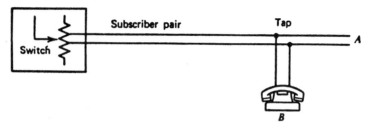

Figure 2.15. Concept of a bridged tap.

A bridged tap appears as a shunting impedance at its point of connection to the loop. This shunt causes a reflection loss that depends on the tap's pair characteristics (gauge, length, etc.). It also depends somewhat on the characteristics of the loop to which it is connected.

For digital service or data rates $\geq$ 9600 bps on a particular subscriber loop, bridged taps and load coils should be removed.

2.7.5.3 Resistance Design (Ref. 14). Resistance design (RD) is a "cookbook" design method for the subscriber loop. If we follow the rule, it assumes that the transmission loss limit will take care of itself if the requisite resistance limit is met. This limit was 1300 Ω. It defined a limit or boundary around a local serving switch called the *resistance design boundary*. For subscribers served outside of the boundary, the long route design (LRD) method was used.

There are three steps in the RD procedure.

1. Determination of the resistance design boundary
2. Determination of the design loop
3. Selection of cable gauges to meet the design objectives

Design loop is defined as the subscriber loop under study for a given distribution area to which the switch design limit (i.e., 1300 Ω) is applied to determine conductor sizes (wire gauges). It is normally the longest loop expected for the period of fill for the cable involved.

The *theoretical design* is used to determine the wire gauge or combination of wire gauges. If more than one gauge is required, Ref. 13 states that the most economical approach, neglecting existing plant, is to use the two consecutive finest (i.e., smallest diameter) gauges that meet a particular switch design limit. The smaller of the two gauges is usually placed outward from the serving switch because it usually has a larger cross section of wire pairs. Since the design loop length has been determined, the resistance per kilofoot (or km) for each gauge may be determined from Tables 2.5, 2.6, 2.7, and 2.9. The theoretical design can now be calculated applying two simultaneous equations with two unknowns. If we call the unknowns X and Y, the $X + Y$ is the total loop length. Allow 9 Ω for each load coil.

The North American design loss limit is 8 dB for a subscriber loop. However, to ensure that this loss limit is met, the following additional rules in RD design should be adhered to:

Use inductive loading on all loops over 18 kft long.

Limit the cumulative length of all bridged taps on nonloaded loops to 6 kft or less.

Reference 13 recommends H88 loading where the first load coil out from the serving exchange is 3000 ft and the separation between subsequent load coils is 6000 ft. The spacing tolerance between load coils is ±120 ft.

2.7.5.4 Long Route Design (LRD). Long route design applies to subscriber loops that had a total dc resistance greater than 1300 Ω. Of course, each LRD loop had to be able to carry out the supervisory signaling function and meet the 8-dB maximum loop attenuation rule for the United States. LRD provides for a specific combination of range extenders and/or fixed-gain device (VF amplifiers) to meet the supervision and loss criteria.

On most long loops, a range extender with gain was employed at the switch. A range extender boosts the standard −48 V up to −84 V, and the gain provided is from 3–6 dB.

2.7.5.5 Current Loop Design Rules for the United States. The present loop design rules for the United States and employed by the Regional Bell Operating Companies are outlined in Table 2.10. Resistance design (RD) described in Section 2.7.5.3 was updated to "Revised Resistance Design" (RRD). Long route design described in Section 2.7.5.4 was changed to modified long-route design (MLRD).

2.7.5.5.1 Revised Resistance Design (RRD). The cornerstone of RRD is that the maximum loop resistance was increased to 1500 Ω, where under RD, it was 1300 Ω. It is applicable for loops that are 24 kft or less in length. It reduces the amount of bridged tap allowed. The length of 18 kft for nonloaded loops is the same for both plans (i.e., RD and RRD) except that with RRD the maximum length includes the length of any bridged tap (Ref. 8). Nearly all modern switches can handle loop resistances of 1600 Ω or more. Thus any reduction of signaling margins brought about by RRD are well compensated for with the increased maximum loop resistance of these modern switches.

For loops where the resistance is greater than 1500 Ω, Ref. 8 recommends the use of a digital loop carrier (DLC) as a first choice, carrier serving area (CSA), then modified long-route design (MLRD), in that order.

TABLE 2.10 Current Loop Design Plans

Design Parameter	Carrier Serving Area	Revised Resistance Design	Modified Long-Route Design
Loop Resistance[a] (Ω)	N/A (limited by loss)	0–18 kft: 1300 max. 18–24 ktf: 1500 max.	1501–2800
Loading	None[b]	Full H88 > 18 kft	Full H88
Cable gauging	Two gauges, except stubs and fuse cables (max. lengths including BT): • 24-, 22-, and/or 19-gauge: 12 kft • 26-gauge: 9 kft[c]	Two-gauge combinations (22-, 24-, 26-gauge) preferred	
Bridged tap (BT) and end section (ES)	Total BT 2.5 kft max. No single BT > 2 kft	Nonloaded cable and BT: 18 kft max. Total BT: 6 kft max. Loaded: ES and BT, 3–12 kft	ES and BT: 3–12 kft
Transmission limitations	None; supports ISDN DSL, 56-kb data, and "despecialized" special services	Compatible with ISDN DSL; no digital services > 18 kft	No digital services Needs range extender with gain if > 1500 Ω

[a]Includes (only) the resistance of the cable and loading coils.
[b]At least one exchange carrier uses an "extended Carrier Serving Area (CSA)" in some rural areas. This variant allows loading but does not accommodate digital services.
[c]Multigauge designs incorporating 26 gauge are restricted in total length to $12 - [3L_{26} / (9 - BT]$ kft, where L_{26} is the total length of the 26-gauge and BT is the sum of bridged taps of all gauges.
Source: Reference 8, Table 7-11; copyright © Bellcore 1994. Reprinted with permission.

2.7.5.5.2 Modified Long-Route Design. MLRD was introduced in the United States in 1980 to provide for the extension of loops on a per-line basis. Under this plan, the 1500- to 2000-Ω range is designated Resistance Zone 18 (RZ 18) with an additional 3 dB gain required. The 2000- to 2800-Ω range (RZ 28) requires 6 dB of gain.

2.7.5.5.3 The Carrier Serving Area (CSA) Concept. The evolution to a network that can readily provide digital services via loop facilities led to the CSA concept. A carrier serving area is an area that is or may be served by DLC. DLC comes in many flavors. One example is the ISDN Basic Rate at 160 kbps in North America and 192 kbps elsewhere. AT & T's SLC-96 is another application of DLC. Many DLC applications are based on DS1 and E1 systems. These are described in Chapter 3. Chapter 4 discusses several other DLC applications.

All loops within a CSA are nonloaded. Loop length in the CSA is limited by attenuation, not dc resistance. Bridged tap lengths are controlled to preserve capability for high-bit rate operation.

2.8   DESIGN OF LOCAL AREA TRUNKS (JUNCTIONS)

Historically, trunks (called junctions in the United Kingdom) used VF transmission over wire-pair cable. In much of North America, these trunk circuits are now digital based on DS1 (T1) formats. In other parts of the world the change over to digital transmission has shown equal progress or is somewhat slower to come about.

There are two choices for transmission media: wire-pair cable or fiber-optic cable. In the case of wire-pair cable, often the older existing VF cable plant used selected pairs, each pair carrying 24 equivalent telephone channels, with one pair required for transmit and another for receive. On each pair, a regenerative repeater is installed at each 6000-ft interval. Often these repeaters just replace load coils (H-type loading), which are also spaced at 6000-ft intervals.

Digital trunks, whether on fiber or metallic pair, are commonly designed as "lossless" facilities.

On wire-pair digital transmission using a common cable for both go and return (i.e., transmit and receive) DS1/DS1C configurations, crosstalk between go and return can degrade error performance. This crosstalk is categorized as NEXT (near-end crosstalk) and FEXT (far-end crosstalk). The main thrust here is that a high-level signal leaving a regenerative repeater will spill into its companion (or other) low-level signal(s) entering the same repeater location.

Digital transmission on wire pairs has distance limitations due to the buildup of jitter (see Chapter 3). The accumulation of jitter, which degrades error performance, is a function of the number of regenerative repeaters in tandem.

When fiber-optic facilities are employed for trunk transmission, a number of advantages accrue. For one thing, fiber optics has nearly unlimited bandwidth. Another advantage is that there are much fewer regenerative repeaters per unit length so that jitter accumulation in most cases can be neglected. There are no crosstalk problems in the light domain. Most new trunk facilities use fiber optics as the transmission medium.

The cost of fiber-optic facilities is continuously decreasing. However, an implementer should carry out a cost trade-off analysis between wire pair or fiber optics to determine the most cost-effective solution for a particular project.

2.9   VF REPEATERS

VF repeaters in telephone terminology imply the use of *uni*directional amplifiers at VF on VF trunks. On a two-wire trunk two amplifiers must be used on each pair with a hybrid in and a hybrid out. A simplified block diagram is shown in Figure 2.16.

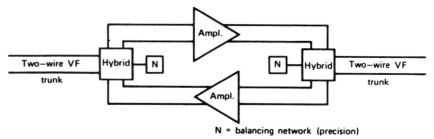

N = balancing network (precision)

Figure 2.16. Simplified block diagram of a VF repeater.

The gain of a VF repeater can be run up as high as 20 or 25 dB, and originally they were used at 50-mi intervals on 19-gauge loaded cable in the long-distance (toll) plant. Today they are seldom found on long-distance circuits, but they do have application on local trunk circuits where the gain requirements are considerably less. Trunks using VF repeaters have the repeater's gain adjusted to the equivalent loss of the circuit minus the 4-dB loss to provide the necessary singing margin. In practice a repeater is installed at each end of the trunk circuit to simplify maintenance and power feeding. Gains may be as high as 6–8 dB.

An important consideration with VF repeaters is the balance at the hybrids. Here precision balancing networks may be used instead of the compromise networks employed at the two-wire–four-wire interface (Section 2.6.3). It is common to achieve a 21-dB return loss; 27 dB is also possible, and, theoretically, 35 dB can be reached.

Another repeater commonly used on two-wire trunks is the negative-impedance repeater. This repeater can provide a gain as high as 12 dB, but 7 or 8 dB is more common in practice. The negative-impedance repeater requires a line build out (LBO) at each port and is a true two-way, two-wire repeater. The repeater action is based on regenerative feedback of two amplifiers. The advantage of negative-impedance repeaters is that they are transparent to dc signaling. On the other hand, VF repeaters require a composite arrangement to pass dc signaling. This consists of a transformer bypass.

2.10 ITU-T INTERFACE

2.10.1 Introduction

To facilitate satisfactory communications between telephone subscribers in different countries, the ITU-T Organization has established specific transmission guidelines in the form of recommendations. According to the ITU-T

Organization, a connection is satisfactory if it meets certain criteria for the following:

- Loudness rating
- Noise (for analog connectivities)
- Echo/singing
- Variation of transmission loss with time
- Attenuation distortion
- Group delay (phase distortion)
- Propagation time and total delay, among others

Specific digital network interface requirements, such as standard formats, slip rate, error performance, jitter/wander, and quantizing distortion, are covered in Chapter 3.

The following subsections outline and highlight these criteria from the point of view of a telephone company or an administration's international interface. This is a point where the national network meets the international system. However, in most cases the performance is related to the end-user (e.g., loudness rating).

2.10.2 Maximum Number of Links in Tandem

As we add links in tandem, transmission impairments tend to deteriorate, typically adding noise (for analog networks), degrading the bit error rate, and accumulating jitter/wander on digital networks. Thus it is prudent to limit the number of links in tandem. Of course, the ITU-T Organization's interest is for international connectivities.

ITU-T Rec. G.101 (Ref. 15) limits the total number of links in tandem to 12. We define a link* as that trunk connectivity between "adjacent" switches. The maximum number of links in tandem is shown in Figure 2.17. The 12 links are allocated as follows:

- Four links in the originating country
- Four links in the international portion
- Four links in the terminating country

Because of the use of direct circuits via satellite and, to a lesser extent, similar terrestrial fiber-optic circuits, nearly all international connectivities require considerably less than 12 links in tandem.

---

*Called a "telephone circuit" in the ITU-T reference document.

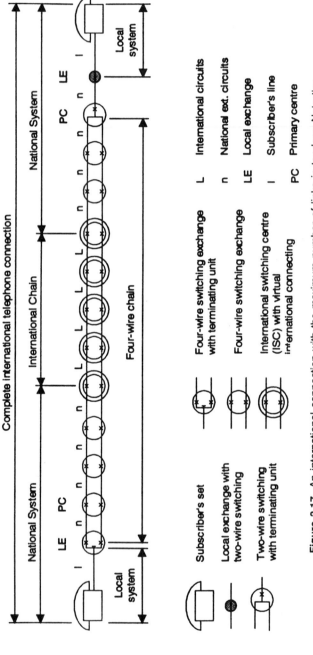

Figure 2.17. An international connection with the maximum number of links in tandem. Note the options given for conversation from two-wire to four-wire. (From Ref. 15, Figure 6/G.101, ITU-T Rec. G.101.)

Complete international telephone connection

National System · International Chain · National System

Local system · Four-wire chain · Local system

LE PC ... PC LE

Subscriber's set

Local exchange with two-wire switching

Two-wire switching with terminating unit

Four-wire switching exchange with terminating unit

Four-wire switching exchange

International switching centre (ISC) with virtual international connecting

L International circuits
n National ext. circuits
LE Local exchange
l Subscriber's line
PC Primary centre

2.10.3 Loudness Rating

A suitable value for OLR is about 10 dB in most cases. The long-term traffic-weighted mean value should lie in the range of 8–12 dB (Ref. 5).

2.10.4 Noise

ITU-T Rec. G.113 (Ref. 16) states that the network performance objective for circuit noise on complete telephone connections—mean value, and taken over a large number of worldwide connections (each including four international circuits [links]), of the distribution of 1-minute mean values of signal-independent noise power of the connections—should not exceed −43 dBm0p referred to the input of the first circuit in the chain of international circuits.

2.10.5 Echo and Singing

In order to minimize the effects of echo on international connections, the echo loss should have a value of $(15 + n)$ dB with a standard deviation not exceeding $\sqrt{9 + 4n}$ dB, where n is the number of analog or mixed analog/digital circuits (links) in tandem (Ref. 7). (*Note*: The 15-dB minimum of loss should also fully protect the circuit from any singing condition.) To avoid excessive loss insertion to counter echo, ITU-T recommends the use of echo control devices on connectivities where the round-trip delay equals or exceeds 50 ms (Ref. 17, Rule M).

2.10.6 Variation of Transmission Loss with Time

The ITU-T Organization recommends the following circuit performance objective, which has been used to assess the stability (see Ref. 17) of international connections (Ref. 18, p. 171):

1. "The standard deviation of the variation of transmission loss of a circuit should not exceed 1 dB. This objective can be achieved already for circuits on a single group link equipped with automatic regulation and should be achieved on each national circuit, whether regulated or not. The standard deviation should not exceed 1.5 dB for other international circuits."
2. "The difference between the mean value and the nominal value of the transmission loss for each circuit should not exceed 0.5 dB."

2.10.7 Attenuation Distortion

The network performance objectives for the variation with frequency of transmission loss in terminal condition of a worldwide four-wire chain of 12 circuits (international plus national extensions), each one routed over a single

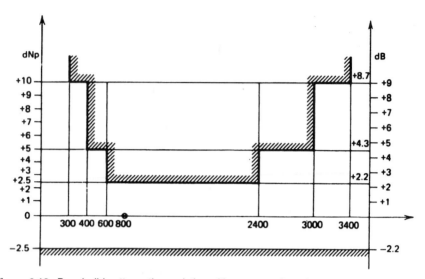

Figure 2.18. Permissible attenuation variation with respect to its value at 800 Hz, objective for a worldwide four-wire chain in terminal service. (From Ref. 19, Figure 1/G.132, CCITT Rec. 132.)

group link, are shown in Figure 2.18. This assumes that no use is made of high-frequency (HF) radio circuits using 3-kHz channel equipment (Ref. 19).

2.10.8 Group Delay (Phase Distortion)

The network performance objectives for the permissible differences for a worldwide chain of 12 circuits (links) in tandem, each on a single 12-channel group link, between the minimum group delay (throughout the transmitted frequency band) and the group delay at the lower and upper limits of the frequency band are shown in Table 2.11. (Limits given in Table 2.11 should be met for both analog circuits and mixed circuits with analog and digital sections. For digital service these values should be improved.)

TABLE 2.11   Group Delay Distortion for 12 Circuits in Tandem

Circuit	Lower Limit of Frequency Band (ms)	Upper Limit of Frequency Band (ms)
International chain	30	15
Each of the national four-wire extensions	15	7.5
On the whole four-wire chain	60	30

Source: Reference 20, Table 1/G.133, CCITT Rec. G.133.

2.10.9   Error in the Reconstituted Frequency

As channels of any international telephone circuit should be suitable for voice-frequency telegraphy, the network performance objective for the accuracy of the virtual carrier frequencies should be such that the difference between an audio frequency applied at one end of the circuit and the frequency received at the other end should not exceed 2 Hz, even when there are intervening modulating and demodulating processes (Ref. 21).

2.10.10   Crosstalk

2.10.10.1   Between Circuits. The circuit performance objective for the near-end or far-end crosstalk ratio (intelligible crosstalk only) measured at audio frequency at trunk exchanges between two complete circuits in terminal service should not be less than 65 dB. However, when a minimum noise level of at lcast 4000 pWp is always present in a system (such as may be the case for satellite circuits), a reduced crosstalk ratio of 58 dB between circuits is acceptable (Ref. 18).

2.10.10.2   Between Go and Return Channels of a Four-Wire Circuit. Since all ordinary telephone circuits may be used as VF telegraph bearers, the circuit performance objective for the near-end crosstalk ratio between the two directions of transmission should be at least 43 dB (Ref. 18).

2.10.11   One-Way Propagation Time

CCITT recommends the following one-way transmission time for connection where echo is adequately controlled:

0 to 150 ms. Acceptable for most user applications.

150 to 400 ms. Acceptable provided that administrations (telephone companies) are aware of the transmission time impact on the transmission quality of user applications. For example, international connections with satellite hops that have transmission times below 400 ms are considered acceptable.

Above 400 ms. Unacceptable for general network planning purposes. However, it is recognized that in some exceptional cases this will be exceeded. Examples where such exceptions are unavoidable are double satellite hops, satellites used to restore terrestrial routes, fixed satellite and digital cellular interconnections, videotelephony over satellite circuits, and very long international connections with two cellular systems connected by long terrestrial facilities.

Planning values for propagation and processing delays are given in Table 2.12.

TABLE 2.12   Planning Values for Propagation and Processing Delays

Transmission Medium	Contribution to One-Way Transmission Time	Remarks
Terrestrial coaxial cable or radio-relay system; FDM and digital transmission	4 μs/km	Allows for delay in repeaters and regenerators
Optical fiber cable system; digital transmission	5 μs/ km[a]	
Submarine coaxial cable system	6 μs/km	
Satellite system		
14,000-km altitude	110 ms	Between earth stations only
36,000-km altitude	260 ms	
FDM channel modulator or demodulator	0.75 ms[b]	Half the sum of propagation times in both directions of transmission
FDM compandored channel modulator or demodulator	0.5 ms[c]	
PCM coder or decoder	0.3 ms[b]	
PCM/ADPCM/PCM transcoding	0.5 ms	
G.728 coder and decoder	2.0 ms	
8 kbps	32 ms[d]	
PLMS (Public Land Mobile System) (objective 40 ms, Rec. G. 173)	80–110 ms	
H.261 video coder and decoder	FS	
G.763 coder and decoder	FS	
G.765 coder and decoder	FS	
Transmultiplexer	1.5 ms[e]	
Digital transit exchange, digital-digital	0.45 ms[f]	
Digital local exchange analog–analog	1.5 ms[f]	Half the sum of propagation times in both directions of transmission
Digital local exchange, analog subscriber line–digital junction	0.975 ms[f]	
Digital local exchange, digital subscriber line–digital junction	0.825 ms[f]	
Echo cancelers	1 ms[g]	

[a]This value is provisional and is under study.

[b]These values allow for group-delay distortion around frequencies of peak speech energy and for delay of intermediate higher order multiplex and through-connecting equipment.

[c]This value refers to FDM equipment designed to be used with a compandor and special filters.

[d]This is a performance requirements value. Hardware is currently not available.

[e]For satellite digital communications where the transmultiplexer is located at the earth station, this value may be increased to 3.3 ms.

[f]These are mean values. Depending on traffic loading, higher values can be encountered: e.g., 0.75 ms (1.950 ms, 1.350 ms, or 1.250 ms) with 0.95 probability of not exceeding. (For details, see Recommendation Q.551.)

[g]Echo cancelers, when placed in service, will add a one-way propagation time of up to 1 ms in the send path of each echo canceler. This delay excludes the delay through any CODEC in the echo canceler. No significant delay should be incurred in the receive path of the echo canceler.

Source: Reference 22, Table A.1/G.114, ITU-T Rec. G.114.

For an average size country, the terrestrial one-way propagation delay will be less than 18 ms.

To estimate the propagation delay on purely digital circuits between local exchanges using fiber-optic transmission, the transmission time will probably not exceed

$$3 + (0.005 \times \text{distance in km}) \text{ ms}$$

The 3-ms constant term makes allowance for one PCM coder–decoder and five digital exchanges (Ref. 22).

REVIEW EXERCISES

1. Where does the frequency response peak for a carbon subset transmitter (e.g., what frequency or frequencies)?

2. A subscriber loop provides a metallic electrical path for four purposes. Name the four.

3. What singular parameter does loudness rating measure at the subscriber subset?

4. In an analog network, why is it difficult to achieve ideal loudness loss values? (This is a conceptual question.)

5. What are the two basic parameters that limit the length of a subscriber loop?

6. What are the three basic components of overall loudness rating (OLR)? (That is, overall loudness rating $= x + y + z$.)

7. At generally what level of psophometric weighted circuit noise (dBmp) does such noise begin to impact speech transmission quality?

8. Wire-pair transmission has notable limitations when used on a geographically extended network. Name five basic transmission methods that can be used to extend the network utilizing the same wire pair.

9. Define two-wire and four-wire transmission.

10. Why is it conceptually important to achieve as high as possible a return loss between the two-wire side of a hybrid and its associated balancing network?

11. Referring to Exercise 10, give a one-sentence reason why it is difficult to achieve such high return loss values. What is the generally accepted median value of such return loss? How will this situation improve once the network from the serving switch outward becomes all digital?

12. What is liable to occur when a hybrid is connected to a two-wire connection with a poor match?

13. Define echo and singing.

14. Why is stability return loss important? Differentiate stability return loss from echo return loss.

15. What is the total *design* loss of a hybrid? This is made up of two components. What are they?

16. When a hybrid demonstrates a poor return loss for a particular two-wire connection, what two impairments may occur?

17. How do we prevent singing?

18. How does delay affect echo as an impairment to a speech subscriber?

19. Echo can be controlled by adding loss. Loss is added as a function of delay. If we continue to add loss, how does it affect overall subscriber satisfaction? At what value(s) of round-trip delay must we use echo suppressors? Discuss this situation, including the effects on overall loudness rating (OLR).

20. When we examine the voice channel, what frequency bands are most susceptible as a cause of singing?

21. What makes via net loss (VNL) different from fixed-loss plans?

22. What is the loss value assigned across the digital toll network for all connections no matter the length? We refer here to the North American transmission plan.

23. In the design of a subscriber loop with inductive loading, besides the wire loop itself, what are the additional resistance values that must be taken into account? Give a budgetary value for each in ohms.

24. Given a cable with 26-gauge wire pairs displaying a loss of 0.51 dB/1000 ft of loop and a resistance of 83.5 Ω per 1000 ft of loop, an exchange that handles up to 1500 Ω of loop resistance, and a transmission plan that allows up to 8.0-dB loop loss, what is the maximum length of loop permitted without conditioning of any sort?

25. Why is fiber-optic cable superior to wire-pair cable in local area trunk design?

26. 24H88 load coils are used on a loop. Explain the three elements of this load coil coding.

27. Resistance design does not treat loop loss per se. With revised resistance design (RRD), a resistance of 1500 Ω is accommodated. How do we know that we do not exceed the maximum loop loss limits?

28. How do standard modified long-route design (MLRD) procedures meet the additional loss and resistance ($> 1500\ \Omega$) requirements found on such loops?

29. What is the one major overriding advantage of using digital loop carrier (DLC) on long-route design?

30. What is the purpose of a range extender?

31. What range of dB values would define excellent OLR performance?

32. What range of *applied* gain values in dB would we expect to find used for VF repeaters?

33. What type of VF repeater lends itself well to two-wire metallic VF circuits? For what reason?

34. Why do we limit the number of links in tandem on an international connection? What are the ramifications of this limitation on internal national connections?

35. Estimate the transmission time of a 1000-km connectivity using fiber-optic cable with five intervening digital exchanges and a coder/decoder at each end.

REFERENCES

1. *Transmission Systems for Communications*, 5th ed., Bell Telephone Laboratories, Holmdel, NJ, 1982.
2. *Handbook on Transmission Planning*, ITU-TS, Geneva, 1993.
3. Roger L. Freeman, *Telecommunication System Engineering*, 3rd ed., Wiley, New York, 1996.
4. *Transmission Systems for Communications*, revised 3rd ed., Bell Telephone Laboratories, Holmdel, NJ, 1964.
5. "Loudness Ratings (LRs) in an International Connection," ITU-T Rec. G.111, ITU, Helsinki, 1993.
6. "Effect of Transmission Impairments," ITU-T Rec. P.11, ITU, Helsinki, Mar. 1993.
7. "Influence of National Systems on Stability and Talker Echo in International Connections," ITU-T Rec. G.122, Helsinki, Mar. 1993.
8. *BOC Notes on the LEC Networks—1994*, Bellcore, Piscataway, NJ, 1994.
9. *Network Performance—Loss Plan for Evolving Digital Networks*, ANSI T1.508-1992, ANSI, New York, 1992, with supplements 1992 and 1993.
10. *Outside Plant*, Telecommunication Planning Documents, ITT Laboratories, Spain, 1973.
11. H. R. Huntley, "Transmission Design of Intertoll Telephone Trunks," *Bell Syst. Tech. J.* Sept. 1953.

12. Whitham Reeves, *Subscriber Loop Signal and Transmission Handbook—Analog*, IEEE Press, New York, 1992.

13. *Telecommunication Transmission Engineering*, 2nd ed., vols. 1–3, American Telephone & Telegraph Co., New York, 1977.

14. *Notes on the BOC Intra-LATA Networks—1986*, TR-NPL-000275, Bellcore, Livingston, NJ, 1986.

15. "The Transmission Plan," ITU-T Rec. G.101, ITU, Helsinki, 1993.

16. "Transmission Impairments," ITU-T Rec. G.113, ITU, Helsinki, Mar. 1993.

17. "Stability and Echo," CCITT Rec. G.131, Fascicle III.1, IXth Plenary Assembly, Melbourne, 1988.

18. "General Performance Objectives Applicable to All Modern International Circuits and National Extension Circuits," CCITT Rec. G.151, Fascicle III.1, IXth Plenary Assembly, Melbourne, 1988.

19. "Attenuation Distortion," CCIT Rec. G.132, Fascicle III.1, IXth Plenary Assembly, Melbourne, 1988.

20. "Group-Delay Distortion," CCITT Rec. G.133, Fascicle III.1, IXth Plenary Assembly, Melbourne, 1988.

21. "Error on the Reconstituted Frequency," CCITT Rec. G.135, Facile III.1, IXth Plenary Assembly, Melbourne, 1988.

22. "One-Way Transmission Time," ITU-T Rec. G.114, ITU, Helsinki, Mar. 1993.

3

MULTIPLEXING TECHNIQUES

3.1 DEFINITION AND SCOPE

Multiplexing deals with the transmission of two or more signals over a common transmission facility such as a wire pair, fiber-optic cable, or radio carrier. When multiplexing is carried out in the frequency domain, we call it frequency division multiplex (FDM). More commonly today it is carried out in the time domain, and we call this time division multiplex (TDM).

With a FDM system, each information-bearing channel is assigned a distinct frequency slot in a band of frequencies. Of course, with telephony, each frequency slot carries a telephone voice channel. Each voice channel amplitude modulates a different carrier frequency, which permits translation of that voice channel to its own frequency slot or segment of a broadband spectrum that is different from all other modulated channels sharing the same spectrum. Such derived signals are combined in an electrical network for transmission to the line at one end of a circuit and then separated electrically at the other end. Because carrier frequencies are used in the multiplexing process, the term *carrier techniques* (or *carrier transmission*) evolved.

With TDM, each information-bearing channel, such as the voice channel, is assigned a time slot. Each channel then occupies the entire frequency spectrum for a very short period of time.

The bulk of this chapter discusses these two multiplexing methods and their associated waveforms. The FDM discussion is short and concise. It is being phased out in favor of TDM. Several of the ITU-T FDM formats are covered.

The subsequent TDM discussion is based on common practice of pulse code modulation (PCM). Both North American and ITU-T standards are covered. There are notable differences between these two. This is followed by short descriptions of two other time division techniques: ADPCM and delta modulation/CVSD.

It should be noted that all conventional multiplexing works on a four-wire basis. The transmit and receive paths are separate. Two-wire and four-wire transmission and the conversion from two-wire to four-wire systems are covered in Section 2.6.3.

3.2 FREQUENCY DIVISION MULTIPLEX

3.2.1 Introduction

In FDM an available bearer bandwidth is divided into a number of nonoverlapping frequency slots. Each slot carries a single information-bearing signal such as a voice channel. We can consider a FDM multiplexer as a frequency translator. At the opposite end of the circuit a demultiplexer filters and translates the channel frequency slots into the original information-bearing channels. This concept is shown in Figure 3.1.

In practice, the "frequency translator" (multiplexer) uses single-sideband modulation of RF carriers. A different RF carrier is used for each channel to be multiplexed. This technique is based on mixing or heterodyning a signal to be multiplexed, typically a voice channel, with a RF carrier. A simplified block diagram of a FDM link is shown in Figure 3.2.

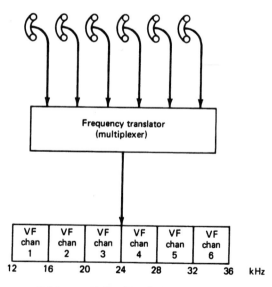

Figure 3.1. Frequency division multiplex (FDM) frequency translation concept. Note the 4-kHz channel spacing.

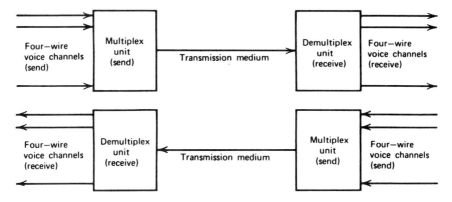

Figure 3.2. A simplified block diagram of a FDM Link.

3.2.2 Mixing

The heterodyning or mixing of signals of frequencies A and B is shown below. What frequencies may be found at the output of the mixer?

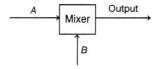

Both the original signals will be present as well as the signals representing their sum and their difference in the frequency domain. Thus at the output of the above mixer we will have present the signals of frequency A, B, $A + B$, and $A - B$. Such a mixing process is repeated many times in FDM equipment.

Let us now look at the boundaries of the nominal 4-kHz voice channel. These are 300 and 3400 Hz. Let us further consider these frequencies as simple tones of 300 and 3400 Hz. Now consider the mixer below and examine the possibilities at its output:

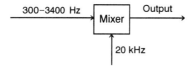

First, the output may be the sum

$$
\begin{array}{r}
20{,}000 \text{ Hz} \\
+ \quad 300 \text{ Hz} \\
\hline
20{,}300 \text{ Hz}
\end{array}
\quad \text{or} \quad
\begin{array}{r}
20{,}000 \text{ Hz} \\
+ \quad 3{,}400 \text{ Hz} \\
\hline
23{,}400 \text{ Hz}
\end{array}
$$

A simple low-pass filter could filter out all frequencies below 20,300 Hz.

Now imagine that instead of two frequencies, we have a continuous spectrum of frequencies between 300 and 3400 Hz (i.e., we have the voice channel). We represent the spectrum as a triangle:

As a result of the mixing process (translation) we have another triangle as follows:

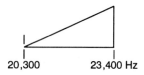

When we take the sum, as we did previously, and filter out all other frequencies, we say we have selected the upper sideband. Therefore we have a triangle facing to the right, and we call this an upright or erect sideband.

We can also take the difference, such that

$$
\begin{array}{rr}
20{,}000 \text{ Hz} & 20{,}000 \text{ Hz} \\
-\quad\ 300 \text{ Hz} & -\ 3{,}400 \text{ Hz} \\
\hline
19{,}700 \text{ Hz} & 16{,}600 \text{ Hz}
\end{array}
$$

and we see that in the translation (mixing process) we have had an inversion of frequencies. The higher frequencies of the voice channel become the lower frequencies of the translated spectrum, and the lower frequencies of the voice channel become the higher when the difference is taken. We represent this by a right triangle facing the other direction:

This is called an inverted sideband. To review, when we take the sum, we get an erect sideband. When we take the difference, frequencies invert and we have an inverted sideband represented by a triangle facing left.

3.2.3 The CCITT Modulation Plan

3.2.3.1 *Introduction.* A modulation plan sets forth the development of a band of frequencies called the line frequency (i.e., ready for transmission on the line or transmission medium). The modulation plan usually is a diagram showing the necessary mixing, local oscillator mixing frequencies, and the sidebands selected by means of the triangles described previously in a step-by-step process from voice channel input to line frequency output. The CCITT has recommended a standardized modulation plan with a common terminology. This allows large telephone networks, on both national and multinational systems, to interconnect. In the following paragraphs the reader is advised to be careful with terminology.

3.2.3.2 *Formation of the Standard CCITT Group.*
The standard *group* as defined by the CCITT occupies the frequency band of 60–108 kHz and contains 12 voice channels. Each voice channel is the nominal 4-kHz channel occupying the 300–3400-Hz spectrum. The group is formed by mixing each of the 12 voice channels, with a particular carrier frequency associated with each channel. Lower sidebands are then selected. Figure 3.3 shows the preferred approach to the formation of the standard CCITT group. It should be noted that in the 60–108-kHz band, voice channel 1 occupies the highest frequency segment by convention, between 104 and 108 kHz. The layout of the standard group is shown in Figure 3.4. The applicable CCITT recommendation is G.232 (Ref. 1).

Single-sideband suppressed carrier (SSBSC) modulation techniques are universally utilized. CCITT recommends that carrier leak be down to at least − 26 dBm0 referred to a 0 relative level point (see Section 1.9.5).

3.2.3.3 *Formation of the Standard CCITT Supergroup.*
A supergroup contains five standard CCITT groups, equivalent to 60 voice channels. The standard supergroup before translation occupies the frequency band of 312–552 kHz. Each of the five groups making up the supergroup is translated in frequency to the supergroup band by mixing with the proper carrier frequencies. The carrier frequencies are 420 kHz for group 1, 468 kHz for group 2, 516 kHz for group 3, 564 kHz for group 4, and 612 kHz for group 5. In the mixing process the difference is taken (lower sidebands are selected). This translation process is shown in Figure 3.5.

3.2.3.4 *Formation of the Standard CCITT Basic Mastergroup and Supermastergroup.*
The basic mastergroup contains five supergroups, equivalent to 300 voice channels, and occupies the spectrum of 812–2044 kHz. It is formed by translating the five standard supergroups, each occupying the 312–552-kHz band, by a process similar to that used to form the supergroup from five standard CCITT groups. This process is shown in Figure 3.6

The basic supermastergroup contains three mastergroups and occupies the band of 8516–12,388 kHz. The formation of the supermastergroup is shown in Figure 3.7.

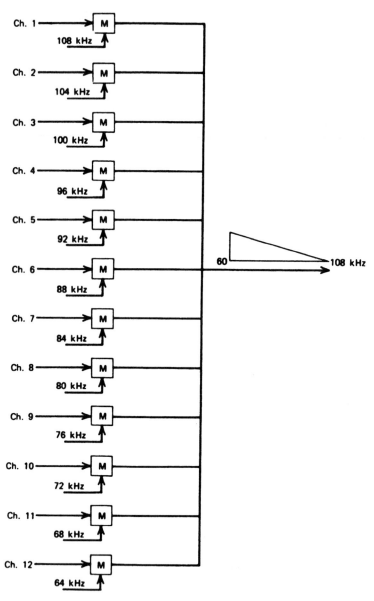

Figure 3.3. Formation of the standard CCITT group. The mixing frequencies are 64, 68, 72, ... , 108 kHz.

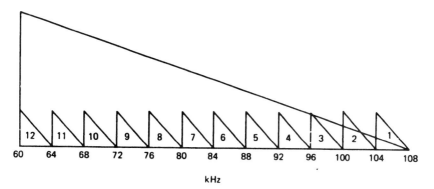

Figure 3.4. The standard CCITT group.

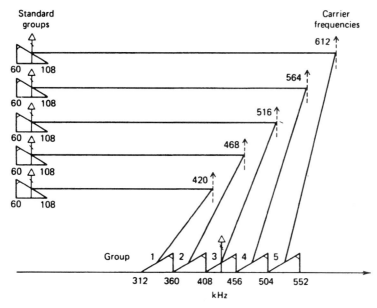

Figure 3.5. Formation of the standard CCITT supergroup. *Note:* Vertical arrows show group level regulating pilot tones; the larger arrow is the supergroup regulating pilot (see Section 3.2.5). (From Ref. 2, CCITT Rec. G.233. Courtesy of ITU-T Organization.)

3.2.3.5   The "Line" Frequency. The band of frequencies that the multiplex applies to the line, whether the line is a radiolink, coaxial cable, wire pair, or open-wire line, is called the line frequency. Another expression often used is HF (or high frequency), not to be confused with high-frequency radio, discussed in Chapter 9.

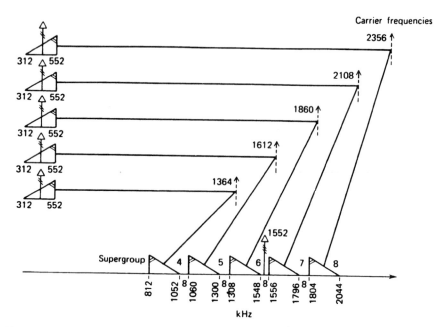

Figure 3.6. Formation of the standard CCITT mastergroup. (From Ref. 2, CCITT Rec. G.233. Courtesy of ITU-CCITT.)

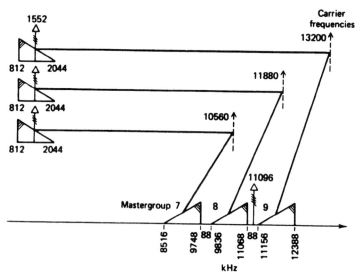

Figure 3.7. Formation of the standard CCITT supermastergroup. (From Ref. 2, CCITT Rec. G.233. Courtesy of ITU-CCITT.)

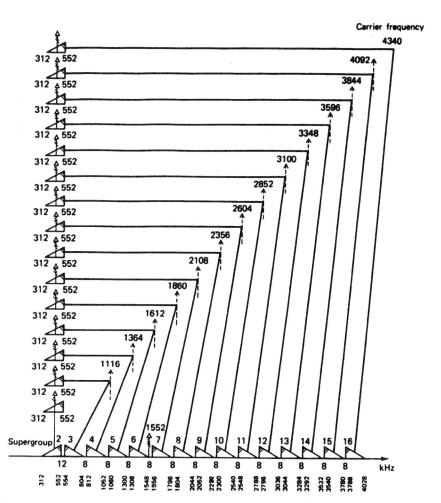

Figure 3.8. Makeup of basic CCITT 15-supergroup assembly. (From Ref. 2, CCITT Rec. G.233. Courtesy of ITU-CCITT.)

The line frequency in this case may be direct application of a group or supergroup to the line. However, more commonly a final translation stage occurs, particularly on high-density systems. One of these line configurations is shown in Figure 3.8 which illustrates the makeup of the basic 15-supergroup assembly.

3.2.4 Loading of Multichannel FDM Systems

3.2.4.1 Basic Loading Concepts. FDM carrier equipment was designed to carry speech telephony traffic, sometimes misnamed "message traffic" in

North America. In this context we refer to full-duplex conversations by telephone between two "talkers." It also carries data traffic, which has distinctly different characteristics.

Our discussion here deals with human speech and how multiple telephone users may load a FDM carrier system. If we load a carrier system too heavily, meaning here that the input levels are too high, intermodulation noise and crosstalk become intolerable. If we do not load the system sufficiently, the signal-to-noise ratio suffers. The problem is fairly complex because speech amplitude varies:

- With talker volume (i.e., the loudness of the talker)
- At a syllabic rate
- At an audio rate
- With varying circuit losses as different loops and trunks are switched into the same channel bank channel input

Also, the loading of a particular system varies with the busy hour* (i.e., the number of channels occupied up to the maximum).

As we are aware, speech is very spurty. Empirically, for a typical talker the peak power is about 18.6 dB higher than the average speech power. Peakiness of speech level means that FDM carrier equipment must be operated at a low average power to withstand voice peaks so as not to overload the system and cause distortion.

The activity factor, T_a, of a FDM speech circuit is an important parameter. We can write an equation for average talker power using the familiar VU of Chapter 2 as follows:

$$P_{\text{dBm}} = \text{VU} - 1.4 + 10 \log T_a \qquad (3.1)$$

If now a second talker is added operating on a different frequency segment on the same equipment, but independent of the first talker, the aggregate level of the output increases 3 dB. If we have N talkers, each on a different frequency segment, the average power developed is

$$P_{\text{dBm}} = \text{VU} - 1.4 + 10 \log N \qquad (3.2)$$

where P_{dBm} = electrical power developed across the frequency band occupied by all the talkers.

Empirically, we have found that the peakiness or peak factor of multitalkers over a multichannel analog system reaches the characteristics of random noise peaks when the number of talkers N reaches 64. When $N = 2$, the peaking factor is 18 dB; for 10 talkers it is 16 dB, for 50 talkers 14 dB, and so forth. Above 64 talkers the peak factor is 12.5 dB (Ref. 3).

---

*The *busy hour* is a term used in traffic engineering and is defined by the ITU-T Organization as "the uninterrupted period of 60 minutes during which the average traffic flow is maximum."

Up to now we have been using an activity factor of 1 (e.g., 100% duty cycle). This value is unrealistic for speech. This means that someone is talking all the time on a circuit. The traditional figure for activity factor accepted by the ITU-T Organization and used in North American practice is 0.25. This figure is derived as follows.

For one thing, the multichannel equipment cannot be designed for N callers and no more. If this were true, a new call would have to be initiated every time a call terminated or calls would have to be turned away for an "all trunks busy" condition. In the real-life situation, particularly for automatic service, carrier equipment, like switches, must have a certain margin by being overdimensioned for busy-hour service. For this overdimensioning we drop the activity factor to 0.70. Other causes reduce the figure even more. For instance, circuits are essentially inactive during call setup as well as during pauses for thinking during a conversation. The factor of 0.70 now reduces to 0.50. This latter figure is divided in half owing to the talk–listen effect. If we disregard isolated cases of "double-talking," it is obvious that on a full-duplex telephone circuit, while one end is talking, the other is listening. Hence a circuit (in one direction) is idle half the time during the "listen" period. The resulting activity factor is 0.25.

3.2.4.2 *Overload.*

In Section 1.9.6, where we discussed intermodulation (IM) products, we showed that one cause of IM products is overload. One definition of overload follows (Ref. 3):

> The overload point, or overload level, of a telephone transmission system is 6 dB higher than the average power in dBm0 of each of two applied sinusoids of equal amplitude and of frequencies a and b, when these input levels are so adjusted that an increase of 1 dB in both their separate levels causes an increase, at the output, of 20 dB in the intermodulation product of frequency $2a - b$.

Up to this point we have been talking about average power. Overload usually occurs when instantaneous signal peaks exceed some threshold. Consider that peak instantaneous power exceeds average power of a simple sinusoid by 3 dB. For multichannel systems the peak factor may exceed that of a sinusoid by 10 dB.

White noise is often used to simulate multitalker situations for systems with more than 64 operative channels.

3.2.4.3 *Loading.*

For loading multichannel FDM systems, CCITT recommends (Ref. 4, CCITT Rec. G.223):

> It will be assumed for the calculation of intermodulation below the overload point that the multiplex signal during the busy hour can be represented by a uniform spectrum [of] random noise signal, the mean absolute power level of

which, at a zero relative level point (in dBm0),

$$P_{av} = -15 + 10 \log N \quad \text{when } N \geq 240 \tag{3.3}$$

and

$$P_{av} = -1 + 4 \log N \quad \text{when } 12 \leq N < 240 \cdots \tag{3.4}$$

where N = numbers of 4-kHz voice channels.

All logs are to the base 10. *Note*: These equations apply only to systems without preemphasis and using independent amplifiers in both directions. Preemphasis is discussed in Section 5.4.2.3. An activity factor of 0.25 is assumed. See Figure 5.10 and the discussion therewith. Examples of the application of these formulas are discussed in the same section. It should also be noted that the formulas above include a small margin for loads caused by signaling tones, pilot tones, and carrier leaks.

Example 1. What is the average power of the composite signal for a 600-voice channel system using CCITT loading? N is greater than 240; thus equation 3.3 is valid.

$$\begin{aligned} P_{av} &= -15 + 10 \log 600 \\ &= -15 + 10 \times 2.7782 \\ &= +12.782 \text{ dBm0} \end{aligned}$$

Example 2. What is the average power of the composite signal for a 24-voice channel system using CCITT loading? N is less than 240, thus equation 3.4 is valid.

$$\begin{aligned} P_{av} &= -1 + 4 \log 24 \\ &= -1 + 4 \times 1.3802 \\ &= +4.5208 \text{ dBm0} \end{aligned}$$

3.2.4.4 *Single-Channel Loading.* A number of telephone administrations have attempted to standardize on -16 dBm0 for single-channel speech input to multichannel FDM equipment. With this input, peaks in speech level may reach -3 dBm0. Tests indicated that such peaks will not be exceeded more than 1% of the time. However, the conventional value of average power per voice channel allowed by the CCITT is -15 dBm0. (Refer to equation 3.3 and Ref. 4, CCITT Rec. G.223.) This assumes a standard deviation of 5.8 dB and the traditional activity factor of 0.25. Average talker level is assumed to be at -11.5 VU. We must turn to the use of standard deviation because we are dealing with talker levels that vary with each talker, and consequently with the mean or average.

3.2.4.5 Loading with Constant-Amplitude Signals. Speech on multi-channel systems has a low duty cycle or activity factor. We established the traditional figure of 0.25. Certain other types of signals transmitted over the multichannel equipment have an activity factor of 1. This means that they are transmitted continuously, or continuously over fixed time frames. They are also characterized by constant amplitude. Examples of these types of signals follow:

- Telegraph tone or tones
- Signaling tone or tones
- Pilot tones
- Data signals (particularly FSK and PSK; see Chapter 13)

Here again, if we reduce the level too much to ensure against overload, the signal-to-noise ratio will suffer, and hence the error rate will suffer.

For typical constant-amplitude signals, traditional (taken from CCITT) transmit levels (input to the channel modulator on the carrier [FDM] equipment) are as follows:

- Data: -13 dBm0
- Signaling (SF supervision), tone-on when idle: -20 dBm0
- Composite telegraph: -8.7 dBm0

For one FDM system now on the market with 75% speech loading and 25% data/telegraph loading with more than 240 voice channels, the manufacturer recommends the following:

$$P_{\text{rms}} = -11 + 10 \log N \qquad (3.5)$$

using -5 dBm0 per channel for the data input levels* and -8 dBm0 for the composite telegraph level.

Table 3.1 shows some of the standard practice for data/telegraph loading on a per-channel basis. Data and telegraph should be loaded uniformly. For instance, if equipment is designed for 25% data and telegraph loading, then voice channel assignment should, whenever possible, load each group and supergroup uniformly. For instance, it is bad practice to load one group with 75% data, while another group carries no data traffic at all. Data should also be assigned to voice channels that will not be near group band edge. Avoid channels 1 and 12 on each group for the transmission of data, particularly medium- and high-speed data. It is precisely these channels that display the

---

*All VF channels may be loaded at -8 dBm0 level, whether data or telegraph with this equipment, but for -5 dBm0 level data, only two channels per group may be assigned this level, the remainder voice, or the group must be "deloaded" (i.e., idle channels).

TABLE 3.1   Voice Channel Loading of Data / Telegraph Signals

Signal Type	CCITT	North American
High-speed data	− 10 dBm0 simplex − 13 dBm0 duplex	− 10 dBm0 switched network − 8 dBm0 leased line − 5 dBm0 occasionally − 8 dBm0 total power
Medium-speed data Telegraph (multichannel)		
≤ 12 channels	− 19.5 dBm0/channel − 8.7 dBm0 total	
≤ 18 channels	− 21.25 dBm0/channel − 8.7 dBm0 total	
≤ 24 channels	− 22.25 dBm0/channel − 8.7 dBm0 total	

Source: Reference 5. Courtesy of Lenkurt Electric Corp.

poorest attenuation distortion and group delay due to the sharp roll-off of group filters (see Chapter 13).

3.2.5   Pilot Tones

3.2.5.1   Introduction. Pilot tones in FDM carrier equipment have essentially two purposes:

- Control of level
- Actuation of alarms when levels are out of tolerance

3.2.5.2   Level-Regulating Pilots. The nature of speech, in particular its varying amplitude, makes it a poor prospect as a reference for level control. Ideally, simple single-sinusoid constant-amplitude signals with 100% duty cycles provide simple control information for level-regulating equipment. Multiplex level regulators operate in the same manner as automatic gain control circuits on radio systems, except that their dynamic range is considerably smaller.

Modern carrier systems initiate a level-regulating pilot tone on each group at the transmit end. Individual level-regulating pilots are also initiated on all supergroups and mastergroups. The intent is to regulate the system level within ±0.5 dB.

Pilots are assigned frequencies that are part of the transmitted spectrum yet do not interfere with voice channel operation. They usually are assigned a frequency appearing in the guard band between voice channels or are residual carriers (i.e., partially suppressed carriers). CCITT has assigned the following as group regulation pilots:

- 84.080 kHz (at a level of − 20 dBm0)
- 84.140 kHz (at a level of − 25 dBm0)

TABLE 3.2   Frequency and Level of CCITT Recommended Pilots

Pilot for	Frequency (kHz)	Absolute Power Level at a Zero Relative Level Point (dB)(Np)
Basic group B	84.080	$-20\,(-2.3)$
	84.140	$-25\,(-2.9)$
	104.080	$-20\,(-2.3)$
Basic supergroup	411.860	$-25\,(-2.9)$
	411.920	$-20\,(-2.3)$
	547.920	$-20\,(-2.3)$
Basic mastergroup	1552	$-20\,(-2.3)$
Basic supermastergroup	11096	$-20\,(-2.3)$
Basic 15-supergroup assembly (No. 1)	1552	$-20\,(-2.3)$

Source: Reference 6, CCITT Rec. G.241. Courtesy of ITU-CCITT.

The Defense Communications Agency of the U.S. Department of Defense recommends 104.08 kHz $\pm$ 1 Hz for group regulation and alarm.

For CCITT group pilots, the maximum level of interference permissible in the voice channel is -73 dBm0p. CCITT pilot filters have essentially a bandwidth at the 3-dB points of 50 Hz (refer to CCITT Rec. G.232).

Table 3.2 presents other CCITT pilot tone frequencies as well as those standard for group regulation. Respective levels are also shown. The operating range of level-control equipment activated by pilot tones is usually about ±4 or 5 dB. If the incoming level of a pilot tone in the multiplex receive equipment drops outside the level-regulating range, then an alarm will be indicated (if such an alarm is included in the system design). CCITT recommends such an alarm when the incoming level varies 4 dB up or down from the nominal (Ref. 6, CCITT Rec. G.241).

3.2.6   Noise and Noise Calculations

3.2.6.1   General. Carrier equipment is the principal contributor of noise on coaxial cable systems and other metallic transmission media. On radio-links it makes up about one-quarter of the total noise. The traditional approach is to consider noise from the point of view of a hypothetical reference circuit. Two methods are possible, depending on the application. The first is the CCITT method, which is based on a 2500-km hypothetical reference circuit. The second is used by the U.S. Department of Defense in specifying communications systems. Such military systems are based on a 12,000-nautical mi reference circuit with 1000-mi links and 333-mi sections.

3.2.6.2   CCITT Approach. CCITT Rec. G.222 (Ref. 7) states:

> The mean psophometric power, which corresponds to the noise produced by all modulating (multiplex) equipment...shall not exceed 2500 pW at a zero relative level point. This value of power refers to the whole of the noise due to

various causes (thermal, intermodulation, crosstalk, power supplies, etc.). Its allocation between various equipments can be to a certain extent left to the discretion of design engineers. However, to ensure a measurement agreement in the allocation chosen by different administrations, the following values are given as a guide to the target values:

Equipment	Maximum Value Contributed by the Send and Receive Side Together	Assumptions About Loading		
Channel modulators	200 pW0p[a]	Adjacent channels loaded with: Other channels loaded with:	-15 dBm0 -6.4 dBm0	(Signal corresponding to Rec. G.227)
Group modulators	80 pW0p	Load in group to be measured: Load in other groups:	$+3.3$ dBm0 -3.1 dBm0 (each)	
Supergroup modulators	60 pW0p	Load in supergroup to be measured: Load in other supergroups:	-6.1 dBm0 $+2.3$ dBm0 (each)	
Mastergroup modulators	60 pW0p	Load in each mastergroup:	$+9.8$ dBm0	
Supermaster-group modulators	60 pW0p	Load in each supermastergroup:	$+14.5$ dBm0	
Basic 15-super-group assembly modulators	60 pW0p	Load in each 15-supergroup assembly	$+14.5$ dBm0	

[a]No account is taken of the values attributed to pilot frequencies and carrier leaks.

Experience has shown that often these target figures can be improved upon considerably. The CCITT notes that they purposely loosened the value for channel modulators. This permits the use of the subgroup modulation scheme shown above as the alternative method for forming the basic 12-channel group.

For instance, one piece of solid-state equipment now on the market, when operated with CCITT loading, has the following characteristics:

1 pair of channel modulators	224 pWp
1 pair of group modulators	62 pWp
1 pair of supergroup modulators	25 pWp
1 (single) line amplifier	30 pWp

If out-of-band signaling* is used, the noise in a pair of channel modulators reduces to 75 pWp.

---

*For a discussion of out-of-band signaling, see Ref. 8.

Using the same solid-state equipment mentioned previously and increasing the loading to 75% voice, 17% telegraph tones, and 8% data, the following noise information is applicable:

1 pair of channel modulators	322 pWp
1 pair of group modulators	100 pWp
1 pair of supergroup modulators	63 pWp
1 (single) line amplifier	51 pWp

If this equipment is used on a real circuit with the heavier loading, the sum for noise for channel modulators, group modulators, and supergroup modulator pairs is 485 pWp. Accordingly, a system would be permitted to demodulate to voice only five times over a 2500-km route (i.e., 5 × 485 = 2425 pWp).

3.2.7 Compandors

The word *compandor* is derived from two words that describe it functions: compressor–expandor, to compress and to expand. A compandor does just that. It compresses a signal on one end of a circuit and expands it on the other. A simplified functional diagram and its analogy are shown in Figure 3.9.

The compressor compresses the intensity range of speech signals at the input circuit of a communication channel by imparting more gain to weak signals than to strong signals. At the far-end output of the communication circuit, the expandor performs the reverse function. It restores the intensity of the signal to its original dynamic range. We cover only syllabic compandors in this discussion.

The three advantages of compandors are that they:

1. Tend to improve the signal-to-noise ratio on noisy speech circuits.
2. Limit the dynamic power range of voice signals, reducing the chances of overload of carrier systems.
3. Reduce the possibility of crosstalk.

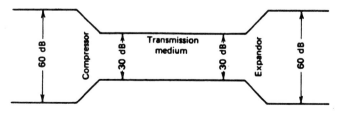

Figure 3.9. Functional analogy of a compandor.

A basic problem in telephony stems from the dynamic range of talker levels. This intensity range can vary 60 dB for the weakest syllables of the weakest talker to the loudest syllables of the loudest talker. The compandor brings this range down to more manageable proportions.

An important parameter of a compandor is its compression–expansion ratio, the degree to which speech energy is compressed and expanded. It is expressed by the ratio of the *input* to the *output* power (dB) in the compressor and expandor, respectively. Compression ratios are always greater than 1 and expansion ratios are less than 1. The most common compression ratio is 2 (2 : 1). The corresponding expansion ratio is thus $\frac{1}{2}$. The meaning of a compression ratio of 2 is that the dynamic range of the speech volume has been cut in half from the input of the compressor to its output. Figure 3.10 is a simplified functional block diagram of a compressor and an expandor.

Another important criterion for a compandor is its companding range. This is the range of intensity levels a compressor can handle at its input. Usually 50–60 dB is sufficient to provide the expected signal-to-noise ratio and reduce the possibility of distortion. High-level signals appearing outside this range are limited without markedly affecting intelligibility.

But just what are the high and the low levels? Such a high or low level is referred to as an "unaffected level" or focal point. CCITT Rec. G.162 (Ref. 9) defines the unaffected level as:

> the absolute level, at a point of zero relative level on the line between the compressor and expandor of a signal at 800 Hz, which remains unchanged whether the circuit is operated with the compressor or not.

CCITT goes on to comment:

> The unaffected level should be, in principle, 0 dBm0. Nevertheless, to make allowances for the increase in mean power introduced by the compressor, and to avoid the risk of increasing the intermodulation noise and overload which might result, the unaffected level may, in some cases, be reduced as much as

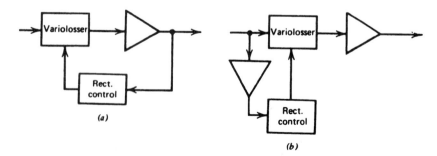

Figure 3.10. Simplified functional block diagram: (a) compressor and (b) expandor.

5 dB. However, this reduction of unaffected level entails a diminution of improvement in signal-to-noise ratio provided by the compandor.... No reduction is necessary, in general, for systems with less than 60 channels.

CCITT recommends a range of level from $+5$ to -45 dBm0 at the compressor input and $+5$ to -50 dBm0 at the nominal output of the expandor. Figure 3.11 shows diagrammatically a typical compandor range of $+10$ to -50 dBm.

A syllabic compandor operates much in the same fashion as any level-control device where the output level acts as a source for controlling input to the device (see Figure 3.10). Automatic gain control (AGC) on radio receivers operates in the same manner. This brings up the third important design parameter of syllabic compandors: attack and recovery times. These are the response to suddenly applied signals such as a loud speech syllable or burst of syllables. Attack and recovery times are a function of design time constants and are adjusted by the designer to operate as a function of the speech envelope (syllabic variations) and *not* with instantaneous amplitude changes (such as used in PCM). If the operation time is too fast, wide bandwidths would be required for faithful transmission. When attack times are too slow, the system may be prone to overload.

CCITT Rec. G.162 (Ref. 9) specifies an attack time equal to or less than 5 ms and a recovery time equal to or less than 22.5 ms.

The signal-to-noise ratio advantage of a compandor varies with the multi-channel loading factor of FDM equipment and thus depends on the voice level into the FDM channel modulation equipment. At best an advantage of 20 dB may be attained on low-level signals.

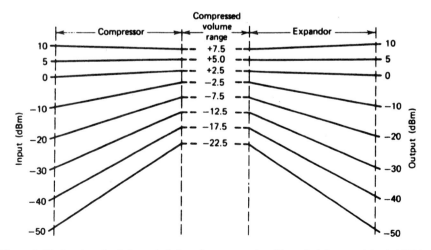

Figure 3.11. Input–output characteristics of a compandor. (From Ref. 3; copyright © 1970 by Bell Telephone Laboratories.)

3.3 TIME DIVISION MULTIPLEX—PCM

3.3.1 What Is PCM?

Pulse code modulation (PCM) is a method of modulating in which a continuous analog wave is transmitted in an equivalent digital mode. The cornerstone of an explanation of the functioning of PCM is the Nyquist sampling theorem, which states (Ref. 10, Section 23):

> If a band-limited signal is sampled at regular intervals of time and at a rate equal to or higher than twice the highest significant signal frequency, then the sample contains all the information of the original signal. The original signal may then be reconstructed by use of a low-pass filter.

As an example of the sampling theorem, the nominal 4-kHz channel would be sampled at a rate of 8000 samples per second (i.e., 4000 × 2).

To develop a PCM signal from one or several analog signals, three processing steps are required: sampling, quantization, and coding. The result is a serial binary signal or bit stream, which may or may not be applied to the line without additional modulation steps.

One major advantage of digital transmission is that signals may be regenerated at intermediate points on links involved in transmission. The price for this advantage is the increased bandwidth required for PCM. Practical systems require 16 times the bandwidth of their analog counterpart (e.g., a nominal 4-kHz analog voice channel requires 16 × 4 or 64 kHz when transmitted by PCM). Regeneration of a digital signal is simplified and particularly effective when the transmitted line signal is binary, whether neutral, polar, or bipolar. An example of bipolar transmission is shown in Figure 3.12.

Binary transmission tolerates considerably higher noise levels (i.e., degraded signal-to-noise ratios) when compared to its analog counterpart (i.e., FDM, Section 3.2). This plus the regeneration capability is a great step forward in transmission engineering. The regeneration that takes place at

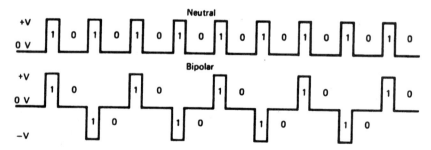

Figure 3.12. Neutral versus bipolar bit streams. *Top:* Alternative 1's and 0's transmitted in a neutral mode. *Bottom:* Equivalent in a bipolar mode.

each repeater by definition recreates a new digital signal. Therefore noise, as we know it, does not accumulate. However, there is an equivalent to noise in PCM systems that is generated in the modulation–demodulation process. This is called quantizing distortion and can be equated with thermal noise in annoyance to the listener.

Error rate is another important factor (see Chapter 13). If we can maintain an end-to-end error rate on the digital portion of the system of 1×10^{-3},* intelligibility will not be degraded. A third factor is important in PCM cable applications. This is crosstalk spilling from one PCM system to another or from the send path to the receive path inside the same cable sheath.

The purpose of the discussion of PCM in this chapter is to provide a background of the problems involved with it and its transmission, including the several PCM formats now in use. Long-distance (toll) transmission via PCM is discussed in Chapter 4.

3.3.1.1 *Clarification.* There are two comparatively different PCM systems in use today worldwide. These are the North American DS1(T1) system and the European E-1 system. Previously the E-1 system was called the CEPT30 + 2 system, where CEPT stands for Conference European Post & Telecommunications. Both use 8-bit time slots.

3.3.2 Development of a PCM Signal

3.3.2.1 *Sampling.* Consider the sampling theorem given previously. If we now sample the standard CCITT voice channel, 300–3400 Hz (a bandwidth of 3100 Hz), at a rate of 8000 samples per second, we will have complied with the theorem and we can expect to recover all the information in the original analog signal. Therefore a sample is taken every 1/8000 s, or every 125 μs. These are key parameters for our future argument.

Another example may be a 15-kHz program channel. Here the lowest sampling rate would be 30,000 times per second. Samples would be taken at 1/30,000-s intervals or every 33.3 μs.

3.3.2.2 *The PAM Wave.* With several exceptions, namely, single-channel systems (e.g., INTELSAT SPADE systems), practical PCM systems involve TDM. Sampling in these cases does not involve just one voice channel, but several. In practice, one system to be discussed samples 24 voice channels in sequence; another one samples 32 channels. The result of the multiple sampling is a pulse amplitude modulation (PAM) wave. A simplified PAM wave is shown in Figure 3.13, in this case a single sinusoid. A simplified diagram of the processing involved to derive a multiplexed PAM wave is shown in Figure 3.14.

---

*This value is a threshold floor, which will be discussed in Section 4.2.

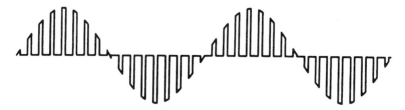

Figure 3.13. PAM wave as a result of sampling a single sinusoid.

If the nominal 4-kHz voice channel must be sampled 8000 times per second and a group of 24 such voice channels are to be sampled sequentially to interleave them, forming a PAM multiplexed wave, this could be done by gating. Open the gate for a 5.2-μs (125/24) period for each voice channel to be sampled successively from channel 1 through channel 24. This full sequence must be done in a 125-μs period ($1 \times 10^6/8000$). We call this 125-μs period a *frame*, and inside the frame all 24 channels are successively sampled once.

3.3.2.3 *Quantization.* The next step, in the process of forming a PCM serial bit stream, is to assign a binary coded sequence to each sample as it is presented to the coder.

The number of bits required to represent a character is called a code length or, more properly, a coding *level.* For instance, a binary code with four discrete elements (a 4-level code) could code 2^4 separate and distinct meanings or 16 characters, not enough for the 26 letters in our alphabet; a 5-level code would provide 2^5 or 32 characters or meanings. The ASCII, the commonly used source code for data transmission, which is described in Chapter 13, is basically a 7-level code, allowing 128 discrete meanings for each code combination ($2^7 = 128$). An 8-level code would yield 256 possibilities.

Another concept that must be kept in mind as the discussion leads into coding is that bandwidth is related to information rate (more exactly, to modulation rate) or, for this discussion, to the number of bits per second transmitted. The goal is to keep some control over the amount of bandwidth necessary. It follows, then, that the coding length (number of levels) must be limited.

As it stands, an infinite number of amplitude levels are being presented to the coder on the PAM highway. If the excursion of the PAM wave is between 0 and +1 V, the reader should ask how many discrete values there are between 0 and 1. All values must be considered, even 0.0176487892 V.

The intensity range of voice signals over an analog telephone channel is on the order of 60 dB (see Section 3.2.7). The 0–1-V range of the PAM highway

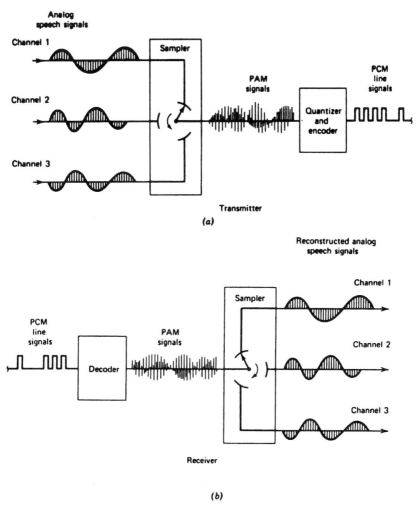

Figure 3.14. Simplified analogy of the formation of a PAM wave and PCM bit stream: (a) transmitter and (b) receiver. (From Ref. 11. Courtesy of GTE Lenkurt Inc., San Carlos, CA.)

at the coder input may represent that 50-dB range. Furthermore, it is obvious that the coder cannot provide a code of infinite length (e.g., an infinite number of coded levels) to satisfy every level in the 50-dB range (or a range from -1 to $+1$ V). The key is to assign discrete levels from -1 V through 0 to $+1$ V (50-dB range).

The assignment of discrete values to the PAM samples is called *quantization*. To cite an example, consider Figure 3.15. Between -1 and $+1$ V,

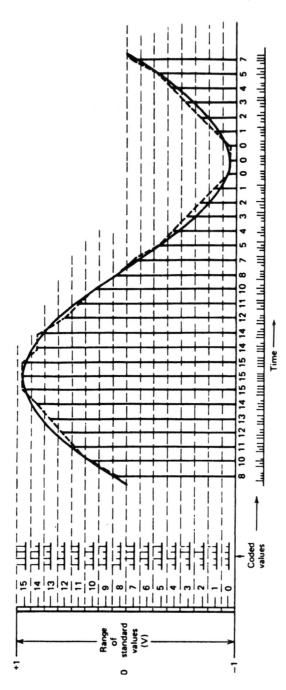

Figure 3.15. Quantization and resulting coding using 16 quantizing steps.

16 quantum steps exist and are coded as follows:

Step	Code	Step	Code
0	0000	8	1000
1	0001	9	1001
2	0010	10	1010
3	0011	11	1011
4	0100	12	1100
5	0101	13	1101
6	0110	14	1110
7	0111	15	1111

Examination of Figure 3.15 shows that step 12 is used twice. Neither time it is used is it the true value of the impinging sinusoid. It is a rounded-off value. These rounded-off values are shown with a dashed line that follows the general outline of the sinusoid. The horizontal dashed lines show the points where the quantum changes to the next higher or next lower level if the sinusoid curve is above or below that value. Take step 14, for example. The curve, dropping from its maximum, is given two values of 14 consecutively. For the first, the curve is above 14, and for the second, below. That error, in the case of 14, for instance, from the quantum value to the true value is called *quantizing distortion*. This distortion is the major source of imperfection in PCM systems.

In Figure 3.15, maintaining the $-1, 0, +1$-V relationship, let us double the number of quantum steps from 16 to 32. What improvement would we achieve in quantization distortion? First determine the step increment in millivolts in each case. In the first case the total range of 2000 mV would be divided into 16 steps, or 125 mV per step. The second case would have 2000/32, or 62.5 mV per step. For the 16-step case, the worst quantizing error (distortion) would occur when the input to be quantized was at the half-step level, or, in this case, 125/2 or 62.5 mV, above or below the nearest quantizing step. For the 32-step case the worst quantizing error (distortion) would again be at the half-step level, or 62.5/2 or 31.25 mV. Thus the improvement in decibels for doubling the number of quantizing steps is

$$20 \log\left(\frac{62.5}{31.25}\right) = 20 \log 2$$

$$\approx 6 \text{ dB}$$

This is valid for linear quantization only (see Section 3.3.2.7). Thus increasing the number of quantizing steps for a fixed range of input values reduces quantizing distortion accordingly. Experiments have shown that if 2048 uniform quantizing steps are provided, sufficient voice signal quality is achieved.

For 2048 quantizing steps a coder will be required to code the 2048 discrete meanings (steps). We find that a binary code with 2048 separate characters or meanings (one for each quantum step) requires an 11-element code, or $2^n = 2048$. Hence $n = 11$.

With a sampling rate of 8000 samples per second per voice channel, the binary information rate per voice channel will be 88,000 bps. Consider that equivalent bandwidth is a function of information rate; the desirability of reducing this figure is therefore obvious.

3.3.2.4 *Coding.* Practical PCM systems use 7- and 8-level binary codes, or

$$2^7 = 128 \text{ quantum steps}$$

$$2^8 = 256 \text{ quantum steps}$$

Two methods are used to reduce the quantum steps to 128 or 256 without sacrificing fidelity. These are nonuniform quantizing steps and companding prior to quantizing, followed by uniform quantizing. Keep in mind that the primary concern of digital transmission using PCM techniques is to transmit speech, as distinct from digital data transmission covered in Chapter 13, which deals with the transmission of data and message information. Unlike data transmission, in speech transmission there is a much greater likelihood of encountering signals of small amplitudes than those of large amplitudes.

A secondary, but equally important, aspect is that coded signals are designed to convey maximum information considering that all quantum steps (meanings, characters) will have an equally probably occurrence. (We obliquely refer to this inefficiency in Chapter 13 because practical data codes assume equiprobability. When dealing with a pure number system with complete random selection, this equiprobability does hold true. Elsewhere, particularly in practical application, it does not. One of the worst offenders is our written language. Compare the probability of occurrence of the letter *e* in written text with that of *y* or *q*.) To get around this problem larger quantum steps are used for the larger amplitude portion of the signal, and finer steps for signals with low amplitudes.

The two methods of reducing the total number of quantum steps can now be labeled more precisely:

- Nonuniform quantizing performed in the coding process.
- Companding (compression) before the signals enter the coder, which now performs uniform quantizing on the resulting signal before coding. At the receive end, expansion is carried out after decoding.

An example of nonuniform quantizing could be derived from Figure 3.15 by changing the step assignment. For instance, 20 steps may be assigned between 0.0 and $+0.1$ V (another 20 between 0.0 and -0.1 V, etc.), 15 between

0.1 and 0.2 V, 10 between 0.2 and 0.35 V, 8 between 0.35 and 0.5 V, 7 between 0.5 and 0.75 V, and 4 between 0.75 and 1.0 V.

Most practical PCM systems use companding to give finer granularity (more steps) to the smaller amplitude signals at the expense of the larger amplitude signals. This is instantaneous companding compared to syllabic companding described in Section 3.2.7. Compression imparts more gain to lower amplitude signals. The compression and later expansion functions are logarithmic and follow one of two laws, the A-law and the μ-law. The logarithmic curve for A-law compression may be plotted from the following formulas:

$$
\begin{aligned}
F_A(x) &= \mathrm{sgn}(x)\left[\frac{A|x|}{1 + \ln(A)}\right] && 0 \le |x| \le \frac{1}{A} \\
&= \mathrm{sgn}(x)\left[\frac{1 + \ln|Ax|}{1 + \ln(A)}\right] && \frac{1}{A} \le |x| \le 1
\end{aligned}
\tag{3.6}
$$

The inverse or expansion characteristic is defined as

$$
\begin{aligned}
F_A^{-1}(y) &= \mathrm{sgn}(y)\frac{|y|[1 + \ln(A)]}{A} && 0 \le |y| \le \frac{1}{1 + \ln(A)} \\
&= \mathrm{sgn}(y)\frac{(e^{|y|[1+\ln(A)]-1})}{A} && \frac{1}{1 + \ln(A)} \le |y| \le 1
\end{aligned}
\tag{3.7}
$$

where $y = F_A(x)$, $\mathrm{sgn}(x)$ = polarity of x, and $A = 87.6$.

The μ-law logarithmic curve may be plotted from

$$
F_\mu(x) = \mathrm{sgn}(x)\frac{\ln(1 + \mu|x|)}{\ln(1 + \mu)}
\tag{3.8}
$$

where x = input signal amplitude ($-1 \le x \le 1$), $\mathrm{sgn}(x)$ = polarity of x, and μ = amount of compression. Current North American and Japanese PCM systems use the value $\mu = 255$. Note that equations 3.7 and 3.8 use the natural logarithm rather than the base 10 logarithm.

A common expression used in dealing with the "quality" of a PCM signal is the signal-to-distortion ratio, expressed in decibels. Parameters A and μ determine the range over which the signal-to-distortion ratio is comparatively constant. This is the dynamic range. Using a μ of 100 can provide a dynamic range of 40 dB of relative linearity in the signal-to-distortion ratio. Bellamy (Ref. 12) states that an 8-bit ($\mu = 255$) codec provides a theoretical signal-to-distortion (S/D) ratio greater than 30 dB across a dynamic range of 48 dB. The 8-bit A-law codec provides similar S/D for about a 42-dB dynamic range.

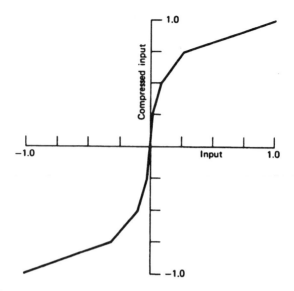

Figure 3.16. Seven-segment linear approximation of the logarithmic curve for the μ-law (μ = 100). (From Ref. 3; copyright © 1970 by Bell Telephone Laboratories. Reprinted with permission.)

In actual PCM systems the companding circuitry does not provide an exact replica of the logarithmic curves shown. The circuitry produces approximate equivalents using a segmented curve, each segment being linear. The more segments the curve has, the more it approaches the true logarithmic curve desired. Such a segmented curve is shown in Figure 3.16.

If the μ-law were implemented using a segment linear approximate equivalent, it would appear as shown in Figure 3.16. Thus, upon coding, the first three coded digits would indicate the segment number (e.g., $2^3 = 8$). Of the seven-digit code, the remaining four digits would divide each segment in 16 equal parts to further identify the exact quantum step (e.g., $2^4 = 16$).

For low-level signals the companding improvement is approximately

A-law	24 dB
μ-law	30 dB

using a 7-level code.*

Coding in PCM systems utilizes a straightforward binary coding. Two good examples of this coding are shown in Figures 3.17 and 3.18.

The coding process is closely connected to the quantizing. In fact, these two steps are carried out together. In practical systems, whether using A-law

---

*Older vintage PCM systems used 7-level (7-bit) coding and the 8th bit was used for supervisory signaling.

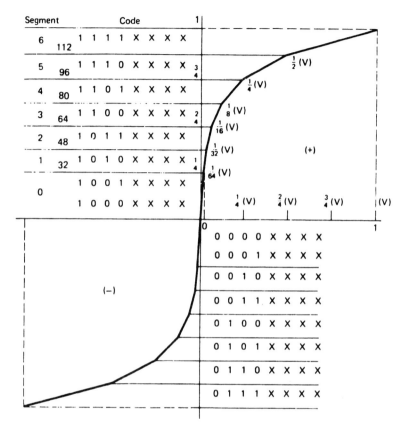

Figure 3.17. Quantization and coding used in the E-1 PCM system.

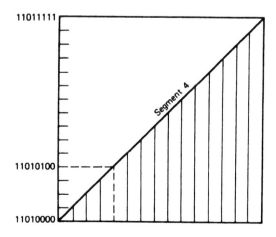

Figure 3.18. E-1 PCM system, coding of segment 4 (positive).

or μ-law, quantizing uses segmented equivalents as discussed previously and shown in Figure 3.16. Rather than use the term *segmented equivalents*, the term "linear approximations" might be more appropriate. In the coding process, it would be helpful to have 16 segments, which would make it compatible for binary coding.

The E-1 PCM system also uses a 13-segment approximation of the A-law, where $A = 87.6$. However, the segment passing through the origin contains four steps, two above and two below the x axis; thus a 16-segment representation is used, which leads us to an 8-level code. The coding for this system is shown in Figure 3.17. Again, if the first code element (bit) is 1, it indicates a positive value (e.g., the quantum step is located above the origin). The following three elements (bits) identify the segment, there being seven segments above the seven segments below the origin (horizontal axis).

As an example consider the fourth positive segment, given as 1101XXXX in Figure 3.17. The first 1 indicates that it is above the horizontal axis (e.g., it is positive). The next three elements indicate the fourth step, or

$$
\begin{aligned}
&0\text{---}1000 \text{ and } 1001 \\
&1\text{---}1010 \\
&2\text{---}1011 \\
&3\text{---}1100 \\
\rightarrow\ &4\text{---}1101 \\
&5\text{---}1110 \text{ etc.}
\end{aligned}
$$

Figure 3.18 shows a "blowup" of the uniform quantizing and subsequent straightforward binary coding of step 4; it illustrates final segment coding, which is uniform, providing 16 ($2^4 = 16$) coded quantum steps.

The North American DS1 PCM system uses a 15-segment approximation of the logarithmic μ-law. Again, there are actually 16 segments. The segments cutting the origin are collinear and are counted as one but are actually two, one above and one below the x axis. This leads to the 16 value. The quantization in the DS1 system is shown in Figure 3.19 for the positive portion of the curve. Segment 5 representing quantizing steps 64–80 is shown blown up in the figure. Figure 3.20 shows the DS1 coding. As can be seen again, the first code element, whether a 1 or a 0, indicates if the quantum step is above or below the horizontal axis. The next three elements identify the segment, and the last four elements (bits) identify the actual quantum level inside that segment.

3.3.2.5 The Concept of Frame. As shown in Figure 3.14, PCM multiplexing is carried out in the sampling process, sampling several sources sequentially. These sources may be nominal 4-kHz voice channels or other information sources, possibly data or video. The final result of the sampling and subsequent quantization and coding is a series of pulses, a serial bit stream that requires some indication or identification of the beginning of a

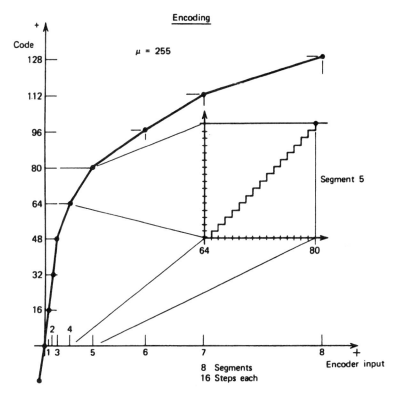

Figure 3.19. Positive portion of the segmented approximation of the μ-law quantizing curve used in the North American DS1 PCM channelizing equipment. (From Ref. 13. Courtesy of ITT Telecommunications, Raleigh, NC.)

scanning sequence. This identification tells the far-end receiver when each full sampling sequence starts and ends; it times the receiver. Such identification is called *framing*. A full sequence or cycle of samples is called a frame in PCM terminology.

CCITT Rec. G.701 (Ref. 14) defines a frame as:

A cyclic set of consecutive digit time slots in which the position of each digit time slot can be identified.

Consider the framing structure of several practical PCM systems. The AT&T D1A system is a 24-channel PCM system using a 7-level code (e.g., $2^7 = 128$ quantizing steps). To every 7 bits representing a coded quantum step, 1 bit is added for signaling. To the full sequence 1 bit is added, called a framing bit. Therefore a D1A-frame consists of

$$(7 + 1) \times 24 + 1 = 193 \text{ bits}$$

Code Level		Digit Number							
		1	2	3	4	5	6	7	8
255	(Peak positive level)	1	0	0	0	0	0	0	0
239		1	0	0	1	0	0	0	0
223		1	0	1	0	0	0	0	0
207		1	0	1	1	0	0	0	0
191		1	1	0	0	0	0	0	0
175		1	1	0	1	0	0	0	0
159		1	1	1	0	0	0	0	0
143		1	1	1	1	0	0	0	0
127	(Center levels)	1	1	1	1	1	1	1	1
126	(Nominal zero)	0	1	1	1	1	1	1	1
111		0	1	1	1	0	0	0	0
95		0	1	1	0	0	0	0	0
79		0	1	0	1	0	0	0	0
63		0	1	0	0	0	0	0	0
47		0	0	1	1	0	0	0	0
31		0	0	1	0	0	0	0	0
15		0	0	0	1	0	0	0	0
2		0	0	0	0	0	0	1	1
1		0	0	0	0	0	0	1	0
0	(Peak negative level)	0	0	0	0	0	0	1*	0

*One digit added to ensure that timing content of transmitted pattern is maintained.

Figure 3.20. Eight-level coding of the North American (AT&T) DS1 PCM system. Note that actually there are only 255 quantizing steps because steps 0 and 1 use the same bit sequence, thus avoiding a code sequence with no transitions (i.e., 0's only). (From Ref. 13. Courtesy of ITT Telecommunications, Raleigh, NC.)

making up a full sequence of frame. By definition, 8000 frames are transmitted, so the bit rate is

$$193 \times 8000 = 1,544,000 \text{ bps}$$

The E-1 system is a 32-channel system where 30 channels transmit speech derived from incoming telephone trunks and the remaining two channels transmit signaling and synchronization information. Each channel is allotted a time slot, and we can speak of time slots 0–31 as follows:

Time Slot	Type of Information
0	Synchronizing (framing)
1–15	Speech
16	Signaling
17–31	Speech

In time slot 0 a synchronizing code or word is transmitted every second frame, occupying digits 2–8 as follows:

$$0011011$$

In those frames without the synchronizing word, the second bit of time slot 0 is frozen at a 1 so that in these frames the synchronizing word cannot be imitated. The remaining bits of time slot 0 can be used for the transmission of supervisory information signals.

The current North American basic 24-channel PCM system, typified by the AT&T D1D channel bank, varies compared with the older D1A system described previously in that all 8 bits of a channel word are used in five out of six frames. In the remaining frame digit 8 is used for signaling. To accommodate these changes, it is necessary to change the framing format so that the specific frames containing signaling information can be identified. By using 8 bits instead of 7 for each channel word, allowing 256 amplitude values to be represented instead of 128 (in five out of six frames), quantizing noise is reduced. The companding characteristic is also different from its older D1A counterpart, D1D uses $\mu = 255$, whereas with D1A $\mu = 100$. This change made a significant difference in the signal-to-noise ratio over a wide range of input signals.

The D1D frame has a similar makeup as the D1A frame in that

$$8 \times 24 + 1 = 193 \text{ bits per frame}$$

producing a line data rate of $193 \times 8000 = 1.544$ Mbps. The frame structure is shown in Figure 3.21. Note that signaling is provided by "robbing" bit 8 from every channel in every sixth frame. For all other frames all bits are used to transmit information coding.

Framing and basic timing should be distinguished. Framing ensures that the PCM receiver is aligned regarding the beginning (and end) of a sequence or frame; timing refers to the synchronization of the receiver clock, specifically, that it is in step with its companion (far-end) transmit clock. Timing at the receiver is corrected via the incoming mark–space (and space–mark) transitions. It is important, then, that long periods without transitions do not occur. This point is discussed later in reference to line codes and digit inversion. (See Section 3.3.4.)

3.3.2.6 *Details of and Enhancements to the DS1 Frame.* A TDM demultiplexer has to identify individual time slot positions so that they can relate to the particular channel inputs of the associated multiplexer or channel bank. A typical demultiplexer uses a counter synchronized to the

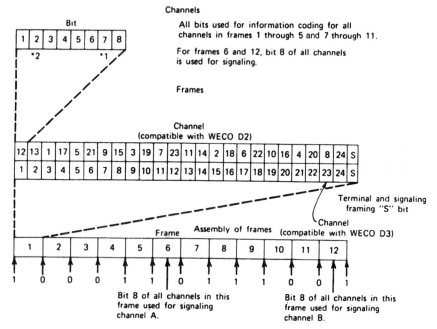

Figure 3.21. Frame structure of the North American (AT&T) D1D (DS1) PCM system for the channel bank. Note the bit "robbing" technique used on each sixth frame to provide signaling information. *Notes:* (1) If bits 1–6 and 8 are 0, then bit 7 is transmitted as 1. (2) Bit 2 is transmitted as 0 on all channels for transmission if end-to-end alarm. (3) Composite pattern 000110111001, etc. (From Ref. 13. Courtesy of ITT Telecommunications, Raleigh, NC.)

frame format of its companion multiplexer. There are at least five regimes available to carry out PCM frame alignment so that time slots can be appropriately identified. Two of these regimes are used in common PCM practice today. E-1 uses *added-channel framing*, as described in Section 3.3.2.5. The North American DS1 PCM system uses *added-bit framing* by means of a framing bit or *F-bit*. This is the first bit of a DS1 193-bit frame. When superframe implementations are used, there are 12 frames in the conventional superframe and thus 12 framing-bit slots of which 6 are used for frame delineation and 6 for superframe delineation (Ref. 28).

AT&T Technical Advisory No. 71 (Ref. 15) outlines two distinct methods of handling the F-bits (framing bits) of the DS1 format. The F-bit is bit 1 of a 193-bit sequence we call a frame. The first method is where a 12-frame superframe is developed. This format is shown in Table 3.3. The second method is where a 24-frame group is used and this is called the *extended superframe* (ESF). This is illustrated in Table 3.4.

The framing-bit sequence (terminal framing F_t in the table) of the superframe is 101010, whereas with the extended superframe it is the binary

TABLE 3.3   The DS1 Superframe Format

Frame Number	Bit Number	S-Bits Terminal Framing F_t	S-Bits Signaling Framing F_s	Traffic	Sig	Signaling Channel
1	0	1	—	1–8	—	
2	193	—	0	1–8		
3	386	0	—	1–8		
4	579	—	0	1–8	—	
5	772	1	—	1–8	—	
6	965	—	1	1–7	8	A
7	1158	0	—	1–8	—	
8	1351	—	1	1–8	—	
9	1544	1	—	1–8	—	
10	1737	—	1	1–8	—	
11	1930	0	—	1–8	—	
12	2123	—	0	1–7	8	B

Notes: Frame 1 transmitted first.
Frames 6 and 12 are denoted signaling frames.
Source: Reference 15, Figure 4.

sequence 001011 found in the FPS column of Table 3.4. It should be noted that there is a distinct incompatibility between systems operating in the superframe regime and those operating with the extended superframe.

Turning to Table 3.3, we make two observations:

1. Terminal framing (F_t) bits identify frame boundaries.
2. Signaling framing (F_s) bits identify superframe boundaries. (When the 192-bit time slots are channelized, the F_s-bits identify the robbed-bit signaling frames and associated signaling channels A and B.)

3.3.2.6.1 *The Extended Superframe.* An extended superframe (ESF) consists of 24 consecutive frames. With an ESF, the F-bits are used for a variety of functions as shown in Table 3.4. Beside the frame alignment discussed above, two additional functions are identified: the cyclic redundancy check (CRC) and the ESF data link.

The rationale for the provision of these important overhead functions was that one did not have to tell the distant end 8000 times a second where a frame started. With modern alignment algorithms and search strategies, only one frame in four provides an FPS (framing pattern sequence) bit. The remainder of the framing bits support those two overhead functions.

CYCLIC REDUNDANCY CHECK (CRC). Two kilobits per second are dedicated to error detection using a CRC-6 code. The CRC-6 bits transmitted in ESF ($n + 1$) and the CRC calculation is based on the generating polynomial

TABLE 3.4   The DS1 Extended Superframe Format

Frame Number	F-Bits				Bit Use int Each Time Slot		Robbed Bit Signaling
	Bit Number	FPS	DL	CRC	Traffic	Sig	
1	0	—	m	—	1–8	—	
2	193	—	—	C1	1–8	—	
3	386	—	m	—	1–8	—	
4	579	0	—	—	1–8	—	
5	772	—	m	—	1–8	—	
6	965	—	—	C2	1–7	8	A
7	1158	—	m	—	1–8	—	
8	1351	0	—	—	1–8	—	
9	1544	—	m	—	1–8	—	
10	1737	—	—	C3	1–8	—	
11	1930	—	m	—	1–8	—	
12	2123	1	—	—	1–7	8	B
13	2316	—	m	—	1–8	—	
14	2509	—	—	C4	1–8	—	
15	2702	—	m	—	1–8	—	
16	2895	0	—	—	1–8	—	
17	3088	—	m	—	1–8	—	
18	3281	—	—	C5	1–7	8	C
19	3474	—	m	—	1–8	—	
20	3667	1	—	—	1–8	—	
21	3860	—	m	—	1–8	—	
22	4053	—	—	C6	1–8	—	
23	4246	—	m	—	1–8	—	
24	4439	1	—	—	1–7	8	D

Notes: Frame 1 transmitted first.
Frames 6, 12, 18, and 24 are denoted signaling frames.
FPS = framing pattern sequence (...001011...).
DL = 4-kbps facility data link (message bits *m*).
CRC = CRC6 cyclic redundancy check (bits C1–C6).
Source: Reference 16, Table 1.

$X^6 + X + 1$. The coefficients of the remainder polynomial are used in order of occurrence, as the ordered set of check bits C1 through C6 for ESF $(n + 1)$. The check bits C1 through C6 contained in an ESF are always those associated with the content of the ESF immediately preceding.*

THE ESF DATA LINK. Beginning with the F-bit of frame 1, every other F-bit of an ESF is used to form a 4-kbps channel, referred to as the "data link" or DL.

Two distinct formats are available for the DL. The first format supports a scheduled message that uses a 15-byte packet to transmit facility perfor-

*See Chapter 13 for a more detailed description of how CRC operates.

mance information. The second format supports a group of unscheduled messages that use repeated 16-bit codewords to send alarms, commands, and responses.

When no other signals are present, the DL sends an idle sequence consisting of repetitions of the pattern 01111110.

SCHEDULED PERFORMANCE REPORT MESSAGE. Once each second, a performance report message is sent by a destination terminal receiving a DS1 signal back to the source terminal originating that DS1 signal. The message informs the source terminal of transmission errors received by the destination on the DS1 facility between them. In this manner, a performance report message transmitted in one direction indicates the quality of transmission in the opposite direction.

In full-duplex operation, performance report messages are sent in both directions of transmission.

Separate counts of transmission error events are taken in each contiguous 1-s interval. The 1-s timing may be derived from the DS1 signal or from a separate, equally accurate (± 32 ppm) source. The beginning of the first 1-s interval does not necessarily correspond to the time of occurrence of any error event.

Transmission error events include CRC errors, frame synchronization bit errors, severely errored framing, line code violations, and controlled slips. (*Note*: Controlled slip reporting depends on the particular application and synchronization procedures. They are detected and reported whenever feasible to do so.)

The performance report also indicates when no errors have occurred.

The five different error events are described below.

1. *CRC Error Event.* This is the occurrence of a received CRC-6 code that is not identical to the corresponding locally calculated code.
2. *Frame Synchronization Bit Error Event.* This is the occurrence of a single bit error within the framing pattern sequence in the terminal synchronization channel.
3. *Severely Errored Framing Event.* This is the occurrence of two or more frame synchronization bit errors within a 3-ms period. Contiguous 3-ms intervals are examined. The 3-ms period may coincide with the extended superframe.
4. *Line Code Violation Event.* A line code violation event for an AMI coded signal is the occurrence of a received bipolar violation (BPV). A line code violation event for a B8ZS (see Section 3.3.4) coded signal is the occurrence of a BPV that is not the associated zero-substitution code. (Binary *N*-zero substitution [BNZS] is described in Section 3.3.4.)
5. *Controlled Slip Event.* This is the occurrence of a replication or deletion of one or more DS1 frames by the receiving terminal. A controlled slip

occurs when the difference in the timing between a synchronous receiving terminal and the received signal is of such magnitude that it exhausts the buffer capacity of the synchronous terminal. (See Chapter 4 for a discussion of slips.)

Table 3.5 shows the scheduled performance report message format.

3.3.2.7 Quantizing Distortion. Quantizing distortion has been defined as the difference between the signal waveform as presented to the PCM multiplex (codec*) and its equivalent quantized value. Quantizing distortion produces a signal-to-distortion ratio S/D given by (Ref. 3, equation 28-10, p. 615)

$$\frac{S}{D} = 6n + 1.8 \text{ dB} \quad \text{(for uniform quantizing)}$$

where n = number of bits used to express a quantizing level. This bit grouping is often referred to as a PCM word. For instance, the AT&T D1A system uses a 7-bit codeword to express a level, and the 30 + 2 and D1D systems use essentially 8 bits.

With a 7-bit codeword (uniform quantizing),

$$\frac{S}{D} = 6 \times 7 + 1.8 = 43.8 \text{ dB}$$

This demonstrates the linear relationship between the number of digits per sample and the signal-to-distortion ratio in decibels. Each added digit increases the signal-to-distortion ratio by 6 dB. Practical signal-to-distortion values 8 range on the order of 33–38 dB[†] for average talker levels.

3.3.2.8 Idle Channel Noise. An idle PCM channel can be excited by the idle Gaussian noise, ac hum (50 or 60 Hz), and crosswalk present on the input channel. A decision threshold may be set that would control idle noise if it remains constant. With a constant-level input there will be no change in codeword output, but any change of amplitude will cause a corresponding change in codeword, and the effect of such noise may be an annoyance to the telephone listener, particularly during pauses in conversation.

One important overall PCM design issue to control idle channel noise is the selection of small signal quantizing values. Of course, these are the values near the origin on the segmented logarithmic companding curve. Typical curves are shown in Figures 3.16–3.19. The question arises as to how to bias these values to minimize idle channel noise. Biasing in the y axis (up and

---

*The term *codec*, meaning coder–decoder, is analogous to *modem* in analog circuits.
[†] Using 8-level coding.

TABLE 3.5 Scheduled Performance Report Message Format

Octet Number	Bit Number 8	7	6	5	4	3	2	1		Octet Content
1	FLAG									01111110
2	SAPI						C/R	EA		00111000 or 00111010
3	TEI							EA		00000001
4	CONTROL									00000011
5	G3	LV	G4	U1	U2	G5	SL	G6	} t_0	
6	FE	SE	LB	G1	R	G2	Nm	NI		
7	G3	LV	G4	U1	U2	G5	SL	G6	} $t_0 - 1$	
8	FE	SE	LB	G1	R	G2	Nm	NI		
9	G3	LV	G4	U1	U2	G5	SL	G6	} $t_0 - 2$	One-Second
10	FE	SE	LB	G1	R	G2	Nm	NI		Report
11	G3	LV	G4	U1	U2	G5	SL	G6	} $t_0 - 3$	
12	FE	SE	LB	G1	R	G2	Nm	NI		
13	FCS									Variable
14										01111110
15	FLAG									

Address	Interpretation
00111000	SAPI = 14, C/R = 0 (CI) EA = 0
00111010	SAPI = 14, C/R = 1 (carrier) EA = 0
00000001	TEI = 0, EA = 1
Control	**Interpretation**
00000011	Unacknowledged information transfer
One-Second Report	**Interpretation**
G1 = 1	CRC error event = 1
G2 = 1	1 < CRC error event ≤ 5
G3 = 1	5 < CRC error event ≤ 10
G4 = 1	10 < CRC error event ≤ 100
G5 = 1	100 < CRC error event ≤ 319
G6 = 1	CRC error event > 320
SE = 1	Severely errored framing event ≥ 1 (FE shall = 0)
FE = 1	Frame synchronization bit error event ≥ 1 (SE shall = 0)
LV = 1	Line code violation event ≥ 1
SL = 1	Controlled slip event ≥ 1
LB = 1	Payload loopback activated
U1, U2 = 0	Under study for synchronization
R = 0	Reserved (default value is 0)
Nm, NI = 00, 01, 10, 11	One-second report modulo 4 counter
FCS	**Interpretation**
Variable	CRC16 frame check sequence

Note: Bit 1 of each octet is transmitted first.

Source: Reference 17, Table 10-11.

down) is called midriser biasing; biasing in the x axis (left and right) is called midtread biasing.

The length of quantizing segments is related to the power of 2. The 15-segment ($\mu = 255$ law) and the 13-segment ($A = 87.6$ law) both use the power of 2 relationship. For the μ-law we have a midtread codec with 8 segments of each polarity with a segment size increasing by a factor of 2 in successive segments. Step size at segment boundaries is adjusted so that the output levels always are halfway between decision thresholds.

In the case of the A-law codec we consider it to have 8 segments of each polarity where the two innermost segments of each are collinear, which leads to its name, a 13-segment codec. Again, the output levels are always midway between decision thresholds. But in this case small signal biasing values near the origin are twice as large as those for the 15-segment μ-law codec. These characteristics cause the 13-segment A-law codec to have poorer idle channel noise characteristics than its μ-law counterpart. On the other hand, the A-law codec exhibits flat signal-to-distortion performance over a somewhat wider range of signal levels.

Midtread quantization characteristics can give better quantization characteristics than midriser (Ref. 12).

CCITT Rec. G.712 (Ref. 18) states that with the input and output ports of the channel terminated in the nominal impedance, idle channel noise should not exceed -65 dBm0p.

3.3.3 Operation of a PCM Codec (Channel Bank)

PCM is four-wire. Voice channel inputs and outputs to and from a PCM multiplex are on a four-wire basis. The term *codec* is used to describe a unit of equipment carrying out the function of PCM multiplex and demultiplex and stands for coder–decoder, even though the equipment carries out more functions than just coding and decoding. A block diagram of a codec is shown in Figure 3.22.

A codec accepts 24 or 30 voice channels, depending on the system used, and digitizes and multiplexes the information. It delivers 1.544 Mbps to the line for the AT&T DS1 channelizing equipment and 2.048 Mbps for the E-1 channelizing equipment. On the decoder side it accepts a serial bit stream at one or the other line modulation rate, demultiplexes the digital information, and performs digital-to-analog conversion. Output to the analog telephone network is 24 or 30 nominal 4-kHz voice channels. Figure 3.22 illustrates the processing of a single analog voice channel through a codec. The voice channel to be transmitted is passed through a 3.4-kHz low-pass filter. The output of the filter is fed to a sampling circuit. The sample of each channel of a set of n channels (n usually equals 24 or 30) is released in turn to the pulse amplitude modulation (PAM) highway. The release of samples is under control of a channel gating pulse derived from the transmit clock. The input to the actual coder is the PAM highway. The coder accepts a sample of each

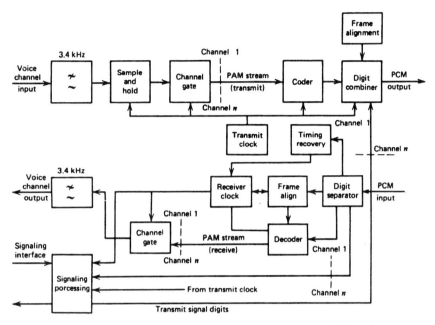

Figure 3.22. Simplified functional block diagram of a typical PCM codec.

channel in sequence and then generates the appropriate signal character (channel word) corresponding to each channel presented. The coder output is the basic PCM signal that is fed to the digit combiner, where framing alignment signals are inserted in the appropriate time slots, as well as the necessary supervisory signaling digits corresponding to each channel (European approach). The output of the digit combiner is the serial PCM bit stream fed to the line. As mentioned previously, supervisory signaling is carried out somewhat differently in the North American (AT&T) approach by robbing 1 bit in frame 6 and in frame 12 for this purpose.

On the receive side the codec accepts the serial PCM bit stream at the line rate at the digit separator. Here the signal is regenerated and split four ways to carry out the following processing functions: (1) timing recovery, (2) decoding, (3) frame alignment, and (4) signaling (supervisory). The timing recovery keeps the receive clock in synchronism with the far-end transmit clock. The receive clock provides the necessary gating pulses for the receive side of the PCM codec. The frame alignment circuit senses the presence of the frame alignment signal at the correct time interval, thus providing the receive terminal with frame alignment. The decoder, under control of the receive clock, decodes the code character (channel word) signals corresponding to each channel. The output of the decoder consists of the reconstituted pulses making up a PAM highway. The channel gate accepts the PAM highway, gating the n-channel PAM highway in sequence under control of the receive clock. The output of the channel gate is fed in turn to each channel filter, therefore enabling the reconstituted analog voice signal to reach each appropriate voice path. Gating pulses extract signaling information in the signaling processor and apply this information to each of the reconstituted voice channels with the supervisory signaling interface as required by the analog telephone system in question.

3.3.4 The Line Code

PCM signals as transmitted to wire-pair cable are in the bipolar mode (biternary), as shown in Figure 3.12. The marks or 1's have only a 50% duty cycle. There are several advantages to this mode of transmission:

- No dc return is required; hence transformer coupling can be used on the line.
- The power spectrum of the transmitted signal is centered at a frequency equivalent to half the bit rate.

It will be noted in bipolar transmission that the 0's are coded as absence of pulses, and the 1's are alternately coded as positive and negative pulses, with the alternation taking place at every occurrence of a 1. This mode of transmission is also called alternate mark inversion (AMI).

One drawback to straightforward AMI transmission is that when a long string of 0's is transmitted (e.g., no transitions), a timing problem may come about because regenerative repeaters and decoders have no way of extracting timing with state transitions. The problem is alleviated by forbidding long strings of 0's. Codes have been developed that are bipolar but with N zeros substitution; they are called BNZS codes. The objective here is to assure a certain 1's density. As bit rates increase, the 1's density requirement becomes more stringent. In our description below, B represents a pulse conforming to the alternating polarity rule, V represents a pulse violating that rule, and 0 represents a binary zero.

3.3.4.1 Bipolar with 8 Zero Substitution (B8ZS).

Bellcore (Ref. 17) recommends B8ZS as the long-term network solution in providing clear channel capability in a DS1 and DS1C.* This coding forbids, any sequence of eight consecutive zeros (00000000). Such code sequence is replaced with (000VB0VB). The polarity of the V (violation) pulses in the fourth and seventh bit positions are the same as the preceding pulses in the bit stream. As a result, if the preceding pulse were positive (+), the B8ZS substitution would be (000 + − 0 − +), while for a preceding negative pulse (−), the substitution would be (000 − + 0 + −). At the receiver, the decoder recognizes the 000VB0VB code and replaces it with the original eight 0's. Table 3.6 illustrates an example of B8ZS coding.

---

*DS1C is described in Section 3.3.7.

TABLE 3.6   Example of B8ZS Coding Used with the DS1, DS1C Rates

Preceding Bit	Next 8 Bits	B8ZS Substitution	Output Bits
+	00000000	000VB0VB	000 + − 0 − +
−	00000000	000VB0VB	000 − + 0 + −

TABLE 3.7   Example of B6ZS Coding Used with the DS2 Rate

Preceding Bit	Next 6 Bits	B6ZS Substitution	Output Bits
+	000000	0VB0VB	0 + − 0 − +
−	000000	0VB0VB	0 − + 0 −

Bipolar with 6 Zero Substitution (B6ZS). B6ZS is used with the DS2 line rate. This code forbids, any sequence of six consecutive zeros. That bit sequence is replaced with (0VB0VB). The bipolar violations occur in the second and fifth bit positions of the substitutions. At the receiver, the decoder recognizes the (0VB0VB) sequence and replaces it with the original six zeros. Table 3.7 illustrates an example of B6ZS coding.

Bipolar with 3 Zero Substitution (B3ZS). B3ZS is used with the DS3 line rate. In this case, any sequence of three consecutive zeros is replaced by (00V) or (B0V). To avoid introducing unnecessary dc component to the signal, the choice of (00V) or (B0V) is made so that the polarity of consecutive V pulses alternates. (This is equivalent to requiring the number of B-pulses between successive V pulses to be odd.) At the receiver, the decoder recognizes the substitution and replaces it with the original three zeros. HDB3 (high density binary three) used with the E1 family of PCM systems is the same as B3ZS; different names for the same thing.

3.3.4.2   *Zero Byte Time Slot Interchange (ZBTSI).* ZBTSI is a signal processing technique that operates with DS1 only. It requires an additional 2 kbps of overhead. Thus the use of ESF is mandatory. Key to ZBTSI is the maintenance of sufficient 1's density to meet timing requirements. It substitutes a variable address word for bytes with all zeros and is referred to as "zero bytes." Bellcore (Ref. 17) states that certain bit sequences that result from the ZBTSI process do not meet the constraints of DS1 1's density requirements. Any violations that occur do not persist long enough to have any impact on performance.

The necessary overhead required for ZBTSI is carried in the ESF DL (data link). Each of these overhead bits is associated with 96 bytes or octets that immediately follow it. The data in the 96-octet block consists of PCM data from four DS1 frames.

The data is scrambled and after scrambling, the octets are buffered at the encoder, and each octet of the 96 is examined, along with the two octets adjacent to it. If, in those three octets (24 bits), there is produced a string of zeros longer than 15 or an insufficient number of 1's on the data (1's density), the all-zero octets are flagged and identified by an address chain. A one-way transmission delay of 0.5 ms is incurred because DS1 frames are buffered for each encode/decode pair encountered.

3.3.5   Signal-to-Gaussian-Noise Ratio on PCM Repeatered Lines

As we mentioned earlier, noise accumulation on PCM systems is not an important consideration. This does not mean that Gaussian noise (nor crosstalk, impulse noise) is not important. Indeed, it does affect the error performance, expressed as error rate (see Chapter 13). The error rate, from one point of view, is cumulative. A decision in error, whether 1 or 0, made anywhere in the digital system, is not recoverable.* Thus such an incorrect decision made by one regenerative repeater adds to the existing error rate on the line, and errors taking place in subsequent repeaters further down the line add in a cumulative manner, tending to deteriorate the received signal.

In a purely binary transmission system, if a 20-dB signal-to-noise ratio is maintained, the system operates nearly error free. In this respect, consider Table 3.8.

As discussed in Section 3.3.4, PCM, in practice, is transmitted online with alternate mark inversion. The marks have a 50% duty cycle, permitting energy concentration at a frequency of half the transmitted bit rate. Consequently, it is advisable to add 1 or 2 dB to the values shown in Table 3.8 to achieve a desired error rate on a practical system.

3.3.5.1   *The Eye Pattern.*   The "eye" pattern provides a convenient method of checking the quality of a digital transmission line. A sketch of a typical eye pattern is shown in Figure 3.23. Any oscilloscope can produce a suitable eye pattern provided it has the proper rise time, which most quality oscilloscopes now available on the market have. The oscilloscope should either terminate or bridge the repeatered line or output of a terminal repeater. The display on the oscilloscope contains all the incoming bipolar pulses superimposed on one another.

Eye patterns are indicative of decision levels. The wider the eye opening vertically the better defined is the decision (whether 1 or 0 in the case of PCM). The opening is often referred to as the decision area (crosshatched in Figure 3.23). Degradations reduce the area. Eye patterns are often measured off in the vertical, giving a relative measure of the margin of decision.

Amplitude degradations shrink the eye in the vertical. Among amplitude degradations can be included echos, intersymbol interference, and decision threshold uncertainties.

---

*Unless some special form of coding is used to correct the errors (see Chapters 5 and 13).

**TABLE 3.8   Error Rate of a Binary Transmission System Versus
Signal-to-RMS Noise Ratio**

Error Rate	S/N (dB)	Error Rate	S/N (dB)
10^{-2}	13.5	10^{-7}	20.3
10^{-3}	16	10^{-8}	21
10^{-4}	17.5	10^{-9}	21.6
10^{-5}	18.7	10^{-10}	22
10^{-6}	19.6	10^{-11}	22.2

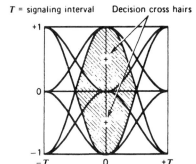

Figure 3.23. Sketch of an eye pattern.

Horizontal shrinkage of the eye pattern is indicative of timing degradations (i.e., jitter and decision time misalignment).

Noise is the other degradation to be considered. Usually noise may be expressed in terms of some improvement in the signal-to-noise ratio to bring the operating system into the bounds of some desired objective (e.g., see Table 3.8). This ratio may be expressed as 20 × log* of the ideal eye opening (in the vertical as read on the oscilloscope's vertical scale) to the degraded reading.

3.3.6   Regenerative Repeaters

As we probably know, pulses passing down a digital transmission line suffer attenuation and are badly distorted by the frequency characteristic of the line. A regenerative repeater amplifies and reconstructs such a badly distorted digital signal and develops a nearly perfect replica of the original at its output. Regenerative repeaters are an essential key to digital transmission in that we could say that the "noise stops at the repeater."

Figure 3.24 is a simplified block diagram of a regenerative repeater and shows typical waveforms corresponding to each functional stage of signal processing. As shown in the figure, the first stage of signal processing is

*Oscilloscopes are commonly used to measure voltage. Thus we can measure the degraded opening in voltage units and compare it to the full-scale perfect opening in the same units. A ratio is developed, and we take 20 log that ratio to determine the signal-to-noise ratio.

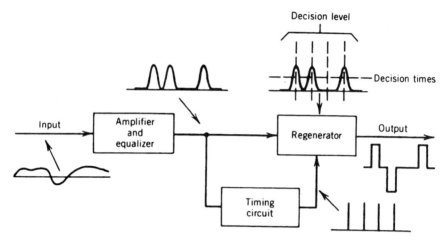

Figure 3.24. Simplified functional block diagram of a regenerative repeater for use on PCM cable systems.

amplification and equalization. Equalization is often a two-step process. The first is a fixed equalizer that compensates for the attenuation–frequency characteristic of the nominal section, which is the standard length of transmission line between repeaters (often 6000 ft). The second equalizer is variable and compensates for departures between nominal repeater section length and the actual length and loss variations due to temperature. The adjustable equalizer uses automatic line build-out (ALBO) networks that are automatically adjusted according to characteristics of the received signal.

The signal output of the repeater must be accurately timed to maintain accurate pulsewidth and space between the pulses. The timing is derived from the incoming bit stream. The incoming signal is rectified and clipped, producing square waves that are applied to the timing extractor, which is a circuit tuned to the timing frequency. The output of the circuit controls a clock-pulse generator that produces an output of narrow pulses that are alternately positive and negative at the zero crossings of the square wave input.

The narrow positive clock pulses gate the incoming pulses of the regenerator, and the negative pulses are used to run off the regenerator. Thus the combination is used to control the width of the regenerated pulses.

Regenerative repeaters are the major source of timing jitter found in a digital transmission system. Jitter is one of the principal impairments in a digital network, giving rise to pulse distortion and intersymbol interference. Jitter is discussed in more detail in Chapter 4.

Most regenerative repeaters transmit a bipolar (AMI) waveform (see Figure 3.12). Such signals can have one of three possible states in any instant in time—positive, zero, or negative—and are often designated $+, 0, -$. The threshold circuits are gated to admit the signal at the middle of the pulse interval. For example, if the signal is positive and exceeds a positive threshold, it is recognized as a positive pulse. If it is negative and exceeds a negative threshold, it is recognized as a negative pulse. If it has a value between the positive and negative thresholds, it is recognized as a 0 (no pulse).

3.3.7 Higher-Order PCM Multiplex Systems

In Section 3.2 a FDM multiplex hierarchy was developed based on the 12-channel group. Five such groups were formed into a supergroup, thence the mastergroup and the supermastergroup. Likewise, in PCM a hierarchy of multiplex is developed based on the 24- or 30-channel group, which is called level 1. Subsequent levels are then developed (i.e., levels 2, 3, 4, and in one system, level 5). Table 3.9 summarizes and compares these multiplex levels for the North American system, Japan, and Europe (based on CCITT). The North American PCM hierarchy is shown in Figure 3.25, giving respective DS line rates and multiplex nomenclature. Regarding this nomenclature, we see from the figure that M12 accepts 1-level input, delivering 2-level to the line. It actually accepts four DS1 inputs, deriving a DS2 output (6.312 Mbps). M13 accepts 1-level inputs, delivering 3-level to the line. In this case 28 DS1 inputs form one DS3 output (44.736 Mbps); the M34 takes six DS3 inputs (level 3) to form one DS4 line rate (274.176 Mbps). DS1C is a special case where two 1.544-Mbps DS1 rates are multiplexed to form a 48-channel group with a line rate of 3.152 Mbps.

By simple multiplication we can see that the higher-order line rate is a multiple of the lower input rate plus some number of bits. The DS1C is an example. Here the line rate is 3152 kbps, which is 2×1544 kbps + 64 kbps. The additional 64 kbps is used for multiplex synchronization and framing. Multiplex (and demultiplex) timing is very important, as one might imagine.

TABLE 3.9   PCM Multiplex Hierarchy Comparison

	Level				
System Type	1	2	3	4	5
North American	1	2	3	4	
Number of voice channels	24	96	672	4032	
Line bit rate (Mbps)	1.544	6.312	44.736	274.176	
Japan					
Number of voice channels	24	96	480	1440	5760
Line bit rate (Mbps)	1.544	6.312	32.064	97.728	397.200
Europe					
Number of voice channels	30	120	480	1920	7680
Line bit rate (Mbps)	2.048	8.448	34.368	139.264	565.148

Sources: References 3, 17, and 20.

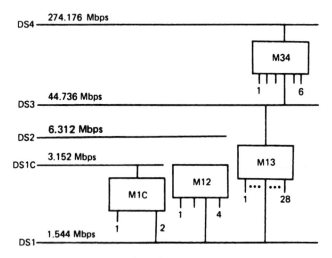

Figure 3.25. North American (AT&T) PCM hierarchy and multiplexing plan.

The two DS1 signal inputs are each 1.544 Mbps plus and minus some tolerance (actually specified as ±130 ppm). The two input signals must be made alike in repetition rate and a rate suitable for multiplexing. This is done by *bit stuffing*. In this process time slots are added to each signal in sufficient quantity to make the signal operate at a precise rate controlled by a common clock circuit in the multiplex. Pulses are inserted (or stuffed) into these time slots but carry no information. Thus it is necessary to code the signal in such a manner that these noninformation bits can be recognized and removed at the receiving terminal (demultiplex).

Of course, in the above example we are dealing with two DS1 sources that are physically separate and controlled by different clocks. If the sources were co-located (no difference in arrival due to delay) and controlled by a common clock, bit stuffing would not be necessary. However, codecs usually operate with free-running clocks (see Table 3.14, tolerance). Suppose we had two DS1 tributary codecs multiplexed by a M1C multiplexer (Figure 3.25) and each codec operated at maximum tolerance on the high side (i.e., 1.544 × 130 ppm × 2), which is 401 bps in excess of 2 × 1.544 bps. If the output of the M1C were just 2 × 1.544 Mbps, its buffers would be accumulating 401 bps more than would be released to the multiplexed line. In some time period, depending on buffer size, the buffers would overflow and a frame would have to be dropped or "slipped" periodically. Such loss of data is called a "slip," which is a highly undesirable impairment. This then is one cogent reason why PCM multiplex line rates must be greater than the aggregate nominal input rates. These added bits to compensate for input tributary variation are called "stuff" bits.

Consider the more general case using CCITT terminology. If we wish to multiplex several lower-level PCM bit streams deriving from separate tributaries into a single PCM bit stream at a higher level, a process of *justification* is required (called *bit stuffing* above). ITU-T (Rec. G.701 defines justification (Ref. 14):

> A process of changing the rate of a digital signal in a controlled manner so that it can accord with a rate different from its own inherent rate usually without loss of information.

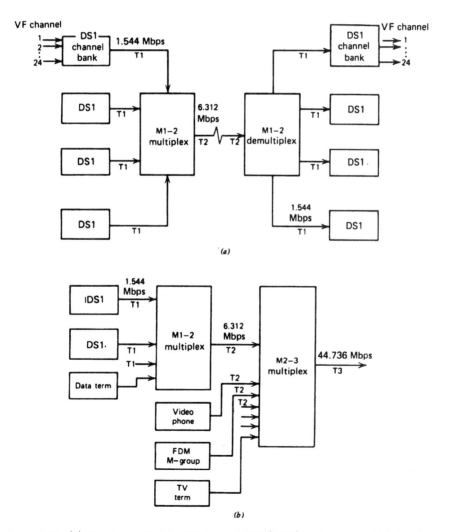

Figure 3.26. (a) Development of the 96-channel DS2 (AT&T) system by multiplexing four 24-channel DS1 channel bank outputs. (b) Development of higher-order PCM (AT&T plan). (From Ref. 3; copyright © 1970 by Bell Telephone Laboratories from 1970 edition. Reprinted with permission.)

Positive justification (as previously noted) adds or stuffs digits; negative justification deletes or slips digits.

In the case of positive justification, normally each separate tributary bit stream is read into a store at its own data rate t, but the store is read out at a rate corresponding to T/n, where T = rate of the multiplex equipment and n = number of tributary signals being multiplexed. T/n is selected relative to t so that $T/n > t$ with a sufficient margin to accommodate the difference in relative data rates of the multiplex and input tributary signals and also to allow for the addition of frame alignment and other service digits.

Under normal operational conditions there will be variations between T/n and t. To provide for these variations, the sequence of time slots at the output of each tributary store has available in it certain designated time slots known as *justifiable digit time slots*. These occur at fixed intervals, and the state of the store fill determines whether or not the justifiable digit time slot has information written into it from the store. If the store is filling, the time slot is used; if not, the time slot is ignored. By this means, a degree of "elasticity" is acquired that enables the relative timing difference to be absorbed.

Figure 3.26*a* is a simplified functional block diagram of the AT&T M12 multiplex, and Figure 3.26*b* of a higher-level multiplex scheme.

3.3.7.1 *Second-Level Frame Structures.* The organization of the DS2 bit stream is shown in Figure 3.27. DS2 is transmitted online at 6.312 Mbps (equivalent of 96 VF channels) and consists of four DS1 bit streams at 1.544 Mbps multiplexed, plus synchronization, framing, and stuff (justification) bits. All of the control information for the far-end demultiplexer is carried within an 1176-bit frame, which is divided into four 294-bit subframes. The control-bit word, disbursed throughout the frame, begins with an M-bit, as shown in Figure 3.27. The four M-bits are transmitted as 011X; the fourth bit X may be a 1 or a 0. This bit may be used as an alarm where 0 indicates an alarm condition and 1 no alarm. The 011 sequence for the first three M-bits identifies (formats) the frame.

Within each of the four subframes two other control sequences are used. Each control bit is followed by a 48-bit block of information of which 12 bits are taken from each of the four DS1 input signals. These are interleaved sequentially in the 48-bit block. The first bit in the third and sixth blocks is designated an F-bit. The F-bits are the sequence 0101 and are used to identify the location of the control bit sequences and the start of each information block.

The stuff (justification) control bits are transmitted at the beginning of each of the 48-bit blocks numbered 2, 4, and 5 within each subframe. These control bits are designated C in Figure 3.27. When a sequence is 000, no stuff pulse is present; when 111, a stuff pulse is added in the stuff position. The stuff bit positions are all assigned to the sixth 48-bit block of each subframe. In subframe 1 the stuff bit is the first bit after the F1 bit; for subframe 2 it is

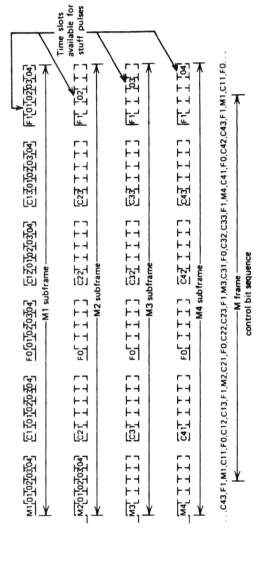

Figure 3.27. Organization of the DS2 signal bit stream. (From Ref. 20, vol. 2; copyright © 1977 by American Telephone and Telegraph Company.)

...C43,F1,M1,C11,F0,C12,C13,F1,M2,C21,F0,C22,C23,F1,M3,C31,F0,C32,C33,F1,M4,C41,F0,C42,C43,F1,M1,C11,F0...

147

TABLE 3.10   The 8448-kbps Multiplexing Frame Structure

Frame Structure	Bit Number
	Set I
Frame alignment signal (1111010000)	1–10
Alarm indication to the remote digital multiplex equipment	11
Bit reserved for national use	12
Bits from tributaries	13–212
	Set II
Justification control bits C_{j1}^a	1–4
Bits from tributaries	5–212
	Set III
Justification control bits C_{j2}^a	1–4
Bits from tributaries	5–212
	Set IV
Justification control bits C_{j3}^a	1–4
Bits from tributaries available for justification	5–8
Bits from tributaries	9–212
Tributary bit rate	2048 kbps
Number of tributaries	4
Frame length	848 bits
Bits per tributary	206 bits
Maximum justification rate per tributary	10 kbps
Nominal justification ratio	0.424

$^a C_{ji}$ indicates the ith justification control bit of the jth tributary.

Source: Reference 21. Courtesy of ITU-CCITT.

the second bit after the F1 bit, and so on through the fourth subframe. The nominal stuffing rate is 1796 bps for each DS1 signal input; the maximum is 5367 bps.

The output of the M12 multiplexer is in the B6ZS format.

The basic European second-order multiplex using positive justification is described in CCITT Rec. G.742. It accepts four tributary inputs, each from the standard E-1 channel bank, which has a nominal line data rate of 2.048 Mbps. The output of the multiplex is a nominal 8.448 Mbps online, an equivalent of 4×30 or 120 VF channels. The multiplex frame structure is described in Table 3.10. CCITT Rec. G.742 (Ref. 21) recommends cyclic bit interleaving in the tributary numbering order and positive justification. Justification should be distributed using C_{jn} bits ($n = 1, 2, 3$). Positive justification is indicated by the signal 111 (stuffing), no stuffing by 000. Two bits per frame are available as service digits. Bit 11 of set I is used to transmit an alarm indication to the remote multiplex equipment. Faults indicated by service digits may be power supply failure, loss of tributary input (2.048 Mpbs), loss of incoming 8.448-Mpbs line signal, and loss of frame alignment.

3.3.7.2   Third-Level Frame Structures

3.3.7.2.1   The North American DS3.   The DS3 format is the most widely used higher-layer multiplex format in North America. The DS3 line rate is

44.736 Mbps and carries 28 DS1 signals bearing a total of 672 digital equivalent voice channels (i.e., 24 × 28). There are two ways to develop a DS3 signal. The first method involves multiplexing 28 DS1 signals in an M13 multiplexer. With the second method, a DS3 signal is developed in an M23 multiplexer where seven DS2 signals are multiplexed to develop the DS3 output. An overview of the DS3 signal format is shown in Figure 3.28.

The DS3 signal is partitioned into M-frames of 4760 bits each. The M-frame format as applied to the M13 multiplex is shown in Figure 3.29. The M13 multiplex provides a digital interface between the DS1 and DS3 levels. The M13 multiplex is totally compatible with multiplexing through M12* and M23 multiplexes connected in tandem. That is, a DS3 signal built up by multiplexing DS1 signals can be demultiplexed by either an M13 multiplex or a combination of the M12 and M23 multiplexers. The compatibility is provided in the M13 by multiplexing in two steps. In the first step, four DS1 signals are multiplexed using pulse stuffing synchronization to reach a 6.312 Mbps transmission rate inside the multiplex. In the second step, seven of the 6.312 Mbps signals are multiplexed using pulse stuffing synchronization to generate the DS3 signal. Demultiplexing is also accomplished using a two-step process. In the first step, a DS3 signal is decomposed into seven 6.312 Mbps signals. In the second step, each of the 6.312 Mbps signals is decomposed into four DS1 signals.

SIGNAL FORMAT. The first multiplexing stage results in the 6.312 Mbps equivalent DS2 signal. Bit multiplexing is used and the bits are interleaved according to the input numbering order. Further, the information from inputs 2 and 4 is inverted logically prior to multiplexing with inputs 1 and 3. This multiplexed format is identical with that created within an M12 multiplex.

The second stage of multiplexing results in a 44.736-Mbps data stream having the format shown in Figures 3.28 and 3.29. Again, bit multiplexing is used and the bits are interleaved according to the input numbering order of the equivalent M23 multiplex. This multiplexed format is identical to that created within the M23 multiplex. In the M13 multiplex, the DS3 clock is used to generate all timing signals. By inserting a stuffed time slot into seven of every 18 M-frames, a clock at 6312016 Hz is created to operate the seven first stages of multiplexing. These dedicated stuffed time slots are still signaled by the stuffing indicator word to provide full compatibility with an M23 multiplex.

As shown in Figure 3.29, the M-frames are divided into 7 M subframes each having 680 bits. Then each subframe is further divided into 8 blocks of 85 bits each. Of these 85 bits, 84 are available for information and 1 bit is used for control.

---

*The M12 multiplexes four DS1 signals to form a DS2 signal at the 6312 kbps transmission bit rate.

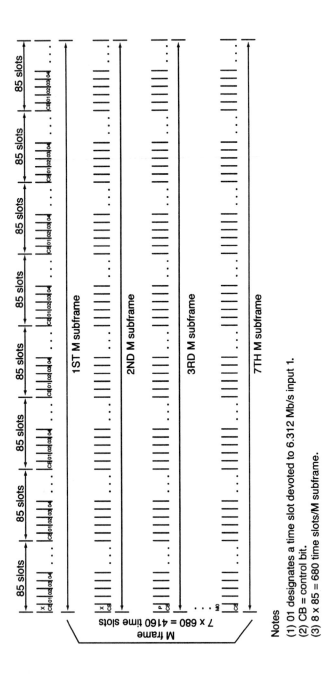

Figure 3.28. Overview of the DS3 signal format. (From Ref. 22, Figure 4. Reprinted with permission.)

Notes

(1) 01 designates a time slot devoted to 6.312 Mb/s input 1.
(2) CB = control bit.
(3) 8 × 85 = 680 time slots/M subframe.

150

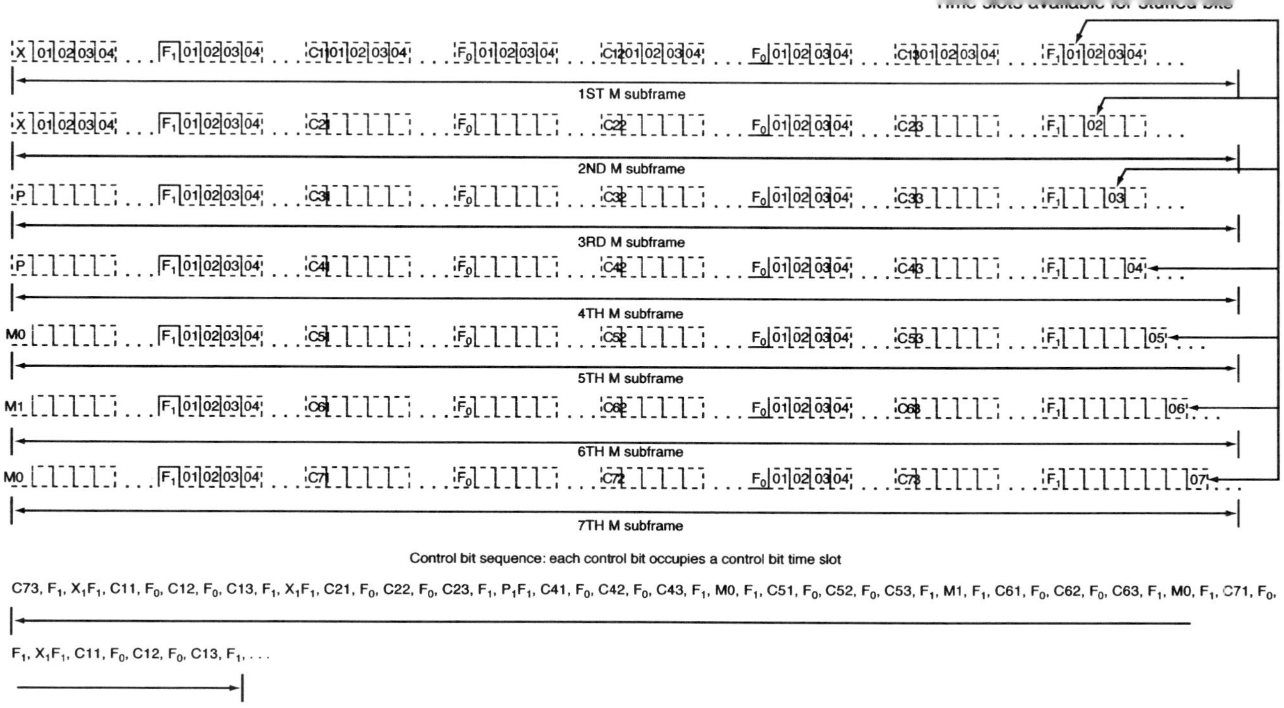

Time slots available for stuffed bits
1ST M subframe
2ND M subframe
3RD M subframe
4TH M subframe
5TH M subframe
6TH M subframe
7TH M subframe
Control bit sequence: each control bit occupies a control bit time slot
C73, F1, X1F1, C11, F0, C12, F0, C13, F1, X1F1, C21, F0, C22, F0, C23, F1, P1F1, C41, F0, C42, F0, C43, F1, M0, F1, C51, F0, C52, F0, C53, F1, M1, F1, C61, F0, C62, F0, C63, F1, M0, F1, C71, F0,
F1, X1F1, C11, F0, C12, F0, C13, F1, . . .

Notes: (1) The frame alignment signal is F0 = F1 = 1. (2) M0, M1 is the multiframe alignment signal and appears in the 5th, 6th, and 7th M subframes. M0-1 and M1-1. (3) PP is parity information taken over all information time slots in the preceding M frame. PP-11 if the digital sum of all the information bits is 1 and the PP-00 if the sum 0. These two parity bits are in the 3rd and 4th M subframes.
(4) The X bits occupy the bit positions at the beginnning of the first and second subframes. In any M frame, the two X bits must be identical (i.e., either 11 or 00). A DS3 source may use the X bits source for asynchronous low-speed signaling. If so, the source shall not change the state of the X bits more than once every second, because transmission equipment may also use the X bits at higher rates. If not used, the X bits shall be set to 11. (5) C11, C12, C13: Stuffing indicator word 6.321 Mb/s input 1. 000 indicates no stuffing and 111 indicates stuffing was done. (6) The time slot available for stuffing 6.312 Mb/s input is the first time slot for input 1, 01. Following F1 in the 1th subframe. (7) The maximum stuffing rates per 6.312 Mb/s input is 9398 bits/sec. (8) The nominal stuffing rate per 6.312 Mb/s input is 3671 bits/sec.

Figure 3.29. Format of the DS3 44.736 Mbps signal. From Ref. 22, Figure 5. Reprinted with permission.

The frame alignment signal ($F_0 F_0 F_1 F_1 \ldots$) is used to identify all control bit time slots and the multiframe alignment signal ($M_0 M_1 M_0$) is used to locate all seven subframes. Within each subframe a single time slot is available for inserting a stuffed bit and a three-bit stuffing indicator word identifies whether or not a stuffed bit has been inserted.* At the beginning of the first and second M subframes is an X-bit that may be used for an alarm service channel. However in any one M-frame, the two X-bits must be identical (i.e., either an 11 or a 00). The source may not change the state of the X-bits more than once per second. This is because the transmission equipment may also use the X-bits.

At the beginning of the third and fourth M subframes is a P-bit that contains CRC (error detection) information. The error detection function covers all information time slots in an M-frame (4704 time slots) and inserts the CRC remainder in the following M-frame. PP = 11 if the digital sum of all information bits is 1 and PP = 00 if the digital sum of all information bits is 0.

Alarm Indication Signal (AIS). The AIS must have correct M-bits, F-bits, and P-bits. All the C-bits in the M-frame are set to 0, and the X-bits are set to 1. The information bits are set to the repeating sequence 1010, with a 1 immediately following any of the control bit positions.

Remote Defect Indication (RDI). This may be sent by a DS3 receiving terminal as soon as it cannot frame-align the incoming DS3 signal or it is receiving an AIS. When this option is implemented, the DS3 receiving terminal, on the return link to the source transmitter, sets both of the X-bits to 0. Otherwise the X-bits are set to 1. The state of the X-bits are not to change more than once a second.

Table 3.11a shows the functions of the framing pattern bits and Table 3.11b details the housekeeping functions of the M subframes. For Table 3.11a, remember that the M-frame consists of seven M subframes. Each subframe contains 680 bits, with the first bit in each subframe making up the framing pattern.

Far-End Alarm and Control (FEAC). The FEAC function is carried in the third C-bit in the first M subframe that provides a communication channel between the distant end multiplex terminal and the near-end multiplexer. This channel transmits to the near-end:

1. Alarm and status information and
2. Loopback commands to initiate and deactive DS3 and DS1 loopbacks at that far-end terminal.

---

*In one version of the M13 multiplexer, the stuffed bit is the same as the previous information bit for the stuffed input.

TABLE 3.11a   Multiframe Framing Pattern, First Bit Functions

Subframe Number	Bit Identifier	Function/Value
1	X	X-bit, see text
2	X	X-bit, see text
3	P	CRC bit
4	P	2nd CRC bit
5	M0	Multiframe align. 0
6	M1	Multiframe align. 1
7	M0	Multiframe align. 0

TABLE 3.11b   M Subframe Housekeeping—Bit 1 of Each 85-Bit Field

Field Number	Bit Identifier	Function/Value
1	Various (See Table 3.11)	Multiframe framing pattern bit
2	F1	Frame align. 1
3	C-bit	Control/service
4	F0	Frame align. 0
5	C-bit	Control/service
6	F0	Frame align. 0
7	C-bit	Control/service
8	F1	Frame align. 1

Note: Remember that an M subframe has eight 85-bit fields. This table outlines the functions of the first bit in each field. The C-bits in M subframe 3 carry out the CRC function and are called CP-bits.

Far-End Bit Error (FEBE) Function. The FEBE bits are the three C-bits in M-subframe 4. If a path error is detected on the incoming signal by CP-bits as described previously, or a framing error is detected, the FEBE bits are sent to indicate an error condition. The three FEBE bits can be set to any binary pattern except 111 to indicate an error. The FEBE bits are set to 111 only if no M-bit or F-bit framing error has occurred and no CP-bit parity error has occurred. The FEAC and FEBE signals can determine the overall performance of the full-duplex DS3 path at either end or at any place along the path.

Terminal-to-Terminal Maintenance Data Link for a Path. The three C-bits in M subframe 5 may be used collectively as a 28.2 kbps data link terminal-to-terminal. If implemented, the data link conforms to the requirements set out below. If not implemented, the three C-bits are set to one. The data link is based on the LAPD protocol outlined in Ref. 36.

3.3.7.2.2 The European E-3 Multiplex Operating at 34.368 Mbps and Using Positive/Zero/Negative Justification. The nominal bit rate at this third-level E-multiplex is 34.368 Mbps ± 20 ppm. It is based on multiplexing four E-2 8.448-Mbps tributaries. Table 3.12 outlines the E-3 format.

TABLE 3.12   The 34.368-Mbps Multiplexing Frame Structure Using Positive/Zero/Negative Justification

Tributary bit rate (kbps)	8448
Number of tributaries	4

Frame Structure	Bit Number
	Set I
Frame alignment signal (111110100000)	1–12
Bits from the secondary tributaries	13–716
	Set II
Justification control bits $(C_{j1})^a$	1–4
Bits for service functions	5–8
Justification control bits $(C_{j2})^a$	9–12
Bits from the secondary tributaries	13–716
	Set III
Justification control bits $(C_{j3})^a$	1–4
Bits reserved for national use	5–8
Bits from tributaries available for negative justification	9–12
Bits from tributaries available for positive justification	13–16
Bits from the tributaries	17–716
Frame length	2148 bits
Frame duration	62.5 μs
Bits per tributary	528
Maximum justification rate per tributary	16 kbps

$^a C_{jn}$ indicates the nth justification control pulse of the jth tributary.
Source: Reference 23, Table 1/G.753, CCITT Rec. G.753.

LOSS AND RECOVERY OF FRAME ALIGNMENT AND CONSEQUENT ACTIONS. The frame alignment system is adaptive to the error rate on the link. Until frame alignment is restored, the frame alignment system retains its position. A new search for frame alignment signal is undertaken when three or more consecutive frame alignment signals have been incorrectly received in their position. Frame alignment is considered recovered when more than two consecutive frame alignment signals have been correctly received in their predicted positions.

MULTIPLEXING METHOD. Cyclic bit interleaving in the tributary numbering order and positive negative justification with command control is recommended by CCITT. The justification control signal is distributed and uses C_{jn}-bits ($n = 1, 2, 3$). See Table 3.12. Correction of one error in a command is possible.

Positive justification is indicated by the signal 111, transmitted in each of two consecutive frames; negative justification is indicated by the signal 000, transmitted in each of two consecutive frames, and no justification by the signal 111 in one frame followed by 000 in the next frame.

Digit time slots 9, 10, 11, 12 (Set III) are used for information-carrying bits (for negative justification), and digit time slots 13, 14, 15, 16 in Set III, when necessary, are used for no information-carrying bits (for positive justification) for tributaries 1, 2, 3, and 4.

Besides, when information from tributaries 1, 2, 3, and 4 is not transmitted, bits 9, 10, 11, and 12 in Set III are available for transmitting information concerning the type of justification (positive or negative) in frames containing commands of positive justification control and intermediate amount of jitter in frames containing commands of negative justification. Table 3.12 gives the maximum justification rate per tributary.

SERVICE DIGITS. Some spare bits per frame are available for service functions (bits 5, 6, and 8 in Set II) for national and international use. Bits 5 and 6 in Set II are available for a digital service channel (using 32-kbps adaptive delta modulation) and bit 8 in Set II is available for ringing on a digital service channel.

FAULT CONDITIONS AND CONSEQUENT ACTIONS. Digital multiplex equipment will detect the following fault conditions:

- Power supply failure
- Loss of incoming signal at 8.448 Mbps at the input of the multiplexer
- Loss of incoming signal at 34.368 Mbps at the input of the demultiplexer (*Note*: The detection of this fault condition is required only when it does not result in an indication of loss of frame alignment.)
- Loss of frame alignment
- Alarm indication received from the remote multiplex equipment at the 34.368-Mbps input of the demultiplexer

CONSEQUENT ACTIONS. After detecting a fault condition appropriate actions are taken as specified in Table 3.13 (Ref. 23).

3.3.7.3 *Line Rates and Codes.*

Table 3.14 summarizes the North American DS series of PCM line rates, tolerances of these rates, and respective line codes. ITU-T Rec. G.703 deals with line interfaces. For the E-1 series PCM and higher-order multiplex derived therefrom, Table 3.15 provides a similar summary as Table 3.14.

3.3.8 Transmitting Data on PCM Links

3.3.8.1 *Introduction.*

Conventional data transmission rates are incompatible with the standard 64-kbps equivalent digital voice channel. These data rates are based on the relationship 75×2^n (e.g., 75, 150, 300, 600, 1200, 2400, 4800, ..., 28,800 bps). CCITT, AT&T, and Bellcore have set forth methods to bring about a compatibility.

TABLE 3.13   Fault Conditions and Consequent Actions

Equipment Part	Fault Conditions	Consequent Actions				
				AIS Applied		
		Prompt Maintenance Alarm Indication Generated	Alarm Indication to the Remote Multiplexer Generated	To All Tributaries	To the Composite Signal	To the Relevant Time Slots of the Composite Signal
Multiplexer and demultiplexer	Failure of power supply	Yes	Yes, if practicable	Yes, if practicable	Yes, if practicable	
Multiplexer only	Loss of incoming signal on a tributory	Yes				Yes
	Loss of incoming signal at 34,368 kbps	Yes	Yes	Yes		
Demultiplexer only	Loss of frame alignment	Yes	Yes	Yes		
multiplexer	AIS received from the remote					

Note: A *Yes* in the table signifies that a certain action should be taken as a consequence of the relevant fault condition. An *open space* in the table signifies that the relevant action should *not* be taken as a consequence of the relevant fault condition, if this condition is the only one present. If more than one fault condition is simultaneously present the relevant action should be taken if, for at least one of the conditions, a *Yes* is defined in relation to this action.

Source: Reference 23, Table 2/G.753, CCITT Rec. G.753.

TABLE 3.14   North American Line Rates, Tolerances, and Line Codes

Signal	Repetition Rate (Mbps)	Tolerance (ppm)[a]	Format	Duty Cycle (%)
DS0	0.064	[b]	Bipolar	100
DS1	1.544	± 130	Bipolar	50
DS1C	3.152	± 30	Bipolar	50
DS2	6.312	± 30	B6ZS	50
DS3	44.736	± 20	B3ZS	50
DS4	274.176	± 10	Polar	100

[a]Parts per million.
[b]Expressed in terms of slip rate.

TABLE 3.15   Summary of E-1 Related Line Rates, Tolerances, and Codes

Level	Line Data Rate (Mbps)	Tolerance (ppm)	Code	Mark Peak Voltage (V)
1	2.048	±50	HDB3	2.37 or 3[a]
2	8.448	±30	HDB3	2.37
3	34.368	±20	HDB3	1.0
4	139.264	±15	CMI[b]	1 ± 0.1 V[c]

[a]2.37 V on coaxial pair; 3 V on symmetric wire pair.
[b]Coded mark inversion.
[c]Peak-to-peak voltage.

One way to force compatibility is to use overhead and stuff bits. For example, we might multiplex six 9600 bit streams deriving an aggregate of 57,600 bps with a remainder of 6400 bits, which could be used for that necessary overhead to identify each of the six channels and to provide some maintenance functions. Some bits may still be required to reach the requisite 64,000 bps and these may be what we would call "stuff bits."

Let us now examine some other approaches that might be a little more elegant.

3.3.8.2   Data Transmission on the Digital Network Based on CCITT Rec. V.110.* CCITT Rec. V.110 provides one approach for the specification of an ISDN TA (terminal adapter) device. A TA adapts incompatible ISDN terminal equipment for ISDN application. ISDN transport is based on the 64-kbps channel,[†] which we may call DS0 or E-0. In ISDN it is called a B-channel.

CCITT Rec. V.110 breaks data transmission down into three categories:

1. Adaptation of synchronous data rates up to 19.2 kbps.
2. Rate adaptation of 48 and 56 kbps to 64 kbps.
3. Adaptation of asynchronous (start–stop) rates up to 19.2 kbps. (Refer to Ref. 24 for this category.)

One-, two- or three-step adaptation is used, depending on the category.

3.3.8.2.1   Synchronous Data Rates Up to 19.2 kbps. The two-step bit rate adaptation function of a TA is shown in Figure 3.30. The function RA1 converts the user data rate to an appropriate intermediate rate expressed by $2^k \times 8$ kbps (where $k = 0, 1,$ or 2). RA2 performs the second conversion from one of the intermediate rates to the requisite 64 kbps. The data rates of 48 and 56 kbps are converted directly into the 64-kbps B-channel rate.

---

*This section is based on Ref. 24, CCITT Rec. V.110. The writeup is abridged and thus the reader is encouraged to review the original V.110 recommendation before committing to a final design.
†In North America, ISDN may be on a 56-kbps channel unless the signaling bit slots are cleared.

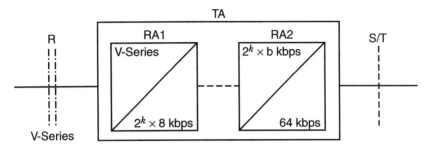

Figure 3.30. Two-step bit rate adaptation. *Note:* The V-series rates are listed in Table 3.16. R, S, and T are ISDN interfaces.

TABLE 3.16   First-Step Rate Adaptation

	Intermediate Rate		
Data Signaling Rate (bit/s)	8 kbps	16 kbps	32 kbps
600	X		
1,200	X		
2,400	X		
4,800	X		
7,200		X	
9,600		X	
12,000			X
14,400			X
19,200			X

Source: Reference 24, Table 1/V.110, CCITT Rec. V.110.

The intermediate rate used with each of the V-series data rates is shown in Table 3.16. The "V-series" refers to the ITU-T/CCITT V recommendations on data transmission.

FRAME STRUCTURE. Table 3.17 shows the first-step adaptation frame structure. The frame has 80 bits. Octet zero contains all binary 0's and octet 5 consists of a binary 1 followed by 7 E-bits (described on page 160). Octets 1–4 and 6–9 contain a binary 1 in bit position number 1, a status bit (S- or X-bit) in bit position number 8, and six data bits (called D-bits) in bit positions 2–7. The order of transmission is from left to right and top to bottom.

FRAME SYNCHRONIZATION AND STATUS BITS. Frame alignment is carried out with a 17-bit frame alignment pattern where all 8 bits of octet 0 are set to binary 0 and bit 1 in the following nine octets is set to binary 1.

The bits S and X convey channel control information associated with the data bits in the data transfer state as illustrated in Table 3.18. The S-bits are placed into two groups denoted SA and SB, to carry the condition of two

TABLE 3.17   Frame Structure

Octet Number	Bit Number							
	1	2	3	4	5	6	7	8
0	0	0	0	0	0	0	0	0
1	1	D1	D2	D3	D4	D5	D6	S1
2	1	D7	D8	D9	D10	D11	D12	X
3	1	D13	D14	D15	D16	D17	D18	S3
4	1	D19	D20	D21	D22	D23	D24	S4
5	1	E1	E2	E3	E4	E5	E6	E7
6	1	D25	D26	D27	D28	D29	D30	S6
7	1	D31	D32	D33	D34	D35	D36	X
8	1	D37	D38	D39	D40	D41	D42	S8
9	1	D43	D44	D45	D46	D47	D48	S9

Source: Reference 24, Table 2/V.110, CCITT Rec. V.110.

TABLE 3.18   General Mapping Scheme

108	S1, S3, S6, S8 = SA	 107
105	S4, S9 = SB	 109
Frame synch .... and 106/IWF	X	 106

Source: Reference 24, Table 3/V.110, CCITT Rec. V.110.

interchange circuits.* The X-bit carries the condition of circuit 106 and, in addition, gives the state of frame synchronization between cooperating TAs. Optionally, the X-bit can carry flow control information between TAs supporting asynchronous terminal equipment.

Table 3.18 shows the mechanism for proper assignment of the control information from the transmitting signal rate adapter interface via these bits to the receiving rate adapter interface.

For S- and X-bits a 0 corresponds to an ON condition and a 1 corresponds to an OFF condition.

Control information conveyed by the S-bits and user data conveyed by the D-bits should not have different transmission delays. The S-bits should therefore transmit control information sampled simultaneously with the D-bits in the positions specified in Table 3.19 and presented in Figure 2/V.110 of Ref. 24.

---

*Interchange circuits are mechanical–electrical interface points and appear as pins on a standard socket. Each pin is an electrical connection and carries out a specific function. The pins are numbered with three-digit numbers (CCITT) (e.g., 105, 106, 107, 108, 109). These pins/sockets are specific to CCITT Rec. V.110.

TABLE 3.19 Coordination Between S-Bits and D-Bits

S-Bit	D-Bit	
	Octet Number	Bit Number
S1	2	3 (D8)
S3	3	5 (D16)
S4	4	7 (D24)
S6	7	3 (D32)
S8	8	5 (D40)
S9	9	7 (D48)

Source: Reference 24, Table 4/V.110, CCITT Rec. V.110.

E-BIT USAGE. The E-bits are used to carry the following information:

1. *Rate Repetition Information.* Bits E1, E2, and E3, in conjunction with the intermediate rate (see Table 3.17) provide the user data rate (synchronous) identification. The coding of these bits is shown in Table 3.20.
2. *Network Independent Clock Information.* Bits E4, E5, and E6 are used to carry network independent clock phase information (i.e., displacement in percentage of nominal R1 clock period and compensation control).
3. *Multiframe Information.* Bit E7 is used as indicated in Table 3.20.

TABLE 3.20 E-Bit Usage

Intermediate Rates[a] (kbps)			E1	E2[b]	E3	E4	R5[c]	E6	E7
8	16	32							
bps	bps	bps							
600			1	0	0	C	C	C	1 or 0[d]
1,200			0	1	0	C	C	C	1
2,400			1	1	0	C	C	C	1
		12,000	0	0	1	C	C	C	1
	7,200	14,400	1	0	1	C	C	C	1
4,800	9,600	19,200	0	1	1	C	C	C	1

[a]The data signaling rates of 600, 2,400, 4,800, and 9,600 bps are also Rec. X.1 user classes of service (see also Rec. X.30/I.461).
[b]Synchronous rate information is carried by bits E1, E2, and E3 as indicated. Asynchronous rate information must be provided with out-of-band signaling (layer 3 messages in the D-channel) or with in-band parameter exchange as described in Appendix I of Ref. 24.
[c]C indicates the use of E4, E5, and E6 for the transport of network independent blocking information. These bits shall be set to 1 when unused.
[d]In order to maintain compatibility with Rec. X.30 (I.461), the 600-bps user rate E7 is coded to enable the 4 × 80 bit multiframe synchronization. To this end, E7 in the fourth 80-bit frame is set to binary 0 (see Table 3.21a).

Source: Reference 24, Table 5/V.110, CCITT Rec. V.110.

Data Bits. Data are conveyed by the D-bits, up to 48 bits per 80-bit frame. CCITT Rec. V.110 states that the octet boundaries of the user's data stream are not defined.

Bit Assignments. Tables 3.21a–3.21f show the adaptation of the V series synchronous rates to the specified intermediate rates (i.e., 8, 16, or 32 kbps).

Frame Synchronization. The following 17-bit alignment pattern is used to achieve frame synchronization:

0 0 0 0 0 0 0 0   1XXXXXXX   1XXXXXXX   1XXXXXXX   1XXXXXXX

1XXXXXXX   1XXXXXXX   1XXXXXXX   1XXXXXXX   1XXXXXXX

To ensure reliable synchronization, at least two 17-bit alignment patterns in consecutive frames should be detected.

TABLE 3.21a   Adaptation of 600-bps User Rate to 8-kbps Intermediate Rate

0	0	0	0	0	0	0	0
1	D1	D1	D1	D1	D1	D1	S1
1	D1	D1	D2	D2	D2	D2	X
1	D2	D2	D2	D2	D3	D3	S3
1	D3	D3	D3	D3	D3	D3	S4
1	1	0	0	E4	E5	E6	E7[a]
1	D4	D4	D4	D4	D4	D4	S6
1	D4	D4	D5	D5	D5	D5	X
1	D5	D5	D5	D5	D6	D6	S8
1	D6	D6	D6	D6	D6	D6	S9

[a]In order to maintain compatibility with Rec. X.30 (I.461), the 600-bps user rate E7 is coded to enable the 4 × 80 bit multiframe synchronization. To this end, E7 in the fourth 80-bit frame is set to binary 0.

Source: Reference 24, Table 6a/V.110, CCITT Rec. V.110.

TABLE 3.21b   Adaptation of 1200-bps User Rate to 8-kbps Intermediate Rate

0	0	0	0	0	0	0	0
1	D1	D1	D1	D1	D2	D2	S1
1	D2	D2	D3	D3	D3	D3	X
1	D4	D4	D4	D4	D5	D5	S3
1	D5	D5	D6	D6	D6	D6	S4
1	0	1	0	E4	E5	E6	E7
1	D7	D7	D7	D7	D8	D8	S6
1	D8	D8	D9	D9	D9	D9	X
1	D10	D10	D10	D10	D11	D11	S8
1	D11	D11	D12	D12	D12	D12	S9

Source: Reference 24, Table 6b/V.110, CCITT Rec. V.110.

TABLE 3.21c   Adaptation of 2400-bps User Rate to 8-kbps Intermediate Rate

0	0	0	0	0	0	0	0
1	D1	D1	D2	D2	D3	D3	S1
1	D4	D4	D5	D5	D6	D6	X
1	D7	D7	D8	D8	D9	D9	S3
1	D10	D10	D11	D11	D12	D12	S4
1	1	1	0	E4	E5	E6	E7
1	D13	D13	D14	D14	D15	D15	S6
1	D16	D16	D17	D17	D18	D18	X
1	D19	D19	D20	D20	D21	D21	S8
1	D22	D22	D23	D23	D24	D24	S9

Source: Reference 24, Table 6c/V.110, CCITT Rec. V.110.

TABLE 3.21d   Adaptation of $N^a \times$ 3600-bps User Rate to the Intermediate Rate

0	0	0	0	0	0	0	0
1	D1	D2	D3	D4	D5	D6	S1
1	D7	D8	D9	D10	F[b]	F	X
1	D11	D12	F	F	D13	D14	S3
1	F	F	D15	D16	D17	D18	S4
1	1	0	1	E4	E5	E6	E7
1	D19	D20	D21	D22	D23	D24	S6
1	D25	D26	D27	D28	F	F	X
1	D29	D30	F	F	D31	D32	S8
1	F	F	D33	D34	D35	D36	S9

[a]N = 2 or 4 only.
[b]F = fill bit.
Source: Reference 24, Table 6d/V.110, CCITT Rec. V.110.

TABLE 3.21e   Adaptation of $N^a \times$ 4800-bps User Rate to the Intermediate Rate

0	0	0	0	0	0	0	0
1	D1	D2	D3	D4	D5	D6	S1
1	D7	D8	D9	D10	D11	D12	X
1	D13	D14	D15	D16	D17	D18	S3
1	D19	D20	D21	D22	D23	D24	S4
1	0	1	1	E4	E5	E6	E7
1	D25	D26	D27	D28	D29	D30	S6
1	D31	D32	D33	D34	D35	D36	X
1	D37	D38	D39	D40	D41	D42	S8
1	D43	D44	D45	D46	D47	D48	S9

[a]N = 1, 2, or 4 only.
Source: Reference 24, Table 6e/V.110, CCITT Rec. V.110.

TABLE 3.21f   Adaptation of 12,000-bps User Rate to 32-kbps Intermediate Rate

0	0	0	0	0	0	0	0
1	D1	D2	D3	D4	D5	D6	S1
1	D7	D8	D9	D10	F[a]	F	X
1	D11	D12	F	F	D13	D14	S3
1	F	F	D15	F	F	F	S4
1	0	0	1	E4	E5	E6	E7
1	D16	D17	D18	D19	D20	D21	S6
1	D22	D23	D24	D25	F	F	X
1	D26	D27	F	F	D28	D29	S8
1	F	F	D30	F	F	F	S9

[a]F = fill.
Source: Reference 24, Table 6f/V.110, CCITT Rec. V.110.

Loss of frame synchronization is not assumed unless at least three consecutive frames, each with at least one framing bit error, are detected.

Adaptation of Intermediate Rates to 64 kbps. In the 64-kbps B-channel octet, it is assumed that the bit positions are numbered 1 through 8 with bit position 1 being transmitted first. The procedure here for step 2 adaptation requires:

1. The 8-kbps stream occupies bit position 1; the 16-kbps stream occupies bit positions 1 and 2; the 32-kbps stream occupies bit positions 1, 2, 3, and 4.
2. All unused bit positions are set to binary 1.

Multiplexing several intermediate rate streams should be considered, if feasible.

Rate Adaptation of 48- and 56-kbps User Rates to 64 kbps. The 48- and 56-kbps user data rates are adapted to the 64-kbps B-channel rate in one step as illustrated in Tables 3.22a–3.22c.

TABLE 3.22a   Adaptation of 48-kbps User Rate to 64 kbps

Octet Number	Bit Number							
	1	2	3	4	5	6	7	8
1	1	D1	D2	D3	D4	D5	D6	S1
2	0	D7	D8	D9	D10	D11	D12	X
3	1	D13	D14	D15	D16	D17	D18	S3
4	1	D19	D20	D21	D22	D23	D24	S4

Note 1. 48 kbps is also a Rec. X.1 user class of service (see also Rec. X.30/I.461).
Note 2. Refer to "frame synchronization" above for the use of status bits and bit X; however, for international operation over restricted 64-kbps bearer capabilities, bit X must be set to binary 1.
Source: Reference 24, Table 7a/V.110, CCITT Rec. V.110.

TABLE 3.22b Adaptation of 56-kbps User Rate to 64 kbps

Octet Number	\multicolumn{8}{c}{Bit Number}
	1	2	3	4	5	6	7	8
1	D1	D2	D3	D4	D5	D6	D7	1
2	D8	D9	D10	D11	D12	D13	D14	1
3	D15	D16	D17	D18	D19	D20	D21	1
4	D22	D23	D24	D25	D26	D27	D28	1
5	D29	D30	D31	D32	D33	D34	D35	1
6	D36	D37	D38	D39	D40	D41	D42	1
7	D43	D44	D45	D46	D47	D48	D49	1
8	D50	D51	D52	D53	D54	D55	D56	1

Source: Reference 24, Table 7b/V.110, CCITT Rec. V.110.

TABLE 3.22c Alternative Frame Structure for the Adaptation of 56-kbps User Rate to 64 kbps

Octet Number	\multicolumn{8}{c}{Bit Number}
	1	2	3	4	5	6	7	8
1	D1	D2	D3	D4	D5	D6	D7	0
2	D8	D9	D10	D11	D12	D13	D14	X
3	D15	D16	D17	D18	D19	D20	D21	S3
4	D22	D23	D24	D25	D26	D27	D28	S4
5	D29	D30	D31	D32	D33	D34	D35	1
6	D36	D37	D38	D39	D40	D41	D42	1
7	D43	D44	D45	D46	D47	D48	D49	1
8	D50	D51	D52	D53	D54	D55	D56	1

Note 1. Refer to "Frame Synchronization and Status Bits" in Section 3.3.8.2.1 for the use of status bits and bit X.
Note 2. Table 3.22c is a permitted option to provide for signaling to enter and to leave the data phase. However, the recommended approach shall be as in Table 3.22b and the responsibility shall be on the user of Table 3.22c to ensure that interworking can be achieved.
Source: Reference 24, Table 7c/V.110, CCITT Rec. V.110.

Frame Synchronization. At the user data rate of 48 kbps, the frame alignment pattern consists of 1011 in bit position 1 of consecutive octets of one frame. To ensure reliable synchronization, at least five 4-bit alignment patterns in consecutive frames should be detected.

At the user data rate of 56 kbps with the alternative frame structure in Table 3.22c, the frame alignment pattern consists of 0YYY1111 in bit 8 of consecutive octets of one frame. Bits marked with Y may be either 0 or 1. To assure reliable synchronization, at least four 5-bit (01111) alignment patterns of the 8-bit sequence of 0YYY1111 in consecutive octets should be detected.

Interchange Circuits. Table 3.23 shows interchange circuits by pin number and function.

TABLE 3.23 Interchange Circuits: Pin Number and Function (See Note 1)

Interchange Circuit		
Number	Description	Notes
102	Signal ground or common return	
102a	DTE common return	2
102b	DCE common return	2
103	Transmitted data	
104	Received data	
105	Request for sending	3
106	Ready for sending	
107	Data set ready	
108/1	Connect data set to line	4
108/2	Data terminal ready	4
109	Data channel received line signal detector	
111	Data signaling rate selector (DTE source)	5
112	Data signaling rate selector (DCE source)	5
113	Transmitter signal element timing (DTE source)	6
114	Transmitter signal element timing (DC source)	
115	Receiver signal element timing (DCE source)	
125	Calling indicator	7
140	Loopback/maintenance test	8
141	Local loopback	8
142	Test indicator	8

Note 1. All essential circuits and any others that are provided shall comply with the functional and operational requirements of Rec. V.24 (see Chapter 13 for a discussion of CCITT Rec. V.24). All interchange circuits provided shall be properly terminated in the data terminal equipment and in the data circuit-terminating equipment in accordance with the appropriate recommendation for electrical characteristics.
Note 2. Interchange circuits 102a and 102b are required where the electrical characteristics defined in Rec. V.10 are used at data signaling rates above 20 kbps.
Note 3. Not required for DTEs that operate with DCEs in the continuous carrier mode.
Note 4. This circuit shall be capable of operating as circuit 108/1 or 108/2, depending on its use (by the associated TE).
Note 5. The use of this circuit is for further study.
Note 6. The use of circuit 113 is for further study, since its application is restricted by the synchronous nature of ISDN.
Note 7. This circuit is used with the automatic answering terminal adapter function.
Note 8. The use for loopback testing is for further study.
Source: Reference 24, Table 9/V.110, CCITT Rec. V. 110.

3.3.8.3 *AT&T Digital Data System (DDS).* The AT&T digital data system (DDS) provides full-duplex point-to-point and multipoint private line digital data transmission at a number of synchronous data rates. This system is based on the standard 1.544-Mbps DS1 PCM line rate discussed in Section 3.3.2, where individual bit streams have data rates that are submultiples of that line rate (i.e., based on the 64-kbps channel). However, pulse slots are reserved for identification in the demultiplexing of individual user bit streams as well as for certain status and control signals to ensure that sufficient line pulses are transmitted for receive clock recovery and pulse regeneration. The

maximum data rate available to a subscriber to the system is 56 kbps, some 87.5% of the 64-kbps theoretical maximum.

The PCM line signal is the standard DS1 clear channel; that is, a signal consisting of 24 sequential 8-bit words (i.e., channel time slots) plus one additional framing bit. The sequence is repeated 8000 times a second, developing the familiar 1.544-Mbps DS1 line rate. This fact offers the advantage of allowing a mix of voice (PCM) and data where the full dedication of a DS1 facility to data transmission would be inefficient.

AT&T calls the basic 8-bit word a byte. One bit of each 8-bit word is reserved for network control and for stuffing to meet nominal line bit rate requirements. The control bit is called a C-bit. With the C-bit removed we see where the standard channel bit rate is derived, namely, 56 kbps or 8000×7. Three subrates or submultiple data rates are also available: 2.4, 4.8, and 9.6 kbps. Newer versions of DDS also offer the 19.2-kbps rate.

When the subrates are implemented, an additional bit must be robbed from the basic byte to establish flag patterns to route each subrate channel to its proper demultiplexer port. This allows only 48 kbps out of the original 64 kbps for the transmission of user data. The 48-kbps composite total may be divided down to five 9.6-kbps channels, or ten 4.8-kbps channels, or twenty 2.4-kbps channels (Ref. 25). The subhierarchy of DDS signals is shown in Figure 3.31.

AT&T also offers a switched-56 version of DDS called ACCUNET™ Switched 56 (Ref. 26).

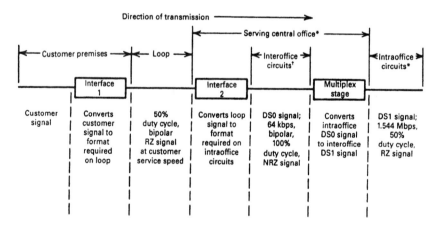

Figure 3.31. Subhierarchy of DDS signals. *Note:* Inverse processing must be provided for the opposite direction of transmission. Four-wire transmission is used throughout. *Exchange (local switch); †PCM trunk. (From Ref. 20, vol. 3; copyright © 1977, American Telephone & Telegraph Company.)

3.4   ADAPTIVE DIFFERENTIAL PCM (ADPCM)

3.4.1   Introduction

Standard ADPCM coders operate at 32 kbps, just half the transmission rate of conventional PCM discussed in Section 3.3. Other transmission rates for ADPCM are 24 and 40 kbps. There is an embedded ADPCM that uses a variable coding rate, which can be reduced at any point in the network without the need for coordination between transmitter and receiver (coder and decoder). The rates of the embedded coder are 16, 24, 32, and 40 kbps.

3.4.2   Fixed-Rate ADPCM

CCITT/ANSI/Bellcore specified ADPCM (Refs. 17, 27, 28) is based on a transcoding operation, meaning one digital code is recoded into another code. In this case the input is conventional 8-bit, either A-law or μ-law, DS1 or E1 PCM at 64 kbps per channel. The output of the transcoder is 32 kbps ADPCM. The advantage here is that the bit rate has been cut in half, to 32 kbps rather than 64 kbps. In one application we are familiar with, a leased E0 channel carries two ADPCM channels, essentially getting two for the price of one. The speech quality is reported equal to that of 8-bit PCM.

3.4.2.1   Background and Concept. In the term ADPCM we have _adaptive_ and _differential_. First, let us consider differential. In this context, differential refers to a "difference."

For analog signals (typically video and speech), the difference in quantization values between consecutive amplitude samples is usually small. With conventional PCM the value of an amplitude sample theoretically could take any one of 256 possibilities. In practice, however, the sample amplitude rarely differs much from the previous sample. A forerunner of ADPCM was differential PCM (DPCM), where only the difference value between successive samples was transmitted. In one version of DPCM, the difference value could be expressed with only 4 bits, rather than the 8 bits of conventional PCM. However, the quality of DPCM was inferior to standard PCM. With such differential devices, particularly at low bit rates, signal distortion occurred because at least some samples would differ by an amount greater than the DPCM coder could handle.

ADPCM operates in a similar manner as DPCM, but it has a further refinement to permit even lower bit rates. In ADPCM a predictive algorithm is incorporated that estimates what the next sample is likely to be. The actual value is then compared to the prediction, and only the bit sequence describing the difference in the two values is transmitted to the line. At the decoder, the same predictive algorithm is used together with the correction value received online to allow the signal to be reconstituted.

3.4.2.2 ADPCM Based on CCITT Rec. G.721. This recommendation describes a transcoding technique where the input to an ADPCM coder is standard (CCITT Rec. G.703) PCM, which converts the signal to an ADPCM format, which is then transmitted to the line. At the distant-end decoder that signal is returned to the original standard PCM waveform.

CCITT reports (in Rec. G.71) that, for the time being, the 32-kbps ADPCM algorithm defined in the recommendation is intended for transmission purposes only since switching applications at this bit rate are subject to further study.

3.4.2.2.1 Brief Description of the Operation of the ADPCM Coder. A simplified block diagram of the ADPCM coder is shown in Figure 3.32.

Subsequent to the conversion of the *A*-law or *μ*-law PCM input signal to uniform PCM, a difference signal is obtained, by subtracting an estimate of the input signal from the input signal itself. An adaptive 15-level quantizer is used to assign four binary digits to the value of the difference signal for transmission to the distant-end decoder. An inverse quantizer produces a quantized difference signal from these same four binary digits. The signal estimate is added to this quantized difference signal to produce the reconstructed version of the input signal. Both the reconstructed signal and the quantized difference signal are operated upon by an adaptive predictor, which produces the estimate of the input signal, thereby completing the feedback loop.

3.4.2.2.2 ADPCM Decoder. Figure 3.33 is a simplified functional block diagram of the ADPCM decoder. The decoder includes a structure identical to the feedback portion of the encoder, together with a uniform PCM to *A*-law or *μ*-law conversion and a synchronous coding adjustment. The synchronous coding adjustment prevents cumulative distortion occurring on

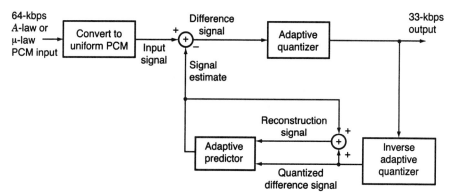

Figure 3.32. Simplified block diagram of the ADPCM encoder. (From Ref. 28, Figure 1/G.721, CCITT Rec. G. 721.)

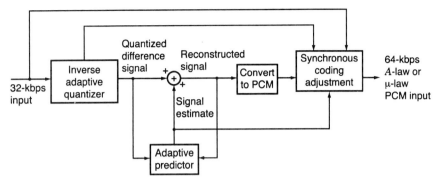

Figure 3.33. Simplified functional block diagram of the ADPCM decoder. (From Ref. 28, Figure 1/G.721, CCITT Rec. G.721.)

synchronous tandem codings (e.g., ADPCN–PCM–ADPCM digital connections) under certain conditions. The synchronous coding adjustment is achieved by adjusting the PCM output codes in a manner that attempts to eliminate quantizing distortion in the next ADPCM encoding stage.

3.4.2.2.3 Input and Output Signals. The input and output signals are given in Table 3.24.

3.4.2.2.4 Network Aspects of ADPCM. Bit robbing for signaling cannot be accommodated on ADPCM, as described previously, without incurring serious performance degradation. In this same vein, bit integrity of the original PCM signal cannot be guaranteed end to end.

The synchronous coding adjustment described above is dependent on 32-kbps paths and on intermediate 64-kbps paths being uncorrupted by other

TABLE 3.24 ADPCM Encoder/Decoder Input Output Signals

	Name	Number of Bits	Description
		Encoder	
Input	S	8	PCM input word
Input	LAW	1	PCM law select, 0 = μ-law, 1 = A-law
Input	R (*optional*)	1	Reset
Output	I	4	ADPCM word
		Decoder	
Input	I	4	ADPCM word
Input	LAW	1	PCM law select, 0 = μ-law, 1 = A-law
Input	R (*optional*)	1	Reset
Output	SD	8	Decoder PCM output word

Source: Reference 28, Table 2/G.721, CCITT Rec. G.721.

digital processes. For example, the use of digital pads, A-law to μ-law converters, echo cancelers, or digital speech interpolation (DSI) at these intermediate points will inhibit the correct functioning of this adjustment. However, the performance will still be better than that achieved when an asynchronous connection is employed.

Voiceband data performance up to 2400 bps using tone-type modems (see Chapter 13) based typically on CCITT Recs. V.22 bis, V.23, and V.26 ter will not be subject to much degradation over 32-kbps ADPCM links. However, voiceband data performance at 4800 bps, typically based on CCITT Rec. V.27 bis, can be accommodated but is subject to additional degradation over and above that expected from standard 64-kbps PCM. Operation at 9600 bps and above will be unsatisfactory if based on CCITT Rec. V.29. Other modulation implementations might alleviate this problem.

The 24-channel voice frequency carrier telegraph cannot be accommodated over ADPCM links.

3.5 DELTA MODULATION

3.5.1 Basic Delta Modulation (DM)

DM is another method of transmitting an audio (analog) signal in a digital format. It is quite different from PCM in that coding is carried out before multiplexing and the code is far more elemental, actually coding at only 1 bit at a time.

The DM code is a one-element code and differential in nature. Of course we mean here that comparison is always made to the prior condition. A 1 is transmitted to the line if the incoming signal at the sampling instant is greater than the immediate previous sampling instant; it is a 0 if it is of smaller amplitude. With DM the derivative of the analog input is transmitted rather than the instantaneous amplitude as in PCM. This is achieved by integrating the digitally encoded signal and comparing it with the analog input to decide which of the two has the larger amplitude. The polarity of the next binary digit placed online is either plus or minus, to reduce the amplitude of the two waveforms (i.e., analog input and integrated digital output [previous digit]). We thus see the delta encoder basically as a feedback circuit, as shown in Figure 3.34.

Let us see how this feedback concept is applied to the delta encoder. Figure 3.35 illustrates the application. The switch is the double NAND gate and flip-flop. The comparator is the amplifier in Figure 3.34, and the feedback network is the integrating network.

The basic delta decoder consists of a current source, integrating network, amplifier, and low-pass filter. Figure 3.36 illustrates a simplified delta decoder.

We have seen that the digital output signal of the delta coder is indicative of the slope of the analog input signal (its derivative is the slope)—a 1 for

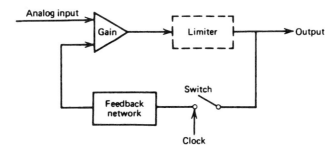

Figure 3.34. Basic electronic feedback circuit employed in DM.

positive slope and 0 for negative slope. But the 1 and 0 give no idea of an instantaneous or even semi-instantaneous slope. This leads to the basic weakness found in the development of the DM system, namely, poor dynamic range or poor dynamic response, given a satisfactory signal-to-quantizing-noise ratio. For delta circuits this limit is about 26 dB. A number greater than 26 dB is generally satisfactory, and a number numerically less than 26 dB is unsatisfactory. The reader is cautioned not to numerically equate quantizing noise in PCM to quantizing noise in DM, although the concept is the same.

One method used to improve the dynamic range of a DM system is to utilize two integrator circuits (we showed just one in Figure 3.35). This is

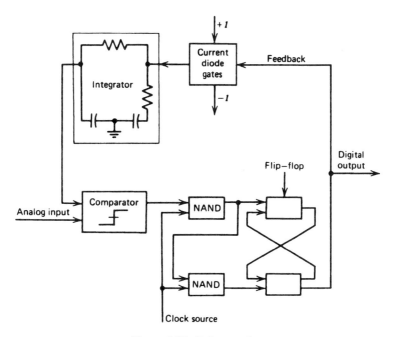

Figure 3.35. Delta encoder.

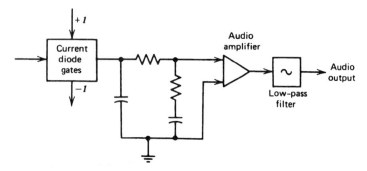

Figure 3.36. Delta decoder.

TABLE 3.25   Dynamic Range: Digital Waveform Systems Compared (Refs. 12, 29)

System	Maximum Signal-to-Quantizing-Noise Ratio (dB)	Dynamic Range for Minimum Signal-to-Quantizing-Noise Ratio of 26 dB (dB)
Basic DM, single integration	34	8
Basic DM, double integration	44	18
Companded DM, single integration	34	15
Companded DM, double integration	44	31
Seven-digit companded PCM	30-dB UVR[a]	At S/N of 31 dB

[a]Useful volume range.

called *double integration.* Companding provides further improvement. Table 3.25 compares several 56-kbps digital systems with a 3-kHz bandwidth input analog signal regarding dynamic range.

As seen from Table 3.25, companded DM has properties equal to PCM. One reason we can take advantage of the good coding nature of companded DM is that voice signals are predictable, unlike band-limited random signals. The predictability can be based on knowledge of the speech spectrum or on the autocorrelation function. Therefore delta coders can be designed on the principle of prediction (Refs. 12, 29).

The advantages of DM over PCM are as follows:

• Multiplexing is carried out by simple digital multiplexers, whereas PCM interleaves analog samples.
• It is essentially more economical for small numbers of channels.

- It has few varieties of building blocks.
- It is less complex, consequently giving improved reliability.
- Intelligibility is maintained with a BER down to 10^{-2}.

The disadvantages remain for the dynamic range and for the multiplexing of many channels.

DM has found wide application in military communications where low-bit-rate digital systems are required. Both the U.S. and NATO forces have fielded large quantities of delta multiplexers and switches based on 16 and 32 kbps. DM is also used in the commercial telephony world on thin-line telephone systems, for rural subscribers, and in certain satellite DAMA applications, such as Canada's TeleSat.

3.5.2 CVSD—A SUBSET OF DELTA MODULATION

Continuous variable-slope DM (CVSD) is a DM scheme that improves the basic weakness of DM, namely, the dynamic range over which the noise level is constant. CSVD is used by the U.S. military forces TRI-TAC field-tactical communication system and by NATO's Eurocom. Both use either 16- or 32-kbps digital transmission rates.

A CVSD coder is shown in Figure 3.37, a decoder in Figure 3.38, and the CVSD waveforms in Figure 3.39. CVSD circuitry provides increased dynamic range capability by adjusting the gain of the integrator. For a given clock frequency and input bandwidth, the additional circuitry increases the delta modulator's dynamic range (i.e., up to 50 dB of range). External to the basic delta modulator is an algorithm that monitors the past few outputs of the delta modulator in a simple shift register. The register is usually 3–4 bits

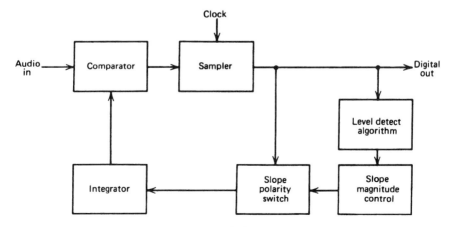

Figure 3.37. A CVSD encoder.

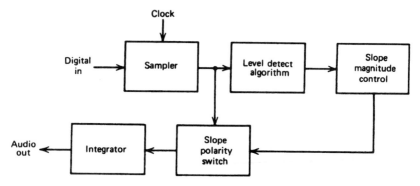

Figure 3.38. A CVSD decoder.

long, depending on the application. A CVSD algorithm monitors the contents of the shift register and indicates whether it contains all 1's or all 0's. This condition is called coincidence. When it occurs, it indicates that the gain of the integrator is too small. In this particular design, the coincidence changes a single-pole low-pass filter. The voltage output of this syllabic filter controls the integrator gain through a pulse amplitude modulator whose other input is the sign bit or up/down control.

The algorithm provides a means of measuring the average power or level of the input signal. The purpose of the algorithm is to control the gain of the integrator and to increase the dynamic range. The algorithm is repeated in the receiver to recover level data in the decoding process. Because the algorithm only operates on past serial data, it changes the nature of the bit stream without changing the channel bit rate.

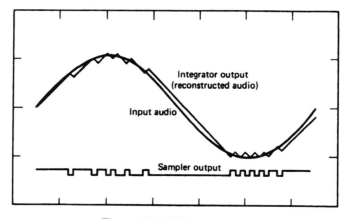

Figure 3.39. CVSD waveforms.

The effect of the algorithm is to compand the input signal. If a CVSD encoder is fed into a basic delta demodulator, the output of that demodulator will reflect the shape of the input signal, but all of the output will be at an equal level. Thus the algorithm at the output is needed to restore the level variations.

3.6 SONET AND SDH

3.6.1 Introduction

SONET (synchronous optical network) and SDH (synchronous digital hierarchy) are digital formats comparatively recently developed to handle the higher PCM bit rates typical of optical fiber links. However, there is no reason that these formats cannot be used on other transmission media such as LOS microwave and satellite circuits if these media can support the wider bandwidths required. SONET was developed in North America, whereas SDH was developed in Europe and is covered by ITU-T recommendations.

It should be pointed out that these formats are optimized for voice operation with 125-μs frames. Both types commonly transport plesiochronous digital hierarchy (PDH) formats such as DS1 and E-1 as well as asynchronous transfer mode (ATM) cells.

In the general scheme of things, the interface from one to the other will take place at North American gateways. In other words, international trunks are SDH equipped, not SONET equipped.

3.6.2 SONET

3.6.2.1 Introduction and Background. The SONET standard was developed by the ANSI T1X1 committee dating back to 1988. The SONET standard defines the features and functionality of a transport system based on the principles of synchronous multiplexing. SONET can be used in the local area, with digital loop carrier (Chapter 4) and with the long-distance network.

Unlike the PDH (e.g., DS1/E-1), individual tributary signals may be multiplexed directly into a higher rate SONET signal without using intermediate stages of multiplexing. SONET provides built-in signal capacity for advanced network management and maintenance capabilities. About 5% of SONET capacity is dedicated to these functions.

SONET is capable of transporting all the tributary signals that have been defined for the digital networks in existence today. This means that SONET can be deployed as an overlay to the existing network and, where appropriate, provide enhanced network flexibility by transporting existing signal types. In addition, SONET has the flexibility to readily accommodate the new types of customer service signals such as SMDS (switched multimegabit data service) and ATM. Actually, it can carry any octet-based binary format such as TCP/IP, SNA, OSI regimes, X.25, frame relay, and various LAN formats, which have been packaged for long-distance transmission.

3.6.2.2 *Synchronous Signal Structure.* SONET is based on a synchronous signal comprised of 8-bit octets, which are organized into a frame structure. The frame can be represented by a two-dimensional map consisting of N rows and M columns, where each box so derived contains one octet (or byte). The upper left-hand corner of the rectangular map representing a frame contains an identifiable marker to tell the receiver it is the start of frame.

SONET consists of a basic, first-level, structure called STS-1 (STS stands for synchronous transport signal), which is discussed below. The definition of the first level also defines the entire hierarchy of SONET signals because high-level SONET signals are obtained by synchronously multiplexing the lower-level modules. When lower-level modules are multiplexed together, the result is denoted as STS-N, where N is an integer. The resulting format then can be converted to an OC-N (OC stands for optical carrier) or STS-N electrical signal. There is an integer multiple relationship between the rate of the basic module STS-1 and the OC-N electrical signals (i.e., the rate of an OC-N is equal to N times the rate of an STS-1). Only OC-1, -3, -12, -24, -48, and -192 are supported by today's SONET.

3.6.2.2.1 *Basic Structure.* The STS-1 frame is shown in Figure 3.40. STS-1 is the basic module and building block of SONET. It is a specific sequence of

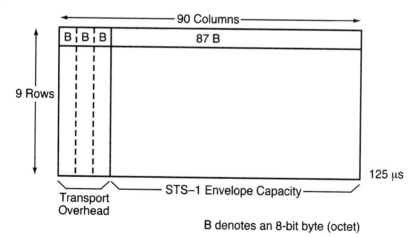

Figure 3.40. The STS-1 frame.

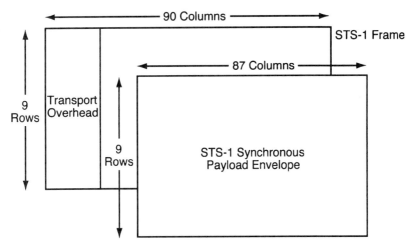

Figure 3.41. STS-1 synchronous payload envelope (SPE).

810 octets* (6480 bits), which includes various overhead octets and an envelope capacity for transporting payloads. STS-1 is depicted as a 90-column by 9-row structure. With a frame period of 125 μs (i.e., 8000 frames per second), STS-1 has a bit rate of 51.840 Mbps. Consider Figure 3.40. The order of transmission of octets is row-by-row, from left to right. In each octet of STS-1, the most significant bit is transmitted first.

As shown in Figure 3.40, the first three columns of the STS-1 frame are the transport overhead. These three columns contain 27 octets, of which nine are overhead for the section layer (i.e., section overhead) and 18 octets are overhead for the line layer (i.e., line overhead). The remaining 87 columns make up the STS-1 envelope capacity as shown in Figure 3.41.

The STS-1 synchronous payload envelope (SPE) occupies the STS-1 envelope capacity. The STS-1 SPE consists of 783 octets and is depicted as an 87-column by 9-row structure. In that structure, column 1 contains 9 octets and is designated as the STS path overhead (POH). In the SPE columns 30 and 59 are not used for payload but are designated as *fixed stuff* columns. The 756 octets in the remaining 84 columns are used for the actual STS-1 payload capacity.

Figure 3.42 shows the fixed stuff columns 30 and 59 inside the SPE. The reference document (Ref. 31) states that the octet in these fixed stuff columns are undefined and are set to binary 0's. However, the values used to stuff these columns of each STS-1 SPE will produce even parity in the calculation of the STS-1 Path BIP-8 (BIP stands for bit-interleaved parity.)

---

*The reference publications use the term byte, meaning an 8-bit sequence. We prefer to use the term *octet* because some argue that byte is ambiguous.

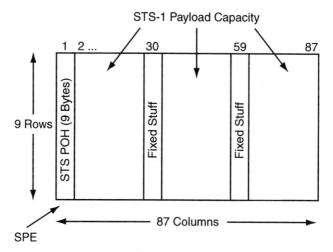

Figure 3.42. POH and STS-1 payload capacity with the STS-1 SPE. Note that the *net* payload capacity in a STS-1 frame is only 84 columns.

The STS-1 SPE may begin anywhere in the STS-1 envelope capacity. Typically the SPE begins in one STS-1 frame and ends in the next. This is shown in Figure 3.43. However, on occasion the SPE may be wholly contained on one frame. The STS payload pointer contained in the transport overhead designates the location of the octet where the SPE begins. Payload pointers are described in a following subsection.

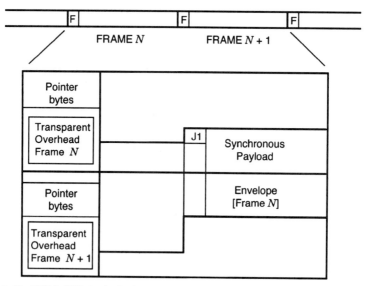

Figure 3.43. STS-1 SPE typically located in STS-1 frames. (From Ref. 30. Courtesy of Hewlett-Packard Co.)

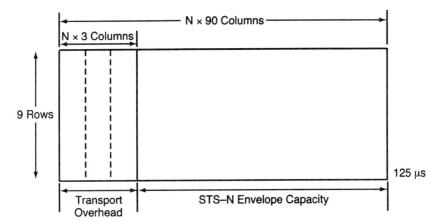

Figure 3.44. STS-N frame.

The STS POH is associated with each payload and is used to communicate various pieces of information from the point where the payload is mapped into the STS-1 SPE to the point where it is delivered.

3.6.2.2.2 STS-N Frames. Figure 3.44 illustrates the structure of a STS-N frame. This frame consists of a specific sequence $N \times 810$ octets. The STS-N frame is formed by octet-interleaving STS-1 and STS-M $(M < N)$ modules. The transport overheads of the individual STS-1 and STS-M modules are frame-aligned before interleaving, but the associated STS SPEs are not required to be aligned because each STS-1 has a payload pointer to indicate the location of the SPE or to indicate concatenation.

3.6.2.2.3 STS Concatenation. Superrate payloads require multiple STS-1 SPEs. Some B-ISDN payloads fall into this category. Concatenation means the linking together. A STS-Nc module is formed by linking N constituent STS-1's together in fixed phase alignment. The superrate payload is then mapped into the resulting STS-Nc SPE for transport. Such STS-Nc SPE requires an OC-N or a STS-N electrical signal. Concatenation indicators contained in the second through the Nth STS payload pointer are used to show that the STS-1's of a STS-Nc are linked together.

There are $N \times 783$ octets in a STS-Nc. Such a STS-Nc arrangement is illustrated in Figure 3.45 and is depicted as an $N \times 87$ column by 9-row structure. Because of the linkage, only one set of STS POH is required in the STS-Nc SPE. Here the STS POH always appears in the first of the N STS-1's that make up the STS-Nc.

Figure 3.46 shows the transport overhead assignment for an OC-3 carrying a STS-3c SPE.

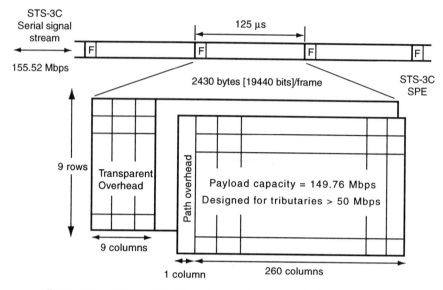

Figure 3.45. STS-Nc SPE. (From Ref. 30. Courtesy of Hewlett-Packard Co.)

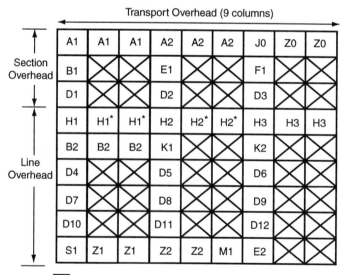

Transport Overhead (9 columns)

A1	A1	A1	A2	A2	A2	J0	Z0	Z0
B1	⊠	⊠	E1	⊠	⊠	F1	⊠	⊠
D1	⊠	⊠	D2	⊠	⊠	D3	⊠	⊠
H1	H1*	H1*	H2	H2*	H2*	H3	H3	H3
B2	B2	B2	K1	⊠	⊠	K2	⊠	⊠
D4	⊠	⊠	D5	⊠	⊠	D6	⊠	⊠
D7	⊠	⊠	D8	⊠	⊠	D9	⊠	⊠
D10	⊠	⊠	D11	⊠	⊠	D12	⊠	⊠
S1	Z1	Z1	Z2	Z2	M1	E2	⊠	⊠

Section Overhead (rows 1–3), Line Overhead (rows 4–9)

⊠ = Undefined overhead byte (all-zeros pattern as an objective)

* = Concentration Indication

H1* = 10010011

H2* = 11111111

Figure 3.46. Transport overhead assignment showing OC-3 carrying a STS-3c SPE. (From Ref. 31, Figure 3-8. Copyright © 1994 Bellcore. Reprinted with permission.)

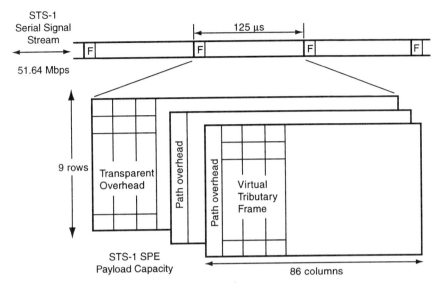

Figure 3.47. The virtual tributary (VT) concept (From Ref. 30. Courtesy of Hewlett-Packard Co.)

3.6.2.2.4 *Structure of Virtual Tributaries (VTs).* The SONET STS-1 SPE with a channel capacity of 50.11 Mbps has been designed specifically to transport a DS3 tributary signal. To accommodate sub-STS-1 rate payloads such as DS1, the VT structure is used. It consists of four sizes: VT1.5 (1.728 Mbps) for DS1 transport, VT2 (2.304 Mbps) for E1 transport, VT3 (3.456 Mbps) for DS1C transport, and VT6 (6.912 Mbps) for DS2 transport. The virtual tributary concept is illustrated in Figure 3.47. The four VT configurations are shown in Figure 3.48. In the 87-column by 9-row structure of the STS-1 SPE, the VTs occupy 3, 4, 6, and 12 columns, respectively.

There are two VT operating modes: *floating mode* and *locked mode.*

The floating mode was designed to minimize network delay and provide efficient cross-connects of transport signals at the VT level within the synchronous network. This is achieved by allowing each VT SPE to float with respect to the STS-1 SPE in order to avoid the use of unwanted slip* buffers at each VT cross-connect point. Each VT SPE has its own payload pointer that accommodates timing synchronization issues associated with the individual VTs. As a result, by allowing a selected VT1.5, for example, to be cross-connected between different transport systems without unwanted network delay, this mode allows a DS1 to be transported effectively across a SONET network.

The locked mode minimizes interface complexity and supports bulk transport of DS1 signals for digital switching applications. This is achieved by

*Slips are discussed in Chapter 4.

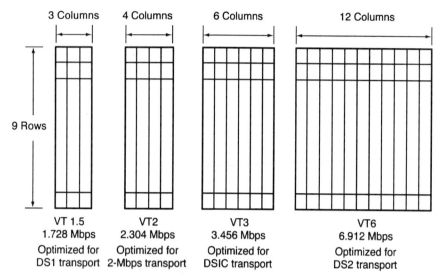

Figure 3.48. The four sizes of virtual tributary frames (From Ref. 30. Courtesy of Hewlett-Packard Co.)

locking individual VT SPEs in fixed positions with respect to the STS-1 SPE. In this case, each VT1.5 SPE is not provided with its own payload pointer. With the locked mode it is not possible to route a selected VT1.5 through the SONET network without unwanted network delay caused by having to provide slip buffers to accommodate the timing/synchronization issues.

3.6.2.2.5 The Payload Pointer. The STS payload pointer provides a method of allowing flexible and dynamic alignment of the STS SPE within the STS envelope capacity, independent of the actual contents of the SPE.

SONET, by definition, is intended to be a synchronous network. As described in Chapter 4, it derives its timing from a master network clock.

Modern digital networks must make provision for more than one master clock. Examples in North America are the several interexchange carriers, which interface with local exchange carriers, each with their own master clock. Each master clock operates independently. Each independent master clock has excellent stability (i.e., better than 1×10^{-11}/month), yet there may still be a very small variance in time among the clocks. Likewise, SONET must take into account loss of master clock or a segment of its timing delivery system. In this case, switches fall back on lower-stability internal clocks. This situation must also be handled by SONET. Therefore synchronous transport must be able to operate effectively under these conditions where network nodes are operating at slightly different rates.

To accommodate these clock offsets, the SPE can be moved (justified) in the positive or negative direction one octet at a time with respect to the transport frame. This is achieved by recalculating or updating the payload pointer at each SONET network node. In addition to clock offsets, updating the payload pointer also accommodates any other timing phase adjustments required between the input SONET signals and the timing reference at the SONET node.

This is what is meant by dynamic alignment where the STS SPE is allowed to float within the STS envelope capacity.

The payload pointer is contained in the H1 and H2 octets in the line overhead and designates the location of the octet where the STS SPE begins. These two octets are viewed as one word, as illustrated in Figure 3.49. Bits 1 through 4 of the pointer word carry the *new data flag*, and bits 7 through 16 carry the pointer value. Bits 5 and 6 are undefined.

Pointer word bits 7 through 16 contain the pointer value, a binary number with a range of 0 to 782. It indicates the offset of the pointer word and the first octet of the STS SPE (i.e., the J1 octet). The transport overhead octets are not counted in the offset. For example, a pointer value of 0 indicates that the STS SPE starts in the octet location that immediately follows the H3 octet, whereas an offset of 87 indicates that it starts immediately after the K2 octet location.

Payload pointer processing introduces a signal impairment known as *payload adjustment jitter*. This impairment appears on a received tributary signal after recovery from a SPE that has been subjected to payload pointer changes. The operation of the network equipment processing the tributary signal immediately downstream is influenced by this excessive jitter. By careful design of the timing distribution for the synchronous network, payload pointer adjustments can be minimized, thus reducing the level of tributary jitter that can be accumulated through synchronous transport.

3.6.2.2.6 The Three Overhead Levels in SONET. The three embedded overhead levels of SONET are:

- Path (POH)
- Line (LOH)
- Section (SOH)

These overhead levels, represented as spans, are illustrated in Figure 3.50. One important function is to support network operation and maintenance (OAM).

The POH consists of 9 octets and occupies the first column of the SPE as we pointed out previously. It is created and included in the SPE as part of the SPE assembly process. The POH provides the facilities to support and maintain the transport of the SPE between path terminations, where the SPE is assembled and disassembled.

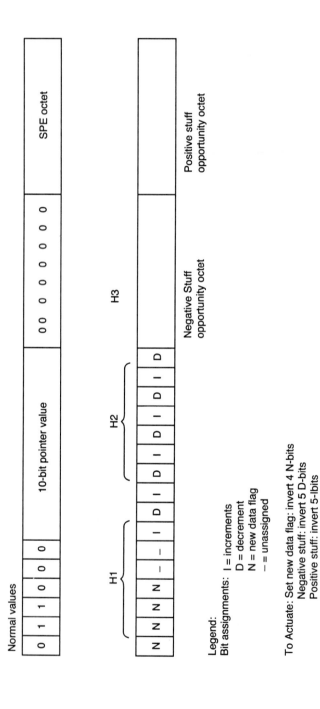

Figure 3.49. STS payload pointer (H1, H2) coding.

Normal values

| 0 | 1 | 1 | 0 | 0 | 0 | 10-bit pointer value | 0 0 0 0 0 0 0 0 | SPE octet |

H1 · H2

N N N N – – I D I D I D I D I D

H3

Negative Stuff opportunity octet

Positive stuff opportunity octet

Legend:
Bit assignments: I = increments
D = decrement
N = new data flag
– = unassigned

To Actuate: Set new data flag: invert 4 N-bits
Negative stuff: invert 5 D-bits
Positive stuff: invert 5-Ibits

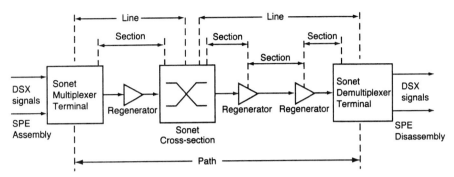

Figure 3.50. SONET section, line, and path definitions.

Among the POH specific functions are:

- An 8-bit wide (octet B3) BIP check calculated over all bits of the previous SPE. The computed value is placed in the POH of the following frame.
- Alarm and performance information (octet G1).
- A path signal label (octet C2) gives details of SPE structure. It is 8 bits wide, which can identify up to 256 structures (2^8).
- One octet (J1) repeated through 64 frames can develop an alphanumeric message associated with the path. This allows verification of continuity of connection to the source of the path signal at any receiving terminal along the path by monitoring the message string.
- An orderwire for network operator communications between path equipment (octet F2).

Facilities to support and maintain the transport of the SPE between adjacent nodes are provided by the line and section overhead. These two overheads share the first three columns of the STS-1 frame (see Figure 3.40). The SOH occupies the top three rows (total of 9 octets) and the LOH occupies the bottom 6 rows (total of 18 octets).

The line overhead functions include:

- Payload pointer (octets H1, H2, and H3) (each STS-1 in a STS-N frame has its own payload pointer)
- Automatic protection switching control (octets K1 and K2)
- BIP parity check (octet B2)
- 576-kbps Data channel (octets D4 through D12)
- Express orderwire (octet E2)

A *section* is defined in Figure 3.50. Among the section overhead functions are:

- Frame alignment pattern (octets A1, A2)
- STS-1 identification (octet C1): a binary number corresponding to the order of appearance in the STS-N frame, which can be used in the framing and deinterleaving process to determine the position of other signals
- BIP-8 parity check (octet B1): section error monitoring
- Data communications channel (octets D1, D2, and D3)
- Local orderwire channel (octet E1)
- User channel (octet F1).

3.6.2.2.7 The SPE Assembly/Disassembly Process. Payload mapping is the process of assembling a tributary signal into a SPE. It is fundamental to SONET operation.

The payload capacity provided for each individual tributary signal is always slightly greater than that required by the tributary signal. The mapping process, in essence, is to synchronize the tributary signal with the payload capacity. This is achieved by adding stuffing bits to the bit stream as part of the mapping process.

An example might be a DS3 tributary signal at a nominal rate of 44.736 Mbps to be synchronized with a payload capacity of 49.54 Mbps provided by a STS-1 SPE. The addition of path overhead completes the assembly process of the STS-1 SPE and increases the bit rate of the composite signal to 50.11 Mbps. The SPE assembly process is shown graphically in Figure 3.51.

At the terminus or drop point of the network, the original DS3 payload must be recovered as in our example. This process of SPE disassembly is shown in Figure 3.52. The term used here is *payload demapping.*

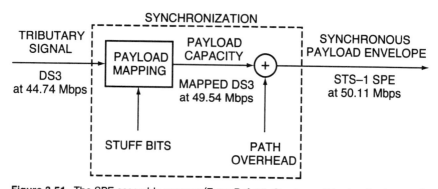

Figure 3.51. The SPE assembly process (From Ref. 30. Courtesy of Hewlett-Packard Co.)

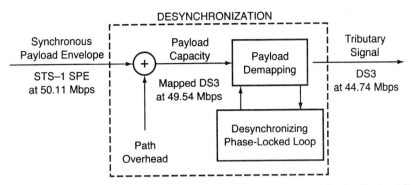

Figure 3.52. The SPE disassembly process (From Ref. 30. Courtesy of Hewlett-Packard Co.)

The demapping process desynchronizes the tributary signal from the composite SPE signal by stripping off the path overhead and the added stuff bits. In the example, a STS-1 SPE with a mapped DS3 payload arrives at the tributary disassembly location with a signal rate of 50.11 Mbps. The stripping process results in a discontinuous signal representing the transported DS3 signal with an average signal rate of 44.74 Mbps. The timing discontinuities are reduced by means of a desynchronizing phase-locked loop, which then produces a continuous DS3 signal at the required average transmission rate (Refs. 30, 31).

3.6.2.2.8 Line Rates for Standard SONET Interface Signals. Table 3.26 shows the standard line transmission rates for OC-N and STS-N.

3.6.2.2.9 Add–Drop Multiplex (ADM) Functions. The SONET ADM multi-plexes one or more DSn signals into the SONET OC-N channel. An ADM can be configured for either the add–drop or the terminal mode. In the ADM mode, it can operate when the low-speed DS1 signals terminating at the SONET ADM derive timing from the same or equivalent source as SONET (i.e., synchronous) or do not derive timing from the same source as SONET (called asynchronous in this case).

TABLE 3.26 Line Rates for Standard SONET Interface Signals

OC-N Level	STS-N Electrical Level	Line Rate (Mbps)
OC-1	STS-1 electrical	51.84
OC-3	STS-3 electrical	155.52
OC-12	STS-12 electrical	622.08
OC-24	STS-24 electrical	1244.16
OC-48	STS-48 electrical	2488.32
OC-192	STS-192 electrical	9953.28

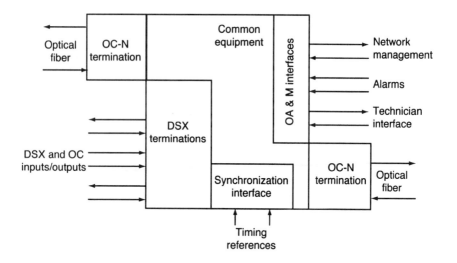

Notes: DSX = DS1–DS3 and OC (optical carrier, SONET)
OA & M = Operations, administration and maintenance

Figure 3.53. SONET ADM add–drop configuration example.

Figure 3.53 shows an example of ADM configured in the add–drop mode with DS1 and OC-N interfaces. A SONET ADM interfaces two full-duplex OC-N signals and one or more full-duplex DS1 signals. It may optionally provide low-speed DS1C, DS2, DS3, or OC-M ($M \leq N$) interfaces. There is nonpath-terminating information payloads from each incoming OC-N signal, which are passed through the SONET ADM and transmitted by the OC-N interface at the other side.

Timing for transmitted OC-N is derived from either an external synchronization source, an incoming OC-N signal, each incoming OC-N signal in each direction (called through-timing), or its local clock, depending on the network application. Each DS1 interface reads data from an incoming OC-N and inserts data into an outgoing OC-N bit stream as required. Figure 3.53 also shows a synchronization interface for local switch application with external timing and an operations interface module (OIM) that provides local technician orderwire, local alarm, and an interface to remote operations systems. A controller is part of each SONET ADM, which maintains and controls the ADM functions, to connect to local or remote technician interfaces, and to connect to required and optional operations links that permit maintenance, provisioning, and testing.

Figure 3.54 shows an example of an ADM in the terminal mode of operation with DS1 interfaces. In this case, the ADM multiplexes up to $N \times (28DS1)$ or equivalent signals into an OC-N bit stream. Timing for this terminal configuration is taken from either an external synchronization source, the received OC-N signal (called loop timing), or its own local clock, depending on the network application.

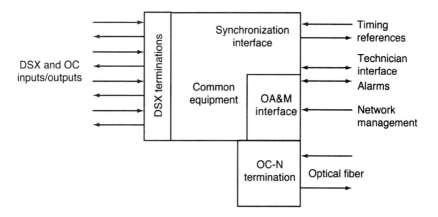

Notes: DSX = DS1–DS3 and OC = optical carrier, SONET
OA & M = Operations, administration and maintenance

Figure 3.54. An ADM in a terminal configuration.

3.6.3 Synchronous Digital Hierarchy (SDH)

3.6.3.1 Introduction. Synchronous digital hierarchy (SDH) was a European/CCITT development, whereas SONET was a North American development. They are very similar. One major difference is that their level 1 rates are dissimilar. For SONET, the basic building block is STS-1 at 51.84 Mbps, and for SDH it is STM-1, which derives the lowest transmission rate at 155.52 Mbps. The STM-1 rate is the same as SONET's STS-3 transmission rate. SDH is generally understood to be more flexible than SONET and was designed more with the PTT (government-owned telecommunication administrations) in mind. The SONET design, on the other hand, has more the flavor of a private network. SDH is several years behind SONET in implementation.

3.6.3.2 SDH Standard Bit Rates. The standard SDH bit rates are shown in Table 3.27. ITU-T Rec. G.707 (Ref. 32) states "that the first level of the digital hierarchy shall be 155,520 kbps ... and ... higher synchronous digital hierarchy rates shall be obtained as integer multiples of the first level bit rate."

TABLE 3.27 SDH Bit Rates with SONET Equivalents

SDH Level[a]	SDH Bit Rate (kbps)	SONET Equivalent Line Rate
1	155,520	STS-3/OC-3
4	622,080	STS-12/OC-12
16	2,488,320	STS-48/OC-48

[a]Two other SDH hierarchical levels are under consideration by CCITT (Ref. 32):
Level 8   1,244,160 kbps
Level 12   1,866,240 kbps

3.6.3.3 *Interface and Frame Structure of SDH.* Figure 3.55 illustrates the relationship between various multiplexing elements subsequently described and shows generic multiplexing structures. Figures 3.56, 3.57, and 3.58 show specific derived multiplexing methods.

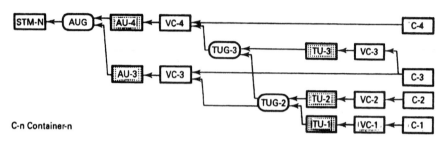

Figure 3.55. Generalized SDH multiplexing structure. (From Ref. 33, Figure 2-1/G.708, ITU-T Rec. G.708.)

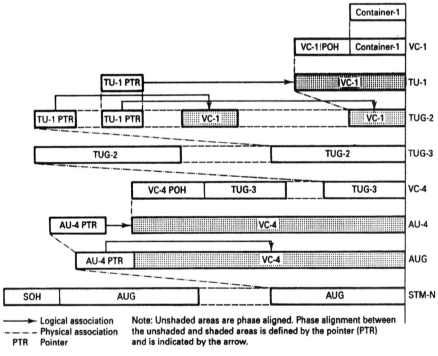

Figure 3.56. Multiplexing method directly from container-1 using AU-4. (From Ref. 33, Figure 2-2/G.708, ITU-T Rec. G.708.)

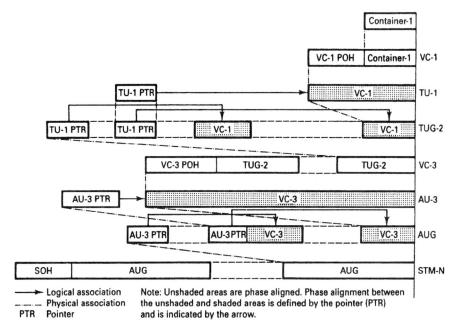

Figure 3.57. SDH multiplexing method directly from container-1 using AU-3. (From Ref. 33, Figure 2-3/G.708, ITU-T Rec. G.708.)

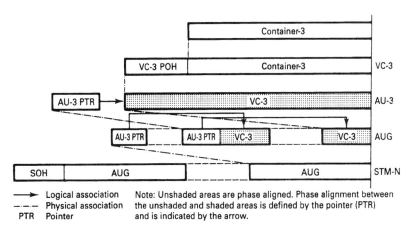

Figure 3.58. Multiplexing method directly from container-3 using AU-3. (From Ref. 33, Figure 2-4/G.708, ITU-T Rec. G.708.)

3.6.3.3.1 Definitions

Synchronous Transport Module (STM). A STM is the information structure used to support section layer connections in the SDH. It is analogous to STS in the SONET regime. STM consists of information payload and section overhead (SOH) information fields organized in a block frame structure that repeats every 125 μs. The information is suitably conditioned for serial transmission on selected media at a rate that is synchronized to the network. A basic STM (STM-1) is defined at 155,520 kbps. Higher-capacity STMs are formed at rates equivalent to N times multiples of this basic rate. STM capacities for $N = 4$ and $N = 16$ are defined, and higher values are under consideration by ITU-T. A STM comprises a single administrative unit group (AUG) together with the SOH. STM-N contains N AUGs together with SOH.

Container, C-n ($n = 1$ to $n = 4$). This element is a defined unit of payload capacity, which is dimensioned to carry any of the bit rates currently defined in Section 3.6.3.2 and may also provide capacity for transport of broadband signals that are not yet defined by CCITT (ITU-T Organization) (Ref. 33).

Virtual Container-n (VC-n). A virtual container is the information structure used to support path layer connection in the SDH. It consists of information payload and path overhead (POH) information fields organized in a block frame that repeats every 125 or 500 μs. Alignment information to identify VC-n frame start is provided by the server network layer. Two types of virtual container have been identified:

1. *Lower-Order Virtual Container-n, VC-n ($n = 1, 2$).* This element consists of a single C-n ($n = 1, 2$), plus the basic virtual container POH appropriate to that level.
2. *Higher-Order Virtual Container-n, to VC-n ($n = 3, 4$).* This element consists of a single C-n ($n = 3, 4$), an assembly of tributary unit groups (TUG-2's), or an assembly of TU-3's, together with virtual container POH appropriate to that level.

Administrative Unit-n, AU-n. An administrative unit is the information structure that provides adaptation between the higher-order path layer and the multiplex section. It consists of an information payload (the higher-order virtual container) and an administrative unit pointer, which indicates the offset of the payload frame start relative to the multiplex section frame start. Two administrative units are defined. The AU-4 consists of a VC-4 plus an administrative unit pointer, which indicates the phase alignment of the VC-4 with respect to the STM-N frame. The AU-3 consists of a VC-3 plus an administrative unit pointer, which indicates the phase alignment of the VC-3 with respect to the STM-N frame. In each case the administrative unit pointer location is fixed with respect to the STM-N frame (Ref. 33).

One or more administrative units occupying fixed, defined positions in a STM payload is termed an *administrative unit group* (AUG). An AUG consists of a homogeneous assembly of AU-3's or an AU-4.

Tributary Unit-*n*, TU-*n*. A tributary unit is an information structure that provides adaptation between the lower-order path layer and the higher-order path layer. It consists of an information payload (the lower-order virtual container) and a tributary unit pointer, which indicates the offset of the payload frame start relative to the higher-order virtual container frame start.

The TU-*n* ($n = 1, 2, 3$) consists of a VC-*n* together with a tributary unit pointer.

One or more tributary units occupying fixed, defined positions in a higher-order VC-*n* payload is termed a *tributary unit group* (TUG). TUGs are defined in such a way that mixed-capacity payloads made up of different-size tributary units can be constructed to increase flexibility of the transport network.

A TUG-2 consists of a homogeneous assembly of identical TU-1's or a TU-2. A TUG-3 consists of a homogeneous assembly of TUG-2's or a TU-3 (Ref. 33).

Container-*n* ($n = 1$–4). A container is the information structure that forms the network synchronous information payload for a virtual container. For each of the defined virtual containers there is a corresponding container. Adaptation functions have been defined for many common network rates into a limited number of standard containers (Ref. 33). These include standard E-1/DS-1 rates defined in ITU-T Rec. G.702 (Ref. 34).

3.6.3.3.2 Frame Structure. The basic frame structure for STM-N is illustrated in Figure 3.59. The three main areas of STM-1 frame are section overhead, AU pointers, and STM-1 payload.

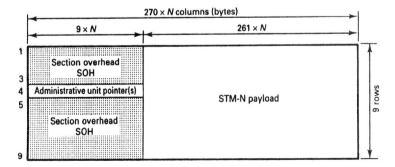

Figure 3.59. STM-N frame structure. (From Ref. 33, Figure 3-1/G.708, ITU-T Rec. G.708.)

Section Overhead (SOH). Section overhead is contained in rows 1–3 and 5–9 of columns 1–9 × N of the STM-N shown in Figure 3.59.

Administrative Unit (AU) Pointers. The AU-*n* pointer (like the SONET pointer) allows flexible and dynamic alignment of the VC-*n* within the AU-*n* frame. Dynamic alignment means that the VC-*n* floats within the AU-*n* frame. Thus the pointer is able to accommodate differences, not only in the phases of the VC-*n* and the SOH but also in the frame rates.

Row 4 of columns 1–9 × N in Figure 3.59 is available for AU pointers. The AU-4 pointer is contained in octets H1, H2, and H3 as shown in Figure 3.60. The three individual AU-3 pointers are contained in three separate H1, H2, and H3 octets as shown in Figure 3.61.

The pointer contained in H1 and H2 designates the location of the octet where the VC-*n* begins. The two octets (or bytes) allocated to the pointer

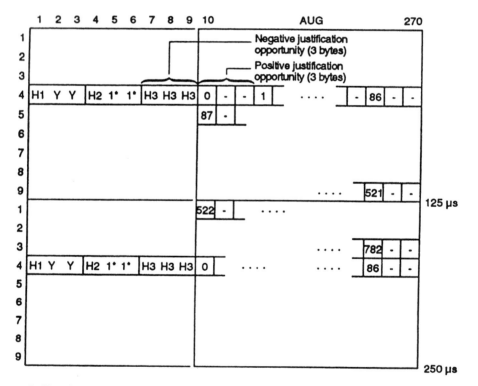

1* All 1s byte
Y 1001SS11 (S bits are unspecified)

Figure 3.60. AU-4 pointer offset numbering. (From Ref. 35, Figure 3-1/G.709, ITU-T Rec. G.709.)

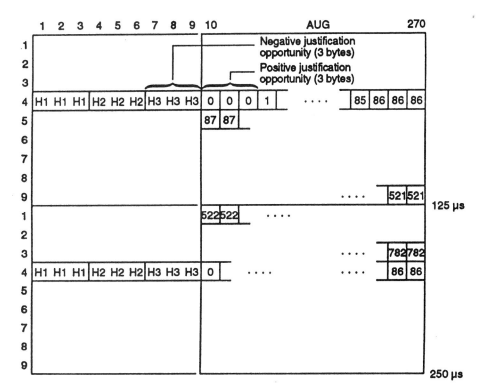

Figure 3.61. AU-3 offset numbering. (From Ref. 35, Figure 3-2/G.709, ITU-T Rec. G.709.)

function can be viewed as one word as shown in Figure 3.62. The last ten bits (bits 7–16) of the pointer word carry the pointer value.

As shown in Figure 3.62, the AU-4 pointer value is a binary number with a range of 0–782, which indicates offset, in three octet increments, between the pointer and the first octet of the VC-4 (see Figure 3.60). Figure 3.62 also indicates one additional valid point, the concatenation indication. This concatenation indication is given by 1001 in bits 1–4; bits 5–6 are unspecified, and there are ten 1's in bits 7–16. The AU-4 pointer is set to concatenation indication for AU-4 concatenation.

There are three AU-3's in an AUG, where each AU-3 has its own associated H1, H2, and H3 octets. As detailed in Figure 3.61, the H octets are shown in sequence. The first H1, H2, H3 set refers to the first AU-3, and the second set to the second AU-3, and so on. For the AU-3's, each pointer operates independently. In all cases the AU-n pointer octets are not counted in the offset. For example, in an AU-4, the pointer value of 0 indicates that the VC-4 starts in the octet location that immediately follows the last H3

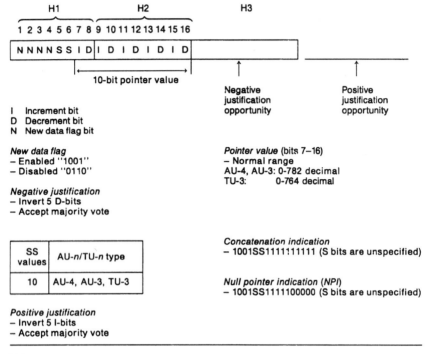

I   Increment bit
D   Decrement bit
N   New data flag bit

New data flag
– Enabled "1001"
– Disabled "0110"

Negative justification
– Invert 5 D-bits
– Accept majority vote

SS values	AU-*n*/TU-*n* type
10	AU-4, AU-3, TU-3

Positive justification
– Invert 5 I-bits
– Accept majority vote

Pointer value (bits 7–16)
– Normal range
AU-4, AU-3: 0-782 decimal
TU-3:   0-764 decimal

Concatenation indication
– 1001SS1111111111 (S bits are unspecified)

Null pointer indication (*NPI*)
– 1001SS1111100000 (S bits are unspecified)

Notes:
1. NPI value applies only to TU-3 pointers.
2. The pointer is set to all "1"s when an AIS occurs.

Source: Ref. 20.

Figure 3.62. AU-*n*/TU-3 pointer (H1, H2, H3) coding. (From Ref. 35, Figure 3-3/G.709, ITU-T Rec. G.709.)

octet, whereas an offset of 87 indicates that the VC-4 starts three octets after the K2 octet (byte) (Ref. 35). (Note the similarity to SONET.)

Frequency Justification. If there is a frequency offset between the frame rate of the AUG and that of the VC-*n*, then the pointer value will be incremented or decremented as needed, accompanied by a corresponding positive or negative justification octet or octets. Consecutive pointer operations must be separated by at least three frames (i.e., every fourth frame) in which the pointer values remain constant.

If the frame rate of the VC-*n* is too slow with respect to that of the AUG, then the alignment of the VC-*n* must periodically slip back in time and the pointer value must be incremented by one. This operation is indicated by inverting bits 7, 9, 11, 13, and 15 (I-bits) of the pointer word to allow 5-bit majority voting at the receiver. Three positive justification octets appear immediately after the last H3 octet in the AU-4 frame containing the inverted I-bits. Subsequent pointers will contain the new offset.

For AU-3 frames, a positive justification octet appears immediately after the individual H3 octet of the AU-3 frame containing inverted I-bits. Subsequent pointers will contain the new offset.

If the frame rate of the VC-n is too fast with respect to that of the AUG, then the alignment of the VC-n must periodically be advanced in time and the pointer value must then be decremented by one. This operation is indicated by inverting bits 8, 10, 12, 14, and 16 (D-bits) of the pointer word to allow 5-bit majority voting at the receiver. Three negative justification octets appear in the H3 octets in the AU-4 frame containing inverted D-bits. Subsequent pointers will contain the new offset.

For AU-3 frames, a negative justification octet appears in the individual H3 octet of the AU-3 frame containing inverted D-bits. Subsequent pointers will contain the new offset.

The following summarizes the rules (Ref. 35) for interpreting the AU-n pointers:

1. During normal operation, the pointer locates the start of the VC-n within the AU-n frame.
2. Any variation from the current pointer value is ignored unless a consistent new value is received three times consecutively or it is preceded by one; see rules 3, 4, or 5 (below). Any consistent new value received three times consecutively overrides (i.e., takes priority over) rules 3 and 4.
3. If the majority of I-bits of the pointer word are inverted, a positive justification operation is indicated. Subsequent pointer values shall be incremented by one.
4. If the majority of D-bits of the pointer word are inverted, a negative justification operation is indicated. Subsequent pointer values shall be decremented by one.
5. If the NDF (new data flag) is set to 1001, then the coincident pointer value shall replace the current one at the offset indicated by the new pointer values unless the receiver is in a state that corresponds to a loss of pointer.

Administrative Units in the STM-N. The STM-N payload can support N AUGs, where each AUG may consist of one AU-4 or three Au-3's. The VC-n associated with each AUG-n does not have a fixed phase with respect to the STM-N frame. The location of the first octet of the VC-n is indicated by the AU-n pointer. The AU-n pointer is in a fixed location in the STM-N frame. This is illustrated in Figures 3.56–3.59, 3.63, and 3.64.

The AU-4 may be used to carry, via the VC-4, a number of TU-n's ($n = 1, 2, 3$) forming a two-stage multiplex. An example of this arrangement is shown in Figures 3.56 and 3.64a. The VC-n associated with each TU-n does not have a fixed phase relationship with respect to the start of the VC-4.

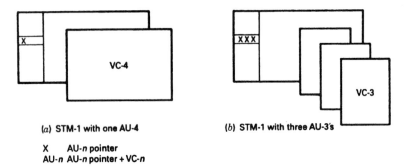

(a) STM-1 with one AU-4 (b) STM-1 with three AU-3's

X   AU-*n* pointer
AU-*n* AU-*n* pointer + VC-*n*

Figure 3.63. Administrative units in a STM-1 frame. (From Ref. 33, Figure 3-2/G.708, ITU-T Rec. G.708.)

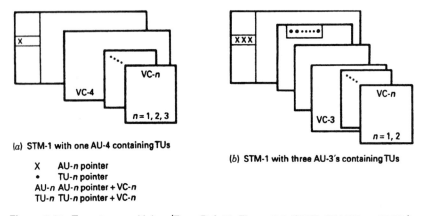

(a) STM-1 with one AU-4 containing TUs

X   AU-*n* pointer
•   TU-*n* pointer
AU-*n* AU-*n* pointer + VC-*n*
TU-*n* TU-*n* pointer + VC-*n*

(b) STM-1 with three AU-3's containing TUs

Figure 3.64. Two-stage multiplex. (From Ref. 33, Figure 3-3/G.708, ITU-T Rec. G.708.)

The TU-*n* pointer is in a fixed location in the VC-4 and the location of the first octet of the VC-*n* is indicated by the TU-*n* pointer.

The AU-3 may be used to carry, via the VC-3, a number of TU-*n*'s ($n = 1, 2$) forming a two-stage multiplex. An example of this arrangement is shown in Figure 3.63 and 3.64*b*. The VC-*n* associated with each TU-*n* does not have a fixed phase relationship with respect to the start of the VC-3. The TU-*n* pointer is in a fixed location in the VC-3 and the location of the first octet of the VC-*n* is indicated by the TU-*n* pointer (Ref. 35).

3.6.3.3.3 Interconnection of STM-1's. SDH has been designed to be universal, allowing transport of a large variety of signals including those specified in ITU-T Rec. G.702 (Ref. 34), such as the North American DS1 hierarchy and the European E-1 hierarchy. However, different structures can be used for the transport of virtual containers. The following interconnection rules are used.

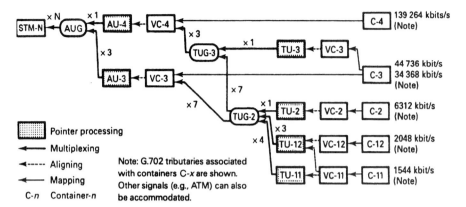

Figure 3.65. Basic SDH multiplexing structure. (From Ref. 35, Figure 1/G.709, ITU-T Rec. G.709.)

1. The rule for interconnecting two AUGs based on two different types of administrative unit, namely, AU-4 and AU-3, is to use the AU-4 structure. Therefore the AUG based on AU-3 is demultiplexed to the TUG-2 or VC-3 level according to the type of payload and is remultiplexed within an AUG via the TUG-3/VC-4/AU-4 route.

2. The rule for interconnecting VC-11's transported via different types of tributary unit, namely, TU-11 and TU-12, is to use the TU-11 structure. VC-11, TU-11, and TU-12 are described in ITU-T Rec. G. 709 (Ref. 35).

3.6.3.3.4 Basic SDH Multiplexing Structure. The SDH multiplexing structure is illustrated in Figure 3.65.

REVIEW EXERCISES

1. Define the two basic generic types of multiplexing covered in the text.

2. The input to a mixer used in FDM equipment is the standard CCITT voice channel, and a local oscillator injects a 64-kHz signal into that mixer. Define the lower sideband boundaries at the output from the mixer.

3. Define the standard CCITT FDM group regarding voice channel capacity and the frequency band it occupies.

4. Define the standard CCITT FDM supergroup regarding voice channel capacity, number of groups contained therein, and the frequency band that the supergroup occupies.

5. List at least three transmission media to which we could apply FDM line configurations.

6. Speech loading on FDM systems is a complex matter. The complexity arises because speech amplitude varies with _____. List at least three parameters here.

7. How does speech level in dBm vary with the number of channels? Expressed another way, suppose the test tone input to a channel was −16 dBm and there were 400 channels. What would the aggregate level value be in dBm?

8. Relate peak power of an average talker to average power. What peaking factor (dB) would there be for 100 talkers (compared to average power)?

9. What is the activity factor (speech) for conventional commercial FDM equipment?

10. Name at least three types of constant-amplitude signals that might be transmitted on FDM equipment? What would the activity factor be here?

11. What is the principal purpose of a FDM pilot tone? A secondary purpose?

12. On long routes, why is it necessary to limit the number of modulation/demodulation (multiplex/demultiplex) steps in FDM systems?

13. Name at least two advantages of syllabic compandors.

14. What is the principal transmission advantage of TDM (PCM) networks over analog networks (FDM)? Name at least one disadvantage.

15. What are the three basic steps involved in the development of a PCM signal from an analog input (e.g., voice)?

16. In accordance with the Nyquist sampling theorem, what is the sampling rate of a standard nominal 4-kHz voice channel? Of a 15-kHz program channel? Of a 4.2-MHz NTSC video TV signal?

17. A 7-level (7-bit) codeword has how many distinct coding sequences? An 8-level (8-bit) codeword? A 12-bit codeword (typical of a CD)?

18. What is the cause of quantization noise developed in a PCM coder/decoder combination?

19. What is the period (time duration) of a standard PCM frame?

20. If we double the number of quantizing steps, based on linear quantization, what is the quantizing noise improvement expressed in dB?

21. Of a PCM codeword, what does the first significant bit tell a decoder? Bits 2, 3, and 4?

22. Approximately what signal-to-distortion ratio may be expected for 8-bit PCM using the A-law? Using the μ-law?

23. Numerically derive the two basic PCM bit rates (i.e., DS1 and E-1).

24. How many bits are in the North American DS1 frame? How many in an E-1 frame?

25. There are two basic framing strategies used in today's PCM systems. Define and discuss these framing strategies (DS1 and E-1).

26. Describe and discuss the two DS1 multiframe (or superframe) strategies. Where are we finding these extra bits, which serve as overhead? What can these overhead bits do for us? (List at least three items.)

27. Can we really derive a bit error rate with the CRC used in an ESF?

28. Discuss and compare supervisory signaling on DS1 and on E-1.

29. What are the causes of idle channel noise on PCM systems? How can it be mitigated?

30. What is a *violation* of an AMI (bipolar) line code?

31. What are the meanings and implications of BNZS? Of HDB3?

32. Key to the principal advantage of PCM systems, and all binary digital systems for that matter, is the regenerative repeater. Explain why.

33. Why does a PCM system operating on a metallic medium use the AMI code rather than just a serial binary bit stream?

34. An M12 multiplexer provides a DS2 line rate of 6.312 Mbps at its output and has four DS1 1.544-Mbps sources at its input. If we multiply 1.544 Mbps by 4, we get 6.176 Mbps. Hence the M12 output is greater than the aggregate of the four inputs by 136 kbps. Why? For what are the extra bits used?

35. Explain the rationale for justification/stuffing?

36. Match B3ZS, B6ZS, and B8ZS to their respective bit rates or DS level.

37. Why would we scramble a PCM bit stream?

38. What is the use of an *eye pattern*?

39. Name at least four differences between the North American (DS1 family) and European (E-1 family) PCM systems.

40. What are the C-bits used for in the DS3 frame?

41. Explain the purpose(s) of the FEBE bits.

42. What is the purpose of channel 0 in E-1? Channel 16?

43. Discuss compatibility issues between standard digital data transmission rates and the 64-kbps digital voice channel.

44. Explain the CCITT two-step bit rate adaptation for data transmission on the digital 64-kbps channel.

45. Why is *bit integrity* so important for data transmission on the digital 64-kbps channel?

46. What is the difference between DS0-A and DS0-B?

47. How does the AT&T DDS system transmit data on the 64-kbps channel? Include the use of DS0 and overhead bits in the discussion.

48. How does ADPCM compress 64 kbps into 32 kbps? Include the words *adaptive* and *differential* in the discussion.

49. Discuss limitations for data modem operation over an ADPCM channel. Up to what bit rate can we be comfortable transmitting data?

50. ADPCM uses a 4-bit code for information, yet 32 is 2^5. For what purpose, then, is the other bit?

51. What kind of coding is used with delta modulation/CVSD?

52. How does CVSD achieve quality of voice transmission?

53. What is the major drawback (besides voice quality at 16 kbps) of delta modulation?

54. Use just one word to describe a major advantage of the DM/CVSD concept and equipment.

REFERENCES

1. "12-Channel FDM Terminal Equipments," CCITT Rec. G.232, Fascicle III.2, IXth Plenary Assembly, Melbourne, 1988.

2. "FDM Translating Equipments," CCITT Rec. G.233, Fascicle III.2, IXth Plenary Assembly, Melbourne, 1988.

3. *Transmission Systems for Communications*, 5th ed., Bell Telephone Laboratories, Holmdel, NJ, 1982.

4. "Assumptions for the Calculation of Noise on Hypothetical Reference Circuits for Telephony," CCITT Rec. G.223, Fascicle III.2, IXth Plenary Assembly, Melbourne, 1988.

5. *Voice Channel Loading of Data and Telegraphy Signals*, Lenkurt Demodulator, San Carlos, CA, July 1968.

6. "Pilots on Groups, Supergroups, Mastergroups and Super Mastergroups," CCITT Rec. G.241, Fascicle III.2, IXth Plenary Assembly, Melbourne, 1988.

7. "Noise Objectives for the Design of FDM Carrier-Transmission Systems on a 2500 km Hypothetical Reference Circuit," CCITT Rec. G.222, Fascicle III.2, IXth Plenary Assembly, Melbourne, 1988.

8. Roger L. Freeman, *Telecommunication System Engineering*, 3rd ed., Wiley, New York, 1996.

9. "Characteristics of Compandors for Telephony," CCITT Rec. G.162, Fascicle III.1, IXth Plenary Assembly, Melbourne, 1988.

10. *Reference Manual for Engineers: Radio, Electronics, Computers & Communications*, 8th ed., Sams Publishing (Prentice Hall), Carmel, IN, 1995.

11. *Introduction to PCM*, GTE Lenkurt Demodulator, San Carlos, CA, 1976.

12. John C. Bellamy, *Digital Telephony*, 2nd ed., Wiley, New York, 1990.

13. *Operations and Maintenance Manual for T324 PCM Cable Carrier System*, ITT Telecommunications, Raleigh, NC, Apr. 1973.

14. "Vocabulary of Digital Transmission and Multiplexing and Pulse Code Modulation (PCM) Terms," ITU-T Rec. G.701, ITU, Helsinki, Mar. 1993.

15. "Digital Access and Cross-Connect System Technical Reference and Compatibility Specification," *Technical Advisory for USITA*, No. 71, Issue 2, AT&T, Basking Ridge, NJ, Apr. 1982.

16. "The Extended Framing Format Interface Specification," *Technical Advisory for USITA*, No. 70, AT&T, Basking Ridge, NJ, Sept. 1981.

17. *Transport Systems Generic Requirements: Common Requirements*, TR-NWT-000499, Issue 5, Bellcore, Piscataway, NJ, Dec. 1993.

18. "Transmission Performance Characteristics of Pulse Code Modulation," CCITT Rec. G.712, ITU, Geneva, Sept. 1992.

19. "Physical/Electrical Characteristics of Hierarchical Digital Interfaces," CCITT Rec. G.703, CCITT, Geneva, 1991.

20. *Telecommunications Transmission Engineering*, 2nd ed., vols. 1–3, American Telephone & Telegraph Co., New York, 1977.

21. "General Requirements on Second Order Multiplex Equipment," CCITT Rec. G.742, Fascicle III.4, IXth Plenary Assembly, Melbourne, 1988.

22. "M13 Multiplex Compatibility Specification," *Technical Advisory for USITA*, No. 51, Issue 4, AT&T, Basking Ridge, NJ, Oct. 1979.

23. "Third Order Digital Multiplex Equipment Operating at 34.368 Mbps and Using Positive/Zero/Negative Justification," CCITT Rec. G.753, Fascicle III.4, IXth Plenary Assembly, Melbourne, 1988.

24. "Support of Data Terminal Equipments with V-Series-Type Interfaces by an Integrated Services Digital Network," CCITT Rec. V.110, CCITT, Geneva, Sept. 1992.

25. *Digital Data System Channel Interface Specification*, Bell System Technical Reference PUB 62310, American Telephone & Telegraph Co., New York, 1983.

26. *Dataphone*™ *II 2600 Switched Digital Data Service ACCUNET*™ *Switched-56 Service, User's Manual*, American Telephone & Telegraph Co., Indianapolis, IN, Mar. 1986.

27. *Algorithm and Line Format for 32 kbps Adaptive Differential Pulse-Code Modulation (ADPCM)*, ANSI T1.301-1987, ANSI, New York, 1987.

28. "32 kbps Adaptive Differential Pulse Code Modulation (ADPCM)," CCITT Rec. G.721, Fascicle III.4, IXth Plenary Assembly, Melbourne, 1988.

29. H. R. Schindler, *Delta Modulation*, "IEEE Spectrum," New York, Oct. 1970.

30. "Introduction to SONET," seminar, Hewlett-Packard Co., Burlington, MA, Nov. 1993.

31. *Synchronous Optical Network (SONET) Transport Systems, Common Generic Criteria*, Bellcore GR-253-CORE, Issue 1, Bellcore, Piscataway, NJ, Dec. 1994.

32. "Synchronous Digital Hierarchy Bit Rates," ITU-T Rec. G.707, ITU, Helsinki, Mar. 1993.

33. "Network Node Interface for the Synchronous Digital Hierarchy," ITU-T Rec. G.708, ITU, Helsinki, Mar. 1993.

34. "Digital Hierarchy Bit Rates," CCITT Rec. G.702, Fascicle III.4, IXth Plenary Assembly, Melbourne, 1988.

35. "Synchronous Multiplexing Structure," ITU-T Rec. G.709, ITU, Geneva, Mar. 1993.

36. "ISDN User-Network Interface-Data Link Layer Specification," CCITT Rec. Q.921, ITU Geneva, Mar. 1993.

4

DIGITAL NETWORKS

4.1 SCOPE AND RATIONALE

The objective of this chapter is to summarize the transmission factors of a typical PSTN digital network. These same factors will apply, perhaps less stringently, in private digital networks. We assume that *digital network* transmission is based on 8-bit PCM, where, in some instances, it will be in the European E-1 format and in others, in the DS1 format. These formats are described in detail in Chapter 3.

What makes digital networks different from their analog counterparts? Noise is a primary issue in analog networks; it is a secondary issue in digital networks. Certainly error performance of the underlying digital bearer is the major consideration in digital networks; it is of a concern in analog networks only when channels in those networks carry data.

Whereas we were concerned about the nominal 4-kHz voice channel in the analog network, here our concern is the 8-bit time slot and the resulting 64-kbps digital voice channel, variously called the B-channel and different groups of these channels to form H-channels.

Up to a certain point in the analog PSTN we were allowed the luxury of compartmentalizing transmission and switching. With the digital network they are deeply intertwined and cannot be compartmentalized. For example, master timing so necessary in transmission derives from a switch. In many cases traffic-bearing digital bit streams disseminate timing to switches. These bit streams are dependent on the timing imparted by a switch at its output ports.

There are underlying impairments in the digital network that were nonexistent or of secondary importance in the analog network. Among these impairments are jitter and wander, slips, and slip rate, and *self-imposed delay* brought about by buffer storage in the switching process.

Testing of digital networks is far different from testing of analog networks. Nearly every digital network has an analog side and we can use typical test tones. However, where we used 800- and 1000-Hz tones on an analog connectivity, they are forbidden if the connectivity is digital. Harmonics of these frequencies can lie on the 8000-Hz sampling rate. Thus we prefer a tone such as 1004 Hz or 1020 Hz, where the harmonics are safely away from the sampling rate. For testing on the digital side, we use specific bit sequences.

Other important subjects covered in this chapter are network synchronization and timing, transmission characteristics of digital switches, the digital subscriber loop, and transmission issues in ISDN BRI and ATM. SONET and SDH are also discussed. These are digital formats primarily used at the higher bit rates more fitting to optical fiber transmission. However, there are radio systems that can accommodate SONET/SDH transmission rates up to 655 Mbps.

4.2 TRANSMISSION FACTORS IN PCM-BASED DIGITAL NETWORKS

4.2.1 Introduction

Error performance of a digital network is paramount. In the PSTN the error performance threshold is a bit error rate (BER) $\leq 1 \times 10^{-3}$. As error performance degrades from this value, supervisory signaling will drop out and a dial tone will be returned at either end of the connection for a circuit in active use. Any PCM circuit (B-channel) should have an end-to-end error performance (ITU-T Rec. G.821) better than 1×10^{-6}. This value is driven by data communication users. Speech on a PCM connection is intelligible with a BER around 1×10^{-2}.

Another impairment is quantization noise or quantization distortion, which was discussed in Chapter 3. If there is only one A/D stage (i.e., a PCM channel bank) in a connectivity, the quantization distortion is determined there and nowhere else. However, in hybrid analog–digital–analog connectivities, each additional conversion from analog-to-digital format further degrades quantization distortion (or noise).

Jitter and wander are important impairments in a digital network, which can seriously limit circuit length. We will relate jitter accumulation as a function of the number of regenerative repeaters.

Slips represent still another digital network impairment. It is more appropriate to discuss slips under a general heading of timing and synchronization. This is covered in Section 4.3.

Delay is still another issue. Delay is an annoyance in human conversation. It also may have serious consequences for some interactive data users. The ITU-T Organization recommends (G.114) total delay in a voice connectivity be limited as follows:

- 0–150 ms: acceptable (except for some interactive users).
- 150–400 ms: acceptable if telephone companies and users are aware of the interactive shortcomings. Delay due to geostationary satellite communications falls into this category.
- above 400 ms: unacceptable.

4.2.2 Error Performance

Error performance is examined from three different perspectives. The first is based on ITU-T Rec. G.821; the second briefly reviews ITU-T Rec. G. 826; and the third perspective is from Bellcore.

4.2.2.1 *Error Performance Based on ITU-T Rec. G.821.* The error performance objectives of Rec. G.821 (Ref. 1) are based on a 64-kbps circuit-switched connection used for voice traffic or as a "bearer channel" for data traffic. It is directly tied to international ISDN connections.

These error performance parameters are defined as follows: the percentage of averaging periods each of time interval $T(0)$ during which the BER exceeds a threshold value. The percentage is assessed over a much longer time interval $T(L)$. A suggested interval for $T(L)$ is 1 mo.

It should be noted that total time $T(L)$ is broken down into two parts:

- Time that the connection is available
- Time that the connection is unavailable

The following BERs and intervals are used in CCITT Rec. G.821 in the statement of objectives:

- A BER of less than 1×10^{-6} for $T(0) = 1$ min
- A BER of less than 1×10^{-3} for $T(0) = 1$ s
- Zero errors for $T(0) = 1$ s

Table 4.1 gives the ITU-T Rec. G.821 error performance objectives. Table 4.2 gives some guidelines for the interpretation of Table 4.1.

4.2.2.2 *Error Performance Objectives Based on ITU-T Rec. G.826.* Rather than consider bit errors, this recommendation deals with the error performance measurement of blocks. A block, in this context, is defined as a

TABLE 4.1   ITU-T Error Performance Objectives for International ISDN Connections[a]

Performance Classification	Objective (Note 3)
(a) (Degraded minutes) (Notes 1, 2)	Fewer than 10% of 1-min intervals to have a bit error ratio worse than 1×10^{-6} (Note 4)
(b) (Severely errored seconds) (Note 1)	Fewer than 0.2% of 1-s intervals to have a bit error ratio worse than 1×10^{-3}
(c) (Errored seconds) (Note 1)	Fewer than 8% of 1-s intervals to have any errors (equivalent to 92% error-free seconds)

[a]See Table 4.2 for interpretation.

Note 1. The terms "degraded minutes," "severely errored seconds," and "errored seconds" are used as a convenient and concise performance objective "identifier." Their usage is not intended to imply the acceptability, or otherwise, of this level of performance.

Note 2. The 1-min intervals mentioned in the table and in the notes (i.e., the periods for $M > 4$ in Annex B [Ref. 1]) are derived by removing unavailable time and severely errored seconds from the total time and then consecutively grouping the remaining seconds into blocks of 60. The basic 1-s intervals are derived from a fixed period.

Note 3. The time interval T_L, over which the percentages are to be assessed, has not been specified since the period may depend on the application. A period of the order of any one moment is suggested as a reference.

Note 4. For practical reasons, at 64 kbps, a minute containing four errors (equivalent to an error ratio of 1.04×10^{-6}) is not considered degraded. However, this does not imply relaxation of the error ratio objective of 1×10^{-6}.

Source: Reference 1, Table 1/G.821, CCITT Rec. G.821.

set of consecutive bits associated with the path; each bit belongs to one and only one block. Thus a frame is a block, but the term *block* takes on a wider meaning as any grouping of consecutive bits. Table 4.3 specifies the recommended range of the number of bits within each block for the various bit rate ranges.

In Chapter 3 we discussed how the superframe and extended superframe permitted in-service monitoring of performance of DS1 and E-1 links. In these monitoring systems we were not able to obtain an accurate BER, but we could get information on errored frames, which in this instance we call blocks.

Recommendation G.826 considers that each block is monitored by means of an *inherent error detection code* (EDC) [e.g., bit-interleaved parity (BIP) or cyclic redundancy check (CRC)]. While carrying out in-service monitoring, the EDC is considered as part of the block even though the EDC bits can physically be separated from the block to which they apply. If an error occurs in the EDC, an *errored block* (EB) will be counted. It should be noted that Rec. G.826 does not define a specific EDC. However, the EDC should have a probability of failing to detect an error of at least 10%. In our discussion in Chapter 3, CRC-4 (E-1) and CRC-6 (DS1) were employed.

TABLE 4.2   Guidelines for the Interpretation of Table 4.1

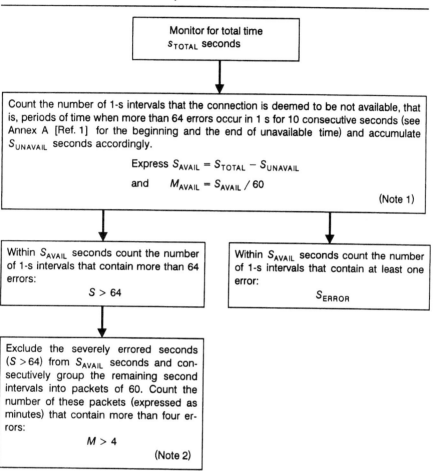

Note 1. The result is rounded off to the next higher integer.
Note 2. The last packet, which may be incomplete, is treated as if it were a complete packet with the same rules being applied.

Performance Classification (see Ref. 1 Table 1/G.821)	Objective
(a)	$\dfrac{M > 4}{M_{AVAIL}} < 10\%$
(b)	$\dfrac{S > 64}{S_{AVAIL}} < 0.2\%$
(c)	$\dfrac{S_{ERROR}}{S_{AVAIL}} < 8\%$

Source: Reference 1, CCITT Rec. G.821.

TABLE 4.3 End-to-End Performance Objectives for a 27,500-km International Digital HRP at or Above the Primary Rate

Rate (Mbps)	1.5-5	> 5-15	> 15-55	> 55-160	> 160-3500	> 3500
Bits/block	2000-8000 (Note 1)	2000-8000	4000-20,000	6000-20,000	15,000-30,000 (Note 2)	For further study
ESR	0.04	0.05	0.075	0.16	(Note 3)	For further study
SESR	0.002	0.002	0.002	0.002	0.002	For further study
BBER	3×10^{-4}	2×10^{-4}	2×10^{-4}	2×10^{-4}	10^{-4}	For further study

Note 1. VC-11 and VC-12 (see ITU-T Rec. G.709) paths are defined with a number of bits/block of 832 and 1120, respectively, that is, outside the recommended range for 1.5-5-Mbps paths. For these block sizes, BBER objective for VC-11 and VC-12 is 2×10^{-4}.

Note 2. Because bit error ratios are not expected to decrease dramatically as the bit rates of transmission systems increase, the block sizes used in evaluating very high bit rate paths should remain within the range 15,000-30,000 bits/block. Preserving a constant block size for very high bit rate paths results in relatively constant BBER and SESR objectives for these paths.

As currently defined, VC-4-4c (see Rec. G.709) is a 601-Mbps path with a block size of 75,168 bits/block. Since this is outside the recommended range for 160-3500-Mbps paths, performance on VC-4-4c paths should not be estimated in-service using this table. The BBER objective for VC-4-4c using the 75,168 bit block size is taken to be 4×10^{-4}. There are currently no paths defined for bit rates greater than VC-4-4c (> 601 Mbps).

Digital sections are defined for higher bit rates and guidance on evaluating the performance of digital sections can be found in Section 5.1 of Rec. G.826a.

Note 3. Due to the lack of information on the performance of paths operating above 160 Mbps, no ESR objectives are recommended at this time. Nevertheless, ESR processing should be implemented within any error performance measuring devices operating at these rates for maintenance or monitoring purposes.

Source: Reference 2, Table 1/G.826, ITU-T Rec. G.826.

4.2.2.2.1 Definitions

Errored Block (EB). A block in which one or more bits are in error.

Errored Second (ES). A 1-second period with one or more errored blocks.

Severely Errored Second (SES). A 1-second period that contains $\geq 30\%$ errored blocks or at least one severely disturbed period (SDP). Recommendation G.826 (Ref. 2) points out that some definitions of SES use a figure different than $\geq 30\%$.

When carrying out out-of-service measurements, a SDP occurs when, over a period of time equivalent to four contiguous blocks or 1 ms, whichever is larger, either all the contiguous blocks are affected by a high bit error density of $\geq 10^{-2}$ or a loss of signal is observed. For the case of in-service monitoring, a SDP is estimated by the occurrence of a network defect. The term defect is defined differently for PDH [plesiochronous digital hierarchy, such as DS1 family or E-1 family of formats), SDH (typically SONET and synchronous digital hierarchy (Section 3.6)], and ATM (asynchronous transfer mode, Section 4.8).

Background Block Error (BBE). An errored block not occurring as part of a SES.

Recommendation G.826 lists the preceding as *events*. The following are *parameters*, which deal directly with error performance. Recommendation G.826 states that error performance should only be evaluated while a path is in the *available state*.

Unavailable Time. This begins at the onset of ten consecutive SES events. These 10 s are considered to be part of unavailable time. A new period of *available time* begins at the onset of ten consecutive non-SES events. These 10 s are considered to be part of available time.

Errored Second Ratio (ESR). The ratio of ES to total seconds in available time during a fixed measurement interval.

Severely Errored Second Ratio (SESR). The ratio of SES to total seconds in available time during a fixed measurement interval.

Background Block Error Ratio (BBER). The ratio of errored blocks to total blocks during a fixed measurement interval, excluding all blocks during SES and unavailable time.

For PDH and SDH systems, SES events can be monitored for both directions at a single path end point. Thus a network provider is able to determine the unavailable state of the path. In some cases, it is possible to monitor the full set of error performance parameters in both directions from one end of the path.

4.2.2.2.2 Recommendation G.826 Error Performance Objectives

END-TO-END OBJECTIVES. The end-to-end objectives are outlined in Table 4.3 for a 27,500-km hypothetical reference path (HRP) in terms of the *parameters* defined above. Recommendation G.826 states that an international digital path at or above the primary rate* should meet its allocated objectives for all parameters concurrently. If a path fails to meet any of these objectives, then the path fails to meet this performance requirement. A hypothetical reference path is illustrated in Figure 4.1.

In the case where a higher-level multiplex signal (e.g., DS3, E-4) meets these performance objectives, it then can be assumed that its constituent parts (e.g., any of the four DS1's making up a DS2 signal) also meet the same objectives.

Primary rate is defined as DS1 or E-1.

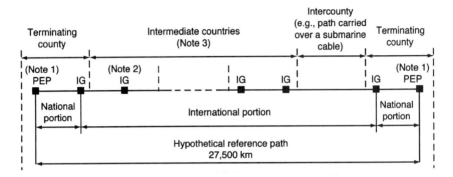

NOTE 1. If a path is considered to terminate at the IG, only the international portion allocation applies.
NOTE 2. One or two IGs (entry or exit) may be defined per indeterminate country.
NOTE 3. Four intermediate countries are assumed.

Figure 4.1. Hypothetical reference path (HRP). (From Ref. 2, Figure 3/G.826, ITU-T Rec. G.826.)

APPORTIONMENT OF END-TO-END OBJECTIVES. The apportionment is based on the HRP illustrated in Figure 4.1. This apportionment allows us to specify the performance expected from the national and international portions of a digital connectivity based on the SRP.

The boundary between the national and international portions is defined to be at an *international gateway* (IG). The actual boundary usually is found in a cross-connect, a higher-order multiplex, or a switch (ISDN or B-ISDN). The IG boundary is found in terrestrially based equipment physically resident in the terminating (or intermediate) country. Higher-order paths relative to the HRP under consideration may be used between IGs. Such paths receive only the allocation corresponding to the international portion between the IGs. In intermediate countries, the IGs are only located in order to calculate the overall length of the international portion of the path in order to calculate the overall allocation.

The allocation methodology described in Rec. G.826 applies to each parameter previously defined and takes into account both the complexity and length of the international path. All paths should be designed to meet the allocated objectives described in the following paragraphs. Of course the reader should also refer to Table 4.3 for overall end-to-end performance objectives.

ALLOCATION TO THE NATIONAL PORTION OF THE END-TO-END PATH. Each national portion (of course there are two) is allocated a fixed block allowance of 17.5% of the end-to-end objective. Furthermore, a distance-based allocation is added to the block allowance. The actual route length between the path end point (PEP) and the IG should be calculated first, if it is known. If not known, then the crow-fly distance should be used with a 1.5 routing factor (i.e., crow-fly distance × 1.5). When both actual and calculated route lengths are known, use the smaller value. The distance is usually rounded to the nearest 500 km. An allocation of 1% per 500 km is then applied to the resulting distance.

Independent of the distance spanned, any satellite hop in the national portion receives a 35% allocation of the objectives given in Table 4.3. If a satellite hop is used in the national portion, apply the 35% allocation, which replaces the distance-based allocation for this portion.

ALLOCATION TO THE INTERNATIONAL PORTION OF THE END-TO-END PATH. The international portion is allocated a block allowance of 2% per intermediate country plus 1% for each terminating country. In addition, a distance-based allocation is added to the block allowance. When an international path passes through one or more intermediate countries, the actual route length between intermediate IGs (one or two per intermediate country) is added to calculate the overall length of the international portion. When international route lengths are not known, then crow-fly distance between intermediate IGs is used and multiplied by the 1.5 routing factor. If both actual and calculated route distances are known, use the smaller value of the two. Next, round off the overall distance to the nearest 500 km. An allocation of 1% per 500 km is then applied to the resulting distance.

For satellite hops in the international portion, apply the same criterion given for the national portion and remove the distance-based allocation.

4.2.2.2.3 *Relationship Between PDH Path Performance Monitoring and Block-Based Parameters.* In-service anomaly conditions are used to determine the error performance of a PDH* path that is not in a fault condition. The two categories of anomalies related to the incoming signal are:

1. An errored frame alignment signal, denoted a_1.
2. An EB as indicated by an EDC, denoted a_2.

For in-service fault conditions (see ITU-T Recs. G.730–G.751) relevant to PDH multiplex equipment, the following three changes of performance state may occur on a path related to the incoming signal:

1. Loss of signal (LOS), denoted d_1.
2. Alarm indication signal (AIS), denoted d_2.
3. Loss of frame alignment (LOF), denoted d_3.

TYPES OF PATHS. An ISM facility carries out in-service monitoring. The number of performance factors to be measured depends on the type of ISM

---

*Again, a PDH path is one that uses either the DS1 family of PCM formats or the E-1 family of PCM formats.

and the PDH path under consideration. Recommendation G.826 classifies four types of paths and their associated ISMs.

> *Type 1: Frame and Block Structured Paths.* Here the full set of fault conditions d_1 through d_3 and anomaly indications a_1 and a_2 are provided by the ISM facilities. Examples include:
> - Primary rate and second-order paths using CRC-4 or CRC-6 as defined in Rec. G.704
> - Fourth-order paths with a parity bit per frame as defined in Rec. G.755.

> *Type 2: Frame Structured Paths.* The full set of fault conditions d_1 through d_3 and the anomaly indication a_1 are provided by the ISM facilities. Examples include:
> - Primary rate up to fourth-order paths in the 2-Mbps hierarchy as defined by Recs. G.732, G.742, and G.751
> - Primary rate paths in the 1.5-Mbps hierarchy as defined in Recs. G.733 and G.734.

> *Type 3: Other Frame Structured Paths.* A limited set of fault conditions d_1 and d_2 and the anomaly indication a_1 are provided by the ISM facilities. In addition, the number of consecutive errored frame alignment signal (FAS) per second is available. An example of this type of path is:
> - Second-order up to fourth-order paths in the 1.5-Mbps hierarchy as defined in Recs. G.743 and G.752.

> *Type 4: Unframed Paths.* A limited set of fault conditions d_1 and d_2 is provided by ISM facilities, which do not include any error check. No FAS control is available. An example of this type of path is:
> - End-to-end path (e.g., leased line) carried over several higher-order paths placed in tandem.

ESTIMATION OF PERFORMANCE PARAMETERS. Table 4.4 gives information on which set of parameters should be estimated and the related measurement criteria according to the type of path considered.

IN-SERVICE MONITORING CAPABILITIES AND CRITERIA FOR DECLARATION OF PERFORMANCE PARAMETERS. Table 4.5 provides guidance on the criteria for declaration of a SES event on PDH paths. This table lists examples of the ISM SES criteria x for signal formats with EDC capabilities implemented prior to ITU-T Rec. G.826.

TABLE 4.4 Performance Parameters and Measurement Criteria by Type of Path

Type	Set of Parameters	Measurement Criteria
1	ESR	An ES is observed when, during 1 s, at least one anomaly a_1 or a_2, or one defect d_1 to d_3 occurs.
	SESR	A SES is observed when, during 1 s, at least X anomalies a_1 or a_2, or one defect d_1 to d_3 occurs (Notes 1 and 2).
	BBER	EBs are accumulated as defined in Section 4.2.2.2.
2	ESR	A ES is observed when, during 1 s, at least one anomaly a_1 or one defect d_1 to d_3 occurs.
	SESR	A SES is observed when, during 1 s, at least X anomalies a_1 or one defect d_1 to d_3 occurs (Note 2).
2	ESR	A ES is observed when, during 1 s, at least one anomaly a_1 or one defect d_1 or d_2 occurs.
	SESR	A SES is observed when, during 1 s, at least X anomalies a_1 or one defect d_1 or d_2 occurs (Note 2).
4	SESR	A SES is observed when, during 1 s, at least one defect d_1 or d_2 occurs (Note 3).

Note 1. If more than one anomaly a_1 or a_2 occurs during the block interval, then only one anomaly has to be counted.
Note 2. Values of X can be found in Section B.4 of Rec. 826.
Note 3. The estimates of the ESR and SESR will be identical since the SES event is a subset of the ES event.
Source: Reference 2, Table B.1/G.826, ITU-T Rec. G.826.

TABLE 4.5 Examples of ISM SES Criteria X

Bit rate (kbps)	1544	2048	44,736
Recommendation	G.704	G.704	G.752
EDC type	CRC-6	CRC-4	Parity
Blocks/second	333	1000	9396
Bits/block	4632	2048	4704
SES threshold used on equipment developed prior to the acceptance of this	$x = 320$	$x = 805$	$x = 45$ or $x = 4698$ as suggested in
Recommendation			Rec. M.2100
ISM threshold based on Rec. G.826 SES (30% errored blocks)	(Note 2)	(Note 2)	$x = 2444$ (Note 3)

Note 1. It is recognized that there are discrepancies between the figures above and those given in Table C.1 of G.826. This requires further study.
Note 2. Due to the fact that there is a large population of systems in-service, the criteria for declaration of a SES will not change for the frame formats of these systems.
Note 3. This figure takes into account the fact that, although 30% of the blocks could contain errors, a lesser number will be detected by the EDC due to the inability of the simple parity code to detect even numbers of errors in a block. It should be noted that such a simple EDC is noncompliant with the intent of this Recommendation.
Note 4. Completion of this table for other bit rates is for further study.
Source: Reference 2, Table B.2/G.826, ITU-T Rec. G.826.

4.2.2.3 Error Performance from a Bellcore Perspective

4.2.2.3.1 Introduction. Bellcore and interexchange carriers in the United States have been tightening up BER requirements. In 1986 Bellcore (Ref. 3) required a BER of 1×10^{-6} for 95% of connections. In 1990 Bellcore (Ref. 4) improved the BER value to 1×10^{-7} for 95% of all connections. During informal discussions with the SPRINT technical staff (Sept. 1996), I was given the value as BER $= 1 \times 10^{-12}$. If we were to assume 100 links in tandem,* the end-to-end (i.e., POP to POP) value would be 1×10^{-10}.

4.2.2.3.2 Bellcore Based on TSGR. The requirements given in this reference assume one-way options and are based on the maximum short-haul design length. The measurement interval consists of a series of 1-s intervals. The following paragraphs (1–4) are based on Ref. 5.

1. The BER at the interfaces DSX-1, DSX-1C, DSX-2, and DSX-3 will be less than 2×10^{-10}, excluding burst errored seconds in the measurement period. During a burst errored second, neither the number of bit errors nor the number of bits is counted.

2. The frequency of burst errored seconds, other than those caused by protection switching induced by hard equipment failures, will average no more than four per day at each of the interface levels DSX-1, DSX-1C, DSX-2, and DSX-3.[†]

3. For systems interfacing at the DS1 level, the long-term percentage of errored seconds (measured at the DS1 rate) will not exceed 0.04%. This equates to 99.96 error-free seconds (EFS). This requirement applies in a normal operating environment *and is also an acceptance criterion.* It is equivalent to no more than 10 errored seconds during a 7-hour, one-way (loopback) test. While no errored second requirements are given for the DS1C and DS2 rates, any DS1 channel embedded in a DS1C, DS2, or higher-rate channel will meet the DS1 rate errored second requirement.

4. For systems interfacing at the DS3 level, the long-term percentage of errored seconds (measured at the DS3 rate) will not exceed 0.4%. This is equivalent to 99.6% error-free seconds. This requirement applies in a normal operating environment and *is also an acceptance criterion.* It is equivalent to no more than 29 errored seconds during a 2-hour, one-way (loopback) test. The reference document points out that if the above BER requirements are met, there is no assurance that the errored second requirements will also be met.

---

*Assuming random noise, where each tandem link degraded uniformly across the network.
[†]This is a long-term average over many days. Due to day-to-day variation, the number of burst errored seconds occurring on a particular day may be greater than the average.

Note: A burst errored second is any errored second containing at least 100 errors.

4.2.3 Timing Jitter and Wander

In a digital network, typically the PSTN, jitter and wander accumulate according to the jitter and wander generation and transfer characteristics of each equipment interconnected. These equipments may be different types of multiplexers/demultiplexers and line systems.

An excessive amount of jitter and wander can adversely affect digital networks by causing the generation of bit errors and uncontrolled slips (see Section 4.3.3).

4.2.3.1 Definitions of Jitter and Wander. CCITT Rec. G.810 defines jitter as "short-term variations of the significant instants of a digital signal from their reference positions in time."

Bellcore (in Ref. 5) defines timing jitter as "the short-term variations of digital significant instants (e.g., optimum sampling instants from their ideal positions in time)." Short-term variations are phase oscillations of frequency greater than a demarcation point that is specified for each interface rate (such as DS1), phase modulations that, after demodulation, pass through a high-pass filter with a cutoff frequency of 10 Hz, and a 20-dB/decade roll-off. Bellcore states that long-term variations (i.e., variations of frequency less than 10 Hz) are defined as *wander*.

CCITT Rec. G.810 (Ref. 6) defines wander as long-term variations of significant instances of a digital signal from its ideal positions in time, where "long-term" implies that these variations are of frequency less than 10 Hz.

In this section we will concentrate on timing jitter.

4.2.3.2 Causes of Jitter * Jitter is a product of timing circuits in regenerative repeaters. A model of a typical timing circuit is shown in Figure 4.2. These timing circuits are key to regeneration. In a typical regenerator the incoming signal is fed to a timing path equalizer. After equalization, the bit stream in the repeater is in a suitable form for regeneration. The timing circuits in the repeater control the regeneration process by providing a clock signal to (1) sample the equalized bit stream near the center of the eye (of the eye pattern), (2) maintain the correct bit spacing, and (3) maintain the proper bit pulsewidth.

The timing signal is extracted from the equalized and amplified bit stream. In general, this signal does not contain a discrete frequency component at the pulse repetition rate (e.g., 1.544 megapulses/second). The bandwidth required for the transmission of the digital information signal alone is only half the bit repetition rate. Addition of a discrete component at the bit repetition

*Section 4.2.3.2 is based on Ref. 7.

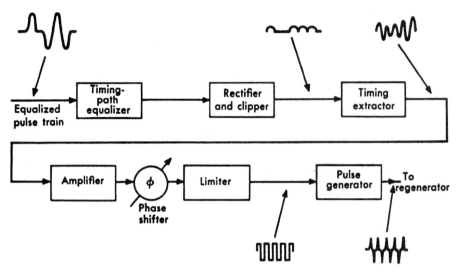

Figure 4.2. A repeater timing-path model. (From Ref. 7, Figure 30-19. Reprinted with permission of AT&T Bell Laboratories.)

rate generally means that the timing path incurs more loss, noise, and crosstalk, as well as causing more crosstalk into other systems. Since the digital signal does not normally contain a suitable timing signal, timing information must be extracted from the equalized bit stream by nonlinear processing.

Consider the block diagram of a typical repeater timing path in Figure 4.2. Since the optimum equalizers for the information and timing signals are not necessarily identical, additional timing-path equalization may be provided. The equalized signal is then rectified, which produces a discrete spectral component at the bit signaling rate. The timing extractor, which may be a high-Q tuned circuit, a crystal filter, or a phase-locked loop, extracts this desired timing component and eliminates most of the other spectral components of the rectified signal. A phase-shifter circuit allows adjustment of the timing pulse phase so that sampling will occur at the time of maximum eye opening. The timing-extractor output is sinusoidal, but its amplitude and phase may fluctuate as the received bit pattern changes. The timing extractor output is fed to a limiting amplifier, which eliminates the amplitude variations, producing an output that is roughly a square wave. However, the phase variations, or *jitter*, remain an impairment. This method of obtaining a clock signal is termed *forward-acting timing*, and the digital repeater is said to be *self-timed*. "Self-timing" refers to the fact that the timing is recovered from the information-bearing bits themselves, rather than from a timing waveform that is added to the signal or from a separate timing path. It is also possible to have *backward-acting timing*. Backward-acting timing is a method in which the timing wave is obtained from the reconstructed bit stream at the regenerator output and then fed back to a point internal to the repeater.

4.2.3.2.1 Sources of Timing Jitter. The random phase modulation or phase jitter introduced at each repeater accumulates in a repeater chain and may lead to crosstalk and distortion in the reconstructed analog signal. In digital switching systems, jitter on the incoming lines is a potential source of slips. The sources of timing jitter may be classified as systematic or nonsystematic according to whether or not they are related to the pulse pattern. Systematic jitter sources lead to jitter that degrades the bit stream in the same way at each repeater in the chain. Systematic sources include intersymbol interference, finite pulsewidth, and clock threshold offsets. Nonsystematic jitter sources such as mistuning and crosstalk result in timing degradations that are random from repeater to repeater. In a long repeater chain, the total accumulated jitter is dominated by components produced by systematic sources.

4.2.3.2.2 Accumulation of Timing Jitter. It can be shown (Ref. 7) that the mean square value of jitter in a long chain of regenerative repeaters increases with N (the number of repeaters in the chain), and the rms value of jitter increases with $\sqrt{N}$. Also, the jitter is proportional to the timing-filter bandwidth, which confirms the fact that higher-Q tuned circuits in a repeater will reduce jitter.

In conventional PCM coding of 4-kHz voice channels, jitter of up to 1.4 μs can be tolerated before the distortion due to jitter approaches the distortion due to quantizing noise.

4.2.3.3 Timing Jitter from an ITU-T Perspective. Our examination of jitter and wander is based on two ITU-T Recommendations: G.824 (Ref. 8) for the 1.544-Mbps hierarchy and G.823 (Ref. 9) for the 2.048-Mbps hierarchy. (SDH is covered in Section 3.6.)

4.2.3.3.1 Timing Jitter at Interfaces of the 1.544-Mbps Hierarchy. The maximum permissible output jitter limits at hierarchical interfaces of a digital network based on the 1.544-Mbps hierarchy are given in Table 4.6. The arrangement for measuring output jitter at a digital interface is illustrated in Figure 4.3. The specific jitter limits and values of filter cutoff frequencies are given in Table 4.6.

NETWORK LIMITS ON WANDER FOR 1.544-Mbps HIERARCHY. Network output wander specifications at synchronous network nodes are necessary to ensure satisfactory network performance (e.g., slips, error bursts). For network nodes the following limits are specified, based on the assumption of a nonideal synchronizing signal (nonideal in that it contains jitter, wander,

TABLE 4.6   Maximum Permissible Output Jitter at 1.544-Mbps Hierarchical Interfaces

Digital rate (kbps)	Network Limit (Unit interval peak-to-peak)[a]		Bandpass Filter Having a Lower Cutoff Frequency f_1 or f_3 and Minimum Upper Cutoff Frequency f_4		
	B_1	B_2	f_1 (Hz)	f_3 (kHz)	f_4 (kHz)
1,544	5.0	0.1[b]	10	8	40
6,312	3.0	0.1[b]	10	3	60
32,064	2.0	0.1[b]	10	8	400
44,736	5.0	0.1	10	30	400
97,728	1.0	0.05	10	240	1000

[a]Consider a unit interval (UI) to be the duration of 1 bit, assuming 1's and 0's of equal duration.
[b]This value requires further study.
Source: Reference 8, Table 1/G.824, ITU-T Rec. G.824.

frequency departure, and other impairments) on the line delivering timing information. The maximum time interval error (MTIE) (see Section 4.3.4 and ITU-T Rec. G.810) over a period of S seconds should not exceed the following:

- $S < 10^4$; region requires further study
- $(10^{-2}S + 10,000)$ ns; applicable to values greater than 10^4

The resulting overall specification is illustrated in Figure 4.4. (*Note:* The full MTIE of 10 μs, superimposed on the average timing, as specified in ITU-T Rec. G.824, may only occur at the output of the last node in a chain of nodes.)

JITTER AND WANDER FROM INDIVIDUAL DIGITAL EQUIPMENT. The principal parameters of importance when considering jitter and wander performance of digital equipment are the following:

- The amount of jitter and wander that can be tolerated at the equipment input
- The proportion of this input jitter and wander that filters through to the equipment output
- The amount of jitter and wander generated by the equipment itself

JITTER AND WANDER TOLERANCE AT INPUT PORTS. Equipment input ports should be capable of accommodating levels of network output jitter up to the maximum given in Table 4.6. Most specifications of equipment input jitter tolerance are in terms of the amplitude of sinusoidal jitter that can be applied at various frequencies without causing a designated degradation of error performance.

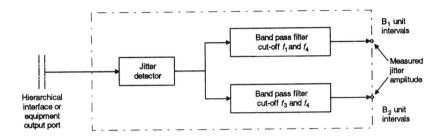

Figure 4.3. Measurement arrangements for output jitter from a hierarchical interface or an equipment output port. (From Ref. 8, Figure 1/G.824, ITU-T Rec. G.824.)

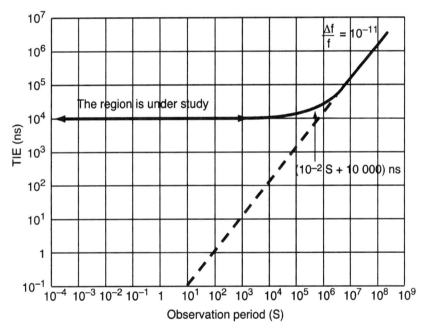

Figure 4.4. Permissible maximum time interval error (MTIE) versus observation period S for the output of a network node. (From Ref. 8, Figure 2/G/824, ITU-T Rec. G.824.)

As a minimum guideline for equipment tolerance, ITU-T Rec. G.824 suggests that all digital input ports of equipments be able to tolerate the sinusoidal jitter and wander defined in Figure 4.5 and Table 4.7. The limits are to be met in an operating environment.

Recommendation G.824 states that in deriving the specifications contained in Table 4.7 for frequencies above f_3, the effects of the amount of

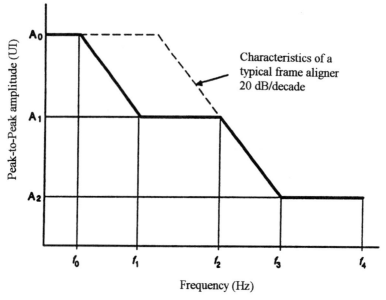

Figure 4.5. Mask of peak-to-peak jitter and wander that must be accommodated at the input of a node in a digital network. (From Ref. 8, Figure 3/G.824, ITU-T Rec. G.824.)

alignment jitter of the equipment clock decision circuit are considered to be predominant.

In deriving these specifications, the wander effects are considered to be predominant at frequencies below f_1, and many transmission equipments, such as digital line systems and asynchronous muldexes using justification (stuffing) techniques, are effectively transparent to these very-low-frequency changes in phase. However, such phase variation does need to be accommodated at the input of certain equipment such as digital exchanges and synchronous muldexes.* The requirement contained in Table 4.7 for frequencies below f_1 is not amenable to simple practical evaluation, but account should be taken of the requirement at the design stage of the equipment.

The network design should be prepared for certain eventualities. One example is that an input synchronizing a node and another input not synchronizing the node may derive their respective timing from the same reference clock, but over different paths, and may therefore, in an extreme case, have opposite phase deviation. The maximum relative phase deviation is 18 μs, which must be accommodated by the equipment.

---

*Muldex is a contraction of multiplex–demultiplex. A channel bank can be considered a muldex and also higher-level multiplex/demultiplex equipment.

TABLE 4.7   Jitter and Wander Tolerance at Input Ports (Provisional Values)

Bit Rates (kbps)	Jitter Amplitude (peak-to-peak)			Frequency					Test Signal
	A_0 (μs)	A_1 (UI)	A_2 (UI)	f_0 (Hz)	f_1 (Hz)	f_2 (Hz)	f_3 (kHz)	f_4 (kHz)	
1,544	18 (Note 2)	5.0	0.1 (Note 2)	1.2×10^{-5}	10	120	6	40	$2^{20}-1$ (Note 3)
6,312	18 (Note 2)	5.0	0.1	1.2×10^{-5}	10	50	2.5	60	$2^{20}-1$ (Note 2)
32,064	18 (Note 2)	2.0	0.1	1.2×10^{-5}	10	400	8	400	$2^{20}-1$ (Note 3)
44,736	18 (Note 2)	5.0	0.1 (Note 2)	1.2×10^{-5}	10	600	30	400	$2^{20}-1$ (Note 2)
97,728	18 (Note 2)	2.0	0.1	1.2×10^{-5}	10	12,000	240	1000	$2^{23}-1$ (Note 2)

Note 1. Reference to individual equipment specifications should always be made to check if supplementary input jitter tolerance requirements are necessary.
Note 2. This value requires further study.
Note 3. It is necessary to suppress long zero strings in the test sequence in networks not supporting 64-kbps transparency.
Note 4. The value A_0 (18 μs) represents a relative phase deviation between the incoming signal and the internal local timing signal derived from the reference clock.
Note 5. The absolute phase deviation requires further study.
Note 6. An example of a reference configuration explaining the A_0 value is given in Annex A of Ref. 8.
Source: Reference 8, Table 2/G.824, ITU-T Rec. G.824.

Equipment wander tolerance must be compatible with network output wander limits given in Figure 4.4. Insufficient wander tolerance at synchronous equipment input ports may result in controlled or uncontrolled slips, depending on the specific slip control strategy employed.

4.2.3.3.2 Jitter and Wander Within Digital Networks Based on the 2.048-Mbps Hierarchy.

Table 4.8 gives the maximum permissible levels of jitter at hierarchical interface within a digital network based on the 2.048-Mbps hierarchy. These limits should be met for all operating conditions and regardless of the amount of equipment preceding the interface.

The arrangements for measuring output jitter at a digital interface are shown in Figure 4.3. The specific jitter limits and values of filter cutoff frequencies for the different hierarchical levels are given in Table 4.8. The frequency response of the filters associated with the measurement equipment should have a roll-off of 20 dB/decade.

NETWORK LIMITS ON WANDER. See Section 4.2.3.3.1.

JITTER LIMITS OF DIGITAL EQUIPMENTS INCORPORATED IN A NETWORK. See Section 4.2.3.3.1.

TABLE 4.8   Maximum Permissible Jitter at a 2.048-Mbps Hierarchical Interface

Parameter Value →	Network Limit		Measurement Filter Bandwidth		
Digit Rate (kbps) ↓	B₁ Unit Interval (peak-to-peak)	B₂ Unit Interval (peak-to-peak)	Bandpass Filter Having a Lower Cutoff Frequency f_1 or f_3 and an Upper Cutoff Frequency f_4		
			f_1	f_3	f_4
64 (Note 1)	0.25	0.05	20 Hz	3 kHz	20 kHz
2,048	1.5	0.2	20 Hz	18 kHz (700 Hz)	100 kHz
8,448	1.5	0.2	20 Hz	3 kHz (80 kHz)	400 kHz
34,368	1.5	0.15	100 Hz	10 kHz	800 kHz
139,264	1.5	0.075	200 Hz	10 kHz	3500 kHz

Note 1. For the codirectional interface only.
Note 2. The frequency values shown in parentheses only apply to certain national interfaces.
Note 3. UI-unit interval:

For 64 kbps	1 UI = 15.6 μs
For 2048 kbps	1 UI = 488 ns
For 8448 kbps	1 UI = 118 ns
For 34,368 kbps	1 UI = 29.1 ns
For 139,264 kbps	1 UI = 7.18 ns

Source: Reference 9, Table 1/G.823, ITU-T Rec. G.823.

JITTER AND WANDER TOLERANCE AT DIGITAL INPUT PORTS. The required tolerance is defined in terms of the amplitude and frequency of sinusoidal jitter, which, when modulating a test pattern, should not cause any significant degradation in the operation of the equipment. It should be recognized that the test condition is not, in itself, intended to be representative of the type of jitter to be found in practice in a network. Nevertheless, the test does ensure that the "*Q*" factor associated with the timing signal recovery of the equipment's input circuitry is not excessive and, where necessary, that an adequate buffer storage capacity has been provided.

With this in mind, all digital input ports of equipment should be able to tolerate a digital signal having electrical characteristics based on ITU-T Rec. G.703 but modulated by sinusoidal wander and jitter having an amplitude–frequency relationship given in Figure 4.6. Table 4.9 shows the appropriate limits for the different hierarchical levels.

In principle, these requirements should be met regardless of the information content of the digital signal. For test purposes, the equivalent binary content of the signal with jitter modulation should be a pseudorandom bit sequence as defined in Table 4.9.

In deriving these limits, the wander effects are considered to be predominant at frequencies below f_1 and many transmission equipments, such as regenerators and asynchronous muldexes using justification techniques (stuff-

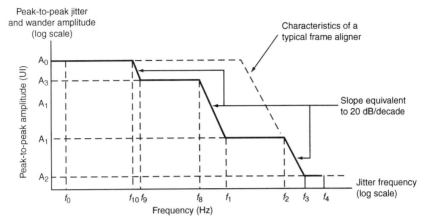

Figure 4.6. Lower limit of maximum tolerable input jitter and wander. (From Ref. 9, Figure 3/G.823, ITU-T Rec. G.8.23.)

ing), are effectively transparent to these very-low-frequency changes in phase. Nevertheless, it does not need to be accommodated at the input of certain equipment such as digital switches and muldexes. The requirement below f_1 is not amenable to simple practical evaluation, but account should be taken of the requirement during the design stage of the equipment.

Unlike that part of the mask between frequencies f_1 and f_4, which reflects the maximum permissible jitter magnitude in a digital network, that part of the mask below frequency f_1 does not aim to represent the maximum permissible wander that might occur in practice. Below frequency f_1, the mask is derived such that, where necessary, the provision of this level of buffer storage at the input of an equipment facilitates the accommodation of wander generated in a large proportion of real connections.

An input synchronizing a node and another input not synchronizing the node may derive their respective timing from the same reference clock but over different paths and may therefore, in an extreme case, have opposite phase deviation. The expected maximum relative phase deviation is 18 μs, which must be accommodated by the equipment.

A short-term reversal of the relative time interval error (TIE) between the incoming signal and the internal timing signal of the terminating equipment shortly after the occurrence of a controlled slip should not cause another slip. In order to prevent such a slip, the equipment should be designed with a suitable hysteresis for the phenomenon. The hysteresis should be at least 18 μs.

4.2.3.3.3 Jitter Accumulation with Cascaded Homogeneous Regenerators. Most digital regenerators currently in use are fully regenerative and self-timed. Here we mean that the output signal is retimed under control of a

TABLE 4.9 Parameter Values for Input Jitter and Wander Tolerance at 2.048-Mbps Hierarchical Levels

Digit Rate (kbps)	Peak-to-Peak Amplitude Unit Interval				Frequency								Pseudorandom Test Signal
	A_0	A_1	A_2	A_3	f_0	f_{10}	f_9	f_8	f_1	f_2	f_3	f_4	
64 (Note 1)	1.15 (18 μs)	a	0.25	0.05	1.2×10^{-5} Hz	a	a	a	20 Hz	600 Hz	3 kHz	20 kHz	$2^{11}-1$
2,048	36.9 (18 μs)	18 (Note 7)	1.5	0.2	1.2×10^{-5} Hz	4.88×10^{-3} Hz (Note 7)	0.01 Hz (Note 7)	1.667 Hz (Note 7)	20 Hz	2.4 kHz (93 Hz)	18 kHz (700 Hz)	100 kHz	$2^{15}-1$ (Rec. O.151)
8,448	152 (18 μs)	a	1.5	0.2	1.2×10^{-5} Hz	a	a	a	20 Hz	400 Hz (10.7 kHz)	3 kHz (80 kHz)	400 kHz	$2^{15}-1$ (Rec. O.151)
34,368	618.6 (18 μs)	a	1.5	0.15	a	a	a	a	100 Hz	1 kHz	10 kHz	800 kHz	$2^{23}-1$ (Rec. O.151)
139,264	2506.6 (18 μs)	a	1.5	0.075	a	a	a	a	200 Hz	500 Hz	10 kHz	3500 kHz	$2^{23}-1$ (Rec. O.151)

^{a}Values under study.

Further Explanatory Notes:

Note 1. For the codirectional interface only.

Note 2. For interfaces within national networks the frequency values (f_2 and f_3) shown in parenthesis may be used.

Note 3. UI-unit interval:

For 64 kbps   1 UI = 15.6 μs
For 2048 kbps   1 UI = 488 ns
For 8448 kbps   1 UI = 118 ns
For 34,368 kbps   1 UI = 29.1 ns
For 139,264 kbps   1 UI = 7.18 ns

Note 4. The value for A_0 (18 μs) represents a relative phase deviation between the incoming signal and the internal timing local signal derived from the reference clock.

Note 5. The absolute phase deviation requires further study.

Note 6. An example of reference configuration explaining the A_0 values is given in Annex C of Ref. 9.

Note 7. These values refer to 2048-kbps interfaces, which are not used for carrying synchronization signals. Specifications for synchronization signals are under study.

Source: Reference 9, Table 2/G.823, ITU-T Rec. G.823.

timing signal derived from the incoming signal (see Section 4.2.3.2). The most significant form of jitter arises from imperfections in the circuitry, which cause jitter that is dependent on the sequence of bits in a digital signal being transmitted, and is termed pattern-dependent jitter.* The mechanisms that generate jitter within a regenerator are principally related to imperfections in the timing-recovery circuit.

Since pattern-dependent (systematic) jitter from regenerator sections is the dominant type of jitter in a network, the manner in which it accumulates must be considered. For jitter purposes, a regenerative repeater acts as a low-pass filter to the jitter present on the input signal. It also generates its own jitter, which can be represented by an additional jitter source at the input. If this added jitter were truly random, as distinct from pattern-dependent, then the total rms jitter, J_N, present on the digital signal after N regenerators is given by the approximate relationship

$$J_N = J \times N^{1/4} \tag{4.1}$$

where J is the rms jitter from a single regenerator due to uncorrelated jitter sources. The equation assumes that the jitter added at each regenerator is uncorrelated.

As we pointed out earlier, most of the jitter added is pattern-dependent (systematic) and, since the pattern is the same at each regenerator, it can be assumed that the same jitter is added at each regenerator in a chain of identical regenerators. In this case it can be shown that the low-frequency components of the jitter add linearly, whereas the high-frequency components are increasingly attenuated by the low-pass filtering effect of successive regenerators. If a random signal is being transmitted, the rms jitter J_N, present on the signal after N regenerators, would be given by the approximate relationship

$$J_N \approx J_1 \times (2N)^{1/2} \quad \text{for large values of } N \tag{4.2}$$

where J_1 is the rms jitter from a single regenerator due to pattern-dependent mechanisms.

ITU-T Rec. G.823 (Ref. 9, Appendix B) goes on to note that, based on operational experience to date (March 1993), values for J_1 in the range of 0.4–1.5% of a unit interval (UI) are achievable using cost-effective designs.

It is also noted that, with the implementation of timing recovery, use of a transversal surface acoustic filter produces a rate of accumulation approaching that obtained for uncorrelated jitter sources.

If timing recovery is based on a phased-locked loop, the rate of accumulation will be marginally greater than that of equation 4.2 and is given by the

---

*It was called "systematic jitter" in Section 4.2.3.2.

approximate relationship

$$J_N = J_1 \times (2NA)^{1/2} \tag{4.3}$$

where A is a factor dependent on both the number of regenerators and the phase-locked loop damping factor. The latter parameter is generally selected, for this application, such that A has an amplitude marginally greater than unity.

With the use of a scrambler/descrambler combination in a digital line, radio relay, or optical fiber system, when such systems are connected in cascade, there is notably reduced jitter accumulation. This is because the jitter contributed to each system is uncorrelated and accumulates with the fourth root (rather than the square root) of the number of cascaded systems.

4.2.4   Quantization Distortion as a Network Impairment*

Quantization or quantizing is defined as "a process in which a continuous range of values is divided into a number of adjacent intervals, and any value within a given interval is represented by a single predetermined value within the interval" (Ref. 12). This was discussed in Chapter 3, Section 3.3.2.

Quantization distortion results from the process of quantizing samples within the working range (Ref. 12). We can think of it as the error between a real voltage sample of an analog signal and its resulting quantized value.

In looking at quantization distortion across a digital network, or hybrid analog–digital network, the *qdu* (quantization distortion unit) is used. This was introduced by CCITT in 1982. A qdu is defined as equivalent to the distortion that results from a single encoding and decoding by an average PCM codec as defined in CCITT Rec. G.711 (Ref. 10). Such a device has a signal-to-distortion ratio of 35 dB when measured in accordance with CCITT Rec. O.132.

4.2.4.1 Transmission Impairments Due to Digital Processes. It is necessary to ensure that the accumulation of impairments due to digital processes across the network does not reach a point where it seriously degrades overall transmission quality. ITU-T Rec. G.113 (Ref. 11) describes "the quantization distortion method." This method is recommended for use where the digital processing is performed using coders that are waveform oriented. The following would come under this category: A-law and μ-law encoders, 32-kbps ADPCM coders, digital loss pads, PCM to ADPCM to PCM conversions, and μ-/A-law converters. The following coders do not lend themselves to this method: LD-CELP, VSELP, RELP, RPE-LTP, and CELP + .

*Section 4.2.4 is based on Refs. 11 and 12.

Specifically, the distortion allocation given to a digital process is given in qdu. Thus a unit assigned a value of 4 qdu is presumed to provide a level of impairment equivalent to four unintegrated 8-bit PCM processes in tandem. It assumes that quantization distortion adds on a $15\log_{10}(n)$ law for n codec pairs in tandem.

First we must assume that a digital network or hybrid analog–digital network meets the basic requirements of ITU-T recommendations for amplitude distortion (i.e., Rec. G.132) and circuit noise (Rec. G.143, i.e., -43 dBm0p referred to the input of the first circuit in the chain of international circuits). Furthermore, a worldwide chain of four-wire circuits with 14 analog-to-digital conversions will keep group delay within acceptable limits.*

Based on these assumptions, Rec. G.113 has established the *5 + 4 + 5 = 14 qdu* planning rule. Under this rule, each of the two national portions of an international telephone connection is permitted to introduce up to a maximum of 5 qdu of transmission impairment and the international portion up to a maximum of 4 qdu. The units of quantization distortion (qdu) tentatively assigned to a number of digital processes are given in Table 4.10.

4.3 NETWORK SYNCHRONIZATION

4.3.1 Introduction

A PCM-based digital network must be synchronized at three levels: bit, time slot, and frame. Bit synchronization refers to the need for the transmitter (coder) and receiver (decoder) to operate at the same bit rate. It also refers to the requirement that the receiver decision point be exactly at the midposition of the incoming bit. Bit synchronization assures that the bits will not be misread by the receiver.

Obviously a digital receiver must also know where a time slot begins and ends. If we can synchronize a frame, time-slot synchronization can be assured. Frame synchronization assumes that bit synchronization has been achieved. We know where a frame begins (and ends) by some kind of marking device. With DS1 it is the framing bit. In some frames it appears as a 1 and in others it appears as a 0. If the 12-frame superframe regime has been adopted, it has 12 framing bits, one in each of the 12 frames. This provides the 000111 framing pattern. In the case of the 24-frame extended superframe (ESF), the repeating pattern is 001011, and the framing bit occurs only once in four frames.

---

*The resultant group delay on an international connection is a function of the number of translations to voiceband that occur within the network. The evolution toward an all-digital network reduces the number of translations and results in a greater number of less complicated connections, which reduces overall group delay distortion.

TABLE 4.10 Planning Values for Quantizing Distortion, Speech Service Only (See Notes 1, 11, and 12)

Digital Process	Quantizing Distortion Units (qdu)	Notes
Processes involving A/D conversion		
8-bit PCM codec pair (according to Rec. G.711, A- or μ-law)	1	2, 3
7-bit PCM codec pair (A- or μ-law)	3	3, 4, 5
Transmultiplexer pair based on 8-bit PCM, A- or μ-law (according to Rec. G.792)	1	3
32 kbps ADPCM (with adaptive predictor) (combination of an 8-bit PCM codec pair and a PCM–ADPCM–PCM tandem conversion according to Recs. G.721, G.726, and G.727)	3, 5	6
Purely digital processes		
Digital loss pad (8-bit PCM, A- or μ-law)	0, 7	7
A/μ-law or μ/A-law converter (according to Rec. G.711)	0, 5	10
$A/\mu/A$-law tandem conversion	0, 5	
$\mu/A/\mu$-law tandem conversion	0, 25	
PCM to ADPCM to PCM conversion (according to Recs. G.721, G.726, and G.727)	2, 5	8, 9
8-7-8 bit transcoding (A- or μ-law)	3	9

Note 1. As a general remark, the number of units of quantizing distortion entered for the different digital processes is that value which has been derived at a mean Gaussian signal level of about -20 dBm0. The cases dealt with in Supplement No. 21, Ref. 25, are in accordance with this approach.

Note 2. By definition.

Note 3. For general planning purposes, half the value indicated may be assigned to either of the send or receive parts.

Note 4. This system is not recommended by the ITU-T but is in use by some administrations in their national networks.

Note 5. The impairment indicated for this process is based on subjective tests.

Note 6. Recommendations G.726 and G.727 perform equivalently at corresponding bit rates, including 24 and 40 kbps. However, qdu values cannot be assigned for 24- and 40-kbps operation, at this time. For evaluation of ADPCM codecs in the context of overall circuit quality, it appears the "equipment impairment factor" method described in Clause 6 of Ref. 11 gives a more accurate description of their subjective effects on speech quality.

Note 7. The impairment indicated is about the same for all digital pad values in the range 1–8 dB. One exception is the 6-dB A-law pad, which introduces negligible impairment for signals down to about -30 dBm0 and thus attracts 0 units for quantizing distortion.

Note 8. The value of 2.5 units was derived by subtracting the value for an 8-bit PCM codec pair from the 3.5 qdu determined subjectively for the combination of an 8-bit PCM code pair and a PCM/ADPCM / PCM conversion. Multiple synchronous digital conversions, such as PCM/ADPCM, and PCM/ADPCM /PCM, are assigned a value of 2.5 qdu. For evaluation of ADPCM codecs in the context of overall circuit quality, it appears that the "equipment impairment factor" method described in Clause 6 gives a more accurate description of their subjective effects on speech quality.

Note 9. This process might be used in a digital speech interpolation system.

Note 10. The qdu contributions made by coding law converters (e.g., μ-law to A-law) are assigned to the international part.

Note 11. The qdu assignments to these digital processes reflect, to the extent possible, only the effect of quantization distortion on speech performance. Other impairments, such as circuit noise, echo, and attenuation distortion, also affect speech performance. The effect of these other impairments must therefore be taken into account in the planning process.

Note 12. The qdu impairments in this table are derived under the assumption of negligible bit error.

Source: Reference 11, Table 1/G.113, ITU-T Rec. G.113.

E-1, as we remember in Chapter 3, has a separate framing and synchronization channel, namely, channel 0. In this case the receiver looks in channel 0 for the framing sequence in bits 2 through 8 (bit 1 is reserved) of every other frame. The framing sequence is 0011011. Once the sequence is acquired, the receiver knows exactly where frame boundaries are. It is also time-slot aligned.

4.3.2   The Network Switch—Responsible for Local Synchronization and Timing

All digital switches have a master clock. Outgoing bit streams from a switch are slaved to the switch's master clock. Incoming bit streams to a switch derive timing from bit transitions of that incoming bit stream. It is mandatory that each and every switch in a digital network generate outgoing bit streams whose bit rate is extremely close to the nominal bit rate. To achieve this, network synchronization is necessary. Network synchronization can be accomplished by synchronizing all switch (node) master clocks so that transmissions from these nodes have the same average line bit rate. Buffer storage devices are judiciously placed at various transmission interfaces to absorb differences between the actual line bit rate and the average rate. Without this network-wide synchronization, *slips* will occur. Slips are a major impairment in digital networks. Slip performance requirements are discussed in Section 4.3.3.1. A properly synchronized network will not have slips (assuming negligible phase jitter and wander). In the next paragraph we explain the fundamental cause of slips.

As we mentioned earlier, timing of an outgoing bit stream is governed by the switch clock. Suppose a switch is receiving a bit stream from a distant source and expects this bit stream to have a transmission rate of $F(0)$ in Mbps. Of course the switch has a buffer of finite storage capacity into which it is streaming these incoming bits. Let us further suppose that this incoming bit stream is arriving at a rate slightly higher than $F(0)$, yet the switch is draining bits from the buffer at exactly $F(0)$. Obviously, at some time, sooner or later, that buffer must overflow. That overflow is a *slip*. Now consider the contrary condition: the incoming bit stream has a bit rate slightly less than $F(0)$. Now we will have an underflow condition. The buffer has been emptied and for a moment in time there are no further bits to be streamed out. This must be compensated for by the insertion of idle bits, false bits, or frame. However, it is more common just to repeat the previous frame. This is also a slip. We may remember the discussion of stuffing* in Chapter 3, Section 3.3.7, in the description of higher-order multiplexers. Stuffing allowed some variance of incoming bit rates without causing slips.

When a slip does occur at a switch–port buffer, it should be controlled to occur at frame boundaries. This is much more desirable than to have an

*Called *justification* by the ITU-T Organization.

uncontrolled slip that can occur anywhere, even inside a time slot. Slips occur for two basic reasons:

1. Lack of frequency synchronization among clocks at various network nodes (switches).
2. Excessive phase wander and jitter on digital bit streams.

Thus, even if all network nodes are operating in the synchronous mode and are properly synchronized to the network master clock, slips can still occur due to transmission impairments. An example of environmental effects that can produce phase wander of bit streams is the daily ambient temperature variation affecting the electrical length of a digital transmission line.

Consider this example. A 1000-km length of coaxial cable carrying 300 Mbps (3×10^8 bps) will have about 1 million bits in transit at any given time, each bit occupying about 1 meter of cable. A 0.01% increase in propagation velocity, as would be produced by a 1°F decrease in temperature, will result in 100 fewer bits in the cable; these bits must be absorbed by the switch's incoming elastic store buffer. This may end up causing an underflow problem, forcing a controlled slip. Because it is underflow, the slip will be manifested by a frame repeat; usually the last frame just before the slip occurs is repeated.

In speech telephony, a slip causes a click in the received speech. For the data user, however, the problem is far more serious. In this case, at least one data frame or packet will be corrupted.

Slips due to jitter and wander can be prevented by adequate buffering.* Therefore adequate buffer size at the digital line interfaces and synchronization of the network node clocks are basic means to achieve a slip-free network, or comply with stringent slip rate criteria.

4.3.3 Methods of Network Synchronization

There are a number of methods that can be employed to synchronize a digital network. Six such methods are shown graphically in Figure 4.7.

Figure 4.7a illustrates plesiochronous operation. In this case each switch clock is free-running (i.e., it is *not* synchronized to the network master clock). Each network nodal switch has identical high-stability clocks operating at the same nominal rate. When we say high-stability, we mean stability in the range from 1×10^{-11} to about 5×10^{-13} per month. This implies an atomic clock, either rubidium or cesium. The accuracy and stability of each clock are such that there is almost complete coincidence in timekeeping, and the phase drift among many clocks is, in theory, avoided or the slip rate between network nodes is acceptably low. This requires that all switching nodes, no matter how

---

*Good design of timing recovery circuits in regenerators and/or the implementation of scrambling all bit streams can vastly reduce or eliminate slips due to such causes.

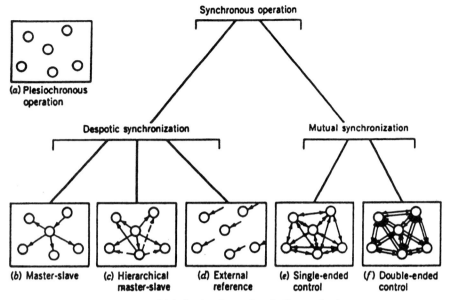

Figure 4.7. Digital network synchronization methods.

small, have such high-precision clocks. For commercial networks, this causes a high cost burden. However, for military networks it is very attractive for survivability because there is no mutual network synchronization. Thus the loss of a node or a clock does not affect the rest of the network synchronization and timing. CCITT (ITU-T) recommends plesiochronous operation on transnational connectivities (i.e., at international switching centers). Refer to CCITT Rec. G.811 (Ref. 13).

Another general synchronization scheme is mutual synchronization, which is shown in Figures 4.7e and 4.7f. Here all nodes in the network exchange frequency references, thereby establishing a common network clock frequency. Each node averages the incoming references and uses the result to correct its locally transmitted clock. After an initialization period, the network aggregate clock normally converges to a single stable frequency.

A number of military networks* as well as a growing number of private networks use external synchronization.[†] This is illustrated in Figure 4.7d. Switch clocks use "disciplined" oscillators slaved to an external radio source. One of the most popular of these external sources is the GPS (geographical

---

*TRI-TAC, a U.S. military digital network, is based on plesiochronous operation using rubidium oscillators located in transmission equipment (e.g., AN/TRC-170 (V); see Chapter 6).

[†]We believe that by the 1999 time frame all local exchange carrier (LEC) switches will change over to this type of synchronization using GPS. Each switch will then have a stratum 1 clock traceability. See Table 4.11.

positioning system), which disseminates universal coordinated time called UTC, an abbreviation from the French equivalent. It is a multiple-satellite system using polar orbits. At any moment in time, at least three or four satellites are in view from a particular location anywhere on the globe. Its time transfer capability is in the 10–100-ns range from UTC.

Other time dissemination systems by radio are also available, such as satellite-based Transit and GOES, terrestrially based Omega (worldwide coverage), and Loran-C at 100 kHz with somewhat spotty coverage. HF radio time transfer techniques, such as WWV, are *not* recommended.

4.3.3.1 *North American Synchronization Plan as Specified by ANSI/ Bellcore.*

The North American network and many other national digital networks use a hierarchical timing distribution system shown in Figure 4.7c. It is based on a four-level hierarchy and these levels are called strata (stratum in the singular). The North American synchronization (Bellcore) arrangement is shown in Figure 4.8.

Timing requirements for each stratum level are shown in Table 4.11. The parameters given in the table are described below.

The stratum levels for synchronized clocks are based on three parameters:

1. *Free-run accuracy.* This is a clock's accuracy when free running, that is, when it has no external reference or has been in holdover for some period of time (i.e., > 3 or 4 days).
2. *Holdover Stability.* When a clock loses its frequency reference, it must depend on its own internal stability. Holdover stability is the drift a clock experiences when it is free running.

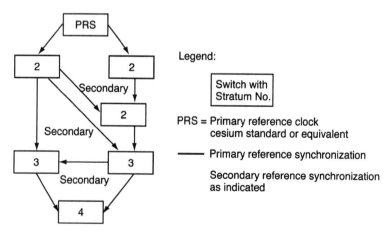

Figure 4.8. North American hierarchical network synchronization. Based on data from Refs. 14 and 24.

TABLE 4.11   Stratum-Level Parameters

Stratum No.	Free-Run Accuracy	Holdover Stability	Pull-in/Hold-in
1	$\pm 10^{-11}$	—	—
2	$\pm 1.6 \times 10^{-8}$	$\pm 1 \times 10^{-10}$ per day	$\pm 1.6 \times 10^{-8}$
3	$\pm 4.6 \times 10^{-6}$	< 255 slips for 1st day of holdover	4.6×10^{-6}
4	$\pm 32 \times 10^{-6}$	no holdover	32×10^{-6}

Note: There is a stratum 3E level:

	$\pm 4.6 \times 10^{-6}$	$\pm \times 10^{-8}$	4.6×10^{-6}

Source: References 14 and 24.

3. *Pull-in / Hold-in.* When a slaved clock is initially synchronized or resynchronized when losing its external reference, it must have the ability to be synchronized from its external reference. To do this it must have a sufficient range to capture the external reference. This range must be at least as wide as its free-running accuracy.

The stratum 1 clock is, by definition, the primary reference source (PRS). Such a clock is required to have a long-term frequency accuracy of 1×10^{-11} and be completely autonomous of other frequency references. To meet this requirement, Refs. 14 and 24 state that the only clock that can meet such a requirement is the cesium beam atomic frequency standard. The PRS must also have the capability of being verified with UTC. Another approach for stratum 1 is to use a broadcast frequency/time standard. Only GPS and Loran C can provide such frequency and time accuracy/stability. In fact, GPS is a recognized time dissemination service for UTC. Some digital networks use a disciplined oscillator slaved to GPS at each switch allowing autonomous plesiochronous operation of the switch.

Stratum 2 clocks can be found in tandem and transit switches. To meet the required stability of $\pm 1.6 \times 10^{-8}$, double-ovened crystal oscillators are employed as the frequency source. Stratum 2 clocks should also have long time constants for averaging their input frequency reference. To improve reliability, these clocks often use dual input frequency references with automatic switch-over.

Stratum 3 clocks are used in local serving switches. They use temperature-compensated crystal oscillators. There is a stratum 3E clock. To clean up wander, a stratum 3E clock filters its incoming references. This creates a timing signal with low levels of wander. These 3E clocks usually provide better holdover performance.

Stratum 4 is the lowest level of the North American hierarchical synchronization plan. These clocks are commonly employed in PCM channel banks.

There is a 4E clock, which uses two input references and is used in digital PABXs.

Building Integrated Timing Supply (BITS). Each switching center has one master clock embodied in a timing supply generator (TSG), which is the basis of the BITS. All other synchronized clocks in the switching center derive timing from the TSG. The BITS clock in the TSG is the same or higher stratum level than all other clocks in the switching center and is the only clock that has an external reference.

With the introduction of DS0 data ports, all channel banks must derive timing from the BITS TSG.

The TSG provides DS1 and composite clock (CC) timing to all synchronized clocks in the switching center. The TSG is externally synchronized by two DS1 signals. The most common method is to bridge off traffic-bearing DS1s entering the switching center. Another method is to terminate two DS1s dedicated to synchronization distribution. Another approach is to use SONET network elements (NEs) as described in Bellcore GR-253 (Ref. 15). Still another method is to employ timing from a DS1 derived from a colocated PRS. We might call the GPS-derived disciplined oscillator an equivalent PRS.

Figure 4.9 is a simplified functional block diagram of a BITS implementation.

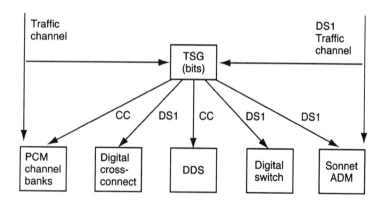

Legend:

DDS: Digital Data Service
CC: Composite Clock
ADM: Add—Drop Multiplex
TSG: Timing Signal Generator

Figure 4.9. Simplified functional block diagram of a BITS implementation.

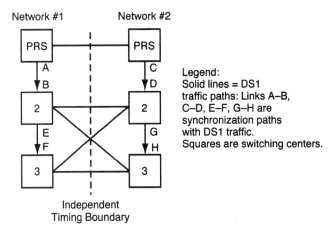

Figure 4.10. A typical two-clock problem. Networks 1 and 2 are shown. Each network has an independent PRS.

Timing Distribution. The synchronization network historically has been tied to the switching network, using traffic-bearing DS1s between digital switches for synchronization distribution. The switching network includes local serving exchanges and tandem switches (and transit switches in the long-distance network). The tandem switches (and transit switches), which are connected to many downstream switches, provide a convenient point to distribute synchronization. In the future, as synchronization distribution evolves to become SONET-based,* SONET facility hubs may become synchronization distribution hubs. Thus Bellcore recommends that synchronization hubs that distribute timing to several downstream switches have stratum 2 BITS clocks, and consideration should be given to making them PRS sites.

4.3.3.2 The Two-Clock Problem. When two disparate networks must interface, we are up against what some call the *two-clock problem.* In the United States, typically this is the interface between a local exchange carrier (LEC) and one or several interexchange carriers (IECs). Each network is autonomous for timing synchronization, and each has a clock traceable to some sort of PRS. A similar situation can arise when a private network interfaces with the PSTN. When two disparate networks have traceability each to their own PRS, it is called *plesiochronous operation.* This concept is shown in Figure 4.10. If both clocks are traceable to stratum 1 sources, Bellcore (Ref. 14) reports that a slip rate of no greater than 1 slip in 72 days can be expected.

---

*SONET is described in Section 3.6.

4.3.4 CCITT/ITU-T Synchronization Plan

CCITT Rec. G.811 (Ref. 13) deals with synchronization of international links. Plesiochronous operation is preferred. (See Section 4.3.3.) The recommendation states the problem at the outset:

> International digital links will be required to interconnect a variety of national and international networks. These networks may be of the following form:
>
> (a) a wholly synchronized network in which the timing is controlled by a single reference clock.
> (b) a set of synchronized subnetworks in which the timing of each is controlled by a reference clock but with plesiochronous operation between the subnetworks.
> (c) a wholly plesiochronous network (i.e., a network where the timing of each node is controlled by a separate reference clock).

Plesiochronous operation is the only type of synchronization that can be compatible with all three types listed. Such operation requires high-stability clocks. Thus Rec. G.811 states that all clocks at network nodes that terminate international links will have a long-term frequency departure of not greater than 1×10^{-11}. This is further described in what follows.

The theoretical long-term mean rate of occurrence of controlled frame or octet (time slot) slips under ideal conditions in any 64-kbps channel is consequently not greater than *1 in 70 days* per international digital link.

Any phase discontinuity due to the network clock or within the network node should result only in the lengthening or shortening of a time signal interval and should not cause a phase discontinuity in excess of one-eighth of a unit interval on the outgoing digital signal from the network node.

Recommendation G.811 also states that when plesiochronous and synchronous operation coexist within the international network, the interfacing nodes will be required to provide both types of operation. It is therefore important that the synchronization control does not cause short-term frequency departure of clocks, which is unacceptable for plesiochronous operation. The magnitude of the short-term frequency departure should meet the requirements specified in Section 4.3.4.1.

4.3.4.1 *Time Interval Error and Frequency Departure.* Time interval error (TIE) is based on the variation of ΔT, which is the time delay of a given timing signal with respect to an ideal timing signal, such as UTC. The TIE over a period of S seconds is defined to be the magnitude of difference between time delay values measured at the end and at the beginning of the period:

$$TIE(S) = |\Delta T(t + S) - \Delta T(t)| \qquad (4.4)$$

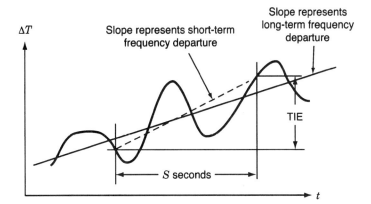

Figure 4.11. Definition of time interval error (TIE). (From Ref. 13, CCITT Rec. G.811.)

where t is the beginning of the time interval, often given as UTC. This is shown diagrammatically in Figure 4.11. The corresponding normalized frequency departure $\Delta f/f$ is the TIE divided by the duration of the period (i.e., S seconds).

The TIE at the output of a reference clock is specified in CCITT Rec. G.811 for three values of frequency as follows:

The TIE over a period of S seconds shall not exceed the following limits:

(a) $(100S)$ ns + 1/8 unit interval. Applicable to S less than 5. These limits may be exceeded during periods of internal clock testing and rearrangements. In such cases the following conditions should be met: TIE over any period up to 2 UI (unit intervals) should not exceed 1/8 of a UI. For periods greater than 2 UI, the phase variation for each interval of 2 UI should not exceed 1/8 UI up to a total maximum TIE of 500 ns.

(b) $(5S + 500)$ ns for values of S between 5 and 500.

(c) $(10^{-2S} + 3000)$ ns for values of S greater than 500.

The allowance in (c) of 3000 ns is for component aging and environmental effects.

For clarification, CCITT defines UI in Rec. G.701 (Ref. 12) as the nominal difference in time between consecutive significant instants of an isochronous signal. With NRZ coding we can think of a UI as the duration of 1 bit or the bit period. The bit period in NRZ is the inverse of the bit rate.

Care must be taken with this definition because in the case of AMI coding the mark (1) and space (0) are usually not of equal duration.

Table 4.12 shows the permissible degradation of a timing network node. The performance category refers to slip rate performance objectives given in Section 4.3.5.

TABLE 4.12 Maximum Permissible Degradation of Timing at a Network Node

Performance Category[a]	Frequency Departure $\left\|\dfrac{\Delta f}{f}\right\|$ of Node Timing[b]		Proportion of Time During Which Degradation May Occur, Referred to Total Time[d]	
	Local[c]	Transit	Local[c]	Transit
Nominal	See Section 5.5.5.1 of Ref. 13	See Section 5.5.5.1 of Ref. 13	$\geq 98.89\%$	$\geq 99.945\%$
(a)	$10^{-11} < \left\|\dfrac{\Delta f}{f}\right\| \leq 10^{-8}$	$10^{-11} < \left\|\dfrac{\Delta f}{f}\right\| \leq 2.0 \times 10^{-9}$	$\leq 1\%$	$\leq 0.05\%$
(b)	$10^{-8} < \left\|\dfrac{\Delta f}{f}\right\| \leq 10^{-6}$	$2.0 \times 10^{-9} < \left\|\dfrac{\Delta f}{f}\right\| \leq 5.0 \times 10^{-7}$	$\leq 0.1\%$	$\leq 0.005\%$
(c)	$\left\|\dfrac{\Delta f}{f}\right\| > 10^{-6}$	$\left\|\dfrac{\Delta f}{f}\right\| > 5.0 \times 10^{-7}$	$\leq 0.01\%$	$\leq 0.0005\%$

[a]The performance categories (b) and (c) correspond to (b) and (c) in Rec. G.822 while category (a) in Rec. G.822 corresponds to "Nominal" and (a) in Rec. G.811, combined.
[b]All values are provisional.
[c]The values for local nodes are given for guidance only, and administrations are free to adopt other performance levels provided the overall controlled slip performance objectives of Rec. G.822 are met.
[d]These values are more stringent than would be strictly required by Rec. G.822 for a 64-kbps connection, to allow for the future introduction of services at higher bit rates that may require a better slip performance. They also allow a margin for possible network effects.
Source: Reference 13, CCITT Rec. G.811.

4.4 DIGITAL REFERENCE SIGNALS AND PROCESSES*

4.4.1 Digital Milliwatt

The digital milliwatt (DMW) is an important concept relating to signals in digital facilities. A DMW is a digital representation of a 0-dBm level, 1000-Hz (analog) sine wave. A representation of eight 8-bit words has been adopted by the ITU-T Organization for a DMW that is encoded according to the $\mu = 255$ encoding law. The following is a DMW pattern taken from AT&T Technical Advisory Note No. 32:

DIGIT NUMBER

1	2	3	4	5	6	7	8
0	0	0	1	1	1	1	0
0	0	0	0	1	0	1	1
0	0	0	0	1	0	1	1
0	0	0	1	1	1	1	0
1	0	0	1	1	1	1	0
1	0	0	0	1	0	1	1
1	0	0	0	1	0	1	1
1	0	0	1	1	1	1	0

*Section 4.4 is based on Refs. 5 and 16.

The DMW is not recommended for transmission over DS1 because the bit pattern duplicates the framing sequence of D-type channel banks. Under certain conditions, the channel bank can frame on the DMW sequence, causing frame misalignment without any accompanying alarm condition. However, the DMW can be used internally in digital switching systems because these systems do not operate on the DS1 rate.

In addition, it is not advisable to inject either an 800- or 1000-Hz signal into a DS1 or E-1 system because of the harmonic relation of the 8000-Hz sampling rate. Interference signals can result in erroneous loss measurements when these frequencies are mixed.

4.4.2 Digital Reference Signal

The digital reference signal (DRS) is an encoded analog signal, which by current international convention is a sine wave in the frequency range of 1013–1022 Hz having a level (power) of 0 dBm ±0.03 dB. The ITU-T Organization and ANSI are considering adopting a specific encoded signal that uses a 797-byte repetitive pattern and has a frequency of approximately 1013.801756 ⋯ Hz. The avoidance of the DMW sequence means that these DRSs do not have the limitations of the DMW and can be transmitted over the network. For reference level purposes, the DRS is equivalent to the analog milliwatt signal in analog facilities.

4.4.3 Digital Pads

Digital pads are used to insert loss before decoding. Such losses may be required as described in Section 2.6.5.3 to control echo and singing. When a signal is in digital form (typically PCM), gain or loss can be applied directly to the bit stream without first converting it to an analog equivalent. A simple look-up table is often used to convert from one code set to the other; or the nonlinear PCM can be converted to a linear code, multiplied by the gain or loss desired, and converted back to nonlinear PCM. This avoids the cost of digital-to-analog conversion but does not avoid signal impairment, since approximately the same amount of quantizing noise is added in either case.

Digital level control always reduces the signal-to-quantizing noise ratio because of round-off errors in mapping between code sets. This differs from analog gain adjustment, which seldom affects signal-to-noise ratio.

To provide nearly constant ratio of signal-to-quantizing noise over a signal range from −40 to +3 dBm0, a nonlinear $\mu = 255$ PCM coding is used (i.e., DS1). This range closely approximates the range of talker levels to be found at local serving switches and at tandem switches and assures optimum quality for nearly all talkers. Digital loss or gain pushes speech power toward one end or the other of this range by the amount of the loss or gain inserted. It reduces the code set used, since the maximum output signal will be x dB below the maximum code (for x-dB loss) or the minimum will be y dB above the minimum code (for y-dB gain). The use of digital pads will cause the loss of bit integrity because bit sequences are being changed enroute. It causes no harm to speech communications but should not be used on channels carrying data where bit integrity is vital (Ref. 5).

4.4.4   North American Digital Transmission Conventions

4.4.4.1   Encoders. Encoders, in this context, convert analog signals into digital signals for transmission through digital facilities such as trunks and switches. If a sinusoidal signal of e dBm at an encoder input results in a DRS at the encoder output, the encoder is said to have an encode level of e and is called an *e-level encoder*. The input side of the encoder is said to be an *e encode level point* (e-ELP).

For $\mu = 255$ encoders, overload (clipping) begins to occur at an input power level that is 3.17 dB higher than the power level of a sinusoid that would produce a DRS at the encoder output. Thus the overload level of an e-level encoder is ($e = 3.17$) dBm.

4.4.4.2   Decoders. Decoders convert digital signals into analog signals. If a DRS at a decoder input results in a sinusoidal signal of d dBm at the decoder output, the decoder is said to have a decode level of d and is called a *d-level decoder*. The output side of the decoder is said to be a *d decode level point* (d-DLP).

4.4.4.3   Equivalent Power Level. Equivalent power level is a term used to denote the power level of an analog signal that is encoded into a digital bit stream. Equivalent power should not be confused with the actual power of the digital signal. The equivalent power can be measured indirectly by mapping the digital signal to the output of a 0-level decoder. The power of the decoded signal can, of course, be measured directly with analog test equipment. The equivalent power is sometimes referred to as *power relative to a 0-level decoder*.

4.4.4.4   Loss Implementation in a Digital Network

4.4.4.4.1   Analog–Digital Combination. In this case a digital tandem switch is connected to an analog serving exchange, as shown in Figure 4.12. Using Bellcore convention, because the interface is digital at the digital switch, the digital transmission facility is considered a *combination trunk* (i.e., digital on one end and analog through a channel bank at the other). Using the concept of equivalent power at the center of the digital switch, a loss of 6 dB is shown in the figure from the tandem to the serving switch line appearance, and 0 dB is shown in the opposite direction (A).

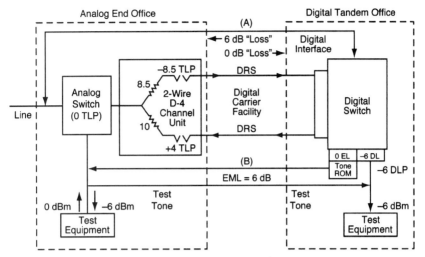

Figure 4.12. Analog end office switch to digital tandem. (From Ref. 16, Figure 7-18. Copyright © 1994 Bellcore. Reprinted with permission.)

With the tandem switch encode and decode levels of 0 and -6, respectively, the test level received at both switches is -6 dBm; thus the EML* (expected measured loss) of this trunk is 6 dB in both directions (B).

4.4.4.4.2 Loss on Digital Trunks. Figure 4.13 shows a digital tandem switch connected to a digital local serving switch (digital end office). The trunk is digital. Using the concept of equivalent power, it can be seen that there is no loss between these two digital switches. Both switches encode at 0 and decode at -6. Thus the loss is 6 dB from the tandem switch to the local serving switch line appearance and 0 dB in the opposite direction (A). The EML of the trunk is 6 dB in both directions (B).

4.5 DELAY IN DIGITAL NETWORKS

4.5.1 Delay Essentials

Round-trip delay affects voice, data, and video-telephony. We gave certain delay guidelines in Section 4.2.1. Some highly interactive data circuits may experience degradation even at delays on the order of 100 ms. The upper limit suggested in ITU-T Rec. G.114 (Ref. 17) is 400 ms for one-way transmission time.

---

*EML, *expected measured loss*, as defined in Ref. 16, is the loss of a trunk at 1004 Hz that is expected to be measured under specified test conditions. This loss is calculated by summing all the losses and gains of the trunk in the specified measuring configuration. It is used as a reference for comparisons with actual measurements.

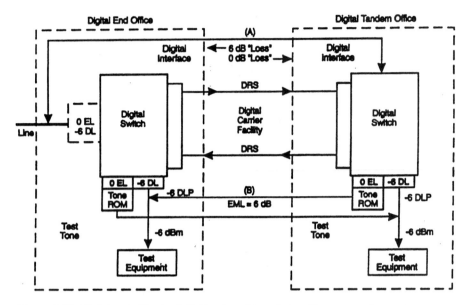

Figure 4.13. Digital end office to digital tandem. (From Ref. 5, Figure 7-19. Copyright © 1994 Bellcore. Reprinted with permission.)

Delay is made up of two components: (1) transmission or propagation delay, which is a function of the velocity of propagation of the media in the connection and their length, and (2) processing delay. Of course, propagation delay makes up the lion's share of the total delay. Satellite communication circuits using geostationary satellites are the worst offenders, where we budget approximately 260 ms of one-way delay for one satellite hop.

ITU-T Rec. G.114 provides guidelines on estimating total delay. These guidelines are given in the next section.

4.5.2   Estimating Delay

1. *Analog Networks.* The transmission time in milliseconds will probably not exceed

$$12 + (0.004 \times \text{distance in km}) \quad \text{ms}$$

The factor 0.004 is based on a comparatively high-velocity plant (about 250/km/ms). The 12-ms constant term makes allowance for terminal equipment and for the probable presence of loaded cable over a portion of the connectivity (e.g., three pairs of channel translating equipments plus about 160 km of H88 loaded cables).

2. *Digital Networks.* The transmission time in ms will probably not exceed

$$3 + (0.004 \times \text{distance in km}) \quad \text{ms}$$

The 3-ms constant term makes allowance for one PCM coder and decoder and five digitally switched exchanges. The value of 0.004 is a mean value for coaxial cable and radio-relay systems; for optical fiber systems, use 0.005.

The following constants may be used to estimate one-way transmission time.

- Terrestrial coaxial cable or line-of-sight (LOS) microwave, whether analog or digital: 4 μs/km. This allows for repeaters and regenerators.
- Optical fiber cable: 5 μs/km.
- Submarine coaxial cable: 6 μs/km.
- PCM coder or decoder: 0.3 ms. This value allows for higher-order multiplex equipment.
- PCM/ADPCM/PCM transcoding: 0.5 ms.
- Digital transit exchange (digital–digital): 0.45 ms. (Because of variants involved, often 1 ms is budgeted per digital exchange.)

These estimations are based on ITU-T Rec. G.114 (Ref. 17).

4.5.3 Delay from a Bellcore Perspective*

To control echo and to minimize the effect on digital throughput, the maximum round-trip delay for steady-state operation of a transport system with no intermediate terminals is as follows:

- The maximum round-trip delay in milliseconds of a system having route mileage (i.e., one-way mileage) M miles shall be no more than

$$\text{RTD}_{\text{max}} = 0.32 + 0.0168 \times M$$

This requirement can be interpreted in terms of separate processing delay and propagation delay requirements as follows:

- The maximum round-trip propagation delay of the transmission facility (including delays of regenerators, repeaters, or equivalent elements) shall be no more than 0.0168 ms per route mile.

*From Ref. 5, Issue 5, Section 8.

- The maximum round-trip processing delay of a digital facility interface (DFI) unit that terminates the trunk at a switching system at each end shall be no more than 0.16 ms. This requirement applies for all interface options provided.

As an example, the required maximum round-trip delay for a system of 100 route-miles is 2.0 ms.

4.6   DIGITAL SUBSCRIBER LOOP

4.6.1   Introduction

A digital subscriber loop (DSL) covers any digital transmission system found in the subscriber plant. In North America, there is a wide scope of these DSL systems. Among them we include: SLC-96,* a 96-channel digital subscriber line and a predecessor of IDLC, developed specifically for North American LECs;[†] IDLC (integrated digital loop carrier) and UDLC (universal digital loop carrier); ISDN basic rate service terminating in residences or very small offices and ISDN primary rate service terminating at PABXs; subrate digital loops (SRDL), principally used for data connectivity; switched 56 loops (56 kbps); and HDSL (high bit-rate digital subscriber line) and ADSL (asymmetric digital subscriber line). The list is not exhaustive.

Few of these systems actually terminate at the subscriber's premises with the exception of ISDN basic rate, SRDL, switched 56, and HDSL when employed to supply CATV[‡]-like service. For example, the SLC-96, IDLC, and UDLC terminate in a "box" near the subscriber premises. This box is variously called a remote terminal (RT) or remote interface. The RT can have as many as 96 lines without concentration (or more) capability and each line is converted to an analog subscriber loop. Each line terminates in a SLIC (subscriber line interface card). The SLIC carries out the following functions, often helpfully abbreviated by *BORSCHT*:

B = battery feed
O = overvoltage protection
R = ringing
S = signaling
C = coding (the A/D and D/A conversion)
H = hybrid
T = test

These functions were addressed in Chapter 2 under the analog subscriber loop description. Figure 4.14 is a block diagram of a typical SLIC.

---

*SLC-96 is copyrighted by AT&T.
[†]LEC = local exchange carrier, a strictly U.S. term.
[‡]CATV = community antenna television, more commonly known as cable television.

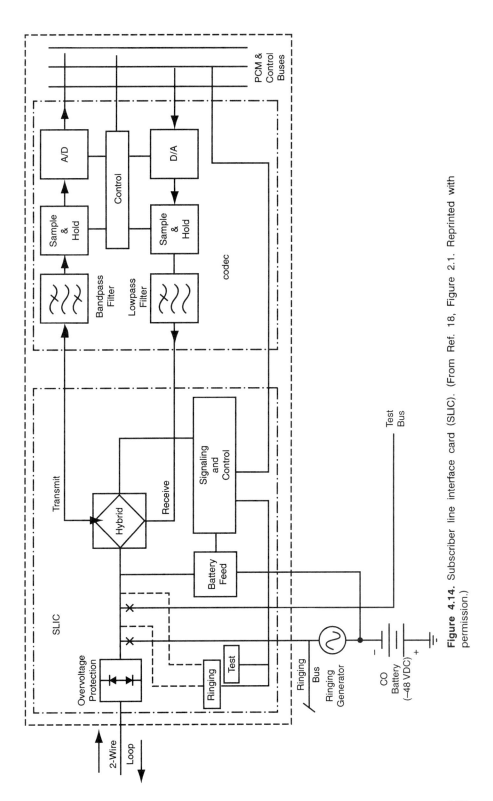

Figure 4.14. Subscriber line interface card (SLIC). (From Ref. 18, Figure 2.1. Reprinted with permission.)

247

The RT may be powered from the COT (central office terminal, the local serving switch) or may be powered locally. If powered locally, some form of backup battery should be provided with a reasonably, yet cost-effective reserve time. DLC recommends an 8-hour reserve time. The battery voltage is a nominal -48 Vdc. If powered by the local serving switch, the voltage drop of the line would be a major consideration.

4.6.2 Generic Considerations for DLC Systems

In North America, the DS1 signal is the most prevalent signal used on digital facilities in the loop plant. The following are requirements placed by Bellcore (Ref. 19) on DS1-based DLC systems.

- Have a DSX-1 cross-connect interface to ensure compatibility with other DS1-hierarchy plant.
- Use superframe (SF) or extended superframe (ESF), with ESF being favored because of improved false framing protection and network maintenance capabilities inherent in the ESF format.
- The COT and RT DS1 interfaces of a DLC system must tolerate 8 unit interval (UI) phase steps of jitter, where 1 UI is about 684 ns at the DS1 transmission rate.
- Use of B8ZS to ensure minimum 1's density and ease of provision of clear channel capability.*
- A RT must accept an input signal from the adjacent DS1 repeater or COT that can have as much as 25-dB loss on the cable at 772 kHz; the design objective is up to 34 dB.
- The COT must operate satisfactorily with a cable pair input signal attenuated up to 25 dB at 772 kHz.
- The COT interface must supply power to the T1 line at a constant current of 60 $\pm$ 3 mA. An option is to supply a nominal 140-mA constant dc current.
- DLC systems providing a DS1 interface must provide the capability of determining the location of a fault on a T1-type (DS1) digital line by tests performed at only one end of the T1-type line using standard T1-type fault-locating methods. This capability must be provided at both the COT and RT locations.
- A DLC system must provide the capability to initiate a switch-over to the protection facility if an alarm indication signal (AIS) is received on the primary transmission facility at either the COT or the RT.

---

*Clear channel means a 64-kbps channel unencumbered with robbed-bit supervisory signaling. Of course DS1 has 24 such 64-kbps channels.

4.6.2.1 Synchronization of DLC Systems. When a DLC system is to connect to other digital network elements, some means of synchronizing the clock rates of the DLC system and connecting network elements are required. The system needs to obtain bit and/or byte synchronization with other connecting network elements. DLC systems that provide digital data services or features must have the capability of obtaining synchronization from the local serving switch timing supply.

A DLC system is a stratum 4 node; thus a DLC system switch-side terminal must meet the stratum 4 node clock requirements (± 32 ppm frequency accuracy).

4.6.2.2 The Carrier Serving Area Concept. The carrier serving area (CSA) is strictly a North American notion. A CSA is a distinct geographic area capable of being served by one theoretical DLC RT (remote terminal) site. CSA is an approach to the planning of outside plant cable. When the CSA concept was first developed, its goal was to sectionalize a cable route into discrete geographical units (CSAs) so that every user along the route could be provided with the most restrictive digital service of that time (e.g., 56-kbps DDS) over an unrepeatered facility when all CSAs were activated. With such a planning approach, the entire cable route beyond 12,000 ft from a local serving switch is segmented into CSAs. When the wire pairs in a cable become exhausted, rather than lay new cable a DLC system is an alternative to provide that relief, the preplanned RT site and its serving area can be activated.

The boundaries of a CSA were determined by the following:

- The maximum unrepeatered transmission range for the worst case digital service, 56-kbps DDS (when CSA was first conceived).
- The distance over which currently available DLC systems could provide acceptable standard voice telephone service on unloaded cable pairs.

Based on these considerations, a CSA cable from the RT of the DLC system to the network interface on the customer's premises must meet the following requirements:

1. Nonloaded cable only.
2. Multigauge cable is restricted to two gauges, excluding short cable sections used for stubbing or fusing.
3. Total bridged-tap length not to exceed 2500 feet. No single bridged tap may exceed 2000 ft.
4. The amount of 26-gauge cable, used alone or in combination with another gauge cable, may not exceed a total length of 9000 ft including bridged tap.

5. For single-gauge or multigauge cables containing only 19-, 22-, or 24-gauge cable, the total cable length including bridged tap may not exceed 12,000 ft.

6. The total length including bridged tap of a multigauge cable that contains 26-gauge cable may not exceed:

$$12,000 - 3(L26)/(9000\text{-}LBTAP) \text{ ft}$$

where L26 is the total length (ft) of 26-gauge cable in the cable, excluding any 26-gauge bridged tap in the cable, and LBTAP is the total length of the bridged tap in the cable. All lengths are in feet (Ref. 19).

4.6.3 High Bit-Rate Digital Subscriber Line

4.6.3.1 Introduction to HDSL. HDSL (high bit-rate digital subscriber line) provides repeaterless DS1 equivalent service over wire pairs that conform to CSA design rules (given previously). HDSL is a DS1 substitute that can deliver 1.544-Mbps services such as high capacity digital service (HCDS), ISDN primary rate access, SMDS (switched multimegabit data service), and switched DS1/switched fractional DS1 transmission rates.

4.6.3.2 Highlights of HDSL. HDSL provides full duplex 1.544-Mbps re-peaterless transmission capability over two-wire pairs. All inductive loading is removed and the pairs are based on CSA design rules. The bit rate on each pair is 784 kbps in each direction. The 1.544-Mbps transmission rate is achieved by combining the 784-kbps signals from both pairs. This type of transmission is called *dual duplex*, and its format is DS1 compatible. The 784-kbps rate includes overhead for performance monitoring, framing, and timing functions.

Figure 4.15 shows the dual duplex HDSL system with two transceivers at both the remote terminal (RT) and at the serving switch terminal. The two transceivers along with DS1 interface circuitry compose an HTU. The HTU located at the serving switch is called an HTU-T, and the one at the remote

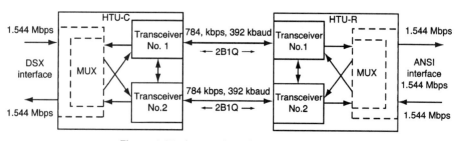

Figure 4.15. An overview of the HDSL system.

terminal is the HTU-R. The HTU-T is connected to the HTU-R by two CSA-compatible wire pair loops.

Each HTU accepts a DS1 signal, adds HDSL overhead for synchronization and maintenance purposes, performs digital signal processing, and generates a line signal that is placed on the loop. The HTU at the opposite end performs digital signal processing on the incoming 782-kbps line signal to minimize the effects of impairments on the loop and recovers the original DS1 signal.

Additional bits are required to support HDSL overhead. Framing, start-up and recovery, and performance monitoring are typical functions that are supported by such overhead.

An echo canceler with hybrid principle is used with HDSL to provide full-duplex signal transmission. Figure 4.16 is a block diagram of the echo canceler. The echo canceler removes echos of the transmitted signal that have mixed with the received signal. The echos are reflections of the transmitted signal from discontinuities in the line, such as bridged taps, gauge changes, impedance mismatches, and hybrid leakage. This echo canceler permits a relatively weak received signal to be accurately detected and is a means of avoiding the use of a separate wire pair in each direction of transmission.

4.6.3.3 HDSL Transmission Considerations

Line Code. The line code is 2B1Q, which is a 4-level code meaning that there are four amplitude levels as shown in Figure 4.17. The 784-kbps

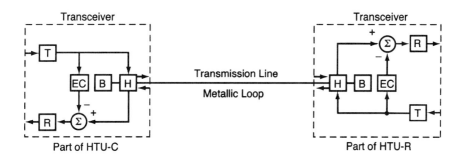

B = Balance impedance
EC = Echo canceler
H = Hybrid
R = Receiver
T = Transmitter
S = Summer

Figure 4.16. Simplified block diagram of the echo canceler with hybrid (ECH) transceiver. (From Ref. 20, Figure 2-2. Copyright 1991 © Bellcore. Reprinted with permission.)

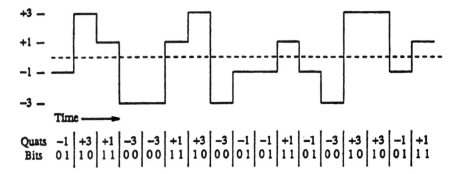

Figure 4.17. Example of 2B1Q quaternary symbols.

data bit stream is made up of 768-kbps DS1 payload, 8-kbps DS1 framing bits, and 8 kbps of HDSL overhead. It is grouped in pairs of bits for conversion to quaternary symbols that are called *quats*. As we look at these bits the first bit is called the sign bit and the second is called the magnitude bit. All data bits, except the 14 bits of the seven-quat sync word and any stuff quat bits, are scrambled before coding.

Each successive pair of scrambled bits in the binary data stream is converted to a quaternary symbol, which is the output of the transmitter. When the first (sign) bit is 1 and the second (magnitude) bit is 0, the quat symbol is $+3$; when the first (sign) bit is 1 and the second (magnitude) bit is 1, the quat symbol is $+1$; when the first (sign) bit is 0 and the second (magnitude) bit is 1, the quat symbol is -1; and when the first (sign) bit is 0 and the second (magnitude) bit is 0, the quat symbol is -3. Figure 4.17 shows a segment of bit stream of binary data and equivalent 2B1Q mapping.

Received Pulses. With HDSL pulses transmitted over the telephone loop plant that conforms to CSA design rules, the HDSL receiver shall receive any random sequence of these pulses with a BER of less than 1×10^{-7}.

Total Power. The average power of the transmitted equiprobable symbols shall be between $+13$ and $+14$ dBm over the frequency band from 0 Hz to 784 kHz.

Baud Rate. The baud rate from the HTU-C is 392 kbaud ± 32 ppm.

Impedance and Return Loss. The nominal driving point impedance of the HTUs, when powered, is 135 $\pm 1\%$. The return loss rises from 0 dB up to 4 kHz; there is a 20-dB/decade rise from 4 to 40 kHz, a 20 dB rise from 40 to 200 kHz, and a drop of 20 dB/decade thereafter.

4.6.3.4 Pictorial View of the HDSL Frame Structure. Figure 4.18 presents a simplified quaternary symbol view of the HDSL frame structure. The frame starts with the seven-symbol synchronization word followed by one quat of other HDSL overhead. This is followed by 12 DS1 blocks of payload. Each DS1 block is composed of a DS1 framing bit (F) (if present, or a HDSL chosen substitute) and 12 bytes of DS0 data, for a total of 97 bits encoded to $48\frac{1}{2}$ quats. For loop 1, the 12 bytes in the block are group 1 of the DS1 frame, bytes 1–12. For loop 2, the 12 bytes are DS1 group 2, bytes 13–24. Another five quats of HDSL overhead follows, and then more payload and more overhead until a total of 48 DS1 blocks and 23 quats of overhead (including the sync word) have been transmitted. At that point, none or two stuffing quat symbols are inserted, depending on whether the DS1 rate is fast or slow relative to the HDSL clock and the status of the data buffers.

The HDSL frame will be either $6 + \frac{1}{392}$ ms ("6 + ") (2353 quats) long if the two stuffing quats are transmitted, or $6 - \frac{1}{392}$ ms ("6 − ") (2351 quats) long when they are omitted. The frame length will tend to average about 6 ms or 2352 quats. The receiver side is able to determine the length of a given incoming frame from the loop by detection of the sync word in the following frame and then adjust the demultiplexing of the data stream.

4.6.4 Asymmetric Digital Subscriber Lines

4.6.4.1 Introduction. ADSL (asymmetric digital subscriber lines), as presented here, is an engineering concept meant to extend the life of metallic subscriber pairs. At the time the Bellcore reference document was prepared (Ref. 21), the architecture of ADSL could provide a CATV (cable television)-like service or "movies on demand." In the current period, it could also be an answer to the consumer demand for high-speed downstream Internet data service. In either case, this high-speed downstream service is provided in addition to simultaneous two-way conventional telephone service (POTS—"plain old telephone service") as well as a slow-speed full-duplex command data channel.

ADSL provides one-way 1.544-Mbps downstream digital transmission capability. In addition, a full-duplex 16-kbps payload signal and a nominal 4-kHz POTS telephone channel can operate simultaneously with the high-speed DS1 channel.

A second, more advanced version, called ADSL-2, will be discussed briefly at the end of this section. Many applications marry ADSL to revised resistance design (RRD) of the copper pair subscriber plant. RRD was described in Chapter 2.

ADSL has a number of potential applications, such as movies-on-demand where an end user can select a movie from a large library of programs. The user selection is carried out on the low-rate (16 kbps) return channel or on the POTS channel. In fact, this technique can be used for an entire range of

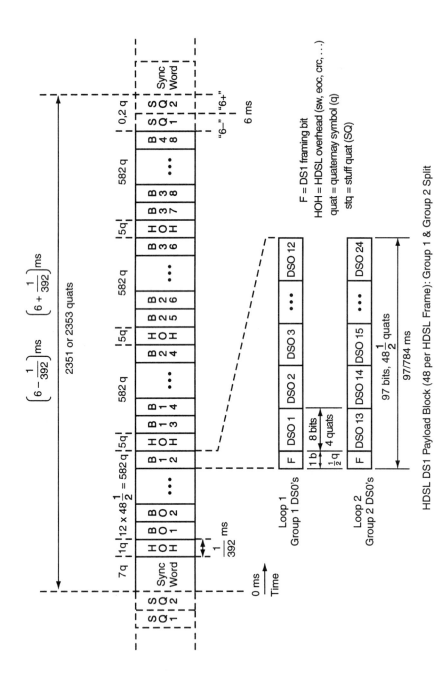

Figure 4.18. A pictorial view of the HDSL frame structure based on quats (not bits). (From Ref. 20, Figure 6-8. Reprinted with permission.)

HDSL DS1 Payload Block (48 per HDSL Frame): Group 1 & Group 2 Split

F = DS1 framing bit
HOH = HDSL overhead (sw, eoc, crc, . . .)
quat = quaternary symbol (q)
stq = stuff quat (SQ)

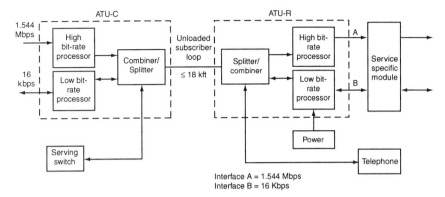

Figure 4.19. Overall generic ADSL-1 architecture providing a full-duplex POTS channel, a full-duplex low-rate data channel, usually 16 kbps, and a one-way 1.544-Mbps channel.

CATV programming services. Other applications are interactive video, tele-education, and transaction services. Figure 4.19 illustrates the overall architecture for ADSL-1. Similar to HDSL, the ATU-C is the ADSL terminal unit that is located at the local serving switch or at a remote node. ATU-R is the remote unit located at or near the customer premises.

The basic reference source (Ref. 21) provides a limited overview of ADS1-1 where few of the requirements are firm. It is conceptual in nature. The following paragraphs must be read with this in mind. ADSL-1 is the baseline version and ADSL-2 is a more advanced design basically offering higher downstream bit rates.

4.6.4.2 ADSL-1 System Requirements. The ADSL-1 system provides transport for a 1.544 Mbps bit stream framed or unframed from the network to the end-user premises. This payload is called *AS1* in the reference 21. The system also provides a bidirectional channel with a nominal synchronous 16-kbps data rate, with an asynchronous 9.6-kbps option. This payload is called *LS1*. Furthermore, the system provides a full-duplex conventional telephone channel on the same metallic pair without degradation of the telephone service.

The one-way transmission delay objective for the high bit-rate channel, from the network end interface through the ADSL-1 system to the service module (SM) interface, should be no greater than 10 ms. The SM interfaces ADSL with customer equipment. The one-way transmission delay for the low-rate channel from the SM interface through the ADSL-1 system to the network-end interface should be less than 1 ms.

BER Requirements. The ADSL-1 system will operate over a single non-loaded metallic twisted pair that conforms to RRD guidelines. If the loop has load coils, they must be permanently removed. The BER requirement at the remote end of the loop is a BER of 1×10^{-7} with a 6-dB margin. Section 8 of Ref. 21 sets up a test scenario to verify this operation using a set of laboratory test loops without injected impulse noise and specified interference.

Transmission Rates to the Line. The ADSL-1 line bit rates must accommodate the downstream channel of 1.544-Mbps AS1 payload plus the LS1 upstream and downstream payloads of either 16 or 9.6 kbps. Bellcore suggests one possible alternative is a line rate of 1.744 Mbps for the high-rate and 28 kbps for the low-rate channel. The additional 200-kbps rate in the downstream includes an allocation for the system overhead function as well as redundancy for forward error correction (FEC) (10% redundancy; see Chapter 5 for a discussion of FEC). The objectives in the selection of bit rates and formats are:

- Minimize bit rate (equivalent bandwidth)
- Select convenient clock rates and
- Provide robust timing recovery

Frame Lengths. The main payload of the ADSL-1 system is the 1.544-Mbps downstream AS1 signal. Either the standard DS1 superframe (SF) or extended superframe (ESF) should be considered.* This signal may be expected to have either a 1.5-ms SF or a 3-ms ESF structure. ADSL frames with multiples of 3 m are logical choices for evaluating alternatives for packaging 3 ms length payloads. Ref. 21 states that frames of 6 ms lengths are probably an optimum compromise considering the number of overhead bits, efficiency, and possible delay for operations information through the system. There should also be a capability to insert dummy bits to replace the FEC redundancy. The proposed line rate of the low-rate channel is 28 kbps including 16-kbps payload and 12-kbps overhead and padding.

Frequency Assignment Overview. ADSL transmits to the wire-pair transmission line a frequency division multiplex signal. In other words, each of the three services is assigned its own frequency slot. The lowest slot might conveniently be assigned to the nominal 4-kHz telephone channel. The midband slot would be assigned to the two-way low speed data channel (LS1), and the highest slot to the downstream high-speed digital channel (AS1). Some form of frequency translation will be required to the LS1 and AS1 segments. Of course, the bandwidths occupied by the three components will depend on the modulation and coding techniques employed.

---

*See Section 3.3.2 for descriptions of SF and ESF.

4.6.4.3   ADSL-2 Architectures. An overriding objective of ADSL-2 is to operate over shorter length loops, conforming to CSA guidelines, in order to achieve higher bit rate service capabilities. Reference 21 states that one objective is to achieve twice the AS1 payload, or about 3.2 Mbps in the downstream direction, so that the downstream channel can be used for either two simultaneous AS1 channels or one composite 3.2-Mbps channel. Recent press articles have stated that a 6.4-Mbps rate is planned for the ADSL-2 high-rate channel.

Another objective is to increase the narrowband full-duplex capability, and to achieve more synergy with ISDN* basic rate access (BRA) transport capabilities.

4.6.4.4   ADSL According to ANSI

4.6.4.4.1   Introduction. ANSI (Ref. 22) has defined a new version of ADSL that is more versatile than that described in Sections 4.6.4.1–4.6.4.3 and with greater capacity, up to 6.144 Mbps downstream and up to 640 kbps upstream as a default data rate and up to 6.944 Mbps for transporting ATM cells downstream. This section provides a brief overview of ANSI ADSL.

4.6.4.4.2   Discrete Multitone (DMT) Modulation. A number of modulation schemes were proposed to increase aggregate bit rates above the 1.544 Mbps proposed by Bellcore and reported in Sections 4.6.4.1 and 4.6.4.2. These are discrete multitone (DMT), carrierless amplitude–phase (CAP) modulation, and quadrature-amplitude modulation (QAM). ANSI T1.413-1995 (Ref. 22) describes DMT modulation as a multitone scheme with up to 255 subcarriers, each separated by 4.3125 kHz and each modulated at 4000 baud. A form of QAM is utilized such that a maximum of 15-bits/Hz bit packing can be achieved. Each of the 255 subcarrier channels is adaptive, where the modulation rate is adjusted to line conditions.

There are two major impairments to ADSL at these higher modulation rates: crosstalk and impulse noise. Thus some modest FEC is provided using Reed–Solomon coding with (optional) interleaving. This allows error bursts up to 500 μs to be corrected but at a latency price of 17 ms, mostly due to the interleaving span. Many line impairments are frequency selective. There are 255 distinct frequency segments used with DMT ADSL. Those frequency segments displaying greater impairments (e.g., crosstalk, impulse noise, group delay) carry lower bit rates extending the baud period. Those segments less affected by these impairments are assigned higher bit rates, decreasing the baud period (Ref. 23).

ADSL can also accommodate many ATM cell rates as well as at least 3 E-1 configurations. The following table (Ref. 23) shows some upstream and

*ISDN is discussed in Chapter 13, Section 13.14.

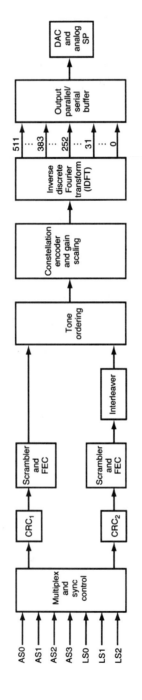

Figure 4.20. Functional diagram of an ATU-C DMT transmitter.

CRC = cyclic redundancy check
SP = signal processing
ASO = Three downstream simplex subchannel
    designators
LSO = Two duplex subchannel designators

258

downstream bit rate combinations:

Downstream Bit Rate (Mbps)	Upstream Bit Rate (kbps)
1.536	64
2.048	160
3.072	176
4.096	384
4.608	576
6.144	640

Figure 4.20 is a functional block diagram of an ATU-C transmitter equipped for DMT operation. The ATU-C can accommodate up to four downstream simplex data channels and up to three duplex data channels. These are synchronized to the 4000-symbol/second ADSL DMT rate and are multiplexed into two separate data buffers, fast and interleaved. Cyclic redundancy check (CRC) and scrambling processing functions are carried out, and FEC is applied to each buffer separately. The data from the interleaved buffer are passed through an interleaving function. The two data streams are now tone-ordered and then combined into a data symbol, which is fed to the constellation encoder. After constellation encoding, the data are modulated to produce an analog signal (DAC—digital-to-analog converter in Figure 4.20) for transmission on the subscriber loop.

REVIEW EXERCISES

1. What is the principal advantage of digital transmission over its analog counterpart? What is a major disadvantage?

2. Give three impairments typical of a digital network that do not exist in its analog counterpart.

3. Test tones used in the analog network are 800 and 1000 Hz. Why is it not a good idea to use these same frequency test tones in the digital network?

4. The 8-bit PCM is intelligible with a BER as poor as 1×10^{-2}. Why then is the BER threshold for a digital network set at 1×10^{-3}?

5. Considering exercise 4, why would we want a BER of 1×10^{-6} at least 80% of the time? What is driving the network operator for such "stringent" BERs?

6. Define *unavailable time*.

7. If the desired BER at the end of a system with ten links is 1×10^{-8} (assuming random errors only), what is the BER we must specify for each link?

8. What are the deleterious effects of excessive jitter and wander?

9. The accumulation of timing jitter is a function of the _____. Express this simple relationship mathematically.

10. Give at least two methods to reduce the accumulation of timing jitter.

11. Why would scrambling/descrambling reduce accumulated jitter?

12. Define quantization distortion.

13. Define a *qdu*.

14. What is the maximum number of qdu permitted on an international connection?

15. What are the three levels of synchronization required for a digital network?

16. How do we know where a frame begins and ends for E-1? For DS1?

17. What device in the digital network is responsible for local timing and synchronization?

18. What is a slip?

19. Carefully explain the cause of slips.

20. If a network is perfectly synchronized, what slip rate can we expect (leaving aside transmission impairments)?

21. What happens to a traffic bit stream in the digital network when there is buffer underflow?

22. What is a good value for timing stability for a high-stability clock?

23. What is plesiochronous operation? Why would this type of operation be used at an international (digital) switch?

24. Name at least three distinct methods of synchronizing a digital network.

25. How is synchronization accomplished for a North American PSTN? What is the means of transport?

26. Describe the typical "two-clock problem." How can it be optimally handled?

27. Why do we wish to avoid the use of the DMW?

28. Define the present digital reference signal.

29. How does a digital pad operate?

30. Why should we avoid using digital pads with channels transporting data?

31. Describe in simple terms the North American digital loss plan.

32. What are the two components of delay in the digital network?

33. Name the eight functions carried out by a SLIC. Associate it with BORSCHT.

34. Define B8ZS.

35. In the case of DS1, define a "clear channel."

36. Define HDSL in three sentences giving the highlights of the system.

37. Name three functions supported by HDSL overhead.

38. For HDSL and similar systems (e.g., ISDN BRI), why not use simple NRZ waveforms rather than 2B1Q?

39. What were the two principal services that designers had in mind for ADSL? What is a third service that is now appearing?

40. Over and above the initial design of ADSL, notably higher bit rates are being achieved. Explain how a wire pair can accommodate bit rates now in excess of 6 Mbps.

41. What are the two basic impairments to ADSL at these higher bit rates?

REFERENCES

1. "Error Performance of an International Digital Connection Forming Part of an Integrated Services Digital Network," CCITT Rec. G.821, Fascicle III.5, IXth Plenary Assembly, Melbourne, 1988.

2. "Error Performance Parameters and Objectives for International, Constant Bit Rate Digital Paths at or Above the Primary Rate," ITU-T Rec. G.826, ITU, Geneva, Nov. 1993.

3. *Notes on the BOC Inter-LATA Networks—1986*, TR-NPL-000275, Bellcore, Piscataway, NJ, 1986.

4. *BOC Notes on the LEC Networks—1990*, SR-TSV-002275, Bellcore, Piscataway, NJ, Mar. 1991.

5. *Transport Systems Generic Requirements (TSGR): Common Requirements*, GR-499-CORE, Issue 1, Bellcore, Piscataway, NJ, Dec. 1995.

6. "Considerations on Timing and Synchronization Issues," CCITT Rec. G.810, Fascicle III.5, IXth Plenary Assembly, Melbourne, 1988.

7. *Transmission Systems for Communications*, 5th ed., Bell Telephone Laboratories, Holmdel, NJ, 1982.

8. "The Control of Jitter and Wander Within Digital Networks Which Are Based on the 1544 kbps Hieararchy," ITU-T Rec. G.824, ITU, Geneva, Mar. 1993.

9. "The Control of Jitter and Wander Within Digital Networks Which Are Based on the 2048 kbps Hierarchy," ITU-T Rec. G.823, ITU, Geneva, Mar. 1993.

10. "Pulse Code Modulation (PCM) of Voice Frequencies," CCITT Rec. G.711, Fascicle III.4, IXth Plenary Assembly, Melbourne, 1988.

11. "Transmission Impairments," ITU-T Rec. G.113, ITU, Geneva, Feb. 1996.

12. "Vocabulary of Digital Transmission and Multiplexing, and Pulse Code Modulation (PCM) Terms," ITU-T Rec. G.701, ITU, Helsinki, Mar. 1993.

13. "Timing Requirements at the Outputs of Primary Reference Clocks Suitable for Plesiochronous Operation of International Digital Links," CCITT Rec. G.811, Fascicle III.5, IXth Plenary Assembly, Melbourne, 1988.

14. "Digital Network Synchronization Plan," GR-436-CORE, Revision 1, June 1996, Bellcore, Piscataway, NJ, 1996.

15. *Synchronous Optical Network (SONET) Transport Systems: Common Generic Criteria*, GR-253-CORE, Issue 1, Bellcore, Piscataway, NJ, Dec. 1994.

16. *BOC Notes on the LEC Networks—1994*, SR-TSV-002275, Bellcore, Piscataway, NJ, Apr. 1994.

17. "One-Way Transmission Time," ITU-T Rec. G.114, Geneva, Feb. 1996.

18. Whitham D. Reeve, *Subscriber Loop Signaling and Transmission Handbook—Digital*, IEEE Press, New York, 1995.

19. *Functional Criteria for Digital Loop Carrier Systems*, TR-NWT-000057, Issue 2, Bellcore, Piscataway, NJ, Jan. 1993.

20. *Generic Requirements for High-Bit-Rate Digital Subscriber Lines*, TA-NWT-001210, Issue 1, Bellcore, Piscataway, NJ, Oct. 1991.

21. *Framework Generic Requirements for Asymmetric Digital Subscriber Lines*, FA-NWT-001307, Issue 1, Bellcore, Piscataway, NJ, Dec. 1992.

22. *Network and Customer Installation Interfaces—Asymmetric Digital Subscriber Line (ADSL) Metallic Interface*, ANSI T1. 413-1995, ANSI, New York, 1995.

23. Kim Maxwell, "Asymmetric Digital Subscriber Line: Interim Technology for the Next Forty Years," *IEEE Commun. Mag.*, Oct. 1996.

24. *Synchronization Interface Standards for Digital Networks*, ANSI T1.101-1994, ANSI, New York, 1994.

25. Supplement 21, "The Use of Quantization Distortion Units in the Planning of International Connections," page 326, Fascicle III.1, Red Book, Geneva, 1985.

5

LINE-OF-SIGHT MICROWAVE

5.1 INTRODUCTION

Line-of-sight (LOS) microwave is a widely employed means of broadband radio transmission in point-to-point service. In Europe, LOS microwave is often referred to as radio relay or radiolinks. A link can be defined as a radio connectivity from a near-end transmitter to a far-end receiver. In this case, a link length is limited by "line-of-sight."

Historically, LOS microwave was the basic transport means for FDM configurations, usually on LOS microwave using frequency modulation. In industrialized nations, the trend is away from analog transmission and all new LOS links now use digital waveforms. Most existing FDM/FM systems have either been taken out of service in favor of fiber optic cable connectivity or have been/are being replaced by digital equivalents. When referring to digital, we mean a digital baseband that modulates an analog RF carrier. The baseband is 8-bit PCM, either of the DS1 hierarchy or the E-1 hierarchy. Some military networks are based on CVSD (Chapter 3).

Among the many applications of LOS microwave are the following:

1. Point-to-point links as a backbone or tails of large networks for common carriers, specialized common carriers, and private and government entities.
2. Point-to-multipoint systems for TV, telephony, data, or various mixes thereof.
3. Transport of TV or other video signals such as community antenna television (CATV) head-end extension, broadcast transport, and studio-to-transmitter links.
4. Specialized digital and digital data networks.
5. Power and pipeline companies for the transport of telemetry, command, and control information.
6. Air traffic control center interconnectivity.

7. Short-haul applications such as linking offices and buildings in congested urban areas; final connectivities for common carrier/specialized common carriers; tails off fiber-optic trunks.

8. Military applications: fixed point-to-point, point-to-multipoint, and transportable point-to-point.

Many of these application implementations are now being tempered by fiber-optic links. Under certain circumstances fiber optics is more cost-effective and provides considerably greater information bandwidth. Fiber, however, is hampered by right-of-way requirements and by cables being severed by construction activities.

5.1.1 Definition

Let us define line-of-sight microwave systems as those that fulfill the following criteria:

1. The signal follows a "straight line" or "LOS" path.*

2. Signal propagation is affected by free-space attenuation, precipitation, and gaseous absorption.

3. Use of frequencies greater than 150 MHz, thereby permitting transmission of more information per RF carrier by the use of wider RF bandwidths.

4. Use of angle modulation (i.e., FM or PM), digital modulation, or spread-spectrum and time-sharing techniques.

Table 5.1 gives frequency assignments for the United States and its territories. A more generalized listing may be found in the ITU *Radio Regulations* (Ref. 1).

A valuable characteristic of LOS transmission is that we can predict the level of a signal arriving at a distant receiver with known accuracy.

5.2 LINK ENGINEERING

Engineering a LOS microwave system involves the following steps:

1. Selection of sites (radio equipment plus tower locations) that are in line-of-sight of each other.

2. Selection of an operational frequency band from those set forth in Table 5.1, considering RF interference environment and legal restraints.

---

*The quotation marks are purposeful. As we will see, in nearly all cases, the ray beam is bent. Line-of-sight must be very carefully defined.

TABLE 5.1 Frequency Bands for LOS Microwave (Point-to-Point) (United States Only)

Frequency Bands (MHz)	Comments[a]
932.5–935.0	Shared with government fixed/private fixed
941.5–944.0	Shared with government fixed/private fixed
2,110–2,130	No TV, 3.5-MHz maximum bandwidth
2,169–2,180	Restrictions
3,700–4,200	Shared with fixed satellite service
5,925–6,425	Shared with fixed satellite, no mobile ES[b]
10,550–10,680	Shared with private fixed microwave service
10,700–11,700	Portion shared with satellite, local TV
13,200–13,250	Shared fixed/mobile stations, other service
17,700–18,220	Shared with fixed satellite service
18,920–19,160	Shared with fixed satellite service
19,260–19,700	Shared with fixed satellite service
21,200–22,000	Shared with government, fixed/mobile other services
22,000–23,600	Shared with government, fixed/mobile other services
27,500–29,500	Shared with fixed-satellite service
31,000–31,300	Shared with auxiliary broadcast, CATV relay, mobile
38,600–40,000	Shared with fixed and mobile and other services

[a]For more definitive restrictions consult Ref. 2.
[b]ES = earth station.
Source: Reference 2, Part 21.701.

3. Development of path profiles to determine radio tower heights. If tower heights exceed a certain economic limit, then step 1 must be repeated, bringing the sites closer together or reconfiguring the path, usually along another route. In making a profile, it must be taken into consideration that microwave energy is

- Attenuated or absorbed by solid objects
- Reflected from flat conductive surfaces such as water and sides of metal buildings
- Diffracted around solid objects
- Refracted or bent by the atmosphere; often the bending is such that the beam may be extended beyond the optical horizon

4. Path calculations. After setting a propagation reliability expressed as a percentage of time, the received signal will be above threshold level. Often this level is the FM improvement threshold of the FM receiver. To this level a margin is set for signal fading under all anticipated climatic conditions.

5. Making a path survey to ensure correctness of steps 1–4. It also provides certain additional planning information vital to the installation project or bid.

6. Establishment of a frequency plan and necessary operational parameters.

7. Equipment configuration to achieve the fade margins set in step 4 most economically.

8. Installation.

9. Beam alignment, equipment lineup, checkout, and acceptance by a customer.

Reference will be made, where applicable, to these steps so that the reader will be exposed to practical radiolink problems.

5.3 PROPAGATION

5.3.1 Free-Space Loss (Spreading Loss)

Consider a signal traveling between a transmitter at A and a receiver at B:

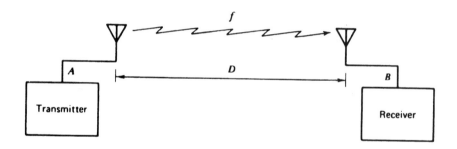

The distance between antennas is D and the frequency of transmission is f. Let D be in kilometers and f in megahertz; then the free-space loss in decibels may be calculated with the following formula:

$$L_{dB} = 32.44 + 20 \log D + 20 \log f \qquad (5.1a)$$

If D is in statute miles, then

$$L_{dB} = 36.58 + 20 \log D + 20 \log f \qquad (5.1b)$$

(all logarithms are to the base 10).

Suppose that the distance separating A and B were 40 km ($D = 40$). What would the free-space path loss be at 6 GHz ($f = 6000$)?

$$L = 32.44 + 20 \log 40 + 20 \log(6 \times 10^3)$$
$$= 32.44 + 20 \times 1.6021 + 20 \times 3.7782$$
$$= 32.44 + 32.042 + 75.564$$
$$= 140.046 \text{ dB}$$

Let us look at it another way. Consider a signal leaving an isotropic antenna.* At one wavelength (1λ) away from the antenna, the free-space attenuation is 22 dB. At two wavelengths (2λ) it is 28 dB; at four wavelengths it is 34 dB. Every time we double the distance, the free-space attenuation (or loss) is 6 dB greater. Likewise, if we halve the distance, the attenuation (or loss) decreases by 6 dB.

5.3.2 Bending of a Radio Wave Ray Beam Above 100 MHz from Straight-Line Propagation

Radio waves traveling through the atmosphere do not follow true straight lines. They are refracted or bent. They may also be diffracted.

The velocity of an electromagnetic wave is a function of the density of the media through which it travels. This is treated by Snell's law, which provides a valuable relationship for an electromagnetic wave passing from one medium to another (i.e., from an air mass with one density to an air mass with another density). It states that the ratio of the sine of the angle of incidence to the sine of the angle of refraction is equal to the ratio of the respective velocities in the media. This is equal to a constant that is the refractive index of the second medium relative to the first medium.

The absolute refractive index of a substance is its index with respect to a vacuum and is practically the same value as its index with respect to air. It is the change in the refractive index that determines the path of an electromagnetic wave through the atmosphere, or how much the wave is bent from a straight line.

If radio waves above 100 MHz traveled a straight line, the engineering of LOS microwave (radiolink) systems would be much easier. We could then accurately predict the height of the towers required at repeater and terminal stations and exactly where the radiating device on the tower should be located (steps 1 and 3, Section 5.2). Essentially what we are dealing with here, then, is a method to determine the height of a microwave radiator (i.e., an antenna or other radiating device) to permit reliable radiolink communication from one location to another.

To determine tower height, we must establish the position and height of obstacles in the path between stations with which we want to communicate by radiolink systems. To each obstacle height, we will add earth bulge. This is the number of feet or meters an obstacle is raised higher in elevation (into the path) owing to earth curvature or earth bulge. The amount of earth bulge in feet at any point in a path may be determined by the formula

$$h = 0.667 d_1 d_2 \qquad (5.2)$$

---

*An isotropic antenna is an ideal antenna with a reference gain of 1 (0 dB). It radiates uniformly in all directions (i.e., is perfectly omnidirectional).

where d_1 = distance from the near end of the link to the point (obstacle location), and d_2 = distance from the far end of the link to the obstacle location.

The equation will become more useful if it is made directly applicable to the problem of ray bending. As the equation is presented, the ray is unbent or a straight line.

Atmospheric refraction may cause the ray beam to be bent toward the earth or away from the earth. If it is bent toward the earth, it is as if we shrank the earth bulge or lowered it from its true location. If the beam is bent away from the earth, it is as if we expanded the earth bulge or raised it up toward the beam above its true value. This lowering or raising is handled mathematically by adding a factor K to the earth bulge equation. It now becomes

$$h_{\text{ft}} = \frac{0.667d_1d_2}{K} \quad (d_1 \text{ and } d_2 \text{ in mi}) \tag{5.3a}$$

$$h_{\text{m}} = \frac{0.078d_1d_2}{K} \quad (d_1 \text{ and } d_2 \text{ in km}) \tag{5.3b}$$

The K-factor can be calculated from the formula

$$K = \frac{\text{effective earth radius}}{\text{true earth radius}} = \frac{r}{r_0} \tag{5.4}$$

The value commonly used for r_0 is 6370 km. We calculate the effective earth radius, r, from the formula

$$r = r_0[1 - 0.04665 \exp(0.005577N_s)]^{-1} \tag{5.5}$$

Note that the use of the notation "exp" means e, the natural number raised to the indicated power, in this case $(0.005577N_s)$.

For example, if the surface refractivity N_s is 301, we substitute that value in equation 5.5 and

$$r = 6370[1 - 0.04665e^{(0.005577 \times 301)}]^{-1}$$

$$= 6370[1 - 0.04665 \times 5.3585]^{-1}$$

$$= 6370(1 - 0.24997)^{-1}$$

$$= 6370 \times 1.333$$

$$= 8493 \text{ km}$$

Note that the K-factor comes out to be 1.33 or $\frac{4}{3}$.

Of course we now need to find a value for surface refractivity N_s. This is the refractivity at the altitude of the LOS microwave site that we selected or the average refractivity of the path. The sea-level refractivity can be obtained for the area from a nearby weather bureau or from a chart (such as in Figure 5.2). To calculate N_s when we are given N_0, the mean sea-level refractivity, we use the following formula:

$$N_s = N_0 \exp(-0.1057 h_s) \tag{5.6}$$

where h_s is the altitude above mean sea level (in kilometers) of the LOS radio site.

For example, the sea-level refractivity is about 312 around the Boston, Massachusetts, area of the United States. A potential site near Boston is 220 m above sea level (i.e., 0.22 km). Using equation 5.6 we have

$$N_s = N_0 X e^{(-0.1057 \times 0.22)}$$

$$= 304.8$$

If the K-factor is greater than 1, the ray beam is bent toward the earth, which essentially allows us to shorten radiolink towers. If K is less than 1, the earth bulge effectively is increased, and the path is shortened or the tower height must be increased. Figure 5.1 gives earth curvature, in feet, for various values of K.

Many texts on radiolinks refer to normal refraction, which is equivalent to a K-factor of $\frac{4}{3}$ or 1.33. It follows a rule of thumb that applies to refraction in

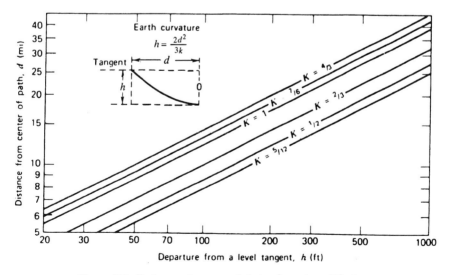

Figure 5.1. Earth curvature or earth bulge for various K-factors.

that a propagated wavefront (or beam) bends toward the region of higher density, that is, toward the region having the higher index of refraction. The preponderance of publications dealing with LOS microwave insist that the K-factor should nearly always be $\frac{4}{3}$. Care should be taken when engineering LOS links that the $\frac{4}{3}$ theory (standard refraction or normal refraction) not be accepted *carte blanche* on many paths likely to be encountered. We recommend going through the exercise of calculating the K-factor for each path based on local weather bureau refractivity information. However, $K = \frac{4}{3}$ may be used for *gross* planning of LOS microwave systems.

Regarding gross planning, there is a short-cut formula that provides distance to the horizon with many provisos:

- Based on $K = \frac{4}{3}$
- Smooth earth (i.e., no obstacles in the path whatsoever)
- Grazing (later we will talk about mandatory clearance criteria)

With antenna height in feet and distance to the horizon in statute miles, the following formula may be used:

$$d = \sqrt{2h} \qquad\qquad (5.7a)$$

If we wish to use meters and kilometers, the following formula is valid:

$$d = 2.9\sqrt{2h'} \qquad\qquad (5.7b)$$

where h' is the antenna height in meters.

For example, an antenna is 200 ft high. What is the distance to the horizon ($K = \frac{4}{3}$)?

$$d = \sqrt{400} = 20 \text{ mi}$$

Keep in mind that formula 5.7 may only be used for gross path estimations.

5.3.3 Fresnel Zone Clearance

In Section 5.3.4 we describe the plotting of a path profile to obtain tower heights. The exercise will involve identifying obstacles in the path with their height above sea level, extending this height for earth curvature, and then adding a further extension for Fresnel zone clearance.

Fresnel was a French scientist and mathematician who did landmark work in optics. The *Fresnel zone clearance* derives from electromagnetic wave theory that a wavefront (which our ray beam is) has expanding properties as it travels through space. These expanding properties are a consequence of reflections and phase transitions as the wave passes over an obstacle. The result is an increase or decrease in the received signal level (RSL). Call it *fading.* In our design of a LOS radiolink we wish to avoid all sources of fading if at all possible.

If we can provide sufficient clearance of the ray beam above an obstacle, this source of fading can essentially be removed. Empirically we have determined that if the clearance exceeds 0.6 of the first Fresnel zone radius, the fading properties are removed for that obstacle. Some link engineers add 10 feet to that 0.6 value as a safety factor.

The first Fresnel zone radius, in feet, may be calculated from the following formula:

$$R_{ft} = 72.1\sqrt{\frac{d_1 d_2}{FD}} \qquad (5.8a)$$

where
F = frequency of signal (GHz)
d_1 = distance from transmitter to path obstacle (statute mi)
d_2 = distance from path obstacle to receiver (statute mi)
$D = d_1 + d_2$, total path length (statute mi)

To determine the first Fresnel zone radius when using metric units, use

$$R_m = 17.3\sqrt{\frac{d_1 d_2}{FD}} \qquad (5.8b)$$

where F = frequency of signal (GHz), d_1, d_2, and D are the same as in equation 5.8a but in kilometers, and R_m is in meters.

As an example, calculate the clearance value (0.6 of the first Fresnel zone) where an obstacle is 5 mi from the transmit antenna; the total path length is 20 mi and the operational frequency is 6 GHz.

$$R_{ft} = 72.1\sqrt{\frac{5 \times 15}{6 \times 20}}$$

$$= 72.1\sqrt{\frac{75}{120}}$$

$$= 72.1 \times 0.7906$$

$$= 57 \text{ ft}$$

Multiply this value by 0.6; then the clearance value is 34 ft. Of course, the more clearance provided, the less chance for fading. However, the more clearance, the higher the towers must be.

5.3.4 Path Profiling—Practical Application

5.3.4.1 Introduction. Path profiles can be done on a PC, where the terrain information for the area of interest is imported on disk or CD-ROM. The method given below is manual. It is tedious and time-consuming. However, we believe that all LOS microwave system designers should be able to do profiles manually. One reason is that PC programs can be fallible. Knowing the manual method allows spot checks of the program and imported topographical information.

5.3.4.2 Plotting Path Profiles Manually. After tentative terminal or repeater sites have been selected, path profiles are plotted on rectangular graph paper. Obstacle information is taken from topographical maps. For the continental United States the best topographical maps available are from the U.S. Geological Survey. They are $7\frac{1}{2}$-min maps of latitude and longitude with a scale $1:24,000$, where 1 in. $= 2000$ ft, and 15-min maps with a scale of $1:62,5000$, where 1 in. $=$ approx. 1 mi. Also, 30-min and 1-deg maps are available, but their scales are not fine enough for path profile application. Many areas of Canada are covered by maps with scales of $1:50,000$ (1 in. $= 0.79$ mi) and $1:63,360$ (1 in. $= 1$ mi).

Profiles are made on available linear graph paper; 10 divisions per inch is suggested. Any convenient combination of vertical and horizontal scales may be used. For paths 30 mi or less in length, 2 mi to the inch plotted on the horizontal scale is suggested. For longer paths, graph paper may be extended by trimming and pasting. For the vertical scale, 100 ft to the inch is satisfactory for fairly flat country, where there is no more than an 800-ft change in altitude along the path; for hilly country, 200 ft to the inch; and for mountainous country, 500 or 1000 ft to the inch are appropriate. One suggestion to preserve a proper relationship between height and distance is that if the distance scale is doubled, the height scale should be quadrupled. All heights should be shown with reference to mean sea level (MSL), except the base (0 elevation on the graph paper) need not be MSL, but the lowest elevation of interest on the profile.

On the topological map draw a straight line connecting the two adjacent radiolink sites. Carefully trace with your eye or thumb down the line from one site to the other, marking all obstacles or obstructions and possible points of reflection, such as bodies of water, marshes, or desert areas, assigning consecutive letters to each obstacle.

Plot the horizontal location of each point on the graph paper. Mark the path midpoint, which is the point of maximum earth bulge and should be marked as an obstacle. Determine the K-factor by one of the following methods:

1. Refer to a sea-level refractivity profile chart such as in Figure 5.2. Select the appropriate N_0 for the area of interest; correct for the altitude of midpath (equation 5.6). Calculate the K-factor from equations 5.4 and 5.5.

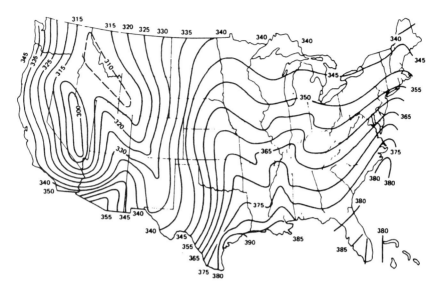

Figure 5.2. Sea-level refractivity index N_0 for the contiguous United States: maximum for worst month, August.

2. Lacking refractivity information, plot the path profile using three K-factors: 1.33, 1.0, and 0.5. A later field survey will help to decide the appropriate K-factor. Table 5.2 will also be a useful guide.

For each obstacle point compute d_1, the distance to one repeater site, and d_2, the distance to the other. Compute the equivalent earth curvature EC for each point with equation 5.3:

$$EC = \frac{0.667 d_1 d_2}{K} \tag{5.9}$$

TABLE 5.2   *K*-Factor Guide[a]

	Propagation Conditions				
	Perfect	Ideal	Average	Difficult	Bad
Weather	Standard atmosphere	No surface layers or fog	Substandard, light fog	Surface layers, ground fog	Fog moisture over water
Typical	Temperate zone, no fog, no ducting, good atmospheric mix day and night	Dry, mountainous, no fog	Flat, temperate, some fog	Coastal water tropical	Coastal
K-factor	1.33	1–1.33	0.66–1.0	0.66–0.5	0.5–0.4

[a]For 99.9–99.99% path reliability.

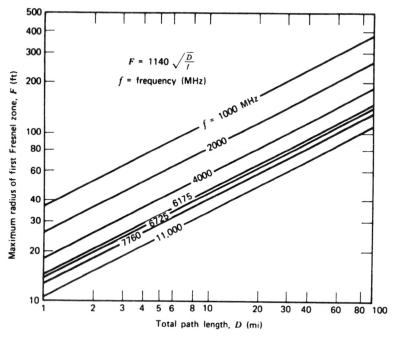

Figure 5.3. Midpath Fresnel zone clearance: 0.6 of this value is used in a path profile. For other than midpath Fresnel zone clearance, the value derived here is used in conjunction with Figure 5.4. (From Ref. 3. Courtesy of GTE Lenkurt, San Carlos, CA.)

Compute the Fresnel zone clearance by equation 5.8 or by using Figures 5.3 and 5.4. Figure 5.3 gives the Fresnel clearance for midpath for various bands of frequencies. For other obstacle points compute the percentage of total path length. For instance, on a 30-mi path, midpoint is 15 mi (50%). However, a point 5 mi from one site is 25 mi from the other, or represents $\frac{5}{30}$ and $\frac{25}{30}$, or $\frac{1}{6}$ and $\frac{5}{6}$. Converted to percentage, $\frac{1}{6} = 16.6\%$ and $\frac{5}{6} = 83\%$. Apply this percentage on the x axis of Figure 5.4. In this case 16.6% or 83% is equivalent to 76% of midpath clearance. If midpath Fresnel clearance were 40 ft, only 30.4 ft of Fresnel clearance would be required at the 5-mi point. Remember, we use 0.6 of R, the first Fresnel zone radius, for the clearance value.

Set up a table at the bottom of the profile chart as follows:

Obstacle Point	Altitude	d_1	d_2	Earth Curvature	Fresnel	Trees and Growth	Total
A.							
B.							
C.							
D.							
E.							

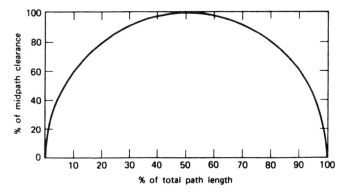

Figure 5.4. Conversion from midpath Fresnel zone clearance for other than midpath obstacle points.

On standard linear graph paper (10 × 10 sq. in.) plot the height above sea level of each obstacle (second column of chart). Remember, d_1 and d_2 are the distances from the transmitter site to a particular obstacle and the distance from the receiver site to that obstacle. From column 5 of the table, add the value of EC (earth curvature); from column 6 add the value of 0.6 of the radius of the first Fresnel zone; from column 7 add a value of the height of trees and growth at the top of the obstacle. If no real value exists, use 40 + 10 ft. Of course, we will neglect this column if there are no trees, such as on bald mountains and ridge tops and the tops of buildings. The all-important path survey will confirm or deny these figures or will permit adjustment.

Minimum tower heights may now be determined by drawing a straight line from transmitter to receiver site through the highest obstacle points. Often these cluster around midpath. Figure 5.5 is a hypothetical path profile exercise.

5.3.5 Reflection Point

From the profile, possible reflection points may be obtained. The objective is to adjust the tower heights such that the reflection point is adjusted to fall on land area where the reflected energy will be broken up and scattered. Bodies of water and other smooth surfaces cause reflections that are undesirable. Figure 5.6 will assist in adjusting the reflection point. It uses a ratio of tower heights h_1/h_2, and the shorter tower height is always h_1. The reflection area lies between a K-factor of grazing ($K = 1$) and a K-factor of infinity. The distance expressed is always from h_1, the shorter tower. By adjusting the ratio h_1/h_2, the reflection point can be moved. The objective is to ensure that the reflection point does not fall on an area of smooth terrain or on

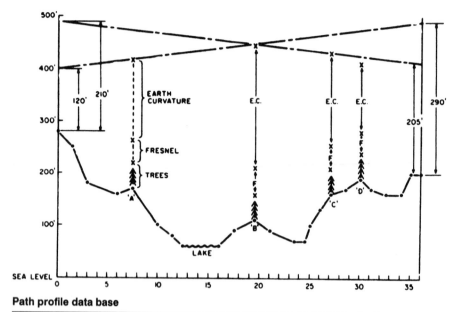

Path profile data base

Obstacle	d_1 (mi)	d_2 (mi)	0.6 Fresnel (ft)	EC (ft)[b]	Vegetation	Total Height Extend. (ft)
A	7.5	28.5	43	152	50	245
B	19.4	16.6	53	233	50	336
C	27.0	9.0	46	176	50	272
D	30.0	6.0	39	130	50	219

[a]Frequency bands; 6 GHz; K factor 0.92; $D = d_1 + d_2 = 36$ mi; vegetation /tree conditions 40 ft plus 10 ft growth.

Figure 5.5. Example path profile.

water, but rather on a land area where the reflected energy will be broken up or scattered (e.g., wooded areas).

For a highly reflective path, space diversity operation may be desirable to minimize the effects of multipath reception (see Section 5.4.3).

5.3.6 Penalty for Not Meeting Obstacle Clearance Criteria

If a LOS microwave path does not meet the obstacle clearance criteria established in Section 5.3.3 (i.e., EC and 0.6 Fresnel zone), a penalty must be invoked of excess attenuation due to diffraction loss.

Diffraction loss depends on the type of terrain and vegetation. For a given path ray clearance, the diffraction loss will vary from a minimum value for a single knife-edge obstruction to a maximum for spherical smooth earth.

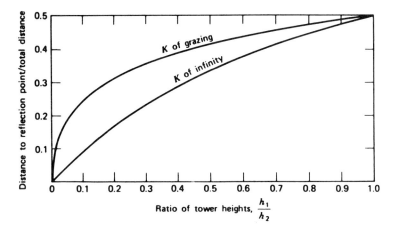

Figure 5.6. Calculation of reflection points.

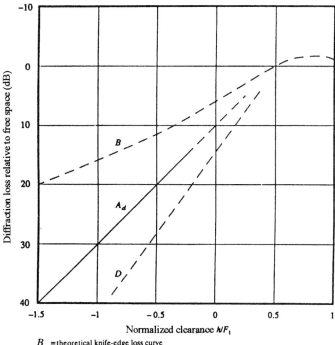

B = theoretical knife-edge loss curve

D = theoretical smooth spherical earth loss curve, at 6.5 GHz and $k_e = 4/3$

A_d = empirical diffraction loss based on equation for intermediate terrain

h = amount by which the radio path clears the earth's surface

F_1 = radius of the first Fresnel zone

Figure 5.7. Diffraction loss for obstructed line-of-sight microwave paths. (From Ref. 4, Figure 1, ITU-R Rec. 530-5.)

Methods of calculating diffraction loss are provided in Chapter 6. These upper and lower limits of diffraction loss are given in Figure 5.7.

The diffraction loss over average terrain can be approximated for losses greater than about 15 dB by the formula

$$A_d = -20h/F_1 + 10 \tag{5.10}$$

where h = height in meters of the most significant path blockage (obstacle) above the path trajectory. The height h is negative if the top of the obstruction of interest is above the virtual line-of-sight. F_1 is the radius of the first Fresnel zone, which is calculated by formula 5.8b. The curve in Figure 5.7 is strictly valid for losses greater than 15 dB. It has been extrapolated up to 6-dB loss as an aid to link designers.

5.3.7 Attenuation Through Vegetation

Figure 5.8 provides data on the excess attenuation of microwave ray beams passing through woodland. The curves represent an approximate average for all types of woodland for frequencies up to about 3 GHz. When the attenuation inside such woodland becomes large (i.e., ≥ 30 dB), the possibility of diffraction or surface modes has to be considered.

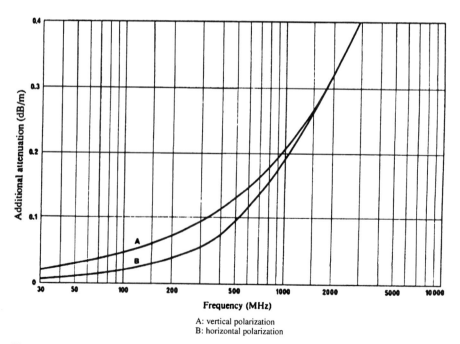

A: vertical polarization
B: horizontal polarization

Figure 5.8. Excess attenuation due to woodland. (From Ref. 5, Figure 1, ITU-R Rec. 833-1.)

5.4 PATH CALCULATIONS (LINK BUDGET)

5.4.1 General

The next step in path engineering is to carry out path calculations. Essentially this entails the determination of equipment parameters and configurations to meet a minimum performance requirement. Such performance requirements are usually related to noise in an equivalent voice channel or they are related to the signal-to-noise ratio, or both. Either way, the requirement is given for a percentage of time. For a single path this may be stated as 99%, 99.9%, or 99.99% of the time said performance exceeds a certain minimum. This is often called propagation reliability. Table 5.3 states reliability percentages.

5.4.2 Basic Path Calculations

5.4.2.1 Introduction. Whereas the path profile provides the required tower heights to achieve the necessary ray beam path clearance, from the path calculation (link budget) we derive parameters for dimensioning the radio equipment. This includes antenna size or aperture, transmitter power output, receiver noise figure, required bandwidth (or an equivalent), whether diversity is required and its type, coding gain for digital systems, performance measured in signal-to-noise ratio, and noise in the derived voice channel for FDM/FM systems or bit error rate (BER) for digital systems. Nearly always the performance is given with a time distribution (e.g., 99.995% of the time).

5.4.2.2 The Far-End Receiver. The far-end receiver is the starting point on a path calculation or link budget. Let us consider a generic receiver. The following question must be answered: What signal level entering the receiver will give the desired link performance?

The receiver will have a noise floor or noise threshold. At this juncture we consider only thermal noise. This was discussed in Section 1.9.6, specifically

TABLE 5.3 Reliability Versus Outage Time

Reliability (%)	Outage Time (%)	Outage Time		
		Per Year	Per Month (avg.)	Per Day (avg.)
0	100	8760 h	720 h	24 h
50	50	4380 h	360 h	12 h
80	20	1752 h	144 h	4.8 h
90	10	876 h	72 h	2.4 h
95	5	438 h	36 h	1.2 h
98	2	175 h	14 h	29 min
99	1	88 h	7 h	14 min
99.9	0.1	8.8 h	43 min	1.4 min
99.99	0.01	53 min	4.3 min	88.6 s
99.999	0.001	5.3 min	26 s	0.86 s

equation 1.22, which is restated here:*

$$P_n = -204\ dBW + NF_{dB} + 10\log(bandwidth_{Hz}) \qquad (1.22)$$

Consider a microwave receiver with a 10-MHz bandwidth and a 3-dB noise figure. Its thermal noise threshold will then be

$$P_n = -204\ dBW + 3\ dB + 10\log(10^7)$$

$$= -204 + 73 = -131\ dBW$$

From this example, we know that the receive signal level (RSL) must be above this noise level or floor by some value in dB. How much above will depend on a number of factors:

- The desired link reliability (time availability)
- The BER or S/N or noise in the voice channel
- The type of modulation
- Fading and fade margin
- The type of coding and other coding factors (for digital systems)
- The interference environment

5.4.2.3 Brief Review of FM Systems

5.4.2.3.1 Background. LOS microwave systems using frequency modulation (FM) were the mainstay of local and long-distance telephony from about 1950 to the 1980s, when the transition to an all-digital network began to occur. Some FM systems using frequency division multiplex (FDM) are still being installed in emerging nations. It is also still being used for a number of studio-to-transmitter links for TV broadcasters. However, even this application is converting to a digital format.

5.4.2.3.2 Required Bandwidth. With broadband FM radio, bandwidth is traded off for signal-to-noise ratio. The bandwidth is defined by Carson's

---

*With modern LOS microwave receivers, noise performance has seen major improvements. Low noise mixers can have a noise figure better than 2 dB and LNAs in the 1 dB range or even less. As a result, we should consider the receiver noise figure more from the view of a receiving system noise figure. The antenna noise contribution should now be taken into account. This methodology is described in Section 7.4.7.1. First calculate antenna noise temperature in K. Convert the receiver noise figure to noise temperature and add that value to the antenna noise temperature. Convert the sum to the noise figure using equation 1.34. This then should be the "noise figure" value used in calculating the receiver thermal noise threshold. The sky noise portion of the antenna noise can be derived from Figure 7.5. In most cases, the antenna elevation angle will be around 0°. In some instances, a microwave antenna may be depressed below 0°, where the sky noise temperature value should be 290 K.

rule, namely,

$$B_{IF} = 2(\text{highest modulation frequency} + \text{peak frequency deviation}) \quad (5.11)$$

The highest modulation frequency is the highest frequency of the FDM configuration used to modulate the FM transmitter (see Chapter 3). For example, if the configuration was a single supergroup (i.e., 312–552 kHz), the highest modulating frequency would be 552 kHz.

Deviation is a measure of instantaneous frequency variation and is a direct function of the level of the modulating signal. The higher the level of modulating signal, the more deviation we can expect. There is a minimum amount of deviation required to achieve FM improvement. This is a point where we are assured of achieving the noise advantage we spoke of earlier. If a transmitter is overdeviated, more bandwidth is required, resulting in increased thermal noise, and we invite intermodulation distortion. Both are undesirable. Equation 1.22 (repeated on previous page) has the term $10 \log B_{IF}$, which simply tells us that if we double the IF bandwidth of the receiver, the thermal noise will increase 3 dB.

Consider an example of applying Carson's rule. A certain FM transmitter has the peak deviation set for 8 MHz and the highest modulating frequency is 5 MHz. What would be the bandwidth of the IF?

$$B_{IF} = (2 \times 8 + 2 \times 5)$$

$$= 26 \text{ MHz}$$

The next step is the calculation of peak deviation, the second term in the Carson's rule formula (equation 5.11). First, however, we must define preemphasis/deemphasis as employed in FM radiolinks.

PREEMPHASIS/DEEMPHASIS FOR FM RADIO SYSTEMS. The output characteristics of a FM receiver with system preemphasis/deemphasis, with an input signal of a modulation of uniform amplitude, are such that it has linearly increasing noise amplitude with increasing baseband frequency. This is shown in Figure 5.9. Note the ramplike or triangular noise of the higher-frequency demodulated baseband components. The result is decreasing signal-to-noise ratio with increasing baseband frequency. The desired receiver output is a constant signal-to-noise ratio across the entire baseband. Preemphasis at the FM transmitter and deemphasis at the FM receiver accomplish this end.

Preemphasis is accomplished by increasing the peak deviation during the FM modulation process for the higher baseband frequencies. This increase of peak frequency deviation is done in accordance with a curve designed to effect compensation for the ramplike noise at the FM receiver output. CCIR, in Rec. 275-3 (Ref. 6), has recommended standardization on the curve shown in Figure 5.10 for multichannel telephony.

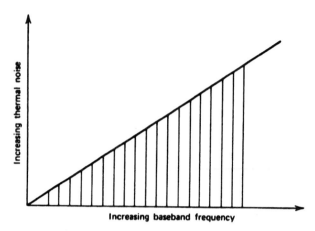

Figure 5.9. Sketch of increasing thermal noise from the output of a FM receiver in a system with preemphasis/deemphasis.

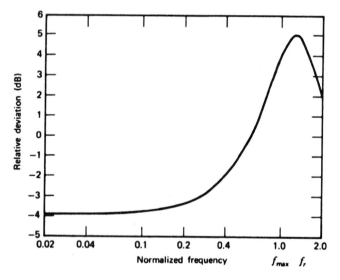

Figure 5.10. Preemphasis characteristic for multichannel telephony. (From Ref. 6, Figure 1, CCIR Rec. 275-3.)

5.4.2.3.3 Calculation of Peak Deviation for FM Transmitter. As discussed earlier, to calculate the bandwidth, using Carson's rule (equation 5.11), we need to know the peak deviation of the transmitted FM signal.

For systems using SSBSC FDM basebands, peak deviation D may be calculated as follows: for $N = 240$ or more,

$$D = 4.47d\left(\log^{-1}\frac{-15 + 10\log N}{20}\right) \qquad (5.12a)$$

TABLE 5.4   Frequency Deviation Without Preemphasis

Maximum Number of Channels	rms Deviation per Channel (kHz)
12	35
24	35
60	50, 100, 200
120	50, 100, 200
300	200
600	200
960	200
1260	140, 200
1800	140
2700	140

Source: Reference 7, CCIR Rec. 404-2.

or for N between 12 and 240,

$$D = 4.47d\left(\log^{-1}\frac{-1 + 4\log N}{20}\right) \qquad (5.12b)$$

where D = peak deviation (kHz)
  d = per-channel rms test tone deviation (kHz)
  N = number of SSBSC voice channels in the system

Table 5.4 may be used to obtain a value for d, the per-channel rms test tone deviation in kHz.

5.4.2.3.4 FM Improvement Threshold. At the FM receiver, FM improvement will be achieved if the carrier-to-noise ratio is at least 10 dB. Just about at this point, we will see the signal-to-noise ratio in a derived FDM voice channel jump some 20 dB. Of course, there remains the proviso that the transmitter has the right deviation.

First, we define RSL. It is the receive level at the input to a receiver's first active stage. On many, if not most, LOS microwave receivers, the first active stage is the mixer. In other cases, it would be the input of the low-noise amplifier (LNA). RSL is commonly measured in dBm or dBW. It could just as well be in milliwatts or watts. RSL is synonymous with carrier level (C). So we can just as well write C/N as RSL/N, where N is the thermal noise threshold, (equation 1.22).

The following is an equation for the FM improvement threshold (in dBW):

FM improvement threshold $_{\text{dBW}}$

$$= -204 \text{ dBW} + NF_{\text{dB}} + 10\log(BW_{\text{Hz}}) + 10 \text{ dB} \qquad (5.13)$$

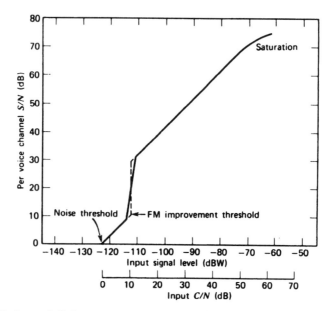

Figure 5.11. Input C/N in dB versus output S/N (per voice channel) for a typical FM microwave receiver.

Equation 5.13 is the same as equation 1.22 with $+10$ dB appended. It just states that, for FM improvement, the carrier level must be at least 10 dB above the noise level at the receiver. Figure 5.11 illustrates a typical curve for C/N versus S/N in a derived FDM voice channel. From a point where the C/N ratio is zero to 10 dB, the derived S/N increases about 1 dB for 1-dB increase in C/N. At about $C/N = 10$ dB, the S/N jumps to 30 dB. For every 1-dB increase in C/N from a $C/N = 10$ dB, the S/N again increases 1 dB. At $C/N = 12$ dB, for example, the S/N is about 32 dB, and so on. The curve is illustrative only.

This FM phenomenon is often called the *FM capture effect.*

5.4.2.3.5 Basic Path Calculations for FM Link. The approach used here will be the same for the remainder of this book when dealing with radio/wireless systems. It is a step-by-step process, which is usually summarized in a table. First we list the steps. Then we describe each step, defining new terms where appropriate. Several example exercises will be worked out and a typical table will be provided. It is assumed that a path profile has been carried out and all obstacle clearance criteria have been met. If the clearance criteria have not been met, the results of the path calculations in the following paragraphs are invalid.

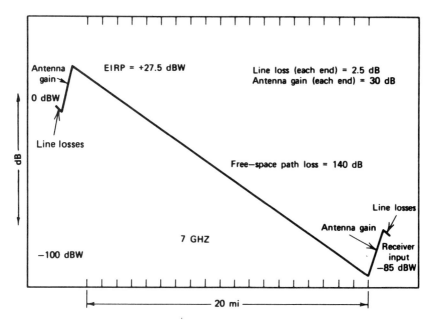

Figure 5.12. LOS microwave link: typical gains and losses. Transmitter output is 1 watt or 0 dBW.

Steps Required in a Link Budget or Path Calculation

Note: All values are in dB or derived dB units.

1. Calculate the effective isotropic* radiated power (EIRP) of the transmitter system.
2. From that value, subtract the free-space loss (equation 5.1). The resulting value is the isotropic receive level (IRL).
3. Add the receive antenna gain and subtract the transmission line losses. The result is the unfaded RSL. Fading and fade margin will be covered later.

A typical illustration of LOS microwave link gains and losses is shown in Figure 5.12.

CALCULATION OF EIRP. The EIRP calculation expresses the power at the peak of the ray beam emanating from the microwave transmit antenna. It is simply the algebraic sum of the power output (at the flange) of the microwave

---

*An isotropic or isotropic antenna is an antenna that radiates uniformly in all directions. Then, by definition, the antenna has a 0-dB gain. Antenna gain is commonly expressed in dBi, where the isotropic antenna is the reference.

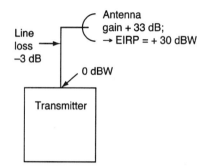

Figure 5.13. The concept of EIRP with illustrative values.

transmitter, less the line losses, plus the antenna gain in the desired direction. Expressed as an equation,

$$EIRP_{dBW} = \text{transmit power output}_{dBW} - \text{line losses} + \text{antenna gain}_{dBi}$$

$$(5.14)$$

Figure 5.13 illustrates the concept of EIRP.

Example. A transmitter has an output of 100 mW, the line losses are 1.3 dB, and the antenna gain is +3 dBi. What is the EIRP?

First convert the 100 mW to dBW. The value is −10 dBW. Add to this the antenna gain and subtract the line losses. EIRP = −8.3 dBW.

CALCULATION OF ISOTROPIC RECEIVE LEVEL (IRL). We place an imaginary isotropic receive antenna in exactly the same location where we would have the microwave receive antenna. The IRL value is simply the EIRP less the free-space loss (FSL). So:

$$IRL_{dBW} = EIRP_{dBW} - FSL_{dB} \qquad (5.15)$$

Example. A LOS microwave link 23 mi long operates at 7.13 GHz. The EIRP is +33 dBW. What is the IRL?

First calculate the FSL using formula 5.1a.

$$FSL_{dB} = 36.58 + 20\log(23) + 20\log(7130)$$

$$= 36.58 + 27.23 + 77.06$$

$$= 140.87 \text{ dB}$$

$$IRL_{dBW} = +33 \text{ dBW} - 140.87 \text{ dB}$$

$$= -107.87 \text{ dBW}$$

CALCULATION OF THE RECEIVE SIGNAL LEVEL (RSL). To the IRL value, add the receive antenna gain and subtract the receive antenna line losses.

$$\mathrm{RSL}_{\mathrm{dBW}} = \mathrm{IRL}_{\mathrm{dBW}} + \text{rec. ant. gain}_{\mathrm{dBi}} - \text{rec. line losses}_{\mathrm{dB}} \quad (5.16)$$

Example. The IRL for a particular link is -107.87 dBW, the receive antenna gain is $+30$ dBi, and the line losses are 2.7 dB. What is the RSL (dBW)?

$$\mathrm{RSL}_{\mathrm{dBW}} = -107.87 \text{ dBW} + 30 \text{ dBi} - 2.7 \text{ dB}$$
$$= -80.57 \text{ dBW}$$

CALCULATION OF C/N. We can rewrite C/N as follows, if we use logarithmic quantities, such as dB.

$$C/N_{\mathrm{dB}} = C_{\mathrm{dBW}} - N_{\mathrm{dBW}} \quad (5.17a)$$

As we have defined $C = \mathrm{RSL}$. Then

$$C/N_{\mathrm{dB}} = \mathrm{RSL}_{\mathrm{dBW}} - N_{\mathrm{dBW}} \quad (5.17b)$$

N is calculated using equation 1.22. RSL is calculated using equation 5.16.

Example. A LOS microwave link has a RSL of -80.57 dBW. The receiver has an 18-MHz bandwidth, and its noise figure is 4 dB. What is the C/N?

Calculate the thermal noise threshold of the receiver (dBW) using formula 1.22:

$$P_{\mathrm{n}} = -204 \text{ dBW} + 4 \text{ dB} + 10\log(18 \times 10^{6})$$
$$= -127.45 \text{ dBW}$$

Calculate C/N using formula 5.17.

$$C/N = -87.07 \text{ dBW} - (-127.45 \text{ dBW})$$
$$= 40.38 \text{ dB}$$

If this were a FM system, subtract the 10 dB required for FM improvement threshold, and we would have 30.38-dB margin.

5.4.2.3.6 Link Budget or Path Calculations in Tabular Form. The most straightforward and shortest way of carrying out a link budget is by using a table. The table should include as much pertinent information as possible. The end point is link performance and margin. The table should also include margin for fading and what it was based upon. The sample table (Table 5.5) has been simplified to illustrate methodology. For instance, we use a 40-dB thermal fade margin, which will provide a 99.99% path reliability, which we

TABLE 5.5 Example Link Budget

Number	Item	Value	Comments
1	Path identifier		Latitude and longitude of each site, tower heights, antenna azimuth, elevation for each site
2	Transmitter power output	0 dBW	
3	Transmitter line losses	3.2 dB	This includes all such losses, e.g., transitions
4	Antenna gain	34.5 dBi	55% antenna efficiency
5	EIRP	+31.3 dBW	Equation 5.14
6	Free-space loss	137.23 dB	Equation 5.1
7	Isotropic receive level (IRL)	−105.93 dBW	Equation 5.15
8	Receive antenna gain	34.5 dBi	
9	Receive line losses	3.2 dB	
10	Receive signal level (RSL)	−74.63 dBW	Add items Nos. 7 and 8, subtract No. 9. Equation 5.16
11	Receive noise threshold	−127.45 dBW	Equation 1.22. NF = 4 dB, BW = 18 MHz
12	C/N	52.82 dB	Equation 5.17
13	FM improvement threshold	−117.45 dBW	Add +10 dB to item No. 11. Equation 5.13
14	Overall margin	42.82 dB	Item No. 10 − item No. 13
15	Required fade margin	40 dB	
16	Remaining margin	2.82 dB	Item No. 14 − item No. 15

prefer to call *time availability*. That value is based on one method,* albeit conservative, to calculate fade margin.

Scenario. A LOS microwave path is 28 mi long and the operating frequency is 3.85 GHz. The transmitter is frequency modulated based on CCIR deviation criterion. The bandwidth is 18 MHz. Obstacle clearance criteria have been met. The parabolic antenna gain is 34.5 dB (at each end). The receiver noise figure is 4 dB. Transmitter output power at the waveguide flange is 1.0 watt, and the transmission line losses are 3.2 dB at each end. Table 5.5 is prepared based on this information.

A link budget, such as this, lends itself to computer application with prompts.

For each link, the path profile and link budget should be kept in archive. Both should be sufficiently clear so that, years hence, the data can be analyzed with ease.

*Based on Rayleigh fading criterion.

The technique used here will be repeated for digital LOS microwave links, with some small changes, with troposcatter/diffraction links, with satellite communications, with HF links, with PCS/cellular with other changes, and, to a certain extent, with fiber-optic links.

5.4.3 The Mechanisms of Fading—An Introductory Discussion

5.4.3.1 General. Up to this point we have considered that the RSL remains constant. In other words, the only path attenuation is free-space loss, assuming path clearance criteria have been fully met. On short paths, < 2 mi, for frequencies below 10 GHz, this will hold true. As a path gets longer, a variation in RSL will be noted. This is fading. Depending on *conditions* and path length, fade depths up to 30 dB or more may be expected. "Conditions" can cover many items such as terrain roughness, climate, reflectivity, and rainfall for frequencies above 10 GHz and so on.

The following sections cover two general types of fading: multipath and power.

5.4.3.2 Multipath Fading. This type of fading is by far the most commonly encountered. It stems from the interference between a direct wave (the peak of the ray beam) and another wave, usually a reflected wave. The reflection may be from the ground or from atmospheric sheets or layers. Direct path interference may also occur. It may be caused by surface layers of strong refractive index gradients or horizontally distributed changes in the refractive index.

Multipath fading may display fades in excess of 30 dB for periods of seconds or minutes. Typically this form of fading will be observed during quiet, windless, and foggy nights, when temperature inversion near the ground occurs and there is not enough turbulence to mix the air. Thus stratified layers, either elevated or ground based, are formed. Two-path propagation may also be due to specular reflections from a body of water, salt beds, or flat desert between transmitting and receiving antennas. Deep fading of this latter type usually occurs in daytime on over-water paths or other such paths with high ground reflection. Vegetation or other "roughness" found on most radiolink paths breaks up the reflected components, rendering them rather harmless. Multipath fading at its worst is independent of obstruction clearance, and its extreme condition approaches a Rayleigh distribution.

5.4.3.3 Power Fading. Dougherty (Ref. 9) defines power fading as follows:

partial isolation of the transmitting and receiving antennas because of

- Intrusion of the earth's surface or atmospheric layers into the propagation path (earth bulge or diffraction fading)
- Antenna decoupling due to variation of the refractive index gradient

- Partial reflection from elevated layers interpositioned between terminal antenna elevations
- "Ducting" formations containing only one of the terminal antennas
- Precipitation along the propagation path

Power fading is characterized by marked decreases in the free-space signal level for extended time periods. Diffraction may persist for several hours with fade depths of 20–30 dB.

5.4.4 Methods of Estimating Fade Margin for Analog Microwave Links

Fading is a random increase in path loss during abnormal propagation conditions. During such conditions path loss may increase 10, 20, 30 dB, or more for short periods. The objective of this subsection is to assist in setting a margin or overbuild in system design to minimize the effects of fading. Remember, margin costs money; thus we want to set the minimum required and no more.

The factors involved in fading phenomena are many and complex. They have been discussed in some detail in Section 5.4.3. The principal type of fading below 10 GHz is multipath fading. CCIR Rep. 784-3 (Ref. 8) states that

> multipath effects due to atmosphere increase slowly with frequency, but much more rapidly with path length (the deep fading probability follows an approximate law $f \cdot d^{3.5}$, where f is the carrier frequency and d is distance).

Determining the fade margin without resorting to live path testing is not an easy matter for the radio system engineer. One approach is simply to assume what is often considered the worst fading condition on a single radiolink hop. This is the familiar Rayleigh fading, which can be summarized as follows:

Single-Hop Propagation Reliability (%)	Required Fade Margin (dB)
90	8
99	18
99.9	28
99.99	38
99.999	48

For other propagation reliability values, simple interpolation can be used. For instance, for a hop where a 99.95% reliability is desired, the fade margin required for Rayleigh would be 33 dB.

This expresses worst-case fading and does not take into account the variables mentioned in the CCIR report, namely, hop length (distance) and frequency. Adding margin costs money, and it would be desirable to reduce the margin yet maintain hop reliability.

Further refinements have been made in the methodology of estimating fade margins. Experience in the design of many radiolink systems has shown that the incidence of multipath fading varies not only as a function of path length and frequency but also as a function of climate and terrain conditions. It has been found that hops over dry, windy, mountainous areas are the most favorable, displaying a low incidence of fading. Worst fading conditions usually occur in coastal areas that are hot and humid, and inland temperate regions are somewhere in between the extremes. Flat terrain along a radiolink path tends to increase the probability of fading, while irregular hilly terrain, especially with vegetation, tends toward lower incidence of fading (e.g., depth and frequency of events).

W. T. Barnett and A. Vigants of Bell Telephone Laboratories (Refs. 3, 10) have developed an empirical method to further refine the estimation of fade margins. The percentage we have been using really expresses an availability A of a radiolink path or hop, sometimes called *time availability*. Subtracting the percentage from 1.000 gives the path unavailability u. As a percentage, $A = 100(1 - u)$. Following Barnett's argument (in general), let U_{ndp} be the nondiversity annual outage probability and r the fade occurrence factor:

$$r = \frac{\text{actual fade probability}}{\text{Rayleigh fade probability}} \tag{5.18}$$

If F is the fade margin in decibels, then

$$r = \frac{\text{actual fade probability}}{10^{-F/10}} \tag{5.19}$$

For the worst month,

$$r_{\text{m}} = a \times 10^{-5} \left(\frac{f}{4} \right) D^3 \tag{5.20}$$

where D = path length (statute mi)
$\quad f$ = frequency (GHz)
$\quad F$ = fade margin (dB)
$\quad a$ = 4 for very smooth terrain, over water, flat desert
$\qquad$ = 1 for average terrain with some roughness
$\qquad$ = 0.25 for mountains, or very rough or very dry terrain

Over a year,

$$r_{yr} = br_m \tag{5.21}$$

where b = 0.5 for hot, humid coastal areas
   = 0.25 for normal, interior temperate, or subarctic areas
   = 0.125 for mountainous or very dry but nonreflective areas

$$U_{ndp} = r_{yr}(10^{-F/10}) = br_m(10^{-F/10})$$

$$= 2.5abfD^3(10^{-F/10})(10^{-6}) \tag{5.22}$$

Example. Given a 25-mi path with average terrain but with some roughness, an inland temperate climate, and a link operating at a frequency of 6.7 GHz with a desired propagation reliability of 99.95%, what fade margin should be assigned to the link?

$$U_{ndp} = 1 - \text{percentage}$$

$$= 1 - 0.9995$$

$$= 0.0005$$

then

$$0.0005 = 2.5abfD^3(10^{-F/10})(10^{-6})$$

$$= 2.5(1)(0.25)(6.7)25^3(10^{-F/10})(10^{-6})$$

$$F = 31.2 \text{ dB}$$

It is good design practice to add a miscellaneous loss margin to the derived fade margin by whatever method the derivation has been made. This additional margin is required to account for minor antenna misalignment and system gain degradation (e.g., waveguide corrosion, and transmitter output and receiver NF degradations due to aging). The Defense Communications Agency of the U.S. Department of Defense recommends 6 dB (Ref. 11) for this additional margin.

We introduce *diversity operation* in the next section. Diversity can reduce the fade margin requirements previously established; it tends to reduce the depth of fades. Figure 5.14 shows how fade margins may be reduced with diversity. For instance, assume a Rayleigh distribution (random fading) with a 35-dB fade margin requirement without diversity. Then using frequency diversity with a frequency separation of 2%, only a 23.5-dB fade margin would be required to maintain the same propagation reliability (time availability, the percentage of time the level is exceeded).

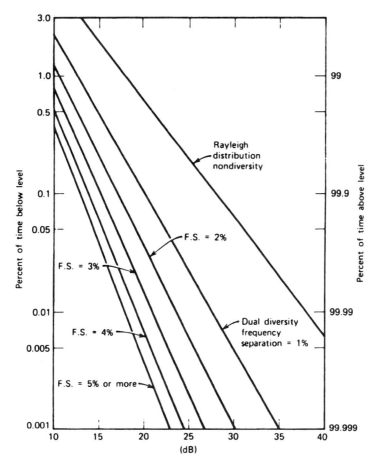

Figure 5.14. Approximate interference fading distribution for a nondiversity system with Rayleigh fading versus frequency diversity systems for various percentages of frequency separation (F.S.).

5.4.5 Diversity Operation

Diversity operation is widely used on many types of radio systems. These systems include tropospheric scatter/diffraction links, cellular radio/PCS, and HF radiolinks. A special form of diversity is used on satellite communication links operating above 10 GHz to mitigate the effects of rainfall fading. Space diversity is becoming mandatory on longer (i.e., > 5 mi) digital microwave links. Diversity operation is attractive for the following reasons:

- It tends to reduce depth of fades on combined output.
- It provides improved equipment reliability (if one diversity path is lost due to equipment failure, other path[s] remain in operation).

- Depending on the type of combiner in use, the combined output signal-to-noise ratio is improved over that of any single signal path.

Diversity reception is based on the fact that radio signals arriving at a point of reception over separate paths may have noncorrelated signal levels. More simply, at one instant of time a signal on one path may be in a condition of fade while the identical signal on another path may not.

First one must consider what are separate paths and how "separate" they must be (CCIR Rep. 376-6; Ref. 12). The separation may be in:

- Frequency
- Space (including angle of arrival and polarization)
- Time (a time delay of two identical signals on parallel paths)
- Path (signals arrive on geographically separate paths)

The most common forms of diversity in radiolink systems are those of frequency and space. A frequency diversity system utilizes the phenomenon that the period of fading differs for carrier frequencies separated by 2–5%. Such a system employs two transmitters and two receivers, with each pair tuned to a different frequency (usually 2–3% separation, since the frequency band allocations are limited). If the fading period at one frequency extends for a period of time, the same signal on the other frequency will be received at a higher level, with the resultant improvement in propagation reliability.

As far as equipment reliability is concerned, frequency diversity provides a separate path, complete and independent, and consequently one whole order of reliability has been added. Besides the expense of the additional equipment, the use of additional frequencies without carrying additional traffic is a severe disadvantage to the employment of frequency diversity, especially when frequency assignments are even harder to get in highly developed areas where the demand for frequencies is greatest.

One of the main attractions of space diversity is that no additional frequency assignment is required. In a space diversity system, if two or more antennas are spaced many wavelengths apart (in the vertical plane), it has been observed that multipath fading will not occur simultaneously at both antennas. Sufficient output is almost always available from one of the antennas to provide a useful signal to the receiver diversity system. The use of two antennas at different heights provides a means of compensating, to a certain degree, for changes in electrical path differences between direct and reflected rays by favoring the stronger signal in the diversity combiner. The antenna separation required for optimum operation of a space diversity system may be calculated using the following formula:

$$S = \frac{3\lambda R}{L} \tag{5.23}$$

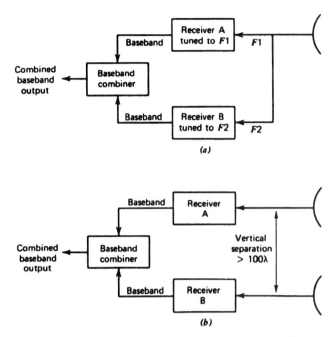

Figure 5.15. (a) Simplified block diagram of a frequency diversity configuration, $F_1 < F_2 + 0.02F_2$. (b) Simplified block diagram of a space diversity configuration. Note the physical separation of antennas.

where S = separation (m)
   R = effective earth radius (m)
   λ = wavelength (m)
   L = path length (m)

However, any spacing between 100λ and 200λ is usually found to be satisfactory. The goal in space diversity is to make the separation of diversity antennas such that the reflected wave travels a half-wavelength further than the normal path. In addition, the CCIR states that for acceptable space diversity operation the spacing should be such that the value of the space correlation coefficient does not exceed 0.6 (CCIR Rep. 376-6; Ref. 12).

Figure 5.15a is a simplified block diagram of a radiolink frequency diversity system and Figure 5.15b is that of a space diversity system.

5.4.5.1 Diversity Combiners

5.4.5.1.1 General. A diversity combiner combines signals from two or more diversity paths. Combining is traditionally broken down into two major categories:

- Predetection
- Postdetection

The classification is made according to where in the reception process the combining takes place. Predetection combining takes place in the IF. However, at least one system performs combining at RF. With the second type, combining is carried out at baseband (i.e., after detection).

For predetection combining, phase control circuitry is required unless some form of path selection is used.

Figure 5.16 shows simplified functional block diagrams of radiolink receiving systems using predetection and postdetection combiners.

5.4.5.1.2 Types of Combiners. Three types of combiners find more common application in radiolink diversity systems. These are

- Selection combiner
- Equal gain combiner
- Maximal ratio combiner (ratio squared)

The selection combiner uses but one receiver at a time. The output signal-to-noise ratio is equal to the input signal-to-noise ratio from the receiver selected for use at the time.

The equal gain combiner simply adds the diversity receiver outputs, and the output signal-to-noise ratio of the combiner is

$$\frac{S_0}{N_0} = \frac{S_1 + S_2}{2N} \tag{5.24}$$

where N = receiver noise.

The maximal ratio combiner uses a relative gain change between the output signals in use. For example, let us assume that the stronger signal has unity output and the weaker signal has an output proportional to gain G. It then can be shown that $G = S_2/S_1$ such that the signal gain is adjusted to be proportional to the ratio of the input signals. We then have

$$\left(\frac{S_0}{N_0}\right)^2 = \left(\frac{S_1}{N}\right)^2 + \left(\frac{S_2}{N}\right)^2 \tag{5.25}$$

where N = receiver noise.

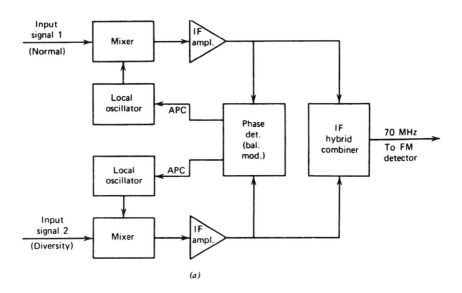

(a)

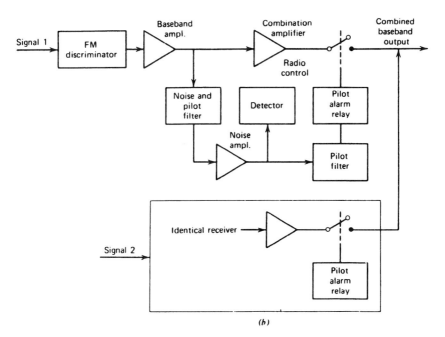

(b)

Figure 5.16. (*a*) Predetection combiner. APC = automatic phase control. (*b*) Postdetection combiner (maximum ratio squared).

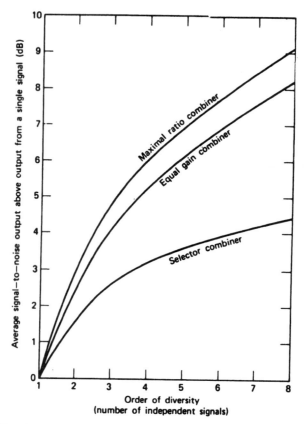

Figure 5.17. Signal-to-noise ratio improvement in a diversity system for various orders of diversity (Ref. 13).

For the signal-to-noise ratio equation for the latter two combiners, we assume the following:

- All receivers have equal gain.
- Signals add linearly; noise adds on a rms basis.
- Noise is random.
- All receivers have equal noise outputs N.
- The output (from the combiner) signal-to-noise ratio S_0/N_0 is a constant.

Figure 5.17 shows graphically a comparison of the three types of combiners. We see the gain that we can expect in output signal-to-noise ratio for various orders of diversity for the three types of combiners discussed, assuming a Rayleigh distribution. The order of diversity refers to the number of independent diversity paths. If we were to use space *or* frequency diversity alone, we would have two orders of diversity. If we used space *and* frequency diversity, we would then have four orders of diversity.

The reader should bear in mind that the efficiency of diversity depends on the correlation of fading of the independent diversity paths. If the correlation coefficient is zero (i.e., there is no relationship in fading for one path to another), we can expect maximum diversity enhancement. The efficiency of a diversity systems drops by half with a correlation coefficient of 0.8, and nearly full efficiency can be expected with a correlation coefficient of 0.3.

5.4.6   Repeaters for Analog LOS Microwave Systems

Repeaters extend LOS microwave systems for several additional miles or all the way across a continent. At a minimum, a repeater receives a signal at frequency F_1, amplifies it, and translates the frequency to F_2. The resulting signal is then amplified and radiated.

There are three types of analog microwave radio repeaters: baseband, IF repeater, and RF repeater. These are shown in Figure 5.18.

Repeater sites that have drop and insert requirements use baseband repeaters. A baseband repeater fully demodulates the incoming RF signal to baseband. In its simplest configuration, the demodulated baseband is used to modulate the transmitter used in the link section. If a switch or concentrator is associated with a repeater, the baseband may be partially or entirely demultiplexed and remultiplexed so that the switch may have access to individual voice channels. A typical baseband repeater is shown in Figure 5.18*a*.

Two other types of repeaters are also available: the IF heterodyne repeater (Figure 5.18*b*) and the RF heterodyne repeater (Figure 5.18*c*). The IF repeater is attractive for use on long backbone systems where noise and/or differential phase and gain should be minimized.

Generally, a system with fewer modulation–demodulation stages or steps is less noisy. The IF repeater eliminates two modulation steps. It simply translates the incoming signal to IF with the appropriate local oscillator and a mixer, amplifies the derived IF, and then up-converts it to a new RF frequency. The up-converted frequency may then be amplified by a traveling wave tube (TWT) or solid-state amplifier.

With a RF heterodyne repeater (Figure 5.18*c*), amplification is carried out directly at RF frequencies. The incoming signal is amplified, up- or down-converted, and amplified again, usually by a TWT, and then reradiated. RF repeaters are troublesome in their design in such things as sufficient selectivity, limiting and automatic gain control, and methods to correct group delay. "Bent pipe" satellite transponders are RF repeaters. Analog LOS microwave favors IF repeaters. Digital LOS microwave necessarily must use baseband repeaters to achieve regeneration of the digital signal.

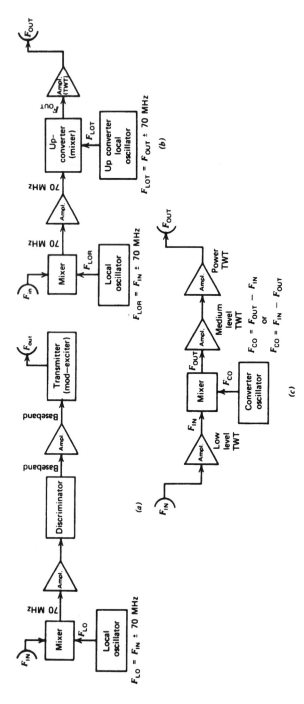

Figure 5.18. Radiolink repeaters. (a) Baseband repeater. (b) IF heterodyne repeater. (c) RF heterodyne repeater. F_{IN} = input frequency to receiver; F_{OUT} = output frequency of the transmitter; F_{CO} = output frequency of the converter local oscillator; F_{LOR} = frequency of the receiver local oscillator; F_{LOT} = frequency of the transmitter local oscillator; F_{LO} = frequency of the local oscillator; TWT = traveling wave tube.

5.4.7 Wideband FM Radio Noise Considerations

The principal concern of the transmission engineer when designing an analog LOS microwave system is noise accumulation. This is the primary factor that separates analog and digital systems: noise accumulation. Noise does not accumulate on digital systems; bit errors do accumulate.

5.4.7.1 *Noise Planning—Requirements.* For noise planning, CCIR established the 2500-km hypothetical reference circuit, where, at the end-point receiver, there shall be no more than 10,000 pWp. This value is broken down into two component parts: the underlying FDM multiplex portion and the radio portion. Thus 2500 pWp is allocated to the FDM part and 7500 pWp to the radio portion. If we divide 10,000 pWp by 2500, we can roughly say that there should not be more than 4 pWp per kilometer of noise accumulation; of this 3 pWp/km is allocated to the radio portion.

CCIR in Rec. 393-4 (Ref. 14) states:

1. The noise power at a point of zero relative level in any telephone channel on a 2500 km hypothetical reference circuit for FDM radio-relay systems should not exceed the values given below, which have been chosen to take account of adverse propagation conditions:

    1.1 7500 pW0p, psophometrically weighted* one-minute mean power for more than 20% of any month;

    1.2 47,500 pW0p psophometrically weighted one-minute power for more than 0.1% of any month;

    1.3 1,000,000 pW0, unweighted (with integrating time of 5 ms) for more than 0.01% of any month.

If we were to connect a psophometer to the end of a 2500-km (reference) circuit made up of homogeneous radio-relay sections, in a derived voice channel we should read values no greater than those shown above plus 2500 pW0p mean noise power due to the FDM equipment contribution. For the first entry above, we would add the 2500 pW0p to the 7500 pW0p value for a total of 10,000 pW0p, which would not be exceeded for more than 20% of any month; and so on for the other two entries.

It should be appreciated that these noise power values are specified statistically. In a fading environment, C/N at a receiver varies with time. As the C/N ratio decreases, noise in the derived voice channel increases. The statistical nature of the specification simply expresses the fading nature of the received signal.

---

*The level of uniform-spectrum noise power in a 3.1-kHz band (the voice channel) must be reduced by 2.5 dB to obtain the psophometrically weighted noise power.

5.4.7.2 Noise in the Derived FDM Voice Channel

5.4.7.2.1 Introduction. On a FM LOS microwave link, noise is examined during conditions of fading and for the unfaded or median noise condition. For the low RSL or faded condition, Ref. 3 states that

> the noise in a derived (FDM) voice channel at FM threshold, falls approximately at, or slightly higher than the level considered to be the maximum tolerable noise for a telephone channel in the public network. By present standards, this maximum is considered to be 55 dBrnc0 (316,200 pWp0). In industrial systems, a value of 59 dBrnc0 (631,000 pWp0) is commonly used as the maximum acceptable noise level.

The FM improvement threshold, therefore, is the usual point where fade margin is added (Section 5.4.2.3) to achieve the unfaded RSL. In many systems, however, the CCIR guideline of 3 pWp/km may not be achieved at the high-signal-level condition, and the unfaded RSL may have to be increased still further.

At low RSL, the primary contributor to noise is thermal noise. At high-signal-level conditions (unfaded RSL), there are three contributors:

- Thermal noise
- Radio equipment IM noise
- IM noise due to antenna feeder distortion

Each is calculated for a particular link for unfaded RSL and each value is converted to noise power in mW, pW, or pWp and summed. The sum is then compared to the noise apportionment for the link from Section 5.4.7.1.

5.4.7.2.2 Calculation of Thermal Noise. The conventional formula (Ref. 15) to calculate signal-to-noise power ratio on FM radiolink (test tone to flat weighted noise for thermal noise) is

$$\frac{S}{N} = \frac{\text{RSL}}{(2ktbF)[\Delta f/f_c]^2} \tag{5.26}$$

where
b = channel bandwidth (3.1 kHz)

Δf = channel test tone peak deviation (as adjusted by preemphasis)

f_c = highest channel (center) frequency in baseband (kHz)

F = noise factor of the receiver (i.e., the numeric equivalent of the noise figure, dB)

$kt = 4 \times 10^{-18}$ mW/Hz

RSL = receive signal level (mW)

S/N = test tone-to-noise ratio (numeric equivalent of S/N, dB)

In the more useful decibel form, equation 5.26 is

$$\left(\frac{S}{N}\right)_{dB} = RSL_{dBm} + 136.1 - NF_{dB} + 20\log\left(\frac{\Delta f}{f_c}\right) \quad \text{(flat)} \quad (5.27)$$

In a similar fashion (Ref. 3) the noise power in the derived voice channel can be calculated as follows:

$$P_{dBrnc0} = -RSL_{dBm} - 48.1 + NF_{dB} - 20\log\left(\frac{\Delta f}{f_c}\right) \quad (5.28)$$

and

$$P_{pWp0} = \log_{10}^{-1}\left[\frac{-RSL_{dBm} - 48.6 + NF_{dB} - 20\log(\Delta f/f_c)}{10}\right] \quad (5.29)$$

Note: Values for Δf, the *channel* test tone *peak* deviation (Table 5.4): 200-kHz rms deviation for 60–960 VF channel loading and 140-kHz rms deviation for transmitters loaded with more than 1200 voice channels. Per-channel peak deviation is 282.8 and 200 kHz, respectively (e.g., use sinusoid peaking factor of 1.414 from rms).

Example. A 50-hop microwave system is designed to meet CCIR noise requirements. The system is 2500 km long, so each hop contributes 150 pWp of noise. The system carries 300 FDM channels and the highest channel center frequency is 1248 kHz. The receiver noise figure is 10 dB. Calculate the receiver RSL at the end of one hop.

Use equation 5.29 and set 150 pWp0 equal to the value of the right-hand side of the equation:

$$150 \text{ pWp0} = \log_{10}^{-1}\left[\frac{-RSL - 48.6 + 10 \text{ dB} - 20\log(\Delta f/f_c)}{10}\right]$$

Calculate the value of

$$20\log\left(\frac{\Delta f}{f_c}\right) = 20\log\left(\frac{282.8}{1248}\right)$$

$$= -12.9 \text{ dB}$$

Then

$$10\log(150) = -RSL - 48.6 + 10 + 12.9$$

$$= -47.4 \text{ dBm}$$

Table 5.6 presents equivalent values of dBrnc, pWp, and signal-to-noise power ratio for the standard voice channel.

TABLE 5.6 Approximate Equivalents for Signal-to-Noise Ratio (S/N) and Common Noise Level Values[a]

dBrnc0	pWp0	dBm0p	S/N	NPR (flat)	dBrnc0	pWp0	dBm0p	S/N	NPR (flat)
0	1.0	−90	88	71.6	30	1,000	−60	58	41.6
1	1.3	−89	87	70.6	31	1,259	−59	57	40.6
2	1.6	−88	86	69.6	32	1,585	−58	56	39.6
3	2.0	−87	85	68.6	33	1,995	−57	55	38.6
4	2.5	−86	84	67.6	34	2,520	−56	54	37.6
5	3.2	−85	83	66.6	35	3,162	−55	53	36.6
6	4.0	−84	82	65.6	36	3,981	−54	52	35.6
7	5.0	−83	81	64.6	37	5,012	−53	51	34.6
8	6.3	−82	80	63.6	38	6,310	−52	50	33.6
9	7.9	−81	79	62.6	39	7,943	−51	49	32.6
10	10.0	−80	78	61.6	40	10,000	−50	48	31.6
11	12.6	−79	77	60.6	41	12,590	−49	47	30.6
12	15.8	−78	76	59.6	42	15,850	−48	46	29.6
13	20.0	−77	75	58.6	43	19,950	−47	45	28.6
14	25.2	−76	74	57.6	44	25,200	−46	44	27.6
15	31.6	−75	73	56.6	45	31,620	−45	43	26.6
16	39.8	−74	72	55.6	46	39,810	−44	42	25.6
17	50.1	−73	71	54.6	47	50,120	−43	41	24.6
18	63.1	−72	70	53.6	48	63,100	−42	40	23.6
19	79.4	−71	69	52.6	49	79,430	−41	39	22.6
20	100	−70	68	51.6	50	100,000	−40	38	21.6
21	126	−69	67	50.6	51	125,900	−39	37	20.6
22	158	−68	66	49.6	52	158,500	−38	36	19.6
23	200	−67	65	48.6	53	199,500	−37	35	18.6
24	252	−66	64	47.6	54	252,000	−36	34	17.6
25	316	−65	63	46.6	55	316,200	−35	33	16.6
26	398	−64	62	45.6	56	398,100	−34	32	15.6
27	501	−63	61	44.6	57	501,200	−33	31	14.6
28	631	−62	60	43.6	58	631,000	−32	30	13.6
29	794	−61	59	42.6	59	794,300	−31	29	12.6

[a]This table is based on the following commonly used relationships, which include some rounding off for convenience: Correlations between columns 3 and 4 are valid for all types of noise. All other correlations are valid for white noise only. dBrnc0 = $10 \log_{10}$ pWp0 = dBm0p + 90 = 88 − S/N (flat) = 7.16 − NPR.

Source: Extracted from Ref. 15. Courtesy of Electronics Industries Association.

5.4.7.2.3 Calculation of Radio Equipment IM Noise. Up to this point we have only dealt with thermal noise in a radiolink. In an operational analog radiolink, a second type of noise can be equally important. This is intermodulation noise (IM noise).

IM noise is caused by nonlinearity when information signals in one or more channels give rise to harmonics or intermodulation products that appear as unintelligible noise in other channels. In a FDM/FM radiolink, nonlinear noise in a particular channel varies as the multiplex signal level and the position of the channel in the multiplex baseband spectrum. For a fixed multiplex signal level and for a specific FDM channel, nonlinear noise is constant.

For low receive signal levels at the far-end FM receiver, such as during conditions of deep fades, thermal noise limits performance. During the converse condition, when there are high signal levels, IM noise may become the limiting performance factor.

In a FM radio system, nonlinear (IM) noise may be attributed to three principal factors: (1) transmitter nonlinearity, (2) multipath effects of the medium, and (3) receiver nonlinearity. Amplitude and phase nonlinearity are equally important in contribution to total noise and each should be carefully considered.

A common method of measuring total noise on a FM radiolink under maximum (traffic) loading conditions consists of applying a "white noise" signal at the baseband input port of the FM transmitter. A white noise generator produces a noise spectrum approximating that produced by the FDM equipment. The output noise level of the generator is adjusted to a desired multiplex composite baseband level (composite noise power). Then a notched filter is switched in to clear a narrow slot in the spectrum of the noise signal, and a noise analyzer is connected to the output of the system. The analyzer can be used to measure the ratio of the composite noise power to the noise power in the cleared slot. The noise power is equivalent to the total noise (e.g., thermal plus IM noise) present in the slot bandwidth. Conventionally, the slot bandwidth is made equal to that of a single FDM voice channel.

The most common unit of noise measurement in white noise testing is noise power ratio (NPR), which is defined as follows (Ref. 17):

NPR is the decibel ratio of the noise level in a measuring channel with the baseband fully loaded ... to ... the level in that channel with all of the baseband noise loaded except the measuring channel.

The notched (slot) filters used in white noise testing have been standardized for radiolinks by CCIR in Rec. 399-3 (Ref. 16) for ten common FDM baseband configurations. Available measuring channel frequencies and high and low baseband cutoff frequencies are shown in Table 5.7. In a NPR test, usually three different slots are tested separately: high frequency, midband, and low frequency.

When the NPR measurement is made at high RF signal levels, such as when the measurement is made in a back-to-back configuration, the dominant noise component is equipment IM noise. This parameter can be used as an approximation of the equipment IM noise contribution. This value together with stated equivalent noise loading should also be available from manufacturers' published specifications on the equipment to be used. Modern, new radiolink terminal equipment (i.e., that equipment that accepts an information baseband for modulation and demodulates a RF signal to

TABLE 5.7 CCIR Measurement Frequencies for White Noise Testing

System Capacity (Channels)	Limits of Band Occupied by Telephone Channels (kHz)	Effective Cutoff Frequencies of Band-Limiting Filters (kHz)		Frequencies of Available Measuring Channels (kHz)
		High Pass	Low Pass	
60	60–300	60 ± 1	300 ± 2	70 270
120	60–552	60 ± 1	552 ± 4	70 270 534
300	60–1,300 / 64–1,296	60 ± 1	1,296 ± 8	70 270 534 1,248
600	60–2,540 / 54–2,660	60 ± 1	2,600 ± 20	70 270 534 1,248 2,438
960	60–4,028 / 64–4,024	60 ± 1	4,100 ± 30	70 270 534 1,248 2,438 3,886
900	316–4,188	316 ± 5	4,100 ± 30	534 1,248 2,438 3,886
1,260	60–5,636 / 60–5,564	60 ± 1	5,600 ± 50	70 270 534 1,248 2,438 3,886 5,340
1,200	316–5,564	316 ± 5	5,600 ± 50	534 1,248 2,438 3,886 5,340
1,800	312–8,120 / 312–8,204 / 316–8,204	316 ± 5	8,160 ± 75	534 1,248 2,438 3,886 5,340 / 7,600
2,700	312–12,336 / 316–12,388 / 312–12,388	316 ± 5	12,360 ± 100	534 1,248 2,438 3,886 5,340 / 7,600 11,700

Source: Reference 16, CCIR Rec. 399-3. Courtesy of ITU-CCIR.

baseband) should display a NPR of at least 55 dB when tested back-to-back. It should be noted, however, that diversity combining can improve NPR. This is because in equal gain and maximal ratio combiners, the signal powers are added coherently, whereas the IM noise contribution, which is similar (for this discussion) to other noise, is added randomly. Reference 18 allows a 3-dB improvement in NPR when diversity combining is used on a link.

Up to this point NPR has been treated for terminal radio equipment or baseband repeaters. If the designer is concerned with heterodyne (IF) repeaters (Section 5.4.6), the white noise test procedure as previously described cannot be carried out per se, and the designer should rely on manufacturers' specifications. Alternatively, about a 4-dB improvement (Ref. 18) in NPR over baseband radio equipment may be assumed, or for a new IF repeater of modern design, a NPR of at least 59 dB should be achieved.

Specifying the noise in a test channel by NPR provides a relative indication of IM noise and crosstalk. An alternative is to express the noise in decibels relative to a specified absolute signal level in a test channel. In this case we can define the signal-to-noise power ratio (S/N) as the decibel ratio of the level of the standard test tone to the noise in a standard channel bandwidth (3100 Hz) within the test channel or

$$\left(\frac{S}{N}\right)_{dB} = NPR + BWR - NLR \qquad (5.30)$$

where NLR = noise load ratio and BWR = bandwidth ratio.

For FDM telephony baseband loading of a FM microwave transmitter following CCIR recommendations, NLR in decibels may be calculated using the following formulas:

$$NLR_{dB} = -1 + 4 \log N \qquad (5.31)$$

for FDM configurations from 12 to 240 voice channels and

$$NLR_{dB} = -15 + 10 \log N \qquad (5.32)$$

for FDM configurations where N is greater than 240. For U.S. military systems, the value

$$NLR_{dB} = -10 + 10 \log N \qquad (5.33)$$

is used.

Following AT&T practice, the following formula is used in lieu of equation 5.32:

$$NLR_{dB} = -16 + 10 \log N \qquad (5.34)$$

The bandwidth ratio

$$BWR_{dB} = \left(\frac{\text{occupied baseband of white noise test signal}}{\text{voice channel bandwidth}}\right) \qquad (5.35)$$

The denominator in equation 5.35 can usually be taken as 3.1 kHz. The numerator can be taken from Table 5.7.

Example. A particular FM radiolink transmitter and receiver back-to-back display a NPR of 55 dB. They have been designed and adjusted for 960 VF channel operation and will use CCIR loading. What is the S/N in a voice channel?

Consult Table 5.7. The baseband occupies 60–4028 kHz.

$$BWR = 10 \log\left(\frac{4028 - 60}{3.1}\right) \text{ (kHz)}$$

$$= 10 \log\left(\frac{3968}{3.1}\right)$$

$$= 31.07 \text{ dB}$$

$$NLR = -15 + 10 \log(960)$$

$$= 14.82 \text{ dB}$$

$$\frac{S}{N} = 55 \text{ dB} + 31.07 \text{ dB} - 14.82 \text{ dB}$$

$$= 71.25 \text{ dB}$$

5.4.7.2.4 IM Noise Due to Antenna Feeder Distortion. Antenna feeder distortion or echo distortion is caused by mismatches in the transmission line connecting the radio equipment to the antenna. These mismatches cause echos or reflections of the incident wave. Similar distortion can be caused by long IF runs; however, in most cases, this can be neglected.

Echo distortion actually results from a second signal arriving at the receiver but delayed in time by some given amount. It should be noted that multipath propagation may also cause the same effect. In this case, though, the delay time is random and continuously varying, thereby making analysis difficult, if not impossible.

The level of the echo signal is an inverse function of the return loss at each end of the transmission line and its terminating device (i.e., the antenna at one end and the communication equipment at the other). An echo signal so generated will be constant, since the variables that established it are constant. Thus the distortion created by the echo will be constant but contingent on modulation. In other words, if the carrier were unmodulated, there would be no distortion due to echo. When the carrier is modulated, echo appears.

Reference 18 provides a method of calculating IM noise due to antenna feeder distortion. Such IM noise makes a contribution to total noise at each end of a link. For noise budgeting purposes, allow about 7-pWp IM feeder noise contribution at each end of a link, for a total of 14 pWp.

5.4.7.2.5 Total Noise in the Derived Voice Channel. To calculate the total noise in the voice channel from each of the three noise contributors, convert to absolute values of noise power—such as mW, pW, or pWp—sum, and reconvert to a decibel unit if so desired. The thermal noise value should be taken for unfaded RSL conditions.

Example. Suppose the thermal noise contribution were 206 pWp, the IM noise contribution 37.5 pWp, and antenna IM feeder noise contribution 12.5 pWp. The total noise would be 256 pWp, or simply the sum of the three contributors.

Some Notes on Noise in the Voice Channel

- A voice channel with psophometric weighting has a 2.5-dB noise improvement over a flat channel.
- dBrnc (dB above reference noise with C-message weighting): The reference frequency/level is a 1000-Hz tone at -90 dBm (Refs. 14 and 31).
- pWp (picowatts of noise power with psophometric weighting): The noise power reference frequency/level is an 800-Hz tone where 1 pWp = -90 dBm (pWp = pW $\times$ 0.56) (Refs. 14 and 31).
- dBmp (psophometrically weighted noise power measured in dBm), where, with an 800-Hz tone 0 dBmp = 0 dBm. For flat noise in the band 300–3400 Hz, dBmp = dBm $-$ 2.5 dB (Refs. 14 and 31).

- 0 dBrnc = −88.5 dBm. Commonly, the −88.5-dBm value is rounded off to −88 dBm; thus 0 dBrnc = −88.0 dBm (Refs. 14 and 31).
- dBrnc = dBmp + 90 dB: It should be noted that C-message weighting and psophometric weighting in fact vary by 1 dB (−1.5 dB and −2.5 dB) and this equivalency has an inherent error of 1 dB. However, the equivalency is commonly accepted in the industry (Refs. 14 and 31).
- dBrnc0 = 10 log pWp0 + 0.8 dB = dBmp + 90.8 dB = 88.3 (dB) − $(S/N)_{dB\,(flat)}$

(Ref. 3)

To calculate flat noise in the test channel, the following expression applies:

$$P_{tcf} = \log^{-1}\left(\frac{90 - (S/N)_{dB}}{10}\right) \quad (\text{pW0}) \tag{5.36}$$

and to calculate psophometrically weighted noise:

$$P_{tcp} = 0.56\log^{-1}\left(\frac{90 - (S/N)_{dB}}{10}\right) \quad (\text{pWp0}) \tag{5.37}$$

Example. If S/N in a voice channel is 71.25 dB, what is the noise level in that channel in pWp0?

Use equation 5.37:

$$P_{tcp} = 0.56\log^{-1}\left(\frac{90 - 71.25}{10}\right)$$

$$= 0.56 \times 74.99$$

$$= 41.99 \text{ pWp0}$$

5.5 DIGITAL LINE-OF-SIGHT MICROWAVE

5.5.1 Definition and Scope

All new LOS microwave systems used for duplex communication are digital systems. The public switched telecommunication network in North America is virtually all-digital. Private, industrial, and government networks are now all-digital. Thus microwave links supporting these networks are all-digital. First, let us define digital radio.

A digital radio for this discussion is defined as a radio set in which one or more properties (amplitude, frequency, and phase) of the RF carrier are quantized by a modulating signal. Digital implies fixed sets of discrete values. A digital radio waveform, then, can assume one of a discrete set of amplitude levels, frequencies, or phases (or hybrids thereof) as a result of the modulating signal. Let us assume that the modulating signal is a serial synchronous bit stream.

For a given system the information in the serial bit stream may be PCM, DM (Chapter 3), or any other form of serial data information (Chapter 13). Although we define data transmission in Chapter 13 as the digital transmission of record traffic, graphics, numbers, and so on, the industry has given *data* a wider meaning in many instances. Depending on the person one is talking to, data can mean any digital bit stream, no matter what its information content. For instance, *data rate* is a common term that is synonymous with bit rate. We shall also be using the term *symbol rate*, meaning the number of transitions or changes in state per second. In a two-state or two-level system the symbol rate measured in bauds is the same as the data rate in bits per second. If there are more than two states or levels, it is called an *M*-ary system, where *M* is the number of states. For example, 8-ary frequency shift keying (FSK) is an eight-frequency system, and at any instant one of the eight frequencies is transmitted.

We must not lose sight of the system aspects of digital radio. A digital radio transmitter does not sit out there alone; it must work with a far-end receiver making up a link. There may be several links in tandem. These links may or may not be part of a larger network. Timing, bit count integrity, justification (slips and stuffing), bit error rate (BER), and special coding may all impact the design of a digital radiolink.

5.5.2 Applications

Digital radio-relay (LOS microwave) systems find application on:

- Trunk routes in the PSTN, private, government, and industrial networks
- Studio-to-transmitter links for TV and other broadcasters
- Specialized links for remote radar, telemetry, and sensor data
- Military networks, both strategic and tactical

Digital radio application is tempered by the advantages of fiber-optic cable.

5.5.3 Basic Digital Radio and Link Design Considerations

5.5.3.1 The Primary Feature. Digital radio system design follows most of the basic procedures outlined in Section 5.4. However, there is one overriding issue that must be resolved by the system design engineer, namely, that of spectrum conservation without significant performance degradation.

A digital radio must transmit a standard 8-bit PCM bit stream described in Chapter 3. An analog VF channel occupies 4 kHz; its digital counterpart occupies 64 kHz, assuming 1-bit/Hz occupancy. Based on this same thinking, we can accommodate 1800 nominal 4-kHz VF channels in approximately 30 MHz of RF bandwidth using conventional FDM/FM, as discussed in Section 5.4. The same number of voice channels using PCM, allowing 1 bit/Hz of bandwidth, would then require 1800 × 64 kHz or 115 MHz of bandwidth.

The RF spectrum is one of our natural resources and it is limited. Thus we should not be wasteful of it. Therefore something has to be done to make digital radio transmission more bandwidth conservative. It is the 1-bit/Hz occupancy assumption that we will be working with in this section. The problem of achieving more bits per hertz will be treated at length further on in the chapter.

The U.S. Federal Communications Commission (FCC) has taken up the issue in Part 21.122 of *Rules and Regulations* (Ref. 2), stating:

- Microwave transmitters employing digital modulation techniques and operating below 15 GHz shall, with appropriate multiplex equipment, comply with the following additional requirements:

On or Before June 1, 1997

1. The bit rate in bits per second shall be equal to or greater than the bandwidth specified by the emission designator in hertz (e.g., to be acceptable, equipment transmitting at a 20-Mbps rate must not require a bandwidth greater than 20 MHz).

2. Equipment to be used for voice transmission shall be capable of satisfactory operation within the authorized bandwidth to encode at least the following number of voice channels:

Frequency Range	Number of Encoded Voice Channels
2,110–2,130 MHz	96
2,160–2,180 MHz	96
3,700–4,200 MHz	1,152
5,925–6,425 MHz	1,152
10,700–11,700 MHz	1,152

After June 1, 1997

In the 4-, 6-, 10-, and 11-GHz bands, Table 5.8 applies.

The FCC rules the following on emission limitation (Part 21.106):

When using transmissions employing digital modulation techniques

(i) For operating frequencies below 15 GHz, in any 4-kHz band, the center frequency of which is removed from the assigned frequency by more than 50% up to and including 250% of the authorized bandwidth: As specified by the following equation but in no event less than 50 dB:

$$A = 35 + 0.8(P - 50) + 10 \log_{10} B \qquad (5.38)$$

TABLE 5.8   U.S. FCC Traffic and Loading Requirements

Nominal Channel Bandwidth (MHz)	Minimum Payload Capacity (Mbps)	Minimum Traffic Loading Payload (as Percent of Payload Capacity)	Typical Utilization[a]
0.400	1.54	N/A	1 DS-1
0.800	3.08	N/A	2 DS-1
1.25	3.08	N/A	2 DS-1
1.60	6.17	N/A	4 DS-1
2.50	6.17	N/A	4 DS-1
3.75	12.3	N/A	8 DS-1
5.0	18.5	N/A	12 DS-1
10.0	44.7	50[b]	1 DS-3/STS-1
20.0	89.4	50[b]	2 DS-3/STS-1
30.0 (11 GHz)	89.4	50[b]	2 DS-3/STS-1
30.0 (6 GHz)	134.1	50[b]	3 DS-3/STS-1
40.0	134.1	50[b]	3 DS-3/STS-1

[a]DS and STS refer to the number of voice circuits a channel can accommodate. 1 DS-1 = 24 voice circuits; 2 DS-1 = 48; 4 DS-1 = 96; 8 DS-1 = 192; 12 DS-1 = 288; 1 DS-3/STS-1 = 672; 2 DS-3/STS-1 = 1344; 3 DS-3/STS-1 = 2016.
[b]This loading requirement must be met within 30 months of licensing. If two transmitters simultaneously operate on the same frequency over the same path, the requirement is reduced to 25%.
Source: Reference 2.

where A = attenuation (dB) below the mean output power level
  P = percent removed from carrier frequency
  B = authorized bandwidth (MHz)
(Attenuation greater than 80 dB is not required.)

(ii) For operating frequencies above 15 GHz, in any 1-MHz band, the center frequency of which is removed from the assigned frequency by more than 50% up to and including 250% of the authorized bandwidth: As specified by the following equation but in no event less than 11 dB:

$$A = 11 + 0.4(P - 50) + 10 \log_{10} B \qquad (5.39)$$

(Attenuation greater than 56 dB is not required.)

(iii) In any 4-kHz band, the center frequency of which is removed from the assigned frequency by more than 250% of the authorized bandwidth: At least $43 + 10 \log_{10}$ (mean power output in watts) dB or 80 dB, whichever is the lesser attenuation.

ITU-R (CCIR) also provides guidance on spectral occupancy for digital radio systems. Reference should be made to ITU-R Recs. F.1101 and SM.1046 (Refs. 19 and 20). However, ITU-R is less specific on spectral occupancy versus bit rate and number of equivalent voice channels. Rather, these recommendations show ways of measuring spectral efficiency. Other ITU-R recommendations give guidance on frequency allocations for specific operational bands.

Analyzing Table 5.8, it would appear that spectral occupancy should be at least 4.5 bits/Hz. We must keep in mind that bandwidth must be set aside for RF filter roll-offs.

5.5.3.2 Other Design Considerations in Contrast to Analog FM Systems.

The performance of a digital radio system is stated in terms of bit error rate (BER). If the digital radiolink is part of the PSTN, error performance should meet the criteria given in Chapter 4. Of course, the BER of a link or system should be stated statistically as a time percentage of a year or worst month.

E_b/N_0 is an expression commonly used on digital radio systems. It is usually related to BER. We first introduced E_b/N_0 in Section 1.11. It expresses "energy per bit per hertz of noise spectral density ratio." We will be using E_b/N_0 exclusively rather than S/N when discussing digital radio performance. If we are given a particular modulation waveform, we can derive a plot of E_b/N_0 versus BER. (Figure 5.24 is a family of such curves.) Since we will be using logarithmic values for E_b/N_0, it is helpful to state the identity:

$$E_b/N_0 - E_{b(\text{dbW or dBm})} - N_{0(\text{dBw or dBm})} \qquad (5.40)$$

In Section 5.7.1 we give a more rigorous mathematical definition of E_b/N_0.

Since we are dealing with synchronous serial data, synchronization is a very important system aspect. It would seem that this goes without saying. On radiolinks over, say, 7 km (4.3 mi) long, we will encounter some sort of fading at least some of the time. Fading on radiolinks (LOS microwave) is more probably caused by multipath, or at least we can say that it is the principal offender. When fades are deep enough, a dropout occurs and synchronization (i.e., the receiver is in synchronization with its companion far-end transmitter) may be lost. A digital radio system must be designed to withstand short periods of dropout without losing synchronization. This capability is expressed as maintenance of bit count integrity (BCI) and is often measured in milliseconds. The system should also have some rapid automatic means of resynchronizing itself once BCI is exceeded.

One result of multipath on microwave radiolinks is dispersion. Dispersion means that an arriving signal at the receiver has been dispersed or "elongated." In the case of a pulse, there would be a main signal component arriving in its proper time slot, this signal having followed the direct and expected path. During multipath conditions, reflected signal energy due to multipath conditions arrives somewhat later, spilling into the time slot of the next pulse. Dispersion is a measure of elongation of smearing, usually given in nanoseconds. Such dispersion causes intersymbol interference, deteriorating link error performance. Digital radiolinks often can operate up to 10 Mbps without concern for dispersion. Those operating at rates greater than 10 Mbps and especially those at rates higher than 20 Mbps should take dispersion into account in the link design and equipment selection.

On digital radio systems there are basically two causes of deteriorated error performance. One is multipath dispersion just discussed. A second is due to Gaussian noise in the receiving system. Basic to any link design is to establish a value of E_b/N_0 to meet the required error performance criteria of the link expressed as BER. Deep fades will degrade this value for the period of the fade, causing an error burst. Another cause of errors is on systems designed to operate near the E_b/N_0 margin, resulting in bursts of errors and periods when the link is comparatively error free. Thus the system designer at the outset must establish whether a digital radiolink is dominated by random errors or by bursts of errors. Earth stations* would probably fall into the random error category, whereas longer LOS radiolinks and tropospheric scatter links would fall into the bursty error category. If coding[†] is used, and we mean channel coding here, to improve error performance, the code selected will much depend on whether a link will expect random errors or bursty errors. Other causes of errors are timing jitter and interference.

5.5.3.3 *Error Performance.* ITU-R Rec. 634-3 (Ref. 21) provides recommended BER performance for digital radio-relay (LOS microwave) links forming part of a high-grade circuit within an ISDN (integrated services digital network) for a link of length L, between 280 and 2500 km.

1.1 BER $> 1 \times 10^{-3}$ for no more than $(L/2500) \times 0.054\%$ of any month, integration time 1 second.

1.2 BER $> 1 \times 10^{-6}$ for more than $(L/2500) \times 0.4\%$ of any month, integration time 1 minute.

1.3 Errored seconds (ES) for no more than $(L/2500) \times 0.32\%$ of any month.

1.4 RBER (residual bit error rate) $\leq (L \times 5 \times 10^{-9})/2500$.

CCIR Rep. 930-2 (Ref. 22) defines RBER as the error ratio in the absence of fading and includes allowance for system inherent errors, environmental and aging effects, and long-term interference. A RBER over a 2500-km high-grade real circuit of 5.0×10^{-9} is specified.

The rationale for selecting a bottoming-out threshold value of BER at 1×10^{-3} is purposeful. For digital systems carrying telephone traffic, error bursts can cause call dropout due to the loss of supervisory signaling. A United Kingdom proposal to the CCIR suggests that with an error rate threshold of 1×10^{-3}, any error burst of 70 ms or less would not cause call dropout.

---

*Earth Stations operating below 10 GHz and with elevation angle $> 7°$.
[†]Coding is discussed in Section 5.6.

In Chapter 4 we discussed much more stringent error performance criteria; for example, end-to-end performance of 5×10^{-10}. This is certainly feasible on fiber-optic links. Improving error performance of radiolinks from the present 1×10^{-6} value is also feasible. A design engineer should take this under advisement. What is driving the push for BER improvement are data transmission techniques that throw out errored frames or cells. In particular, we include frame relay and ATM (asynchronous transfer mode) transmission. The better the error performance, the lower the probability of discarded frames.

5.5.3.4 Modulation Techniques and Spectral Efficiency*

5.5.3.4.1 Introduction. There are three generic modulation techniques available: AM, FM, and PM (i.e., amplitude, frequency, and phase modulation, respectively). Terminology of the industry appends the letters "SK" to the first letter of the modulation type, as in ASK, meaning amplitude shift keying; FSK, meaning frequency shift keying (classified as digital FM); and PSK, meaning phase shift keying. Any of the three modulation types may be two-level or multilevel. For the two-level case, one state represents the binary "1" and the other state a binary "0." For multilevel or *M*-ary systems there are more than two levels or states, usually a multiple of 2, with a few exceptions, such as partial response systems, duo-binary being an example. Four-level or 4-ary systems are in common use, such as QPSK. In this case, each level or state represents two information bits or coded symbols. For 8-level systems such as 8-ary FSK or 8-ary PSK, 3 bits are transmitted for each transition or change of state, and for a 16-level system (16-ary) 4 bits are transmitted for each transition or change of state. Of course, in *M*-ary systems some form of coding or combining is required prior to modulation and decoding after demodulation to recover the original bit stream. Conceptually, a typical QPSK modulator is shown in Figure 5.19. The divide-by-2 box does the coding; the Σ box does the combining. With a little imagination we can see that the modulator is really made up of two BPSK modulators shifted 90° in phase. One is the I for in-phase modulator; the other is the Q, for quadrature modulator. In fact, BPSK and QPSK systems have the same error performance.

Let us consider that each of the three basic modulation techniques may be represented by a modulated sinusoid. At the far-end receiver, some sort of detection must be carried out. Coherent detection requires a sinusoidal reference signal extremely closely matched in both frequency and phase to the received carrier. This phase frequency reference may be obtained from a transmitted pilot tone or from the modulated signal itself. Noncoherent detection, being based on waveform characteristics independent of phase

*Section 5.5.3.4 is based on Refs. 19, 23, and 24.

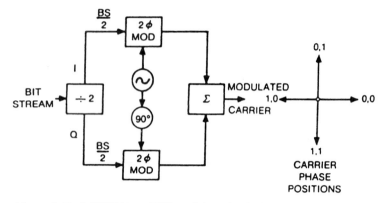

Figure 5.19. A QPSK (4-ary PSK) modulator: I = in-phase; Q = quadrature.

(e.g., energy or frequency), does not require a phase frequency. FSK commonly uses noncoherent detection.

After detection in the receiver, there is usually some device that carries out a decision process, although in less sophisticated systems this process may be carried out in the detector itself. Some decision circuits make decisions on a baud-by-baud basis. Others obtain some advantage by examining the signal over several baud intervals before making each "baud" decision. The observation interval is the portion of the received waveform examined by the decision device.

5.5.3.4.2 Selection of a Spectrally Efficient Modulation Technique. There are three aspects to be considered in selecting a particular modulation technique:

1. To meet the spectral efficiency (confinement) requirements (i.e., a certain number of megabits/second for a given bandwidth, as described in Section 5.5.3.1).
2. To meet realizable performance requirements (Section 5.5.3.3).
3. Economic constraints that are a function of equipment complexity.

The Nyquist bandwidth is 2 bits/Hz. Filters for this bandwidth must have a near vertical rise-time for a pulse, which is not easy to achieve. For practical roll-off containing the Nyquist bandwidth, it is more realizable to apportion 1 bit/Hz. We then say, in theory, that a system transmitting 10 Mbps in the binary domain will require 10 MHz of bandwidth.*

---

*We recommend that the reader not fall into the quagmire of equating bit rate and bandwidth as is commonly done in the trade press. Bandwidth is measured in hertz; bit rate in bits per second. They are *not* the same.

ITU-R Rec. F.1101 (Ref. 19) gives some excellent guidance on the selection of digital modulation schemes. The recommendation states in part that different modulation techniques may be compared on the basis of their bandwidth (B) and BER versus C/N behavior.

Three different concepts of power level may be defined:

- W_{in}: Received maximum steady-state signal power (the value of the carrier mean level related to the highest state of the modulation format
- W_{av}: Received average signal power
- W_p: Received absolute peak of the mean power (peak of the signal envelope)

$$W_{(dB)} = 10\log[W_{in}/(W_n \times f_n)] \qquad (5.41)$$

$$S/N_{(dB)} = 10\log[W_{av}/(W_n \times b_n)] \qquad (5.42)$$

where W_n = noise power density at the receiver input
  F_n = a bandwidth numerically equal to the bit rate (B) of a binary signal before the modulation process
  $b_n = f_n/n$, the Nyquist bandwidth

The bit rate B is the gross bit rate along the radio system and takes into account the redundancy possibly introduced for transmission of service and supervisory channels, error control, and so on, but not the redundancy introduced by coded modulation schemes.

The above carrier-to-noise definitions are related through simple scaling factors:

$$W = S/N - 10\log n + 10\log(W_{in}/W_{av}) \qquad (5.43)$$

Table 5.9 compares several modulation schemes regarding W, which is E_b/N_0, bandwidth B as defined, and is based on a BER of 1×10^{-6}. The BER is based completely on thermal noise.

Bit packing is a term used when discussing spectrally efficient modulation techniques. The packing ratio is the ratio of bits to hertz of bandwidth. For instance, the theoretical packing ratio of BPSK and BFSK is 1 bit/Hz. In practice, it is something like 0.8 bit/Hz. Thus we must distinguish theoretical and practical packing ratios. The theoretical packing ratio of QPSK is 2 bits per hertz of bandwidth; 32-QAM is 5 bits/Hz ($32 = 2^5$) and so forth. Thus the theoretical bit packing value in bits per hertz is the square root of the number of states. For example, for $M = 16$ (such as 16-QAM), $\sqrt{16} = 4$.

Digital modulation schemes are often represented by space diagrams, usually called signal constellations. Typical signal space diagrams are shown in Figure 5.20. On the left side of the figure is 8-ary PSK providing 3 bits/Hz theoretical bit packing; to the right, 16-QAM providing 4 bits/Hz.

TABLE 5.9   Comparison of Different Modulation Schemes

System	Variants	W^a (dB)	S/N^a (dB)	Nyquist Bandwidth (b_n)
	Basic Modulation Schemes			
FSK	2-state FSK with discriminator detection	13.4	13.4	B
	3-state FSK (duo-binary)	15.9	15.9	B
	4-state FSK	20.1	23.1	$B/2$
PSK	2-state PSK with coherent detection	10.5	10.5	B
	4-state PSK with coherent detection	10.5	13.5	$B/2$
	8-state PSK with coherent detection	14.0	18.8	$B/3$
	16-state PSK with coherent detection	18.4	24.4	$B/4$
QAM	16-QAM with coherent detection	17.0	20.5	$B/4$
	32-QAM with coherent detection	18.9	23.5	$B/5$
	64-QAM with coherent detection	22.5	26.5	$B/6$
	128-QAM with coherent detection	24.3	29.5	$B/7$
	256-QAM with coherent detection	27.8	32.6	$B/8$
	512-QAM with coherent detection	28.9	35.5	$B/9$
QPR^b	9-QPR with coherent detection	13.5	16.5	$B/2$
	25-QPR with coherent detection	16.0	20.8	$B/3$
	49-QPR with coherent detection	17.5	23.5	$B/4$
	Basic Modulation Schemes with Forward Error Correction			
QAM with block codes^c	16-QAM with coherent detection	13.9	17.6	$B/4*(1+r)$
	32-QAM with coherent detection	15.6	20.6	$B/5*(1+r)$
	64-QAM with coherent detection	19.4	23.8	$B/6*(1+r)$
	128-QAM with coherent detection	21.1	26.7	$B/7*(1+r)$
	256-QAM with coherent detection	24.7	29.8	$B/8*(1+r)$
	512-QAM with coherent detection	25.8	32.4	$B/9*(1+r)$

[a]Theoretical W and S/N values at 10^{-6} BER; calculated values may differ slightly due to different assumptions.
[b]Quadrature partial response.
[c]As an example, BCH error correction with a redundancy of 6.7% ($r = 6.7\%$) is used for calculations in this table.
Source: Reference 19, Table 1a, ITU-R Rec. F.1101.

5.5.3.4.3   *M-QAM: A Hybrid Modulation Scheme.*   QAM stands for quadrature amplitude modulation. It is a hybrid modulation scheme where both ASK and PSK are used in conjunction with one another.

One example is QPSK using four amplitude levels as well. This is 16-QAM. Figure 5.21 illustrates signal state diagrams for 4-, 8-, 16-, 32-, and 64-QAM. For 16-QAM the amplitude levels are $+3$, $+1$, -1, and -3. 16-QAM provides the same bit packing as 16-ary PSK but is more robust because signal states are derived from two distinct modulation techniques: ASK and PSK. Also, it requires less E_b/N_0 for a given error performance than 16-ary PSK.

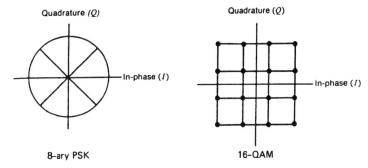

Figure 5.20. Two signal space diagrams.

Adding more amplitude levels can be carried still further: to 32-QAM and 64-QAM, shown in Figure 5.21. 32-QAM can be derived using six amplitude levels, and the corner states are removed, achieving a theoretical bit packing of 5 bits/Hz.

In a similar fashion 128-QAM, 256-QAM, and 512-QAM have evolved. 1024-QAM is on the drawing boards.

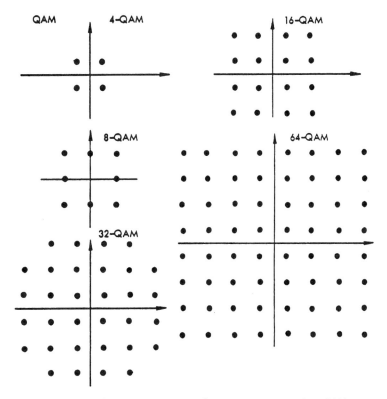

Figure 5.21. Signal state diagrams for 4-, 8-, 16-, 32-, and 64-QAM.

Practical packing ratios are less optimistic. Binary FSK and PSK achieve about 0.8 bit/Hz, QPSK achieves 1.9 bits/Hz, 8-ary PSK 2.6 bits/Hz, 16-ary PSK 2.9 bits/Hz, 16-QAM 3.1 bits/Hz, 32-QAM 4 bits/Hz, and 64-QAM about 4.9 bits/Hz. These practical values will vary depending on filter design.

The following is a relationship between the transmitted bit rate (R_b), baud rate (symbol rate) ($1/T$), and the value of M:

$$R_b = (1/T)10 \log_2 M \quad \text{(bps)} \tag{5.44}$$

This formula shows that the bit rate increases linearly with the baud rate (symbol rate) and logarithmically with M.

5.5.3.4.4 Spectral Efficiency.

Let us assume that a radio frequency assignment had a bandwidth of W. Common bandwidths in the 4-, 6-, 7-, and 11-GHz bands are 20, 30, or 40 MHz. For example, one of these bandwidths may be the value of W.

Let η be the spectral efficiency; then

$$\eta = R_b/W \tag{5.45}$$

Substituting from equation 5.44 we have

$$\eta = (1/WT)\log_2 M \tag{5.46}$$

Noguchi et al. (Ref. 23) state that, in theory, WT can be as low as 1 without adjacent channel interference. Consider now that we will use a raised cosine filter. With cosine roll-off shaping, we can achieve the WT value of 1 or the Nyquist bandwidth with a roll-off factor of $\alpha = 0$. α defines the excess bandwidth, meaning in excess of the Nyquist bandwidth, usually expressed as $1/T$. The Nyquist bandwidth plus the excess bandwidth is commonly expressed as $(1 + \alpha)/T$. Figure 5.22 shows some typical transmitted spectra for cosine roll-off shaping.

Selecting the value of α is very important. If α is 1, we compromise spectral efficiency because we require double the Nyquist bandwidth. Manufacturing is made more difficult if α is 0, as well as having a system more vulnerable to impairments. In modern digital radios, α is commonly selected at 0.5. It can be shown that, with $1/T = 3W/4$ and α near 0.5, FCC mask requirements can be met. The resulting η from equation 5.46 is $\frac{3}{4}\log_2 M$, so that systems using $M = 4, 16, 64,$ and 256 have spectral efficiencies of 1.5, 3.0, 4.5, and 6.0 bits/Hz, respectively.

5.5.3.4.5 Power Amplifier Distortion.

Digital microwave transmitter power amplifiers are peak-power-limited devices. This brings about a serious distortion problem because these devices become increasingly nonlinear as they approach saturation.

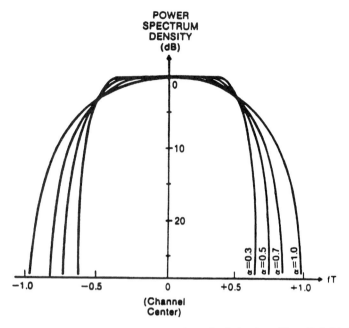

Figure 5.22. Transmitted spectra for typical cosine roll-off shaping. (From Ref. 23, Figure 5. Reprinted with permission.)

The problem here is peak power. However, at the receiver we are interested in the average power to achieve a specified bit error rate. Figure 5.23 shows peak instantaneous power relative to QPSK with square pulses as a function of the cosine roll-off factor, α. The figure is based on a BER of 1×10^{-6}.

For $M > 8$, as M increases so does the ratio of peak to average power. To accommodate a given M-ary waveform and roll-off factor, the saturation power of the transmitter power amplifier must be sufficiently large that the peak input power lies in its linear range. If power peaks push the transmitter into its nonlinear range, then nonlinear distortion results. It causes spectral spreading of the transmitter output. Predistortion and postdistortion of the transmitter signal help.

5.5.4   Performance of Typical *M*-QAM Waveforms

Figure 5.24 gives detailed theoretical symbol error rate (SER) performance for ideal 16-QAM, 64-QAM, and 256-QAM waveforms. Conversion of SER to BER is described in the caption to the figure.

We must distinguish between theoretical and practical values of E_b/N_0. For a given BER, practical values are always higher than theoretical values.

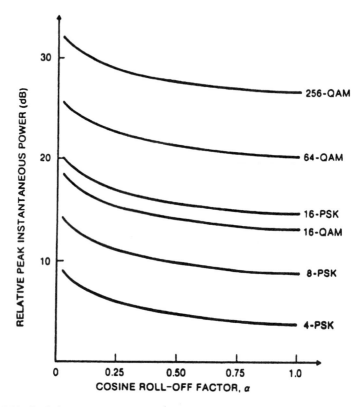

Figure 5.23. Peak instantaneous power (relative to QPSK with square pulses) for *M*-ary waveforms for several values of *M*. (From Ref. 23, Figure 8. Reprinted with permission.)

The amount that they are higher, in dB, is called modulation implementation loss. This loss takes into account degradation parameters not included in an ideal system.

Items that may contribute to the modulation implementation loss budget are:

1. The effect of nonideal sharp cutoff spectral shaping filters.
2. The effect of additional linear distortion created by realistic filters.
3. The effect of signaling with a peak-power-limited amplifier.
4. The relative efficiency between preamplifier filtering and postamplifier filtering.
5. The effect of practical nonlinearities encountered with real amplifiers on system performance.
6. Techniques required for adapting realistic amplifier nonlinearities in order to render the amplifier linear for the purpose of supporting highly bandwidth-efficient digital radio transmission.

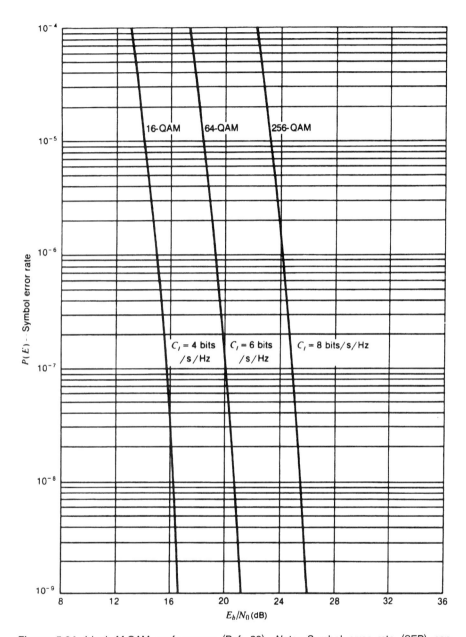

Figure 5.24. Ideal *M*-QAM performance (Ref. 26). *Note:* Symbol error rate (SER) can be converted to bit error rate (BER) assuming Gray-coded state assignment by: BER = $(1/\log_2 M) \times$ SER. Thus for 16-QAM, BER = SER/4; for 64-QAM, BER = SER/6; and for 256-QAM, BER = SER/8. For example, for 64-QAM where SER = 1×10^{-9}, the BER = 1.667×10^{-10} (Ref. 27).

7. The effect on performance of practical imperfections in an implementation of a bandwidth-efficient modem including baseband equalizer for counteracting linear distortion caused by realistic filter characteristics.

Values for modulation implementation loss can be as high as 5 dB or more (Ref. 25); my own experience suggests values of 2 dB or less on a well-designed system.

5.5.4.1 Impairments Peculiar to Digital Radio Operation.

One particular concern we must be aware of is signal space distance, briefly described in Section 5.5.3.4. Here we showed that as the value of M increases, the distance between points on the signal constellation (state diagram, see Figures 5.20 and 5.21) decreases. For example, 16-QAM has 16 points in its constellation, 64-QAM has 64 points, and 256-QAM has 256 points on a 16×16 grid; a 512-QAM system has twice as many signal points. The real vulnerability lies in the receiver because it must resolve which signal point was transmitted. Outside disturbances tend to deteriorate the system by masking or otherwise confusing the correct signal point, degrading error performance.

One such outside disturbance is additive white Gaussian noise (AWGN). Consider Figure 5.25. A bandwidth-efficient modem consequently requires a higher E_b/N_0 for a given symbol rate as the number of bits per second per hertz is increased. This is aptly shown in Figure 5.24.

Another type of outside disturbance is created by cochannel and/or adjacent-channel interference. Here the bandwidth-efficient modem is more vulnerable to interference since less interference power is required to push the transmitted signal point selection (at the receiver) to an adjacent point, thus resulting in a *hit* causing error(s).

For the system engineer, one of the most perplexing impairments is caused by the transmission medium itself. This is multipath fading causing signal dispersion. The problem is dealt with, in part, in the next subsection.

5.5.4.2 Mitigation Techniques for the Effects of Multipath Fading.

In analog radiolink systems, multipath fading results in an increase in thermal noise (AWGN) as the RSL drops. In digital radio systems, however, there is also a degradation in BER during periods of fading that is usually caused by intersymbol interference (ISI) due to multipath. Even rather shallow fades can cause relatively destructive amounts of ISI. The interference results from frequency-dependent amplitude and group delay changes. The degradation depends on the magnitude of in-band amplitude and delay distortion. This, in turn, is a function of fade depth and time delay between the direct and reflected signals.

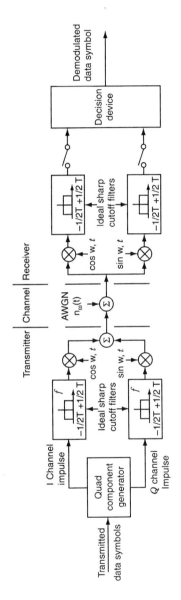

Figure 5.25. An ideal radio system using quadrature modulation.

325

Five of the most common methods to mitigate the effects of multipath fading on digital radiolinks are:

- System configuration (i.e., adjusting antenna height to avoid ground reflection; implementation of space and/or frequency diversity)
- Use of IF combiners in diversity configurations
- Use of baseband switching combiners in a diversity configuration
- Adaptive IF equalizers
- Adaptive transversal equalizers

System configuration techniques have been described earlier in this chapter, such as sufficient clearance to avoid obstacle diffraction, high–low antennas to place a reflection point on "rough" ground, and particularly the use of diversity. In the following paragraphs we will concentrate on space diversity. We now expand on several of the items listed above for the mitigation of the effects of multipath fading.

An *optimal IF combiner* for digital radio receiving systems can be designed to adjust adaptively to path conditions. One such combining technique, the maximum power IF combiner, vectorially adds the two diversity paths to give maximum power output from the two input signals. This is done by conditioning the signal on one path with an endless phase shifter, which rotates the phase on this path to within a few degrees of the signal on the other diversity path prior to combining. The output of this type of combiner can display in-band distortion that is worse than the distortion on either diversity path alone but functions well to keep the signal at an acceptable level during deep fades on one of the diversity paths.

A minimum-distortion IF combiner operation is similar in most respects to the maximum power IF combiner but uses a different algorithm to control the endless phase shifter. The output spectrum of the combiner is monitored for flatness such that the phase of one diversity path is rotated and, when combined with the second diversity path, produces a comparatively flat spectral output. The algorithm also suppresses the polarity inversion on the group delay, which is present during nonminimum phase conditions. One disadvantage of this combiner is that it can cancel two like signals such that the signal level can be degraded below threshold.

Reference 28 suggests a dual algorithm combiner that functions primarily as a maximum power combiner and automatically converts to a minimum-distortion combiner when signal conditions warrant. Using space diversity followed by a dual-algorithm combiner can give improvement factors better than 150.

Adaptive IF equalizers attempt to compensate directly at IF for multipath passband distortion. Digital radio transmitters emit a transmit spectra of relatively fixed shape. Thus various points on the spectrum can be monitored, and when distortion is present, corrective action can be taken to restore spectral fidelity. The three most common types of IF adaptive equalizers are shape-only equalizers, slope and fixed-notch equalizers, and tracking-notch equalizers.

Another equalizer is the adaptive transversal equalizer, which is efficient at canceling intersymbol interference due to signal dispersion caused by multipath. The signal energy dispersion can be such that energy from a digital transition or pulse arrives both before and after the main bang of the pulse. The equalizer uses a cascade of baud delay sections that are analog elements to which the symbol or baud sequence is inputted. The "present" baud or symbol is defined as the output of the Nth section. Sufficient sections are required to encompass those symbols or bauds that are producing the distortion. These transversal equalizers provide both feedforward and feedback information. There are both linear and nonlinear versions. The nonlinear version is sometimes called a decision feedback equalizer. Reference 28 reports that both the IF and transversal equalizers show better than three times improvement in error rate performance over systems without such equalizers.

5.5.4.3 ITU-R Guidelines for Combating Propagation Effects*

5.5.4.3.1 Space Diversity. Space diversity is one of the most effective methods of combating multipath fading. For digital radio systems, where performance objectives are difficult to meet owing to waveform distortion caused by multipath effects, system designs must often be based on the use of space diversity.

By reducing the effective incidence of deep fading, space diversity can reduce the effects of various types of interference. In particular, it can reduce the short-term interference effects from cross-polar channels on the same or adjacent channel frequencies and the interference from other systems and from within the same system.

Linear amplitude dispersion (LAD) is an important component of waveform distortion and quadrature crosstalk effects and can be reduced by the use of space diversity. Diversity combining used specifically to minimize LAD is among the methods that are particularly effective in combating this type of distortion.

The improvement derived from space diversity depends on how the two signals are processed at the receiver (combiner). Two examples of techniques are "hitless" switching and variable phase combining. The "hitless" switch selects the receiver with the greatest eye opening or the lowest error rate, and the combiners use either cophase or various types of dispersion-minimizing control algorithms. "Hitless" switching and cophase combining provide very similar improvements.

*Section 5.5.4.3 is based on Ref. 29 (Section 3.2/3.3).

5.5.4.3.2 Adaptive Channel Equalization. Some form of receiver equalizing is usually necessary in the radio channel(s). As propagation conditions vary, an equalizer must be adaptively controlled to follow the variations in transmission characteristics. Such equalizers work in either the frequency domain or the time domain.

FREQUENCY DOMAIN EQUALIZATION. Equalizers operating in the frequency domain comprise one or more linear networks, which are designed to produce amplitude and group delay responses. They compensate for transmission impairments, which are most likely to produce system performance degradation during periods of multipath fading. Table 5.10 shows several alternative equalizer structures that may be considered by the system engineer.

TIME DOMAIN EQUALIZATION. Time domain equalizers combat intersymbol interference directly. With these equalizers control information is derived by correlating the interference that appears at the instant of decision with the various adjacent symbols producing the control information, and is used to adjust tapped delay line networks to provide appropriate cancellation signals.

TABLE 5.10   Comparison of Adaptive Equalizers

| | | | | Fade Characteristic and Position of Maximum Attenuation[a] | | | |
| | | | | Minimum Phase | | Nonminimum Phase | |
Generic Type		Description of Equalizer	Complexity of Implementation	Out-of-Band	In-Band	Out-of-Band	In-Band
Frequency-domain equalizers	F1	Amplitude tilts	Simple	2	1	2	1
	F2	F1 + parabolic amplitude	Simple	2	2	2	2
	F3	F2 + group delay tilt	Complex (moderately complex)	3	2	3(1)	2(1)
	F4	F3 + parabolic group delay (For F3 and F4 ratings in brackets apply "minimum phase" control assumptions)	Complex (moderately complex)	3	3	3(1)	3(0)
	F5	Single tuned circuit ("agile notch")	Simple	3	3	1	0
Time domain equalizers	T1	Two-dimensional linear transversal equalizer	Moderately complex/complex	3	2	3	2
	T2	Cross-coupled decision feedback equalizer	Moderately complex	3	3	2	1
	T3	T1 + T2 full time domain equalization	Complex	3	3	3	2

[a]Effectiveness of equalization: 3, produces a well-equalized response; 2, produces a moderately equalized response; 1, produces a partially equalized response; 0, not effective.
Source: Reference 29, Table 1, ITU-R Rec. F.1093.

Such an equalizer is able to handle, simultaneously and independently, distortion that arises from amplitude and group delay deviations in the fading channel. It therefore can provide either minimum-phase or non-minimum-phase characteristic compensation.

Quadrature distortion compensation is very important in QAM-type systems. Significant destructive effects are associated with crosstalk generated by channel asymmetries for which a time domain equalizer must correct.

ADAPTIVE EQUALIZERS USED IN COMBINATION WITH SPACE DIVERSITY COMBINING. By far the best method of combating performance deterioration and outage due to multipath effects is to use adaptive equalization in conjunction with space diversity. There are synergistic effects between the two. It has been shown that the total improvement usually exceeds the product of the corresponding individual improvements of space diversity and adaptive equalization considered separately. The product of the space diversity improvement and the square of the equalizer improvement equals the total improvement of the combination of space diversity and adaptive equalization.

SUMMARY. There are three principal degradations that can cause deteriorated performance or system outages on digital radio systems: interference, thermal noise, and waveform distortion. Equalization is really only effective against waveform distortion. Adaptive equalizer performance must be judged in this light, especially where link degradation or outage is traced principally to waveform distortion, especially dispersion.

5.5.5 Fading and Fade Margins on Digital LOS Microwave Links

5.5.5.1 Introduction.* For a well-designed LOS microwave path that is not subject to diffraction fading or surface reflections, multipath conditions provide the dominant factor for fading below 10 GHz. Above about 10 GHz, the effects of rainfall (Chapter 9) determine maximum path lengths through system availability criteria. The necessary reduction in path length due to rainfall loss limitations tends to reduce the severity of multipath fading. These two principal causes of fading are mutually exclusive.[†] Given the split between availability and error performance objectives, rainfall effects contribute mainly to unavailability and multipath propagation contributes mainly to error performance. Another result of rainfall is backscatter from the raindrops. This may influence the selection of RF channel arrangements.

Propagation effects from rainfall tend not to be frequency dispersive, while multipath propagation caused by tropospheric layers can be, which may cause severe distortion and ISI on digital signals. Rainfall fading is covered in Chapter 8.

---

*Section 5.5.5.1 is based on Ref. 29.
[†]This means that we do not add fade margins (i.e., multipath and rainfall). The fade margin to be implemented is the larger of the two, not their sum.

5.5.5.2 Other Views on the Calculation of Fade Margins on Digital LOS Microwave Links.[*] Digital radio systems react to fading differently than their analog counterparts. Earlier in this chapter we were concerned with thermal fade margin (Section 5.4.4). Here we are more concerned with the dispersiveness of a path. The concept of "net" or "effective" fade margin is used for digital radio systems. The "net" fade margin is defined as a single-frequency fade depth in dB that is exceeded for the same number of seconds as that selected for the BER threshold, usually 1×10^{-3}.

A composite fade margin accounts for the dispersiveness of the fading on a hop by using dispersion ratios, which can be used as a parameter to compare the dispersiveness of different hops in relation to single-frequency fading. This net fade margin is considered as the composite of the effects of thermal noise, ISI due to multipath dispersion, and interference from other radio systems. At the detector of a radio receiver during fading these three sources will give three voltage components, which add on a power basis since each is independent. Thus the total outage time is the sum of the contributions due to single-frequency fading, dispersion, and interference.

The dispersive fade margin, which we call DFM, may be determined from the measured net fade margin by correcting it for any thermal noise or interference contributions as necessary. Because the dispersive fade margin reflects the impact of multipath dispersion on a radio system, its value must depend on the fading and on the radio equipment. The first step is to determine the dispersive fade margin of a radio system on a path with a known dispersion ratio of DR_0. This value in dB is taken as a reference dispersive fade margin, DFMR. Then the dispersive fade margin that would be measured or predicted on a path with a dispersion ratio of DR is given by

$$DFM = DFMR - 10\log(DR/DR_0) \tag{5.47}$$

The referenced ITU-R recommendation (Ref. 29) states that calculations based on this procedure have shown good agreement with measured radio performance in the field in the presence of interference, as well as detailed estimates based on propagation models.

DR is given by the following relationship:

$$DR = \frac{T_{IBPD}}{T_{SFF} \times BF^2} \tag{5.48}$$

where T_{IBPD} = amount of time that a chosen in-band power difference IBPD (i.e., the amount of dispersion on a hop) value is exceeded

T_{SFF} = amount of time that a chosen single-frequency fade value is exceeded

BF = bandwidth correction factor, which is the ratio of 22 MHz to the measurement bandwidth

---

[*]Section 5.5.5.2 is based on Ref. 29.

Modern digital radio systems (e.g., 64-QAM) equipped with adaptive time domain equalizers experience outage time (i.e., BER $> 10^{-3}$) due to IBPD distortion in the region of 10–15 dB. Thus a suitable threshold for comparing dispersion would be 10 dB. The values of dispersion ratio measured on a number of hops in North America and Europe are in the range of 0.09–8.1 for hop lengths in the range of 38–112 km. This is based on values of 10 dB and 30 dB for IBPD and single-frequency fade, respectively.

5.5.5.3 *Multipath Fade Margin Calculations Based on TIA TSB 10-F.* *

On digital microwave radiolinks, the fade margin consists of four factors that are power added and constitute the composite fade margin (CFM). These four factors are defined below:

TFM. Thermal fade margin (dB) (sometimes called the flat fade margin) is the fade margin discussed in Section 5.4.4. TFM is the algebraic difference between the nominal RSL and the 1×10^{-3} BER outage threshold for flat (i.e., nondispersive) fades. Since interference affects unfaded baseband noise, TFM is the only fade margin that needs to be considered on analog LOS links.

DFM. Dispersive fade margin (dB), also to the 1×10^{-3} BER. DFM is defined by the radio equipment manufacturer. It is determined by the type of modulation, the effectiveness of equalization employed in the receive path, and the multipath signal's delay time. This is standardized on manufacturers' data sheets as 6.3 ns. DFM characterizes the digital radio's robustness to dispersive (spectrum-distorting) fades. One means to improve DFM on some paths is to increase the antenna discrimination to reduce the level of longer-delay multipath signals, which can unacceptably degrade a link's DFM. According to TIA (Ref. 30), a DFM greater than 50 dB is a good baseline criterion. *(Note the difference between DFM defined here and in Section 5.5.5.2).*

EIFM. External interference fade margin (dB) is a receiver threshold degradation due to interference from a total of the three (MEA[†] factor) external systems (usually 1 dB, but depends on CFM objective). In the absence of adjacent channel interference (AIFM), EIFM is simply IFM.

AIFM. Adjacent-channel interference fade margin (dB). Receiver threshold degradation is due to interference from adjacent-channel transmitters on the same path due to transmitters in one's own system. This is normally a negligible parameter except in cases of frequency diversity and multiline hot-standby systems.

---

*Section 5.5.5.3 was extracted from Ref. 30 (Section 4.2.3).

[†] MEA = multiple exposure allowance (dB).

These four fade margins just defined are power added to derive the composite fade margin (CFM) as follows:

$$CFM = 10\log(10^{-TFM/10} + 10^{-DFM/10} + 10^{-EIFM/10} + 10^{-AIFM/10})$$

(5.49)

The outage time due to multipath fading in a nondiversity link is calculated by

$$T = rT_0\, 10^{-(CFM/10)}/I_0$$

(5.50)

where T = outage time in seconds

$\quad r$ = fade occurrence factor

$\quad T_0 = (t/50)\,(8 \times 10^6)$ = length of fade season in seconds

$\quad t$ = average annual temperature in °F

$\quad CFM$ = composite fade margin

$\quad I_0$ = space diversity improvement factor ($I_0 = 1$ for nondiversity; $I_0 > 1$ for space diversity)

The fade occurrence factor, r, is calculated from the basic outage equation for atmospheric multipath fading:

$$r = c\left(\frac{f}{4}\right)\left(\frac{D}{1.6}\right)^3 \times 10^{-5}$$

$$= x\left(\frac{15}{w}\right)^{1.3}\left(\frac{f}{4}\right)\left(\frac{D}{1.6}\right)^3 \times 10^{-5} \quad \text{(metric)}$$

(5.51a)

or

$$r = c\left(\frac{f}{4}\right)D^3 \times 10^{-5}$$

$$= x\left(\frac{50}{w}\right)^{13}\left(\frac{f}{4}\right)D^3 \times 10^{-5} \quad \text{(English)}$$

(5.51b)

where c = climate–terrain factor (see Figure 5.26)

$\quad x$ = climate factor (see Figure 5.27)

$\quad w$ = terrain roughness

$\qquad 6 \leq w \leq 42$ m—average 15 m or

$\qquad 20 \leq w \leq 140$ ft—average 50 ft

$\quad f$ = frequency (GHz)

$\quad D$ = path length (km or mi)

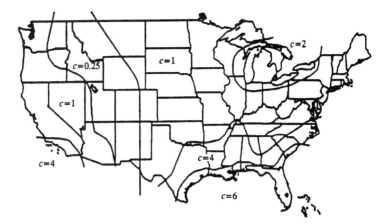

Figure 5.26. Values of climate–terrain factor c.

Hawaii/Caribbean	Alaska
$c = 4/\ x = 2$	$c = 0.25/x = 0.5$: costal and mountainous areas
	$c = 1/x = 1$: flat permafrost tundra areas in west and north Alaska

Figure 5.27. Values of climate factor, x.

To calculate I_0 in equation 5.50 above, use the following procedure (equation 5.52):

$$I_0 = 1.2 \times 10^{-3} s^2 \left(\frac{f}{D}\right) \times 10^{CFM/10}, \quad s \leq 15 \text{ m} \quad \text{(metric)} \quad (5.52a)$$

$$= 7 \times 10^{-5} s^2 \left(\frac{f}{D}\right) \times 10^{CFM/10}, \quad s \leq 50 \text{ ft} \quad \text{(English)} \quad (5.52b)$$

where fade margins on both antennas are about equal, and s = vertical antenna separation in meters (feet), center to center.

The Space diversity improvement factor (I_0) may underestimate diversity improvements for small antenna spacings and overestimate diversity improvement for large antenna spacings on "flat land" microwave links.

For the purposes of this text, average climate ($x = 1$), temperature [10°C (50°F)], and terrain roughness [15 m (50 ft)] conditions may usually be assumed. This simplifies the outage time equation to

$$T = \frac{5fD^3 \times 10^{-CFM/10}}{I_0} \quad \text{(metric)}$$

$$= \frac{20fD^3 \times 10^{-CFM/10}}{I_0} \quad \text{(English)} \quad (5.53)$$

It is seen from the above equations that nondiversity multipath outage increases directly as a function of the path length cubed (D^3). Therefore short digital paths can usually meet outage objectives with less composite fade margin (more interference) since the outage probability of fading is low.

Since the total number of seconds in one year equals 31.5×10^6, the annual path reliability is computed from

$$\text{Path reliability } (\%) = \left(\frac{31.5 \times 10^6 - T}{31.5 \times 10^6}\right) \times 100 \quad (5.54)$$

The nondiversity outage equations can be rearranged to derive the analog radio fade margin (FM) or digital radio composite fade margin (CFM) required for a given outage time:

$$\text{FM or CFM (nondiversity)} = -10 \log\left(\frac{T}{5fD^3}\right) \quad \text{(metric)}$$

$$= -10 \log\left(\frac{T}{20fD^3}\right) \quad \text{(English)} \quad (5.55)$$

where T = outage time objective (s/yr)
   f = frequency (GHz)
   D = path distance (km or mi)

Space diversity improvement plays such a significant role in increasing path reliability that it often allows higher interference levels that degrade (reduce) the composite fade margin of many digital links. By combining the nondiversity outage equation and the space diversity improvement factor equations, we arrive at the following equation for the annual outage in a space diversity path:

$$T_{SD} = \frac{4 \times 10^3 D^4 \times 10^{-CFM/5}}{s^2} \quad \text{(metric)}$$

$$= \frac{3 \times 10^5 D^4 \times^{-CFM/5}}{s^2} \quad \text{(English)}$$

(5.56)

Note that the frequency term has disappeared from the space diversity outage equation and the annual outage now varies as a function of D^4. Rearranging this equation to solve for the required fade margin or composite fade margin for a given outage time with space diversity gives

$$\text{FM or CFM (space diversity)} = -5\log\left(\frac{2.5 \times 10^{-4} Ts^2}{D^4}\right) \quad \text{(metric)}$$

$$= -5\log\left(\frac{3.5 \times 10^{-6} Ts^2}{D^4}\right) \quad \text{(English)}$$

(5.57)

Calculation of the required fade margins for nondiversity or space diversity links with the above equations may provide improved spectrum utilization (efficiency) by permitting higher interference levels without overly degrading the required reliability for many short and diversity links. For example, if the required fade margin (above) is 25 dB, and the path calculations with no interference show 33 dB, an interference level 7 dB above the value calculated on the basis of threshold degradation (by equation 13.10 in Ref. 42, for instance) would probably not cause the hop outage to exceed objectives.

Since analog radios are nonregenerative, the baseband noise is additive on N tandem hops (typically per-hop noise plus $13\log N$). Fading on different hops is noncorrelated, so the outage time (probability of outage) of a digital or analog radio system is equivalent to the sum of the outage times (probabilities of outage) of the individual hops. While the above outage and fade margin calculations are applicable to both analog and digital radio hops, analog radio noise buildup poses a more complex problem. With analog systems, one must consider the overall system noise objectives in parallel with the system reliability (outage) objectives. Most analog systems require significant increases in RSL above FM improvement threshold just to achieve acceptable baseband S/N.

5.5.5.4 Simple Calculations of Path Dispersiveness. Multipath delay for LOS microwave paths can be as high as 20 μs. Ideally this dispersion should be less than half a symbol period to avoid destructive ISI. For STM-1/STS-3 with a bit rate of 155 Mbps, using a binary waveform, this value should be < 0.0019 μs. If we use 64-QAM, the symbol period is six times as long or 0.0116 μs. Thus a path with dispersion on the order of 10 μs would be highly destructive for such symbol rates. Many paths, however, even with hop lengths on the order of 50 mi, display median dispersion in the very low nanosecond range with maxima in the range of 20 or 30 ns.

Kolton (Ref. 31) suggests two formulas to calculate maximum dispersion. The first depends entirely on path length:

$$\tau_{\mathrm{m}} = 3.7(D/20)^3 \quad (\mathrm{ns}) \tag{5.58}$$

where D is the path length in miles. The second formula is based on path length in kilometers and half-power beamwidth of the antenna in degrees:

$$\tau_{\mathrm{m}} = 1668 D \tan^2(\theta/2) \quad (\mathrm{ns}) \tag{5.59}$$

Suppose a path is 60 km (37.5 mi) long and the half-power beamwidth is 1°. By the first formula the maximum dispersion is 24.39 ns; by the second formula, it is 7.62 ns.

5.6 FORWARD ERROR CORRECTION (FEC)

5.6.1 Objective and Background

Forward error correction (FEC) and its related *coding gain* provide yet another technique to improve digital performance on radio systems. It does not give dramatic performance improvements such as, say, doubling the diameter of an antenna at both ends (some 12 dB). But it does provide some 2–6-dB improvement as a gain in a link budget. It also has some hidden advantages such as with scintillation and rainfall fading. Ordinarily we have to pay some price for the added redundancy required for FEC with additional bandwidth.

Error rate is a principal design factor for digital transmission systems. The digital public switched telecommunication network (PSTN) is based on 8-bit PCM, which, for speech transmission, can tolerate a BER degraded to 1×10^{-2} and still be intelligible. CVSD (continuous variable slope delta modulation) remains intelligible with a BER of nearly 1×10^{-1}. For conventional telephony on the PSTN the gating error rate is determined by supervisory signaling.* This value of BER should be $\leq 1 \times 10^{-3}$. This is the threshold for the PSTN. Computer data users of the network drive the BER value to at least 1×10^{-6} at the receive end as recommended in CCITT Rec. G.821 (Ref. 32). For digital LOS microwave a single hop should display a BER of 1×10^{-9}.

---

*Supervisory signaling informs a circuit switch if a line is idle or busy.

To design a transmission system to meet a specified error performance, we must first consider the cause of errors. For this initial discussion, let us assume that intersymbol interference (ISI) is negligible. The errors derive from insufficient E_b/N_0, which results in bit mutilation by thermal noise peaks (additive white Gaussian noise or AWGN). Error rate is a function of signal-to-noise ratio or E_b/N_0. As we have seen in this chapter, we can achieve a desired BER on a link by specifying the requisite E_b/N_0. On many satellite and tropospheric scatter links, there may be a more economically feasible way.

Errors derive from insufficient E_b/N_0 (for this initial discussion), because of deficient system design, equipment deterioration, or fading. On unfaded links or during unfaded conditions, these errors are random in nature; while during fading, errors are predominantly bursty in nature. The length of a burst can be related to fade duration.

There are several tools available to the design engineer to achieve a desired link error rate. Obviously the first is to specify sufficient E_b/N_0 for the waveform selected, adding margin for link deterioration due to equipment aging and a margin allowance for fading.

Another approach is to specify a lower E_b/N_0 for a BER, perhaps in the range of 1×10^{-4}, and implement an ARQ (automatic repeat request) regime. ARQ requires a return channel. It is usually implemented on an end-to-end basis or a section-by-section basis depending on the protocol used and whether it is a network layer or link-layer protocol. With ARQ, data messages are broken down into blocks, frames, or packets at the originating end. Each block, frame, or packet has appended a parity tail, often referred to as a frame check sequence (FCS). This tail is generated at the originating end of the link by a processor that determines the parity characteristics of the message or uses a cyclic redundancy check (CRC), and the tail is the remainder of that check, often 16 bits in length. At the receiving end a similar processing technique is used, and the locally derived remainder is compared to the remainder received from the distant end. If they are the same, the message is said to be error-free; if not, the block or packet is in error.

There are several ARQ methods. One is called stop-and-wait ARQ. In this case, if the frame is error-free, the receiver transmits an ACK (acknowledgment) to the transmitting end, which in turn sends the next frame or packet. If the frame is in error, the receiver sends a NACK (negative acknowledgment) to the transmitter, which now repeats the frame just sent. It continues to repeat it until the ACK signal is received.

A second type of ARQ is variously called "continuous ARQ" or "selective ARQ." At the transmitting end, in this case, sending of frames is continuous. There is accounting information, which involves some sort of sequential frame or packet numbering in each frame or packet header. Similar parity checking is carried out as before. When the receive end encounters a frame in error, it identifies the errored frame to the transmitter, which intersperses the repeated frames, each with a proper sequence number, with its regular continuous frame transmissions. Obviously, continuous ARQ is a more complex implementation than stop-and-wait ARQ, but no valuable circuit time is lost stopping and waiting.

Go-back-n ARQ is similar to continuous ARQ but rather than intersperse a repeated frame with the normal sequence of frames, the receiver tells the transmit end to "go-back-n" frames. The transmitter then repeats all frames from the errored frame forward. One can imagine that ARQ is completely unfeasible on voice circuits and other connectivities requiring a constant bit rate, such as conference TV. The far-end voice listener will not tolerate the ARQ delays or interspersed frames, which also imply delay.

On satellite circuits the delay problem is magnified because of the long propagation delays in addition to ARQ delay, on the order of 0.5 s. It is also bad on data circuits with connectivity through a geostationary satellite, particularly with stop-and-wait ARQ. Here the wasted circuit time is multiplied. Continuous ARQ mitigates much of that time waste.

Another approach is to use error-correction coding, often called channel coding. It offers a number of advantages and at least one major disadvantage. To meet BER requirements in a most cost-effective manner, it is then up to the design engineer to trade off system overbuild with error correction techniques, of which there are two types: ARQ and FEC. A brief description of the former was given earlier. The latter is described in this section.

5.6.2   Basic Forward Error Correction

The IEEE (Ref. 33) defines a forward error correcting system as

> a system employing an error-correcting code and so arranged that some or all of the signals detected as being in error are automatically corrected at the receiving terminal before delivery to the data sink or to the telegraph receiver.

Our definition may help clarify the matter. FEC is a method of error control that employs the adding of systematic redundancy at the transmit end of a link such that many or all of the errors caused by the medium can be corrected at the receiver by means of a decoding algorithm. The price we pay for this is the redundancy.

Figure 5.28 shows a digital communication system with FEC. The binary data source generates information bits at R_s bits per second. An encoder encodes these information bits for FEC at a code rate R. The output of the encoder is a binary sequence of R_c symbols per second. The coder output is related to the input bit rate by the following expression:

$$R_c = R_s/R \qquad (5.60)$$

R, the code rate, is the ratio of the number of information bits to the number

Figure 5.28. Simplified functional block diagram of a binary communication system employing FEC.

of encoded symbols for binary transmission. For example, if the information bit rate were 2400 bps and the code rate (R) were $\frac{1}{2}$, then the symbol rate (R_c) would be 4800 coded symbols per second. As we can see, the transmission rate is doubled without any increase in the information rate.

The encoder output sequence is then modulated and transmitted over the transmission medium or channel. Demodulation is then performed at the receive end. The output of the demodulator is R_c symbols per second, which is fed to the decoder. The decoder output to the data sink is the recovered 2400 bps (R_s). The entire system is assumed to be synchronous.

The major advantages of a FEC system are:

- No feedback channel required as with ARQ systems
- Constant information throughput (e.g., no stop-and-wait gaps)
- Generally small and constant decoding delay
- Significant coding gain achieved for an AWGN channel

There are two disadvantages to a FEC system. To effect FEC, for a fixed information rate, the bandwidth must be increased* because, by definition, the symbol rate on the transmission channel is greater than the information bit rate. There is also the cost of added complexity of the coder and decoder.

5.6.2.1 Coding Gain. For a given information bit rate and modulation waveform (e.g., QPSK, 8-ary FSK, 16-QAM), the required E_b/N_0 for a specified BER with FEC is less than the E_b/N_0 required without FEC. The coding gain is the difference in E_b/N_0 values.

FEC can be used on any digital transmission system. There is a difference in application depending on the medium employed. Wire-pair, coaxial cable, and satellite radio systems (operating below 10 GHz) generally do not suffer from fading. LOS microwave, tropospheric scatter/diffraction, HF, and satellite systems operating above 10 GHz can be prone to fading.

Let us consider a satellite system operating below 10 GHz, where we can generally neglect rainfall losses. Here we can effect major savings by simply adding FEC processors at each end of a link. Suppose we can achieve 3 dB of net coding gain. Thus, keeping the link BER unchanged from the uncoded to

*Trellis coded systems are an exception.

coded condition, the 3-dB coding gain can be used to reduce satellite EIRP by 3 dB without affecting error performance. This can provide many economic savings. First, we can reduce the size of the high-power amplifier (HPA) stage on the satellite transponder because only half the power output is required when compared to the uncoded link. It also reduces the weight of the transponder payload, possibly by 75%. The principal reason for this is that the transmitter power supply size can be decreased. Battery weight can also be reduced, possibly up to 50% [batteries are used to power the transponder during eclipse (darkness)], with the concurrent reduction in solar cells. It is not only the savings in the direct cost of these items, but also the savings in lifting weight of the satellite to place it in orbit, whether by space shuttle or rocket booster. With unfaded conditions, assuming random errors, coding gains of 2–7 dB can be realized. The amount of gain achievable under these conditions is a function of the modulation type (waveform), the code employed, the coding rate (equation 4.1), the constraint length, the type of decoder, soft- or hard-decision decoding, and the demodulation approach. Coding gain values will be provided later in this section.

A FEC system can use one of two broad classes of error-correcting codes: block or convolutional. These are briefly described later. However, we first discuss demodulators using hard decisions or soft decisions and these could provide a form of synergism in the decoding process.

5.6.2.2 *Hard- and Soft-Decision Demodulators.*

In this section, the outputs of a demodulator are defined on a continuum. Before these demodulator outputs can be processed with digital circuits, some form of amplitude quantization* is required. We consider the quantizer as part of the demodulator.

With binary modulation and no coding, a demodulator produces an output defined over a continuum for each bit transmitted. It makes a hard (irrevocable) decision as to which information bit was transmitted (i.e., a 1 or a 0) by determining the polarity of the demodulator output. We could say a 1-bit quantizer is used. Such a 1-bit quantization is also referred to as *hard quantization*. Without coding, providing additional amplitude information about the demodulator is of no help in the improvement of error performance.

With coding a decoding decision on a particular bit is based on several demodulator outputs. Retaining some amplitude information rather than just the sign of the demodulator outputs can be very helpful. If, for example, the output of the demodulator is very large, there is confidence that a polarity decision on that demodulator output is correct. On the other hand, if the output is very small, there is a high probability that the output would be in error. We can design a decoder that uses this amplitude information, which,

---

*Quantization in this context means converting from an output defined by a continuum to an output with discrete values, usually based on the power of 2.

in effect, weighs the contributions of demodulator outputs to the decoding decision. Such decoders can perform better than a similar decoder that only uses the polarity information. A quantizer that retains some amplitude information (i.e., more than one bit is retained) is called a soft quantizer.

No quantization refers to the ideal situation where no quantizer is used at the demodulator output. That is, all of the amplitude information is retained.

An interesting example is 8-ary PSK where several methods of quantizing demodulator outputs are suggested. One method is to quantize the in-phase and quadrature outputs so that the signal space, consisting of signal components every 45° on a circle of radius $\sqrt{E_s}$, is divided into small squares as shown in Figure 5.29a, where E_s is the energy per symbol. Another method divides the receive signal space into pie-shaped wedges depending on the angle of the received signal component as shown in Figure 5.29b. The particular quantization technique depends largely on implementation considerations (Ref. 34).

5.6.3 FEC Codes

5.6.3.1 Block Codes. Block codes were the first error-correction coding techniques to be used. For this class of codes, the data are transmitted in blocks of symbols. In the binary case for every block of K information bits input to the coder, $N - K$ redundant parity check bits are generated as linear (modulo-2) combinations of the information bits and transmitted along with the information bits at a code rate of K/N bits per binary channel symbol. N represents the total number of binary symbols transmitted. Thus the code rate (R) is given by

$$R = K/N \tag{5.61}$$

And as we see, the code rate is the ratio of the information bits to the total bits (binary symbols) transmitted: this is also the inverse of the bandwidth expansion factor. The more successful block-coding techniques have centered about finite-field algebraic concepts, culminating in various classes of codes that can be generated by means of a linear feedback register.

Linear block codes can be described by a $K \times N$ generator matrix **G**. If the K-symbol encoder input is represented by a K-dimensional column vector **x**, and the encoder output by a K-dimensional column vector **y**, the encoder input–output relationship is given by

$$\mathbf{y} = \mathbf{xG} \tag{5.62}$$

So the N-symbol encoder output blocks are linear algebraic combinations of the rows of the generator matrix. In the binary symbol case, the output blocks are bit-by-bit modulo-2 sums* of the appropriate rows of **G**.

---

*Modulo-2 addition is binary addition without carries. The sign of modulo-2 addition is $\oplus$. For example, $0 \oplus 0 = 0$; $0 \oplus 1 = 1$; $1 \oplus 0 = 1$; and $1 \oplus 1 = 0$.

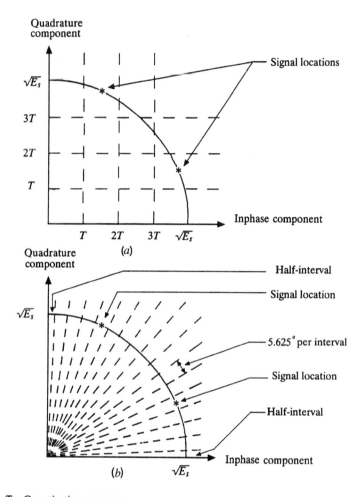

Figure 5.29. Diagrams of the first quadrant signal space quantization intervals for two possible 6-bit quantization techniques for 8-ary PSK. (From Ref. 34.)

Usually block codes are decoded using algebraic techniques that require the demodulator to make a hard decision on each received symbol. Hard quantization reduces the potential performance of a coding system. For BPSK and QPSK modulation, based only on an AWGN channel, the potential coding gain of a finely quantized coding system is about 2 dB more than that of a hard-quantized system. However, block codes are now starting to use soft quantization and some of the 2 dB that was lost is being recovered. However, the implementation complexity of such systems is usually greater than that of a hard-quantized system.

Another disadvantage of block codes as compared to convolutional codes is that with block codes the receiver must resolve an n-way ambiguity to determine the start of a block, whereas with Viterbi- or feedback-decoded convolutional codes, a much smaller ambiguity needs to be resolved.

Block codes are often described with notation such as $(7, 4)$ meaning $N = 7$ and $K = 4$. In this case the information bits are stored in $K = 4$ storage devices and the device is made to shift $N = 7$ times. The first K symbols of the block output are the information symbols, and the last $N - K$ symbols are a set of check symbols that form the whole N-symbol word. A block code may also be identified with the notation (N, K, t), where t corresponds to the number of errors in a block of N symbols that the code will correct.

In the following subsections several specific block codes are reviewed.

5.6.3.1.1 Extended Golay Code. A widely used block code is the extended Golay code where $N = 24$, $K = 12$ [i.e., $(24, 12)$]. It is formed by adding an overall parity bit to the perfect $(23, 12)$ Golay code. The parity bit increases the minimum distance of the code from 7 to 8, producing a rate $\frac{1}{2}$ code, which is easier to work with than the rate $\frac{12}{23}$ of the $(23, 12)$ code.

Extended Golay codes are considerably more powerful than earlier codes such as the Hamming codes. To achieve the improved performance, a more complex decoder at a lower rate is required, involving a larger bandwidth expansion. Decoding algorithms that make use of soft-decision demodulator outputs are available for these codes. When such soft-decision decoding algorithms are used, the performance of an extended Golay code is similar to that of a simple Viterbi-decoded convolutional coding system of constraint length 5 ($K = 5$). While it is difficult to compare the implementation complexity of two different coding systems, it can be concluded that when hard-decision demodulator outputs are available, extended Golay coding systems are of the same approximate complexity as similar performance convolutional coding schemes. However, when soft decisions are available, convolutional coding is superior.

Figure 5.30 gives block, bit, and undetected error probabilities versus E_b/N_0 for BPSK or QPSK modulation, an AWGN channel, and extended Golay coding. Table 5.11 provides E_b/N_0 ratios required to obtain a 1×10^{-5} BER with extended Golay coding and several different modulation/demodulation techniques for AWGN and Rayleigh fading channels.

5.6.3.1.2 Bose–Chaudhuri–Hocquenghem (BCH) Codes. Binary BCH codes are a large class of block codes with a wide range of code parameters. The so-called primitive BCH codes, which are the most common, have codeword lengths of the form $2^m - 1$, $m \geq 3$, where m describes the degree of the generating polynomial (i.e., the highest exponent value of the polynomial). For BCH codes, there is no simple expression for the N, K, t parameters. Table 5.12 gives these parameters for all binary BCH codes of length

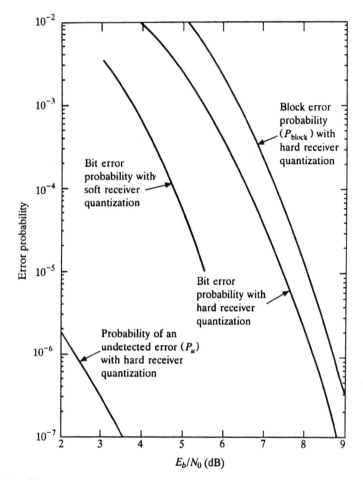

Figure 5.30. Block, bit, and undetected error probabilities versus E_b/N_0 for BPSK and QPSK modulation, an AWGN channel, and extended Golay coding. (From Ref. 34.)

255 and less. In general, for any m and t, there is a BCH code of length $2^m - 1$ that corrects any combination of t errors and that requires no more than m/t parity check symbols.

BCH codes are cyclic and are characterized by a generating polynomial. A selected group of primitive generating polynomials is given in Table 5.13. The encoding of a BCH code can be performed with a feedback shift register of length K or $N - K$ (Ref. 36).

Figure 5.31 gives the BER versus channel error rate performance of several block-length 127 BCH codes. Channel error rate performance can be taken as the uncoded error performance of the modulation type given.

TABLE 5.11 E_b/N_0 Required to Achieve a 1×10^{-5} BER with Extended Golay Coding and Several Modulation/Demodulation Techniques

Type of Interference	Modulation/Demodulation Technique	E_b/N_0 (in dB) Required to Achieve a 10^{-5} Bit Error Rate
AWGN	BPSK or QPSK	7.5 (hard decision), 5.6 (soft decision)
AWGN	Octal-PSK with interleaving	10.4 (hard decision)
AWGN	DBPSK with interleaving	9.0 (hard decision)
AWGN	DQPSK with interleaving	10.0 (hard decision)
Independent Rayleigh fading	Noncoherent binary FSK with optimum diversity (i.e., $L = 8$ or 16 channel bits per information bit with rate $\frac{1}{2}$ coding)	15.9 (hard decision)

Source: Reference 34.

TABLE 5.12 BCH Code Parameters for Primitive Codes of Length 255 and Less

N	K	t	N	K	t	N	K	t
7	4	1	127	106	3	255	179	10
				99	4		171	11
15	11	1		92	5		163	12
	7	2		85	6		155	13
				78	7		147	14
31	26	1		71	9		139	15
	21	2		64	10		131	18
	16	3		57	11		123	19
	11	5		50	13		115	21
	6	7		43	14		107	22
				36	15		99	23
63	57	1		29	21		91	25
	51	2		22	23		87	26
	45	3		15	27		79	27
	39	4		8	31		71	29
	36	5					63	30
	30	6	255	247	1		55	31
	24	7		239	2		47	42
	18	10		231	3		45	43
	16	11		223	4		37	45
	10	13		215	5		29	47
	7	15		207	6		21	55
				199	7		13	59
127	120	1		191	8		9	63
	113	2		187	9			

Source: Reference 36.

TABLE 5.13   Selected Primitive Binary Polynomials

m	$p(x)$
3	$1 + x + x^3$
4	$1 + x + x^4$
5	$1 + x^2 + x^5$
6	$1 + x + x^6$
7	$1 + x^3 + x^7$
8	$1 + x^2 + x^3 + x^4 + x^8$
9	$1 + x^4 + x^9$
10	$1 + x^3 + x^{10}$
11	$1 + x^2 + x^{11}$
12	$1 + x + x^4 + x^6 + x^{12}$
13	$1 + x + x^3 + x^4 + x^{13}$
14	$1 + x + x^6 + x^{10} + x^{14}$
15	$1 + x + x^{15}$

Source: Reference 36.

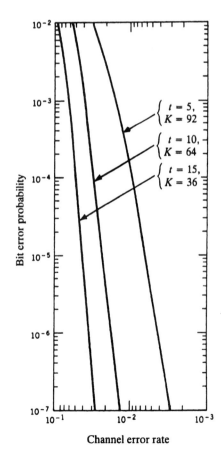

Figure 5.31. Bit error rate probability versus channel error rate performance of several BCH codes with block length of 127. (From Ref. 34.)

5.6.4   Binary Convolutional Codes

Viterbi (Ref. 35) defines a convolutional coder as the following:

> a linear finite state machine consisting of a K-stage shift register and n linear algebraic function generators. The input data which are usually, but not necessarily always, binary, are shifted along the register b bits at a time.

Convolutional codes are defined by a code rate, R, and a constraint length, K. The code rate is the information bit rate into the coder divided by the coder's output symbol rate. The constraint length K is defined as the total number of binary register stages in the coder.

Let us consider a convolutional encoder with $R = \frac{1}{2}$ and $K = 3$. This is shown in Figure 5.32, which indicates the outputs for a particular binary input sequence assuming the state (i.e., the previous two data bits into the shift register) as 0. Modulo-2 addition is used. With the input and output sequences defined from right to left, the first three input bits 0, 1, and 1 generate the code outputs 00, 11, and 01, respectively. The outputs are shown multiplexed into a single code sequence. Of course, the coded sequence has twice the bit rate as the data sequence. We will return to this simple model again.

A tree diagram is often used instructively to show a convolutional code. A typical tree diagram is illustrated in Figure 5.33.

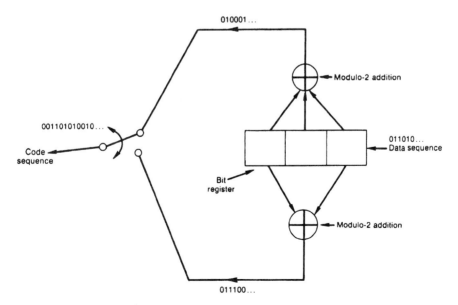

Figure 5.32. Rate $\frac{1}{2}$, constraint length 3, convolutional encoder. (From Ref. 34.)

In Figure 5.33, if the first input bit is a zero, the code symbols are shown on the first upper branch; while if it is a one, the output symbols are shown on the first lower branch. Similarly, if the second input bit is a zero, we can trace the tree diagram to the next upper branch, while if it is a one, the diagram is traced downward. In such a way, all 32 possible outputs may be traced for the first five inputs.

From the tree diagram in Figure 5.33 it also becomes clear that after the first three branches the structure becomes repetitive. In fact, we readily recognize that beyond the third branch the code symbols on branches emanating from the two nodes labeled "a" are identical, and so on, for all the similarly labeled pairs of nodes. The reason for this is obvious from examination of the encoder. As the fourth bit enters the coder at the right, the first data bit falls off on the left and no longer influences the output code symbols. Consequently, the data sequences $100xy \cdots$ and $000xy \cdots$ generate the same code symbols after the third branch and, as is shown in the tree diagram, both nodes labeled "a" can be joined together.

This leads to redrawing the tree diagram as shown in Figure 5.34. This is called a trellis diagram, since a trellis is a treelike structure with remerging branches. A convention is adopted here where the code branches produced by a zero-input bit are shown as solid lines and code branches produced by a one-input bit are shown as dashed lines.

The completely repetitive structure of the trellis diagram suggests a further reduction in the representation of the code to the state diagram in Figure 5.35. The "states" of the state diagram are labeled according to the nodes of the trellis diagram. However, since the states correspond merely to the last two input bits to the coder, we may use these bits to denote the nodes or states of this diagram.

It can be seen that the state diagram can be drawn directly observing the finite-state machine properties of the encoder and, particularly, the fact that a four-state directed graph can be used to represent uniquely the input–output relation of the finite-state machine, since the nodes represent the previous two bits, while the present bit is indicated by the transition branch. For example, if the encoder (synonymous with finite-state machine) contains the sequence 011, this is represented in the diagram by the transition from state $b = 01$ to state $d = 11$ and the corresponding branch indicates the code symbol outputs 01.

This example will be used when we describe the Viterbi decoder.

5.6.4.1 *Convolutional Decoding.* Decoding algorithms for block and convolutional codes are quite different. Because a block code has a formal structure, advantage can be taken of the known structural properties of the words or algebraic nature of the constraints among the symbols used to represent an information sequence. One class of powerful codes with well-defined algorithms is the BCH codes described in Section 5.6.3.1.2.

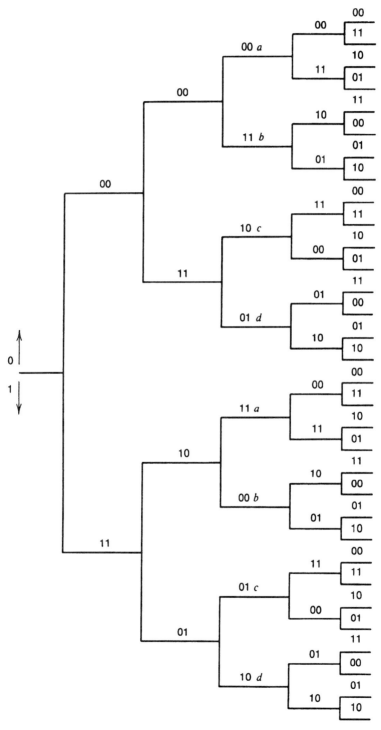

Figure 5.33. Tree representation of the coder shown in Figure 5.32. (From Ref. 34.)

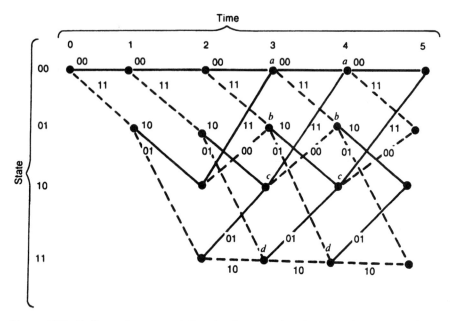

Figure 5.34. Trellis code representation for the encoder shown in Figure 5.32. (From Ref. 34.)

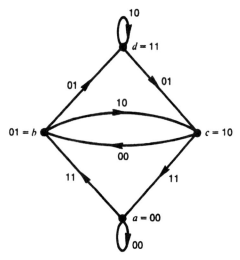

Figure 5.35. State diagram representation for the encoder shown in Figure 5.32. (From Ref. 34.)

The decoding of convolutional codes can be carried out by a number of different techniques. Among these are the simpler threshold and feedback decoders and those more complex decoders with improved performance (coding gain) such as the Viterbi decoder and sequential decoder. These techniques depend on the ability to home in on the correct sequence by designing efficient search procedures that discard unlikely sequences quickly. The sequential decoder differs from most other types of decoders in that when it finds itself on the wrong path in the tree, it has the ability to search back and forth, changing previously decoded information bits until it finds the correct tree path. The frequency with which the decoder has to search back and the depth of the backward searches are dependent on the value of the channel BER.

An important property of the sequential decoder is that, if the constraint length is large enough, the probability that the decoder will make an error approaches zero (i.e., a BER better than 1×10^{-9}). One cause of error is overflow, being defined as the situation in which the decoder is unable to perform the necessary number of computations in the performance of the tree search. If we define a computation as a complete examination of a path through the decoding tree, a decoder has a limit on the number of computations that it can make per unit time. The number of searches and computations is a function of the number of errors arriving at the decoder input, and the number of computations that must be made to decode one information bit is a random variable. An important parameter for a decoder is the average number of computations per decoded information bit. As long as the probability of bit error is not too high, the chances of decoder overflow will be low, and satisfactory performance will result.

For the previous discussion it has been assumed that the output of a demodulator has been a hard decision. By "hard" decision we mean a firm, irrevocable decision. If these were soft decisions instead of hard decisions, additional improvement in error performance (or coding gain) on the order of several decibels can be obtained. By a "soft" decision we mean that the output of a demodulator is quantized into four or eight levels (e.g., 2- or 3-bit quantization, respectively), and then certain decoding algorithms can use this additional information to improve the output BER. Sequential and Viterbi decoding algorithms can use this soft-decision information effectively, giving them an advantage over algebraic decoding techniques, which are not designed to handle the additional information provided by the soft decision (Ref. 34).

The soft-decision level of quantization is indicated conventionally by the letter Q, which indicates the number of bits in the quantized decision sample. If $Q = 1$, we are dealing with a hard-decision demodulator output; $Q = 2$ indicates a quantization level of 4; $Q = 3$ a quantization level of 8; and so on.

5.6.4.1.1 Viterbi Decoding. The Viterbi decoder is one of the more common decoders on links using convolutional codes. The Viterbi decoding algorithm is a path maximum-likelihood algorithm that takes advantage of the remerging path structure (see Figure 5.34) of convolutional codes. By path maximum-likelihood decoding we mean that of all the possible paths through the trellis, a Viterbi decoder chooses the path, or one of the paths, most likely in the probabilistic sense to have been transmitted. A brief description of the operation of a Viterbi decoder using a demodulator giving hard decisions is now given.

For this description our model will be a binary symmetric channel (i.e., BPSK, BFSK). Errors that transform a channel code symbol 0 to 1 or 1 to 0 are assumed to occur independently from symbols with a probability of p. If all input (message) sequences are equally likely, the decoder that minimizes the overall path error probability for any code, block or convolutional, is one that examines the error-corrupted received sequence, which we may call $y_1, y_2, \ldots, y_j \ldots$, and chooses the data sequence corresponding to that sequence that was transmitted or $x_1, x_2, \ldots, x_j \ldots$, which is closest to the received sequence as measured by the Hamming distance. The Hamming distance can be defined as the transmitted sequence that differs from the received sequence by the minimum number of symbols.

Consider the tree diagram (typically Figure 5.33). The preceding statement tells us that the path to be selected in the tree is the one whose code sequence differs in the minimum number of symbols from the received sequence. In the derived trellis diagram (Figure 5.34) it was shown that the transmitted code branches remerge continually. Thus the choice of possible paths can be limited in the trellis diagram. It is also unnecessary to consider the entire received sequence at any one time to decide upon the most likely transmitted sequence or minimum distance. In particular, immediately after the third branch (Figure 5.34) we may determine which of the two paths leading to node or state "a" is more likely to have been sent. For example, if 010001 is received, it is clear that this is a Hamming distance 2 from 000000, while it is a distance 3 from 111011. As a consequence, we may exclude the lower path into node "a." For no matter what the subsequent received symbols will be, they will affect the Hamming distances only over subsequent branches after these two paths have remerged and, consequently, in exactly the same way. The same can be said for pairs of paths merging at the other three nodes after the third branch. Often, in the literature, the minimum distance path of the two paths merging at a given node is called the "survivor." Only two things have to be remembered: the minimum distance path from the received sequence (or survivor) at each node and the value of that minimum distance. This is necessary because at the next node level we must compare two branches merging at each node level, which were survivors at the previous level for different nodes. This can be seen in Figure 5.34 where the comparison at node "a" after the fourth branch is among the survivors of the comparison of nodes "a" and "c" after the third branch. For example, if the received sequence over the first four branches is 01000111, the survivor at the third node level for node "a" is 000000 with distance 2, and at node "c" it is 110101, also with distance 2. In going from the third node level to the fourth, the received sequence agrees precisely with the survivor from "c" but has distance 2, from the survivor from "a." Hence the survivor at node "a" of the fourth level is the data sequence 1100 that produced the code sequence 11010111, which is at minimum distance 2 from the received sequence.

In this way we may proceed through the received sequence and at each step preserve one surviving path and its distance from the received sequence, which is more generally called a "metric." The only difficulty that may arise is the possibility that, in a given comparison between merging paths, the distances or metrics are identical. In this case we may simply flip a coin, as is done for block code words at equal distances from the received sequence. For even if both equally valid contenders were preserved, additional received symbols would affect both metrics in exactly the same way and thus not further influence our choice.

Another approach to the description of the algorithm can be obtained from the state diagram representation given in Figure 5.35. Suppose we sought that path around the directed state diagram arriving at node "a" after the kth transition, whose code symbols are at a minimum distance from the received sequence. But, clearly, this minimum distance path to node "a" at time k can only be one of two candidates: the minimum distance path to node "a" at time $k - 1$ and the minimum distance path to node "c" at time $k - 1$. The comparison is performed by adding the new distance accumulated in the kth transition by each of these paths to their minimum distances (metrics) at time $k - 1$.

Thus it appears that the state diagram also represents a system diagram for this decoder. With each node or state, we associate a storage register that remembers the minimum distance path into the state after each transition as well as a metric register that remembers its (minimum) distance from the received sequence. Furthermore, comparisons are made at each step between the two paths that lead into each node. Thus four comparators must also be provided.

We will expand somewhat on the distance properties of convolutional codes following the example given in Figure 5.32. It should be noted that, as with linear block codes, there is no loss in generality in computing the distance from the all-zeros codeword to all other codewords, for this set of distances is the same as the set of distances from any specific codeword to all the others.

For this purpose we may again use either the trellis diagram or state diagram. First we redraw the trellis diagram in Figure 5.34, labeling the branches according to their distances from the all-zeros path. Now consider all the paths that merge with the all-zeros path for the first time at some arbitrary node "j." From the redrawn trellis diagram (Figure 5.36), it can be seen that of these paths there will be just one path at distance 5 from the all-zeros path and this diverged from it three branches back. Similarly, there

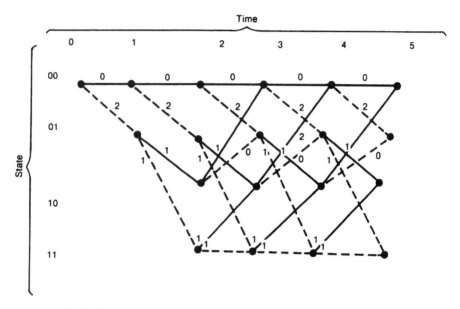

Figure 5.36. Trellis diagram labeled with distances from the all-zeros path. (From Ref. 34.)

are two at distance 6 from it, one that diverged four branches back and the other that diverged five branches back, and so forth. It should be noted that the input bits for the distance 5 path are $00 \cdots 01000$ and thus differ in only one input bit from the all-zero path. The minimum distance, sometimes called the "minimum free distance," among all paths is thus seen to be 5. This implies that any pair of channel errors can be corrected, for two errors will cause the received sequence to be at a distance 2 from the transmitted (correct) sequence, but it will be at least at distance 3 from any other possible code sequence. In this way the distances of all paths from all-zeros (or any arbitrary) path can be determined from the trellis diagram.

5.6.4.2 Systematic and Nonsystematic Convolutional Codes. The term
"systematic" convolutional code refers to a code on each of whose branches the uncoded information bits are included in the encoder output bits generated by that branch. Figure 5.37 shows an encoder for rate $\frac{1}{2}$ and $K = 2$ that is systematic.

For linear block codes, any nonsystematic code can be transformed into a systematic code with the same block distance properties. This is not the case for convolutional codes because the performance of a code on any channel depends largely on the relative distance between codewords and, particularly, on the minimum free distance. Making the code systematic, in general, reduces the maximum possible free distance for a given constraint length and code rate. For example, the maximum minimum-free-distance systematic

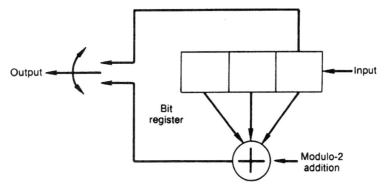

Figure 5.37. Systematic convolutional encoder, $K = 3$, $R = \frac{1}{2}$. (From Ref. 34.)

code for $K = 3$ is that of Figure 5.36 and this has $d = 4$, while the nonsystematic $K = 3$ code of Figure 5.32 has a minimum free distance of $d = 5$. Table 5.14 shows the maximum free distance for $R = \frac{1}{2}$ systematic and nonsystematic codes for $K = 2$ through 5. It should be noted that for large constraint lengths the results are even more widely separated.

5.6.5 Channel Performance of Uncoded and Coded Systems

5.6.5.1 Uncoded Performance. For uncoded systems a number of modulation implementations are reviewed in the presence of additive white Gaussian noise (AWGN) and with Rayleigh fading. The AWGN performance of BPSK, QPSK, and 8-ary PSK is shown in Figure 5.38. AWGN is typified by thermal noise. The demodulator for this system requires a coherent phase reference.

Another similar implementation is differentially coherent phase shift keying. This is a method of obtaining a phase reference by using the previously received channel symbol. The demodulator makes its decision based on the change in phase from the previous to the present received channel symbol.

TABLE 5.14   Comparison of Systematic and Nonsystematic $R = \frac{1}{2}$ Code Distances

	Maximum, Minimum Free Distance	
K	Systematic	Nonsystematic
2	3	3
3	4	5
4	4	6
5	5	7

Source: Reference 34.

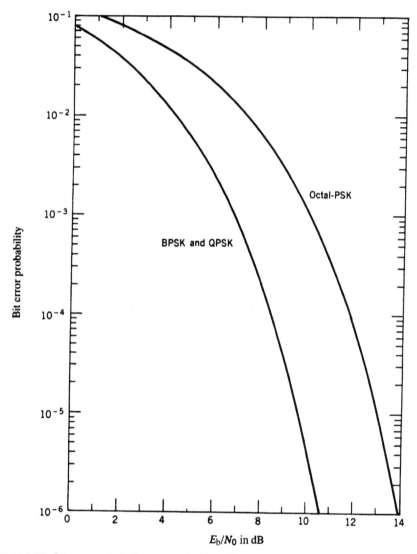

Figure 5.38. Bit error probability versus E_b/N_0 performance of BPSK, QPSK, and octal PSK (8-ary PSK). (From Ref. 34.)

Figure 5.39 gives the performance of DBPSK and DQPSK with values of BER versus E_b/N_0.

Figure 5.40 gives performance for M-ary FSK.

Independent Rayleigh fading can be assumed during periods of heavy rainfall on satellite links operating above about 10 GHz (see Chapter 8). Such fading can severely degrade error rate performance. The performance with this type of channel can be greatly improved by providing some type of

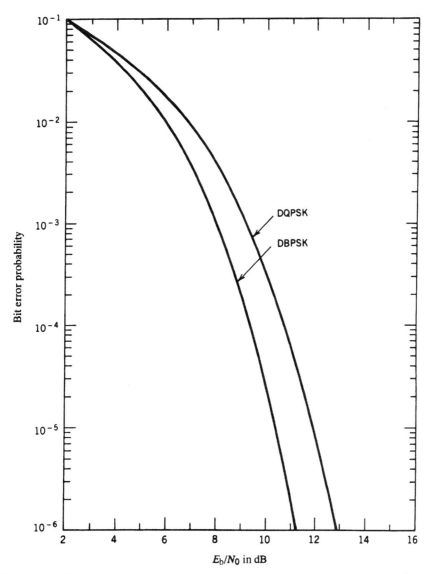

Figure 5.39. Bit error probability versus E_b/N_0 performance of DBPSK and DQPSK. (From Ref. 34.)

diversity. Here we mean providing several independent transmissions for each information symbol. In this case we will restrict the meaning to some form of time diversity that can be achieved by repeating each information symbol several times and using interleaving/deinterleaving for the channel symbols. Figure 5.41 gives binary bit error probability for several orders of diversity (L = order of diversity; $L = 1$, no diversity) for the mean bit

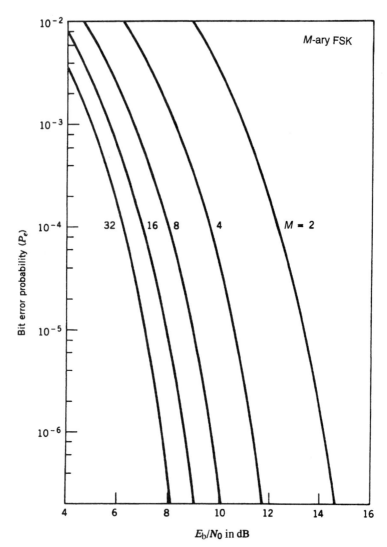

Figure 5.40. Bit error probability versus E_b/N_0 for M-ary FSK. $m = 2$ for BFSK. (From Ref. 34.)

energy-to-noise ratio $(\overline{E}_b/N_0)$. This figure shows that for a particular error rate there is an optimum amount of diversity. The modulation is binary FSK.

Table 5.15 recaps error performance versus E_b/N_0 for the several modulation types considered. The reader should keep in mind that the values for E_b/N_0 are theoretical values. A certain modulation implementation loss should be added for each case to derive practical values. The modulation implementation loss value in each case is equipment driven.

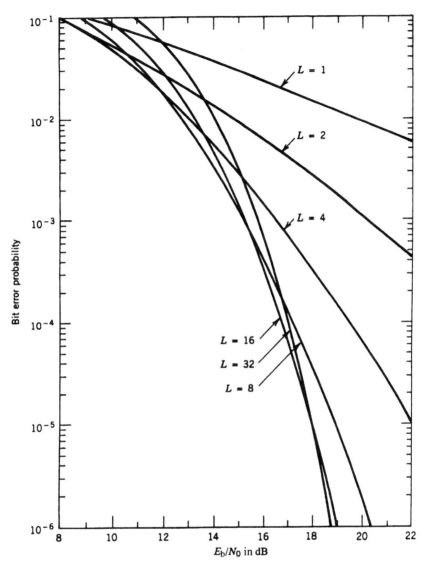

Figure 5.41. Bit error probability versus E_b/N_0 performance of binary FSK on a Rayleigh fading channel for several orders of diversity. L = order of diversity. (From Ref. 34.)

5.6.5.2 Coded Performance. Figures 5.42, 5.43, and 5.44 show the BER performance for $K = 7$, $R = \frac{1}{2}$; $K = 7$, $R = \frac{1}{3}$; and $K = 9$, $R = \frac{3}{4}$ convolutional coding systems on an AWGN channel with hard and 3-bit soft quantization. Figures 5.42 and 5.44 again illustrate the advantages of soft quantization discussed in Section 5.6.2.2.

TABLE 5.15   Summary of Uncoded System Performance

Channel	Modulation/ Demodulation	E_b/N_0 (dB) Required for Given Bit Error Rate						
		10^{-1}	10^{-2}	10^{-3}	10^{-4}	10^{-5}	10^{-6}	10^{-7}
Additive white Gaussian noise	BPSK and QPSK	−0.8	4.3	6.8	8.4	9.6	10.5	11.3
	Octal-PSK	1.0	7.3	10.0	11.7	13.0	13.9	14.7
	DBPSK	2.1	5.9	7.9	9.3	10.3	11.2	11.9
	DQPSK	2.1	6.8	9.2	10.8	12.0	12.9	13.6
	Noncoherently demodulated binary FSK	5.1	8.9	10.9	12.3	13.4	14.2	14.9
	Noncoherently demodulated 8-ary MFSK	2.0	5.2	7.0	8.2	9.1	9.9	10.5
Independent Rayleigh fading	Binary FSK, $L = 1$	9.0	19.9	30.0	40.0	50.0	60.0	70.0
	Binary FSK, $L = 2$	7.9	14.8	20.2	25.3	30.4	35.4	40.4
	Binary FSK, $L = 4$	8.1	13.0	16.5	19.4	22.1	24.8	27.3
	Binary FSK, $L = 8$	8.7	12.8	15.3	17.2	18.9	20.5	22.0
	Binary FSK, $L = 16$	9.7	13.2	15.3	16.7	18.0	19.1	20.0
	Binary FSK, $L = 32$	10.9	14.1	15.8	17.1	18.1	18.9	19.7

Source: Reference 34.

Reference 34 reports that simulation results have been run for many other codes (from those discussed) on an AWGN channel. The results show that, for rate $\frac{1}{2}$ codes, each increment increase in the constraint length in the range of $K = 3$ to $K = 8$ provides an approximate 0.4–0.5 dB improvement in E_b/N_0 at a BER of 1×10^{-5}.

Figure 5.45 shows the 3-bit quantization coding gain for the codes shown in Figures 5.42, 5.43, and 5.44.

Figure 5.46 shows bit error probability versus E_b/N_0 performance of a $K = 7$, $R = \frac{1}{2}$ convolutional coding system with DBPSK modulation and an AWGN channel. The performance is notably degraded compared to a coherent PSK system as shown in Figure 5.38.

Table 5.16 recapitulates coding gain information for several modulation types using $K = 7$, $R = \frac{1}{2}$ convolutional coding and a Viterbi decoder assuming a BER of 1×10^{-5}. Table 5.17 summarizes E_b/N_0 requirements of several coded communication systems with BPSK modulation and a BER of 1×10^{-5}.

5.6.6   Coding with Bursty Errors

In our entire discussion so far we have limited our communication channels to additive white Gaussian noise (AWGN) and *only* AWGN. As we mentioned, this would suffice for well-behaved propagation media. Typically by "well-behaved" media we list wire-pair, coaxial cable, fiber-optic cable, very

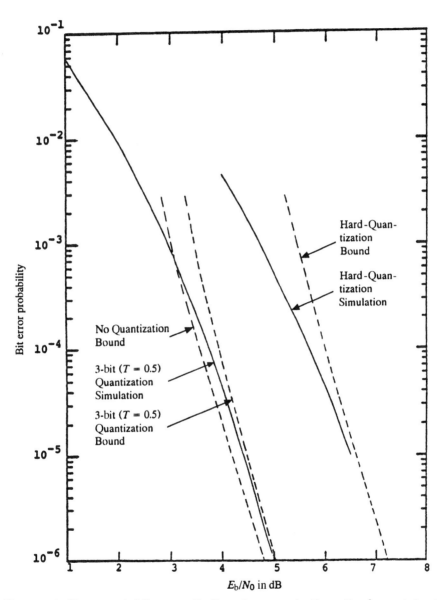

Figure 5.42. Bit error probability versus E_b/N_0 performance of a $K = 7$, $R = \frac{1}{2}$ convolutional coding system with BPSK modulation and an AWGN channel. (From Ref. 34.)

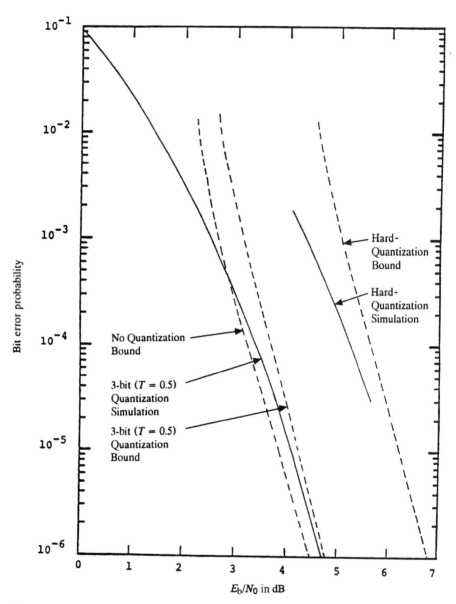

Figure 5.43. Bit error probability versus E_b/N_0 performance of $K = 7$, $R = \frac{1}{3}$ convolutional coding system with BPSK modulation and an AWGN channel. (From Ref. 34.)

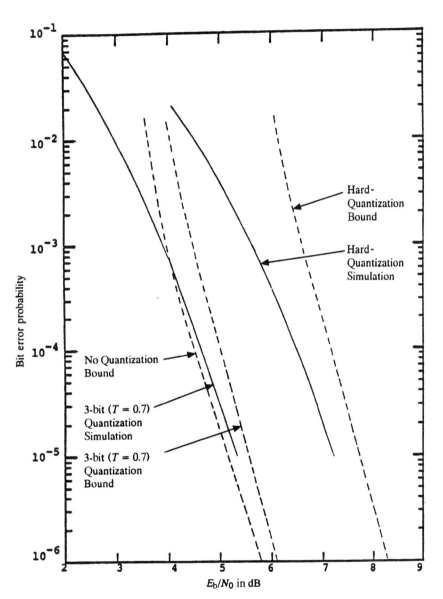

Figure 5.44. Bit error probability versus E_b/N_0 performance of a $K = 9$, $R = \frac{3}{4}$ convolutional coding system with BPSK modulation and an AWGN channel. (From Ref. 34.)

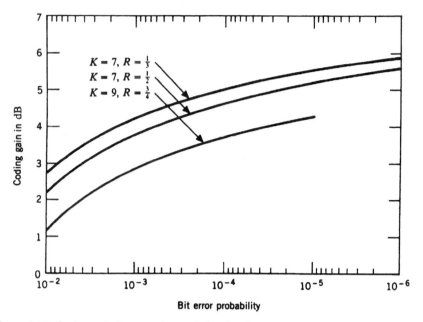

Figure 5.45. Coding gain for several convolutional codes with BPSK modulation, AWGN, and 3-bit receiver quantization. (From Ref. 34.)

short LOS microwave links [e.g., <1 mi (1.6 km)], and satellite links above a 5° elevation angle and operating below 10 GHz.*

Poorly behaved (fading) media are MF/HF radio, all LOS microwave on links over 1-mi (1.6-km) range, tropospheric scatter and diffraction circuits, and satellite communication paths above 10 GHz. There is an easy "fix," but beware of added delay.

There are basically two causes of error bursts (vis-à-vis random errors): fading and impulse noise. The characteristics of each are quite different, but the result is the same—error bursts. We will make an assumption here that during the period when there are no error bursts, the signal is benign with only a very few random errors.

It should be noted that impulse noise could be found on any of the media. However, the chances are very slight that impulse noise occurred on fiber-optic circuits. It could "invade" these circuits when still in the electrical domain.

The objective here is to randomize the bit stream. To do this we use a device called an *interleaver*. An interleaver takes blocks of bits and pseudo-randomizes the data in each block. It is the interleaver that "blocks" the

---

*Purists will note certain exceptions in the low UHF and VHF bands operating in the vicinity of the magnetic equator.

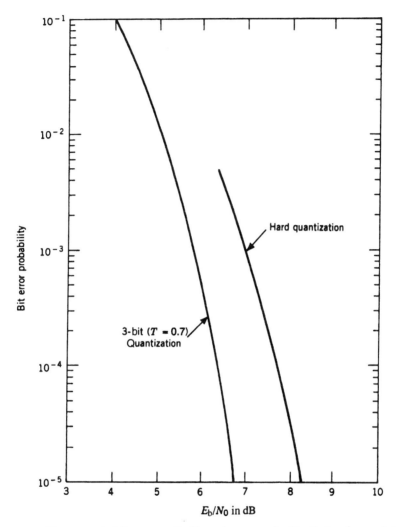

Figure 5.46. Bit error probability versus E_b/N_0 performance of a $K = 7$, $R = \frac{1}{2}$ convolutional coding system with DBPSK modulation and an AWGN channel. (From Ref. 34.)

data, and we will call each block a *span*. We use the term span because it is a time span. This span must be greater than the duration of a fade.

Suppose we found that, statistically, fades seldom had durations greater than 0.9 s. We might make the span have a 1-s duration. What would be 1 s of data? At 9600 bps, it would be 9600 bits or binary coded symbols. So the interleaver would break up the binary symbol stream into a span of at least 9600 symbols. Suppose it was a bit stream with DS1 (T1) at 1.544 Mbps and the coding rate was $\frac{1}{2}$. Thus the symbol rate out of the coder would be

TABLE 5.16 Summary of E_b/N_0 Requirements and Coding Gains of $K = 7$, $R = \frac{1}{2}$ Viterbi-Decoded Convolutional Coding Systems with Several Modulation Types, BER $= 1 \times 10^{-5}$

Modulation	Number of Bits of Receiver Quantization per Binary Channel Symbol	E_b/N_0 (dB) Required for $P_b = 10^{-5}$	Coding Gain (dB)
Coherent biphase			
BPSK or QPSK	3	4.4	5.2
BPSK or QPSK	2	4.8	4.8
BPSK or QPSK	1	6.5	3.1
Octal-PSK[a]	1	9.3	3.7
DBPSK[a]	3	6.7	3.6
DBPSK[a]	1	8.2	2.1
Differentially[a] coherent QPSK	1	9.0	3.0
Noncoherently demodulated binary FSK	1	11.2	2.1

[a]Interleaving/deinterleaving assumed.

Source: Reference 34.

2×1.544 Mbps or 3.088 Mbps, which would be the length of the span or 1 s worth of binary symbols.

The danger here is the added delay of the interleaver. The span time is directly additive to the total delay. This might be quite acceptable on circuits carrying data but may be highly unacceptable on circuits carrying voice or even conference television. If the span time is only a few milliseconds, it might not alter the delay budget enough to affect speech telephony circuits. A block diagram of a typical interleaver implementation is shown in Figure 5.47.

One important aspect of interleaving is that the interleaver and companion deinterleaver must be time synchronized.

5.7 LINK CALCULATIONS FOR DIGITAL LOS MICROWAVE SYSTEMS

5.7.1 Review of E_b / N_0

N_0 is simply the thermal noise in 1 Hz of bandwidth or

$$N_0 = -204 \text{ dBW/Hz} + \text{NF}_{\text{dB}} \tag{5.63}$$

for receivers operating at room temperature.

TABLE 5.17   Summary of E_b/N_0 Requirements of Several Coded Communication Systems for a BER $= 1 \times 10^{-5}$ with BPSK Modulation

	Coding Type	Number of Bits Receiver Quantization	Coding[a] Gain (dB)
$K = 7,\ \ R = \frac{1}{2}$	Viterbi-decoded convolutional	1	3.1
$K = 7,\ \ R = \frac{1}{2}$	Viterbi-decoded convolutional	3	5.2
$K = 7,\ \ R = \frac{1}{3}$	Viterbi-decoded convolutional	1	3.6
$K = 7,\ \ R = \frac{1}{3}$	Viterbi-decoded convolutional	3	5.5
$K = 9,\ \ R = \frac{3}{4}$	Viterbi-decoded convolutional	1	2.4
$K = 9,\ \ R = \frac{3}{4}$	Viterbi-decoded convolutional	3	4.3
$K = 24,\ R = \frac{1}{2}$	Sequential-decoded convolutional 20 kbps,[b] 1000-bit blocks	1	4.2
$K = 24,\ R = \frac{1}{2}$	Sequential-decoded convolutional 20 kbps,[b] 1000-bit blocks	3	6.2
$K = 10,\ L = 11,\ R = \frac{1}{2}$	Feedback-decoded convolutional	1	2.1
$K = 8,\ \ \ L = 8,\ \ R = \frac{2}{3}$	Feedback-decoded convolutional	1	1.8
$K = 8,\ \ \ L = 9,\ \ R = \frac{3}{4}$	Feedback-decoded convolutional	1	2.0
$K = 3,\ \ \ L = 3,\ \ R = \frac{3}{4}$	Feedback-decoded convolutional	1	1.1
(24, 12) Golay		3	4.0
(24, 12) Golay		1	2.1
(127, 92) BCH		1	3.3
(127, 64) BCH		1	3.5
(127, 36) BCH		1	2.3
(7, 4) Hamming		1	0.6
(15, 11) Hamming		1	1.3
(31, 26) Hamming		1	1.6

[a]9.6 dB required for uncoded system.
[b]The same system at a data rate of 100 kbps has 0.5 dB less coding gain.

Notation

K Constraint length of a convolutional code defined as the number of binary register stages in the encoder for such a code. With the Viterbi-decoding algorithm, increasing the constraint length increases the coding gain but also the implementation complexity of the system. To a much lesser extent the same is also true with sequential and feedback decoding algorithms.

L Look-ahead length of a feedback-decoded convolutional coding system defined as the number of received symbols, expressed in terms of the corresponding number of encoder input bits, that are used to decode an information bit. Increasing the look-ahead length increases the coding gain but also the decoder implementation complexity.

(n, k) Denotes a block code (Golay, BCH, or Hamming here) with n decoder output bits for each block of k encoder input bits.

Receiver   Quantization describes the degree of quantization of the demodulator outputs. Without coding and biphase ($0°$ or $180°$) modulation the demodulator output (or intermediate output if the quantizer is considered as part of the demodulator) is quantized to one bit (i.e., the sign if provided). With coding, a decoding decision is based on several demodulator outputs, and the performance can be improved if in addition to the sign the demodulator provides some magnitude information.

Source: Reference 34.

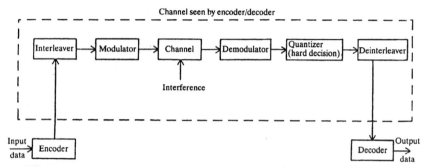

Figure 5.47. Block diagram of a coding system using interleaving and deinterleaving.

Example 1. A certain receiver has a 3-dB noise figure. What is the value of N_0?

$$N_0 = -204 \text{ dBW} + 3 \text{ dB}$$

$$= -201 \text{ dBW/Hz}$$

E_b is the signal energy per bit (*not* symbol or baud). Restated:

$$E_b = \text{RSL}_{dBW} - 10\log(\text{bit rate}_{bps}) \tag{5.64}$$

Example 2. The RSL of a certain receiver is -87 dBW and the bit rate is 1.544 Mbps. What is E_b?

$$E_b = -87 \text{ dBW} - 10\log(1.544 \times 10^6)$$

$$= -87 \text{ dBW} - 61.89 \text{ dB}$$

$$= -148.89 \text{ dBW}$$

We can write an equation for E_b/N_0 by combining equations 5.63 and 5.64:

$$E_b/N_0 = E_{b(dBW)} - N_{0(dBW)} \tag{5.65}$$

Then

$$E_b/N_0 = \text{RSL}_{dBW} - 10\log(\text{bit rate}) - (-204 \text{ dBW} + \text{NF}_{dB})$$

$$= +204 \text{ dBW} + \text{RSL}_{dBW} - 10\log(\text{bit rate}) - \text{NF}_{dB} \tag{5.66}$$

Calculate E_b/N_0 from the values of E_b and N_0 from the preceding examples:

$$E_b/N_0 = -148.9 \text{ dBW} + 201 \text{ dBW/Hz}$$

$$= 52.1 \text{ dB}$$

5.7.2 Link Calculation Procedures

The procedure for digital radiolink analysis is very similar to its analog counterpart described in Section 5.4 or simply:

- Calculate EIRP.
- Algebraically add FSL and other losses due to the medium (P_L) such as gaseous absorption loss, if applicable.
- Add receiving antenna gain (G_r).
- Algebraically add the line losses (L_{Lr}).

Inserting minus signs for losses, we have the following familiar equation to calculate RSL (receive signal level):

$$\text{RSL}_{dBW} = \text{EIRP}_{dBW} - \text{FSL}_{dB} - P_L + G_r - L_{Lr} \qquad (5.67)$$

In Section 5.5.3.2 we introduced an expression for E_b/N_0. We will use equipment manufacturers' error performance curves to determine a value for E_b/N_0 given a certain required BER. The manufacturer should also provide a modulation implementation loss or that loss is included in the E_b/N_0 value. A typical curve is shown in Figure 5.24 or the values in the "W" column may be used from Table 5.9. The following example may be useful.

Example. Assume a 15-mi (24-km) path with an operating frequency at 6 GHz on a link designed to transmit STS-3 (155.520 Mbps). The 64-QAM modulation will be employed requiring a 30-MHz bandwidth. The required BER is 1×10^{-6}; thus the theoretical E_b/N_0 is 21.4 dB. The modulation implementation loss is 2.0 dB, resulting in a practical E_b/N_0 of 23.4 dB. Assume no fading and a zero margin; the receiver noise figure is 5.0 dB, and the waveguide losses at each end are 1.5 dB. Find a reasonable transmitter output power and antenna aperture to meet these conditions.

Calculate N_0 using equation 5.63:

$$N_0 = -204 \text{ dBW} + 5 \text{ dB}$$

$$= -199 \text{ dBW}$$

We can now calculate E_b, because we know that its level must be 23.4 dB higher than the N_0 level (i.e., $E_b = 23.4 + N_0$). Then

$$E_b = -175.6 \text{ dBW}$$

Calculate RSL_{dBW}:

$$\begin{aligned}
\text{RSL}_{\text{dBW}} &= E_b + 10\log(\text{bit rate}) \\
&= E_b + 10\log(155{,}520{,}000) \\
\text{RSL} &= -175.6 + 81.92 \text{ dB} \\
&= -93.68 \text{ dBW}
\end{aligned}$$

Calculate FSL:

$$\begin{aligned}
\text{FSL} &= 36.58 + 20\log 15 + 20\log 6000 \\
&= 135.66 \text{ dB}
\end{aligned}$$

Apply equation 5.16:

$$\text{RSL}_{\text{dBW}} = \text{EIRP}_{\text{dBW}} - 135.66 \text{ dB} + G_r - 1.5 \text{ dB}$$

Simplifying this expression, we find

$$\text{RSL}_{\text{dBW}} = \text{EIRP}_{\text{dBW}} - 137.36 + G_r$$

G_r is the gain of the receive antenna; let G_t be the gain of the transmit antenna. Let the aperture of each antenna be 2 ft (use equation 5.71a).

$$\begin{aligned}
G_t = G_r &= 20\log 2 + 20\log 6 + 7.5 \\
&= 29.08 \text{ dB}
\end{aligned}$$

Calculate the EIRP (use equation 5.14):

$$\text{EIRP}_{\text{dBW}} = P_0 + L_L + G_t$$

Let the transmitter output be 1 watt or 0 dBW; then

$$\begin{aligned}
\text{EIRP} &= 0 - 1.5 \text{ dB} + 29.08 \text{ dB} \\
&= +27.58 \text{ dBW}
\end{aligned}$$

Now make a "trial" run for RSL (equation 5.16):

$$\begin{aligned}
\text{RSL}_{\text{dBW}} &= +27.58 \text{ dBW} - 137.36 \text{ dB} - 1.5 \text{ dB} + 29.08 \text{ dB} \\
&= -82.20 \text{ dBW}
\end{aligned}$$

The required RSL, with no margin, was -94.58 dBW. We end up with a 12.38-dB margin. To bring the link budget down 12.38 dB, we can reduce antenna sizes and/or reduce transmit power. Reducing the antennas to 1-ft dishes at each end would allow us to subtract 6.02×2 or 12.04 dB (we multiply by 2 because there are two antennas, a transmit and a receive antenna) and we are left with 0.34 dB in excess of what we wanted. We could reduce the transmitter output power by 0.34 dB. The resulting output power would be 0.92 watts.

Of course, in practice, we would want a margin, depending on the time availability desired.

5.8   DIGITAL LOS MICROWAVE TERMINAL SYSTEMS

5.8.1   Transmitter / Receiver Subsystem

Figure 5.48 is a functional block diagram of a digital microwave transmitter/receiver. Starting from the bottom of the figure on the transmit side (left side), there are a number of baseband processing functions. The line code converter takes the standard PCM line code (AMI, HDB3), which has been implemented for good baseband transmission properties, and converts the code, usually to a nonreturn-to-zero (NRZ) format. Then the resulting code is scrambled using typical scrambling (pseudorandomizing) techniques with a generating polynomial. This action tends to remove internal correlation among symbols such as long strings of 0's or 1's. The output waveform of the scrambler has a comparatively constant power spectrum.

The signal may then be channel coded (optional) for forward error correction, although this is not commonly done on terrestrial radiolinks. Differential coding/decoding is one method used to remove phase ambiguity at the receive end at a cost of 1 or 2 dB of E_b/N_0. A coherent system would not require this function. However, a phase coherence base would be required.

A serial-to-parallel converter divides the serial bit stream into two components for the I and Q inputs of the phase modulator (see Figure 5.19). Multilevel coders and decoders are required for 8-ary PSK and QAM modulation schemes. In the cases of 16-QAM, 32-QAM, and above, more than two amplitude levels are used per orthogonal modulation. Therefore the digital data have to be converted to multilevel logic. For instance, 64-QAM requires 8-level modulation signals.

The predistorter is designed to compensate for the distortion imparted on the signal by the power amplifier.

The digital modulator, of course, carries out the modulation function. Modulation to achieve spectral efficiency was discussed in Section 5.5.3.4. The output of the modulator is then amplified, filtered, and passed to the upconverter. The upconverter translates this signal to the operating frequency of the terminal. The output of the upconverter is then fed to the high-power amplifier (HPA), which amplifies the signal to the desired output

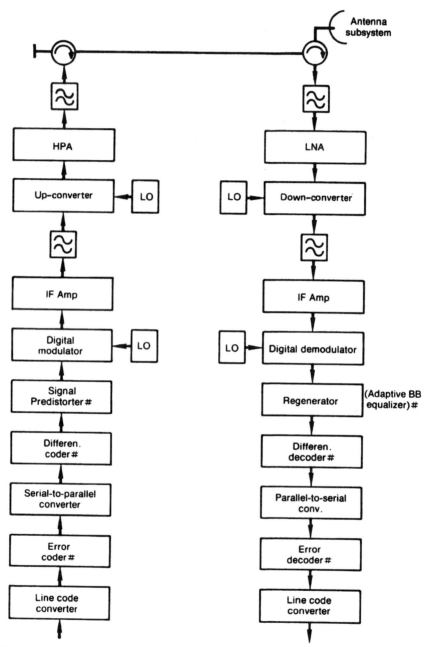

Figure 5.48. Typical functional block diagram of a microwave digital radio terminal. Typical PCM line codes: AMI, polar, BNZS, and HDB3/CMI. BB = baseband, # = optional.

level. The HPA may be a TWT or a solid-state amplifier (SSA). Commonly, power outputs of HPAs for microwave LOS operation are 0.1, 1, or, in some cases, 10 W. The output of the HPA generally incorporates a bandpass filter to reduce spurious and harmonic out-of-band signals. The signal is then fed to the antenna subsystem for radiation to the distant end.

The received signal, starting from the antenna downward in Figure 5.48 (right side), is fed from the antenna subsystem through a bandpass filter through the low-noise amplifier (LNA) to a downconverter. In many installations the LNA is omitted. The decision is driven by economics. If the link can tolerate additional thermal noise, some savings can be made on the installation. Common downconverters display noise figures from 5 to 10 dB, whereas GaAs field-effect transistor (FET) or pseudomorphic high electron mobility (PHEMT) LNAs display noise figures from 0.75 to 2.5 dB to 20 GHz and, above 20 GHz, 2.5–4 dB.

The downconverter translates the incoming signal to the IF, which is commonly 70 or 140 MHz. These are the more common IF frequencies used in the industry, and it does not mean that other IF frequencies cannot be used, such as 300, 600, and 700 MHz.

The output of the downconverter is then fed through a filter to an IF amplifier prior to inputting the digital demodulator. Excellent linearity throughout the receive chain is extremely important, particularly on high-bit-rate equipments using higher-order modulation schemes such as 64-QAM.

Demodulation and regeneration are probably the most important elements in the digital microwave receiver. The principal purpose, of course, is to achieve a demodulator serial bit stream output that is undistorted and comparatively free of intersymbol interference.

Coherent demodulation is commonly used. Reference phase has to be maintained. The vector status of the modulated signal is compared with that of the carrier. However, since the carrier is not available in the received modulation signal, it has to be reproduced from it. One method that can be used for an AM-SSB signal is to employ a squaring process, where a discrete spectral component is available at twice the frequency of the carrier from which a carrier can then be derived. Another method is to use a Costas loop for carrier regeneration.

After demodulation, the data clock at the incoming symbol rate is regenerated from the baseband digital bit stream. The clock is usually derived in a similar manner as the carrier. A phase-locked loop (PLL) is synchronized to the spectral component occurring at the clock rate. Here, timing jitter is a major impairment that should be prevented or minimized. Specifications on jitter are called out in CCITT Rec. G.703 (Ref. 37).

The demodulated signal is then regenerated by means of the regenerated clock and a sample-and-hold circuit is applied to regenerate the actual signal. For modulation schemes using M-PSK, the I and Q channel sequences are regenerated separately. The regenerative repeater may also incorporate a baseband equalizer to reduce signal distortion. Methods of reducing distortion on a digital radiolink are discussed in Section 5.5.4.

For the case of I and Q bit streams, the combining of these two bit streams into a single serial bit stream is accomplished in the parallel-to-serial converter. If FEC is implemented, the signal is then error decoded and the resulting signal is then conditioned for line transmission in the line code converter.

5.8.2   The Antenna Subsystem

5.8.2.1   General. For conventional LOS microwave links, the antenna subsystem offers more room for trade-off to meet system requirements than any other subsystem. Basically, the antenna group looking outward from a transmitter must have:

- Transmission line (waveguide or coaxial line)
- Antenna: a reflecting surface or device
- Antenna: a feed horn or other feeding device

In addition, the antenna system may have:

- Circulators
- Directional couplers
- Phasers
- Passive reflectors
- Radome

5.8.2.2   Antennas. Below 700 MHz, antennas used for point-to-point microwave systems are often a form of yagi and are fed with coaxial transmission lines. Above 700 MHz some form of parabolic reflector-feed arrangement is used. However, 700 MHz is no hard and fast dividing line. Above 2000 MHz the transmission line is usually waveguide. The same antenna is used for transmission and reception.

An important antenna parameter is its radiation efficiency. Assuming no losses, the power radiated from an antenna would be equal to the power delivered to the antenna. Such power is equal to the square of the rms current flowing on the antenna times a resistance, called the radiation resistance:

$$P = I_{rms}^2 R \qquad (5.68)$$

where P = radiated power (W)

R = radiation resistance (Ω)

I = current (A)

In practice, all the power delivered to the antenna is *not* radiated in space. The radiation efficiency is defined as the ratio of the power radiated to the total power delivered to an antenna.

To derive a more realistic equation to express power, we divide the resistance, which we shall call the terminal resistance, into two component parts, namely, R = radiation resistance and R_1 = equivalent terminal loss resistance, so that

$$P = I_{rms}^2 (R + R_1) \tag{5.69}$$

and

$$\text{Radiation efficiency }(\%) = \frac{R}{R + R_1} \times 100 \tag{5.70}$$

Antenna gain is a fundamental parameter in radiolink engineering. Gain is conventionally expressed in decibels and is an indication of the antenna's concentration of radiated power in a given direction. Antenna gain expressed anywhere in this work is gain over an isotropic (dBi). An isotropic is a theoretical antenna with a gain of 1 (0 dB). In other words, it is an antenna that radiates equally in all directions.

For parabolic reflector-type antennas, gain is a function of the diameter D of the parabola and the frequency f. The theoretical gain is expressed by the formula

$$G_{dB} = 20 \log F_{GHz} + 20 \log D_{ft} + 7.5 \tag{5.71a}$$

For metric units where D is in meters,

$$G_{dB} = 20 \log F_{GHz} + 20 \log D_m + 17.8 \tag{5.71b}$$

Note: In practice we assume a surface efficiency of 55% for LOS microwave systems. This efficiency is incorporated in equations 5.71a and 5.71b. Chapter 7 discusses antennas with improved efficiencies.

For LOS microwave applications, we do not recommend antenna diameters over 12 ft (3.7 m). There are two reasons for this recommendation. The first deals with "sail area." Larger diameter dishes present more sail area to the wind, causing greater tower movement and increasing chances of pulling the antenna ray beam out of the capture area of the distant receive antenna. To reduce this effect, towers have to be notably more stiffened, costing considerably more investment. The second reason is beamwidth, which we discuss below.

As the antenna diameter increases, beamwidth decreases. A relationship to calculate approximate beamwidth, ϕ, is given below.

$$\phi = 70/(F_{GHz} \times D_{ft}) \quad \text{(degrees)} \tag{5.72a}$$

TABLE 5.18 Antenna Beamwidths as a Function of Gain

Antenna Gain (dB)	Antenna Beamwidth
30	5°
33	3.75°
36	2.5°
40	1.75°
44	1.0°
47	0.73°
50	0.5°

In metric units where D is in meters,

$$\phi = 22/(F_{GHz} \times D_m) \quad \text{(degrees)} \tag{5.72b}$$

In practice, absolutely uniform illumination is not used so that sidelobe levels may be reduced using a tapered form of illumination in keeping with FCC and ITU *Radio Regulations* (Ref. 1) requirements. Expect practical beamwidths to be 0.1°–0.2° wider.

Beamwidths are narrow for LOS microwave applications. Table 5.18 illustrates their narrowness. From this it is evident that considerable accuracy is required in pointing the antenna at the distant end. It also underlines the importance of maintaining antenna tower twist and sway within certain recommended bounds. This problem is addressed in the next section.

Polarization. In radiolink systems antennas use linear polarization and, depending on the feed, may be horizontally or vertically polarized, or both. That is, an antenna radiating and receiving several frequencies at once often will radiate adjacent frequencies with opposite polarizations, or the received polarization is opposite to the transmit. Isolation of 26 dB or better may be expected between polarizations, and on well-designed installations 35 dB, allowing closer interworking at installations using multi-RF carrier operation.

5.8.2.3 Transmission Lines and Related Devices

5.8.2.3.1 Waveguide and Coaxial Cable. Waveguides may be used on installations operating above 2 GHz to carry the signal from the radio equipment to the antenna (and vice versa). For those systems operating above 4 GHz it is mandatory from a transmission efficiency point of view. Coaxial lines are used on those systems operating below 2 GHz.

From a systems engineering aspect, the concern is loss with regard to a transmission line. Figures 5.49 and 5.50 identify several types of commonly used waveguides and coaxial cable expressing loss versus frequency.

Waveguide installations are always maintained under dry air or nitrogen pressure to prevent moisture condensation within the guide. Any constant

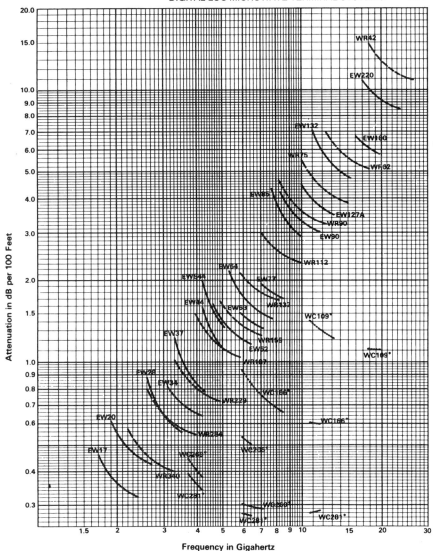

Figure 5.49. Waveguide attenuation. Add 0.3 dB for top and bottom transitions. (From Ref. 38. Courtesy of the Andrew Corp.)

positive pressure up to 10 lb/in.^{2} (0.7 kg/cm^2) is adequate to prevent "breathing" during temperature cycles.

Waveguides may be rectangular, elliptical, or circular. Nearly all older installations used rectangular waveguide exclusively. Today its use is limited to routing in tight places, where space is limited. However, bends are troublesome and joints add 0.06-dB loss each to the system. Optimum electrical performance is achieved by using the minimum number of compo-

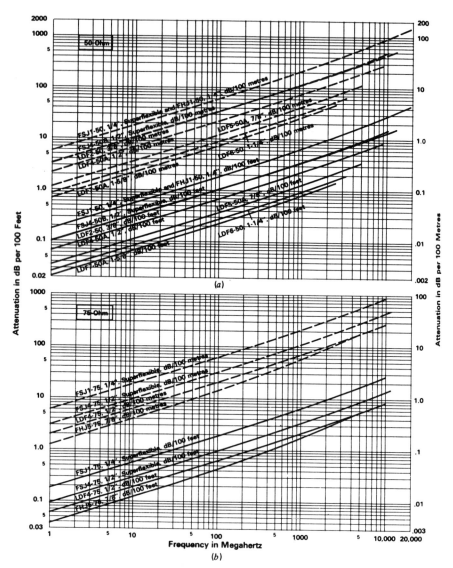

Figure 5.50. Attenuation per unit length for certain types of coaxial cable. Attenuation curves based on VSWR 1.0 : 1, 50-Ω copper conductors. For 75-Ω cables, multiply by 0.95. (a) Air dielectric. (b) Foam dielectric. (From Ref. 38. Courtesy of the Andrew Corp.)

nents. Therefore it has become the practice that wherever a single length is required, elliptical waveguide is used from the antenna to the radio equipment without the addition of miscellaneous flex-twist or rigid sections that are used for rectangular waveguide.

Circular waveguide offers generally lower loss plus dual polarized capability such that only one waveguide run up the tower is necessary for dual polarized installations. Circular waveguide is used when the run is long so that excessive loss is not introduced.

As Figure 5.49 demonstrates, as operational frequencies increase, attenuation increases. For frequencies above 15 GHz, an interesting and economic alternative is to mount the transmitting upconverter and HPA, the LNA, and downconverter into the antenna mount, and bring the IF up/down by coaxial cable to the transmitting modulator and receiving demodulator. Still another possibility is to use fiber-optic connectivity between the tower-mounted equipment and the shelter-mounted equipment. The IFs amplitude modulate a light signal.

5.8.2.3.2 Transmission Line Devices. A ferrite load isolator is a waveguide component that provides isolation between a single source and its load, reducing ill effects of voltage standing wave ratio (VSWR) and often improving stability as a result. Most commonly, load isolators are used with transmitting sources absorbing much of the reflected energy from high VSWR. Owing to the ferrite material with its associated permanent magnetic field, ferrite load isolators have a unidirectional property. Energy traveling toward the antenna is relatively unattenuated, whereas energy traveling back from the antenna undergoes fairly severe attenuation. The forward and reverse attenuations are on the order of 1 and 40 dB, respectively.

A waveguide circulator is used to couple two or three microwave radio equipments to a single antenna. The circulator shown in Figure 5.51 is a

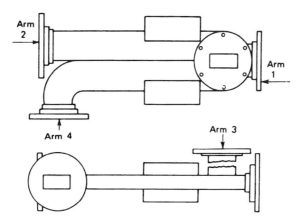

Figure 5.51. Waveguide circulator (four ports).

four-port device. It consists essentially of three basic waveguide sections combined into a single assembly. The center section is a ferrite nonreciprocal phase shifter. An external permanent magnet causes the ferrite material to exhibit phase shifting characteristics. Normally an antenna transmission line is connected to one arm and either three radio equipments are connected to the other three arms, or two equipments and a shorting plate. Attenuation in a clockwise direction from arm to arm is low, on the order of 0.5 dB, whereas in the counterclockwise direction from arm to arm it is high, on the order of 20 dB.

A power splitter is a simple waveguide device that divides the power coming from or going to the antenna. A 3-dB power split divides power in half; such a device could be used, for instance, to radiate the power from a transmitter in two directions. A 20-dB power split, or 30 dB, has an output that serves to sample the power in a transmission line. Often such a device has directional properties (therefore called a directional coupler) and is used for VSWR measurements allowing measurement of forward and reverse power.

Magic tees or hybrid tees are waveguide devices used to connect several equipments to a common waveguide run.

5.9 SYSTEM AND LINK PROPAGATION RELIABILITY (TIME AVAILABILITY)

Reliability is a principal concern of the microwave design engineer, whether a design involves a LOS microwave system of links in tandem or only just one link. There are two facets to reliability: outage due to propagation and outage due to equipment malfunction or failure. Availability is a term we can expect to encounter. We define availability as the percentage of time a system or link meets performance requirements; unavailability is the percentage of time the system or link does not meet requirements.

Time availability is a term that has become synonymous with propagation reliability. We use the two terms interchangeably.

CCIR Rep. 445-3 (Ref. 39) gives five general causes of interruption:

1. Equipment and power failure.
2. Propagation.
3. Interference.
4. Support facilities (e.g., collapse of a tower).
5. Human activity (human error).

Previous editions of CCIR Rep. 557 allocated about half of the unavailability time to propagation (deep fades) and half to the remaining four items on the list above. CCIR Rec. 557-1 (XVIth Plenary Assembly, 1986) does not take this liberty.

Suppose we did assume that half the outage time was due to propagation and we allot the remainder to "other." If a system availability is stated as 99.7%, then the unavailability would be 0.3% (i.e., 1.000 − 0.997 = 0.003 or 0.3%). For what time availability would we design the system? We divide the unavailability in half and subtract that value from 1.0000. In this case 0.15% or 0.0015; 1.0000 − 0.0015 = 99.85%.

Suppose there were 10 hops in tandem on a certain LOS microwave system and the system time availability was 99.85%. What would the per hop time availability be? Calculate the system unavailability (time). This is 0.0015. Divide this value by 10 and subtract from 1.0000; 1.0000 − 0.00015 = 99.985%.

This method we show is extremely conservative because it assumes that fading is well correlated along the microwave route or that all paths fade nearly simultaneously. This is highly unlikely. Nevertheless, many system engineers use this conservative technique.

One guideline is that for long systems of many hops in tandem (e.g., more than 10) only one-third should be permitted to fade at once. Such an assumption is fairly safe, erring on the conservative side, and can provide considerable relief for per hop time availability values.

CCIR Rep. 445-3 (Ref. 39) gives the equation shown below to calculate availability and assumes a two-way circuit:

Overall availability is defined by the following formula:

$$A = 100 - (2500 \times 100/L) \times [(T_1 + T_2 - T_b)/T_s] \qquad (5.73)$$

where A = percentage availability based on 2500 km
   L = length of one-way radio channel under consideration (km)
   T_1 = total interruption time for at least 10 consecutive seconds for one direction of transmission (s)
   T_2 = total interruption time for at least 10 consecutive seconds for the other direction of transmission (s)
   T_b = bidirectional interruptions for at least 10 consecutive seconds (s)
   T_s = period of study (s)

Note: For short circuits, the unavailability is unlikely to add in a linear manner and it may be necessary to define a minimum value for the parameter L. For higher grade circuits, the value of 280 km quoted in Report 930 might be appropriate.

For unidirectional transmission set $T_2 = 0$ and $T_b = 0$ in equation 5.73.

"Fundamental system availability" is defined by CCIR (Ref. 39) as a system without built-in redundancy (e.g., hot-standby) and is taken to be a function of the mean time between failure(s) (MTBF) and of the mean time to repair (MTTR). We use the familiar formula in this case as

$$A = \frac{\text{MTBF}}{\text{MTBF} + \text{MTTR}} \qquad (5.74)$$

The unit for MTBF and MTTR is hours. CCIR uses 10 for MTTR. The number is large because it assumes no technician on duty. The U.S. Department of Defense may use 20 min because it assumes a technician on duty and spare parts are available on site.

Example. What is the availability of a system where the MTBF is 40,000 and the MTTR is 10 h?

$$A = \frac{40,000}{40,000 + 10}$$
$$= 0.99975 \text{ or } 99.975\%$$

If we change MTTR to 20 min or 0.33 h, then we get an availability of 99.999%.

5.10 HOT-STANDBY OPERATION

LOS microwave radiolinks commonly provide transport of multichannel telephone service and/or point-to-point broadcast television on high-priority backbone routes. A high order of route reliability is essential. Route reliability depends on path reliability or link time availability (propagation) and equipment/system reliability. We have discussed path reliability and equipment reliability in Section 5.9. Redundancy is one way to achieve equipment reliability to minimize downtime and maximize link availability regarding equipment degradation or failure.

One straightforward way to achieve redundancy effectively is to provide a parallel terminal/repeater system. Frequency diversity effectively does just this. With this approach all equipment is active and operated in parallel with two distinct systems carrying the same traffic. This is expensive but necessary if a high order of link reliability is desired. Here we mean route reliability.

Often the additional frequency assignments to permit operation in frequency diversity are not available. When this is the case, the equivalent equipment reliability may be achieved by the use of a hot-standby configuration. Figure 5.52 illustrates the hot-standby concept.

The equipment marked *A* and *B* in Figure 5.52 could be single modules, shelves or groups of modules, or whole equipment racks. On a complex equipment, such as a radiolink terminal, whether digital or analog, more than one set of protection can exist.

On a radiolink terminal such as shown in Figure 5.53 the sections marked "baseband" and "RF" are both hot-standby protected. The operation of the

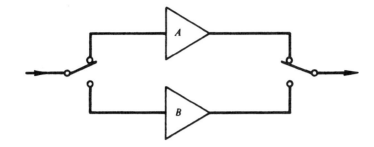

Figure 5.52. The concept of hot-standby protection.

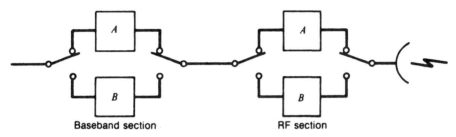

Baseband section   RF section

Figure 5.53. Hot-standby radio.

protection system of these two sections, however, is independent. The protection system is broken down further in that the protections for the transmit and receive paths operate independently, as shown in Figure 5.54. It should be noted that the switches ahead of each set of modules have been replaced by a signal splitter. This technique allows the signal to be fed into each set of modules simultaneously.

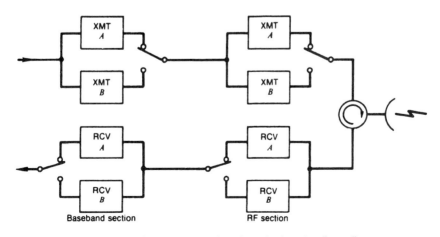

Baseband section   RF section

Figure 5.54. Separate transmit and receive hot-standby radio.

As the expression indicates, hot standby is the provision of parallel redundant equipment such that this equipment can be switched in to replace the operating on-line equipment nearly instantaneously when there is a failure in the operating equipment. The switchover can take place in the order of microseconds or less. The changeover of a transmitter and/or receiver line can be brought about by a change, over/under a preset amount, in one of the following values: for a transmitter,

- Frequency
- RF power
- Demodulated baseband (radio) pilot level

and for a receiver,

- Automatic gain control (AGC) voltage
- Squelch
- Received pilot level
- Degraded bit error rate for a digital system
- Loss of frame alignment for a digital system

(*Note:* Digital systems do not use pilot tones.)

Hot-standby-protection systems provide sensing and logic circuitry for the control of waveguide switches (or coaxial switches where appropriate), in some cases IF switches as well as baseband switches on transmitters, and IF and baseband output signals on receivers. The use of a combiner on the receiver side is common with both receivers on line at once.

There are two approaches to the use of protection equipment. These are called one-for-one and one-for-*n*. One-for-one operation provides one full line of standby equipment for each operational system (see Figure 5.55).

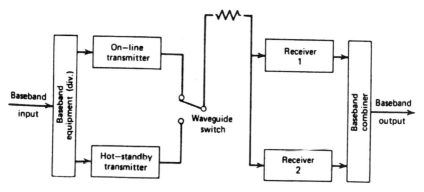

Figure 5.55. Functional block diagram of a typical one-for-one hot-standby configuration.

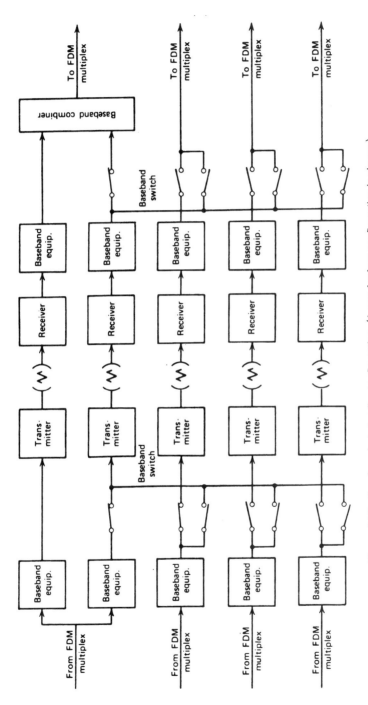

Figure 5.56. One-for-*n* hot-standby configuration. (A one-for-four configuration is shown.)

385

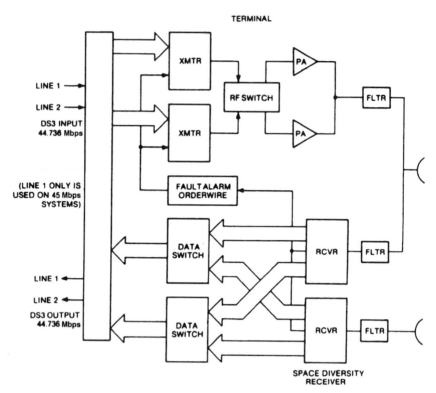

Figure 5.57. A typical hot-standby (one-for-one) digital implementation using space diversity. (From Ref. 40. Courtesy of Rockwell International, Collins Transmission Division.)

One-for-n provides only one full line of equipment for n operational lines of equipment, where n is greater than one. Figure 5.56 illustrates a typical one-for-four configuration.

One-for-one is more expensive but provides a higher order of reliability. Its switching system is comparatively simple. One-for-n is more economic, with only one line of spare equipment for several operational lines. It is less reliable (i.e., suppose there were equipment failures in two lines of operational equipment of the n lines), and switching is considerably more complex.

Figure 5.57 shows a typical digital hot-standby configuration with space diversity.

5.11   SERVICE CHANNELS

Service channels are separate facilities from the information baseband (on analog radiolinks) but transmitted on the same carrier. Service channels operate in frequency slots (on analog microwave) below the information baseband for FDM telephony operation and above the video baseband for those links transmitting video (and associated but separated aural channels). Service channels are used for maintenance, link, and network coordination, and, in these cases, may be called "orderwire(s)." They may also be assigned to carry network status data and fault information from unmanned radio relay sites. For FDM telephony links, service channels commonly occupy the band from 300 Hz to 12 kHz of the transmitted baseband, allowing three nominal 4-kHz channel operation.

Digital microwave radiolinks utilize specific time slots in the digital bit stream for service channels. At each terminal and repeater, local digital service channels are dropped and inserted. An express orderwire can also be accommodated. This, of course, requires a reconstitution of the information bit stream at the terminal locations to permit the insertion of service channel information. In most instances the transmitter develops its own frame with overhead assigned for service functions, such as service channels, and supervisory functions, and to fulfill other operational requirements. Another option is to utilize spare overhead in SDH and SONET formats set aside for this function.

5.12 ALARM AND SUPERVISORY SUBSYSTEMS

Many LOS microwave sites are unattended, especially repeater sites. To ensure improved system availability, it is desirable to know the status of unattended sites at a central or manned location. This is accomplished by means of a fault-reporting system. Commonly, such fault alarms are called status reports. The microwave sites originating status reports are defined as reporting stations. A site that receives and displays such reports is defined as a supervisory location. This is the standard terminology of the industry. Normally, supervisory locations are those terminals that terminate a LOS microwave section. Status reports may also be required to be extended over a wire circuit to a remote location, often a maintenance center.

The following functions at a microwave site, which is a reporting location (unmanned), are candidate functions for status reports:

Equipment Alarms

Loss of receive signal
Loss of pilot (at receiver) (analog systems only)
High noise level (at receiver)
Power supply failure
Loss of modulating signal
TWT overcurrent
Low transmitter output

Off-frequency operation

Hot-standby actuation

Site Alarms

Illegal entry

Commercial power failure

Low fuel supply

Standby power unit failure

Standby power unit online

Tower light status

Additional Fault Information for Digital Systems

Loss of bit count integrity (BCI)

Loss of sync and alignment

Excessive BER

Often alarms are categorized into "major" (urgent) and "minor" (nonurgent) in accordance with their importance. For instance, a major alarm would be one where the fault would cause the system or link to go down (cease operation) or seriously deteriorate performance. A major alarm may be audible as well as visible on the status panel. A minor alarm may then show only as an indication on the status panel. On military equipment, alarms are a part of built-in-test equipment (BITE).

The design intent of alarm or BITE systems is to make all faults binary: a tower light is either on or off; the RSL (receive signal level) has dropped below a specified level, -100 dBW, for example; the transmitter power output is 3 dB below its specified output; or the noise on a derived analog channel is above a certain level in picowatts. By keeping all functions binary, using relay closure (or open) or equivalent solid-state circuitry, the job of coding alarms for transmission is made much easier. Thus all alarms are of a "go/no-go" nature.

5.12.1 Transmission of Fault Information

On analog systems, common practice today is to transmit fault information in a voice channel associated with the service channel groupings of voice channels (Section 5.11). Binary information is transmitted by VF telegraph equipment using frequency shift keying (FSK) or tone-on, tone-off (see Chapter 13). Depending on the system used, 16, 18, or 24 tone channels may occupy the voice channel assigned. A tone channel is assigned to each reporting location (i.e., each reporting location will have a tone transmitter operating on the specific tone frequency assigned to it). The supervisory location will have a tone receiver for each reporting (unmanned site under its supervision) location.

At each reporting location the fault or BITE points previously listed are scanned every so many seconds, and the information from each monitor or scan point is time division multiplexed in a simple serial bit stream code. The data output from each tone receiver at the supervisory location represents a series of reporting information on each remote unmanned site. The coded sequence in each case is demultiplexed and displayed on the status panel.

A simpler method is the tone-on, tone-off method. Here the presence of a tone indicates a fault in a particular time slot; in another method it is indicated by the absence of a tone. A device called a fault-interrupter panel is used to code the faults so that different faults may be reported on the same tone frequency.

On digital microwave radiolinks, fault information in a digital format is stored and then inserted into one of the service channel time slots, with one or more time slots reserved for each reporting location. Each bit in a time slot is assigned a function. For example, bit 1 is assigned to transmit power, bit 2 to loss of receive sync, and so on.

5.12.2 Remote Control

Through a similar system to that previously described, which operates in the opposite direction, a supervisory station can control certain functions at reporting locations via a voice frequency (VF) telegraph tone line (for analog systems) with a tone frequency assigned to each separate reporting location on the span. If only one condition is to be controlled, such as turning on tower lights, then a mark condition could represent lights on and a space could represent lights off. If more than one condition is to be controlled, then coded sequences are used to energize or deenergize the proper function at the remote reporting location.

There is an interesting combination of fault reporting and remote control that particularly favors implementation on long spans. Here only summary status is normally passed to the supervisory location; that is, a reporting location is either in a "go" or "no-go" status. When a "no-go" is received, that reporting station in question is polled by the supervisory location, and detailed fault information is then released. Polling may also be carried out on a periodic basis to determine detailed minor alarm fault data.

On the digital radiolink equipment of one manufacturer (Ref. 40) the following operational support system maintenance functions and performance monitoring are listed:

Maintenance Functions

Alarm Surveillance
   Alarm reporting
   Status reporting

Alarm conditioning

Alarm distribution

Attribute report

Control Functions

Allow–inhibit local alarms

Operate alarm cutoff

Allow–inhibit protection switching

Operate release protection switch

Remove–restore service

Restart processor

Preemptive switching–override an existing protective switch

Activate/restore lockout (lockout prevents switching)

Local/remote control (i.e., inhibits local operation)

Operate–release loopback

Command (control) verification (i.e., set status point)

Completion acknowledgment (i.e., completed the command received earlier)

Performance Monitoring

Report performance monitoring data such as BER, sync error, and error second

Inhibit–allow performance monitoring data (i.e., collect but do not send unless asked)

Start–stop performance monitoring data

Initialize (reset) performance monitoring data storage registers

5.13   FREQUENCY PLANNING AND THE INTERFERENCE PROBLEM

5.13.1   General

To derive maximum performance from a microwave radiolink system, the systems engineer must set out a frequency usage plan that may or may not have to be approved by the local administration.

The problem has many aspects. First, the useful RF spectrum is limited, from above direct current to about 150 GHz. The frequency range of discussion for microwave radiolinks is essentially from the VHF band at 150 MHz (overlapping) to the millimeter region of 23 GHz. Second, the spectrum from 150 MHz to 23 GHz must be shared with other services such as radar, navigational aids, research (i.e., space), meteorological, and broadcast. For point-to-point communications, we are limited by international agreement to those bands shown in Table 5.1.

Although many of the allocated bands are wide, some up to 500 MHz in width, FM by its very nature is a wideband form of emission. It is not uncommon to have a RF bandwidth $B_{RF} = 25$ or 30 MHz for just one emission. Guard bands must also be provided. These are a function of the frequency drift of transmitters as well as "splatter" or out-of-band emission, which in some areas is not well specified.

Occupied bandwidth is discussed in Sections 5.4.2.3 and 5.5.3.4.

5.13.2   Spurious Emission

For a digital LOS microwave transmit terminal, we discussed spectral confinement of the digital waveform in accordance with FCC Regulations, Part 21 (21.106). This FCC paragraph is titled "emission limitations." The FCC is very specific of signal attenuation as a function of frequencies removed from the carrier center frequency. Portions of the FCC regulation is quoted in Section 5.5.3.4.

The CCIR discusses the issue of spurious emission in Rep. 937-2 (Ref. 41). Both the FCC and CCIR give power density values in a 4-kHz bandwidth. CCIR references spurious signal levels at the antenna feeder input:

For an analog signal, the maximum spurious power density is -70 dB (W/4 kHz).

For digital transmission, spurious is broken down into two parts:

- Equal or less than -40 dB (W/4 kHz) for carrier wave (CW) emissions
- Equal or less than -70 dB (W/4 kHz) for noise spectrumlike emissions

The reference point again is at the antenna feeder input.

5.13.3   Radio Frequency Interference (RFI)

On planning a new radiolink system or on adding RF carriers to an existing installation, careful consideration must be given to the radio frequency interference (RFI) of the existing (or planned) emitters in the area. Usually the governmental authorizing agency has information on these and their stated radiation limits. Typical limits have been given previously. Equally important as those limits are those of antenna directivity and sidelobe radiation. Not only must the radiation of other emitters be examined from this point of view, but also the capability of the planned antenna to reject unwanted signals. The radiation pattern of all licensed emitters should be known. Convert the lobe level in the direction of the planned installation to EIRP in dBW. This should be done for all interference candidates within interference frequency range. For each emitter's EIRP, compute a path loss to the planned installation to determine interference. Such a study could well affect a frequency plan or antenna design.

Nonlicensed emitters should also be looked into. Many such emitters may be classified as industrial noise sources such as heating devices, electronic ovens, electric motors, or unwanted radiation from your own and other microwave installations (i.e., radar harmonics). In the 6-GHz band a coordination contour should be carried out to verify interference from earth stations (see Ref. 44). For a general discussion on the techniques for calculating interference noise in radiolink systems, see Refs. 30 and 42.

Another RFI consideration is the interference that can be caused by radiolink transmitting systems with the fixed satellite service in those frequency bands that are shared with that service. Guidance on this potential problem is provided by CCIR Rec. 406-7 (Ref. 43). Some of the major points of the CCIR recommendation are summarized below:

1. In those shared bands between 1 and 10 GHz the maximum EIRP of a radiolink transmitting system should not exceed +55 dBW, and the input power to the transmitting antenna should not exceed +13 dBW. As far as practicable, sites for new radiolink transmitting stations where the EIRP exceeds +35 dBW should be selected so that the direction of maximum radiation of any antenna will be at least 2° away from a geostationary satellite orbit. In special situations the EIRP should not exceed +47 dBW for any antenna directed within 0.5° of a geostationary orbit and +47 to +55 dBW on a linear dB scale (8 dB per angular degree) for any antenna beam directed between 0.5° and 1.5° of a geostationary satellite orbit.

2. In those shared bands between 10 and 15 GHz, the maximum EIRP of a radiolink transmitting system should not exceed +55 dBW, and the input power to the antenna system should not exceed +10 dBW. As far as practicable, transmitting stations where the EIRP exceeds +45 dBW should be selected so that the direction of maximum radiation of any antenna will be at least 1.5° away from the geostationary satellite orbit.

5.13.4   Overshoot

Overshoot interference may occur when radiolink hops in tandem are in a straight line. Consider stations A, B, C, and D in a straight line, or that a straight line on a map drawn between A and C also passes through B and D. Link A–B has frequency F_1 from A to B. F_1 is reused in the direction C to D. Care must be taken that some of the emission F_1 on the A–B hop does not spill over into the receiver at D. Reuse may even occur on an A, B, and C combination, so F_1 at A to B may spill into a receiver at C tuned to F_1. This can be avoided, provided stations are removed from the straight line. In this case the station at B should be moved to the north of a line A to C, for example.

5.13.5   Transmit–Receive Separation

If a transmitter and receiver are operated in the same frequency band at a radiolink station, the loss between them must be at least 120 dB. One way to

TABLE 5.19   Frequency Frogging at Radiolink Repeaters

Number of Voice Channels	Minimum Separation (MHz)	
	2000–4000 MHz	6000–8000 MHz
120 or less	120	161
300 or more	213	252

Source: Reference 45.

assure the 120 figure is to place all "go" channels in one-half of an assigned band and all "return" channels in the other. The terms *go* and *return* are used to distinguish between the two directions of transmission.

5.13.6   Basis of Frequency Assignment

"Go" and "return" channels are assigned as in the preceding section. For adjacent RF channels in the same half of the band, horizontal and vertical polarizations are used alternately. To carry this out we may assign, as an example, horizontal polarization H to the odd-numbered channels in both directions on a given section and vertical polarization V to the even-numbered channels. The order of isolation between polarizations is on the order of 26 dB, but often specified as 35 dB or more.

In order to prevent interference between antennas at repeaters between receivers on one side and transmitters in the same chain on the other side of the station, each channel shall be shifted in frequency (called "frequency frogging") as it passes through the repeater station. Recommended separations or shifts of frequency are shown in Table 5.19.

5.13.7   IF Interference

Care must be taken when assigning frequencies of transmitter and receiver local oscillators as to whether these are placed above or below the desired operating frequency. Avoid frequencies that emit the received channel frequency F_R and check those combinations of $F_R \pm 70$ MHz for equipment with 70-MHz IFs or $F_R \pm 140$ MHz when 140-MHz IFs are used. Often plots of all station frequencies are made on graph paper to assure that forbidden combinations do not exist. When close frequency stacking is desired, and/or nonstandard IFs are to be used, the system designer must establish rules as to minimum adjacent channel spacing and receive–transmit channel spacing.

5.14   ANTENNA TOWERS AND MASTS

5.14.1   General

Two types of towers are used for LOS microwave systems: guyed and self-supporting. However, other natural or constructed towers should also be considered or at least taken advantage of. Radiolink engineers should consider mountains, hills, and ridges so that tower heights may be reduced. They should also consider office buildings, hotels, grain elevators, high-rise apartment houses, and other steel structures (e.g., the sharing of a TV broadcast tower) for direct antenna mounting. For tower heights of 30–60 ft, wooden masts are often used.

One of the most desirable construction materials for a tower is hot-dipped galvanized steel. Guyed towers are usually preferred because of overall economy and versatility. Although guyed towers have the advantage that they can be placed closer to a shelter or building than self-supporting types, the fact that they need a larger site may be a disadvantage where land values are high. The larger site is needed because additional space is required for installing guy anchors. Table 5.20 shows approximate land areas needed for several tower heights.

Tower foundations should be reinforced concrete with anchor bolts firmly embedded. Economy or cost versus height trade-offs usually limit tower heights to no more than 300 ft (188 m). Soil bearing pressure is a major consideration in tower construction. Increasing the foundation area increases soil bearing capability or equivalent design pressure. Wind loading under no-ice (i.e., normal) conditions is usually taken as 30 lb/ft^2 for flat surfaces. A design guide (EIA RS-222-E, Ref. 46) indicates that standard tower foundations and anchors for self-supporting and guyed towers should be designed for a soil pressure of 4000 lb/ft^2 acting normal to any bearing area under specified loading.

5.14.2 Tower Twist and Sway

Just like any other structure, a radiolink tower tends to twist and sway due to wind loads and other natural forces. Considering the narrow beamwidths referred to in Section 5.8.2.2 (Table 5.18), with only a little imagination we can see that only a very small deflection of a tower or antenna will cause a radio ray beam to fall out of the reflection face of an antenna on the receive side or move the beam out on the far-end transmit side of a link.

Twist and sway, therefore, must be limited. Table 5.21 sets certain limits. The table has been taken from EIA RS-222-E (Ref. 46). From the table we can see that angular deflection and tower movement are functions of wind velocity. It should also be noted that the larger the antenna, the smaller the beamwidth, besides the fact that the sail area is larger. Hence the larger the antenna (and the higher the frequency of operation), the more we must limit the deflection.

To reduce twist and sway, tower rigidity must be improved. One generality we can make is that towers that are designed to meet required wind load or ice load specifications are sufficiently rigid to meet twist and sway tolerances. One way to increase rigidity is to increase the number of guys, particularly at the top of the tower. This is often done by doubling the number of guys from three to six.

TABLE 5.20 Minimum Land Area Required for Guyed Towers

Tower Height (ft)	Area Required[a] (ft)				
	80% Guyed	75% Guyed	70% Guyed	65% Guyed	60% Guyed
60	87 × 100	83 × 96	78 × 90	74 × 86	69 × 80
80	111 × 128	105 × 122	99 × 114	93 × 108	87 × 102
100	135 × 156	128 × 148	120 × 140	113 × 130	105 × 122
120	159 × 184	150 × 174	141 × 164	132 × 154	123 × 142
140	183 × 212	178 × 200	162 × 188	152 × 176	141 × 164
160	207 × 240	195 × 226	183 × 212	171 × 198	159 × 184
180	231 × 268	218 × 252	204 × 236	191 × 220	177 × 204
200	255 × 296	240 × 278	225 × 260	210 × 244	195 × 226
210	267 × 304	252 × 291	236 × 272	220 × 264	204 × 236
220	279 × 322	263 × 304	246 × 284	230 × 266	213 × 246
240	303 × 350	285 × 330	267 × 308	249 × 288	231 × 268
250	315 × 364	296 × 342	278 × 320	254 × 282	240 × 277
260	327 × 378	308 × 356	288 × 334	269 × 310	249 × 288
280	351 × 406	330 × 382	309 × 358	288 × 332	267 × 308
300	375 × 434	353 × 408	330 × 382	308 × 356	285 × 330
320	399 × 462	375 × 434	351 × 406	327 × 376	303 × 350
340	423 × 488	398 × 460	372 × 430	347 × 400	321 × 372
350	435 × 502	409 × 472	383 × 442	356 × 411	330 × 381
360	447 × 516	420 × 486	393 × 454	366 × 424	339 × 392
380	471 × 544	443 × 512	414 × 478	386 × 446	357 × 412
400	495 × 572	465 × 536	425 × 502	405 × 468	375 × 434
420	519 × 599	488 × 563	456 × 527	425 × 490	393 × 454
440	543 × 627	510 × 589	477 × 551	444 × 513	411 × 475

Tower Height (ft)	Area Required[a] (acre)				
	80% Guyed	75% Guyed	70% Guyed	65% Guyed	60% Guyed
60	0.23	0.21	0.19	0.17	0.15
80	0.38	0.34	0.30	0.26	0.23
100	0.56	0.50	0.44	0.39	0.34
120	0.77	0.69	0.61	0.53	0.46
140	1.03	0.91	0.80	0.70	0.61
160	1.31	1.16	1.03	0.90	0.77
180	1.63	1.45	1.27	1.11	0.96
200	1.99	1.76	1.55	1.35	1.16
210	2.18	1.93	1.70	1.48	1.27
220	2.38	2.11	1.85	1.61	1.39
240	2.81	2.49	2.18	1.90	1.63
250	3.04	2.69	2.36	2.05	1.76
260	3.27	2.89	2.54	2.21	1.90
280	3.77	3.33	2.92	2.54	2.18
300	4.30	3.80	3.33	2.89	2.49
320	4.87	4.30	3.77	3.27	2.81
340	5.48	4.84	4.24	3.65	3.15
350	5.79	5.11	4.48	4.88	3.33
360	6.12	5.40	4.73	4.10	3.52
380	6.79	5.99	5.25	4.55	3.90
400	7.50	6.62	5.79	5.02	4.30
420	8.24	7.27	6.36	5.52	4.73
440	9.03	7.96	6.96	6.03	5.17

[a]Preferred area is a square using the larger dimension of minimum area. This will permit orienting the tower in any desired position.

TABLE 5.21   Allowable Twist and Sway Values for Parabolic Antennas, Passive Reflectors, and Periscope System Reflectors

A	B	C	D	E	F	G	H	I
		Parabolic Antennas		Passive Reflectors		Periscope System Reflectors		
3-dB Beamwidth 2Θ HP for Antenna Only (Note 8) (Degrees)	Deflection Angle at 10-dB Points (Notes 1 and 7) (Degrees)	Limit of Antenna Movement with Respect to Structure (Degrees)	Limit of Structure Movement Twist or Sway at Antenna Attachment Point (Degrees)	Limit of Passive Reflector Sway (Notes 4 and 5) (Degrees)	Limit of Passive Reflector Twist (Note 4) (Degrees)	Limit of Reflector Movement with Respect to Structure (Degrees)	Limit of Structure Twist at Reflector Attachment Point (Degrees)	Limit of Structure Sway at Reflector Attachment Point (Degrees)
5.6	5.0	0.4	4.6	3.5	2.5	0.2	4.8	2.3
5.6	4.8	0.4	4.4	3.3	2.4	0.2	4.6	2.2
5.4	4.6	0.4	4.2	3.2	2.3	0.2	4.4	2.1
5.1	4.4	0.4	4.0	3.0	2.2	0.2	4.2	2.0
4.9	4.2	0.4	3.8	2.9	2.1	0.2	4.0	1.9
4.7	4.0	0.3	3.7	2.8	2.0	0.2	3.8	1.8
4.4	3.8	0.3	3.5	2.6	1.9	0.2	3.6	1.7
4.2	3.6	0.3	3.3	2.5	1.8	0.2	3.4	1.6
4.0	3.4	0.3	3.1	2.3	1.7	0.2	3.2	1.5
3.7	3.2	0.3	2.9	2.2	1.6	0.2	3.0	1.4
3.5	3.0	0.3	2.7	2.1	1.5	0.2	2.8	1.4
3.4	2.9	0.2	2.7	2.0	1.45	0.1	2.8	1.3
3.3	2.8	0.2	2.6	1.9	1.4	0.1	2.7	1.3
3.1	2.7	0.2	2.5	1.8	1.35	0.1	2.6	1.25
3.0	2.6	0.2	2.4	1.8	1.3	0.1	2.5	1.2
2.9	2.5	0.2	2.3	1.7	1.25	0.1	2.4	1.15
2.8	2.4	0.2	2.2	1.6	1.2	0.1	2.3	1.1
2.7	2.3	0.2	2.1	1.6	1.15	0.1	2.2	1.05
2.6	2.2	0.2	2.0	1.5	1.1	0.1	2.1	0.1
2.5	2.1	0.2	1.9	1.4	1.05	0.1	2.0	0.95
2.3	2.0	0.2	1.8	1.4	1.0	0.1	1.9	0.9
2.2	1.9	0.2	1.7	1.3	0.95	0.1	1.8	0.85
2.1	1.8	0.2	1.6	1.2	0.9	0.1	1.7	0.8
2.0	1.7	0.2	1.5	1.1	0.85	0.1	1.6	0.75
1.9	1.6	0.2	1.4	1.1	0.8	0.1	1.5	0.7
1.7	1.5	0.2	1.3	1.0	0.75	0.1	1.4	0.65
1.6	1.4	0.2	1.2	0.9	0.7	0.1	1.3	0.6
1.5	1.3	0.1	1.2	0.9	0.65	0.1	1.2	0.55
1.4	1.2	0.1	1.1	0.8	0.6	0.1	1.1	0.5
1.3	1.1	0.1	1.0	0.7	0.55	0.1	1.0	0.45
1.2	1.0	0.1	0.9	0.7	0.5	0.1	0.9	0.4
1.1	0.9	0.1	0.8	0.6	0.45	0.1	0.8	0.35
0.9	0.8	0.1	0.7	0.5	0.4	0.1	0.7	0.3
0.8	0.7	0.1	0.6	0.4	0.35	0.1	0.6	0.25
0.7	0.6	0.1	0.5	0.4	0.3	0.1	0.5	0.2
0.6	0.5	0.1	0.4	0.3	0.25	0.1	0.4	0.15
0.5	0.4	0.1	0.3	0.2	0.2	0.07	0.3	0.13
0.3	0.3	0.05	0.25	0.2	0.15	0.05	0.25	0.10
0.2	0.2			0.14	0.1			
0.1	0.1			0.07	0.05			

Only for configuration where antenna is directly under the reflector

TABLE 5.21 (*Continued*)

Note 1. If values for columns "A" and "B" are not available from the manufacturer(s) of the antenna system or from the user of the antenna system, then values shall be obtained from Figure C1, C2, or C3 of the reference document.

Note 2. Limits of beam movement for twist or sway (treated separately in most analyses) will be the sum of the appropriate figures in columns C and D, G and H, and G and I. Columns G, H, and I apply to a vertical periscope configuration.

Note 3. It is not intended that the values in this table imply an accuracy of beamwidth determination or structural rigidity calculation beyond known practicable values and computational procedures. For most microwave structures it is not practicable to require a calculated structural rigidity of less than 0.25 twist or sway with a 50 mi/h (22.4 m/s) basic wind speed.

Note 4. For passive reflectors the allowable twist and sway values are assumed to include the effects of all members contributing to the rotation of the face under wind load. For passives not elevated far above ground, [approximately 5–20 ft (1.5–6 m) clearance above ground] the structure and reflecting face supporting elements are considered an integral unit. Therefore separating the structure portion of the deflection is only meaningful when passives are mounted on conventional microwave structures.

Note 5. The allowable sway for passive reflectors is considered to be 1.4 times the allowable twist to account for the amount of rotation of the face about a horizontal axis through the face center and parallel to the face compared to the amount of beam rotation along the direction of the path as it deviates from the plane of the incident and reflected beam axis.

Note 6. Linear horizontal movement of antennas and reflectors in the amount experienced for properly designed microwave antenna system support structures is not considered a problem (no significant signal degradation attributed to this movement).

Note 7. For systems using a frequency of 450 MHz, the half-power beamwidths may be nearly 2Θ degrees for some antennas. However, structures designed for microwave relay systems will usually have an inherent rigidity less than the maximum 5° deflection angle shown on the chart.

Note 8. The 3-dB beamwidths, 2Θ HP in column "A," are shown for convenient reference to manufacturers' published antenna information. The minimum deflection reference for this standard is the allowable total deflection angle Θ at the 10-dB points.

Source: Reference 46, Appendix C, EIA/TIA-222-E. Reprinted with permission.

5.15 PLANE REFLECTORS AS PASSIVE REPEATERS

A plane reflector as a passive repeater offers some unique advantages. Suppose that we wish to provide multichannel service to a town in a valley, and a mountain is nearby with poor access to its top. The radiolink engineer should consider the use of a passive repeater as an economic alternative. A prime requirement is that the plane reflector be within line-of-sight of the terminal antenna in town as well as line-of-sight of the distant microwave radiolink station. Such a passive repeater installation may look like the following example, where a = the net path loss in decibels:

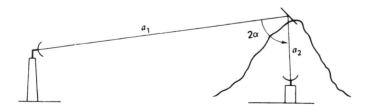

$$a = G_t + G_r + G_A - a_1 - a_2 \qquad (5.75)$$

where a_1 = path loss on path 1 (dB)

a_2 = path loss on path 2 (dB)

G_t = transmitting antenna gain

G_r = receiving antenna gain

G_A = passive reflector gain

Let us concern ourselves with G_A for the moment. The gain of a passive reflector results from the capture of a ray beam of RF energy from a distant antenna emitter and the redirection of it toward a distant receiving antenna. The gain of the passive reflector is divided into two parts: (1) incoming energy and (2) redirected or reflected energy. The gain for (1) and (2) is

$$G_A = 20 \log \frac{4\pi A \cos \alpha}{\lambda^2} \tag{5.76}$$

where α = one-half the included horizontal angle between incident and reflected wave and A = the surface area (ft^2 or m^2) of the passive reflector. If A is in square feet, then λ must also be in feet.

The reader will find the following relationship useful:

$$\text{Wavelength (ft)} = \frac{985}{F} \tag{5.77}$$

where F is in megahertz. Passive reflector path calculations are not difficult. The first step is to determine if the shorter path (a_2 path in the figure) places the passive reflector in the near field of the nearer parabolic antenna. To determine whether near field or far field, solve the following formula:

$$\frac{1}{K} = \frac{\pi \lambda d'}{4A} \tag{5.78}$$

If the ratio $1/K$ is less than 2.5, a near-field condition exists; if $1/K$ is greater than 2.5, a far-field condition exists. d' = length of the path in question (i.e., the shorter distance).

For the far-field condition, consider paths 1 and 2 as separate paths and sum their free-space path losses. Determine the gain of the passive plane reflector using equation 5.76. Sum this gain with the two free-space path losses algebraically to obtain the net path loss.

For the near-field condition, where $1/K$ is less than 2.5, the free-space path loss will be that of the longer hop (e.g., a_1 above). Algebraically add the repeater gain (or loss), which is determined as follows. Compute the parabola/reflector coupling factor l:

$$l = D'\sqrt{\frac{\pi}{4A}} \tag{5.79}$$

where D' = diameter of parabolic antenna (ft)
       A = effective area of passive reflector (ft^2)

Figure 5.58 is now used to determine near-field gain or loss. The $1/K$ value is on the abscissa, and l is the family of curves shown.

The examples below are from Ref. 47.

Example 1. *Far-field:* A plane passive reflector 10 ft × 16 ft, or 160 ft^2, is erected 21 mi from one active site and only 1 mi from the other. $2\alpha = 100°$ and $\alpha = 50°$. The operating frequency is 2000 MHz. By formula the free-space loss for the longer path is 129.5 dB, and for the shorter path it is 103 dB.
    Calculate the gain G_A of the passive plane reflector, equation 5.76.

$$G_A = \frac{20 \log 4 \times 160 \cos 50°}{(985/2000)^2} = 20 \log 5.340$$

$$= 74.6 \text{ dB}$$

$$\text{Net path loss} = -129.5 - 103 + 74.6 = -157.9 \text{ dB}$$

Example 2. *Near-field:* The passive reflector selected in this case is 24 ft × 30 ft. The operating frequency is 6000 MHz. The long leg is 30 mi and the short leg 4000 ft. A 10-ft parabolic antenna is associated with the active site on the short leg. (Note that 6 GHz is approximately equivalent to 0.164 ft.) Determine $1/K$:

$$\frac{1}{K} = \frac{\pi \lambda d}{4A} = \frac{\pi(0.164)(4000)}{4 \times 720} = 0.717$$

Note that this figure is less than 2.5, indicating the near-field condition. Calculate l:

$$l = D'\sqrt{\pi/4A} = 10\sqrt{\pi/4 \times 720} = 0.33$$

Using these two inputs, the values of l and $1/K$, we go to Figure 5.58 and find the net gain of the system to be +0.2 dB. The net free-space loss is then $+0.2 - 142.3 = -142.1$ dB. The free-space loss of the 30-mi leg is 142.3 dB.

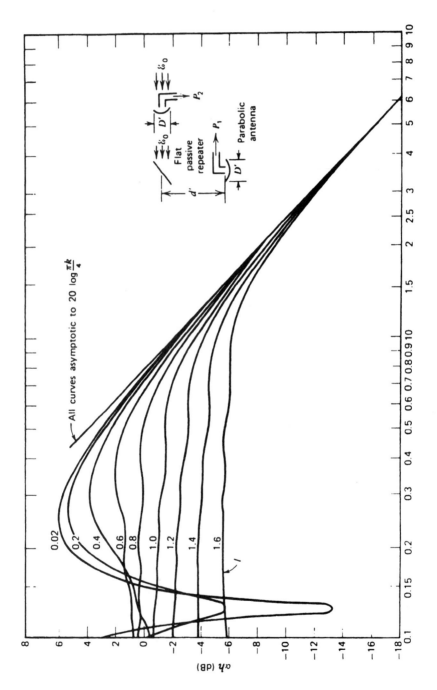

Figure 5.58. Antenna–reflector efficiency curves. (From Ref. 47. Courtesy of Microflect, Inc.)

REVIEW EXERCISES

1. Give at least five applications of LOS microwave systems for telecommunications.

2. What are the principal competing transmission media for LOS microwave? Give some advantages and disadvantages of each.

3. Give the four most important steps in the design of a LOS microwave link.

4. Calculate the free-space loss (FSL) in dB for the following LOS microwave paths: 1 GHz, 22 mi; 3800 MHz, 45 mi; 12.27 GHz, 7.3 mi; and an Andean path 127 mi long at 2.1 GHz.

5. What is the smooth earth distance to the horizon ($K = \frac{4}{3}$) for an antenna mounted on the top of a tower 50 ft high? 450 ft high?

6. In a particular geographic area and altitude, the effective earth radius is found to be 7800 km. Calculate the equivalent K-factor.

7. What do we derive from a path profile? (What does it tell us?)

8. When drawing a path profile, we calculated the appropriate K-factor as 0.85. How will such a K-factor impact tower height compared to a K-factor of 1.0?

9. On a particular path 30 km long, there is an obstacle just 14 km from the "transmitter." The K-factor is 1.2. Calculate the additional height to compensate for earth curvature at that obstacle point.

10. What is the standard Fresnel zone clearance for a microwave ray beam to clear an obstacle?

11. A Fresnel zone clearance is really a function of only two variables. What are they? Where in a LOS microwave path is the Fresnel zone clearance maximum?

12. What is the primary purpose of the path profile? What is the primary purpose of the path calculation (link budget) exercise?

13. Calculate the thermal noise floor of a LOS microwave receiver with a bandwidth of 20 MHz and a noise figure of 4 dB.

14. A microwave transmitter has an output of 2 W, line losses are 2 dB, and the antenna gain is 37 dB. What is the EIRP in dBW?

15. Calculate the unfaded RSL of a microwave link operating at 6 GHz, 35 km long; antenna diameters are 1.6 m at each end; line losses are 3.5 dB at each end; transmitter output is 0.5 W and the receiver noise figure is 7 dB.

16. In exercise 15, give the value for C/N_0.

17. We designed a LOS microwave radio hop where we find, late in the acceptance test period, that it does not meet noise or BER requirements. If it is a digital link, we assume, in this case, that measures have been taken to mitigate the effects of dispersion. Now list, in order of economic impact (least to most costly), measures that we can take to bring the performance up to specifications.

18. Calculate the EIRP in dBW of a microwave transmitting installation where the transmitter output is 0.5 W, the total transmission line loss is 5 dB, and the antenna gain is 33 dB.

19. Determine the IRL at the receive end of a 4-GHz link 25 mi long where the transmitter has 1-W output to the flange, the transmission line loss is 4 dB, and the transmit antenna has a gain of 37 dB. There is a radome over the antenna with a transmission loss of 1 dB.

20. What are the two types of fading covered in the text?

21. Calculate the Rayleigh fade margin where the desired time availability on a LOS microwave link is 99.93%.

22. Using the Vigants and Barnett method to calculate fade margin or time availability, a certain link has a fade margin of 37 dB. What is the time availability of the link? The path length is 28 km, average terrain, interior temperate climate. The frequency is 7 GHz.

23. Name four different ways in which we can achieve independent diversity paths.

24. What is the most common type of diversity used with LOS microwave links? Why not use frequency diversity?

25. Define a digital radio.

26. What is the overriding feature required for digital radios when operating with standard 8-bit PCM (Chapters 3 and 4), particularly for application in North America?

27. Calculate N_0 for a digital radio receiver with a noise figure of 8 dB; of 3 dB.

28. What is the meaning of bit count integrity (BCI)?

29. Following U.S. FCC rules, what is the minimum practical bit packing required?

30. Give a short definition of dispersion.

31. How many bits per hertz can theoretically be achieved with QPSK? With 8-ary PSK? With 64-QAM? With 512-QAM?

32. What is the principal cause of bit packing degradation from theoretical bit packing to practical values? For example, 256-QAM theoretically provides 8 bits/Hz, but the practical value is more like 6 bits/Hz.

33. As M increases for M-QAM systems, what is the effect on E_b/N_0 assuming the same BER for all values of M?

34. If a RSL is -76 dBW and the bit rate is 20 Mbps, what is the value of E_b in dBW?

35. Name at least three of the contributors of modulation implementation loss?

36. Name at least four measures we can take to mitigate the effects of multipath fading (and dispersion).

37. Given a value of dispersion on a path (in nanoseconds), relate this to bit rate. This, of course, should deal with pulsewidth. Assume NRZ waveform and simple binary PSK.

38. For digital radios, name the four factors that make up composite fade margin.

39. Calculate path dispersiveness for a path 50 km long and the half-power beamwidth of the antenna is 1.5°.

40. Give two ways in which FEC can improve link performance and/or the cost of implementing the link.

41. Differentiate between random and bursty errors and state the causes of each.

42. What are the two generic types of FEC coding? Name some advantages and disadvantages of each.

43. Define code rate. Give a simple formula for code rate.

44. What is the principal advantage of coding gain? Give a disadvantage.

45. What is the difference between a hard- and soft-decision demodulator? What type gives improved coding gain?

46. Block codes are often defined with the notation (N, K, t). What parameter is given for each letter of the notation?

47. Name three types of block codes.

48. A rate $\frac{3}{4}$ convolutional coder has a 10-Mbps input. What is the symbol rate at the output?

49. List the three types of decoders used with convolutional coding. Make the list in ascending order of performance.

50. Coding gain is a function of at least four factors. What are they?

51. Using QPSK modulation, hard-decision demodulator, convolutional coding, Viterbi decoder, and a BER = 1×10^{-5}, the coding gain is 4.8 dB and the required E_b/N_0 = 4.8 dB. What would the required E_b/N_0 be for the same system uncoded?

52. FEC systems can only handle random errors. How can we design a FEC system to handle bursty errors? What must we know about the fading characteristics of the link in the design of this "system"?

53. A digital microwave link is 22 mi long with an operating frequency at 6 GHz. The bit rate is 155.520 Mbps. The required BER is 1×10^{-6} and the modulation is 64-QAM; thus the theoretical E_b/N_0 is 21.4 dB. The modulation implementation loss is 1.8 dB. Assume no fading and zero link margin. The receiver noise figure is 6 dB and the waveguide losses at each end are 2.1 dB. Find a reasonable transmitter output power and antenna aperture to meet these conditions.

54. Why not use microwave antennas with apertures greater than 12 ft?

55. Calculate the beamwidth of a 12-ft antenna operating at 12 GHz.

56. Why would we prefer to use an elliptical waveguide over the other two types (rectangular and circular)?

57. A 2-m microwave parabolic dish antenna is being operated at 6 GHz. What is its gain?

58. Give at least four possible causes of a microwave link service interruption.

59. Argue why the value of MTTR is so critical in the availability equation. What might be a reasonable value in hours?

60. What are the two types of hot-standby operation?

61. Why is meeting antenna tower twist and sway requirements so important in LOS microwave installations?

62. There are ten LOS microwave links in tandem. At the last receiver in the system we desire a time availability (due to propagation alone) of 99.97%. What time availability must we assign to each link? Use the most conservative method.

REFERENCES

1. *Radio Regulations*, edition of 1990, revised 1994, International Telecommunication Union, Geneva, 1994.

2. (U.S.) "Code of Federal Regulations—Title 47, Telecommunication," *FCC Rules and Regulations*, Part 21, revised Oct. 1994, Special Edition of the *Federal Register*, Washington, DC, 1994.

3. *Engineering Considerations for Microwave Communication Systems*, GTE Lenkurt, San Carlos, CA, 1975.

4. "Propagation Data and Prediction Methods Required for the Design of Terrestrial Line-of-Sight Systems," ITU-R Rec. 530-5, 1994 PN Series Volume, ITU, Geneva, 1994.

5. "Attenuation in Vegetation," ITU-R Rec. 833-1, 1994 PN Series Volume, ITU, Geneva, 1994.

6. "Pre-emphasis Characteristics for Frequency Modulation Radio-Relay Systems for Telephony Using Frequency-Division Multiplex," CCIR Rec. 275-3, Vol. IX, Part 1, XVIIth Plenary Assembly, Dusseldorf, 1990.

7. "Frequency Deviation for Analogue Radio-Relay Systems for Telephony Using Frequency-Division Multiplex," CCIR Rec. 404-2, Vol. IX, Part 1, XVIIth Plenary Assembly, Dusseldorf, 1990.

8. "Effects of Propagation on the Design and Operation of Line-of-Sight Radio-Relay Systems," CCIR Rep. 784-3, Annex to Vol. IX, Part 1, XVIIth Plenary Assembly, Dusseldorf, 1990.

9. H. T. Dougherty, *A Survey of Microwave Fading Mechanisms, Remedies, and Applications*, ESSA Tech. Rep. ERL-69-WPL-4, Boulder, CO, Mar. 1968.

10. A. Vigants, *The Number of Fades and Their Durations on Microwave Line-of-Sight Links With and Without Diversity*, ICC-1969, IEEE, New York, 1969.

11. *Design Objectives of DCEC LOS Digital Radio Links*, Eng. Pub. DCEC EP-27-77, Defense Communications Engineering Center, Washington, DC, 1977.

12. "Diversity Techniques for Radio-Relay Systems," CCIR Rep. 376-6, Annex to Vol. IX, Part 1, XVIIth Plenary Assembly, Dusseldorf, 1990.

13. A. P. Barkhausen et al., *Equipment Characteristics and Their Relationship to Performance for Troposcatter Communication Circuits*, NBS Tech. Note 103, Boulder, CO, Jan. 1963.

14. "Allowable Noise Power in the Hypothetical Reference Circuit for Radio-Relay Systems for Telephony Using Frequency-Division Multiplex," CCIR Rec. 393-4, 1994 F Series Volume, Part 1, ITU, Geneva, 1994.

15. *Standard Microwave Transmission Systems*, EIA RS-252-A, Electronic Industries Assoc., Washington, DC, Sept. 1972.

16. "Measurement of Noise Using a Continuous Uniform Spectrum Signal on Frequency-Division Multiplex Telephony Radio-Relay Systems," CCIR Rec. 399-3, Vol. IX, Part 1, XVIIth Plenary Assembly, Dusseldorf, 1990.

17. W. Oliver, *White Noise Loading of Multichannel Communication Systems*, Marconi Instruments Ltd., St. Albans, UK, Sept. 1964.

18. *Design Handbook for Line-of-Sight Microwave Communication Systems*, MIL-HDBK-416, U.S. Department of Defense, Washington, DC, 1977.

19. "Characteristics of Digital Radio-Relay Systems Below About 17 GHz," ITU-R Rec. F.1101, 1994 F Series Volume, Part 1, ITU, Geneva, 1994.

20. "Definition of Spectrum Use and Efficiency of a Radio System," ITU-R Rec. SM.1046, 1994 SM Series Volume, ITU, Geneva, 1994.

21. "Error Performance Objectives for Real Digital Radio-Relay Links Forming Part of a Higher-Grade Circuit Within an Integrated Services Digital Network," ITU-R Rec. F.634-3, 1994 F Series Volume, Part 1, ITU, Geneva, 1994.

22. "Performance Objectives of Digital Radio-Relay Systems," CCIR Rep. 930-2, Annex to Vol. IX, Part 1, XVIIth Plenary Assembly, Dusseldorf, 1990.

23. T. Noguchi, Y. Daido, and J. A. Nossek, "Modulation Techniques for Microwave Digital Radio," *IEEE Commun. Mag.*, vol. 24, no. 10, Oct. 1986.

24. "Principles of Digital Transmission," ER79-307, under U.S. Govt. contract MDA-904-79-C-0470, Raytheon Co., Sudbury, MA, 1979 (limited circulation).

25. K. Feher, *Digital Communications Microwave Applications*, Prentice Hall, Englewood Cliffs, NJ, 1981.

26. "Linear Modulation Techniques for Digital Microwave," Harris Corp., RADC-TR-79-C-56, U.S. Govt. contract F30602-77-C-0039, USAF RADC, Rome, NY, Aug. 1979.

27. Private communication, W. P. Norris, Harris Electronic Systems Sector, Palm Bay, FL, Sept. 13, 1995.

28. E. W. Allen, "The Multipath Phenomenon in Line-of-Sight Digital Transmission Systems," *Microwave J.*, May 1984.

29. "Effects of Multipath Propagation on the Design and Operation of Line-of-Sight Digital Radio-Relay Systems," ITU-R Rec. F.1093, 1994 F Series Volume, Part 1, ITU, Geneva, 1994.

30. *Interference Criteria for Microwave Systems*, TIA Telecommunications Systems Bulletin, TSB 10-F, Telecommunication Industry Assoc., Washington, DC, 1994.

31. Eli Kolton, "Results and Analysis of Static and Dynamic Multipath in a Severe Atmospheric Environment," NTIA Contractor Report 86-37, Boulder, CO, Sept. 1986.

32. "Error Performance of an International Digital Connection Forming Part of an Integrated Services Digital Network," CCITT Rec. G.821, Fascicle III.5, IXth Plenary Assembly, Melbourne, 1988.

33. *The New Standard Dictionary of Electrical and Electronic Terms*, 5th ed., IEEE Std. 100-1992, IEEE, New York, 1992.

34. *Error Control Handbook*, Linkabit Corp., San Diego, CA, July 1976, under USAF contract F44620-76-C-0056.

35. A. J. Viterbi, "Convolutional Codes and Their Performance in Communication Systems," *IEEE Trans. Commun. Tech.*, vol. Com-19, pp. 751–792, Oct. 1971.

36. *Error Protection Manual, with Summary and Supplement*, NTIS AD-759 836, Computer Sciences Corporation, Falls Church, VA, 1973.

37. "Physical/Electrical Characteristics of Hierarchical Digital Interfaces," CCITT Rec. G.703, Geneva, 1991.

38. *Andrew Catalog 34*, Andrew Corporation, Orland Park, IL, 1988.

39. "Availability and Reliability of Radio-Relay Systems," CCIR Rep. 445-3, Annex to Vol. IX, Part 1, XVIIth Plenary Assembly, Dusseldorf, 1990.

40. "Transmission Systems Engineering Symposium," Rockwell International, Collins Transmission Systems Division, Dallas, TX, Sept. 1985.

41. "Spurious Emissions of Radio-Relay Systems," CCIR Rep. 937-2, Annex to Vol. IX, Part 1, XVIIth Plenary Assembly, Dusseldorf, 1990.

42. Roger L. Freeman, *Radio System Design for Telecommunications*, 2nd ed., Wiley, New York, 1997.

43. "Maximum Equivalent Isotropically Radiated Power of Line-of-Sight Radio-Relay System Transmitters Operating in the Frequency Bands Shared with the Fixed Satellite Service," CCIR Rec. 406-7, RFS Series, ITU, Geneva, 1992.

44. "Determination of the Interference Potential Between Earth Stations of the Fixed-Satellite Service and Stations in the Fixed Service," ITU-R Rec. SF.1006, ITU-R SF Series, ITU, Geneva, 1993.

45. *Microwave Radio-Relay Systems*, USAF Tech. Order TO 31R5-1-9, U.S. Department of Defense, Washington, DC, Apr. 1965.

46. *Structural Standards for Steel Antenna Towers and Antenna Supporting Structures*, ANSI/EIA/TIA-222-E-1991, Electronic Industries Assoc., Washington, DC, Feb. 1991.

47. *Microflect Passive Repeater Engineering Manual, No. 161*, Microflect, Inc., Salem, OR, 1962.

6

BEYOND LINE-OF-SIGHT: TROPOSPHERIC SCATTER AND DIFFRACTION LINKS

6.1 INTRODUCTION

Tropospheric scatter and diffraction are methods of propagating microwave energy beyond LOS or "over the horizon." Communication systems utilizing the tropospheric scatter/diffraction phenomena handle from 12 to 240 multiplexed telephone channels. Well-planned tropospheric scatter diffraction links may have propagation reliabilities on the order of 99.9% or better. These reliabilities are comparable to those of radiolink (LOS microwave) discussed in Chapter 5. In fact, the discussion of tropospheric scatter is a natural extension of that chapter.

Tropospheric scatter takes advantage of the refraction and reflection phenomena in a section of the earth's atmosphere called the troposphere. This is the lower portion of the atmosphere from sea level to a height of about 11 km (35,000 ft). UHF/SHF signals are scattered in such a way as to follow reliable communications on hops up to 640 km (400 mi). Long distances of many thousands of kilometers may be covered by operating a number of hops in tandem. The North Atlantic Radio System (NARS) of the U.S. Air Force is an example of a lengthy tandem system. It extended from Canada to Great Britain via Greenland, Iceland, the Faeroes, and Scotland. A mix of radiolinks (LOS microwave) and tropospheric scatter became fairly common. The Canadian National Telephone Company (CNT) operated such a system in the Northwest Territories. The Bahama Islands are interconnected for communications by a mix of LOS microwave, tropospheric scatter, and HF.

Tropospheric scatter systems generally use transmitter power outputs of 1 or 10 kW, parabolic-type antennas with diameters of 4.5 m (15 ft), 9 m (30 ft), or 18 m (60 ft), and sensitive (uncooled) broadband FM receivers with front-end NFs on the order of 1.0–4.0 dB. A tropospheric scatter installation is obviously a bigger financial investment than a LOS microwave installation. Tropospheric scatter/diffraction, however, has many advantages for commercial application that could well outweigh the issue of high cost. These advantages are summarized as follows:

1. It reduces the number of stations required to cover a given large distance when compared to radiolinks. Tropospheric scatter may require from one-third to one-tenth the number of stations as a radiolink system over the same path.
2. It provides reliable multichannel communication across large stretches of water (e.g., over inland lakes, to offshore islands, between islands) or between areas separated by inaccessible terrain.
3. It may be ideally suited to meet toll-connecting requirements of areas of low population density.
4. It is useful when radio waves must cross territories of another political administration.
5. It requires less maintenance staff per route-kilometer than conventional LOS microwave systems over the same route.
6. It allows multichannel communication with isolated areas, especially when intervening territory limits or prevents the use of repeaters.
7. It is desirable for multichannel communications in the tactical military field environment for links from 30 to 200 mi long (50–340 km).
8. It can be used in thin-line (e.g., ≤ 16-kbps) military systems with links up to 800 mi long (1480 km).

There is a tradeoff with these eight terms and other transmission media, such as fiber optics and satellite communications

6.2   THE PHENOMENON OF TROPOSPHERIC SCATTER

There are a number of theories explaining over-the-horizon communications by tropospheric scatter. One theory postulates atmospheric air turbulence, irregularities in the refractive index, or similar homogeneous discontinuities capable of diverting a small fraction of the transmitted radio energy toward a receiving station. This theory accounts for the scattering of radio energy in a way much as fog or moisture seems to scatter a searchlight on a dark night. Another theory is that the air is stratified into discrete layers of varying thickness in the troposphere. The boundaries between these layers become partially reflecting surfaces for radio waves and thereby scatter the waves downward over the horizon.

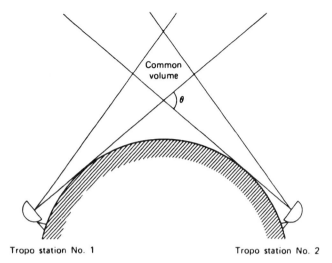

Figure 6.1. Tropospheric scatter model. θ = scatter angle.

Figure 6.1 is a simple diagram of a tropospheric scatter link showing two important propagation concepts:

- Scatter angle, which may be defined as either of two acute angles formed by the intersection of the two portions of the tropospheric scatter beam (lower boundaries) tangent to the earth's surface. Keeping the angle small effectively reduces the overall path attenuation.
- Scatter volume or "common volume," which is the common enclosed area where the two beams intercept.

6.3 TROPOSPHERIC SCATTER FADING

Fading is characteristic of tropospheric scatter. It is handy to break fading in tropospheric scatter systems into two types, slow and fast fading. Expressed another way, these are long-term (slow) and short-term variations in the received signal level.

When referring to tropospheric scatter received signal level, we usually use the median received level as reference. In general the hourly median and minute median are the same. In Chapter 5, the reference level was the unfaded signal level, which turned out to approximate sufficiently the calculated level under no-fade conditions. Such a straightforward reference signal level is impossible in tropospheric scatter because a tropospheric scatter signal is in a constant condition of fade. Thus for path calculations and path loss, we refer to the long-term median, usually extended over the whole year.

TABLE 6.1   Time Block Assignments

Time Block	Month	Hours
1	Nov.–Apr.	0600–1300
2	Nov.–Apr.	1300–1800
3	Nov.–Apr.	1800–2400
4	May–Oct.	0600–1300
5	May–Oct.	1300–1800
6	May–Oct.	1800–2400
7	May–Oct.	0000–0600
8	Nov.–Apr.	0000–0600

At any one moment a received tropospheric scatter signal will be affected by both slow and fast fading. It is believed that fast fading is due to the effects of multipath (i.e., due to a phase incoherence at various scatter angles). Fast fading is treated statistically "within the hour" and has a Rayleigh distribution with a sampling time of 1–7 min, although in some circumstances it has been noted up to 1 h. The fading rate depends on both frequency and distance or length of hop.

The U.S. National Bureau of Standards (NBS) describes long-term variations in signal level as variations of *hourly median* values of transmission loss. This is the level of transmission loss that is exceeded for a total of one-half of a given hour. A distribution of hourly medians gives a measure of long-term fading. Where these hourly medians are considered over a period of 1 month or more, the distribution is log-normal.

In studying variations in tropospheric scatter transmission path loss (fading), we have had to depend on empirical information. The signal level varies with the time or day, the season of year, and the latitude, among other variables. To assist in the analysis and prediction of long-term signal variation, the hours of the year have been broken down into eight time blocks, given in Table 6.1.

Most commonly we refer to time block 2 for a specific median path loss. Time block 2 may be thought of as an average winter afternoon in the temperate zone of the Northern Hemisphere.

It should be noted that signal levels average 10 dB lower in winter than in summer, and that morning or evening signals are at least 5 dB higher than midafternoon signals. Slow fading is believed due to changes in path conditions such as atmospheric changes (e.g., a change in the index of refraction of the atmosphere).

6.4   PATH LOSS CALCULATIONS

Tropospheric scatter paths typically display considerably larger losses when compared to radiolink (LOS microwave) paths. Losses up to 260 dB are *not* uncommon. There are a number of acceptable methods of estimating path losses for tropospheric scatter systems. One such method is outlined in CCIR Rep. 238-3 (Ref. 1), and another is described in NBS Tech. Note 101 (Ref. 2). These are more commonly known as the CCIR method and the NBS method. Their approach to the problem is somewhat similar. In the following paragraphs we will summarize some of the more important aspects of the NBS method as described in Ref. 3. The procedure has been considerably simplified and abbreviated for this discussion.

The objective is to predict a long-term path loss that will not be exceeded for specified time availabilities, such as 50%, 90%, 99%, 99.9%, or 99.99% of the time. In Chapter 5 on LOS microwave we referred to this as propagation reliability.

The NBS method describes how to calculate these losses with a 50% probability (confidence level) of being the correct prediction for a path in question. Then it shows how to systematically add margin to assure improved probability that the prediction will be correct or more than the minimum necessary on a particular path. This probability of prediction is called *service probability*. It is common with tropospheric scatter path calculations to show a 50% and a 95% service probability. Service probability indicates the confidence level of the prediction.

6.4.1 Mode of Propagation

An over-the-horizon microwave path can and often does display two modes of propagation—diffraction and tropospheric scatter. In most cases either one or the other will predominate. Particularly on shorter paths the possibility of the diffraction mode should be investigated. Engineers experienced in over-the-horizon systems can often identify which propagation mode can be expected during the path profile and path survey phases of the link engineering effort. In the absence of this expertise, the following criteria may be used as an aid to identify the principal propagation mode (Ref. 3):

1. The distance at which diffraction and tropospheric scatter losses are approximately equal is $65(100/f)^{1/3}$ km, where f = radio frequency in MHz. For path lengths less than this value, diffraction will be the predominant mode, and vice versa.
2. For paths having angular distances of 20 mrad or more, the diffraction mode may be neglected and the path can be considered to be operating in the troposcatter mode. (Angular distance is explained in the following.)

6.4.2 Basic Long-Term Tropospheric Scatter Transmission Loss

Following the NBS method, we determine the basic long-term tropospheric scatter transmission loss L_{bsr}:

$$L_{bsr} = 30 \log f - 20 \log d + F(\theta d) - F_0 + H_0 + A_a \qquad (6.1)$$

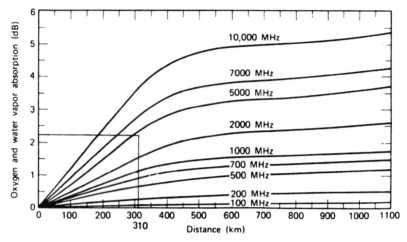

Figure 6.2. Determination of median oxygen and water vapor absorption in decibels for various operating frequencies when path length is given (for August, Washington, DC). (From Ref. 3.)

where f = operating frequency (MHz)
          d = great-circle path length (km)
  $F(\theta d)$ = attenuation function (dB)
       F_0 = scattering efficiency correction factor
       H_0 = frequency gain function
       A_a = atmospheric absorption factor from Figure 6.2

We have simplified our procedure by neglecting the frequency gain function and the scattering efficiency correction factor.

   The numerical values of the first two terms of equation 6.1 are determined by substituting the assigned frequency in megahertz of the radio system to be installed (see Section 6.9) and the great-circle distance in kilometers. The third term requires some detailed discussion, which follows.

6.4.3 Attenuation Function

The attenuation function $F(\theta d)$ is derived from Figure 6.3. θd is the product of the angular distance (scatter angle) in radians and the great-circle path length in kilometers. The following is an abbreviated method of approximating the scatter angle θ. We assume that a path profile has been carried out (see Section 5.3.4). Arbitrarily, one site is denoted the transmitter site t and the other site the receiver site r. From the profile the horizon location in the direction of the distant site and its altitude above mean sea level (MSL) are determined as well as its distance from its corresponding site. For all further

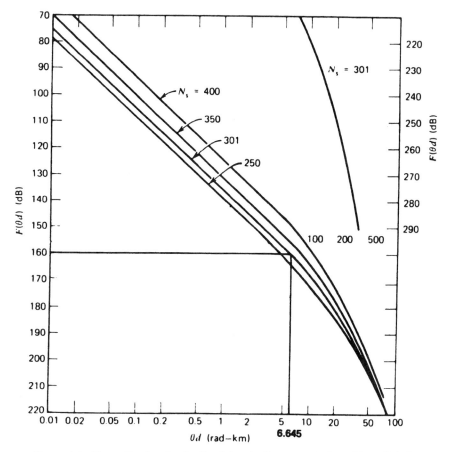

Figure 6.3. Attenuation function for the determination of scatter loss. (From Ref. 3).

calculations, distances and altitudes (elevations) are measured in kilometers and angles in radians. It is important to use only these units throughout. Figure 6.4 will assist in identifying the following distances, elevations, and angles. Let

d = great-circle distance between transmitter and receiver sites

d_{Lt} = distances from transmitter site to transmitter horizon

d_{Lr} = distance from receiver to receiver horizon

h_{ts} = elevation above MSL to center of transmitting antenna (km)

h_{rs} = elevation above MSL to center of receiving antenna (km)

h_{Lt} = elevation above MSL of transmitter horizon point (km)

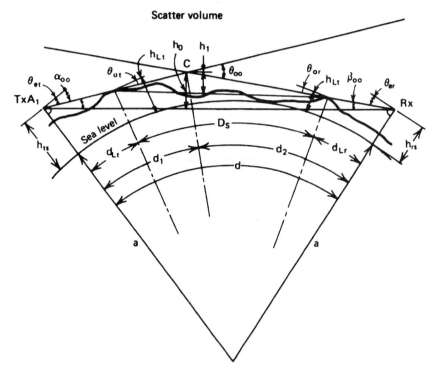

Figure 6.4. Tropospheric scatter path geometry.

h_{Lr} = elevation above MSL of receiver horizon point (km)

N_0 = surface refractivity corrected for MSL (for the continental United States use Figure 5.2, and for other locations use Ref. 4, Figures 1.4 and 1.5)

Adjust the surface refractivity N_0 for the elevation of each site by the following formula:

$$N_{ts,rs} = N_0 \exp(-0.1057 h_{ts}, h_{rs}) \qquad (6.2)$$

Compute N_s:

$$N_s = \frac{N_{ts} + N_{rs}}{2} \qquad (6.3)$$

Calculate the effective earth radius by the following formula:

$$a = a_0 (1 - 0.04665 \exp^{0.005577 N_s})^{-1} \qquad (6.4)$$

where a_0 = true earth radius, 6370 km. If N_s = 301 then a = 8500 km, which is the familiar $\frac{4}{3}$ earth radius case alluded to in Chapter 5.

Calculate the antenna takeoff angles at each site by the following equations:

$$\theta_{et} = \frac{h_{Lt} - h_{ts}}{d_{Lt}} - \frac{d_{Lt}}{2a} \tag{6.5a}$$

$$\theta_{er} = \frac{h_{Lr} - h_{rs}}{d_{Lr}} - \frac{d_{Lr}}{2a} \tag{6.5b}$$

Calculate the scatter angle components α_0 and β_0 by the following formulas:

$$\alpha_0 = \frac{d}{2a} + \theta_{et} + \frac{h_{ts} - h_{rs}}{d} \tag{6.6a}$$

$$\beta_0 = \frac{d}{2a} + \theta_{er} + \frac{h_{rs} - h_{ts}}{d} \tag{6.6b}$$

Calculate the scatter angle (often called *angular distance*) θ_0 by the following equation:

$$\theta_0 = \alpha_0 + \beta_0 \text{ (rad)} \tag{6.7}$$

Multiply θ_0 in radians by the path length in kilometers. This is θd.

Determine $F(\theta d)$ from Figure 6.3 using the product θd calculated above, and interpolate, if necessary, for the value of N_s taken from equation 6.3.

L_{bsr} is now calculated by equation 6.1, neglecting terms F_0 and H_0.

6.4.4 Basic Median Transmission Loss

The predicted median long-term transmission loss $L_n(0.5, 50)$, abbreviated $L_n(0.5)$, for the appropriate climatic region n is related to L_{bsr} by the following formula:

$$L_n(0.5) = L_{bsr} - V_n(0.5, d_e) \tag{6.8}$$

where $L_n(0.5)$ = predicted transmission loss (in dB) exceeded by half of all hourly medians, and hence the yearly median value, and $V_n(0.5, d_e)$ = variability of the median value about the basic long-term transmission loss L_{bsr} for the appropriate climatic region n and the effective distance d_e. NBS has established eight climatic regions for the world as follows (Ref. 3):

1. Continental temperate. Large land mass, $30°-60°$ N latitude, $30°-60°$ S latitude.

2. Maritime temperate overland. In this region, prevailing winds, unobstructed by mountains, carry moist maritime air inland. Latitudes 20°–50° N and 20°–50° S, typified by United Kingdom, the west coasts of North America and Europe, and northwestern coastal areas of Africa.

3. Maritime temperate oversea. Fully over-water paths in temperate regions.

4. Maritime subtropical overland. Latitudes 10°–30° N, 10°–30° S, near the sea with defined rainy and dry seasons.

5. Maritime subtropical. Latitudes same as region 4. Over-water paths. However, valid curves are not available due to lack of empirical data for this region. Use region 3 or region 4, whichever is more applicable.

6. Desert, Sahara. Regions with year-round semiarid conditions.

7. Equatorial. Latitude ±20° from the equator, characterized by monotonous heavy rains and high average summer temperatures.

8. Continental subtropical. Usually 20°–40° N latitude, an area of monsoons with seasonal extremes of summer rainfall and winter drought.

Select the most appropriate region (n) for the path in question, then compute the effective distance d_e. To calculate d_e, effective antenna heights are required, namely, h_{te} and h_{re}. These heights are functions of the average elevation of the terrain between each antenna and its respective radio horizon in the direction of the distant end of the path. For smooth earth condition (i.e., typically an over-water path or a hypothetical path with no obstacles except central earth bulge), h_{te} and h_{re} are the effective elevations of each site above MSL. Under real overland conditions, the effective height is the average height above MSL of the central 80% between the antenna and its radio horizon.

Calculate d_L by the following formula:

$$d_L = 3\sqrt{2h_{te} \times 10^3} + 3\sqrt{2h_{re} \times 10^3} \qquad (6.9)$$

Determine d_{sl} by the following formula:

$$d_{sl} = 65\left(\frac{100}{f}\right)^{1/3} \qquad (6.10)$$

where f = frequency (MHz).

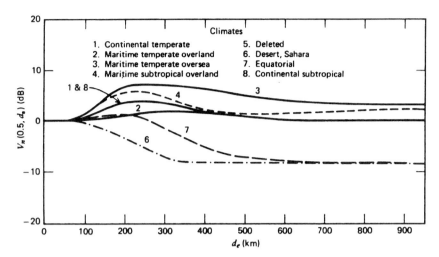

Figure 6.5. Function $V_n(0.5, d_e)$ for eight climatic regions. (From Ref. 3.)

There are two cases to calculate d_e. Use whichever is applicable.

1. If $d \leq d_L + d_{sl}$, then

$$d_e = \frac{130d}{(d_L + d_{sl})} \qquad (6.11a)$$

2. If $d > d_L + d_{sl}$, then

$$d_e = 130 + d - (d_L + d_{sl}) \qquad (6.11b)$$

With the climatic region n and the effective distance d_e determined, derive V_n in decibels from Figure 6.5. Calculate $L_n(0.5)$ using equation 6.8. The value $L_n(0.5, 50)$ represents the long-term median path loss for a 50% time availability and 50% service probability.

6.4.5 The 50% Service Probability Case

The next step is to extend the time availability to the specified or desired value for the tropospheric scatter path in question. Often it is convenient to state time availability for a number of percentages as follows:

Time Availability q (%)	Path Loss
50	$L_n(0.5, 50)$
90	A
99	B
99.9	C
99.99	D

Values for A, B, C, and D in the path loss column are determined by adding a factor to $L_n(0.5, 50)$ called $Y_n(q, 50, d_e)$, where $q = 0.9$, 0.99, 0.999, and 0.9999. Y_n values are derived from curves for the appropriate climatic region and frequency band. One example family of curves is presented in Figure 6.6, where n is region 1, the continental temperate region, and for frequencies above 1 GHz. Y_n is derived for several values of q using the appropriate effective distance d_e of the tropospheric scatter path under study.

6.4.6 Improving Service Probability

Under the values of path loss calculated in the previous section, only half of the paths installed for a specific set of conditions would have a measured long-term path loss equal to or less than those calculated. By definition, this is a service probability of 50%. To extend the service probability (i.e., improve the confidence level), the following procedures should be followed. Again we are dealing only with long-term power fading. The basic data required are the values obtained for $Y_n(q, d_e)$ and the standard normal deviate Z_{mo} for the service probability desired. Several standard normal deviates and their corresponding service probabilities are provided below:

Service Probability (%)	Standard Normal Deviate Z_{mo}
50	0
60	0.25
75	0.75
80	0.85
90	1.28
95	1.65
99	2.35

Calculate the path-to-path variance $\sigma_c^2(q)$, where q is the corresponding (or desired) time availability:

$$\sigma_c^2(q) = 12.73 + 0.12Y_n^2(q, d_e) \qquad (6.12)$$

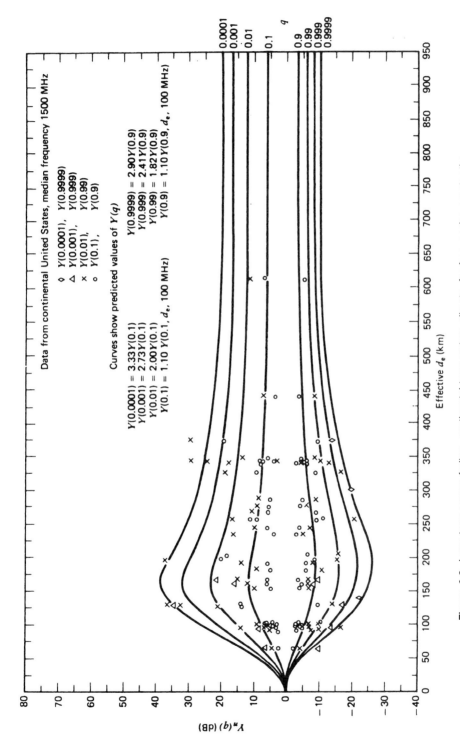

Figure 6.6. Long-term power fading, continental temperature climate, for frequencies greater than 1000 MHz. (From Ref. 3.)

Determine the prediction error $\sigma_{rc}(q)$ by the following equation:

$$\sigma_{rc}(q) = \sqrt{\sigma_c^2(q) + 4} \tag{6.13}$$

Calculate the product of Z_{mo} and $\sigma_{rc}(q)$. This value is now added to the path loss value for the corresponding time availability q given in the previous section.

6.4.7 Example Problem

Assume smooth earth condition (i.e., no intervening obstacles besides earth bulge) and calculate the path loss from Newport, NY, to Bedford, MA (U.S.). The great-circle distance between the sites is 310.5 km.

Site elevation, Newport, $h_{ts} = 2000$ ft (0.61 km)
Site elevation, Bedford, $h_{rs} = 100$ ft (0.031 km)
$N_0 = 310$
Operating frequency 4700 MHz
$d_{Lt} = 102$ km (smooth earth); $d_{Lr} = 23$ km (smooth earth) (see equation 5.7)
$h_{Lt}, h_{Lr} = 0.0$ km (smooth earth, by definition)
$N_{st} = 291$; $N_{sr} = 309$
$N_s = 300$
$a = 8500$ km
$\theta_{et} = 0.0119$ rad; $\theta_{er} = -0.0027$ rad
$\alpha_0 = 0.008$ rad; $\beta_0 = 0.0137$ rad
$\theta = 0.0217$ rad
$\theta d = 0.0217 \times 310.5 = 6.784$ km-rad

From Figure 6.3, $F(\theta d) = 160$ dB.
Determine L_{bsr}:

$$
\begin{array}{ll}
30 \log f = 110.16 \text{ dB} & \text{where } f = 4700 \text{ MHz} \\
-20 \log d = -49.84 \text{ dB} & \\
F(\theta d) = 160 \quad \text{dB} & \\
\underline{A_a = 2.2 \quad \text{dB}} & \text{(from Figure 6.2)} \\
L_{bsr} = 222.52 \text{ dB} &
\end{array}
$$

Calculate d_L and d_{sL}:

$h_{te} = 0.609$ km $= h_{ts}$; $h_{re} = 0.031 \, km = h_{rs}$

$f = 4700$ MHz; $d_{sL} = 65(100/f)^{1/3} = 18$ km

(from equation 6.10)

$$d_{\mathrm{L}} = 3(2h_{\mathrm{te}} \times 10^3)^{1/2} + 3(2h_{\mathrm{re}} \times 10^3)^{1/2} = 128 \text{ km}$$

(from equation 6.9)

$$d_{\mathrm{e}} = 130 + 310.5 - 128 - 18 \text{ km} = 294.5 \text{ km (from equation 6.11b)}$$

Determine V_n from Figure 6.5:

$$V_n = 3.5 \text{ dB}, \quad n = \text{region 1}$$

Then

$$L_n(0.5) = 222.52 - 3.5 = 219.02 \text{ dB}$$

This is the predicted path loss for a 50% time availability and 50% service probability. All further path loss calculations are based on this value. We now make up a path loss/time distribution table similar to that shown in Section 6.4.5 for the 50% service probability case:

Time Availability $q(\%)$	$Y_1(q)$ (dB)	Transmission Loss (dB)
50	0	219
90	7	226
99	13	232
99.9	17	236
99.99	20	239

The values for $Y_1(q)$ were taken from Figure 6.6.

We now prepare a similar table for the 95% service probability case:

Time Availability $q(\%)$	$Z_{\mathrm{mo}} \sigma_{\mathrm{rc}}(q)$ (dB)	Transmission Loss (dB)
50	6.7	226
90	7.8	234
99	10.0	242
99.9	11.8	245
99.99	13.3	252

6.5 APERTURE-TO-MEDIUM COUPLING LOSS

Some tropospheric scatter link designers include aperture-to-medium coupling loss as another factor in the path loss equation, and others subtract the values from the antenna gains. In any event the loss must be included somewhere.

Aperture-to-medium coupling loss has sometimes been called the *antenna gain degradation*. It occurs because of the very nature of tropospheric scatter in that the antennas used are not doing the job we would expect them to do. This is evident if we use the same antenna on a LOS (or radiolink) path. The problem stems from the concept of the common volume. High-gain parabolic antennas used on tropospheric scatter paths have very narrow beamwidths (see Section 6.10.1). The tropospheric scatter loss calculations consider a larger common volume than would be formed by these beamwidths. As the beam becomes more narrow due to the higher gain antennas, the received signal level does not increase in the same proportion as it would under free-space (LOS) propagation conditions. The difference between the free-space expected gain and its measured gain on a tropospheric scatter hop is called the aperture-to-medium coupling loss. This loss is proportional to the scatter angle θ and the beamwidth Ω. The beamwidth may be calculated from

$$\Omega = \frac{70}{F \times D_r} \tag{6.14}$$

where F = carrier frequency (GHz) and D_r = antenna reflector diameter (ft).

The ratio θ/Ω is computed, and from this ratio the aperture-to-medium coupling loss may be derived from Table 6.2.

Example. Calculate the aperture-to-medium coupling loss for two 30-ft antennas, one at each end of the path, with a 2° scatter angle and a 900-MHz

TABLE 6.2   Aperture-to-Median Coupling Loss

Antenna Beamwidth Ratio, θ/Ω	Coupling Loss (dB)	Antenna Beamwidth Ratio, θ/Ω	Coupling Loss (dB)
0.3	0.18	1.4	2.95
0.4	0.40	1.5	3.22
0.5	0.60	1.6	3.55
0.6	0.90	1.7	3.80
0.7	1.10	1.8	4.10
0.75	1.20	1.9	4.25
0.8	1.40	2.0	4.63
0.9	1.70	2.1	4.90
1.0	1.95	2.2	5.20
1.1	2.2	2.3	5.48
1.2	2.42	2.4	5.70
1.3	2.75	2.5	6.00

Source: Reference 8.

operating frequency. The beamwidth Ω is

$$\Omega = \frac{70}{30 \times 0.9}$$

$$= 2.7°, \text{ equivalent to 47 mrad}$$

The scatter angle is

$$\theta = 2°, \text{ equivalent to 35 mrad}$$

Thus the ratio $\theta/\Omega = 35/47$, or approximately 0.75. From Table 6.2 this is equivalent to a loss of 1.2 dB.

CCIR Rep. 238-3 (Ref. 1) suggests another approach to the calculation of aperture-to-medium coupling loss using solely antenna gains G_t and G_r:

$$\text{Aperture-to-medium coupling loss (dB)} = 0.07 \exp[0.055(G_t + G_r)]$$

$$(6.15)$$

where G_t = gain of the transmitting antenna (dB) and G_r = gain of the receiving antenna (dB).

Using the previous example of two 30-ft antennas and 900-MHz operating frequency, $G_t = G_r = 36$ dB and $G_t + G_r = 72$ dB, and

$$\text{Aperture-to-medium coupling loss (dB)} = 0.07 \exp(0.055 \times 72)$$

$$= 3.67 \text{ dB}$$

The difference in value between the two methods should be noted.

6.6   TAKEOFF ANGLE (TOA)

The TOA is probably the most important factor under control of the engineer who selects a tropospheric scatter site in actual path design. The TOA is the angle between a horizontal ray extending from the radiation center of an antenna and a ray extending from the radiation center of the antenna to the radio horizon. Figure 6.7 illustrates the definition.

The TOA is computed by means of path profiling several miles out from the candidate site location. It then can be verified by means of transit siting. Path profiling is described in Section 5.3.4.

Figure 6.8 shows the effect of TOA on transmission loss. As the TOA is increased, about 12 dB of loss is added for each degree increase in TOA. This loss shows up in the scatter loss term of the equation for computing the

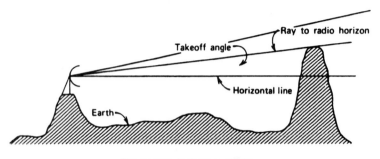

Figure 6.7. Definition of TOA.

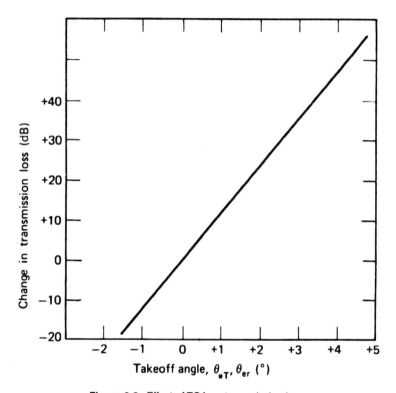

Figure 6.8. Effect of TOA on transmission loss.

median path loss (equation 6.1). This approximation is valid at $0°$ in the range of $+10°$ to $-10°$.

The benefit of siting a tropospheric scatter station on as high a site as possible is obvious. The idea is to minimize obstructions to the horizon in the direction of the "shot." As we shall see later, every decibel saved in median path loss may represent a savings of many thousands of dollars. Therefore

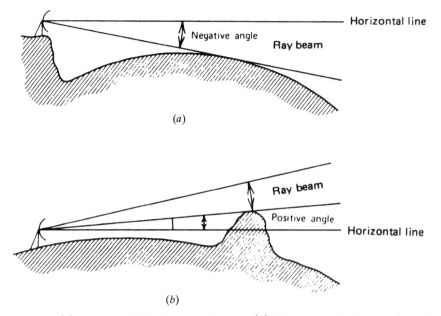

Figure 6.9. (a) A more desirable site regarding TOA. (b) A less desirable site regarding TOA (note that the beam should actually just clear the obstacle).

the more we can minimize the TOA, the better. Negative TOAs are very desirable. Figure 6.9 illustrates this criterion.

6.7 OTHER SITING CONSIDERATIONS

6.7.1 Antenna Height

Increasing the antenna height decreases the TOA, in addition to the small advantage of getting the antenna up and over surrounding obstacles. Raising an antenna from 20 ft above the ground to 100 ft above the ground provides something on the order of less than a 3-dB improvement in median path loss at 400 MHz and about a 1-dB improvement at 900 MHz (Ref. 5).

6.7.2 Distance to the Radio Horizon

The radio horizon may be considered one more obstacle that the tropospheric scatter ray beam must get over. Varying the distance to the horizon varies the TOA. If we maintain a constant TOA, the distance to the horizon can vary widely with insignificant effect on the overall transmission loss.

6.7.3   Other Issues

If we vary the path length with constant TOA, the median path loss varies about 0.1 dB/mi. The primary effect of increasing the path length is to change the TOA, which will notably affect the total median path loss. This is graphically shown in Figure 6.8.

6.8   PATH CALCULATIONS

6.8.1   Basis

New troposcatter/diffraction links will be digital. Troposcatter/diffraction paths are highly dispersive. Whereas with LOS microwave links we might expect to find maximum dispersion values at about 10 ns, tropo paths often exceed 200 ns with maximum values on the order of 500 ns. This puts some severe limitations on digital transmission by diffraction/tropo means. Waveforms must be robust. Thus simple modulation schemes such as BPSK/QPSK are in order.

The dispersion generally results from multipath propagation. Thus some form of compensation for these effects must be employed. One approach is to use effective adaptive equalizers in each diversity branch as suggested in Section 5.5.4.3.2 or a more revolutionary approach used in Raytheon's AN/TRC-170(V) described in Section 6.12.

On troposcatter/diffraction paths, dual diversity is a requirement, not just an option as it is with LOS microwave. Quadruple diversity is desirable. Diversity can compensate for much of the Rayleigh fading due to multipath; it can do nothing for long-term (log-normal) fading.

6.8.2   Overview of Procedure

The link budget or path calculation procedure is very similar to that described in Section 5.7 for digital LOS microwave. Again we will be dealing with E_b/N_0. We select the type of modulation and a value for BER and apply this to the appropriate E_b/N_0 curve such as one of the curves shown in Figure 5.38 or 5.39.

At this juncture we have calculated the transmission loss from Section 6.4.6, which includes an additional loss value for long-term fading calculated for the desired time availability and service probability. BER, modulation type, and target E_b/N_0 have been selected.

The link budget can be used as a tool to solve many specific design alternatives. A transmission engineer may be asked what maximum link length a given equipment configuration can support or what equipment configuration may be used for a given specific link. A situation that the author has faced involved a given certain link and specific equipment configuration that did not meet time availability requirements given a specific

BER threshold. The problem was to determine how to modify the configuration so that time availability requirements were complied with.

A warning to the potential tropo link engineer: tropo/diffraction paths are extremely susceptible to siting and topology. As we mentioned in Section 6.6, for every 1° increase in takeoff angle, we add about 12 dB to transmission loss. The reverse is also true; for every 1° decrease in takeoff angle, we can subtract 12 dB from the transmission loss value. One equipment configuration may meet requirements with margin to spare on a 250-mi path, yet on another path only 70 mi long with the same configuration, the time availability may be unacceptable.

One good approach is to set up a model configuration for a particular path and adjust for any shortfall or overage.

6.8.2.1 Candidate Model Configuration. Any reasonable model will do. The following was selected because it seems reasonable and applicable:

Frequency: 4700 MHz

Dual diversity, 15-ft parabolic dish antennas; 65% efficient; antenna gain, 45 dB

Transmission line loss: 2 dB

Aperture-to-medium coupling loss: 10 dB (CCIR method)

Transmitter power output: 1 kW ($+30$ dBW)

Receiver noise figure: 1.2 dB; $N_0 = -202.8$ dBW/Hz

Bit rate: 1.544 Mbps; BER $= 1 \times 10^{-6}$; $E_b/N_0 = 10.6$ dB $+ 1.4$ dB modulation implementation loss for a total of 12.0 dB

Transmission loss: 220 dB, continental temperate climate; time availability 99.95% and service probability 95%

We assume that the dispersion problem has been solved by means of adaptive equalization on each diversity path and that there is a hitless maximal ratio square combiner. This will give us 3-dB diversity improvement on the RSL.

We go through the following steps (see Section 5.7.2):

- Calculate EIRP: $+30$ dBW $- 2$ dB $+ 45$ dB $= +73$ dBW
- Add the aperture-to-medium coupling loss to the path loss: Total loss $= 230$ dB
- Compute the IRL: $+73$ dBW $- 230$ dB $= -157$ dBW
- Calculate the RSL: -157 dBW $+ 45$ dB $- 2$ dB $= -114$ dBW
- Compute E_b: -114 dBW $- 10\log(1.544 \times 10^6) = -175.89$ dBW
- Calculate E_b/N_0: -175.89 dBW $- (-202.8$ dBW$) = 26.9$ dB
- Calculate margin: $26.9 - 12$ dB $= 14.9$ dB

The configuration meets the path performance requirements plus a margin of 14.9 dB. Add to this the diversity improvement factor of 3 dB and the total margin then is 17.9 dB, which would be left to compensate for residual Rayleigh fading.

If additional margin is deemed necessary, an economic alternative would be to use FEC (Section 5.6) because there were no bandwidth constraints imposed in the model.

Suppose, now, that we wanted to extend the link and improve the time availability and/or the service probability. What alternative actions are open to us? Moving from least to most expensive, we list:

1. FEC coding (3–6 dB of coding gain).
2. Increase antenna aperture. For larger aperture antennas, cost varies approximately by the exponent 3.5.
3. Use better grade transmission line (possibly 1–2 dB improvement).
4. Increase output power of transmitter (1 to 10 kW, 10 dB).
5. Change over to quadruple diversity. Reduces required Rayleigh fading margin and adds net about 2.4-dB diversity gain.
6. Resite to a more favorable location for takeoff angle.

It should be noted that little can be done to improve noise performance. One could also argue the order of items (items 1–6 above) regarding cost.

A typical quadruple-diversity troposcatter installation is shown in Figure 6.10.

6.9 TROPOSCATTER/DIFFRACTION OPERATIONAL FREQUENCY BANDS

The following frequency bands have been used for troposcatter/diffraction operation:

350–450 MHz
755–985 MHz
1700–2400 MHz
4400–5000 MHz

6.10 ANTENNAS, TRANSMISSION LINES, DUPLEXERS, AND OTHER RELATED TRANSMISSION LINE DEVICES

6.10.1 Antennas

The antennas used in tropospheric scatter installations are broadband high-gain parabolic reflector devices, usually ground-mounted. The antennas

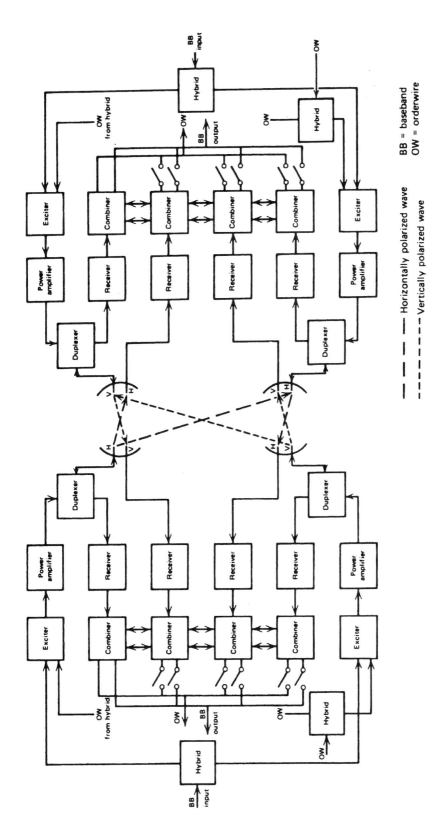

Figure 6.10. Simplified functional block diagram of a quadruple-diversity tropospheric scatter configuration.

BB = baseband
OW = orderwire

— — — Horizontally polarized wave
- - - - - Vertically polarized wave

covered here are similar in many respects to those discussed in Section 5.8.2 but have higher gain and consequently are larger and considerably more expensive. As we discussed in that section, the gain of this type of antenna is a function of the reflector diameter, frequency, and efficiency. Beamwidth is a function of gain; the higher the gain, the narrower the beamwidth. Table 6.3 gives some typical gains with beamwidths for several frequency bands and several standardized diameters. In this case, we have assumed a 55% efficiency. Improved feed methods illuminate the reflector more uniformly and reduce spillover, with consequent improvement of antenna efficiency. For example, for a 30-ft reflector operating at 2 GHz, improving the efficiency from 55% to 61% will increase the net antenna gain by about 0.5 dB.

It is desirable, but not always practical, to have the two antennas (as shown in Figure 6.10) spaced not less than 100 wavelengths apart to ensure proper space diversity operation. Antenna spillover (i.e., radiated energy in the sidelobes and back lobes) must be reduced to improve radiation efficiency and to minimize interference with simultaneous receiver operation and with other services.

The first sidelobes should be down (attenuated) at least 23 dB (from the main lobe) and the remainder of the unwanted lobes down at least 40 dB from the main lobe. Antenna alignment with the distant end is extremely important because of narrow beamwidths. Examples of typical half-power beamwidths are given in Table 6.3. A good impedance match between the transmission line and antenna, and between the equipment and transmission line, is important. The match is expressed by voltage standing wave ratio (VSWR), where a VSWR of 1:1 is a perfect match (i.e., an infinite return loss). A good VSWR improves system efficiency; with a poor VSWR, resulting reflected power may damage components further back in the transmis-

TABLE 6.3 Some Typical Antenna Gains and Beamwidths

Reflector Diameter (ft)	Frequency (GHz)	Gain (dB)	Beamwidth (degrees)
15	0.4	23	11.5
	1.0	31	4.6
	2.0	37	2.4
	4.0	43	1.15
30	0.4	29	5.4
	1.0	37	2.4
	2.0	43	1.15
	4.0	49	0.57
60	0.4	35	2.8
	1.0	43	1.15
	2.0	49	0.57
	4.0	55	0.28
120	4.0	61	0.14

sion system. Often, in tropo installations, load isolators are required to minimize the damaging effects of reflected power. In very high-power installations, these devices are often cooled.

A load isolator is a ferrite device with approximately 0.5-dB insertion loss. The forward wave (the energy radiated toward the antenna) is attenuated 0.5 dB; the reflected wave (the energy reflected back from the antenna) is attenuated more than 20 dB.

Another important consideration in planning a tropospheric scatter antenna system is polarization (see Figure 6.10). For a common antenna the transmit wave should be orthogonal to the receive wave. This means that if the transmitted signal is horizontally polarized, the receive signal should be vertically polarized. The polarization is established by the feeding device, usually a feed horn. The primary reason for using opposite polarization is to improve isolation, although the correlation of fading on diversity paths may be reduced. A figure commonly encountered for isolation between polarizations on a common antenna is 30 dB. However, improved figures may be expected in the future.

6.10.2 Transmission Lines

In selecting and laying out transmission lines for tropospheric scatter installations, it should be kept in mind that losses must be kept to a minimum. That additional fraction of a decibel is much more costly in tropospheric scatter than in radiolink installations. The tendency therefore is to use waveguide on most tropospheric scatter installations because of its lower losses than coaxial cable. Waveguide is universally used above 1.7 GHz.

Transmission line runs should be less than 200 ft (60 m). Ideally, the attenuation of the line should be kept under 1 dB from the transmitter to the antenna feed and from the antenna feed to the receiver, respectively. To minimize reflective losses, the VSWR of the line should be 1.05:1 or better when terminated in its characteristic impedance. Figures 5.49 and 5.50 show several types of transmission lines commercially available.

6.10.3 The Duplexer

The duplexer is a transmission line device that permits the use of a single antenna for simultaneous transmission and reception. For tropospheric scatter application a duplexer is a three-port device (see Figure 6.11) so tuned that the receiver leg appears to have an admittance approaching (ideally) zero at the transmitting frequency. At the same time the transmitter leg has an admittance approaching zero at the receiving frequency. To establish this, sufficient separation in frequency is required between the transmitted and received frequencies. Figure 6.11 is a simplified block diagram of a duplexer. The insertion loss of a duplexer in each direction should be less than 0.5 dB. Isolation between the transmitter port and the receiver port should be better

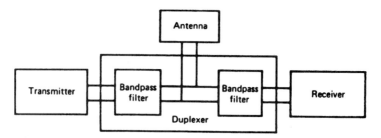

Figure 6.11. A simplified block diagram of a duplexer.

than 30 dB. High-power duplexers are usually factory tuned. It should be noted that some textbooks call the duplexer a diplexer.

6.11 ISOLATION

An important factor in tropospheric scatter installation design is the isolation between the emitted transmit signal and the receiver input. Normally we refer to the receiver sharing a common antenna feed with the transmitter.

A nominal receiver input level for military tropospheric scatter systems is -80 dBm (Ref. 9) for design purposes. If a transmitter has an output power of 10 kW or $+70$ dBm and transmission line losses are negligible, then isolation must be greater than 150 dB.

To achieve overall isolation such that the transmitted signal interferes in no way with receiver operation when the equipment is operating simultaneously, the following items aid the required isolation when there is sufficient frequency separation between transmitter and receiver:

- Polarization
- Duplexer
- Receiver preselector
- Transmit filters
- Normal isolation from receiver conversion to IF

6.12 OVERCOMING DISPERSION—ANOTHER APPROACH

6.12.1 Introduction

Dispersion is the principal cause of degradation of BER on digital transhorizon links. With conventional waveforms such as BPSK, MPSK, BFSK, and MFSK, dispersion may be such, on some links, that BER performance is unacceptable.

Dispersion is simply the result of some signal power from an emitted pulse that is delayed, with that power arriving later at the receiver than other power components. The received pulse appears widened or smeared or what we call dispersed. These late-arrival components spill over into the time slots of subsequent pulses. The result is intersymbol interference (ISI), which deteriorates BER.

Expected values of dispersion on transhorizon paths vary from 30 to 380 ns. The cause is multipath. The delay is a function of path length, antenna beamwidth, and the scatter angle components α_0 and β_0.

6.12.2 Overcoming Dispersion by Gated Pulses

One simple method to avoid overlapping pulse energy is to time-gate the transmitted energy, which allows a resting time after each pulse. Suppose we were transmitting a megabit per second and we let the resting time be half a pulsewidth. Then we would be transmitting pulses of 500 ns of pulse width, and there would be a 500-ns resting time after each pulse, time enough to allow the residual delayed energy to subside. The cost in this case is a 3-dB loss of emitted power.

A two-frequency approach taken to reach the same objective in the design of the Raytheon AN/TRC-170 DAR modem, which is the heart of this digital troposcatter radio terminal, is to transmit on two separate frequencies, alternatively gating each. The two-frequency pulse waveform is simply the time interleaving of two half-duty cycle pulse waveforms, each on a separate frequency. This technique offers two significant advantages over the one-frequency waveform. First, the two signals (subcarriers) are interleaved in time and are added to produce a composite transmitted signal with nearly constant amplitude, thereby nearly recovering the 3 dB of power lost due to time-gating. The operation of this technique is shown in Figure 6.12.

The second advantage is what is called intrinsic or implicit diversity. This can be seen as achieved in two ways. First, the residual energy of the "smear" can be utilized, whereas in conventional systems it is destructive (i.e., it causes intersymbol interference). Second, on lower bit rate transmission, where the bit rate is R, R is placed on each subcarrier, rather than $R/2$ for the higher bit rates. The redundancy at the lower bit rates gives an order of in-band diversity. The modulation on the AN/TRC-170 is QPSK on each subcarrier. The maximum data rate is 4.608 Mbps, which includes a digital orderwire and service channel.

The AN/TRC-170 operates in the 4.4–5.0-GHz band with a transmitter output power of 2 kW. The receiver noise figure is 3.1 dB. In its quadruple-diversity configuration with 9.5-ft antennas and when operating at a trunk bit rate of 1.024 Mbps, the terminal can support a path loss typically of 240 dB (BER = 1×10^{-5}). This value is based on an implicit diversity advantage for a multipath delay spread typical of profile B (see Figure 6.13). On a less dispersive path based on profile A, the performance would be degraded by

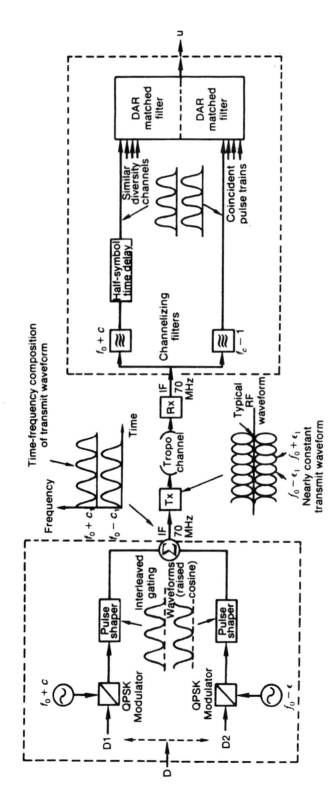

Figure 6.12. Operation of the two-frequency AN / TRC-170 modem. (From Ref. 7.)

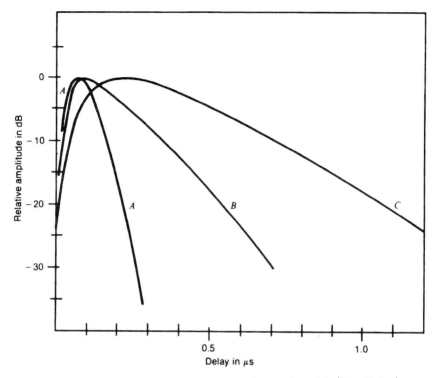

Figure 6.13. Characterization of multipath profiles *A*, *B*, and *C*. (From Ref. 6.)

about 2.6 dB or a transmission loss of 237.4 dB. With a more dispersive path based on profile *C*, we would expect a 1.1-dB improvement over that of profile *B* (Ref. 6).

The more dispersive the path, the better the equipment operates up to about 1 μs of rms dispersion. This maximum value would be shifted upward or downward depending on the data rate.

Three multipath profiles are shown in Figure 6.13; rms values for the multipath delay spread of each profile are

Profile *A*: 65 ns
Profile *B*: 190 ns
Profile *C*: 380 ns

REVIEW EXERCISES

1. List at least five applications of troposcatter/diffraction communication links.
2. Compare a typical troposcatter terminal to a typical microwave LOS terminal. There should be at least five key items compared.

3. Describe the two types of fading usually encountered on a tropo link. Give the type of time distribution associated with each.

4. Differentiate between time availability and service probability when describing the calculated performance of a tropo/diffraction link.

5. Define θd as used in the tropo transmission loss equation.

6. With microwave LOS links, fading is a function of path length. How does long-term fading vary with distance in the case of a troposcatter link?

7. What is the primary cause of aperture-to-medium coupling loss?

8. We are forced to increase the takeoff angle by 1.5° of the antenna on one side of a tropo link. Give a quantitative value of how this will affect the transmission loss.

9. Calculate the aperture-to-medium coupling loss using the CCIR method when the antennas at each end of the link display a gain of 41 dB.

10. Determine the C/N (long term) on a 2-GHz tropo link with the following parameters: transmission loss, 212.5 dB; transmitter output, 5 kW; antenna gain, 43 dB each end; line losses at each end, 1.5 dB; receiver noise figure, 1.5 dB; and IF bandwidth of the receiver, 2 MHz.

11. What is the function of a duplexer in a tropo terminal?

12. Discuss trade-offs of using the tropo frequency band of 900 MHz versus 4 GHz, considering range and varying antenna size but keeping all other parameters constant.

13. Discuss the importance of isolation on a quadruple-diversity terminal. How do we achieve the isolation? Use some dB values.

14. What is the value in dB of E_b/N_0 when the RSL is -121 dBW, the bit rate is 2.048 Mbps, and the receiving system noise figure is 2.2 dB?

15. Explain the cause of dispersion on a digital tropo path. It can bring about a serious impairment unless we design to mitigate its effects. What is the principal impairment that results from dispersion?

16. What range of dispersion in nanoseconds might be encountered on tropo links?

REFERENCES

1. "Propagation Data Required for Trans-Horizon Radio-Relay Systems," CCIR Rep. 238-3, XIIIth Plenary Assembly, Kyoto, 1978.
2. P. L. Rice et al., *Transmission Loss Predictions for Tropospheric Scatter Communication Circuits*, Tech. Note 101 as revised, National Bureau of Standards, Boulder, CO, Apr. 1959.

3. *General Engineering—Beyond Horizon Radio Communications*, USAF Tech. Order TO 31Z-10-13, U.S. Department of Defense, Washington, DC, Oct. 1971.

4. Roger L. Freeman, *Radio System Design for Telecommunications*, 2nd ed., Wiley, New York, 1997.

5. K. O. Kornberg, *Siting Criteria for Tropospheric Scatter Propagation Circuits*. Memo. Rep. PM-85-15, National Bureau of Standards, Boulder, CO, Apr. 1959.

6. W. J. Connor, "AN/TRC-170(V)—A Digital Troposcatter System," IEEE ICC'78 Conference Record, 1978.

7. T. E. Brand, W. J. Connor, and R. J. Sherwood, "AN/TRC-170—Troposcatter Communication System," NATO Conference on Digital Troposcatter, Brussels, Mar. 1980.

8. "Forward Propagation Tropospheric Scatter Communications Systems," *Handbook for Planning and Siting*, USAF Tech. Order TO 31R5-1-11, U.S. Department of Defense, Washington DC, as revised Nov. 30, 1959.

9. "Long-Haul Communications Transversing Microwave LOS Radio and Tropospheric Scatter Radio," MIL-STD-188-313, U.S. Department of Defense, Washington DC, Dec. 19, 1973.

7

SATELLITE
COMMUNICATIONS

7.1 BACKGROUND AND INTRODUCTION

Satellite communication created a quantum leap forward in long-distance communication. It competed with undersea cable for transoceanic voice channel connectivity. It brought reliable, high-quality communication to countries and rural areas that previously depended on high-frequency (HF) radio and/or ground return single-wire telegraph. For instance, in Argentina there was a 500% multiplier on international traffic when the first INTEL-SAT facility was installed. Canada's TeleSat ANIK satellites brought automatic telephony and TV to its far arctic regions.

Ships at sea are provided almost instantaneous connectivity to the international public switched telecommunication network by means of INMARSAT (International Maritime Satellite [Organization]). INMARSAT earth stations are found in nearly all major maritime nations. The United States and Russia have a joint venture for search and rescue (SAR) alert and location by means of satellites. The U.S. geographical positioning system (GPS) consisting of constellations on several polar orbits provides universal coordinated time (UTC time) with several microseconds' accuracy and position within ± 10 m in three dimensions anywhere on earth to platforms equipped with the appropriate receiver.

India, Brazil, Indonesia, the Arab States, and Europe have regional or domestic satellite systems. The United States, however, is probably the leader in this area. In the United States and Canada, TV programming relay is one area of business activity that truly has mushroomed. TV relay by satellite is being widely used by

- Broadcasters
- Cable TV

- Hotels/motels
- Industrial/education users
- Direct-to-home TV subscribers such as Direct Broadcast Satellite

The U.S. and NATO armed forces rely heavily on satellite communication for strategic, tactical, and support communications. Typical systems are the U.S. Navy's FLTSAT (fleet satellite [series]), DSCS (Defense Satellite Communication System), MILSTAR, and NATO satellites.

Telephone circuit trunking on national and international routes is another major application. Private industrial networks use satellites to provide long-haul connectivity. Bypass has become the byword. Large- and medium-size corporations have found it cost-effective to bypass local exchange carriers (LECs) and interLATA carriers by using satellite transponder space allowing direct access to their own local PBX facilities.

VSAT (very small aperture terminal) satellite systems are a vastly expanding subset of bypass. Hotels, fast-food chains, chain stores, and other commercial entities that are spread far and wide across a geographical expanse connect individual low-bit-rate terminals through a large hub facility. This facility is often associated with the company's large mainframe computer. Such systems turn out to be very inexpensive alternatives to using the public switched telecommunication network (PSTN).

Satellite communication has grown so much and so fast that the Western Hemisphere equatorial orbit is full for the 6/4-GHz band and nearly full for the 14/12-GHz. The 30/20-Gz band is only now beginning to be exploited. The notable advantages of terrestrial fiber-optic links are tempering the further growth of satellite communications in the geostationary earth orbit (GEO). Low earth orbit (LEO) satellite systems are covered in Chapter 11.

This chapter sets forth methods of design of satellite communication links. Setting aside satellite communication being used for entertainment (e.g., CATV), satellite links transporting voice and data often have an international flavor (e.g., INTELSAT). The public switched networks in many emerging nations remain analog. Thus connectivities with these nations must necessarily be analog using FDM/FM configurations. The emphasis of this chapter is digital, but we start the introductory discussion with conventional analog FDM/FM links.

The primary thrust of this chapter is system design and terminal dimensioning and its rationale. The chapter also provides standard interface information for several example systems. Regarding propagation issues, this chapter deals with satellite systems operating below 10 GHz. We have arbitrarily made 10 GHz a dividing line below which excess attenuation due to rainfall and gaseous absorption can be neglected. Chapter 8 addresses the problems of propagation above 10 GHz for satellite and terrestrial links.

7.2   AN INTRODUCTION TO SATELLITE COMMUNICATION

7.2.1   Two Broad Categories of Communication Satellites

This chapter deals with two broad categories of communication satellites. The first is the repeater satellite, affectionately called the "bent pipe" satellite. The second is the processing satellite, which is used exclusively on digital circuits, where, as a minimum, the satellite demodulates the uplink signal to baseband and regenerates that signal. Analog circuits use exclusively "bent pipe" techniques; digital circuits may use either variety.

The bent pipe satellite is simply a frequency translating RF repeater. Figure 7.1 is a simplified functional block diagram of a transponder payload of such a satellite.

7.2.2   Satellite Orbits

There are three types of satellite orbits:

- Polar
- Equatorial
- Inclined

The figure a satellite defines in orbit is an ellipse. Of course, a circle is a particular class of ellipse. A Molinya orbit is a highly inclined elliptical orbit.

The discussion here will dwell almost entirely on geostationary satellites. A geostationary satellite has a circular orbit. Its orbital period is one sidereal day (23 h, 56 min, 4.091 s) or nominally 24 h. Its inclination is 0°, which

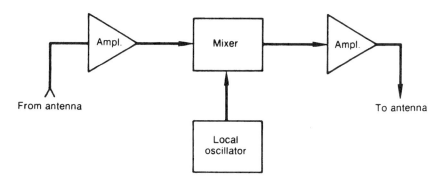

Figure 7.1. Simplified functional block diagram of a transponder payload of a conventional translating RF repeater or bent pipe satellite.

means that the satellite is always directly over the equator. It is geostationary: it appears stationary over any location on earth that is within optical view.

Geostationary satellites are conventionally located with respect to the equator (0° latitude) and a subsatellite point, which is given in longitude at the earth's surface. The satellite's range at this point, and only at this point, is 35,784 km. Table 7.1 gives details and parameters of the geostationary satellite.

The table also outlines several of the advantages and disadvantages of this satellite. Most of these points are self-explanatory. For satellites not at geosynchronous altitude and not over the equator, there is the appearance of movement. The movement with relation to a point on earth will require some form of automatic tracking on the earth station antenna to keep it always pointed at the satellite. If a satellite system is to have full earth coverage using a constellation of geostationary satellites, a minimum of three satellites would be required to be separated by 120°. As one moves northward or southward from the equator, the elevation angle to a geostationary satellite decreases (see Section 7.2.3). Elevation angles below 5° are generally undesirable because of fading and increase in antenna noise. This is the rationale in Table 7.1 for "area of no coverage." Handover refers to the action taken by a satellite earth station antenna when a nongeostationary (often misnamed "orbiting satellite") disappears below the horizon (or below 5° elevation angle) and the antenna slews to a companion satellite in the system that is just appearing above the opposite horizon. It should be pointed out here that geostationary satellites do have small residual relative motions. Over its

TABLE 7.1   The Geostationary Satellite Orbit

For the special case of a synchronous orbit—satellite in prograde circular orbit over the equator:	
Altitude	19,322 nautical mi; 22,235 statute mi; 37,784 km
Period	23 h, 56 min, 4.091 s (one sidereal day)
Orbit inclination	0°
Velocity	6879 statute mi / h
Coverage	42.5% of earth's surface (0° elevation)
Number of satellites	Three for global coverage with some areas of overlap (120° apart)
Subsatellite point	On the equator
Area of no coverage	Above 81° north and south latitude
Advantages	Simpler ground station tracking
	No handover problem
	Nearly constant range
	Very small Doppler shift
Disadvantages	Transmission delay
	Range loss (free-space loss)
	No polar coverage

Source: Reference 1.

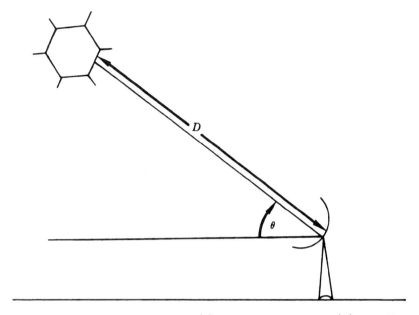

Figure 7.2. Definition of elevation angle (θ) or "look angle" and range (D) to satellite.

subsatellite point, a geostationary satellite carries out a small apparent suborbit in the form of a figure eight because of higher space harmonics of the earth's gravitation and tidal forces from the sun and the moon. The satellite also tends to drift off station because of the gravitational attraction of the sun and the moon as well as solar winds. Without correction the inclination plane drifts roughly 0.86° per year (Ref. 1, Section 13.4.1.9).

7.2.3   Elevation Angle Definition

The elevation angle or "look angle" of a satellite terminal antenna is the angle measured from the horizontal to the point on the center of the main beam of the antenna when the antenna is pointed directly at the satellite. This concept is shown in Figure 7.2. Given the elevation angle of a geostationary satellite, we can define the range. We will need the range, D in Figure 7.2, to calculate the free-space loss or spreading loss for the satellite radiolink.

7.2.4   Calculation of Azimuth, Elevation, and Range of a Geostationary Satellite

Figure 7.3 is a nomogram from which we can calculate from a known earth station location, azimuth, elevation angle, and range of a geostationary

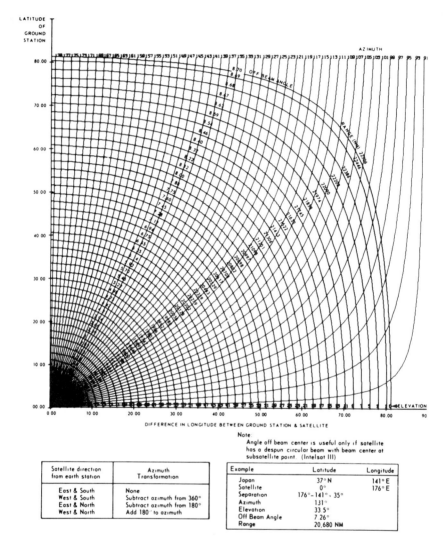

Figure 7.3. Determination of range to a geostationary satellite, azimuth, and elevation angles. (From Ref. 1. Courtesy of COMSAT, Washington, DC.)

satellite if the subsatellite point of that satellite is known. Reference 2 provides a mathematical method of calculating these important parameters.

7.3 FREQUENCY BANDS AVAILABLE FOR SATELLITE COMMUNICATIONS

When making a general reference to frequency bands available for satellite communication, we speak of "frequency pairs." For instance, we refer to the

6/4-GHz band. In all cases the first number represents the uplink band or the band of frequencies available for an earth station to transmit to a satellite. The second number represents the downlink band or the band of frequencies available for a satellite to transmit to an earth station. It will be noted that the uplink frequency is always higher in frequency than the downlink. This is purposeful. The higher frequency suffers greater spreading or free-space loss than its lower frequency counterpart, and an earth station aims upward with well-controlled antenna sidelobes. Obviously, an earth station has better transmitting assets than a satellite. It has unlimited prime power available and can use large aperture antennas and large power amplifiers. A satellite spews a signal to earth that must be limited in flux density so as not to interfere with terrestrial services that share the same band. A satellite does not have the transmission assets of an earth station. For example, it derives its prime power from solar cells backed up by secondary cells. One small advantage we can give the satellite is that it transmits on the lower frequency of the frequency pair with lower path loss. Much of this thinking derived, of course, when earth coverage antennas were nearly universally used.

The following lists commonly used frequency bands for satellite communication:

6/4 GHz	5925–6425 MHz	Uplink	Commercial
	3700–4200 MHz	Downlink	
8/7 GHz	7900–8400 MHz	Uplink	Military
	7250–7750 MHz	Downlink	
14/11 Gz	14.0–14.5 GHz	Uplink	Commercial
	11.7–12.2 GHz	Downward	
30/20 GHz	27.5–30.5 GHz	Uplink	Commercial
	17.7–20.2 GHz	Downlink	
30/20 GHz	30.0–31.0 GHz	Uplink	Military
	20.2–21.2 GHz	Downlink	
44/20 GHz	43.5–45.5 GHz	Uplink	Military
	20.2–21.2 GHz	Downlink	

7.4 LINK DESIGN PROCEDURES—THE LINK BUDGET

7.4.1 Introduction

To size or dimension a satellite communication terminal correctly, we will carry out a link budget analysis. The methodology is very similar to the path analyses described in Chapters 5 and 6. There are also certain legal constraints of which we should be aware.

At the outset, a certain satellite link must meet performance requirements. If the link is analog FDM/FM, the requirements will be expressed as noise in the derived voice channel; if TV, a certain value of signal-to-noise ratio S/N; if a digital link, a bit error rate. To meet these requirements, we will establish a value of carrier-to-noise ratio C/N or C/N_0 at the receiver input of the downlink. The C, of course, is the RSL measured in dBW or dBm; the N or N_0 value is based on the receiving system thermal noise threshold. Such a procedure should now be thoroughly familiar to the reader from Chapters 5 and 6.

Link budgets are carried out in a tabular manner. Table 5.5 shows a tabular method of doing a path calculation for a line-of-sight (LOS) microwave link. A path calculation is just another name for a link budget.

The first item in a link budget table is the EIRP expressed in dBW or dBm. If we are dealing with an uplink, it is the satellite terminal EIRP; if a downlink, it is the satellite EIRP. The next item in the table is free-space loss. Several other miscellaneous loses are then included. We then calculate the IRL (isotropic receive level) by algebraically summing the column.

To the IRL value in dBm or dBW we algebraically add the G/T. Here we encounter a difference in methodology with the link budget's LOS microwave counterpart. G/T will be new to those not familiar with satellite communication. G/T is the receiving system figure of merit. It quantifies the receiving system sensitivity.

To this sum we subtract Boltzmann's constant, -228.6 dBW. The result is C/N_0 or, if you will, RSL/N_0. From this value we can derive E_b/N_0 or S/N and link margin.

In the following subsections we calculate free-space loss to get some idea of the range of loss values with which we have to deal. Then we show why noise is a driving factor in satellite link design by introducing limitations introduced by CCIR on satellite EIRP, resulting in very low signal levels at the satellite terminal. The reason for limiting the flux density on the earth's surface is that the satellite communication frequency bands are shared by terrestrial communication.

G/T is then introduced and example link budgets are worked.

7.4.2   Path Loss or Free-Space Loss

We introduced free-space loss in Section 5.3.1. Free-space loss formulas are repeated below for convenience. Unfortunately, we do not have a worldwide (or even nationwide) standard unit of distance. We find that range to a satellite (distance) may be given in statute miles (sm), kilometers (km), or nautical miles (nm),. Accordingly, we give three equations:

Range (D) unit sm:

$$L_{dB} = 36.58 + 20 \log D_{sm} + 20 \log f \text{ (MHz)} \tag{7.1a}$$

Range (D) unit km:

$$L_{dB} = 32.4 + 20 \log D_{km} + 20 \log f \text{ (MHz)} \tag{7.1b}$$

Range (D) unit nm:

$$L_{dB} = 37.80 + 20 \log D_{nm} + 20 \log f \text{ (MHz)} \qquad (7.1c)$$

where D is the distance to the satellite in the unit indicated and f is the operating frequency in megahertz.

Often we are given the elevation angle of the satellite of interest. If this is the case, use Figure 7.3, which gives range (distance) to the satellite in nautical miles for various elevation angles.

Example. The elevation angle to a geostationary satellite is 23° and the transmitting frequency is 3840 MHz. What is the free-space loss in dB?

Turn to Figure 7.3 and we find the range (distance) to the satellite is 21,201 nm. Use this value for D in equation 7.1c. Thus

$$L_{dB} = 37.8 + 20 \log 21{,}201 + 20 \log 3840$$

$$= 196.01 \text{ dB}$$

7.4.3 Isotropic Receive Level—Simplified Model

Consider the example in the previous subsection where there was a downlink operating at 3840 MHz and the range to the satellite was 21,201 nm, producing a free-space loss of 196.01 dB. Assume the satellite has an EIRP of +31 dBW. If we neglect all other link losses, what is the IRL at the earth station (satellite terminal) antenna?

$$\text{IRL (dBW)} = \text{EIRP (dBW)} + \text{FSL (dB)}$$

$$= +31 \text{ dBW} + (-196.01) \text{ dB}$$

$$= -165.01 \text{ dBW} \qquad (7.2)$$

Some important losses have been left out. Although each may seem small in value, when totaled they will make a considerable contribution to the total link loss. Among these losses are pointing losses, polarization loss, random loss (if radome is used), gaseous absorption loss, and excess attenuation due to rainfall.

7.4.4 Limitation of Flux Density on the Earth's Surface

The satellite communication frequency bands shown in Section 7.3 are shared with terrestrial services such as point-to-point LOS microwave. The flux density of satellite downlink signals on the earth's surface must be limited so as not to interfere with terrestrial radio services that share these same frequency bands. ITU-R Rec. SF.358-4 recommends the following flux density limits (Ref. 3):

1. That, in frequency bands in the range 2.5 to 23 GHz shared between systems in the fixed-satellite service and line-of-sight radio-relay systems, the maximum power flux-density produced at the surface of the Earth by emissions from a satellite, including those from a reflecting satellite, for all conditions and methods of modulation, should not exceed:

   1.1 in the band 2.5 to 2.690 GHz, in any 4 kHz band:

   -152 dB(W/m^2) for $\theta \leq 5°$

   $-152 + 0.75(\theta - 5)$ dB(W/m^2) for $5° < \theta \leq 25°$

   -137 dB(W/m^2) for $25° < \theta \leq 90°$

   1.2 in the band 3.4 to 7.750 GHz, in any 4 kHz band:

   -152 dB(W/m^2) for $\theta \leq 5°$

   $-152 + 0.5(\theta - 5)$ dB(W/m^2) for $5° < \theta \leq 25°$

   -142 dB(W/m^2) for $25° < \theta \leq 90°$

   1.3 in the band 8.025 to 11.7 GHz, in any 4 kHz band:

   -150 dB(W/m^2) for $\theta \leq 5°$

   $-150 + 0.5(\theta - 5)$ dB(W/m^2) for $5° < \theta \leq 25°$

   -140 dB(W/m^2) for $25° < \theta \leq 90°$

   1.4 in the band 12.2 to 12.75 GHz, in any 4 kHz band:

   -148 dB(W/m^2) for $\theta \leq 5°$

   $-148 + 0.5(\theta - 5)$ dB(W/m^2) for $5° < \theta \leq 25°$

   -138 dB(W/m^2) for $25° < \theta \leq 90°$

   1.5 in the band 17.7 to 19.7 GHz, in any 1 MHz band:

   -115 dB(W/m^2) for $\theta \leq 5°$

   $-115 + 0.5(\theta - 5)$ dB(W/m^2) for $5° < \theta \leq 25°$

   -105 dB(W/m^2) for $25° < \theta \leq 90°$

   Where θ is the angle of arrival of the radio-frequency wave (degrees above the horizontal);

2. That the aforementioned limits relate to the power flux-density and angles of arrival which would be obtained under free-space propagation conditions.

Note 1: Definitive limits applicable in shared frequency bands are laid down in Nos. 2561 to 2580.1 of Article 28 of the Radio Regulations [Ref. 4]. The CCIR is continuing its study of these problems, which may lead to changes in the recommended limits.

Note 2: Under Nos. 2581 to 2585 of the Radio Regulations, the power flux-density limits in the band 17.7 to 19.7 GHz shall apply provisionally to the band 31.0 to 40.5 GHz until such time as the CCIR has recommended definitive values, endorsed by a competent Administrative Conference (No. 2582.1 of the Radio Regulations).

7.4.5   Thermal Noise Aspects of Low-Noise Systems

We deal with very low signal levels in space communication systems. Down-link signal levels are in the approximate range of -154 to -188 dBW. The objective is to achieve sufficient S/N or E_b/N_0 at demodulator outputs.

There are two ways of accomplishing this:

- Increasing system gain, usually with antenna gain
- Reducing system noise

In this section we will give an introductory treatment of thermal noise analytically, and later the term G/T will be discussed.

Around noise threshold, thermal noise predominates. To set the stage, we quote from Ref. 3:

> The equipartition law of Boltzmann and Maxwell (and the works of Johnson and Nyquist) states that the available power per unit bandwidth of a thermal noise source is
>
> $$p_n(f) = kT \text{ watts/Hz} \tag{7.3}$$
>
> where k is Boltzmann's constant $(1.3806 \times 10^{-23}$ joule/K) and T is the absolute temperature of the source in kelvins.

Looking at a receiving system, all active components and all components displaying ohmic loss generate noise. In LOS radiolinks, system noise temperatures are in the range of 1500–4000 K, and the noise of the receiver front end is by far the major contributor. In the case of space communication, the receiver front end may contribute less than one-half the total system noise. Total receiving system noise temperatures range from as low as 70 up to 1000 K (for those types of systems considered in this chapter).

In Chapters 5 and 6, receiving system noise was characterized by noise figure expressed in decibels. Here, where we deal often with system noise temperatures of less than 290 K, the conventional reference of basing noise at room temperature is awkward. Therefore noise figure is not useful at such low noise levels. Instead, it has become common to use effective noise temperature T_e (equation 7.6).

It can be shown that the available noise power at the output of a network in a band B_w is (Ref. 5)

$$p_n = g_a(f)(T + T_e)B_w \tag{7.4}$$

where g_a = the network power gain at frequency f
 T = the noise temperature of the input source
 T_e = the effective input temperature of the network

For an antenna–receiver system, the total effective system noise temperature T_{sys}, conventionally referred to the input of the receiver, is

$$T_{sys} = T_{ant} + T_r \tag{7.5}$$

where T_{ant} is the effective input noise temperature of the antenna subsystem and T_r is the effective input noise temperature of the receiver subsystem. The ohmic loss components from the antenna feed to the receiver input also generate noise. Such components include waveguide or other types of transmission lines, directional couplers, circulators, isolators, and waveguide switches.

It can be shown that the effective input noise temperature of an attenuator is (Ref. 5)

$$T_e = \frac{p_a}{kB_w g_a} + T_s = \frac{T(1 - g_a)}{g_a} \tag{7.6}$$

where T_s is the effective noise temperature of the source, the lossy elements have a noise temperature T, k is Boltzmann's constant, and g_a is the gain (available loss); p_a is the noise power at the output of the network.

The loss of the attenuator l_a is the inverse of the gain or

$$l_a = \frac{1}{g_a} \tag{7.7}$$

where l_a and g_a are numeric equivalents of the respective decibel values. Substituting into equation 7.6 gives

$$T_e = T(l_a - 1) \tag{7.8}$$

It is accepted practice (Ref. 5) that

$$n_f = l_a + \frac{T_e}{T_0} \tag{7.9}$$

where in Ref. 5 n_f is called the noise figure. Other texts call it the noise factor* and

$$NF_{dB} = 10 \log_{10} n_f \tag{7.10}$$

and T_0 is standard temperature of 290 K. NF_{dB} is the conventional noise figure discussed in Section 1.12.

From equation 7.8 the noise figure (factor) is

$$n_f = 1 + \frac{(l_a - 1)T}{T_0} \tag{7.11}$$

---

* We like to distinguish the two. Noise figure is measured in decibels and noise factor is in decibel units (e.g., 3 dB = 2, 10 dB = 10, 13 dB = 20).

If the attenuator lossy elements are at standard temperature (i.e., 290 K), the noise figure equals the loss (the noise factor equals the numeric of the loss):

$$n_f = l_a \qquad (7.12a)$$

or expressed in decibels,

$$NF_{dB} = 10 \log l_a = L_{a(dB)} \qquad (7.12b)$$

For low-loss (i.e., ohmic loss) devices whose loss is less than about 0.5 dB, such as short waveguide runs, which are at standard temperature, equation 7.8 reduces to a helpful approximation:

$$T_e \approx 66.8 L \qquad (7.13)$$

where L is the loss of the device in decibels.

The noise figure in decibels may be converted to effective noise temperature by

$$NF_{dB} = 10 \log_{10} \left(1 + \frac{T_e}{290} \right) \qquad (7.14)$$

Example. If a noise figure were given as 1.1 dB, what is the effective noise temperature?

$$1.1 \text{ dB} = 10 \log \left(1 + \frac{T_e}{290} \right)$$

$$0.11 = \log \left(1 + \frac{T_e}{290} \right)$$

$$1 + \frac{T_e}{290} = \log^{-1}(0.11)$$

$$1 + \frac{T_e}{290} = 1.29$$

$$T_e = 84.1 \text{ K}$$

7.4.6 Calculation of C/N_0

We present two methods to carry out this calculation. The first method follows the rationale given in Section 5.7. C/N_0 is measured at the input of the first active stage of the receiving system. For space receiving systems this is the low-noise amplifier (LNA) or other device carrying out a similar function. Figure 7.4 is a simplified functional block diagram of such a receiving system.

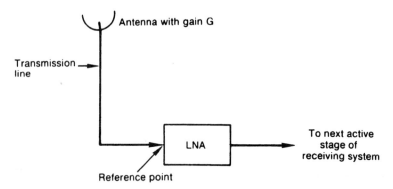

Figure 7.4. Simplified block diagram of a space receiving system.

C/N_0 is simply the carrier-to-noise ratio, where N_0 is the noise density in 1 Hz of bandwidth, and C is the receive signal level (RSL). Restating equation 7.3,

$$N_0 = kT \tag{7.15}$$

where k is Boltzmann's constant and T is the effective noise temperature, in this case of the space receiving system. We can now state this identity:

$$\frac{C}{N_0} = \frac{C}{kT} \tag{7.16}$$

Turning to Figure 7.4, we see that if we are given the signal level impinging on the antenna, which we call the isotropic receive level (IRL), the receive signal level (RSL or C) at the input to the LNA is the IRL plus the antenna gain minus the line losses, or, stated in equation form,

$$C_{\text{dBW}} = \text{IRL}_{\text{dBW}} + G_{\text{ant}} - L_{\text{L(dB)}} \tag{7.17}$$

where L_{L} represents the line losses in decibels. These losses will be the sum of the waveguide or other transmission line losses, antenna feed losses, and, if used, directional coupler loss, waveguide switch loss, power split low, bandpass filter loss (if not incorporated in LNA), circulator/isolator losses, and so forth.

To calculate N_0, equation 7.15 can be restated as

$$N_0 = -228.6 \text{ dBW} + 10 \log T_{\text{sys}} \tag{7.18}$$

where -228.6 dBW is the theoretical value of the noise level in dBW for a perfect receiver (noise factor of 1) at absolute zero in 1 Hz of bandwidth, and T_{sys} is the receiving system effective noise temperature, often just called system noise temperature.

Example 1. Given a system (effective) noise temperature of 84.1 K, what is N_0?

$$N_0 = -228.6 \text{ dBW} + 10 \log 84.1$$
$$= -228.6 + 19.25$$
$$= -209.35 \text{ dBW}$$

To calculate C or RSL, consider example 2.

Example 2. The IRL from a satellite is -155 dBW; the earth station receiving system (space receiving system) has an antenna gain of 47 dB, an antenna feed loss of 0.1 dB, a waveguide loss of 1.5 dB, a directional coupler insertion loss of 0.2 dB, and a bandpass filter loss of 0.3 dB; the system noise temperature T_{sys} is 117 K. What is C/N_0?

Calculate C (or RSL):

$$C = -155 \text{ dBW} + 47 \text{ dB} - 0.1 \text{ dB} - 1.5 \text{ dB} - 0.2 \text{ dB} - 0.3 \text{ dB}$$
$$= -110.1 \text{ dBW}$$

Calculate N_0:

$$N_0 = -228.6 \text{ dBW} + 10 \log 117 \text{ K}$$
$$= -207.92 \text{ dBW}$$

Thus

$$\frac{C}{N_0} = C_{\text{dBW}} - N_{0(\text{dBW})} \qquad (7.19)$$

In this example, substituting:

$$\frac{C}{N_0} = -110.1 \text{ dBW} - (-207.92 \text{ dBW})$$

$$= 97.82 \text{ dB}$$

The second method of determining C/N_0 involves G/T, which is discussed in the next section.

7.4.7   Gain-to-Noise Temperature Ratio G/T

G/T can be called the "figure of merit" of a radio receiving system. It is most commonly used in space communication. It not only gives an experienced engineer a "feel" of a receiving system's capability to receive low-level signals effectively, it is also used quite neatly as an algebraically additive factor in space system link budget analysis.

G/T can be expressed by the following identity:

$$\frac{G}{T} = G_{dB} - 10 \log T \qquad (7.20)$$

where G is the receiving system antenna gain and T (better expressed as T_{sys}) is the receiving system noise temperature. Now we offer a word of caution. When calculating G/T for a particular receiving system, we must stipulate where the reference plane is. In Figure 7.4 it was called the "reference point." It is at the reference plane where the system gain is measured. In other words, we take the gross antenna gain and subtract all losses (ohmic and others) up to that plane or point. This is the net gain at that plane.

System noise is treated in the same fashion. Equation 7.5 stated

$$T_{sys} = T_{ant} + T_r$$

The antenna noise temperature T_{ant}, coming inward in the system, includes all noise contributors, including sky noise, up to the reference plane. Receiver noise T_r includes all noise contributors from the reference plane to the baseband output of the demodulator.

In most commercial space receiving systems, the reference plane is taken at the input to the LNA, as shown in Figure 7.4. In many military systems it is taken at the foot of the antenna pedestal. In one system, it was required to be taken at the feed. It can be shown that G/T will remain constant as long as we are consistent regarding the reference plane.

Calculation of the net gain G_{net} to the reference plane is straightforward. It is the gross gain of the antenna minus the sum of all losses up to the reference plane. Determining T_{sys} is somewhat more involved. We use equation 7.5. The calculation of T_{ant} is described in Section 7.4.7.1, and that of T_r in Section 7.4.7.2.

7.4.7.1 Calculation of Antenna Noise Temperature T_{ant}.

The term T_{ant}, or antenna noise, includes all noise contributions up to the reference plane. Let us assume for all further discussion in this chapter that the reference plane coincides with the input to the LNA (Figure 7.4). There are two "basic" contributors of noise: sky noise and noise from ohmic losses.

Sky noise is a catchall for all external noise sources that enter through the antenna, through its main lobe, and through its sidelobes. External noise is largely due to extraterrestrial sources and thermal radiation from the atmosphere and the earth. Cosmic noise is extraterrestrial radiation that seems to come from all directions.

The sun is an extremely strong source of noise, and it can interrupt satellite communication when it passes behind the satellite being used and thus lies in the main lobe of an earth station's antenna receiving pattern. The moon is a much weaker source, which is relatively innocuous to satellite communication. Its radiation is due to its own temperature and reflected radiation from the sun.

The atmosphere affects external noise in two ways. It attenuates noise passing through it, and it generates noise because of the energy of its constituents. Ground radiation, which includes radiation of objects of all kinds in the vicinity of the antenna, is also thermal in nature.

For our discussion we will say that sky noise T_{sky} varies with frequency, elevation angle, and surface water-vapor concentration. Figure 7.5 gives values of sky noise for elevation angles θ of 0°, 5°, 10°, 20°, 30°, 60°, and 90° for water-vapor concentration 7.5 g/m³.* These figures do not include ground radiation contributions.[†]

Antenna noise T_{ant} is the total noise contributed to the receiving system by the antenna up to the reference plane. It is calculated by the formula (Ref. 2)

$$T_{ant} = \frac{(l_a - 1)290 + T_{sky}}{l_a} \qquad (7.21)$$

where l_a is the numeric equivalent of the system ohmic losses (in decibels) up to the reference plane. Then, l_a may be expressed as

$$l_a = \log_{10}^{-1} \frac{L_a}{10} \qquad (7.22)$$

where L_a is the sum of the losses to the reference plane.

Example. Assume an earth station with an antenna at an elevation angle of 10°, clear sky, 7.5 g/m³ water-vapor concentration, and ohmic losses as follows: waveguide loss of 2 dB, feed loss of 0.1 dB, directional coupler insertion loss of 0.2 dB, and a bandpass filter insertion loss of 0.4 dB. These are the losses up to the reference plane, which is taken as the input to the LNA (Figure 7.6). What is the antenna noise temperature T_{ant}? The operating frequency is 12 GHz.

Determine the sky noise from Figure 7.5. For an elevation angle of 10° and a frequency of 12 GHz, the value is 22 K.

Sum the ohmic losses up to the reference plane:

$$L_a = 0.1 \text{ dB} + 2 \text{ dB} + 0.2 \text{ dB} + 0.4 \text{ dB} + 2.7 \text{ dB}$$

$$l_a = \log^{-1}\left(\frac{2.7}{10}\right)$$

$$= 1.86$$

---

* For other water-vapor concentrations consult Ref. 6.
[†] Warm ground (earth) is considered to be a noise source with a noise temperature of 290 K.

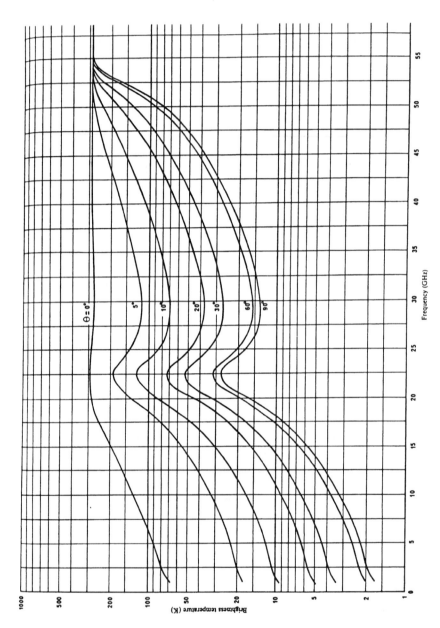

Figure 7.5. Brightness temperature (sky noise) for clear air for 7.5 g/m³ of water-vapor concentration; θ is the elevation angle. (From Ref. 6, Figure 5, CCIR Rep. 720-2. Courtesy of ITU-CCIR.)

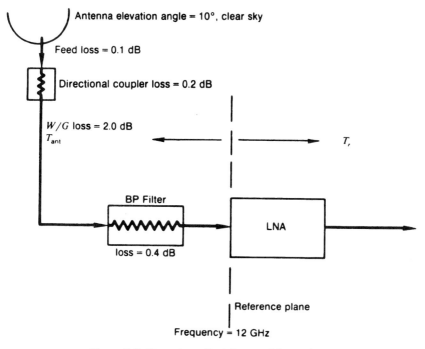

Figure 7.6. Example earth station receiving system.

Substitute into equation 7.21:

$$T_{\text{ant}} = \frac{(1.86 - 1)290 + 22}{1.86}$$

$$= 145.9 \text{ K}$$

7.4.7.2 Calculation of Receiver Noise Temperature T_r.

A receiver will probably consist of a number of stages in cascade, as shown in Figure 7.7. The effective noise temperature of the receiving system, which we will call T_r,

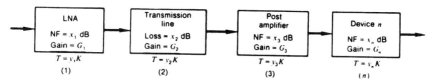

Figure 7.7. Generalized cascaded amplifiers/attenuators for noise temperature calculation.

is calculated from the traditional cascade formula

$$T_r = y_1 + \frac{y_2}{G_1} + \frac{y_3}{G_1 G_2} + \cdots + \frac{y_n}{G_1 G_2 \cdots G_{n-1}} \qquad (7.23)$$

where y is the effective noise temperature of each amplifier or device and G is the numeric equivalent of the gain (or loss) of the device.

Example. Compute T_r for the first three stages of a receiving system. The first stage is a LNA with a noise figure of 1.1 dB and a gain of 25 dB. The second stage is a lossy transmission line with 2.2-dB loss. The third and final stage is a postamplifier with a 6-dB noise figure and a gain of 30 dB.

Convert the noise figures to equivalent noise temperatures, using equation 7.14:

$$1.1 \text{ dB} = 10 \log\left(1 + \frac{T_e}{290}\right)$$

$$T_e = 83.6 \text{ K}$$

$$6.0 \text{ dB} = 10 \log\left(1 + \frac{T_e}{290}\right)$$

$$T_e = 864.5 \text{ K}$$

Calculate the noise temperature of the lossy transmission line, using equation 7.4. First determine l_a:

$$l_a = \log^{-1}\left(\frac{L_a}{10}\right)$$

$$= 1.66$$

$$T_e = (1.66 - 1)290$$

$$= 191.3 \text{ K}$$

Calculate T_r, using equation 7.23:

$$T_r = 83.6 + \frac{191.3}{316.2} + \frac{864.5}{316.2 \times 1/166}$$

$$= 83.6 + 0.605 + 4.53$$

$$= 88.735 \text{ K}$$

It should be noted that in the second and third terms, we divided by the numeric equivalent of the gain, not the gain in decibels. In the third term, of course, it was not a gain, but a loss (e.g., 1/1.66, where 1.66 is the numeric equivalent of a 2.2-dB loss). It will also be found that in cascaded systems the loss of a lossy device is equivalent to its noise figure.

7.4.7.3 *Example Calculation of G/T.* A satellite downlink operates at 21.5 GHz. Calculate the G/T of a terminal operating with this satellite. The reference plane is taken at the input to the LNA. The antenna has a 3-ft aperture displaying a 44-dB gross gain. There is 2 ft of waveguide with 0.2 dB/ft of loss. There is a feed loss of 0.1 dB, a bandpass filter has 0.4-dB insertion loss, and a radome has a loss of 1.0 dB. The LNA has a noise figure of 5.0 dB and a 30-dB gain. The LNA connects directly to a downconverter/IF amplifier combination with a single sideband noise figure of 13 dB.

Calculate the net gain of the antenna to the reference plane:

$$G_{\text{net}} = 44 \text{ dB} - 1.0 \text{ dB} - 0.1 \text{ dB} - 0.4 \text{ dB} - 0.4 \text{ dB}$$

$$= 42.1 \text{ dB}$$

This will be the value for G in the G/T expression.

Determine the sky noise temperature value at the 10° elevation angle; assume clear sky with dry conditions at 21.5 GHz. Use Figure 7.5. The value is 145 K.

Calculate L_A, the sum of the losses to the reference plane. This will include, of course, the radome loss:

$$L_A = 1.0 \text{ dB} + 0.1 \text{ dB} + 0.4 \text{ dB} + 0.4 \text{ dB} = 1.9 \text{ dB}$$

Determine l_a, the numeric equivalent of L_A from the 1.9-dB value (equation 7.22):

$$l_a = \log^{-1}\left(\frac{1.9}{10}\right)$$

$$= 1.55$$

Calculate T_{ant}, using equation 7.21:

$$T_{\text{ant}} = \frac{(1.55 - 1)290 + 145}{1.55}$$

$$= 196.45 \text{ K}$$

Calculate T_r, using equation 7.23. First convert the noise figures to equivalent noise temperatures using equation 7.14. The LNA has a 5.0-dB noise figure, and its equivalent noise temperature is 627 K; the downconverter/IF amplifier has a noise figure of 13 dB and an equivalent noise temperature of 5496 K.

$$T_r = 627 + \frac{5496}{1000}$$

$$= 632.5 \text{ K}$$

Determine T_{sys} using equation 7.5:

$$T_{\text{sys}} = 196.45 + 632.5$$

$$= 828.95 \text{ K}$$

Calculate G/T using equation 7.20:

$$\frac{G}{T} = 42.1 \text{ dB} - 10\log(828.95)$$

$$= +12.91 \text{ dB/K}$$

The following discussion, taken from Ref. 7, further clarifies G/T analysis.

Figure 7.8 shows a satellite terminal receiving system and its gain/noise analysis. The notation used is that of the reference document.

The following are some observations of Figure 7.8 (using notation from the reference):

1. The value of T_S is different at every junction.
2. The value of G/T (where $T = T_S$) is the same at every junction.

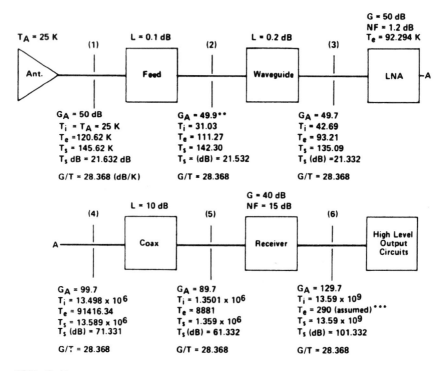

*$G/T = G_A/T_s$

**Dimensions are same as at position (1)

***T_e for the output circuits is likely to be inconsequential in comparison with T_i at (6) and can usually be neglected.

Figure 7.8. Example of noise temperature and G/T calculations for cascade of two ports. Note that G/T is independent of position in cascade. (From Ref. 7, Paper 2A, "Noise Temperature and G/T of Satellite Receiving System." Reprinted with permission.)

3. The system noise temperature at the input to the LNA is influenced largely by the noise temperature of the components that precede the LNA and the LNA itself. The components that follow the LNA have a negligible contribution to the system noise temperature at the LNA input junction (reference plane) if the LNA gain is sufficiently high.

The parameters that have significant influence on G/T are the following:

1. The antenna gain and the antenna noise temperature.
2. The antenna elevation angle. The lower the angle, the higher the sky noise; thus the higher the antenna noise and, hence, the lower the G/T for a given antenna gain.
3. Feed and waveguide insertion losses. The lower the insertion loss of these devices, the higher the G/T.
4. LNA. The lower the noise temperature of the LNA, the higher the G/T. The higher the gain of the LNA, the less the noise contribution of the stages following the LNA. For instance, in Figure 7.8, if the gain of the LNA were reduced to 40 dB, the value of T_S would increase to 144.1 K. This means that the value of G/T would be reduced by about 0.26 dB. For a LNA with a gain of only 30 dB, the G/T would then drop by an additional 1.96 dB.

7.4.8 Calculation of C/N_0 Using the Link Budget Technique

The link budget is a tabular method of calculating space communication system parameters. It is a similar approach used in Chapters 5 and 6, where it was called link analysis. In the method presented here the starting point of a link budget is the platform EIRP. The platform can be a terminal or a satellite. In an equation, it would be expressed as follows:

$$\frac{C}{N_0} = \text{EIRP} - \text{FSL}_{\text{dB}} - (\text{other losses}) + \frac{G}{T_{\text{dB/K}} - k} \qquad (7.24)$$

where FSL is the free-space loss, k is Boltzmann's constant expressed in dBW or dBm, and the "other losses" may include

- Polarization loss
- Pointing losses (terminal and satellite)
- Off-contour loss
- Gaseous absorption losses
- Excess attenuation due to rainfall (as applicable)

The off-contour loss refers to spacecraft antennas that are not earth coverage, such as spot beams, zone beams, and multiple beam antenna (MBA), and the contours are flux density contours on the earth's surface. This loss expresses in equivalent decibels the distance the terminal platform is from a contour line. Satellite pointing loss, in this case, expresses contour line error or that the contours are not exactly as specified.

7.4.8.1 *Some Link Loss Guidelines.* Free-space loss is calculated using equations 7.1a–c. Care should be taken in the use of units for distance (range) and frequency.

When no other information is available, use 0.5 dB as an estimate for polarization loss and pointing losses (e.g., 0.5 dB for satellite pointing loss and 0.5 dB for terminal pointing loss).

For off-contour loss, the applicable contour map should be used, placing the prospective satellite terminal in its proposed location on the map and estimating the loss. Figure 7.9 is a typical contour map.

Atmospheric absorption losses are comparatively low for systems operating below 10 GHz. These losses vary with frequency, elevation angle, and altitude. For the 7-GHz downlink, 0.8 dB is appropriate for a 5° elevation angle, dropping to 0.5 dB for 10° and 0.25 dB for 15°; all values are for sea level. For 4 GHz, recommended values are 0.5 dB for 5° elevation angle and 0.25 dB for 10°; all values are for sea level. Atmospheric absorption losses are treated more extensively in Chapter 8.

Excess attenuation due to rainfall is rigorously treated in Chapter 8. This attenuation also varies with elevation angle and altitude. Suggested estimates are 0.5 dB at 5° elevation angle for a 4-GHz downlink band, 0.25 at 10°, and 0.15 dB at 15° with similar values for the 6-GHz uplink. For the 7-GHz military band: 3 dB for 5°, 1.5 dB at 10°, and 0.75 dB at 15°, all values at sea level. Use similar values for the 8-Gz uplink.

7.4.8.2 *Link Budget Examples.* The link budget is used to calculate C/N_0 when other system parameters are given. It is also used when C/N_0 is given when it is desired to calculate one other parameter such as G/T or EIRP of either platform in the link.

Example 1. We have a 4-GHz downlink, FDM/FM, 5° elevation angle. Satellite EIRP is +30 dBW. Range to satellite (geostationary) is 22,208 nautical mi or 25,573 statute mi from, Figure 7.3 Terminal $G/T = +20$ dB/K. Calculate C/N_0.

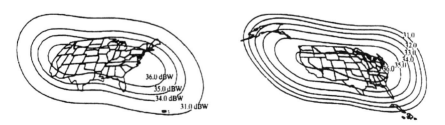

Figure 7.9. GE SPACENET 4-GHz EIRP contours for satellite at 70° W (left) and 119° W (right). (From Ref. 1. Courtesy of GE SpaceNet.)

Calculate free-space Loss (L_{dB}) from equation (7.1c):

$$L_{dB} = 36.58 + 20\log(4 \times 10^3) + 20\log(25{,}573)$$

$$= 196.78 \text{ dB}$$

EIRP of satellite	$+30$ dBW
Free-space loss	-196.78 dB
Satellite pointing loss	-0.5 dB
Off-contour loss	-0.5 dB
Atmospheric absorption loss	-0.5 dB
Rainfall loss	-0.5 dB
Polarization loss	-0.5 dB
Terminal pointing loss	-0.5 dB
Isotropic receive level	-169.78 dBW
Terminal G/T	$+20$ dB/K
Sum	-149.78 dBW
Boltzmann's constant	$-(-228.6$ dBW$)$
C/N_0	78.82 dB

Example 2. Calculate the required satellite G/T, where the uplink frequency is 6.0 GHz and the terminal EIRP is $+70$ dBW, and with a 5° terminal elevation angle. The required C/N_0 at the satellite is 102.16 dB (typical for an uplink video link). The free-space loss is determined as in Example 1 but with a frequency of 6.0 GHz. This loss is 200.3 dB. Call the satellite G/T value X.

Terminal EIRP	$+70$ dBW	
Terminal pointing loss	-0.5 dB	
Free-space loss	-200.3 dB	
Polarization loss	-0.5 dB	
Satellite pointing loss	-0.5 dB	
Atmospheric absorption loss	-0.5 dB	
Off-contour loss	-0.5 dB	
Rainfall loss	-0.5 dB	
Isotropic receive level at satellite	-133.3 dBW	
G/T of satellite	X dB/K	(initially let $X = 0$ dB/K)
Sum	-133.3 dBW	
Boltzmann's constant	$-(-228.6$ dBW$)$	
C/N_0 (as calculated)	95.3 dB	
C/N_0 (required)	102.16 dB	
G/T	$+6.86$ dB/K	(difference)

This G/T value, when substituted for X, will derive a C/N_0 of 102.16 dB. It is not advisable to design a link without some margin. Margin will compensate for link degradation as well as errors of link budget estimation. The more margin (in decibels) that is added, the more secure we are that the link will work. On the other hand, each decibel of margin costs money. Some compromise between conservatism and economic realism should be met. For this link, 4 dB might be such a compromise. If it were all allotted to satellite G/T, then the new G/T value would be $+10.86$ dB/K. Other alternatives to build in a margin would be to increase transmitter power output, thereby increasing EIRP, or to increase terminal antenna size, among other possibilities. The power of using the link budget can now be seen easily. The decibels flow through on a one-for-one basis, and it is fairly easy to carry out trade-offs.

It should be noted that when the space platform (satellite) employs an earth coverage (EC) antenna, satellite pointing loss and off-contour loss are disregarded. The beamwidth of an EC antenna is usually accepted as 17° or 18° when the satellite is in geostationary orbit.

7.4.8.3 Calculating System C/N_0.

The final C/N_0 value we wish to know is that at the terminal receiver. This must include, as a minimum, the C/N_0 for the uplink and C/N_0 for the downlink. If the satellite transponder is shared (e.g., simultaneous multicarrier operation on one transponder), a value for C/N_0 for satellite intermodulation noise must also be included. The basic equation to calculate C/N_0 for the system is given as

$$\left(\frac{C}{N_0}\right)_{\text{sys}} = \frac{1}{1/(C/N_0)_{\text{u}} + 1/(C/N_0)_{\text{d}}} \tag{7.25}$$

Example. Consider a bent pipe satellite system where the uplink $C/N_0 = 105$ dB and the downlink $C/N_0 = 95$ dB. What is the system C/N_0?

Convert each value of C/N_0 to its numeric equivalent:

$$\log^{-1}\left(\frac{105}{10}\right) = 3.16 \times 10^{10}; \qquad \log^{-1}\left(\frac{95}{10}\right) = 0.316 \times 10^{10}$$

Invert each value and add. Invert this value and take 10 log:

$$\left(\frac{C}{N_0}\right)_{\text{sys}} = 94.6 \text{ dB/Hz}$$

Many satellite transponders simultaneously permit multiple carrier access, particularly when operated in the FDMA or SCPC regimes. FDMA (frequency division multiple access) and SCPC (single channel per carrier) systems are described subsequently in Section 7.5.2.

When transponders are operated in the multiple-carrier mode, the down-link signal is rich in intermodulation (IM) products. Cumulatively, this is IM noise, which must be tightly controlled, but cannot be eliminated. Such noise must be considered when determining $(C/N_0)_{sys}$ when two or more carriers are put through the same transponder simultaneously. The following equation now applies to calculate $(C/N_0)_{sys}$:

$$\left(\frac{C}{N_0}\right)_{sys} = \frac{1}{(C/N_0)_u^{-1} + (C/N_0)_d^{-1} + (C/N_0)_i^{-1}} \quad (7.26)$$

where the subscript sys, u, d, and i refer to system, uplink, downlink, and intermodulation, respectively.

The principal source of IM noise (products) in a satellite transponder is the final amplifier, often called HPA (high-power amplifier). With present technology, this is usually a TWT (traveling-wave tube) amplifier, although SSAs (solid-state amplifiers) are showing increasing implementation on space platforms.

Figure 7.10 shows typical IM curves for a bent pipe type transponder using a TWT transmitter. The lower curve in the figure (curve A) is the IM performance for a large number of equally spaced carriers within a single

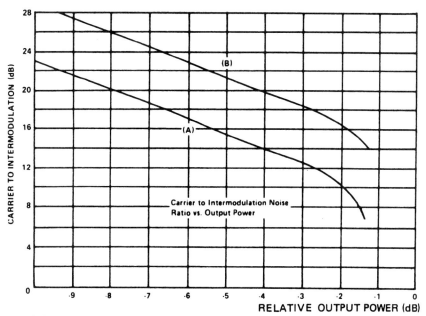

(A) Large number of uniformly spaced carriers.
(B) Two Carriers.

Figure 7.10. TWT power transfer characteristics. (From Ref. 7. Courtesy of Scientific-Atlanta.)

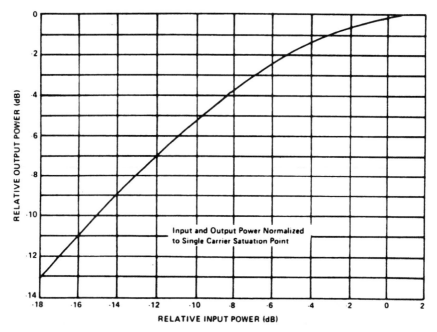

Figure 7.11. Output power normalized to single-carrier saturation point. (From Ref. 7. Courtesy of Scientific-Atlanta.)

transponder. The upper curve (curve B) shows the IM characteristics of two carriers.

In order to bring the IM products to an acceptable level and thus not degrade the system C/N_0 due to poor C/IM performance, the total power of the uplink must be "backed off" or reduced, commonly to a C/IM ratio of 20 dB or better. The result, of course, is a reduction in downlink EIRP. Figure 7.11 shows the effect of input (drive) power reduction versus output power of a TWT. To increase the C/IM ratio to 24 dB for the two-carrier case, we can see from Figure 7.10 that the input power must be backed off by 7 dB. Figure 7.11 shows that this results in a total downlink power reduction of 3 dB. It should also be noted that the resulting downlink power must be shared among the carriers actually being transmitted. For example, if two equal-level carriers share a transponder, the power in each carrier is 3 dB lower than the backed-off value.

7.4.9 Calculating S/N—FM Analog Service

7.4.9.1 In a VF Channel of a FDM/FM Configuration. To calculate S/N and psophometrically weighted noise (pWp0) in a voice channel when C/N

in the IF is given, use the following formula (Ref. 7):

$$\frac{S}{N} = \frac{C}{N} + 20\log\left(\frac{\Delta F_{TT}}{f_{ch}}\right) + 10\log\left(\frac{B_{IF}}{B_{ch}}\right) + P + W \qquad (7.27)$$

where B_{IF} = IF noise bandidth
   ΔF_{TT} = rms test tone deviation
   f_{ch} = highest baseband frequency
   B_{ch} = voice channel bandwidth (3.1 kHz)
   P = top VF channel emphasis improvement factor
   W = psophometric weighting improvement factor (2.5 dB)

Once the voice channel S/N has been calculated, the noise in the voice channel in picowatts may be determined from

$$\text{Noise} = \log^{-1}\left\{\frac{90 - (S/N)}{10}\right\} \quad (\text{pWP0}) \qquad (7.28)$$

Table 7.2 lists the transmission parameters of INTELSAT V, VA, and VI, which we will use below in an example problem. These are typical bent pipe satellite system parameters for communication links.

Example. Using Table 7.2 calculate S/N and psophometrically weighted noise in a FDM/FM derived voice channel for a 972-channel system. First use equation 7.27 to calculate S/N and then equation 7.28 to calculate the noise in pWp. Use the value f_r for rms test tone deviation, f_m for the maximum baseband frequency, and B_{IF} for the allocated satellite bandwidth. The emphasis improvement was discussed in Chapter 5. In this case, C/N is 17.8 dB:

$$\frac{S}{N} = 17.8\text{ dB} + 20\log\left(\frac{802 \times 10^3}{4028 \times 10^3}\right) + 10\log\left(\frac{36 \times 10^6}{3.1 \times 10^3}\right)$$

$$+ 2.5\text{ dB} + 4.5\text{ dB}$$

$$= 51.43\text{ dB}$$

$$\text{Noise} = \log^{-1}\left(\frac{90 - 51.43}{10}\right)$$

$$= 7194.5\text{ pWp}$$

TABLE 7.2 INTELSAT V, VA, VA(IBS), VI, VII, VIIA, and K Transmission Parameters (Regular FDM/FM Carriers)

Carrier Capacity (Number of Channels)	Top Baseband Frequency (kHz)	Allocated Satellite BW Unit (MHz)	Occupied Bandwidth (MHz)	Deviation (rms) for 0-dBm0 Test Tone (kHz)	Multichannel rms Deviation (kHz)	Carrier-to-Total Noise Temperature Ratio at Operating Point 8000 ± 200 pW0p from RF Sources (dBW/K)	Carrier-to-Noise Ratio in Occupied BW (dB)	Ratio of Unmodulated Carrier Power to Maximum Carrier Power Density Under Full Load Conditions
n	f_m	B_a	B_o	f_r	f_{mc}	C/T	C/N	(dB/4 kHz)
12	60	1.25	1.125	109	159	−154.7	13.4	20.0
24	108	2.5	2.00	164	275	−153.0	12.7	22.3
36	156	2.5	2.25	168	307	−150.0	15.1	22.8
48	204	2.5	2.25	151	292	−146.7	18.4	22.6
60	252	2.5	2.25	136	276	−144.0	21.1	22.4
60	252	5.0	4.0	270	546	−149.9	12.7	25.3
72	300	5.0	4.5	294	616	−149.1	13.0	25.8
96	408	5.0	4.5	263	584	−145.5	16.6	25.6
132	552	5.0	4.4	223	529	−141.4	20.7	$24.2^a(X = 1)$
96	408	7.5	5.9	360	799	−148.2	12.7	27.0
132	552	7.5	6.75	376	891	−145.9	14.4	27.5
192	804	7.5	6.4	297	758	−140.6	19.9	$25.8^a(X = 1)$
132	552	10.0	7.5	430	1020	−147.1	12.7	28.0
192	804	10.0	9.0	457	1167	−144.4	14.7	28.6
252	1052	10.0	8.5	358	1009	−139.9	19.4	$27.0^a(X = 1)$

252	1052	15.0	12.4	577	1627	−144.1	13.6	30.0
312	1300	15.0	13.5	546	1716	−141.7	15.6	30.2
372	1548	15.0	13.5	480	1646	−138.9	18.4	30.1
432	1796	15.0	13.0	401	1479	−136.2	21.2	27.6[a](X = 1)
312	1300	17.5	15.75	663	2081	−143.4	13.2	31.2
372	1548	17.5	15.75	583	1999	−140.8	15.9	31.0
432	1796	17.5	15.75	517	1919	−138.5	18.2	30.8
432	1796	20.0	18.0	616	2276	−139.9	16.1	31.5
492	2044	20.0	18.0	558	2200	−137.8	18.2	31.4
552	2292	20.0	18.0	508	2121	−136.0	20.0	30.2[a](X = 1)
432	1796	25.0	20.7	729	2688	−141.4	14.1	32.2
492	2044	25.0	22.5	738	2911	−140.3	14.8	32.6
552	2292	25.0	22.5	678	2833	−138.5	16.6	32.5
612	2540	25.0	22.5	626	2755	−136.9	18.1	32.4
612	2540	36.0	32.4	983	4325	−141.0	12.5	34.3
792	3284	36.0	32.4	816	4085	−137.0	16.5	34.1
972	4028	36.0	32.4	694	3849	−133.8	19.7	32.8[a](X = 1)
792	3284	36.0	36.0	930	4653	−138.3	14.8	34.7
972	4028	36.0	36.0	802	4417	−135.2	17.8	34.5

[a]This value is X dB lower than the value calculated according to the normal formula used to derive this ratio:

$$10 \log_{10}(f_{mc}\sqrt{2\pi}/4)$$

where X is the value in brackets in the last column and f_{mc} is the rms multichannel deviation in kHz. This factor is necessary in order to compensate for low modulation index carriers that are not considered to have a Gaussian power density distribution.

Source: Reference 8, Table 3, INTELSAT IESS-301.

7.4.9.2 For a Typical Video Channel. As suggested by Ref. 7,

$$\frac{S}{N_v} = \frac{C}{N} + 10\log 3\left(\frac{\Delta f}{f_m}\right)^2 + 10\log\left(\frac{B_{IF}}{2B_v}\right) + W + CF \qquad (7.29)$$

where S/N_v = peak-to-peak luminance signal-to-noise ratio
Δf = peak composite deviation of the video
f_m = highest baseband frequency
B_v = video noise bandwidth (for NTSC systems this is 4.2 MHz)
B_{IF} = IF noise bandwidth
W = emphasis plus weighting improvement factor (12.8 dB for NTSC North American systems)
CF = rms to peak-to-peak luminance signal conversion factor (6.0 dB)

For many satellite systems, without frequency reuse, a 500-MHz assigned bandwidth is broken down into twelve 36-MHz segments.* Each segment is then assigned to a transponder. For video transmission either a half or full transponder is assigned.

Example. A video link is related through a 36-MHz (full) transponder bent pipe satellite where the peak composite deviation is 11 MHz and the C/N is 14.6 dB. What is the weighted S/N? Assume NTSC standards.
Using equation 7.29,

$$\frac{S}{N} = 14.6\ dB + 10\log\left[3\left(\frac{11}{4.2}\right)^2\right] + 10\log\left(\frac{36}{8.4}\right) + 12.8 + 6$$

$$= 52.9\ dB$$

In this case the video noise bandwidth is 4.2 MHz, which is also the highest baseband frequency. European systems would use 5 MHz.
Equation 7.29 is only useful for C/N above 11 dB. Below that C/N value, impulse noise becomes apparent and the equation is not valid.

7.4.10 System Performance Parameters

Table 7.3 gives some typical performance parameters for video, aural channel, and FDM/FM for a 1200-channel configuration.

---

*Many new systems coming online use wider bandwidths.

TABLE 7.3 Some Typical Satellite Link Performance Parameters (Analog)

System Parameters	Units	FDM/FM	Video
TV Video			
C/N	dB		14.6
Maximum video frequency	MHz		4.2
Overdeviation	dB		
Peak operating deviation	MHz		10.7
FM improvement	dB		13.2
BW improvement	dB		6.3
Weighting/emphasis improvement	dB		12.8
P-rms conversion factor	dB		6.0
Total improvement	dB		38.3
S/N (peak-to-peak/rms-luminance signal)	dB		52.9
TV Program Channel (Subcarrier)			
Peak carrier deviation	MHz		2.0
Subcarrier frequency	MHz		7.5
FM improvement	dB		11.5
BW improvement	dB		14.0
Total improvement	dB		2.5
C/N_{sc} (subcarrier)	dB		16.9
Peak subcarrier deviation	kHz		75
Maximum audio frequency	kHz		15
FM improvement	dB		18.8
BW improvement	dB		13.8
Emphasis improvement	dB		12.0
Total improvement	dB		44.6
S/N (audio)	dB		59.2
FDM/FM			
Number of channels		1,200	
Test tone deviation (rms)	kHz	650	
Top baseband frequency	kHz	5,260	
FM improvement	dB	18.2	
BW improvement	dB	40.7	
Weighting improvement	dB	2.5	
Emphasis (top slot) improvement	dB	4.0	
Total improvement	dB	29.0	
TT$/N$ (test tone to noise ratio)	dB	49.0	
Noise	pWp0	12,589	

Source: Reference 7. Courtesy of Scientific-Atlanta.

7.5 ACCESS TECHNIQUES

7.5.1 Introduction

Access refers to the way a communication system uses a satellite transponder. There are three basic access techniques:

- FDMA (frequency division multiple access)—analog or digital operation
- TDMA (time division multiple access)—digital operation exclusively
- CDMA (code division multiple access)—digital operation exclusively

With FDMA a satellite transponder is divided into frequency band segments where each segment is assigned to a user. The number of segments can vary from one, where an entire transponder is assigned to a single user, to literally hundreds of segments, which is typical of SCPC (single channel per carrier) operation. For analog telephony operation, each segment is operated in a FDM/FM mode. In this case FDM group(s) and/or supergroup(s) are assigned for distinct distant location connectivity.

TDMA works in the time domain. Only one user appears on the transponder at any given time. Each user is assigned a time slot to access the satellite. System timing is crucial with TDMA. It lends itself only to digital transmission, typically PCM.

CDMA is especially attractive to military users due to its antijam and low probability of intercept (LPI) properties. With CDMA, the transmitted signal is spread over part or all of the available transponder bandwidth in a time–frequency relationship by a code transformation. Typically, the modulated RF bandwidth is ten to hundreds of times greater than the information bandwidth.

CDMA had previously only been attractive to the military user because of its antijam properties. Since 1980 some interest in CDMA has been shown in the commercial sector for demand access for large populations of data circuit/network users with bursty requirements in order to improve spectral utilization. For further discussion of spread spectrum systems, refer to Ref. 9.

Figure 7.12 depicts the three basic types of satellite multiple access. The horizontal axis represents time and the vertical axis spectral bandwidth.

7.5.2   Frequency Division Multiple Access (FDMA)

FDMA has been the primary method of multiple access for satellite communication systems since around 1965 and will probably remain so for quite some time. In the telephone service, it is most attractive for permanent direct or high-usage (HU) routes, where during the busy hour, trunk groups to a particular, distinct location require 5 or more Erlangs (e.g., one or more FDM groups).

For this most basic form of FDMA, a single earth station transmits a carrier to a bent pipe satellite. The carrier contains a FDM configuration in its modulation envelope consisting of groups and supergroups for distinct distant locations. Each distant location receives and demodulates the carrier, but demultiplexes only those FDM groups and supergroups pertaining to it. For full duplex service, a particular earth station receives, in return, one carrier from each distant location with which it provides connectivity. So, on its downlink, it must receive, and select by filtering, carriers from each distant location and demultiplex only that portion of each derived baseband that contains FDM channelization destined for it.

Another form of FDMA is SCPC, where each individual telephone channel independently modulates a separate radio-frequency carrier. Each carrier

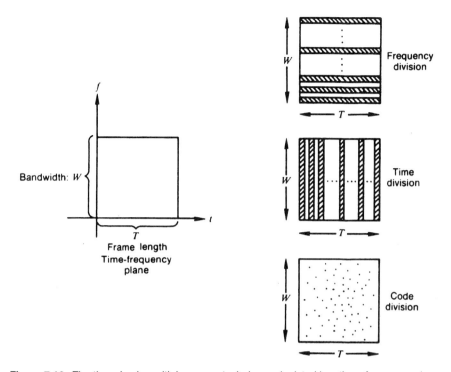

Figure 7.12. The three basic multiple access techniques depicted in a time–frequency plane.

may be frequency modulated or digitally modulated, often PSK. SCPC is useful on low-traffic-intensity routes (e.g., less than 5 Erlangs) or for overflow from FDM/FDMA. SCPC will be discussed in Section 7.5.2.1.

FDMA is a mature technology. Its implementation is fairly straightforward. Several constraints must be considered in system design. Many of these constraints center around the use of TWT as HPA in satellite transponders. Depending on the method of modulation employed, amplitude and phase nonlinearities must be considered to minimize IM products, intelligible crosstalk, and other interfering effects by taking into account the number and size (bandwidth) of carriers expected to access a transponder. These impairments are maintained at acceptable levels by operating the transponder TWT at the minimum input backoff necessary to ensure that performance objectives are met. However, this method of operation results in less available channel capacity when compared to a single access mode.

TWT amplifiers operate most efficiently when they are driven to an operating point at or near saturation. When two or more carriers drive the TWT near its saturation point, excessive IM products are developed. These can be reduced to acceptable levels by reducing the drive, which, in turn, reduces the amplifier efficiency. This reduction in drive power is called backoff.

In early satellite communication systems, IM products created by TWT amplitude nonlinearity were the dominant factor in limiting system operation. However, as greater power became available and narrow bandwidth transponders capable of operation with only a few carriers near saturation became practical, maximum capacity became dependent on a carefully evaluated trade-off analysis of a number of parameters, which include the following:

1. Satellite TWT impairments, including in-band IM products produced by both amplitude and phase nonlinearities, and intelligible crosstalk caused by FM−AM−PM conversion during multicarrier operation.
2. FM transmission impairments not directly associated with the satellite transponder TWT, such as adjacent carrier interference caused by frequency spectrum overlap between adjacent carriers, which gives rise to convolution and impulse noise in the baseband; dual path distortion between transponders; interference due to adjacent transponders' IM; earth station out-of-band emission; and frequency reuse cochannel interference.
3. General constraints, including available power and allocated bandwidth; uplink power control; frequency coordination; and general vulnerability to interference.

Backoff was discussed briefly in Section 7.4.8.3. From Figures 7.10 and 7.11 we can derive a rough rule of thumb that tells us that for approximately every decibel of backoff, IM products drop 2 dB for the multicarrier case (i.e., more than three carriers on a transponder). Also for every decibel of backoff on TWT driving power, TWT output power drops 1 dB. Of course, this causes inefficiency in the use of the TWT. As the number of carriers is increased on a transponder, the utilization of the available bandwidth becomes less efficient. If we assume 100% efficient utilization of a transponder with only 1 carrier, then with 2 carriers the efficiency drops to about 90%, with 4 to 60%, with 8 to about 50%, and with 14 carriers to about 40%.

Crosstalk is a significant impairment in FDMA systems. It can result from a sequence of two phenomena:

• An amplitude response that varies with frequency-producing amplitude modulation coherent with the original frequency modulation of a RF carrier or FM−AM transfer
• A coherent amplitude modulation that phase-modulates all carriers occupying a common TWT amplifier due to AM−PM conversion

As a carrier passes through a TWT amplifier, it may produce amplitude modulation from gain−slope anomalies in the transmission path. Another carrier passing through the same TWT will vary in phase at the same rate as the AM component and thereby pick up intelligible crosstalk from any carrier sharing the transponder. Provisions should be made in specifying TWT amplifier characteristics to ensure that AM–PM conversion and gain–slope variation meet system requirements. Intelligible crosstalk should be 58 dB down or better, and with modern equipment this goal can be met quite easily.

Out-of-band RF emission from an earth station is another issue; 500 pWp0 is commonly assigned in a communication satellite system voice channel noise budget for earth station RF out-of-band emission. The problem centers on the earth station HPA TWT. When a high-power, wideband TWT is operated at saturation, its full output power can be realized, but it can also produce severe IM RF products to the up-path of other carriers in the FDMA system. To limit such unwanted RF emission, we must again turn to the backoff technique of the RF drive to the TWT. Some systems use as much as 7-dB backoff. For example, a 12-kW HPA with 7-dB backoff is operated at about 2.4 kW. This is one good reason for overbuilding earth station TWTs and accepting the inefficiency of use.

Uplink power control is an important requirement for FDM/FM/FDMA systems. Sufficient power levels must be maintained on the uplink to meet signal-to-noise ratio requirements on the derived downlinks for bent pipe transponders. On the other hand, uplink power on each carrier accessing a particular transponder must be limited to maintain IM product generation (IM noise) within specifications. This, of course, is the backoff discussed above. Close control is required of the power level of each carrier to keep transmission impairments within the total noise budget.

One method of meeting these objectives is to study each transponder configuration on a case-by-case basis before the system is actually implemented on the satellite. Once the proper operating values have been determined, each earth station is requested to provide those values of uplink carrier levels. These values can be further refined by monitoring stations that precisely measure resulting downlink power of each carrier. The theoretical power levels are then compared to both the reported uplink and the measured downlink levels.

Energy dispersal is yet another factor required in the design of a bent pipe satellite system. Energy dispersal is used to reduce the spectral energy density of satellite signals during periods of light loading (e.g., off busy hour). The reduction of maximum energy density will also facilitate efficient use of the geostationary satellite orbit by minimizing the orbital separation needed between satellites using the same frequency band and multiple-carrier operation of broadband transponders. The objective is to maintain spectral energy density the same for conditions of light loading as for busy-hour loading. Several methods of implementing energy dispersal are described in ITU-R Rec. S.446-4 (Ref. 10).

One method of increasing satellite transponder capacity is by *frequency reuse*. As the term implies, an assigned frequency band is used more than once. The problem is to minimize mutual interference between carriers operating on the same frequency but accessing distinct transponders. There are two ways of avoiding interference in a channel that is used more than once:

- By orthogonal polarization
- By multiple exclusive spot-beam antennas

The use of opposite-hand circular or crossed-linear polarizations may be used to effect an increase in bandwidth by a factor of 2. Whether the potential increase in capacity can be realized depends on the amount of cross-polarization discrimination that can be achieved for the cochannel operation. Cross-polarization discrimination is a function of the quality of the antenna systems and the effect of the propagation medium on polarization of the transmitted signal. The amount of polarization "twisting" or distortion is a function of the elevation angle for a given earth station. The lower the angle, the more the twist.

Isolation between cochannel transponders should be greater than 25 dB. INTELSAT VI specifieds 27-dB minimum isolation between polarizations.

Whereas a satellite operating in a 500-MHz bandwidth might have only 12 transponders (36- or 40-MHz bandwidth each), with frequency reuse, 24 transponders can be accommodated with the same bandwidth. Thus the capacity has been doubled.

Table 7.2 shows a typical transponder allocation for FDMA operation.

7.5.2.1 Single Channel per Carrier (SCPC) FDMA.

Many SCPC systems have been implemented and are now in operation, and many more are coming online. There are essentially two types of SCPC systems: preassigned and demand assigned (DAMA, which stands for demand assignment multiple access). The latter requires some form of control system. Some systems occupy an entire transponder, whereas others share a transponder with video service, leaving a fairly large guard band between the video portion of the transponder passband and the SCPC portion. Such transponder sharing is illustrated in Figure 7.13.

Channel spacings on a transponder vary depending on the system. INTEL-SAT systems commonly use 45-kHz spacing. Others use 22.5, 30, and 36 kHz as well as 45 kHz. Modulation is FM or BPSK/QPSK. The latter lends itself well to digital systems, whether PCM or CVSD (continuous variable-slope delta modulation).

If we were to divide a 36-MHz bandwidth on a transponder into uniform 45-kHz segments, the total voice channel capacity of the transponder would be 800 VF channels. These are better termed half-channels, because, for telephony, we always measure channel capacity as full-duplex channels. Therefore 400 of the 800 channels would be used in the "go" direction, and

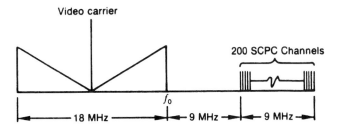

Figure 7.13. Satellite transponder frequency plan, video plus SCPC.

the other 400 would be used in the reverse direction. A SCPC system is generally designed for a 40% activity factor. Hence, statistically, only 320 of the channels can be expected to be active at any instant during the busy hour. This has no effect on the bandwidth, only on the loading. Most systems use voice-activated service. In this case, carriers appear on the transponder for the activated channels only. This provides probably the worst case for IM noise for all conventional bent pipe systems during periods of full activation. A simplified drawing of a SCPC system is shown in Figure 7.14.

SCPC channels can be preassigned or DAMA. The preassigned technique is only economically feasible in situations where source/destination locations

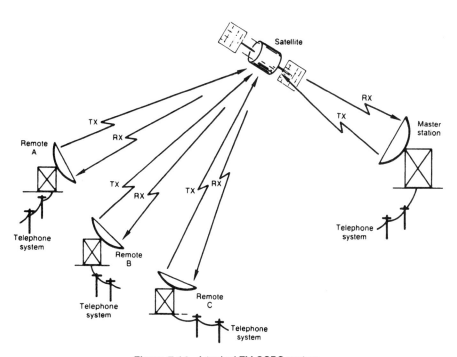

Figure 7.14. A typical FM SCPC system.

have very low traffic density during the busy hour (e.g., less than several Erlangs).

DAMA SCPC systems are efficient for source/sink locations of comparatively low traffic intensity, especially under situations with multiple location community of interest (e.g., trunks to many distinct locations but under 12 channels to any one location). This demarcation line at 12 channels, of course, is where the designer should look seriously at establishing fixed-assignment FDMA/FDM channel group allocation.

The potential improvement in communication system capacity and efficiency is the primary motivation for DAMA systems. Our primary consideration in this section is voice traffic. Call durations average several minutes. The overhead time to connect and disconnect is 1 or 2 s in an efficient channel assignment system. Some of the issues of importance in demand assignment systems are the following:

- User requirements such as traffic intensity in Erlangs or ccs, number of destinations, and grade of service (probability of blocked calls)
- Capacity improvements due to implementation of DAMA scheme
- Assignment algorithms (centralized versus distributed control)
- Equipment cost and complexity

Consider an example to demonstrate the potential improvement due to demand assignment (see Table 7.4). An earth station is required to communicate with 40 destinations, and traffic intensity to each destination is 0.5 Erlang, with a blocking probability of $P_B = 0.01$. Assume that the call arrivals have a Poisson distribution and the call holding times have an exponential distribution. Then for a trunk group with n channels in which blocked calls are cleared, based on an Erlang B distribution (Ref. 1, Section

TABLE 7.4   Comparison of Preassigned Versus DAMA Channel Requirements Based on a Grade of Service of $P_B = 0.01$

Number of Destinations	Channel Requirements[a]					
	A = 0.1		A = 0.5		A = 1.0	
	Preassigned	DAMA	Preassigned	DAMA	Preassigned	DAMA
1	2	2	4	4	5	5
2	4	3	8	5	10	7
4	8	3	16	7	20	10
8	16	4	32	10	40	15
10	20	5	40	11	50	18
20	40	7	80	18	100	30
40	80	10	160	30	200	53

[a]A = Erlang traffic intensity.

Source: Reference 11. Reprinted with permission of IEEE Press.

1-2), A, in Table 7.4, gives the traffic intensity in Erlangs. If the system is designed for preassignment, each destination will require four channels to achieve $P_B = 0.01$. For 40 destinations, 160 channels will be required for the preassigned case. In the DAMA case, the total traffic is considered, since any channel can be assigned to any destination. The total traffic load is 20 Erlangs (40 × 0.5) and the Erlang B formula gives a requirement of only 30 channels. The efficiency of the system with demand assignment is improved over the preassigned by a factor of 5.3 (160/30). Table 7.4 further expands on this comparison using various traffic loadings and numbers of destinations. Reference 12, Chapter 1, provides good introductory information on traffic engineering.

There are essentially three methods for controlling a DAMA system:

- Polling
- Random-access central control
- Random-access distributed control

The polling method is fairly self-explanatory. A master station (Figure 7.14) "polls" all other stations in the system sequentially. When a positive reply is received, a channel is assigned accordingly. As the number of earth stations in the system increases, the polling interval becomes longer, and the system tends to become unwieldy because of excessive postdial delay as a call attempt waits for the polling interval to run its course.

With random-access central control, the status of channels is coordinated by a central computer located at the master station. Call requests (called *call attempts* in switching) are passed to the central computer via a digital orderwire (i.e., digitally over a radio service channel), and a channel is assigned, if available. Once the call is completed and the subscriber goes on-hook, the speech path is taken down and the channel used is returned to the demand-access pool of idle channels. According to system design, there are a number of methods to handle blocked calls (all trunks busy [ATB] in telephone switching), such as queueing and second attempts.

The distributed control random-access method utilizes a processor controller at each earth station accessing the system. All earth stations in the network monitor the status of all channels via continuous updating of channel status information by means of the digital orderwire circuit. When an idle channel is seized, all users are informed of the fact, and the circuit is removed from the pool. Similar information is transmitted to all users when circuits are returned to the idle condition. One problem, of course, is the possibility of double seizure. Also the same problems arise regarding blockage (ATB) as in the central control system. Distributed control is more costly and complex, particularly in large systems with many users. It is attractive in the international environment because it eliminates the "politics" of a master station.

7.5.3 Time Division Multiple Access (TDMA)

7.5.3.1 Introduction. TDMA operates in the time domain and is applicable only to digital systems because information storage is required. With TDMA, use of a satellite transponder is on a time-sharing basis. At any given moment in time, only one earth station accesses the satellite. Individual time slots are assigned to earth stations operating with that transponder in a sequential order. Each earth station has full and exclusive use of the transponder bandwidth during its time-assigned segment. Depending on bandwidth of the transponder and type of modulation used, bit rates of 10–100 Mbps are used.

If only one carrier appears on the transponder at any time, then intercarrier IM products cannot be developed and the TWT-based HPA of the transponder may be operated at full power. This means that the TWT requires no backoff in drive and may be run at or near saturation for maximum efficiency. With multicarrier FDMA systems, backoffs of 5–10 dB are not uncommon.

TDMA also utilizes the transponder bandwidth more efficiently. Reference 11 compares approximate channel capacities of an INTELSAT IV global beam transponder operating with normal INTELSAT Standard A (30-m antenna) earth stations using FDMA and TDMA, respectively. Assuming 10 accesses, typical capacity using FM/FDMA is approximately 450 one-way voice channels. With TDMA, using standard 64-kbps PCM voice channels, the capacity of the same transponder is 900 voice channels. If digital speech interpolation (DSI) is now implemented on the TDMA system, the voice channel capacity is increased to approximately 1800 channels.

Still another advantage of TDMA is flexibility. Flexibility is not only a significant benefit to large systems but is often the key to system viability in smaller systems. Nonuniform accesses pose no problem in TDMA because time-slot assignments are easy to adjust. This applies to initial network configuration, assignments, reassignments, and demand assignments. This is ideal for a long-haul system, where traffic adjustments can be made dynamically as the busy hour shifts from east to west. Changes can also be made for growth or additional services.

Disadvantages of TDMA are timing requirements and complexity. Accesses may not overlap. Obviously, overlapping causes interference between sequential accesses. Guard times between accesses must be made minimal for efficient operation. The longer the guard times, the shorter the burst length and/or number of accesses. Typically, 5–15 accesses can be accommodated per transponder with guard times on the order of 50–200 ns.

As the world's telecommunication network evolves from analog to digital and as frequency congestion increases, there will be more and more demand to shift to TDMA operation in satellite communication.

Table 7.5 compares FDMA and TDMA.

TABLE 7.5   FDMA Versus TDMA

Advantages	Disadvantages
FDMA	
Mature technology	IM in satellite transponder output
No network timing	Requires careful uplink power control
Easy FDM interface	Inflexible to traffic load
TDMA	
Maximum use of transponder power	Still emerging technology
No careful uplink power control	Network timing
Flexible to dynamic traffic loading	Major digital buffer considerations
Straightforward interface with digital network	Difficult to interface with FDM
More efficient transponder bandwidth usage	
Digital format compatible with	
Forward error control	
Source coding	
Demand access algorithms	
Applicable to switched satellite service	

7.6   DIGITAL OPERATION ON A BENT PIPE SATELLITE

7.6.1   Digital Advantages

For the analog network the transmission engineer was primarily a *noise fighter*. Every modulation step in the analog network is a noise generator, and the noise accumulates from source to destination. Not so in the digital network.

At every point of regeneration in the digital network is a point where the noise accumulation stops, and the digital signal is regenerated. This is the principal point of demarcation between a bent pipe satellite and a processing satellite. There is digital signal regeneration in the processing satellite, but none in the bent pipe satellite.

7.6.2   Digital Operation of a Bent Pipe Satellite

There are three approaches to digital bent pipe operation and access:

- FDMA mode
- TDMA mode
- CDMA mode

We will discuss the first two in this chapter. CDMA or spread spectrum is covered in Chapter 11.

7.6.3 Digital FDMA Operation

Digital FDMA operation is very similar to analog FDMA operation discussed in Section 7.5.2. Rather than place a FDM/FM signal in a preassigned transponder frequency slot, a digital waveform is transmitted in that segment using modulation techniques covered in Section 5.5.3.4. Quadrature phase shift keying (QPSK) is commonly employed.

Because there is no signal regeneration in the satellite transponder, the downlink signals suffer a certain amount of distortion due to the satellite transponder. For one thing, there is additive thermal noise deriving from the satellite receiver–transmitter. When there is multichannel activity on the same transponder, IM noise can be a major impairment. A primary source of IM noise is the transponder final amplifier, which is usually a TWT (traveling-wave tube). Solid-state amplifiers (SSAs), which are now being implemented in some satellite transponder designs, tend to reduce IM products because of improved linearity over their TWT counterparts. Both types of amplifiers are inefficient. Noise and distortion result in degraded error performance.

INTELSAT is probably the most extensively used worldwide satellite service provider offering across-the-board telecommunication services. As a good example of digital operation in the FDMA mode, we provide a brief description of INTELSAT's intermediate data rate (IDR) operation in the next subsection.

7.6.3.1 *INTELSAT Intermediate Data Rate (IDR) Carrier System.* IDR digital carriers in the INTELSAT system utilize coherent QPSK modulation operating at information rates ranging from 64 kbps to 44.736 Mbps. The information rate is defined as the bit rate entering the channel unit prior to the application of any overhead or FEC. Any information rate from 64 kbps to 44.736 Mbps inclusive may be transmitted. INTELSAT has, however, defined a set of recommended information rates that are based on ITU-T hierarchical rates (see ITU-T Rec. G.703) as well as ITU-T ISDN rates covered in the I-series recommendations. These recommended bit rates are given in Table 7.6. Bit error rate performance is provided in Table 7.7. Figure 7.15 shows an IDR channel unit.

For certain information rates—namely, 1.544, 2.048, 6.312, 8.448, 32.064, 34.368, and 44.736 Mbps—an overhead structure has also been defined to provide engineering service circuits (ESCs) and maintenance alarms. This overhead increases the data rate of these carrier sizes by 96 kbps. The overhead adds its own frame alignment signal and thus passes the information data stream transparently. For the provision of ESC and alarms on IDR carrier sizes larger than 1.544 Mbps, an overhead of 96 kbps must be used. For other information rates larger than 1.544 Mbps that are not listed above, INTELSAT will determine the overhead framing structure on a case-by-case basis.

TABLE 7.6   Recommended IDR Information Rates and Associated Overhead

		Type of Overhead[a]		
Number of 64-kbps Bearer Channels (n)	Information Rate (n × 64 kbps)	No Overhead (Note 1)	With 96-kbps IDR (Note 2)	With 6.7% IBS Overhead (Notes 2 and 3)
1	64	X		X
2	128			X
3	192	X		
4	256			X
6	384	X		X
8	512			X
12	768			X
16	1024			X
24	1536			X
24	1544 (4)		X	
30 (31)	2048 (4)		X	
90 (93)	6312 (4)		X	
120 (124)	8448 (4)		X	
480	32064 (4)		X	
480 (496)	34368 (4)		X	
630 (651) or 672	44736 (4)		X	

[a]X = Recommended rate corresponding to the type of overhead.
Note 1. For rates less than 1544 kbps, it is possible to use any n × 64 kbps information rate without overhead, but the only INTELSAT-recommended rates are 64 kbps, 192 kbps and 384 kbps. The use of the optional Reed-Solomon outer coding is not defined for any information rate less than 1.544 Mbps which does not use overhead.
Note 2. The optional Reed-Solomon outer coding can be used with the information rates shown with an "x".
Note 3. The carriers in this column are small IDR carriers that can be used with the circuit multiplication concept described in Appendix B of the reference document. (For a definition of the IBS overhead framing, see IESS-309 (IBS).]
Note 4. These are standard ITU-T hierarchical bit rates. Other n × 64 kbps information rates above 1.544 Mbps are also possible and must have an ESC overhead of at least 96 kbps (the overhead framing will be defined on a case-by-case basis by INTELSAT).
Source: Reference 13, Table 1, INTELSAT IESS-308.

The occupied satellite bandwidth unit for IDR carriers is approximately equal to 0.6 times the transmission rate.* To provide guardbands between adjacent carriers, the nominal satellite bandwidth unit is equal to 0.7 times

---

* The transmission rate (R) is defined as the bit rate entering the QPSK modulator at the earth station (i.e., after the addition of any overhead or FEC) and is equal to twice the symbol rate at the output of the QPSK modulator.

TABLE 7.7   IDR Performance INTELSAT VII, VIIA, and K

Weather Condition	Minimum BER Performance (% of year)	Typical BER Performance (% of year)
Clear sky	$\leq 2 \times 10^{-8}$ for $\geq 95.90\%$	$\leq 10^{-10}$ for $\geq 95.90\%$
Degraded	$\leq 2 \times 10^{-7}$ for $\geq 99.36\%$	$\leq 10^{-9}$ for $\geq 99.36\%$
Degraded	$\leq 7 \times 10^{-5}$ for $\geq 99.96\%$	$\leq 10^{-6}$ for $\geq 99.96\%$
Degraded		$\leq 10^{-3}$ for $\geq 99.98\%$

Source: Reference 13, Table 3, INTELSAT IESS-308.

the transmission rate. The actual carrier spacing may be larger and is determined by INTELSAT, based on the particular transponder frequency plan. In particular, for the case of two nominal 45-Mbps rate $\frac{3}{4}$ FEC IDR carriers within a 72-MHz C-band transponder, the allocated RF bandwidth is 36 MHz for each carrier.

EIRP Values. The maximum 6- and 14-GHz example EIRP values and associated parameters for INTELSAT VII are given in Table 7.8 for $R = \frac{3}{4}$ FEC coding.

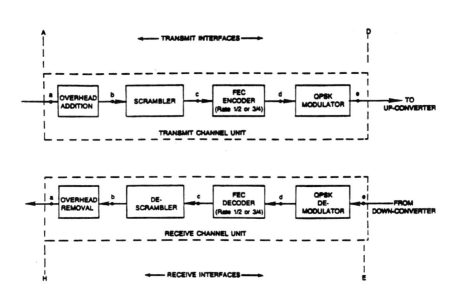

Figure 7.15. An IDR channel unit.

TABLE 7.8   Example Maximum EIRP for Rate $\frac{3}{4}$ FEC INTELSAT VII

Rx Station	EIRP (dBW)						
	A (Previous)	A (Revised)	B	F-3	F-2	F-1	
Off-Axis EIRP	[6.6	4.9	3.1	0.9	−1.5	−6.3]	(Rate $\frac{3}{4}$)
Indicator[a] (dB)	[9.5	7.8	6.0	3.8	1.4	−3.4]	(Rate $\frac{1}{2}$)
Information Rate (kbps)							
64	52.2	53.9	55.7	57.9	60.3	65.1	
192	57.0	58.7	60.5	62.7	65.1	69.9	
384	60.0	61.7	63.5	65.7	68.1	72.9	
512	61.5	63.2	65.0	67.2	69.6	74.4	
1,024	64.5	66.2	68.0	70.2	72.6	77.4	
1,544	66.3	68.0	69.8	72.0	74.4	79.2	
2,048	67.5	69.2	71.0	73.2	75.6	80.4	
6,312	72.2	73.9	75.7	77.9	80.3	85.1	
8,448	73.5	75.2	77.0	79.2	81.6	86.4	
32,064	79.2	80.9	82.7	84.9	87.3	b	
34,368	79.5	81.2	83.0	85.2	87.6	b	
44,736	80.7	82.4	84.2	86.4	b	b	
X (for other information rates)	4.1	5.8	7.6	9.8	12.2	17.0	

Note 1. These maximum EIRP values have been computed using a saturation flux density of −75.0 dBW/m^2 for full transponder loading conditions. Other saturation flux densities may be used depending on the actual traffic loading.

Note 2. For 36-MHz Hemi/Zone Transponder 9, a saturation flux density of −79 dBW/m^2 is used and the EIRP values are 1.5 dB lower for Revised Standard A, 2.5 dB lower for Standard B and F-3, and 3.5 dB lower for Standard F-2 and F-1.

Note 3. The above maximum EIRP values are applicable for the following uplink/downlink beam connections, ocean regions and spacecraft locations:

Uplink		Downlink			D/L EIRP (dBW)	U/L Margin (dB/%)	D/L Margin (dB/%)	S/C Location (°E)
Beam	Direction	Beam	Direction	Region				
H/Z	All	H/Z	All	All	33.0	3/0.02	4/0.02	All
Global	All	Hemi	All	All	33.0	3/0.02	4/0.02	All

Note 4. To obtain the maximum EIRP for rate $\frac{1}{2}$ FEC, a correction factor of 1.1 dB must be subtracted from the EIRP values shown for rate $\frac{3}{4}$ FEC.

[a]Off-axis EIRP density indicator (dB) applies to all values in the column below the number.

[b]Denotes carriers not intended for this connection.

Source: Reference 13, Table 5, INTELSAT IESS-308.

The EIRP values given in these tables are for a 10° elevation angle and located at beam edge. For other angles (i.e., >10°) and for earth station locations other than at satellite antenna beam edge, see IESS-402 for correction factors K_1 and K_2. "High gain" and "low gain" refer to the power attenuator setting.

Table 7.9 gives the QPSK characteristics and transmission parameters for IDR carriers using $R = \frac{3}{4}$ FEC coding, INTELSAT VII, VIIA, and K.

TABLE 7.9   QPSK Characteristics and Transmission Parameters for IDR Carriers, Rate $\frac{3}{4}$ FEC, INTELSAT VII, VIIA, and K

Parameter	Requirement				
1. Information rate, IR	64 kbps to 44.736 Mbps				
2. Overhead data rate for carriers with IR $\geq$ 1.544 Mbps	96 kbps				
3. Forward error correction encoding	Rate $\frac{3}{4}$ convolutional encoding/Viterbi decoding				
4. Energy dispersal (scrambling)	As per Figures 15 and 16 (main text) of Ref.13				
5. Modulation	4-Phase coherent PSK				
6. Ambiguity resolution	Combination of differential encoding (180°) and FEC (90°)				
7. Clock recovery	Clock timing must be recovered from the received data stream				
8. Minimum carrier bandwidth (allocated)[a]	0.7R Hz or [0.933 (IR + overhead)]				
9. Noise bandwidth (and occupied bandwidth)	0.6R Hz or [0.8 (IR + overhead)]				
10. Composite rate E_b/N_0 at BER (rate $\frac{3}{4}$ FEC)	10^{-3}	10^{-6}	10^{-7}	10^{-8}	10^{-10}
a. Modems back-to-back	5.3 dB	7.6 dB	8.3 dB	8.8 dB	10.3 dB
b. Through satellite channel[a]	5.7 dB	8.0 dB	8.7 dB	9.2 dB	11.0 dB
11. Transmission rate E_b/N_0 at BER (rate $\frac{3}{4}$ FEC)	10^{-3}	10^{-6}	10^{-7}	10^{-8}	10^{-10}
a. Modems back-to-back	4.05 dB	6.35 dB	7.05 dB	7.55 dB	9.05 dB
b. Through satellite channel[a]	4.45 dB	6.75 dB	7.45 dB	7.95 dB	9.75 dB
12. C/T^a at typical operating point ($< 10^{-10}$ BER)	$-217.6 + 10\log_{10}$(IR + OH), dBW/K				
13. C/N^a in noise bandwidth at typical operating point	12.0 dB				
14. Typical bit error rate at operating point	Less than 1×10^{-10}				
15. C/T^a at threshold[b] (BER = 10^{-6})	$-220.6 + 10\log_{10}$(IR + OH), dBW/K				
16. C/N^a in noise bandwidth at threshold (BER = 10^{-6})	9.0 dB				
17. C/T^a at BER = 10^{-3}	$-222.9 + 10\log_{10}$(IR + OH), dBW/K				
18. C/N^a in noise bandwidth at BER = 10^{-3}	6.7 dB				

[a]In the special case of two 45-Mbps carriers operating in an INTELSAT VII/VIIA 72-MHz transponder, the E_b/N_0, C/T, and C/N values shown are increased by 0.4 dB (10^{-3}), 0.7 dB (10^{-6}), 1.0 dB (10^{-7}), 1.3 dB (10^{-8}), and 2.0 dB (10^{-10}). The allocated bandwidth is 36.0 MHz.

[b]A BER of 10^{-6} is used as the threshold point for link availability with the fade margin allocation tables.

Note 1. IR is the information rate in bits per second.

Note 2. R is the transmission rate in bits per second and equals (IR + OH) times $\frac{4}{3}$ for carriers employing rate $\frac{3}{4}$ FEC.

Note 3. Composite data rate = (information data rate + overhead).

Note 4. The allocated bandwidth will be equal to 0.7 times the transmission rate, rounded up to the next higher odd integer multiple of 22.5 kHz (for information rates less than or equal to 10 Mbps) or integer multiple of 125 kHz (for information rates greater than 10 Mbps).

Source: Reference 13, Table D.2, INTELSAT IESS-308.

TABLE 7.10 Transmission Parameters for INTELSAT VII, VIIA, and K Recommended IDR Carrier, FEC Rate = $\frac{3}{4}$

Information rate (bit/s)	Overhead Rate (kbps)	Data Rate (IR + OH) (bit/s)	Transmission Rate (bit/s)	Occupied Bandwidth (Hz)	Allocated Bandwidth (Hz)	C/T (dBW/K) 10^{-10}	C/N_0 (dB – Hz) 10^{-10}	C/N (dB) 10^{-10}
64 k	0	64 k	85.33 k	51.2 k	67.5 k	−169.5	59.1	12.0
192 k	0	192 k	256.00 k	153.6 k	202.5 k	−164.8	63.8	12.0
384 k	0	384 k	512.00 k	307.2 k	382.5 k	−161.8	66.8	12.0
512 k[a]	34.1[a]	546.1 k[a]	728.18 k[a]	436.9 k[a]	517.5 k[a]	−160.2 [a]	68.4[a]	12.0 [a]
1.024 M[a]	68.3[a]	1.092 k[a]	1.456 M[a]	873.8 k[a]	1057.5 k[a]	−157.2 [a]	71.4[a]	12.0 [a]
1.544 M	96	1.640 M	2.187 M	1.31 M	1552.5 k	−155.5	73.1	12.0
2.048 M	96	2.144 M	2.859 M	1.72 M	2002.5 k	−154.3	74.3	12.0
6.312 M	96	6.408 M	8.544 M	5.13 M	6007.5 k	−149.5	79.1	12.0
8.448 M	96	8.544 M	11.392 M	6.84 M	7987.5 k	−148.3	80.3	12.0
32.064 M	96	32.160 M	42.880 M	25.73 M	30125.0 k	−142.5	86.1	12.0
34.368 M	96	34.464 M	45.952 M	27.57 M	32250.0 k	−142.2	86.4	12.0
44.736 M	96	44.832 M	59.776 M	35.87 M	41875.0 k	−141.1	87.5	12.0
44.736 M[(Note 4)]	96	44.832 M	59.776 M	35.87 M	36000.0 k	−142.4	86.2	10.7

[a]See Appendix B of Ref. 1. The approach described in Appendix B for 512 kbps may also be applied to other IDR carrier sizes ($n \times 64$ kbps) equal to or smaller than 1.536 Mbps, where n may be equal to 1, 2, 4, 6, 8, 12, 16, or 24.

Note 1. The above table illustrates parameters for recommended carrier sizes. However, any other information rate between 64 kbps and 44.736 Mbps may be used.

Note 2. C/T, C/N_0, and C/N values have been calculated to provide a clear-sky link BER of better than 10^{-10} and assume the use of rate $\frac{3}{4}$ FEC.

Note 3. For carrier information rates of 10 Mbps and below, carrier frequency spacings will be odd integer multiples of 22.5 kHz. For rates greater than 10 Mbps, they will be any integer multiple of 125 kHz.

Note 4. In the special case of transmissions of two 45 Mbps carriers in an INTELSAT VII/VIIA 72-MHz transponder with Standard A or B earth stations, the maximum uplink EIRP values can be maintained at the present levels (however, the nominal clear-sky BER = 10^{-7}). Operation of two 45-Mbps carriers with the Reed–Solomon outer code is provisional. The uplink EIRP values are the same with and without the Reed–Solomon outer coding.

Source: Reference 13, Table D.3, INTELSAT IESS-308. The reader should refer to Ref. 2 for a description of INTELSAT's other digital FDMA offering, IBS (international business services).

Table 7.10 provides one example of transmission parameters for INTELSAT VII, VIIA, and K recommended IDR carriers, FEC code rate = $\frac{3}{4}$.

7.6.4 TDMA Operation on a Bent Pipe Satellite

In Section 7.5.3 TDMA operation was introduced. The following description gives greater breadth and depth on the subject. It is placed here because TDMA lends itself well to digital bent pipe satellite operation.

7.6.4.1 General. With a time division multiple access (TDMA) arrangement on a bent pipe satellite, each earth station accessing a transponder is assigned a time slot for its transmission, and all uplinks use the same carrier frequency on a particular transponder. We recall that a major limitation of FDMA systems is the required backoff of drive in a transponder to reduce

IM products developed in the TWT final amplifier owing to simultaneous multicarrier operation. With TDMA, on the other hand, only one carrier appears at the transponder input at any one time, and, as a result, the TWT can be run to saturation minus a small backoff to reduce waveform spreading. The result is notably more efficient use of a transponder and permits greater system capacity. In some cases capacity can be doubled when compared to an equivalent FDMA counterpart. Another advantage of a TDMA system is that the traffic capacity of each access can be modified on a nearly instantaneous basis. The resulting spare traffic capacity can be assigned to another access with a higher traffic volume. The traffic loading of a long-haul system can be varied as the busy hour moves across it, assuming that accesses are located in different time zones. This is difficult to achieve on a conventional FDMA system.

7.6.4.2 Simplified Description of TDMA Operation. An important requirement of TDMA is that transmission bursts do not overlap. To ensure nonoverlap, bursts are separated by a guard time, which is analogous to a guardband in FDMA. The longer we make the guard time, the greater assurance of nonoverlap. However, this guard time reduces the efficiency of the system. Ideally guard time should be reduced to zero, but due to imperfections of even the best timing systems, some guard time must be assigned between each access. Typical guard times for operating systems are on the order of 50–200 ns.

Figure 7.16 shows a typical TDMA frame. A frame is a complete cycle of bursts, usually with one burst per access. The burst length per access need not be of uniform duration; in fact, it is usually not. Burst length can be made a function of the traffic load of a particular access at a particular time. This nonuniformity is shown in the figure. The frame period is the time required to sequence the bursts through a frame.

The number of accesses on a transponder can vary from 3 to over 100. Obviously, the number of accesses is a function of the traffic intensity of each access, assuming a full-capacity system. It is also a function of the transponder bandwidth and the digital modulation employed (e.g., the packing ratio or the number of bits per hertz). For example, more bits can be packed per unit bandwidth with an 8-ary PSK signal of fixed duration than a BPSK signal of equal duration. For high-capacity systems, frame periods vary from 100 μs to over 2 ms. As an example, the INTELSAT TDMA system has a frame period of 2 ms.

Figure 7.16 also shows a typical burst format, which we call an access subframe. The first segment of the subframe is called the CR/BTR (carrier recovery/bit timing recovery). This symbol sequence is particularly necessary on a coherent PSK system, where the CR is used by the PSK demodulator in each receiver to recover local carrier and the BTR to synchronize (sync) the local clock. In the INTELSAT system CR/BTR is 176 symbols long. Other systems may use as few as 30 symbols. We must remember that CR/BTR is

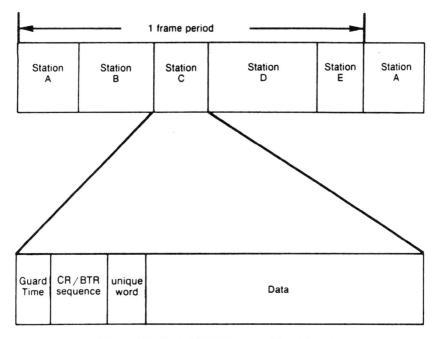

Figure 7.16. Typical TDMA frame and burst formats.

overhead, and thus it would be desirable to shorten its duration as much as possible.

Generally, the minimum number of bits (or symbols—in QPSK systems one symbol or baud carries two bits of information) required in a CR/BTR sequence is only roughly a function of system bit or symbol rate. Carrier recovery and bit timing recovery at the receiver must be accomplished by realizable phase-lock loops and/or filters that have a sufficiently narrow bandwidth to provide a satisfactory output signal-to-noise ratio S/N. There is a trade-off in system design between acquisition time (implying a wider bandwidth) and S/N (implying a narrower bandwidth). Reference 14 suggests that CR/BTR bandwidths of 0.5–2% of the bit or symbol rate provide a good compromise between acquisition time and bit error rate performance resulting from a finite-loop-output S/N. Adaptive phase-lock loops that acquire in a wide-band mode and track with a narrower bandwidth can be used to reduce CR/BTR overhead.

The next bit sequence in the burst subframe (Figure 7.16) is the unique word (UW), which establishes an accurate time reference in the received burst. The primary purpose of the UW is to perform the clock alignment function. It can also be used as a transmit station identifier. Alternatively, the UW can be followed by a transmit station identifier sequence (SIC—station identification code).

The loss of either the CR/BTR or the UW is fatal to the receipt of a burst. For voice traffic, a lost burst causes clicks, and sounds like impulse noise to the listener. In the case of a data bit stream, large blocks of data can be lost due to a "skew" or slip of alignment. TDMA systems are designed for a probability of miss or false detection of 1×10^{-8} or better per burst to maintain a required threshold bit error rate of 1×10^{-4}. A major guideline we should not lose sight of is the point where supervisory signaling will be lost. This value is a BER of approximately 1×10^{-3}. The design value of 1×10^{-8} threshold will provide a mean time to miss or false detection of several hours with a frame length on the order of 1 ms (Ref. 14).

TDMA system design usually allows for some errors in the UW without loss of alignment. One approach to reduce chances of misalignment suggests a change in waveform during the UW interval. For instance, on an 8-ary PSK system, we might transmit BPSK during this interval, providing more energy per bit, ensuring an improved error performance during this important interval and thus improving the threshold performance.

Other overhead or housekeeping functions are inserted in the burst subframe between the UW and the data text. These may include voice and teleprinter orderwires, BERT (bit error rate test) and other sequences, alarms, and a control and delay channel. Preambles or burst overhead usually require between 100 and 600 bits. INTELSAT uses 288 symbols or 576 bits (with QPSK, 1 symbol is equivalent to 2 information bits).

The efficiency of a TDMA system depends largely on how well we can amortize system overhead and reduce guard times, in both length and number. Frame length affects efficiency. As the length increases, the number of overhead bits per unit time decreases. Also, as the length increases, the receiving system buffer memory size must increase, increasing the complexity and cost of terminals.

7.6.4.3 TDMA System Clocking, Timing, and Synchronization.
It was previously stressed that an efficient TDMA system must have no burst overlap on one hand, and as short a guard time as possible between bursts on the other hand. We are looking at guard times in the nanosecond regime. The satellites under discussion here are geostationary. For a particular TDMA system the range to a satellite can vary from 23,000 to 26,000 statute miles. We can express these range values in time equivalents by dividing by the velocity of propagation in free space, or 186,000 mi/s. These values are 23,000/186,000 and 26,000/186,000 or 123.469 and 139.573 ms. The time difference for a signal to reach a geostationary satellite from a very-low-elevation-angle earth station and a very-high-elevation-angle earth station is 16.104 ms (e.g., 139.573 − 123.469) or 16,104 μs or 16,104,335 ns. Of course, this is a worst case, but still feasible. The TDMA system must be capable of handling these orders of time differences among the accessing earth stations. How do we do it and meet the guidelines set out previously? We must also

keep in mind that geostationary satellites actually are in motion in a suborbit, causing an additional time difference and Doppler shift, both varying dynamically with time.

There are two generic methods used to handle the problem: "open loop" and "closed loop." Open-loop methods are characterized by the property that an earth station's transmitted burst is not received by that station. We mean here that an earth station does not monitor the downlink of its own signal for sync and timing purposes. By not using its delayed receiving signal for timing, the loop is not closed; hence, it is open.

Closed loop covers those synchronization techniques in which the transmitted signals are returned through the bent pipe transponder repeated to the transmitting station. This permits nearly perfect synchronization and high-precision ranging. The term *closed loop* derives from the looping back through the satellite of the transmitted signal, permitting the transmitting TDMA station to compare the time of the transmitted-burst leading edge to that of the same burst after passing through the satellite repeater. The TDMA transmitter is then controlled by the result, an early or late arrival relative to the transmitting station's time base. (*Note*: These definitions of open loop and closed loop should be taken in context and not confused with open-loop and closed-loop satellite tracking.)

One open-loop method uses no active form of synchronization. It is possible to achieve accuracies from 5 μs to 1 ms (Ref. 15) through what can be termed "coarse sync." The system is based on very stable free-running clocks, and an approximation is made of the orbit parameters where burst positioning can be done to better than 200-μs accuracy. The method was used on some early TDMA trial systems and on some military systems and will probably be employed on satellite-based data networks, especially with long frames.

One of the most common methods used to synchronize a family of TDMA accesses is by a reference burst. A reference burst is a special preamble only, and its purpose is to mark the start of frame with a burst codeword. The station transmitting the reference burst is called a reference station. The reference bursts are received by each member of the family of TDMA accesses, and all transmissions of the family are locked to the time base of the reference station. This, of course, is a form of open-loop operation. Generally, a reference burst is inserted at the beginning of frame. Since reference bursts pass through the bent pipe repeater and usually occupy the same bandwidth as traffic bursts, they provide each station in the family with information on Doppler shift, time-delay variations due to satellite motion, and channel characteristics. However, the reference burst technique has some drawbacks. It can only serve one repeater and a specific pair of uplinks and downlinks. There are difficulties with this technique in transferring such results accurately to other repeaters, other beams, or stations in different locations. The reference bursts also add to system overhead by using a bandwidth/time product not strictly devoted to the transfer of useful, revenue-bearing data/information. However, we must accept that some satellite capacity must be devoted to achieve synchronization.

7.7   PROCESSING SATELLITES

Processing satellites, as distinguished from "bent pipe" satellites, operate in the *digital* mode and, as a minimum, demodulate the uplink signal to baseband for regeneration. As a maximum, at least as we envision today, they operate as digital switches in the sky. In this section we will discuss satellite systems that demodulate and decode in the transponder and then we will present some ideas on switching schemes suggested in the NASA 30/20-GHz system. This will be followed by a section on coding gain and a section on link analysis for processing satellites.

7.7.1   Primitive Processing Satellite

The most primitive form of processing satellite is the implementation of on-board regenerative repeaters. This only requires that the uplink signal be demodulated and passed through a hard limiter or a decision circuit. The implementation of regenerative repeaters accrues the following advantages:

- Isolation of the uplink and downlink by on-board regeneration prevents the accumulation of thermal noise and interference. Cochannel interference is a predominant factor of signal degradation because of the measures taken to augment communication capacity by such means as frequency reuse.
- Isolating the uplink and downlink makes the optimization of each link possible. For example, the modulation format of the downlink need not be the same as that for the uplink.
- Regeneration at the satellite makes it possible to implement various kinds of signal processing on board the satellite. This can add to the communication capacity of the satellite and provide a more versatile set of conveniences for the user network.

Reference 16 points out that it can be shown on a PSK system that a regenerative repeater on board can save 6 dB on the uplink budget and 3 dB on the downlink budget over its bent pipe counterpart, assuming the same BER on both systems.

Applying this technique to a digital TDMA system requires carrier recovery and bit timing recovery. Although the carrier frequencies and clocking are quite close among all bursts, coherency of the carrier and the clocking recovery may not be anticipated between bursts. To regenerate baseband signals on TDMA systems effectively, carrier and bit timing recovery are done in the preamble of each burst and must be done very rapidly to maintain a high communication efficiency. There are two methods that can be implemented to resolve the correct phase of the recovered carrier. One uses reference codewords in the transmitted bit stream at regular intervals. The other solution is to use differential encoding on the transmitted bit stream. Although this latter method is simpler, it does degrade BER considering equal C/N_0 of each approach.

An ideal regenerative repeater for a satellite transponder is shown in Figure 7.17 for PSK operation. It will carry out the following functions:

- Carrier generation
- Carrier recovery
- Clock recovery
- Coherent detection
- Differential decoding
- Differential encoding
- Modulation
- Signal processing
- Symbol/bit decision

The addition of forward error correction (FEC) coding/decoding on the uplink and on the downlink carries on-board processing one step further. FEC coding and decoding are discussed subsequently. In a fading environment such as one might expect with satellite communication systems operating above 10 GHz during periods of heavy rainfall, an interleaver would be added after the coder and a deinterleaver before the decoder. Fading causes burst errors, and conventional FEC schemes handle random errors. Interleavers break up a digital bit stream by shuffling coded symbols such that symbols in error due to the burst appear to the decoder as random errors. Of course, interleaving intervals or spans should be significantly longer than the fade period expected.

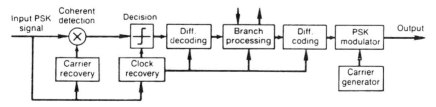

Figure 7.17. Configuration of ideal regenerative repeater.

7.7.2   Switched-Satellite TDMA (SS/TDMA)

We now carry satellite processing one step further by employing antenna beam switching in conjunction with TDMA. This technique provides bulk trunk routing, increasing satellite capacity by frequency reuse. Figure 7.18 depicts this concept. TDMA signals from a geographical zone are cyclically interconnected to other beams or zones so that a set of transponders appears to have beam-hopping capability. A sync window or reference window is usually required to synchronize the TDMA signals from earth terminals to the on-board switch sequence.

Sync window is a generic method to allow earth stations to synchronize to a switching sequence being followed in the satellite. A switching satellite, as described here, consists of a number of transmitters cross-connected to receivers by a high-speed time division switch matrix. The switch matrix connections are changed throughout the TDMA frame to produce the required interconnections of earth terminals. A special connection at the beginning of the frame is the sync window, during which signals from each spot-beam zone are looped back to their originating spot-beam zone, thus forming the timing reference for all zones. This establishes closed-loop synchronization.

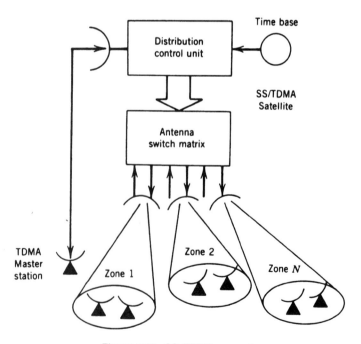

Figure 7.18. SS/TDMA concept.

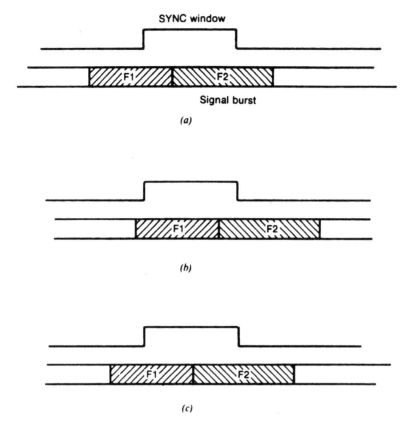

Figure 7.19. Sync window for SS/TDMA and use of bursts of two frequencies. (a) Signal burst too early; (b) signal burst too late; (c) synchronized to sync window. (From Ref. 15. Reprinted with permission.)

Figure 7.19 shows a scheme for locking and tracking a sync window in a satellite switching sequence. A burst of two tones, F_1 and F_2, is transmitted by a single access station. Only the portion that passes through the sync window is received back at the access station.

The basic concept is to measure and compare the received subbursts F_1 and F_2 as shown in the figure. Although a very narrow bandwidth and full RF power are used, digital averaging over many frames still is required. The difference is used to control F_1/F_2 burst to a resolution of one symbol, and the process is continually repeated in closed loop. The sync bursts to the TDMA network were slaved, and the network is therefore synchronized to the sync window and the switching sequence on the satellite.

With SS/TDMA the network connectivity and the traffic volume between zones can be adapted to changing needs by reprogramming the control processor of the antennas. Also, of course, the narrow beams (e.g., higher-gain antennas) increase the uplink C/N and the downlink EIRP for a given transponder HPA power output. However, the applicability of the system must be analyzed carefully. Since there is a limit to the speed at which the antenna beam can be switched as the data symbol rate increases, guard times become an increasing fraction of the message frame, resulting in a loss of efficiency. This can be offset somewhat by the use of longer frames, but this requires increased buffering and increases the transmission time.

The scheme becomes very inefficient if a large proportion of the traffic is to be broadcast to many zones. SS/TDMA is not applicable to channelized satellites where multiple transponders share common transmit and receive antennas and the signals in each transponder cannot be synchronized for transmission between common terminal zones.

Figure 7.20 is a block diagram of a beam-switching processor in which narrow scanning beams are implemented by the use of a processor-controlled multiple-beam antenna (MBA) for both uplinks and downlinks. The output of the address selector is modified by the memory to provide control signals to the beam-switching network, which, in turn, controls the antenna-weighting networks. This process steers the uplink and downlink antennas to the selected zones for a time interval that matches the time interval of a data frame to be exchanged between the selected zones. Initial synchronization of the system is obtained from a synchronizing signal transmitted to the processor from the TDMA master station. The memory can be updated at any time

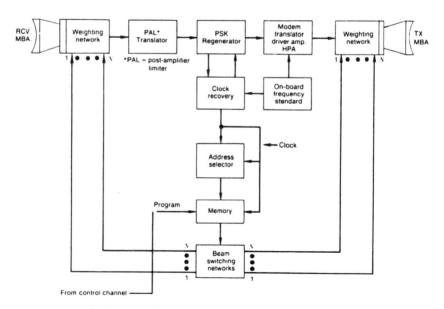

Figure 7.20. Block diagram, beam-switching processor.

by signals from the master terminal via a control channel. Often, in practice, two memories are provided so that one can be updated while the other is on-line, thus preserving traffic continuity.

7.7.3 IF Switching

When the principal requirement for switching is for traffic routing rather than just for antenna selection, one method is to convert all uplink signals to a common IF, followed by a switching matrix and upconverting translators. This provides greater switching capability and flexibility. It allows for a wider choice of switching components in the design, since many more device types show good performance at the lower IF frequencies than at the higher uplink and downlink RF frequencies. Table 7.11 lists four switch architectures and their advantages and disadvantages. Of these, Ref. 17 states that the coupler crossbar offers the best performance for a large switching matrix (e.g., on the order of 20 × 20—20 inlets, 20 outlets). Figure 7.21 is a block diagram of such a crossbar matrix. In this case the crosspoints could be PIN diodes or dual-gate field effect transistors (FETs).

TABLE 7.11   Comparison of Switch Matrix Architectures

Description	Advantages	Disadvantages
Fan-out/fan-in	Broadcast mode capability	High-input VSWR[a] High insertion loss Redundancy difficult to implement
Single pole/multiple throw	Low insertion loss	Poor reliability
Rearrangeable switch	Low insertion loss	Poor reliability Random interruptions Control algorithm complicated
Coupler crossbar	Planar construction (minimum size, weight, and volume)	Difficult feedthroughs Difficult broadbanding
	Broadcast mode capability	Isolation hard to maintain
	Redundancy easy to implement	
	Good input/output VSWR[a]	
	Signal output level independent of the path	
	Enhanced reliability	

[a]VSWR = voltage standing wave ratio.
Source: Reference 17.

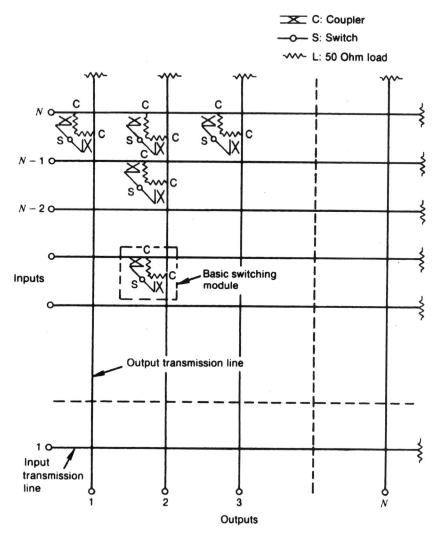

Figure 7.21. Coupler crossbar switching matrix for satellite IF switching. (From Ref. 17.)

7.7.4 Intersatellite Links

Carrying the on-board processing concept still further, we now consider intersatellite links that can greatly extend the coverage area of a network. An intersatellite link (ISL) is a full-duplex link between two satellites, and other similar links can be added, providing intersatellite connectivity among satellites in a large constellation. A number of military satellite constellations (e.g., MILSTAR) have cross-linking capability. The 58–62-GHz band has been assigned for this purpose by the ITU. Laser cross-links are actively under consideration by the U.S. Department of Defense.

An ISL capability can have a significant impact on system design. The following three system design characteristics are most directly affected:

1. Connectivity. The ISL can be used to provide connectivity among users served by regional satellites, still retaining the required level of interconnectivity within a community of users. It is assumed that each satellite in the system will serve a region with a high intraregion community of interest. The community of interest among regions will be lower. Thus each satellite will handle the high-traffic-intensity intraregion, and cross-links will serve as tandem relays for the lower-intensity traffic between regions.

2. Capacity. As described previously, uplinks and downlinks will have a high fill factor, and with cross-links implemented, low-traffic-intensity uplinks and downlinks will not be required to serve other regions with low interregional community of interest.

3. Coverage. Many satellite systems are designed for worldwide coverage where users of one satellite footprint require connectivity to users not in view of that satellite. The use of cross-links eliminates the need of earth relay at a dual-antenna earth station. The cross-link saves money and reduces propagation delay. Generally, an ISL is more economic than adding the additional uplink and downlink to the next satellite.

The cross-link concept incurs other advantages:

- Not affected by climatic conditions (e.g., the link does not pass through the atmosphere, assuming geostationary satellites)
- Low antenna noise temperature
- Low probability of earth-based intercept

7.8 LINK BUDGETS FOR DIGITAL SATELLITES

A link budget is a tool used to dimension or size a satellite communication system. In Section 7.4.8 link budgets for analog systems were described. The approach described here will be similar, especially for digital access via FDMA, such as the INTELSAT IDR system.

For digital operation using FDMA on a bent pipe satellite, we calculate C/N_0 for the uplink, then C/N_0 for the downlink. The final step is to calculate C/N_0 for the system combining uplink and downlink. With this value for C/N_0, we calculate E_b/N_0 for the modulation type in use subtracting an appropriate value for modulation implementation loss. This net E_b/N_0 value is then applied to a BER curve to obtain the equivalent BER for the link endpoint earth station receiver.

The transmission engineer might be called upon to size or dimension a link given a required BER. In this case we work backward from the equivalent E_b/N_0 value for that BER. Given E_b/N_0, we calculate C/N_0, which we remember is synonymous with RSL/N_0. Thus

$$E_b/N_0 = C/N_0 - 10\log(\text{bit rate}) \tag{7.30}$$

We can also obtain a target C/N_0 value from Table 7.10 if we are dimensioning a system to work with INTELSAT IDR service.*

Example 1. An earth station has a G/T of $+25$ dB/K. Its uplink is at 6.0 GHz; its downlink at 4.0 GHz. The bit rate is 45 Mbps and the satellite EIRP is $+34$ dBW; the satellite G/T is $+0.5$ dB/K. The uplink free-space loss is 199.6 dB and the downlink free-space loss is 196.2 dB. The required E_b/N_0 is 12.0 dB, which includes modulation implementation loss. The earth station EIRP is $+67$ dBW. From this information, set up an uplink budget and a downlink budget.

uplink budget

Earth station EIRP	$+67$ dBW
Pointing loss	0.5 dB
Polarization loss	0.5 dB
Satellite pointing loss	0.5 dB
Free-space loss	199.6 dB
IRL satellite	-134.1 dBW
G/T satellite	$+0.5$ dB/K
Sum	-133.6 dBW
Boltzmann's constant	$-(-228.6$ dBW/Hz$)$
C/N_0	95 dB/Hz

downlink budget

Satellite EIRP	$+34$ dBW
Satellite pointing loss	0.5 dB
Polarization loss	0.5 dB
Earth station pointing loss	0.5 dB
Free-space loss	196.2 dB
IRL at earth station	-163.7 dBW
G/T earth station	25.0 dB/K
Sum	-138.7 dBW/K
Boltzmann's constant	$-(-228.6$ dBW/Hz$)$
C/N_0	89.9 dB/Hz

---

*For INTELSAT services, standard earth stations are employed that have already been "dimensioned." See Section 7.10.

Turn back to Section 7.4.8.3 and use equation 7.25. Convert our C/N_0 values to their equivalent numerical values.

$$89.9 \text{ dB} = 0.977 \times 10^9 \qquad 95.0 \text{ dB} = 3.16 \times 10^9$$

$$(C/N_0)_{\text{sys}} = 1/\left[1/(0.977 \times 10^9) + 1/(3.16 \times 10^9)\right]$$

$$= 88.73 \text{ dB/Hz}$$

Apply equation 7.30:

$$E_b/N_0 = 88.73 - 10\log(45 \times 10^6)$$

$$= 12.19 \text{ dB}$$

The required E_b/N_0 was 12.0 dB, leaving only a 0.19-dB margin.

Link budget for processing satellites, uplinks, and downlinks are treated separately, carrying the link budget to the calculated E_b/N_0 value. These values are adjusted by increasing or decreasing terminal parameters such as G/T and EIRP to meet the required BER. The following examples will serve to illustrate these points.

TABLE 7.12   Example 2, Uplink Power Budget

Terminal EIRP	+65 dBW	
Terminal pointing loss	0.5 dB	
Free-space loss	199.2 dB	
Satellite pointing loss	0.0 dB	(earth coverage)
Polarization loss	0.5 dB	
Atmospheric losses	0.5 dB	
Rainfall (excess attenuation)	0.25 dB	(10° elevation angle)
Isotropic receive level	− 135.95 dBW	
G/T	0.0 dB	
Sum	− 135.95 dBW	
Boltzmann's constant	− (− 228.6 dBW)	
C/N_0	92.65 dB	
− 10 log(bit rate)	− 70.00 dB	
E_b/N_0	+ 22.65 dB	
Required E_b/N_0	− 9.6 dB	
Implementation loss	− 2.0 dB	
Margin	11.05 dB	
Allowable margin	− 4 dB	
Excess margin	7.05 dB	

The satellite G/T can be − 7.05 dB for the uncoded system.
For the coded system with a 3.1-dB coding gain, the satellite G/T can be degraded to − 10.15 dB [− 7.05 dB + (− 3.1 dB)].

Example 2. A specific uplink at 6 GHz working into a processing satellite is to have a BER of 1×10^{-5}. The modulation is QPSK and the data rate is 10 Mbps. The terminal EIRP is $+65$ dBW. The free-space loss to the satellite is 199.2 dB. What satellite G/T will be required without coding? What G/T will be required when FEC is employed? The satellite will use an earth coverage antenna. See Table 7.12.

Select the required E_b/N_0 first for no coding and then with convolutional coding rate $\frac{1}{2}$ and $K = 7$, Viterbi-decoded. Use a modulation implementation loss of 2.0 dB in both cases. From Figure 5.38, lower left curve, the E_b/N_0 is 9.6 dB for QPSK (uncoded), and from Table 5.15 for the coded system use 6.5 dB. Allow 4 dB of margin. Initially set the satellite G/T at 0 dB/K.

Example 3. A satellite has a $+30$-dBW EIRP downlink at 7.3 GHz. The desired BER is 1×10^{-6}; the modulation is coherent BPSK; FEC is implemented with convolutional encoding and 3-bit receiver quantization; the bit rate is 45 Mbps; and the free-space loss is 202.0 dB. What is the terminal G/T assuming a 5-dB margin? See Table 7.13. Assume that the satellite uses a spot beam, and a rate $\frac{1}{2}$, $K = 7$ convolutional code.

The required G/T is the sum of $26.18 + 4.9 + 2.0 + 5.0$ dB or $+38.08$ dB/K. If that value is now substituted for the initial G/T of 0 dB/K, the last entry in the table or "sum" would drop to 0. The value for the required E_b/N_0 was derived first from Figure 5.38, lower left curve, using BER = 1 $\times$

TABLE 7.13   Example 3, Downlink Power Budget

Satellite EIRP	$+30$ dBW
Satellite pointing loss	0.5 dB
Footprint error	0.25 dB
Off-footprint center loss	1.0 dB
Terminal pointing loss	0.5 dB
Polarization loss	0.5 dB
Atmospheric losses	0.5 dB
Free-space loss	202.0 dB
Rainfall loss	3.0 dB
Isotropic receive level	-178.25 dBW
Terminal G/T	0.0 dB/K
Sum	-178.25 dBW
Boltzmann's constant	$-(-228.6)$ dBW
C/N_0	50.35 dB
10 log(bit rate)	-76.53
Difference	-26.18
Required	-4.9 dB
Modulation implementation loss	-2.0 dB
Margin	-5.0 dB
Sum	38.08 dB

10^{-6}, thence to Figure 5.45 where we found that the coding gain for $K = 7$, $R = \frac{1}{2}$ is 5.6 dB, and we then subtracted. The rainfall loss value was, in a way, arbitrary. As we will show in Chapter 8, 3 dB in this band, at a 10° elevation angle, would provide this performance 99.9% of the time for central North America. However, with interleaving using a sufficient interleaving interval (about 4 or 5 s) and the coding employed, we can decrease the excess attenuation due to rainfall to zero. The value would be accommodated in the coding gain. The coding gain used was for AWGN conditions only. It will also provide a fading improvement. Heavy rain manifests itself in fading, which, in the worst case, can be considered a Rayleigh distribution. The incidental losses (i.e., polarization, pointing errors) are good estimates and probably can be improved upon with a real system where firm values can be used.

7.9 VSAT (VERY SMALL APERTURE TERMINAL)

7.9.1 Definitions of VSAT

A very small aperature terminal (VSAT) is a digital satellite terminal where *economy* is the key word. The term *very small aperture*, of course, refers to the size of the terminal antenna. The diameters of VSAT parabolic antennas vary from 0.6 m (2 ft) to 2.4 m (7.8 ft), depending a great deal on the capabilities desired from the terminal. These can vary from a data connectivity (inbound) of 1200 bps up to DS1 or E-1.

In most cases, the definition denotes a family of modest "out terminals" and a comparatively large "hub" terminal. This implies a wheel made up of a hub and spokes. In fact, most VSAT architectures can be seen as hub and spokes, the spokes being the connectivities to the VSAT outstations, most often in a star network configuration as shown in Figure 7.22. The larger hub, in theory, compensates for the smaller handicapped VSAT outstations.

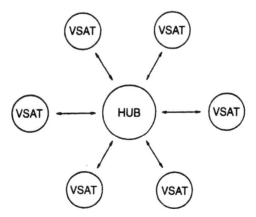

Figure 7.22. The conventional VSAT star network configuration.

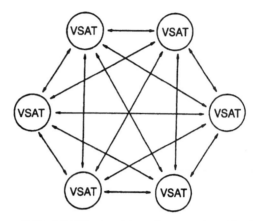

Figure 7.23. A VSAT network based on a mesh architecture.

With the VSAT star network, traffic can be one-way or two way. VSAT networks have extended the definition to include mesh-connected networks with no hub, as shown in Figure 7.23. The star configuration does not lend itself well to VSAT-to-VSAT voice communication because of the added delay, whereas the mesh architecture does lend itself to voice interconnectivity among VSATs. Some sources even call any network of small satellite terminals a VSAT network.

VSAT networks operate in the 6/4-GHz band. However, the 14/12/11-GHz band is the more popular operational band because a higher satellite EIRP is permitted, allowing for smaller VSAT outstations. Excess attenuation due to rainfall becomes a propagation issue in this higher-frequency band. It is expected that the wider 30/20-GHz bands will become well exploited by VSAT when commercial space segments become available.

7.9.2 VSAT Network Applications

VSATs are commonly implemented in private networks. Why they are attractive depends largely on a country's telecommunication infrastructure. In the United States and Canada, economy is the driving factor. Such VSAT networks bypass the local and long-distance telephone companies and ostensibly save money by doing so.

In many other countries, national governments permitting, well-engineered VSAT networks can provide sterling quality service, whereas the local telecommunication administrations cannot. There is a third category that includes countries with a poor infrastructure and where many communities are afforded no electrical communications whatsoever.

As we mentioned, there are one-way and two-way applications.

7.9.2.1   One-Way Applications. This application basically involves data distribution from a hub outward to VSAT receive-only terminals. These applications include:

- Press releases, news from press agencies or the like
- Stocks, bonds, commodity information
- Remote loading of computer programs
- Weather information from meteorological agencies, typically to airports
- Video distribution utilizing compressed video, typically 1.544 or 2.048 Mbps

Another data application is in the direction VSAT-to-hub for data collection purposes. This application may involve remote sensors on oil pipelines, environmental monitoring, or electrical power utilities' remote facilities. However, with many of these applications, some form of network control is necessary, making two-way communication desirable.

7.9.2.2   Two-Way Applications. The most widely used application of VSAT communications is for diverse types of two-way data communications. Such a network provides complete flexibility for file transfer and all types of interactive data exchanges such as inquiry/response. In most configurations, the hub is colocated with corporate headquarters. Typical two-way applications are:

- Point-of-sale operations
- Financial, banking, and insurance from field branches to central headquarters
- Credit card verification
- ATM (automatic teller machine) operations
- Hotel/motel reservations and all other types of reservations
- Support for shipping and freight-handling facilities
- Inventory control; cash flow
- Technical support network, manufacturer to manufacturer's representatives
- Supervisory control and data acquisition (SCADA), pipelines, railroads
- Extending local area networks

TELEPHONE CONNECTIVITIES. If sufficient capacity is built into a VSAT network, telephone operation is feasible from outstation to headquarters and vice versa. It is not feasible because of added delay from outstation to other outstation via the hub. It would be feasible on VSAT mesh networks.

In some emerging nations, VSAT-like networks provide rural telephone connectivity.

With current video compression techniques, video conferencing may also be feasible (Ref. 1).

7.9.3 Technical Description of VSAT Networks

7.9.3.1 Introduction. The most common network topology of a VSAT network is the star configuration illustrated in Figure 7.22. The hub is the centerpiece and is almost always located at the corporate headquarters or state capital or national capital, for that matter. The hub may have an antenna 5 m (16 ft) to 11 m (36 ft) in diameter, whereas a VSAT may have an antenna diameter in the range of 0.6 m (1 ft) to 2.4 m (8 ft). The RF output at the hub will vary from 100 to 1000 W, whereas a VSAT will be in the range of 1–10 W.

To reduce space segment recurring charges, it is incumbent on the system designer to use as small a bandwidth as possible on a satellite transponder. Outbound traffic is usually carried on a TDM bit stream, 56 or 64 kbps, 128, 256, 384 kbps (etc.). Inbound traffic depends greatly on the traffic profile and will use some type of demand assignment process or contention, polling, or other protocol with bit rates ranging from 1200 bps to 64 kbps or greater.

Definition. Inbound—traffic or circuit(s) in the direction of VSAT to hub; outbound—traffic or circuits in the direction of hub to VSAT.

7.9.3.2 Summary of VSAT RF Characteristics. VSAT operation commonly uses either a 6/4-GHz band (C-band) or 14/12-GHz band (Ku-band).* As operational frequencies increase, receiver noise performance degrades. At C-band, we can expect an LNA with 50-K noise temperature; at Ku-band, 100 K. Antenna noise temperature (T_{ant}) at C-band (5° elevation angle) is 100 K and at Ku-band (10° elevation angle) it is 106 K. Thus the typical T_{sys} for C-band VSAT operation is 150 K while for Ku-band it is 206 K. Line losses for both bands are taken at 1.5 dB for this particular model. In the case of Ku-band, the LNA is placed as close as practical to the feed to reduce line losses.

We will now construct Table 7.14, which will give typical G/T values for several discrete antenna diameters for both C-band and Ku-band operation. The table is based on the T_{sys} figures given above. The antenna gains are based on 65% aperture efficiency (η). Formula 7.40 was used to calculate the parabolic antenna gains.

---

*There is nascent VSAT activity in Ka-band (30/20-Gz band) with the advent of the NASA ACTS satellite. Surely other satellites will follow, and in time there will be equal or greater VSAT activity in the Ka-band.

TABLE 7.14   Typical VSAT G/T Values

Antenna Aperture (m)	Gain, C-Band	G/T, C-Band	Gain, Ku-Band	G/T, Ku-Band
0.5 m (1.625 ft)	23.46 dB	+1.7 dB/K	33 dB	+9.86 dB/K
0.75 m (2.44 ft)	27.0 dB	+5.23 dB/K	36.53 dB	+13.39 dB/K
1.0 m (3.25 ft)	29.49 dB	+7.73 dB/K	39.03 dB	+15.89 dB/K
1.5 m (4.875 ft)	33.01 dB	+11.24 dB/K	42.55 dB	+19.41 dB/K
2.0 m (6.5 ft)	35.51 dB	+13.75 dB/K	45.05 dB	+21.91 dB/K
2.5 m (8.125 ft)	37.45 dB	+15.69 dB/K	47.0 dB	+23.86 dB/K
3.0 m (9.75 ft)	39.03 dB	+17.54 dB/K	48.57 dB	+25.43 dB/K

The reference frequencies used for antenna gain calculations are 4000 and 12,000 MHz, respectively. The table includes 1-dB line loss for both bands.

7.9.4   VSAT Access Techniques

The most common VSAT architecture is the interactive network based on star topology (hub and spokes, Figure 7.22). Reference 2 describes a VSAT network with as many as 16,000 outstations. The author is familiar with networks with up to 2500 outstations interoperating with one hub. There are many access techniques, and the type selected will be fairly heavily driven by the traffic profile. The access technique will often determine the efficiency of usage of the space segment. For example, a completely assigned FDMA regime would prove to be a very ineffective use of transponder bandwidth if there were hundreds of outstations, each interchanging short, interactive messages with the hub with a medium to low activity factor.

In selecting the type of channel assignment technique for a VSAT network the following factors should be considered:

- Statistical properties of the traffic
- The permissible delay in transmission, including channel setup and propagation delay
- Efficiency of channel sharing, throughput performance
- Complexity, equipment and implementation cost
- Operations and maintenance

For example, for credit card verifications and transactions, delay is probably the most important factor, throughput much less so; whereas with file transfer and batch transactions, throughput is more important than delay (comparatively).

We will discuss three categories of access: random, demand assigned, and fixed assigned. Here, of course, we refer to inbound channels. Outbound service is assumed to be a TDM bit stream and is discussed in Section 7.9.4.5.

7.9.4.1 Random Access

7.9.4.1.1 Pure Aloha. Random access schemes lend themselves well to short and bursty traffic. In this case, the inbound channel is shared by several or many VSAT terminals. This is really a contention scheme. When a VSAT has traffic, it bursts the traffic on the inbound channel, taking a chance that another VSAT is not transmitting at the same time. If there is a collision—in other words, another VSAT is transmitting traffic at the same time—the traffic is corrupted, and both VSATs must try again. Each VSAT has a random backoff algorithm. In theory, the backoff time will be different for each terminal, and the second attempts will be successful. A VSAT knows if an attempt is successful because it will receive an acknowledgment from its associated hub on the TDM outbound channel.

This scheme works out well when traffic volume is small. Delay is normally short because most transmissions only require one exchange with the hub.

As traffic volume picks up, the system becomes more and more unwieldy. The point where this occurs, according to Ref. 20, is when throughput approaches 25–30%.* As we increase loading above the 30% value for throughput, the probability of collision increases as do transaction delays. When the traffic volume exceeds a further limiting value (some argue 50%), the system will tend to "crash," meaning that throughput begins to approach zero because of nearly continuous collisions, backoffs, and reattempts. Flow control mechanisms can help prevent this from happening.

The type of access described here is called pure Aloha. The principal advantage of pure Aloha is its simplicity. There is no time synchronization required, and the hub and VSATs do not need precise timing control (Refs. 18–21).

7.9.4.1.2 Slotted Aloha. Slotted Aloha is more complex than pure Aloha in that it requires time synchronization among VSATs. In this case users can transmit only in discrete time slots. With such a scheme two (or more) users can collide with each other only if they start transmitting exactly at the same time. One disadvantage of slotted Aloha is the wasted periods of time when a message or packet does not use up the total slot time allowed. Slotted Aloha has about twice the efficiency of pure Aloha or about 34% throughput versus 18% for pure Aloha. These percentage values are points where throughput begins to level off or decrease because collisions begin to increase (Refs. 18, 22).

---

* The CCIR (Ref. 18) places it at an 18% throughput value. As further traffic is offered, actual throughput drops.

7.9.4.1.3 Selective Reject (SREJ) Aloha. With SREJ Aloha, messages or packets are broken down into subpackets. These subpackets have fixed length and can be received independently. Each subpacket has its own acquisition preamble and header. Generally, there is no total collision. Some subpackets may collide, but not whole messages; some of the message or main packet gets through successfully. In SREJ Aloha, only the subpackets that are corrupted are retransmitted. This reduces retransmission because the only retransmission required is a smaller subpacket, not the whole message. SREJ Aloha does not need time synchronization for messages or subpackets. It can achieve higher throughput than pure Aloha and is well suited for variable length messages (Refs. 18, 20).

7.9.4.1.4 R-Aloha or Aloha with Capacity Reservation. This is still another version of Aloha. It is useful when there are a few high-traffic-intensity users and other low-intensity, sporadic users. The high-intensity users have reserved slots and the remainder of the slots are for low-intensity users. This latter group operates on a contention basis similar to pure Aloha. There are many variants of this reservation system (Ref. 21).

One efficient derivative is called slot reservation or demand assignment TDMA (DA/TDMA). When a VSAT has data packets to be transmitted, a request is sent to the hub, which then replies with TDMA slot assignments. The request specifies the length and number of data packets to be sent. Upon receipt of the slot assignment, the VSAT transmits the data packets in the assigned slots without any risk of collision because the hub informs all participating VSATs that particular slots have been reserved. Although requiring more response time since a delay time of two satellite round trips is spent before the actual packets are transmitted, the reservation mode is very effective for the case of longer messages. It is pointed out that this reservation scheme is different from conventional DAMA schemes in that the reservations are made on a packet-by-packet (or for a group of packets) basis and not for a continuous channel.

The reservation request can be made on a dedicated request channel or on a traffic channel, which operates on a random access (pure Aloha) Scheme. Of course, there is the added overhead for reservation requests. Figure 7.24 is a conceptual drawing of reservation mode transmission. Figure 7.25 compares throughput versus delay performance for three Aloha access schemes (Ref. 18).

7.9.4.2 Demand Assignment Multiple Access. Demand assignment multiple access (DAMA) is a satellite access method based on the concept of a pool of traffic channels that can be assigned on demand. When a VSAT user has traffic, the hub is petitioned for a channel. If a channel is available, the hub assigns the channel to the VSAT, which then proceeds to transmit its traffic. When the traffic transaction is completed, the channel is turned back into the pool of available channels.

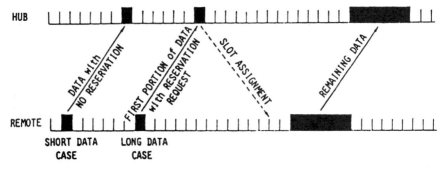

Figure 7.24. Reservation mode transmission. (From Ref. 18.)

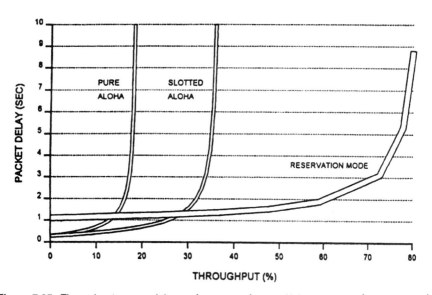

Figure 7.25. Throughput versus delay performance of some Aloha schemes. (From Ref. 18.)

These channels that are assigned can be on a FDMA basis or on a TDMA basis. In other words, the pool of channels may consist of a group of frequency slots or a group of time slots. If the frequency domain is employed, it is called FDMA/SCPC, where SCPC stands for single channel per carrier. DAMA VSAT systems are attractive for voice operation in mesh networks or for voice connectivities VSAT to hub. Of course, they are also useful when there is intense data traffic that is more or less continuous (Refs. 18, 21).

7.9.4.3 *Fixed-Assigned FDMA.* When a nearly continuous traffic flow is expected from a VSAT to a hub, SCPC operation may be an attractive alternative. In this case, each VSAT is assigned a frequency slot on a full

TABLE 7.15   Comparison of Multiple Access Techniques—Inbound

Multiple Access Technique	Maximum Throughput	Practical Message Delay	Suitable Application
Random access			
Pure Aloha	18.4%	< 0.5 s	Interactive data
Slotted Aloha	36.8%	< 0.5 s	Interactive data
SREJ Aloha	20–30%	< 0.5 s	Interactive data
Slot reservation	60–90%	< 2 s	Batch data
Demand assigned			
FDMA, TDMA	High	0.25 s	Batch data, voice
Fixed assigned			
FDMA, TDMA	High	0.25 s	Multiplexed data, voice

Source: Reference 18, Table 3.1.1.

period basis. The bandwidth of the slot should be sufficient to accommodate the traffic flow. Another alternative is TDMA, where a time slot is assigned full period for the connectivity.

7.9.4.4 Summary of Access Techniques. Table 7.15 presents a comparison of the several access techniques covered in Section 7.9.4. The choice of which technique to adopt depends on the traffic and delay requirements of the proposed VSAT system. System complexity may also be an issue. Complexity not only can be costly but may also impact reliability.

7.9.4.5 Outbound TDM Channel. Besides a vehicle for outbound traffic, the outbound TDM channel may have other functions. Among these functions are:

- Provide timing to slave VSAT clocks, typically for slotted Aloha
- Channel assignment, typically for DAMA schemes, reservation Aloha
- Acknowledgment of incoming packets (frames)
- Other control functions

The TDM channel sends a continuous series of frames, where data packets are inserted in the frame's information or data field. In many cases, as we pointed out earlier, the frames are formatted following the HDLC* link layer protocol (see Ref. 23, Chapter 5). The HDLC control field often is modified to carry out VSAT control functions; the acknowledgment technique can also be patterned after HDLC. If there is no outbound message traffic to be sent, the info field can be filled with null data or supervisory frames can be sent. In some implementations, a frame timing and control field, including preamble and unique word (UW), are used.

---

* HDLC = high-level data link control, a link layer protocol developed by the ISO. ADCCP, LAPB, and LAPD are direct derivatives of HDLC.

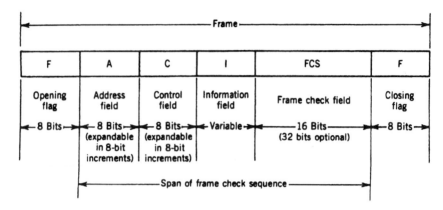

Figure 7.26. The HDLC frame format. It should be noted that there are three types of HDLC frames: information, supervisory, and unnumbered. In the latter two frame types there is no information field, and the other variant is the control field.

A typical HDLC frame is shown in Figure 7.26, unmodified. It should be pointed out that the address field must have sufficient length to accommodate addresses for all VSAT terminals in the system as well as group and broadcast addresses. In some systems, an OSI layer 3 (the network layer) can be employed. Often this is based on ITU-T Rec. X.25 (see Ref. 23).

7.10 INTERFERENCE ISSUES WITH VSATs

VSATs by definition have small antennas. As a result, they have comparatively wide beamwidths. For aperture antennas, beamwidth is related to gain. Jasik and Johnson (Ref. 24) provide the following relationship for estimating beamwidth:

$$BW_{3\text{-}dB} = 70°(\lambda/D) \qquad (7.31)$$

where λ is the wavelength and D is the diameter of the antenna. Both D and λ must be expressed in the same units.

If we use the downlink Ku-band frequency of 12,000 MHz, its equivalent wavelength is 0.025 m. If we use a 1-m dish, the beamwidth will be 1.75°. This gives rise to one interference problem in installations with such small antennas. Satellites in geostationary orbit are now placed 2° apart. A beamwidth of 1.75° with a VSAT antenna pointed to one satellite will be prone to interference on the downlink from a neighboring satellite, just 2° away in the orbital equatorial plane.

For the 12-GHz frequency (i.e., $\lambda = 0.025$ m), we develop Table 7.16 for various diameters of parabolic dish antennas that may be employed in a VSAT installation. The table is based on formula 7.31.

TABLE 7.16   Antenna Beamwidths for Various Diameter Dishes (F = 12 GHz)

Antenna Diameter (m)	Antenna Diameter (ft)	Beamwidth (degrees)
0.5	1.625	3.5
0.75	2.44	2.33
1.0	3.25	1.75
1.5	4.875	1.166
2.0	6.5	0.875
2.5	8.125	0.70
5.0	9.75	0.35

From the table we see that the larger the antenna, the narrower the beamwidth and the less the possibility of overlap from one satellite to an adjacent satellite in the geostationary orbit.

The ITU-R Organization has established some interference guidelines among various telecommunication services offered by satellite relay. These guidelines are summarized in Table 7.17 based on carrier-to-interference ratios (C/I). For digital systems it is more convenient to use E_b/I_0 or energy per bit to interference spectral density ratio. The C/I and E_b/I_0 values in the table apply to the combined uplink and downlink values of the link budget process. Ku-band antenna discrimination values in dB are given in Figure 7.27.

In a homogeneous arrangement of adjacent satellite systems providing narrowband digital services in which carrier power flux densities are approximately the same (Figure 7.27 and Table 7.17) in order to obtain a single entry C/I ratio (or antenna discrimination value) of about 20 dB, the VSAT

TABLE 7.17   Examples of Single Entry Interference Protection Ratios for Typical Satellite Carrier Services

Fixed Satellite Service Carrier Type	Single Entry Interference Protection Ratio
A Frequency-modulated television (FM/TV)	
Studio quality	C/I = 28 dB
Good quality	C/I = 22 dB
B Digital data channels	
Wideband, full transponder bandwidth	E_b/I_0 = 25 dB
Narrowband, SCPC, T1 (1.544 Mbps)	E_b/I_0 = 20 dB
Narrowband, SCPC, 56 kbps	E_b/I_0 = 20 dB
C Spread spectrum channels	E_b/I_0 = 20 dB
D Frequency-modulated SCPC,	1000 pW maximum in worst
voice interference contribution	baseband channel
E Frequency modulated SCPC, audio	C/I = 24 dB
program	

Note 1. E_b/I_0 refers to the ratio of energy per hertz of interference.
Note 2. C/I refers to the ratio of carrier power to interference power.
Source: Reference 18, Table 3.5.1.

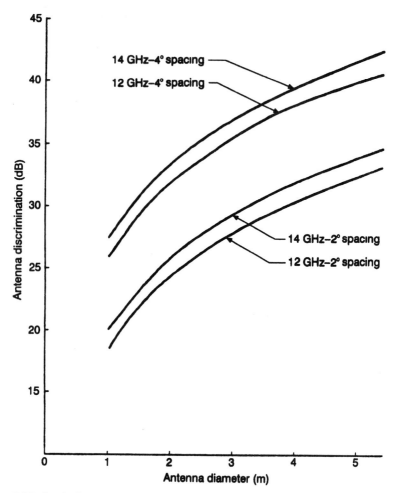

Figure 7.27. Parabolic antenna discrimination at the 14/11/12-Gz band for 2° and 4° satellite orbit spacing. (From Ref. 18, Figure 3.5.1.)

antenna diameter must exceed 1.2 m for satellite separations of 2° or about 0.8 m for satellite separations of 3°. In this example it is assumed that the VSAT system employs a star network with a hub station of at least 4 m in diameter so that the hub-to-satellite link has at least 30 dB of antenna discrimination and contributes less than 0.5 dB to the overall link C/I. On the other hand, if the system is composed of VSATs only (typically with a mesh network, Figure 7.23), and neither uplink nor downlink is controlling with regard to interference, then the antenna sizes must be larger than the VSATs given in the above example. In this case, if the antenna are of the same size, the antenna discrimination value needed will more likely be on the order of 23 dB. From Figure 7.27, this results in an antenna diameter requirement of 1.7 m at 2° of satellite spacing or 1.1 m at 3° of satellite spacing.

Besides antenna discrimination, interference from adjacent satellite emission can be reduced by:

- Using channelization plans in which carrier center frequencies in adjacent satellite systems are offset from each other
- Employing cross-polarization techniques
- Using FEC techniques that can reduce receiver sensitivity to interference

Unfortunately, VSAT systems often are bunched together, because many VSAT applications are found in urban areas. Thus many interference scenarios are set up such as:

- VSAT to VSAT
- VSAT to large earth station (and vice versa)
- Line-of-sight microwave to VSAT and vice versa

7.11 INTELSAT INTERFACES AND STANDARDS

7.11.1 Introduction

INTELSAT has issued six categories of earth station standards, which are broken down by function as follows:

IESS-101 Introduction and Approved Document List
IESS-200 series: Antenna and RF Equipment Characteristics
IESS-300 series: Modulation and Access Characteristics
IESS-400 series: "Supplementary" (deals with such subjects as IM products, engineering orderwires, and satellite characteristics)
IESS-500 series: Baseband Processing Including DCME and Digital TV
IESS-600 series: Generic Earth Station Standards

Highlights of some of these standards are covered as follows.

7.11.2 Standard A—Antenna and Wideband RF Performance Characteristics

Frequency band: 6/4 GHz

$$G/T \geq 35.0 + 20\log(f/4) \quad \text{dB/K} \tag{7.32}$$

where f is the receive frequency in gigahertz. The reader should note that there is an extension to the 6/4-Gz band as follows:

Previous: 3.700–4.200-GHz downlink
           5.925–6.425-GHz uplink
Extended: 3.625–4.200-GHz downlink
           5.850–6.425-GHz uplink

Bandwidth: The transmitter shall be able to transmit one or more RF carriers anywhere in the specified 6-Gz band; the receiver shall be able to receive one or more RF carriers anywhere in the specified 4-GHz band.

Table 7.18 gives earth station polarization requirements in the 6/4-GHz bands for INTELSAT V, VA, VII, VII, and VIIA satellites. The applicable standard is IESS-207.

TABLE 7.18 Earth Station Polarization Requirements to Operate with INTELSAT V, VA, VI, VII, and VIIA (6/4 GHz)

Coverage	INTELSAT V		INTELSAT VA/VA (IBS)		INTELSAT VI		INTELSAT VII/VIIA	
	Earth Station Transmit	Earth Station Receive	Earth Station Transmit	Earth Station Receive	Earth Station Transmit	Earth Station Receive	Earth Station Transmit	Earth Station Receive
1. Global A	LHCP	RHCP	LHCP	RHCP	LHCP	RHCP	LHCP	RHCP
2. Global B	N/A	N/A	RHCP	LHCP	RHCP	LHCP	RHCP	LHCP
3. Western Hemisphere (Hemi 1)[a,c]	LHCP	RHCP	LHCP	RHCP	LHCP	RHCP	LHCP	RHCP
4. Eastern Hemisphere (Hemi 2)[a,c]	LHCP	RHCP	LHCP	RHCP	LHCP	RHCP	LHCP	RHCP
5. NW Zone (Z1)[b] ZA[c]	RHCP	LHCP	RHCP	LHCP	RHCP	LHCP	RHCP	LHCP
6. NE Zone (Z3)[b] ZB[c]	RHCP	LHCP	RHCP	LHCP	RHCP	LHCP	RHCP	LHCP
7. SW Zone (Z2)[b] ZC[c]	N/A	N/A	N/A	N/A	RHCP	LHCP	RHCP	LHCP
8. SE Zone (Z4)[b] ZD[c]	N/A	N/A	N/A	N/A	RHCP	LHCP	RHCP	LHCP
9. C-Spot A	N/A	N/A	N/A	RHCP	N/A	N/A	LHCP	RHCP
10. C-Spot B	N/A	N/A	N/A	LHCP	N/A	N/A	RHCP	LHCP

[a]Hemi 1, Hemi 2, ZA, ZB, ZC, ZD nomenclature applies to INTELSAT VII and VIIA only.
[b]Z1, Z2, Z3, Z4 nomenclature applies to INTELSAT VI only.
[c]This indicates the normal mode of operation for INTELSAT VII and VIIA; the inverted mode implies different beams in the East and West, as illustrated in IESS-409.
Note: LHCP = left-hand circularly polarized
      RHCP = Right-hand circularly polarized
      N/A = Not applicable to this spacecraft
Source: Reference 25, Table 2(a), IESS-207.

7.11.3   Standard B—Antenna and Wideband
RF Performance Characteristics

Frequency band: 6/4 GHz

$$G/T \geq 31.7 + 20\log(f/4)\ dB/K \qquad (7.33)$$

where f is the receive frequency in gigahertz.

For transmitting and receiving bandwidths, refer to Table 7.21.
The polarization requirements are shown in Table 7.18.
The applicable standard is IESS-207 (Ref. 25).

7.11.4   Standard C—Antenna and Wideband
RF Performance Characteristics

Frequency band: 14/12 and 14/11 GHz

$$G/T \geq 37.0 + 20\log(f/11.2) \quad dB/K$$

where f is the frequency in gigahertz. The G/T value from the above equation applies only to clear sky conditions. For degraded weather conditions (i.e., heavy rain storms; see Chapter 8), the following G/T requirement should be met:

$$G/T \geq 37.0 + 2\log\left(\frac{f}{11.2}\right) + X\ dB \quad (dB/K) \qquad (7.34)$$

where X is the excess (attenuation) of the downlink degradation predicted by local rain statistics over the reference downlink degradation margin given in Table 7.19 for the same time percentage. Downlink degradation is defined in the sum of the precipitation attenuation in decibels and the increase in receiving system noise temperature in decibels for the given time percentage.

Minimum earth station transmitting and receiving bandwidths are given in Table 7.20.

The applicable standard is IESS-208 (Ref. 26).

7.11.5   Standard D—Antenna and Wideband
RF Performance Characteristics

Frequency band: 6/4 GHz

There are two versions of the Standard D earth station: D-1 and D-2.

$$G/T \geq 22.7 + 20\log(f/4) \quad dB/K \quad D\text{-}1 \qquad (7.35)$$

$$G/T \geq 31.7 + 20\log(f/4) \quad dB/K \quad D\text{-}2 \qquad (7.36)$$

where f is the receiving frequency in gigahertz.

TABLE 7.19   Reference Downlink Degradation Margin for G/T Determination (INTELSAT, V, VA, VA(IBS) and VI)

Satellite Orbital Location (°E)	Earth Station Location in Satellite Beam	Reference Downlink Degradation Margin (dB)	Percentage of Year Margin Can Be Exceeded (%)
325.5 to 341.5	West spot	13	0.03
	East spot	11	0.01
174 to 180	West spot	13	0.03
	East spot	11	0.01
307 to 310	West spot	13	0.02
	East spot	11	0.02

Note: Margins apply to clear sky conditions.
Source: Reference 26, Table 1, IESS-208.

Polarization requirements:

Polarization A: uplink LHCP; downlink, RHCP

Polarization B: uplink, RHCP; downlink, LHCP

where LHCP = left-hand circularly polarized and RHCP = right-hand circularly polarized. Simultaneous operation in both polarization senses is not normally required.

The applicable standard is IESS-207 (Ref. 25).

7.11.6   Standard E—Antenna and Wideband RF Performance Characteristics

Frequency bands: 14/12 and 14/11 GHz

There are three versions of the Standard E earth station: E-1, E-2, and E-3.

$$G/T \geq 25.0 \text{ (but } < 29.0) + 20\log(f/11) \quad \text{dB/K} \quad \text{E-1} \quad (7.37a)$$

$$G/T \geq 29.0 \text{ (but } < 34.0) + 20\log(f/11) \quad \text{dB/K} \quad \text{E-2} \quad (7.37b)$$

$$G/T \geq 34.0 + 20\log(F/11) \quad \text{dB/K} \quad \text{E-3} \quad (7.37c)$$

where f is the frequency in gigahertz. The G/T values shown above are for clear sky conditions. For degraded conditions (i.e., heavy rainfall), the following G/T values are used:

$$G/T \geq 25.0 + 20\log(f/11) + X \quad \text{dB/K} \quad \text{E-1} \quad (7.38a)$$

$$G/T \geq 29.0 + 20\log(f/11) + X \quad \text{dB/K} \quad \text{E-2} \quad (7.38b)$$

$$G/T \geq 34.0 + 20\log(f/11) + X \quad \text{dB/K} \quad \text{E-3} \quad (7.38c)$$

TABLE 7.20   Minimum Bandwidth Requirements for Standard C and E Earth Stations

Satellite	ITU Region (Note 8)	Earth Station Transmit Frequency (GHz)	Earth Station Receive Frequency (GHz)
V, VA	All	14.00–14.50	10.95–11.20 and 11.45–11.70
VA(IBS) (Notes 3 and 7)	All	14.00–14.50	10.95–11.20 and 11.45–11.70
	2 (Note 4)	14.00–14.25 (Note 5)	11.70–11.95
	1 & 3 (Note 4)	14.00–14.25 (Note 5)	12.50–12.75
VI	All	14.00–14.50	10.95–11.20 and 11.45–11.70
VII, VIIA, VIII (Note 7)	All	14.00–14.50	10.95–11.20 and 11.45–11.70
	2 (Notes 4 and 6)	14.00–14.25	11.70–11.95
	1 & 3 (Note 4)	14.00–14.25	12.50–12.75
VIIIA (806)	2	14.00–14.25	11.70–11.95
VIIIA (805)	1 & 3	14.00–14.25	12.50–12.75
K	1 & 2	14.25–14.50	11.45–11.70
	1	14.00–14.25	12.50–12.75
	2	14.00–14.25	11.70–11.95

Note 1. Users are referred to the INTELSAT IESS-400 series modules for details of the channelization of the various INTELSAT spacecraft.

Note 2. For Standard C, E-2, and E-3 earth stations, the minimum bandwidth requirements apply to the antenna feed elements and RF electronics utilizing the 14/11-GHz and 14/12-GHz bands. These requirements also apply to the RF electronics and antenna feed elements of Standard E-1 earth stations using the 14/12-GHz band. For Standard E-1 earth stations using the 14/11-GHz band, these requirements apply to the antenna feed elements only (for the minimum bandwidth requirements of the RF electronics of Standard E-1 earth stations, see paragraph 2.2.1(b) of reference document).

Note 3. If INTELSAT VA(IBS) satellites are configured to operate at 11 GHz, the earth station receive frequency band will be 10.95–11.70 GHz.

Note 4. In INTELSAT VA(IBS) and INTELSAT VII, the receive band segments of 11.70–11.95 GHz and 12.50–12.75 GHz are interchangeable between the East and West spot beams, so that these spacecraft series can be operated in any ocean region.

Note 5. Earth stations intending to operate in the Spot-to-Hemi multibeam mode available in transponder 7-8 on INTELSAT VA (IBS) should have this transmit bandwidth extended to 14.35 GHz.

Note 6. Earth station users should consider in their design the possibility of extending their usable bandwidth to 14.35 GHz in the transmit band and to 11.45 GHz in the receive band.

Note 7. Consideration should be given to designing the RF system with a receive bandwidth of 10.95–12.75 GHz and a transmit bandwidth of 14.0–14.5 GHz. This will simplify conversion from the 11-GHz band to the 12-GHz band and provide maximum flexibility for operation with any spacecraft series.

Note 8. The world is divided into three ITU Regions separated by meridians. See ITU Radio Regulations (Ref. 4) for a more detailed description.

Source: Reference 26, Table 7.20, IESS-208.

where X is the excess of the downlink degradation predicted by local rain statistics over the reference downlink degradation margin shown in Table 7.19.

The minimum bandwidth requirements for Standard E earth stations are given in Table 7.20.

Polarization is linear and the transmit beam is orthogonal to the receive beam.

The applicable standard is IESS-208 (Ref. 26).

7.11.7   Standard F—Antenna and Wideband RF Performance Characteristics

Frequency band: 6/4 GHz

There are three versions of the Standard F earth station: F-1, F-2, and F-3.

$$G/T \geq 22.7 + 20\log(f/4) \quad \text{dB/K} \quad \text{F-1} \qquad (7.39a)$$

$$G/T \geq 27.0 + 20\log(f/4) \quad \text{dB/K} \quad \text{F-2} \qquad (7.39b)$$

$$G/T \geq 29.0 + 20\log(f/4) \quad \text{dB/K} \quad \text{F-3} \qquad (7.39c)$$

where f is the frequency in gigahertz.

Polarization: An F-type earth station shall be capable of LHCP (heft-hand circular polariation) and RHCP (right-hand circular polarization) on the transmit and receive beams. However, this is not required simultaneously on the uplink nor on the downlink. Polarization requirements are shown in Table 7.18.

The minimum bandwidth requirements for the F-type earth station are shown in Table 7.21; the minimum tracking requirements are shown in Table 7.22.

The applicable standard is IESS-207 (Ref. 25).

7.12   SATELLITE EARTH STATIONS (TERMINAL SEGMENT)

7.12.1   Functional Operation of a Standard Earth Station

7.12.1.1   The Communication Subsystem. Figure 7.28 is a simplified functional block diagram of an earth station showing the communication subsystem only. We shall use this figure to trace a signal through the equipment chain from antenna to baseband. Typically, large earth stations transmit and receive more RF carriers than shown in the figure. By "standard" we can assume typically INTELSAT A, B, or C service or high-

TABLE 7.21   Minimum Bandwidth Requirements for Standard A, B, D, and F Earth Stations

Earth Station Standard	Satellite	Earth Station Type (Note 1)	Earth Station Transmit Band (MHz)	Earth Station Receive Band (MHz)	Tx & Rcv Bandwidth (MHz)	Notes
A, B, D-2	All	Existing	5925–6425	3700–4200	500	
	V, VA, VA(IBS), VII, VIIA	New	5925–6425	3700–4200	500	
	VI, VIII	New	5850–6425	3625–4200	575	2 and 3
	VIIIA	New	5850–6650	3400–4200	800	2 and 3
D-1	All	Existing and new	5925–6425	3700–4200	500	
F-1, F-2, F-3	All	Existing	5925–6256 (band segment 1)	3700–4031 (band segment 1)	331	4
			or			
			6094–6425 (band segment 2)	3869–4200 (band segment 2)	331	4
	All	New	5925–6425	3700–4200	500	

Note 1. Existing earth stations are defined to be those with a RFP, or similar contractual document, dated prior to 1996. New earth stations are those with a RFP dated after 1995.

Note 2. Users are encouraged to utilize the extended transmit and receive bandwidth available in the INTELSAT VI, VIII, and VIIIA hemispheric beams. Users intending to buy new earth stations or antennas should consult with INTELSAT in the event they have any difficulty in utilizing these additional bandwidth segments. The extended bandwidths are as follows:

> INTELSAT VI and VII   Receive: 3625–4200 MHz   Transmit: 5850–6425 MHz
> INTELSAT VIIIA   Receive: 3400–4200 MHz   Transmit: 5850–6650 MHz

Note 3. Administrations not permitting operation across the full extent of the (5850–6650)/(3400–4200) MHz bands are only required to equip for operation in the (5925–6425)/(3700–4200) MHz bands.

Note 4. Existing Standard F earth stations are required to be capable of operating over the full extent of one of the 331-MHz bandwidth segments listed.

Source: Reference 25, Table 4, IESS-207.

traffic capacity regional/national domestic satellite service. A VSAT would be "nonstandard."

The operation of an earth station communication subsystem in the FDMA mode really varies little from that of a LOS radiolink system. The variances are essentially these:

- Universal use of low-noise front ends on receiving systems; GaAs or PHEMT-based LNAs.
- HPA with capability of 200–8000-watt output. The higher level output is used for standard broadcast quality TV transmission.
- Larger high-efficiency antennas and feeds.
- Careful design to achieve as low a noise as possible.
- Use of forward error correction (FEC) on many digital systems and, above 10 GHz, possibly with interleaving to mitigate rainfall fading.

TABLE 7.22   Minimum Tracking Requirements for 6/4-GHz Earth Stations

Earth Station Standard	V, VA, and VA(IBS)	VI, VII, VIIA, VIII, and VIIIA (Note 2)
A (Note 3)	Manual and autotrack	Manual and autotrack
B (Note 3)	Manual and autotrack	Manual and autotrack
D-2 (Notes 3, 6, 10)	Manual and autotrack	Manual and autotrack
D-1 (Note 4)	"Fixed" antenna (Note 5)	"Fixed" antenna
F-3 (Note 7)	Autotrack	"Fixed" antenna (Notes 4, 8, 9)
F-2 (Note 7)	Manual E/W only (Note 5) (weekly peaking)	"Fixed" antenna (Note 4)
F-1 (Notes 4, 7)	"Fixed" antenna (Note 5)	"Fixed" antenna

Note 1. The minimum tracking requirements are subject to the earth station meeting the axial ratio requirements of paragraph 1.3.2 of the reference document.

Note 2. Antenna tracking requirements are based on provisional INTELSAT VIIA, VIII, and VIIIA station-keeping tolerances. These tolerances will be reviewed after operational experience is gained with these new series of spacecraft.

Note 3. Users are urged to include in their designs a provision to add program steering.

Note 4. "Fixed" antenna amounts will still require the capability to be steered from one satellite position to another, as dictated by operational requirements (typically once or twice every 2–3 years). These antennas should also be capable of being steered at least over a range of $\pm 5°$ from beam center for the purpose of verifying that the antenna pointing is correctly set toward the satellite, and for providing a means of verifying the sidelobe characteristics in the range.

Note 5. Standard D-1, F-1, and F-2 users should consider the possible need to upgrade the earth station with manual or autotrack systems in the event it becomes necessary to operate with satellites having higher than nominal inclination.

Note 6. The use of autotracking is recommended for Standard D-2.

Note 7. These minimum requirements apply to Standard F earth stations that either transmit or transmit and receive.

Note 8. Standard F-3 users are encouraged to consider autotrack designs. Earth stations using fixed antennas must meet all specifications of the reference document irrespective of the satellite position within the box defined by nominal station-keeping limits.

Note 9. Standard F-3 users should take into consideration their tracking requirements under contingency operation with another satellite series.

Note 10. The tracking requirements shown for Standard D-2 earth stations is only mandatory for earth stations with a RFP issued after 1994. (Prior to the release of INTELSAT IESS-207 autotrack was only recommended).

Source: Reference 25, Table 3, IESS-207.

Of course, another vital difference is that there is no fading on satellite links, with the following exceptions:

- Operating below the minimum elevation angle.
- Rainfall fading. This is generally not an issue below 10 GHz.

Now let us trace a signal through the communication subsystem typical of Figure 7.28. We will assume the signal is digital, E-1 or DS1 based. From the baseband patch panel, the signal is fed to the modulator. In the modulator, the signal is converted to a NRZ waveform from conventional AMI or HDB3 and then fed to a FEC coder, thence through an interleaver (optional).

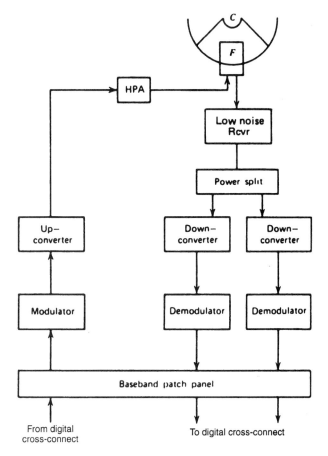

Figure 7.28. Simplified functional block diagram of an earth station communication subsystem. F = feed; HPA = high-power amplifier; C = Cassegrain subreflector.

The digital modulation is commonly QPSK. Thus the next step is to split the digital bit stream into two: one is fed to the in-phase channel and one is fed to the quadrature channel. This is the same approach as in Chapter 5 in digital LOS microwave. The signals (there are two) are then modulated in a QPSK modulator that can be viewed as two BPSK modulators. The quadrature modulator lags the in-phase by 90°. Thus the modulator output is just one IF carrier. This output is filtered, amplified, and up-converted to the operational frequency. The output of the upconverter is fed to the HPA, filtered again, and fed to the antenna subsystem where it is radiated.

Transmit and receive IFs are trending more from the conventional 70 MHz to 140 MHz to accommodate the broader bandwidth transponders in the space segment.

On the receive side, the signal derives from the antenna feed and is fed to a low-noise receiver or LNA. In the case of an INTELSAT Standard A earth station, the LNA looks at the entire 3700–4200 MHz (or, in some cases, at the even wider bandwidth permitted; see Table 7.21). The LNA amplifies the receive signal 20–40 dB. The more gain it imparts to the receive signal, the less noise subsequent signal stages contribute to the total receive system noise temperature. LNAs use either GaAs or PHEMT technology, where we can expect noise figures in the range of 0.4 dB at 4 GHz and 0.7 dB at 11–12 GHz. G/T requirements dictate the investment in low-noise operation.

When there are long waveguide runs from the antenna to the equipment building, the signal may be amplified still further by a low-level TWT or solid-state amplifier (SSA). The LNA should be placed as close as possible to the antenna to reduce ohmic loss in front of the LNA. This is vital to higher-frequency receive systems (e.g., 11–12 GHz and 20 GHz) where waveguide loss is considerably greater per unit length than for C-band systems.

The comparatively high-level broadband receive signal is then fed to a power split. There is one output from the power split for every downconverter–demodulator chain. In addition, there is often a test receiver available as well as one or several redundant receivers in case of failure of an operational receiver chain. It should be kept in mind that every time the broadband incoming signal is split into two equal-level paths, there is a 3-dB loss due to the split, plus an insertion loss of the splitter. A splitter with eight outputs will incur a loss of something on the order of 10 dB.

A downconverter is required for each receive carrier, and there will be at least one receive carrier from each distant end. Each downconverter is tuned to its appropriate carrier and converts the signal to the 70- or 140-MHz IF. In some instances dual conversion is used.

The 70- or 140-MHz IF is then fed to the demodulator on each receive chain developing an I and a Q bit stream that are combined to form a normal serial bit stream of NRZ data. This bit stream is then deinterleaved and decoded and the HRZ data are converted to the appropriate PCM line signal such as AMI, BNZS, or HDB3.

7.12.1.2 The Antenna Subsystem. The antenna subsystem is one of the most important component parts of an earth station, since it provides the means of radiating signal to the satellite and/or collecting signals from the satellite. The antenna not only must provide the gain necessary to allow proper transmission and reception, but also must have the radiation characteristics that discriminate against unwanted signals and minimize interference into other satellite and/or terrestrial systems.

Earth stations most commonly use parabolic reflector antennas or derivatives thereof. Dish diameters range from 1 to 25 m.

The sizing of the antenna and its design are driven more by the earth station required G/T than the EIRP. The gain is basically determined by the

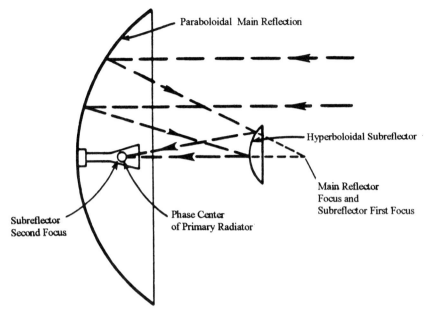

Figure 7.29. Cassegrain antenna functional operation.

aperture (e.g., diameter) of the dish, but improved efficiency can also add to the gain on the order of 0.5–2 dB. For this reason the Cassegrain feed technique is almost always used on larger earth terminal installations. Smaller military terminals also resort to the use of Cassegrain. In some cases efficiencies as high as 70% and more have been reported. The Cassegrain feed working into a parabolic dish configuration (Figure 7.29) permits the feed to look into cool space as far as the spillover from the subreflector is concerned.

Antennas of generally less than 30-ft diameter often use a front-mounted feed horn assembly in the interest of economy. The most common type is the prime focus feed antenna (Figure 7.30). This is a more lossy arrangement, but, since the overall requirements are more modest, it is an acceptable one.

In order to keep interference levels on both the uplink and downlink to acceptable levels, antenna sidelobe envelope limits (in dB) of $32\text{-}25 \log \theta$ relative to the main beam maximum level have been internationally adopted; θ is the angular distance in degrees from the main beam lobe maximum.

Figure 7.29 shows the functional operation of a Cassegrain-fed antenna. Such an antenna consists of a parabolic main (prime) reflector and a hyperbolic subreflector. Here, of course, we refer to truncated parabolic and hyperbolic surfaces. The subreflector is positioned between the focal point and the vertex of the prime reflector. The feed system is situated at the focus of the subreflector, which also determines the focal length of the system.

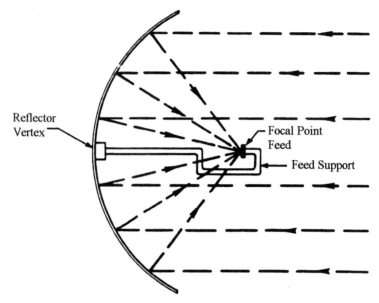

Figure 7.30. Prime focus antenna functional operation.

Spherical waves emanating from the feed are reflected by the subreflector. The wave then appears to be emanating from the virtual focus. These waves are then, in turn, reflected by the primary reflector into a plane wave traveling in the direction of the axis of symmetry. The size of the aperture (diameter) of the prime reflector determines the gain.

The gain of a uniformly illuminated parabolic dish antenna, which has a diameter of D ft operating at a frequency of F MHz, can be expressed by

$$G_{dB} = 20 \log F_{MHz} + 20 \log D_{ft} + 10 \log \eta - 49.92 \text{ dB} \qquad (7.40)$$

where η is the aperture efficiency expressed as a percentage (a decimal). The value of η is usually in the range of 0.50–0.70.

It can be shown that a uniform field distribution over the reflector gives the highest gain of all constant-phase distributions. The aperture efficiency can be shown to be a function of:

- Phase error loss
- Illumination loss due to a nonuniform amplitude distribution over the aperture
- Spillover loss
- Cross-polarization loss
- Blockage loss due to the feed, struts, and subreflector
- Random errors over the surface of the reflector (e.g., surface tolerance)

Nearly all very high gain reflector antennas are of the Cassegrain type. Within the main beam the antenna behaves essentially like a long-focal-length front-fed parabolic reflector. Slight shaping of the two reflector surfaces can lead to substantial gain enhancement. Such a design also leads to more uniform illumination of the main reflector and less spillover. Typically, efficiencies of Cassegrain-type antennas are from 65% to 70%, which is at least 10% above most front-fed designs. It also permits the LNA to be placed close to the feed, if desired. The ability of the antenna to achieve these characteristics rests largely with the feed horn design. The feed horn radiation pattern has to provide uniform aperture illumination and proper tapering at the edges of the aperture.

The simplest type of feed for the antenna is a waveguide, which can be either open ended or terminated with a horn. Both rectangular and circular waveguides have been used, with the circular considered superior, since it produces a more uniform illumination pattern over the aperture and provides better cross-polarization loss characteristics. The illumination pattern should taper to a value of -10 dB at an angle corresponding to the edge of the reflector (subreflector). This results in an asymmetric feed radiation pattern, causing a loss in efficiency, increased cross-polarization losses, and moderate reflector spillover. Cross-polarization loss can be reduced by careful selection of waveguide radius, while improvement in efficiency can be achieved by using a corrugated horn. The positive aspect to 10-dB taper is to reduce sidelobe levels.

Smaller earth stations with less stringent G/T requirements often resort to using the less expensive prime-focus-fed parabolic reflector antenna. The functional operation of this antenna is shown in Figure 7.30. For intermediate aperture sizes, this type of antenna has excellent sidelobe performance in all angular regions except the spillover region around the edge of the reflector. Even in this area a sidelobe suppression can be achieved that will satisfy FCC/CCIR pattern requirements. The aperture efficiency for apertures greater than about 100 wavelengths is around 60%. Hence it represents a good compromise choice between sidelobes and gain. For aperture sizes less than about 40 wavelengths, the blockage of the feed and feed support structure raises sidelobes with respect to the peak of the main beam such that it becomes exceedingly difficult to meet the FCC/CCIR sidelobe specification. However, the CCIR specification can be met, since it contains a modifier that is dependent on the aperture size.

Polarization. By use of suitable geometry in the design of an antenna feed horn assembly, it is possible to transmit a plane electric wavefront in which the E and H fields have a well-defined orientation. For linear polarization of the wavefront, the convention of vertical or horizontal electric (E) field is adopted.The generation of linearly polarized signals is based on the ability of a length of square section waveguide to propagate a field in the $TE_{1,0}$ and $TE_{0,1}$ modes, which are orthogonally oriented. By exciting a short length of square waveguide in one mode by the transmitted signal and extracting signals in the orthogonal mode for the receiver, a means is provided for cross-polarizing the transmitted and received signals. The square waveguide is flared to form the antenna feed horn. Similarly, by exciting orthogonal modes in a section of circular waveguide, a left- and right-hand rotation is imparted to the wavefront, providing two orthogonal circular polarizations.

The waveguide assembly used to obtain this dual polarization is known as a diplexer or an "orthomode junction." In addition to generating the polarization effect, the diplexer has the advantage of providing isolation between the transmitter and the receiver ports that could be in the range of 50 dB if the antenna presented a true broadband match. In practice, some 25–30 dB or more of isolation can be expected.

For linear polarization, discrimination between received horizontal and vertical fields can be as high as 50 dB, but diurnal effects coupled with precipitation can reduce this value to 30 dB. To maintain optimum discrimination, large antennas are equipped with a feed-rotating device driven by a polarization-sensing servo loop. Circularly polarized fields do not provide much more than 30-dB discrimination, but they tend to be more stable and do not require polarity tracking. This makes circular polarization more suitable for systems that include mobile terminals. One of the important parameters of antenna performance is how well cross-polarization is preserved across the operating spectrum.

Polarization discrimination can be used to obtain frequency reuse. Alternate transponder channel spectra are allowed to overlap symmetrically and are provided with alternate polarization. The channelization schemes of INTELSAT V and VI are typical. The number of transponders in the satellite is effectively doubled by this process. Polarization discrimination can also provide a certain degree of interface isolation between satellite networks if the nearest satellite (in terms of angular orbit separation) uses orthogonal uplink and downlink polarizations (Ref. 17).

Antenna Pointing and Tracking. Satellites orbiting the earth are in motion. They can be in geostationary orbits and inclined orbits. Those in geostationary orbits appear to be stationary with respect to a point on earth. Those that are in inclined orbits are in motion with respect to a point on earth. All satellite terminals working with this latter class of satellite require a tracking capability.

Even though we have said that geostationary satellites appear stationary relative to a point on earth, they do tend to drift in small suborbits (figure eights). However, even with improved satellite station-keeping, the narrow beamwidths encountered with large earth station antennas, such as the INTELSAT Standard A 15-m (50-ft), require precise pointing and subsequent tracking by the earth station antenna to maximize the signal to the satellite and from the satellite. The basic modes of operation to provide these capabilities are:

- Manual pointing
- Programmed tracking (open-loop tracking)
- Automatic tracking (closed-loop tracking)

Pointing deals with "aiming" the antenna initially on the satellite. Tracking keeps it that way. Programmed tracking (open-loop tracking) may assume both duties. With programmed tracking, the antenna is continuously pointed by interpolation between values of a precomputed time-indexed ephemeris. With adequate information as to the actual satellite position and true satellite terminal position, pointing resolutions are on the order of 0.03–0.05°.

Manual pointing may be effective for initial satellite acquisition or "capture" for later active tracking (closed-loop tracking). It is also effective for wider beamwidths antennas, where the beamwidth is sufficiently wide to accommodate the entire geostationary satellite suborbit. Midsized installations may require a periodic trim up, and some smaller installations need never be trimmed up, assuming, of course, that the satellite in question is keeping good station-keeping.

There are three types of active or closed-loop tracking: monopulse, conscan, and step-track. Monopulse tracking is the most accurate for automatic tracking. It is expensive and will be found in some military installations and on TT & C (telemetry, tracking, and command) facilities. Conscan, which derives from WWII radar technology, is popular on some military systems.

Of the three automatic tracking techniques, step-track is the simplest and most economical. It is cost effective where there are low dynamic tracking requirements. With today's excellent station-keeping of geostationary satellites, step-tracking can even be used on large installations, such as INTELSAT Standard A. Step-tracking control is based on the automatic gain control (AGC) voltage or other dc signal proportion to the receive RF signal level, such as a signal level indication from a communications demodulator or from a beacon receiver. The output of the step-track processor algorithm can be as simple as a periodic step function or as complex as a pseudorandom sinusoid. This type of output applied to the antenna servo drive subsystem results in smooth, continuous antenna motion. Figure 7.31 is a simplified functional block diagram of a step-track system.

With the step-track techniques, the antenna is periodically moved a small amount along each axis, and the level of the received signal is compared to its previous level. A microprocessor, or part of the terminal control processor, provides processing to convert these level comparisons into input signals for the antenna servo system, which will drive the antenna in directions that maximize the received signal level.

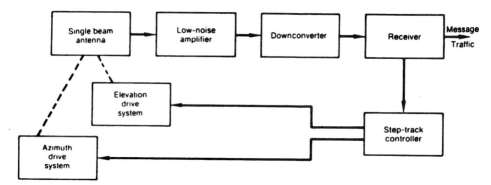

Figure 7.31. Functional block diagram of a step-track system.

7.12.2 VSAT System Terminals

In Section 7.9.2 VSAT networks are described as specialized networks principally used for data traffic. A typical network consists of a hub and a series of small outstation satellite terminals, which are called VSAT terminals.

Small VSATs have parabolic dish antennas with diameters ranging from 0.5 m for a very modest installation to up to 2.4 m for a much higher capacity installation. Hubs are much larger facilities to compensate somewhat for the small disadvantaged VSATs. One hub may serve more than 2000 VSAT outstations. Hub antennas range from 3 to 10 m in diameter. Their HPAs range from 20 to 200 watts of output power for 6/4-GHz operation and 3 to 100 watts for 14/12-GHz operation.

VSAT outstations are optimized for cost because there are so many, in one system over 2000. Thus, in that system, the cost multiplier is 2000. For example, a variance in cost of $1000 multiplied by 2000 would become an increment in cost of $2 million.

Each VSAT and hub consists of an indoor unit and an outdoor unit. A block diagram of a typical hub is shown in Figure 7.32 and of a typical VSAT in Figure 7.33. For the VSAT the outdoor unit is miniaturized and jammed up into the parabolic reflector shell to make waveguide runs as short as possible—1 or 2 inches. This reduces loss and noise due to ohmic noise; it also can save money.

Hubs with larger diameter antennas will require tracking. Step-tracking is a cost-effective alternative. The tracking requirement will depend largely on the station-keeping capabilities of the satellite accessed.

Figure 7.32 shows a hub terminal configured for a very widely used application, that is, to interexchange data messages with VSAT outstations. The outdoor unit will probably be located in the antenna pedestal or in an adjacent shelter; the indoor unit in a standard rack. The terminals in Figures

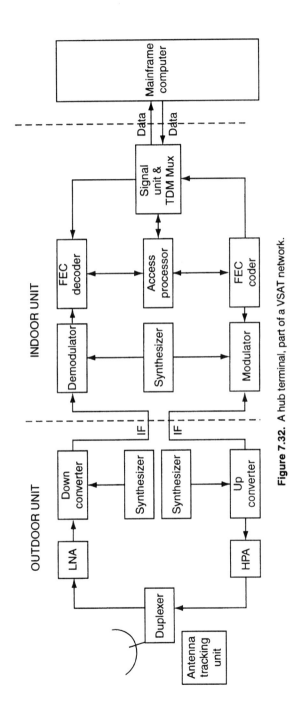

Figure 7.32. A hub terminal, part of a VSAT network.

533

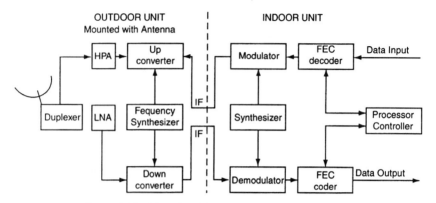

Figure 7.33. A functional block diagram of a typical VSAT.

7.32 and 7.33 are configured for the outbound link as a single-thread high-rate TDM bit stream. The inbound link may use any one of the several access protocols, including DAMA, as described in Section 7.9.4.

REVIEW EXERCISES

1. Give at least six major applications of satellite communication.

2. What transmission technique/medium is now tempering the growth of satellite communication?

3. What are the two broad generic categories of communication satellites? (*Hints*: Consider function. One is fully mature and the second is just beginning to emerge.)

4. Draw a simplified block diagram of a "bent pipe" communication satellite (transponder).

5. The text listed three types of satellite orbits. Which is almost universally used by Western nations? Give at least two advantages of using this type of orbit; give one disadvantage.

6. Give one reason why satellite downlinks are always in a lower frequency band than their companion uplinks.

7. Calculate the EIRP in dBW of an earth terminal with an antenna with a gain of 40 dB, transmitter output of 200 W, and transmission line losses of 2.3 dB.

8. What is the free-space loss on a 12-GHz downlink for a geostationary satellite where the earth terminal uses a 23° elevation angle?

9. What is the isotropic receive level (IRL) at an earth terminal antenna where the free-space loss to the associated satellite is 196.4 dB and other link losses are 2.6 dB? The satellite EIRP is +34 dBW.

10. Why must the flux density of a satellite signal at the earth's surface be limited?

11. What is the receiving system noise temperature (T_{sys}) if the antenna noise temperature is 300 K and the receiver noise temperature is 100 K?

12. Why is receiving system noise temperature so important in satellite communication, when it was a very secondary issue in the design of LOS microwave systems?

13. The noise figure of a low-noise amplifier (LNA) is 1.2 dB. What is its equivalent noise temperature in kelvins?

14. List at least three contributors to receiving system noise temperature.

15. Calculate N_0 of a receiving system with a noise temperature of 100 K.

16. Determine C/N_0 at an earth terminal of a satellite downlink where the RSL is -139 dBW and the system noise temperature is 400 K.

17. State G/T as a mathematical identity (equation). Where would we measure G in dB?

18. Calculate the antenna noise temperature (T_a) for an earth terminal receiving system where the sum of the ohmic losses from the antenna feed to the input of the LNA is 1.7 dB. The sky noise contribution is 15 K.

19. Determine G/T for a satellite downlink receiving system where the antenna gain is 42 dB, ohmic losses from the antenna feed to the input to the LNA are 2.1 dB, sky noise is 12 K, and the LNA noise figure is 1.5 dB and has 30 dB of gain.

20. Consider exercise 19. Why, especially in this case, can we generally neglect noise contributions from receiver stages that follow the LNA? Suppose the LNA only had a 10-dB gain. Explain why, in this case, where latter stages may make some considerable noise contribution to the system noise temperature, we now must take these contributions into consideration.

21. Name at least four parameters of an earth terminal that will affect G/T.

22. Besides free-space loss, name at least four other loss contributors to a link budget.

23. Calculate the required geostationary satellite G/T where the uplink frequency is 14 GHz, the terminal EIRP is +76 dBW, and the terminal antenna elevation angle is 10°. The required C/N_0 at the satellite is 90

dB. Allow 6 dB for excess attenuation due to rainfall (margin), terminal pointing loss of 0.2 dB, satellite pointing loss of 0.5 dB, polarization loss of 0.5 dB, and atmospheric absorption loss of 0.5 dB. (Use Figure 7.3 in the text to compute range to satellite.)

24. Determine system C/N_0 where the uplink C/N_0 is 90 dB and the downlink C/N_0 is 85 dB.

25. What is the major drawback to the use of TWT-based high-power amplifiers in a satellite using multicarrier operation?

26. List the three basic access techniques of a satellite transponder (SCPC does not count).

27. Why is uplink power control necessary for multicarrier FDMA operation?

28. Give at least two important major applications of a SCPC/DAMA system.

29. List three methods of control for a DAMA system.

30. Compare TDMA and FDMA. List at least three advantages and three disadvantages of each.

31. Why must TDMA operate exclusively in a digital mode?

32. Discuss TDMA guard time between bursts. Why is it desirable to shorten the guard time as much as possible and how does this impact total system cost?

33. What is the function of the CR/BTR in TDMA frame overhead?

34. Differentiate open-loop and closed-loop timing on TDMA systems.

35. What is the most primitive form of a processing satellite. What advantages accrue for the system? What functions are required that are not required in a bent pipe satellite?

36. List at least four advantages of a processing satellite.

37. What are the two transmission media that are candidates for satellite cross-links?

38. The most common type of VSAT network is what type of network? What other (synonymous) name is used to describe such a network?

39. For common VSAT operation, where is a common (and desirable) place to locate the hub?

40. Give an overriding reason why a hub facility is so much larger than each VSAT it serves.

41. If a VSAT network is to provide voice operation among VSATs, what type of architecture is advisable to use? Why?

42. For typical hub and spoke VSAT networks, there is one-way operation and two-way operation. Give at least three applications of one-way operation, and give at least five applications for two-way operation VSAT networks.

43. In the case of conventional VSAT networks, the outbound link is nearly always in what type of format?

44. Name at least four possible access methods applicable to the inbound link.

45. Conventional VSATs often have a transmitter with output power in the range of _____ (watts or dBW)? Antenna sizes range from what to what?

46. What factors and parameters basically determine the size of a VSAT?

47. Size a hub and its related VSATs for transmit power, receiver LNA, and antenna apertures. Use EIRP and G/T in the analysis. Outbound transmission rate is 56 kbps TDM; inbound bursty at 9600 bps. Apply reasonable BERs. Select the modulation type and FEC coding, if deemed advantageous (argue these advantages or disadvantages). Select the inbound access mode. Assign transponder bandwidth. Operate at Ku-band.

48. Why is Ku-band often more desirable than C-band operation for VSAT terminals? What is the principal disadvantage of Ku-band when compared to C-band?

49. Why can we eliminate increased free-space loss as a factor in exercise 48? Think! *Hint*: Increasing frequency affects more than free-space loss.

50. What is the principal driver in the selection of the inbound access technique? Name two other driving factors in that selection.

51. Describe how pure Aloha operates. When does it become unwieldy and why?

52. Compare pure Aloha to slotted Aloha.

53. Describe how demand assignment TDMA works.

54. Assume a processing satellite. Calculate the terminal G/T under the following conditions: digital downlink at 14 GHz; 10-Mbps information data rate; BER $= 1 \times 10^{-6}$; convolutional coding $R = \frac{1}{2}$; $K = 7$; soft-decision sequential decoder where $Q = 3$; elevation angle 19°; total margin 10 dB, including rainfall margin. Modulation is QPSK with a modulation implementation loss of 1.5 dB. Other miscellaneous link losses are 2.5 dB. Radome loss is 1.0 dB. The satellite EIRP is $+44$ dBW.

55. Draw a functional block diagram of a generic communications subsystem of an earth station. The earth station will be communicating with two distant ends, full-duplex.

56. Give the principal reason why we might use a Cassegrain antenna despite the additional cost.

57. Why must we pay attention to antenna sidelobe characteristics? There is one principal reason and an important secondary reason.

58. Antenna aperture efficiency is a function of _____ (provide at least four items).

59. Calculate the gain of a parabolic dish antenna with an aperature diameter of 4 m. The frequency is 12 GHz.

60. How much polarization isolation may we expect in practice using conventional satellite terminal antennas?

61. List at least six ways a satellite earth terminal differs from its LOS counterpart.

62. What equipment components might we expect to find in an outdoor unit of a VSAT?

63. When using Ku- or Ka-band,* why must we try to keep waveguide runs as short as possible?

REFERENCES

1. Roger L. Freeman, *Reference Manual for Telecommunication Engineering*, 2nd ed., Wiley, New York 1994.

2. Roger L. Freeman, *Radio System Design for Telecommunications*, 2nd ed., Wiley, New York, 1997.

3. "Maximum Permissible Values of Power Flux-Density at the Surface of the Earth Produced by Satellites in the Fixed-Satellite Service Using the Same Frequency Bands Above 1 GHz as Line-of-Sight Radio-Relay Systems," ITU-R Rec. SF.358-4, ITU-R SF Series, ITU, Geneva, 1993.

4. *Radio Regulations*, Edition of 1990, Revised 1994, ITU, Geneva, 1994.

5. *Transmission Systems for Communications*, 5th ed., Bell Telephone Laboratories, Holmdel, NJ, 1982.

6. "Radio Emission from Natural Sources in the Frequency Range Above About 50 MHz," CCIR Rep. 720-2, Annex to Vol. V, XVIIth Plenary Assembly, Dusseldorf, 1990.

7. *Satellite Communications Symposium*, Scientific-Atlanta, Atlanta, GA, 1982.

---

* Ka-band for satellite communication is the 30/20-GHz band.

8. "Performance Characteristics for Frequency Division Multiplex/Frequency Modulation (FDM/FM) Telephony Carriers," INTELSAT IESS-301, Rev. 3, INTELSAT, Washington, DC, May 1994.

9. R. C. Dixon, *Spread Spectrum Systems*, 3rd ed., Wiley, New York, 1994.

10. "Carrier Energy Dispersal for Systems Employing Angle Modulation by Analogue Signals or Digital Modulation in the Fixed Satellite Service," ITU-R Rec. S.446-4, 1994 S Series Volume, ITU, Geneva, 1994.

11. H. L. Van Trees, Section 3.6.5 in "Demand Assignment," *Satellite Communications*, IEEE Press, New York, 1979.

12. Roger L. Freeman, *Telecommunication System Engineering*, 3rd ed., Wiley, New York, 1996.

13. "Performance Characteristics for Intermediate Data Rate (IDR) Digital Carriers," INTELSAT IESS-308, Rev. 7A, INTELSAT, Washington, DC, Aug. 1994.

14. H. L. Van Trees, Section 3.6.3.1 in "Time Division Multiple Access," *Satellite Communications*, IEEE Press, New York, 1979.

15. V. K. Bargava et al., *Digital Communications by Satellite*, Wiley, New York, 1981.

16. K. Koga, T. Muratani, and A. Ogawa, "On-Board Regenerative Repeaters Applied to Digital Satellite Communications," *Proc. IEEE*, Mar. 1977.

17. N. R. Richards, *Satellite Reference Data Handbook*, Computer Science Corp., Falls Church, VA, 1988, under DCA contract DCA-100-81-C-0044.

18. "VSAT Systems and Earth Stations," Supplement 3 to *Handbook of Satellite Communications*, ITU–Radio Communications Bureau, Geneva, 1994.

19. Edwin B. Parker and Joseph Rinde, "Transaction Network Applications with User Premises Earth Stations," *IEEE Commun. Mag.*, vol. 26, no. 9, Sept. 1988.

20. Dattakumar M. Chitre and John McCoskey, "VSAT Networks: Architectures, Protocols and Management," *IEEE Commun. Mag.*, vol. 26, no. 7, July 1988.

21. D. Raychaudhuri and K. Joseph, "Channel Access Protocols for Ku-Band VSAT Networks: A Comparative Evaluation," *IEEE Commun. Mag.*, vol. 26, no. 5, May 1988.

22. D. Chakraborty, "VSAT Communication Networks: An Overview," *IEEE Commun. Mag.*, vol. 26, no. 5, May 1988.

23. Roger L. Freeman, *Practical Data Communications*, Wiley, New York, 1995.

24. Henry Jasik and Richard C. Johnson, *Antenna Engineering Handbook*, 2nd ed., McGraw-Hill, New York, 1984.

25. "Antenna and Wideband RF Performance Characteristics of C-band Earth Stations Accessing the INTELSAT Space Segment," INTELSAT IESS-207, Rev. 1, INTELSAT, Washington, DC, Aug. 1994.

26. "Antenna and Wideband RF Performance Characteristics of Ku-Band Earth Stations Accessing the INTELSAT Space Segment for Standard Services," INTELSAT IESS-208, Rev. 1, INTELSAT, Washington, DC, Nov. 1994.

8

RADIO SYSTEM DESIGN
ABOVE 10 GHz

8.1 THE PROBLEM—AN INTRODUCTION

There is an ever-increasing demand for radio-frequency spectrum in the industrialized nations of the world. This is due to the information transfer explosion in our society, resulting in a rapid increase in telecommunication connectivity, and the links satisfying that connectivity are required to have ever-greater capacity.

The most desirable spectrum to satisfy these needs is the band between 1 and 10 GHz. It is called the "noise window," where galactic and man-made noises are minimum. Atmospheric absorption may generally be neglected in this region.

Congestion in the 1–10-GHz region has forced us to look above 10 GHz for operational frequencies. By careful engineering we have found that frequencies above 10 GHz can give equivalent performance or nearly equivalent performance to those below 10 GHz.

We have arbitrarily selected 10 GHz as a demarcation line. Generally, below 10 GHz, in radiolink design, we can neglect excess attenuation due to rainfall and atmospheric absorption. For frequencies above 10 GHz, excess attenuation due to rainfall and atmospheric absorption can have an overriding importance in radiolink design. In fact, certain frequency bands display so much gaseous absorption that they are unusable for many applications.

The principal thrust of this chapter is to describe techniques for band selection and link design for line-of-sight (LOS) microwave and earth–space–earth links for frequencies above 10 GHz. The chapter also describes low-elevation-angle propagation effects on space links.

541

8.2   THE GENERAL PROPAGATION PROBLEM ABOVE 10 GHz

Propagation of radio waves through the atmosphere above 10 GHz involves not only free-space loss but several other important factors. As expressed in Ref. 1, these are:

1. The gaseous contribution of the homogeneous atmosphere due to resonant and nonresonant polarization mechanisms.
2. The contribution of inhomogeneities in the atmosphere.
3. The particular contributions due to rain, fog, mist, and haze (dust, smoke, and salt particles in the air).

Under item 1 we are dealing with the propagation of a wave through the atmosphere under the influence of several molecular resonances, such as water vapor (H_2O) at 22 and 183 GHz, oxygen with lines around 60 GHz, and a single oxygen line at 119 GHz. These points with their relative attenuation are shown in Figure 8.1.

Other gases display resonant lines as well, such as N_2O, SO_2, O_3, NO_2, and NH_3, but because of their low density in the atmosphere, they have negligible effect on propagation.

The major offender is precipitation attenuation (under items 2 and 3). It can exceed that of all other sources of attenuation in the atmosphere above 10 GHz. Rainfall and its effect on propagation are covered at length in this chapter.

We will first treat total loss due to absorption and scattering. It will be remembered that when an incident electromagnetic wave passes over an object that has dielectric properties different from the surrounding medium, some energy is absorbed and some is scattered. That which is absorbed heats the absorbing material; that which is scattered is quasi-isotropic and relates to the wavelength of the incident wave. The smaller the scatterer, the more isotropic it is in direction with respect to the wavelength of the incident energy.

We can develop a formula from equation 5.1 to calculate total transmission loss for a given link:

$$\text{Attenuation (dB)} = 92.45 + 20\log f_{GHz} + 20\log D_{km} + a + b + c + d + e$$

$$(8.1)$$

where f is in gigahertz and D is in kilometers. Also,

a = excess attenuation (dB) due to water vapor
b = excess attenuation (dB) due to mist and fog
c = excess attenuation (dB) due to oxygen (O_2)
d = sum of the absorption losses (dB) due to other gases
e = excess attenuation (dB) due to rainfall

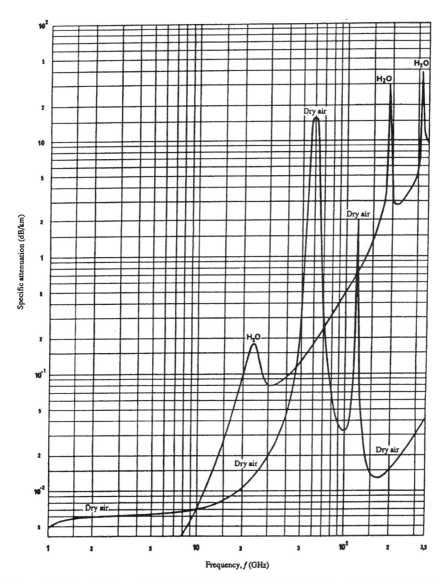

Figure 8.1. Specific attenuation due to atmospheric gases. Pressure = 1013 mb; temperature = 15°C; water vapor = 7.5 g/m^3. (From Ref. 2, Figure 1, ITU-R Rec. 676-1.)

Notes and Comments on Equation 8.1

1. a varies with relative humidity, temperature, atmospheric pressure, and altitude. The transmission engineer assumes that the water-vapor content is linear with these parameters and that the atmosphere is homogeneous (actually horizontally homogeneous but vertically stratified).

There is a water-vapor absorption band about 22 GHz caused by molecular resonance.

2. c and d are assumed to vary linearly with atmospheric density, thus directly with atmospheric pressure, and are a function of altitude (e.g., it is assumed that the atmosphere is homogeneous).

3. b and e vary with the density of the rainfall cell or cloud and the size of the rainfall drops or water particles such as fog or mist. In this case the atmosphere is most certainly not homogeneous. (Droplets less than 0.01 cm in diameter are considered mist/fog, and more than 0.01 cm, rain.) Ordinary fog produces about 0.1-dB/km excess attenuation at 35 GHz, rising to 0.6 dB/km at 75 GHz.

In equation 8.1 terms b and d can often be neglected; terms a and c are usually lumped together and called "atmospheric attenuation." If we were to install a 10-km LOS link at 22 GHz, in calculating transmission loss, 1.6 dB would have to be added for what is called atmospheric attenuation but is predominantly water-vapor absorption, as shown in Figure 8.1.

It will be noted in Figure 8.1 that there are frequency bands with relatively high levels of attenuation per unit distance; some are rather narrow bands and some fairly wide. For example, the O_2 absorption band covers from about 58 to 62 GHz and with the skirts down to 50 and up to 70 GHz. At its peak at about 60 GHz, the sea-level attenuation is about 15 dB/km. One could ask: "Of what use are these bands?" Actually, the 58–62-GHz band is appropriately assigned for satellite cross-links. These links operate out in space far above the limits of the earth's atmosphere, where the terms a through e may be completely neglected. It is particularly attractive on military cross-links having an inherent protection from earth-based enemy jammers by that significant atmospheric attenuation factor. It is also useful for very-short-haul military links such as a ship-to-ship secure communication system. Again, it is the atmospheric attenuation that offers some additional security for signal intercept (low probability of intercept, LPI) and against jamming.

On the other hand, Figure 8.1 shows a number of bands that are relatively open. These openings are often called windows. Three such windows are suggested for point-to-point service in Table 8.1.

TABLE 8.1   Windows for Point-to-Point Service

Band (GHz)	Excess Attenuation Due to Atmospheric Absorption (dB/km)
20 <	0.08 <
28–42	0.13
75–95	0.4
125–140	1.8

8.3   RAINFALL LOSS

8.3.1   Basic Rainfall Considerations

Of the factors a through e in equation 8.1, factor e, excess attenuation due to rainfall, is the principal one affecting path loss. For instance, even at 22 GHz, the water-vapor line, excess attenuation due to atmospheric gases accumulates at only 0.165 dB/km, and for a 10-km path only 1.65 dB must be added to free-space loss to compensate for water vapor loss. This is negligible when compared to free-space loss itself, such as 119.3 dB for the first kilometer at 22 GHz, accumulating thence roughly 6 dB each time the path length is doubled (i.e., add 6 dB for 2 km, 12 dB for 4 km, etc.). Accordingly, a 10-km path would have a free-space loss of 139.3 dB plus 1.65 dB added for excess attenuation due to water vapor (22 GHz), or a total of 140.95 dB.

Excess attenuation due to rainfall is another matter. It has been common practice to express rainfall loss as a function of precipitation rate. Such a rate depends on the liquid water content and the fall velocity of the drops. The velocity, in turn, depends on raindrop size. Therefore our interest in rainfall boils down to drop size and drop-size distribution for point rainfall rates. All this information is designed to lead the transmission engineer to fix an excess attenuation due to rainfall on a particular path as a function of time and time distribution. This is a method similar to that used in Chapter 5 for overbuilding a link to accommodate fading.

An earlier approach dealt with rain on a basis of rainfall given in millimeters per hour. Often this was done with rain gauges, using collected rain averaging over a day or even periods of days. For path design above 10 GHz such statistics are not sufficient where we may require path availability better than 99.9% and do not wish to resort to overconservative design procedures (e.g., assign excessive link margins).

As we mentioned, there is a fallacy in using annual rainfall rates as a basis for calculation of excess attenuation due to rainfall. For instance, several weeks of light drizzle will affect the overall long-term path availability much less than several good downpours that are short lived (e.g., 20-min duration). It is simply this downpour activity for which we need statistics. Such downpours are cellular in nature. How big are the cells? What is the rainfall rate in the cell? What are the size of the drops and their distribution?

Hogg (Ref. 3) suggests the use of high-speed rain gauges with outputs readily available for computer analysis. These gauges can provide minute-by-minute analysis of the rate of fall, something lacking with conventional types of gauges. Of course, it would be desirable to have several years' statistics for a specific path to provide the necessary information on fading caused by rainfall that will govern system parameters such as LOS repeater spacing, antenna size, and diversity.

Some such information is now available and is indicative of a great variation in short-term rainfall rates from one geographical location to another. As an example, in one period of measurement it was found that Miami, Florida, has maximum rain rates about 20 times greater than those of heavy showers occurring in Oregon, the region of heaviest rain in the United States. In Miami a point rainfall rate may exceed 700 mm/h. The effect of 700 mm/h on 70- and 48-GHz paths can be extrapolated from Figure 8.2. In the figure the rainfall rate in millimeters per hour extends to 100, which at 100 mm/h provides an excess attenuation of from 25 to 30 dB/km.

When identical systems were compared (Ref. 3) at 30 GHz with repeater spacings of 1 km and equal desired signals (e.g., producing a 30-dB signal-to-noise ratio), 140 min of total time below the desired level was obtained at Miami, Florida; 13 min at Coweeta, North Carolina; 4 min at Island Beach, New Jersey; 0.5 min at Bedford, United Kingdom; and less than 0.5 min at Corvallis, Oregon. Such outages, of course, can be reduced by increasing transmitter output power, improving receiver noise figure (NF), increasing antenna size, implementing a diversity scheme, and so on.

One valid approach to lengthen repeater sections (space between repeaters) and still maintain performance objectives is to use path diversity. This is the most effective form of diversity for downpour rainfall fading. Path diversity is the simultaneous transmission of the same information on paths separated by at least 10 km, the idea being the rain cells affecting one path will have a low probability of affecting the other at the same time. A switch would select the better path of the two. Careful phase equalization between the two paths would be required, particularly for the transmission of high-bit-rate information.

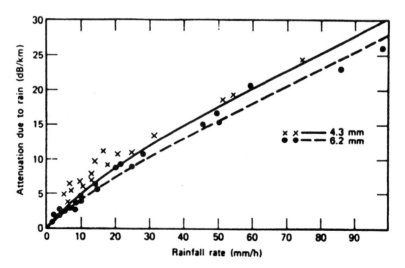

Figure 8.2. Measurements made by Bell Telephone Laboratories of excess attenuation due to rainfall at wavelengths of 6.2 and 4.2 mm (48 and 70 GHz) compared with calculated values. (From Ref. 3; copyright © 1968 by American Telegraph & Telephone Company.)

8.3.2 Calculation of Excess Path Attenuation Due to Rainfall for LOS Paths

When designing terrestrial LOS microwave links (or satellite links), a major problem in link engineering is to determine the excess path attenuation due to rainfall. The adjective "excess" is used to denote path attenuation in *excess* of free-space loss (i.e., the terms a, b, c, d, and e in equation 8.1 are in excess of free-space attenuation [loss]).

Before treating the methodology of calculation of excess rain attenuation, we will review some general link engineering information dealing with rain. When discussing rainfall here, all measurements are in millimeters per hour of rain and are point rainfall measurements. From our previous discussion we know that heavy downpour rain is the most seriously damaging to radio propagation above 10 GHz. Such rain is cellular in nature and has limited coverage. We must address the question as to whether the entire hop is in the storm for the whole period of the storm. Light rainfall (e.g., less than 2 mm/h), on the other hand, is usually widespread in character, and the path average is the same as the local value. Heavier rain occurs in convective storm cells, which are typically 2–6 km across and are often embedded in larger regions measured in tens of kilometers (Ref. 4). Thus for short hops (2–6 km) the path-averaged rainfall rate will be the same as the local rate, but for longer paths it will be reduced by the ratio of the path length on which it is raining to the total path length.

This concept is further expanded upon by CCIR Rep. 563-1 (Ref. 5), where rain cell size is related to rainfall rate as shown in Figure 8.3. This concept of rain cell size is very important, whether engineering a LOS link or a satellite link, in particular, when the satellite link has a low elevation angle.

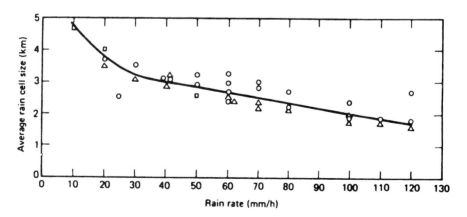

Figure 8.3. Average rain cell size as a function of rain rate. (From Ref. 5, CCIR Rep. 563-1. Courtesy of ITU-CCIR.)

CCIR Rep. 338-3 (Ref. 6) is quoted in part below:

> Measurements in the United Kingdom over a period of two years... at 11, 20
> and 37 GHz on links of 4–22 km in length show that the attenuation due to rain
> and multipath, which is exceeded for 0.01% of the time and less, increased
> rapidly with path length up to 10 km, but further increase up to 22 km
> produced a small additional effect.

The application of specific rain cell size was taken from CCIR 1978. We will
call this method the "liberal method" because when reviewing CCIR Vol. V
of 1982, CCIR Rep. 563-2 (Ref. 7), we can see that the CCIR has turned
more cautious and conservative. Our Figure 8.3 does not appear in the
report. Let us partially quote from this report:

> For attenuation predictions the situation is generally more complex (than that
> of interference scattering by precipitation). Volume cells are known to cluster
> frequently within small mesoscale areas.... Terrestrial links exceeding 10 km
> may therefore traverse more than one volume cell within a mesoscale cluster. In
> addition, since the attenuating influence of the lower intensity rainfall
> surrounding the cell must be taken into account, any model used to calculate
> attenuation must take these larger rain regions into account. The linear extent
> of these regions increases with decreasing rain intensity and may be as large as
> several tens of kilometers.

One of the most accepted methods of dealing with excess path attenuation A
(dB) due to rainfall is an empirical procedure based on the approximate
relation between A and the rain rate R:

$$A = aR^b \qquad (8.2)$$

where a and b are functions of frequency f and rain temperature T.
Allowing a rain temperature of 20°C and allowing for the Laws and Parsons
drop-size distribution, Table 8.2 gives the regression coefficients for estimating specific attenuations from equation 8.2.

We note that horizontally polarized waves suffer greater attenuation than
vertically polarized waves because large raindrops are generally shaped as
oblate spheroids and are aligned with a vertical rotation axis. Hence we use
the subscript notation h and v for horizontal and vertical polarizations in
Table 8.2 for the values a and b. A is the specific attenuation in dB/km. We
obtain a and b from Table 8.2. Next we must obtain a value for the rain rate.
This is obtained from local data sources, and we need a value of rain
intensity exceeded 0.01% of the time with an integration time of 1 min. If this
information cannot be obtained from local sources, an estimate can be
obtained by identifying the region of interest from the maps appearing in
Figures 8.4–8.6, then selecting the appropriate rainfall intensity for the
specified time percentage from Table 8.3, which gives the 15 regions in the

TABLE 8.2   Regression Coefficients for Estimating Specific Attenuations in Equation 8.2[a]

Frequency (GHz)	a_h	a_v	b_h	b_v
1	0.0000387	0.0000352	0.912	0.880
2	0.000154	0.000138	0.963	0.923
4	0.000650	0.000591	1.121	1.075
6	0.00175	0.00155	1.308	1.265
7	0.00301	0.00265	1.332	1.312
18	0.00454	0.00395	1.327	1.310
10	0.0101	0.00887	1.276	1.264
12	0.0188	0.0168	1.217	1.200
15	0.0367	0.0335	1.154	1.128
20	0.0751	0.0691	1.099	1.065
25	0.124	0.113	1.061	1.030
30	0.187	0.167	1.021	1.000
35	0.263	0.233	0.979	0.963
40	0.350	0.310	0.939	0.929
45	0.442	0.393	0.903	0.897
50	0.536	0.479	0.873	0.868
60	0.707	0.642	0.826	0.824
70	0.851	0.784	0.793	0.793
80	0.975	0.906	0.769	0.769
90	1.06	0.999	0.753	0.754
100	1.12	1.06	0.743	0.744
120	1.18	1.13	0.731	0.732
150	1.31	1.27	0.710	0.711
200	1.45	1.42	0.689	0.690
300	1.36	1.35	0.688	0.689
400	1.32	1.31	0.683	0.684

[a]Raindrop size distribution (Laws and Parsons, 1943); terminal velocity of raindrops (Gunn and Kinzer, 1949); index of refraction of water at 20°C (Ray, 1972). Values of a and b for spheroidal drops (Fedi, 1979; Maggiori, 1981) based on regression for the range 1–150 mm/h. (*Note*: These references are from Ref. 8.)

Source: Reference 8, Table 1, CCIR Rec. 838.

maps. This gives a value for R in equation 8.2. Knowing the frequency and polarization, we calculate A in dB/km from values of a and b from Table 8.2 using equation 8.2. Figure 8.7 and/or the nomogram in Figure 8.8 may also be used to calculate A.

We now turn again to the problem of effective path length (Ref. 11). This is obtained by multiplying the actual path length L by a reduction factor r. A first estimate to calculate r is given by

$$r = 1/(1 + L/L_0) \qquad (8.3)$$

where

$$L_0 = 35 \exp(-0.015 R_{0.01}) \qquad (8.4)$$

From Table 8.3, $R_{0.01}$ is the rain rate with an exceedance 0.01% of the time.

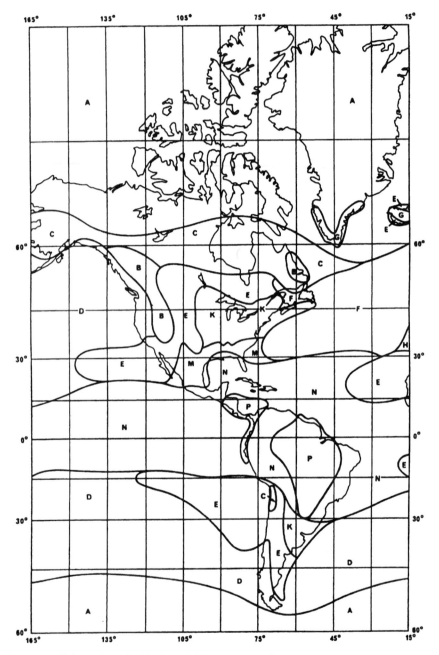

Figure 8.4. Rain regions for North and South America. (From Ref. 9, Figure 1, ITU-R Rec. PN.837-1.)

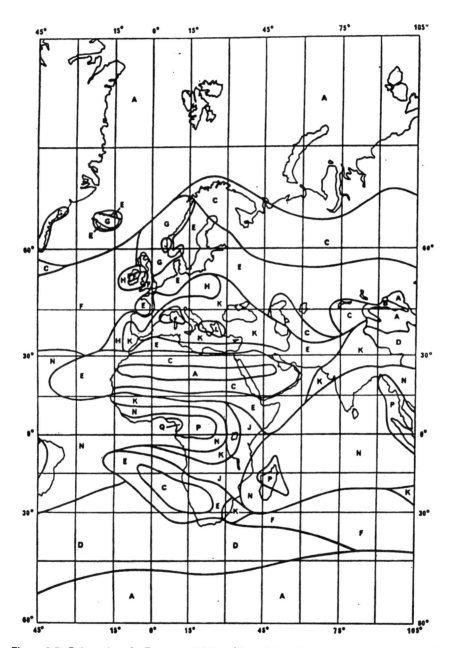

Figure 8.5. Rain regions for Europe and Africa. (From Ref. 9, Figure 2, ITU-R Rec. PN.837-1.)

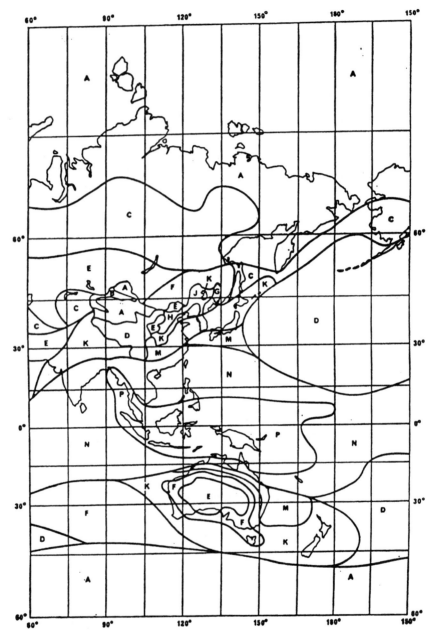

Figure 8.6. Rain regions for Asia, Oceana, and Australia. (From Ref. 9, Figure 3, ITU-R Rec. PN.837-1.)

**TABLE 8.3   Rain Climatic Zones—Rainfall Intensity Exceeded (mm/h)
(Reference to Figures 8.4–8.7)**

Percentage of Time (%)	A	B	C	D	E	F	G	H	J	K	L	M	N	P	Q
1.0	< 0.5	1	2	3	1	2	3	2	8	2	2	4	5	12	24
0.3	1	2	3	5	3	4	7	4	13	6	7	11	15	34	49
0.1	2	3	5	8	6	8	12	10	20	12	15	22	35	65	72
0.03	5	6	9	13	12	15	20	18	28	23	33	40	65	105	96
0.01	8	12	15	19	22	28	30	32	35	42	60	63	95	145	115
0.003	14	21	26	29	41	54	45	55	45	70	105	95	140	200	142
0.001	22	32	42	42	70	78	65	83	55	100	150	120	180	250	170

Source: Reference 9, Table I, ITU-R Rec. PN.837-1.

The path attenuation per kilometer exceeded for 0.01% of the time is found from (Ref. 8)

$$A_{0.01} = aR_{0.01}^b \tag{8.5}$$

The effective excess (A_{eff}) attenuation due to rainfall for a path with a rainfall time availability of 99.99% (exceedance = 0.01%) is

$$A_{\text{eff}} = A_{0.01} \times L \times r \tag{8.6}$$

where A is the value derived from equation 8.5 and r, the reduction factor, from equation 8.3. Attenuation exceedance for other percentages P can be found by the following power law:

$$A_p = A_{\text{eff}(0.01)}[0.12\ P^{-(0.546+0.0743\ \log P)}] \tag{8.7}$$

This formula has been determined to give factors of 0.12, 0.39, 1, and 2.14 for 1%, 0.1%, 0.01%, and 0.001%, respectively (Ref. 11).

Example 1. Consider a path near Madrid, Spain, 10-km long and operating at 30 GHz. Design the path for a 99.9% time availability, horizontal polarization. This then would be an exceedance of 0.1%. From Figure 8.5 we see the path is in zone H. Turn now to Table 8.3. Under column H, the value for 0.01% of the time is 32 mm/h. This is the value for R. Extrapolate values for a and b for 38 GHz, which we will apply to equation 8.5. Use Table 8.2.

$$a_h = 0.315 \qquad b_H = 0.963$$

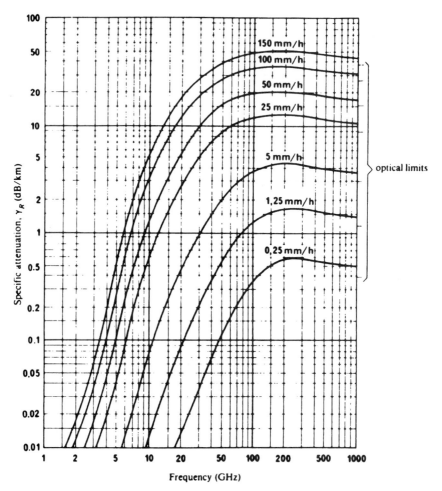

Figure 8.7. Specific attenuation *A* due to rain. Raindrop-size distribution (Laws and Parsons, 1943); terminal velocity of raindrops (Gunn and Kinzer, 1949); index of refraction of water at 20°C (Ray, 1972); spherical drops. (From Ref. 10, Figure 1, CCIR Rep. 721-3.)

Calculate $A_{0.01}$ with equation 8.5:

$$A_{0.01} = 0.315(32 \text{ mm/h})^{0.963}$$

$$= 8.87 \text{ dB/km}$$

Calculate the reduction factor, r, using equation 8.3:

$$r = 1/(1 + 10/L_0)$$

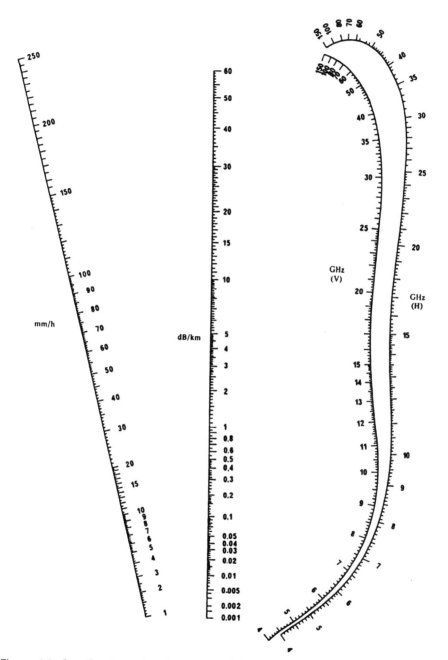

Figure 8.8. Specific attenuation due to rain. (H) = horizontal polarization; (V) = vertical polarization. It is recommended that this nomogram not be used for frequencies above 406 Hz. (From Ref. 10, Figure 2, CCIR Rep. 721-3.)

Calculate L_0 using equation 8.4:

$$L_0 = 21.66$$

$$r = 0.684$$

Calculate the effective excess attenuation due to rainfall for the path using equation 8.6:

$$A_{eff} = 8.87 \text{ dB/km} \times 0.684 \times 10$$

$$= 60.67 \text{ dB}$$

Apply equation 8.7 for the time availability 99.9%, exceedance 0.1%; or multiply the 0.01% value by 0.39.

$$A_{eff(0.1)} = 23.66 \text{ dB}$$

Example 2. Calculate the excess attenuation due to rainfall for a 13-km path operating at 18 GHz. Calculate excess attenuation for exceedance values for 0.1%, 0.01%, and 0.001%. Design for vertical polarization. The path will be in the Caracas area of Venezuela.

From Figure 8.4, the climatic area is N. From Table 8.3 derive the rainfall rate for 0.01% exceedance. The value is 95 mm/h. Interpolate the regression coefficients from Table 8.2 for 18 GHz, vertical polarization.

$$a_v = 0.0548 \qquad b_v = 1.0902$$

Calculate $A_{0.01}$ from equation 8.5:

$$A_{0.01} = 0.0548 \times 95^{1.0902}$$

$$= 7.85 \text{ dB/km}$$

Calculate the reduction factor, r, using equation 8.3. However, first calculate L_0 using equation 8.4:

$$L_0 = 8.42$$

$$r = 0.393$$

Calculate the 0.01% exceedance value for the 13-km path:

$$A_{eff(0.01)} = 13 \times 7.85 \times 0.393$$

$$= 40.1 \text{ dB}$$

Adjust this value for 0.1% and 0.001% exceedances (99.9% and 99.999% time availabilities). For 0.1%, $40.1 \times 0.39 = 15.64$ dB; for 0.001%, $40.1 \times 2.14 = 85.814$ dB.

8.3.3 Calculation of Excess Attenuation Due to Rainfall for Satellite Paths

Rainfall attenuation on satellite paths is a function of frequency and elevation angle. The calculation of excess attenuation due to rainfall for uplinks and downlinks is very similar to the exercise for terrestrial line-of-sight radiolinks described in Section 8.3.2. The methodology presented here is suggested by CCIR in Rec. 564-4 (Ref. 12) and repeated in ITU-R Rec. PN.618-3 (Ref. 13). It is based on the path geometry shown in Figure 8.9. The reference document advises that the procedure provides estimates of long-term statistics of the slant-path attenuation at a given location for frequencies up to 30 GHz. The calculations are only valid for elevation angles $\geq 5°$.

The following parameters are defined:

$R_{0.01}$ point rainfall rate for the location for 0.01% of an average year (mm/h)

h_s height above mean sea level of the earth station (km)

θ elevation angle

ϕ absolute value of latitude of the earth station (degrees)

f frequency (GHz)

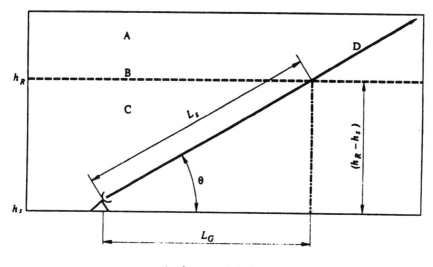

A: frozen precipitation
B: rain height
C: liquid precipitation
D: Earth-space path

Figure 8.9. Schematic presentation of an earth–space path giving parameters that will be inputs to the attenuation prediction process. (From Ref. 13, Figure 1, ITU-R Rec. PN.618-3.)

The procedure is described in eight steps as follows:

Step 1. Calculate the effective rain height, h_R, for the latitude of the station ϕ:

$$h_R(\text{km}) = \begin{cases} 3.0 + 0.028\phi & \text{for } 0 \le \phi < 36° \quad (8.8a) \\ 4.0 - 0.075(\phi - 36) & \text{for} \quad \phi \ge 36° \quad (8.8b) \end{cases}$$

Step 2. For $\theta \ge 5°$ compute the slant-path length, L_S, below the rain height from

$$L_s = \frac{(h_R - h_s)}{\sin\theta} \quad \text{km} \tag{8.9}$$

For $\theta < 5°$, the following formula is used:

$$L_s = \frac{2(h_R - h_s)}{\left(\sin^2\theta + \dfrac{2(h_R - h_s)}{R_e}\right)^{1/2} + \sin\theta} \quad \text{km} \tag{8.10}$$

Step 3. Calculate the horizontal projection, L_G, of the slant-path length from

$$L_G = L_s \cos\theta \quad \text{km} \tag{8.11}$$

We have now calculated the horizontal projection of the slant path. This is analogous to L, the path length used in Section 8.2. It is called L_G here. The steps from here onward are the same as we used in Section 8.2 with equations 8.3, 8.4, 8.5, 8.6, and 8.7. The equations used here will have the same numbers used in Section 8.2, but we append an S at the end of each formula number.

Step 4. Obtain the rain rate intensity, $R_{0.01}$, exceeded for 0.01%. If this information cannot be obtained from the local weather bureau, consult Figures 8.4, 8.5, and 8.6 and select the appropriate rain region. With that information, consult Table 8.3 using the exceedance row 0.01% for that rain region. This will be the value for $R_{0.01}$, the rain rate for an exceedance of 0.01%.

Step 5. Calculate the reduction factor, $r_{0.01}$, for 0.01% of the time for $R_{0.01} \le 100$ mm/h. Use equations 8.3 and 8.4.

$$r_{0.01} = 1/(1 + L_G/L_0) \tag{8.3S}$$

If $R_{0.01} > 100$ mm/h, use the 100-mm/h value in place of $R_{0.01}$. First, however, calculate L_0 using equation 8.4:

$$L_0 = 35 \exp^{(-0.015 \times R_{0.01})} \tag{8.4S}$$

Substitute the value of L_0 in equation 8.3S.

Step 6. Calculate the effective attenuation, $A_{\text{eff}(0.01)}$, using formula 8.5:

$$A_{\text{eff}(0.01)} = aR_{0.01}^b \quad \text{dB/km} \tag{8.5S}$$

Obtain the value for R from Table 8.3 using the 0.01% exceedance row and the appropriate climatic zone column. The values for a and b, using the appropriate polarization, are taken from Table 8.2. Interpolate, if required.

Step 7. The predicted attenuation for the satellite path exceedance for 0.01% of an average year is obtained by

$$A_{\text{eff}(\text{path}, 0.01)} = A_{\text{eff}(0.01)} \times r \times L_s \tag{8.6S}$$

where L_s is taken from Step 2, formula 8.9.

Step 8. Use formula 8.7 for other time availabilities (i.e., other exceedances). Or use the interpolation formula factors of 0.12, 0.38, 1, and 2.14 for 1%, 0.1%, 0.01%, and 0.001%, respectively (Ref. 13).

Example. An earth station near London (UK) has its downlink at 12 GHz; its elevation angle to the satellite is 20°; its latitude is 48° N; its height above mean sea level is 1 km; horizontal polarization. The time availability desired is 99.999%. Calculate the excess attenuation due to rainfall.

Step 1. Calculate the effective rain height for 48° N latitude. Use formula 8.8b:

$$h_R = 3.1 \text{ km}$$

Step 2. Calculate the slant-path length, L_s, below the rain height from equation 8.9:

$$L_s = 2.1/0.342$$
$$= 6.14 \text{ km}$$

Step 3. Calculate the horizontal projection, L_G, of the slant-path length using equation 8.11:

$$L_G = L_s \cos 20°$$
$$= 6.14 \times 0.94$$
$$= 5.77 \text{ km}$$

Step 4. Obtain the rain intensity (rain rate) for 0.01% exceedance for the London (UK) area. Use Figure 8.4, thus rain region E.

$$R_{0.01} = 22 \text{ mm/h}$$

Step 5. Calculate the reduction factor, $r_{0.01}$, by equations 8.3S and 8.4S. First calculate L_0:

$$L_0 = 25.16$$

$$r_{0.01} = 0.813$$

Step 6. Obtain the specific attenuation, $A_{\text{eff}(0.01)}$, using the regression coefficients from Table 8.2. $R_{0.01} = 22$ mm/h.

$$a_h = 0.0188 \qquad b_h = 1.217$$

Use equation 8.5S:

$$A_{\text{eff}(0.01)} = 0.0188(22)^{1.217}$$

$$= 0.81 \text{ dB/km}$$

Step 7. Calculate the predicted excess attenuation for 0.01% of the time. Use equation 8.6S:

$$A_{\text{eff}(\text{path}, 0.01)} = 0.81 \text{ dB/km} \times 0.813 \times 5.77 \text{ km}$$

$$= 3.79 \text{ dB}$$

Step 8. Use a factor of 2.14 to calculate excess attenuation for 0.001% of the time.

$$A_{0.001} = 3.79 \times 2.14$$

$$= 8.11 \text{ dB}$$

8.3.4 Utilization of Path Diversity to Achieve Performance Objectives

Excess attenuation due to rainfall often degrades satellite uplinks and downlinks operating above 10 GHz so seriously that the requirements of optimum economic design and reliable performance cannot be achieved simultaneously. Path diversity can overcome this problem at some reasonable cost compromise. Path diversity advantage is based on the hypothesis that rain cells and, in particular, the intense rain cells that cause the most severe fading are rather limited in spatial extent. Furthermore, these rain cells do not occur immediately adjacent to one another. Therefore the probability of simultaneous fading on two paths to spatially separated earth stations would be less than that associated with either individual path. The hypothesis has been borne out experimentally (Ref. 14).

Let us define two commonly used terms: diversity gain and diversity advantage. Diversity gain is defined (in this context) as the difference between the rain attenuation exceeded on a single path and that exceeded jointly on separated paths for a given percentage of time. Diversity advantage is defined (in this context) as the ratio of the percentage of time exceeded on a single path to that exceeded jointly on separated paths for a given rain attenuation level.

Diversity gain may be interpreted as the reduction in the required system margin at a particular percentage of time afforded by the use of path diversity. Alternatively, diversity advantage may be interpreted as the factor by which the fade time is improved at a particular attenuation level due to the use of path diversity.

The principal factor to achieve path diversity to compensate for excess attenuation due to rainfall is separation distance. The diversity gain increases rapidly as the separation distance d is increased over a small separation distance, up to about 10 km. Thereafter the gain increases more slowly until a maximum value is reached, usually between about 10 and 30 km. This is shown in Figure 8.10.

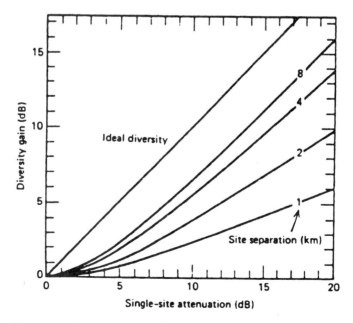

Figure 8.10. Diversity gain for various site separations. (From Ref. 15.)

The uplink/downlink frequencies seem to have little effect on diversity gain up to about 30 GHz (Ref. 16). This same reference suggests that for link frequencies above 30 GHz attenuation on both paths simultaneously can be sufficient to create an outage. Hence extrapolation beyond 30 GHz is not recommended, at least with the values given in Figure 8.10.

8.4   EXCESS ATTENUATION DUE TO ATMOSPHERIC GASES ON SATELLITE LINKS

The zenith one-way attenuations for a moderately humid atmosphere (e.g., 7.5-g/m^3 surface water-vapor density) at various starting heights above sea level are given in Figure 8.11 and in Table 8.4. These curves were computed by Crane and Blood (Ref. 17) for temperate latitudes assuming the U.S. standard atmosphere, July, 45° N latitude. The range of values shown in Figure 8.11 refers to the peaks and valleys of the fine absorption lines. The range of values for starting heights above 16 km is even greater.

Figure 8.11 also shows the standard deviation of the clear-air zenith attenuation as a function of frequency. The standard deviation was calculated from 220 measured atmosphere profiles spanning all seasons and geographical locations (by Crane, Ref. 18). The zenith attenuation is a function of frequency, earth terminal altitude above sea level, and water-vapor content. Compensating for earth terminal altitudes can be done by interpolating between the curves in Figure 8.11.

The water-vapor content is the most variable component of the atmosphere. Corrections should be made to the values derived from Figure 8.11 and Table 8.4 in regions that notably vary from the 7.5-g/m^3 value given. Such regions would be arid or humid, jungle or desert. This correction to the total zenith attenuation is a function of the water-vapor density at the surface p_0 as follows:

$$\Delta A_{c1} = b_p \left(p_0 - 7.5 \text{ g/m}^3 \right) \qquad (8.12)$$

where ΔA_{c1} is the additive correction to the zenith clear-air attenuation that accounts for the difference between the actual surface water-vapor density and 7.5 g/m^3. The coefficient b_p is frequency dependent and is given in Figure 8.12. To convert from the more familiar relative humidity or partial pressure of water vapor, refer to Section 8.4.2.

The surface temperature T_0 also affects the total attenuation because it affects the density of both the wet and dry components of the gaseous attenuation. This relation is (Ref. 17)

$$\Delta A_{c2} = C_T \left(21° - T_0 \right) \qquad (8.13)$$

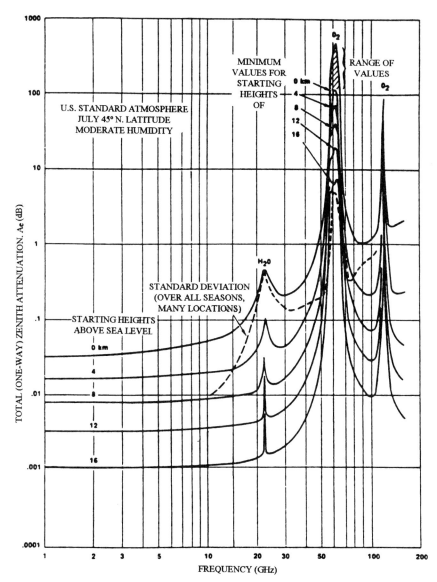

Figure 8.11. Total zenith attenuation versus frequency. (From Ref. 16.)

where ΔA_{c2} is an additive correction to the zenith clear-air attenuation. Figure 8.12 gives the frequency-dependent values for C_T.

The satellite earth terminal elevation angle has a major impact on the gaseous attenuation value for a link. For elevation angles greater than about 5°, the zenith clear-air attenuation value A_c is multiplied by the cosecant of the elevation angle θ. The total attenuation for an elevation angle θ is

**TABLE 8.4   Typical One-Way Clear-Air Total Zenith Attenuation Values
(7.5 g/m³ H₂O, July, 45° N Latitude, 21°C)**

Frequency (GHz)	Altitude				
	0	0.5	1.0	2.0	4.0
10	0.055	0.05	0.043	0.035	0.02
15	0.08	0.07	0.063	0.045	0.023
20	0.30	0.25	0.19	0.12	0.05
30	0.22	0.18	0.16	0.10	0.045
40	0.40	0.37	0.31	0.25	0.135
80	1.1	0.90	0.77	0.55	0.30
90	1.1	0.92	0.75	0.50	0.22
100	1.55	1.25	0.95	0.62	0.25

Source: Reference 16.

given by

$$A_c = A'_c \csc \theta \qquad (8.14)$$

8.4.1   Example Calculation of Clear-Air Attenuation—Hypothetical Location

For a satellite downlink, we are given the following information: frequency, 20 GHz; altitude of earth station, 600 m; relative humidity (RH), 50%; temperature (surface, T_0), 70°F (21.1°C); and elevation angle, 25°. Calculate clear-air attenuation.

Obtain total zenith attenuation A'_c from Table 8.4, and interpolate value for altitude: $A'_c = 0.24$ dB.

Find the water-vapor density p_0. From Figure 8.13, the saturated partial pressure of water vapor at 70°F is $e_s = 2300$ N/m². Apply formula 8.15 (Section 8.4.2) and

$$p_0 = \frac{(0.5)2300}{(0.461)(294.1)}$$

$$= \frac{1150}{135.6}$$

$$= 8.48 \text{ g/m}^3$$

Calculate the water-vapor correction factor ΔA_{c1}. From Figure 8.12, for a frequency of 20 GHz, correction coefficient $b_p = 0.05$. Then use equation 8.12:

$$\Delta A_{c1} = (0.05)(8.48 - 7.5) = 0.05 \text{ dB}$$

Compute the temperature C_T using Figure 8.12. At 20 GHz, $C_T = 0.0015$. As can be seen, this value can be neglected in this case.

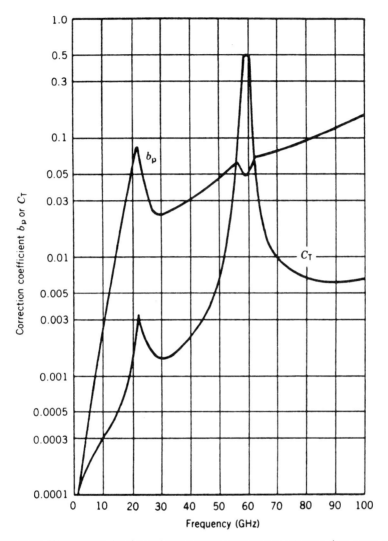

Figure 8.12. Water-vapor density and temperature correction coefficients. (From Ref. 16.)

Calculate the clear-air zenith attenuation corrected A'_c:

$$A'_c = 0.24 \text{ dB} + 0.05 \text{ dB} + 0 \text{ dB}$$
$$= 0.29 \text{ dB}$$

Compute the clear-air slant attenuation using equation 8.14:

$$A_c = 0.29 \csc 25°$$
$$= 0.29 \times 2.366$$
$$= 0.69 \text{ dB}$$

This value would then be used in the link budget for this hypothetical link.

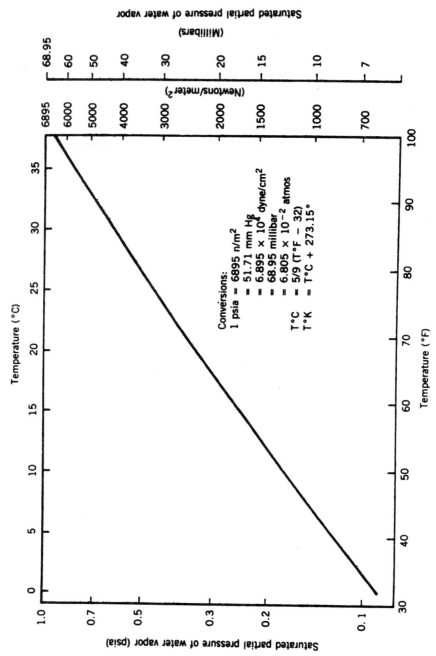

Figure 8.13. The saturated partial pressure of water vapor versus temperature. (From Ref. 16.)

8.4.2   Conversion of Relative Humidity to Water-Vapor Density

The surface water-vapor density p_0 (g/m^3) at a given surface temperature T_0 may be calculated from the ideal gas law:

$$p_0 = \frac{(\text{RH})e_s}{0.461 \text{ J} \cdot \text{g}^{-1}/\text{K}^{-1}(T_0 + 273)} \tag{8.15}$$

where RH = relative humidity and e_s (N/m^2) = the saturated partial pressure of water vapor that corresponds to the surface temperature T_0 (°C). See Figure 8.13. The relative humidity corresponding to 7.5 g/m^3 at 20°C (68°F) is RH = 0.42 or 42% (Ref. 16).

8.5   ATTENUATION DUE TO CLOUDS AND FOG

Water droplets that constitute clouds and fog are generally less than 0.01 cm in diameter (Ref. 16). This allows a Rayleigh approximation to calculate the attenuation due to clouds and fog for frequencies up to 100 GHz. The specific attenuation a_c is, unlike the case of rain, independent of drop-size distribution. It is a function of liquid water content p_1 and can be expressed by

$$a_c = K_c p_1 \quad (\text{dB/km}) \tag{8.16}$$

where p_1 is normally expressed in g/m^3. K_c is the attenuation constant, which is a function of frequency and temperature and is given in Figure 8.14. The curves in Figure 8.14 assume pure water droplets. The values for salt-water droplets, corresponding to ocean fogs and mists, are higher by approximately 25% at 20°C and 5% at 0°C (Ref. 20).

The liquid water content of clouds varies widely. Stratiform or layered clouds display ranges of 0.05–0.25 g/m^3 (Ref. 16). Stratocumulus, which is the most dense of this cloud type, has shown maximum values from 0.3 to 1.3 g/m^3 (Ref. 16). Cumulus clouds, especially the large cumulonimbus and cumulus congestus that accompany thunderstorms, have the highest values of liquid content. Fair-weather cumulus clouds generally have a liquid water content of less than 1 g/m^3. Reference 21 reported values exceeding 5 g/m^3 in cumulus congestus and estimates an average value of 2 g/m^3 for cumulus congestus and 2.5 g/m^3 for cumulonimbus clouds.

Care must be exercised in estimating excess attenuation due to clouds when designing uplinks and downlinks. First, clouds are not homogeneous masses of air containing uniformly distributed droplets of water. Actually, the liquid water content can vary widely with location in a single cloud. Even sharp differences have been observed in localized regions on the order of 100 m across. There is a fairly rapid variation with time as well, owing to the complex patterns of air movement taking place within cumulus clouds.

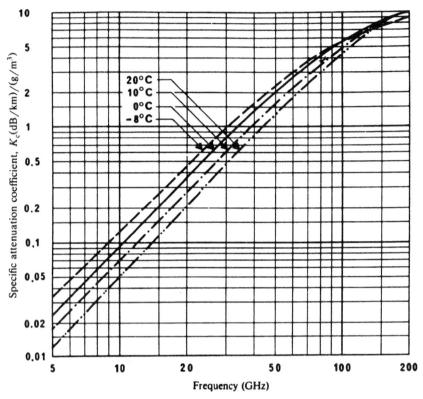

Figure 8.14. Attenuation coefficient K_c due to water-vapor droplets. (From Ref. 19, Figure 1, ITU-R Rec. PN.840-1.)

Typical path lengths through cumulus congestus clouds roughly fall between 2 and 8 km. Using equation 8.16 and the value given for water-vapor density and the attenuation coefficient K_c from Figure 8.14, an added path loss at 35 GHz from 4 to 16 dB would derive. Fortunately, for the system designer, the calculation grossly overestimates the actual attenuation that has been observed through this type of cloud structure. Table 8.5 provides values that seem more dependable. In the 35- and 95-GHz bands, cloud attenuation, in most cases, is 40% or less of the gaseous attenuation values. One should not lose sight of the fact, in these calculations, of the great variability in the size and state of development of the clouds observed. Data from Table 8.5 may be roughly scaled in frequency, using the frequency dependence of the attenuation coefficient from Figure 8.14.

Fog results from the condensation of atmospheric water vapor into water droplets that remain suspended in air. The water-vapor content of fog varies from less than 0.4 up to as much as 1 g/m^3.

TABLE 8.5 Atmospheric Attenuation in the Vertical Direction for 95 and 150 GHz, Slough, UK, October 1975–May 1976

Frequency	95 GHz	150 GHz
Attenuation (dB) in clear air; water-vapor content at ground-level		
4–11 g/m^3	0.7–1	1–3
Additional attenuation (dB) due to clouds:		
Stratocumulus	0.5–1	0.5–1
Small, fine-weather cumulus	0.5	0.5
Large cumulus	1.5	2
Cumulonimbus	2–7	3–8
Nimbostratus (rain cloud)	2–4	5–7

Source: Reference 10, Table 2, CCIR Rep. 721-3.

The attenuation due to fog in dB/km can be estimated using the curves in Figure 8.14. The 10°C curve is recommended for summer, and the 0°C curve should be used for the other seasons. Typical liquid water content values for fog vary from 0.1 to 0.2 g/m^3. Assuming a temperature of 10°C, the specific attenuation would be about 0.08–0.16 dB/km at 35 GHz and 0.45–0.9 dB/km for 95 GHz. In a typical fog layer 50 m thick, a path at a 30° elevation angle would have only 100-m extension through fog, producing less than 0.1-dB excess attenuation at 95 GHz. In most cases, the result is that fog attenuation is negligible for satellite links.

8.6 CALCULATION OF SKY NOISE TEMPERATURE AS A FUNCTION OF ATTENUATION

The effective sky noise (see Section 7.4.7.1 and Figure 7.5) due to the troposphere is primarily dependent on the attenuation at the frequency of observation. References 13 and 22 show the derivation of an empirical equation relating specific attenuation A to sky noise temperature:

$$T_s = T_m[1 - 10^{(-A/10)}] \qquad (8.17)$$

where T_s is the sky noise, T_m the mean absorption temperature of the attenuating medium (e.g., gaseous, clouds, rainfall), and A the specific attenuation that has been calculated in the previous subsections. Temperatures are in kelvins. The value

$$T_m = 1.12(\text{surface temperature in K}) - 50 \text{ K} \qquad (8.18)$$

has been empirically determined by Refs. 13 and 22.

TABLE 8.6 Cumulative Statistics of Sky Temperature Due to Rain for Rosman, North Carolina, at 20 GHz (T_m = 275 K)

Percentage of Year	Point Rain Rate Values (mm/h)	Average Rain Rate (mm/h)	Total Rain Attenuation[a] (dB)	Sky Noise Temperature[b] (K)
0.001	102	89	47	275
0.002	86	77	40	275
0.005	64	60	30	275
0.01	49	47	23	274
0.02	35	35	16	269
0.05	22	24	11	252
0.1	15	17	7	224
0.2	9.5	11.3	4.6	180
0.5	5.2	6.7	2.6	123
1.0	3.0	4.2	1.5	82
2.0	1.8	2.7	0.93	53

[a]At 20 GHz the specific attenuation $A = 0.06 R_{av}^{1.12}$ dB/km and for Rosman, NC, the effective path length is 5.1 km to ATS-6.
[b]For a ground temperature of 17°C = 63°F, the T_m is 275 K.
Source: Reference 16.

Some typical values taken in Rosman, North Carolina (Ref. 16) are given in Table 8.6 for rainfall. Also, CCIR Rep. 564-4 (Ref. 12), Section 3, should be consulted.

Example. From Table 8.6 with a total rain attenuation of 11 dB, what is the sky noise at 20 GHz? Assume T_m = 275 K.
Use equation 8.17.

$$T_s = 275(1 - 10^{-11/10})$$

$$= 253.16 \text{ K}$$

8.7 THE SUN AS A NOISE GENERATOR

The sun is a white-noise jammer of an earth terminal when aligned with the downlink terminal beam. This alignment occurs, for a geostationary satellite, twice a year near the equinoxes, and in the period of the equinox will occur for a short period each day. The sun's radio signal is of sufficient level to nearly saturate the terminal's receiving system, wiping out service for that period. Figure 8.15 gives the power flux density of the sun as a function of frequency. Above about 20 GHz the sun's signal remains practically constant at −188 dBW/Hz · m² for "quiet sun" conditions.

Reception of the sun's signal or any other solar noise source can be viewed as an equivalent increase in a terminal's antenna noise temperature by an

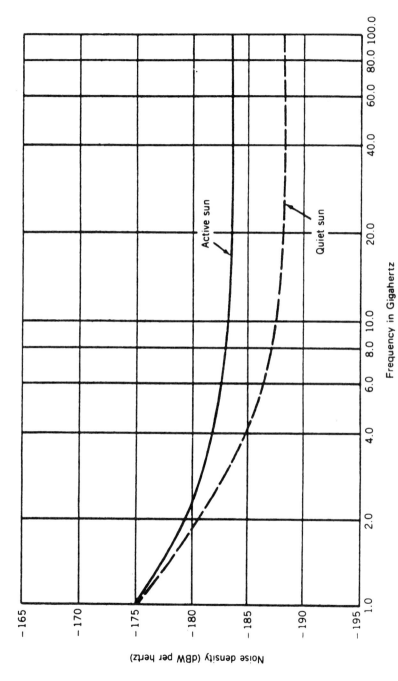

Figure 8.15. Values of noise from quiet and active sun. Sun fills entire antenna beam. (From Ref. 16.)

amount T_s. T_s is a function of terminal antenna beamwidth compared to the apparent diameter of the sun (e.g., 0.48°C), and how close the sun approaches the antenna boresight. The following formula, taken from Ref. 16, provides an estimate of T_s when the sun or any other extraterrestrial noise source is aligned in the antenna beam:

$$T_s = \frac{1 - \exp\left[-(D/1.2\theta)^2\right]}{f^2 D^2} \left(\log^{-1} \frac{S + 250}{10}\right) \qquad (8.19)$$

where D = apparent diameter of the sun or 0.48°
      f = frequency (GHz)
      S = power flux density, dBW/Hz · m²
      θ = half-power beamwidth of the terminal antenna (deg)

Example. An earth station operating with a 20-GHz downlink has a 2-m antenna (beamwidth of 0.5°). What is the maximum increase in antenna noise temperature that would be caused by a quiet sun transit?
    Use formula 8.19

$$T_s = 8146 \text{ K}$$

8.8 PROPAGATION EFFECTS WITH A LOW ELEVATION ANGLE

As the elevation angle of an earth station is lowered, the ray beam traverses an ever-increasing amount of atmosphere. Below about 10°, fading on the downlink signal must be considered. Fading and signal fluctuations apply only to the ground terminal downlink because its antenna is in close proximity to a turbulent medium. The companion uplink satellite path will suffer uplink fluctuation degradation only due to scattering of energy out of the path (Ref. 13). Because of the large distance traversed by the uplink signal since leaving the troposphere, the signal arrives at the satellite as a plane wave and with only a small amount of angle-of-arrival effects.
    The ITU-R Organization makes the following comments (Ref. 13) in the design of earth–space links. The effects of the nonionized atmosphere need to be considered at all frequencies but become critical above about 1 GHz and for low elevation angles. The effects include:

1. Absorption in atmospheric gases; absorption, scattering, and depolarization by hydrometers (water and ice droplets in precipitation, clouds, etc.); and emission noise from absorbing media; all of which are especially important at frequencies above about 10 GHz.
2. Loss of signal due to beam divergence of earth station antenna, due to the normal refraction in the atmosphere.

3. A decrease in effective antenna gain, due to phase decorrelation across the antenna aperture, caused by irregularities in the refractive index structure.

4. Relatively slow fading due to beam-bending caused by large-scale changes in the refractive index; more rapid fading (scintillation) and variations in angle-of-arrival caused by small-scale variations in the refractive index.

5. Possible limitation in bandwidth due to multiple scattering or multi-path effects, especially in high-capacity digital systems.

6. Attenuation by the local environment of the earth station (buildings, trees, etc.).

7. Short-term variations of the ratio of attenuations at the uplink and downlink frequencies, which may affect the accuracy of adaptive fade countermeasures.

8. Rotation of the plane of polarization (Faraday rotation), particularly if frequency reuse and linear polarization are employed.

9. Dispersion, which results in differential time delay across the bandwidth of the signal.

10. Excess time delay.

11. Scintillation, which affects amplitude, phase, and angle-of-arrival of the received signal.

Several of these items are commented on in the sections that follow.

8.8.1 Defocusing Loss

The regular decrease of refractive index with height causes ray-bending and hence a defocusing effect at low elevation angles. The magnitude of the defocusing loss of the antenna beam is independent of frequency and is less than 0.4 dB at a 3° elevation angle, even for high values of refractivity at ground level. It is therefore seldom important in system design in comparison with the attenuation due to other causes.

8.8.2 Decrease in Antenna Gain Due to Wavefront Incoherence

Incoherence of a wavefront of a wave incident on a receiving antenna is caused by small-scale irregularities of the refractive index structure of the atmosphere. Apart from the rapid signal fluctuations discussed in Section 8.8.3, they cause an antenna-to-medium coupling loss that can be described as a decrease in the antenna gain.

The effect increases both with increasing frequency and decreasing elevation angle. Measurements (Ref. 13) with a 22-m antenna at a 5° elevation angle showed losses of about 0.2 and 0.4 dB at 4 and 6 GHz, respectively.

Measurements with a 7-m antenna at a 5° elevation angle at 15.5 and 31.6 GHz gave losses of 0.3 and 0.6 dB, respectively. In practice, the effect is only likely to be significant for narrow-beamwidth antennas, higher frequencies, and elevation angles below 5°.

8.8.3 Scintillation

Scintillations are variations in amplitude, phase, polarization, and angle-of-arrival produced when radio waves pass through electron density irregularities in the ionosphere. Ionospheric scintillations present themselves as fast fluctuations of signal level with peak-to-peak amplitude fluctuations from 1 dB to over 10 dB and lasting for several minutes to several hours. The phenomena are caused by one of two types of ionospheric irregularities:

- Sufficiently high electron density fluctuations at scale sizes comparable to the Fresnel zone dimension of the propagation path
- Sharp gradients of ambient electron density, especially in the direction transverse to the direction of propagation

Either type of irregularity is known to occur in the ionosphere under certain solar, geomagnetic, and upper atmospheric conditions, and the scintillations can become so severe that they represent a practical limitation for telecommunication systems. Scintillations have been observed at frequencies from about 10 MHz to 12 GHz.

For systems applications, scintillations can be characterized by the depth of fading and fading period. A useful index to quantify the severity of scintillation is the scintillation index, S_4, which is defined as the standard deviation of received power divided by the mean value of the received power, or

$$S_4^2 = (\langle I^2 \rangle - \langle I \rangle^2)/\langle I \rangle^2 \qquad (8.20)$$

where I is the carrier intensity, and $\langle \, \rangle$ denotes ensemble average.

The fading period of scintillation varies over a large range from less than 0.1 s to several minutes, as the fading period depends both on the apparent motion of the irregularities relative to the ray path and, in the case of strong scintillation, on its severity. The fading period of gigahertz scintillation ranges from approximately 1 to 10 s. Long-period (of the order of tens of seconds) components of saturated scintillation (S_4 approaches 1) at VHF and UHF have also been observed.

The frequency dependence of S_4 is on the order of $f^{-1.5}$.

8.8.4   Cross-Polarization Effects

Frequency reuse by means of orthogonal polarization is often employed to increase the capacity of space telecommunication systems. This technique is restricted, however, by depolarization on atmospheric propagation paths. Various depolarization mechanisms, especially hydrometer effects, are important in the troposphere. Hydrometer effects are typically caused by ice, rain, and snow.

Faraday rotation of the plane of polarization of the ionosphere is yet another aspect. As much as 1° of rotation may be encountered at 10 GHz, and greater rotations at lower frequencies. Typical maximum rotations are 1.2 rotations at 500 MHz, 108° at 1 GHz, and 12° at 3 GHz. The frequency dependence of Faraday rotation is approximated by $1/f^2$. As seen from an earth station, the planes of polarization rotate in the same direction on the uplinks and downlinks. It is therefore impossible to compensate for Faraday rotation by rotating the feed system of the antenna, if the same antenna is used for both transmitting and receiving (Refs. 13, 23).

REVIEW EXERCISES

1. Give at least two causes of excess attenuation on satellite paths that must be taken into account for satellite systems operating above 10 GHz.

2. Identify the two frequency bands between 10 and 100 GHz where excess attenuation due to atmospheric gases is high, one of which is excessive.

3. List some uses one might make of these high-attenuation bands.

4. Argue why cumulative annual rainfall rates may not be used for calculation of excess attenuation due to rainfall and why we must use point rainfall rates.

5. Name at least four ways a system design engineer can build in a link margin.

6. In early attempts to build in sufficient margin on satellite and LOS link operation above 10 GHz, it was found that the required margins were excessively large because excess attenuation per kilometer was integrated along the entire path (the entire path in the atmosphere for satellite links). Describe how statistics on rain cell size assisted to better estimate excess attenuation due to rainfall.

7. Calculate the specific attenuation per kilometer for a path operating at 30 GHz on a LOS basis with a time availability for the path of 99.9%. Neglect path length considerations, of course. The path is located in

northeastern United States. Carry out the calculation for both horizontal and vertical polarizations.

8. Determine the excess attenuation due to rainfall for a LOS path operating at 50 GHz in central Australia. The path length is 20 km and the desired time availability 99.99%. Assume vertical polarization.

9. Calculate the excess attenuation due to rainfall for each polarization for a LOS path 25 km long with an operating frequency of 18 GHz; the desired path availability (propagation reliability) is 99.99%. The path is located in Massachusetts.

10. Name five ways to build a rainfall margin for the path in exercise 9.

11. Calculate the excess attenuation due to rainfall for a satellite path with a 20° elevation angle for a 21-GHz downlink. The earth station is located in southern Minnesota and the desired time availability for the link is 99%.

12. An earth station is to be located near Bonn, Germany, and will operate at 14 GHz. The desired uplink time availability is 99.95% and the subsatellite point is 10° W. Assume an elevation angle of 20°. What is the excess attenuation due to rainfall?

13. An earth station is to be installed on Diego Garcia, an island in the Indian Ocean, with an uplink at 44 GHz. The elevation angle is 15° and the desired path (time) availability is 99%. What value of excess attenuation due to rainfall should be used in the link budget?

14. There is an uplink at 30 GHz and the required excess attenuation due to rainfall is 15 dB. Path diversity is planned. Show how the value of excess attenuation due to rainfall for a single site can be reduced for site separations of 1, 2, 4, and 8 km.

15. For an earth station, calculate the excess attenuation due to atmospheric gases for a site near sea level. The site is planned for 30/20-GHz operation. The elevation angle is 15°. The relative humidity is 60% and the surface temperature is 27°C (70°F).

16. Determine the sky noise contribution for the attenuation of gases calculated in exercise 15. Calculate the sky noise temperature due to the excess attenuation due to rainfall from exercise 13.

17. A certain digital LOS hop operates at 23 GHz. The fade margin is 35 dB and the excess attenuation due to rainfall is 16 dB. By how many decibels must we overbuild the link?

REFERENCES

1. H. J. Liebe, *Atmospheric Propagation Properties in the* 10 *to* 75 *GHz Region: A Survey and Recommendations*, ESSA Tech. Report ERL 130-ITS, NBS, Boulder, CO, 1969.

2. "Attenuation by Atmospheric Gases," ITU-R (CCIR) Rec. 676-1, 1994 PN Series Volume, ITU, Geneva, 1994.

3. D. C. Hogg, "Millimeter Wave Propagation Through the Atmosphere," *Science*, Mar. 1968.

4. R. K. Crane, "Prediction of the Effects of Rain on Satellite Communication Systems," *Proc. IEEE*, vol. 65, 456–474, 1977.

5. "Radiometeorological Data," CCIR Rep. 563-1, Vol. V, XIVth Plenary Assembly, Kyoto, 1978.

6. "Propagation Data and Prediction Methods Required for Line-of-Sight Radio Relay Systems," CCIR Rep. 338-3, Vol. V, XVIth Plenary Assembly, Geneva, 1982.

7. "Radiometeorological Data," CCIR Rep. 563-2, Vol. V, XVth Plenary Assembly, Geneva, 1982.

8. "Specific Attenuation Model for Rain for Use in Prediction Methods," CCIR Rec. 838, 1994 PN Series Recommendations, ITU, Geneva, 1994.

9. "Characteristics of Precipitation for Propagation Modeling," ITU-R Rec. PN.837-1, 1994 PN Series Volume, ITU, Geneva, 1994.

10. "Attenuation by Hydrometers, in Particular Precipitation, and Other Atmospheric Particles," CCIR Rep. 721-3, Annex to Vol. V, XVIIth Plenary Assembly, Dusseldorf, 1990.

11. "Propagation Data and Prediction Methods Required for the Design of Terrestrial Line-of-Sight Systems," ITU-R Rec. PN.530-5, 1994 PN Series Volume, ITU, Geneva, 1994.

12. "Propagation Data and Prediction Methods Required for Earth–Space Telecommunication Systems," CCIR Rep. 564-4, Annex to Vol. V, XVIIth Plenary Assembly, Dusseldorf, 1990.

13. "Propagation Data and Prediction Methods Required for the Design of Earth–Space Telecommunication Systems," ITU-R Rec. PN.618-3, 1994 PN Volume, ITU, Geneva, 1994.

14. D. B. Hodge, "The Characteristics of Millimeter Wavelength Satellite-to-Ground Space Diversity Links," IEE Conf. No. 98, London, Apr. 1978.

15. Roger L. Freeman, *Reference Manual for Telecommunications Engineering*, 2nd ed., Wiley, New York, 1994.

16. R. Kaul, R. Wallace, and G. Kinal, *A Propagation Effects Handbook for Satellite Systems Design: A Summary of Propagation Impairments on 10–100 GHz Satellite Links, with Techniques for System Design*, NTIS N80-25520, ORI Inc., Silver Spring, MD, 1980 (U.S. Govt.—DoD, contract).

17. R. K. Crane and D. W. Blood, *Handbook for the Estimation of Microwave Propagation Effects—Link Calculations for Earth–Space Paths*, Environmental Research and Technology Report No. 1, Doc. No. P-7376-TRI, U.S. Department of Defense, Washington, DC, 1979.

18. R. K. Crane, *An Algorithm to Retrieve Water Vapor Information from Satellite Measurements*, NEPRF Tech. Report 7-76, Project No. 1423 (U.S. DoD contract), Environmental Research and Technology, Inc., Concord, MA, 1976.

19. "Attenuation Due to Clouds and Fog," ITU-R Rec. PN.840-1, 1994 PN Series Volume, ITU, Geneva, 1994.

20. K. L. Koester and L. H. Kosowsky, "Millimeter Wave Propagation in Ocean Fogs and Mists," *Proc. IEEE Antenna Propagation Symp.*, AP-26, 1978.

21. Roger L. Freeman, *Radio System Design for Telecommunications*, Wiley, New York, 1997.

22. K. H. Wulfsberg, "Apparent Sky Temperatures at Millimeter-Wave Frequencies," Physical Science Research Paper No. 38, Air Force Cambridge Research Laboratories, No. 64-590, 1964.

23. "Ionospheric Effects Influencing Radio Systems Involving Spacecraft," ITU-R Rec. PI.531-3, ITU, Geneva, 1994.

HIGH-FREQUENCY RADIO

9.1 GENERAL

Radio-frequency (RF) transmission between 3 and 30 MHz by ITU convention is called high frequency (HF) or shortwave. Many in the industry extend the HF band downward to just above the standard broadcast band from 2.0 to 30 MHz. This text holds with the ITU convention.

HF communication is a unique method of radio transmission because of its peculiar characteristics of propagation. This propagation phenomenon is such that many radio amateurs at certain times carry out satisfactory communication better than halfway around the world with 1–2 W of radiated power.

9.2 APPLICATIONS OF HF RADIO COMMUNICATION

HF is probably the most economic means of low information rate transmission over long distances (e.g., > 200 mi). It might be argued that meteor burst communication is yet more economic in some circumstances. Performance makes the difference. As we will learn in Chapter 10, meteor burst transmission links have the disadvantage of waiting time between short data packets. HF does not.

Traditionally, since the 1930s, HF has been the mainstay of ship–shore–ship communication. It still is. Satellite communication offered by INMARSAT (International Marine Satellite [Organization]) certainly provides a more reliable service, but HF continues to hold sway. We will not argue rationale one way or the other.

Ship–shore HF communication service includes:

- CW or continuous wave, which traditionally connotes keying a RF carrier on and off, forming the dots and dashes of the international Morse code.

- Selective calling teleprinter service (CCIR Rec. 493-3 [Ref. 1]) using frequency shift keying (FSK) with a wide frequency shift (see Chapter 13).
- Simplex teleprinter service with narrow and wide shift FSK as used by many of the world's navies such as the U.S. Navy's Fox broadcast, merchant marine broadcast (MERCAST), and Hydro broadcasts (Refs. 31 and 38).
- Single-sideband (SSB) voice telephony. This is often used for access to a national public switched telephone network.

HF is also used on ground–air–ground circuits for secondary, backup, and primary communication on some transoceanic paths.

HF is widely employed for propaganda broadcasts such as the U.S. Voice of America, Radio Free Europe, and the Voice of Moscow, to only mention a few. Many of these HF installations are very large, with effective isotropically radiated power (EIRPs) in excess of a megawatt.

HF also can provide very inexpensive point-to-point radio teleprinter service: simplex, half-duplex, full-duplex; four-channel time division multiplex (TDM); and 16-channel frequency division multiplex (FDM) using voice frequency carrier techniques (see Chapter 13).

Weather maps to ships and other entities are broadcast on HF by facsimile transmission using narrowband frequency modulation. It is used for the transmission of time signals such as WWV on 5, 10, 15, and 20 MHz. HF also remains one of the principal means of military communication, both tactical and strategic, for voice, record traffic, data, and facsimile.

HF has many disadvantages. Some of these are:

- Low information rate. The maximum bandwidth by radio regulation is four 3-kHz independent sideband voice channels in a quasi-frequency division configuration with low data rates (e.g., about 2400 bps per 3-kHz voice channel).
- Degraded link time availability when compared to satellite, fiber-optic, wire pair, troposcatter, and line-of-sight (LOS) microwave communication. HF link time availabilities vary from 80% up to better than 95% for some new spread spectrum wideband adaptive systems.
- Impairments include dispersion in both the time and frequency domains. Fading is endemic on skywave links. Atmospheric, galactic, and man-made interference noise are among the primary causes of low availability besides the basic propagation phenomenon itself.

9.3 COMPOSITION OF TYPICAL HF TERMINALS

9.3.1 Basic Equipment

A HF installation for two-way communication consists of one or more transmitters and one or more receivers. The most common type of modulation/waveform is single-sideband suppressed carrier (SSBSC). The operation of this equipment may be half- or full-duplex. We define half-duplex as the operation of a link in one direction at a time. In this case, the near-end transmitter transmits and the far-end receiver receives; then the far-end transmitter transmits and the near-end receiver receives. There are several advantages with this type of operation. A common antenna may be shared by a transmitter and receiver. Under many circumstances, both ends of the link use the same frequency. A simplified diagram of half-duplex operation with a shared antenna is shown below.

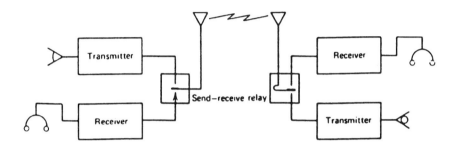

Full-duplex is when there is simultaneous two-way operation. At the same time the near end is transmitting to the far end, it is receiving from the far end. Usually, a different antenna is used for transmission than for reception. Likewise, the transmit and receive frequencies must be different. This is to prevent the near-end transmitter from interfering with its own receive frequency. Typical full-duplex operation is shown below.

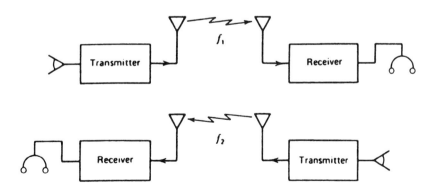

Being interfered with by one's own transmitter is called *cosite interference*. Care must be taken to assure sufficient isolation. There are many measures to be taken to mitigate cosite interference. Among those that should be considered are frequency separation, receiver selectivity, use of separate antennas and their sufficient separation, shielding, filtering transmit output, power amplifier linearity and grounding, bonding, and shielding. The worst environments for cosite interference and other forms of electromagnetic interference (EMI) include airborne and shipboard situations, particularly on military platforms.

When there is multiple transmitter and receiver operation required from a common geographical point, we may have to resort to two- or three-site operation. With this approach, isolation is achieved by physical separation of transmitters from receivers, generally by 2 km or more. However, such separations may even reach 20–30 km. At more complex installations, a third site may serve as an operational center, well isolated from the other two sites. In addition, we may purposely search for a "quiet" site to locate the receiver installation; "quiet" in this context means quiet regarding man-made noise. Both the transmitter site and receiver site may require many acres of cleared land for antennas.

The sites in question are interconnected by microwave LOS links and/or fiber-optic cable. Where survivability and reliability are very important, such as at key military installations, at least two distinct transmission media may be required. Generally, one is LOS microwave and the other some type of cable.

9.3.2 Basic Single-Sideband (SSB) Operation

Figures 9.1*a* and 9.1*b* are simplified functional block diagrams of a typical HF SSB transmitter and receiver, respectively.

Transmitter Operation. A nominal 3-kHz voice channel amplitude modulates a 100-kHz* stable intermediate-frequency (IF) carrier; one sideband is filtered and the carrier is suppressed. The resultant sideband couples the RF spectrum of $100 + 3$ kHz or $100 - 3$ kHz, depending on whether the upper or lower sideband is to be transmitted. This signal is then up-converted in a mixer to the desired output frequency, F_0. Thus the local oscillator output to the mixer is $F_0 + 100$ kHz or $F_0 - 100$ kHz. Note that frequency inversion takes place when lower sidebands are selected. The output of the mixer is fed to a linear power amplifier and then radiated by an antenna. Transmitter power outputs can vary from 10 W to 100 kW or more.

Receiver Operation. The incoming RF signal from the distant end, consisting of a suppressed carrier plus a sideband, is received by the antenna and filtered by a bandpass filter, often called a preselector. The signal is then amplified and mixed with a stable oscillator to produce an IF (assume again

---

*Other common IF frequencies are 455 and 1750 kHz.

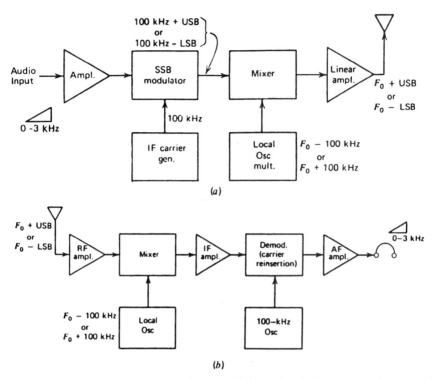

Figure 9.1. Simplified functional block diagram of (a) a typical SSB transmitter (100-kHz IF) and (b) a typical SSB receiver (100-kHz IF). USB = upper sideband; LSB = lower sideband; F_0 = operating frequency; IF = intermediate frequency.

that the IF is 100 kHz*). Several IF amplifiers increase the signal level. Demodulation takes place by reinserting the carrier at IF. In this case, it would be from a stable 100-kHz source, and detection is usually via a product detector. The output of the receiver is the nominal 3-kHz voice channel.

9.3.3 SSB System Considerations

One of the most important considerations in the development of a SSB signal at the near-end transmitter and its demodulation at the far-end receiver is accurate and stable frequency generation for use in frequency sources.

Generally, SSB circuits can maintain tolerable intelligibility when the transmitter and companion far-end receiver have an operating frequency difference no greater than 50 Hz. Narrow shift FSK operation such as voice frequency carrier telegraph (VFTG) will not tolerate frequency differences greater than 2 Hz.

*Other common IF frequencies are 455 and 1750 kHz.

Prior to 1960, end-to-end frequency synchronization for high-quality SSB circuits was by means of a pilot carrier. At the far end, the receiver would lock on the semisuppressed transmitted carrier and slave its local oscillators to this circuit.

Today, HF transmitters and receivers use synthesizers as frequency sources. A synthesizer is a tunable, highly stable oscillator. It gives one or several simultaneous sinusoidal RF outputs on *discrete* frequencies in the HF range. In most cases it provides the frequency supply for all RF carrier needs in SSB applications. For example, it will supply:

- Transmitter IF carrier
- IF carrier reinsertion supply
- Transmitter local oscillator supply
- Receiver local oscillator supplies

The following are some of the demands we must place on a HF synthesizer:

- Frequency stability
- Number of frequency increments
- Spectral purity of RF outputs
- Frequency accuracy
- Supplementary outputs
- Capability of being slaved to a frequency standard

Synthesizer RF output stability should be better than 1.5×10^{-8}. Frequency increments may be in 1-kHz steps for less expensive equipment and can range down to every 10 Hz for higher quality operational equipment.

9.3.4 Linear Power Amplifiers

The power amplifier in a SSB transmitter raises the power of a low-level signal with minimum possible added distortion. That is, the envelope of the signal output must be as nearly as possible an exact replica of the signal input. Therefore, by definition, a linear power amplifier is required. Such power amplifiers used for HF communication display output powers in the range of 10 W to 100 kW or more. More commonly, we would expect to find this range narrowed to 100 to 10,000 W. Power output of the transmitter is one input used in link calculations and link prediction computer programs described later in Section 9.5.1.2).

The trend in HF power amplifiers today, even for high-power applications, is the use of solid-state amplifier modules. A single module may be used for the lower power applications and groups of modules with combiners for higher power use. One big advantage of using solid state is that it eliminates the high voltage required for vacuum tube operation. Another benefit is improved reliability and graceful degradation. Here we mean that solid-state devices, especially when used as modular building blocks, tend to degrade rather than suffer complete failure. This latter is more the case for vacuum tubes.

9.3.4.1   *Intermodulation Distortion.* Nonlinearity in a HF transmitter results in intermodulation (IM) distortion when two or more signals appear in the waveform to be transmitted. Intermodulation noise or intermodulation products are discussed in Chapter 1.

IM distortion may be measured in two different ways:

- Two-tone test
- Tests using white noise loading

The two-tone test is carried out by applying two tones simultaneously at the audio input of the SSB transmitter. A $3:5$ frequency ratio between the two tones is desirable so that the IM products can be identified easily. For a 3-kHz input audio channel, a $3:5$ frequency ratio could be tones of 1500 and 2500 Hz.

The test tones are applied at equal amplitude, and their gains are increased to drive the transmitter to full power output. Exciter or transmitter output is sampled and observed on a spectrum analyzer. The amplitudes of the undesired products (see Section 1.9.1) and the carrier products are measured in terms of decibels below either of the equal-amplitude test tones as they appear in the exciter or transmitter output. The decibel difference is the signal-to-distortion ratio (S/D). This should be 40 dB or better. As one might expect, the highest level product is the third-order product. As discussed in Section 1.9.6, this product is two times the frequency of one tone minus the frequency of the second tone. For example, if the two test tones are 1500 and 2500 Hz, then

$$2 \times 1500 - 2500 = 500\ \text{Hz} \quad \text{or} \quad 2 \times 2500 - 1500 = 3500\ \text{Hz}$$

and, consequently, the third-order products are 500 and 3500 Hz. The presence of IM products numerically lower than 40 dB indicates maladjustment or deterioration of one or several transmitter stages, or overdrive.

The white noise test for IM distortion more nearly simulates operating conditions of a complex signal such as voice. The approach here is similar to that used to determine noise power ratio (described in Chapter 5). The 3-kHz audio channel is loaded with uniform-amplitude white noise and a slot is cleared, usually the width of a VFTG channel (e.g., 170 Hz). The signal-to-distortion ratio is the ratio of the level of the white noise signal outside the slot to the level of the distortion products in the slot.

9.3.5   HF Configuration Notes

In large HF communication facilities, space diversity reception is the rule. A rule of thumb for space diversity is that the antennas must be separated by a distance greater than 6λ. Each antenna terminates in its own receiver. The outputs of the receivers are combined either at the receiver site or at the operational center.

One receive antenna can serve many receivers by means of a multicoupler. A multicoupler, in this case, is a broadband amplifier with many output ports to which we can couple receivers.

A major impairment at HF receiver sites is man-made noise. For this reason, we select receiver sites that are comparatively quiet or quiet rural, as described in CCIR Rep. 258-5 (Ref. 25).

9.3.6   HF Link Operation

Figure 9.2 illustrates the operation of a HF link carrying teleprinter traffic. The upper side of the drawing shows the transmit side of the link. One-hop skywave operation is portrayed. The lower part of the drawing shows the receive side. The receive site is at the right. Its several noise impairments are shown, such as cosmic noise, man-made noise, atmospheric noise propagated into the site from long distances, local thunderstorm noise, and interference from other users.

In the figure the basic elements of a transmitter and of a receiver are shown as described in Sections 9.3.1–9.3.4. Antennas are described in Section 9.14.

The drawing shows a modem that converts the binary bit stream from a teleprinter keyboard to a signal compatible with the transmitter audio input. The modem develops an audio tone that is frequency shifted (FSK). A higher frequency tone represents a mark (binary 1) and the lower frequency tone represents a space (binary 0). This is further discussed in Chapter 13.

Another method is to offset the transmit synthesizer frequency by one and the other tone frequency. However, the first method is by far the most common. Several other tone formats are discussed in Section 9.12.

A companion modem is used on the receive side, which converts the receiver audio tone output to a binary serial bit stream compatible with the teleprinter. The modem need not be a separate entity but may be incorporated in the transmit exciter and receiver after the demodulator.

9.4   BASIC HF PROPAGATION

9.4.1   Introduction

A HF wave emitted from an antenna is characterized by a groundwave and a skywave component. The groundwave follows the surface of the earth and

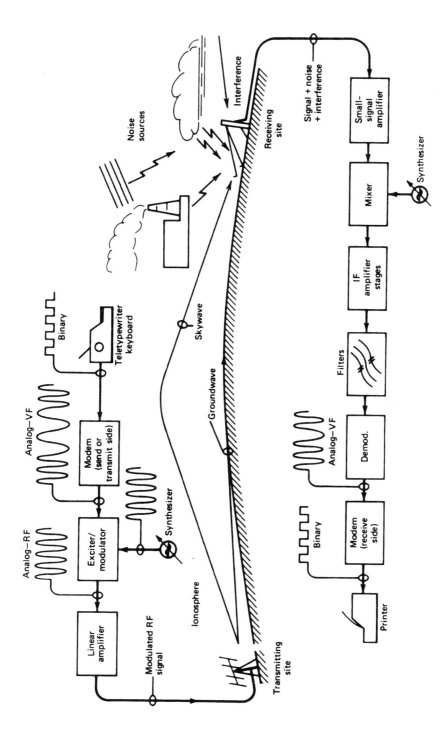

Figure 9.2. A typical HF radiolink providing teleprinter service. (Courtesy of Radio General Company.)

can provide useful communication over salt water up to about 650 mi (1000 km) and over land from some 25 (40 km) to 100 mi (160 km) or more, depending on RF power, antenna type and height off the ground, atmospheric noise, man-made noise, and ground conductivity. Generally, the lowest part of the HF band is the most desirable for groundwave communication. However, as we progress lower in frequency, depending on geographical location, atmospheric noise becomes a serious impairment. Well-designed groundwave links, with their shorter range, will achieve a better time availability than skywave links.

Skywave links are used for long circuits from about 100 mi (160 km) to over 8000 mi (12,800 km). We have had successful connectivity by continuous wave (CW) from a ship in the Ross Sea (Antarctica) to WCC (Chatham, MA) in the early morning hours (local time) for over 8 consecutive weeks of the Austral summer, and SSB voice communication from Frobisher Bay (Canada NWT) to Little America (Antarctica) in 1958 for periods of several hours a day for about $1\frac{1}{2}$ weeks. In both cases, modest antennas were used with transmitter outputs on the order of several hundred watts.

9.4.2 Skywave Transmission

The skywave transmission phenomenon of HF depends on ionospheric refraction. Transmitted radio waves hitting the ionosphere are bent or refracted. When they are bent sufficiently, the waves are returned to earth at a distant location. Often at the distant location they are reflected back to the sky again, only to be returned to earth still again, even further from the transmitter.

The ionosphere is the key to HF skywave communication. Look at the ionosphere as a layered region of ionized gas above the earth. The amount of refraction varies with the degree of ionization. The degree of ionization is primarily a function of the sun's ultraviolet (UV) radiation. Depending on the intensity of the UV radiation, more than one ionized layer may form (see Figure 9.3). The existence of more than one ionized layer in the atmosphere is explained by the existence of different UV frequencies in the sun's radiation. The lower frequencies produce the upper ionospheric layers, expending all their energy at high altitude. The higher frequency UV waves penetrate the atmosphere more deeply before producing appreciable ionization. Ionization of the atmosphere may also be caused by particle radiation from sunspots, cosmic rays, and meteor activity.

For all practical purposes four layers of the ionosphere have been identified and labeled as follows:

D Region or D-Layer. Not always present, but when it does exist, it is a
   daytime phenomenon. It is the lowest of the four layers. When it exists,
   it occupies an area between 50 and 90 km above the earth. The D
   region is usually highly absorptive due to its high collision frequency.

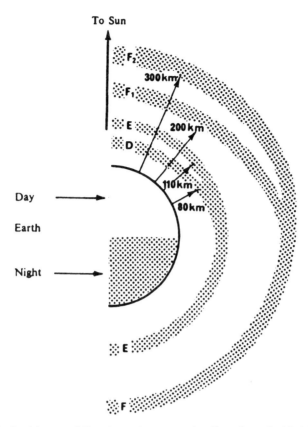

Figure 9.3. Ionized layers of the atmosphere as a function of nominal height above the earth's surface.

E Region or E-Layer. A daylight phenomenon, existing between 90 and 140 km above the earth. It depends directly on the sun's UV radiation and hence it is most dense directly under the sun. The layer all but disappears shortly after sunset. Layer density varies with seasons owing to variations in the sun's zenith angle with seasons.

F1-Layer. A daylight phenomenon existing between 140 and 250 km above the earth. Its behavior is similar to that of the E-layer in that it tends to follow the sun (i.e., most dense under the sun). At sunset the F1-layer rises, merging with the next higher layer, the F2-layer.

F2-Layer. This layer exists day and night between 150 and 250 km (night) and 250 and 300 km above the earth (day). During the daytime in winter, it extends from 150 to 300 km above the earth. Variations in height are due to solar heat. It is believed that the F2-layer is also strongly influenced by the earth's magnetic field. The earth is divided into three magnetic zones representing different degrees of magnetic intensity called east, west, and intermediate. Monthly F2 propagation predictions are made for each zone.* The north and south auroral zones are also important for F2 propagation, particularly during high sunspot activity.

Consider these layers as mirrors or partial mirrors, depending on the amount of ionization present. Thus, transmitted waves striking an ionospheric layer, particularly the F-layer, may be refracted directly back to earth and received after their first hop, or they may be reflected from the earth back to the ionosphere again and repeat the process several times before reaching the distant receiver. The latter phenomenon is called multihop transmission. Single-hop and multihop transmission are illustrated diagrammatically in Figure 9.4.

To obtain some idea of the estimated least possible number of F-layer hops as related to path length, the following may be used as a guide:

Number of Hops	Path Length (km)
1	< 4000
2	4000–7000
3	7000–12,000

(see Section 9.10.2.4).

HF propagation above about 8 MHz encounters what is called a *skip zone*. This is an "area of silence" or a zone of no reception extending from the outer limit of groundwave communication to the inner limit of skywave communication (first hop). The skip zone is shown graphically in Figure 9.5.

The region of coverage from a HF transmitter can be extended through the "skip zone" by a subset of skywave transmission called "near-vertical incidence" (NVI) or "quasi-vertical incidence" (QVI). The objective is to launch a wave nearly directly upward from the antenna. Therefore special antennas that have high takeoff angles are required (see Section 9.14). By radiating the HF RF energy nearly vertically, we can achieve "reflection" from the F-layer almost overhead. NVI propagation provides HF connectivity from 20 mi or less to over 500 mi. The best performance for this type of operation is to use frequencies from 3 to 6 MHz. Receive signals using this mode of propagation will fade rapidly (1 Hz), producing time dispersion on the order of 100 μs and with some Doppler spread. With proper signal processing at the receiver, these impairments can be mitigated. Using some of the new wideband HF techniques that are just now emerging can resolve the multipath and much of the dispersion.

---

*Monthly propagation forecasts are made by the Central Radio Propagation Laboratory (CRPL), U.S. National Bureau of Standards, now called NIST (National Institutes of Standards and Technology).

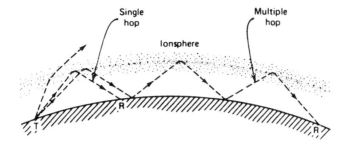

Figure 9.4. Single-hop and multihop HF skywave transmission. T = transmitter; R = receiver.

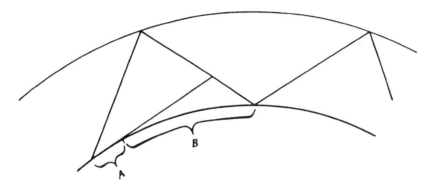

Figure 9.5. Skip zone. *A* = limit of groundwave communication; *B* = skip zone.

9.5 CHOICE OF OPTIMUM OPERATING FREQUENCY

One of the most important elements for the successful operation of a HF link using skywave transmission is the selection of an operating frequency that will assure a link with 100% time availability, day, night, and year-round. Seldom, if ever, can this be achieved.

Optimum HF propagation between points X and Y anywhere on the earth varies with the following:

- Location, in particular, latitude
- Season
- Diurnal variations (time of day)
- Cyclical variations (relating to the sunspot number)
- Abnormal (disturbed) propagation conditions

Location. The intensity of ionizing radiation that strikes the atmosphere varies with latitude. The intensity is greatest in equatorial regions, where the sun is more directly overhead than in the higher latitudes.

We find that the critical frequencies for E and F1 regions vary directly with the sun's elevation, being highest in equatorial regions and decreasing as a function of the increase in latitude. The *critical frequency* is the highest frequency, using vertical incidence (transmitting a wave directly overhead, at a 90° elevation angle), at a certain time and location, at which RF energy is reflected back to earth.

The F2-layer also varies with latitude. Here the issue is more complex. It is postulated that the variance in ionization is caused by other sources, such as X-rays, cosmic rays, and the earth's magnetic field. However, F2-layer critical frequency does not have a strong variance as a function of latitude but is more associated with longitude. Critical frequencies related to the F2-layer are generally higher in the Far East than in Africa, Europe, and the Western Hemisphere.

HF transmission engineers consider regions (location) in a more general sense, again related to latitude. The most attractive region that displays relatively low values of dispersion and Doppler spread is the temperate region, especially overwater paths in that region. The second region covers HF paths that are transequatorial. These paths encounter *spread F* propagation, which causes relatively high time dispersion because the F-layer over the equator is more diffuse, resulting in much greater multipath effects.

The third difficult region covers those HF paths that cross the auroral oval. There are two auroral ovals, one in the Southern Hemisphere and one in the Northern Hemisphere. Each is centered on its respective magnetic pole. Transauroral paths are difficult paths in terms of meeting performance requirements and are even more difficult during magnetic storms and other solar flare disturbances.

The most difficult of all HF paths are those that cross the polar cap, where Doppler spread has been measured in excess of 10 Hz and time dispersion over 1–2 ms (Refs. 37, 40).

Figure 9.6 shows an ionogram of a benign one-hop path in the evening. It shows the groundwave return appearing vertically on the far left, the one-hop F2 return(s) consisting of an O-ray and an X-ray. O stands for "ordinary" and the X for "extraordinary." How these rays are generated is discussed in Section 9.6.2.1. The ionogram also shows a vestige of multimode transmission appearing as a two-hop mode return on the far right side of the figure. Figure 9.7 shows a typical auroral oval, in this case, north of the equator. It should be kept in mind that the oval is indeed centered on the magnetic pole but that its boundary can shift; it is not fixed (Ref. 2).

Seasonal Factors. Our earth orbits the sun with an orbital period of 1 yr. It is this orbit that brings about our seasons: spring, summer, autumn, and winter. The sun is a major controlling element on the behavior of the ionosphere. For instance, E-layer ionization of sufficient magnitude to support skywave propagation depends almost entirely on the sun's elevation in the sky; it is stronger in summer than in winter.

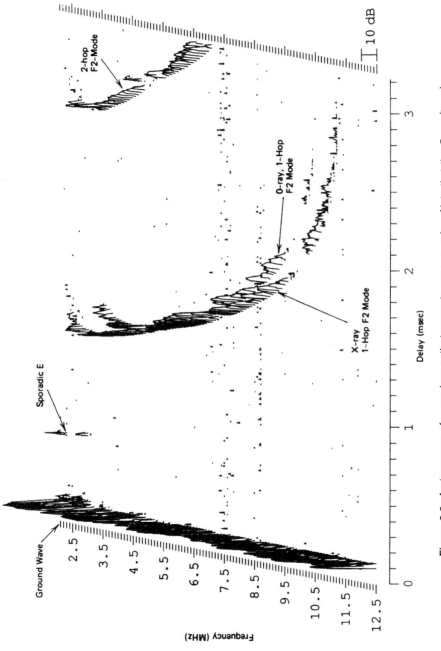

Figure 9.6. An ionogram of a one-hop path, temperate zone, evening. Note that the O-ray (trace) and X-ray (trace) are clearly visible. (Ionogram courtesy of Dr. L. S. Wagner, Naval Research Laboratories, Washington, DC, Ref. 39.)

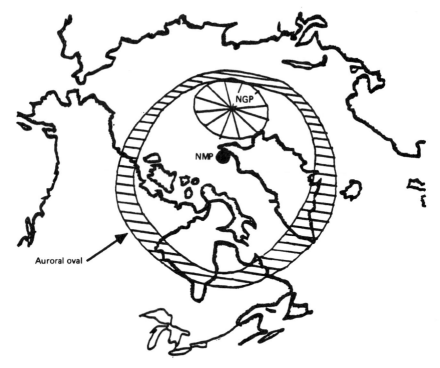

Figure 9.7. The auroral oval. NMP = northern magnetic pole; NGP = northern geographical pole.

The F1-layer, in general, exists only during daylight hours. During the winter its critical frequency varies in a similar manner as the E-layer, being dependent on the sun's elevation. Ionization, as one might imagine, is greater in the summer months than in the winter months, when there is less sunlight. Often, in the winter months, the F1-layer merges with the F2-layer and cannot be specifically identified except in equatorial regions.

The F2-layer is the reverse; daytime ionization is very intense and critical frequencies are higher than its F1 counterpart. Because of the extended periods of darkness in winter months, the ionosphere has more time to lose its electrical charge (recombination), resulting in nighttime critical frequencies dipping to very low values.

During the summer with its extended daylight hours, the F2-layer heats and expands, resulting in notably lower ionization density than in winter. This makes summer daytime F2 critical frequencies lower than winter values. The general assumption is that the F1- and F2-layers combine at night. This certainly is true in the winter months. In summer months, because of the longer daylight hours, this recombination does not occur to the extent that it does in winter. This results in nighttime F2 critical frequencies that are significantly higher in summer than in winter. Also, the difference between day and night critical frequencies is much smaller in summer than in winter.

Diurnal Variations (Time of Day). We break the 24-h day into three generalized time periods:

1. Day
2. Transitions
3. Night

The transition periods occurs twice: once around sunrise and again around sunset.

Again we are dealing with the intensity and duration of sunlight through the daytime hours. It is primarily the UV radiation from the sun that ionizes the atmosphere. During the daylight hours this radiation reaches a maximum intensity; during the hours of darkness, there is little UV radiation impinging the atmosphere and the E and F regions decrease to a relatively weak single layer through which propagate clouds with greater electron densities, which gives rise to sporadic E (E_s) propagation modes.

The F2 region critical frequency rises steeply at sunrise, reaching its maximum after the sun has reached its zenith. It starts to drop off sharply when the sun is starting to set. The transition periods, around sunrise and again around sunset, are when we encounter rapidly changing critical frequencies.

Cyclical Variations. The results of sunspots lead to the one phenomenon that affects atmospheric behavior more than any other. Sunspot activity is cyclical and is therefore referred to as the solar cycle. We characterize solar activity (i.e., number, intensity, and duration of sunspots) by the sunspot number. Sunspots appear on the sun's surface and are tremendous eruptions of whirling electrified gases. These gases are cooled to temperatures below those of the sun's surface, resulting in darkened areas appearing to us as spots on the sun. The whirling gases are accompanied by extremely intense magnetic fields. At times there are sudden flare-ups on the sun that some call *magnetic storms*. Magnetic storms will be discussed later.

Sunspots last from several days to several months. The sun has a rotational period of 27 days. Larger sunspots can remain visible for several rotations of the sun, giving rise to propagation anomalies with 27-day cycles.

Galileo was the first to report sunspots in modern recorded history. Some sources credit the Chinese, long before the birth of Christ, for being the first to observe sunspots.

The Zurich observatory began reporting sunspot data on a regular basis in 1749. Here Rudolf Wolf devised a standard method of measure of relative sunspot activity. This is called the Wolf sunspot number, or the Zurich sunspot number or just the sunspot number. The sunspot number is more of an index of solar activity than a measure of the number of sunspots observed.

The official sunspot number is derived from daily recording of sunspot numbers at a solar observatory in Locarno, Switzerland. The raw sunspot data are processed and disseminated by the Sunspot Index Data Center in Brussels. There has been a continuity of these measurements since 1749 when they were begun in Zurich. The sunspot cycle is roughly 11 years. A cycle is the time in years between contiguous solar activity minima. A minimum sunspot number has an average value of 5 and an average maximum value of 109 (Ref. 3). Solar cycles are numbered, with cycle No. 1 starting in 1749; cycle 21 began in June 1976 and cycle 22 in September 1986, reaching a maximum value of about 190 in January 1990. Sunspot cycles are not of uniform duration. Some are as short as 9 years and others as long as 14 years (Ref. 3). Figure 9.8 shows observed and predicted sunspot numbers up through 1991 and extrapolated until 1999.

Sunspots have a direct bearing on UV radiation intensity of the sun, which, as mentioned previously, is the principal cause of ionization of the upper atmosphere, which we call the ionosphere. It has a direct relation to the critical frequency and consequently to the *maximum usable frequency* (MUF). During periods of low sunspot numbers, the daytime MUF may not reach 17 MHz; during periods of sunspot maximum, the MUF has been known to far exceed 50 MHz. In sunspot cycle 19, state police radios operating in the 50-MHz region in Massachusetts and Rhode Island were

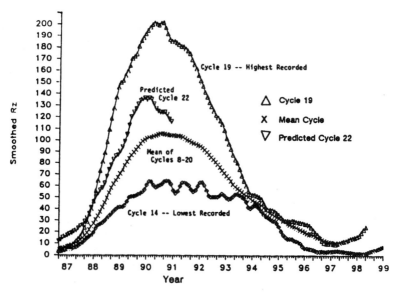

Figure 9.8. Observed and predicted sunspot numbers from McNish–Lincoln analysis. R_z is relative sunspot number (smoothed). (From Ref. 4.)

made virtually useless by interfering signals from the West Coast, some 3000 mi away. This is one reason most dispatching systems, especially in public safety, are now using 150 MHz and some low ultrahigh frequencies (UHF). Thus they avoid the cyclical problem of those periods when sunspot numbers are high.

Ionospheric Disturbances. We have classified four factors dealing with sky-wave HF transmission. If we neglect interference, there is some predictability for the design of a HF path. The computer has helped us along the way with such excellent programs as IONCAP and PROPHET (Refs. 5–7). They take into account sunspot number, latitude/longitude, time of day, season, and system characteristics. However, the transmission engineer feels nearly help-less when faced with the unexpected, the ionospheric disturbance.

We have classified three types of ionospheric disturbances:

1. Ionospheric storm (magnetic storm)
2. Sudden ionospheric disturbance (SID)
3. Polar blackout

The author was in the Arctic during cycle 19 peak (1958), aboard ship. Whenever there was a good display of aurora borealis, the HF band would black out and would remain that way from 12 to 72 h. All one would hear is a high level of thermal noise. Curiously, it seemed that signals in the low medium-frequency (MF) and low-frequency (LF) bands were enhanced in these periods of HF blackout.

There is fair correlation with the number and intensity of SIDs, iono-spheric storms, and blackouts with the sunspot cycle. The higher the sunspot number, the greater the probability of occurrence and the greater the intensity. Again, we believe the cause to be solar flare activity, a sudden change in the magnetic field around sunspots (Ref. 3). These produce high levels of X-rays, UV radiation, and cosmic noise. It takes about 8 min for the effects of this radiation to degrade the ionosphere. This is a SID. The result is heavy HF skywave signal attenuation.

Slower-traveling charged particles arrive some 18–36 h after the sudden flare-up on the sun. The result of these particles hitting the ionosphere is often blackout, especially in the higher latitudes. In other areas we notice a remarkable lowering of the MUF.

WWV transmitting from Boulder, Colorado, on 5, 10, 15, and 20 MHz is normally used for time and frequency calibration. At 18 min after each hour, WWV of the U.S. National Bureau of Standards (NBS) transmits timely solar and geomagnetic activity with updates every 6 h. WWV transmits the K index and the A index. The K index is a single digit from 0 through 9 and is a measure of current geomagnetic activity; 0 is the lowest activity factor and 9 the highest. The A index is a measure of solar flux activity.

9.5.1 Frequency Management

9.5.1.1 Definitions. In the context of this chapter, frequency management is the art/technology of selecting an optimum frequency for HF communication at a certain time of day between any two points on the earth. One important factor is omitted in the process, that is, the interference that may be present in the path, affecting one or both ends of the path.

We will be using the terms *MUF*, *LUF*, and *FOT* (*OWF*). The MUF and LUF are the upper and lower limiting frequencies for skywave communication between points X and Y. The MUF is the maximum usable frequency on an oblique incidence path; the LUF is the lowest usable frequency.

The concept of MUF is most important for the HF link design engineer for skywave links. We will find that the MUF is related to the critical frequency by the secant law. The optimum working frequency (OWF; original derivation from a French term), sometimes called FOT (from the French, *fréquence optimum de travail*), is often taken as $0.85 \times$ MUF. Little is mentioned in the literature about the LUF, since it is so system sensitive.

The MUF is a function of the sunspot number, time of day, latitude, day of the year, and so on, which are things completely out of our control. The LUF, on the other hand, is somewhat under our control. If we hold a transmitting station EIRP constant, as the operating frequency decreases, the available power at the distant receiver normally decreases owing to increased ionospheric absorption. Furthermore, the noise power increases so that the signal-to-noise ratio deteriorates and the circuit reliability decreases. The minimum frequency below which the reliability is unacceptable is called the lowest usable frequency (LUF). The LUF depends on transmitter power; antenna gain at the desired takeoff angle (TOA); factors that determine transmission loss over the path, such as frequency, season, sunspot number, and geographical location; and the external noise level. One of the primary factors is ionospheric absorption and hence, since this varies with the solar zenith angle, the LUF peaks at about noon. Consequently, in selecting a frequency, it is necessary to ascertain whether the LUF exceeds this frequency.

When applying computer prediction programs such as IONCAP, under certain situations we will find the LUF exceeds the MUF. This tells us, given the input parameters to the program, that the link is unworkable. We may be able to shift the LUF downward in frequency by relaxing link reliability (time availability), reducing signal-to-noise ratio requirements, increasing EIRP, and increasing receive antenna directional gain.

One factor in the transmission loss formula for HF is D-layer absorption. It varies as $1/f^2$. For this reason we want to operate at the highest possible frequency to minimize D-layer absorption. Suppose we operated at the MUF. It is a boundary limit and would be unstable with heavy fading. We choose a frequency as close as possible to the MUF yet stay out of boundary conditions. The operating frequency goes under two names: FOT or OWF, both defined earlier. The OWF is usually 0.85 the value of the MUF for F2 operation. Our objective is to keep our transmitter frequency as close to the

OWF as possible. This is done to minimize atmospheric absorption but yet not too close to the MUF to reduce ordinary–extraordinary ray fading. As the hours of the day pass, the OWF will move upward or downward and through the cycle of our 24-h day. It could move one or even two octaves from maximum to minimum frequency. How do we know when and where to move to in frequency?

9.5.1.2 Methods. There are six general methods in use today to select the best frequency. We might call these methods *frequency management*. The six methods are:

1. By experience.
2. Use of CRPL (Central Radio Propagation Laboratory) predictions.
3. Carrying out one's own predictions by one of several computer programs available.
4. Use of ionospheric sounders.
5. Use of distant broadcast facilities.
6. Self- and embedded sounding.

Experience. Many old-time operators still rely on experience. First they listen to their receivers, then they judge if an operating frequency change is required. It is the receive side of a link that commands a distant transmitter. An operator may well feel that "yesterday I had to QSY* at 7 p.m. to 13.7 MHz, so today I will do the same." Listening to his/her receiver will confirm or deny this belief. When he/she hears his/her present operating frequency (from the distant end) start to take deep fades and/or the signal level starts to drop, it is time to start searching for a new frequency. The operator checks other assigned frequencies on a spare receiver and listens for other identifiable signals near these frequencies to determine conditions to select a better frequency. If conditions are found to be better on a new frequency, the transmitter operator is ordered to change frequency (QSY at the distant end). This, in essence, is the experience method.

CRPL Predictions. Here, of course, we are dealing with the use of predictions issued by the U.S. Institute of Telecommunication Sciences, located in Boulder, Colorado. The predictions are published monthly, three months in advance of their effective dates (Ref. 8).

Consider a HF circuit designed for 95% time availability[†] (propagation reliability) and assume a minimum signal-to-noise ratio of 12 dB. The median

---

*A "Q" signal is an internationally recognized three-letter operational signal used among operators. A typical Q signal is QSY, which means "change your frequency to _____." Q signals were used almost exclusively on CW (Morse) circuits and their use has extended to teleprinter and even voice circuits.

[†]We define "time availability" as the percentage of time a certain performance objective (e.g., BER) is met.

receive signal level (RSL) must be increased on the order of 14 dB to overcome slow variations of skywave field intensity and atmospheric noise, and 11 dB to overcome fast variations of skywave field intensity. Therefore a rough order of magnitude value of signal-to-median-noise ratio with sufficient margin for 90% of the days is 37 dB for M-ary frequency shift keying (MFSK) data/telegraph transmission for a bit error rate (BER) of 1×10^{-4}. See CCIR Rec. 339-6 (Ref. 9).

Predictions by PC. Quite accurate propagation predictions for a HF path can be carried out on a personal computer such as the IBM/IBM clone PC-AT. A very widely accepted program is IONCAP, which stands for Ionospheric Communication Analysis and Prediction Program. It is written in Fortran (ANSI) and is divided into seven largely independent subsections (Refs. 5, 6):

1. Input subroutines.
2. Path geometry subroutines.
3. Antenna pattern subroutines.
4. Ionospheric parameter subroutines.
5. Maximum usable frequency (MUF) subroutines.
6. System performance subroutines.
7. Output subroutines.

Table 9.1 is a listing of 30 available output methods. The IONCAP computer program performs four basic analysis tasks. These tasks are summarized as follows. Note that E_s indicates highest observed frequency of the ordinary component of sporadic E and HPF means highest probable frequency.

1. *Ionospheric Parameters.* The ionosphere is predicted using parameters that describe four ionospheric regions: E, F1, F2, and E_s. For each sample area, the location, time of day, and all ionospheric parameters are derived. These may be used to find an electron density profile, which may be integrated to construct a predicted ionogram. These options are specified by methods 1 and 2 in Table 9.1.

2. *Antenna Patterns.* The user may precalculate the antenna gain pattern needed for the system performance predictions. These options are specified by methods 13–15 in Table 9.1. If the pattern is precalculated, then the antenna gain is computed for all frequencies (1–30 MHz) and elevation angles. If the pattern is not precalculated, then the gain value is determined for a particular frequency and elevation angle (takeoff angle) as needed.

3. *Maximum Usable Frequency (MUF).* The maximum frequency at which a skywave mode exists can be predicted. The 10% (FOT), 50% (MUF), and 90% (HPF [highest probable frequency]) levels are calculated for each of the

TABLE 9.1   IONCAP—Available Output Methods

Method	Description of Method
1	Ionospheric parameters
2	Ionograms
3	MUF–FOT lines (nomogram)
4	MUF–FOT graph
5	HPF–MUF–FOT graph
6	MUF–FOT–E_s graph
7	FOT–MUF table (full ionosphere)
8	MUF–FOT graph
9	HPF–MUF–FOT graph
10	MUF–FOT–ANG graph
11	MUF–FOT–E_s graph
12	MUF by magnetic indices, K (not implemented)
13	Transmitter antenna pattern
14	Receiver antenna pattern
15	Both transmitter and receiver antenna patterns
16	System performance (S.P.)
17	Condensed system performance, reliability
18	Condensed system performance, service probability
19	Propagation path geometry
20	Complete system performance (C.S.P.)
21	Forced long-path model (C.S.P.)
22	Forced short-path model (C.S.P.)
23	User-selected output lines (set by TOPLINES and BOTLINES)
24	MUF–REL table
25	All modes table
26	MUF–LUF–FOT table (nomogram)
27	FOT–LUF graph
28	MUF–FOT–LUF graph
29	MUF–LUF graph
30	Create binary file of variables in "COMMON/MUFS/" (allows the user to save MUFs–LUFs for printing by a separate user-written program)

Source: Reference 6.

four ionospheric regions predicted. These numbers are a description of the state of the ionosphere between two locations on the earth and not a statement of the actual performance of any operational communication circuit. These options are specified by methods 3–12 in Table 9.1.

4. *Systems Performance.* A comprehensive prediction of radio system performance parameters (up to 22) is provided. Emphasis is on the statistical performance over a month's time. A search to find the lowest usable frequency (LUF) is provided. These options are specified by methods 16–29 in Table 9.1.

Table 9.2 shows a typical run for method 16. The path is between Santa Barbara, California, and Marlborough, Massachusetts.

TABLE 9.2 Samples of an IONCAP Run, Method 16

```
                    METHOD  16  IONCAP  PC.20  PAGE 2
        OCT 1990              SSN = 192.
SANTA BARBARA TO MARLBOROUGH           AZIMUTHS    N. MI.      KM
34.50 N  119.00 W − 42.50 N  71.50 W   63.19   273.90  2252.5    4171.2
                    MINIMUM ANGLE 1.0 DEGREES

ITS − 1 ANTENNA PACKAGE
XMTR   2.0  TO  30.0  CONST. GAIN H  .00 L  .00 A  .0  OFF AZ  .0
RCVR   2.0  TO  30.0  CONST. GAIN H  .00 L  .00 A  .0  OFF AZ  .0
POWER = 1.000 KW 3 MHZ NOISE = −150.0 DBW REQ. REL = .80 REQ. SNR = 48.0
MULTIPATH POWER TOLERANCE = 6.0 DB    MULTIPATH DELAY TOLERANCE = .500 MS
```

UT MUF												
5.0	18.9	2.0	3.4	4.2	6.3	8.5	10.7	12.8	15.0	17.2	19.4	21.5 FREQ
	1F2	4 E	1F2	2ES	2F2	2F2	2F2	2F2	2F2	2F2	1F2	1F2 MODE
	3.8	8.7	3.2	1.3	11.1	11.4	12.2	14.0	19.0	19.0	3.5	3.5 ANGLE
	14.8	14.3	14.7	14.1	14.8	14.8	14.9	15.1	15.7	15.7	14.8	14.8 DELAY
	513.	103.	487.	110.	302.	309.	326.	363.	475.	475.	500.	500. V HITE
	.50	1.00	1.00	1.00	1.00	1.00	1.00	.99	.93	.74	.41	.10 F DAYS
	151.	158.	153.	157.	143.	142.	142.	145.	177.	235.	153.	173. LOSS
	16.	−15.	0.	2.	12.	15.	16.	14.	−16.	−73.	10.	−9. DBU
	−116	−128	−117	−117	−110	−111	−111	−114	−147	−204	−123	−143 S DBW
	−171	−136	−143	−145	−151	−158	−163	−166	−168	−170	−172	−173 N DBW
	55.	9.	25.	28.	40.	47.	51.	51.	22.	−34.	49.	30. SNR
	10.	48.	28.	24.	12.	5.	2.	8.	43.	99.	16.	34. RPWRG
	.64	.00	.00	.01	.16	.45	.68	.59	.11	.00	.52	.20 REL
	.00	.00	.00	.01	.03	.00	.00	.00	.00	.00	.00	.00 MPROB
11.0	13.7	2.0	3.5	5.1	6.6	8.2	8.6	9.7	11.2	12.8	14.3	15.9 FREQ
	1F2	3 E	2ES	2F2	2F2	1F2	1F2	2F2	1F2	1F2	1F2	1F2 MODE
	4.9	4.1	1.3	16.7	15.0	4.9	3.7	17.9	1.2	2.5	4.9	4.9 ANGLE
	14.9	14.1	14.1	15.4	15.2	14.9	14.8	15.6	14.5	14.7	14.9	14.9 DELAY
	557.	89.	110.	422.	385.	559.	510.	450.	407.	457.	557.	557. V HITE
	.50	1.00	.91	1.00	1.00	1.00	1.00	.98	.90	.68	.35	.10 F DAYS
	148.	172.	168.	147.	144.	142.	141.	150.	139.	142.	153.	169. LOSS
	12.	−28.	−6.	7.	10.	17.	18.	8.	19.	17.	8.	−8. DBU
	−117	−140	−123	−114	−113	−107	−107	−118	−109	−112	−122	−139 S DBW
	−167	−144	−148	−151	−154	−157	−158	−161	−164	−166	−168	−169 N DBW
	49.	4.	25.	36.	41.	49.	50.	42.	55.	54.	45.	30. SNR
	15.	54.	31.	19.	15.	6.	5.	22.	2.	8.	19.	35. RPWRG
	.53	.00	.01	.09	.18	.56	.61	.32	.74	.64	.43	.20 REL
	.00	.00	.00	.02	.00	.00	.00	.00	.00	.00	.00	.00 MPROB
17.0	42.3	2.0	4.6	7.1	9.7	12.2	14.8	17.4	23.0	28.7	34.3	40.0 FREQ
	1F2	3 E	2ES	2ES	2ES	2ES	4F2	2F2	2F2	2F2	2F2	1F2 MODE
	3.9	3.3	1.3	1.3	1.3	1.3	23.8	9.4	9.2	11.2	18.3	8.4 ANGLE
	14.8	14.1	14.1	14.1	14.1	14.1	15.8	14.6	14.6	14.8	15.6	15.4 DELAY
	517.	79.	110.	110.	110.	110.	261.	267.	264.	304.	459.	715. V HITE
	.50	1.00	1.00	1.00	1.00	.92	1.00	1.00	.99	.95	.83	.61 F DAYS
	152.	626.	348.	253.	213.	195.	186.	159.	152.	151.	188.	160. LOSS
	22.	* * * *	* * * *	−96.	−46.	−26.	−26.	5.	13.	16.	−21.	12. DBU
	−117	−355	−317	−220	−172	−154	−156	−126	−120	−119	−158	−127 S DBW
	−181	−145	−155	−160	−163	−165	−167	−169	−174	−176	−179	−180 N DBW
	63.	* * * *	* * * *	−60.	−9.	11.	11.	42.	53.	56.	21.	54. SNR
	1.	262.	221.	118.	67.	48.	50.	17.	8.	9.	44.	11. RPWRG
	.78	.00	.00	.00	.00	.00	.00	.25	.62	.66	.10	.61 REL
	.00	.00	.00	.00	.00	.00	.00	.00	.00	.00	.00	.00 MPROB

TABLE 9.2 (*Continued*)

24.0	38.2	2.0	4.0	6.1	10.3	14.6	18.8	23.0	27.3	31.5	35.8	40.0	FREQ
	1F2	3 E	2ES	4F2	2F2	2F2	2F2	2F2	2F2	2F2	1F2	1F2	MODE
	3.7	4.1	1.3	24.4	9.3	9.3	9.8	11.0	14.1	18.9	9.2	3.7	ANGLE
	14.8	14.1	14.1	15.9	14.6	14.6	14.7	14.8	15.1	15.7	15.5	14.8	DELAY
	510.	89.	110.	269.	266.	265.	276.	300.	366.	473.	749.	510.	V HITE
	.50	1.00	.89	1.00	1.00	1.00	1.00	.99	.97	.87	.66	.31	F DAYS
	157.	227.	177.	175.	151.	148.	148.	149.	155.	210.	169.	164.	LOSS
	11.	−84.	−27.	−22.	8.	13.	15.	15.	11.	−43.	2.	5.	DBU
	−127	−196	−146	−145	−119	−116	−117	−119	−123	−179	−136	−134	S DBW
	−180	−144	−149	−153	−161	−167	−171	−174	−176	−178	−179	−180	N DBW
	53.	−53.	3.	9.	42.	50.	54.	55.	52.	−2.	43.	46.	SNR
	12.	111.	55.	47.	13.	6.	3.	3.	13.	67.	22.	18.	RPWRG
	.60	.00	.00	.00	.22	.58	.72	.71	.57	.01	.38	.46	REL
	.00	.00	.00	.00	.00	.00	.00	.00	.00	.00	.00	.00	MPROB

Notes: Header information largely supplied by user.

Line 1: Month, year, and sunspot number.

Line 2: Label as supplied by user and headings for next line.

Line 3: Transmitter location, receiver location, the azimuth of the transmitter to the receiver in degrees east of north; path distance in nm and km.

Line 4: Minimum radiation angle in degrees.

Line 5: Antenna subroutine used (use IONCAP antenna subroutine).

Line 6: Transmitter antenna data.

Line 6A: Receiver antenna data.

Line 7: System line, which has transmitter power in kW, man-made noise level at 3 MHz in dBW, required reliability, and required S/N in dB.

System Performance Lines That Are Repeated for Each Hour

Line 0, FREQ: Time and frequency line as associated with each column. The first four lines always refer to the most reliable mode (MRM). The system performance parameter usually comes from the sum of all six modes.

Line 1, MODE: E is E-layer, F1 is F1-layer, F2 is F2-layer, E_s is E_s-layer; N is a one-hop E_s with n F1 or F2 hops (MRM).

Line 2, ANGLE: The radiation angle in degrees (MRM).

Line 3, DELAY: Time delay in ms (MRM).

Line 4, V HITE: Virtual height in km (MRM).

Line 5, F DAYS: The probability that the operating frequency will exceed the predicted MUF.

Line 6, LOSS: Median system loss in dB for the most reliable mode (MRM).

Line 7, DBU: Median field strength expected at the receiver location in dB above 1 μV/m.

Line 8, S DBW: Median signal power expected at the receiver input terminals in dB above a watt.

Line 9, N DBW: Median noise power expected at the receiver in dB above a watt.

Line 10, SNR: Median signal-to-noise ratio in dB.

Line 11, RPWRG: Required combination of transmitter power and antenna gains needed to achieve the required reliability.

Line 12, REL: Reliability. The probability that the SNR exceeds the required SNR. Note this applies to all days of the month and includes the effect of all mode types: E, F1, F2, E_s, and over-the-MUF modes.

Line 13, MPROB: The probability of an additional mode within the multipath tolerances (short paths only).

Source: Radio General Company, Stow, MA.

There are several other programs that can be run on a PC. One is Minimuf, which predicts only the MUF in midlatitudes (Ref. 10). PROPHET and Advanced PROPHET are two HF prediction codes for military application (Refs. 7, 11). The accuracy is not as good as IONCAP, but they are a useful tool for propagation prediction and for the specific military applications for which they are designed. They provide area coverage easily, which IONCAP does not, and also allow interaction for various details of electronic warfare.

Ionospheric Sounders. Ionospheric oblique sounders give real-time data on the MUF, transmission modes propagating (F1, F2, one-, two- or three-hop), and the LUF between two points that are sounder-equipped. One location has a sounder transmitter installed that sweeps the entire HF band using a FM/CW signal. Transmitter power output usually is in the range of 1–10 W. A sounder receiver and display is operated at the distant end. The receiver is time synchronized with the distant transmitter. It displays a time history of the sweep of the band. The *x* axis of the display is the HF band 2 (3) to 30 MHz and the *y* axis is time delay measured in milliseconds. Here received power is displayed as a function of frequency and time delay. Of course, the shortest delay is that power reflected off the E-layer, the next shortest the F-layer (F1 and F2 in daytime, in that order), then multihop power from the F-layers. A multihop signal is delayed even more. Figure 9.9 shows a conceptual diagram of oblique incidence sounder operation.

One system designed and manufactured by BR Communications, is the AN/TRQ-35. This unit includes a low-power transmitter and, at the far end of the circuit, a receiver with a spectrum monitor that compiles channel occupancy statistics for 9333 channels spaced 3 kHz from 2 to 30 MHz. BR Communications calls their spectrum monitor a "signal-to-noise ratio" analyzer that determines the background noise level—usually atmospheric noise —and then compiles statistics on the percentage of time that level is exceeded in each channel. The monitor has memory and can display any channel usage up to 30 min (Ref. 12).

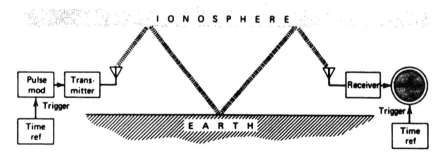

Figure 9.9. Synchronized oblique sounding.

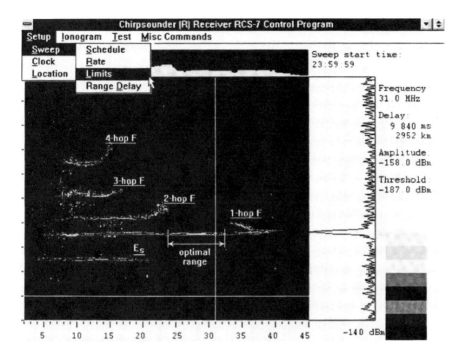

Figure 9.10. Typical HF ionogram from an oblique sounding system. (Courtesy of TCI International, Inc., Sunnyvale, CA.)

A typical oblique ionospheric display (ionogram) is shown in Figure 9.10. Another type of sounder is the backscatter or vertical incidence sounder, where the receiver and transmitter are collocated. In this case, the HF spectrum is swept as before, but the RF energy is launched vertically for refraction by the ionosphere directly overhead. Backscatter sounding provides data on the critical frequency, from which we can derive the MUF for any one-hop path. This is done with the following formula, which is based on Snell's law:

$$F_0 = F_n \sec \phi \tag{9.1}$$

where F_n = the maximum frequency (critical frequency) that will be reflected back at vertical incidence

F_0 = the maximum usable frequency (MUF) at oblique angle ϕ

ϕ = the angle of incidence (i.e., the angle between the direction of propagation and a line perpendicular to the earth)

Backscatter techniques can be used on any one-hop path where one point is where the backscatter sounder is located. On the other hand, oblique incidence sounders provide information on the single, equipped path, which may be one hop or multiple hops.

Another approach that can be used on some occasions is to know the locations of other operating oblique sounder transmitters near the desired location for HF connectivity. "Near" can mean several hundred miles either side of the north–south direction and about 150 mi either side in the east–west direction of the desired distant end. Naturally, synchronization and interface data on these other sounders are necessary if we wish to make use of their signals.

Use of Distant Broadcast Facilities. We can obtain rough order-of-magnitude propagation data derived from making comparative signal-level measurements on distant HF broadcast facilities "near" the desired distant point with which we wish to establish HF connectivity. Many facilities broadcast simultaneously on multiple HF bands, so a comparison can be made from one band to another on signal level, fade rate, and fade depth to determine the optimum band for operation. Many HF receivers are equipped with S-meters* from which we can derive some crude comparative level measurements.

For example, if we wish to communicate with an area in or near the U.S. state of Colorado, we can use WWV, which has excellent time and frequency dissemination service transmitting on 5, 10, 15, and 20 MHz simultaneously 24 h a day.

International broadcast such as the BBC and the Voice of America are other candidates. With present state-of-the-art synthesized HF receivers, we can tune directly to them. If one can read CW (copy international Morse code), there are hundreds of marine coastal stations that will have continuous transmissions on 4, 6, 8, 12, 17, and 22 MHz (the marine bands), depending on which bands each facility considers to be "open" (useful) for the service that facility provides. These facilities will drop off automatic operation and into manual CW operation when they are "in traffic" (operating with a distant ship station). We may call this method "poor man's sounding."

Embedded Sounding, Early Versions. These earlier devices, still very widely used, are based on the idea that a facility or facility grouping has only a limited number of assigned operating frequencies. These devices transmit low RF energy, usually using FSK at a low bit rate, to sample each frequency to derive link quality assessment (LQA) data and, in some instances, the same device may be used for selective calling. LQA involves the exchange of link quality information on a one-on-one basis or networkwide. SELSCAN, a Rockwell–Collins trademark, is a good example of such a system. It measures

---

*An S-meter gives a comparative measurement of received signal level.

signal-to-noise ratio and multipath delay distortion on up to 30 stored preset channels for use by automatic frequency selection algorithms. SELSCAN also has an ALE (automatic link establishment) feature (Ref. 13).

It is suggested that the reader consult MIL-STD-188-141A (Ref. 14) for some excellent methods of ALE and LQA. We are really dealing here with an OSI layer-2 data link access protocol (see Chapter 13). SELCAN, however, has been designed as an adjunct to SSB voice operation on HF for setting up a link on an optimum frequency. Harris (United States) and Tadiran (Israel), among others, have similar ALE/LQA devices.

More Advanced Embedded Sounding. These are self-sounding systems designed for HF data/telegraph circuits that are half- or full-duplex. We will consider such circuits, for this argument, to have as near as possible to 100% duty cycle (i.e., they are active nearly 100% of the time). Such systems dedicate pure overhead, perhaps 5% or 10%, for self-sounding. Self-sounding, in this context, means that no separate sounding equipment is used and that a cooperative distant end is required.

Two cooperating facilities tied by a HF link operate on a fully synchronous basis and have their clocks, and hence time of day (TOD), slaved to a common system time or to universal coordinated time (UCT). At intervals, every 5 min, for example, both facilities take time out from exchanging traffic and search, first above their operating frequency and then below. Such systems work well with wideband HF, which operates over a full 1 MHz or 500 kHz with very low level PN spread phase shift keying (PSK) signals. Here then each station moves up 1 MHz, for instance, then down 1 MHz in synchronism, each exchanging LQA data with the other regarding new frequencies.

LQA, for instance, can be a 3-bit group contained in the header information. A 1 1 1 exchange from each would indicate to move to the higher frequency, and a 1 1 0 to move to the lower frequency. Thus both stations are constantly homing in on the optimum working frequency (OWF). During transition periods, the search is carried out more frequently during calm periods, such as from, say, 10 a.m. to 3 p.m. At midpath (local time) the search may only be required every 10 min. Such systems often exchange only short message blocks and LQA data are updated in every header. As the binary number starts to drop, an algorithm kicks off a retimer to shorten the period between searches. If the first two digits hold in at binary 1 for long periods (e.g., 15 min.), the timer is extended. The LQA three-digit word can be controlled by BER measurement or signal-to-noise ratio monitor (S/N). The problem with the use of S/N only is that it may not be indicative of multipath where BER or automatic repeat request (ARQ) negative acknowledgments (NACKs) (see Chapter 13 for ARQ) will be indicative of BER, which often can be traced to intersymbol interference, a typical effect of multipath.

9.6 PROPAGATION MODES

There are three basic modes of HF propagation:

1. Groundwave.
2. Skywave (oblique incidence).
3. Near-vertical incidence (NVI). This is a distinct subset of the skywave mode.

9.6.1 Basic Groundwave Propagation

The spacewave (not skywave) intensity decreases with the inverse of the distance, the groundwave with the inverse of the distance squared. Therefore at long distances and nonzero elevation angles, the intensity of the spacewave exceeds that of the groundwave. The groundwave is diffracted somewhat to follow the curvature of the earth. The diffraction increases as frequency decreases. Diffraction is also influenced by the imperfect conductivity of the ground. Energy is absorbed by currents induced in the earth so that energy flow takes place from the wave downward. The loss of energy dissipated in the earth leads to attenuation dependent on conductivity and dielectric constant. With horizontal polarization the wave attenuation is greater than with vertical polarization due to the different behavior of Fresnel reflection coefficients for both polarizations.

To summarize, groundwave is an excellent form of HF propagation where we can, during daylight hours, achieve 99% or better time availability. Groundwave propagation decreases with increasing frequency and with decreasing ground conductivity. As we go down in frequency, atmospheric noise starts to limit performance in daytime, and skywave interference at night is a basic limiter of groundwave performance on whichever of the lower frequencies we wish to use, providing we are not operating above the MUF.

9.6.2 Skywave Propagation

A wave that has been reflected from the ionosphere is commonly called a skywave. The reflection can take place at the E region and the F1 and/or F2 regions. In some circumstances RF energy can be reflected back from any two or all three regions at once. Figure 9.3 shows the three regions or layers of interest as well as the D-layer, which is absorptive. Figure 9.4 illustrates skywave communication.

HF skywave communications can be by one-hop, two-hop, or three-hop, depending on path length and ionospheric conditions. Figure 9.11 shows eight possible skywave modes of propagation. On somewhat longer paths ($\approx > 1000$ km) we can receive RF energy from two or more modes at once, giving rise to multipath reception, which causes signal dispersion. Dispersion

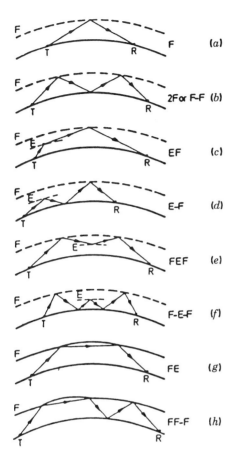

F   F   (a)

F   2F or F-F (b)

F   E   EF   (c)

F   E   E-F   (d)

F   E   FEF   (e)

F   E   F-E-F   (f)

F   FE   (g)

F   FF-F   (h)

Figure 9.11. Examples of possible skywave multihop/multimode paths.

results in intersymbol interference (ISI) on digital circuits. Multimode reception is shown a bit later in Figure 9.13.

In general, we can say that multipath propagation can arise from the following:

- Multihop, especially when transmit and receive antennas have low gain and low takeoff angles
- Low and high angle paths (the low ray and the high ray)
- Multilayer propagation
- Ordinary (O) and extraordinary (X) rays from one or more paths

These effects, more often than not, are existent in combination. An example of the high ray and low ray as they might be seen on an ionogram is shown in Figure 9.12. It also shows the MUF.

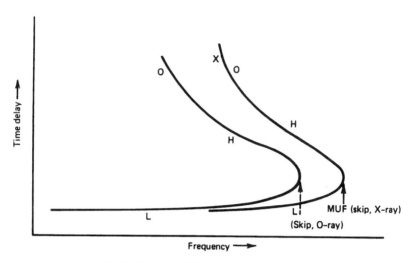

Figure 9.12. Ionogram idealized sketch. Oblique propagation along magnetic field for a fixed distance (point-to-point), F region. High ray (H) and low ray (L) and splitting into X-ray and O-ray are shown.

Typical one-hop ranges are 2000, 3400, and 4000 km for E-, F1-, and F2-layer reflections, respectively. These limits depend on the layer height of maximum electron density for rays launched at grazing incidence. The range values take into account the poor performance of HF antennas at low elevation angles.

Distances beyond the values given previously can be achieved by utilizing consecutive reflections between the ionosphere and the earth's surface (see Figure 9.4 and Figure 9.11*b–h*). For each ground reflection the signal must pass through the absorptive D-layer twice, adding significantly to signal attenuation. The ground reflection itself is absorptive. It should also be noted that the elevation angle increases as a function of the hop number, which results in the lowering of the path MUF.

The skywave propagation based on F-layer modes is identified by a three-character notation such as 1F1, 1F2, 2F2, and 3F2. The first digit is the hop number and the second two characters identify the dominant mode (i.e., F1 or F2 reflection). Accordingly, 2F2 means two hops where the dominant mode is F2 reflection.

A HF receiving installation will commonly receive multiple modes simultaneously, typically 1F2 and 2F2 and at greater distances 2F2 and 3F2. The strongest mode on a long path is usually the lowest order F2 mode unless the antenna discriminates against this. It can be appreciated that higher order F2 modes suffer greater attenuation due to absorption by D-layer passage and ground reflection. The result is a lower level signal than the lower order F2 propagating modes. In other words, a 3F2 mode is considerably more attenuated at a certain location than a 2F2 mode if it can be received at the same location (assuming isotropic antennas). It has traveled through the D-layer two more times than its 2F2 counterpart and been absorbed one more time by ground reflection.

E-layer propagation is rarely of importance beyond one hop. Reflections from the F1-layer occur only under restricted conditions, and the 1F1 mode is less common than the 1E and 1F2 modes. The 1F1 mode is more common at high latitudes at ranges of 2000–2500 km. Multiple-hop F1 modes are very rare (Ref. 15).

Long-distance HF paths ($\approx > 4000$ km) involve multiple hops and a changing ionosphere as a signal traverses a path. This is especially true on east–west paths, where much of the time some portion of the path will be in transition, part in daylight, part in darkness. At the equator, a 4000-km path extends across nearly three time zones, further north or south, four and five time zones. Mixed-mode operation is a common feature of transequatorial paths. Figure 9.11 illustrates single-mode paths (*a* and *b*) and mixed-mode paths (*d–f*). Figure 9.11*c* shows a path with asymmetry. This occurs when a wave frequency exceeds the E-layer MUF only slightly, so that the wave does not penetrate the layer along a rectilinear path but will be bent downward, resulting in the asymmetry.

9.6.2.1 *Ray or Wave Splitting.* Because of the presence of the geomagnetic field, the ionosphere is a doubly refracting medium. Magnetoionic theory shows that a ray entering the ionosphere is split into two separate waves owing to the influence of the earth's magnetic field. One ray of the split rays is called the ordinary wave (O-wave) and the other the extraordinary wave (X-wave). The O-wave is refracted less than its X-wave counterpart and this is reflected at a greater altitude; the O-wave will have a lower critical frequency, hence a lower MUF. Both waves experience different amounts of refraction and thus travel independently along different ray paths displaced in time at the receiver, typically from 1 to 10 μs. The O-wave suffers less absorption and is usually the stronger and therefore the more important. Figure 9.12 is a conceptual sketch of O- and X-wave propagation as well as the high ray and the low ray. Figure 9.13 is a simplified diagram of the arrival at a receiver of the X-ray and O-ray (or wave).

So even if we have a one-mode path, say, somewhat under 1000 km, the received signal will still be impaired by multipath arising from the O-ray and X-ray; and if we shift down in frequency somewhat from the MUF, there is further multipath impairment of the high ray and the low ray, which are both split into O and X.

Simple multipath fading is shown in Figure 9.14, which is a snapshot sketch of 1 MHz of HF spectrum showing the effects of multipath on an emitted signal. At certain frequencies, the multipath components are in phase and the signal level constructs and builds to a higher level than predicted by calculation. At other frequencies the multipath energy components are out of phase and the signal amplitude resultant shows the out-of-

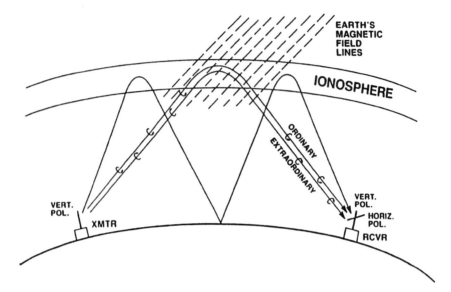

Figure 9.13. The skywave channel, showing the formation of the X- and O-waves, and simultaneous reception of one hop and two hops at receiver.

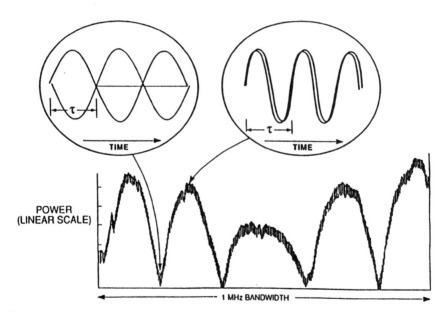

Figure 9.14. A snapshot sketch of a 1-MHz bandwidth, showing the frequency dependence of multipath effects on a transmitted wideband HF (WBHF) signal.

phase destruction. Where multipath components are just 180° out of phase, there is full destruction and no signal present at that snapshot moment in time at that frequency. There are five such nulls shown in the snapshot picture (Figure 9.14).

9.6.3 Near-Vertical Incidence (NVI) Propagation

NVI propagation is used for short-range HF communication. It can fill the *skip zone* or *zone of silence*; here groundwave propagation is no longer effective, to the point where one-hop skywave, using oblique incidence propagation, may be used. NVI utilizes the same skywave principles of propagation previously discussed. The key factor in NVI operation is the antenna. For effective HF communication using the NVI mode, the antenna must radiate its main beam energy at a very high angle (TOA), near vertical, if you will.

NVI circuits suffer the same impairments as oblique skywave circuits, but in the case of NVI the fading is more severe, particularly polarization fading. One rough rule of thumb is that if an equivalent oblique path experiences, say, 5 μs of dispersion, the NVI path will experience about 10 times as much, or 50 μs. NVI paths in a temperate zone have been known to suffer as much as 100 μs of dispersion. In equatorial areas such paths may experience even greater dispersion. We can also have dispersion from a second hop where the principal signal power is being derived from a one-hop dominant mode. Figure 9.15 shows diagrammatically the operation of NVI propagation. The letter A in the figure shows the extent of useful communication by means of a groundwave component if we were to transmit with a low-elevation-angle antenna with vertical polarization, such as a whip.

We will generally favor the lower frequencies for NVI operation, from 2 to 7 MHz. Higher-elevation-angle circuits tend to lower MUFs to start with. However, atmospheric noise limitations may drive us up in frequency as it does with groundwave operation. We will discuss atmospheric noise in Section 9.9.3. Atmospheric and man-made noise levels decrease with increasing operating frequencies.

9.6.4 Reciprocal Reception

Many HF system engineers assume reciprocity on a HF path. This means that if there is a good path from point X to point Y, the path is equally good from point Y to point X. This serves as the basis of one-way sounding and we do not quarrel with the assumption. There are several points that should be considered. The interference levels (including atmospheric noise) may be different at X than at Y. Also, propagation loss may differ owing to the influence of the earth's magnetic field. These two points may be particularly true on long paths. Using proper link quality assessment (LQA) procedures in both directions can alleviate such asymmetric situations by each far-end

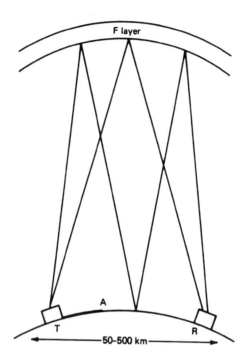

Figure 9.15. A near-vertical incidence (NVI) path showing one and two hops from the F-layer. The letter A shows the maximum effective range of groundwave.

receiver optimizing its companion near-end transmitter operation both in frequency management and RF output power.

9.7 HF COMMUNICATION IMPAIRMENTS

9.7.1 Introduction

There are a number of important impairments on a HF channel that affect received signal quality. Different signals are affected in different ways and in severity. Digital data transmission can be severely affected by dispersion, both in the time domain and in the frequency domain. Analog transmission is much less affected, as is CW. FSK is much more tolerant than coherent PSK. Signal-to-noise ratio requirements are a function of signal type. Table 9.3 reviews some typical HF waveforms (signal modulation types), required bandwidths, and signal-to-noise ratios.

Much of the thrust in this section is on digital transmission. At the end of the section we will briefly discuss medium-related impairments to analog transmission.

TABLE 9.3 Emission Types, Bandwidths, and Required Signal-to-Noise Ratios

Class of Emission	Predetection Bandwidth of Receiver (Hz)	Postdetection Bandwidth of Receiver (Hz)	Grade of Service	Audio Signal-to-Noise Ratio (1) (dB)	RF Signal-to-Noise Density Ratio (2, 3) (dB) Stable Condition	Fading Condition (4) Nondiversity	Fading Condition (5) Dual Diversity
A1A telegraphy 8 bauds	3000	1500	Aural reception (6)	−4	31	38	
A1B telegraphy 50 bauds, printer	250	250	Commercial grade (7)	16	40		58
A1B telegraphy 120 bauds, undulator	600	600		10	38		49
A2A telegraphy 8 bauds	3000	1500	Aural reception (6)(19)	−4	35	38	
A2B telegraphy 24 bauds	3000	1500	Commercial grade (7)(19)	11	50	56	
F1B telegraphy 50 bauds, printer 2D = 200–400 Hz	1500	100	(8) $P_C = 0.01$ / $P_C = 0.001$ / $P_C = 0.0001$		45 / 51 / 56 (9)	53 / 63 / 74 (9)	45 / 52 / 59 (9)
F1B telegraphy 100 bauds, printer 2D = 170 Hz, ARQ	300	300	(10)		43	52	
F7B telegraphy 200 bauds, printer 2D = …, ARQ			(10)				
F1B telegraphy MFSK 33-tone ITA2 10 characters/s	400	400	(8) $P_C = 0.01$ / $P_C = 0.001$ / $P_C = 0.0001$		23 / 24 / 26	37 / 45 (25) / 52 (25)	29 / 34 / 39
F1B telegraphy MFSK 12-tone ITA5 10 characters/s	300	300	(8) $P_C = 0.01$ / $P_C = 0.001$ / $P_C = 0.0001$		26 / 27 / 29	42 / 49 (25) / 56	32 / 36 / 42
F1B telegraphy MFSK 6-tone ITA2 10 characters/s	180	180	(8) $P_C = 0.01$ / $P_C = 0.001$ / $P_C = 0.0001$		25 / 26 / 28	41 / 48 (25) / 55	31 / 35 / 41
F7B telegraphy							

615

TABLE 9.3 (Continued)

Class of Emission	Predetection Bandwidth of Receiver (Hz)	Postdetection Bandwidth of Receiver (Hz)	Grade of Service		Audio Signal-to-Noise Ratio (1) (dB)	RF Signal-to-Noise Density Ratio (2, 3) (dB)		
						Stable Condition	Fading Condition (4), (5)	
							Nondiversity	Dual Diversity
R3C phototelegraphy 60 rpm	3000	3000				50	59	
F3C phototelegraphy 60 rpm	1100	3000	Marginally commercial	(22)	15	50	58	
			Good commercial	(22)	20	55	65	
A3E telephony double-sideband	6000	3000	Just usable	(11)	6	50	51	48
			Marginally commercial	(12)	15 (18)	59	64 (20)	60 (15)
			Good commercial	(13)	33	67(14)	75(14)	70(14) (20)
H3E telephony single-sideband full carrier	3000	3000	Just usable	(11)	6	53	54	51
			Marginally commercial	(12)	15 (18)	62 (23)	67 (20)	63 (15)
			Good commercial	(13)	33	70(14)	78(14)	73(14) (20)
R3E telephony single-sideband reduced carrier	3000	3000	Just usable	(11)	6	48	49	46
			Marginally commercial	(12)	15 (18)	57 (24)	62 (20)	58 (15)
			Good commercial	(13)	33	64(14)	73(14)	68(14) (20)
J3E telephony single-sideband suppressed carrier	3000	3000	Just usable	(11)	6	47	48	45
			Marginally commercial	(12)	15 (18)	56	61 (20)	57 (15)
			Good commercial	(13)	33	64(14)	72(14)	67(14) (20)
B8E telephony independent-sideband 2 channels	6000	3000 per channel	Just usable	(11)	6	49	50	47
			Marginally commercial	(12)	15 (18)	58	63 (20)	59 (20)
			Good commercial	(13)	33	66(14)	74(14)	69(14)
B8E telephony independent-sideband 4 channels	12000	3000 per channel	Just usable	(11)	6	50	51	48
			Marginally commercial	(12)	15 (18)	59	64 (20)	60 (15)
			Good commercial	(13)	33	67(14)	75(14)	70(14) (20)
J7B multichannel VF telegraphy 16 channels 75 bauds each	3000	110 per channel	$P_C = 0.01$	(8)		59 (21)	67 (21)	59 (21)
			$P_C = 0.001$			65	77	66
			$P_C = 0.0001$			69	87	72

616

Footnotes to TABLE 9.3

Note 1. Noise bandwidth is equal to postdetection bandwidth of receiver. For an independent-sideband telephony, noise bandwidth is equal to the postdetection bandwidth of one channel.

Note 2. The figures in this column represent the ratio of signal peak envelope power to the average noise power in a 1-Hz bandwidth except for double-sideband A3E emission, where the figures represent the ratio of the carrier power to the average noise power in a 1-Hz bandwidth.

Note 3. The values of the radio-frequency signal-to-noise density ratio for telephony listed in this column apply when conventional terminals are used. They can be reduced considerably (by amounts as yet undetermined) when terminals of the type using linked compressor–expanders (Lincompex) are used. A speech-to-noise (rms voltage) ratio of 7 dB measured at audio frequency in a 3-kHz band has been found to correspond to just marginally commercial quality at the output of the system, taking into account the compandor improvement.

Note 4. The values in these columns represent the median values of the fading signal power necessary to yield an equivalent grade of service and do not include the intensity fluctuation factor (allowance for day-to-day fluctuation), which may be obtained from Report 252-2 + Supplement (published separately) in conjunction with Report 322 (published separately). In the absence of information from these reports, a value of 14 dB may be added as the intensity fluctuation factor to the values in these columns to arrive at provisional values for the total required signal-to-noise density ratios, which may be used as a guide to estimate required monthly-median values of hourly-median field strength. This value of 14 dB has been obtained as follows:

The intensity fluctuation factor for the signal, against steady noise, is 10 dB, estimated to give protection for 90% of the days. The fluctuations in intensity of atmospheric noise are also taken to be 10 dB for 90% of the days. Assuming that there is no correlation between the fluctuations in intensity of the noise and those of the signal, a good estimate of the combined signal and noise intensity fluctuation factor is

$$\sqrt{10^2 + 10^2} = 14 \text{ dB}$$

Note 5. In calculating the radio-frequency signal-to-noise density ratios for rapid short-period fading, a log-normal amplitude distribution of the received fading signal has been used (using 7 dB for the ratio of median level to level exceeded for 10% or 90% of the time) except for high-speed automatic telegraphy services, where the protection has been calculated on the assumption of a Rayleigh distribution. The following notes refer to protection against rapid or short-period fading.

Note 6. For protection 90% of the time.

Note 7. For A1B telegraphy, 50-baud printer: for protection 99.99% of the time. For A2B telegraphy, 24 bauds: for protection 98% of the time.

Note 8. The symbol P_C stands for the probability of character error.

Note 9. Atmospheric noise ($V_d = 6$ dB) is assumed (see Report 322).

Note 10. Based on 90% traffic efficiency.

Note 11. For 90% sentence intelligibility.

Note 12. When connected to the public service network: based on 80% protection.

Note 13. When connected to the public service network: based on 90% protection.

Note 14. Assuming 10-dB improvement due to the use of noise reducers.

Note 15. Diversity improvement based on a wide-spaced (several kilometers) diversity.

Note 16. Transmitter loading of 80% of the rated peak envelope power of the transmitter by the multichannel telegraph signal is assumed.

Note 17. Required signal-to-noise density ratio based on performance of telegraphy channels.

Note 18. For telephony, the figures in this column represent the ratio of the audio-frequency signal, as measured on a standard VU-meter, to the rms noise, for a bandwidth of 3 kHz. (The corresponding peak signal power, i.e., when the transmitter is 100% tone-modulated, is assumed to be 6 dB higher.)

Note 19. Total sideband power, combined with keyed carrier, is assumed to give partial (two-element) diversity effect. An allowance of 4 dB is made for 90% protection (8 bauds), and 6 dB for 98% protection (24 bauds).

Note 20. Used if Lincompex terminals will reduce these figures by an amount yet to be determined.

Note 21. For fewer channels these figures will be different. The relationship between the number of channels and the required signal-to-noise ratio has yet to be determined.

Note 22. Quality judged in accordance with article 23.1 of ITU publication "Use of the Standardized Test Chart for Facsimile Transmissions."

Note 23. For class of emission H3E the levels of sideband signals and pilot-carrier corresponding to 100% modulation are each -6 dB relative peak envelope power (PEP). SSB receiver used for reception.

Note 24. For class of emission R3E the pilot-carrier level of -20 dB relative to PEP is applied and the level of the sideband signal corresponding to 100% modulation is 1 dB lower than the PEP.

Note 25. Dependent on fading rate, typical values shown.

Source: Reference 9, Table 1, CCIR Rep. 339-6 (see also Ref. 17).

9.7.2 Fading

HF skywave (and NVI) signals suffer from fading, a principal impairment on HF. It results from the characteristics of the ionosphere. The amplitude and phase of skywave signals fluctuate with reference to time, space, and frequency. These effects collectively are described as fading and have a decisive influence on the performance of HF radio communication systems.

We consider four types of fading here (Ref. 16):

1. *Interference Fading.* This is the most common type of fading encountered on HF circuits. It is caused by mixing of two or more signal components propagating along different paths. This is multipath fading, which may arise from multiple-mode and multiple-layer propagated rays, high- and low-angle modes, groundwaves, and skywaves. This latter phenomenon is usually encountered during transition periods and at night.

2. *Polarization Fading Due to Faraday Rotation.* This is brought about by the earth's magnetic field and the split into the O-ray and the X-ray, which become two elliptically polarized components. Both components can interfere to yield an elliptically polarized resultant wave. The major axis of the resulting ellipse will have continuous changes in direction due to changes in the electron density encountered along its propagation paths. HF antennas are ordinarily linearly polarized. However, when an elliptically polarized wave with a constantly changing electric field vector shifts from perpendicular to the receiving HF antenna to parallel to the HF receiving antenna, the input signal voltage to the receiver will vary from maximum (perpendicular case) to zero (parallel case). Consequently, the receiver input voltage will vary according to the spatial rotation of the ellipse of polarization. Fading periods vary from a fraction of a second to seconds.

3. *Focusing and Defocusing Due to Atmospheric Irregularities.* These deformed layers can focus or defocus a signal wave if they encounter deformities that are concave or convex, respectively. The motion of these structures can cause fades with periods up to some minutes.

4. *Absorption Fading Caused by Solar Flare Activity.* This type of fading particularly affects the lower frequencies, and fades may last from minutes to more than an hour.

Of interest to the communication engineer are fading depth, duration, and frequency (fade rate). For short-term fading, the fading depth is the difference in decibels between the signal levels exceeded for 10% and 90% of the time. Measurements have confirmed that we may expect about 14 dB for a Rayleigh distribution short-term fading, which is the most common form of such fading. This value is valid for paths 1500–6000 km long and does not vary much with the time of day or season.

For long-term variations in signal level (i.e., variations of hourly median signal values of a month), the log-normal distribution provides a best fit. A good value is an 8-dB margin for HF paths below 60° geomagnetic latitude and 11 dB for paths above 60° and especially over the polar cap region (Ref. 16). This suggests that a 22-dB fade margin should provide better than a 99% time availability (for fading, not frequency management) assuming a 3σ point and for temperate zone paths. This decibel value is referenced to the median signal level.

Some fade rate values on typical HF circuits are provided in Figure 9.16. Figure 9.17 gives a sampling of typical fade durations on a medium-to-long path.

9.7.3 The Effects of Impairments at the HF Receiver

9.7.3.1 General. In Section 9.7.2 we discussed fade rate and depth. This is amplitude fading. It is accompanied by associated group path delay and phase path delay. Doppler shift arises from ionospheric movement as well as movement if one or both ends of the path are mobile (i.e., in motion).

9.7.3.2 Time Dispersion. On skywave paths the primary cause of time dispersion is multipath propagation, which derives from differences in transit time between different propagation paths, as discussed in Section 9.4. The multipath spread causes amplitude and phase variations in the signal spectrum owing to interference of the multipath wave components. When these fluctuations are correlated within the signal bandwidth (e.g., 3 kHz) and all the spectral components behave more or less in the same manner, we then call this *flat fading*. When these fluctuations have little correlation, the fading is called *frequency selective fading*. Time dispersion is characterized by a delay power spectrum and is measured as multipath delay spread in microseconds or milliseconds.

Time dispersion is an especially serious and destructive impairment to digital communication signals on HF. One rule of thumb that is useful is that if the delay spread exceeds half the time width (period) of a signal element (baud), the error rate becomes intolerable. This is one rationale for extending the width of a signal element (e.g., by lowering the *baud* rate), to combat time dispersion. For instance, if a serial bit stream is transmitted at 100 bauds, a baud period is 0.01 s (10 ms). In this case, the circuit will remain operational, although with a degraded BER, if the time dispersion remains under 5 ms (i.e., half the period of a signal element or bit). At 200 bauds the half-baud period value drops to 2.5 ms and at 50 bauds it is 10 ms.

Multipath has been shown to be a function of the operating frequency relative to the MUF. Multipath delay tends to approach zero as the operating frequency approaches the MUF value. Turn to Figure 9.12 and we see that as we approach the MUF moving upward in frequency, we will receive only one signal power component. (*Note*: There are really two MUFs, one for the O-ray and one for the X-ray trace. The X is usually below the O in signal strength.)

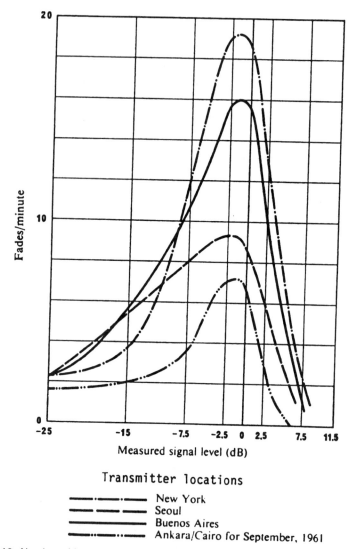

Figure 9.16. Number of fades per minute as a function of the signal level for various circuits terminating in Frankfurt, Germany. (From Ref. 18, Figure 3, CCIR Rep.197-4.)

9.7.3.3 Frequency Dispersion. Experience on operating HF circuits has shown that frequency dispersion (Doppler spread) nearly always is present when there is time dispersion. But the converse does not necessarily hold true. The Doppler shift and spread are due to a drifting ionosphere. As the signal encounters elemental surfaces of the ionosphere, each with a different velocity vector, the result is a Doppler spread. Such Doppler shifts and spreads can have disastrous effects, particularly on narrowband FSK systems,

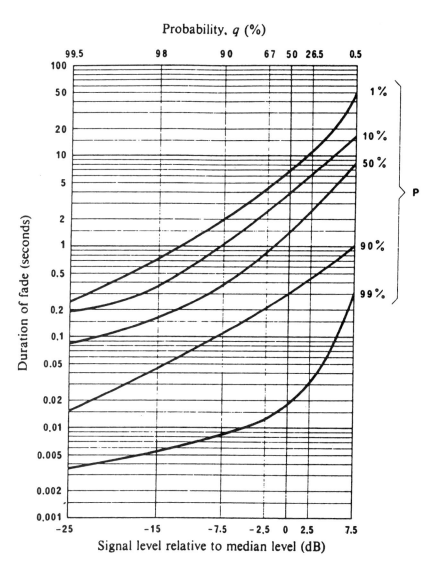

Figure 9.17. Duration of fades as a function of the level of the test signal. Circuit: New York–Frankfurt, Germany; September 14, 1961; 1100 h Central European Time; frequency 13.79 MHz. The figures on the right-hand side of the curves represent the percentage *p* of the number of fades for which a given duration of fade is exceeded. The measured values of signal levels are shown together with the probability *q* that these levels will be exceeded. (From Ref. 18, Figure 4, CCIR Rep. 197-4.)

which can tolerate no more than ± 2-Hz total frequency departure. If the 2-Hz value is exceeded, the BER will approach 5×10^{-1}. Typical values of Doppler spread on midlatitude paths are 0.1–0.2 Hz.

9.8 MITIGATION OF PROPAGATION-RELATED IMPAIRMENTS

The general approach microwave engineers use to overcome propagation impairments is to increase link margin (i.e., S/N). Generally, this approach has drawbacks with HF propagation because many of the impairments may be difficult to overcome with increasing power. Other measures also merit consideration.

One measure commonly used at HF installations is to employ diversity. The basis of diversity is to take advantage of signals that are not correlated. There are three basic types of diversity: space, time, and frequency. Other types of diversity are variants of the three basic types. Several types of diversity are discussed below.

Space Diversity. The same signals are received by at least two antennas separated in space. Separation should be greater than six wavelengths.

Polarization Diversity. Signals are received on antennas with different polarizations.

Frequency Diversity. Information is transmitted simultaneously on different frequencies. Some refer to this as in-band diversity, where subcarrier tones are separated by from 300 to 7000 Hz. CCIR recommends at least 400-Hz separation between redundant FSK tones in ITU-R Rec. F436-3 (Ref. 19). MIL-STD-188-342 (Ref. 20) suggests a voice frequency (VF) carrier modulation plan shown in Table 9.4. The tone pairing plan is given at the bottom of the table. In this case redundant data are transmitted on pairs of FSK tones separated by 1360 Hz.

Time Diversity. The signal is transmitted several times. Some techniques are available that can dramatically improve error performance using time diversity. These techniques not only take advantage of the decorrelation of fading with time, but also the simple redundancy.

Channel coding with interleaving is only now being exploited on HF. We discussed some typical channel coding and interleaving techniques in Section 5.6. This coupled with automatic repeat request (ARQ) schemes using short message blocks can bring error performance on HF links into manageable bounds.

One method of mitigation of the dispersive effects of multipath propagation by pulse width extension was discussed in Section 9.7.3.2.

TABLE 9.4   A Voice Frequency Carrier Telegraph (VFCT) Modulation Plan for HF

Channel Designation	Mark Frequency (Hz)	Center Frequency (Hz)	Space Frequency (Hz)
1	382.5	425	467.5
2	552.5	595	637.5
3	722.5	765	807.5
4	892.5	935	977.5
5	1062.5	1105	1147.5
6	1232.5	1275	1317.5
7	1402.5	1445	1487.5
8	1572.5	1615	1657.5
9	1742.5	1785	1827.5
10	1912.5	1955	1997.5
11	2082.5	2125	2167.5
12	2252.5	2295	2337.5
13	2422.5	2465	2507.5
14	2592.5	2635	2677.5
15	2762.5	2805	2847.5
16[a]	2932.5	2975	3017.5
17[b]	3012.5	3145	3187.5
18[b]	3272.5	3315	3357.5

Diversity

Pair 1(9), 2(10), 3(11), 4(12), 5(13), 6(14), 7(15), and 8(16)
Note: Connect loop to lower numbered channel of each pair.

[a]Marginal over HF (nominal 3-kHz) channels.
[b]Not usable over HF channels.

Source: Reference 20, MIL-STD-188-342.

9.9   HF IMPAIRMENTS—NOISE IN THE RECEIVING SYSTEM

9.9.1   Introduction

In previous chapters on radio systems, the primary source of noise was thermal noise generated in the receiver front end. Except under certain special circumstances, this is not the case for HF receiving systems. External noise is by far dominant for HF receivers. In declining importance, we categorize this noise as follows:

1. Interference from other emitters.
2. Atmospheric noise.
3. Man-made and galactic noise.
4. Receiver thermal noise.

9.9.2   Interference

The HF band has tens of thousands of users who are assigned operating frequencies by national authorities, such as the FCC in the United States, and by the International Frequency Registration Board (IFRB), a subsidiary organization of the ITU. The emitters operate in frequency bands in accordance with the service they perform and the region of the world they are in. The ITU has divided the world into three regions. We must also take into consideration noise from ionospheric sounder transmitters, harmonics, and spurious emissions from licensed emitters. Of primary importance in establishing a HF link is a "clear" frequency. Clear means interference-free, and the next consideration is how interference-free. On several tests of HF data modems it was found that interference even 15 dB or more below desired signals corrupted BERs. If we were to try to use a 3-kHz channel, we could well find that an emitter 5000 km away can place a signal 30 dB, 40 dB, or more over our desired signal. With nominal 1-MHz wideband HF systems, it is estimated that interference from in-band emitters may contribute an integrated noise level over 30 dB above atmospheric noise (Ref. 21). A typical spectrum analyzer snapshot of a European interference spectrum is shown in Figure 9.18.

One can look at the HF spectrum between the LUF and MUF as a picket fence in which the pickets are the interferers. Unfortunately, by the very nature of HF propagation, these "pickets" randomly appear and disappear,

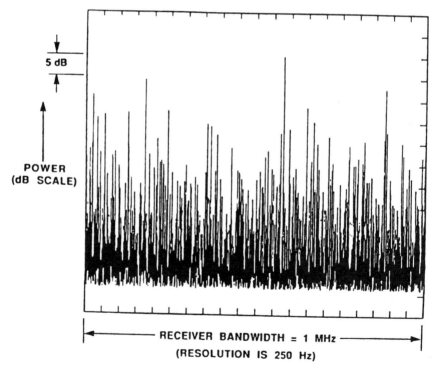

Figure 9.18. Spectrum analyzer snapshot of European HF interference spectrum.

their location changes, and their amplitudes increase and decrease not only owing to fading but also to constant changes in the ionosphere as the MUF changes with time, sporadic E propagation, transition F-layer changes, and so on.

Reference 22 gives congestion values in Manchester, United Kingdom, in July 1982 for the period between about noon and midnight local time. The HF band was scanned by stepping a receiver in 1-kHz increments and with a bandwidth of 1 kHz. It shows that the broadcast bands are the most congested and have the highest interference levels. In this study receive levels starting at −117 dBm were recorded in decades up to −77 dBm. Several examples of their measurements are given in Table 9.5, which shows the percentage of congestion values at defined thresholds.

There are several techniques to overcome or mitigate interference. The most desirable, of course, is to operate on a clear frequency with no interference at the distant-end receiver at the optimum working frequency (OWF). This is not often fully achievable.

In nearly all cases a facility is assigned a group of frequencies, usually with one of the following bandwidths, dependent on the service: 1, 3, 6, 9, or 12 kHz. Also see Table 9.3. Broadcast installations have other bandwidths. Broadcast is not a concern of this text. Distant-end receivers usually "command" near-end transmitters. The receiver site checks a new frequency for propagation and occupancy. More modern systems use advanced oblique

TABLE 9.5 Percentage Congestion Values at Defined Threshold[a]

Frequency Band (kHz)	Service (dBm)	Day				
		− 117	− 107	− 97	− 87	− 77
5950–6200	Broadcast	63	41	23	9	2
8195–8500	Maritime mobile	24	11	5	1	1
10150–10600	Fixed mobile	70	33	17	10	4
12050–12230	Fixed	82	50	24	11	5
13800–14000	Fixed/mobile	60	30	17	6	2
15000–15100	Aeronautical mobile	57	31	10	4	1
15100–15600	Broadcast	96	80	55	33	16
17410–17550	Fixed	44	21	8	5	1
		Night				
5950–6200	Broadcast	100	100	93	75	51
8195–8500	Maritime mobile	100	92	52	29	13
10150–10600	Fixed/mobile	100	91	49	24	11
12050–12230	Fixed	100	88	60	35	16
13800–14000	Fixed/mobile	72	15	6	2	1
15000–15100	Aeronautical mobile	97	63	32	7	5
15100–15600	Broadcast	96	78	52	33	16
17410–17550	Fixed	53	32	17	7	2

[a] Note that − 107 dBm is equivalent here to a received field strength of 2 μV/m.

Source: Reference 22.

sounders such as the AN/TRQ-35 or SELSCAN, Autolink, or MESA (Rockwell, Harris, and Tadiran trademarks, respectively), which also provide occupancy data. Thus modern systems have the propagation and occupancy check automated.

We can see from Table 9.5 that there is a fair probability that an optimum propagating frequency may be occupied by other users. There are several means to overcome this problem:

1. Find a clear frequency with suboptimum propagation.
2. Increase transmit power to achieve the desired signal-to-noise ratio at the distant end.
3. Use directional antennas where the interferer is in a sidelobe and hence attenuated compared to the desired signal.
4. Use antenna nulling. This is a form of electronic beam steering that creates a null in the direction of the interferer.
5. Use sharp front-end preselection on receiver(s).

Some military circuits use FEC coding rich in redundancy. Rather than using code rates such as $\frac{7}{8}$, $\frac{3}{4}$, $\frac{2}{3}$, or $\frac{1}{2}$, we resort to code rates of $\frac{1}{8}$, $\frac{1}{16}$, $\frac{1}{32}$, or $\frac{1}{64}$. This highly redundant bit stream phase shifts a RF signal that is frequency hopped (spread spectrum) over a fairly wide frequency band in the vicinity of the OWF (for the desired connectivity). Typical spreads are 100 kHz, 500 kHz, up to 2 MHz. In theory, some of the hopped energy must get through the holes in the picket fence. Such coding multiplies the symbol rate of transmission. For example, if we wish to maintain 75-bps information rate, at rate $\frac{1}{16}$ we will have to transmit 16×75, or 1200 symbols a second. If we do not increase our RF power output, the energy per symbol is reduced by 16, or 6% of the energy if we had no coding. To equate energy per bit in this situation, if we use a transmitter with a 100-W output at 75 bps, we then would require a 1.6-kW transmitter output to maintain the same energy per symbol (bit) as when rate $\frac{1}{16}$ was employed.

On the other hand, the emerging wideband HF (WBHF) systems excise the high-level interferers with only a loss of several tenths of a decibel, up to 2 dB of equivalent receive power. For instance, with 30% excision, only about 1.5 dB of receive power is lost. A 1-MHz PN spread system using a chip rate of 512 kchips in effect has 512,000 pieces of redundant RF-energy-carrying information spread across 1 MHz of HF spectrum. With so much redundancy, certainly we can effectively assure some portion of the desired transmit power to successfully get through the picket fence. Such systems have very low transmitted spectral energy density and are compatible with conventional narrowband HF systems. Processing gains are on the order of 60–66 dB when information bit rates are on the order of 75–150 bps. We should not equate processing gain to power gain. In fact there is no power gain. Processing gain describes how well it will work against a jammer.*

---

*Whether intentional or unintentional.

9.9.3 Atmospheric Noise

Atmospheric noise is a result of numerous thunderstorms occurring at various points on the earth, but concentrated mainly in tropical regions. These electrical disturbances are transmitted long distances via the ionosphere in the same manner as HF skywaves. Because the resulting field intensities of the noise decrease with the distance traveled, the level of atmospheric noise encountered from this source becomes progressively smaller in the higher latitudes of the temperate zones and in the polar regions. As we are aware, skywave propagation varies with time of day and season. Hence the intensity of atmospheric noise varies with both location and time.

Since the major portion of atmospheric noise is traced to thunderstorm activity, the atmospheric noise level at a particular location is due to contributions from both local and distant sources. During a local thunderstorm, the average noise level is about 10 dB higher than the average noise of the same period in the absence of local thunderstorm activity (Ref. 23). From this it can be seen that the atmospheric noise level is related directly to weather conditions. The position of the equatorial weather front greatly affects atmospheric noise at all locations. This front varies in position from day to day and its general location seasonally moves north and south with the sun.

The degree of activity varies from time to time and from place to place, being much greater over land than over sea. The main areas of thunderstorm activity lie in equatorial regions, notably the East Indies, equatorial Africa, equatorial South America, and Central America. Thunderstorms are present about 50% of the days at locations in these equatorial belts and this activity is the principal source of long-distance atmospheric noise. It has been estimated that there are about 2000 thunderstorms in progress at each instant throughout the world (Ref. 23).

Thunderstorm activity is located over tropical land masses during local summer seasons and is more active over land, usually between 1200 and 1700 local time. Thunderstorm activity over the sea generally occurs at night and can last for more than a day.

Atmospheric noise from local sources shows discrete crashes similar to impulse noise, while long-distance atmospheric noise consists of rapid and irregular fluctuations with a frequency of 10 or 20 kHz per second and a damped wave train of oscillations. The amplitude of lightning disturbances varies approximately inversely with the frequency squared and is propagated in all directions both for groundwaves and skywaves.

9.9.3.1 *Calculating Atmospheric Noise Using CCIR Rep. 322-2.* CCIR Rep. 322-2 (Ref. 24) is the most widely accepted data source and methodology on atmospheric noise. It has been derived from years of noise-monitoring data taken from monitoring stations around the world. The data are presented graphically as atmospheric noise contours at 1 MHz and then may be scaled graphically or by formula to the desired frequency.

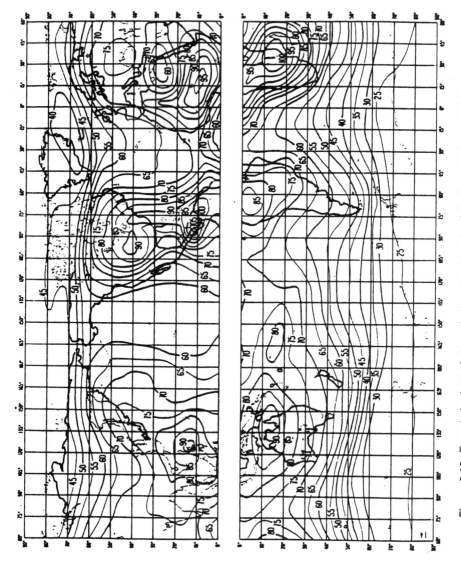

Figure 9.19. Expected values of atmospheric radio noise F_{am} (dB above $kT_0 b$ at 1 MHz), summer, 0000–0400 universal time (UT). (From Ref. 24, Figure 14a, CCIR Rep. 322-2. Courtesy of ITU-CCIR.)

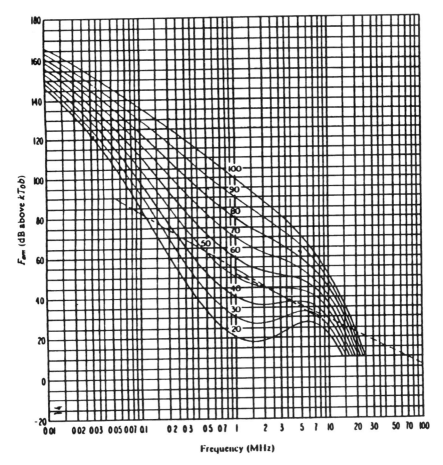

Figure 9.20. Variation of radio noise with frequency (summer, 0000–0400 h): —, expected values of atmospheric noise; ⋯⋯, expected values of man-made noise at a quiet receiving location; and ---, expected values of galactic noise. (From Ref. 24, Figure 14b, CCIR Rep. 322-2. Courtesy of ITU-CCIR.)

CCIR divides a 24-h day into six time blocks of 4 h each starting at 12 midnight. For seasonal variations in atmospheric noise, there are charts for the four seasons. Thus there are 24 charts (6 × 4), each of which is followed by a frequency scaling chart accompanied by a noise variability chart. A sample of these is given in Figure 9.19. Figure 9.20 is for frequency scaling and Figure 9.21 for noise variability and they accompany Figure 9.19.

It is noted in Figure 9.19 that the noise contours given represent *median* noise levels and are measured in decibels above kT_0b at 1 MHz; their notation is F_{am}. We will derive a formula for F_{am}.

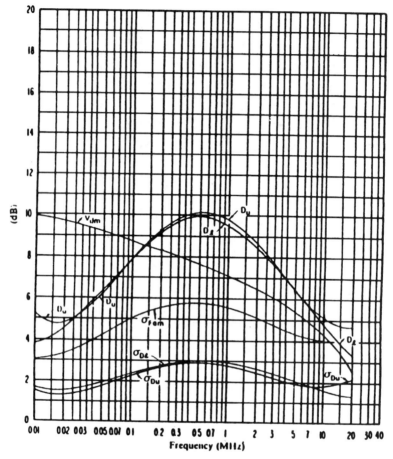

Figure 9.21. Data on noise variability and character (summer, 0000–0400 h): $\sigma_{F_{am}}$, standard deviation of values of F_{am}; D_u, ratio of upper decile to median value F_{am}; σ_{D_u}, standard deviation of values of D_u; D_l, ratio of median value F_{am} to lower decile; σ_{D_l}, standard deviation of value of D_l; and, V_{dm}, expected value of median deviation of average voltage. The values shown are for a bandwidth of 200 Hz. (From Ref. 24, Figure 14c, CCIR Rep. 322-2. Courtesy of ITU-CCIR.)

We can express the antenna noise factor f_a, which relates the noise power received from sources *external* to the antenna:

$$f_a = \frac{P_n}{kT_0b} = \frac{T_a}{T_0} \tag{9.2}$$

where P_n = the noise power available from an equivalent loss-free antenna (W)

k = Boltzmann's constant = -228.6 dBW/Hz

b = the effective noise bandwidth (Hz)

T_0 = 290 K (we have used 290 K in previous chapters)

T_a = the effective antenna temperature in the presence of external noise

$$10 \log kT_0 = -204 \text{ dBW}$$

$$F_a = 10 \log f_a$$

Both f_a and T_a are independent of bandwidth because the available noise power from all sources may be assumed to be proportional to bandwidth, as is the reference power level.

The antenna noise factor F_a, in decibels, is for a short vertical antenna over a perfectly conducting ground plane. This parameter is related to rms noise field strength along the antenna by

$$E_n = F_a - 65.5 - 20 \log F_{MHz} \tag{9.3}$$

where E_n is the rms field strength for a 1-kHz bandwidth (dB[μV/m]) and F_a is the noise factor for the frequency f in question (dB). F_{MHz} is the frequency.

The value of field strength for any bandwidth B_{Hz} other than 1 kHz can be derived by adding ($10 \log b - 30$) to E_n. For instance, to derive E_n in 1 Hz of bandwidth, we subtract 30 dB. CCIR Rep. 322-2 cautions that E_n is the vertical component of the noise field at the antenna.

In predicting the expected noise levels, the systematic trends with time of day, season, frequency, and geographical location are taken into account explicitly. There are other variations that must be taken into account statistically. The value of F_a for a given hour of the day varies from day to day because of random changes in thunderstorm activity and propagation conditions. The median of the hourly values within a time block (the time block median) is designated F_{am}. Variations of the hourly values within the time block can be represented by values exceeded for 10% and 90% of the hours, expressed as deviations D_u and D_l from the time block median plotted on normal probability graph paper (level in dB), the amplitude distribution of the deviations, D, above the median can be represented with reasonable accuracy by a straight line through the median and upper decile values, and a corresponding line through the median and the lower decile values can be used to represent values below the median. Extrapolation beyond D_u and D_l, however, yields only very approximate values of noise.

The value of F_a is average noise power. For digital circuits, it is useful to have knowledge of the amplitude probability distribution (APD) of the noise. This will show the percentage of time for which any level is exceeded, which usually is the noise envelope described. The APD is dependent on the short-term characteristics of the noise and, as a result, cannot be deduced from the hourly values of F_a alone. The reader should consult CCIR Rep. 322-2 if more detailed analysis is desired in this area.

From equation 9.2 we can state the noise threshold (noise power) of a receiver (P_n):

$$P_n = F_{am} + X\sigma + 10 \log B_{Hz} \qquad (9.4)$$

If

$$X = 1 \,(\text{one standard deviation}), 68\% \,(\text{of the time})$$

$$X = 2 \,(\text{two standard deviations}), 95\% \,(\text{of the time})$$

$$X = 3 \,(\text{three standard deviations}), 99\% \,(\text{of the time})$$

Example. Calculate the receiver noise threshold for 95% time availability where the receiver is connected to a lossless quarter-wavelength whip antenna equipped with a good ground plane. The receiver is tuned to 10 MHz with a 1-kHz bandwidth. The receiving facility is located in Massachusetts, and we are interested in the summer time block of 0000–0400. From Figure 9.19 we see that the F_{am} noise contour is 85 dB at 1 MHz. We turn to Figure 9.20 and F_{am} (10 MHz) scales down to 48 dB. Figure 9.21 gives $\sigma_{F_{am}}$ as 4 dB and we wish two standard deviations (95%), which is 8 dB for the variability.

$$P_n = 48 + 8 + 10 \log 1000 - 204 \text{ dBw}$$

$$= -118 \; dBW \; in \; 1\text{-}kHz \; bandwidth$$

We have shown much of the methodology given in CCIR Rep. 322-2, enough for an initial system design. However, several comments are in order. F_{am} decreases rapidly with increasing frequency. Besides atmospheric noise sources, other noise sources affecting total receiver noise power must start to be considered as frequency increases. These are:

f_c, the noise factor of the antenna (function of its ohmic loss)
f_t, the noise factor of the transmission line (function of its ohmic loss)
f_r, noise factor of the receiver

The operating noise factor f, then, equals

$$f_a - 1 + f_c f_t f_r \quad (\text{numerics}) \qquad (9.5)$$

At low frequencies (e.g., <10 MHz) atmospheric noise predominates and will determine the value of f. Because f_a decreases with increasing frequency, transmission line noise and receiver thermal noise become more important and f_c tends to approach unity. The values of f_t and f_r can be determined from calculations involving design features of the transmission line and receiver or by direct measurement.

The effective noise factor of the antenna, insofar as it is determined by atmospheric noise, may be influenced in several ways. If the noise sources were distributed isotropically, the noise factor would be independent of directional properties. In practice, however, the azimuthal direction of the beam may coincide with the direction of an area where thunderstorms are prevalent, and the noise factor will be measured accordingly compared with an omnidirectional antenna. On the other hand, the converse may be true. The directivity in the vertical plane may be such as to differentiate in favor or against reception of noise from a strong source.

Unfortunately, CCIR Rep. 322-2 does not take into account the sunspot number and, as mentioned previously, horizontal polarization.

9.9.4 Man-Made Noise

Above about 10 MHz we will often find that man-made noise is predominant. Man-made noise can be generated by many sources, such as electrical machinery, automobile ignitions, all types of electronic processors/computers, high-power electric transmission lines, and certain types of lighting. As a result, man-made noise is a function of industrialization and habitation density.

The most recognized reference on man-made noise is CCIR Rep. 258-5 (Ref. 25). Figure 9.22 gives median values of man-made noise, expressed in decibels, above -204 dBW/Hz. The figure shows five curves:

A. Business
B. Residential
C. Rural
D. Quiet rural
E. Galactic (extending upward in frequency from 10 MHz)

Business areas are defined as any area used predominantly for any type of business, such as stores, offices, industrial parks, large shopping centers, and main streets and highways lined with various business enterprises. Residential areas are those predominantly used for single- or multiple-family dwellings of at least two single-family units per acre (five per hectare) and with no large or busy highways. Rural areas are defined as areas in which dwellings number no more than one per 5 acres (2 hectares) and where there are no intense noise sources. Minimum man-made noise may be found in "quiet" rural areas. It is in these areas where galactic noise predominates above about 10 MHz. This is the dashed line (E) shown in Figure 9.22.

We can also calculate the median man-made noise from the following expression:

$$F_{am} = c - d \log f \qquad (9.6)$$

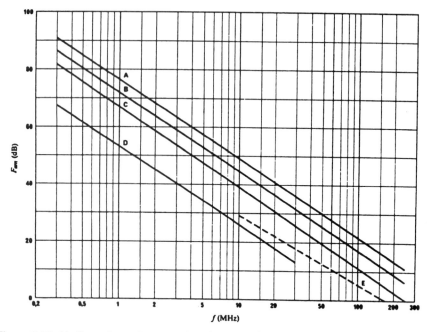

Figure 9.22. Median values of man-made noise power for a short vertical lossless grounded monopole antenna. Environmental category: A, business; B, residential; C, rural; D, quiet rural; E, galactic. (From Ref. 25, Figure 1, CCIR Rep. 258-5. Courtesy of ITU-CCIR.)

where f is the operating frequency in megahertz. The values of the constants c and d may be taken from Table 9.6. Table 9.7 provides some selected values of F_{am} and deviations of the median value. CCIR Rep. 258-5 advises that equation 9.6 may give erroneous values for quiet rural (D) and galactic noise (E) environments.

Example. A receiver operates at 15 MHz with a 1-kHz bandwidth. The receiver is located in a rural environment and uses a lossless whip antenna

TABLE 9.6 Values of Constants c and d (Equation 9.6)

Environmental Category	c	d
Business (curve A)	76.8	27.7
Interstate highways	73.0	27.7
Residential (curve B)	72.5	27.7
Parks and university campuses	69.3	27.7
Rural (curve C)	67.2	27.7
Quiet rural (curve D)	53.6	28.6
Galactic noise (curve E)	52.0	23.0

Source: Reference 25, Table I, CCIR Rep. 258. Courtesy of ITU-CCIR.

TABLE 9.7   Representative Values of Selected Measured Noise Parameters for Business, Residential, and Rural Environmental Categories[a]

Frequency (MHz)	Business				Residential				Rural			
	F_{am} (dB$[kT_0]$)	D_u (dB)	D_l (dB)	σ_{NL} (dB)	F_{am} (dB$[kT_0]$)	D_u (dB)	D_l (dB)	σ_{NL} (dB)	F_{am} (dB$[kT_0]$)	D_u (dB)	D_l (dB)	σ_{NL} (dB)
0.25	93.5	8.1	6.1	6.1	89.2	9.3	5.0	3.5	83.9	10.6	2.8	3.9
0.50	85.1	12.6	8.0	8.2	80.8	12.3	4.9	4.3	75.5	12.5	4.0	4.4
1.00	76.8	9.8	4.0	2.3	72.5	10.0	4.4	2.5	67.2	9.2	6.6	7.1
2.50	65.8	11.9	9.5	9.1	61.5	10.1	6.2	8.1	56.2	10.1	5.1	8.0
5.00	57.4	11.0	6.2	6.1	53.1	10.0	5.7	5.5	47.8	5.9	7.5	7.7
10.00	49.1	10.9	4.2	4.2	44.8	8.4	5.0	2.9	39.5	9.0	4.0	4.0
20.00	40.8	10.5	7.6	4.9	36.5	10.6	6.5	4.7	31.2	7.8	5.5	4.5
48.00	30.2	13.1	8.1	7.1	25.9	12.3	7.1	4.0	20.6	5.3	1.8	3.2
102.00	21.2	11.9	5.7	8.8	16.9	12.5	4.8	2.7	11.6	10.5	3.1	3.8
250.00	10.4	6.7	3.2	3.8	6.1	6.9	1.8	2.9	0.8	3.5	0.8	2.3

[a]*Key:* F_{am}, median value; D_u, D_l, upper, lower decile deviations from the median value within an hour at a given location; and σ_{NL}, standard deviation of location variability.

Source: Reference 25, Table II, CCIR Rep. 258-5. Courtesy of ITU-CCIR.

one-quarter wavelength long with a ground plane. Calculate the noise power threshold of a receiver under these circumstances. Assume a lossless transmission line.

Use equation 9.4 and replace $X\sigma_{am}$ with the expression ($D_u + \sigma_{NL}$, whose values we take from Table 9.7.

$$P_n = 34\ dB + (8.2\ dB + 4.2\ dB) + 10\log 1000 - 204\ dBW$$

$$= -127.6\ dBW$$

9.9.5   Receiver Thermal Noise

Only under very special circumstances does receiver thermal noise become a consideration under normal HF operation. If we consider that atmospheric, man-made, and cosmic noise have a nonuniform distribution around an antenna (i.e., it tends to be directional), then the antenna gain at a certain frequency will generally favor signal and discriminate against noise.

Some HF antennas are designed with a very low gain; actually, they have a directional loss. The gain is expressed in negative dBi units. If we couple this antenna to a long or otherwise lossy transmission line, the noise at the receiver will have a significant thermal noise component. Seldom, however, is receiver-generated thermal noise a significant contributor to total HF receiving system noise. External noise is the dominant noise in determining the noise floor in the case of HF when calculating signal-to-noise ratio or E_b/N_0. In the design of HF receivers, thermal noise is not an overriding issue, and we find noise figures for these receivers are on the order of 12–16 dB.

9.10 NOTES ON HF LINK TRANSMISSION LOSS CALCULATIONS

9.10.1 Introduction

To predict the performance of a HF link or to size (dimension) link terminal equipments to meet some performance objectives, a first step is to calculate the link transmission loss. The procedure is similar to the transmission loss calculations for line-of-sight (LOS) microwave or troposcatter links. In other words, we must determine the signal attenuation in decibels between the transmitting antenna and its companion far-end receiving antenna. For a LOS link our concern was essentially free-space loss. The calculation for a HF link is considerably more involved.

9.10.2 Transmission Loss Components

The unfaded net transmission loss (L_{TL}) of a HF link can be expressed by

$$L_{TL} = L_{FSL} + L_D + L_B + L_M \quad \text{(dB)} \tag{9.7}$$

where
L_D = the D-layer absorption losses
L_B = the ground reflection losses
L_M = miscellaneous losses
L_{FSL} = the free-space loss expressed by the familiar formula

$$L_{FSL} = 32.45 + 20 \log d_{km} + 20 \log F_{MHz} \quad \text{(dB)} \tag{9.8}$$

where d is the distance in kilometers and F the frequency in megahertz.

9.10.2.1 Free-Space Loss. When we calculate L_{FSL} (equation 9.8), d_{eff} is the total distance a signal travels from transmitter A to its far-end receiver B. For short near-vertical incidence (NVI) paths, d_{eff} will be notably greater than the great circle distance from A to B. As the distance from A to B increases, and we start to use "normal" skywave modes, the great circle and the effective distance (d_{eff}) between A and B start to converge. Thus for short paths (e.g., <1000-km great circle distance), we will substitute d_{eff} in equation 9.8 in place of d. For paths greater than 1000 km, we can just use the great circle distance between transmitter A and distant receiver B (Ref. 26).

A geometric representation of a one-hop HF path is shown in Figure 9.23, where we see d_{eff} is the path APB and P is the reflection point h_i' above the earth's surface at midpoint. d_{eff} can be calculated:

$$d_{eff} = 2\sqrt{8.115 \times 10^7 + 12{,}740 h_i' + h_i'^2 - \cos\frac{\alpha}{2} \times \left(8.115 \times 10^7 + 12{,}740 h_i'\right)}$$

$$\tag{9.9}$$

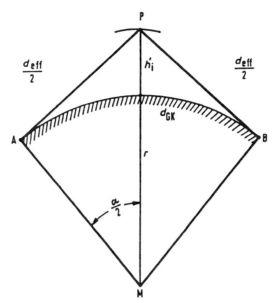

Figure 9.23. Geometric representation of a one-hop HF path where d_{eff} is the distance APB.

Equation 9.9 assumes r_i, the radius of the earth in Figure 9.23, to be 6370 km, α is the great circle arc from A to B, and h'_i is the virtual reflection height. We can calculate α from the great circle equation:

$$\cos \alpha = \sin A \sin B + \cos A \cos B \cos \Delta L \qquad (9.10)$$

where α = angle of the great circle arc (see Figure 9.23)

   A = latitude of station A

   B = latitude of station B

   ΔL = difference in longitude between stations A and B

d_{eff} can also be derived from Figure 9.24, where h' is the reflection point height above the earth's surface. The height of the reflection point h' can be derived from Figures 9.28a and 9.28b.

9.10.2.2   *D-Layer Absorption Losses.* D-layer absorption is a daytime phenomenon. The D-layer disappears at night. D-layer absorption varies with the zenith angle of the sun, the sunspot number, the season, and the operating frequency. In fact, it varies as the inverse of the square of the operating frequency. This is one reason we are driven to use higher frequencies for skywave links, to reduce D-layer absorption.

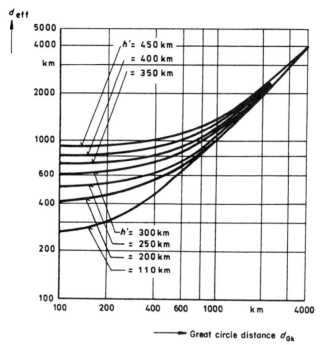

Figure 9.24. Effective distance d_{eff} when given the great circle distance and reflection height h'. (From Ref. 26.)

To calculate D-layer absorption on a particular skywave path, we first compute the absorption index I:

$$I = (1 + 0.0037R)(\cos 0.881\chi)^{1.3} \qquad (9.11)$$

where R is the sunspot number and χ the solar zenith angle of the sun. If χ is greater than 100°, it is nighttime and we can neglect D-layer absorption.

From Ref. 37, we calculate the solar zenith angle with the following formula:

$$\cos \chi = \sin \phi \sin \varepsilon + \cos \phi \cos \varepsilon \cos h \qquad (9.12)$$

where ϕ = geographical latitude

ε = solar declination

h = the local hour angle of the sun measured westward from apparent noon, which is mean noon corrected for the equation of time and the standard time used at the location of interest

Tables of hourly values of cos χ from sunrise to sunset for the 15th day of each month for most of the ionosphere vertical incidence sounding stations are given in the URSI Ionosphere Manual (See Ref. 26, p. 19).

From Ref. 27, the hour angle of the sun, h, is calculated:

$$h = (\text{LST} - \text{right ascension})\left(\frac{15.0}{\pi}\right) \tag{9.13}$$

where the local sidereal time, LST, is

$$\text{LST} = \{[(\text{DN} \times 24 + \text{GMT})(1.002737909) - (\text{DN} \times 24)] - \phi\} + S \tag{9.14}$$

where ϕ = longitude of the point (in this case) (rad)

DN = the number of the day in the year

GMT = Greenwich mean time in decimal hours

If the local sidereal time is less than 0, then 24 h is added to it. If it is more than 24, then 24 h is subtracted from it. S is a correction factor.

Figures 9.25a and 9.25b can also be used to determine the solar zenith angle (Ref. 26). We now can calculate I, the absorption index, using equation 9.11. This value is then corrected for the winter anomaly, if required. These correction factors are given in Table 9.8. To use the nomogram in Figure 9.26 effectively, we need to know the gyrofrequency for the region of interest. We can derive this value from Figure 9.27.

Example. A certain HF path operates during a sunspot number 100, the operating frequency is 18 MHz, and the elevation or takeoff angle (TOA) is 10°. Calculate the D-layer absorption value for 12 noon local time (at midpath) for the month of January at 40° north latitude. First we must derive the solar zenith angle from Figure 9.25a, 70°. Now we can substitute numbers in equation 9.11:

$$I = (1 + 0.0037 \times 100)(\cos 0.881 \times 70°)^{1.3}$$

$$= 0.52$$

Multiply this value by the factor 1.5 taken from Table 9.8 to correct for the winter anomaly. I (corrected) is 0.78. Derive the gyrofrequency (f_H) for the latitude from Figure 9.27; it is 1.4 MHz. The value $f + f_\text{H}$ is 18 + 1.4 MHz, or 19.4 MHz. This value we use in Figure 9.26 to derive the D-layer absorption. We do this by first entering the takeoff angle (elevation angle) of the antenna, which is 10°. Now draw a line vertically to the corrected absorption index (I). We now have a reference point at the intersection of

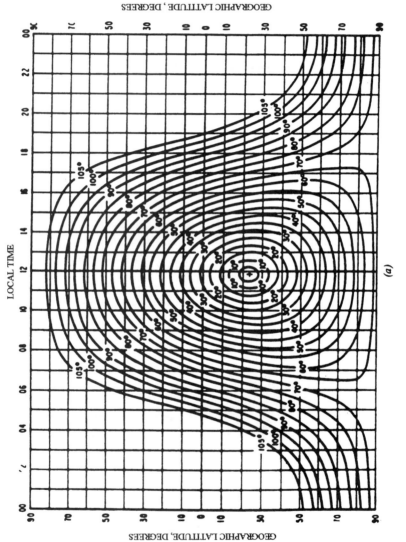

LOCAL TIME

GEOGRAPHIC LATTITUDE, DEGREES

(a)

Figure 9.25a. Solar zenith angle for December. (From Ref. 26.)

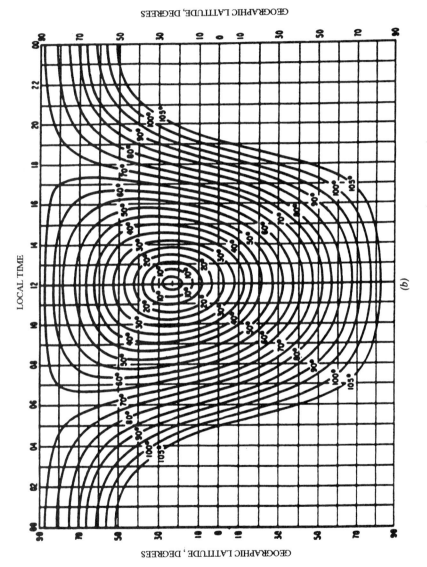

Figure 9.25b. Solar zenith angle for June. (From Ref. 26.)

641

TABLE 9.8   Correction Factors of the Absorption Index to Account for the Winter Anomaly

Month		
Northern Hemisphere	Southern Hemisphere	Factor
November	May	1.2
December	June	1.5
January	July	1.5
February	August	1.2

Source: Reference 26.

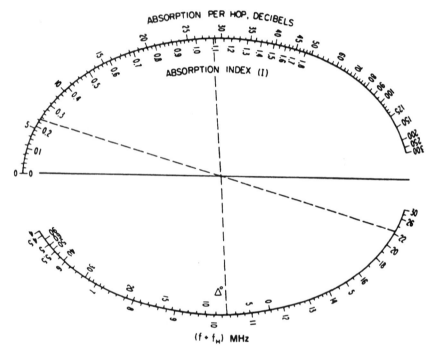

Figure 9.26. Nomogram for obtaining the ionospheric absorption per hop from the absorption index *I*, the effective wave frequency $f + f_H$, and the angle of elevation Δ. Example: (1) Enter with the absorption index *I* (1 = 1.09) and the elevation angle Δ (= 8°). (2) Mark the reference point of intersection with the center line. (3) With reference point and effective frequency $f + f_H$ (= 22.4), draw a straight line as far as the curve marked absorption per hop (= 6 dB). (From Ref. 26.)

the solid horizontal lines. Through this reference point draw another line (right to left) from the frequency value of 19.4 MHz; it intersects "Absorption per Hop, Decibels" at 6 dB. We have a one-hop path, so the absorption value is 6 dB. If it were a two-hop path with these parameters, it would be approximately 6 × 2, or 12 dB for D-layer absorption. Figures 9.28*a* and 9.28*b* can be used to determine the F2-layer height *h'*.

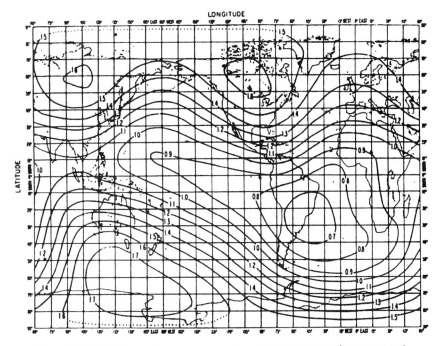

Figure 9.27. World map of gyrofrequency for a height of 100 km. (From Ref. 26.)

9.10.2.3 Ground Reflection Losses and Miscellaneous Skywave Propagation Losses. The following guidelines may be used to calculate ground reflection losses (Ref. 28):

One-hop = 0 dB
Two-hop = 2 dB
Three-hop = 4 dB (etc.)

For miscellaneous skywave propagation losses, if no other information is available, use the value 7.3 dB. We will use this value for L_M in equation 9.7.

9.10.2.4 Guidelines to Determine Dominant Hop Mode. We only consider the lower order E- and F-layer modes (Ref. 28):

For path lengths up to 2000 km: 1E, 1F2, and 2F2
For path lengths between 2000 and 4000 km: 2E, 1F2, and 2F2
For path lengths between 4000 and 7000 km: 2F2 and 3F2
For path lengths between 7000 and 9000 km: 3F2

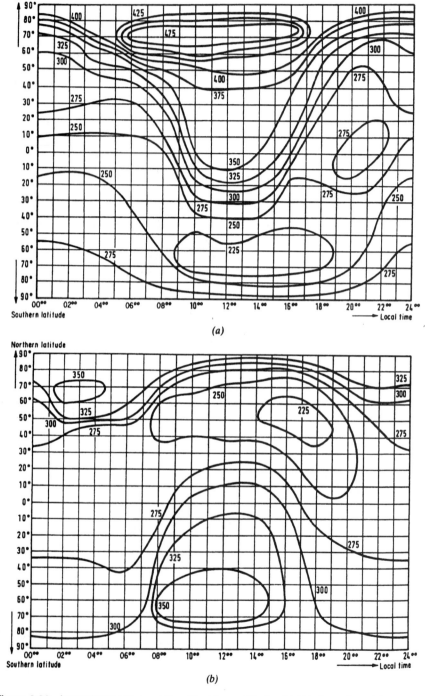

Figure 9.28. Approximate values of the virtual layer altitude of h' F2 (in km) for (a) July and (b) January. (From Ref. 29.)

9.10.3 A Simplified Example of Transmission Loss Calculation

Consider a 1500-km path with 1F2 dominant mode in the temperate zone (40° N) operating over land in June with a sunspot number of 100 (R12) at 12 noon local time (midpath). This is an application of equation 9.7. Use Figure 9.29 to determine the radiation angle when given the great circle distance and virtual (F2) layer height. We can derive the layer height from Figures 9.28a and 9.28b. In this case, of course, we use Figure 9.28a for a value of 375 km. The elevation angle is ~17°. We now calculate the free-space loss using equation 9.8:

$$\text{FSL}_{dB} = 32.45 + 20\log F_{\text{MHz}} + 20\log d_{\text{km}}$$

The optimum working frequency (OWF) is determined to be 17 MHz. We now have the distance and frequency, so we can proceed to calculate FSL, which is 120.57 dB.

We now want to calculate the D-layer absorption loss, L_D in equation 9.7. Use Figure 9.25b to calculate the solar zenith angle. This is 15°. Now we use

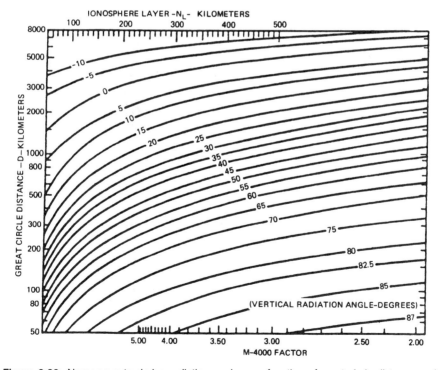

Figure 9.29. Nomogram to derive radiation angle as a function of great circle distance and ionospheric layer height. (From Ref. 26.)

formula 9.11 to calculate the absorption index (I):

$$I = (1 + 0.0037 \times 100)(\cos 0.881 \times 15)^{1.3}$$

$$= 1.37(0.97)^{1.3}$$

$$= 1.45$$

Because the month of interest is June, we can neglect the winter anomaly.

The next step is to derive the gyrofrequency using Figure 9.27. The value is 1.5 MHz, so $f + f_H = 18.5$ MHz. We now derive the D-layer absorption from the nomogram, Figure 9.26. It is 8 dB. This is a one-hop path, so ground reflections are zero. L_M is 7.3 dB (see Section 9.10.2.3). Therefore the total transmission loss (unfaded) from equation 9.7 is

$$L_{TL} = 120.57 \text{ dB} + 8 \text{ dB} + 0 \text{ dB} + 7.3 \text{ dB}$$

$$= 135.87 \text{ dB}$$

9.10.4 Groundwave Transmission Loss

9.10.4.1 *Introduction.* Groundwave or surface wave transmission by HF is particularly effective over "short" ranges. The range or distance for effective transmission decreases with increasing frequency and decreases with decreasing ground conductivity. Seawater is highly conductive and we find that ranges from 500 to 800 mi (800 to 1200 km) can be achieved during daytime. Skywave interference at night, man-made and atmospheric noise, can reduce the range. The link design engineer must find an optimum frequency trading off atmospheric noise with frequency and range. We must also note that vertical polarization is more effective than horizontal polarization for groundwave paths.

Overland groundwave path transmission range is a function of ground conductivity and roughness. The performance prediction of such paths is a rather imperfect art.

9.10.4.2 *Calculation of Groundwave Transmission Loss.* We use a very simple method to calculate groundwave transmission loss. We first turn to CCIR Rec. 368-5 (Ref. 30) and use the appropriate curves. Three such curves are given in Figures 9.30*a*–9.30*c*. Figure 9.30*a* is for a path over seawater; Figures 9.30*b* and 9.30*c* are for overland paths with different ground conductivities. Table 9.9 will help in the selection process for the appropriate curve (ground conductivity).

We then select the operational frequency and distance, from which we will derive a field strength E, in decibels above a microvolt. To be consistent with other chapters dealing with radiolinks, we will want to calculate the transmission loss. Note 4 to CCIR Rec. 368-5 (Ref. 30) provides the following

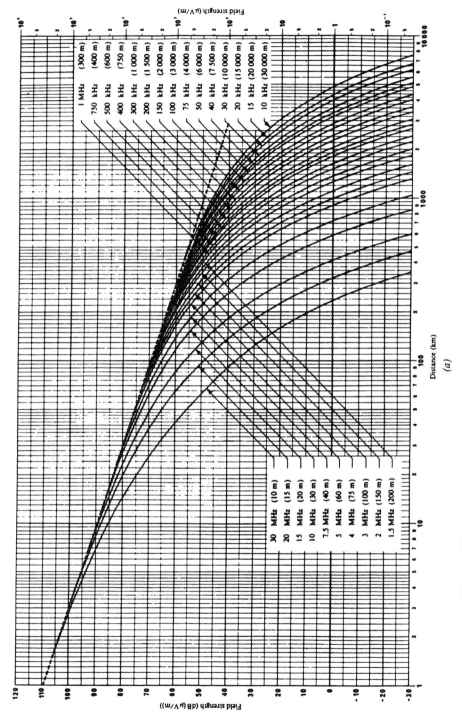

Figure 9.30a. Groundwave propagation curves: seawater, average salinity, 20°C, $\sigma = 5$ S/m, $\varepsilon = 70$. ---, Inverse distance curve. (From Ref. 30, Figure 1, CCIR Rec. 368-5.)

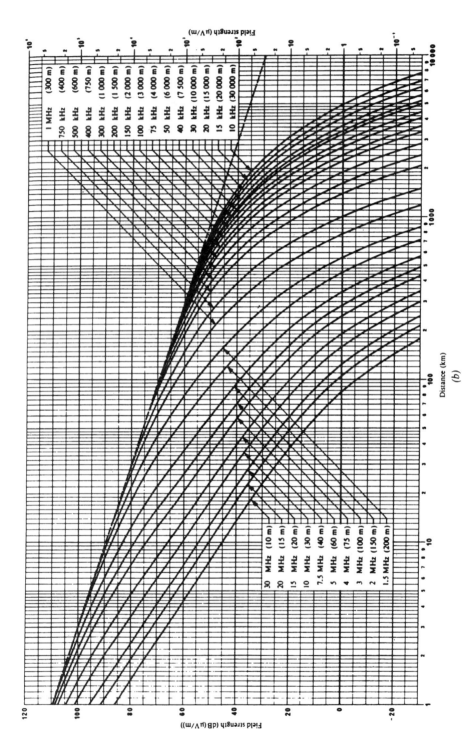

Figure 9.30b. Groundwave propagation curves: land, $\sigma = 3 \times 10^{-2}$ S/m, $\varepsilon = 30$. ----, Inverse distance curve. (From Ref. 30, Figure 2, CCIR Rec. 368-5.)

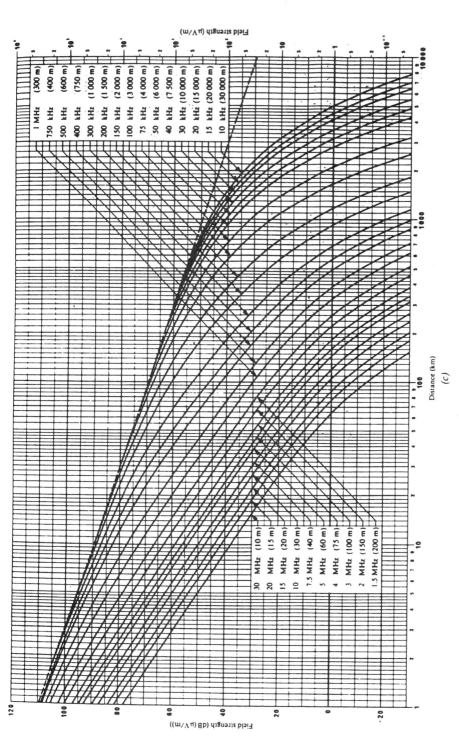

Figure 9.30c. Groundwave propagation curves: medium dry ground, $\sigma = 10^{-3}$ S/m, $\varepsilon = 15$. ——, Inverse distance curve. (From Ref. 30, Figure 5, CCIR Rec. 368-5.)

649

TABLE 9.9   Guidelines in the Selection of Ground Conductivity and Dielectric Constant for Various Soil Types

Type of Terrain[a]	Conductivity σ (S/m^{-1})	Dielectric Constant ε
Seawater of average salt content	4	80
Fresh water (20°C)	3×10^{-3}	80
Moist ⎫	10^{-2}	30
Medium ⎬ soil	10^{-3}	15
Dry ⎪	10^{-4}	4
Very dry ⎭	10^{-5}	4

[a]These types of terrain apply to the following areas:

Seawater	Practically all oceans
Fresh water	Inland waters, such as large lakes, wide rivers, and river estuaries
Moist soil	Marsh and fenland, regions with high groundwater level, floodplains, and so on
Medium soil	Agricultural areas, wooded land, typical of countries in the temperate zones
Dry soil	Dry, sandy regions such as coastal areas, steppes, and also arctic regions
Very dry soil	Deserts, industrial regions, large towns, and high mountains

Source: Reference 26.

equation to derive transmission loss given the value for E:

$$L_b = 137.2 + 20 \log F_{\text{MHz}} - E \qquad (9.15)$$

We illustrate this procedure by example. Suppose we have a 10-km link and have selected 3 MHz for our operational frequency over "normal" land (Figure 9.30*b*). We first determine the value E from the figure. It is 82.5 dB (μV/m). We apply this value of E to the formula:

$$L_b = 137.2 + 20 \log 3 - 82.5 \text{ dB } (\mu\text{V/m})$$

$$= 137.2 + 9.5 - 82.5$$

$$= 64.24 \text{ dB}$$

Hence the transmission loss over *smooth* earth with *homogeneous* ground conductivity and permittivity is 64.24 dB. Because the intervening terrain is not smooth, we add a terrain roughness degradation factor of 6 dB, for a total transmission loss of 70.24 dB.

Some notes of caution: If we were to measure transmission loss on this circuit, we would probably find a much greater loss as measured at the input port of the far-end receiver. The following lists some variables we may encounter on our path and in our equipment, which may help explain such disparities:

1. The CCIR Rec. 368-5 curves are for vertical polarization. A horizontally polarized wave will have a notably higher transmission loss, on the order of 50 dB higher* (Ref. 26, p. 269).
2. With the exception of overwater paths, we will deal with ground conductivities/permittivities that can vary by more than an order of magnitude as we traverse the path.
3. The CCIR Rec. 368-5 (Ref. 30) curves are for a short vertical monopole at the earth's surface. The earth is assumed to be flat and *perfectly conducting* in the immediate area of the antenna but not along the path.

Thus the curves are idealized. Unless an excellent ground screen is installed under the antenna at both ends of the path, a degradation factor should be added for less than perfect ground. Degradation factors must also be added for antenna matching and coupling losses.

9.11 LINK ANALYSIS FOR EQUIPMENT DIMENSIONING

9.11.1 Introduction

In this section we will determine such key equipment/system design parameters as transmitter output power, type of modulation, impact of modulation selection on signal-to-noise ratio, bandwidth, fading and interference, use of FEC coding/interleaving, and type of diversity and diversity improvement. We assume in all cases that we are using the optimum working frequency (OWF) and have calculated transmission loss for that frequency, time, season, and sunspot number. We also assume that we have calculated the appropriate atmospheric noise level for time, season, and frequency.

We will use Table 9.3 as a guide for required bandwidths and S/N requirements for some of the more common types of modulation used for HF point-to-point operation. Section 9.12 will deal with more sophisticated modulation schemes and how they can improve performance.

9.11.2 Methodology

The signal-to-noise ratio at the distant receiver can be expressed by

$$\frac{S}{N} = S_{dBm} - N_{dBm} \qquad (9.16)$$

Substitute RSL_{dBm} for S_{dBm}, where RSL is the receive signal level. This assumes, of course, that $S/N = C/N$ (carrier-to-noise ratio). To permit this

---

*In theory—in practice, the gap between vertical and horizontal polarization may be much smaller.

equality, some value k is added to our transmission loss. For lack of another term, we may call this modulation implementation loss, where we sum up all the inefficiencies in the modulation–demodulation process (including coding and interleaving) to come up with a value for k. Unless we have other values, we will use 2 dB for the value of k.

In equation 9.16 N is the noise floor of the receiver, which is probably dominated by atmospheric noise and/or man-made noise rather than thermal noise. Refer to Sections 9.9.3 and 9.9.4.

RSL is calculated in the conventional manner:

$$\text{RSL}_{\text{dBm}} = \text{EIRP}_{\text{dBm}} - L_{\text{TL}} + G_{\text{rec}} \qquad (9.17)$$

where L_{TL} = total transmission loss in decibels, including k dB of modulation implementation loss

  EIRP = effective isotropically radiated power from the transmit antenna

  G_{rec} = net antenna gain at the receiver

Remember that the antenna gains used in the calculation of EIRP and the receiver antenna gain must be that gain at the proper azimuth and the elevation of the takeoff angle (TOA) (or radiation angle) (Figure 9.29).

Try the following example (refer to Section 9.10.3). There is a 1500-km path with 1F2 as the predominant mode, in the temperate zone (40° N), operating over land in June with a sunspot number of 100 ($R12$), and the time of day is 12 noon. The receiver has a 15-dB noise figure and a 3-kHz bandwidth and operates in a rural setting. The OWF is 17 MHz and the TOA is about 17°. The FSL is 120.57 dB, D-layer absorption is 9 dB, miscellaneous losses are 7.3 dB, and the modulation implementation loss is 2 dB; thus L_{TL} is 138.87 dB.

The atmospheric noise at 17 MHz is on a 10-dB contour for summer, 12 noon (Figures 9.19 and 9.20) with a variability of 8 dB for 95% of the time. Man-made noise has a value F_{am} of 42 dB with a variability of 8 dB. The man-made noise component of the noise floor is by far the dominant noise and it is

$$p_{\text{mm}} = 42 + 8 + 10 \log 3000 - 174 \text{ dBm}$$

$$= -89.23 \text{ dBm}$$

We will use this value for N_{dBm} in equation 9.16.

Let us assume that the transmitter has 1-kW RF output power and the net antenna gain at the angles of interest is 10 dB (including transmission line losses). The net receiver antenna gain is also assumed to be 10 dB. The EIRP is then

$$\text{EIRP}_{\text{dBm}} = +60 \text{ dBm} + 10 \text{ dB}$$

$$= +70 \text{ dBm}$$

$$L_{\text{TL}} = 138.87 \text{ dB}$$

$$\text{RSL} = +70 \text{ dBm} - 138.87 \text{ dB} + 10 \text{ dB}$$

$$= -58.87 \text{ dBm}$$

$$\frac{S}{N} = -58.87 - (-138.87 \text{ dB})$$

$$= 80 \text{ dB}/3\text{-kHz}$$

Suppose the link we are analyzing was going to use single-sideband (SSB) operation. Now turning to Table 9.3, we find for J3E (SSB) operation, stable condition, a S/N of 64 dB is required in 1 Hz of bandwidth. Our S/N value is for 3000-Hz bandwidth. The Table 9.3 (CCIR) value becomes 29.2 dB for a 3000-Hz bandwidth. We arrive at this value by subtracting $10 \log 3000$ from the CCIR value in 1-Hz bandwidth. Our margin is 80 dB − 29.22 dB, or 50.78 dB (unfaded condition).

Still on the J3E line in Table 9.3 we now look to the column that indicates good commercial service with fading and now we require 72 dB S/N in a 1-Hz bandwidth, or 37.2 dB in a 3-kHz bandwidth. Our new margin is 42.8 dB.

We can trade off some of the margin for transmit power or antenna gain. For the case of transmit power, we can reduce it by 20 dB and still have a 22.8-dB margin. The link now operates with 10 W of power. We also can reduce the antenna gain at each end by 5 dB (total of 10 dB for both ends) and the margin would now be 12.8 dB.

The assumption is made, of course, that the HF facility has a frequency assignment near 17 MHz and that it is an "interference-free channel." This latter statement, in practice, is highly problematical.

9.12 SOME ADVANCED MODULATION AND CODING SCHEMES

9.12.1 Two Approaches

There are two approaches to the transmission of binary message traffic or data on a conventional HF link: serial tone transmission and parallel tone transmission. Today it is still probably more common to encounter links using binary frequency shift keying (FSK) on a single channel or multiple channels in the 3-kHz passband with voice frequency carrier telegraph (VFCT). (See Table 9.4.) On single-channel systems, when transmitting 150 bps or less, the center tone frequency is at 1275 Hz with a frequency shift of ± 42.5 Hz

(Ref. 31). For 600-bps operation the center frequency is at 1500 Hz with a frequency shift of ± 200 Hz; for 1200 bps, the center tone frequency is at 1700 Hz shifted ± 400 Hz.

We will briefly describe two parallel tone systems and two serial tone systems. The first parallel tone system transmits 16 simultaneous tones and each tone is differentially phase shifted. The second method uses 39 tones. In either case these tones are contained in the nominal 3-kHz passband envelope. Of the two serial tone techniques, one uses 8-ary PSK and the other 8-ary FSK modulation.

9.12.2 Parallel Tone Operation

The first technique referenced here (MIL-STD-188-110 [Ref. 31]) operates from 75 to 2400 bps. The modulator accepts serial binary data (see Chapter 13) and converts the bit stream into 16 parallel data streams. The signal element interval on each data bit stream is 13.33 ms and its modulation rate is 75 bauds. The modulator provides a separate tone combination for initial synchronization and, if required, a separate tone for Doppler correction. Tone frequencies and bit locations are shown in Table 9.10. For data

TABLE 9.10   Data Tone Frequencies and Bit Locations for HF Data Modems

Tone Frequency (Hz)	Function	Even and Odd Bit Locations of Serial Binary Bit Stream, Encoded and Phase Modulated on Each Data Tone Employing:					
		Quadrature-Phase Modulation		Biphase Modulation			
		2400 bps	1200 bps	600 bps	300 bps	150 bps	75 bps
605	Continuous Doppler Tone		⟵	In-Band Diversity		⟶	
825[a]	Synchronization Slot						
935	Data tone 1	1st and 2nd	1st and 2nd	1st	1st	1st	1st
1045	Data tone 2	3rd and 4th	3rd and 4th	2nd	2nd	2nd	1st
1155	Data tone 3	5th and 6th	5th and 6th	3rd	3rd	1st	1st
1265	Data tone 4	7th and 8th	7th and 8th	4th	4th	2nd	1st
1375	Data tone 5	9th and 10th	9th and 10th	5th	1st	1st	1st
1485	Data tone 6	11th and 12th	11th and 12th	6th	2nd	2nd	1st
1595	Data tone 7	13th and 14th	13th and 14th	7th	3rd	1st	1st
1705	Data tone 8	15th and 16th	15th and 16th	8th	4th	2nd	1st
1815	Data tone 9	17th and 18th	1st and 2nd	1st	1st	1st	1st
1925	Data tone 10	19th and 20th	3rd and 4th	2nd	2nd	2nd	1st
2035	Data tone 11	21st and 22nd	5th and 6th	3rd	3rd	1st	1st
2145	Data tone 12	23rd and 24th	7th and 8th	4th	4th	2nd	1st
2255	Data tone 13	25th and 26th	9th and 10th	5th	1st	1st	1st
2365	Data tone 14	27th and 28th	11th and 12th	6th	2nd	2nd	1st
2475	Data tone 15	29th and 30th	13th and 14th	7th	3rd	1st	1st
2585	Data tone 16	31st and 32nd	15th and 16th	8th	4th	2nd	1st

[a]No tone is transmitted at this frequency.

Source: Reference 31, MIL-STD-188-110.

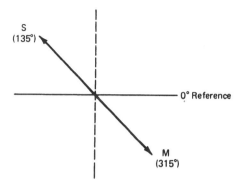

A. For data signaling rates of 75, 150, 300, or 600 bps.

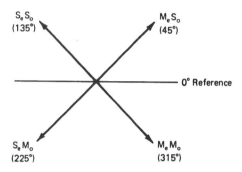

B. For data signaling rates of 1200 or 2400 bps.

Notes:
1. M = logic sense of mark; S = logic sense of space.

2. The subscripts refer to the even (e) or odd (o)
   bit locations of the serial binary bit stream.
   (see Table 9.10)

Figure 9.31. Phase modulation vectors for HF data modem. (From Ref. 31, MIL-STD-188-110.)

signaling rates of 75, 150, 300, and 600 bps at the modulator input, each data tone signal element is biphase modulated, as shown in the upper part of Figure 9.31. Each bit of the serial binary input signal is encoded, depending on the mark or space logic sense of the bit, into a phase change of the data tone signal element as listed in Table 9.11. For data signaling rates of 1200 and 2400 bps at the modulator input, each data tone signal is four-phase modulated (QPSK), as shown in the lower part of Figure 9.31. In this case, each dibit of the serial binary input signal is encoded, depending on the mark or space logic sense and the even or odd bit location of each bit, into a phase change of the data tone signal element as listed in Table 9.11. The phase change of a data tone signal element is relative to the phase of the immedi-

TABLE 9.11   Modulation Characteristics for the 16 Parallel Tone HF Modem

Input Data Signaling Rate (bps)	Degree of In-Band Diversity Combining	Type of Modulation	Logic Sense of Dibits or Bits in Serial Binary Bit Stream Depending on:		Phase of Data Tone Signal Element Relative to Phase of Preceding Signal Element (deg)
			Even Bit Locations	Odd Bit Locations	
2400	N/A		Mark	Space	+45
		Four-phase	Space	Space	+135
			Space	Mark	+225
1200	2		Mark	Mark	+315
600	2		Mark[a]		+315
300	4	Two-phase			
150	8				
75	16		Space[a]		+135

[a]Regardless of even or odd bit locations.
Source: Reference 31, MIL-STD-188-110.

ately preceding signal element. This is called differential phase shift keying. In-band diversity (see Section 9.8) is provided for all but the 2400-bps data rate.

There are two arguments in favor of the parallel tone approach. The first is that the transmitted signal element is long (13.33 ms) for all user data rates. Consequently, in theory, a system using this technique can withstand up to 6 ms of multipath dispersion (see Section 9.7.3.2). The second argument is that equalization across the 3-kHz audio passband becomes a moot point because the band is broken down into 110-Hz segments and each segment carries a very low modulation rate, so that delay and amplitude equalization are unnecessary.

The second technique uses 39 parallel tones each with quadrature differential phase-shifted modulation. A 40th unmodulated tone is added for Doppler correction. In-band diversity is available for all data rates below 1200 bps. Forward error correction (FEC) coding employing a shortened Reed–Solomon (15, 11) block code with appropriate interleaving is incorporated in the modem. A means is provided for synchronization of the signal element and interleaved block timing.

Each of the 39 tones is assigned a 52.25-Hz channel. The lowest frequency data tone is at 675.00 Hz and the highest is at 2812.50 Hz. The Doppler correction tone is at 393.75 Hz. The frequency accuracy of a data tone must be maintained within ± 0.05 Hz. For operation below 1200 bps both time and in-band frequency diversity are used.

Table 9.12 gives bit error probabilities versus signal-to-noise ratio for the 39-tone modem based on a HF baseband simulator, assuming two independent, equal, average-power Rayleigh fading paths with a 2-Hz fading bandwidth and 2-ms multipath dispersion (Ref. 31).

TABLE 9.12   Probability of Error Versus Signal-to-Noise Ratio[a]

Signal-to-Noise Ratio (dB in 3-kHz Bandwidth)	Probability of Bit Error	
	2400 bps	1200 bps
5	8.6 E-2	6.4 E-2
10	3.5 E-2	4.4 E-3
15	1.0 E-2	3.4 E-4
20	1.0 E-3	9.0 E-6
30	1.8 E-4	2.7 E-6

Signal-to-Noise Ratio (dB in 3-kHz Bandwidth)	Probability of Bit Error	
	300 bps	75 bps
0	1.8 E-2	4.4 E-4
2	6.4 E-3	5.0 E-5
4	1.0 E-3	1.0 E-6
6	5.0 E-5	1.0 E-6
8	1.5 E-6	1.0 E-6

[a]Two independent, equal, average-power Rayleigh fading paths, with 2-Hz fading bandwidth and 2-ms multipath spread.

Source: Reference 31, MIL-STD-188-110.

9.12.3   Serial Tone Approaches

The first technique we describe is covered in MIL-STD-188-110 (Ref. 31) and employs M-ary phase shift keying (PSK) on a single carrier frequency (or tone). The modem accepts serial binary data at 75×2^n up to 2400 bps and converts this bit stream into an 8-ary PSK modulated output signal. The serial bit stream, before modulation, is coded with a convolutional code with a constraint length of 7 ($K = 7$). Table 9.13 gives the coding rates for the various input data rates. An interleaver is used with storage of 0.0-, 0.6-, and 4.8-s block storage (interleaving interval). The carrier tone frequency is 1800 Hz $\pm$ 1 Hz. Figure 9.32 shows the modulation state diagram for M-ary PSK as employed by this modem.

TABLE 9.13   Error Correction Coding, Fixed-Frequency Operation

Data Rate (bps)	Effective Code Rate	Method for Achieving the Code Rate
2400	$\frac{1}{2}$	Rate $\frac{1}{2}$ code
1200	$\frac{1}{2}$	Rate $\frac{1}{2}$ code
600	$\frac{1}{2}$	Rate $\frac{1}{2}$ code
300	$\frac{1}{4}$	Rate $\frac{1}{2}$ code repeated 2 times
150	$\frac{1}{8}$	Rate $\frac{1}{2}$ code repeated 4 times
75	$\frac{1}{2}$	Rate $\frac{1}{2}$ code

Source: Reference 31, MIL-STD-188-110.

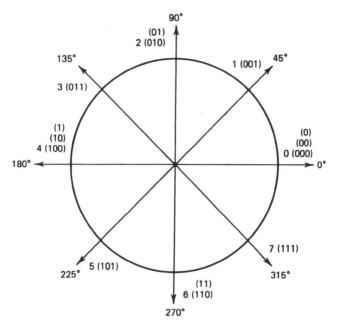

Legend:
0°... 315°= Phase (degrees)
0 ... 7 = Tribit numbers
(000) ... (111) = Three–bit channel symbols
(00) ... (11) = Two–bit channel symbols
(0) ... (1) = One–bit channel symbols

Figure 9.32. State constellation diagram. (From Ref. 31, MIL-STD-188-110.)

The waveform (signal structure) of the modem has four functionally distinct sequential transmission phases:

1. Synchronization preamble phase
2. Data phase
3. End-of-message (EOM) phase
4. Coder and interleaver flush phase

The length of the preamble depends on the interleaver setting. For the 0.0 setting, it is 0.0-s duration; for the 0.6 setting, 0.6-s duration; and for the 4.8-s setting, 4.8-s duration. This known preamble sequence allows the distant-end receive modem to achieve time and frequency synchronization.

During the data transmission phase, the desired data message is interspersed with known data sequences to train the far-end modem for channel equalization. We are aware, of course, that on a HF medium channel group delay and amplitude distortion are constantly changing with time. For instance, at 2400 bps 16 symbols of a known sequence (called known data) are followed by 32 symbols of user data (called unknown data). After the last unknown data bit has been transmitted, a special 32-bit sequence is sent to the coder, which performs the EOM function. It informs the distant receiving modem of end of message. The EOM sequence consists of the hexadecimal number 4B65A5B2. The final transmission phase is used to flush (reset) the far-end FEC decoder.

Table 9.14 gives bit error probabilities versus signal-to-noise ratio for various bit rates. This performance is based on the Waterson simulator described in CCIR Rep. 549-2 (Ref. 32). Here the modeled multipath spread values and fading (2σ) bandwidth values derive from two independent but equal average-power Rayleigh paths (Ref. 31).

The second serial tone approach is taken from Annex A of MIL-STD-188-141A (Ref. 14). The primary purpose of the standard is to automate a low-data-rate HF system, whether a single link, a star network with polling, or a large grid network. A fully detailed open system interconnection (OSI) layer 2 (data link layer; see Chapter 13) is outlined. Two important functions are incorporated: automatic link establishment (ALE) and a method of self-sounding, including a method of exchanging link quality assessment (LQA) data among network members.

The modulation is 8-ary FSK and the eight tones are as follows: 750 Hz (000), 1000 Hz (001), 1250 Hz (011), 1500 Hz (010), 1750 Hz (110), 2000 Hz (111), 2250 Hz (101), and 2500 Hz (100). It will be appreciated that just one tone is transmitted at a time. The tone transitions are phase continuous with a baud rate of 125 bauds, which is a transmission rate of 375 coded symbols per second (1 baud = 3 symbols).

TABLE 9.14   Performance Characteristics for Serial Tone Modem Using Waterson Simulator (CCIR Rep. 549-2)

User Bit Rate	Channel Paths	Multipath (ms)	Fading[a] Bandwidth (Hz)	SNR[b] (dB)	Coded Bit Error Rate
2400	1 fixed			10	1.0 E-5
2400	2 fading	2	1	18	1.0 E-5
2400	2 fading	2	5	> 30	1.0 E-3
2400	2 fading	5	1	> 30	1.0 E-5
1200	2 fading	2	1	11	1.0 E-5
600	2 fading	2	1	7	1.0 E-5
300	2 fading	5	5	7	1.0 E-5
150	2 fading	5	5	5	1.0 E-5
75	2 fading	5	5	2	1.0 E-5

[a]Per CCIR Rep. 549-2 (Ref. 32).
[b]3-kHz bandwidth.
Source: Reference 31, MIL-STD-188-110.

TABLE 9.15 Probability of Linking

Probability of Linking (%)	Signal-to-Noise Ratio (dB)		
	Gaussian Noise Channel	CCIR Good Channel	CCIR Poor Channel
≥ 25	−2.5	+0.5	+1.0
≥ 50	−1.5	+2.5	+3.0
≥ 85	−0.5	+5.5	+6.0
≥ 95	0.0	+8.5	+11.0

Source: Reference 14, MIL-STD-188-141A.

The system uses block coding FEC with the Golay $(24, 12, 3)$ rate $\frac{1}{2}$ code. In other words, 1 data bit is represented by two coded symbols. In the data text mode (DTM), automatic message display (AMD) mode, and the basic ALE mode, an auxiliary coding is also employed: redundant $\times 32$ with $\frac{2}{3}$ majority voting (with 49 transmitted symbols).

The uncoded data rate (user data rate) is 61.22 bps in the DTM, the AMD mode, and the ALE modes. In the data block mode (DBM), the uncoded data rate is 187.5 bps $(375/2)$. The throughput maximum data rate is 53.57 bps in the DTM, AMD, and basic ALE modes. The source coding is a subset of ASCII (see Chapter 13).

The LQA is built into the protocol header. LQA information includes BER, SINAD (signal + noise + distortion to noise + distortion ratio), and multipath value (MP). The LQA field consists of 24 bits, including 11 overhead bits, 3 multipath bits, 5 SINAD bits, and 5 BER bits. The system incorporates a sounding probe for frequency optimization/management.

Table 9.15 gives performance data regarding the probability of linking (ALE) using the standard HF simulator described in CCIR Rec. 549-2 (Ref. 32).

9.13 IMPROVED LINCOMPEX FOR HF RADIO TELEPHONE CIRCUITS

Lincompex is an acronym for *link compression and expansion*. It is a technique that provides uniquely controlled companding (compression–expansion) function on single-sideband (SSB) voice circuits. Performance improvement on links employing Lincompex is reported to be 22 dB minimum across time for typical speech signals under varying propagation conditions with a maximum improvement of 47 dB (Refs. 35, 36).

On systems using Lincompex, speech is compressed to a comparatively constant amplitude and the compressor control current is utilized to frequency modulate an oscillator in a separate control channel carried in a frequency lot just above the voice channel. The speech channel, which contains virtually all the frequency information of the speech signal, and the control channel, which contains the speech amplitude information, are combined for transmission on the nominal HF 3-kHz channel. As each speech syllable is individually compressed, the transmitter is more effectively loaded than with current SSB practice (without Lincompex). On reception, both the speech and control signals are amplified to constant level, the demodulated control signal being used to determine the expander gain and thus restore the original amplitude variations to the speech signal. Because the output level at the receiving end depends solely on the frequency of the control signal, which is itself directly related to the input level at the transmitting end, the overall system gain and loss can be maintained at a constant level. Operation with a slight loss (two-wire to two-wire) eliminates the need for singing suppressors, although echo suppressors will still be needed on long delay circuits.

Lincompex equipment based on ITU-R Rec. F.1111 (Ref. 35) can accommodate four different HF speech channels: 250–2500 Hz, 250–2380 Hz, 250–2000 Hz, and 250–1575 Hz; each with a control tone at 2900, 2580, 2200, or 1775 Hz, respectively. Calibrated frequency deviation on a tone is ± 60 Hz. Change in frequency for each 1-dB change in level is 2 Hz. There is a modulator frequency calibrate format that can be activated on command. It will correct errors of ± 80 Hz and bring the end-to-end frequency error down to ± 2 Hz. A Lincompex transmit unit can accommodate input levels from $+5$ dBm0 to -55 dBm0.

The maximum end-to-end frequency error of each Lincompex channel should be within ± 2 Hz if no Lincompex frequency calibrate option is available.

Performance data in this section were derived from ITU-R Rec. F.1111 (Ref. 35). Operational information was provided by Link Plus of Columbia, Maryland.

9.14 HF ANTENNAS*

9.14.1 Introduction

The HF antenna installation can impact link performance more than any other system element of a HF system. It is also often the least understood and appreciated.

The antenna subsystem can be a "force-multiplier" if you will. A 10-dB net gain antenna system can make a 1-kW RF power output behave like 10 kW. It can attenuate interfering signals entering sidelobes. With an arrayed antenna system using advanced interference nulling techniques, interference rejection can be even more effective.

---

*The material in Section 9.14 is primarily based on Ref. 33.

TABLE 9.16   Antenna Applications

Application	Antenna Type	Notes
Point-to-point skywave one-hop	Horizontal LP[a]	
	Terminated long wire	
	Dipole and doublet	
Point-to-point skywave, multihop	Vertical LP[a]	
	Rhombic	
	Sloping vee	
Point-to-point groundwave	Whip ⎫	Vertical polarization
	Tower ⎭	generally preferred
NVI	Doublet ⎫	Reasonable gain
	Frame ⎭	at high TOAs[b]
Multipoint skywave, one-hop	Conical monopole ⎫	
	Discone	
	Cage antenna	Broadband
Multipoint skywave, multihop	LP[a] rosette ⎬	omnidirectional
	Array of rhombics	(azimuth)
Multipoint groundwave	Whip with counterpoise	
	Tower ⎭	

[a]Log periodic.
[b]Takeoff angles.

The selection of a particular type of HF antenna is application driven. We consider three generalized applications: (1) point-to-point, (2) near-vertical incidence (NVI), and (3) multipoint and subsets of skywave and groundwave. Table 9.16 reviews these applications and some antenna types appropriate to the application.

Important parameters in the selection of antennas are:

- Radiation patterns: azimuthal and vertical (elevation)
- Directive gain (receiver)
- Power gain (tuner efficiency, transmission line loss—transmitter)
- Polarization
- Impedance
- Ground effects
- Instantaneous bandwidth

Other important factors are size and cost.

In previous chapters we have used the isotropic radiator as the reference antenna, where gain is expressed in dBi (dB related to an isotropic). The isotropic antenna has a gain of 1 or 0 dB and radiates uniformly in all directions in free space.

HF engineers often use other antennas as reference antennas. Table 9.17 compares some of these with the isotropic.

TABLE 9.17   Reference Antennas

Antenna Type	Gain (dBi)	Notes
Isotropic	0	In free space
Half-wave dipole	2.15	In free space
Full-wave dipole	3.8	In free space
Short vertical	4.8	In free space
Vertical	5.2	Quarter-wave on perfectly conducting flat ground; lossless tuning device

Source: Reference 33, DCAC 330-175-1.

9.14.2 Antenna Parameters

9.14.2.1 Radiation Patterns. Radiation patterns are usually provided by the antenna manufacturer for a specific model antenna. These are graphical plots, one for the horizontal plane or azimuthal and one for the vertical plane (elevation). Figures 9.33*a* and 9.33*b* show typical plots. The vertical plane should have some or total correspondence with the calculated elevation angle(s) or TOAs (takeoff angles).

One quantity determined uniquely by the radiation pattern is the antenna *directivity* that an antenna can attain. It is defined as the ratio of the maximum radiated power density to the average radiated power density. Whether a gain is in fact attained depends on losses with the antenna system.

9.14.2.2 Polarization. Polarization of the radiation from an antenna is produced by virtue of the fact that the current flow direction in an antenna is a vector quantity to which the spatial orientations of the electric and magnetic fields are related. Single linear antennas in free space produce linear polarized waves in the far field, with the electric vector in a plane parallel to and passing through the axis of the radiating element. Antennas containing radiating elements with different spatial orientations and time phases produce elliptically polarized waves, with the ellipticity being a function of direction.

A linear antenna in free space used for receiving responds most strongly to another antenna with the same polarization and not at all to one polarized at 90° to the first antenna. Right-hand circularly (RHC) polarized antennas produce maximum receiving response when receiving transmission from a right-hand polarized antenna, and none from a left-hand circularly (LHC) polarized antenna.

Several other effects are also important. A circularly polarized antenna receiving a signal from a linear polarized antenna delivers 3 dB less power for a given transmitting power than would be received from a circular transmitted wave of the correct rotation sense. Conversely, of course, the same received signal loss applies to a linear receiving antenna and a circularly polarized transmitting antenna.

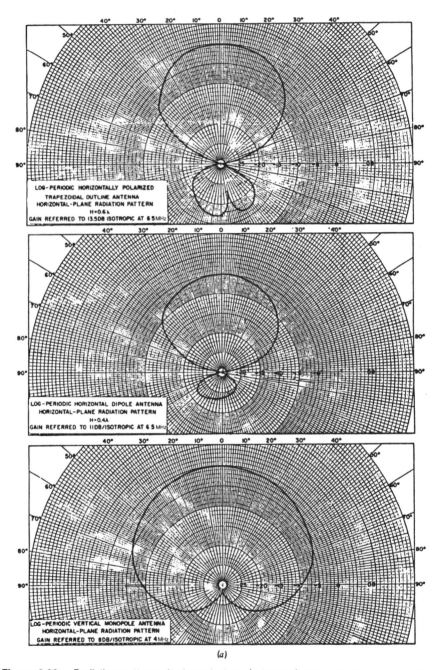

(a)

Figure 9.33a. Radiation patterns, horizontal plane (azimuthal), log periodic antennas. For directive values (dB): upper curve, add 13.5 dB; middle curve, add 11 dB; and lower curve, add 8 dB to readings. (From Ref. 33, DCAC 330-175-1.)

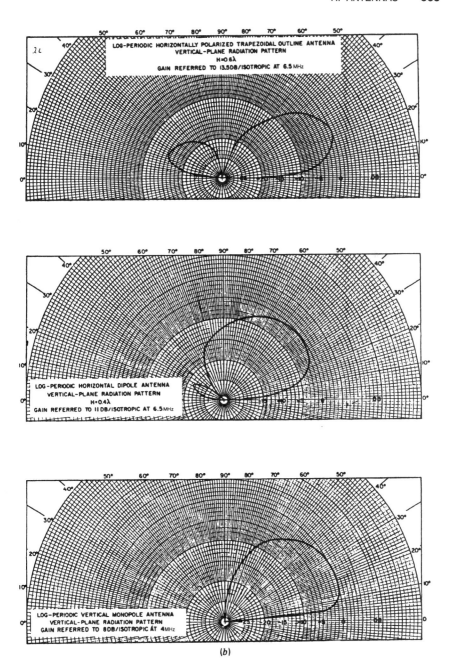

(b)

Figure 9.33b. Radiation patterns, vertical plane (elevation or TOA), log periodic antennas. For directive gain values (dB): upper curve, add 13.5 dB; middle curve, add 11 dB; and lower curve, add 8 dB to the readings. (From Ref. 33, DCAC 330-175-1.)

Vertically polarized waves radiated by a vertical antenna over imperfect ground are elliptically polarized because of the presence of a small electric field along the ground in space quadrature and a different time phase from the major electric component.

For HF skywave propagation, the received signal is usually elliptically polarized even when the transmitting antenna produces linear polarization. This condition arises from the effect of the earth's magnetic field in the ionized layers, which splits the incident linearly polarized signal into the ordinary and extraordinary wave. The two waves travel at different velocities and experience different polarization rotations. The resulting received signal is elliptically polarized.

9.14.2.3 *Impedance.*

The impedance of an antenna depends on the radiation resistance, the reactive storage field, antenna conductor losses, and coupled impedance effects from nearby conductors. For simple antennas such as dipoles, integration of the power pattern over a spherical surface will yield a quantity called the radiation resistance, which appears as a multiplying factor for the antenna current in the integrated power pattern. The radiation resistance obtained from the antenna impedance can be measured by a RF bridge placed at the current reference terminals of the antenna.

The antenna impedance as measured with a bridge will contain the radiation resistance, a resistance proportional to antenna conductor losses and, if the antenna is near other conducting objects, a component representing an interaction between the antenna and the nearby conductors. This latter component may be either positive or negative resistance and will usually have a reactive component.

The reactive storage field of an antenna depends to a considerable extent on the antenna diameter and physical shape. The storage field represents energy that flows into the antenna and then is returned to the generator during each cycle.

Antennas that have small diameters in terms of wavelength have large storage fields, whereas thicker antennas have lesser storage energy. Since the storage field represents circulating energy, which is analogous to that flowing in a physical reactance, the storage field contributes a reactive component to the antenna impedance.

Dipole antennas and other balanced standing wave antennas have reactive behavior similar to that of an open-circuited transmission line with a length equal to half the antenna length. For example, the reactive component of the impedance of a dipole is capacitive for lengths shorter than half-wavelength, is zero at about half-wavelength, and is inductive for lengths between one-half and one wavelength. The antenna reactance continues to alternate cyclically in a half-wavelength period for longer lengths. For convenience in matching the standing wavelength antenna, it is usual at high frequencies to use antennas at a frequency where resonance occurs, that is, where the reactance is zero.

Traveling-wave antennas, such as rhombics and terminated vees, have wide-impedance bandwidths. In fact, for these antennas, impedance bandwidths usually exceed radiation pattern bandwidths. Log-periodic (LP) antennas achieve wide-impedance bandwidth by selectively choosing only elements that are nearly resonant at a specific frequency.

Nominal impedance values of some common HF antennas are as follows (Ref. 33):

Horizontal rhombic and terminated vee	600 Ω
Horizontal LP	50/300 Ω
Vertical (dipole and monopole)	50 Ω
Yagi and half-wave dipole	50 Ω
Conical monopole, discone, and inverted discone	50 Ω
Vertical tower	50 Ω

9.14.2.4 Gain and Bandwidth. The gain of an antenna is defined as the ratio of the maximum power density radiated by the antenna to the maximum power density radiated by a reference antenna (see Table 9.17). The directivity of an antenna, which is sometimes confused with antenna gain, is the ratio of the maximum power density radiated by the antenna to the *average power* radiated by the antenna. The distinction between the two terms arises from the fact that directivity will exceed antenna gain. Since all antennas have some losses, the directivity will exceed antenna gain. The directivity of an antenna can be obtained from the antenna radiation patterns alone, without consideration of antenna circuit losses. Because directivity is a ratio, absolute power values are not required in determination of directivity, and convenient normalizing factors can be applied to radiation power values.

Rhombic and vee antennas, which dissipate portions of the antenna input power in terminators, will have lower gain values than directivity values by about the termination losses. Cophased dipole arrays have low conductor losses, and directivity, as a result, only slightly exceeds gain values for such antennas. LPs and Yagis can have 1-dB or more difference between directivity and gain. Table 9.18 compares power gain, usable radiation angles, nominal bandwidth, horizontal beamwidth, and sidelobe suppression.

9.14.2.5 Ground Effects. The free-space radiation pattern efficiency and impedance of an antenna are modified when the antenna is placed near ground. The impedance change is small for antennas placed at least one wavelength above ground, but the change becomes increasingly greater as that height is reduced. Since the ground appears as a lossy dielectric medium for HF, location of the antenna near ground may increase the losses a considerable extent unless a ground plane (a wire mesh screen or ground radials) is used to reduce ground resistance. Vertical monopole antennas, which are often fed at the ground surface, require a system of ground radials extending from the antenna to a sufficient distance to provide a

TABLE 9.18 Antenna Comparison

Type	Power Gain (dB)	Usable Radiation Angle (deg)	Bandwidth Ratio	Horizontal Beamwidth	Sidelobe Suppression
Horizontal rhombic	8–23	4–35	≥ 2 : 1	6–26°	> 6 dB
Terminated vee	4–10	4–35	≥ 2 : 1	8–36°	> 6 dB
Horizontal LP	10–17	5–45	≥ 8 : 1	55–75°	> 12 dB
Vertical LP (dipole)	6–10	3–25	≥ 8 : 1	90–140°	> 12 dB
Horizontal half-wave dipole	5–7	5–80	≥ 5%	80–180°/lobe	N/A
Discone	2–5	4–40	≥ 4 : 1	N/A	N/A
Conical monopole	−2–+2	3–45	≥ 4 : 1	N/A	N/A
Inverted monopole	1–5	5–45	≥ 4 : 1	N/A	N/A
Vertical tower	−5–+2	3–30	≥ 4 : 1	N/A	N/A

Note 1. Typical power gains are gains over good earth for vertical polarization and over poor earth for horizontal polarization.
Note 2. Usable radiation angles are typical over good earth for vertical polarization and over poor earth for horizontal polarization.
Note 3. Normal bandwidth is the ratio of the two frequencies within which the specified voltage standing wave ratio (VSWR) will not be exceeded or within which the desired pattern will not suffer more than 3-dB degradation.
Source: Reference 33, DCAC 330-175-1.

low-resistance ground return path for the ground currents produced by the induction fields. For short vertical antennas, the radial length should be approximately $\lambda/2\pi$ long; for longer antennas the length should be approximately the antenna height. In addition, near the antenna base a wire mesh ground screen is recommended to reduce I^2R losses. For medium- or long-distance HF links, horizontal antennas should be mounted greater than one-half wavelength above ground and usually do not require a ground screen or ground plane.

In addition to losses, the presence of ground causes a change in antenna impedance. The change is brought about by the interaction between the antenna fields and fields of the ground currents.

Ground-reflected energy combines with the direct radiation of an antenna to modify the significantly vertical radiation pattern of the antenna. The magnitude of the image antenna current is equal to the magnitude of the real antenna current multiplied by the ground-reflection amplitude coefficient. An additional phase difference between the direct and the ground-reflected field at a distant point results from the difference of the two path lengths.

9.14.2.6 Bandwidth.
Antenna bandwidth is specified as the frequency band over which the voltage standing wave ratio (VSWR) criteria are met and the radiation pattern provides the required performance. The frequency band is, to some extent, determined by the application, since greater deterioration of antenna characteristics can be tolerated for receiving use than for transmitting use. Antenna bandwidth is usually limited by the change of either or both the radiation shape and impedance with change of frequency.

Simple antennas, such as dipole, when in free space, have a figure-eight pattern broadside to the antenna. For dipoles, the pattern maxima are oriented normal to the dipole with lengths up to approximately $1\frac{1}{4}$ wavelengths. With greater lengths, the pattern maxima may not be oriented in this direction.

Linear conductors with traveling-wave distributions have a continuous shift of the distribution in the direction of radiation pattern maximum, and the maximum approaches the axis of the conductor with increasing frequency. For this type of antenna and all other linear antennas, the pattern maximum can never fall exactly along the conductor axis because of the inherent null of the conductor in the direction of current flow, except when located near the earth.

When simple antennas are arrayed with other like antennas, the pattern changes increase with frequency because the changing electrical spacing between the arrayed elements also becomes a factor in the determination of the radiation pattern. In an array with discrete elements, an additional complication is produced by grating lobes. These lobes can occur whenever the interelement spacing exceeds a half-wavelength. If the element spacing equals one wavelength, the grating lobes can have intensities equal to the main radiation lobe. These lobes arise because the wide interelement spacing allows all element radiations to combine in phase in more than one direction.

Another factor in pattern change with frequency is produced by antenna energy, which is reflected from the ground and combined with the direct radiated energy. Since the electrical path difference between the direct and the ground-reflected energy is proportional to the frequency, the elevation pattern of an antenna above ground is a function of frequency.

In a communication circuit, two degrees of severity in pattern change with frequency can occur. In one the pattern shape changes, but the maximum is in the required radiation direction. In the second the maximum is deflected from the desired direction with frequency. The first can be tolerated at the expense of increased circuit interference and noise. The second can cause complete circuit outages with frequency change.

Considerations of pattern change with frequency usually limit the use of broadband monopoles or dipoles to no more than two octaves. This usable frequency range is reduced to no more than one octave when these elements are arrayed. Rhombic antennas have a tolerable pattern shift, if used within a frequency range of less than one octave. LP antennas escape the directivity limitations of antennas using conductors of fixed physical length by selectively changing the active conductor length with frequency. However, unless the LP antenna is specially designed for use over ground, the antenna will have a pattern shift caused by ground reflection.

Input impedance change of an antenna may limit the coverage of an antenna to a smaller frequency range than would be expected from the antenna radiation pattern. This is particularly true of transmitting antennas, when the antenna supplies the load for the transmission line and transmitter, and any inability to load the transmitter. Input impedance requirements for receiving antennas are not so severe because the receiver is the termination for the line and the antenna VSWR causes only a reduction in received power transfer.

9.15   REFERENCE FIELDS—THEORETICAL REFERENCES*

In most propagation problems, the methods of solution are based on a transmitting antenna that will produce some standard value of field intensity at a standard distance. This standard field may be one of several values produced by one of several types of reference antenna. Three of the most commonly used reference antennas are the omnidirectional radiator (isotropic source) in free space, the omnidirectional radiator over perfectly conducting earth, and the short lossless vertical antenna over perfect earth. Each field is expressed in reference to 1-kW power input at a standard distance of 1 mi or 1 km.

The most often used reference antenna is the omnidirectional radiator, a theoretical antenna that radiates equally well in all directions. The field intensity produced by this antenna may be found by equating the radiated power to the surface integral of a uniform field over a spherical surface surrounding the radiator:

$$E^2 = \frac{P_t \eta}{4 \pi d^2} \quad \text{mV/m} \tag{9.18}$$

where P_t = the transmit power (W)
   η = the impedance of free space
   d = the distance (km)

Thus if the transmit power P_t is 1 kW and the distance is 1 km, the field intensity is found by

$$E = \left(\frac{1000 \eta}{4 \pi} \right)^{1/2} \quad \text{mV/m} \tag{9.19}$$

and since $\eta = 120 \pi$ (impedance of free space),

$$E = [1000(30)]^{1/2}$$
$$= 173.2 \text{ mV/m at 1 km}$$

If this omnidirectional antenna is placed on the surface of perfectly conducting earth, we obtain

$$E = 173.2 \sqrt{2}$$
$$= 245.0 \text{ mV/m at 1 km}$$

*The following material is extracted from Ref. 33.

as the field intensity, since all the power is radiated in a hemisphere and, therefore, the power per unit area is doubled.

The field of a short vertical antenna over perfect earth may be found by

$$E_\Delta = \frac{60\pi I}{d}\left(\frac{L}{\lambda}\right)\cos \Delta(1 + e^{-j4\pi h/\lambda \sin \Delta}) \qquad (9.20)$$

where l/λ = length in wavelengths

Δ = vertical radiation angle

I = input current

h/λ = effective height in wavelengths

d = the distance in kilometers

But since the height of the differential element for a ground-based antenna is zero, the field intensity in the ground plane over perfect earth is

$$E_\Delta = \frac{120\pi I}{d}\left(\frac{l}{\lambda}\right) \qquad (9.21)$$

The radiation resistance R_r is

$$R_r = 160\pi^2\left(\frac{l}{\lambda}\right)^2 \qquad (9.22)$$

and since $I = \sqrt{P_t/R_r}$ then

$$I = \frac{\sqrt{P_t}}{\sqrt{160\pi^2(l/\lambda)^2}} - \frac{\sqrt{P_t}}{4\sqrt{10\pi l/\lambda}} \qquad (9.23)$$

and the field intensity becomes

$$E_\Delta = \frac{120\pi}{d}\left(\frac{l}{\lambda}\right)\frac{\sqrt{P_t}\cdot\lambda}{4\sqrt{10}\,\pi l} \qquad (9.24)$$

and

$$E_\Delta = \frac{30}{d}\sqrt{\frac{P_t}{10}} \qquad (9.25)$$

where d is in kilometers, P is in watts, and E_Δ is in mV/m.

Solving for a power input of 1 kW at a distance of 1 km in the ground plane ($\Delta = 0°$),

$$E_\Delta = 30\sqrt{\frac{1000}{10}} = 300 \text{ mV/m}$$

9.16 CONVERSION OF RADIO FREQUENCY (RF) FIELD STRENGTH TO POWER*

Many radio engineers are accustomed to working in the power domain (e.g., dBm, dBW). We may wish to know the receive signal level (RSL) at the input to the first active stage of a HF receiver. In the power domain, the characteristic impedance is not a consideration by definition.

HF engineers traditionally work with field strength usually expressed in microvolts per meter (μV/m). When we convert μV/m to dBm, characteristic impedance becomes important. We remember the familiar formula

$$
\text{Power}_{(W)} = \frac{E^2}{R} = I^2 R \tag{9.26}
$$

where E is expressed in volts and I in amperes, and we can consider R to be the characteristic impedance.

Carrying this one step further,

$$
\text{Power}_{(W)} = \frac{[E(V/m)]^2 [\text{effective antenna area } (m^2)]}{(\text{impedance of free space})} \tag{9.27}
$$

The impedance of free space is 120π or 377 Ω.

The effective antenna area is

$$
A_{(m^2)} = \frac{G\lambda^2}{4\pi} \tag{9.28}
$$

where A = effective antenna area (m^2)
G = antenna gain (numeric, *not* dB)
λ = wavelength (m)

If $G = 1$ (0 dBi), then

$$
P_{(W)} = \frac{E^2\lambda^2}{377 \times 4\pi} \tag{9.29}
$$

We rewrite equation 9.29 expressed in frequency rather than wavelength:

$$
P_{(W)} = \frac{\dot{E}^2(c^2)}{4737.5(f^2)} \tag{9.30}
$$

where c = velocity of propagation in free space, or 3×10^8 m/s. Convert to

*The following material is extracted from Ref. 34.

more useful units: express P in milliwatts, E in microvolts per meter, and f in megahertz. Then

$$P_{(mW)} = \frac{1.89972 \times 10^{-8}(E)^2}{f^2_{(MHz)}} \qquad (9.31)$$

Here E is expressed in microvolts per meter. We now derive

$$P_{(dBm)} = -77 + 20\log(E) - 20\log(f) \qquad (9.32)$$

where E is the field strength in microvolts per meter and f is expressed in megahertz. To convert back to field strength

$$E = 10^{[P+77+20 \ \log(f)]/20} \qquad (9.33)$$

REVIEW EXERCISES

1. Give three advantages and three disadvantages of HF as a communication medium.

2. Identify at least five applications of HF for telecommunication that are used very widely.

3. Give at least three of the variables that affect range of HF groundwave communication links.

4. In the text, four ionospheric layers are described that affect HF propagation. What are they? One of these layers disappears at night. It affects a HF link in several ways. What are these effects and how can they be mitigated?

5. Describe two completely different ways in which we can communicate short distances by HF, say, out to 100–200 km.

6. Classify by region on the earth's surface where it is most desirable and least desirable to communicate by HF.

7. Describe how season affects HF propagation.

8. Considering transition periods, why are north–south paths better behaved than east–west paths?

9. How does sunspot activity affect HF propagation?

10. Describe the average length of a sunspot cycle and variation in sunspot number (average) over the period.

11. List the five methods given in the text to carry out frequency management.

12. Using Table 9.2, at 1100 universal time (UT), what frequency would be recommended?

13. How can we derive the maximum usable frequency (MUF) when given the critical frequency?

14. How can one get comparative signal strength data for a particular connectivity without the use of an ionospheric sounder or by the "cut and try" method?

15. Describe how an embedded ionospheric sounding system operates. Some trademark sounding devices give other important data. Explain what some of the other data are and how they can be used to aid path performance.

16. Describe an advanced self-sounding system. What does link quality assessment (LQA) mean and how can it be used?

17. What is the weakness in the approach of basing LQA entirely on received signal strength?

18. Identify the three basic HF transmission modes. Distinguish each from the other and give their application.

19. What detracts from groundwave operation at night?

20. What three effects give rise to multipath propagation?

21. What is the effect of D-layer absorption on multihop propagation?

22. Justify why we would be better off choosing the lowest order F2 mode on multimode reception.

23. What is the meaning of 3F2?

24. Explain the basic impairments due to propagation on transequatorial skywave paths, especially if they have only small latitude changes.

25. Explain the cause of the formation of the O-ray and X-ray. Which arrives first at the distant receiver?

26. Suppose we operate well below the MUF on a 1F2 path. How many modes could we receive? (Assume 1F2 only.)

27. What sort of antenna would we require for optimum near-vertical incidence (NVI) operation?

28. What frequency band would we favor (approximately) for NVI operation?

29. Discuss the *reciprocity* of HF operation.

30. What fade margin is recommended by CCIR on a HF skywave circuit for 99% time availability?

31. How much dispersion will a 50-baud FSK signal tolerate (i.e., binary FSK, or BFSK)?

32. Give at least three mitigation techniques that we might use to combat the effects of HF multipath.

33. Identify four types of diversity we may consider using for HF operation. Describe each in one or two sentences.

34. Why is receiver noise figure a secondary concern on HF systems?

35. There is a fair probability that the optimum operational frequency may be occupied by at least one other user. Name six methods to overcome this problem.

36. What is the cause of atmospheric noise and how does it propagate? How does it vary with frequency?

37. Calculate the receiver noise threshold at 1900 MHz, summer at 7 p.m. local time in Massachusetts for a receiver with a 3-kHz bandwidth, valid for 95% of the time.

38. Above 20 MHz, what types of external noise may be predominant?

39. Man-made noise is a function of what two items?

40. What are the components (contributors) of HF link transmission loss?

41. When we calculate the free-space loss for a skywave path 100 km long, what is the most significant range component? For a 1500-km path?

42. How does D-region absorption vary with frequency? It also varies as a function of what? Give at least five items.

43. For a three-hop path, what is the estimated value of ground-reflection losses?

44. Using the text, calculate D-layer absorption for the following path: 1F2 mode, sunspot number 10, operating frequency 14.5 MHz, 12-noon midpath at 45° north latitude with an elevation angle of 9°.

45. Determine the transmission loss for a groundwave path over the ocean 100 km long at 4 MHz.

46. Calculate the optimum groundwave frequency over medium dry land in Colombia at 1400 local time assuming BFSK service, a signal-to-noise ratio (S/N) of 15 dB, a receiver noise figure of 15 dB, 75-bps operation, 1-kW EIRP, and a receiving antenna of 0 dBi, which includes line losses to receiver front end, for a time availability of 95%.

47. On long groundwave paths at night, leaving aside atmospheric noise, for what must we be particularly watchful? What are some compensating methods for this impairment?

48. Give two basic arguments in favor of parallel tone transmission (versus serial tone transmission) for 2400-bps operation. Give at least one argument against the parallel tone approach.

49. The text describes a serial tone system for synchronous transmission of digital traffic. Why is "known data" interspersed with "unknown data"?

50. What is LQA and what is its purpose? Describe how one might use a LQA sequence for a self-sounding system. Argue both sides: Using LQA, is a (PN) spread wideband system easier or more difficult than a conventional narrowband system for self-sounding?

51. Explain in six sentences or less the operation of Lincompex. Where is it applied and what approximate improvement (dB) does it provide?

52. Why is the control of out-of-band emission so important? Give two related answers.

53. Give two reasons why a high-gain, low sidelobe antenna can, if you will, be a force-multiplier on a point-to-point HF link.

54. Name at least five of the six important antenna parameters given in the text.

55. Differentiate between gain and directivity of an antenna.

56. Select an optimum antenna for (a) a groundwave path, multipoint; (b) a NVI path; (c) a short one-hop skywave path (e.g., 1200 km); (d) a very long (> 5000 km) path; and (e) a HF star network operating half-duplex over one- and two-hop paths.

57. Identify the reference value of field strength for the 1-km reference field with a 0° takeoff angle (groundwave).

58. Give the equivalent values in dBm of the following field strengths: 300 mV/m, 300 μV/m, 55μV/m, and 10 μV/m. How many microvolts per meter are equivalent to -114 dBm?

REFERENCES

1. "Digital Selective Calling System for Use in the Maritime Mobile Service," CCIR Rec. 493-3, vol. VIII-2, XVIth Plenary Assembly, Dubrovnik, 1986.
2. L. S. Wagner, J. S. Goldstein, and W. D. Meyers, *Wideband Probing of the Trans-auroral HF Channel, Solar Minimum*, Naval Research Laboratories, Washington, DC, 1987.
3. George Jacobs and Theodore Cohen, *The Shortwave Propagation Handbook*, CQ Publishing, Hicksville, NY, 1979.
4. *Solar Geophysical Data, Prompt Reports*, National Geophysical Data Center, Boulder, CO, Mar. 1990.

5. John L. Lloyd et al., *Estimating the Performance of Telecommunication Systems Using the Ionospheric Transmission Channel—Techniques for Analyzing the Ionospheric Effects upon HF Systems*, Institute of Telecommunication Sciences, NTIA, Boulder, CO, 1983.

6. Larry Teters et al., *Estimating the Performance of Telecommunication Systems Using the Ionospheric Transmission Channel—Ionospheric Analysis and Prediction Program User's Manual*, Institute of Telecommunication Sciences, NTIA, Boulder, CO, 1983.

7. PROPHET Software Description Document, IWG Corp., San Diego, CA, 1984.

8. *CRPL Predictions*, NBS Circular 462, National Bureau of Standards, Washington, DC, 1948.

9. "Bandwidths, Signal-to-Noise Ratios and Fading Allowances in Complete Systems," CCIR Rec. 339-6, 1994 F Series Volume, Part 2, ITU, Geneva, 1994.

10. P. H. Levine, R. B. Rose, and J. N. Martin, *Minimuf-3—A Simplified HF MUF Prediction Algorithm*, IEE Conference on Antennas and Propagation, Pub. no. 78, 1978.

11. *Operator's Manual for the USCG Advanced PROPHET System*, Naval Ocean Systems Command, San Diego, CA, 1987.

12. *Real-time Frequency Management for Military HF Communications*, BR Communications Tech. Note no. 2, BR Communications, Sunnyvale, CA, June 1980.

13. Product Sheet (SELSCAN registered trademark), Collins 309L-2 & 4 and 514A-12 SELSCAN Adaptive Communications Processor, Rockwell International, Cedar Rapids, IA, 1981.

14. "Interoperability and Performance Standards for MF and HF Radio Equipment," MIL-STD-188-141A (Fed. Std. 1045), U.S. Department of Defense, Washington, DC, Sept. 1988.

15. Klaus-Juergen Hortenbach, *HF Groundwave and Skywave Propagation*, AGARD-R-744, Cologne, Germany, Oct. 1986.

16. "Ionospheric Propagation Characteristics Pertinent to Terrestrial Radio Communication Systems Design (Fading)," CCIR Rep. 266-7, Annex to vol. VI, XVIIth Plenary Assembly, Dusseldorf, 1990.

17. *Radio Regulations*, ITU, Geneva, 1982; revised 1988.

18. "Factors Affecting the Quality of Performance of Complete Systems in the Fixed Service," CCIR Rep. 197-4, vol. III, XVIth Plenary Assembly, Dubrovnik, 1986.

19. "Arrangement of Voice-Frequency Frequency-Shift Telegraph Channels on HF Radio Circuits," ITU-R Ref. F.436-3, 1994 F Series Volume, ITU, Geneva, 1996.

20. "Equipment Technical Design Standards for Voice Frequency Carrier Telegraph (FSK)," MIL-STD-188-342, U.S. Department of Defense, Washington, DC, Feb. 1972.

21. B. D. Perry and L. G. Abraham, *Wideband HF Interference and Noise Model Based on Measured Data*, Rep. M-88-7, MITRE Corp., Bedford, MA, 1988.

22. P. J. Laycock et al., "A Model for HF Spectral Occupancy," *Fourth International Conference on HF Systems and Technology*, IEE, London, 1988.

23. *Electrical Communication System Engineering, Radio*, Department of the Army, Washington, DC, Aug. 1956.

24. "Characteristics and Applications of Atmospheric Radio Noise," CCIR Rep. 322-2 (issued separately from other CCIR documents), ITU, Geneva, 1983.

25. "Man-made Radio Noise," CCIR Rep. 258-5, Annex to Vol. VI, XVIIth Plenary Assembly, Dusseldorf, 1990.

26. K. Davies, *Ionospheric Radio Propagation*, NBS Monograph 80, U.S. Department of Commerce, National Bureau of Standards, Boulder, CO, 1965.

27. "Technical Description of the Communication Assessment Program (CAP)," Rev. 3.0, Defense Communications Agency, Arlington Hall Station, VA, Jan. 24, 1986.

28. "Simple HF Propagation Prediction Method for MUF and Field Strength," CCIR Rep. 894-1, vol. V, XVIth Plenary Assembly, Dubrovnik, 1986.

29. Technical Report No. 9, U.S. Army Signal Corps., Radio Propagation Agency, Ft. Monmouth, NJ, 1956.

30. "Groundwave Propagation Curves for Frequencies Between 10 kHz and 30 MHz," CCIR Rec. 368-5, vol. V, XVIth Plenary Assembly, Dubrovnik, 1986.

31. "Equipment Technical Design Standards for Common Long-Haul/Tactical Data Modems," MIL-STD-188-110, through notice 2, U.S. Department Defense, Washington, DC, Nov. 1988.

32. "HF Ionospheric Channel Simulators," CCIR Rep. 549-2, Vol. III, XVIth Plenary Assembly, Dubrovnik, 1986.

33. "MF/HF Communications Antennas," DCAC 330-175-1, Addendum no. 1 to *DCS Engineering—Installation Standards Manual*, Defense Communications Agency, Washington, DC, May 1966.

34. *Technical Issues #89-1* (Dave Adamy), Association of Old Crows, Alexandria, VA, 1989.

35. "Improved Lincompex System for HF Radio Telephone Circuits," ITU-R Rec. F.1111, 1994 F Series Volume, Part 2, ITU, Geneva, 1994.

36. "Improved Transmission Systems for Use over HF Radiotelephone Circuits," CCIR Rep. 354-5, vol. III, XVIth Plenary Assembly, Dubrovnik, 1986.

37. Gerhard Braun, *Planning and Engineering of Shortwave Links*, Siemens–Heyden, London, 1982.

38. "Voice Frequency Telegraphy on Radio Circuits," CCITT Rec. R.39, Fascicle VII.1, IXth Plenary Assembly, Melbourne, 1988.

39. Private communication. Dr. L. Wagner, Naval Research Laboratories, Washington, DC, Jan. 5, 1991.

40. Roy F. Basler et al., "Ionospheric Distortion of HF Signals," Final Rept., Contract DNA-008-85-C-0155, SRI International, Menlo Park, CA, 1987.

10

METEOR BURST
COMMUNICATION

10.1 INTRODUCTION

Meteor burst communication (MBC) can provide inexpensive very low data rate connectivity for links up to about 1000 sm (1600 km) long. Very low data rates in this context are in the throughput range of tens to hundreds of bits per second.

MBC utilizes the phenomenon of scattering of a radio signal from the ionization trails caused by meteors entering the atmosphere. The trails are of short duration from tens of milliseconds to several seconds. A meteor trail must have some form of common geometry between one end of a link and the other. Thus a particular link can use this trail, common to both ends, for a very short period of time and then the users will have to wait for another trail entering the atmosphere with similar common geometry characteristics. On a particular link a transmitter bursts data when a common trail is discovered, waits for another trail, and bursts data again. The time between useful trails is called *waiting time*.

The useful radio frequency range for meteor burst operation is between about 20 and 120 MHz. The lower frequencies are ideal and provide the best performance. As we have seen in the previous chapter, receivers are externally noise limited up to about 40–50 MHz. The types of noise that concern us are man-made and galactic noise. The intensity of both noise types varies inversely with frequency. It is the presence of this noise that drives MBC system designers to use frequencies in the range of 40–50 MHz.

MBC transmitters have output powers ranging from 100 W to 5 kW or more. Antennas for fixed-frequency operation are usually yagis; and horizontally polarized log periodics (LPs) if we wish to cover a frequency range of more than 5 MHz. Figure 10.1 shows the concept of MBC.

The implementation of MBC systems is attractive, especially from the standpoint of economy. The low data rate and the waiting times are

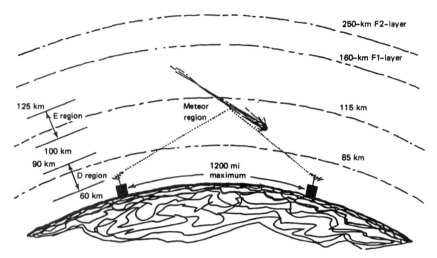

Figure 10.1. The concept of operation of a meteor burst communication link.

disadvantages. One common application is remote sensing of meteorological conditions and/or seismic conditions. One large MBC system is installed in the Rocky Mountains in the United States to provide data on snowfall and accumulated snow. It is called the SnoTel system. MBC networks can also serve as orderwires for larger networks, particularly for the military.

10.2   METEOR TRAILS

10.2.1   General

Billions of meteors enter the earth's atmosphere every day. One source (Ref. 1) states that each day the earth sweeps up some 10^{12} objects that, upon entering the atmosphere, produce sufficient ionization to be potentially useful for reflecting/scattering radio signals.

Meteors enter the atmosphere and cause trails at altitudes of 70–140 km. The trails are long and thin, generating heat that causes the ionization. They sometimes emit visible light. The forward scatter of radio waves from these trails can support communication. The trails quickly dissipate by diffusion into the background ionization of the earth's atmosphere.

Meteor trails are classified into two categories, underdense and overdense, depending on the line density of free electrons. The dividing line is 2×10^{14} electrons per meter. Trails with a line density less than the value are *underdense*; those with a line density greater than 2×10^{14} electrons per meter are termed *overdense*. The dividing line of about 2×10^{14} electrons per meter corresponds to the ionization produced by a meteor whose weight is about 1×10^{-3} g. When averaged over 24 h, the number of meteors is almost inversely proportional to weight. As a result, we would expect that the number of underdense trails would far exceed the number of overdense trails. However, the signals reflected from underdense trails fall off roughly in proportion to the square of the weight, whereas signals from overdense trails increase only a little with weight. In practice, though, we find perhaps only 70% of received MBC signals are from underdense trails. Even so, the mainstay of a MBC system is the underdense trail.

Another interesting and useful fact about sporadic meteors is their mass distribution. This distribution is such that the total masses of each size of particle are approximately equal (i.e., there are 10 times as many particles of mass 10^{-4} g as there are particles of mass 10^{-3} g). Table 10.1 lists the approximate relationship between mass, size, electronic density, and number (Ref. 2).

10.2.2 Distribution of Meteors

At certain times of the year meteors occur as showers and may be prolific over durations of hours. Typical of these are the Quadrantids (early January), Arietids (May, June), Perseids (July, August), and Geminids (December)

TABLE 10.1   Estimate of Properties of Sporadic Meteors

Notes		Mass (g)	Radius (cm)	Number Swept Up by Earth per Day	Electron Line Density (Electrons/Meter)
Particles that survive passage through atmosphere		10^4	8	10	
	Overdense	10^3	4	10^2	
	visual	10^2	2	10^3	
		10	0.8	10^4	10^{18}
		1	0.4	10^5	10^{17}
Particles totally disintegrated in upper atmosphere		10^{-1}	0.2	10^6	10^{16}
		10^{-2}	0.08	10^7	10^{15}
	Underdense	10^{-3}	0.04	10^8	10^{14}
	nonvisual	10^{-4}	0.02	10^9	10^{13}
		10^{-5}	0.008	10^{10}	10^{12}
		10^{-6}	0.004	10^{11}	10^{11}
		10^{-7}	0.002	10^{12}	10^{10}
Particles that cannot be detected by radio means		10^{-8} to 10^{-13}	0.004 to 0.0002	Total about 10^{20}	Practically none

Source: Reference 2. Courtesy of Meteor Communications Corporation. Reprinted with permission.

(Ref. 3). However, for MBC planning purposes, we use the type of meteors discussed in Section 10.2.1, which CCIR calls "sporadic meteors."

Meteor intensity (i.e., the number of usable meteor trails per hour) varies with latitude, becoming less in number but more uniform diurnally for the high latitudes. With the midlatitude case, there is roughly a sinusoidal diurnal variation of incidence, with the maximum intensity about 0600 local time and the minimum some 12 hours later, or 1800 local time. The ratio of maximum to minimum is about 4:1. There is a seasonal variation of similar magnitude with a minimum in February and a maximum in July. Considerable day-to-day variability exists in the incidence of sporadic and shower meteors (Ref. 4).

These variabilities are explained by the earth's rotation and its orbit around the sun. The diurnal variability is a consequence of the earth's motion in its orbit around the sun. All meteor particles are in some form of orbit about the sun. At about 0600 local time a MBC site in question is on the forward side of the earth and at that time it sweeps up slower moving particles as well as a random number of particles colliding with the earth. At 1800 local time the same MBC site is on the back side of the earth, where the slower moving particles cannot catch up with the earth. Thus 1800 is the least productive time of the day. There is always the value of random meteor counts, with the earth's velocity modulating the mean, increasing the morning count and decreasing the evening count. Sometimes at midday we see a day-to-day variation of 5:1, and this is due to useful returns from sporadic E (layer) rather than just meteor intensity variation.

The season maximum in the July/August period is due to the fact that the earth's orbit takes it through a region of more dense solar orbit material.

10.2.3   Underdense Trails

MBC links basically depend on reflections from underdense trails. An underdense trail does not actually *reflect* energy; instead, radio waves excite individual electrons as they pass through the trail. These excited electrons act as small dipoles, reradiating the signal at an angle equal, but opposite, to the incident angle of the trail.

Signals received from an underdense trail rise to a peak value in a few hundred microseconds, then tend to decay exponentially (Figure 10.2). Decay times from a few milliseconds to a few seconds are typical. The decay in signal strength results from the destructive phase interference caused by radial expansion or diffusion of the trail's electrons.

In Section 10.5.6 of this chapter we will show that the received power at a MBC terminal is proportional $1/f^3$, which limits the maximum useful frequency to below 80 MHz. Commonly used frequencies are in the range of 30–50 MHz.

From equation 10.8 in Section 10.5.6.1 an amplitude–range relationship can also be found. An approximate normalized plot (Ref. 2) of range versus

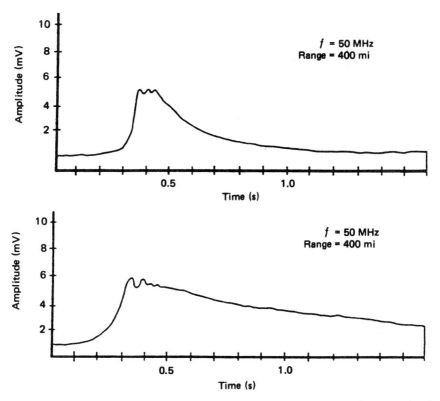

Figure 10.2. Typical underdense trails/returns where received signal intensity versus time is plotted. (From Ref. 2. Courtesy of Meteor Communications Corporation. Reprinted with permission.)

amplitude is shown in Figure 10.3. Reference 2 shows that range and frequency also affect time duration. Such a plot is shown in Figure 10.4 for four different frequencies.

10.2.4 Overdense Trails

We defined overdense meteor trails as those with electron line densities greater than 2×10^{14} electrons/meteor. In this case the line density is so great that signal penetration is impossible and reflection occurs rather than reradiation. Donich (Ref. 2) reports that there are no distinctive patterns for overdense trails except that they may reach a higher amplitude of signal level and last longer. With long-lasting trails we can expect fading of received signals. Some of the fading may be attributed to destructive interference of reflections off different parts of the trail and the breakup of a trail during late periods because of ionospheric winds. The period of fade is over several

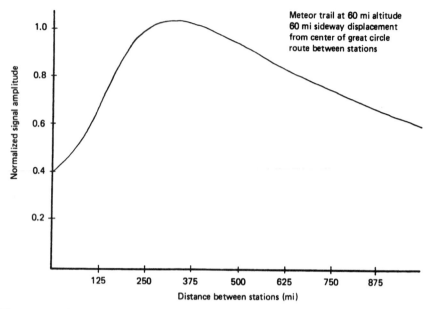

Figure 10.3. Normalized underdense reflected signal power. (From Ref. 2. Courtesy of Meteor Communications Corporation. Reprinted with permission.)

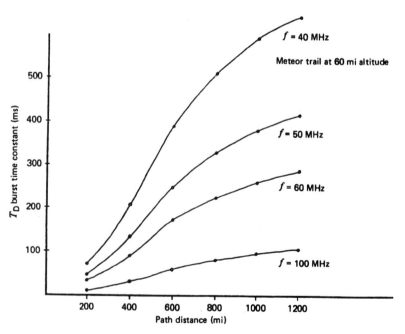

Figure 10.4. Burst time constant versus range. $T_D = \lambda^2 \sec^2 \phi / 32\pi^2 D$. (From Ref. 2. Courtesy of Meteor Communications Corporation. Reprinted with permission.)

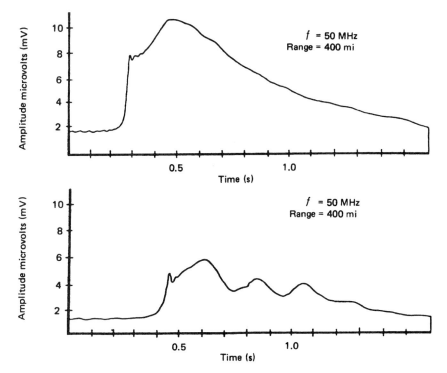

Figure 10.5. Overdense meteor reflections. (From Ref. 2. Courtesy of Meteor Communications Corporation. Reprinted with permission.)

hundred milliseconds, permitting a MBC system to operate between nulls. Figure 10.5 shows plots of received power level versus time for overdense trails.

10.3   TYPICAL METEOR BURST TERMINALS AND THEIR OPERATION

A meteor burst terminal consists of a transmitter, receiver, and modem/processor. Such a terminal is shown in Figure 10.6. For half-duplex operation, one antenna will suffice. However, a fast-operating T/R (transmit/receive) switch is required. We will appreciate why the switch should be *fast* in a moment.

The most efficient modulation is phase shift keying, either binary (BPSK) or quaternary (QPSK). Fixed-frequency (assigned-frequency) facilities commonly use yagi antennas optimized for the frequencies of interest. If the transmit and receive frequencies are separated by more than 3 or 4 MHz, a second, optimized yagi may be desirable, because a yagi is a narrowband

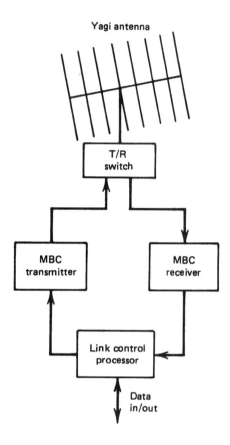

Figure 10.6. A typical meteor burst terminal (half-duplex).

antenna. The transmitter and receiver in Figure 10.6 are conventional. The receiver noise figure may be in the range of 1–3 dB. In nearly all situations, external noise is dominant. In the quiet rural condition, we would expect galactic noise to predominate; in other locations, man-made noise.

The link control processor contains the software and firmware necessary to make a MBC link work. It detects the presence of the probe from the distant end, identifies that it is its companion probe, and then releases the message. Several methods of MBC link operation are described as follows.

There are several ways a meteor burst system can operate. One of the most common is the "master–remote" technique. Here a master station, usually more robust than the remote, sends a continuous probe in the direction of the remote. A probe is a transmitted signal on frequency F_1, usually with identification information. The remote is continuously monitoring frequency F_1, and when it hears the probe, properly authenticated with the identification information, it knows that a common meteor trail exists between the two and transmits a burst of data at frequency F_2.

Another method is quasi-half-duplex, where we transmit and receive on the same frequency. In this case a probe is sent in bursts, with sufficient resting time between bursts to accommodate the propagation time from the distant end. When the remote terminal hears a probe burst, it replies with its message, and the poll bursts from the master station cease.

Of course, we can expand this concept to a polling regime where a master station polls remotes, one at a time, from a large family of remotes. Such an operation is carried out on the SnoTel system mentioned earlier. One master has 600 remotes under its jurisdiction. One can carry this concept to full-duplex operation. In this case the transmitter and receiver each have separate antennas separated and isolated as much as possible. Cosite interference can be a major problem with this type of operation.

Another method of operation is broadcast. In this case a master station broadcasts traffic to "silent" remotes. Such broadcasts take advantage of the statistics of a MBC channel. A short message is continuously repeated over a comparatively long period of time. It is expected that the recipient will receive various pieces of the message and will have to reconstruct the message when the last piece is received.

"Message piecing" is a common technique used on MBC systems. If a message has any length at all—let us say a burst rate of 8 kbps and a burst duration of 100 ms or $\frac{1}{10}$ s, thus a maximum message length of 800 bits (this includes all overhead)—then message piecing must occur if a message is longer than 800 bits. This is the same concept as packet transmission. Each message piece is a "packet" and these packets must be reassembled at the receive location to reconstruct the originator's complete message.

10.4   SYSTEM DESIGN PARAMETERS

10.4.1   Introduction

MBC link performance is defined as the "waiting time" required to transfer a message with a specified reliability. The principal parameters affecting performance are operating frequency, data burst rate (bps), transmitter power, antenna gain, and receiver sensitivity threshold.

10.4.2   Operating Frequency

Meteor trails will reradiate or reflect very high frequency (VHF) radio signals in the 20–200-MHz frequency range. However, since the reflected signal amplitude is proportional to $1/f^3$ and its time duration to $1/f^2$, the message waiting time increases sharply as frequency is increased. Frequencies in the 20–50-MHz range are most practical for minimum waiting time. The lower limit exists, as we mentioned previously, due to external noise conditions.

External noise in this region consists of man-made and galactic noise types. In fact, in many instances, the lower limit of 20 MHz must be increased to 30 or 40 MHz in typical urban noise environments.

10.4.3 Data Rate

Here we refer to the burst rate or burst data rate. A MBC terminal transmits data in high-rate bursts, ideally throughout the duration of the usable portion of the meteor trail event, from 0.2 to 2 s, typically. Data burst rates generally are in the range of 2–16 kbps. The upper limit may be constrained by legal bandwidth considerations dictated by national regulatory authorities such as the FCC. Each data burst must contain overhead information, and the amount of overhead will restrict useful data throughput. Typical modulation is coherent PSK, either BPSK or QPSK.

10.4.4 Transmit Power

The higher the transmit power, the shorter the waiting time. Many MBC terminals with moderate performance features operate with RF power output in the range of 150–200 W. Larger facilities operate at 0.5, 1, 5, or even 10 kW. Naturally, there is a trade-off between performance and economy.

10.4.5 Antenna Gain

As we are aware, antenna gain and beamwidth are inversely proportional. We want our MBC antennas to encompass a large portion of the sky to take advantage of as many meteor trails as possible. As we increase antenna gain, we decrease beamwidth and decrease the "slice of the sky." The trade-off between amount of sky encompassed by an antenna ray beam and antenna gain seems to be in the region of +13 dBi gain. Donich (Ref. 2) gives +16 dBi for short links (400–600 mi) and +21 to +24 dBi for long-range links (600–1200 mi).

10.4.6 Receiver Threshold

Receiver threshold can be defined as the receive signal level (RSL) required to achieve a certain bit error rate (BER). It is a function of the type of modulation used, bandwidth, and the receiver noise, which is usually externally limited. Of course, the lower the receiver threshold, the lower the waiting time. Often a receiver is noise limited by man-made noise. Methodology for calculating threshold using man-made noise as the overriding noise contributor is given in Section 9.9.4. Donich (Ref. 2) shows that a MBC system operating at 40 MHz using coherent BPSK at 2 kbps would have a noise threshold at -121 dBm for a BER of 1×10^{-3}.

10.5   PREDICTION OF MBC LINK PERFORMANCE

10.5.1   Introduction

The starting point in the methodology we present for predicting link performance is the calculation of receiver noise threshold and a threshold for the required BER.

We next provide a basic set of MBC relationships, taken from CCIR Rec. 843 (Ref. 4), used to calculate MBC link transmission loss. Several generalized shortcuts are also presented. Methods of calculating other prediction parameters, such as meteor rate, burst duration time, and waiting time probability, are also described.

10.5.2   Receiver Threshold

We use a similar method to calculate MBC receiver noise threshold as a high-frequency (HF) receiver. The receiver is externally noise limited; receiver thermal noise is of secondary importance. In most cases the type of noise will be man-made. Only in a quiet rural environment will we find a receiver galactic noise limited. Therefore we turn to the methodology given in CCIR Rep. 258, which we discuss in Section 9.9.4. Use equation 9.4 modified.

$$P_n = F_{am} + (D_u + \sigma_{a1}) + 10 \log B_{Hz} - 204 \text{ dBW} \qquad (10.1)$$

Values for D_u and σ_{a1} are taken from Table 9.7.

Example. A MBC receiver operates in a residential environment at 48 MHz and its bandwidth is 8 kHz. Find the noise threshold of the receiver. Neglect transmission line losses. From Table 9.7 we find $F_{am} = 25.9$ dB, $D_u = 12.3$ dB, and $\sigma_{a1} = 4$ dB. Hence

$$P_n = 25.9 + 12.3 + 4 + 10 \log 8000 - 204 \text{ dBW}$$

$$= -148.6 \text{ dBW or } -118 \text{ dBm}$$

If we were to assume BPSK modulation and a 2-dB modulation implementation loss, we would require an E_b/N_0 of $9 + 2 = 11$ dB for a BER of 1×10^{-4}. (See Figure 5.38.) From the example, we can calculate N_0 by subtracting $10 \log 8000$ from the P_n value or -157 dBm. Then the threshold for a BER of 1×10^{-4} must be 11 dB above the noise threshold in 1 Hz of bandwidth or -157 dBm $+ 11$ dB $= -146$ dBm, or the RSL threshold value for a BER $= 1 \times 10^{-4}$ is -146 dBm $+ 39$ dB or -107 dBm.

10.5.3 Positions of Regions of Optimum Scatter

The scattering of straight meteor ionization trails is strongly aspect sensitive. To be effective, it is necessary for the trails approximately to satisfy a specular reflection condition. This requires the ionized trail to be tangential to a prolate spheroid whose foci are at the transmitter and receiver terminals (Figure 10.7). The fraction of incident meteor trails that are expected to have usable orientations is about 5% in the area of the sky that is most effective (Ref. 5). Figure 10.8 shows the estimated percentages of useful trails for a terminal separation of 1000 km. This figure aptly shows that the optimum scattering regions ("hot spots") are situated about 100 km to either side of the great circle, independent of path length.

This feature, together with the fact that the trails lie mainly in the height range of 85–110 km, serves to establish the two hot-spot regions toward which both antennas should be directed. The two hot spots vary in relative importance according to time of day and path orientation (Ref. 5). Generally, antennas used in practice have beams broad enough to cover both hot spots. Thus the performance is not optimized, but on the other hand the need for beam swinging does not arise.

10.5.4 Effective Length, Average Height, and Radius of Meteor Trails*

Consider the ray geometry for a meteor burst propagation path as shown in Figure 10.7 between transmitter T and receiver R. P represents the tangent point and P' a point further along the trail such that $(R'_1 + R'_2)$ exceeds

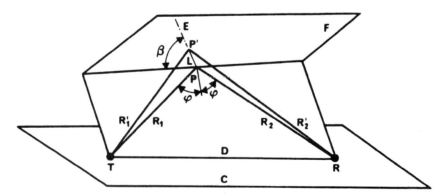

Figure 10.7. Ray geometry for a meteor burst propagation path. C, earth's surface; D, plane of propagation; E, trail; F, tangent plane; β, angle between the trail axis and the plane of propagation; T, transmitter; R, receiver. For the terms L, P, P', R_1, R_2, R'_1, R'_2, and φ, see Section 10.5.4. (From Ref. 4, Figure 2, Rec. 843.)

*This section is adapted from CCIR Rec. 843 (Ref. 4).

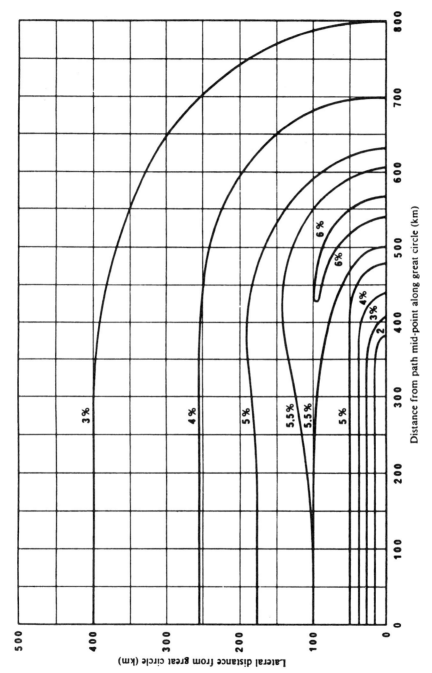

Figure 10.8. Estimated percentages of useful trails as a function of scattering position for a terminal separation of 1000 km. (From Ref. 4, Figure 5, CCIR Rec. 843.)

Distance from path mid-point along great circle (km)

Lateral distance from great circle (km)

691

$(R_1 + R_2)$ by half a wavelength. Thus PP' (of length L) lies within the principal Fresnel zone and the total length of the trail within this zone is $2L$. Provided R_1 and R_2 are much greater than L, it follows that

$$L = \left[\frac{\lambda R_1 R_2}{(R_1 + R_2)(1 - \sin^2 \varphi \cos^2 \beta)} \right]^{1/2} \tag{10.2}$$

where φ = angle of incidence
β = angle between the trail axis and the plane of propagation
λ = wavelength

In order to evaluate the scattering cross section of the trail it is usual to assume that ambipolar diffusion causes the radial density of electrons to have a Gaussian distribution and that the volume density is reduced while the line density remains constant. These assumptions lead to an equation for the volume density N_v in electrons per cubic meter as a function of radius r and time from the instant of formation t, which is

$$N_v(r, t) = \frac{q}{\pi(4Dt + r_0^2)} \exp\left[\frac{-r^2}{(4Dt + r_0^2)} \right] \tag{10.3}$$

where q = electron line density per meter
D = ambipolar diffusion coefficient in m^2/s
r_0 = initial radius of trail in meters

Both D and r_0 are marked functions of height. From experimental results the following empirical formula for evaluating the average height of trails, which is a function of frequency, can be derived:

$$h = -17 \log f + 124 \tag{10.4}$$

where h = average trail height (km)
f = wave frequency (MHz)

The average trail height is a function of other system parameters in addition to frequency. However, equation 10.4 is a good approximation.

Various empirical relationships have been derived between the initial trail radius and the meteor height. An average expression is

$$\log r_0 = 0.035h - 3.45 \tag{10.5}$$

10.5.5   Ambipolar Diffusion Constant*

A good estimate of the ambipolar diffusion constant is provided by the expression

$$\log D = 0.067h - 5.6 \tag{10.6}$$

---

*This section is adapted from CCIR Rec. 843 (Ref. 4).

The ratio of the ambipolar diffusion constant D to the velocity of the meteor V (required in the evaluation of received power) can be approximated by

$$\frac{D}{V} = \left[0.0015h + 0.035 + 0.0013(h - 90)^2\right]10^{-3} \qquad (10.7)$$

where V = velocity of the meteor (m/s).

10.5.6 Received Power*

10.5.6.1 Underdense Trails

$$p_R(t) = \frac{p_T g_T g_R \lambda^2 \sigma a_1 a_2(t) a_2(t_0) a_3}{64\pi^3 R_1^2 R_2^2} \qquad (10.8)$$

where
λ = wavelength (m)

σ = echoing area of the trail (m^2)

a_1 = loss factor due to finite initial trail radius

$a_2(t)$ = loss factor due to trail diffusion

a_3 = loss factor due to ionospheric absorption

t = time measured from the instant of complete formation of the first Fresnel zone(s)

t_0 = half the time taken for the meteor to traverse the first Fresnel zone

p_T = transmitter power (W)

$p_R(t)$ = power available from the receiving antenna (W)

g_T = transmit antenna gain relative to an isotropic antenna in free space

g_R = receive antenna gain relative to an isotropic antenna in free space

R_1, R_2 = see Figure 10.7

(Lossless transmitting and receiving antennas are assumed.)

The echoing area σ is given as

$$\sigma = 4\pi r_e^2 q^2 L^2 \sin^2 \alpha \qquad (10.9)$$

where r_e = effective radius of the electron = 2.8×10^{-15} m

α = angle between the incident electric vector at the trail and the direction of the receiver from that point

*Note that the next three subsections are adapted from CCIR Rec. 843 (Ref. 4).

Since L^2 is directly proportional to λ, the echoing area σ is also proportional to λ and hence the received power for underdense trails varies as λ^3. Horizontal polarization normally is used at both terminals. The $\sin^2 \alpha$ term in equation 10.9 is then nearly unity for trails at the two hot spots.

The loss factor a_1 is given by

$$a_1 = \exp\left(-\frac{8\pi^2 r_0^2}{\lambda^2 \sec^2 \varphi}\right) \tag{10.10}$$

It represents losses arising from interference between the reradiation from the electrons wherever the thickness of the trail at formation is comparable with the wavelength.

The factor $a_2(t)$ allows for the increase in radius of the trail by ambipolar diffusion. It may be expressed

$$a_2(t) = \exp\left(-\frac{32\pi^2 Dt}{\lambda^2 \sec^2 \varphi}\right) \tag{10.11}$$

For angle φ, see Figure 10.7.

The increase in radius due to ambipolar diffusion can be appreciable even for as short a period as is required for the formation of the trail. The overall effect with regard to the reflected power is equal to that which would arise if the whole trail within the first Fresnel zone had expanded to the same extent as at its midpoint. Since this portion of trail is of length $2L$ the midpoint radius is that arising after a time lapse of L/V s. Calling the time lapse t_0 gives, for trails near the path midpoint ($R_1 \approx R_2 \approx R$):

- For trails at right angles to the plane of propagation ($\beta = 90°$):

$$t_0 \simeq \left(\frac{\lambda R}{2V^2}\right)^{1/2} \tag{10.12}$$

- For trails in the plane of propagation ($\beta = 0$):

$$t_0 \simeq \left(\frac{\lambda R}{2}\right)^{1/2} \times \frac{\sec \varphi}{V} \tag{10.13}$$

Substituting t_0 from equation 10.12 into equation 10.11 gives for the $\beta = 90°$ case

$$a_2(t_0) = \exp\left[-\frac{32\pi^2}{\lambda^{3/2}}\left(\frac{D}{V}\right)\left(\frac{R}{2}\right)^{1/2}\frac{1}{\sec^2 \varphi}\right] \tag{10.14}$$

For $\beta = 0°$ the exponent in this expression is $\sec^2 \varphi$ times greater.

The term $a_2(t)$ is the only time-dependent term and gives the decay time of the reflected signal power. Defining a time constant T_{un} for the received power to decay by a factor e^2 (i.e., 8.7 dB) leads to

$$T_{un} = \frac{\lambda^2 \sec^2 \varphi}{16\pi^2 D} \qquad (10.15)$$

With reflection at grazing incidence, $\sec^2 \varphi$ will be large and hence so is the echo-time constant. The echo-time constant is also increased by the use of lower frequencies.

10.5.6.2 *Overdense Trails.* The formula for the received power in the case of overdense meteor trails is usually based on the assumption of reflection from a metallic cylinder whose surface coincides with the region for which the dielectric constant is zero. The effect of refraction in the underdense portion of the trail is usually ignored. As in the underdense case, the received power varies as λ^3 and again the echo duration varies as λ^2. However, the maximum received power is now proportional to $q^{1/2}$, in contrast to q^2 for underdense trails. Thus the increase in received signal power with ionization intensity is more modest.

10.5.6.3 *Typical Values of Basic Transmission Loss.* Since any practical meteor burst communication system will rely mainly on underdense trails, the overdense formulas are of less importance. Satisfactory performance estimates can be made using formulas for the underdense case with assumed values of q in the range of 10^{13}–10^{14} electrons per meter according to the prevailing system parameters.

Basic transmission loss curves derived from equation 10.8 with $q = 10^{14}$ electrons per meter are given in Figure 10.9. As the angle β can take any value between $0°$ and $90°$ only these two extreme cases are shown. The advantage of lower propagation loss at the lower frequencies is clearly seen. Average meteor heights given from equation 10.4 have been used in deriving the curves. It should be noted that the prediction of system performance depends critically on the heights assumed.

10.5.7 Meteor Rate

The meteor rate or meteor burst per unit time (M_c) is related to system parameters through the following expression (Ref. 6):

$$M_c = \left(P_T \times G_T \times G_R / F_c^3 \times T_R \right)^{1/2} \qquad (10.16)$$

where P_T = the transmitter power (W)

T_R = the receiver threshold (W)

G_T = the transmit antenna gain relative to an isotropic

F_c = the carrier frequency (MHz)

G_R = the receiver antenna gain relative to an isotropic

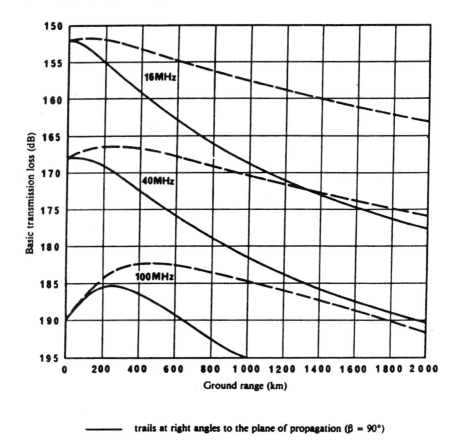

Figure 10.9. Basic transmission loss for underdense trails derived from equation 10.8 with electron density 1×10^{14} electrons/m and horizontal polarization. (From Ref. 4, Figure 3, CCIR Rec. 843.)

From this relationship and a known meteor rate M_T of a test system, operating at a known frequency and a known power level, an expression can be derived to calculate the meteor rate M_c at different values and parameters.

We now let $P = P_T G_T G_R / T_R$, where P is a power ratio. Now the meteor rate ratio of the desired system to the test system becomes

$$\frac{M_c}{M_T} = \left(\frac{P_c}{P_T} \right)^{1/2} \left(\frac{f_T}{f_c} \right)^{3/2} \tag{10.17}$$

This method advises that the higher the system power factor (PF), where $PF = 10 \log P$, and/or the lower the frequency, the higher the observed meteor rate. When power factor is expressed in terms of decibels,

$$M_c/M_T = 10^{(PF_c - PF_T)} + 10^{1.5 \log(f_T/f_c)} \qquad (10.18)$$

where PF_c = power factor of the unknown system (dB)
   PF_T = power factor of the test system (dB)
   f_c = operating frequency of the unknown system
   f_T = operating frequency of the test system
   M_c = meteor rate of the desired system
   M_T = meteor rate of the test system (see Table 10.2)

Values of M obtained from a test system operated for over 1 yr are given in Table 10.2. The test system PF was 180 dB and its operating frequency f_T was 47 MHz. The values given in Table 10.2 show the diurnal and seasonal variations. Plots M_c/M_T versus the power factor of the desired system are given in Figure 10.10 for various operating frequencies. Thus, if the operating frequency and the power factor are known, the meteor rate M_c can be obtained. Conversely, if a desired M_c and operating frequency are given, the required power factor can be defined.

TABLE 10.2   Test System Data[a]

	Meteor Bursts[b] / h	
	Daily Average[c]	Daily Minimum
January	50	19
February	50	19
March	50	19
April	50	19
May	50	19
June	55	20
July	65	22
August	70	25
September	70	25
October	70	25
November	65	25
December	70	25
Yearly	60	23

[a]System power factor $(PF)_T$ = 180 dB; frequency f_T = 47 MHz.
[b]A burst is defined as the recognition of the coded synchronization signal from the master station.
[c]The daily average is defined by the 12 hours per day the average will be exceeded and the 12 hours per day the performance will be less than average.

Source: Reference 6. Courtesy of Meteor Communications Corporation. Reprinted with permission.

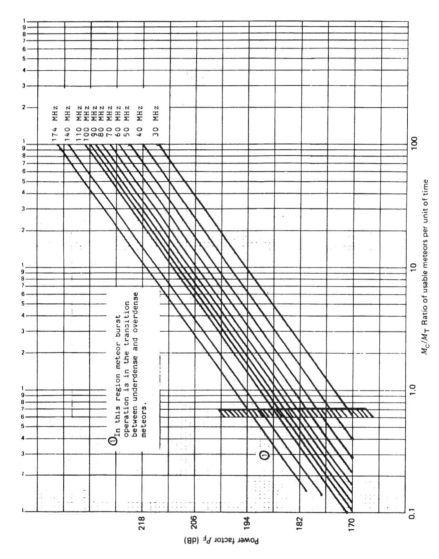

Figure 10.10. Power factor versus meteor rate. M_T established from empirical data; P_F for test data = 180 dB; frequency for test data = 47 MHz. (From Ref. 6. Courtesy of Meteor Communications Corp. Reprinted with permission.)

10.5.8 Burst Time Duration

Signals from underdense trails rise to an initial peak value in a few hundred microseconds, then decay exponentially in amplitude. Decay times from a few milliseconds to a few seconds are typical. This time variation must be taken into account in order to predict message waiting times; accordingly, the next step is to determine the average burst decay times (time above a threshold) for the specific MBC system in question. Eshleman's (Ref. 7) analytic expression for this varying signal strength is given as

$$V(t) = V_D e^{-(t/T_D)} \tag{10.19}$$

where V_D is the peak value of the signal strength and T_D the average time constant in seconds.

T_D is related to both frequency and range between the two MBC stations, as given by

$$T_D = \lambda^2 \sec^2 \phi / 32^2 D \tag{10.20}$$

where D = the diffusion coefficient, which is $8M^2/s$

$\phi = \frac{1}{2}$ the forward scattering angle as a function of range

λ = the wavelength

The value of D given above was obtained from test data given in Ref. 6. It should be noted that D exhibits a diurnal variation, with a maximum in the afternoon and a minimum in the morning. The value of $8M^2/s$ is a daily average and is the value used for all T_D calculations in this section. Plots of T_D versus range with frequency as a parameter are given in Figure 10.11. Thus, once a specific operating frequency is selected and an operating range established, the average burst time can be defined (Ref. 6).

10.5.9 Burst Rate Correction Factor

The number of bursts per hour obtained from Figure 10.10 will require a correction factor where we can calculate the number of bursts that are sufficiently long to transfer a complete message. Of course, we will have to stipulate a certain message length. The distribution of burst durations is an exponential function and can be expressed as

$$M = M_1 e^{-(t/T_D)} \tag{10.21}$$

where M = the number of meteors exceeding the specified threshold for t seconds

M_1 = the total number of meteors exceeding the specified threshold

T_D = the burst time constant

t = the time to transfer the complete message

A normalized plot of M/M_1 is shown in Figure 10.12. Therefore M/M_1 becomes a scaling factor for M_c derived from Figure 10.10 as a function of message transaction time. The value of M_c is reduced to remove bursts that have insufficient time duration to complete a message transmission by setting the value of t in equation 10.21.

10.5.10   Waiting Time Probability

Underdense meteor burst occurrences are random in nature and follow a Poisson distribution as a function of time. The fundamental Poisson equation is

$$P = 1 - e^{-Mt} \tag{10.22}$$

where P is the probability of a meteor occurrence in time t. M is the meteor density or number of bursts per hour and t is the time in hours. Equation 10.22 provides the probability relationship to derive message waiting time. If time t is given in minutes, equation 10.22 becomes

$$P = 1 - e^{-Mt/60} \tag{10.23}$$

In the preceding paragraphs, the meteor burst communication performance

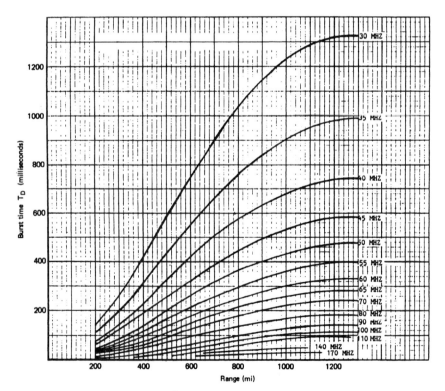

Figure 10.11. Burst time constant. (From Ref. 6. Courtesy of Meteor Communications Corp. Reprinted with permission.)

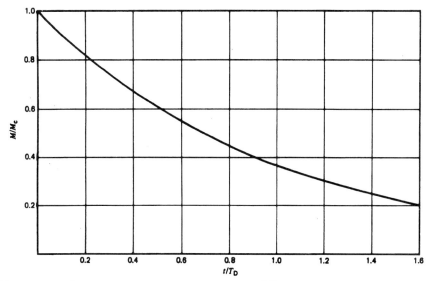

Figure 10.12. Burst duration distribution. $M / M_c = e^{-t/T_D}$. M = total number of bursts/h exceeding a specified threshold for t s; M_c = total number of bursts/h exceeding a specified threshold (the threshold is defined as the recognition of the synchronization code from the transmitting or probing station); T_D = burst time constant; t = time (s). (From Ref. 6. Courtesy of Meteor Communications Corp. Reprinted with permission.)

prediction approach resulted in a value M. Using the Poisson distribution given previously, a family of curves can be generated for a broad set of values for M. Figures 10.13–10.15 show these relationships, with primary interest, of course, where the probability is greater than 0.9 (Ref. 6).

10.6 DESIGN / PERFORMANCE PREDICTION PROCEDURE

Donich (Ref. 6) offers the following step-by-step procedure to calculate the performance of a meteor burst link operating in a single-burst mode. One can either predict the performance given the power factor parameters (Section 10.5.7) or, in reverse, specify a desired performance level and work to the desired power factor to obtain that performance. If the power factor parameters are defined, the procedure below may be followed to predict performance within those given parameters:

1. Calculate the minimum signal levels for a BER of 1×10^{-3} by the methodology shown in Section 10.5.2 given the data rate, modulation type, receiver noise figure, and ambient external noise.
2. Determine the average M for the defined power factor parameters, including line losses, related to long-term test data. Use Section 10.5.7, Table 10.2, and Figure 10.10.

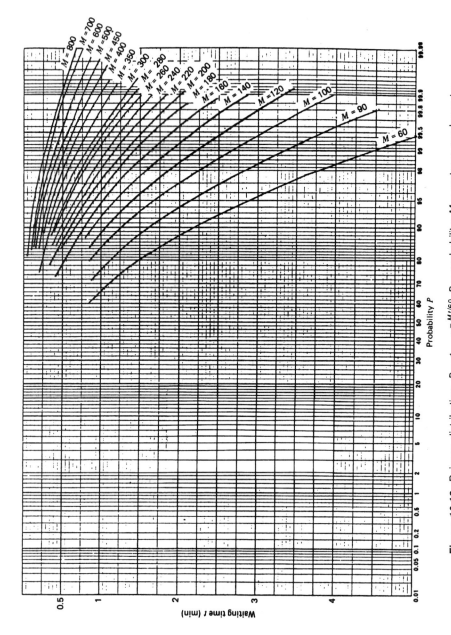

Figure 10.13. Poisson distribution. $P = 1 - e^{-Mt/60}$. P = probability; M = meteors per hour; t = waiting time. (From Ref. 6. Courtesy of Meteor Communications Corp. Reprinted with permission.)

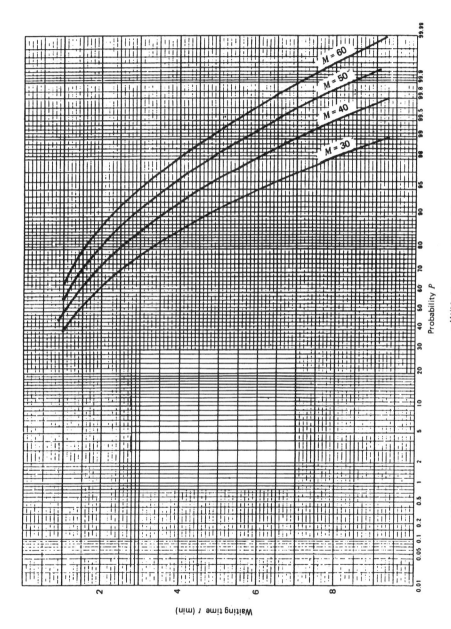

Figure 10.14. Poisson distribution. $P = 1 - e^{-Mt/60}$. P = probability; M = meteors per hour; t = waiting time. (From Ref. 6. Courtesy of Meteor Communications Corp. Reprinted with permission.)

703

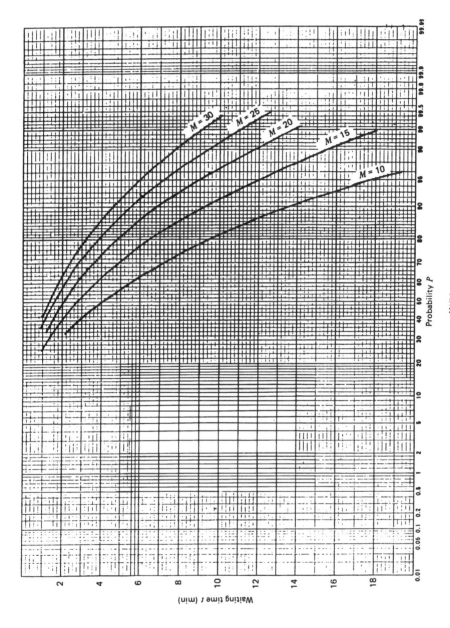

Figure 10.15. Poisson distribution. $P = 1 - e^{-Mt/60}$. P = probability; M = meteors per hour; t = waiting time. (From Ref. 6. Courtesy of Meteor Communications Corp. Reprinted with permission.)

3. Calculate the average burst time constant, given the operating frequency and range. Use Section 10.5.8 and Figure 10.11.
4. Determine the message transfer time required to complete the message reception, including propagation time and message overhead requirements.
5. Remove from the value of *M*, obtained from step 2, the number of bursts of insufficient duration, using the message transfer time, obtained from step 4, and correct *M*. Use Figure 10.12.
6. Use the value of *M*, obtained from step 5, and determine the waiting times for required reliabilities. Use Figures 10.13–10.15. Reference 2 reports that these derived waiting times are for the average time of the year and time of day. For the worst case, multiply by three.

10.7 NOTES ON MBC TRANSMISSION LOSS

Some insight is given in Ref. 8 regarding transmission loss at 40 MHz. Table 10.3 shows link distance; the two components of MBC transmission loss, MBC scatter loss and free-space loss; and the total transmission loss. We see that the values range around Donich's 180-dB reference model (180 dB) (Ref. 6) and the 40-MHz values of Figure 10.9.

How will a typical link operate with such loss values when a simple link budget technique is applied? We use the receiver model described in Section 10.5.2, where the RSL threshold is -107 dBm, and there is an 8-kHz bandwidth, BPSK modulation, and a BER of 1×10^{-4}. Antenna gains at each end are 13 dB and the transmitter output power is 1 kW ($+60$ dBm). Line losses are neglected.

EIRP	$+73$ dBm
MBC transmission loss	-180 dB
Receiver antenna gain	$+13$ dB
RSL	-94 dBm
Threshold	-107 dBm
Margin	13 dB

If we assume the parameters given, once the return from the trail falls below -107 dBm, the link performance is unacceptable. We also see that with this system the maximum transmission loss is $180 + 13$ dB, or 193 dB.

Forward error correction (FEC) coding as well as its attendant coding gain has been offered as one means to extend a trail's useful life. FEC requires symbol redundancy and thus more bandwidth. Of course, the more we open up a receiver's bandwidth, the more thermal noise, degrading operation. Therefore coding gain derived from FEC (in decibels) must be greater than the required noise bandwidth increase (in decibels) due to redundancy (see Ref. 9).

TABLE 10.3   MBC Transmission Loss at 40 MHz[a]

Distance (km)	MBC Scatter Loss (dB)	Free-Space Loss (dB)	Total Loss (dB)
300	52	114.03	166
500	53	118.46	171.5
1000	57	124.49	181.5
1500	58.5	128.01	186.5
2000	61.5	130.5	192

[a]The free-space loss column uses great circle distance. For links under 300 km, R_1 and R_2 should be used (the distance up and down from the reflection height). The MBC scatter loss is taken from Ref. 8.

Ince (Ref. 8) also aptly points out that MBC link maximum range is often more constrained by antenna gain degradation due to low takeoff angles than MBC transmission loss per se. For a 1200-km link, the takeoff angle is 5°, but for a 2000-km link the optimum takeoff angle is less than 1°. If we wish to implement such long links, we must use very elevated antennas to avoid ground reflections, which cause the gain degradation. The optimum range of MBC systems is really in the range of 400–800 km.

MBC links will often experience scatter from the E-layer and such scatter is very sporadic. With E scatter, continuous connectivity can last from minutes to hours (Ref. 15).

Another phenomenon that MBC links can experience is multipath effects (fading/dispersion) during late trail conditions. This is likely to occur on more intense trails, such as some returns from overdense trails. Such trails still give useful returns even when solar winds start to break up the trail. Multipath derives from returns from different trail segments after breakup, occurring, of course, toward the end of a trail's useful life (Ref. 14). Coding with appropriate interleaving is one method of mitigating these multipath effects.

D-layer absorption often is neglected in link budget analyses of meteor burst links. We discuss D-layer absorption and its calculation in Chapter 9. Another source is CCIR Rep. 252-2 (Ref. 10). D-layer absorption, a daytime-only phenomenon, may exceed 3 dB at 40 MHz at midlatitudes during comparatively high sunspot number periods. It is about half this value at 60 MHz.

Yet another path loss is due to Faraday rotation. The D region and the earth's magnetic field cause a linearly polarized VHF wave to be rotated both before and after meteor trail reflection. These rotations result in an overall end-to-end loss due to polarization mismatching between an incident wave and a linearly polarized receiving antenna. An excellent paper dealing with Faraday rotation effects on meteor burst communication links was published by Cannon in 1985 (Ref. 11).

As in the case of D-layer absorption, polarization rotation loss is also affected by path length (secant of the takeoff angle), the sun's zenith angle, and the sunspot number. Like D-layer absorption, Faraday rotation losses also disappear at night. Faraday rotation losses decrease rapidly as frequency increases. Cannon (Ref. 11) shows Faraday rotation losses varying from 1 dB to over 15 dB at 40 MHz (very path dependent) and dropping to a maximum of 1.6 dB at 60 MHz on the same paths.

10.8 MBC CIRCUIT OPTIMIZATION

Obviously, to help optimize MBC link operation, we want to take as much advantage of trail duration as possible. Among the details we should observe is turnaround time. Here we mean the time from receiving a valid probe to the time transmission begins. At the probe transmitter there may be a short period of receiver desensitization after transmitting. This can be minimized in the receiver design and by having a separate transmit and receive antenna. T/R (transmit/receive) switches must be fast operating. Message overhead must be as short as possible.

Adaptive trail operation has also been suggested (Refs. 12, 13, 16, 17). Underdense meteor trails are idealized with an exponential decay. One method measures the signal-to-noise ratio (S/N) of the probe to provide a measure of trail intensity. Burst rate is adjusted for the initial intensity. More intense trails can support higher burst rates. Another system, operating in the full-duplex mode, makes periodic measures of probe intensity (which operates throughout the message transmission), periodically adjusting the burst rate accordingly. Some of the economy of MBC systems is being given up for better throughput performance. However, some argue that the performance improvement is marginal.

10.9 METEOR BURST NETWORKS

Since MBC links are limited to 1000 miles per hop, longer range communications have been achieved using several master stations "chained" together to form a network. Message piecing software has been expanded to provide routing and relay functions. Message piecing is similar to the packet relay function described in Chapter 13. For example, seven MBC master stations have been chained together to provide a network stretching from Tampa, Florida (USA), to Anchorage, Alaska (USAF NORAD-SAC network—Ref. 18).

REVIEW EXERCISES

1. Describe the basis of meteor burst communication (MBC) link operation.
2. What kind of average data throughput in bits per second can we expect from a MBC link?

3. "Waiting time" describes exactly what?

4. Describe the operating radio-frequency relationship with MBC path loss. Why, then, do we go higher in frequency on most operational links when returns are more intense the lower we go in frequency?

5. For fixed-frequency operation, what is the most commonly used antenna? What is one type of antenna we might use to cover a broad band of frequencies for MBC link operation? Remember polarization.

6. What is the principal advantage of selecting MBC over other means of communication? Name at least two major disadvantages.

7. Identify the two categories of meteor trails. How are they described (i.e., in what units)?

8. Give the seasonal and daily variation in performance of a MBC link regarding time of day and months of the year. How do we explain these variations?

9. What is the range of useful time duration of an underdense meteor trail?

10. What transmission impairment can we expect from trails with comparatively long time durations?

11. What type of modulation is commonly used on MBC links?

12. Describe one of the most commonly used techniques of MBC link operation, typically from a remote sensor. (*Hint*: Consider how we know the presence of a trail and then transmit traffic.)

13. MBC links often carry very short, often "canned" messages that can be accommodated by one common trail. How are longer messages handled?

14. In urban and suburban environments, what type of external noise is predominant? In quiet, rural environments?

15. What are the two major constraints on MBC bandwidth?

16. What is the ideal gain of a MBC link antenna? Why not more gain? Include the concept of "hot spots" in the discussion.

17. Calculate the BER threshold of a MBC receiver with a 10-kHz bandwidth, BER $= 1 \times 10^{-3}$, and BPSK modulation, operating in a quiet, rural environment. Use a modulation implementation loss of 1 dB.

18. What is the range of altitude of effective meteor burst trails?

19. What is the "two hot spot" theory? How can we practically accommodate both?

20. For a MBC link operating at 40 MHz, determine the average height of useful meteor burst trails.

21. Meteor rate is a function of what parameters? It is directly proportional to three parameters and inversely proportional to two. Give all five parameters.

22. Meteor rate arrival can be defined by what (mathematical) type of distribution?

23. One simple way of viewing MBC link transmission loss is that it is made up of two loss components. What are they? Roughly what range of decibel loss values would we expect?

24. Discuss the use of forward error correction (FEC) to extend the useful life of a meteor trail. Offer some trade-offs.

25. What is a cause of multipath on a MBC link? When is multipath most likely to occur?

26. A MBC link can take advantage of other radio transmission phenomena. Name three (only one is covered in the text).

27. Name at least two additional link losses we can expect.

28. Give at least four ways to increase MBC link throughput.

29. Why would we wish to raise antennas either on towers or high ground (or both) for long MBC links?

30. What complications are seen in the design of a full-duplex MBC link besides economic ones? What advantages?

31. How does Faraday rotation loss vary with frequency? Time of day?

REFERENCES

1. D. W. Brown and W. P. Williams, "The Performance of Meteor-Burst Communications at Different Frequencies," *Aspects of Electromagnetic Wave Scattering in Radio Communications*, AGARD Conf. Proceedings 244, 24–1 to 24–26, Brussels, 1978.

2. Thomas G. Donich, *Theoretical and Design Aspects for a Meteor Burst Communications System*, Meteor Communications Corp., Kent, WA, 1986.

3. D. W. R. McKinley, *Meteor Science and Engineering*, McGraw-Hill, New York, 1961.

4. "Communication by Meteor-Burst Propagation," CCIR Rec. 843, RPI Series, ITU Geneva, Oct. 1992.

5. V. R. Eshleman and L. L. Manning, "Meteors in the Ionosphere," *Proc. IRE*, vol. 49, Feb. 1959.

6. Thomas G. Donich, *MCBS Design/Performance Prediction Method*, Meteor Communications Corp., Kent, WA, 1986.

7. V. R. Eshleman, *Meteors and Radio Propagation*, Stanford University Rept. no. 44, contract N60NR-25132, Feb. 1955.

8. E. Nejat Ince, "Communications through EM-wave Scattering," *IEEE Commun. Mag.*, May 1982.

9. Scott L. Miller and Laurence B. Millstein, "Performance of a Coded Meteor Burst System," IEEE MILCOM'89, Boston, MA, Oct. 1989.

10. "CCIR Interim Report for Estimating Sky-wave Field Strength and Transmission Loss at Frequencies between the Approximate Limits of 2 and 30 MHz," CCIR Rep. 252-2, 1970 (out of print).

11. P. S. Cannon, *Polarization Rotation in Meteor Burst Communication Systems*, Royal Aircraft Establishement Tech. Report 85082 (TR85082), London, Sept. 1985.

12. W. B. Birkemeier et al., "Feasibility of High Speed Communications on the Meteor Scatter Channel," University of Wisconsin, Madison, 1983.

13. *Efficient Communications Using the Meteor Burst Channel*, STS Telecom, Port Washington, NY, 1987 (NSF IS1-8660079).

14. G. R. Sugar, "Radio Propagation by Reflection from Meteor Trails," *Proc. IRE*, Feb. 1964.

15. Michael R. Owen, "VHF Meteor Scatter—An Astronomical Perspective," *QST*, June 1986.

16. David W. Brown, "A Physical Meteor-Burst Propagation Model and Some Significant Results for Communication System Design," *IEEE J. Selected Areas Commun.*, vol. SAC-3, no. 5, Sept. 1985.

17. Dale K. Smith and Thomas G. Donich, *Variable Data Rate Applications in Meteor Burst Communications*, Meteor Communications Corp., Kent, WA, 1989.

18. Dale K. Smith and Richard J. Fulthorp, *Transport, Network and Link Layer Considerations in Medium and Large Meteor Burst Communications Networks*, Meteor Communications Corp., Kent, WA, 1989.

FIBER-OPTIC
COMMUNICATION LINKS

11.1 OVERVIEW

Fiber-optic transmission has grown from a nascent technology to the transmission medium of choice over a 20-year period (1978–1998) for trunk traffic both in the PSTN and in the cable TV plant. In our technology of transmission the commodity is bandwidth. Compared to other transmission media, fiber optics can be considered to have an infinite bandwidth. References 1 and 2 describe bandwidths in excess of 25 THz. The useful band for radio transmission is 100 GHz (i.e., 3 kHz to 100 GHz); for coaxial cable, the useful bandwidth is some 800 or 1000 MHz, and for wire pair, around 200–300 MHz with drastic distance limitations. Considering these values of competing media, fiber does indeed have an infinite bandwidth.

Fiber lends itself particularly well to digital transmission, which will be emphasized in this chapter. For instance, coaxial cable and wire-pair transmission require far more repeaters than their glass fiber cable counterpart with a ratio from 20:1 to 100:1. We commonly transmit 10 Gbits (SONET OC-192) in a single bit stream. Using wave division multiplex (WDM), there is a ten or more multiplier, with aggregates reaching 100 Gbits and above.

A simplified functional block diagram of a typical fiber-optic communication link is shown in Figure 11.1. The optical source (transmitter) may be a light-emitting diode (LED) or laser diode (LD) coupling light to a fiber strand consisting of a glass core covered with a cladding. Two or more such strands are bundled into a cable, each with its own light source. Each strand is connected at the far end to an optical detector or receiver. Such detectors may be a PIN diode or avalanche photodiode (APD).

Fiber-optic links may be as short as several feet or provide intercontinental connectivity. Fiber-optic cable is a transmission medium used on LANs (local area networks), MANs (metropolitan area networks), and WANs (wide area networks). The cable television industry uses fiber on trunk circuits

711

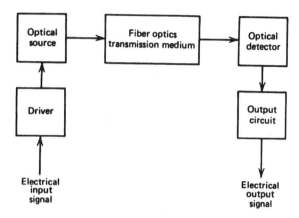

Figure 11.1. A fiber-optic communication link.

bringing multichannel TV to distribution points.

Fiber-optic cable is much smaller and lighter than its metallic counter-parts. Rather than a pair or tube, it uses a strand of glass, slightly larger than a human hair. Fiber is immune to electromagnetic interference (EMI) and the light signal is impossible to detect unless the glass strand is penetrated. Thus it is secure. It does not require equalization. When compared to coaxial cable, it has a flat response over the small band of interest.

In all previous chapters of this book we have used frequency (Hz) to describe where a radio-frequency emission is located in the electromagnetic spectrum or to describe bandwidth. When working with light transmission, such as fiber optics, we use wavelength (m). The reason for this, it is said, is that fiber and other types of light transmission were developed by physicists. The region in which fiber-optic transmission systems operate is in the infrared band, which is shown in Figure 11.2.

Figure 11.3 shows the loss per kilometer achievable across the near-infrared from 800- to 1700-nm (nanometer) band and three fiber-optic transmission windows. The comparatively low loss per kilometer in the wavelength bands of these windows is the reason why we choose these regions of the spectrum to operate fiber-optic links. The windows are in the following wavelength bands:

810–850 nm	Nominal wavelength: 820 nm
1220–1340 nm	Nominal wavelength: 1330 nm
1540–1610 nm	Nominal wavelength: 1550 nm

It will be noted in Figure 11.3 that these windows occur at absorption minima. The derived wavelength units we use are nanometers (nm), or 10^{-9} m, and micrometers or microns (μm), 10^{-6} m. Each window has a usable bandwidth of about 25,000 GHz (Ref. 3).

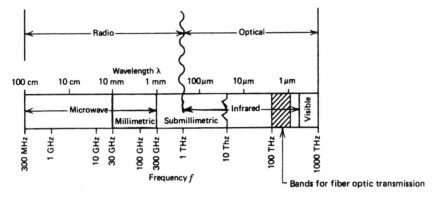

Figure 11.2. The frequency spectrum above 300 MHz.

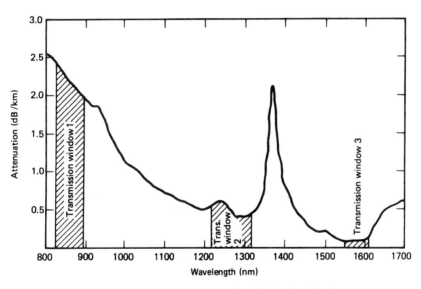

Figure 11.3. Optical fiber attenuation versus wavelength showing the three common transmission windows.

11.2 INTRODUCTION TO OPTICAL FIBER AS A TRANSMISSION MEDIUM

The practical propagation of light through an optical fiber might best be explained using ray theory and Snell's law. Simply stated, we can say that when light passes from a medium of higher refractive index into a medium of lesser refractive index, the refracted ray is bent away from the normal. For instance, a ray traveling in water and passing into an air region is bent away

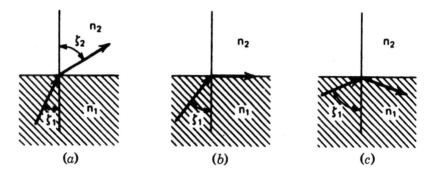

Figure 11.4. Ray paths for several angles of incidence, $n_1 > n_2$.

from the normal to the interface between the two regions. As the angle of incidence becomes more oblique, the refracted ray is bent more until finally the refracted ray emerges at an angle of 90° with respect to the normal and just grazes along the surface. Figure 11.4 shows various incidence angles. Figure 11.4b illustrates what is called the critical angle, where the refracted ray just grazes along the surface. Figure 11.4c is an example of total reflection. This is when the angle of incidence exceeds the critical angle. A glass fiber, when utilized as a medium for the transmission of light, requires total internal reflection.

For effective transmission of light a glass fiber is made up of a glass fiber core that is covered with a jacket called *cladding*. If we let the core have a refractive index of n_1 and the cladding a refractive index of n_2, the structure (core with cladding) will act as a light waveguide when $n_1 > n_2$.

Another property of the fiber for a given wavelength λ is the normalized frequency V; then

$$V = \frac{2\pi a}{\lambda}\sqrt{n_1^2 - n_2^2} \approx n_1\sqrt{2\Delta} \qquad (11.1)$$

where a = the core radius, n_2 for unclad fiber = 1, and $\Delta = (n_1 - n_2)/n_1$.

The term $\sqrt{n_1^2 - n_2^2}$ in equation 11.1 is called the numerical aperture (NA). In essence the numerical aperture is used to describe the light-gathering ability of a fiber. In fact, the amount of optical power accepted by a fiber varies as the square of its numerical aperture. It is also interesting to note that the numerical aperture is independent of any physical dimension of the fiber.

As shown in Figure 11.1, there are three basic elements in an optical fiber transmission system: the source, the fiber link, and the optical detector. Regarding the fiber link, there are two basic design parameters that limit the length of a link without repeaters, or limit the distance between repeaters. These most important parameters are loss, usually expressed in dB/km, and dispersion, which is often expressed as an equivalent bandwidth–distance product in MHz/km. A link may be power limited (loss limited) or it may be dispersion limited.

Dispersion, manifesting itself with intersymbol interference at the far end, is brought about by two factors. One is material dispersion and the other is modal dispersion. Material dispersion is caused by the fact that the refractive index of the material changes with frequency. If the fiber waveguide supports several modes, we have a modal dispersion. Since the different modes have different phase and group velocities, energy in the respective modes arrives at the detector at different times. Consider that most optical sources excite many modes, and if these modes propagate down the fiber waveguide, delay distortion (dispersion) will result. The degree of distortion depends on the amount of energy in the various modes at the detector input.

One way of limiting the number of propagating modes in the fiber is in the design and construction of the fiber waveguide itself. Return again to equation 11.1. The modes propagated can be limited by increasing the radius a and keeping the ratio n_1/n_2 as small as practical, often 1.01 or less.

We can approximate the number of modes N that a fiber can support by applying formula 11.1. If $V = 2.405$, only one mode will propagate (HE_{11}). If V is greater than 2.405, more than one mode will propagate, and when a reasonably large number of modes propagate,

$$N = \left(\tfrac{1}{2}\right)V^2 \tag{11.2}$$

Dispersion is discussed in more detail later in Section 11.3.3.

11.3 TYPES OF FIBER

There are three categories of fiber as distinguished by their modal and physical properties:

- Single mode (monomode)
- Step index (multimode)
- Graded index (multimode)

Single-mode fiber is designed such that only one mode is propagated. To do this, $V < 2.405$. Such a fiber exhibits little modal dispersion. Typically, we might encounter a fiber with indices of refraction on $n_1 = 1.48$ and $n_2 = 1.46$. If the optical source wavelength is 820 nm, for single-mode operation the maximum core diameter would be 2.6 μm, a very small diameter indeed.

Step-index fiber is characterized by an abrupt change in refractive index, and graded-index fiber is characterized by a continuous and smooth change in refractive index (i.e., from n_1 to n_2). Figure 11.5 shows the fiber construc-

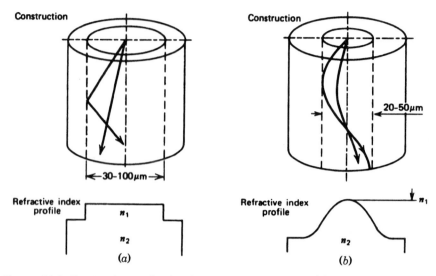

Figure 11.5. Construction and refractive index properties for (a) step-index fiber and (b) graded-index fiber.

tion and refractive index profile for step-index fiber (Figure 11.5a) and graded-index fiber (Figure 11.5b).

Step-index multimode fiber is more economical than graded-index fiber. For step-index fiber the multimode bandwidth–distance product, the measure of dispersion discussed previously, is on the order of 10–100 MHz/km. With repeater spacings on the order of 10 km, only a few megahertz of bandwidth is possible.

Graded-index fiber is more expensive than step-index fiber, but it is one alternative for improved distance–bandwidth products. When a laser diode source is used, values of from 400 to 1000 MHz/km are possible. But if a LED source is used with its much broader emission spectrum, distance–bandwidth products with graded-index fiber can be achieved up to 300 MHz/km or better. Material dispersion in this case is what principally limits the usable bandwidth.

Figure 11.6 illustrates index profiles and modes of propagation for each of the three types of common optical fiber.

11.3.1 Causes of Attenuation in Optical Fiber

There are four causes of loss in optical fiber:

1. Intrinsic loss.
2. Impurity loss.
3. Rayleigh scattering.
4. Loss due to bends and microbends.

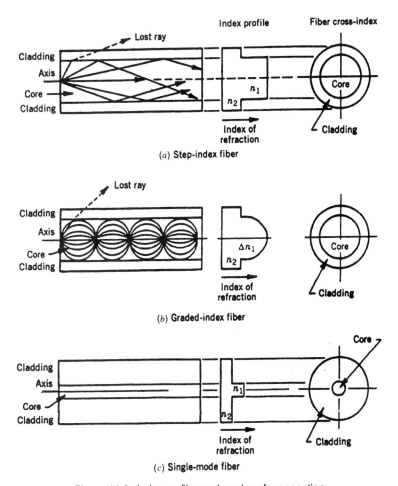

Figure 11.6. Index profiles and modes of propagation.

Intrinsic Loss. There are two intense absorption bands located at 200 and 10,000 nm. There is no way we can reduce these losses (the two bands) because it is a basic property of the silica material from which the fiber is fabricated. The resulting losses on operating fiber-optic transmission systems can be very low if the operating wavelengths are selected far enough away from those absorption bands.

Impurity Losses. These losses can be minimized in the manufacturing process. The losses are due to the residual hydroxyl ion (OH) with the principal absorption line at 2800 nm and its harmonics at 1400, 900, and 700 nm, often mistakenly called the "water line" or water absorption. However, the cause of these losses is due to the water left in the fiber during the fabrication

process. By reducing the water content, absorption line intensities are reduced.

Rayleigh Scattering. This is intrinsically due to minute density and concentration fluctuations by bulk imperfections in the fiber such as bubbles, inhomogeneities, and cracks or by waveguide imperfections due to core and cladding
interfacial irregularities.

Losses Due to Bends and Microbends. Such losses are caused by imperfections
at the core–cladding interface, which can contribute to the net fiber loss. If
care is taken to ensure that the core radius does not vary significantly along
the fiber length during manufacture, variations can be kept under 1% and
the resulting scattering loss is typically under 0.03 dB/km.

Another source of scattering loss is due to bends in the fiber. To experience total internal reflection a guided ray must hit the fiber at an angle
greater than the critical angle (Figure 11.4). Near a bend the angle decreases
and may become smaller than the critical angle at tight bends. This results in
the ray escaping out of the fiber. For single-mode fiber, if the bending radius
is >5 mm, the bending loss is negligible. Since most macroscopic bends well
exceed 5 mm, macrobending losses can be neglected in practice.

Microbending losses can be major sources of high attenuation (e.g., > 100
dB). These losses are related to the random axial distortions, which invariably
occur during cabling when the fiber is pressed against a surface that is not
perfectly smooth. In the case of single-mode (monomode) fibers, these losses
can be minimized by choosing the value of V as close as possible to the cutoff
value of practical range for V between 2 and 2.4 (Ref. 1).

11.3.2 Fiber Strength

The materials used in the manufacturing of optical fiber—silica or composite
glass—have a brittle nature and an elastic behavior until the failure point is
reached. The theoretical failure stress of flawless silica is very high, about
1.6×10^4 N/mm^2. However, this value may be drastically reduced by the
presence of surface flaws produced during the manufacturing process of the
fiber and successive handling operations. For this reason, protective plastic
coatings are applied to freshly drawn fibers (Ref. 10).

Figure 11.7 shows some typical fiber-optic cable layup cross sections.
Figure 11.7*a* shows generic ribbon cable, 11.7*b* generic loose tube cable, and
11.7*c* generic fiber bundle cable.

11.3.3 Dispersion in Optical Fiber Transmission

Dispersion is the broadening of a light pulse as it traverses a fiber. The pulse
is smeared or *dispersed*. Ideally a pulse should be of finite duration. In
operating systems one pulse is followed directly by another with no resting

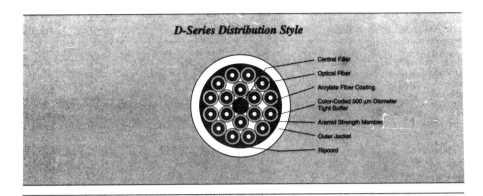

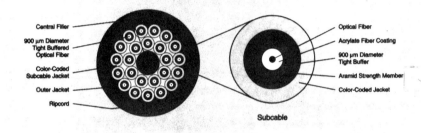

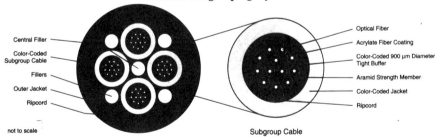

Figure 11.7. Fiber-optic cable cross section: (*a*) distribution cable, (*b*) breakout style cable, and (*c*) subgrouping style cable. (From Ref. 21. Courtesy of Optical Cable Corporation. Reprinted with permission.)

time between pulses. A pulse train progresses down a fiber. The first pulse is a 1 and the second pulse is a 0, where the 1 is the "active state" and the 0 is the "passive state."* When a pulse is broadened or smeared due to delayed signal energy, some signal energy from pulse one will spill into the pulse two position. This could confuse the light detector where it will declare pulse two to be a 1 rather than a 0. This is the typical situation of intersymbol interference. Symbol one energy has corrupted symbol two.

The key here is the delay of energy arrival, where some signal energy related to a pulse position arrives later than the principal, main pulse energy. There are several causes for such delay, and they go under different names in the world of fiber optics. Consider the following dispersion types:

- Intermodal dispersion
- Polarization mode dispersion (PMD)
- Waveguide dispersion
- Chromatic dispersion, material dispersion

Intermodal Dispersion. In multimode fiber, as its name implies, multiple modes are launched down the fiber from the light source. As we can see from Figures 11.6a and 116b, different rays, due to the different modes, travel along paths of different lengths. Thus there is a time delay between wavefront arrivals on the shortest and longest paths. This represents the pulse broadening. As we see in Figure 11.6c for monomode fiber, only one mode (HE_{11}) is launched down the fiber, thus eliminating the cause of intermodal dispersion.

Polarization Mode Dispersion (PMD). As we saw in Figure 11.6c, in monomode (single-mode) fiber only one mode propagates. This is not exactly true. There are actually two independent modes that propagate, which are orthogonally polarized. These two modes are degenerate and have identical group delays (i.e., travel at the same velocity) only when the fiber profile is circularly symmetric. Any local changes to that symmetry such as a slight ovality, which may be introduced during fabrication or by the application of stress to the fiber, will remove the degeneracy and cause the fiber to be slightly birefringent.[†] One consequence is that the polarization state of the optical wave will evolve as it propagates down the fiber and that a group delay difference exists between the two polarization states of a single-mode fiber. This is known as polarization mode dispersion (PMD).

Waveguide Dispersion. The IEEE (Ref. 5) defines *waveguide dispersion* (fiber optics) as follows: "For each mode in an optical waveguide, the term used to

---

*This scenario has been simplified for this discussion.

[†]The IEEE (Ref. 5) defines a *birefringent medium* as "a material that exhibits different indices of refraction for orthogonal linear polarizations of light. The phase velocity of a wave in a birefringent medium thus depends on the polarization of a wave."

describe the process by which an electromagnetic signal is distorted by virtue of dependence of the phase and group velocities on wavelength as a consequence of the geometric properties of the waveguide. In particular, for circular waveguides, the dependence is on the ratio a/λ, where a is the core radius and λ is wavelength."

As we discussed earlier, there are two intense absorption bands, one at 200 nm and the other at 10,000 nm. The goal is to operate as far as possible from these absorption bands, in the range of 800–1600 nm. Nevertheless, these absorption bands influence the entire range between the two peaks such that the index of refraction n varies with wavelength λ.

As some of the readers may appreciate, there is a particular wavelength where there is zero dispersion. We will call this λ_{ZD}. This point occurs where the rate of change of n versus λ reaches a maximum. In a sense we could say that this point is about midway between the two absorption band centers. For pure silica, the group delay at this point reaches a minimum value of approximately 4.87 μs/km, which occurs at 1270 nm. The corresponding dispersion has a zero value at the same wavelength. For optical fiber that is not pure silica but purposefully contains dopants, the point of zero dispersion is 1310 nm for conventional step-index fibers (note that monomode fiber is step-index fiber). This shift of about 40 nm is caused by waveguide dispersion.

Waveguide dispersion is negative and the material dispersion is positive, when added together they give us a value for chromatic dispersion. It thus displaces the resultant zero crossing point to slightly longer wavelengths.

We find that the slope of the waveguide dispersion curve is inversely proportional to the square of the core radius. If we make the core radius smaller, the waveguide dispersion can be made more negative. By doing this, we can shift λ_{ZD}, the point of zero dispersion, to longer wavelength, such as 1550 nm, the point of minimum fiber loss. This is the basis of dispersion-shifted fiber, where the loss is < 0.22 dB/km at 1550 nm. The notation in the literature (Ref. 1) for waveguide dispersion is D_W.

Chromatic Dispersion. Chromatic dispersion arises because group velocity varies at different wavelengths. Thus there are different delays for different frequencies (or wavelengths). This results in an energy component on one frequency arriving before an energy component on another frequency, which is dispersion giving rise to ISI and a degraded BER. Green (Ref. 3) states that material dispersion, D_M, and chromatic dispersion are synonymous for monomode fiber. Material dispersion in multimode fiber when using LED (light-emitting diode) sources is particularly dramatic. This is because of the broad spectral width of the LED emission (e.g., around 30–40 nm).

Total dispersion D for monomode fiber is given by

$$D = D_M + D_W \qquad (11.3)$$

Values of dispersion are usually expressed as ps/km-nm. Let $\delta(\lambda)$ be the dispersion in single-mode fiber. Suppose δ were 15 ps/km-nm. This would mean that two light signal frequencies 1 nm apart would suffer 15 ps dispersion when traveling just 1 km down a monomode fiber (Ref. 1).

11.3.4 Notes on Monomode Transmission

Single-mode (monomode) fiber is cheaper than multimode fiber because of the quantities produced of single mode. Dispersion can often be neglected on monomode fiber-optic links operating at 622 Mbps or below. Above this bit rate, chromatic dispersion may become a limiting factor.

Standard cladding diameter for either multimode or single-mode fiber is 125 μm. The core diameter for multimode fiber is 50 μm.

For single-mode fiber, rather than core diameter, the *mode field diameter** is specified. If single-mode fiber has set $V = 2.405$, then r_0, the mode field radius, is 1.1 times a (the core radius) (Ref. 6). ITU-T Rec. G.652 (Ref. 7) gives the mode field diameter for monomode fiber in the range of 9–10 μm at 1310 nm operating wavelength. Then with these parameters fixed, a (the core radius) is in the range of 8.18/2 to 9.09/2 μm or 4.09–4.545 μm.

For dispersion-shifted single-mode fiber at 1550 nm, the mode field diameter according to ITU-R Rec. G.653 (Ref. 8) should be in the range of 7.0–8.3 μm. Then a would be in the range 6.36/2 to 7.77/2 μm or 3.18–3.88 μm.

ITU-T Rec. G.652 recommends maximum chromatic dispersion coefficients (ps/km-nm) as shown in the following table:

Wavelength (nm)	Maximum Chromatic Dispersion Coefficient (ps / km-nm)
1288–1339	3.5
1271–1360	5.3
1550	20 (approx.)

Based on 1550-nm operation for dispersion-shifted single-mode optical fiber, ITU-T Rec. G.653 specifies $D_{0(max)}$ as 3.5 ps/km-nm between 1525 and 1575 nm. ITU-T Rec. G.554 (Ref. 9) describes the characteristics of wavelength-loss-minimized single-mode optical fiber for 1550-nm operation. It specifies the chromatic dispersion coefficient as 20 ps/km-nm with a slope of 0.06 ps/nm²-km.

---

*Mode field diameter is a measure of the width of the guided optical power's intensity distribution in a single-mode (monomode) fiber.

11.4 SPLICES AND CONNECTORS

Optical fiber cable is commonly available in 1-km sections, although longer sections can be purchased. There are two methods of connecting these sections in tandem and connecting the light source at one end and the optical detector at the other. These are splicing or using connectors. The objective in either case is to transfer as much light as possible through the coupling. Good fusion splices generally couple more light than do connectors.

A good fusion splice can have an insertion loss as low as 0.04 dB, whereas connectors, depending on the type and how well they are installed, can have an insertion loss as low as 0.3 dB and some as high as 1 dB or more. Expect about 0.2-dB insertion loss for butt splices.

An optical fiber splice requires highly accurate alignment and an excellent end finish to the fibers. There are two causes of loss at a splice:

- Lateral displacement of fiber axes
- Angular misalignment

The actual melting at the fusion splice can also contribute to splice loss. Most field-usable semiautomatic fusion splicers now use TV imaging and pattern-recognition algorithms to microposition the two cores correctly before firing the fusing arc.

Connectors should be used at the optical source interface with the fiber and at the far-end optical detector interface. This will speed replacement procedures when one or the other fails or shows deteriorated performance.

11.5 LIGHT SOURCES

A light source, perhaps more properly called a photon source, has the fundamental function in a fiber-optic communication system to convert electrical energy (current) into optical energy (light) efficiently, in a manner that permits the light output to be launched into the optical fiber effectively. The light signal so generated must also accurately track the input electrical signal so that noise and distortion are minimized. Figure 11.8 is a simplified functional block diagram of an optical transmitter showing the actual light source.

The two most widely used light sources for fiber optic communication systems are the LED (light-emitting diode) and the LD (laser diode). LEDs and LDs are fabricated from the same basic semiconductor compounds and have similar heterojunction structures. They do differ considerably in their performance characteristics. LEDs are less efficient than LDs but are cheaper. In practical applications, a LED is limited to ≤ 100 Mbps operation due to

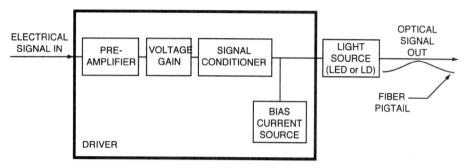

Figure 11.8. A simplified functional block diagram of a fiber-optic transmitter.

its wide spectral linewidth (i.e., 40–80 nm). LDs have much narrower linewidths, ≤2 nm, and are operated with bit rates in excess of 2.4 Gbps.

With present technology, a LED is capable of launching about 100 μW (−10 dBm) of optical power into the core of a fiber with a numerical aperture of 0.2 or greater. A laser diode with the same input power can couple up to 7 mW or more into the same cable. However, LDs usually are operated with outputs of 1 or 2 mW (0 or +3 dBm) to preserve longevity. The coupling efficiency for a surface-emitting LED is about 1%, and for an edge-emitting LED (ELED) it is about 10%. Some improvement is possible in both cases by using fibers that are tapered or have a lensed tip. An external lens also improves coupling efficiency at the expense of reduced mechanical tolerance.

Butt coupling of a laser diode only provides about 10% efficiency, as there is no attempt at matching mode sizes of the LD and the fiber. The mode size of an index-guided InGaAsP LD is about 1 μm. The mode size of its companion fiber is on the order of 6–8 μm. Again, the coupling can be improved by tapering the fiber end and forming a lens at the fiber tip.

Using a lens-coupling approach to optical transmitter design can improve efficiencies to around 70%. This is a confocal design in which a sphere is used to collimate the laser light and focus it onto the fiber core.

Optical feedback can damage or destroy a laser diode. Feedback can be reduced by improving return loss at transitions (e.g., at splices, connectors, and splitters). Even a relatively small amount of feedback can destabilize a laser and affect its performance. Antireflection coatings reduce feedback. Another measure to reduce the effects of feedback is to cut the fiber tip at a slight angle so reflections do not hit the active region of the laser. As a final resort, an optical isolator may be considered. Feedback can be reduced over 30 dB by this method. Most optical isolators base their performance on the Faraday effect, which describes the rotation of the plane of polarization of an optical beam in the presence of a magnetic field (Ref. 1).

LEDs are more reliable than their LD counterparts. Meantime between failures (MTBFs) of LEDs can exceed 1×10^7 hours at 25°C. InGaAsP

TABLE 11.1   ELED Characteristics Summary

Characteristic	Value
Output power into single-mode fiber (25°C, 150-mA drive)	2–10 μW
Rise/fall times	3 ns max.
Half-power linewidth (25°C)	40–60 nm
Output power temperature coefficient	1.2%/°C typical
Center wavelength variation with temperature	0.5–0.8 nm/°C
Spectral broadening	0.4 nm/°C typical

Sources: ITU's *Optical Fibre Planning Guide* (Ref. 10) and MIL-HDBK-415 (Ref. 11).

LEDs have MTBFs that approach 1×10^9 hours. The story is quite different for InGaAsP lasers, where the MTBFs approach 1×10^6 hours at 25°C. Both devices are temperature sensitive such that most transmitters use a thermo-electric cooler to maintain the source temperature around 20°C even when outside ambient displays high temperatures.

One measure that can be taken for the temperature-dependent LD is to resort to a sort of AGC (automatic gain control) as used with radio receivers. Rather than attempt to control the device's temperature, a negative feedback circuit is used whereby a portion of the emitted light is sampled, detected, and fed back to control the drive current.

Table 11.1 summarizes typical ELED performance, and Table 11.2 compares LED and LD performance and other parameters.

Of particular interest are the spectral linewidths of optical transmitters. Table 11.1 shows the half-power linewidth of a LED in the range of 40–60 nm. The LD can have a linewidth < 1 nm. The effect on dispersion of each becomes obvious. The LED with its very wide linewidth has a multitude of frequencies widely spread apart that are transmitted. The LD has fewer frequencies in its spectral line that are well confined to a narrow range. For fiber-optic systems with bit rates in excess of 500 Mbps, the LD is an excellent candidate. Figure 11.9 is an idealized sketch of the spectral line concept.

TABLE 11.2   LED and LD Compared

Characteristic	LED	LD
Coupled power, maximum	− 10 dBm	+ 7 dBm
Radiant power (mW, max.)	20	20
Spectral linewidth	40–60 nm	< 1 nm to 3 nm
Wavelength	1550 nm	1550 nm
Drive current (mA)	10–200	10–200
Modulation bandwidth (max.)	1 GHz	15 GHz
Life (MTBF, hours)	1×10^{-7} to 1×10^9	1×10^6
Cost	Lower	Higher

Sources: MIL-HDBK-415 (Ref. 11), ITU's *Optical Fibre Planning Guide* (Ref. 10), and Ref. 1.

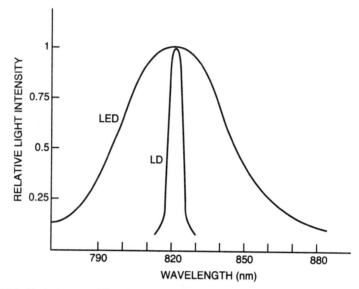

Figure 11.9. Typical spectral linewidths for LED and LD. Note that the peak intensities have been normalized to the same value; the actual peak intensity of the LD is much greater than the LED.

The LD spectral line has been idealized in Figure 11.9. Figure 11.10 shows a more realistic output spectrum of a laser diode (Fabry-Perot). Note the side modes in the figure to the left and right of the principal or dominant mode. For gigabit operation, these side modes should be reduced as much as possible. They represent a spectral broadening inviting increased chromatic dispersion. Enter "single-longitudinal-mode (SLM) operation." Laser cavities

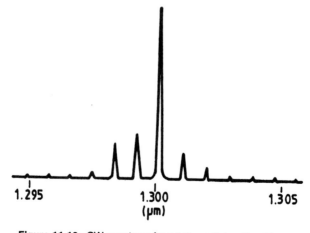

Figure 11.10. CW spectrum for a 1.3-μm Fabry-Perot laser.

are such that different modes have different losses where the longitudinal mode with the least loss becomes the dominant mode. Side mode suppression in these types of lasers can exceed 30 dB.

One of the most common "single-mode" lasers is the DFB (distributed feedback) laser. Similar to the SLM lasers, DFB lasers have structures with built-in wavelength selectivity. These lasers contain a periodic grating on the boundary between two of the layers in the laser structure as shown in Figure 11.11. This corresponds to a periodic variation in the refractive index. As the refractive index varies, a small fraction of the optical power is reflected. The idea is that a wavelength where these small reflections interfere constructively will be favored. DFB lasers can have adjacent mode rejection ratios as high as 40 dB.

If a DFB laser has simple cleaved end facets, the discrete reflections from the facets will interfere with the distributed reflections from the grating. This interference may upset mode selectivity, depending on the positions of the facets relative to the grating shape. As these positions cannot be controlled in the fabrication process, not all the produced lasers will be single mode. The

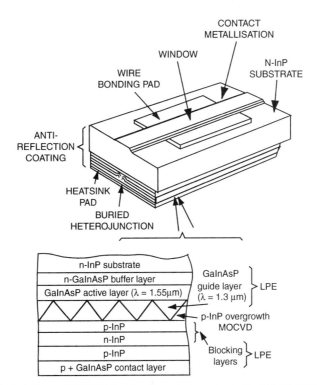

Figure 11.11. A 1.55-μm DCPBH (double-channel planar buried heterostructure) distributed feedback (DFB) laser. MOCVD = metal organic chemical vapor deposition; LPE = liquid phase epitaxy. (From Ref. 10, Figure 3.10. Courtesy of ITU.)

interference problem is, of course, not present if the facet reflections are suppressed by an antireflection coating. In this case, the distributed reflections alone provide the necessary feedback. A more detailed analysis of the operation of such a device shows that two modes spaced about 1–2 nm will be equally favored, thus single-mode operation will not be achieved. An elegant solution to this problem is the use of so-called phase jump gratings. Such a grating contains a discontinuity in the grating shape and, as a result, only one favored mode will exist.

The ITU comments (Ref. 10) that the use of DFB lasers does not eliminate dispersion effects in high data rate* systems completely. This is due to a transient change in the optical frequency when the laser is modulated, the so-called chirping effect. Penalties arising therefrom depend on laser design, modulation index, and drive waveform. They can represent a significant impairment on 1550-nm systems based on ITU-T Rec. G.652 (Ref. 7) at gigabit data rates when fiber dispersion is high (e.g., >20 ps/nm-km). Reference 12 states that a DFB laser with these characteristics, where the central mode is typically <0.1 nm, will have a bit rate–distance product in excess of 60 Gbps/km before this impairment begins to affect performance.

11.6   LIGHT DETECTORS

The most commonly used detectors (receivers) for fiber-optic communication systems are photodiodes, either PIN or avalanche photodiode (APD). The terminology *PIN* derives from the semiconductor construction of the device where an intrinsic (I) material is used between the p–n junction of the diode.

A photodiode can be considered a photon counter. The photon energy E is a function of frequency and is given by

$$E = h\nu \qquad (11.4)$$

where h = Planck's constant (W/s^2) and ν = frequency (Hz). E is measured in watt-seconds or kilowatt-hours.

The receiver power in the optical domain can be measured by counting, in quantum steps, the number of photons received by a detector per second. The power in watts may be derived by multiplying this count by the photon energy, as given in equation 11.4.

The efficiency of the optical-to-electrical power conversion is defined by a photodiode's *quantum efficiency* η, which is a measure of average number of electrons released by each incident photon. A highly efficient photodiode would have a quantum efficiency near 1, and decreasing from 1 indicates progressively poorer efficiencies. The quantum efficiency, in general, varies with wavelength and temperature.

---

*In this context "high data rate" means >1000 Mbps.

For the fiber-optic communication system engineer, *responsivity* is a most important parameter when dealing with photodiode detectors. Responsivity is expressed in amperes per watt or volts per watt and is sometimes called sensitivity. Responsivity is the ratio of the rms value of the output current or voltage of a photodetector to the rms value of the amount of electrical power.

In other words, responsivity is a measure of the amount of electrical power we can expect at the output of a photodiode, given a certain incident light power signal input. For a photodiode the responsivity R is related to the wavelength λ of the light flux and to the quantum efficiency η, the fraction of the incident photons that produce a hole–electron pair. Thus

$$R = \frac{\eta\lambda}{1234} \quad (A/W) \tag{11.5}$$

with λ measured in nanometers.

Responsivity can also be related to electron charge Q by the following:

$$R = \frac{\eta Q}{h\nu} \tag{11.6}$$

where $h\nu$ = photon energy (equation 11.4) and Q = electron charge, 1.6×10^{-19} coulombs (C).

Figure 11.12 plots typical responsivities for four photodetectors. The two upper curves are for semiconductor photodiodes shown with material codes Ge and Si. Curves S-1 and S-20 are for the two photodiode materials AgOCs

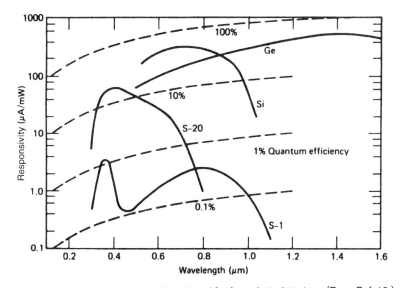

Figure 11.12. Typical responsivities plotted for four photodetectors. (From Ref. 13.)

and Na_2KsbCs, respectively. The curves are plotted with quantum efficiency η as a parameter. The dashed lines are for comparative purposes, where η is assumed to be constant with wavelength.

As in all transmission systems, noise is a most important consideration. Just as in the other systems treated in this text, the noise analysis of a fiber-optic system is centered on the receiver. We must treat noise for these types of systems in both the optical and the electrical domain. In the optical domain, as shown in Figure 11.13, quantum noise is dominant. Quantum noise manifests itself as shot noise on the primary photocurrent from the detector. There is also shot noise on the dark current, and in the case of an APD there is excess noise from avalanche multiplication. Because the optical detector converts light energy into electrical energy, thermal noise must be treated on the electrical side of the detector. In most cases this will be the noise generated in the preamplifier that follows the light detector. In fact in PIN detectors, because their gain is very nearly unity, thermal noise is the principal contributor to total noise.

$$\text{Thermal noise} = \frac{4kT_{\text{eff}}B}{R_{\text{eq}}} = i^2_{\text{NA}} \qquad (11.7)$$

Figure 11.14a is a functional block diagram of a fiber-optics receiver and Figure 11.14b is a simplified model of that receiver. In Figure 11.14b the optical detector is shown by the two intertwining circles at the left.

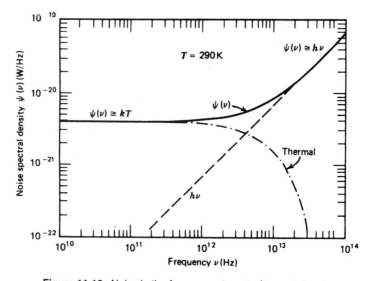

Figure 11.13. Noise in the frequency domain. (From Ref. 13.)

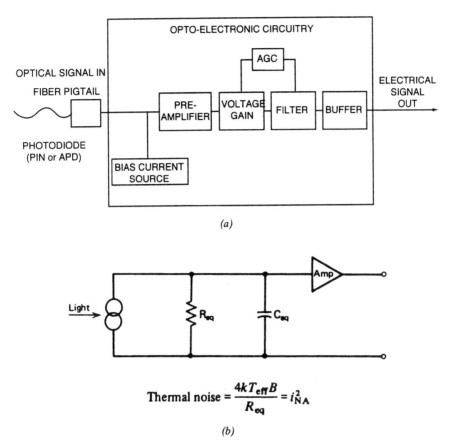

Figure 11.14. (a) Functional block diagram of a fiber-optics receiver. (b) An optical receiver model.

where R_{eq} = equivalent resistance of driver amplifier

T_{eff} = effective noise temperature of load resistor (K)

B = bandwidth (Hz)

k = Boltzmann's constant

Noise is related, in this case, to the mean-square value of the current of the load resistor i_{NA}^2.

Shot noise can also be related to the mean-square value of current and consists of two parts, one from the fluctuations in the signal, i_{NS}^2,

$$i_{NS}^2 = 2q\left(2P_{opt}\frac{\eta q}{h\nu}\right)B \tag{11.8}$$

where P_{opt} = optical power

$\quad q$ = electron charge $(1.6 \times 10^{-19}$ C)

$\quad \eta$ = quantum efficiency

$\quad h\nu$ = photon energy (equation 11.4)

$\quad B$ = system bandwidth

and the second from dark current in the detector, i_{ND}^2,

$$i_{ND}^2 = 2qi_D B \qquad (11.9)$$

where i_D = photodetector dark current.

$$\text{Signal-to-noise ratio}\ \frac{S}{N} = \frac{\text{signal power}}{\text{shot noise power} + \text{amplifier noise power}}$$

$$(11.10)$$

Signal power is a function of the mean-square value of the detector photocurrent, i_S^2,

$$i_S^2 = \frac{1}{2}\left(2P_{opt}\frac{\eta q}{h\nu}\right)^2 \qquad (11.11)$$

Noise power, the two terms in the denominator in equation 11.10, was discussed earlier. The thermal noise power in this case is the preamplifier noise power. The signal-to-noise ratio equation 11.10 can now be written (Ref. 14):

$$\frac{S}{N} = \frac{2\left[P_{opt}(\eta q/h\nu)\right]^2}{\left[2qi_D + 4qP_{opt}(\eta q/h\nu) + 4kT_{eff}/R_L\right]B} \qquad (11.12)$$

where R_L = load resistance and B has been factored out in the denominator. A bit error rate (BER) of 1×10^{-9}, a standard for fiber-optic systems, requires a signal-to-noise ratio of 21.5 dB. The signal-to-noise ratio of the optical power incident on the detector is the square root of this figure, or 10.75 dB. For an APD the signal-to-noise ratio is given by (Ref. 14)

$$\frac{S}{N} = \frac{2\left[P_{opt}(\eta q/h\nu)^2\right]M^2}{\left[\left[2qi_D + 4qP_{opt}(\eta q/h\nu)\right]M^2F(M) + 4kT_{eff}/R_L\right]B} \qquad (11.13)$$

where M = avalanche gain of photocurrent and $F(M)$ = excess noise introduced by avalanche gain.

An APD has what is called optimum gain M_{opt}. The gain M of an APD cannot be increased indefinitely; there is a point where, as M is increased, the signal-to-noise ratio begins to degrade, and we have passed the point of M_{opt}. Reference 15 describes a group of 53 APDs produced by Bell Telephone Laboratories with a calculated theoretical gain of 140 (equivalent to a power gain of about 21.5 dB), whereas the practical M_{opt} for the group, as measured, turned out to be around 80 (19 dB). Other sources specify no more than 15 dB of practical gain M_{opt} for an APD, and another, only 7 dB (Ref. 12).

In the noise analysis presented earlier we have seen that signal-to-noise ratios are a function of bandwidth B, which, in turn, is a function of the bit rate (see Section 11.9). Table 11.3 summarizes detector thresholds in dBm for a standard BER of 1×10^{-9} for many of the bit rates common to the telecommunications industry. It is becoming evident that this standard BER is becoming obsolete as we discussed in Chapter 4. A BER of 1×10^{-12} would be a better value.

Noise equivalent power (NEP) is commonly used as the figure of merit of a photodiode. NEP is defined as the rms value of the optical power required to produce a unity signal-to-noise ratio (i.e., $S/N = 1$) at the output of a light-detecting device. Typical values of NEP range from 1 to 10 $pW/Hz^{1/2}$ (Ref. 1).

TABLE 11.3 Summary of Experimental Receiver Diode Sensitivities (Average Received Optical Power, P, in dBm; BER $= 1 \times 10^{-9}$)

Bit Rate (Mbps)	InGaAs PIN 1.3 μm	InGaAs PIN 1.55 μm	InGaAs SAM/SAGM[a] APD 1.3 μm	InGaAs SAM/SAGM[a] APD 1.55 μm	Ge APD 1.3 μm	Ge APD 1.55 μm	InGaAs Photoconductor 1.3 μm	InGaAs Photoconductor 1.55 μm
34	−52.5				−46	−55.8		
45	−49.9		−51.7		−51.9			
100					−40.5			
140	−46				−45.2	−49.3		
274	−43		−45	−38.7		−36		
320	−43.5							
420			−43	−41.5				
450			−42.5		−39.5	−40.5		
565					−33			
650	−36							
1000			−38	−37.5	−28		−34.4	
1200	−33.2	−36.5						
1800			−31.3		−30.1			
2000				−36.6		−31		−28.8
4000				−32.6				

[a]SAGM = separated absorption, grating, and multiplication regions. SAM = separated absorption and multiplication regions.
Source: Reference 16.

Of the two types of photodiodes discussed here, the PIN is cheaper and requires less complex circuitry than its APD counterpart. The PIN diode has peak responsivity from about 800 to 900 nm for silicon devices. These responsivities range from about 300 to 600 μA/mW. The overall response time for the PIN is good for about 90% of the transient, but sluggish for the remaining 10%, which is a "tail." The poorer response of the tail portion of a pulse may limit the net bit rate on digital systems.

The PIN detector does not display gain, whereas the APD does. The response time of the APD is far better than that of the PIN, but the APD displays certain temperature instabilities where responsivity can change significantly with temperature. Compensation for temperature is usually required in APD detectors and is often accomplished by a feedback control of bias voltage. It should be noted that bias voltages for APDs are much higher than for PIN diodes, and some APDs require bias voltage as high 200 V or more. Both the temperature problem and the high-voltage bias supply complicate repeater design.

11.7 OPTICAL AMPLIFIERS

Optical amplifiers were in the developmental stages from the 1970s until 1989, when rare-earth-doped amplifiers began to be fielded. The technology has come a long way since then.

Amplifiers of light offer several major advantages to the light wave system designer. It has given us a major advance toward repeaterless fiber-optic links. Consider this example. Suppose we had available in a fiber-optic link budget 40 dB for fiber plus splice loss at 1550 nm. The monomode fiber loss plus splice loss was 0.2 dB/km. We then could have 200-km extension to our system without repeaters.* Install a light amplifier after the fiber-optic transmitter with a 30-dB gain[†]; install another with the same gain just ahead of the far-end fiber-optic receiver. Now 60 dB has been added to the link budget, extending the link (without repeaters) another 300 km. We now have a fiber-optic link extending 500 km (i.e., 200 + 300 km).

Advantages in this situation are (1) the elimination of the cost of repeaters, (2) the reduction of jitter (jitter is a function of the number of repeaters in tandem, thus reduction of jitter), and (3) improvement in the availability of the system by reducing active components. For undersea cable the improved availability issue would be overriding.

We will consider two types of fiber-optic amplifiers: semiconductor laser amplifiers (SLAs) and erbium-doped fiber amplifiers (EDFAs). Both types of

---

Repeater is defined here as a device that demodulates an incoming light signal from a particular section to its equivalent electrical signal; the signal is regenerated and then remodulates a light signal in a laser diode for transmission through the next section.
[†]The 30 dB is problematical where amplifier output power is limited to some +10 dBm.

amplifier operate through the process of stimulated emission and with the attendant processes of spontaneous emission and absorption. Here signal photons incident on the amplifier's active material cause generation of additional photons, which are of the same frequency, phase, and polarization as the incident photons. Stimulated emission requires the achievement of population inversion in active material, in which the population of atomic energy states is altered from its normal distribution such that the higher energy states are more populated. The population inversion is achieved by the injection of current in the SLA and the optical pumping in the EDFA. Gain in these amplifiers is determined by the degree of population inversion. In the SLA gain may be increased by increasing bias current; in an EDFA by increasing the optical pump power (Ref. 2).

11.7.1 Semiconductor Laser Amplifiers

There are three types of semiconductor laser amplifiers (SLAs): injection-locked, Fabry-Perot, and traveling wave (TW). When a conventional semiconductor laser is biased above lasing threshold and is used as an amplifier (not a laser diode optical source), it is an injection type. When a SLA is biased below the lasing threshold, it is called a Fabry-Perot type; and when both facets of the laser are given antireflective coatings, it is called a traveling wave (TW) type. Reference 17 states that the TW amplifier has recently become the most important of the three types because of its superior performance. These types of amplifiers display gains on the order of 10–18 dB. However, Agrawal (Ref. 1) reports that TW amplifiers can achieve a 30-dB gain. Some of this gain is reduced (8–10 dB) due to coupling losses. However, if the amplifier is to be used as a postamplifier (i.e., to boost power after a LD optical source), its power output is limited to the 5–10-mW range (Ref. 1).

SLAs also are comparatively noisy, displaying noise figures in the range of 5–6 dB, which detracts from their application as a preamplifier (i.e., as an amplifier placed in front of the PIN or APD detector). The noise in this case is spontaneous-emission noise. Agrawal (Ref. 1) reports that in one experiment with an 8-Gbps system, the SLA extended the sensitivity threshold of an APD by 3.7 dB. When operating in the TW mode, the bandwidth (3 dB) is about 70 nm (9 THz). The SLA would be a good choice for multichannel amplification (i.e., amplifying WDM signals). However, several nonlinear phenomena typical of the SLA cause crosstalk among the channels.

11.7.2 Erbium-Doped Fiber Amplifiers (EDFAs)

EDFAs operate in the 1.55-μm band, which is propitious because it is the band of lowest loss. The amplifier bandwidth is in the 100's to 1000's of GHz range, which can well support WDM operation. Spirit and O'Mahoney (Ref. 2) report a 35-nm bandwidth with peak gain at 1541 nm. However, the

TABLE 11.4 Typical Characteristics of EFDAs

Parameter	Performance
Wavelength band	1.55 μm
Gain	30 to 50 dB
Power output	+10 to +20 dBm
Conversion efficiency	40–80%
Noise figure	3 to 5 dB
Bandwidth	Several 100's to several 1000's GHz
Polarization	No polarization dependence
Connection	Simple coupling to other fiber
Connection loss	Around 0.5 dB
Pump wavelengths	0.98 and 1.48 μm most popular

Sources: References 1 and 17.

response across 35 nm is not completely flat. Table 11.4 summarizes the characteristics of an erbium-doped fiber amplifier.

EDFA amplification is based on a pumping principle, where the most common pumping wavelengths are 1480 or 980 nm. Gains from 30 to 40 dB are possible with pump power in the range of 50–100 mW. Figure 11.15 illustrates five approaches to pumping. The figures show a *certain length* of erbium-doped optical fiber. This "certain length" can vary from 25 to 100 m with the 100-m length displaying the highest gain. The gain tails off after 100–150 m, and where the curve starts to drop off is more or less a function of the pump power. For instance, at about 90 mW of pump power, the knee of the curve is at about 135 m of doped-fiber length.

The noise performance of an EDFA tends to deteriorate at the higher gains. One way to improve noise performance, when high gain and high output power are required, is to use a two-stage amplifier incorporating an isolator and a bandpass filter (about 1-nm bandwidth). When used as a power amplifier, optimum output is around +10 dBm. The better noise performance is achieved at the 980-nm pump wavelength. Also, performance gain degrades as the ambient temperature is increased where the 1.552-μm operating wavelength is less temperature sensitive than the 1.535-μm wavelength (Ref. 17).

11.8 MODULATION AND ELECTRICAL WAVEFORM

The most widely used type of modulation of a light carrier is a form of amplitude modulation (AM) called *intensity modulation*. Both types of light sources, the LED and the LD, can be conveniently modulated in the intensity mode. The detectors (PIN and APD) discussed earlier each respond directly to intensity modulation, producing a photocurrent proportional to the incident light intensity. Today's optical fiber communication systems are more suitable to digital than to analog operation.

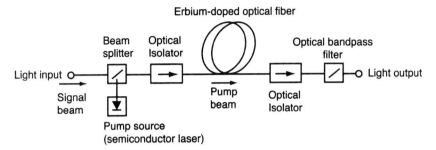

(a) Forward Pumping

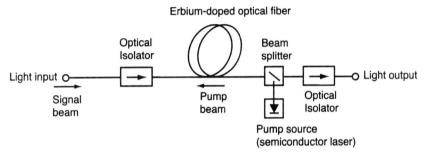

(b) Backward Pumping

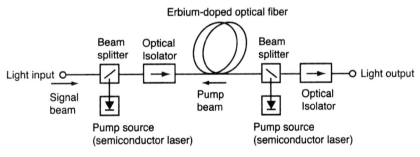

(c) Bidirectional Pumping

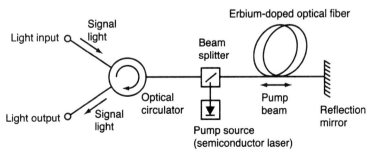

(d) Reflection Pumping

Figure 11.15. Construction of erbium-doped fiber amplifiers. (From Ref. 17, Figure 5.7. Reprinted with permission.)

Considering the previous chapters, let us review some advantages and disadvantages of optical fiber communication systems, particularly for digital operation:

- We are dealing with much higher frequencies.
- Coherence bandwidths are much greater.
- Component response times are inherently faster than anything discussed previously.
- The linearity of optical fiber systems is comparatively poor.

As mentioned before, both the LED and the LD light sources are semiconductor devices that, in most cases, are directly modulated. Biasing of the source and the adjustment of the quiescent operational point are most important considerations. In view of this, the following guidelines should be followed:

1. The intensity of the driving source should vary directly with the bias current either in the lasing region (for the LD, see Figure 11.16) or in the spontaneous emission region (for the LED).
2. In the case of continuous analog modulation the quiescent bias current must be established at a point such that the modulating signal causes an equal plus and minus swing about the quiescent value: it should also be in the most linear range of the intensity characteristic.
3. For digital modulation of the several variants of pulse amplitude modulation (PAM) the quiescent bias current is adjusted, given a specific source, as follows. In the case of a LED, it should be either near zero or at a quiescent point optimum for noise and/or response time. For the LD case it should be adjusted at a point near or below threshold (Figure 11.16) providing that the transition noise is tolerable;

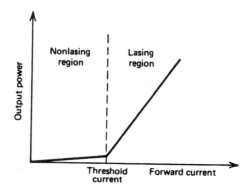

Figure 11.16. Power–current relationship for laser diode (LD) source.

slightly above threshold if transition noise must be reduced; or near zero if transition response time is adequate for the design modulation rate and if threshold transition noise is tolerable.

4. For continuous AM of a LD, consideration must be given to the laser's life when operating in this mode.

An important digital system design consideration is the pulse format. For the discussion that follows, only binary digital systems will be covered (see Chapters 3, 4, and 13). By definition, then, information such as on-line data, PCM, or delta modulation (DM) is transmitted serially as 1's and 0's. The manner (format) in which the 1's and 0's are presented to the modulator is important for a number of reasons.

First, amplifiers for fiber-optic receivers are usually ac coupled. As a result, each light pulse that impinges on the detector produces a linear electrical output response with a low-amplitude negative tail of comparatively long duration. At high bit rates, tails from a sequence of pulses may tend to accumulate, giving rise to a condition known as baseline wander, and such tails cause intersymbol interference. If the number of "on" pulses and "off" pulses can be kept fairly balanced for periods that are short compared to the tail length, the effect of ac coupling is then merely to introduce a constant offset in the linear output of the receiver, which can be compensated for by adjusting the threshold of the regenerator. A line-signaling format can be selected that will provide such a balance. The selection is also important on synchronous systems for self-clocking at the receiver.

Figure 11.17 illustrates five commonly used binary formats. Each is briefly discussed in the following:

1. *NRZ (Nonreturn to Zero)*. This signal format is discussed in Chapter 13, where by convention a 1 represents the active state and a 0 the passive state. A change of state only occurs when there is a 1-to-0 or 0-to-1 transition. A string of 1's is a continuous pulse or "on" condition, and a string of 0's is a continuous "off" condition. In NRZ information is extracted from transitions or lack of transitions in a synchronous format, and a single pulse completely occupies the designated bit interval.

2. *RZ (Return to Zero)*. In this case there is a transition for every bit transmitted, whether a 1 or a 0, as shown in Figure 11.17, and, as a result, a pulse width is less than the bit interval to permit the return-to-zero condition.

3. *Bipolar NRZ*. This is similar to NRZ except that binary 1's alternate in polarity.

4. *Bipolar RZ*. This is the same as bipolar NRZ, but in this case there is a return-to-zero condition for each signal element, and again the pulse width is always less than the bit interval.

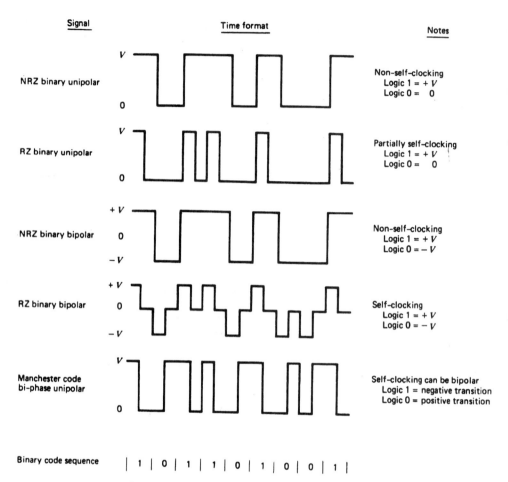

Figure 11.17. Five examples of binary signal formats.

5. *Manchester Code.* This code format is commonly used in digital fiber-optic systems. Here the binary information is carried in the transition, which occurs at midpulse. By convention a logic 0 is defined as a positive-going transition and a logic 1 as a negative-going transition. The signal can be either unipolar or bipolar.

The choice of code format is important in fiber-optic communication system design, and there are a number of trade-offs to be considered. For instance, RZ formats assist in reducing baseline wander. To extract timing on synchronous systems the Manchester code and the RZ bipolar codes are good candidates because of their self-clocking capability. However, it will be appreciated from Figure 11.17 that they will require at least twice the bandwidth as the NRZ unipolar code format. An advantage of the Manchester code is that it can be unipolar, which adapts well to direct intensity modulation of LED or LD sources and also provides at least one transition per unit interval (i.e., per bit) for self-clocking.

With the NRZ format we can attain the highest power per information bit if we wish to tolerate baseline wander. Achieving this power is especially desirable with LED sources. On the other hand, LD sources can be driven to high power levels for short intervals, hence conserving the life of the laser diode, making the RZ format attractive. Longer life with the shorter duty cycle can be traded off for higher modulation rate, which will result in greater bandwidth requirements of the system. As we have seen, RZ systems require twice the bandwidth of NRZ systems for a given bit rate.

11.9 SYSTEM DESIGN

11.9.1 General Application

The design of a fiber-optic communication system involves several steps. Certainly, the first consideration is to determine the feasibility of such a transmission system for a desired application. There are two aspects to this decision, economic and technical. Analog applications are for wideband transmission of such information as video, particularly for CATV trunks, studio-to-transmitter links, and multichannel FDM mastergroups. Throughout this chapter, however, we stress digital application for fiber-optic transmission and in particular for the following:

- On-premises data bus; LANs, particularly FDDI (fiber distributed data interface)
- Higher level PCM or DM multiplex configurations for telephone trunks
- Radar data links
- Conventional data links well in excess of 9.6 kbps
- Digital video, uncompressed and compressed

The present trend is for the cost of fiber cable and components to go down, even with the impact of inflation. Fiber-optic repeaters are more costly than their metallic PCM counterparts, but fewer repeaters are required for a given distance, and the powering of these repeaters is more involved, especially if the power is to be taken off the cable itself. In this case a metallic pair (or pairs) must be included in the cable sheath if the power is to be provided from trunk terminal points. Another approach is to supply power locally at the repeater site with a floating battery backup. Consider the following features that make fiber-optic transmission systems attractive to the telecommunication system design:

1. *Low Transmission Loss.* This is compared to wire pairs or coaxial cable for broadband transmission (i.e., in excess of several megahertz [or Mbps]), allowing a much greater distance between repeaters.

2. *Wide Bandwidth.* With present systems in the 1550-nm region using monomode transmission, the bit rate–distance products are up to 60 Gbps/km. Future systems may see bit rate–distance products up to 1000 Gbps/km.

3. *Small Bending Radius.* With proper cable design, a bending radius on the order of a few centimeters can have negligible effect on transmission.

4. *Nonradiative, Noninductive, Nonconductive, Low Crosstalk.* Complete isolation from nearby electrical systems, whether telecommunication or power, avoiding such problems as groundloops, radiative or inductive interference, and reducing lightning-induced interference.

5. *Lighter Weight, Smaller Diameter.* This is in comparison to its metallic equivalents.

6. *Growth Potential.* Low-bit-rate systems can easily be expanded in the future on fiber to much higher bit-rate systems using the same cable.

11.9.2 Design Procedure

The first step in designing a fiber-optic communication system is to establish the basic input system parameters. Among these we would wish to know:

- Signal to be transmitted
- Link length
- Growth requirements (i.e., additional circuits, increased bit rates)
- Tolerable signal impairment levels stated as signal-to-noise ratio or BER at the output of the terminal detector

Throughout the design procedure, when working with trade-offs, the designer must establish whether he/she is working in a power-limited domain or in a bandwidth- (bit-rate) limited* domain. For instance, with low bit rates such as the T1 carrier (1.544 Mbps), we would expect to be power limited in almost all circumstances. Once we get into higher bit rates, say, above 45 Mbps using multimode fiber or monomode fiber operating above 600 Mbps, the proposed system must be tested (by calculation) to determine if the system rise time is not bit-rate limited. This test will be described in detail (see Section 11.9.4). The designer then selects the most economic alternatives

---

*Perhaps more properly this should be called *dispersion-limited.*

and trade-offs among the following:

- Fiber parameters: single mode or multimode; if multimode, step index or graded index; if monomode, dispersion-shifted
- Cable parameters: loose or tight protection, strength members
- Source type: LED or LD; if LD, type (e.g., DFB)
- Wavelength band: 850, 1330, 1550 nm
- Use of light amplifiers
- Detector type: PIN or APD
- Repeaters, if required; their powering (prime power)
- Modulation type and waveform
- Use of wave division multiplexing (WDM)

With the exception of special fiber-optic links (possibly military applications), joining fiber will be by fusion splices to take advantage of low insertion losses. As fiber links get longer, the fiber itself becomes the cost driver. The break point may be in the range of 2–4 km. On shorter links, the source and detector become cost drivers. As an example, for on-premises data bus systems, low-cost components and cable can be selected such as LED sources, plastic, high-loss, fiber, and PIN detectors.

As in the design of any telecommunication system or subsystem, the bottom line is cost over the life of the system. The question then arises: What is the most economic fiber-optic transmission subsystem that will meet requirements over the system life? One example might be factory-installed connectors versus field fusion splicing. We can argue for factory-installed connectors because the insertion loss difference is so small compared to the savings of labor and field splicing equipment.

Another trade-off is WDM versus adding extra fibers to a particular cable to accommodate the additional bit streams. Route diversity significantly adds to system cost. Here the question is: What is the cost of survivability? PSTN operators and power companies widely use route diversity (geographical diversity) to overcome "back-hoe fades." Here we mean the accidental cutting of a fiber cable by a back-hoe or any other possible way of damaging or destroying a portion of a fiber cable connectivity. Route diversity provides an alternative route for the traffic when such a mishap occurs.

Selecting a *reasonable* link margin is another cost trade-off.

11.9.3 General Approach

11.9.3.1 Introduction. A fiber-optic transmission link can be power limited or dispersion limited. In most practical cases with present technology, using monomode step-index fiber, links will be power limited. Dispersion limitations should be examined when transmitting over 500 Mbps on long

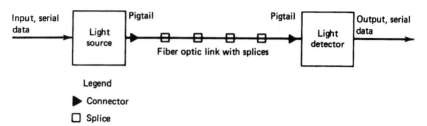

Figure 11.18. Fiber-optic link model.

monomode links. The type of dispersion will be chromatic dispersion, not modal dispersion. If we employ multimode fiber using laser diode sources, modal dispersion will predominate.

11.9.3.2 Power-Limited Operation.
Figure 11.18 is a model of a fiber-optic link. This may be a terminal-to-terminal repeaterless link or a terminal-to-first-repeater link. As shown in the figure, the link consists of a light source coupled to a pigtail fiber with a connector on the end, and fiber sections with connectors or splices and a light detector at the distant end, connecting to the fiber link via a connector/pigtail combination.

Conventionally, the output power of the optical source is specified at the end of the pigtail and the threshold of the detector at a BER of 1×10^{-9},* referenced at the detector pigtail.

Suppose the power output of a LD is $+3$ dBm and the threshold of the detector is -41 dBm. The link power degradation from source to detector is 44 dB and we have to budget for this value. That is,

$$\text{Budget}_{dB} = \text{output}_{dBm} - \text{threshold}_{dBm}$$

$$= +3 \text{ dBm} - (-41 \text{ dBm})$$

$$= 44 \text{ dB} \tag{11.14}$$

The 44 dB is budgeted to the following:

- Connector insertion losses, at the source and at the detector
- Fiber loss
- Intervening connector or splice losses
- Extinction ratio penalty at light detector (in dB), see directly below
- Margin

If we measured the output power of a LD with a photometer, the reading would be average power. Average power must be expressed in terms of the

---

*As we mentioned earlier, a BER of 1×10^{-12} may be more appropriate. See Section 4.2.2.3.

LD duty cycle and we relate these in the two equations that follow:

$$P_{avg} = \frac{P_{max} + P_{min}}{M} \tag{11.15}$$

$$r_{ex} = \frac{P_{min}}{P_{max}} \tag{11.16}$$

where P_{avg} = average power
  P_{max} = maximum or peak power
  P_{min} = minimum power
  r_{ex} = extinction ratio
  M = 2 for a NRZ waveform and 4 for a RZ waveform

We must understand that a LD is biased so that there is a P_{min} with a no-signal condition and P_{max} is the peak power during the signal condition.

From the value of r_{ex} we derive an extinction ratio penalty in decibels for the system based on using a PIN detector (Ref. 18):

r_{ex}	Penalty (dB)
0.5	3
0.4	2.2
0.3	1.7
0.2	1.0
0.1	0.5
0.07	0.3
0.05	0.2
0.02	0.1

Example. The bit rate of a particular fiber-optic link is 622 Mbps. The optical source is a LD coupling 0 dBm into single-mode fiber at 1550 nm. The optical detector is an APD, and the 1×10^{-12} BER threshold is -37.0 dBm. Ahead of the detector is an EDFA with a 30-dB gain. Fusion splicing is used, and the fiber is in 1-km reels. What is the maximum distance we can have between source and the first regenerative repeater?

Derive the dB difference between the source and detector. The total link budget should sum to this number. The value is 67 dB (i.e., 37 + 30 dB). We assign a 6-dB margin for the link. The following are fixed losses:

Connector losses at both ends	1.5 dB
Margin	6 dB
Extinction ratio penalty	0.5 dB
Subtotal	8.0 dB

Calculate the remaining loss (dB) to be assigned to the link:

$$67 \text{ dB} - 8.0 \text{ dB} = 59.0 \text{ dB}$$

Assign a 0.2-dB/km fiber loss and 0.1-dB per kilometer section for fusion splice loss: total = 0.3 dB/km. Divide 59.0 dB by 0.3 dB/km: result = 196 km.

11.9.4 Dispersion-Limited Domain and System Bandwidth

To be assured that a particular fiber-optic link is not dispersion limited, the following calculations should be carried out. We can calculate the system bandwidth or maximum bit rate that can be supported for a certain link distance by calculating the system rise-time, T_r. System rise-time is defined as the time for the voltage to rise from 0.1 to 0.9 of its final value. Furthermore, system rise-time is a function of the square root of the sum of the squares of the component rise-times. As a minimum, the following components must be considered:

- Source S
- Detector d
- Fiber rise-time
    Multimode dispersion F_{mm}
    Material dispersion (when LED is used) F_{md}
    Chromatic dispersion (sometimes called material dispersion), when using a LD transmitter and single-mode fiber, F_{chr}

Rise-time, T_r, must be considered in view of the electrical waveform entering the optical source, whether it is in a RZ or NRZ format. For a given pseudorandom sequence of bits, RZ has twice as many transitions as a NRZ format. We can now set up a simple equation for system rise-time, T_r (Ref. 1):

$$T_r^2 = S^2 + d^2 + F_{mm}^2 + F_{md}^2 + F_{chr}^2 \qquad (11.17)$$

Then

$$T_r = \left(S^2 + d^2 + F_{mm}^2 + F_{md}^2 + F_{chr}^2 \right)^{1/2} \qquad (11.18)$$

Smith (Ref. 19) adds a constant k_t as a factor on the right side of equation 11.18 to account for pulse degradation at the light detector. If no other value is available, use 1.1 for k_t. With this modification, equation 11.18 becomes

$$T_r = 1.1 \left(S^2 + d^2 + F_{mm}^2 + F_{md}^2 + F_{chr}^2 \right)^{1/2} \qquad (11.19)$$

Relating T_r to bit rate, B (Ref. 1),

$$T_r \leq 0.35/B \quad \text{for RZ format} \qquad (11.20a)$$

$$T_r \leq 0.70B \quad \text{for NRZ format} \qquad (11.20b)$$

Then

$$B \geq 0.35/T_r \quad \text{for RZ format} \qquad (11.21a)$$

$$B \geq 0.70/T_r \quad \text{for NRZ format} \qquad (11.21b)$$

Example. Determine the feasibility, regarding dispersion-limited operation, for a fiber-optic link 50 km long, using single-mode, dispersion-shifted fiber operating at 1550 nm. The fiber displays 3.5-ps/nm-km chromatic dispersion.* The bit rate on the link is 1.24416 Gbps (OC-24). Transmitter (source, DFB) rise-time is 200 ps; receiver (detector) rise-time is 300 ps (ADP). All other dispersion can be neglected because of the use of (1) DFB (LD) and (2) single-mode fiber.

The total fiber dispersion is then 50×3.5 ps or 175 ps. Set up the following table:

Parameter/Item	Rise-Time (ps)	(Rise-Time)^2
Source (transmitter) S	200 ps	40,000 ps
Fiber F_{chr}	175 ps	30,625 ps
Detector (receiver) d	300 ps	90,000 ps
Sum		160,625 ps

$$T_r = 400.78 \text{ ps} \times 1.1 = 400.858 \text{ ps}$$

Maximum bit rate for the link is $0.7/(440.858 \times 10^{-12}) = 1.587$ Gbps (formula 11.21b).

11.10 WAVELENGTH DIVISION MULTIPLEXING (WDM)

Wavelength division multiplexing (WDM) is just another name for frequency division multiplexing (FDM). We will hold with the conventional term, WDM. The WDM concept is illustrated in Figure 11.19.

The 1550-nm band is the most attractive for WDM applications because the aggregate wavelengths of a WDM signal can be amplified in the aggregate by an EDFA. This is a great advantage. On the other hand, if we wish to

---

*Actually this is the sum of waveguide and chromatic dispersion. Some references call this material dispersion. I like to think of material dispersion as that occurring due to the very wide spectral linewidth of the LED.

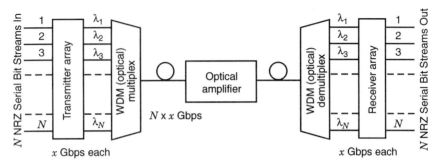

Figure 11.19. The WDM concept.

regenerate a WDM signal in a repeater, the aggregate must be broken down into its components as shown in Figure 11.20.

The cogent question is: How many individual wavelength signals can we multiplex on a single fiber? It depends on the multiplexing/demultiplexing approach employed, the number of EDFAs that are in tandem, available bandwidth, channel separation, and the impairments peculiar to light systems and the WDM techniques used.

As mentioned earlier, the number of WDM wavelength channels depends on the fiber bandwidth available and the channel spacing among other parameters. For instance, in the 1330-nm band about 80 nm is available and in the 1550-nm band, there is about 100 nm. Including the operational bandwidth and guardbands, the capacity on a single fiber would be 40 and 50 WDM channels, respectively, at 2.5 Gbps each. There then would be 100 Gbps and 120 Gbps, respectively, on each fiber. At present (1996), 10-channel systems are available, off-the-shelf (Ref. 20).

If SONET OC-192 (10 Gbps) were placed on each WDM channel, the total capacity on a single fiber strand would increase fourfold.

Spirit and O'Mahoney (Ref. 2) provide the following relationship to calculate frequency bandwidth given wavelength bandwidth (and vice versa). We are familiar with the common formula to calculate frequency, given

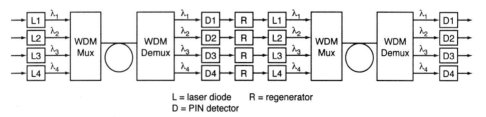

L = laser diode   R = regenerator
D = PIN detector

Figure 11.20. A regenerator/repeater in a WDM fiber link.

wavelength (and vice versa) or 3×10^8 m/s $= f\lambda$. On differentiation we get

$$\Delta f = 3 \times 10^8 \Delta \lambda / \lambda^2 \qquad (11.22)$$

where λ = operating wavelength (e.g., 850, 1330, or 1550 nm).

11.10.1 Methods of Wave Division Multiplexing

There are two broad categories of WDM mechanisms: active and passive. The latter is far more prevalent. In the passive device category there are angularly dispersive elements, optical filters, and wavelength-selective directional couplers. In the angularly dispersive element category include prisms and diffraction gratings. Each of these disperses light into its wavelength components.

The multiplexing task is simpler than that of demultiplexing WDM signals. One of the most elementary WDM multiplexers is the passive combiner or wavelength-selective combiner. A typical two-branch passive combiner is shown in Figure 11.21. One problem with passive combiners is splitting/combining loss. This loss is expressed as

$$\text{Loss}_{dB} = 10 \log_{10} N \qquad (11.23)$$

where N is the number of branches (assuming equal split for all branches). For instance, if there were 10 branches, the loss would be 10 dB. One-hundred-branch couplers are feasible.

To demultiplex a WDM light wave signal, generally a two-step process is required. First the incoming signal must be power split. The amount of splitting is, of course, a function of the number of WDM channels. The next step is a signal-selection step using, typically, tunable filters. This is called wavelength-selective demultiplexing (or multiplexing).

Another approach is to use a grating for multiplexing or demultiplexing. It is comprised of an array of fibers, a lens, and an optical grating. Two types of grating demultiplexers are shown in Figure 11.22. The multiplexed light signal enters one fiber, where it is focused by the lens on the grating. The grating returns the incident light beam in which the different wavelength

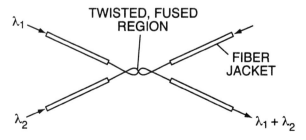

Figure 11.21. A two-branch single-mode fiber passive coupler.

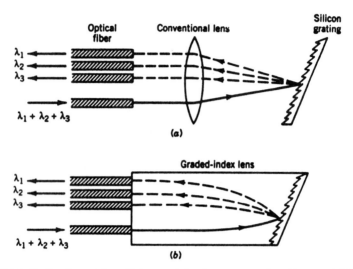

Figure 11.22. Grating-based demultiplexers: (*a*) using a conventional lens and (*b*) using a graded-index lens. (From Ref. 1, Figure 7.8. Reprinted with permission.)

components of the light beam are spatially separated. The geometry of the device is set such that each individual light channel illuminates a particular fiber. Thus the group of output fibers each carry a light signal corresponding to the wavelength of one of the multiplexed input signals, effectively carrying out the demultiplex function.

Agrawal (Ref. 1) points out that a problem exists with grating demultiplexers in that their bandpass characteristics depend on the dimensions of the input and output fibers. In particular, the core size of the output must be large enough to ensure a comparatively flat passband and low insertion losses. For this reason, most grating demultiplexers must use the larger core multimode fiber. Monomode fiber grating demultiplexers are being developed using a microlens array to solve the problem.

The optical filter is a basic building block for many WDM systems. There are three different types of optical filters: fixed interference filters, tunable Fabry-Perot (FP) filters, and grating-based filters. There are two important characteristics of these filters when used for WDM devices. First is the filter passband and its roll-off to ensure minimum crosstalk. The second characteristic is evident when these filters are cascaded. Often, when there are many filter stages, the passband of the aggregate will be less than the passband of an individual filter. FP filters can select one channel from an aggregate of over 100 WDM channels (Ref. 1).

11.10.2 WDM Impairments

The most serious impairment in WDM systems is crosstalk. Crosstalk is essentially the transfer of power from one channel to another. One cause of crosstalk is due to the nonlinear effects in optical fiber. Accordingly, such crosstalk is called *nonlinear crosstalk*. Even the perfectly linear channel can produce crosstalk, which, of course, is called *linear crosstalk*. One source of linear crosstalk is the nonideal filtering associated with the demultiplexing process. Crosstalk degrades bit error rate.

There are two principal causes of nonlinear crosstalk. These are *stimulated Raman scattering* (SRS) and *four-wave mixing* (FWM). SRS is a nonlinear effect in which light in the fiber interacts with molecular vibrations (optical photons) and scattered light is generated at a wavelength longer than that of the incident light. If another signal is present at this longer wavelength, then it will be amplified, with energy transferred from the first wavelength to the second. The longer the light wave WDM link without regeneration, the more prone it is to SRS. Thus SRS tends to limit the length of links carrying WDM signals (Ref. 2).

Four-wave mixing arises from the dependence of the fiber refractive index on the intensity of the optical wave propagating along the fiber. In this interaction, new optical waves of frequencies f_{ijk} are generated through mixing three waves of frequencies f_i, f_j, and f_k. These obey the relationship $f_{ijk} = f_i + f_j - f_k$ required for the energy conservation condition. These newly generated waves can cause crosstalk between channels in a WDM transmission system if the channels are equally spaced, since the mixing product could fall within another of the signal channels. The efficiency of the FWM process is dependent on the degree of optical phase matching, which in turn is dependent on the fiber dispersion and channel spacing. Due to the presence of dispersion in monomode fiber, exact phase matching is difficult to achieve, but approximate phase matching may be obtained in a low-dispersion fiber of significant link lengths. It should be noted that FWM increases with decreasing wavelength separation and the maximum distance limitation becomes more severe (Ref. 2).

A power penalty is invoked to compensate for SRS and FWM.

There is considerable insertion loss (and splitting loss) in the WDM process (e.g., 5–20 dB and more). Thus it is attractive, or even mandatory, to use fiber amplifiers in conjunction with WDM to compensate for these losses. There is a word of caution. EDFAs have a typical bandwidth of 30 nm or less, and when these amplifiers are cascaded, the bandwidth decreases. Therefore advantage may not be taken for the entire 100-nm bandwidth at 1550 nm, only a small portion of it.

11.11 SOLITONS

We may define a soliton as a single light pulse of very special shape. A soliton can propagate indefinitely along a fiber with dispersive properties without being broadened. Not only must the soliton pulse be of special shape, it must also have sufficient amplitude launch power in the hundreds of

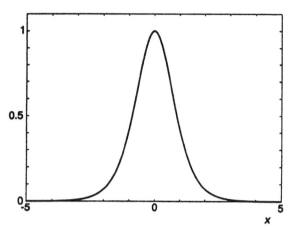

Figure 11.23. A soliton pulse. Based on $P(x) = 1/\cosh^2 x$.

milliwatts range (Ref. 1).* Such launch power is achievable with today's technology. This also implies periodic amplification along the path.

By now we see that the shorter a pulse (i.e., the higher the bit rate), the more it is affected by dispersion. Not so with a soliton because it is not affected by dispersion. We can have soliton transmission with picosecond pulses.

Another interesting aspect of solitons is that if a soliton pulse is launched and it is not exactly the right shape, as it travels down a fiber it will change into the right shape.

The concept of the soliton deals with the trading off of two impairing characteristics of optical fiber transmission: the dispersion of the fiber versus the nonlinearity of glass refractive index with light power. Both of these effects lead to *chirping* of the pulse, which is a change of optical frequency with time. The nonlinearity of the index of refraction occurs at that part of a pulse with the highest amplitude.

It has been found that the special pulse shape of a soliton corresponds to the reciprocal hyperbolic cosine as shown in Figure 11.23. With the pulse shown in the figure, the portion richest in harmonics lies near the pulse peak. Another requirement for soliton transmission is that each active soliton pulse needs to be separated from its adjacent neighbors. This separation can be achieved by having the soliton pulsewidth occupy only a fraction of the bit slot. If a soliton is too close to its neighbor, *soliton interaction* occurs. This interaction tends to pull a soliton from its normal position. Of course, this spacing will put a limit on the effective bit rate of a soliton system.

---

*Agrawal (Ref. 1, p. 400) speaks of an experiment where the required peak power was rather modest, on the order of 3 mW.

Fiber loss works against maintaining a pulse with soliton characteristics because of amplitude reduction as the pulse travels down a fiber. To maintain the proper amplitude, EDFAs or other amplifiers may be used at periodic intervals. One guideline on amplifier spacing is to keep the spacing considerably below the soliton period (Ref. 1), on the order of every 20 or 30 km. Solitons are quite stable and are not easily destroyed by small propagation imperfections, as long as the power level is regenerated frequently along the path (Ref. 3).

11.12   NOTES ON OUTSIDE PLANT INSTALLATION*

11.12.1   System Gain

In the procurement process of an optical fiber system, it is assumed that competitive bids will be solicited. Different manufacturers will submit proposals including power budgets. The designer has to know how the various gain and loss values were derived.

Extreme conditions, such as laser operating wavelength for which the system gain value is *not* optimized, provide a more realistic picture of the fiber system over time. Here is a good example. A system is designed for 1300-nm operation, and the laser manufacturing tolerances allow the central wavelength to range between 1275 and 1325 nm. The system gain measurement might be taken while the laser is operating at 1275 or 1325 nm. If the laser is not operating at the optimized wavelength, there is less available system gain. The highest loss over the wavelength range may better characterize system operation.

Especially for link budget purposes, the point at which the terminal ends and the link span begins must be specified. Does it start at the laser output and include the pigtail or does it start at the fiber end of the pigtail at the connector? There is similar reasoning at the light detector (receiver).

11.12.2   Comments on Outside Plant (OSP) Loss

Maximum OSP path loss = available system gain − equipment margin

Furthermore,

Maximum OSP path loss > fiber loss + interconnection allowance

+ joint loss + growth loss

+ restoration/reroute loss + safety margin

*Section 11.12 is based on Ref. 10. See also Ref. 4.

Definitions and Discussions of Terms

Equipment Margin. This is an allowance for end-of-life eye degradation, environmental factors, modal variation, and optical feedback loss. This can total to 2 or 3 dB.

Interconnection Allowance. This is the loss for jumper cable and connectors used to connect the terminal equipment to the fiber cable. Included in each connector loss is the loss of 30 m of jumper cable. Hence the interconnection allowance is the loss of each connector times the number of connectors.

Fiber Loss. Specified for a particular wavelength in dB per unit length. Is the "particular wavelength" the actual wavelength of the emitted signal? It is recommended that an additional 0.05 dB/km be added to compensate for wavelength variations. If the fiber is exposed to temperature extremes, an extra loss should be added as specified by the cable manufacturer.

Splice Loss. Splices use up some of the available gain. Each splice has an average loss given by manufacturers,* but actual measured losses could vary significantly from these given values. Furthermore, an additional loss may have to be included in the model if the joints are subjected to extreme temperature conditions.

$$\text{Joint loss} = (L_j) \times (N_j) + (L_{je}) \times (N_{je}) \qquad (11.24)$$

where L_j = average joint loss

N_j = total number of joints

L_{je} = added loss due to extreme temperatures

N_{je} = number of joints exposed to extreme temperatures

Growth. If WDM is anticipated in the future to handle traffic growth, the loss of this equipment must be subtracted from the overall power budget.

Restoration / Reroute Loss. Joints are required to repair breaks, and some margin (i.e., 1 dB) should be added to the overall margin for these eventualities. For automatic restoration, switch loss must be included.

Safety Margin. A safety margin is usually added to the loss budget calculations to take care of unexpected losses and possible net deviation above average losses for system elements. One example is reel-to-reel loss variation running to the positive side of tolerance. A safety margin of 3 dB is recommended.

---

*Manufacturers of splicing equipment.

11.12.3 Planning an Installation

Introduction. Planning the installation of optical fiber cable can be based on many of the procedures used for metallic cable but special consideration must be given to certain particular aspects. Among these are:

- The effect of splices and connectors on span length.
- The different construction of optical fiber cables and their more critical physical parameters such as very low stress limits, bending characteristics, and ambient temperature range. The fiber itself is very brittle when compared to wire pair.
- The longer length of cable that can be installed.
- Route construction, conditions, and access both in terms of installation and for later in-service factors and the value of local information on conditions.

Route Layout. The importance of a route survey cannot be overemphasized. The geometry and condition of the existing plant must be taken into account and, with longer lengths, access arrangements must be considered carefully. Regarding ducts, the generally smaller diameter of optical fiber cables provides the opportunity of considering subduct systems to give better duct utilization, a clear and clean installation track, and better maintenance procedures. In large duct systems containing several cables, a midposition is preferable. In aerial systems it is important to minimize in-service movement, and cable strain and to maximize the stability of the pole route. The greater traffic carrying capacity of optical fibers indicates that they should be placed at the top position of telecommunication cables on poles. Where new underground or overhead structures need to be provided, the special requirements of optical fiber cable in terms of small diameter, long lengths, critical bending limits, large splice configurations, and movement and strain limitations should be kept in mind.

Total Cable Length. This is the link length taken from planning documents. To this should be added an additional length for each splice position. This is to allow, if needed, one complete turn of spare cable surrounding the splice position manhole plus spare fiber in the splice housing. The total cable length will also need to include the building lead-in cable from the first (or last) external splice location to the optical fiber distribution frame located near the terminal equipment, or the in-house cable from the cable distribution room to the fiber distribution frame.

Reel or Drum Length. The amount of cable on a reel is controlled by the continuous length a manufacturer can make and by the size and weight of the reel for easy field handling. It is interesting to note that when we buy fiber in reels longer than 1–2 km, the cost per kilometer of fiber tends to increase because of lower yields. For example, if a 10-km reel does not meet specifications, a whole 10-km reel is lost. Whereas if a 1-km reel does not meet specifications, only 1 km is lost. The maximum separation of splices depends on the physical characteristics of the route and the maximum available cable length (drum or reel capacity). Plowed, direct-buried cable or cables pulled into separate pipes thus require splices at the spacing determined on the basis just given. Cable length required to be pulled into ducts will be shorter because of limited pull-in tension and the specified position of the manholes.

When calculating a desired run (reel) length, keep the following in mind once the splicing locations have been ascertained:

Distance between splices, from plans or measured	x (m)
Length allowance	2% of x (m)
Splicing + measurement allowance, 10 m each end	20 m

The drum length for a duct length of x meters is thus $(1 + 0.02)x + 20$ m.

It is important, especially for long regenerator spans in ducts, that the minimization of the number of splices should be consistent with the ability to install the resultant cable lengths. The allowance given above should be sufficient for an extra splice if the cable installation proves more difficult than anticipated. For plowed, direct-buried routes of cables pulled into separate pipes, drum (reel) lengths should be determined as follows, once splicing locations have been ascertained:

Distance between splices, from plans or measured	x (m)
Splicing + measurement allowance, 10 m each end	20 m

The drum (reel) length for a direct-buried route of x meters is thus $x + 20$ m.

Note: This included the additional allowance of 8 m per splice covered by the total cable length.

Number and Location of Splices. The number of splices is a function of the length of the cable that is available on a reel and the location of the planned cable route. Generally, the longer the length of cable on a reel, the fewer splices will be required in a link. However, the condition and location of a duct system may be such that long lengths of cables cannot be placed. Similarly for direct-burial routes, there may be obstacles in the route that require cutting the cable and placing the cable from two directions to the obstacles. Typically splices will occur every 1–2 km.

Splices are located in manholes in duct systems and in handholes (i.e., small buried boxes) in direct-burial plant and are attached to the support strand in aerial plant. In the distribution plant the splices may be located in junction boxes aboveground at sites where network rearrangements are planned.

Right-of-Way. The actual choice of a route depends on a number of factors including getting a right-of-way. Other factors could be environmental impact, meeting state or local zoning laws, town ordinances, and permission to cross private land. Often, optical fiber cables can be placed and included with the same right-of-way as metallic pair cable. These are generally along the sides of streets and highways. Advantage can be taken by acquiring blanket right-of-way from a railroad, power company, or natural gas and petroleum product lines. These lines will always run the risk of *back-hoe fades.**

At the conclusion of the planning phase, a site/route survey should be carried out to determine:

- If public highways/streets are involved
- If private property is involved
- If facilities of other entities are involved such as utilities (power, telecommunications, gas, etc.)
- If cable ducts belonging to other entities may be used (for some rental consideration)
- If buildings, towers, or private residences might be endangered by the installation of the cable system

Ducts, Subducts, Pipes, and Manholes. With existing ductlines where there are insufficient spare pipes to allow the fiber cable to soley occupy one of them, subducting should be considered. All subducts should be installed at the same time and in such a way that the configuration does not tend to spiral as it is pulled in. Sometimes cables cannot be buried directly for one or more of the following reasons:

1. Mechanical protection required.
2. The presence of roads or other obstacles.
3. Possible expansion at a later date.
4. Protection required against rodents.

In this case one or more spare pipes may be installed (plowed or directly buried) as a preparatory measure. The spare pipes consist of individual sections 2500 m (8125 ft) in length, which are joined together by fittings to yield a single pipe spanning the entire length of the route. The optical fiber cable is pulled in later in conformity with the maximum drum (reel) length or pull-in length. To this end the pipe is opened up at the appropriate position and subsequently sealed.

If the joints are to be mounted inside existing manholes, it will be necessary to verify during the site/route survey that there is sufficient space to accommodate the facilities needed for the mounting (e.g., mounting table, light).

---

*A familiar expression meaning accidental cutting of the cable. The "cut" is often done by a back-hoe.

Plowed or Direct-Buried Cable (Spare Pipe) Sections. Optical fiber is commonly direct-buried by trenching or plowing. Trenching is less demanding and provides a more gentle cable lay installation but is more costly than plowing. Trenching is required in rocky soil and where access is difficult. Large and sharp rocks should be removed from the refill soil to avoid damaging the cable.

Plowing is recommended where the terrain is gentle and along prepared rights-of-way. If the soil is hard, it is advisable to first rip the route with an empty plow ahead of the plow containing the cable. It is also possible to plow more than one cable at a single pass or to plow in a cable and a spare duct simultaneously. The spare duct can be used later for placing a second cable.

Aerial Cable. When considering whether to select aerial or buried cable for a certain route, the following features of the aerial cable solution should be taken into account.

Advantages

- Economic
  Use of existing pole lines
  Independence of soil conditions
- Speed of installation
- Possibilities of long-length cabling
- Ease of maintenance when cables run along roads

Disadvantages

- Shortened lifetime because of environmental stresses
- Danger from excessive cable strain in special conditions such as wind or ice loading and over-length spans
- Susceptibility to certain conditions such as storms and vandalism
- Aesthetic considerations

Installation in Ducts. The small size and relatively low strength of optical fiber cable require some planning and care in installation. Factors that limit the length that can be pulled into a duct include the following:

- Number and degree of bends
- Level changes between manholes
- Configuration changes at manholes
- Duct misalignment, damaged/repaired sections and the general condition of the duct, and duct material

The following techniques may be used to minimize these factors so that the distance between splicing locations can be maximized:

- Pulling-in from a midpoint in both directions after fleeting or figure-eighting the cable before the second pull
- Pulling-in in one direction, looping out at intermediate manholes where the ductline makes a sharp turn or on each side of a known *difficult* section

(The above two techniques can also be used very effectively in combination.)

- Pulling downhill rather than uphill
- Using adequate lubrication
- Thoroughly cleaning and checking clearance in each duct
- Using intermediate pulling points
- Using adequate cable guide equipment

If the cable is to be installed in the same ducts as other big cables such as power cables, the optical fiber cables will have a minimum diameter to avoid wedging.

Optical cable pulling-in winches are available that are capable of continuous monitoring of length and pulling tension (measured at the pulling eye itself). These are trailer-mounted and fully self-contained and should be available where and when required.

It should be ensured, with appropriate choice of guides and bearing surfaces, that the bend radius of the cable under tension is not reduced beyond the specified minimum.

REVIEW EXERCISES

1. What is the basic commodity of transmission (i.e., just what are we selling)?

2. If we compare optical fiber transmission to wire pair, coaxial cable, or radio transmission, describe the bandwidth available.

3. Name the basic elements (a block diagram) we have to deal with in a fiber-optic link. (*Note*: Some are active and some are passive.)

4. What is the basic, overriding advantage of optical fiber transmission over other media? Give at least four more advantages of fiber-optic transmission.

5. Give the nominal wavelength band for each of the three minimum attenuation fiber-optic transmission windows.

6. An optical fiber consists of a _____ and a _____ .

7. The notation we use for refractive index is n. If we let the core have a refractive index of n_1 and the cladding, n_2, what relationship of n_1 to n_2 is required for the optical fiber to act as a light waveguide? Can you give some idea of the typical difference between n_1 and n_2?

8. Define numerical aperture (NA) mathematically. What does NA mean in a practical sense?

9. What two very basic parameters can limit the length of a repeaterless fiber-optic section (link)? (This question is aimed at understanding terminology.)

10. Dispersion is brought about by what four factors (dispersion types)? What is the result of excessive dispersion on a derived digital bit stream?

11. Name the three types of optical fiber given in the text. Which type shows minimum dispersive properties?

12. Why use dispersion-shifted optical fiber?

13. Define chromatic dispersion.

14. Why are the deleterious effects of dispersion a function of the bit rate? Consider pulsewidth and allow 20-ps/km dispersion.

15. In general, we neglect dispersion with monomode fiber as a transmission impairment. However, on very high bit rate links, dispersion can corrupt performance. Discuss this type of dispersion and one method to mitigate it.

16. Give the approximate diameter in microns of a monomode optical fiber strand.

17. Compare LED and LD light sources. Include at least six items in the comparison.

18. Compare hot-fusion splices with factory-installed connectors. The comparison should include at least three items.

19. Name and compare the two types of light detectors given in the text. The comparison should include at least four items.

20. Responsivity is comparable to _____ in radio receivers. What is (are) its unit(s) of measure?

21. What are the benefits of using a distributed feedback laser?

22. How much gain can be expected from an APD? From a PIN diode?

23. Give four advantages of using EFDAs.

24. Where are EFDAs usually installed in a fiber-optic link?

25. Compare RZ and NRZ waveforms in regard to transitions per second. Why would we resort to using RZ or Manchester coding on digital fiber-optic links? What might bode against the selection of a RZ waveform over a NRZ?

26. As a first step in the design of a fiber-optic link, name at least four necessary input parameters.

27. Identify at least six technical alternatives and trade-offs dealing with the most cost-effective design of a digital optical fiber link.

28. Why would we want to resort to route diversity (or ring networks) when designing a fiber-optic transmission system?

29. How large a link margin is required for a fiber-optic link?

30. A laser diode has an output of $+1$ dBm and the far-end detector has a threshold (BER $= 1 \times 10^{-9}$) of -39 dBm. What total dB value will we have for the link budget?

31. Calculate the maximum bit rate that a certain fiber-optic link can sustain, accepting that the link is dispersion limited. The link is 60 km long, using single-mode, dispersion-shifted fiber operating at 1550 nm. The fiber displays 4.0-ps/km-nm chromatic dispersion. Transmitter source rise-time is 200 ps and the detector (APD) rise-time is 300 ps.

32. How does the use of EDFAs limit the number of wavelengths that can be wavelength division multiplexed?

33. What are the two basic methods of wave division multiplexing?

34. Why would we use optical filters in a WDM system? Name three important characteristics of these filters.

35. What is the most serious WDM impairment?

36. Give a general explanation of how solitons work.

REFERENCES

1. Govind P. Agrawal, *Fiber Optic Communication Systems*, Wiley, New York, 1992.
2. D. M. Spirit and M. J. O'Mahoney (eds.), *High Capacity Optical Transmission Explained*, Wiley, Chichester, UK, 1995.
3. Paul E. Green, *Fiber Optic Networks*, Prentice-Hall, Englewood Cliffs, NJ, 1993.
4. *Generic Requirements for Optical Fiber and Optical Fiber Cable*, Tech. Ref. TR-NWT-000020, Issue 5, Bellcore, Piscataway, NJ, Dec. 1992.
5. *The New IEEE Standard Dictionary of Electrical and Electronic Terms*, ed., IEEE Std. 100-1992, IEEE Press, New York, 1993.
6. John A. Buck, *Fundamentals of Optical Fibers*, Wiley, New York, 1995.

7. "Characteristics of a Single-Mode Optical Fibre Cable," ITU-T Rec. G.652, ITU, Helsinki, Mar. 1993.

8. "Characteristics of a Dispersion-Shifted Single-Mode Optical Fiber Cable," ITU-T Rec. G.653, ITU, Helsinki, Mar. 1993.

9. "Characteristics of a 1550 nm Wavelength Loss-Minimized Single-Mode Optical Fiber," ITU-T Rec. G.654, ITU, Helsinki, Mar. 1993.

10. *Optical Fibres Systems Planning Guide*, International Telecommunication Union (CCITT), Geneva, 1989.

11. *Design Handbook for Fiber Optic Communication Systems*, MIL-HDBK-415, U.S. Department of Defense, Washington, DC, 1 Feb. 1985.

12. Patrick R. Trischitta and Dixon T. S. Chinn, "Repeaterless Lightwave Systems," *IEEE Commun. Mag.*, Mar. 1989.

13. R. L. Galawa, "Lecture Notes on Optical Communication Via Glass Fiber Waveguides," IEEE Course Series, given at the Raytheon Co., Equipment Division, Wayland, MA, Apr. 1979.

14. M. J. Howes and D. V. Morgan (eds.), *Optical Fibre Communications*, Wiley, Chichester, UK, 1980.

15. R. G. Smith, C. A. Brackett, and H. W. Reinbold, "Optical Detector Package," *Bell Syst. Tech. J.*, vol. 57, 1809–1822, July–Aug. 1978.

16. *Telecommunication Transmission Engineering*, 3rd ed., Bellcore, Piscataway, NJ, 1991.

17. S. Shimada and H. Ishio (eds.), *Optical Amplifiers and Their Applications*, Wiley, Chichester, UK, 1994.

18. E. E. Basch (ed.), *Optical-Fiber Transmission*, Howard W. Sams, Indianapolis, IN, 1987.

19. David R. Smith, *Digital Transmission Systems*, 2nd ed., Van Nostrand-Reinhold, New York, 1993.

20. *LIGHTWAVE 1996 Buyer's Guide Issue*, Pennwell Publishing Co., Nashua, NH, Mar. 1996.

21. *Fiber Optic Cables for Quality Networking*, 1995 Catalog, Optical Cable Corporation, Roanoke, VA, 1995.

12

CELLULAR RADIO AND PCS

12.1 INTRODUCTION

The cellular radio business has expanded explosively since 1980 and continues to expand rapidly. There are several explanations for this popularity. It adds a new dimension to wired PSTN services. In our small spheres of everyday living, we are never away from the telephone, no matter where we are. Outside of industrialized nations, there are long waiting lists for conventional (wired) telephone installations. Go down to the local cellular radio store, and you will have telephone service within the hour. We have found that cellular service augments local telephone service availability. When our local service failed for several days, our cellular telephone worked just fine, although air time was somewhat expensive.

Enter PCS (personal communications services). Does it supplement/complement cellular radio or is it a competitor? It is an extension of cellular certainly in concept. It uses much lower power and has a considerably reduced range. Rappaport (Ref. 4) points out that cellular is hierarchical in nature when connecting to the PSTN; PCS is not. It is hierarchical in that a MTSO (mobile telephone switching office) controls and interfaces up to hundreds of base stations, which connect to mobile users. According to the reference, PCS base stations connect directly to the PSTN. However, a number of PCS strategies have a hierarchy similar to cellular where a MSC (mobile switching center) provides the connectivity to the PSTN. Cellular radio systems operate in the 800- and 900-MHz band; in the United States narrowband PCS operates in the 900-MHz band, and wideband PCS operates in the band 1850–1975 MHz.

Sprint has established a coast-to-coast PCS network, according to a press release (Ref. 34). It states under the heading *Sprint PCS Services Features* that "Sprint PCS provides better connection [to the PSTN], better service and better value using its state-of-the-art, 100-percent digital wireless network. A Sprint Phone is a versatile communications device, incorporating voice mail, caller ID and other enhanced services in a simple, easy-to-use package. Sprint PCS Phones will also have short messaging services and wireless data communications in the near future." Leaving aside the advertising hype, this tells us that PCS varies little from standard cellular radio services. In fact, it appears to be competitive with cellular.

Other PCS operations are specialized such as a wireless PABX, wireless LAN (WLAN), and wireless local loop (WLL) with no competitive cellular counterpart.

12.1.1 Background

The earliest application of radio was mobile communications. As we turned into the 20th century, the Marconi Company* was formed, providing ships with radio communication. In World War I aircraft were provided with radio installations, and after the war mobile radio moved into the civilian arena, particularly in the public safety sector. Taxis and other service vehicles have been using dispatching radio in the VHF band since the 1950s.

In 1979, an entire issue of the *Bell System Technical Journal* was devoted to a new concept—cellular radio—and the issue outlined a new system called AMPS (advanced mobile phone system). This is still the underlying system in the United States and in some Hispanic-American countries. It uses FM radio allocating 30 kHz per channel per user.

AMPS set the scene for explosive cellular radio growth and usage. What set AMPS apart from previous mobile radio systems is that it was designed to interface with the PSTN. It was based on an organized scheme of adjoining cells and had a unique capability of handoff when a vehicle moves through one cell to another, or when another cell receives a higher signal level from the vehicle, it will then take over the call. Vehicles can *roam* from one service area to another with appropriate handoffs.

In the late 1980s there was pressure to convert cellular radio from the bandwidth-wasteful AMPS to some sort of digital regime. As the reader reviews this chapter, he/she will see that digital is also bandwidth-wasteful, even more so than analog FM. Various ways are described on how to remedy the situation: first by reducing the bandwidth of a digital voice channel, and second by the access/modulation scheme proposed. Of this latter proposal, two schemes are on the table in North America: a TDMA scheme and a CDMA scheme. They are radically different and competing.

Meanwhile, the Europeans critiqued our approaches and came up with a better mousetrap. It is called GSM,[†] and there is some pressure that it be adopted in the United States. GSM is a TDMA scheme, fairly different from the U.S. TIA standards (IS-627, 628, 629; see Ref. 36).

---

*The actual name of the company was Marconi International Marine Communications Company, which was created in London in 1900.
†GSM: Groupe Special Mobile is the name of the specifying committee set up by European PTTs.

As we mentioned earlier, PCS is an outgrowth of cellular radio; it uses a cellular concept. The cells, however, are much smaller, under 1-km diameter. RF power is much lower. As with cellular radio, TDMA and CDMA are vying for the national access/modulation method. Unlike the North American popular press, CCIR/ITU-R takes a mature and reasonable view of the affair called FPLMTS (future public land mobile telecommunication system). The FPLMTS concept breaks down into three terrestrial operational areas: indoor environments (range to 100 m), outdoor environments (range from 100 m to 35 km for rural settings), and outdoor-to-indoor environments, where building penetration is a major theme. They also describe satellite environments.

Personal communication service radio terminals are becoming smaller. There is the potential of becoming wristwatch size. However, the human interface requires input/output devices that have optimum usefulness. A wristwatch/size keyboard or keypad is rather difficult to operate. A hard copy printer requires some minimum practical dimensions, and so forth. Also, battery technology has yet to meet these dimension/weight requirements if we want any reasonable battery life before recharging.

12.1.2 Scope and Objective

This chapter presents an overview of "mobile and personal communications." Much of the discussion deals with cellular radio and extends this thinking inside buildings. The coverage most necessarily includes propagation for the several environments, propagation impairments, methods to mitigate those impairments, access techniques, bandwidth limitations, and ways around this problem. It will cover several mobile radio standards and compare a number of existing and planned systems. The chapter objective is to provide an appreciation of mobile/personal communications. Space limitations force us to confine our discussion to what might loosely be called "land mobile systems."

12.2 SOME BASIC CONCEPTS OF CELLULAR RADIO

Cellular radio systems connect a mobile terminal to another user, usually through the PSTN. The "other user" most commonly is a telephone subscriber of the PSTN. However, the other user may be another mobile terminal. Most of the connectivity is extending "plain old telephone service" (POTS) to mobile users. Data and facsimile services are in various stages of implementation. Some of the terms used in this section have a strictly North American flavor.

Figure 12.1 shows a conceptual layout of a cellular system. The heart of the system for a specific serving area is the MTSO. The MTSO is connected by a trunk group to a nearby telephone exchange providing an interface to, and connectivity with, the PSTN.

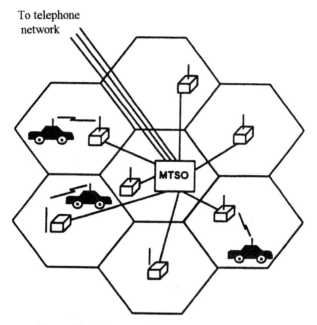

To telephone
network

MTSO

Figure 12.1. Conceptual layout of a cellular system.

The area to be served by a *cellular geographic serving area* (CGSA)* is divided into small geographic cells, which ideally are hexagonal. Cells are initially laid out with centers spaced about 4–8 miles (6.4–12.8 km) apart. The basic system components are the cell sites, the MTSO, and mobile units. These mobile units may be hand-held or vehicle-mounted terminals.

Each cell has a radio facility housed in a building or shelter. The facility's radio equipment can connect and control any mobile unit within the cell's responsible geographic area. Radio transmitters located at the cell site have a maximum effective radiated power (ERP[†]) of 100 watts. Combiners are used to connect multiple transmitters to a common antenna on a radio tower, usually between 50 and 300 ft (15 and 92 m) high. Companion receivers use a separate antenna system mounted on the same tower. The receive antennas are often arranged in a space diversity configuration.

The MTSO provides switching and control functions for a group of cell sites. A method of connectivity is required between the MTSO and the cell site facilities. The MTSO is an electronic switch and carries out a fairly complex group of processing functions to control communications to and from mobile units as they move between cells as well as to make connections with the PSTN. Besides making connectivity with the public network, the MTSO controls cell site activities and mobile actions through command and

---

*CGSA is a term coined by the U.S. FCC. We do not believe it is used in other countries.
[†]Care must be taken with terminology. In this instance ERP and EIRP are *not* the same. The reference antenna in this case is the dipole, which has a 2.15-dBi gain.

control data channels. The connectivity between cell sites and the MTSO is often via DS1 on wire pairs or on microwave facilities, the latter being the most common.

A typical cellular mobile unit consists of a control unit, a radio transceiver, and an antenna. The control unit has a telephone handset, a pushbutton keypad to enter commands into the cellular/telephone network, and audio and visual indications for customer alerting and call progress. The transceiver permits full-duplex transmission and reception between mobile and cell sites. Its ERP is nominally 6 watts. Hand-held terminals combine all functions into one small package that can easily be held in one hand. The ERP of a hand-held is a nominal 0.6 watts.

In North America, cellular communication is assigned a 25-MHz band between 824 and 849 MHz for mobile unit-to-base transmission and a similar band between 869 and 894 MHz for transmission from base to mobile.

The first and most widely implemented North American cellular radio system was called AMPS (advanced mobile phone system). The original system description was contained in an entire issue of the *Bell System Technical Journal* (*BSTJ*) of January 1979. The present AMPS is based on 30-kHz channel spacing using frequency modulation. The peak deviation is 12 kHz. The cellular bands are each split into two to permit competition. Thus only 12.5 MHz is allocated to one cellular operator for each direction of transmission. With 30-kHz spacing, this yields 416 channels. However, nominally 21 channels are used for control purposes with the remaining 395 channels available for cellular end-users.

Common practice with AMPS is to assign 10–50 channel frequencies to each cell for mobile traffic. Of course the number of frequencies used depends on the expected traffic load and the blocking probability. Radiated power from a cell site is kept at a relatively low level with just enough antenna height to cover the cell area. This permits frequency reuse of these same channels in nonadjacent cells in the same CGSA with little or no cochannel interference. A well-coordinated frequency reuse plan enables tens of thousands of simultaneous calls over a CGSA.

Figure 12.2 illustrates a frequency reuse method. Here four channel frequency groups are assigned in a way that avoids the same frequency set used in adjacent cells. If there were uniform terrain contours, this plan could be applied directly. However, real terrain conditions dictate further geographic separation of cells that use the same frequency set. Reuse plans with 7 or 12 sets of channel frequencies provide more physical separation and are often used depending on the shape of the antenna pattern employed.

With user growth in a particular CGSA, cells may become overloaded. This means that grade of service objectives are not being met due to higher than planned traffic levels during the busy hour (BH).* In these cases, congested cells can be subdivided into smaller cells, each with its own base

---

*BH or busy hour is the 1-hour period during a workday with the most telephone traffic calling activity. See Ref. 31 for definitions of the busy hour.

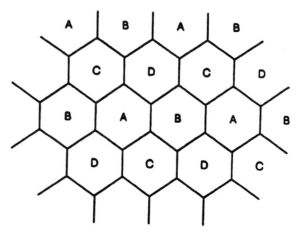

Figure 12.2. Cell separation with four different sets of frequencies.

station, as shown in Figure 12.3. With smaller cells lower transmitter power and antennas with less height are used, thus permitting greater frequency reuse. These subdivided cells can be split still further for even greater frequency reuse. However, there is a practical limit to cell splitting, often with cells with a 1-mile (1.6-km) radius. Under normal, large-cell operation, antennas are usually omnidirectional. When cell splitting is employed, 60° or 120° directional antennas are often used to mitigate interference brought about by increased frequency reuse.

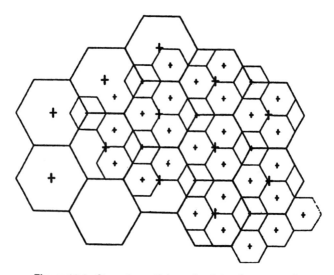

Figure 12.3. Staged growth by cell splitting (subdividing).

Radio system design for cellular operation differs from that used for LOS microwave operation. For one thing, mobility enters the picture. Path characteristics are constantly changing. Mobile units experience multipath scattering, reflection, and/or diffraction by obstructions and buildings in the vicinity. There is shadowing, often very severe. The resulting received signal under these conditions varies randomly as the sum of many individual waves with changing amplitude, phase, and direction of arrival. The statistical autocorrelation distance is on the order of one-half wavelength (Ref. 21). Space diversity at the base station tends to mitigate these impairments.

In Figure 12.1, the MTSO is connected to each of its cell sites by a voice trunk for each of the radio channels at the site. Also, two data links (AMPS design) connect the MTSO to each cell site. These data links transmit information for processing calls and for controlling mobile units. In addition to its "traffic" radio equipment, each cell site has installed signaling equipment, monitoring equipment, and a "setup" radio to establish calls.

When a mobile unit becomes operational, it automatically selects the setup channel with the highest signal level. It then monitors that setup channel for the incoming calls destined for it. When an incoming call is sensed, the mobile terminal in question again samples signal levels of all appropriate setup channels so it can respond through the cell site offering the highest signal level, and then tunes to that channel for response. The responsible MTSO assigns a vacant voice channel to the cell in question, which relays this information via the setup channel to the mobile terminal. The mobile terminal subscriber is then alerted that there is an incoming call. Outgoing calls from mobile terminals are handled in a similar manner.

While a call is in progress, the serving cell site examines the mobile's signal level every few seconds. If the signal level drops below a prescribed level, the system seeks another cell to handle the call. When a more appropriate cell site is found, the MTSO sends a command, relayed by the old cell site, to change frequency for communication with the new cell site. At the same time, the landline subscriber is connected to the new cell site via the MTSO. The periodic monitoring of operating mobile units is known as *locating*, and the act of changing channels is called *handover*. Of course, the functions of locating and handover are to provide subscribers satisfactory service as a mobile unit traverses from cell to cell. When cells are made smaller, handovers are more frequent.

The management and control functions of a cellular system are quite complex. Handover and locating are managed by signaling and supervision techniques, which take place on the setup channel. The setup channel uses a 10-kbps data stream that transmits paging, voice channel designation, and overhead messages to mobile units. In turn, the mobile unit returns page responses, origination messages, and order confirmations.

Both digital messages and continuous supervision tones are transmitted on the voice radio channel. The digital messages are sent as a discontinuous "blank-and-burst" in-band data stream at 10 kbps and include order and handover messages. The mobile unit returns confirmation and messages that contain dialed digits. Continuous positive supervision is provided by an out-of-band 6-kHz tone, which is modulated onto the carrier along with the speech transmission.

Roaming is a term used for a mobile unit that travels such distances that the route covers more than one cellular organization or company. The cellular industry is moving toward technical and tariffing standardization so that a cellular unit can operate anywhere in the United States, Canada, and Mexico.

12.2.1   N-AMPS Increases Channel Capacity Threefold

N-AMPS, developed by Motorola, increases channel capacity three times that of its AMPS counterpart. Rather than 30-kHz segments of AMPS, N-AMPS assigns 10 kHz per voice channel. Its signaling and control are exactly the same as AMPS, except the signaling is one using subaudible data streams.

Of course, to accommodate the narrower FDMA channel width of only 10 kHz, the FM deviation had to be reduced. As a result voice quality was also reduced. To counteract this, N-AMPS uses voice companding to provide a "synthetic" voice channel quieting.

12.3   RADIO PROPAGATION IN THE MOBILE ENVIRONMENT

12.3.1   The Propagation Problem

Line-of-sight microwave and satellite communications covered in Chapters 5 and 7 dealt with fixed systems. Such systems were and are optimized. They are built up and away from obstacles. Sites are selected for best propagation.

This is not so with mobile systems. Motion and a third dimension are additional variables. The end-user terminal often is in motion; or the user is temporarily fixed, but that point can be anywhere within a serving area of interest. Whereas before we dealt with point-to-point, here we deal with point-to-multipoint.

One goal in line-of-sight microwave design was to stretch the distance as much as possible between repeaters by using high towers. In this chapter there are some overriding circumstances where we try to limit coverage extension by reducing tower heights, what we briefly introduced in Section 12.2. Even more importantly, coverage is area coverage, where shadowing is frequently encountered. Examples are valleys, along city streets with high buildings on either side, in verdure such as trees, and inside buildings, to name a few situations. Such an environment is rich with multipath scenarios. Paths can be highly dispersive, as much as 10 μs of delay spread (Ref. 1). Due to a user's motion, Doppler shift can be expected.

The radio-frequency bands of interest are UHF, especially around 800 and 900 MHz and 1700–2000 MHz. Some 400-MHz examples will also be covered.

12.3.2 Several Propagation Models

We concentrate on cellular operation. There is a fixed station (FS) and mobile stations (MS) moving through the cell. A cell is the area of responsibility of the fixed station, a cell site. It usually is pictured as a hexagon in shape, although its propagation profile is more like a circle with the fixed station in its center. Cell radii vary from 1 km (0.6 mi) in heavily built-up urban areas to 30 km (19 mi) or somewhat more in rural areas.

12.3.2.1 *Path Loss.*

We recall the free-space loss (FSL) formula discussed in Chapter 5. It simply stated that FSL was a function of the square of the distance and the square of the frequency plus a constant. It is a very useful formula if the strict rules of obstacle clearance are obeyed. Unfortunately, in the cellular situation, it is impossible to obey these rules. Then to what extent must this free-space loss formula be modified by atmospheric effects, the presence of the earth, and the effects of trees, buildings, and hills that exist in, or close to, the transmission path?

12.3.2.1.1 *CCIR Formula.*

CCIR developed a simple path loss formula (CCIR Rec. 370-5, Ref. 12) for radio and television broadcasting where the frequency term has been factored out. It states that the path loss (L_{dB}) is given by

$$L_{dB} = 40 \log d_m - 20 \log(h_T h_R) \qquad (12.1)$$

Note that the equation is an inverse fourth power law and is fundamental to terrestrial mobile radio. Distance d is in meters; h_T is the height of the transmit antenna above plane earth, again in meters, and h_R is the height of the receive antenna above plane earth in meters.

Suppose $d = 1000$ m (1 km) and the product of $h_T h_R$ is 10 m^2 (low antennas); then $L = 100$ dB. Now suppose that $d = 25$ km and $h_T h_R = 100$ m^2, where the base station is on high ground. The loss L is 136 dB.

The CCIR formula only brings in two new, but important, variables: h_T and h_R.

12.3.2.1.2 *The Amended CCIR Equation.*

This amended model takes the following into account:

- Surface roughness
- Line-of-sight obstacles
- Buildings and trees

The resulting path loss equation is

$$L_{dB} = 40 \log d - 20 \log(h_T h_R) + \beta \qquad (12.2)$$

where β represents the additional losses listed above lumped together (Ref. 2).

12.3.2.1.3 British Urban Path Loss Formula. The following formula was proposed by Allesbrook and Parsons (Ref. 13):

$$L_{dB} = 40 \log d_m - 20 \log(h_T h_R) + 20 + f/40 + 0.18L - 0.34H \quad (12.3)$$

where f = frequency in MHz

   L = land usage factor, a percentage of the test area covered by buildings of any type, 0–100%

   H = terrain height difference between Tx and Rx (i.e., Tx terrain height − Rx terrain height)

Example. Let $d = 2000$ m, $h_T = 30$ m, and $h_R = 3.3$ m; $f = 900$ MHz, $L = 50\%$, and $H = 27$ m.

$$h_T h_R = 100 \text{ m}^2$$

$$L_{dB} = 132 - 40 + 20 + 22.5 + 5.4 - 9.18$$

$$= 130.76 \text{ dB} \qquad (\text{Refs. 3 and 5})$$

12.3.2.1.4 The Okumura Model. Okumura et al. (Ref. 14) carried out a detailed analysis for path predictions around Tokyo for mobile terminals. Hata (Ref. 15) published an empirical formula based on Okumura's results to predict path loss:

$$L_{dB} = 69.55 + 26.16 \log f - 13.82 \log h_t - A(h_r)$$
$$+ (44.9 - 6.55 \log h_t) \log d \qquad (12.4)$$

where f is between 150 and 1500 MHz
   h_t is between 30 and 300 m
   d is between 1 and 20 km

$A(h_r)$ is the correction factor for mobile antenna height and is computed as follows:
   For a small- or medium-size city,

$$A(h_r) = (1.1 \log f - 0.7)h_r - (1.56 \log f - 0.8) \quad (\text{dB}) \quad (12.5a)$$

where h_r is between 1 and 10 m.
   For a large city,

$$A(h_r) = 3.2[\log(11.75h_r)]^2 - 4.97 \quad (\text{dB}) \qquad (12.5b)$$

$(f \geq 400$ MHz$)$.

Example. Let $f = 900$ MHz, $h_t = 40$ m, $h_r = 5$ m, and $d = 10$ km. Calculate $A(h_r)$ for a medium-size city.

$$A(h_r) = 12.75 - 3.8 = 8.95 \text{ dB}$$

$$L_{dB} = 69.55 + 72.28 - 22.14 - 8.95 + 34.4$$

$$= 145.15 \text{ dB}$$

12.3.2.2 *Building Penetration.* For a modern multistory office block at 864 and 1728 MHz, path loss (L_{dB}) includes a value for clutter loss $L(v)$ and is expressed as follows:

$$L_{dB} = L(v) + 20\log d + n_f a_f + n_w a_w \qquad (12.6)$$

where the attenuation in dB of the floors and walls was a_f and a_w, and the number of floors and walls along the line d were n_f and n_w, respectively. The values of $L(v)$ at 864 and 1728 MHz were 32 and 38 dB, with standard deviations of 3 and 4 dB, respectively (Ref. 1).

Another source (Ref. 16) provided the following information. At 1650 MHz the floor loss factor was 14 dB, while the wall losses were 3–4 dB for double plasterboard and 7–9 dB for breeze block or brick. The parameter $L(v)$ was 29 dB. When the propagation frequency was 900 MHz, the first floor factor was 12 dB and $L(v)$ was 23 dB. The higher value for $L(v)$ at 1650 MHz was attributed to a reduced antenna aperture at this frequency compared to 900 MHz. For a 100-dB path loss, the base station and mobile terminal distance exceeded 70 m on the same floor, was 30 m for the floor above, and 20 m for the floor above that, when the propagation frequency was 1650 MHz. The corresponding distances at 900 MHz were 70 m, 55 m, and 30 m. Results will vary from building to building depending on the type of construction of the building, the furniture and equipment it houses, and the number and deployment of the people who populate it.

12.3.3 Microcell Prediction Model According to Lee

For this section a microcell is defined as a cell with a 1-km or less radius. Such cells are used in heavily urbanized areas where demand for service is high and where large cell coverage would be spotty at best. With this model, line-of-sight conditions are seldom encountered; shadowing is the general rule, as shown in Figure 12.4. The major contribution to loss in such situations is due to the dimensions of intervening buildings.

Lee's model (Ref. 6) also includes an antenna height-gain function. Lee (Ref. 6) reports 9 dB/oct or 30 dB/dec for an antenna height change. This would mean that if we doubled the height of an antenna, 9-dB transmission loss improvement would be achieved.

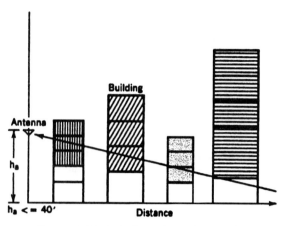

Figure 12.4. A propagation model typical of an urban microcell. (From Ref. 6, Figure 2.24.)

The Lee model for a microcell breaks the prediction process down into a received signal level (dBm) for the line-of-sight component and then the attenuation due to the building blockage component. Figure 12.5 provides a microcell scenario that we will use to understand Lee's model. The receive signal level P_r is equal to the receive signal level for LOS conditions (P_{LOS}) minus the blockage loss due to buildings, α_B. B is the blockage distance in feet. From Figure 12.5, $B = a + b + c$, the sum of the distances (in feet) through each building.

$$P_{LOS} = P_t - 77 \text{ dB} - 21.5 \log(d/100 \text{ ft}) + 30 \log(h_1/20 \text{ ft})$$

for $100 \leq B < 200$ ft   (12.7a)

$$= P_t - 83.5 \text{ dB} - 14 \log(d/200 \text{ ft}) + 30 \log(h_1/20 \text{ ft})$$

for $200 \leq d < 1000$ ft   (12.7b)

$$= P_t - 93.3 \text{ dB} - 36.5 \log(d/1000 \text{ ft}) + 30 \log(h_1/20 \text{ ft})$$

for $1000 \leq d < 5000$ ft   (12.7c)

$\alpha_B = 0$ $1 \text{ ft} \leq B$ (12.8a)

$\alpha_B = 1 + 0.5 \log(B/10 \text{ ft})$ $1 \leq B < 25$ ft   (12.8b)

$\alpha_B = 1.2 + 12.5 \log(B/25 \text{ ft})$ $25 \leq B < 600$ ft   (12.8c)

$\alpha_B = 17.95 + 3 \log(B/600 \text{ ft})$ $600 \leq B < 3000$ ft   (12.8d)

$\alpha_B = 20 \text{ dB}$ $3000 \text{ ft} \leq B$

where P_t is the ERP* in dBm, d is the total distance in feet, and h is the antenna height in feet.

*ERP = effective radiated power (over a dipole).

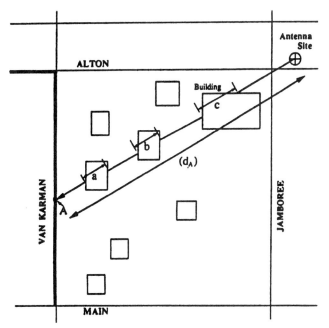

Figure 12.5. Path loss model for a typical microcell in an urban area. Mobile terminal is at location A. With the antenna site so situated, there is blockage by buildings *a*, *b*, and *c*. Thus $B = a + b + c$. (From Ref. 6, Figure 2.26.)

Example 1. The mobile terminal is 500 ft from the cell site antenna, which is 30 ft high. There are three buildings in line between the mobile terminal and the cell site antenna with cross-section (in line with the ray beam) distance of 50, 100, and 150 ft, respectively. Thus $B = 300$ ft. The ERP = +30 dBm (1 watt).

First use equation 12.7b.

$$P_{LOS} = +30 \text{ dBm} - 83.5 \text{ dB} - 14 \log(500 \text{ ft}/200 \text{ ft}) + 30 \log(30 \text{ ft}/20 \text{ ft})$$

$$= +30 \text{ dBm} - 83.5 \text{ dB} - 5.57 \text{ dB} + 5.28 \text{ dB}$$

$$= -53.79 \text{ dBm}$$

Next, use equation 12.8c.

$$\alpha_B = 1.2 + 12.5 \log(300 \text{ ft}/25 \text{ ft})$$

$$= 1.2 \text{ dB} + 13.5 \text{ dB}$$

$$= 14.7 \text{ dB}$$

The receive signal level, P_r, is

$$P_r = -53.79 \text{ dBm} - 14.7 \text{ dB}$$
$$= -68.47 \text{ dBm}$$

Of course, we must assume that the sum of the gain of the receive antenna and the transmission line loss equals 0 dB so that P_r is the same as isotropic receive level.

Example 2. There is 4000 ft separating the cell site transmit antenna from the receive terminal. The transmit antenna is 40 ft high. There are four buildings causing blockage of 150 ft, 200 ft, 140 ft, and 280 ft. These distances are measured along the ray beam line. Thus $B = 770$ ft. The ERP is $+20$ dBm. What is the receive signal level (P_r) assuming no gain or loss in the receive antenna system?

Use equation 12.7c for P_{LOS}.

$$P_{LOS} = +20 \text{ dBm} - 93.3 \text{ dB} - 36.5 \log(4000/1000) + 30 \log(40/20)$$
$$= -73.3 \text{ dBm} - 21.4 \text{ dB} + 9 \text{ dB}$$
$$= -85.67 \text{ dBm}$$

Now use equation 12.8d to calculate α_B.

$$\alpha_B = 17.95 + 3 \log(770/600)$$
$$= 18.28 \text{ dB}$$
$$P_r = -85.67 \text{ dBm} - 18.28 \text{ dB}$$
$$= -103.94 \text{ dBm}$$

12.4   IMPAIRMENTS—FADING IN THE MOBILE ENVIRONMENT

12.4.1   Introduction

Fading in the mobile situation is quite different from the static line-of-sight (LOS) microwave situation discussed in Chapter 5. Radio paths are not optimized as in the LOS environment. The mobile terminal may be fixed throughout a telephone call, but it is more apt to be in motion. Even the hand-held terminal may well have micromotion. When a terminal is in motion, the path characteristics are constantly changing.

Multipath propagation is the rule. Consider the simplified multipath pictorial model in Figure 12.6. Commonly, multiple rays reach the receive antenna, each with its own delay. The constructive and destructive fading can become quite complex. We must deal with both reflection and diffraction. Energy will arrive at the receive antenna reflected off sides of buildings,

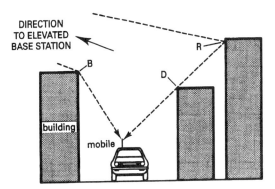

Figure 12.6. Mobile terminal in an urban setting. R = reflection, D = diffraction.

streets, lakes, and so on. Energy will also arrive diffracted from knife edges (e.g., building corners) and rounded obstacles (e.g., water tanks, hilltops) (Ref. 7).

Because the same signal arrives over several paths, each with a different electrical length, the phases of each path will be different, resulting in constructive and destructive amplitude fading. Fades of 20 dB are common, and even 30-dB fades can be expected.

On digital systems, the deleterious effects of multipath fading can be even more severe. Consider a digital bit stream to a mobile terminal with a transmission rate of 1000 bps. Assuming NRZ coding, the bit period would be 1 ms (bit period = 1/bit rate). We find the typical multipath delay spread may be on the order of 10 μs. Thus delayed energy will spill into a subsequent bit (or symbol) for the first 10 μs of the bit period and will have no negative effect on the bit decision. If the bit stream is 64,000 bps, then the bit period is 1/64,000 or 15 μs. Destructive energy from the previous bit (symbol) will spill into the first two-thirds of the bit period, well beyond the midbit sampling point. This is typical intersymbol interference (ISI), and in this case there is a high probability that there will be a bit error. The bottom line is that the destructive potential of ISI increases as the bit rate increases (i.e., as the bit period decreases).

12.4.2 Classification of Fading

We consider three types of channels to place bounds on radio system performance:

- Gaussian channel
- Rayleigh channel
- Rician channel

12.4.2.1 *The Gaussian Channel.* The Gaussian channel can be considered the ideal channel, and it is only impaired by additive white Gaussian noise (AWGN) developed internally by the receiver. Our objective is to achieve a BER typical of a Gaussian channel when we have done everything we can to mitigate fading and its results. Such mitigation measures could be diversity, equalization, and FEC coding with interleaving. The ideal Gaussian channel is very difficult to achieve in the mobile radio environment.

12.4.2.2 *The Rayleigh Channel.* The Rayleigh channel is at the other end of the line, often referred to as a worst-case channel. Remember, in Chapter 5, we treated fading on LOS microwave as Rayleigh fading and it gave us the very worst-case fading scenario. Figure 12.7 shows a channel approaching

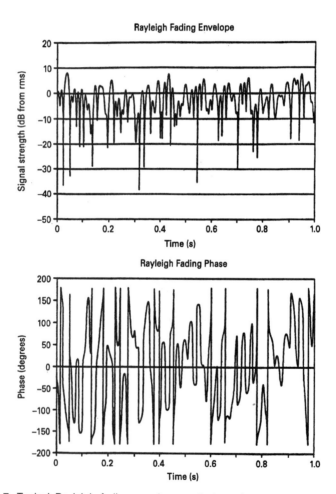

Figure 12.7. Typical Rayleigh fading envelope and phase in a mobile scenario. Vehicle speed is about 30 mph; frequency 900 MHz. (From Ref. 1, Figure 1.1.)

Rayleigh fading characteristics. Of course, we are dealing with multipath here. We showed that in the mobile radio scenario multipath reception commonly had many components. Thus if each multipath component is independent, the PDF (power distribution function) of its envelope is Rayleigh.

12.4.2.3 *The Rician Channel.* The characteristics of a Rician channel are in between those of a Gaussian channel and those of a Rayleigh channel. The channels can be characterized by a function K (not to be confused with the K-factor in Chapter 5). K is defined as follows:

$$K = \frac{\text{power in the dominant path}}{\text{power in the scattered paths}} \tag{12.9}$$

As cells get smaller, the LOS component becomes more and more dominant. There are many cases, in fact nearly all cases where there is no full shadowing, in which there is a LOS component and scattered components. This is a typical multipath scenario. Turning now to equation 12.9, when $K = 0$, the channel is Rayleigh (i.e., the numerator is 0 and all the received energy derives from scattered paths). When $K = \infty$, the channel is Gaussian and the denominator is zero. Figure 12.8 gives BER values for some typical values of K. It shows that those intermediate values of K provide a superior BER than for the Rayleigh channel where $K = 0$. For a microcell mobile scenario, values of K vary from 5 to 30 (Ref. 1). Larger cells tend more toward low values of K.

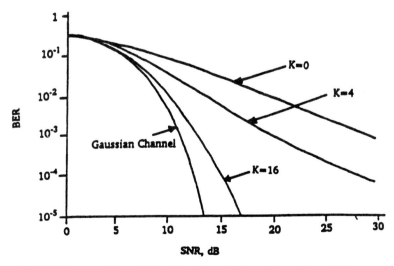

Figure 12.8. BER versus channel signal-to-noise ratio for various values of K; noncoherent FSK. (From Ref. 1, Figure 1.7.)

There is also advantage for Rician fading with higher values of K regarding cochannel interference performance for a desired BER. The smaller the cell, the more fading becomes Rician, approaching the higher values of K.

12.4.3 Diversity—A Technique to Mitigate the Effects of Fading and Dispersion

12.4.3.1 Scope. We discuss diversity to reduce the effects of fading and to mitigate dispersion. Diversity was briefly covered in Chapters 5 and 6, where we dealt with LOS microwave radio system and troposcatter. In those chapters we discussed frequency and space diversity. There is a third diversity scheme called time diversity, which can be applied to digital cellular radio systems. In principle, such techniques can be employed either at the base station and/or at the mobile unit, although different problems have to be solved for each. The basic concept behind diversity is that when two or more radio paths carrying the same information are relatively uncorrelated, when one path is in a fading condition, often the other path is not undergoing a fade. These separate paths can be developed by having two channels, separated in frequency. The two paths can also be separated in space and in time.

When the two (or more) paths are separated in frequency, we call this frequency diversity. However, there must be at least some 2% or greater frequency separation for the paths to be comparatively uncorrelated. Because, in the cellular situation, we are so short of spectrum, using frequency diversity (i.e., using a separate frequency with redundant information) is essentially out of the question, and it will not be discussed further except for its implicit use in CDMA.

12.4.3.2 Space Diversity. Space diversity is commonly employed at cell sites, and two separate receive antennas are required, separated in either the horizontal or vertical plane. Separation of the two antennas vertically can be impractical for cellular receiving systems. Horizontal separation, however, is quite practical. The space diversity concept is illustrated in Figure 12.9.

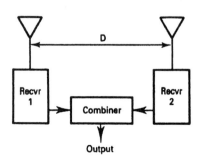

Figure 12.9. The space diversity concept.

One of the most important factors in space diversity design is antenna separation. There are a set of rules for the cell site, and another set of rules for the mobile unit.

Space diversity antenna separation, shown as distance D in Figure 12.9, varies not only as a function of the correlation coefficient but also as a function of antenna height, h. The wider the antennas are separated, the lower the correlation coefficient and the more uncorrelated the diversity paths are. Sometimes we find that by lowering the antennas as well as adjusting the distance between the antennas, we can achieve a very low correlation coefficient. However, we might lose some of the height-gain factor.

Lee (Ref. 6) proposes a new parameter η, where

$$\eta = \frac{\text{antenna height}}{\text{antenna separation}} = \frac{h}{d} \tag{12.10}$$

In Figure 12.10 we relate the correlation coefficient (ρ) with η. α is the orientation of the antenna regarding the incoming signal from the mobile unit. Lee recommends a value of $\rho = 0.7$. Lower values are unnecessary because of the law of diminishing returns. There is much more fading advantage achieved from $\rho = 1.0$ to $\rho = 0.7$ than $\rho = 0.7$ to $\rho = 0.1$.

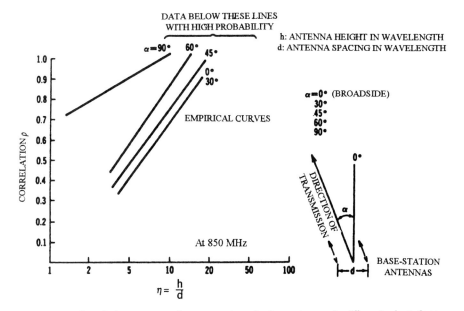

Figure 12.10. Correlation ρ versus the parameter η for two antennas in different orientations. (From Ref. 6, Figure 6.4. Reprinted with permission.)

Based on $\rho = 0.7$, $\eta = 11$, from Figure 12.10, we can calculate antenna separation values (for 850-MHz operation). For example, if $h = 50$ ft (15 m), we can calculate d using formula 12.10:

$$d = h/\eta = 50/11 = 4.5 \text{ ft } (1.36 \text{ m})$$

For an antenna 120 ft (36.9 m) high, we find that $d = 120/11 = 10.9$ ft or 3.35 m (from Ref. 6).

12.4.3.2.1 *Space Diversity on a Mobile Platform.* Lee (Ref. 6) discusses both vertically and horizontally separated antennas on a mobile unit. For the vertical case, 1.5λ is recommended and for the horizontal separation case, 0.5λ. At 850 MHz, $\lambda = 35.29$ cm. Then $1.5\lambda = 1.36$ ft or 52.9 cm. For 0.5λ, the value is 0.45 ft or 17.64 cm.

12.4.3.3 *Frequency Diversity.* We pointed out that conventional frequency diversity was not a practical alternative in cellular systems because of the shortage of available bandwidth. However, with CDMA (direct sequence spread spectrum), depending on the frequency spread, two or more frequency diversity paths are available, and in most CDMA systems we have what is called *implicit diversity*; and multipath can be resolved with the use of a RAKE filter. This is one of the many advantages of CDMA.

12.4.3.4 *Forward Error Correction—A Form of Time Diversity.* Forward error correction (FEC) (see Chapter 5) can be used on digital cellular systems not only to improve bit error rate but to reduce fading. To reduce the effects of fading, a FEC system must incorporate an *interleaver.*

An interleaver pseudorandomly shuffles bits. It first stores a span of bits and shuffles them using a generating polynominal. The span of bits can represent a time period. The rule is that for effective operation against burst errors,* the interleaving span must be much greater than the typical fade duration. The deinterleaver used at the receive end is time synchronized to the interleaver incorporated at the transmit end of the link.

12.4.4 Cellular Radio Path Calculations

Consider the path from the fixed cell site to the mobile platform. There are several mobile receiver parameters that must be considered. The first to be derived are signal quality minima from EIA/TIA IS-19B (Ref. 19).

The minimum SINAD (Signal + Interference + Noise and Distortion to Interference + Noise + Distortion ratio) is 12 dB. This SINAD equates to a threshold of -116 dBm or 7 μV/m. This assumes a cellular transceiver with an antenna with a net gain of 1 dBd (dB over a dipole). The gross antenna

---

*Burst errors are bunched errors due to fading.

gain is 2.5 dBd with a 1.5-dB transmission line loss. A 1-dBd gain is equivalent to a 3.16-dBi gain (i.e., 0 dBd = 2.16 dBi). Furthermore, this value equates to an isotropic receive level of −119.16 dBm (Ref. 19).

One design goal for a cellular system is to more or less maintain a cell boundary at the 39-dBμ contour (Ref. 20). Note that 39 dBμ = −95 dBm (based on a 50-ohm impedance at 850 MHz). Then at this contour, a mobile terminal would have a 24.16-dB fade margin.

If a cellular transmitter has a 10-watt output per channel and an antenna gain of 12 dBi and 2-dB line loss, the EIRP would be +20 dBW or +50 dBm. The maximum path loss to the 39-dBμ contour would be +50 dBm − (−119.16 dBm) or 169 dB.

12.5 THE CELLULAR RADIO BANDWIDTH DILEMMA

12.5.1 Background and Objectives

The present cellular radio bandwidth assignment in the 800- and 900-MHz band cannot support the demand for cellular service, especially in urban areas in the United States and Canada. AMPS, widely used in the United States, Canada, and many other nations of the Western Hemisphere, requires 30 kHz per voice channel. This system can be called a FDMA (frequency division multiple access) system, much like the FDMA/DAMA system described in Chapter 7. We remember that the analog voice channel is a nominal 4 kHz, and 30 kHz is about seven times that value.

The trend is to convert to digital. Digital transmission is notorious for being wasteful of bandwidth, when compared to the 4-kHz analog channel. We can show that PCM has a 16-times multiplier of the 4-kHz analog channel. In other words, the standard digital voice channel occupies 64 kHz (assuming 1 bit per hertz of bandwidth).

One goal of system designers, therefore, is to reduce the required bandwidth of the digital voice channel without sacrificing too much voice quality. As we will show, they have been quite successful.

The real objective is to increase the ratio of users to unit bandwidth. We will describe two distinctly different methods, each claiming to be more bandwidth conservative than the other. The first method is TDMA (time division multiple access) and the second is CDMA (code division multiple access). The former was described in Chapter 7 and the latter was briefly mentioned.

12.5.2 Bit Rate Reduction of the Digital Voice Channel

It became obvious to system designers that conversion to digital cellular required some different technique for coding speech other than conventional

PCM found in the PSTN. The following lists some of the techniques that may be considered:

1. ADPCM (adaptive differential PCM). Good intelligibility and good quality; 32-kbps data transmission over the channel may be questionable.
2. Linear predictive vocoders (voice coders); 2400 bps. Adopted by U.S. Department of Defense. Good intelligibility, poor quality, especially speaker recognition.
3. Subband coding (SBC). Good intelligibility even down to 4800 bps. Quality suffers below 9600 bps.
4. RELP (residual excited linear predictive) type coder. Good intelligibility down to 4800 bps and fair to good quality. Quality improves as bit rate increases. Good quality at 16 kbps.
5. CELP (codebook-excited linear predictive). Good intelligibility and surprisingly good quality even down to 4800 bps. At 8 kbps, near-toll quality speech (Ref. 17).

12.6   NETWORK ACCESS TECHNIQUES

12.6.1   Introduction

The objective of a cellular radio operation is to provide a service where mobile subscribers can communicate with any subscriber in the PSTN, where any subscriber in the PSTN can communicate with any mobile subscriber, and where mobile subscribers can communicate among themselves via the cellular radio system. In all cases the service is full duplex.

A cellular service company is allotted a radio bandwidth segment to provide this service. Ideally, for full-duplex service, a portion of the bandwidth is assigned for transmission from a cell site to mobile subscriber, and another portion is assigned for transmission from a mobile user to a cell site. Our goal here is to select an "access" method to provide this service given our bandwidth constraints.

We will discuss three generic methods of access: FDMA, TDMA, and CDMA. It might be useful for the reader to review our discussion of satellite access in Chapter 7 where we describe FDMA and TDMA. However, in this section, the concepts are the same, but some of our constraints and operating parameters will be different.

12.6.2   Frequency Division Multiple Access (FDMA)

With FDMA, our band of frequencies is divided into segments and each segment is available for one user access. Half of the contiguous segments are

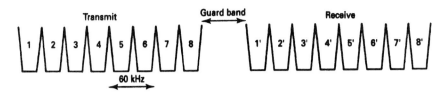

Figure 12.11. A conceptual drawing of FDMA.

assigned to cell site outbound (i.e., to mobile users) and the other half to inbound. A guardband is usually provided between outbound and inbound contiguous channels. This concept is shown in Figure 12.11.

Because of our concern to optimize the number of users per unit bandwidth, the key question is the actual width of one user segment. The North American AMPS system was described in Section 12.2, where each segment width was 30 kHz. The bandwidth of a user segment is greatly determined by the information bandwidth and modulation type. With AMPS, the information bandwidth was a single voice channel with a nominal bandwidth of 4 kHz. The modulation was FM and the bandwidth was then determined by Carson's rule (Chapter 5). As we pointed out, AMPS is not exactly spectrum-conservative. On the other hand, it has a lot of the redeeming features that FM provides such as the noise and interference advantage (FM capture).

Another approach to FDMA would be to convert the voice channel to its digital equivalent using CELP, for example (Section 12.5.2), with a transmission rate of 4.8 kbps. The modulation might be BPSK using a raised cosine filter where the bandwidth could be 1.25% of the bit rate or 6 kHz per voice channel. This alone would increase voice channel capacity five times over AMPS with its 30 kHz per channel. It should be noted that a radio carrier is normally required for each frequency slot.

12.6.3 Time Division Multiple Access (TDMA)

With TDMA we work in the time domain rather than the frequency domain of FDMA. Each user is assigned a time slot rather than a frequency segment and during the user's turn, the full frequency bandwidth is available for the duration of the user's assigned time slot.

Let us say that there are n users and so there are n time slots. In the case of FDMA, we had n frequency segments and n radio carriers, one for each segment. For the TDMA case, only one carrier is required. Each user gains access to the carrier for $1/n$ of the time and there is generally an ordered sequence of time slot turns. A TDMA frame can be defined as cycling through n users' turns just once.

A typical TDMA frame structure is shown in Figure 12.12. One must realize that TDMA is only practical with a digital system such as PCM or any of those discussed in Section 12.5.2. TDMA is a store and burst system.

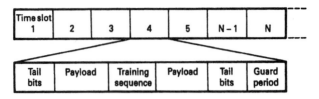

Figure 12.12. A simplified TDMA frame.

Incoming user traffic is stored in memory, and when that user's turn comes up, that accumulated traffic is transmitted in a digital burst.

Suppose there are ten users. Let each user's bit rate be R, then a user's burst must be at least $10R$. Of course, the burst will be greater than $10R$ to accommodate a certain amount of overhead bits as shown in Figure 12.12.

We define downlink as outbound, base station to mobile station(s), and uplink as mobile station to base station. Typical frame periods are:

North American IS-54	40 ms for six time slots
European GSM	4.615 ms for eight time slots

One problem with TDMA, often not appreciated by a novice, is delay. In particular, this is the delay on the uplink. Consider Figure 12.13, where we set up a scenario. A base station receives mobile time slots in a circular pattern and the radius of the circle of responsibility of that base station is 10 km. Let the velocity of a radio wave be 3×10^8 m/s. The time to traverse 1 km is 1000 m/(3×10^8) or 3.333 μs. Making up an uplink frame is a mobile right on top of the base station with essentially no delay and another mobile right at 10 km with 10×3.33 μs or 33.3 μs delay. A GSM time slot is about 576 μs in duration. The terminal at the 10-km range will have its time slot arriving 33.3 μs late compared to the terminal with no delay. A GSM bit

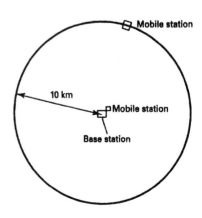

Figure 12.13. A TDMA delay scenario.

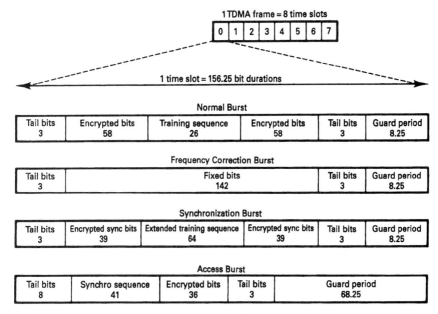

Figure 12.14. GSM frame and burst structures. (From Ref. 1, Figure 8.7. Reprinted with permission.)

period is about 3.69 μs so that the late arrival eats up roughly 10 bits, and unless something is done, the last bit of the burst will overlap the next burst (Refs. 1 and 10).

Refer now to Figure 12.14, which illustrates GSM burst structures. Note that the access burst has a guard period of 68.25-bit durations or a "slop" of 3.69 × 68.25 μs, which will well accommodate the late arrival of the 10-km mobile terminal of only 33.3 μs.

To provide the same long guard period in the other bursts is a waste of valuable "spectrum."* The GSM system overcomes this problem by using adaptive frame alignment. When the base station detects a 41-bit random access synchronization sequence with a long guard period, it measures the received signal delay relative to the expected signal from a mobile station with zero range. This delay, called the timing advance, is transmitted to the mobile station using a 6-bit number. As a result, the mobile station advances its time base over the range of 0–63 bits (i.e., in units of 3.69 μs). By this process the TDMA bursts arrive at the base station in their correct time slots and do not overlap with adjacent ones. As a result, the guard period in all other bursts can be reduced to 8.25 × 3.69 μs or approximately 30.46 μs, the equivalent of 8.25 bits only. Under normal operations, the base station

---

*We are equating bit rate or bit durations to bandwidth. One could assume 1 bit/Hz as a first-order estimate.

continuously monitors the signal delay from the mobile station and thus instructs the mobile station to update its time advance parameter. In very large traffic cells there is an option to actively utilize every second time slot only to cope with the larger propagation delays. This is spectrally inefficient but, in large, low-traffic rural cells, permissible (from Ref. 1).

12.6.3.1 Some Comments on TDMA Efficiency.
Multichannel FDMA can operate with a power amplifier for every channel, or with a common wideband power amplifier for all channels. With the latter, we are setting up a typical generator of intermodulation products as these carriers mix in a comparatively nonlinear common power amplifier. To reduce the level of IM products, just like in satellite communications discussed in Chapter 7, backoff of the power amplifier drive is required. This backoff can be on the order of 3–6 dB.

With TDMA (downlink), only one carrier is present on the power amplifier, thus removing most of the causes of IM noise generation. Thus with TDMA, the power amplifier can be operated to full saturation, a distinct advantage. FDMA required some guardband between frequency segments; there are no guardbands with TDMA. However, as we saw above, a guard time between uplink time slots is required to accommodate the following situations:

- Timing inaccuracies due to clock instabilities
- Delay spread due to propagation*
- Transmission time delay due to propagation distance (see Section 12.6.3)
- Tails of pulsed signals due to transient response

The longer guard times are extended, the more inefficient a TDMA system is.

12.6.3.2 Advantages of TDMA.
The introduction of TDMA results in a much improved system signaling operation and cost. Assuming a 25-MHz bandwidth, up to 23.6 times capacity can be achieved with North American TDMA compared to FDMA (AMPS) (see Ref. 24, Table II).

A mobile station can exchange system control signals with the base station without interruption of speech (or data) transmission. This facilitates the introduction of new network and user services. The mobile station can also check the signal level from nearby cells by momentarily switching to a new time slot and radio channel. This enables the mobile station to assist with handover operations and thereby improve the continuity of service in response to motion or signal fading conditions. The availability of signal

---

*Lee (Ref. 6) reports that a typical urban delay spread is 3 μs.

strength information at both the base and mobile stations, together with suitable algorithms in the station controllers, allows further spectrum efficiency through the use of dynamic channel assignment and power control.

The cost of base stations using TDMA can be reduced if radio equipment is shared by several traffic channels. A reduced number of transceivers leads to a reduction of multiplexer complexity. Outside the major metropolitan areas, the required traffic capacity for a base station may, in many cases, be served by one or two transceivers. The saving in the number of transceivers results in a significantly reduced overall cost.

A further advantage of TDMA is increased system flexibility. Different voice and nonvoice services may be assigned a number of time slots appropriate to the service. For example, as more efficient speech codecs are perfected, increased capacity may be achieved by the assignment of a reduced number of time slots for voice traffic. TDMA also facilitates the introduction of digital data and signaling services as well as the possible later introduction of such further capacity improvements as digital speech interpolation (DSI).

Table 12.1 compares three operational/planned digital TDMA systems.

12.6.4 Code Division Multiple Access (CDMA)

CDMA means code division multiple access implying spread spectrum. There are two types of spread spectrum: frequency hop and direct sequence (sometimes called pseudonoise). In the cellular environment, CDMA means direct sequence spread spectrum (Ref. 1). However, the GSM system uses frequency hop, but not as an access technique.

Using spread spectrum techniques accomplishes just the opposite of what we were trying to accomplish in Chapter 5. Bit packing is used to conserve bandwidth by packing as many bits as possible in 1 Hz of bandwidth. With spread spectrum we do the reverse by spreading the information signal over a very wide bandwidth.

Conventional AM* requires about twice the bandwidth of the audio information signal with its two sidebands of information (i.e., approximately ± 4 kHz). On the other hand, depending on its modulation index, frequency modulation could be considered a type of spread spectrum in that it produces a much wider bandwidth than its transmitted information requires. As with all other spread spectrum systems, a signal-to-noise advantage is gained with FM, depending on its modulation index. For example, with AMPS, a typical FM system, 30 kHz is required to transmit the nominal 4-kHz voice channel.

If we are spreading a voice channel over a very wide frequency band, it would seem that we are defeating the purpose of frequency conservation. With spread spectrum, with its powerful antijam properties, multiple users can transmit on the same frequency with only some minimal interference one

*AM denotes for "toll-quality" telephony.

TABLE 12.1 Three TDMA Systems Compared

Feature	GSM	North America	Japan
Class of emission			
Traffic channels	271KF7W	40K0G7WDT	tbd
Control channels	271KF7W	40K0G1D	tbd
Transmit frequency bands (MHz)			
Base stations	935–960	869–894	810–830 (1.5 GHz tbd)
Mobile stations	890–915	824–849	940–960 (1.5 GHz tbd)
Duplex separation (MHz)	45	45	130 48(1.5 GHz)
RF carrier spacing (kHz)	200	30	25 interleaved 50
Total number of RF duplex channels	124	832	tbd
Maximum base station ERP (W)			
Peak RF carrier	300	300	tbd
Traffic channel average	37.5	100	tbd
Nominal mobile station transmit power (W): peak and average	20 and 2.5 8 and 1.0 5 and 0.625 2 and 0.25	9 and 3 4.8 and 1.6 1.8 and 0.6 tbd and tbd	tbd
Cell radius (km)			
Minimum	0.5	0.5	0.5
Maximum	35 (up to 120)	20	20
Access method Traffic channels/RF carrier	TDMA	TDMA	TDMA
Initial	8	3	3
Design capability	16	6	6
Channel coding	Rate one-half convolutional code with inter-leaving plus error detection	Rate one-half convolutional code	tbd
Control channel structure			
Common control channel	Yes	Shared with AMPS	Yes
Associated control channel	Fast and slow	Fast and slow	Fast and slow
Broadcast control channel	Yes	Yes	Yes
Delay spread equalization capability (μs)	20	60	tbd
Modulation	GMSK (BT = 0.3)	$\pi/4$ diff. encoded QPSK (roll-off = 0.25)	$\pi/4$ diff. encoded QPSK (roll-off = 0.5)
Transmission rate (kbps)	270.833	48.6	37 – 42

TABLE 12.1   (Continued)

Feature	GSM	North America	Japan
Traffic channel structure			
Full-rate speech codec			
Bit rate (kbps)	13.0	8	6.5–9.6
Error protection	9.8 kbps FEC + speech processing	5 kbps FEC	~ 3 kbps FEC
Coding algorithm	RPE-LTP	CELP	tbd
Half-rate speech codec			
Initial	tbd	tbd	tbd
Future	Yes	Yes	Yes
Data			
Initial net rate (kbps)	Up to 9.6	2.4, 4.8, 9.6	1.2, 2.4, 4.8
Other rates (kbps)	Up to 12	tbd	8 and higher
Handover			
Mobile assisted	Yes	Yes	Yes
Intersystem capability with existing analog system	No	Between digital and AMPS	No
International roaming capability	Yes > 16 countries	Yes	Yes
Design capability for multiple system operators in same area	Yes	Yes	Yes

Notes: GMSK = Gaussian minimum shift keying
  tbd = to be defined
  diff. = differentially

Source: Reference 24, Table 1, CCIR Rep. 1156.

to another. This assumes that each user is employing a different key variable (i.e., in essence, using a different time code). At the receiver, the CDMA signals are separated using a correlator that accepts only signal energy from the selected key variable binary sequence (code) used at the transmitter, and then despreads its spectrum. CDMA signals with unmatching codes are not despread and only contribute to the random noise.

CDMA reportedly provides an increase in capacity 15-times that of its analog FM counterpart. It can handle any digital format at the specified input bit rate such as facsimile, data, and paging. In addition, the amount of transmitter power required to overcome interference is comparatively low when utilizing CDMA. This translates into savings on infrastructure (cell site) equipment and longer battery life for hand-held terminals. CDMA also provides so-called soft handoffs from cell site to cell site that make the transition virtually inaudible to the user (Ref. 25).

Dixon (Ref. 18) develops from Claude Shannon's classical relationship an interesting formula to calculate the spread bandwidth given the information rate, signal power, and noise power:

$$C = W \log_2(1 + S/N) \qquad (12.11)$$

where C = capacity of a channel in bits per second
$\quad\quad W$ = bandwidth in Hz
$\quad\quad N$ = noise power
$\quad\quad S$ = signal power

This equation shows the relationship between the ability of a channel to transfer error-free information, compared with the signal-to-noise ratio existing in the channel, and the bandwidth used to transmit the information.

If we let C be the desired system information rate and we change the logarithm base to the natural base (e), the result is

$$C/W = 1.44 \log_e(1 + S/N) \tag{12.12}$$

and, for a S/N that is very small (e.g., ≤ 0.1), which would be used in an antijam system,* we can say

$$C/W = 1.44(S/N) \tag{12.13}$$

From this equation we find that

$$N/S = 1.44W/C \approx W/C \tag{12.14}$$

and

$$W \approx NC/S \tag{12.15}$$

This exercise shows that for any given S/N we can have a low information error rate by increasing the bandwidth used to transfer that information.

Suppose we had a cellular system using a data rate of 4.8 kbps and a S/N of 20 dB (numeric of 100). Then the bandwidth for this 4.8-kbps channel would be

$$W = 100 \times 4.8 \times 10^3/1.44$$

$$= 333.333 \text{ kHz}$$

There are two common ways that information can be embedded in the spread spectrum signal. One way is to add the information to the spectrum-spreading code before the spreading modulation stage. It is assumed that the information to be transmitted is binary because modulo-2 addition is involved in this process. The second method is to modulate the RF carrier with the desired information before spreading the carrier. The modulation is usually PSK or FSK or other angle modulation scheme (Ref. 18).

---

*If we think about it, a cellular scenario where one user is transmitting right on top of others on the same frequency plus adjacent channel interference, we are indeed in an antijam situation.

Dixon (Ref. 18) lists some advantages of the spread spectrum:

1. Selective addressing capability.
2. Code division multiplexing is possible for multiple access.
3. Low-density power spectra for signal hiding.
4. Message security.
5. Interference rejection.

Of most importance for the cellular user (Ref. 18), "when codes are properly chosen for low cross correlation, minimum interference occurs between users, and receivers set to use different codes are reached only by transmitters sending the correct code. Thus more than one signal can be unambiguously transmitted at the same frequency and at the same time; selective addressing and code-division multiplexing are implemented by the coded modulation format."

Figure 12.15 shows a direct sequence (pseudonoise) spread spectrum system with waveforms.

Processing gain is probably the most commonly used parameter to describe the performance of a spread spectrum system. It quantifies the

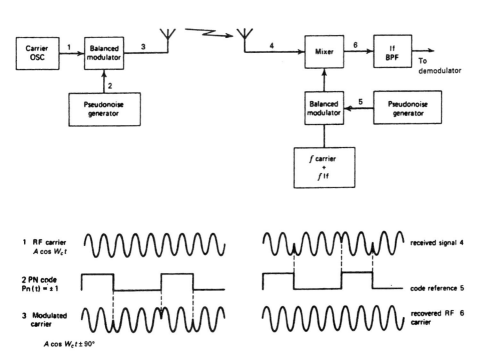

Figure 12.15. A direct sequence spread spectrum system showing waveforms. (From Ref. 18, Figure 2.3. Reprinted with permission.)

signal-to-noise ratio improvement when a spread signal is passed through a "processor." For instance, a certain processor had an input S/N of 12 dB and an output S/N of 20 dB; the processing gain, then, is 8 dB.

Processing gain is expressed by the following:

$$G_p = \frac{\text{spread bandwidth in Hz}}{\text{information bit rate}} \tag{12.16}$$

More commonly, processing gain is given in a dB value; then

$$G_{p(dB)} = 10 \log\left(\frac{\text{spread bandwidth in Hz}}{\text{information bit rate}}\right) \tag{12.17}$$

Example. A certain cellular system voice channel information rate is 9.6 kbps and the RF spread bandwidth is 9.6 MHz. What is the processing gain?

$$G_{p(dB)} = 10 \log(9.6 \times 10^6) - 10 \log 9600$$

$$= 69.8 - 39.8 \text{ (dB)}$$

$$= 30 \text{ dB}$$

It has been pointed out by Steele (Ref. 1) that the power control problem held back the implementation of CDMA for cellular application. If the standard deviation of the received power from each mobile at the base station is not controlled to an accuracy of approximately ± 1 dB relative to the target receive power, the number of users supported by the system can be significantly reduced. Other problems to be overcome were synchronization and sufficient codes available for a large number of mobile users (Ref. 1).

Qualcomm, a North American company, has a CDMA design that overcomes these problems and has fielded a cellular system based on CDMA. It operates at the top of the AMPS band using 1.23 MHz for each uplink and downlink. This is the equivalent of 41 AMPS channels (i.e., 30 kHz × 41 = 1.23 MHz) deriving up to 62 CDMA channels (plus one pilot channel and one synchronization channel) or some 50% capacity increase. The Qualcomm system also operates in the 1.7–1.8-GHz band (Ref. 1). EIA/TIA IS-95 (Ref. 37) is based on the Qualcomm system. Its processing gain, when using the 9600-bps information rate, is $1.23 \times 10^6/9600$ or about 21 dB.

12.6.4.1 Correlation—Key Concept in Direct Sequence Spread Spectrum.

In direct sequence (DS) spread spectrum systems, the chip rate is equivalent to the code generator clock rate. Simplistically, a chip can be considered an element of RF energy with a certain recognizable binary phase characteristic. A chip (or chips) is (are) a result of direct sequence spreading by biphase modulating a RF carrier. Being that each chip has a biphase modulated characteristic, we can identify each with a binary 1 or binary 0.

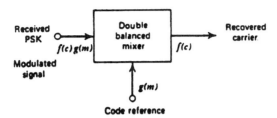

Figure 12.16. In-line correlator.

These chips derive from biphase modulating a carrier where the modulation is controlled by a pseudorandom sequence. If the sequence is long enough, without repeats, it is considered pseudonoise. The sequence is controlled by a key that is unique to our transmitter and its companion remote receiver. Of course the receiver must be time aligned and synchronized with its companion transmitter. A block diagram of this operation is shown in Figure 12.16. It is an in-line correlator.

Let us look at an information bit divided into seven chips coded by a PN sequence $- - - + - + +$ and shown in Figure 12.17a. Now replace the in-line correlator with a matched filter. In this case the matched filter is an electrical delay line tapped at delay intervals, which correspond to the chip time duration. Each tap in the delay line feeds into an arithmetic operator matched in sign to each chip in the coded sequence. If each delay line tap has the same sign (phase shift) as the chips in the sequence, we have a match. This is shown in Figure 12.17b. As shown here, the short sequence of seven chips is enhanced with desired signal seven times. This is the output of the modulo-2 adder, which has an output voltage seven times greater than the input voltage of one chip.

In Figure 12.17c we show the correlation process collapsing the spread signal spectrum to that of the original bit spectrum when the receiver reference signal, based on the same key as the transmitter, is synchronized with the arriving signal at the receiver.

Of overriding importance is that only the desired signal passes through the matched filter delay line (adder). Other users on the same frequency have a different key and do not correlate. These "other" signals are rejected. Likewise, interference from other sources is spread; there is no correlation and those signals also are rejected.

Direct sequence spread spectrum offers two other major advantages for the system designer. It is more forgiving in a multipath environment than conventional narrowband systems, and no intersymbol interference (ISI) will be generated if the coherent bandwidth is greater than the information symbol bandwidth.

If we use a RAKE receiver, which optimally combines the multipath components as part of the decision process, we do not lose the dispersed

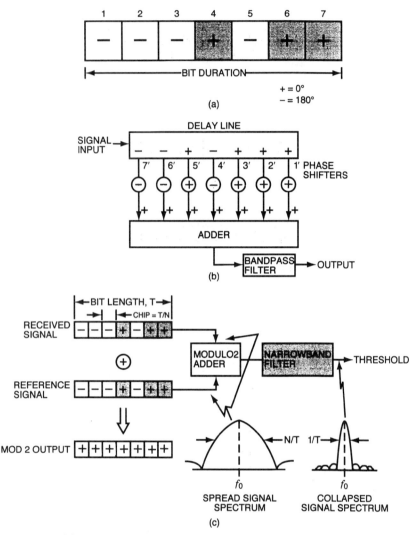

Figure 12.17. (a) An information chip divided into chips coded by a PN sequence. (b) Matched filter for 7-chip PN code. (c) The correlation process collapses the spread signal spectrum to that of the original bit spectrum.

multipath energy. Rather, the RAKE receiver turns it into useful energy to help in the decision process in conjunction with an appropriate combiner. Some texts call this implicit diversity or time diversity.

When sufficient spread bandwidth is provided (i.e., where the spread bandwidth is greater or much greater than the correlation bandwidth), we can get two or more independent frequency diversity paths as well using a RAKE receiver with an appropriate combiner such as a maximal ratio combiner. Figure 12.18 shows a RAKE receiver.

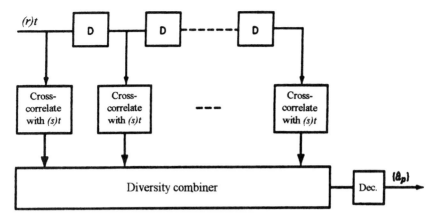

Figure 12.18. A typical RAKE receiver used for direct sequence spread spectrum reception.

12.7 FREQUENCY REUSE

Because of the limited bandwidth allocated in the 800-MHz band for cellular radio communications, frequency reuse is crucial for its successful operation. A certain level of interference has to be tolerated. The major source of interference is cochannel interference from a "nearby" cell using the same frequency group as the cell of interest. For the 30-kHz bandwidth AMPS system, Ref. 5 suggests that C/I be at least 18 dB. The primary isolation derives from the distance between the two cells with the same frequency group. In Figure 12.2 there is only one cell diameter for protection.

Refer to Figure 12.19 for the definition of the parameters R and D. D is the distance between cell centers of repeating frequency groups and R is the "radius" of a cell. We let

$$a = D/R$$

The D/R ratio is a basic frequency reuse planning parameter. If we keep the D/R ratio large enough, cochannel interference can be kept to an acceptable level.

Lee (Ref. 6) calls a the cochannel reduction factor and relates path loss from the interference source to R^{-4}.

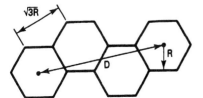

Figure 12.19. Definitions of R and D.

A typical cell in question has six cochannel interferers, one on each side of the hexagon. So there are six equidistant cochannel interference sources. The goal is $C/I \geq 18$ dB or a numeric of 63.1. So

$$C/I = C/\Sigma I = C/6I = R^{-4}/6D^{-4} = a^4/6 \geq 63.1$$

Then

$$a = 4.4$$

This means that D must be 4.4 times the value of R. If R is 6 mi (9.6 km) then $D = 4.4 \times 6 = 26.4$ mi (42.25 km).

Lee (Ref. 6) reports that cochannel interference can be reduced by other means such as directional antennas, tilted beam antennas, lowered antenna height, and an appropriately selected site.

If we consider a 26.4-mi path, what is the height of earth curvature at midpath? From Chapter 5, $h = 0.667(d/2)^2/1.33 = 87.3$ ft (26.9 m). Providing the cellular base station antennas are kept under 87 ft, the 40-dB/decade rule of Lee holds. Of course, we are trying to keep below line-of-sight conditions.

The total available (one-way) bandwidth is split up into N sets of channel groups. The channels are then allocated to cells, one channel set per cell on a regular pattern, which repeats to fill the number of cells required. As N increases, the distance between channel sets (D) increases, reducing the level of interference. As the number of channel sets (N) increases, the number of channels per cell decreases, reducing the system capacity. Selecting the optimum number of channel sets is a compromise between capacity and quality. Note that only certain values of N lead to regular repeat patterns without gaps. These are $N = 3, 4, 7, 9$, and 12, and then multiples thereof. Figure 12.20 shows a repeating 7 pattern for frequency reuse. This means that $N = 7$ or there are 7 different frequency sets for cell assignment.

Cell splitting can take place especially in urban areas in some point in time because the present cell structure cannot support the busy hour traffic load. Cell splitting, in effect, provides more frequency slots for a given area. Marcario (Ref. 5) reports that cells can be split as far down as a 1-km radius.

Cochannel interference tends to increase with cell splitting. Cell sectorization can cut down the interference level. Figure 12.21 shows a three- and six-sector plan. Sectorization breaks a cell into three or six parts each with a directional antenna. With a standard cell, cochannel interference enters from six directions. A six-sector plan can essentially reduce the interference to just one direction. A separate channel set is allocated to each sector.

The three-sector plan is often used with a seven-cell repeating pattern resulting in an overall requirement for 21 channel sets. The six-sector plan with its improved cochannel performance and the rejection of secondary interferers allows a four-cell repeat plan (Figure 12.2) to be employed. This results in an overall 24-channel set requirement. Sectorization requires a

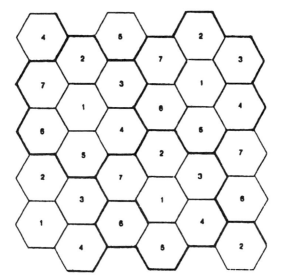

Figure 12.20. A cell layout based on $N = 7$.

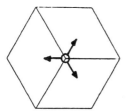 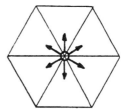

Figure 12.21. Breaking a cell up into three sectors (left) and six sectors (right).

larger number of channel sets and fewer channels per sector. Outwardly it appears that there is less capacity with this approach; however, the ability to use much smaller cells results in actually a much higher capacity operation (Ref. 5).

12.8   PAGING SYSTEMS*

12.8.1   What Are Paging Systems?

Paging is a one-way radio alerting system. The direction of transmission is from a fixed paging transmitter to an individual. It is a simple extension of the PSTN. Certainly paging can be classified as one of the first PCS (personal

*Sections 12.8.2–12.8.8 are based on Refs. 22 and 23.

communications system) operations. The paging receiver is a small box, usually carried on a person's belt. As a minimum, a pager alerts the user that someone wishes to reach him/her by telephone. The person so alerted goes to the nearest telephone and calls a prescribed number. Some pagers have a digital readout, which provides the calling number; whereas others give the number and a short message.

Most paging systems now operate in the VHF and UHF bands with a 3-kHz bandwidth. Transmitters have 1–5-watts output and paging receivers have sensitivities in the range of 10–100 μV/m (Ref. 5).

12.8.2   Radio-Frequency Bands for Pagers

All three ITU regions have some or all of the following frequency bands allocated to mobile services:

26.1–50 MHz	68–88 MHz
146–174 MHz	450–470 MHz
806–960 MHz	

12.8.3   Radio Propagation into Buildings

Measurement results submitted to the CCIR (Ref. 21) have indicated that frequencies in the range of 80–460 MHz are suitable for personal radio paging in urban areas with high building densities. It is possible that frequencies in the bands allocated around 900 MHz may also be suitable but that higher frequencies are less suitable. From measurements made in Japan, median values of propagation loss suffered by signals in the penetration of buildings (building penetration loss) have been derived. These results are summarized in Table 12.2.

12.8.4   Techniques Available for Multiple Transmitter Zones

To cover a service area effectively, it is often necessary to use a number of radio-paging transmitters. When the required coverage area is small, a single

TABLE 12.2   Propagation Loss Suffered by Signals in Penetrating Buildings

Frequency	150 MHz	250 MHz	400 MHz	800 MHz
Building penetration loss[a]	22 dB	18 dB	18 dB	17 dB[b]

[a]The loss is given as the ratio between the median value of the field strengths measured over the lower floors of buildings and the median value of the field strengths measured on the street outside. Similar measurements made in other countries confirm the general trend, but the values of building penetration loss vary about those shown. For instance, measurements made in the United Kingdom indicate that building penetration loss at 160 MHz is about 14 dB and about 12 dB at 460 MHz.
[b]Somewhat less accurate than the other results.
Source: Reference 23, Table 1, CCIR Rep. 499-5.

RF channel should be used so as to avoid the need for multichannel receivers. In these circumstances, the separate transmitters may operate sequentially or simultaneously. In the latter case, the technique of offsetting carrier frequencies, by an amount appropriate to the coding system employed, is often used. It is also necessary to compensate for the differences in the delay to the modulating signals arising from the characteristics of the individual landlines to the paging transmitters. One way to do this is to carry out synchronization of the code bits via the radio-paging channel. Of course, information will be required about the bit rates, which this synchronization method would permit. It is preferable that the frequency offset of the transmitter carrier frequencies in a binary digital radio-paging system be at least twice the signal fundamental frequency. It is also preferable that delay differences between modulation of the transmitters in a binary digital paging system should be less than a quarter of the duration of a bit if direct FSK, NRZ modulation is employed. For subcarrier systems, the corresponding limit should be less than one-eighth of a cycle of the subcarrier frequency (Ref. 23).

12.8.5 Paging Receivers

Built-in antennas can be designed for 150-MHz operation with reasonable efficiency. A typical radio-paging receiver antenna using a small ferrite rod exhibits a loss factor of about 16 dB relative to a half-wave dipole.

The majority of wide-area paging systems use some form of angle modulation.

Repeated transmission of calls can be used to improve the paging success rate of tone alert pagers. If p is the probability of receiving a single call, then $1 - (1 - p)^n$ is the probability of receiving a call transmitted n times, provided that the calls are uncorrelated. Correlations under Rayleigh fading conditions can largely be removed by spacing the calls more than 1 s apart. Longer delays between subsequent transmission (about 20 s) are required to improve the success rate under shadowing conditions.

Receivers with numeric or alphanumeric message displays can only take advantage of call repetitions if the supplementary messages are used to detect and correct errors.

12.8.6 System Capacity

The capacity of any paging system is affected by the following:

- Number and characteristics of the radio channels used
- Number of times each channel is reused within the system
- Actual paging location requirements of the individual users
- Peak information (address and message) requirement in a location(s)

- Tolerable paging delay
- Data transmission rate
- Code efficiency
- Method of using the total code capacity throughout the system (this may also affect the system's capabilities for roaming)
- Any inefficiency introduced by battery-saving provisions
- Possible telephone system input restrictions

12.8.7 Codes and Formats for Paging Systems

The U.S. paging system broadly used across the country is popularly called a Golay code referring to the Golay $(23, 12)$ cyclic code with two codewords representing the address. Messages are coded using a BCH $(15, 7)$ code. The code and format provide queueing and numeric/alphanumeric message flexibility and the ability to operate in a mixed-mode transmission with other formats. The single address capacity of this system is up to 400,000 with noncoded battery saving and up to 4,000,000 with coded preamble.

Japan uses a BCH $(31, 16)$ codeword with a Hamming distance of 7. The format gives approximately 65,000 addresses, 15 groups for battery economy, and a total cycle length of 4185 bits. Each group contains 8 address codewords headed by a 31-bit synchronizing and group-indicating signal.

The U.K. paging system employs a BCH $(31, 21)$ code plus even parity codeword with a Hamming distance of 6. The code format can handle over 8 million addresses and can be expanded. It can also handle any type of data message such as hexadecimal and CCITT Alphabet No. 5. It is designed to share a channel with other codes and to permit mixed simultaneous and sequential multitransmitter operation at the normal 512-bps transmission rate. This code is sometimes referred to as POCSAG and has been adopted as CCIR Radio-Paging Code No. 1 (RPC1).

12.8.8 Considerations for Selecting Codes and Formats

- Number of subscribers to be served
- Number of addresses assigned to each subscriber
- Expected calling rate including that from any included message facility
- Zoning arrangement
- Data transmission rates possible over the linking network and radio channel(s), taking into account the propagation factors of the radio frequencies to be used
- Type of service: vehicular or personal, urban or rural

Once the data are provided from the preceding listing of topics, codes may be compared by their characteristics with respect to:

- Code address capacity
- Number of bits per address
- Codeword Hamming distance
- Code efficiency, such as number of information bits compared to the total number of bits per codeword
- Error-detecting capability; error-correcting capability
- Message capability and length
- Battery-saving capability
- Ability to share a channel with other codes
- Capability of meeting the needs of paging systems, which vary with respect to size and transmission mode (e.g., simultaneous versus sequential)

12.9 PERSONAL COMMUNICATIONS SERVICES (PCS)

12.9.1 Defining Personal Communications

Personal communications services (PCS) are wireless. This simply means that they are radio based. The user requires no *tether*. The conventional telephone is connected by a wire pair through to the local serving switch. The wire pair is a tether. We can only walk as far with that telephone handset as the "tether" allows.

Both of the systems we have dealt with in the previous sections of this chapter can be classified as PCS. Cellular radio, particularly with the handheld terminal, gives the user tetherless telephone communication. Paging systems provide the mobile/ambulatory user a means of being alerted that someone wishes to talk to him/her on the telephone or of receiving a short message.

The cordless telephone is certainly another example, which has extremely wide use around the world. By the end of 1994, it was estimated that there were 60 million cordless telephones in use in the United States (Ref. 30). We will provide a brief overview of cordless telephone developments in the following.

New applications are either on the horizon or going through field tests (1995). One that seems to offer great promise in the office environment is the wireless PABX. It will almost eliminate the telecommunication manager's responsibilities with office rearrangements. Another is the wireless LAN (WLAN).

Developments are expected such that PCS cannot only provide voice communications but facsimile, data, messaging, and possibly video. GSM provides all but video. Cellular digital packet data (CDPD) will permit data services over the cellular system in North America.

Donald Cox (Ref. 26) breaks PCS down into what he calls "high tier" and "low tier." Cellular radio systems are regarded as high-tier PCS, particularly when implemented in the new 1.9-GHz PCS frequency band. Cordless telephones are classified as low-tier.

Table 12.3 summarizes some of the more prevalent wireless technologies.

12.9.2 Narrowband Microcell Propagation at PCS Distances

The microcells discussed here have a radial range of <1 km. One phenomenon is the Fresnel break point, which is illustrated in Figure 12.22. Signal level varies with distance R as A/R^n. For distances greater than 1 km, n is typically between 3.5 and 4. The parameter A describes the effects of environmental features in a highly averaged manner (Ref. 11).

Typical PCS radio paths can be of a LOS nature, particularly near the fixed transmitter where $n = 2$. Such paths may be down the street from the transmitter. The other types of paths are shadowed paths. One type of shadowed path is found in highly urbanized settings, where the signal may be reflected off high-rise buildings. Another is found in more suburban areas, where buildings are often just two stories high.

When a signal at 800 MHz is plotted versus R on a logarithmic scale, as in Figure 12.22, there are distinctly different slopes before and after the Fresnel break point. We call the break distance (from the transmit antenna) R_B. This is the point for which the Fresnel ellipse about the direct ray just touches the ground. This model is illustrated in Figure 12.23. The distance R_B is approximated by

$$R_B = 4h_1h_2/\lambda \tag{12.18}$$

For $R < R_B$, N is less than 2, and for $R > R_B$, n approaches 4.

It was found that on non-LOS paths in an urban environment with low base station antennas and with users at street level, propagation takes place down streets and around corners rather than over buildings. For these non-LOS paths the signal must turn corners by multiple reflections and diffraction at vertical edges of buildings. Field tests reveal that signal level decreases by about 20 dB when turning a corner.

In the case of propagation inside buildings where the transmitter and receiver are on the same floor, the key factor is the clearance height between the average tops of furniture and the ceiling.

Bertoni et al. (Ref. 11) call this clearance W. Here building construction consists of drop ceilings of acoustical material supported by metal frames. That space between the drop ceiling and the floor above contains light fixtures, ventilation ducts, pipes, support beams, and so on. Because the acoustical material has a low dielectric constant, the rays incident on the

TABLE 12.3 Wireless PCS Technologies

	High-Power Systems				Low-Power Systems			
	Digital Cellular (HighTier PCS)				Low-Tier PCS		Digital Cordless	
System	IS-54	IS-95(DS)	GSM	DCS-1800	WACS/PACS	Handi-Phone	DECT	CT-2
Multiple access	TDMA/FDMA	CDMA/FDMA	TDMA/FDMA	TDMA/FDMA	TDMA/FDMA	TDMA/FDMA	TDMA/FDMA	FDMA
Frequency band (MHz) Uplink (MHz) Downlink (MHz)	869–894 824–849 (USA)	869–894 824–849 (USA)	935–960 890–915 (Europe)	1710–1785 1805–1880 (UK)	Emerging Technology [a] (USA)	1895–1907 (Japan)	1880–1900 (Europe)	864–868 (Europe and Asia)
RF channel spacing Downlink (kHz) Uplink (kHz)	30 30	1250 1250	200 200	200 200	300 300	300	1728	100
Modulation	π/4 DQPSK	BPSK/QPSK	GMSK	GMSK	π/4 QPSK	π/4 DQPSK	GFSK	GFSK
Portable txmit power, max./avg.	600 mW/ 200 mW	600 mW	1 W/ 125 mW	1 W/ 125 mW	200 mW/ 25 mW	80 mW/ 10 mW	250 mW/ 10 mW	10 mW/ 6 mW
Speech coding	VSELP	QCELP	RPE-LTP	RPE-LTP	ADPCM	ADPCM	ADPCM	ADPCM
Speech rate (kbps)	7.95	8 (var.)	13	13	32/16/8	32	32	32
Speech channel/RF channel	3	—	8	8	8/16/32	4	12	1
Channel bit rate (kbps) Uplink (kbps) Downlink (kbps)	48.6 48.6		270.833 270.833	270.833 270.833	384 384	384	1152	72
Channel coding	1/2 rate conv.	1/2 rate fwd 1/3 rate rev.	1/2 rate conv.	1/2 rate conv.	CRC	CRC	CRC (control)	None
Frame (ms)	40	20	4.615	4.615	2.5	5	10	2

[a]Spectrum is 1.85–2.2 GHz allocated by the FCC for emerging technologies; DS is direct sequence.

Source: Reference 26, Table 1, Reprinted with permission of the IEEE.

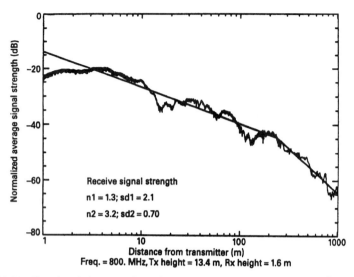

Figure 12.22. Signal variation on a line-of-sight path in a rural environment. (From Ref. 11, Figure 3.)

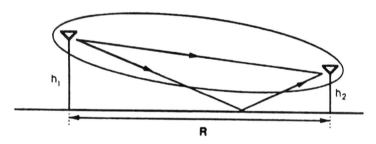

Figure 12.23. Direct and ground reflected rays, showing the Fresnel ellipse about the direct ray. (From Ref. 11, Figure 18.)

ceiling penetrate the material and are strongly scattered by the irregular structure, rather than undergoing specular reflection. Floor-mounted building furnishings such as desks, cubicle partitions, filing cabinets, and workbenches scatter the rays and prevent them from reaching the floor, except in hallways. Thus it is concluded that propagation takes place in the clear space, W.

Figure 12.24 shows a model of a typical floor layout. When both the transmitter and receiver are located in the clear space, path loss can be related to the Fresnel ellipse. If the Fresnel ellipse associated with the path

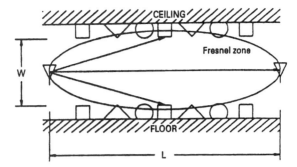

Figure 12.24. Fresnel zone for propagation between transmitter and receiver in clear space between building furnishings and ceiling fixtures. (From Ref. 11, Figure 35.)

lies entirely in the clear space, the path loss has LOS properties $(1/L^2)$. Now as the separation between the transmitter and receiver increases, the Fresnel ellipse grows in size so that scatterers lie within it. This is shown in Figure 12.25. Now the path loss is greater than free space.

Bertoni et al. (Ref. 11) report on one measurement program where the scatterers have been simulated using absorbing screens. It was recognized that path loss will be highly dependent on nearby scattering objects. Figure 12.25 was developed from this program. The path loss in excess of free space calculated at 900 and 1800 MHz for $W = 1.5$ is plotted in Figure 12.25 as a function of path length L. The figure shows that the excess path loss (over LOS) is small at each frequency out to distances of about 20 and 40 m, respectively, where it increases dramatically.

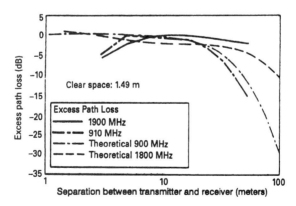

Figure 12.25. Measured and calculated excess path loss at 900 and 1800 MHz for a large office building having head-high cubical partitions, but no floor-to-ceiling partitions. (From Ref. 11, Figure 36.)

Propagation between floors of a modern office building can be very complex. If the floors are constructed of reinforced concrete or prefabricated concrete, transmission loss can be 10 dB or more. Floors constructed of concrete poured over steel panels show much greater loss. In this case (Ref. 11), signals may propagate over other paths involving diffraction rather than transmission through the floors. For instance, signals can exit the building through windows and reenter on higher floors by diffraction mechanisms along the face of the building.

12.10   CORDLESS TELEPHONE TECHNOLOGY

12.10.1   Background

Cordless telephones began to become widely used in North America around 1981. By late 1994, there were over 60 million such units in use and an estimated sales rate of 5 million per year (Ref. 30).

12.10.2   North American Cordless Telephones

These telephones operate using FM with a 20-kHz bandwidth. Their ERP is on the order of 20 μW and they operate on ten frequency pairs in the bands 46.6–47.0 MHz (base transmit) and 49.6–50.0 MHz (handset transmit). Reference 30 suggests that this analog technology will continue on for some time into the future because of the telephone's low cost. It is expected that the FCC will make another 15 frequency pairs available in 1997 near 44 and 49 MHz.

12.10.3   European Cordless Telephones

The first-generation European cordless telephone provided for eight channel pairs near 1.7 MHz (base unit transmit) and 47.5 MHz (handset transmit). Most of these units could only access one or two channel pairs. Some called this "standard" CT0.

This was followed by another analog cordless telephone based on a standard known as CEPT/CT1. CT1 has forty 25-kHz duplex channel pairs operating in the bands 914–915 MHz and 959–960 MHz. There is also a CT1+ in the bands 885–887 MHz and 930–932 MHz, which do not overlap the GSM allocation. CT1 is called a coexistence standard (not a compatible standard) such that cordless telephones from different manufacturers do not interoperate. The present embedded base is about 9 million units and some 3 million units are expected to be sold in 1997.

Two digital standards have evolved in Europe: the CT2 Common Air Interface and DECT (digital European cordless telephone). In both standards, speech coding uses ADPCM (adaptive differential PCM). The ADPCM speech and control data are modulated onto a carrier at a rate of 72 kbps using Gaussian-filtered FSK (GFSK) and are transmitted in 2-ms frames. One base-to-handset burst and one handset-to-base burst are included in each frame.

The frequency allocation for CT2 consists of 40 FDMA channels with 100-kHz spacing in the band 864–868 MHz. The maximum transmit power is 10 mW, and a two-level power control supports prevention of desensitization of base station receivers. As a by-product, it contributes to frequency reuse. CT2 has a call reestablishment procedure on another frequency after 3 seconds of unsuccessful attempts on the initial frequency. This gives a certain robustness to the system when in an interference environment. CT2 supports up to 2400 bps of data transmission and higher rates when accessing the 32 kbps underlying bearer channels.

CT2 also is used for wireless pay telephones. When in this service it is called *Telepoint*. CT2 seems to have more penetration in Asia than in Europe.

Canada has its own version of CT2 called CT2+. It is more oriented toward the mobile environment, providing several of the missing mobility functions in CT2. For example, with CT2+, 5 of the 40 carriers are reserved for signaling, where each carrier provides 12 common channel signaling channels (CSCs) using TDMA. These channels support location registration, updating, and paging and enable Telepoint subscribers to receive calls. The CT2+ band is 944–948 MHz.

DECT takes on more of the cellular flavor than CT2. It uses a picocell concept and TDMA with handover, location registration, and paging. It can be used for Telepoint, radio local loop (RLL), and cordless PABX besides conventional cordless telephony. Its speech coding is similar to CT2, namely, ADPCM. For its initial implementation, 10 carriers have been assigned in the band 1880–1900 MHz.

There are many areas where DECT will suffer interference in the assigned band, particularly from "foreign" mobiles. To help alleviate this problem, DECT uses two strategies: *interference avoidance* and *interference confinement*. The avoidance technique avoids time/frequency slots with a significant level of interference by handover to another slot at the same or another base station. This is very attractive for the uncoordinated operation of base stations because in many interference situations there is no other way around a situation but to change in both the time and frequency domains. The "confinement" concept involves the concentration of interference to a small time–frequency element even at the expense of some system robustness.

Base stations must be synchronized in the DECT system. A control channel carries information about access rights, base station capabilities, and paging messages. The DECT transmission rate is 1152 kbps. As a result of this and a relatively wide bandwidth, either equalization or antenna diversity is typically needed for using DECT in the more dispersive microcells.

Japan has developed the personal handyphone system (PHS). Its frequency allocation is 77 channels, 300 kHz in width, in the band 1895–1918.1 MHz. The upper-half of the band, 1906.1–1918.1 MHz (40 frequencies), is used for public systems. The lower-half of the band, 1895–1906.1 MHz, is reserved for home/office operations. An operational channel is autonomously selected by measuring the field strength and selecting a channel on which it meets certain level requirements. In other words, fully dynamic channel assignment is used. The modulation is $\pi/4$ DQPSK; average transmit power at the handset is 10 mW (80-mW peak power) and no greater than 500 mW (4-W peak power) for the cell site. The PHS frame duration is 5 ms. Its voice coding technique is 32-kbps ADPCM (Ref. 8).

In the United States, digital PCS was based on the wireless access communication system (WACS), which has been modified to an industry standard called PACS (personal access communications services). It is intended for the licensed portion of the new 2-GHz spectrum. Its modulation is $\pi/4$ QPSK with coherent detection. Base stations are envisioned as shoebox-size enclosures mounted on telephone poles, separated by some 600 m. WACS/PACS has an air interface similar to other digital cordless interfaces, except it uses frequency division duplex (FDD) rather than time division duplex (TDD) and more effort has gone into optimizing frequency reuse and the link budget. It has two-branch polarization diversity at both the handset and base station with feedback. This gives it an advantage approaching four-branch receiver diversity. The PACS version has eight time slots and a corresponding reduction in channel bit rate and a slight increase in frame duration over its predecessor, WACS.

Table 12.4 summarizes these several types of digital cordless telephones.

TABLE 12.4   Digital Cordless Telephone Interface Summary

	CT2 CT2+	DECT	PHS	PACS
Region	Europe Canada	Europe	Japan	United States
Duplexing	TDD	TDD	TDD	FDD
Frequency band (MHz)	864–868 944–948	1800–1900	1895–1918	1850–1910/1930–1990 [a]
Carrier spacing (kHz)	100	1728	300	300/300
Number of carriers	40	10	77	16 pairs/10 MHz
Bearer channels/carrier	1	12	4	8/pair
Channel bit rate (kbps)	72	1152	384	384
Modulation	GFSK	GFSK	$\pi/4$ DQPSK	$\pi/4$ QPSK
Speech coding	32 kbps	32 kbps	32 kbps	32 kbps
Average handset transmit power (mW)	5	10	10	25
Peak handset transmit power (mW)	10	250	80	200
Frame duration (ms)	2	10	5	2.5

[a]General allocation to PCS; licensees may use PACS.
Source: Reference 30, Table 2.

12.11   WIRELESS LANs

Wireless LANs, much as their wired counterparts, operate in excess of 1 Mbps. Signal coverage runs from 50 to less than 1000 ft. The transmission medium can be radiated light (around 800 or 900 nm) or radio frequency, unlicensed. Several of these latter systems use spread spectrum with transmitter outputs of 1 watt or less.

WLANs (wireless LANs) using radiated light do not require FCC licensing, a distinct advantage. They are immune to RF interference but are limited in range by office open spaces because their light signals cannot penetrate walls. Shadowing can also be a problem.

One type of radiated-light WLAN uses a directed light beam. These are best suited for fixed terminal installations because the transmitter beams and receivers must be carefully aligned. The advantages for directed beam systems is improved S/N and less problems with multipath. One such system is fully compliant with IEEE 802.5 token ring operation offering 4- and 16-Mbps transmission rates.

Spread spectrum WLANs use the 900-MHz, 2- and 5-GHz industrial, scientific, and medical (ISM) bands. Both direct sequence and frequency hop operation can be used. Directional antennas at the higher frequencies provide considerably longer range than radiated light systems, up to several miles or more. No FCC license is required. A principal user of these higher-frequency bands is microwave ovens. CSMA and CSMA/CD (IEEE 802.3) protocols are often employed.

There is also a standard microwave WLAN (nonspread spectrum) that operates in the band 18–19 GHz. FCC licensing is required.

Building wall penetration loss is high. The basic application is for office open spaces.

12.12   FUTURE PUBLIC LAND MOBILE TELECOMMUNICATIONS SYSTEM (FPLMTS)

12.12.1   Introduction

FPLMTS is a cellular/PCS concept proposed by the ITU-R Organization (previously CCIR). It proposes the bands 1885–2025 MHz and 2210–2200 MHz for FPLMTS service. The system includes both a terrestrial and a satellite component. Figure 12.26 shows a FPLMTS scenario for PCS, terrestrial component; Figure 12.27 shows the satellite component.

12.12.2   Traffic Estimates

CCIR states that the maximum demand for PCS is in large cities where different categories of traffic will traverse the system, such as that generated

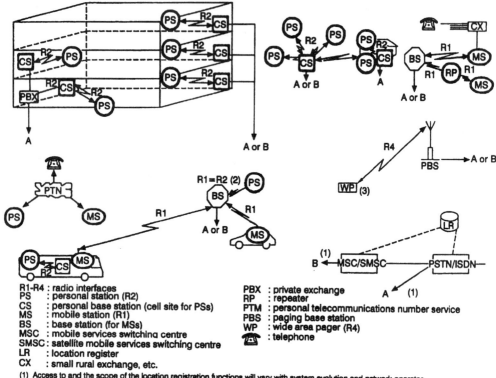

R1-R4 : radio interfaces
PS : personal station (R2)
CS : personal base station (cell site for PSs)
MS : mobile station (R1)
BS : base station (for MSs)
MSC : mobile services switching centre
SMSC: satellite mobile services switching centre
LR : location register
CX : small rural exchange, etc.

PBX : private exchange
RP : repeater
PTM : personal telecommunications number service
PBS : paging base station
WP : wide area pager (R4)
☎ : telephone

(1) Access to and the scope of the location registration functions will vary with system evolution and network operator requirements. This is reflected in network interfaces A and B.

(2) In some implementation scenarios R1 may equal R2.

(3) Can be co-located/integrated with PS.

Figure 12.26. Scenario for PCS within FPLMTS, terrestrial component. (From Ref. 35, Figure 1, CCIR Rec. 687-1.)

by mobile stations (MS) and vehicle-mounted or portable stations (PS), outdoor and indoor.

The worst-case scenario is during a vehicular traffic jam, where the number of vehicles per kilometer along a street is around 600 if they are stationary or 350 if they are moving slowly. Assuming a mean value of 400 vehicles per kilometer of street length, with 50% of these vehicles being equipped with mobile stations, each generating 0.1 E, the traffic density will be 20 E/km of street length leading to 300 E/km^2 based on typical urban street density. Adding about the same amount of traffic for portable mobile stations carried by pedestrians, the combined traffic for MS would be around 500 E/km^2 in the more dense city areas.

The peak traffic for personal stations (PS) is estimated at 1500 E/km^2, assuming 3000 pedestrians per kilometer of street length, 80% penetration(s) of the personal station, and 0.04 E/station. It is estimated that the peak to

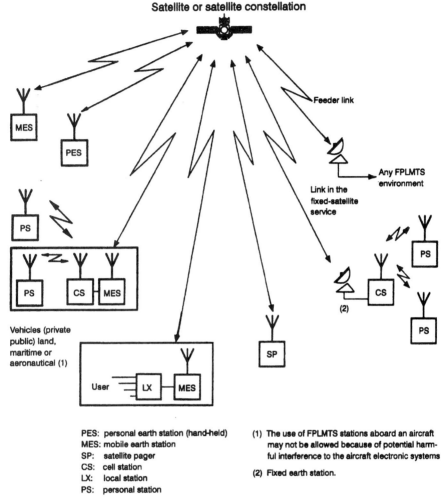

Figure 12.27. Configuration examples for the satellite component of FPLMTS. (From Ref. 35, Figure 2, CCIR Rec. 687-1.)

mean ratio of traffic for pedestrians on busy streets of large cities has a value of around 3.

For PS indoors the traffic may increase by a factor of 10 or more in a multifloor building. CCIR estimates one station every 10×10 m^2 active floor area with a traffic intensity of 0.2 E per station. This corresponds to 20,000 E/km^2/floor.

Nonvoice Traffic. Nonvoice services will constitute an increasing proportion of total traffic. Some data traffic needs more bandwidth than a full-duplex voice channel, increasing spectrum estimates. If nonvoice traffic is handled by queueing procedures, less bandwidth is required, improving spectrum utilization.

One example is facsimile using circuit-switched services, which is mainly relevant for vehicle-mounted MS. If we assume 3000 terminals/km^2, of which 15% are equipped with facsimile terminals, and a call-holding time of 6 min/h, per terminal, the estimated traffic amounts to 45 E/km^2.

CCIR believes that interactive data services are likely to employ packet transmission. The assumption used here is 15 s/h (15 seconds per hour) for a hand-portable (10 pages per hour with 8 kbps per page at a data rate of 4800 bps) and 30 s/h for a vehicle-mobile (4.5 and 9 mE, respectively). If we assume 5000 terminals per km^2 (i.e., 3000 vehicular and 2000 hand-portable), the estimated traffic amounts to 37 E/km^2.

PCS Outdoors. Here the assumption is a cumulative channel occupancy of 5 s/h (10 pages per hour with 8 kbps per page at a data rate of 16 kbps), which corresponds to 1.4 mE/station.

Assuming 2400 stations/km of street length (37,500 stations/km^2), as in the case for voice, the amount of traffic would be 50 E/km^2. To account for other data services, this estimation is increased by a factor of 3 to 150 E/km^2. The nonvoice traffic is then 10% of the voice traffic.

Traffic generated by circuit-switched services (e.g., facsimile) is considered insignificant. Thus only short, interactive data communications are considered in this scenario.

PCS Indoors. The traffic forecast for facsimile is based on 25% of stations having a fax capability and 6 minutes call-holding time per hour, per fax terminal: thus the estimated traffic is 25 mE per station (i.e., one-eighth of the voice traffic, or 2500 E/km^2).

For interactive applications where we assume all stations are using the same application and with the assumption of 20 interactive sessions per hour, with a cumulative channel occupancy of approximately 2 s per station, the estimate is 0.01 E per station, which is equivalent to 1000 E/km^2.

Taking into account contention due to packet transmission, the Erlang value is doubled to 2000 E/km^2.

The grand total for nonvoice service indoors, taking into account batch data application and database retrievals with 10% overhead, a total of 5000 E/km^2 is assumed.

12.12.3   Estimates of Spectrum Requirements

Using the traffic estimates developed in Section 12.12.2, the minimum spectrum bandwidth required for voice and nonvoice services is approximately 230 MHz. The key parameters on which these estimates are based are

TABLE 12.5 General Characteristics of PCS (High-Density Area) Voice Service Traffic Demands and Spectrum Requirements

Specifications	MS R1	PS R2 Outdoor	PS R2 Indoor
Radio coverage (%)	90	>90	99
Base station antenna height (m)	50	10	<3[a]
Base station installed: indoor/outdoor	No/Yes	Yes/Yes[b]	Yes/Yes[b]
Traffic density (E/km^2)	500	1500	20,000[a]
Cell area (km^2)	0.94	0.016	0.0006
Blocking probability (%)	2	1	0.5
Cluster size (cell sites × sectors/site)	9	16	21 (3 floors)
Duplex bandwidth per channel (kHz)	25	50	50
Traffic per cell (E)	470	24	12
Number of channels per cell	493	34	23
Bandwidth (MHz)	111	27	24
Station[c]	Vehicle mounted or portable		
Volume (cm^3)		<200	<220
Weight (g)		<200	<200
Highest power	5 W	50 mW	10 mW

[a]Per floor.
[b]Usual case.
[c]A range of terminal types will be available to suit operational and user requirements.
Source: Reference 35, Table 3, CCIR Rec. 687-1.

given in Tables 12.5 and 12.6. The total requirement at the radio interface for R1 (mobile station—cellular) is 167 MHz and 60 MHz for R2 (PCS).

An important assumption used in Tables 12.5 and 12.6 was the speech coding rate: 8 kbps was assumed for the cellular environment (mobile station). Lower rates are available but the quality of transmission and delay suffer accordingly, as reported by CCIR. For the inexpensive personal

TABLE 12.6 Spectrum Estimation for Nonvoice Services

	MS Outdoor Interface R1		PS Outdoor Interface R2		PS Indoor Interface R2	
	Circuit Switched	Packet Switched	Circuit Switched	Packet Switched	Circuit Switched	Circuit Switched
Traffic density (E/km^2)	45	37	[a]	150	2000[b]	2500[b]
Duplex bandwidth per Channel (kHz)	100	50	50	50	50	50
Bandwidth (MHz)	56		3		6	

[a]Insignificant.
[b]Per floor.
Source: Reference 35, Table 4, CCIR Rec. 687-1.

stations (PCS), higher bit rates have been assumed ranging from 32 kbps down to 10 kbps.

The choice of network access scheme does not substantially affect the spectrum estimates given above.

The data provided in Tables 12.5 and 12.6 are based on densely populated metropolitan areas. CCIR reports that the following items were recognized but not considered in the spectrum estimates provided in the tables:

- Additional signaling traffic for system operation is expected to be significant in FPLMTS due to system complexity and quality objectives.
- Road traffic management and control applications may generate additional nonvoice traffic.
- Sharing of spectrum between several operators may result in less efficient spectrum use.

Taking these items into account, the 230-MHz estimate for FPLMTS may turn out as a lower bound.

12.12.4 Sharing Considerations

The goal is that one service does not interfere with another in excess of the CCIR limits. The basic parameters for sharing with FPLMTS are pfd* (power/km^2/Hz) and the minimum carrier to total noise plus interference that is required. The pfd is derived from the number of terminals per square kilometer and the power for each category of station. Table 12.7 provides some estimates for an urban area.

The level of interference to FPLMTS that can be tolerated has been estimated using a link budget, which shows that PCS systems are expected to be interference-limited rather than noise-limited. If we assume an allocation of 10% of the total interference budget to external interference sources, then we derive a corresponding aggregate interference power level of -117 dBm for indoor PCS and -119 dBM for outdoor PCS. These levels are the maximum permissible without significantly degrading service quality.

12.12.5 Sharing Between FPLMTS and Other Services

Sharing may not be feasible between FPLMTS and other services such as the fixed service, mobile-satellite services, and certain satellite TT&C[†] services. Operational sharing of an allocation common to FPLMTS and other services requires suitable geographic separation between services, or where neither

*pfd = power flux density.
[†]TT&C = tracking, telemetry, and command.

TABLE 12.7   Power Flux Densities for FPLMTS in an Urban Area

Stations	Base and Mobile	Personal
EIRP	10 W (base)	3 mW (indoor)
	1 W (mobile)	20 mW (outdoor)
Traffic density	582 E/km^2	25,000 E/km^2 (indoor)[a]
		1,650 E/km^2 (outdoor)
Assumed bandwidth	167 MHz	60 MHz
Estimated pfd	38 μW/km^2/Hz	1.5 μW/km^2/Hz
	-68 dB(W/m^2/4 kHz)	-82 dB(W/m^2/4 kHz)

[a]This takes into account the vertical frequency reuse of FPLMTS in buildings.
Source: Reference 35, Table 5, CCIR Rec. 687-1.

service requires the total allocated band. If FPLMTS uses adaptive channel assignment, sharing will be greatly facilitated and will simplify the introduction of FPLMTS into bands currently used by other services.

Sharing is not feasible between R1 and R2 interfaces of FPLMTS and the SRS, SOS, and EESS satellite TT & C services in the 2025–2110-MHz and the 2200–2290-MHz bands.

12.13   MOBILE SATELLITE COMMUNICATIONS

12.13.1   Background and Scope

In our earlier discussions on cellular mobile radio and PCS, there seemed to be no clear demarcation where one ended and the other began. Cellular hand terminals certainly are used inside all types of buildings with some fair success—granted some of the connections are marginal. Often we speak of PCS in the bigger picture of cellular mobile radio. Even CCIR (ITU-R Organization) describes FPLMTS as an integrated system where there is no dividing line between PCS and cellular.

In this section we review satellite services that provide PCS and cellular mobile radio on a worldwide basis. We present a short overview and then discuss Motorola's IRIDIUM system in some detail.

12.13.2   Overview of Satellite Mobile Services

12.13.2.1   Existing Systems. INMARSAT [International Maritime Satellite (consortium)] has been providing worldwide full-duplex voice, data, and record traffic service with ships since the mid-1970s. It extended its service to a land-mobile market and to aircraft. At present there are over 25,000 INMARSAT terminals. About 30% of these are land-transportable. INMARSAT satellites are in geostationary earth orbit (GEO).

INMARSAT-M systems provide service to ships and mobile land terminals. By the year 2005 some 600,000 INMARSAT-M terminals are expected to be in operation. The uplinks and downlinks for mobile terminals are in the 1.5- and 1.6-GHz bands. Services are low bit rate voice (5–8 kbps) and data operations. INMARSAT-P is a program specifically directed to the PCS market and is expected to be operational around 1998 (Ref. 9).

American Mobile Satellite Corporation (AMSC) is also providing voice, data, and facsimile service to the Americas, targeting customers in regions not served by conventional terrestrial cellular systems and terrestrial cellular subscribers who have problems "roaming." These satellites are in GEO and provide uplinks and downlinks in the L-band, much like INMARSAT.

12.13.2.2 Post-1997 Systems.
Several satellite communication system companies are launching satellites for low rate services in the VHF band (148–149 MHz uplink and 137–138 MHz downlink). These systems use satellites in low earth orbit (LEO). All provide two-way message services and do not offer voice service. Some orbits are polar and some are inclined.

TRW expects to have its ODYSSEY system in operation by 1998. The satellites will be in LEOs and provide voice, data, facsimile, and paging services worldwide. Serving mobile platforms, uplinks will be in the 1610.5–1626.5-MHz band and the companion downlinks will be in the band 2483.5–2500-MHz band using channelized CDMA access. The orbits will be MEOs (medium earth orbits, about 10,400 km). Twelve satellites will be in three orbital planes.

Loral and Qualcomm have joined forces to offer a LEO system called GLOBALSTAR, consisting of a network of 48 satellites in eight orbital planes. GLOBALSTAR specifically is targeting the hand-held terminal market for interconnection with the PSTN. Access will be by channelized CDMA, and services offered are voice, data, facsimile, and position location. Initial operation is expected in 1997 (Ref. 9).

Constellation Communications of Herndon, Virginia (U.S.A.), plans a large LEO system called ARIES, which will provide users with voice, data, facsimile, and position location services. These LEO satellites will be in four orbital planes at an average altitude of 1020 km. CDMA access is envisioned using a 16.5-MHz segment of L-band around 1.6 GHz for uplinks and another, similar segment for downlinks around 2.5 GHz. Ten or eleven fixed earth stations are planned for connectivity to the PSTN. These facilities will use standard 30/20-GHz uplinks and downlinks. The system became operational in 1996.

ELLIPSO is another planned system employing 15 satellites in elliptical inclined orbits in three planes and up to 9 satellites in equatorial orbit at maximum altitudes of about 7800 km. L-band connectivity is planned for the mobile user and C-band for the feeder uplinks and downlinks. The services offered to customers will be voice, data, facsimile, and paging. Access will be via channelized CDMA now operational.

12.13.3   System Trends

The low earth orbit offers a number of advantages over the geostationary earth orbit, and at least one serious disadvantage.

Delay. One-way delay to a GEO satellite is budgeted at 125 ms; one-way up and down is double this value, or 250 ms. Round-trip delay is about 0.5 s. Delay to a typical LEO satellite is 2.67 ms and round-trip delay is 4×2.67 ms or about 10.66 ms. Calls to/from mobile users of such systems may be relayed still again by conventional satellite services. Data services do not have to be so restricted on the use of "hand-shakes" and stop-and-wait automatic repeat request (ARQ) as with similar services via a GEO system.

Higher Elevation Angles and "Full Earth Coverage". The GEO satellite provides no coverage above about 80° latitude and gives low-angle coverage of many of the world's great population centers because of their comparatively high latitude. Typically, cities in Europe and Canada face this dilemma. LEO satellites, depending on orbital plane spacing, can all provide elevation angles $>40°$. This is particularly attractive in urban areas with tall buildings. Coverage would only be available on the south side of such buildings in the Northern Hemisphere with a clear shot to the horizon. Properly designed LEO systems will not have such drawbacks. Coverage will be available at any orientation.

Tracking, a Disadvantage of LEOs and MEO* Satellites. At L-band quasi-omnidirectional antennas for the mobile user are fairly easy to design and produce. Although such antennas display only modest gain of several dB, links to a LEO satellite can be easily closed with hand-held terminals. However, large feeder, fixed earth terminals will require a good tracking capability as LEO satellites pass overhead. Handoff is also required as a LEO satellite disappears over the horizon and another satellite just appears over the opposite horizon. The handoff should be seamless.

The quasi-omnidirectional user terminal antennas will not require tracking, and the handoff should not be noticeable to the mobile user.

12.13.4   IRIDIUM†

12.13.4.1   Overview. The IRIDIUM system is a LEO satellite system that will provide cellular/personal communication services worldwide. It is being developed and will be operated by the Motorola Satellite Communications Division for the system owner, Iridium, Inc. There is also participation by a

---

*MEO—medium earth orbit (i.e., 5000–13,000 km).
†IRIDIUM is a registered trademark and service mark of Iridium, Inc.

number of other large North American, Asian, and European companies such as Sprint, STET, BCE, and the Raytheon Company.

Subscribers to this system will use portable and mobile terminals with low-profile antennas to reach a constellation of 66 satellites. These satellites will be interconnected by crosslinks as they circle the earth in highly inclined polar orbits about 485 statute miles above the earth. The deployment of the satellites will be in six orbital planes about $31.6°$ apart. However, planes 1 and 6 will be only $22°$ apart. The delay to a satellite varies from 2.5 ms to 11 ms versus about 125 ms to a GEO satellite.

The IRIDIUM system offers a wide range of mobile radio services including voice, data, and facsimile. The subscriber communication services are interconnected with the PSTN through regional gateways. The satellite–subscriber links are at L-band (1.6 GHz); gateway connectivity is via feeder links in the 30/20-GHz band, and the satellite crosslinks operate in the 23-GHz band. Figure 12.28 gives a pictorial overview of the IRIDIUM system.

12.13.4.2 Space Segment.
The space segment includes the 66 satellites in low earth orbit (LEO), which are networked together as a switched-digital communication system utilizing cellular techniques to achieve maximum capability of frequency reuse. The subscriber uplinks and downlinks occupy

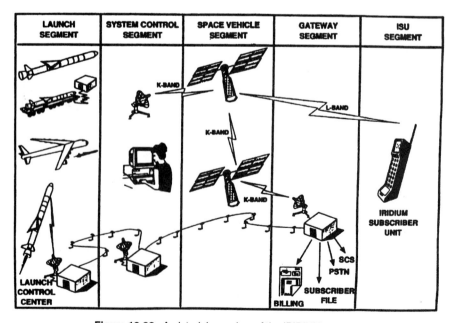

Figure 12.28. A pictorial overview of the IRIDIUM system.

the band 1616–1625.5 MHz. Each satellite will use up to 48 spot beams to form small cells on the surface of the earth. These numerous, relatively narrow beams result in high satellite antenna gains.

Taking advantage of the spatial separation of the beams allows increased spectral efficiency by means of time/frequency/spatial reuse over multiple cells, enabling many simultaneous user traffic connectivities over the same frequency channel. The constellation of satellites and its projection of cells on the earth's surface are analogous to a terrestrial cellular telephone system. With conventional cellular operations, a static set of cells services a large number of mobile/portable users. However, with the IRIDIUM system, the user's motion is relatively slow compared to that of the spacecraft.

Each of the satellites operates cross-links to support internetting. The cross-links operate in the 23-GHz band and include forward- and backward-looking links to two adjacent satellites in the source orbital plane that are normally at a fixed angle about 2100 nautical miles away as well as cross-plane links to adjacent satellites. Up to four interplane links are maintained.

Each satellite communicates with earth-based gateways either directly or through other satellites by means of the crosslink network. Initially, there will be one gateway in the United States and about 3–20 gateways in other parts of the world. The system can accommodate up to 32 gateway stations. A gateway provides the interface between the IRIDIUM system and the PSTN. Table 12.8 gives a summary of major IRIDIUM satellite characteristics.

12.13.4.3 Cell Organization and Frequency Reuse

Satellite L-Band Antenna Pattern. Each satellite has the capability of projecting 48 L-band spot beams to form a continuous overlapping cell pattern on the earth. On a global basis, the entire constellation's beam pattern is projected on the surface of the earth. This results in approximately 2150 active beams with a frequency reuse of about 180 times. Within the contiguous United States, the system achieves up to five times frequency reuse. Each satellite is in view of a individual subscriber unit (ISU) for approximately 9 minutes.

Each satellite has three multiple-beam phased-array antennas. The phased-array antennas are located on side panels of the satellite, each of which forms 16 cellular beams. Active transmit/receive (T/R) modules are used to provide power amplification for the transmit function, low-noise amplification for the receive function, switch selection between transmit and receive, and digital phase control for active beam steering of both the transmit and receive beams in the phased arrays. The 16-cell pattern of each phased array is repeated for each of the three panels.

Cellular Pattern. IRIDIUM operates with a frequency reuse distance designed to minimize cochannel interference. A typical pattern is shown in Figure 12.29. The cells shown as A through G are covered by satellite arrays

TABLE 12.8 Major IRIDIUM Satellite Characteristics

Orbits (nominal)	
Number of operational satellites	66
Number of orbital planes	6
Inclination of orbital planes	86.4°
Orbital period	100 min and 28 s
Apogee	787 km
Perigee	768 km
Argument of perigee	90° ± 10°
Active service arc	360°
RAAN	0°, 31.6°, 63.2°, 94.8°, 126.3°, 157.9°
Earth coverage	5.9 million square (statute) miles per satellite
Maximum number of channels per satellite	
L-band service links	About 3840
Intersatellite links	About 6000
Gateway/TT&C links	About 3000
Frequency bands[a]	
L-band service links	1616–1626.5 MHz
Intersatellite links	23.18–23.38 GHz
Gateway/TT&C links	
Downlinks	19.4–19.6 GHz
Uplinks	29.1–29.3 GHz
Polarization	
L-band service links	Right-hand circular
Intersatellite links	Horizontal
Gateway/TT&C links	
Downlinks	Left-hand circular
Uplinks	Right-hand circular
Transmit EIRP[b]	
L-band service links	7.5–27.7 dBW
Intersatellite links	38.4 dBW
Gateway/TT&C links	13.5–23.2 dBW
Final amplifier output power capability[a]	
L-band service links	0.1–3.5 watts per carrier (burst)[c]
Intersatellite links	3.4 watts per carrier (burst)
Gateway/TT&C links	0.1–1.0 watts per channel
Satellite G/T	
L-band service links	−10.6 to −3.1 dBi/K
Intersatellite links	8.1 dBi/K
Gateway/TT&C links	−1.0 dBi/K
Receiving system noise temperature	
L-band service links	500 K
Intersatellite links	720 K (1188 K with sun)
Gateway/TT&C links	1295 K
Gain of each L-band channel	N.A. (not a transponder)

[a]Does not include circuit losses.
[b]At edge of coverage.
[c]Equipment combined power from phased-array antenna.

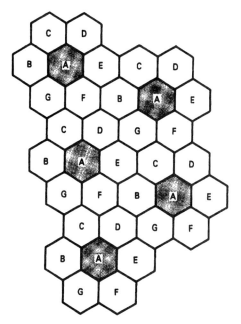

Figure 12.29. A typical seven-cell frequency reuse pattern.

in accordance with the TDMA timing pattern and sequence shown in Figure 12.30. During each TDMA frame, satellite transmission and reception from subscriber units (ISUs) may occur during respective transmit and receive intervals. Each satellite has multiple-beam antennas with fixed boresights to provide contiguous cell coverage. Figure 12.31 shows a typical seven-cell pattern on a satellite and how it is integrated with satellites whose antenna patterns are contiguous.

Frequency Plans. The frequency plan for L-band operation is shown in Figure 12.32. The total peak capacity for a satellite is 1100 channels of which 960 are full duplex. As shown in Figure 12.30, each frequency slot has TDMA

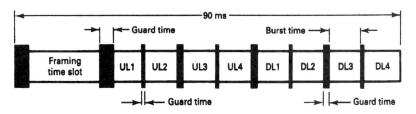

Figure 12.30. TDMA format.

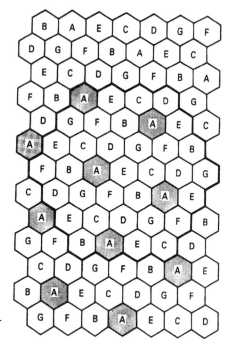

Figure 12.31. A typical frequency reuse pattern with more than one satellite.

capability for four full-duplex voice channels. Without frequency reuse, the contiguous United States are covered by approximately 59 beams, which yield a maximum capacity of 3300 channels of which 2880 are full-duplex voice channels.

12.13.4.4 Communications Subsystem. The IRIDIUM communications system provides L-band communications between each satellite and individual subscriber units, K_a-band communications between each spacecraft and ground-based facilities, which could be either gateway or system control facilities (TT&C), and K_a-band crosslinks from satellite to satellite. The transfer of telemetry, tracking, and control (TT&C) information between the

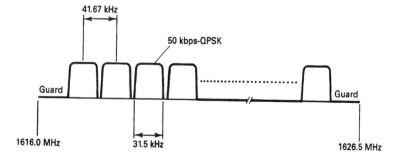

Figure 12.32. L-band uplink/downlink RF plan.

system control facilities and each satellite is generally provided via K_a-band communication links, with a dedicated link (via an omni-antenna) as backup.

L-Band Subscriber Terminal Links. The L-band communication subsystem is supported by an antenna complex consisting of three antenna panels, which form 48 cellular beams. The 48 beams can be formed simultaneously (16 from each antenna panel). The L-band communication system is sized to provide transmit and receive capability for up to five fully loaded cells.

Each active beam supports an average of 236 channels. As shown in Figure 12.32, each FDMA frequency slot is 41.67 kHz wide including guard-bands. Each frequency slot supports four full-duplex voice channels in a TDMA arrangement as shown in Figure 12.31. The modulation is QPSK, the

TABLE 12.9A   L-Band Link Budget, Uplink with Shadow Conditions

	Representative Cells			
	Cell 1	Cell 6	Cell 12	Cell 16
Azimuth angle (deg)	32.4	38.3	40.5	60.0
Ground range (km)	2215.3	1424.9	957.3	528.8
Nadir angle (deg)	61.9	56.4	48.2	33.4
Elevation angle (deg)	8.2	20.8	33.2	51.9
Slant range (km)	2461.7	1696.2	1278.5	960.0
IRIDIUM Subscriber Unit				
HPA burst power (W)	3.7	3.6	1.9	3.7
(dBW)	5.7	5.6	2.9	5.7
Circuit loss (dB)	0.7	0.7	0.7	0.7
Antenna Gain (dBi)	1.0	1.0	1.0	1.0
EIRP (dBW)	6.0	5.9	3.2	6.0
Uplink EIRP density (dBW/4 kHz)	−3.0	−3.1	−5.8	−3.0
Propagation				
Space loss (dB)	164.5	161.3	158.8	156.3
Propagation losses (dB)	15.7	15.7	15.7	15.7
Total Propagation loss (dB)	180.2	177.0	174.5	172.0
Space Vehicle				
Received signal strength (dBW)	−174.2	−171.1	−171.3	−166.0
Effective EOC antenna gain (dBi)	23.9	22.6	22.8	16.4
Signal level (dBW)	−150.3	−148.5	−148.5	−149.6
Required $E_b/(N_0 + I_0)$ (dB)	5.8	5.8	5.8	5.8
E_b/I_0 (dB)	18.0	18.0	18.0	18.0
Required E_b/N_0 (dB)	6.1	6.1	6.1	6.1
T_s (K)	500.0	500.0	500.0	500.0
Signal level required (dBW)	−148.5	−148.5	−148.5	−148.5
Link margin (dB)	−1.8	0.0	0.0	−1.1
G/T (dBi/K)	−3.1	−4.4	−4.2	−10.6

Sources: References 28 and 29.

coded data rate is 50 kbps per channel, and the TDMA frame period is 90 ms.

The L-band communication subsystem is designed to support a bit error rate of 2×10^{-2} end-to-end for voice operation. The improved BER required for data transmission is supported through the use of processing hardware installed in the subscriber units to apply more robust protocols and coding in order to counter deep fading experienced with L-band links in the mobile environment.

Tables 12.9A and 12.9B are typical L-band link budgets for uplink and downlink under shadow conditions.

TABLE 12.9B   L-Band Link Budget, Downlink with Shadow Conditions

	Representative Cells			
	Cell 1	Cell 6	Cell 12	Cell 16
Azimuth angle (deg)	32.4	38.3	40.5	60.0
Ground range (km)	2215.3	1424.9	957.3	528.8
Nadir angle (deg)	61.9	56.4	48.2	33.4
Elevation angle (deg)	8.2	20.8	33.2	51.9
Slant range (km)	2461.7	1696.2	1278.5	960.0
Space Vehicle				
HPA burst power (W)	3.5	2.2	1.3	3.0
(dBW)	5.5	3.5	1.2	4.8
Transmitter circuit loss (dB)	2.1	2.1	2.1	2.1
Effective EOC antenna gain (dB)	24.3	23.1	22.9	16.8
EIRP (dBW)	27.7	24.5	22.0	19.5
Propagation				
Space loss (dB)	164.5	161.3	158.8	156.3
Propagation losses (dB)	15.7	15.7	15.7	15.7
Total propagation loss (dB)	180.2	177.0	174.5	172.0
IRIDIUM Subscriber Unit				
Received signal strength (dBW)	−152.5	−152.5	−152.5	−152.5
Antenna gain (dBi)	1.0	1.0	1.0	1.0
Signal level (dBW)	−151.5	−151.5	−151.5	−151.5
Required $E_b/(N_0 + I_0)$ (dB)	5.8	5.8	5.8	5.8
E_b/I_0 (dB)	18.0	18.0	18.0	18.0
Required E_b/N_0 (dB)	6.1	6.1	6.1	6.1
T_s(K)	250.0	250.0	250.0	250.0
Signal level required (dBW)	−151.5	−151.5	−151.5	−151.5
Link margin (dB)	0.0	0.0	0.0	0.0
G/T_s (dBi/K)	−23.0	−23.0	−23.0	−23.0
SPFD at ISU (dBW/m^2/4 kHz)	−135.8	−135.8	−135.8	−135.8

Sources: References 28 and 29.

Gateway Links. The 30/20-GHz gateway links provide full-duplex operation with two ground-based gateways or system support facilities per satellite. Satellite beam center gains for maximum range are 26.9 dBi on the downlink and 30.1 dBi on the uplink. These transmission links remain operational even with rainfall loss as high as 13 dB on the downlink and 26 dB on the uplink. Multiple earth terminal antennas spaced 34 nautical miles apart provide spatial diversity to avoid sun interference and to mitigate the effects of rainfall attenuation. Time availabilities on these links are expected to be in the range of 99.8%.

Table 12.10 is a typical link budget for gateway 30/20-GHz operation. Each of the two full-duplex gateway links supports 600 simultaneous voice

TABLE 12.10   Link Budget for Gateway Operation

Item	Units	Downlink Rain	Downlink Clear	Uplink Rain	Uplink Clear
Range	km	2326.0	2326.0	2326.0	2326.0
Transmitter					
Power	dBW	0.0	−9.7	13.0	−11.8
Antenna gain	dB	26.9	26.9	56.3	56.3
Circuit loss	dB	−3.2	−3.2	−1.0	−1.0
Pointing loss	dB	−0.5	−0.5	−0.3	−0.3
EIRP	dBWi	23.2	13.5	68.0	43.2
System					
Margin	dB	3.2	3.2	2.1	2.1
Space loss	dB	−185.8	−185.8	−189.1	−189.1
Propagation loss	dB	−14.2	−1.5	−30.0	−1.5
Polarization loss	dB	−0.2	−0.2	−0.2	−0.2
Total propagation loss	dB	−203.4	−190.7	−221.4	−192.9
Receiver					
Received signal strength	dBWi	−180.2	−177.2	−153.4	−149.7
Pointing loss	dB	−0.2	−0.2	−0.8	−0.8
Antenna gain	dB	53.2	53.2	30.1	30.1
Received signal	dBW	−127.2	−124.2	−124.1	−120.4
T_s	K	731.4	731.4	1295.4	1295.4
Noise density	dBW/Hz	−200.0	−200.0	−197.5	−197.5
Noise bandwidth	dB/Hz	64.9	64.9	64.9	64.9
Noise	dBW	−135.1	−135.1	−132.6	−132.6
Link E_b/N_0	dB	7.9	10.9	8.5	12.2
E_b/I_0	dB	25.0	25.0	16.0	16.0
Computed $E_b/(N_0 + I_0)$	dB	7.8	10.7	7.8	10.7
Required $E_b/(N_0 + I_0)$	dB	7.7	7.7	7.7	7.7
Excess margin	dB	0.1	3.0	0.1	3.0
SPFD at GW	dBW/m²/1 MHz	−134.3	−131.3		

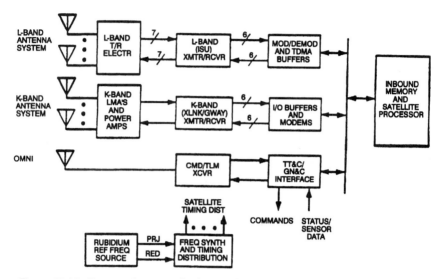

Figure 12.33. Spacecraft communications payload, simplified functional block diagram.

circuits. The frequency plan calls for the allocation of six frequencies each for uplink and downlink operation. The modulation rate in each direction is 12.5 Mbps and the channels are spaced at 15-MHz intervals. Each link supports a BER of better than 1×10^{-7}.

Figure 12.33 is a simplified functional block diagram of an IRIDIUM satellite communication payload.

12.13.4.5 Transmission Characteristics. Digital speech operation on IRIDIUM uses the AMBE* compression technique. Conventional PCM requires 64 kbps for a standard voice channel primarily to support speech operation. The AMBE compression technique used by IRIDIUM cuts this value to 2400 bps or about 27 : 1 compression.

The modulation and multiple access techniques used in the IRIDIUM system are patterned after the GSM terrestrial cellular system. A combined FDMA and TDMA access format is used along with data or vocoded voice and digital modulation techniques.

Each subscriber unit operates in a burst mode using a single carrier transmission. The bursts are controlled to occur at the proper time in the TDMA frame. The TDMA format is shown in Figure 12.30. The frame has four user time slots, both uplink and downlink. Each subscriber unit will burst so that its transmission is received by the satellite in the proper time slot, Doppler corrected. The subscriber unit is similar to a conventional hand-held unit in size and shape. It has a quadrifilar helix antenna with

*AMBE is a registered trademark of Digital Voice Systems, Inc.

nearly hemispherical coverage with a gain of +1 dBi. Its peak EIRP is +8.5 dBW. The G/T is −23 dB/K. Its required E_b/N_0 is 6.3 dB.

12.13.4.5.1 Performance Objectives. The IRIDIUM system is designed to provide service to virtually 100% of the earth. However, it is recognized that it is economically and, at times, physically impossible to provide service to every single point on earth. There are practical limitations to the total number of locations that will physically be within line-of-sight of the satellites. End-to-end bit error rates will be better than 2×10^{-2} for voice transmission. The more typical bit error rates will be in the range of 1×10^{-3} to 1×10^{-4}. Coding and processing will improve these BERs to better than 1×10^{-5} for data connectivities.

Section 12.13 was based on Ref. 27, the original Motorola petition to the FCC for a 77-satellite configuration with seven orbital planes. It was subsequently modified by two amendments to the original petition for a 66-satellite configuration in six orbital planes (Refs. 28, and 29). The writeup was further modified by a Motorola review (Ref. 32) and a second Motorola review (Ref. 33).

REVIEW EXERCISES

1. What is the principal drawback in cellular radio considering its explosive user growth over these past several years?

2. Why is transmission loss so much greater on a cellular path compared to a LOS microwave path on the same frequency and covering the same distance?

3. What is the function of the MTSO or MSC in a cellular network?

4. What is the channel spacing in kilohertz on the North American AMPS system? On the N-AMPS system?

5. Why are antenna heights limited for cells with just sufficient height to cover the cell boundaries?

6. Why do we do cell splitting? What is approximately the minimum practical diameter of a cell (this limits splitting)?

7. When is *handover* necessary?

8. Define *roaming*.

9. Cellular transmission loss varies with what four factors besides distance and frequency?

10. Building penetration varies with what four factors?

11. Using the Okumura/Hata model, calculate the transmission loss for a 900-MHz cellular link where the cellular mobile terminal antenna height is 3 m, the base transmitter antenna height is 35 m, and the distance is 5 km, in a medium-size city.

12. Calculate the building blockage component (dB) according to Lee. The blockage distance is 250 ft, ERP = 1 watt. There are three buildings through which the transmitted wave traverses, 50 ft, 70 ft, and 80 ft. The transmit antenna height is 50 ft.

13. What are some of the fade ranges (dB) we might expect on a cellular link?

14. If the delay spread on a cellular link is about 10 μs, up to about what bit rate will there be little deleterious effects due to multipath dispersion?

15. Space diversity is commonly used at cell base sites. Like LOS microwave, antenna separation is key. This separation varies with two parameters. What are they?

16. For effective space diversity operation, there is a law of diminishing returns when we lower the correlation coefficient below what value?

17. What is the gain of a standard dipole over an isotropic radiator? What is the difference between ERP, as used in this text, and EIRP?

18. Cellular designers use a field strength contour of ___ dBμ, which is equivalent to − ___ dBm.

19. What would be the transmission loss to the 39-dBμ contour line if a cell site EIRP was +52 dBm?

20. A cell site antenna has a gain of +14 dBd. What is the equivalent gain in dBi?

21. There is not enough bandwidth assigned for cellular operation in the United States. The trend is to convert to digital operation. Why at first blush does it seem that we are defeating our purpose (i.e., have more users per unit bandwidth)?

22. What are the three access techniques that might be considered for digital cellular operation?

23. Provide the principal difference between TDMA as used in cellular service and that described in Chapter 7 (digital communications by satellite).

24. Discuss propagation delay as encountered at a TDMA cell site—the problem and its solution.

25. Analyze burst size, guard time, overhead, and training bits for a cellular TDMA system with eight users and 20-ms frame duration. Draw the

frame showing user slot duration and then draw a typical user slot to show how the several fields are assigned.

26. What power amplifier advantage do we have in a TDMA system that we do not generally have with a FDMA system?

27. List four advantages of TDMA when compared to AMPS or other FDMA schemes.

28. Cellular radio, particularly in urban areas, is gated by extreme interference conditions, especially cochannel from frequency reuse. In light of this, describe how we achieve an interference advantage when using CDMA.

29. A CDMA cellular system operates at 4800 bps. It spreads this signal 10 MHz. What is the processing gain in dB? What is the processing gain of IS-95 North American CDMA when the information rate is 9600 bps and the bandwidth is 1.23 MHz?

30. Correlation is the key to CDMA operation. Describe how it works permitting only the wanted signal to pass and rejecting other signals and interference.

31. For effective frequency reuse, the value of D/R must be kept large enough. Define D and R. What value of D/R is "large enough"?

32. In congested urban areas and where cell diameters are small, what measure do we take to reduce C/I?

33. How can paging systems use such low transmit power and achieve comparably long range?

34. PCS cells have diameters under 1 km. Give a generalized expression of path loss for such a system. Discuss the expression parameters and assign some rough values, leaving aside environmental factors.

35. For PCS (and for cellular) links, which are usually tilted in that the transmitter often is higher than the receiver, if we plot the loss curve, two slopes can be identified as we progress from transmitter to receiver. Discuss this phenomenon and its causes.

36. Name at least four different devices in a typical home that may be classified as PCS.

37. Such standards as DECT, CT1, CT2, PHS, and WACS/PACS apply to what basic application of PCS?

38. Typify transmitter output for cellular compared to PCS. There should be at least two values for each.

39. Speech coding is less stringent in the several PCS scenarios (typically 32 kbps). Explain why it is much more stringent for cellular.

40. Why would CDMA be so attractive for PCS, especially in an indoor scenario?

41. WLANs use two different types of transmission media (either one or the other). What are they?

42. FPLMTS provides a skeletal guideline for PCS. In the terrestrial area, which are the three scenarios described and quantified?

43. Give two decided advantages of LEO mobile satellite systems over GEO satellites. Give one advantage of GEO satellites.

44. LEO systems for mobile service have even more constrained allocated bandwidths than terrestrial cellular. How are they overcoming this problem to meet user demand and maintain an acceptable grade of service (i.e., blocking probability)?

45. Describe satellites and orbits of the IRIDIUM system.

46. Trace a call through the IRIDIUM system. Describe its access techniques.

47. What is the bit rate of the new voice compression technique used with IRIDIUM?

REFERENCES

1. Raymond Steele (ed.), *Mobile Radio Communications*, IEEE Press, New York, and Pentech Press, London, 1992.
2. Dennis Bodson et al. (eds.), *Land Mobile Communications Engineering*, IEEE Press, New York, 1983.
3. J. D. Parsons and J. G. Gardiner, *Mobile Communication Systems*, Blackie, London, and Halsted Press, New York, 1989.
4. Theodore E. Rappaport, *Wireless Communications: Principles and Practice*, Prentice-Hall/IEEE Press, New York, 1996.
5. R. C. V. Macario (ed.), *Personal and Mobile Radio Systems*, IEE/Peter Peregrinus, London, 1991.
6. William C. Y. Lee, *Mobile Communications Design Fundamentals*, 2nd ed., Wiley, New York, 1993.
7. A. Jagoda and M. de Villepin, *Mobile Communications*, Wiley, Chichester, UK, 1991/1992.
8. Kaveh Pahlavan and Allen H. Levesque, *Wireless Information Networks*, Wiley, New York, 1995.
9. William W. Wu et al., "Mobile Satellite Communications," *Proc. IEEE*, vol. 82, no. 9, Sept. 1994.
10. Woldemar F. Fuhrmann and Volker Brass, "Performance Aspects of the GSM System," *Proc. IEEE*, vol. 89, no. 9, Sept. 1984.

11. Henry L. Bertoni et al., "UHF Propagation Prediction for Wireless Personal Communication," *Proc. IEEE*, vol. 89, no. 9, Sept. 1994.

12. "VHF and UHF Propagation Curves for the Frequency Range from 30 MHz to 1000 MHz," CCIR Rec. 370-5, XVIIth Plenary Assembly, Dusseldorf, 1990.

13. K. Allesbrook and J. D. Parsons, "Mobile Radio Propagation in British Cities at Frequencies in the VHF and UHF Bands," *Proc. IEE*, vol. 124, no. 2, 1977.

14. Y. Okumura et al., "Field Strength and Its Variability in VHF and UHF Land Mobile Service," *Rev. Electr. Commun. Lab.*, 16, 1968.

15. M. Hata, "Empirical Formula for Propagation Loss in Land-Mobile Radio Services," *IEEE Trans. Veh. Technol.*, vol. VT-20, 1980.

16. F. C. Owen and C. D. Pudney, "In-Building Propagation at 900 MHz and 1650 MHz for Digital Cordless Telephones," *6th International Conference on Antennas and Propagation*, ICCAP'89, Pt. 2: Propagation, Conf. Pub. No. 301, 1989.

17. M. R. Schroeder and B. S. Atal, "Code-Excited Linear Prediction, High Quality Speech at Low Bit Rates," *IEEE Proc. ICASSP*, 1985.

18. Robert C. Dixon, *Spread Spectrum Systems with Commercial Applications*, 3rd ed., Wiley, New York, 1994.

19. "Recommended Minimum Standards for 800-MHz Cellular Subscriber Units," EIA Interim Standard EIA/IS-19B.

20. "Cellular Radio Systems," a seminar given at the University of Wisconsin–Madison by Andrew H. Lamothe, Consultant, Leesburg, VA, 1993.

21. *Telecommunications Transmission Engineering*, 2nd ed., vol. 2, Bellcore, Piscataway, NJ, 1992.

22. "Radio-Paging Systems," CCIR Rep. 900-2, Vol. VIII.1, XVIIth Plenary Assembly, Dusseldorf, 1990.

23. "Radio-Paging Systems," CCIR Rep. 499-5, Vol. VIII.1, XVIIth Plenary Assembly, Dusseldorf, 1990.

24. "Digital Cellular Public Land Mobile Telecommunication Systems (DCPLMTS)," CCIR Rep. 1156, Vol. VIII.1, XVIIth Plenary Assembly, Dusseldorf, 1990.

25. Morris Engelson and Jim Hebert, "Effective Characterization of CDMA Signals," *Wireless Rep.*, Jan. 1995.

26. Donald C. Cox, "Wireless Personal Communications: What is It?" *IEEE Personal Commun.*, vol. 2, no. 2, Apr. 1995.

27. Application of Motorola Satellite Communications, Inc. for "IRIDIUM, A Low Earth Orbit Mobile Satellite System," before the Federal Communications Commission, Dec. 1990.

28. Minor Amendment before the Federal Communications Commission (amending Motorola's original IRIDIUM application), Aug. 1992, by Motorola Satellite Communications, Inc.

29. Minor Amendment before the Federal Communications Commission (amending Motorola's IRIDIUM system application), Nov. 1992 by Motorola Satellite Communications, Inc.

30. Jay C. Padgett, Cristoph G. Gunter, and Takeshi Hattori, "Overview of Wireless Personal Communications," *IEEE Commun. Mag.*, Jan. 1995.

31. Roger L. Freeman, *Telecommunication System Engineering*, 3rd ed., Wiley, New York, 1996.

32. Private communication, Steve Clark, Motorola Chandler, Arizona, Mar. 17, 1996.

33. Private communication, Stephanie Nowack and J. Hoyt, Motorola Chandler, July 22, 1996.

34. Promotional material. "SPRINT PCS^{TM}: Facts-at-a-Glance," Tom Murphy, Sprint, Kansas City, MO (undated), 1996.

35. "Future Public Mobile Land Telecommunication Systems (FPLMTS)," CCIR Rec. 687-1, 1994 M Series Volume, Part 2, ITU, Geneva, 1994.

36. "TDMA Radio Interface: 800 MHz Cellular," TIA/EIA IS-627/628/629, EIA/TIA, Washington, DC, 1996.

37. "Mobile Staion–Base Station Compatibility Standard for Dual-Mode Wideband Spread Spectrum Cellular System," TIA/EIA/IS-95-A EIA/TIA, Washington, DC, 1995.

13

THE TRANSMISSION OF DIGITAL DATA

13.1 INTRODUCTION

We must distinguish again between analog and digital transmission. An analog transmission system has an output at the far end, which is a continuously variable quantity representative of the input. With analog transmission there is continuity (of waveform); with digital transmission there is discreteness.

The simplest form of digital transmission is binary, where an information element is assigned one of two possibilities. There are many binary situations in real life where only one of two possible values can exist; for example, a light may be either on or off, an engine is running or not, and a person is alive or dead.

An entire number system has been based on two values, which by convention have been assigned the symbols 1 and 0. This is the binary system, and its number base is 2. Our everyday number system has a base of 10 and is called the decimal system. Still another system has a base of 8 and is called the octal system.

The basic information element of the binary system is called the bit, which is an acronym for *b*inary dig*it*. The bit, as we know, may have the values 1 or 0.

A number of discrete bits can identify a larger piece of information, and we may call this larger piece a character. A code is defined by the IEEE as "a plan for representing each of a finite number of values or symbols as a particular arrangement or sequence of discrete conditions or events" (Ref. 1).

Binary coding of written information and its subsequent transmission have been with us for a long time. An example is teleprinter service (i.e., the transmission of a telegram).

The greater number of computers now in use operate in binary languages; thus binary transmission fits in well for computer-to-computer communication and the transmission of data.

This chapter introduces the reader to the transmission of binary information or data over telephone networks. It considers data on an end-to-end basis and the effects of the variability of transmission characteristics on the final data output. There is a review of the basics of the makeup of digital data signals and their application. Therefore the chapter endeavors to cover the entire field of digital data and its transmission over telephone network facilities. It includes a discussion of the nature of digital signals, coding, information theory, constraints of the telephone channel, modulation techniques, and the dc nature of data transmission.

13.2 THE BIT AND BINARY CONVENTION

In a binary transmission system the smallest unit of information is the bit. As we know, either one of two conditions may exist, the 1 or the 0. We call one state a mark, the other a space. These conditions may be indicated electrically by a condition of current flow and no current flow. Unless some rules are established, an ambiguous situation would exist. Is the 1 condition a mark or a space? Does the no-current condition mean that a 0 is transmitted, or a 1? To avoid confusion and to establish a positive identity to binary conditions, CCITT Rec. V.1 recommends equivalent binary designations. These are shown in Table 13.1. If the table is adhered to universally, no confusion will exist as to which is a mark, which is a space, which is the active condition, which is the passive condition, which is 1, and which is 0. It defines the *sense* of transmission so that the mark and space, the 1 and 0, will not be inverted. Data transmission engineers often refer to such a table as a table of *mark–space convention.*

TABLE 13.1   Equivalent Binary Designations

Binary 1	Binary 0
Mark or marking	Space or spacing
Perforation (paper tape)	No perforation
Negative voltage[a]	Positive voltage
Condition Z	Condition A
Tone on (amplitude modulation)	Tone off
Low frequency (frequency shift keying [FSK])	High frequency
Opposite to the reference phase	Reference phase (phase modulation)
No phase inversion (differential two-phase modulation)	Inversion of the phase

[a]CCITT Recs. V.10 and V.11 (Refs. 3 and 4, respectively) and EIA/TIA-232E (Ref. 5).
Source: Reference 2, CCITT Rec. V.1. Courtesy of ITU-CCITT.

13.3 CODING

13.3.1 Introduction to Binary Coding Techniques

Written information must be coded before it can be transmitted over a digital system. The following discussion of coding covers only binary codes. But before launching into coding itself, the term *entropy* is introduced.

Operational telecommunication systems transmit information. We can say that information has the property of reducing the uncertainty of a situation. The measurement of uncertainty is called entropy. If entropy is large, then a large amount of information is required to clarify a situation; if entropy is small, then only a small amount of information is required for clarification. Noise in a communication channel is a principal cause of uncertainty. From this we now can introduce Shannon's noisy channel coding theorem, stated approximately (Ref. 6, pp. 41–42):

> If an information source has an entropy H and a noisy channel capacity C, then provided $H < C$, the output from the source can be transmitted over the channel and recovered with an arbitrarily small probability of error. If $H > C$, it is not possible to transmit and recover information with an arbitrarily small probability of error.

Entropy is a major consideration in the development of modern codes. Coding can be such as to reduce transmission errors (uncertainties) due to the transmission medium and even correct the errors at the far end. This is done by reducing the entropy per bit (adding redundancy). We shall discuss errors and their detection in greater detail in Section 13.4. Channel capacity is discussed in Section 13.9.

Now the question arises: How big a binary code? The answer involves yet another question: How much information is to be transmitted?

One binary digit (bit) carries little information; it has only two possibilities. If 2 binary digits are transmitted in sequence, there are four possibilities,

00 10
01 11

or four pieces of information. Suppose 3 bits are transmitted in sequence. Now there are eight possibilities:

000 100
001 101
010 110
011 111

We can now see that for a binary code the number of distinct information characters available is equal to 2 raised to a power equal to the number of elements or bits per character. For instance, the last example was based on a three-element code, giving eight possibilities or information characters.

Another more practical example is the Baudot teleprinter code. It has five bits or information elements per character. Hence the different or distinct graphics or characters available are $2^5 = 32$. The American Standard Code for Information Interchange (ASCII) has seven information elements per character, or $2^7 = 128$; so it has 128 distinct combinations of marks and spaces that are available for assignment as characters or graphic symbols.

The number of distinct characters for a specific code may be extended by establishing a code sequence (a special character assignment) to shift the system or machine to uppercase (as is done with a conventional typewriter). Uppercase is a new character grouping. A second distinct code sequence is then assigned to revert to lowercase. As an example, the CCITT International Telegraph Alphabet (ITA) No. 2 code (Figure 13.1) is a five-unit code with 58 letters, numbers, graphics, and operator sequences. The additional characters (additional above $2^5 = 32$) come from the use of uppercase. Operator sequences appear on a keyboard as *space* (spacing bar), *figures* (uppercase), *letters* (lowercase), *carriage return*, *line feed* (spacing vertically), and so on.

When we refer to a 5-unit, 6-unit, 7-unit, or 12-unit code, we refer to the number of information units or elements that make up a single character or symbol; that is, we refer to those elements assigned to each character that carry information and that make it distinct from all other characters or symbols of the code.

13.3.2 Some Specific Binary Codes for Information Interchange

In addition to the ITA No. 2 code, some of the more commonly used codes are, the IBM data transceiver code (Figure 13.2), the American Standard Code for Information Interchange (ASCII) (Figure 13.3 and Table 13.2), and the extended binary-coded decimal interchange code (EBCDIC) (Figure 13.4).

Parity checks are one way to determine if a character contains an error after transmission. We speak of even parity and odd parity. On a system using an odd-parity check, the total count of 1's or marks has to be an odd number per character (or block) (e.g., it carries 1, 3, 5, or 7 marks of 1's). Some systems, such as the field data code, use even parity (i.e., the total number of marks must be an even number, such as 2, 4, 6, or 8).

To explain parity and parity checks a little more clearly, let us look at some examples. Consider a 7-level* code with an extra parity bit. By system convention, even parity has been established. Suppose that a character is transmitted as 1111111. There are seven marks, so to maintain even parity we would need an even number of marks. Thus an eighth bit is added and must

*Level and bits are often used interchangeably.

Characters				Code Elements[a]						
Letters Case	Communications	Weather	CCITT #2[b]	START	1	2	3	4	5	STOP
A	—	↑			■	■				■
B	?	⊕			■			■	■	■
C	:	○				■	■	■		■
D	$	↗	WRU		■			■		■
E	3	3			■					■
F	1	→	Unassigned		■		■	■		■
G	&	↘	Unassigned			■		■	■	■
H	STOP[c]	↓	Unassigned				■		■	■
I	8	8				■	■			■
J	′	↯	Audible signal		■	■		■		■
K	(	←			■	■	■	■		■
L	)	↖				■			■	■
M	.	.					■	■	■	■
N	,	⊕					■	■		■
O	9	9						■	■	■
P	θ	θ				■	■		■	■
Q	1	1			■	■	■		■	■
R	4	4				■		■		■
S	BELL	BELL			■		■			■
T	5	5							■	■
U	7	7			■	■	■			■
V	;	⊕	=			■	■	■	■	■
W	2	2			■	■			■	■
X	/	/			■		■	■	■	■
Y	6	6			■		■		■	■
Z	″	+	+		■				■	■
BLANK		—								■
SPACE							■			■
CAR. RET.								■		■
LINE FEED						■				■
FIGURE					■	■		■	■	■
LETTERS					■	■	■	■	■	■

[a] Blank, spacing element; crosshatched, marking element.

[b] This column shows only those characters which differ from the American "communications" version.

[c] Figures case H(COMM) may be stop or +.

Figure 13.1. Communication and weather codes, CCITT International Telegraph Alphabet No. 2. (From Ref. 7, CCITT Rec. S.1. Courtesy of ITU-CCITT.)

be a mark (1). Look at another bit pattern, 1011111. Here there are six marks, even; then the eighth (parity) bit must be a space. Still another example would be 0001000. To get even parity, a mark must be added on transmission, and the character transmitted would be 00010001, maintaining even parity. Suppose that, owing to some sort of signal interference, one signal element was changed on reception. No matter which element was changed, the receiver would indicate an error because we would no longer have even parity. If two elements were changed, though, the error could be

Bit Number	Code Assignments							
	0	0	0	0	1	1	1	1
	0	0	1	1	0	0	1	1
	.0	1	0	1	0	1	0	1
X O N R 7 4 2 1								
↓ ↓ ↓ ↓ ↓ ↓ ↓ ↓								
0 0 0 0 0								
0 0 0 0 1								TPH/TGR
0 0 0 1 0								@
0 0 0 1 1				(NA)		(NA)	Space	
0 0 1 0 0								#
0 0 1 0 1				(NA)		(NA)	9	
0 0 1 1 0				(NA)		(NA)	8	
0 0 1 1 1		G	P		X			
0 1 0 0 0								(NA)
0 1 0 0 1				(NA)		(NA)	6	
0 1 0 1 0				(NA)		(NA)	5	
0 1 0 1 1		D	M		U			
0 1 1 0 0				(NA)		(NA)	3	
0 1 1 0 1		B	K		S			
0 1 1 1 0		A	J		/			
0 1 1 1 1	0							
1 0 0 0 0								Restart
1 0 0 0 1				SOC/EOC		$\overset{+}{0}$	EOT	
1 0 0 1 0				•		%		
1 0 0 1 1		&	—		Ø			
1 0 1 0 0				$		.	.	
1 0 1 0 1		I	R		Z			
1 0 1 1 0		H	Q		Y			
1 0 1 1 1	7							
1 1 0 0 0				(NA)		(NA)	(NA)	
1 1 0 0 1		F	O		W			
1 1 0 1 0		E	N		V			
1 1 0 1 1	4							
1 1 1 0 0		C	L		T			
1 1 1 0 1	2							
1 1 1 1 0	1							
1 1 1 1 1								

Figure 13.2. IBM data transceiver code. TPH/TGR = telephone/telegraph. SOC/EOC = start or end of card; EOT = end of transmission; (NA) = valid but not assigned; $\overset{+}{0}$ = plus zero; $\overline{0}$ = minus zero. Transmission order; bit X → bit 1.

b7 b6 b5 →				Column → Row ↓	0 0 0 / 0	0 0 1 / 1	0 1 0 / 2	0 1 1 / 3	1 0 0 / 4	1 0 1 / 5	1 1 0 / 6	1 1 1 / 7
b4	b3	b2	b1									
0	0	0	0	0	NUL	DLE	SP	0	@	P	`	p
0	0	0	1	1	SOH	DC1	!	1	A	Q	a	q
0	0	1	0	2	STX	DC2	"	2	B	R	b	r
0	0	1	1	3	ETX	DC3	#	3	C	S	c	s
0	1	0	0	4	EOT	DC4	$	4	D	T	d	t
0	1	0	1	5	ENQ	NAK	%	5	E	U	e	u
0	1	1	0	6	ACK	SYN	&	6	F	V	f	v
0	1	1	1	7	BEL	ETB	'	7	G	W	g	w
1	0	0	0	8	BS	CAN	(	8	H	X	h	x
1	0	0	1	9	HT	EM	)	9	I	Y	i	y
1	0	1	0	10	LF	SUB	*	:	J	Z	j	z
1	0	1	1	11	VT	ESC	+	;	K	[	k	{
1	1	0	0	12	FF	FS	,	<	L	\	l	|
1	1	0	1	13	CR	GS	—	=	M	]	m	}
1	1	1	0	14	SO	RS	.	>	N	^	n	~
1	1	1	1	15	SI	US	/	?	O	_	o	DEL

Figure 13.3. American Standard Code for Information Interchange FIDS-1. (From Ref. 78, MIL-STD-188-100, 1972; also see Ref. 8.)

masked. This would happen in the case of even or odd parity if two marks were substituted for two spaces or vice versa at any element location in the character.*

The IBM data transceiver code (see Figure 13.2) is used for the transfer of digital data information recorded on perforated cards. It is an 8-bit code providing a total of 256 mark–space combinations. Only those combinations or patterns having a "fixed count" of four 1's (marks) and four 0's (spaces) are made available for assignment as characters. Therefore only 70 bit patterns satisfy the fixed count condition, and the remaining 186 combinations are invalid. The parity (fixed count) or error checking advantage is obvious. Of the 70 valid characters, 54 are assigned to alphanumerics and a limited number of punctuation signs and other symbols. In addition, the code includes special bit patterns assigned to control functions peculiar to the transmission of cards such as "start card" and "end card."

The ASCII (see Figure 13.3) is the latest effort on the part of the U.S. industry and common carrier systems, backed by the American National Standards Institute, to produce a universal common language code. ASCII is

---

*This type of parity is seldom used today. However, the eighth bit remains, making it an 8-bit code and improving its characteristics for 8-bit (16-bit, 32-bit, etc.) processors.

TABLE 13.2   Definitions for Figure 13.3: Control Characters and Format Effectors

Nul: All zero characters. A control character used for fill.

SOH (Start of heading): A communication control character used at the beginning of a sequence of characters which constitute a heading. A heading contains address or routing information. An STX character has the effect of terminating a heading.

STX (Start of text): A communication control character that terminates a heading and indicates the start of text or the information field.

ETX (End of text): A communication control character used to terminate a sequence of characters started with STX and transmitted as an entity.

EOT (End of transmission): A communication control character used to indicate the conclusion of transmission which may have contained one or more texts and associated headings.

ENQ (Enquiry): A communication control character used as a request for response from a remote station. It may be used as a "Who are you" (WRU) to obtain identification, or may be used to obtain station status, or both.

ACK (Acknowledge): A communication control character transmitted by a receiver as an affirmative response to a sender.

BEL: A character for use when there is a need to call human attention. It may control alarm or attention devices.

BS (Backspace): A format effector which controls the movement of the active position one space backward.

HT (Horizontal tabulation): A format effector which controls the movement of the active position forward to the next character position.

LF (Line feed): A format effector which controls the movement of the active position advancing it to the corresponding position of the next line.

VT (Vertical tabulation): A format effector which controls the movement of the active position to advance to the corresponding character position on the next predetermined line.

FF (Form feed): A format effector which controls the movement of the active position to its corresponding character position on the next page or form.

CR (Carriage return): A format effector which controls the movement of the active position to the first character position on the same line.

SO (Shift out): A control character indicating the code combination which follow shall be interpreted as outside of the character set of the standard code table until a Shift In (SI) character(s) is (are) reached. It is also used in conjunction with the (ESC) escape character.

SI (Shift in): A control character indicating that the code combinations which follow shall be interpreted according to the standard code table (Table 13.3).

DLE (Data link escape): A communication control character which will change the meaning of a limited number of contiguously following bit combinations. It is used exclusively to provide supplementary control functions in data networks. DLE is usually terminated by a Shift In (SI) character(s).

DC1, DC2, DC3, DC4 (Device controls): Characters for the control of ancillary devices associated with data or telecommunication networks. It switches these devices "on" or "off." DC4 is preferred for turning a device off.

NAK (Negative acknowledgment): A communication control character transmitted by a receiver as a negative response to a sender.

SYN (Synchronous idle): A communication control character used by a synchronous transmission system in the absence of any other character to provide a signal from which synchronism may be achieved or retained.

ETB (End of transmission block): A communication control character used when block transmission is employed. It indicates the end of a block of data.

CAN (Cancel): A control character used to indicate that the data preceding it is in error or is to be disregarded.

TABLE 13.2 *(Continued)*

EM (End of medium): A control character associated with the sent data which may be used to identify the physical end of the medium, or the end of the used or wanted portion of information recorded on a medium. It should be noted that the position of the character does not necessarily correspond to the physical end of the medium.

SUB (Substitute): A character that may be substituted for a character which is determined to be invalid or in error.

ESC (Escape): A control character intended to provide code extension (supplementary characters) in general information interchange. The ESC character itself is a prefix affecting the interpretation of a limited number of contiguously following characters. ESC is usually terminated by a SI (shift in) character(s).

FS (File separator), GS (Group separator), RS (Record separator) and US (Unit separator): These information separators may be used within data in optional fashion, except that their hierarchical relationship shall be: FS is the most inclusive, then GS, then RS and US is the least inclusive. (The content and length of a File, Group, Record or Unit are not specified).

DEL (Delete): This character is used primarily to "erase" or "obliterate" erroneous or unwanted characters in perforated tape. (In the strict sense DEL is not a control character). DEL characters may also be used for fill without affecting the meaning or information content of the bit stream.

Note: SO, ESC and DLE are all characters which can be used, at the discretion of the system designer, to indicate the beginning of a sequence of digits having special significance.

SP (Space): A nonprinting character used to separate words or sequences. It is a format effector that causes the active position to advance one character position.

Source: Reference 78, pages B-4 through B-6, updated.

a seven-unit code with all 128 combinations available for assignment. Here again, the 128 bit patterns are divided into two groups of 64. One of the groups is assigned to a subset of graphic printing characters. The second subset of 64 is assigned to control characters. An eighth bit is added to each character for parity check. ASCII is widely used in North America and has received considerable acceptance in Europe and Hispanic America.

The reader's attention is called to what are known as computable codes, such as ASCII. Computable codes have the letters of the alphabet plus all other characters and graphics assigned values in continuous binary sequence. Thus these codes are in the native binary language of today's common digital computers. The CCITT ITA No. 2 is not, and when used with a computer, it often requires special processing.

The extended binary-coded decimal interchange code (EBCDIC) is similar to the ASCII, but it is a true 8-bit code. The eighth bit is used as an added bit to "extend" the code, providing 256 distinct code combinations for assignment. Figure 13.4 illustrates the EBCDIC code.

13.3.3 Hexadecimal Representation and the BCD Code

The hexadecimal system is a numeric representation in the number base 16. The number base uses 0 through 9 as in the decimal base, and the letters A through F to represent the decimal numbers 10 through 15. The hexadecimal

B, I T S	4	0	0	0	0	0	0	0	0	1	1	1	1	1	1	1	1
	3	0	0	0	0	1	1	1	1	0	0	0	0	1	1	1	1
	2	0	0	1	1	0	0	1	1	0	0	1	1	0	0	1	1
	1	0	1	0	1	0	1	0	1	0	1	0	1	0	1	0	1
8 7 6 5																	
0 0 0 0		NUL				PF	HT	LC	DEL								
0 0 0 1						RES	NL	BS	IL								
0 0 1 0						BYP	LF	EOB	PRE			SM					
0 0 1 1						PN	RS	UC	EOT								
0 1 0 0		SP										¢	.	<	(	+	|
0 1 0 1		&										!	$	*	)	;	¬
0 1 1 0			/									∧	,	%	—	>	?
0 1 1 1												⁄:	#	("	'	=	"
1 0 0 0			a	b	c	d	e	f	g	h	i						
1 0 0 1			j	k	l	m	n	o	p	q	r						
1 0 1 0				s	t	u	v	w	x	y	z						
1 0 1 1																	
1 1 0 0			A	B	C	D	E	F	G	H	I						
1 1 0 1			J	K	L	M	N	O	P	Q	R						
1 1 1 0				S	T	U	V	W	X	Y	Z						
1 1 1 1		0	1	2	3	4	5	6	7	8	9						¤

PF – Punch Off   RES – Restore   BYP – Bypass
HT – Horiz. Tab   NL – New Line   LF – Line Feed
LC – Lower Case   BS – Backspace   EOB – End of Block
DEL – Delete   IL – Idle   PRE – Prefix
SP – Space   PN – Punch On   RS – Reader Stop
UC – Upper Case   EOT – End of Transmission SM – Start Message

Figure 13.4. Extended binary-coded decimal interchange code (EBCDIC).

numbers can be translated to the binary base as follows:

Hexadecimal	Binary	Hexadecimal	Binary
0	0000	8	1000
1	0001	9	1001
2	0010	A	1010
3	0011	B	1011
4	0100	C	1100
5	0101	D	1101
6	0110	E	1110
7	0111	F	1111

Two examples of the hexadecimal notation are as follows:

Number Base 10	Number Base 16
21	15
64	40

The BCD is a compromise code assigning 4-bit binary numbers to the digits between 0 and 9. The BCD equivalents to decimal digits appear as follows:

Decimal Digit	BCD Digit	Decimal Digit	BCD Digit
0	1010	5	0101
1	0001	6	0110
2	0010	7	0111
3	0011	8	1000
4	0100	9	1001

To cite examples, consider the number 16; it is broken down into 1 and 6. Therefore its BCD equivalent is 0001 0110. If it were written in straight binary notation, it would appear as 10000. The number 25 in BCD combines the digits 2 and 5 above as 0010 0101.

13.4 ERROR DETECTION AND ERROR CORRECTION

13.4.1 Introduction

In the transmission of data the most important goal in design is to minimize the error rate. Error rate may be defined as the ratio of the number of bits incorrectly received to the total number of bits transmitted. According to CCITT the design objective is an error rate no poorer than one error in 1×10^6 and in North America many circuits display an error rate better than one error in 10^{10}.

One method to minimize the error rate is to provide a "perfect" transmission channel, one that will introduce no errors in the transmitted information by the receiver; unfortunately, the engineer designing a data transmission system can never achieve that perfect channel. Besides improvement of the channel transmission parameters themselves, the error rate can be reduced by forms of systematic redundancy. In old-time Morse code on a bad circuit words often were sent twice; this is redundancy in its simplest form. Of course, it took twice as long to send a message. This is not very economical if useful words per minute received is compared to channel occupancy.

This brings up the point of channel efficiency. Redundancy can be increased such that the error rate could approach zero. Meanwhile the information transfer or *throughput* across the channel also approaches zero. Hence unsystematic redundancy is wasteful and merely lowers the rate of useful communication. Maximum efficiency or throughput could be obtained in a digital transmission system if all redundancy and other code elements, such as start and stop elements, were removed from the code, and, in addition, if advantage were taken of the statistical phenomenon of our written language by making high-usage letters, such as E, T, and A, short in code length and low-usage letters, such as Q and X, longer.

13.4.2   Throughput

The *throughput* of a data channel is the expression of how much data are put through. In other words, throughput is an expression of channel efficiency. The term gives a measure of useful data put through the data communication link. These data are directly useful to the computer or data terminal equipment (DTE).

Therefore on a specific circuit, throughput varies with the raw data rate; is related to the error rate and the type of error encountered (whether burst or random); and varies with the type of error detection and correction system used, the message handling time, and the block length, from which we must subtract the "nonuseful" bits such as overhead bits. Among overhead bits we have parity bits, flags, and cyclic redundancy checks.

13.4.3   The Nature of Errors

In data/telegraph transmission an error is a bit that is incorrectly received. For instance, a 1 is transmitted in a particular time slot and the element received in that slot is interpreted as a 0. Bit errors occur either as single random errors or as bursts of error. In fact, we can say that every transmission channel will experience some random errors, but on a number of channels burst errors may predominate. For instance, lightning or other forms of impulse noise often cause bursts of errors, where many contiguous bits show a very high number of bits in error. The IEEE defines error burst as "a group of bits in which two successive bits are always separated by less than a given number of correct bits" (Ref. 1).

13.4.4   Error Detection and Error Correction Defined

The data transmission engineer differentiates between error detection and error correction. Error detection identifies that a symbol, character, block,* packet,* or frame* has been received in error. As discussed earlier, parity is primarily used for error detection. Parity bits, of course, add redundancy and thus decrease channel efficiency or throughput.

Error correction corrects the detected error. Basically, there are two types of error correction techniques: forward-acting (FEC) and two-way error correction (automatic repeat request [ARQ]). The latter technique uses a return channel (backward channel). When an error is detected, the receiver signals this fact to the transmitter over the backward channel, and the block of information containing the error is transmitted again. FEC utilizes a type of coding that permits a limited number of errors to be corrected at the receiving end by means of special coding and software (or hardware) implemented at both ends of a circuit.

---

*A block, packet, or frame is a group of digits or data characters transmitted as a unit over which a coding procedure is usually applied for synchronization and error control purposes.

Error Detection. There are various arrangements or techniques available for the detection of errors. All error detection methods involve some form of redundancy, those additional bits or sequences that can inform the system of the presence of error or errors. Parity, discussed earlier, was character parity, and its weaknesses were presented. Commonly, the data transmission engineer refers to such parity as *vertical redundancy checking* (VRC). The term *vertical* comes from the way characters arranged on paper tape (i.e., hole positions).

Another form of error detection utilizes longitudinal redundancy checking (LRC), which is used in block transmission where a data message consists of one or more blocks. Remember that a block is a specific group of digits or data characters sent as a "package" (not to be confused with *packet*). In such circumstances a LRC character, often called a block check character (BCC), is appended at the end of each block. The BCC modulo-2 sums the 1's and 0's in the columns of the block (vertically). The receiving end also modulo-2 sums the 1's and 0's in the block, depending on the parity convention for the system. If that sum does not correspond to the BCC, an error (or errors) exists in the block. The LRC ameliorates much of the problem of undetected errors that could slip through with VRC if used alone. The LRC method is not foolproof, however, as it uses the same thinking as VRC. Suppose that errors occur such that two 1's are replaced by two 0's in the second and third bit positions of characters 1 and 3 in a certain block. In this case the BCC would read correctly at the receive end and the VRC would pass over the errors as well. A system using both LRC and VRC is obviously more immune to undetected errors than either system implemented alone.

A more powerful method of error detection involves the use of the cyclic redundancy check (CRC). It is used with messages that are data blocks, frames, or packets. Such a data message can be simplistically represented as

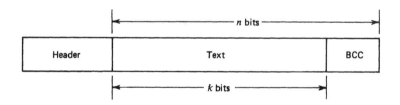

We let n equal the number of bits in the total message after the "header," and k is the number of bits in the message text over which we wish to check for errors. $n - k$ is the number of bits in the BCC or frame check sequence (FCS). For most WANs (wide area networks) it is 16 bits long (2 bytes) and for most LANs (local area networks) it is 32 bits long (4 bytes).

The bit sequence (value) of the BCC is derived from two polynomials: the generating polynomial $P(X)$ and the message polynomial $G(X)$. To develop the 16-bit BCC (FCS) there are three standardized generating polynomials:

CRC-16(ANSI) $X^{16} + X^{15} + X^2 + 1$

CRC(CCITT) $X^{16} + X^{12} + X^5 + 1$

CRC-12 $X^{12} + X^{11} + X^2 + X + 1$ (12-bit BCC)

In binary form we treat a polynomial as follows. A 1 is placed in each position that has a term; absence of a term is indicated by a 0. Of course, if the last term is a 1, that is X^0 power. For example, if a polynomial is given as $X^4 + X + 1$, its binary representation is 10011. The two consecutive zeros tell us that the second and third terms are not present. Let us restate the problem with several more examples.

Mathematically, a message block can be treated as a function such as

$$a_n X^n + a_{n-1} X^{n-1} + a_{n-2} X^{n-2} + \cdots + a_1 X + a_0$$

where coefficients a are set to represent a binary number. Consider the binary number 11011, which is represented by the polynomial

$$\begin{array}{ccccc} 1 & 1 & 0 & 1 & 1 \\ a_4 & a_3 & a_2 & a_1 & a_0 \end{array}$$

and then becomes

$$X^4 + X^3 + X + 1$$

Or consider another example,

$$\begin{array}{ccccc} 0 & 1 & 1 & 0 & 1 \\ a_4 & a_3 & a_2 & a_1 & a_0 \end{array}$$

which then becomes

$$X^3 + X^2 + 1$$

Given now a message polynomial $G(X)$ and a generating polynomial $P(X)$, we wish to construct a code message polynomial $F(X)$ that is evenly divided by $P(X)$. This can be accomplished as follows:

1. Multiply the message $G(X)$ by X^{n-k}, where $n - k$ is the number of bits in the BCC.
2. Divide the resulting product $X^{n-k}[G(X)]$ by the generating polynomial $P(X)$.

3. Disregard the quotient and add the remainder $C(X)$ to the product to yield the code message polynomial $F(X)$, which is represented as $X^{n-k}[G(X)] + C(X)$.

The division is binary division without carries and borrows. In this case, the remainder is always 1 bit less than the divisor. The remainder is the BCC and the divisor is the generating polynomial; hence the bit length of the BCC is always one less than the number of bits in the generating polynomial.

The code message polynomial as represented by the BCC is transmitted to the distant-end receiver. The receiving station divides it by the same generating polynomial. The division will produce no remainder (i.e., all zeros) if there is no error; if there is an error in the n bits of message text, there will be a remainder.

In Ref. 9 it is stated that CRC-12 provides error detection of bursts of up to 12 bits in length. Additionally, 99.955% of error bursts up to 16 bits in length can be detected. CRC-16 provides detection of bursts up to 16 bits in length, and 99.955% of error bursts greater than 16 bits can be detected.

Forward-Acting Error Correction (FEC). FEC uses certain binary codes that are designed to be self-correcting for errors introduced by the intervening transmission media. In this form of error correction the receiving station has the ability to reconstitute messages containing errors.

FEC uses *channel encoding*, whereas the encoding covered previously in this chapter is generically called *source encoding*. The channel encoding used in FEC can be broken down into two broad categories, block codes and convolutional codes. These are discussed at considerable length in Section 5.6.

Key to this type of coding is the modulo-2 adder. Modulo-2 addition is denoted by the symbol $\oplus$. It is binary addition without the "carry," or $1 + 1 = 0$, and we do *not* carry the 1. Summing 10011 and 11001 in modulo-2, we get 01010.

A measure of the error detection and correction capability of a code is given by the Hamming distance. The distance is the minimum number of digits in which two encoded words differ. For example, to detect E digits in error, a code of a minimum Hamming distance of $(E + 1)$ is required. To correct E errors, a code must display a minimum Hamming distance of $(2E + 1)$. A code with a minimum Hamming distance of 4 can correct a single error *and* detect two digits in error (Ref. 10).

Error Correction with Feedback Channel. Two-way or feedback error correction is used very widely today on data and some telegraph circuits. Such a form of error correction is called ARQ, which derives from the old Morse and telegraph signal, "automatic repeat request."

In most modern data systems block transmission is used, and the block is a convenient length of characters sent as an entity. There are two aspects in that "convenience" of length. One relates to the material that is being sent. For instance, the standard "IBM" card has 80 columns. With 8 bits per column, a block of 8 × 80, or 640 bits, would be desirable as data text so that we could transmit an IBM card in each block. In fact, one such operating system, Autodin, bases block length on that criterion, with blocks 672 bits long. The remaining bits, those in excess of 640, are overhead and check bits. On packet networks blocks are typically of uniform length.

The second aspect of block length is the trade-off for optimum length between length and error rate, or the number of block repeats that may be expected on a particular circuit. Longer blocks tend to amortize overhead bits better but are inefficient regarding throughput when an error rate is high. Under these conditions long blocks tend to tie up a circuit with longer retransmission periods.

ARQ, as we know, is based on the block transmission concept. There are three types of ARQ in use today:

- Stop-and-wait
- Continuous, sometimes called selective ARQ
- Go-back-n

Stop-and-wait ARQ is the most straightforward to implement and the least costly from an equipment standpoint. With stop-and wait ARQ, a block (or frame or packet) is transmitted to the distant end. At completion of transmission of the block, transmission ceases. The receiving end of the link receives the block, stores the entire block, and then runs the block through CRC processing. If the CRC remainder is zero, an acknowledgment signal is sent to the far end, and the far-end transmitter sends the next data block. If an error is found (i.e., there is a CRC remainder at the receiver), the receiving station requests retransmission. The transmitter knows that it is the previous block sent that is wanted and retransmits that block. The drawback of stop-and-wait ARQ is the delay of waiting while the receiver does the processing and then transmits back the appropriate signal, which may be an acknowledgment (ACK) or a negative acknowledgment (NACK). It is particularly wasteful of expensive circuit time if propagation delays are long, such as on satellite circuits. It is, however, ideal for half-duplex operation.

Selective or continuous ARQ eliminates the quiescent nonproductive time of waiting and sending the appropriate signal after each block. With this type of ARQ, the transmit end sends a continuous string of blocks, each with an identifying number in the header of each block (or frame or packet). When the receive end detects a block in error, it requests that the distant-end transmitter repeat that block. We assume here, of course, full-duplex, four-wire operation. The request that the receiver sends to the transmitter includes the identifying number of the block in error. The transmit end pulls that block out of its buffer storage and retransmits it as another block in the continuous string with its appropriate identifying number. The receive end is responsible for rerunning a CRC check and then placing that block in its proper sequential order. Obviously, selective or continuous ARQ is more efficient regarding circuit usage. On the other hand, it requires more buffer memory size and a slight increase in overhead to accommodate message sequence numbering. It also requires full-duplex operation, although the return circuit may be only a slow 75-bps channel.

Go-back-*n* ARQ also permits continuous block transmission. When an errored block is discovered, the receiving end informs the far-end transmitter, which then transmits the errored block and all subsequent blocks, even though they have been transmitted before. This approach alleviates the problem at the receive end of inserting the errored block in its proper slot.

13.4.5 Error Performance

From the point of view of the transmission engineer, error performance is the principal measure of quality of service (QoS) of a data circuit. As we discussed in Chapter 4, error performance is measured as BER (bit error rate or ratio). In the 1970s the error performance requirement on a data circuit was 1×10^{-5} or one error in 100,000 bits. By about the mid-1980s, the requirement was improved one order of magnitude to 1×10^{-6}. This is essentially the requirement of CCITT Rec. G.821 (Ref. 21), which is still operative.

Considering that many long-distance data circuits are transported by an underlying digital network, one must not lose sight of the fact that we cannot improve on its error performance. Of course, if we use FEC (forward error correction) with its redundancy overhead, error performance can indeed be improved either one or two orders of magnitude. But note the cost, increased complexity, and additional bandwidth required.

As we discussed in Section 4.2.2, some remarkable improvements have been achieved in the North American digital PSTN, where the underlying transport displays BERs of 1×10^{-10} or better. Particularly if we are using a mapping regime such as AT&T DDS or ITU-T Rec. V.110 described in Chapter 3, we can expect similar error rates for data riding on these digital channels. This assumes, of course, that no errors are introduced by the mapping action itself. ISDN requires no mapping, it being based on the 64-kbps channel, which is used for data or voice communications.

Digital bit streams transported on LANs are protected in an isolated environment, in the floor of a building, across several floors, and so on. A LAN is far less complex than the PSTN, covering very limited distances, measured in feet or meters rather than miles or kilometers. There is just one person responsible for this small network. Thus we can expect error performance with BERs around 1×10^{-11} for those systems on wire pair or coaxial cable, and at least one order of magnitude improvement for LANs on fiber-optic cable (e.g., FDDI).

For a more in-depth examination of error performance, we encourage readers to review Section 4.2.2, in particular, our discussion of ITU-T Rec. G.826.

13.5   THE dc NATURE OF DATA TRANSMISSION

13.5.1   Loops

Binary data are transmitted on a dc loop. More correctly, the binary data end instrument delivers to the line and receives from the line one or several dc loops. In its most basic form a dc loop consists of a switch, a dc voltage source, and a termination. A pair of wires interconnects the switch and termination. The voltage source in data and telegraph work is called the battery, although the device is usually electronic, deriving the dc voltage from an ac power line source. The battery is placed in the line such that it provides voltage(s) consistent with the type of transmission desired. A simplified dc loop is shown in Figure 13.5.

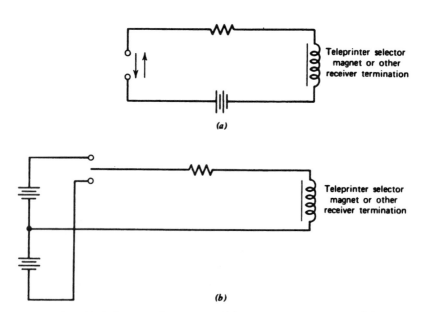

Figure 13.5. Simplified diagram of a dc loop: (a) with neutral keying and (b) with polar keying.

13.5.2 Neutral and Polar dc Transmission Systems

Nearly all dc data and telegraph systems functioning today are operated in either a neutral or a polar mode. However, the neutral mode has pretty much been phased out. The words *neutral* and *polar* describe the manner in which the battery is applied to the dc loop. On a neutral loop, following the mark–space convention in Table 13.1, the battery is applied during marking (1) conditions and is switched off during spacing (0). Current therefore flows in the loop when a mark is sent and the loop is closed. Spacing is indicated on the loop by a condition of no current. Thus we have the two conditions for binary transmission—an open loop (no current flowing) and a closed loop (current flows). Keep in mind that we could reverse this, change the convention (Table 13.1), and, say, assign spacing to a condition of current flowing or closed loop and marking to a condition of no current or open loop. This is sometimes done in practice and is called changing the sense. Either way, a neutral loop is a dc loop circuit where one binary condition is represented by the presence of voltage flow of current, and the other by the absence of voltage/current. Figure 13.5a shows a simplified neutral loop.

Polar transmission approaches the problem a little differently. Two batteries are provided. One is called negative battery and the other positive. During a condition of marking, a positive battery is applied to the loop, following the convention of Table 13.1, and a negative battery is applied to the loop during the spacing condition. In a polar loop, current is always flowing. For a mark or binary 1 it flows in one direction and for a space or binary 0 it flows in the opposite direction. Figure 13.5b shows a simplified polar loop.

13.5.3 Some Common Digital Data Waveforms or Line Signals

In this section we discuss several basic concepts of "electrical" coding to develop line signals or specific line waveforms. Figure 13.6 graphically illustrates several line coding techniques.

Figure 13.6a shows what is still called by many today "neutral transmission." This was the principal method of transmitting telegraph signals until about 1960. First, this waveform is a nonreturn-to-zero (NRZ) format in its simplest form. "Nonreturn-to-zero" simply means that if a string of 1's (marks) is transmitted, the signal remains in the mark state with no transitions. Likewise, if a string of 0's is transmitted, there is no transition and the signal remains in the 0 state until a 1 is transmitted. As we can now see, with NRZ transmission we can transmit information without transitions.

Figures 13.6b and 13.6d show the typical "return-to-zero" (RZ) waveform, where, when a continuous string of marks (or spaces) is transmitted, the signal level (amplitude) returns to the zero-voltage condition at each element or bit. Obviously, RZ transmission is much richer in transitions than NRZ.

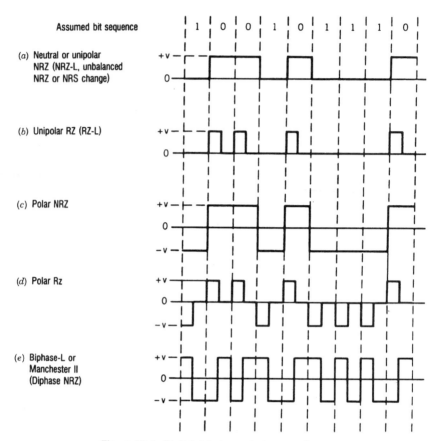

Figure 13.6. Digital data transmission waveforms.

In Section 13.5.2 we discussed neutral and polar dc transmission systems. Figure 13.6a shows a typical neutral waveform where the two state conditions are 0 V for the mark or 1 condition and some positive voltage for the space or 0 condition. On the other hand, in polar transmission, as shown in Figures 13.6c and 13.6d, a positive voltage represents a space, and a negative voltage, a mark. With NRZ transmission, the pulsewidth is the same as the duration of a unit interval or bit. Not so with RZ transmission, where the pulsewidth is less than the duration of a unit interval. This is because we have to allow time for the pulse to return to the zero condition.

Biphase-L or Manchester coding (Figure 13.6e) is a code format that is being used ever more widely on digital systems such as wire pair, coaxial cable, and fiber optics. Here the binary information is carried in the transition. By convention a logic 0 is defined as a positive-going transition and a logic 1 as a negative-going pulse. Note that Manchester coding has a signal transition in the middle of each unit interval (or bit); Manchester coding is a form of phase coding.

The reader should be cognizant of and be able to differentiate between two sets or ways of classifying binary digital waveforms. The first set is *neutral* and *polar*. The second set is *NRZ* and *RZ*. Manchester coding is still another way to represent binary digital data where the transition takes place in the middle of the unit interval. In Chapter 3 one other class of waveform was introduced: alternate mark inversion (AMI).

13.6 BINARY TRANSMISSION AND THE CONCEPT OF TIME

13.6.1 Introduction

Time and timing are most important factors in digital transmission. For this discussion consider a binary end instrument sending out in series a continuous run of marks and spaces. Those readers who have some familiarity with Morse code will recall that the spaces between dots and dashes told the operator where letters and words ended. With the sending device or transmitter delivering a continuous series of characters to the line, each consisting of 5, 6, 7, 8, or 9 elements (bits) to the character, let the receiving device start its print cycle when the transmitter starts sending. If the receiver is perfectly in step with the transmitter, ordinarily one could expect good printed copy and few, if any, errors at the receiving end.

It is obvious that when signals are generated by one machine and received by another, the speed of the receiving machine must be the same or very close to that of the transmitting machine. When the receiver is a motor-driven device, timing stability and accuracy are dependent on the accuracy and stability of the speed of rotation of the motors used.

Most simple data receivers sample at the presumed center of the signal element. It follows, accordingly, that whenever a receiving device accumulates a timing error of more than 50% of the period of 1 bit, it will print in error.

The need for some sort of synchronization is shown in Figure 13.7a. A 5-unit code is employed, and three characters transmitted sequentially are shown. Sampling points are shown in the figure as vertical arrows. Receiver timing begins when the first pulse is received. If there is a 5% timing difference between transmitter and receiver, the first sampling at the receiver will be 5% away from the center of the transmitted pulse. At the end of the 10th pulse or signal element the receiver may sample in error. The 11th signal element will indeed be sampled in error, and all subsequent elements are errors. If the timing error between transmitting machine and receiving machine is 2%, the cumulative error in timing would cause the receiving device to print all characters in error after the 25th bit.

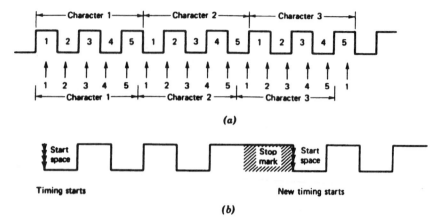

Figure 13.7. (*a*) Five-unit synchronous bit stream with timing error. (*b*) Five-unit start-and-stop stream of bits with a 1.5-unit stop element.

13.6.2   Asynchronous and Synchronous Transmission

In the earlier days of printing telegraphy, start–stop transmission or asynchronous operation was developed to overcome the problem of synchronism. Here timing starts at the beginning of a character and stops at the end. Two signal elements are added to each character to signal the receiving device that a character has begun and ended.

As an example consider a five-element code, such as CCITT ITA No. 2 (see Figure 13.1). In the front of a character an element is added called a start space and at the end of each character a stop mark is inserted. Send the letter A in Figure 13.1. The receiving device starts its timing sequence on receiving element number 1, a space or 0, then a 11000 is received; the A is selected, then the stop mark is received, and the timing sequence stops. On such an operation an accumulation of timing errors can take place only inside each character.

Suppose that the receiving device is again 5% slower or faster than its transmitting counterpart; now the fifth information element will be no more than 30% displaced in time from the transmitted pulse and well inside the 50% or halfway point for correct sampling to take place.

In start–stop transmission the *information* signal elements are each of the same duration, which is the duration or pulsewidth of the start element. The stop element has an indefinite length or pulsewidth beyond a certain minimum.

If a steady series of characters are sent, then the stop element is always the same width or has the same number of unit intervals. Consider the transmission of two A's, 0110001011000111111 → 11111. The start space (0) starts the timing sequence for six additional elements, which are the five code elements in the letter A and the stop mark. Timing starts again on the mark-to-space transition between the stop mark of the first A and the start space of the second. Sampling is carried out at pulse center for most asynchronous systems. One will note that at the end of the second A a continuous series of marks is sent. Hence the signal is a continuation of the stop element or just a continuous mark. It is the mark-to-space transition of the start element that tells the receiving device to start timing a character.

Minimum lengths of stop elements vary. The example discussed previously shows a stop element of 1-unit interval duration (1 bit). Others are of 1.5- and 2-unit interval duration. The proper semantics of data transmission would describe the code of the previous paragraph as a 5-unit start–stop code with a 1-unit stop element.

A primary objective in the design of data systems is to minimize errors received or to minimize the error rate. There are three prime causes of errors. These are noise, intersymbol interference, and improper timing relationships. With start–stop systems a character begins with a mark-to-space transition at the beginning of the start space. Then 1.5-unit intervals later the timing causes the receiving device to sample the first information element, which simply is a mark or space decision. The receiver continues to sample at 1-bit intervals until the stop mark is received. In start–stop systems the last information bit is most susceptible to cumulative timing errors. Figure 13.7b is an example of a 5-unit start–stop bit stream with a 1.5-unit stop element.

Another problem in start–stop systems is that of mutilation of the start element. Once this happens the receiver starts a timing sequence on the next mark-to-space transition it sees and thence continues to print in error until, by chance, it cycles back on a proper start element.

Synchronous data systems do not have start and stop elements but consist of a continuous stream of information elements or bits as shown in Figure 13.7a. The cumulative timing problems eliminated in asynchronous (start–stop) systems are present in synchronous systems. Codes used on synchronous systems are often 7-unit codes with an extra unit added for parity, such as the ASCII code. Timing errors tend to be eliminated by virtue of knowing the exact rate at which the bits of information are transmitted.

If a timing error of 1% were to exist between transmitter and receiver, no more than 50 bits could be transmitted; then the synchronous receiving device would be 50% apart in timing from the transmitter and all bits received would be in error. Even if timing accuracy were improved to 0.05%, the correct timing relationship between transmitter and receiver would exist for only the first 2000 bits transmitted. It follows that no timing error at all can be permitted to accumulate, since anything but absolute accuracy in timing would cause eventual malfunctioning. In practice the receiver is provided with an accurate clock, which is corrected by small adjustments as explained in the following.

13.6.3 Timing

All currently used data transmission systems are synchronized in some manner. Start–stop synchronization has been discussed. Fully synchronous transmission systems all have timing generators or clocks to maintain stability. The transmitting device and its companion receiver at the far end of the data circuit must be a timing system. In normal practice the transmitter is the master clock of the system. The receiver also has a clock, which in every case is corrected by one means or another to its transmitter equivalent at the far end.

Another important timing factor that must also be considered is the time it takes a signal to travel from the transmitter to the receiver. This is called propagation time. With velocities of propagation as low as 20,000 mi/s, consider a circuit 200 mi in length. The propagation time would then be 200/20,000 s or 10 ms, which is the time duration of 1 bit at a data rate of 100 bps. Thus the receiver in this case must delay its clock by 10 ms to be in step with its incoming signal.

Temperature and other variations in the medium may affect this delay. One can also expect variations in the transmitter master clock as well as other time distortions due to the medium.

There are basically three methods of overcoming these problems. One is to provide a separate synchronizing circuit to slave the receiver to the transmitter's master clock. This wastes bandwidth by expending a voice channel or subcarrier just for timing. A second method, which was used fairly widely up to 25 years ago, was to add a special synchronizing pulse for groupings of information pulses, usually for each character. This technique was similar to start–stop synchronization and lost its appeal largely owing to the wasted information capacity for synchronizing. The most prevalent system in use today is one that uses transition timing. With this type of timing the receiving device is automatically adjusted to the signaling rate of the transmitter, and adjustment is made at the receiver by sampling the transitions of the incoming pulses. This offers many advantages, most important of which is that it automatically compensates for variations in propagation time. With this type of synchronization the receiver determines the average repetition rate and phase of the incoming signal transition and adjusts its own clock accordingly.

In digital transmission the concept of a transition is very important. The transition is what really carries the information. In binary systems the space-to-mark and mark-to-space transitions (or lack of transitions) placed in a time reference contain the information. Decision circuits regenerate and retime in sophisticated systems and care only *if* a transition has taken place. Timing cares *when* it takes place. Timing circuits must have memory in case a long series of marks or spaces is received. These will be periods of no transition, but they carry meaningful information. Likewise, the memory must maintain timing for reasonable periods in case of circuit outage. Keep in mind that synchronism pertains to both frequency and phase and that the usual error in high-stability systems is a phase error (i.e., the leading edges of the received pulses are slightly advanced or retarded from the equivalent clock pulses of the receiving device).

High-stability systems once synchronized need only a small amount of correction in timing (phase). Modem internal timing systems may be as stable a 1×10^{-8} or better at both the transmitter and the receiver. Before a significant error condition can build up owing to a time rate difference at 2400 bps, the accumulated time difference between transmitter and receiver must exceed approximately 2×10^{-4} s. This figure neglects phase. Once the transmitter and receiver are synchronized and the circuit is shut down, then the clock on each end must drift apart by at least 2×10^{-4} s before significant errors take place. Again this means that the leading edge of the receiver clock equivalent timing pulse is 2×10^{-4} in advance of or retarded from the leading edge of the received pulse from the distant end. Often an idling signal is sent on synchronous data circuits during conditions of no traffic to maintain timing. Other high-stability systems need to resynchronize only once a day.

Bear in mind that we are considering dedicated circuits only, not switched synchronous data. The problems of synchronization of switched data immediately come to light. Two such problems are that:

- No two master clocks are in perfect phase synchronization.
- The propagation time on any two paths may not be the same.

Consequently, such circuits will need a time interval for synchronization at each switching event before traffic can be passed.

To sum up, synchronous data systems use high-stability clocks, and the clock at the receiving device is undergoing constant but minuscule corrections to maintain an in-step condition with the received pulse train from the distant transmitter by looking at mark–space and space–mark transitions.

13.6.4 Distortion

It has been shown that the key factor in data transmission is timing. The signal must be either a mark or a space, but that alone is not sufficient. The marks and spaces (or 1's and 0's) must be in a meaningful sequence based on a time reference.

In the broadest sense distortion may be defined as any deviation of a signal in any parameter, such as time, amplitude, or wave shape, from that of the ideal signal. For data and telegraph binary transmission, distortion is defined as a displacement in time of a signal transition from the time that the receiver expects to be correct. In other words, the receiving device must make a decision whether a received signal element is a mark or a space. It makes the decision during the sampling interval, which is usually at the center of where the received pulse or bit should be. Therefore it is necessary for the transitions to occur between sampling times and preferably halfway between them. Any displacement of the transition instants is called distortion. The degree of distortion a data signal suffers as it traverses the transmission medium is a major contributor in determining the error rate that can be realized.

Telegraph and data distortion is broken down into two basic types, systematic and fortuitous. Systematic distortion is repetitious and is broken down into bias distortion, cyclic distortion, and end distortion (more common in start–stop systems). Fortuitous distortion is random in nature and may be defined as a distortion type in which the displacement of the transition from the time interval in which it should occur is not the same for every element. Distortion caused by noise spikes in the medium or other transients may be included in this category. Characteristic distortion is caused by transients in the modulation process, which appear in the demodulated signal.

Figure 13.8 shows some examples of distortion. Figure 13.8*a* is an example of a binary signal without distortion, and Figure 13.8*b* shows the sampling instants, which occur ideally in the center of the pulse to be sampled. From this we can see that the displacement tolerance is nearly 50%. This means that the point of sample could be displaced by up to 50% of a pulsewidth and still record the mark or space condition present without error. However, the sampling interval does require a finite amount of time so that in actual practice the displacement permissible is somewhat less than 50%. Figures 13.8*c* and 13.8*d* show bias distortion. An example of spacing bias is shown in Figure 13.8*c*, where all the spacing impulses are lengthened at the expense of the marking impulses. Figure 13.8*d* shows the reverse of this; the marking impulses are elongated at the expense of the spaces. This latter is called marking bias. Figure 13.8*e* shows fortuitous distortion, which is a random type of distortion. In this case the displacement of the signal element is not the same as the time interval in which it should occur for every element.

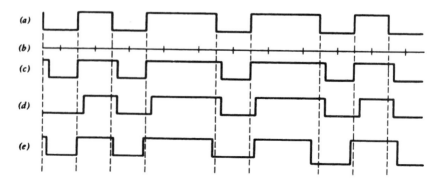

Figure 13.8. Three typical distorted data signals.

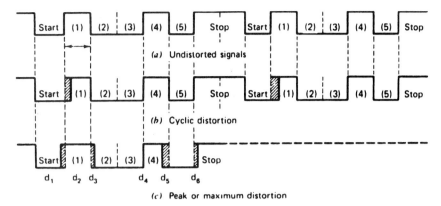

Figure 13.9. Distorted telegraph signals illustrating cyclic and peak distortion. The peak distortion in the character appears at transition d_5.

Figure 13.9 shows distortion that is more typical of start and stop transmission. Figure 13.9a is an undistorted start and stop signal. Figure 13.9b shows cyclic or repetitive distortion typical of mechanical transmitters. In this type of distortion the marking elements may increase in length for a period of time, and then the spacing elements will increase in length. Figure 13.9c shows peak distortion. Identifying the type of distortion present on a signal often gives a clue to the source or cause of distortion. Distortion measurement equipment measures the displacement of the mark-to-space transition from the ideal of a digital signal. If a transition occurs too near to the sampling point, the signal element is liable to be in error. Standards dealing with distortion are published by the Electronic Industries Association and CCITT (Refs. 11–14).

13.7 DATA INTERFACE

13.7.1 Introduction

To interface two or more data nodes satisfactorily is a fairly complex matter. One can argue that the transmission engineer should only be concerned with the electrical (physical) interface to assure compatibility of the electrical signals on a data link. Indeed, in this section, we will emphasize the electrical compatibility aspects. However, it is very shortsighted not to consider and appreciate the complexity and many facets of the complete interface. Many of these "facets" will impact transmission engineering directly or indirectly. To interface means to make compatible so interworking can be effected.

Some interface issues have already been discussed. Table 13.1 is just one example. Consider these other examples for a simple data link connecting two nodes or terminals. There will be no compatibility if one end is using the ASCII code and the other is using EBCDIC. Likewise, there will be incompatibility if the CRC used on one end is CRC-CCITT and on the other,

CRC-ANSI; nor will there be compatibility if one end operates in the synchronous mode and the other expects to receive in the start–stop mode.

In this section we will first present a generalized picture of data communication compatibility and introduce "protocols." Then we will discuss how protocols are implemented on a layer basis using the OSI (open system interconnection) model. The remainder of the section deals with the electrical (physical) interface.

13.7.2 Initial Considerations for Data Network Access and Control Procedures

The term *communications protocol* defines a set of procedures by which communication is accomplished within standard constraints. As shown in the following outline of protocol topics, protocol deals with control functions:

1. Framing: frame makeup (format); block, message, or packet makeup.
2. Error control (note that this is an interface characteristic as well).
3. Sequence control: the numbering of messages (or blocks) to eliminate duplication, maintain proper sequence in the case of packet networks, and maintain a proper record of identification of messages, especially in ARQ systems or for message servicing.
4. Transparency of the communication links, link control equipment, multiplexers, concentrators, modems, and so on. Transparency allows the use of any bit pattern the user wishes to transmit, even though these patterns resemble control characters or prohibited bit sequences such as long series of 1's or 0's.
5. Line control: determination, in the case of half-duplex or multipoint line, of which station is going to transmit and which station is going to receive.
6. Idle patterns to maintain network synchronization.
7. Time-out control: which procedures to allow if message (or block or packet) flow ceases entirely.
8. Start-up control: getting a network into operation initially or after some period of remaining idle for one reason or another.
9. Sign-off control: under normal conditions, the process of ending communication or transaction before starting the next transaction or message exchange.

At this juncture we distinguish data link control from user device control. The data link is defined as the configuration of equipment enabling end terminals in two different stations to communicate directly. The data link includes the paired data terminal equipment (DTE), modems, or other signal converters and the interconnecting facilities. The user device may be a central processing unit (CPU), workstation, or other data peripheral. Pictorially, we can illustrate the difference between data link control and user

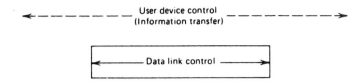

device control by the following diagram:

The current literature distinguishes *circuit connection* from *link connection.* Circuit connection is simply the establishment of an electrical path between two points (or multipoints) that want to communicate. We know from our previous discussions that the connection may be metallic (i.e., wire or cable), fiber optic, and/or radio in either the frequency or time domain (e.g., time slot in a frame). The mere establishment of an electrical connection does not mean that data communication can take place. Link establishment is a group of procedures that prepares the source to send data and the destination to receive that data.

In conventional telephony there are three distinct phases to a telephone call: (1) call setup; (2) information transfer, where the subscribers at each end of the connection carry on their conversation; and (3) call termination. A data link must essentially go through the same three procedures. The user control is analogous to the information transfer portion of the telephone call. Thus we introduce protocols.

13.7.3 Protocols

13.7.3.1 Basic Protocol Functions. Stallings (Ref. 15) lists some basic protocol functions. Typical among these functions are:

- Segmentation and reassembly
- Encapsulation
- Connection control
- Ordered delivery
- Flow control
- Error control

In many respects these are really a restatement of the protocol topics listed in the previous section. A short description of each function follows.

Segmentation and Reassembly. Segmentation refers to breaking up the data into blocks with some bounded size. Depending on the semantics or system, these blocks may be called frames or packets. Reassembly is the counterpart of segmentation, that is, putting the blocks or packets back into their original order. Another name used for a data block is *protocol data unit* (PDU).

Encapsulation. Encapsulation is the adding of control information on either side of the data *text* of a block. Typical control information is the *header*, which contains address information, sequence numbers, and error control.

Connection Control. There are three stages of connection control:

1. Connection establishment.
2. Data transfer.
3. Connection termination.

Some of the more sophisticated protocols also provide connection interrupt and recovery capabilities to cope with errors and other sorts of interruptions.

Ordered Delivery. PDUs are assigned sequence numbers to ensure an ordered delivery of the data at the destination. In a large network, especially if it operates in the packet mode, PDUs (packets) can arrive at the destination out of order. With a unique PDU numbering plan using a simple numbering sequence, it is a rather simple task for a long data file to be reassembled at the destination in its original order.

Flow Control. Flow control refers to the management of data flow from source to destination such that buffers do not overflow but maintain full capacity of all facility components involved in the data transfer. Flow control must operate at several peer layers of a protocol.

Error Control. Error control is a technique that permits recovery of lost or errored PDUs. There are three possible functions involved in error control:

1. Acknowledgment of each PDU or string PDUs.
2. Sequence numbering of PDUs (e.g., missing numbers).
3. Error detection.

Acknowledgment may be carried out by returning to the source the source sequence number of a PDU. This ensures delivery of all PDUs to the destination. Error detection initiates retransmission of errored PDUs (Ref. 15).

13.7.3.2 *Open Systems Interconnection (OSI)*

Rationale. Interfacing data systems can be a complex matter. This is especially true when dealing with CPUs and workstations from different vendors as well as variations in software and operating systems. to accommodate the multitude of processing-related equipment and software, the International Standards Organization (ISO) developed the open systems interconnection (OSI) seven-layer model. This means that there are seven layers of interface starting at the communication input–output ports of a data device. These seven layers are shown in Figure 13.10.

The purpose of the model is to facilitate communication among data entities. It takes at least two to communicate. Therefore we consider the model in twos, one entity on the left of the figure and one on the right. We use the term *peers*. Peers are corresponding entities on each side of Figure 13.10. A peer on one side (system A) communicates with its peer on the other side (system B) by means of a common protocol. For example, the transport layer system A communicates with its peer transport layer at system B. It is important to note that there is no *direct* communication between peer layers except at the physical layer (layer 1). That is, above the physical layer, each protocol entity sends data *down* to the next lower layer to get the data *across* to its peer entity on the other side. Even the physical layer need not be directly connected, as in packet communication. Peer layers must share a common protocol to interface.

There are seven OSI layers, as shown in Figure 13.10. Any layer may be referred to as an N layer. Within a given system there are one or more active entities in each layer. An example of an entity is a process in a multiprocessing system. It could simply be a subroutine. Each entity communicates with entities above and below it across an interface. The interface is at a service access point (SAP). An $(N-1)$ entity provides services to an N entity by use

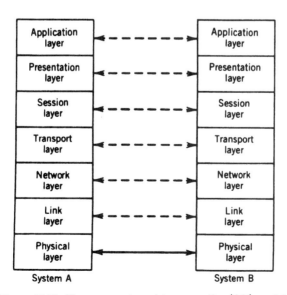

Figure 13.10. The open systems interconnection (OSI) model.

of primitives. A primitive (Ref. 15) specifies the function to be performed and is used to pass data and control information.

CCITT Rec. X.210 (Ref. 16) describes four types of primitive used to define the interaction between adjacent layers of the OSI architecture. A brief description of each of these primitives is given below:

Request. A primitive issued by a service user to invoke some procedure and to pass parameters needed to specify the service fully.

Indication. A primitive issued by a service provider either to invoke some procedure or to indicate that a procedure has been invoked by a service user at the peer service access point.

Response. A primitive issued by a service user to complete at a particular SAP some procedure invoked by an *indication* at that SAP.

Confirm. A primitive issued by a service provider to complete at a particular SAP some procedure previously invoked by a request at that SAP. CCITT Rec. X.210 (Ref. 16) adds this note: Confirms and responses can be positive or negative depending on the circumstances.

The data that pass between entities are a bit grouping called a *data unit*. We discussed protocol data units (PDUs) earlier. Data units are passed downward from a peer entity to the next OSI layer, called the $(N - 1)$ layer. The lower layer calls the PDU a *service data unit* (SDU). The $(N - 1)$ layer adds control information, transforming the SDU into one or more PDUs. However, the identity of the SDU is preserved to the corresponding layer at the other end of the connection. This concept is shown in Figure 13.11.

When we discussed throughput in Section 13.4.2, it became apparent that throughput must be viewed from the eyes of the user. With OSI some form of encapsulation takes place at every layer above the physical layer. To a greater or lesser extent OSI is used on every and all data connectivities. The concept of encapsulation, the adding of overhead, from layers 2 through 7 is shown in Figure 13.12.

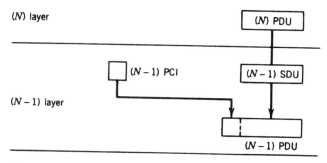

PCI = protocol control information
PDU = protocol data unit
SDU = service data unit

Figure 13.11. An illustration of mapping between data units in adjacent layers.

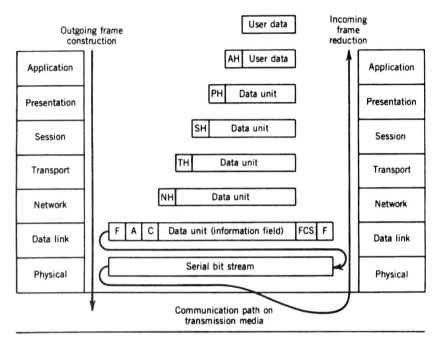

F = unique field
A = address
C = control
FCS = frame check sequence (BCC)
AH = application header
PH = presentation header

Figure 13.12. Buildup and breakdown of a data message following the OSI model. OSI encapsulates at every layer adding significant overhead.

Functions of OSI Layers

PHYSICAL LAYER. The physical layer is layer 1 and the lowest OSI layer. It provides the physical connectivity between two data end-users who wish to communicate. The services it provides to the data link layer are those required to connect, maintain, and disconnect the physical circuits that form the physical connectivity. The physical layer represents the traditional interface between data terminal equipment (DTE) and data communication equipment (DCE). (See Section 13.7.4.)

The physical layer has four important characteristics:

1. Mechanical.
2. Electrical.
3. Functional.
4. Procedural.

The mechanical aspects include the actual cabling and connectors necessary to connect the communication equipment to the media. Electrical characteristics cover voltage and impedance, balanced and unbalanced. Functional characteristics include connector pin assignments at the interface and the precise meaning and interpretation of the various interface signals and data set controls. Procedures cover sequencing rules that govern the control functions necessary to provide higher layer services such as establishing a connectivity across a switched network.

Some applicable standards for the physical layer are the following:

- EIA/TIA-232E, 422-B, 423-B, 530-A, EIA/TIA-561, EIA/TIA-562, and EIA/TIA-574
- CCITT Recs. V10, V.11, V.24, V.28, X.20, X.21, and X.21 bis
- ISO 2110, 2593, 4902, and 4903
- U.S. Fed. Stds. 1020A, 1030A, and 1031
- U.S. MIL-STD-188-114B

DATA LINK LAYER. The data link layer provides services for reliable interchange of data across a data link established by the physical layer. Link layer protocols manage the establishment, maintenance, and release of data link connections. These protocols control the flow of data and supervise error recovery. A most important function of this layer is recovery from abnormal conditions. The data link layer services the network layer or logical link control (LLC; in the case of LANs) and inserts a data unit into the INFO portion of the data frame or block. A generic data frame generated by the link layer is shown in Figure 13.13.

Some of the more common data link layer protocols are the following:

- ISO HDLC, 3309, and 4375
- CCITT LAP-B and LAP-D
- IBM BSC and SDLC
- DEC DDCMP
- ANSI ADCCP (also U.S. government)

NETWORK LAYER. The network layer moves data through the network. At relay and switching nodes along the traffic route, layering concatenates. In

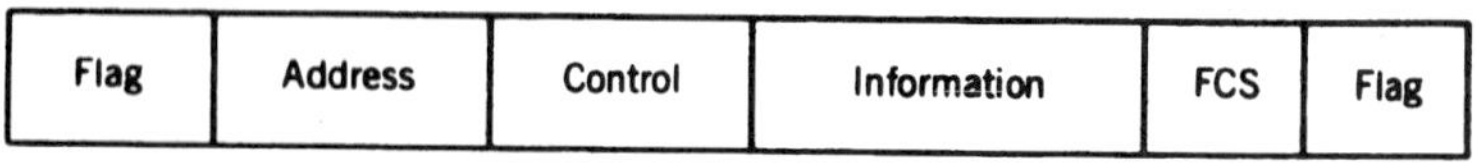

Flag	Address	Control	Information	FCS	Flag

Figure 13.13. Generalized data link layer frame. FCS = frame check sequence.

other words, the higher layers (above layer 3) are not required and are utilized only at user endpoints.

The network layer carries out the functions of switching and routing, sequencing, logical channel control, flow control, and error recovery functions. We note the duplication of error recovery in the data link layer. However, in the network layer, error recovery is networkwide, whereas on the data link layer, error recovery is concerned only with the data link involved.

The network layer also provides and manages logical channel connections between points in a network such as virtual circuits across the public switched network (PSN). It will be appreciated that the network layer concerns itself with the network switching and routing function. On simpler data connectivities, where a large network is not involved, the network layer is not required and can be eliminated. Typical of such connectivities are point-to-point circuits, multipoint circuits, and LANs. A packet-switched network is a typical example where the network layer is required.

The best known standard for layer 3 is the CCITT Rec. X.25 layer 3 standard for packet operation. CCITT Rec. X.21 provides a standard for physical layer functions for circuit-switched operation (Ref. 28).

TRANSPORT LAYER. The transport layer (layer 4) is the highest layer of the services associated with the provider of communication services. One can say that layers 1–4 are the responsibility of the communication system engineer. Layers 5–7 are the responsibility of the data end-user. However, we believe that the telecommunication system engineer should have a working knowledge of all seven layers.

The transport layer has the ultimate responsibility for providing a reliable end-to-end data delivery service for higher-layer users. It is defined as an end system function, located in the equipment using network service or services. In this way its operations are independent of the characteristics of all the networks that are involved. Services provided by a transport layer are as follows:

- *Connection Management.* This includes establishing and terminating connections between transport users. It identifies each connection and negotiates values of all needed parameters.
- *Data Transfer.* This involves the reliable delivery of transparent data between the users. All data are delivered in sequence, with no duplication or missing parts.
- *Flow Control.* This is provided on a connection basis to ensure that data are not delivered at a rate faster than the user's resources can accommodate.

The TCP (transmission control protocol) was the first working version of a transport protocol and was created by DARPA for DARPANET. All the features in TCP have been adopted in the ISO version. TCP is often lumped with the internet protocol and referred to as TCP/IP.

The ISO has developed a "transport protocol specification" for the connection-oriented transfer of data and control information from one transport entity to a peer transport entity. The ISO transport protocol messages are called transport protocol data units (TPDUs). There are connection management TPDUs and data transfer TPDUs. The applicable ISO references are ISO 8073 OSI ("Transport Protocol Specification") and ISO 8072 OSI ("Transport Service Definition").

SESSION LAYER. The purpose of the session layer is to provide the means for cooperating presentation entities to organize and synchronize their dialogue and to manage the data exchange. The session protocol implements the services that are required for users of the session layer. It provides the following services for users:

1. The establishment of session connection with negotiation of connection parameters between users.
2. The orderly release of connection when traffic exchanges are completed.
3. Dialogue control to manage the exchange of session user data.
4. A means to define activities between users in a way that is transparent to the session layer.
5. Mechanisms to establish synchronization points in the dialogue and, in case of error, resumption from a specified point.
6. Interruption of a dialogue and the resumption of it later at a specified point, possibly on a different session connection.

Session protocol messages are called session protocol data units (SPDUs). The session protocol uses the transport layer services to carry out its function. A session connection is assigned to a transport connection. A transport connection can be reused for another session connection if desired. Transport connections have a maximum TPDU size. The SPDU cannot exceed this size. More than one SPDU can be placed on a TPDU for transmission to the remote session layer.

Reference standards for the session layer are ISO 8327 ("Session Protocol Definition"[CCITT Rec. X.225]) and ISO 8326 ("Session Services Definition" [CCITT Rec. X.215]).

PRESENTATION LAYER. The presentation layer services are concerned with data transformation, data formatting, and data syntax. These functions are required to adapt the information handling characteristics of one application process to those of another application process.

The presentation layer services allow an application to interpret properly the data being transferred. For example, there are often three syntactic versions of the information to be exchanged between end-users A and B as follows:

- Syntax used by the originating application entity A
- Syntax used by the receiving application entity B
- Syntax used between presentation entities (this is called the transfer syntax) (Ref. 15)

Of course, it is possible that all three or any two of these may be identical. The presentation layer is responsible for translating the representation of information between the transfer syntax and each of the other two syntaxes as required.

The following standards apply to the presentation layer:

- ISO 8822 "Connection-Oriented Presentation Service Definition"
- ISO 8823 "Connection-Oriented Presentation Service Specification"
- ISO 8824 "Specification of Abstract Syntax Notation One"
- ISO 8824 "Specification of Basic Encoding Rules for Abstract Syntax Notation One"
- CCITT Rec. X.409 "Message Handling Systems: Presentation Transfer Syntax and Notation"

APPLICATION LAYER. The application layer is the highest layer of the OSI architecture. It provides services to the application processes. It is important to note that the applications do not reside in the application layer. Rather, the layer serves as a window through which the application gains access to the communication services provided by the model.

This highest OSI layer provides to a particular application all services related to communication in such a format that easily interfaces with the user application and is expressed in concrete quantitative terms. These include identifying cooperating peer partners, determining the availability of resources, establishing the authority to communicate, and authenticating the communication. The application layer also establishes requirements for data syntax and is responsible for overall management of the transaction.

Of course, the application itself may be executed by a machine, such as a CPU in the form of a program, or by a human operator at a workstation.

The following standards apply to the application layer (Ref. 17):

- ISO 8449/3 "Definition of Common Application Service Elements"
- ISO 8650 "Specification of Protocols for Common Application Service Elements"

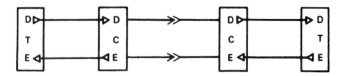

▷   Interface generator
▷   Interface load
≫   Telecommunication channel

Figure 13.14. Digital interface circuit illustrating DTE, DCE, generator, and load. DTE = data terminal equipment; DCE = data communication equipment.

13.7.4   Interfacing the Physical Layer

Figure 13.14 illustrates the physical layer interfaces. The DTE (data terminal equipment) may be an I/O device such as a personal computer (PC) or workstation, a computer, mass storage device, a server, or other peripheral equipment. The DCE (data communication equipment) is some type of device that conditions the data signal for transmission. It is a modem if it interfaces the conventional analog network.

We identify two interfaces in Figure 13.14.

- Between DTE and DCE
- Between the DCEs

The first interface has been well standardized; the second interface is less well standardized.

The U.S. EIA/TIA* has been an *ad hoc* world leader in establishing standards for the DTE–DCE interface (see Figure 13.14). The ITU-T Organization generally followed by issuing "recommendations" for this important area of data communication. Many of the ITU-T standards are similar to (but not exactly identical to) the EIA/TIA standards. There are U.S. military and U.S. federal standards that must be considered if we are working in those types of environments. It should be noted that even these standards are traceable to their EIA/TIA counterparts or are composites of those standards.

At present, the leading DTE–DCE interface standard is still EIA/TIA-232 (now in E-version) (Ref. 5). CCITT Rec. V.24 (Ref. 22) is its counterpart for that international agency. Each of these two standards define electrical, functional, and mechanical interfaces and, with several related standards, define signal levels, conditions, and polarity at each interface connection. It

---

*EIA/TIA = Electronic Industries Association/Telecommunication Industry Association.

should be noted that EIA/TIA-232 can stand on its own and does not require any supporting standards.

This interface as defined by EIA/TIA-232 is a 25-pin plug/socket. Each of the utilized 25 interface pins can be placed in one of four categories, based on the dedicated function the particular pin performs:

1. Electrical ground.
2. Data (interchange).
3. Control.
4. Clock/timing.

Table 13.3 gives EIA/TIA-232 interchange circuits by category, circuit mnemonics, and functions for each of the 25 pins with CCITT Rec. V.24 equivalence. Table 13.4 provides pin assignment information.

The following are several of the pertinent CCITT/ITU-T recommendations:

V.10   "Electrical Characteristics of Unbalanced Double-Current Interchange Circuits for General Use with Integrated Circuit Equipment in the Field of Data Communications" (Ref. 3)

V.11   "Electrical Characteristics for Balanced Double-Current Interchange Circuits for General Use with Integrated Circuit Equipment in the Field of Data Communications" (Ref. 4)

V.12   "Electrical Characteristics for Balanced Double-Current Interchange Circuits for Interfaces with Data Signaling Rates up to 52 Mbps" (Ref. 18)

V.24   "List of Definitions for Interchange Circuits Between Data-Terminal Equipment (DTE) and Data Circuit-Terminating Equipment" (Ref. 22)

V.28   "Electrical Characteristics for Unbalanced Double-Current Interchange Circuits" (Ref. 23)

V.31   "Electrical Characteristics for Single-Current Interchange Circuits Controlled by Contact Closure" (Ref. 26)

V.230   "General Data Communication Interface Layer 1 Specification" (Ref. 24)

X.20   "Interface Between Data Terminal Equipment (DTE) and Data Circuit-Terminating Equipment (DCE) for Start-Stop Transmission Services on Public Data Networks" (Ref. 27)

X.21   "Interface Between Data Terminal Equipment (DTE) and Data Circuit-Terminating Equipment (DCE) for Synchronous Operation on Public Data Networks" (Ref. 28)

TABLE 13.3 EIA/TIA-232 Interchange Circuits by Category

Circuit Mnemonics	CCITT Number	Circuit Name	Circuit Direction	Circuit Type
AB	102	Signal Common	—	Common
BA	103	Transmitted Data	To DCE	Data
BB	104	Received Data	From DCE	Data
CA	105	Request to Send	To DCE	Control
CB	106	Clear to Send	From DCE	Control
CC	107	DCE Ready	From DCE	Control
CD	108/1,/2	DTE Ready	To DCE	Control
CE	125	Ring Indicator	From DCE	Control
CF	109	Received Line Signal Detector	From DCE	Control
CG	110	Signal Quality Detector	From DCE	Control
CH	111	Data Signal Rate Selector (DTE)	To DCE	Control
CI	112	Data Signal Rate Selector (DCE)	From DCE	Control
CJ	133	Ready for Receiving	To DCE	Control
RL	140	Remote Loopback	To DCE	Control
LL	141	Local Loopback	To DCE	Control
TM	142	Test Mode	From DCE	Control
DA	113	Transmitter Signal Element Timing (DTE)	To DCE	Timing
DB	114	Transmitter Signal Element Timing (DCE)	From DCE	Timing
DD	115	Receiver Signal Element Timing (DCE)	From DCE	Timing
SBA	118	Secondary Transmitted Data	To DCE	Data
SBB	119	Secondary Received Data	From DCE	Data
SCA	120	Secondary Request to Send	To DCE	Control
SCB	121	Secondary Clear to Send	From DCE	Control
SCF	122	Secondary Received Line Signal Detector	From DCE	Control

Source: Reference 5. Copyright © 1991 by Electronic Industries Association / Telecommunications Industry Association. Reprinted with permission.

X.32 "Interface Between Data Terminal Equipment (DTE) and Data Circuit-Terminating Equipment (DCE) for Terminals Operating in the Packet Mode and Accessing a Packet Switched Telephone Network or an Integrated Services Digital Network or a Circuit Switched Public Data Network" (Ref. 29)

Note: Where CCITT/ITU-T refers to *double-current* we use the term *polar keying* or *polar transmission*; for *single-current* we use the term *neutral keying* or *neutral transmission*.

The relevant U.S. military standard is MIL-STD-188-114B, "Electrical Characteristics of Digital Interface Circuits" (Ref. 25).

TABLE 13.4 EIA/TIA-232 Pin Assignments

Pin Number	CCITT Number	Circuit	Description
1	—	—	Shield
2	103	BA	Transmitted Data
3	104	BB	Received Data
4	105/133	CA/CJ (Note 1)	Request to Send/Ready for Receiving
5	106	CB	Clear to Send
6	107	CC	DCE Ready
7	102	AB	Signal Common
8	109	CF	Received Line Signal Detector
9	—	—	(Reserved for Testing)
10	—	—	(Reserved for Testing)
11	126	(Note 4)	Unassigned
12	122/112	SCF/CI (Note 2)	Secondary Received Line Signal Detector/Data Signal Rate Selector (DCE Source)
13	121	SCB	Secondary Clear to Send
14	118	SBA	Secondary Transmitted Data
15	114	DB	Transmitter Signal Element Timing (DCE Source)
16	119	SBB	Secondary Received Data
17	115	DD	Receiver Signal Element Timing (DCE Source)
18	141	LL	Local Loopback
19	120	SCA	Secondary Request ιo Send
20	108/1,/2	CD	DTE Ready
21	140/110	RL/CG	Remote Loopback/Signal Quality Detector
22	125	CE	Ring Indicator
23	111/112	CH/CI (Note 2)	Data Signal Rate Selector (DTE/DCE Source)
24	113	DA	Transmit Signal Element Timing (DTE Source)
25	142	TM	Test Mode
26		(Note 3)	No Connection

Note 1: When hardware flow control is required Circuit CA may take on the functionality of Circuit CJ.

Note 2: For designs using interchange circuit SCF, interchange circuits CH and CI are assigned to pin 23. If SCF is not used, CI is assigned to pin 12.

Note 3: Pin 26 is contained on the Alt A connector only. No connection is to be made to this pin.

Note 4: Pin 11 is unassigned. It will not be assigned in future versions of EIA/TIA-232. However, in international standard ISO 2110, this pin is assigned to CCITT Circuit 126, Select Transmit Frequency.

Source: Reference 5. Copyright © 1991 by Electronic Association/Telecommunications Industry Association. Reprinted with permission.

EIA/TIA-232 will probably remain in force for quite some time, as well as its CCITT/ITU-T counterparts. However, perhaps by the year 2000 EIA/TIA-232 may be phased out slowly and replaced by EIA/TIA-530, supplemented by EIA/TIA-422 and -423. Essentially, EIA/TIA-530 specifies the functional and mechanical characteristics of the DTE–DCE interface, and EIA/TIA-422/423 specify the electrical characteristics of that interface. EIA/TIA-422 deals with a balanced voltage interface, and EIA/TIA-423 deals with an unbalanced interface (Refs. 14, 17).

TABLE 13.5 EIA/TIA-530 Connector Contact Assignments

Contact Number	Circuit	CCITT Number[a]	Interchange Points	Circuit Category[b]	Direction
1	Shield	—	—		
2	BA	103	A–A'	I	To DCE
3	BB	104	A–A'	I	From DCE
4	CA/CJ (Note 1)	105/133	A–A'	I	To DCE
5	CB	106	A–A'	I	From DCE
6	CC (Note 3)	107	A–A'	II	From DCE
7	AB	102A	C–C'	—	
8	CF	109	A–A'	I	From DCE
9	DD	115	B–B'	I	From DCE
10	CF	109	B–B'	I	From DCE
11	DA	113	B–B'	I	To DCE
12	DB	114	B–B'	I	From DCE
13	CB	106	B–B'	I	From DCE
14	BA	103	B–B'	I	To DCE
15	DB	114	A–A'	I	From DCE
16	BB	104	B–B'	I	From DCE
17	DD	115	A–A'	I	From DCE
18	LL	141	A–A'	II	To DCE
19	CA/CJ (Note 1)	105/133	B–B'	I	To DCE
20	CD	108/1,/2	A–A'	II	To DCE
21	RL	140	A–A'	II	To DCE
22	CE	125	A–A'	II	From DCE
23	AC	102B	C–C'		
24	DA	113	A–A'	I	To DCE
25	TM	142	A–A'	II	From DCE
26	(Note 2)	—	—	—	—

[a]CCITT number refers to CCITT/ITU-T Rec. V.24 circuit numbers.
[b]There are two categories: I and II. Category I includes data and timing, plus request to send, clear to send, received line signal detector, and ready for receiving. Category II includes remainder of control/indicator, test, and loopback circuits.
Note 1. When hardware flow control is required Circuit CA may take on the functionality of Circuit CJ.
Note 2. Contact 26 is contained on the Alt A connector only. No. connection is to be made to this contact.
Source: Reference 30, Figure 3.9, EIA/TIA-530-A. Reprinted with permission.

   EIA/TIA-232 and -530 (Ref. 30) define a 25-pin plug/socket, and both have an alt A with a 26-pin plug/socket. Table 13.5 lists connector contact assignments for EIA/TIA-530 and Table 13.6, the interchange circuits for the same standard.

   Figures 13.15 and 13.16 illustrate the digital interface circuits for EIA/TIA-422 and -423, respectively. As we discussed earlier EIA/TIA-422 specifies a balanced interface and EIA/TIA-423, an unbalanced interface. The unbalanced circuit, of course, uses a common return (ground), which is shown in Figure 13.16. In either figure, the load may be considered to be one or more receivers. Also note how the generator and the load are configured in Figures 13.15 and 13.16. Turning back to Figure 13.14, when receiving a signal *from* the line, the DCE is the generator and the DTE is the load (receiver). When transmitting signals *to* the line, the DTE is the generator and the DCE is the load.

   Some basic guidelines for the application of balanced and unbalanced electrical connections DTE–DCE are presented as follows. EIA/TIA states

TABLE 13.6   EIA/TIA-530 Interchange Circuits

Circuit Mnemonic	CCITT Number	Circuit Name	Circuit Direction	Circuit Type
AB	102	Signal Common		COMMON
AC	102B	Signal Common		
BA	103	Transmitted Data	To DCE	DATA
BB	104	Received Data	From DCE	
CA	105	Request to Send	To DCE	
CB	106	Clear to Send	From DCE	
CF	109	Received Line Signal Detector	From DCE	
CJ	133	Ready for Receiving	To DCE	
CE	125	Ring Indicator	From DCE	CONTROL
CC	107	DCE Ready	From DCE	
CD	108/1,/2	DTE Ready	To DCE	
DA	113	Transmit Signal Element Timing (DTE Source)	To DCE	
DB	114	Transmit Signal Element Timing (DCE Source)	From DCE	TIMING
DD	115	Receiver Signal Element Timing (DCE Source)	From DCE	
LL	141	Local Loopback	To DCE	
RL	140	Remote Loopback	To DCE	
TM	142	Test Mode	From DCE	

Note: The classification of circuits is similar to EIA/TIA-232: signal common (ground), data, control, and timing.

Source: Reference 30, Figure 4.1, EIA/TIA-530-A. Reprinted with permission.

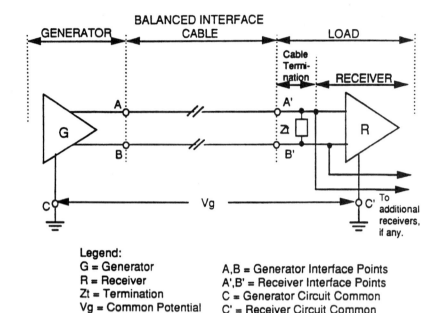

Figure 13.15. Balanced voltage digital interface circuit, EIA/TIA-422-B. (From Ref. 19, Figure 2. Courtesy of EIA/TIA.)

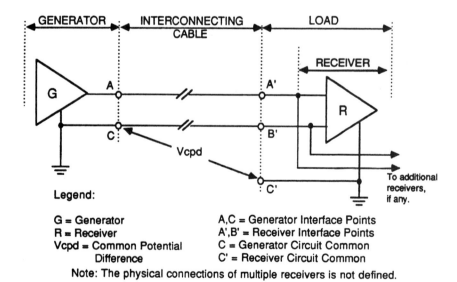

Figure 13.16. Unbalanced voltage digital interface circuit, EIA/TIA-423-B. (From Ref. 20. Courtesy of EIA/TIA.)

in RS-422 the following:

> While the balanced interface is intended for use at the higher modulation rates, it may, in preference to the unbalanced interface circuit, generally be required where any of the following conditions prevail:

- The interconnecting cable (i.e., DTE–DCE) is too long for effective unbalanced operation.
- The interconnecting cable is exposed to extraneous noise sources that may cause an unwanted voltage in excess of plus or minus 1 V measured between the signal conductor and the circuit common at the load end of the cable with a 50-Ω resistor substituted for the generator.
- It is necessary to minimize interference with other signals.
- Inversion of signals may be required, e.g., PLUS MARK to MINUS MARK, may be obtained by inverting the cable pair.

EIA/TIA-422 states, in essence (Figure 13.15), that a generator, as defined in the standard, results in a low-impedance (100 Ω or less) balanced voltage source that will produce a differential voltage applied to the interconnecting cable in the range of 2 V. The signaling sense of the voltages appearing across the interconnection cable is defined as follows:

1. The A terminal of the generator shall be negative with respect to the B terminal for a binary 1 (MARK or OFF state)
2. The A terminal of the generator shall be positive with respect to the B terminal for a binary 0 (SPACE or ON state)

Note: Compare this signaling convention to that in Table 13.1.

EIA/TIA-423 states, in essence (Figure 13.16), that a generator circuit, as defined in the standard, results in a low-impedance (50 Ω or less) unbalanced voltage source that will produce a voltage range of 4–6 V. The signaling sense of the voltage appearing across the interconnecting cable is defined as follows:

1. The A terminal of the generator shall be negative with respect to the C terminal for a binary 1 (MARK or OFF state)
2. The A terminal of the generator shall be positive with respect to the C terminal for a binary 0 (SPACE or ON state)

The test load termination for a balanced circuit is two 50-Ω resistors in series between the two signal leads (A and B in Figure 13.15). The test lead termination for the unbalanced circuit is 450 Ω between the generator output terminal and "common."

For the balanced or unbalanced case, load characteristics during real operating conditions result in a differential receiver having a high input impedance (greater than 4 kΩ), a small input threshold transition region between -0.2 and $+0.2$ V, and an allowance for an internal bias voltage not to exceed 3 V in magnitude.

EIA/TIA-422 and -423 provide guidance on maximum cable length DTE–DCE. For the case of balanced signal lines (EIA/TIA-422): up to 90 kbps, 4000 ft; to 1 Mbps, 380 ft; and to 10 Mbps, 40 ft. For the unbalanced signal lines (EIA/TIA-423): up to 900 bps, 4000 ft; to 10 kbps, 380 ft; and to 100 kbps, 40 ft. EIA states that these lengths are on the conservative side.

EIA/TIA-530-A (Ref. 30) is another in a series of EIA DTE–DCE interface standards whose electrical characteristics are specified in EIA/TIA-422 for balanced circuits or -423 for unbalanced circuits. EIA/TIA-530 has been formulated for data circuits operating at rates from 20 kbps to 20 Mbps. The EIA/TIA-530 connector is a 25-pin connector similar to that of EIA/TIA-232D; however, it is D-shaped. Circuit names for the first eight pins in both standards are the same, but differ for pins 9–11, which are not used with EIA/TIA-232.

The following provides a cross-reference of interface standards:

EIA-423A	U.S. Fed. Std. 1030A	CCITT V.10(X.26)
EIA-422A	U.S. Fed. Std. 1020A	CCITT V.11(X.27)
EIA-449	U.S. Fed. Std. 1031	CCITT V.24/V.10/V.11, ISO 4902
EIA-232D		CCITT V.24/V.28, ISO 2110
EIA-530		CCITT V.230

13.8 DIGITAL DATA TRANSMISSION ON AN ANALOG CHANNEL

13.8.1 Introduction

There is a basic incompatibility between baseband data transmission (i.e., an electrical signal) and the analog voice channel. We defined the analog voice (VF) channel as a frequency band between 300 and 3400 Hz (ITU-T definition). A serial RZ or NRZ data signal contains a major energy content at 0 Hz (dc), which the VF channel will not pass. A *data modem* converts this essentially dc signal to an audio tone usually in the center of the VF channel, at 1700 or 1800 Hz for more advanced modems. *Modem* is an acronym for modulator–demodulator.

13.8.2 Modulation–Demodulation Schemes

A modem modulates and demodulates. There are three generic types of modulation that a modem may use and include one or a combination of the following:

- Amplitude modulation (AM)
- Frequency modulation (FM)
- Phase modulation (PM)

Note: Data transmission terminology derives from its direct predecessor, telegraph transmission. From the language of telegraph we say that a transmitter is *keyed*. Thus we derive the word *keying* and *shift*, meaning a change of state. As a result we have ASK (amplitude shift keying), FSK (frequency shift keying), and PSK (phase shift keying).

Amplitude Modulation (ASK). With this modulation technique, binary states are represented by the presence or absence of an audio tone or carrier. More often it is referred to as on−off telegraphy or on−off keying. The use of ASK (alone) is deprecated because it is susceptible to sudden gain changes and it is inefficient in modulation and spectrum utilization, particularly at the higher data rates.

Frequency Modulation (FSK). For data transmission rates of 1200 bps and below, FSK is almost universally used. The two binary states are represented by two different frequencies and are detected by using two frequency-tuned sections, one tuned to each of the two bit frequencies. The demodulated signals are then integrated over the duration of a bit, and upon the result a binary decision is based.

Digital transmission using frequency shift keying (FSK) has the following advantages: (1) the implementation is not much more complex than an AM system; and (2) since the received signals can be amplified and limited at the receiver, a simple limiting amplifier can be used, whereas the AM system requires sophisticated automatic gain control (AGC) in order to operate over a wide level range. Another advantage is that FSK can show a 3- or 4-dB improvement over AM in most types of noise environment, especially at distortion threshold (i.e., at the point where the distortion is such that good printing is about to cease). As the frequency shift becomes greater, the advantage over AM improves in a noisy environment.

Another advantage of FSK is its immunity from the effects of nonselective level variations even when extremely rapid. Thus it is used almost exclusively on worldwide HF radio transmission, where rapid fades are a common occurrence. In the United States it has nearly universal application for the transmission of data at the lower data rates (i.e., 1200 bps and below).

Phase Modulation (PSK). For systems using higher data rates (i.e., above 1200 bps), a type of phase modulation called phase shift keying (PSK) becomes more attractive. Various forms are used, such as two-phase, relative phase, quadrature-phase, and eight- phase. A two-phase system uses one phase of the carrier frequency for one binary state and the other phase for the other binary state. The two phases are 180° apart and are detected by a synchronous detector using a reference signal at the receiver that is of known phase with respect to the incoming signal. This known signal is at the same frequency as the incoming signal carrier and is arranged to be in phase with one of the binary signals. This is called coherent detection.

In the relative (differential) phase system, a binary 1 is represented by sending a signal burst of the same phase as that of the previous signal burst sent. A binary 0 is represented by a signal burst of a phase opposite to that of the previous signal transmitted. The signals are demodulated at the receiver by integrating and storing each signal burst of one bit period for comparison in phase with the next signal burst.

In the quadrature-phase system, two binary channels (2 bits) are phase multiplexed onto one tone by placing them in phase quadrature as shown in the sketch below. An extension of this technique places two binary channels on each of several tones spaced across the voice channel of a typical telephone circuit.

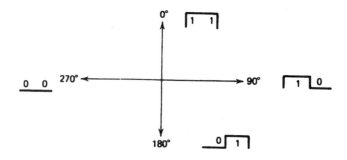

Some of the advantages of phase modulation are that:

- All available power is utilized for intelligence conveyance.
- The demodulation scheme has good noise rejection capability.
- The system yields a smaller noise bandwidth.

13.8.3   Critical Parameters

A most important consideration are the effects of the various telephone circuit parameters on the capability of a circuit to transport data at an acceptable error rate. The following discussion is to familiarize the reader with the problems most likely to be encountered in the transmission of data over analog circuits or the analog end-portion of the digital network and to make some generalizations in some cases, which can help in planning the implementation of data transmission systems.

Phase Distortion (Delay Distortion). Phase distortion or "delay distortion constitutes the most limiting impairment to data transmission, particularly over telephone voice channels" (Ref. 29). When specifying phase distortion, the terms *envelope delay distortion* and *group delay* are often used. The IEEE standard dictionary (Ref. 1) states that "envelope delay is often defined the same as group delay, that is, the rate of change, with angular frequency, of the phase shift between two points in a network" (see Section 1.9.4).

The problem is that in a band-limited system, such as the typical telephone voice channel, not all frequency components of a signal will propagate to the receiving end of a circuit in exactly the same elapsed time. Filters are the principal culprit. Inductive loading looks and acts like a filter. Digital channel banks (codecs) have input and output low-pass filters. FDM equipment has many filters. Figure 13.17 shows a typical frequency-delay response curve in milliseconds for a voice channel end-to-end. For the voice channel (or any symmetrical passband for that matter), delay increases toward band edges and is minimum around the center portion of the passband, around 1700–1800 Hz. It is for this reason that a modem tone operates at 1800 Hz for data rates above 1200 bps. As the bit rate increases (i.e., pulsewidths become shorter and shorter) some of the received signal energy will spill into subsequent bit positions, confusing the decision of the demodulator. This is intersymbol interference (ISI).

In essence, therefore, we are dealing with the phase linearity of a circuit. If the phase–frequency relationship over the passband is not linear, distortion will occur in the transmitted signal. This distortion is best measured by a parameter called *envelope delay distortion* (EDD). Mathematically, envelope delay is the derivative of the phase shift with respect to frequency. The maximum difference in the derivative over any frequency interval is called EDD. Hence EDD is always the difference between the envelope delay at one frequency and that at another frequency of interest in a passband. The EDD unit of measurement is milliseconds or microseconds.

When transmitting data, the shorter the pulse (or symbol) width (in the case of binary systems this would be the width of 1 bit), the more critical the EDD constraints become. As a rule of thumb, delay distortion in the passband should be below the period of 1 bit (or symbol).

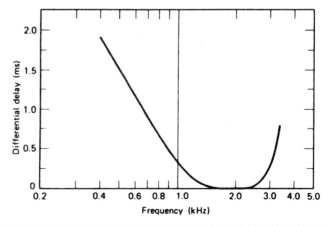

Figure 13.17. Typical differential delay across a voice channel. Note that the central point of maximum flatness is centered at about 1800 Hz.

Attention Distortion (Amplitude Response). Another parameter that seriously affects the transmission of data and that can place very definite limits on the modulation rate is that of amplitude response. Ideally, all frequencies across the passband of the channel of interest should suffer the same attenuation. Place a −10-dBm signal at any frequency between 300 and 3400 Hz, and the output at the receiving end of the channel may be −23 dBm, for example, at any and all frequencies in the band; we would then describe a fully flat channel. Such a channel has the same loss or gain at any frequency within the band. This type of channel is ideal but would be unachievable in a real, working system. In Rec. G.132, CCITT recommends no more than 9 dB of amplitude distortion relative to 800 Hz between 400 and 3000 Hz. This figure of 9 dB describes the maximum variation that may be expected from the reference level at 8000 Hz. This variation of amplitude response often is called attenuation distortion. A conditioned channel, such as a Bell System C-4 channel, will maintain a response of −2 to +3 dB from 500 to 3000 Hz and −2 to +6 dB from 300 to 3200 Hz. Channel conditioning and equalization are discussed in Section 13.11.

Considering tandem operation, the deterioration of amplitude response is arithmetically summed when sections are added. This is particularly true at band edge in view of channel unit transformers and filters, which account for the upper and lower cutoff characteristics.

Amplitude response is also discussed in Section 1.9.3. Figure 13.18 illustrates a typical example of amplitude response across carrier equipment (see Chapter 3) connected back-to-back at the voice channel input–output.

Tables 13.7 and 13.8 give the AT&T requirements for leased lines covering attenuation distortion and EDD.

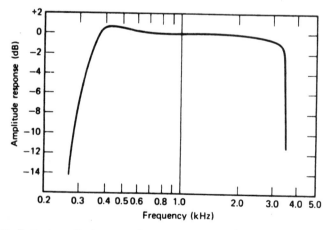

Figure 13.18. Typical amplitude versus frequency response (attenuation distortion) across a voice channel. Carrier equipment is connected back-to-back.

TABLE 13.7   AT&T Requirements for Two-Point or Multipoint Channel Attenuation Distortion

	Frequency Band (Hz)	Attenuation[a] (dB)
Basic requirements	500–2500	−2 to +8
	300–3000	−3 to +12
C1 conditioning	1000–2400	−1 to +3
	300–2700	−2 to +6
	2700–3000	−3 to +12
C2 conditioning	500–2800	−1 to +3
	300–3000	−2 to +6
C4 conditioning	500–3000	−2 to +3
	300–3200	−2 to +6
C5 conditioning	500–2800	−0.5 to +1.5
	300–3000	−3 to +3

[a]Relative to 1000 Hz.
Source: Reference 31.

TABLE 13.8   AT&T Requirements for Two-Point or Multipoint Channel Envelope Delay Distortion

	Frequency Band (Hz)	EDD (μs)[a]
C1 conditioning	800–2600	1750
	1000–2400	1000
C2 conditioning	1000–2600	500
	600–2600	1500
	500–2800	3000
C4 conditioning	1000–2600	300
	800–2800	500
	600–3000	1500
	500–3000	3000
C5 conditioning	1000–2600	100
	600–2600	300
	500–2800	600

[a]Maximum in-band envelope delay difference.
Source: Reference 31.

Noise. Another important consideration in the transmission of data is that of noise. All extraneous elements appearing at the voice channel output that were not due to the input signal are considered to be noise. For convenience noise is broken down into four categories:

- Thermal
- Crosstalk
- Intermodulation
- Impulse

Thermal noise, often called resistance noise, white noise, or Johnson noise, is of a Gaussian nature or fully random. Any system or circuit operating at a temperature above absolute zero will inherently display thermal noise. It is caused by the random motions of discrete electrons in the conduction path.

Crosstalk is a form of noise caused by unwanted coupling from one signal path into another. It may be due to direct inductive or capacitive coupling between conductors or between radio antennas (see Section 1.9.6).

IM noise is another form of unwanted coupling, usually caused by signals mixing in nonlinear elements of a system. Carrier and radio systems are highly susceptible to IM noise, particularly when overloaded (see Section 1.9.6).

Impulse noise* is a primary source of errors in the transmission of data over telephone networks. It is sporadic and may occur in bursts or discrete impulses called *hits*. Some types of impulse noise are natural, such as that from lightning. However, man-made impulse noise, such as from automobile ignition systems or power lines, is ever-increasing. Impulse noise may be of high level in analog telephone switching centers due to dialing, supervision, and switching impulses, which may be induced or otherwise coupled into the data transmission channel.

For our discussion of data transmission, only two forms of noise will be considered: random or Gaussian noise and impulse noise. Random noise measured with a typical transmission measuring set appears to have a relatively constant value. However, the instantaneous value of the noise fluctuates over a wide range of amplitude levels. If the instantaneous noise voltage is of the same magnitude as the received signal, the receiving detection equipment may yield an improper interpretation of the received signal and an error or errors will occur. For a proper analytical approach to the data transmission problem, it is necessary to assume a type of noise that has an amplitude distribution that follows some predictable pattern. Thermal noise or random noise has a Gaussian distribution and is considered representative of the noise encountered on the analog telephone channel (i.e., the voice channel). From the probability distribution curve of Gaussian noise shown in Figure 13.19, we can make some accurate predictions. It may be noted from this curve that the probability of occurrence of noise peaks that have amplitudes 12.5 dB above the rms level is 1 in 10^5. Accordingly, if we wish to ensure an error rate of 10^{-5} in a particular system using binary polar modulation, the rms noise should be at least 12.5 dB below the signal level (Ref. 79, p. 114). This simple analysis is valid for the type of modulation used, assuming that no other factors are degrading the operation of the system and that a cosine-shaped receiving filter is used. If we were to interject EDD, for example, into the system, we could translate the degradation into an equivalent signal-to-noise ratio improvement necessary to restore the desired error

*Called *impulsive noise* in the United Kingdom and by the ITU-T Organization.

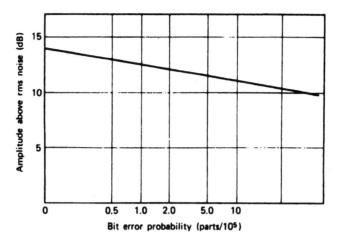

Figure 13.19. Probability of bit error in Gaussian noise, binary polar transmission.

rate. For instance, if the delay distortion were the equivalent of one pulsewidth, the signal-to-noise ratio improvement required for the same error rate would be about 5 dB, or the required signal-to-noise ratio would now be 17.5 dB.

For reasons that will be discussed later in Section 13.10, let us assume that the signal level is -10 dBm at the zero transmission level point of the system. Then the rms noise measured at the same point would be -27.5 dBm to retain the error rate of 1 in 10^5.

In order for the above figure to have any significance, it must be related to the actual noise found in a channel. CCITT recommends no more than 50,000 pW of noise psophometrically weighted on an international connection made up of six circuits in a chain. However, CCITT states (Recs. G.142A and 142D) that, for data transmission at as high a modulation rate as possible without significant error rate, a reasonable circuit objective for maximum random noise would be -40 dBm0p for leased circuits (impulse noise not included) and -36 dBm0p for switched circuits without compandors. This figure obviously appears quite favorable when compared to the -27 dBm0 (-29.5 dBm0p) required in the example above. However, other factors that will be developed later in Section 13.10 will consume much of the noise margin that appears available.

Whereas random noise has a rms value when we measure level, impulse noise is another matter entirely. It is measured as the number of "hits" or "spikes" per interval of time over a certain threshold. In other words, it is a measurement of the recurrence rate of noise peaks over a specified level. The word *rate* should not mislead the reader. The recurrence is not uniform per unit time, as the word *rate* may indicate, but we can consider a sampling and convert it to an average.

AT&T (Ref. 31) states the following:

The impulse noise objective is specified in terms of the rate of occurrence of the impulse voltages above a specified magnitude. The objective is expressed as the threshold in dBrnc0 at which no more than 15 impulses in 15 minutes are measured by an impulse counter with a maximum counting rate of 7 counts per second. The overall objective of 71 dBrnc0 implies a 6-dB signal-to-impulse noise threshold in the presence of a -13-dBm0 signal.

CCITT states (Rec. Q.45) that

in any four-wire international exchange the busy hour impulsive noise counts should not exceed 5 counts in 5 minutes at a threshold level of -35 dBm0.

Remember that random noise has a Gaussian distribution and will produce peaks at 12.5 dB over the rms value (unweighted) 0.001% of the time on a data bit stream for an equivalent error rate of 1×10^{-5}. It should be noted that some references use 12 dB, some 12.5 dB, and others 13 dB. The 12.5 dB above the rms random noise floor should establish the impulse noise threshold for measurement purposes. We should assume in a well-designed data transmission system traversing the telephone network that the signal-to-noise ratio of the data signal will be well in excess of 12.5 dB. Thus impulse noise may well be the major contributor to degrade the error rate.

Care must be taken when measuring impulse noise. A transient such as an impulse noise spike in a band-limited system (which our telephone network most certainly is) tends to cause "ringing." Here the initial impulse noise spike causes what we might call a main bang or principal spike followed by damped subsidiary spikes. If we are not careful, these subsidiary spikes, that ringing effect, may also be counted as individual hits in our impulse noise count total. To avoid this false counting, impulse noise meters have a built-in dead time after each count. It is a kind of damping. The Bell System, for example, specifies a 150-ms dead time after each count. This limits the counting capability of the meter to no more than 6 or 7 counts per minute (cpm).

In this damping or dead time period, missed (real) impulse noise hits may seem to be a problem. For instance, the Bell System suggests that the average improved (increased) sensitivity to measure "all" hits is only 0.9 dB, with a standard deviation of 0.76 dB (Ref. 32).

The period of measurement is also important. How long should the impulse noise measurement set remain connected to a line under test to give an accurate count? It appears empirically that 30 min is sufficient. However, a good estimate of error can be made and corrected for if that period of time is reduced to 5 min. This is done by reducing the threshold (on paper) of the measuring set. From Ref. 32, the standard deviation for a 5-min period is about 2.2 dB. Therefore 95% of all 5-min measurements will be within ± 3.6 dB of a 30-min measurement period.

To clarify this, remember that impulse noise distributions are log-normal and impulse noise level distributions are normal. With this in mind we can relate count distributions, which can be measured readily, to level distributions. The mean of the level distribution is the threshold value of which the impulse noise level meter was set to record the count distribution (in dBm, dBmp, dBrnC, or whatever unit). The set has a count associated with that threshold, which is simply the median of the observed count distribution. The sigma σ_1 standard deviation of the impulse noise level distribution is estimated by the expression

$$\sigma_1 = m\sigma_D \tag{13.1}$$

where m = inverse slope of the peak amplitude distribution in decibels per decade of counts, and averages 7.0 dB, and σ_D = standard deviation of the log-normal count distribution, which is the square root of the $\log_{10}$ of the ratio of the average number of counts to the median count, or

$$\sigma_D = \sqrt{\log_{10}\frac{\text{average count}}{\text{median count}}} \tag{13.2}$$

where the median is not equal to zero. For instance, if we measured 10 cpm at a given threshold, the 1-cpm threshold would be 7 dB above the 10-cpm threshold.

When an unduly high error rate has been traced to impulse noise, there are some methods for improving conditions. Noisy areas may be bypassed, repeaters may be added near the noise source to improve the signal-to-impulse noise ratio, or in special cases pulse smearing techniques may be used. This latter approach uses two delay distortion networks that complement each other such that the net delay distortion is zero. By installing the networks at opposite ends of the circuit, impulse noise passes through only one network and is hence smeared because of the delay distortion. The signal is unaffected because it passes through both networks.

Frequency Translation Errors. Total end-to-end frequency translation errors on a voice channel being used for data or telegraph transmission must be limited to 2 Hz (CCITT Rec. G.135). This is an end-to-end requirement. For systems using FDM carrier equipment, frequency translations occur mostly owing to that carrier equipment's modulation and demodulation steps. FDM carrier equipment widely uses single-sideband suppressed carrier (SSBSC) techniques. Nearly every case of error can be traced to errors in frequency translation (we refer here to deriving the group, supergroup, mastergroup, and its reverse process; see also Chapter 3) and carrier reinsertion frequency offset, the frequency error being exactly equal to the error in translation and offset or the sum of several such errors. Frequency locked (e.g., synchronized) or high-stability master carrier generators (1×10^{-7} or 1×10^{-8}, depending on the system), with all derived frequency sources slaved to the master source, usually are employed to maintain the required stability.

In the digital network frequency translation errors are a secondary issue.

Although 2 Hz seems to be a very rigid specification, when added to the possible back-to-back error of the modems themselves, the error becomes more appreciable. Much of the trouble arises with modems that employ sharply tuned filters. This is true of telegraph equipment in particular. But for the more general case, high-speed data modems can be designed to withstand greater carrier shifts than those that will be encountered over good telephone circuits.

13.8.4 Special-Quality International Leased Circuits

13.8.4.1 Introduction. CCITT offers two recommendations dealing with line conditioning for international leased circuits, Recs. M.1020 and M.1025 (Refs. 34, 35). The two recommendations are similar except for amplitude and phase distortion. The latter is given as group delay. The following subsections give the highlights of the two recommendations. For Rec. M.1025, only amplitude and phase distortion parameters are given.

13.8.4.2 With Special Bandwidth Conditioning. The following requirements are based on CCITT Rec. M.1020. Figure 13.20 shows the limits for

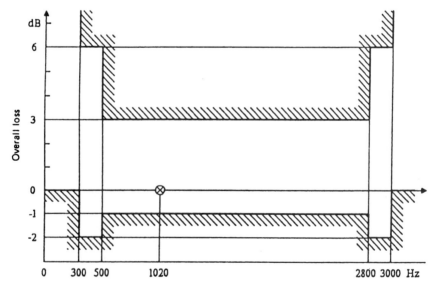

Figure 13.20. Limits of overall loss of the circuit relative to that at 1020 Hz. (From Ref. 34, Figure 1/M.1020, CCITT Rec. M.1020.)

overall loss for a voice channel relative to that at 1020 Hz. The figure, of course, also represents the limits of amplitude distortion, which can vary from -1 to $+3$ dB from 500 to 2500 Hz.

Figure 13.21 gives the limits of group delay (distortion) relative to the minimum measured group delay in the band from 500 to 2800 Hz.

The following summarizes other requirements of the recommendation:

- Amplitude hits greater than ± 2 dB should not exceed 10 in any 15-min measuring period.
- Impulse noise: the number of impulse noise peaks exceeding -21 dBm0 should not exceed 18 in 15 min.
- Phase jitter should not exceed $10°$ peak-to-peak under normal circumstances and $15°$ on particularly complex circuits.
- Frequency error shall not exceed ± 5 Hz.
- Total distortion (including quantizing distortion): The signal-to-total-distortion ratio should be better than 28 dB using a sine wave signal at -10 dBm0.

13.8.4.3 *With Basic Bandwidth Conditioning.* This subsection is based on CCITT Rec. M.1025. Figure 13.22 shows the limits of overall loss for a voice channel relative to 1020 Hz. As in the previous subsection, we can also interpret the figure for amplitude distortion. Figure 13.23 shows the limits of group delay relative to the minimum measured group delay in the band 600–2800 Hz.

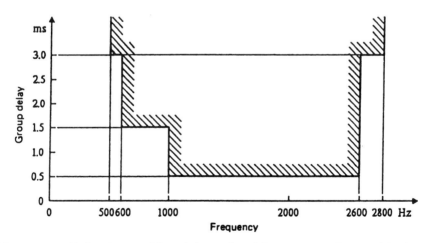

Figure 13.21. Limits of group delay relative to the minimum measured group delay in the 500–2800-Hz band. (From Ref. 34, Figure 2/M.1020, CCITT Rec. M.1020.)

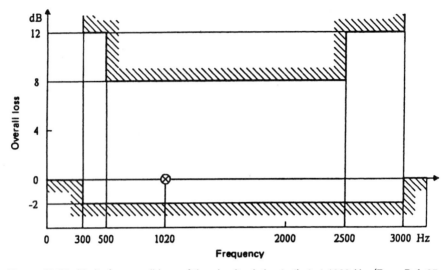

Figure 13.22. Limits for overall loss of the circuit relative to that at 1020 Hz. (From Ref. 35, Figure 1/M.1025, CCITT Rec. M.1025.)

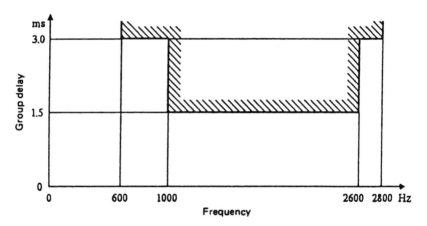

Figure 13.23. Limits for group delay relative to the minimum measured group delay in the 600–2800-Hz band. (From Ref. 35, Figure 2/M.1025, CCITT Rec. M.1025.)

13.9 CHANNEL CAPACITY

A leased or switched voice channel represents a financial investment. The goal of the system engineer is to derive as much benefit as possible from the money invested. For the case of digital transmission this is done by maximizing the information transfer across the system. This subsection discusses how much information in bits can be transmitted, relating information to bandwidth, signal-to-noise ratio, and error rate. An empirical discussion of these matters is carried out in Section 13.10.

First, looking at very basic information theory, Shannon stated in his bandwidth paper (Ref. 33) that if the input information rate to a band-limited channel is less than C (bps), a code exists for which the error rate approaches zero as the message length becomes infinite. Conversely, if the input rate exceeds C, the error rate cannot be reduced below some finite positive number.

The usual voice channel approximates to a Gaussian band-limited channel (GBLC) with additive Gaussian noise. For such a channel consider a signal wave of a mean power of S watts applied at the input of an ideal low-pass filter having a bandwidth of W Hz and containing an internal source of mean Gaussian noise with a mean power of N watts uniformly distributed over the passband. The capacity in bits per second is given by

$$C = W \log_2\left(1 + \frac{S}{N}\right) \tag{13.3}$$

Applying Shannon's "capacity" formula (GBLC) to some everyday voice channel criteria, $W = 3000$ Hz and $S/N = 1023$, then

$$C = 30,000 \text{ bps}$$

(Remember that bits per second and baud are interchangeable in binary systems.)

Nether S/N nor W is an unreasonable value. Seldom, however, can we achieve a modulation rate greater than 3000 bauds. The big question in advanced design is how to increase the data rate and keep the error rate reasonable.

One important item that Shannon's formula did not take into consideration is intersymbol interference. A major problem of a pulse in a band-limited channel is that the pulse tends not to die out immediately, and a subsequent pulse interfered with by "tails" from the preceding pulse. This is shown in Figure 13.24.

Nyquist provided another approach to the data rate problem, this time using intersymbol interference (the tails in Figure 13.24) as a limit (Ref. 36). This resulted in the definition of the so-called Nyquist rate, which is $2W$ elements per second. W is the bandwidth in hertz of a band-limited channel as shown in Figure 13.24. In binary transmission we are limited to $2W$ bps. If we let $W = 3000$ Hz, the maximum data rate attainable is 6000 bps. Some refer to this as "the Nyquist 2-bit rule."

The key here is that we have restricted ourselves to binary transmission and we are limited to $2W$ bps no matter how much we increase the signal-to-noise ratio. The Shannon GBLC equation indicates that we should be able to increase the information rate indefinitely by increasing the

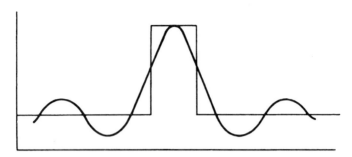

Figure 13.24. Pulse response through a Gaussian band-limited channel (GBLC).

signal-to-noise ratio. The way to attain a higher value of C is to replace the binary transmission system with a multilevel system, often termed an M-ary transmission system with $M > 2$. An M-ary channel can pass $2W \log_2 M$ bps with an acceptable error rate. This is done at the expense of the signal-to-noise ratio. As M increases (as the number of levels increases), so must the signal-to-noise ratio increase to maintain a fixed error rate.

13.10 VOICE CHANNEL DATA MODEMS VERSUS CRITICAL DESIGN PARAMETERS

The critical parameters that affect data transmission have been discussed. They are amplitude–frequency response (sometimes called amplitude distortion), envelope delay distortion (EDD), and noise. Now we relate these parameters to the design of data modems to establish some general limits or "boundaries" for equipment of this type. The discussion that follows purposely avoids HF radio considerations.

As stated earlier in the coverage of EDD, it is desirable to keep the transmitted pulse (bit) length equal to or greater than the residual differential EDD. Since about 1.0 ms is assumed to be a reasonable residual delay after equalization (conditioning), the pulse length should then be no less than approximately 1 ms. This corresponds to a modulation rate of 1000 pulses per second (binary). In the interest of standardization (CCITT Rec. V.22 [Ref. 41]), this figure is modified to 1200 bps.

The next consideration is the usable bandwidth required for the transmission of 1200 bps. This figure is about 1800 Hz, using modulation methods such as PSK, FSK, or DSB-AM, and somewhat less for VSB-AM. Since delay distortion of a typical voice channel is at its minimum between 1700 and 1900 Hz, the required band, when centered about these points, extends from 800 to 2600 Hz or from 1000 to 2800 Hz. From the previous discussion we can see from Figure 13.17 and Table 13.8 that the EDD requirement is met easily over the range of 800–2800 Hz.

Bandwidth limits modulation rate (as will be discussed in the following). However, the modulation rate in bauds and the data rate in bits per second may not necessarily be the same. This is a very important concept.

Suppose a modulator looked at the incoming serial bit stream 2 bits at a time rather than the conventional 1 bit at a time. Now let four amplitudes of a pulse be used to define each of the four possible combinations of 2 consecutive bits, such that

$$A_1 = 00$$

$$A_2 = 01$$

$$A_3 = 11$$

$$A_4 = 10$$

where A_1, A_2, A_3, and A_4 represent the four pulse amplitudes. This form of treating 2 bits at a time is called dibit coding (see Section 13.8.2).

Similarly, we could let eight pulse levels cover all the possible combinations of 3 consecutive bits so that with a modulation rate of 1200 bauds it is possible to transmit information at a rate of 3600 bps. Rather than vary amplitude to 4 or 8 levels, phase can be varied. A four-phase system (PSK) could be coded as follows:

$$F_1 = 0° = 00$$

$$F_2 = 90° = 01$$

$$F_3 = 180° = 11$$

$$F_4 = 270° = 10$$

Again, with a four-phase system using dibit coding, a tone with a modulation rate of 1200-baud PSK can be transmitting 2400 bps. An eight-phase PSK system at 1200 bauds could produce 3600 bps of information transfer. Obviously, this process cannot be extended indefinitely. The limitation comes from channel noise. Each time the number of levels or phases is increased, it is necessary to increase the signal-to-noise ratio to maintain a given error rate. Consider the case of a signal voltage S and a noise voltage N. The maximum number of increments of signal (amplitude) that can be discerned is S/N (since N is the smallest discernible increment). Add to this the no-signal case, and the number of discernible levels now becomes $S/N + 1$ or $S + N/N$, where S and N are expressed as power. The number of levels becomes the square root of this expression. This formula shows that in going from 2 to 4 levels or from 4 to 8 levels, a roughly 6-dB noise penalty is incurred each time we double the number of levels.

A similar analysis is carried out for the multiphase case; the penalty in going from four phases to eight phases is 3 dB, for example. See Chapter 5.

Sufficient background has been developed to appraise the data modem for the voice channel. Now consider a data modem for a data rate of 2400 bps. By using quaternary phase shift keying (QPSK), as described previously, 2400 bps is transmitted with a modulation rate of 1200 bauds. Assume that the modem uses differential phase detection wherein the detector decisions are based on the change in phase between the last transition and the preceding one.

Assume the bandwidth to be present for the data modem under consideration (for most telephone networks the minimum bandwidth discussed for the sample case is indeed present—1800 Hz). It is now possible to determine if the noise requirements can be satisfied. Figure 13.19 shows that a 12.5-dB signal-to-noise ratio (Gaussian noise) is required to maintain an error rate of 1×10^{-5} for a binary polar (AM) system. As is well established, PSK systems have about a 3-dB improvement. In this case only a 9.6-dB signal-to-noise ratio would be needed, all other factors held constant (no other contributing factors).

Assume the input from the line to be -10 dBm0 in order to satisfy loading conditions. To maintain the proper signal-to-noise ratio, the channel noise must be down to -19.6 dBm0.

To improve the modulation rate without the expense of increased bandwidth, QPSK (four-phase) is used. Allow a 3-dB general noise degradation factor, bringing the required noise level down to -22.6 dBm0.

Consider now the effects of EDD. It has been found that for a four-phase differential system, this degradation will amount to 6 dB if the permissible delay distortion is one pulsewidth. This impairment brings the noise requirement down to -28.6 dBm0 of average noise power in the voice channel. Allow 1 dB for frequency translation error or other factors, and the noise requirement is now down to -29.6 dBm0.

If the transmit level were -13 dBm0 instead of -10 dBm0, the numbers for noise must be adjusted another 3 dB such that it is now down to -32.6 dBm0. Therefore it can be seen that to achieve a certain error rate for a given modulation rate, several modulation schemes should be considered. It is safe to say that in the majority of these schemes the noise requirement will fall somewhere between -25 and -40 dBm0. This is safely inside the CCITT ITU-T figure of -43 dBm (see subsection entitled "Noise" under Section 13.8.3). More discussion on this matter may be found in Ref. 29.

13.11 EQUALIZATION

Of the critical circuit parameters mentioned in Section 13.8.3, two that have severe deleterious effects on data transmission can be reduced to tolerable limits by circuit conditioning, or equalization. These two are amplitude–frequency response (distortion) and EDD.

There are several methods of performing equalization. The most common is to use one or several networks in tandem. Such networks tend to flatten response. In the case of amplitude, they add attenuation increasingly toward channel center and less toward its edges. The overall effect is one of making the amplitude response flatter. The delay equalizer operates fairly similarly. Delay increases toward channel edges parabolically from the center. Delay is added in the center, much like an inverted parabola, with less and less delay added as the band edge is approached. Thus the delay response is flattened at some small cost to absolute delay, which, in most data systems, has no effect. However, care must be taken with the effect of a delay equalizer on an amplitude equalizer and, conversely, the amplitude equalizer on the delay equalizer. Their design and adjustment must be such that the flattening of the channel for one parameter does not entirely distort the channel for the other.

Another type of equalizer is the transversal type of filter. It is useful where it is necessary to select among or to adjust several attenuation (amplitude) and phase characteristics. The basis of the filter is a tapped delay line to which the input is presented. The output is taken from a summing network, which adds or sums the outputs of the taps. Such a filter is adjusted to the desired response (equalization of both phase and amplitude) by adjusting the tap contributions.

If the characteristics of a line are known, another method of equalization is predistortion of the output signal of the data set. Some devices use a shift register and a summing network. If the equalization needs to be varied, then a feedback circuit from the receiver to the transmitter would be required to control the shift register. Such a type of active predistortion is valid for binary transmission only.

A major drawback of all the equalizers discussed (with the exception of the last with a feedback circuit) is that they are useful only on dedicated or leased circuits where the circuit characteristics are known and remain fixed. Obviously, a switched circuit would require a variable automatic equalizer, or conditioning would be required on every circuit in the switched system that would be transmitting data.

Circuits are usually equalized on the receiving end. This is called post-equalization. Equalizers must be balanced and must present the proper impedance to the line. Administrations* may choose to condition (equalize) trunks and attempt to eliminate the need to equalize station lines; the economy of considerably fewer equalizers is obvious. In addition, each circuit that would possibly carry high-speed data in the system would have to be equalized, and the equalization must be good enough that any possible combination will meet the overall requirements. If equalization requirements become greater (i.e., parameters more stringent), then consideration may have to be given to the restriction of the maximum number of circuits (trunks) in tandem.

---

*Telephone companies.

Equalization to meet amplitude–frequency response requirements is less exacting on the overall system than envelope delay. Equalization for envelope delay and its associated measurements are time-consuming and expensive. Envelope delay in general is arithmetically cumulative. If there is a requirement of overall EDD of 1 ms for a circuit between 1000 and 2600 Hz, then in three links in tandem, each link must be better than 333 μs between the same frequency limits. For four links in tandem, each link would have to be 250 μs or better. In practice, accumulation of delay distortion is not entirely arithmetical, resulting in a loosening of requirements by about 10%. Delay distortion tends to be inversely proportional to the velocity of propagation. Loaded cables display greater delay distortion than nonloaded cables. Likewise, with sharp filters a greater delay is experienced for frequencies approaching band edge than for filters with a more gradual roll-off.

In carrier multiplex systems, channel banks contribute more to the overall EDD than any other part of the system. Because channels 1 and 12 of the standard CCITT modulation plan, those nearest the group band edge, suffer additional delay distortion owing to the effects of group and, in some cases, supergroup filters, the system engineer should allocate channels for data transmission near group and supergroup centers. On long-haul critical data systems the data channels should be allocated to through-groups and through-supergroups, minimizing as much as possible the steps of demodulation back to voice frequencies (channel demodulation).

Automatic equalization for both amplitude and delay is widely used, especially for switched data systems. Such devices are self-adaptive and require a short adaptation period after switching, on the order of 1–2 s or less. This can be carried out during synchronization. Not only is the modem clock being "averaged" for the new circuit on transmission of a synchronous idle signal, but the self-adaptive equalizer adjusts for optimum equalization as well. The major drawback of adaptive equalizers is their expense.

Many texts use the term "conditioning" as a synonym for equalization. We like to differentiate the two terms. Conditioning is carried out by the common carrier or administration on a dedicated line to reduce its variabilities such as EDD and attenuation distortion. Equalization is carried out at the user data set as described previously.

13.12   PRACTICAL MODEM APPLICATIONS

13.12.1   Voice Frequency Carrier Telegraph (VFCT)

Narrow shifted FSK transmission of digital data goes under several common names. These are VFTG and VFCT, which stand for voice frequency telegraph and voice frequency carrier telegraph, respectively.

In practice, VFCT techniques handle data rates up to 1200 bps by a simple application of FSK modulation. The voice channel is divided into segments or frequency-bounded zones or bands. Each segment represents a data or telegraph channel, each with a frequency-shifted subcarrier.

For proper end-to-end system interface it is convenient to use standardized modulation plans, particularly on international circuits. In order for the far-end demodulator to operate with the near-end modulator, it must be tuned to the same center frequency and accept the same shift. Center frequency is the frequency in the center of the passband of the modulator–demodulator. The shift is the number of hertz that the center frequency is shifted up and down in frequency for the mark and space condition. From Table 13.1, by convention, the mark condition is the center frequency shifted downward, and the space upward. For modulation rates below 80 bps, bandpasses have either 170-* or 120-Hz bandwidths with frequency shifts of ± 42.5 or ± 30 Hz, respectively. CCITT recommends (Rec. R.70 bis [Ref. 37]) the 120-Hz channels for operating at 50 bps and below. However, some administrations operate these channels at higher modulation rates.

The number of tone telegraph or data channels that can be accommodated on a voice channel depends for one thing on the usable voice channel bandwidth. For HF radio with a voice channel limit on the order of 3 kHz, 16 channels may be accommodated using 170-Hz spacing (170 Hz between center frequencies). Twenty-four VFCT channels may be accommodated between 390 and 3210 Hz with 120-Hz spacing, or 12 channels with 240-Hz spacing. This can easily meet standard telephone channels of 300–3400 Hz.

13.12.2 Medium-Data-Rate Modems

In a normal practice, FSK is used for the transmission of data rates up to 1200 bps. The 120-Hz channel is nominally modified as described in Section 13.12.1 such that one 240-Hz channel replaces two 120-Hz channels. Administrations use the 240-Hz channel for modulation rates up to 150 bps. The same process can continue using 480-Hz channels for 300-bps FSK, and 960-Hz channels for 600 bps (Ref. 39).

In the following paragraphs a synopsis is given for CCITT recommendations dealing with modems that operate with the nominal analog 4-kHz voice channel.

CCITT Rec. V.21 (Ref. 40) defines a 300 bps duplex modem standardized for use in the general switched telephone networks and recommends the following:

1. Frequency shift ± 100 Hz.
2. Center frequency of channel 1, 1080 Hz.
3. Center frequency of channel 2, 1750 Hz.
4. In each case, space (0) is the higher frequency.

*CCITT Rec. R.39 (Ref. 38).

It also provides for a disabling tone on echo suppressors, a very important consideration on long circuits.

CCITT Rec. V.22 (Ref. 41) defines a 1200-bps duplex modem for use on the general switched telephone network and on point-to-point two-wire leased-type circuits. In the full-duplex mode, channel separation is by frequency division. The modem uses QPSK modulation and the carrier frequencies are 1200 and 2400 Hz. There is a guard tone at 1800 Hz, which may be disabled at the discretion of users. The modem employs a fixed compromise equalizer and can operate at 1200 or 600 bps in the start–stop or synchronous modes. A scrambler is also included.

CCITT Rec. V.22 bis (Ref. 42) defines a modem for 2400-bps duplex operation using the frequency division technique standardized for use on the general switched telephone network (GSTN) and on point-to-point two-wire leased circuits. It uses quadrature amplitude modulation. The carrier frequencies are the same as used in Rec. V.22 described above. It can operate either in the start–stop or synchronous modes at 2400 or 1200 bps. The signal constellation for this modem is shown in Figure 13.25. The scrambling algorithm uses the generating polynomial $1 + X^{-14} + X^{-17}$.

CCITT Rec. V.23 (Ref. 43) recommends a 600/1200-baud modem standardized for use in the general switched telephone network for application to

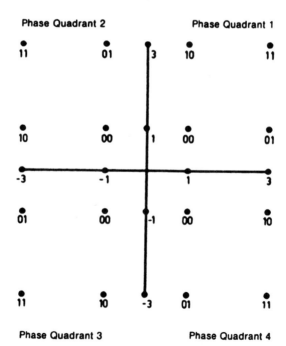

Figure 13.25. Signal constellation used with the CCITT Rec. V.22 bis modem. (From Ref. 42, Figure 2/V.22, CCITT Rec. V.22 bis.)

synchronous or asynchronous systems. Provision is made for an optional backward channel for error control.

For the forward channel the following modulation rates and characteristic frequencies are presented:

	F_0 (Hz)	F_Z (Hz)	F_A (Hz)
Mode 1, up to 600 bps	1500	1300	1700
Mode 2, up to 1200 bps	1700	1300	2100

The backward channel for error control is capable of modulation rates up to 75 bps. Its mark and space frequencies are

F_Z	F_A
390 Hz	450 Hz

Refer to Table 13.1 for the mark–space convention (F_Z = mark or binary 1, F_A = space or binary 0).

CCITT Rec. V.26 (Ref. 44) defines a 2400 bps modem standardized for use on four-wire leased telephone-type circuits and gives the following guidelines:

1. Carrier frequency 1800 Hz ± 1 Hz, four-phase modulation, synchronous mode of operation.
2. Dibit coding as follows:

	Phase Change (deg)	
CDibit	**Alternative A**	**Alternative B**
00	0	+45
01	+90	+135
11	+180	+225
10	+270	+315

3. Data signaling rate 2400 bps ±0.01%.
4. Modulation rate 1200 bauds ±0.01%.
5. Maximum frequency error at receiver, ±7 Hz (allowing ±1 Hz for modulator [transmitter] error).
6. Backward channel 75 bps (see Rec. V.23 [Ref. 43]).

CCITT Rec. V.26 bis (Ref. 45), defining a 2400/1200-bps modem for use in the general switched telephone network, is generally the same as Rec. V.26. At 1200-bps line operation, differential PSK is recommended where

$$0 = +90°$$
$$1 = +270°$$

CCITT Rec. V.27 (Ref. 46), for a 4800-bps modem with manual equalizer for use on leased telephone-type circuits, gives the following principal characteristics of this modem:

1. Full-duplex or half-duplex operation.
2. Differential eight-phase modulation, synchronous mode of operation.
3. Possibility of backward (supervisory) channel with modulation rates to 75 bauds in each direction of transmission.
4. Inclusion of a manually adjustable equalizer.
5. Carrier frequency 1800 Hz ± 1 Hz.
6. Tribit coding as follows:

Tribit Values	Phase Change (deg)
001	0
000	45
010	90
011	135
111	180
110	225
100	270
101	315

7. Frequency tolerance same as Recs. V.26 and V.26 bis.
8. Line signal characteristics: a 50% raised cosine energy spectrum equally divided between transmitter and receiver.
9. A self-synchronizing scrambler/descrambler having a generating polynomial $1 + X^{-6} + X^{-7}$ with additional guards against repeating patterns of 1, 2, 3, 4, 6, 9, and 12 bits to be included in the modem.

The basic characteristics of the modem in CCITT Rec. V.27 bis (Ref. 47), a 4800/2400-bps modem with automatic equalizer standardized for use on leased telephone-type circuits, are the same as for that of Rec. V.27 except for the following:

1. A fallback rate of 2400 bps with V.26 Alternative A characteristics.
2. An automatic equalizer with two types of turn-on sequences and a turn-off sequence. The first type of turn-on sequence is short in duration for comparatively good circuits, meeting CCITT Rec. M.1020 (Ref. 34), and a second is a longer sequence for relatively poor circuits, below the Rec. M.1020 standard. The sequences, which are transmitted online prior to traffic, provide for equalizer training and serve to set descrambler synchronization into proper operation.

CCITT Rec. V.27 ter (Ref. 48) defines a 4800/2400-bps modem standardized for use in the general switched telephone network and is similar to Rec. V.27 bis except for the turn-on sequence, which, in this case, includes the capability to protect against talker echo.

In CCITT Rec. V.29 (Ref. 49), for 9600-bps modem standardized for use on point-to-point four-wire telephone-type circuits, the main characteristics of the modem are as follows:

1. A fallback to rates of 7200 and 4800 bps.
2. Full-duplex and half-duplex operation.
3. Combined amplitude and phase modulation with synchronous mode operation.
4. Automatic equalizer.
5. Optional inclusion of a multiplexer for combining data rates of 7200, 4800, and 2400 bps.
6. Line signal 1700 Hz $\pm$ 1 Hz.
7. Signal space coding. At 9600 bps the scrambled data stream to be transmitted is divided into groups of four consecutive data bits (quadbits). The first bit Q_1 in time of each quadbit is used to determine the signal element amplitude to be transmitted. The second Q_2, third Q_3, and fourth Q_4 bits are encoded as a phase change relative to the phase of the immediately preceding element (Table 13.9). The phase encoding is identical to that of Rec. V.27.

The relative amplitude of the transmitted signal element is determined by the first bit Q_1 of the quadbit and the absolute phase of the signal element. (Table 13.10.) The absolute phase is initially established by the synchronizing signal. The four possible signaling elements Q_1, Q_2, Q_3, and Q_4 represent 16 phase–amplitude possibilities ($2^4 = 16$), which are shown in Figure 13.26.

CCITT Rec. V.32 (Ref. 50) is a standard for a family of two-wire duplex modems operating at data signaling rates of up to 9600 bps for use on the general switched telephone network (GSTN) and on leased-type telephone

TABLE 13.9 Phase Encoding

Q_2	Q_3	Q_4	Phase Change (deg)
0	0	1	0
0	0	0	45
0	1	0	90
0	1	1	135
1	1	1	180
1	1	0	225
1	0	0	270
1	0	1	315

TABLE 13.10 Amplitude–Phase Relationships

Absolute Phase (deg)	Q_1	Relative Signal Element Amplitude
0, 90, 180, 270	0	3
	1	5
45, 135, 225, 315	0	$\sqrt{2}$
	1	$3\sqrt{2}$

circuits. Some highlights of the V.32 standard are listed below:

1. Duplex mode of operation on the GSTN and two-wire, point-to-point leased circuits.
2. Channel separation by echo cancellation techniques.
3. Quadrature amplitude modulation for each channel with synchronous line transmission at 2400 bauds.
4. Any combination of the following data signaling rates may be implemented: 9600-, 4800-, and 2400-bps synchronous.
5. At 9600 bps, two alternative modulation schemes are provided. One uses 16-carrier states and the other uses trellis coding with 32-carrier states. However, modems providing the 9600-bps data signaling rate shall be capable of interworking using the 16-state alternative.
6. There is an exchange of rate sequences during start-up to establish the data rate, coding, and any other special facilities.

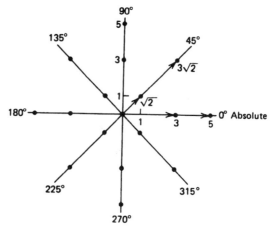

Figure 13.26. Signal space diagram, 9600-bps operation. From Ref. 49, Figure 1/V.29, CCITT Rec. V.29.

**TABLE 13.11   Differential Quadrant Coding for 4800-bps Operation
and for Nonredundant Coding at 9600-bps Operation**

Inputs		Previous Outputs		Phase Quadrant Change	Outputs		Signal State
$Q1_n$	$Q2_n$	$Y1_{n-1}$	$Y2_{n-1}$	(deg)	$Y1_n$	$Y2_n$	for 4800 bps
0	0	0	0	+90	0	1	B
0	0	0	1		1	1	C
0	0	1	0		0	0	A
0	0	1	1		1	0	D
0	1	0	0	0	0	0	A
0	1	0	1		0	1	B
0	1	1	0		1	0	D
0	1	1	1		1	1	C
1	0	0	0	+180	1	1	C
1	0	0	1		1	0	D
1	0	1	0		0	1	B
1	0	1	1		0	0	A
1	1	0	0	+270	1	0	D
1	1	0	1		0	0	A
1	1	1	0		1	1	C
1	1	1	1		0	1	B

Source: Reference 50, Table 1/V.32, CCITT Rec. V.32.

7. There is an optional provision for an asynchronous (start–stop) mode of operation in accordance with CCITT Rec. V.14.

8. Carrier frequency: 1800 Hz ± 1 Hz. The receiver can operate with frequency offsets up to ±7 Hz.

The recommendation offers two alternatives for signal element coding at the 9600-bps data rate: (1) nonredundant and (2) trellis coding.

With nonredundant coding, the scrambled data stream to be transmitted is divided into groups of 4 consecutive bits. The first 2 bits in time $Q1_n$ and $Q2_n$ in each group, where the subscript n designates the sequence number of the group, are differentially encoded into $Y1_n$ and $Y2_n$ in accordance with Table 13.11. Bits $Y1_n$, $Y2_n$, $Q3_n$, and $Q4_n$ are then mapped into coordinates of the signal state to be transmitted according to the signal space diagram shown in Figure 13.27 and as listed in Table 13.13.

When using the second alternative with trellis coding, the scrambled data stream to be transmitted is divided into two groups of 4 consecutive data bits. As shown in Figure 13.28, the first 2 bits in time $Q1_n$ and $Q2_n$ in each group, where the subscript n designates the sequence number of the group, are first differentially encoded into $Y1_n$ and $Y2_n$ in accordance with Table 13.12. The two differentially encoded bits $Y1_n$ and $Y2_n$ are used as input to a systematic convolutional coder, which generates redundant bit $Y0_n$. The redundant bit and the 4 information-carrying bits $Y1_n$, $Y2_n$, $Q3_n$, and $Q4_n$, are then

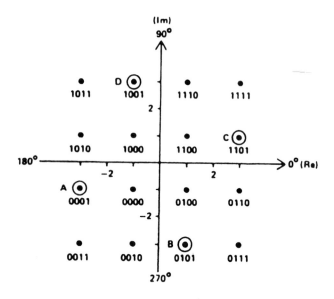

The binary numbers denote Y1$_n$ Y2$_n$ Q3$_n$ Q4$_n$

Figure 13.27. A 16-point signal constellation with nonredundant coding for 9600-bps operation and subset A, B, C, and D states used for 4800-bps operation and for training. (From Ref. 50, Figure 1/V.32, CCITT Rec. V.32.)

mapped into the coordinates of the signal element to be transmitted according to the signal space diagram shown in Figure 13.29 and as listed in Table 13.13.

For 4800-bps operation the data stream to be transmitted is divided into groups of 2 consecutive data bits. These bits, denoted $Q1_n$ and $Q2_n$, where $Q1_n$ is the first in time and the subscript n designates the sequence number of the group, are differentially encoded into $Y1_n$ and $Y2_n$, according to Table 13.11. Figure 13.27 shows the subset A, B, C, and D of signal states used for 4800-bps transmission.

There are two scrambling generating polynomials:

- Call mode modem = $1 + X^{-18} + X^{-23}$
- Answer mode modem = $1 + X^{-5} + X^{-23}$

CCIT Rec. V.33 (Ref. 51) defines a 14,400-bps modem standardized for use on point-to-point four-wire leased telephone-type circuits. The modem is intended to be used primarily on special-quality leased circuits that are typically defined in CCITT Recs. M.1020 and M.1025 (see Section 13.8.4).

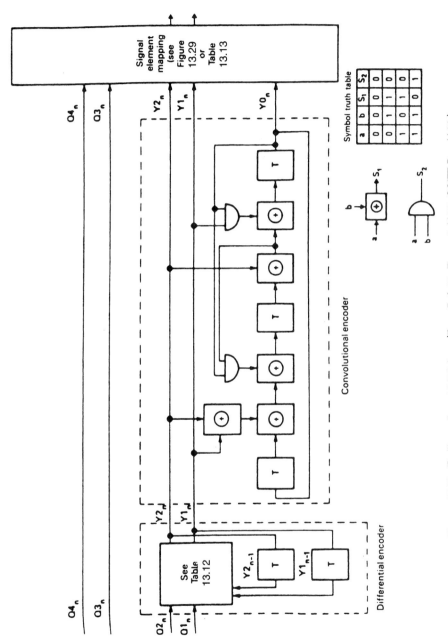

Figure 13.28. Trellis coding at 9600 bps. (From Ref. 50, Figure 2/V.32, CCITT Rec. V.32.)

TABLE 13.12   Differential Encoding for Use with Trellis-Coded Alternative at 9600-bps Operation

Inputs		Previous Outputs		Outputs	
$Q1_n$	$Q2_n$	$Y1_{n-1}$	$Y2_{n-1}$	$Y1_n$	$Y2_n$
0	0	0	0	0	0
0	0	0	1	0	1
0	0	1	0	1	0
0	0	1	1	1	1
0	1	0	0	0	1
0	1	0	1	0	0
0	1	1	0	1	1
0	1	1	1	1	0
1	0	0	0	1	0
1	0	0	1	1	1
1	0	1	0	0	1
1	0	1	1	0	0
1	1	0	0	1	1
1	1	0	1	1	0
1	1	1	0	0	0
1	1	1	1	0	1

Source: Reference 50, Table 2/V.32, CCITT Rec. V.32.

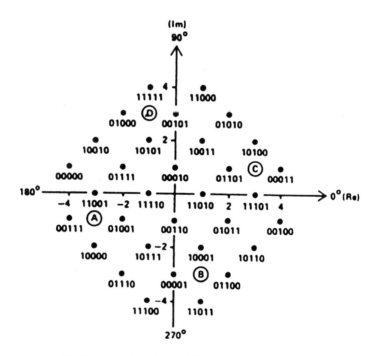

The binary numbers denote $Y0_n$ $Y1_n$ $Y2_n$ $Q3_n$ $Q4_n$

Figure 13.29. A 32-point signal constellation with trellis coding for 9600-bps operation and states A, B, C, and D used at 4800-bps operation and for training. (From Ref. 50, Figure 3/V.22, CCITT Rec. V.32.)

TABLE 13.13   Two Alternative Signal State Mappings for 9600-bps Operation

(Y0)	Y1	Y2	Q3	Q4	Nonredundant Coding Re	Im	Trellis Coding Re	Im
0	0	0	0	0	−1	−1	−4	1
	0	0	0	1	−3	−1	0	−3
	0	10	1	0	−1	−3	0	1
	0	0	1	1	−3	−3	4	1
	0	1	0	0	1	−1	4	−1
	0	1	0	1	1	−3	0	3
	0	1	1	0	3	−1	0	−1
	0	1	1	1	3	−3	−4	−1
	1	0	0	0	−1	1	−2	3
	1	0	0	1	−1	3	−2	−1
	1	0	1	0	−3	1	2	3
	1	0	1	1	−3	3	2	−1
	1	1	0	0	1	1	2	−3
	1	1	0	1	3	1	2	1
	1	1	1	0	1	3	−2	−3
	1	1	1	1	3	3	−2	1
1	0	0	0	0			−3	−2
	0	0	0	1			1	−2
	0	0	1	0			−3	2
	0	0	1	1			1	2
	0	1	0	0			3	2
	0	1	0	1			−1	2
	0	1	1	0		3		−2
	0	1	1	1		−1		−2
	1	0	0	0			1	4
	1	0	0	1			−3	0
	1	0	1	0			1	0
	1	0	1	1			1	−4
	1	1	0	0			−1	−4
	1	1	0	1			3	0
	1	1	1	0			−1	0
	1	1	1	1			−1	4

[a]See Tables 13.11 and 13.12 and Figure 13.28.
Source: Reference 50, Table 3/V.32, CCITT Rec. V.32.

The modem's principal characteristics are the following:

1. A fallback of 12,000 bps.
2. A capability of operating in a duplex mode with continuous carrier.
3. Combined amplitude and phase modulation with synchronous mode of operation.
4. Inclusion of an eight-state trellis-coded modulation.
5. Optional inclusion of a multiplexer for combining data rates of 12,000, 9600, 7200, 4800, and 2400 bps.
6. A carrier frequency of 1800 ± 1 Hz.

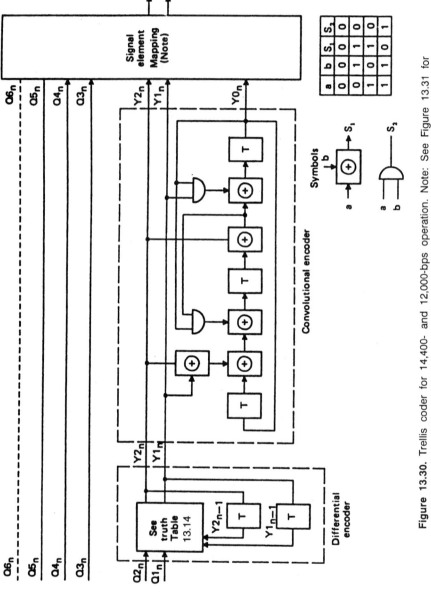

Figure 13.30. Trellis coder for 14,400- and 12,000-bps operation. Note: See Figure 13.31 for 14,400-bps rate and Figure 13.32 for 12,000-bps rate. (From Ref. 51, Figure 1/V.33, CCITT Rec. V.33.)

TABLE 13.14   Differential Encoding for Use with Trellis Coding

Inputs		Previous Outputs		Outputs	
$Q1_n$	$Q2_n$	$Y1_{n-1}$	$Y2_{n-2}$	$Y1_n$	$Y2_n$
0	0	0	0	0	0
0	0	0	1	0	1
0	0	1	0	1	0
0	0	1	1	1	1
0	1	0	0	0	1
0	1	0	1	0	0
0	1	1	0	1	1
0	1	1	1	1	0
1	0	0	0	1	0
1	0	0	1	1	1
1	0	1	0	0	1
1	0	1	1	0	0
1	1	0	0	1	1
1	1	0	1	1	0
1	1	1	0	0	0
1	1	1	1	0	1

Source: Reference 51, Table 1A/V.33, CCITT Rec. V.33.

At the 14,400-bps data rate, the scrambled data stream to be transmitted is divided into groups of 6 consecutive data bits. As illustrated in Figure 13.30, the first 2 bits in time $Q1_n$ and $Q2_n$ in each group are first differentially coded into $Y1$ and $Y2$ in accordance with Table 13.14. The two differentially encoded bits $Y1_n$ and $Y2_n$ are used as input to a systematic convolutional encoder, which generates a redundant bit $Y0_n$. This redundant bit and the 6 information-carrying bits $Y1_n$, $Y2_n$, $Q3_n$, $Q4_n$, $Q5_n$, and $Q6_n$ are then mapped into the coordinates of the signal element to be transmitted in accordance with the signal space diagram shown in Figure 13.31. Figure 13.32 shows the signal constellation and mapping for trellis-coded modulation at 12,000 bps. The self-synchronizing scrambler generating polynomial is $1 + X^{-18} + X^{-23}$.

13.12.3   Notes on Trellis-Coded Modulation (TCM)

In Chapter 5 we introduced forward error correction (FEC) coding and "coding gain." With FEC we can improve error rate performance, or, given a fixed bit error rate (BER), we can reduce power (level) because of coding gain. Depending on the coding/decoding design parameters, for a fixed BER, a lower E_b/N_0 will achieve the same performance if we used the same modulation scheme without coding. For example, with QPSK without coding, to achieve a BER of 1×10^{-5}, we would require an E_b/N_0 of 9.6 dB. Using FEC, we may only need an E_b/N_0 of 5 dB for the same BER. The result is a coding gain of $9.6 - 5$, or 4.6 dB. We pay for this by an increase in symbol rate (ergo bandwidth) due to the redundancy added for FEC.

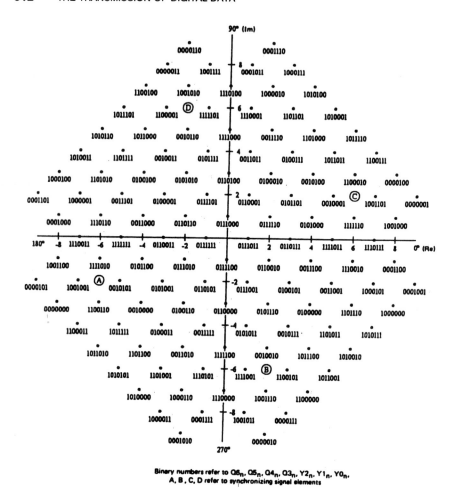

Figure 13.31. Signal constellation and mapping for trellis-coded modulation at 14,400 bps. (From Ref. 51, Figure 2/V.33, CCITT Rec. V.33.)

When working with the voice channel we have severe bandwidth limitations, nominally 3 kHz. When implementing FEC and maintaining a high data rate, our natural tendency is to use bit packing schemes turning from binary modulation (two states) to M-ary modulation (M-states). Such bit packing schemes were described in Section 5.5.3.4. Typical types of M-ary modulation are QPSK, 8-ary, PSK, and 4-, 8-, 16-,... quadrature amplitude modulation (QAM). QAM schemes, such as used in CCITT Rec. V.29 described previously, have two-dimensional signal constellations; one dimension is phase and the other amplitude. The signal constellation shown in Figure 13.27 has 16 points. As the number of points increases, the receive

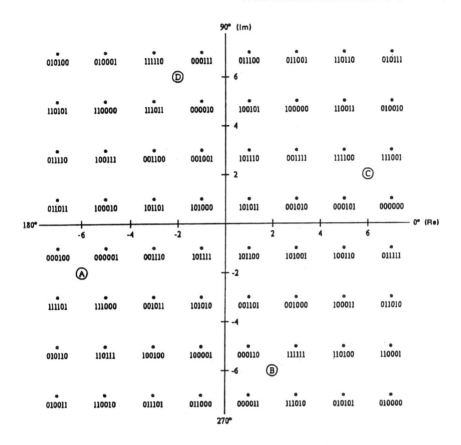

Binary numbers refer to Q5$_n$, Q4$_n$, Q3$_n$, Y2$_n$, Y1$_n$, Y0$_n$,
A, B, C, D refer to synchronizing signal elements

Figure 13.32. Signal constellation and mapping for trellis-coded modulation at 12,000 bps. (From Ref. 51, Figure 3V.33, CCITT Rec. V.33.)

modem becomes more prone to errors induced by Gaussian noise hits and intersymbol interference. This problem can be mitigated by the design of the M-ary signal set.

In our discussions in Chapter 5, coding and modulation were carried out separately. With trellis-coded modulation (TCM), there is a unified concept. Here coding gain is achieved without bandwidth expansion by expanding the signal set.

In conventional, multilevel (amplitude and/or phase) modulation systems, during each modulation interval, the modulator maps m binary symbols (bits) into one of $M = 2^m$ possible transmit signals, and the demodulator recovers the m bits by making an independent M-ary nearest-neighbor decision on each signal received. Previous codes were designed for maximum Hamming free distance. Hamming distance refers to the number of symbols in which two codes symbols or blocks differ regardless of how these symbols differ (Ref. 52).

TCM broke with this tradition and maximized free Euclidean distance (distance between maximum likely neighbors in the signal constellation) rather than Hamming distance. The redundancy necessary for TCM would have to come from expanding the signal set to avoid bandwidth expansion. Ungerboeck (Ref. 52) states that, in principle, TCM can achieve coding gains of about 7–8 dB over conventional uncoded multilevel modulation schemes. Most of the achievable coding gain would be obtained by expanding the signal sets only by a factor or 2. This means using signal set sizes of 2^{m+1} for transmission of m bits per modulation interval.

All TCM systems (Ref. 52) achieve significant distance gains with as few as 4, 8, and 16 states. Roughly speaking, it is possible to gain 3 dB with 4 states, 4 dB with 8 states, nearly 5 dB with 16 states, and up to 6 dB with 128 or more states. Doubling the number of states does not always yield a code with larger free distance. Generally, limited distance growth and increasing numbers of nearest neighbors and neighbors with next-largest distances are the two mechanisms that prevent realizing coding gains from exceeding the ultimate limit set by channel capacity. This limit can be characterized by the signal-to-noise ratio at which the channel capacity of a modulation system with $(2m + 1)$-ary signal set equals m bps/Hz (Ref. 52; also see Ref. 53). Practical TCM systems can achieve coding gains of 3–6 dB at spectral efficiencies equal to or larger than 2 bits/Hz.

13.12.4 Data Rate at 28,800 bps—The V.34 Modem

13.12.4.1 General. The V.34 modem (Ref. 54) is designed to operate in the PSTN and on point-to-point two-wire leased telephone-type circuits. The following are some of the salient characteristics of a V.34 modem:

1. Duplex and half-duplex modes.
2. Channel separation by echo cancellation techniques.
3. Quadrature amplitude modulation (QAM) for each channel with synchronous line transmission at selectable symbol rates including the mandatory rates of 2400, 3000, and 3200 symbols per second and optional rates of 2743, 2800, and 3429 symbols per second.
4. Synchronous primary channel data rates of:

28,800 bps	26,400 bps
24,000 bps	21,600 bps
19,200 bps	16,800 bps
14,400 bps	12,000 bps
9600 bps	7200 bps
4800 bps	2400 bps

5. Trellis coding for all data rates.

6. An optional auxiliary channel with a synchronous data rate of 200 bps, a portion of which may be provided to the user as an asynchronous secondary channel.

7. Adaptive techniques that enable the modem to achieve close to the maximum data rate the channel can support in each direction.

8. Exchange of rate sequences during start-up to establish the data rate.

9. Automoding to V-Series modems supported by ITU-T Rec. V.32 bis automode procedures and Group 3 facsimile machines.

13.12.4.2 *Selected Definitions*

Constellation Shaping. A method for improving noise immunity by introducing a nonuniform two-dimensional probability distribution for transmitted signal points. The degree of constellation shaping is a function of the amount of constellation expansion.

Data Mode Modulation Parameters. Parameters determined during start-up and used during data mode transmission.

Frame Switching. A method for sending a fractional number of bits per mapping frame, on average, by alternating between sending an integer $b - 1$ bits per mapping frame and b bits per mapping frame according to a periodic switching pattern.

Line Probing. A method for determining channel characteristics by sending periodic signals, which are analyzed by the modem and used to determine data mode modulation parameters.

Nonlinear Encoding. A method of improving distortion immunity near the perimeter of a signal constellation by introducing a nonuniform two-dimensional (2D) signal point spacing.

Precoding. A nonlinear equalization method for reducing equalizer noise enhancement caused by amplitude distortion. Equalization is performed at the transmitter using precoding coefficients provided by the remote modem.

Preemphasis. A linear equalization method where the transmit signal spectrum is shaped to compensate for amplitude distortion. The preemphasis filter is selected using a filter index provided by the remote modem.

Shell Mapping. A method of mapping data bits to signal points in a multidimensional signal constellation, which involves partitioning a two-dimensional signal constellation into rings containing an equal number of signal points.

Trellis Encoding. A method of improving noise immunity using a convolutional coder to select a sequence of subsets in a partitioned signal constellation. The trellis encoders employed in the V.34 modem are all four-dimensional (4D) and they are used in a feedback structure where the inputs to the trellis encoder are derived from the signal points.

13.12.4.3 An Overview of Selected Key Areas of Modem Operation

13.12.4.3.1 Line Signals. The primary channel supports synchronous data rates of 2400–28,800 bps in multiples of 2400 bps. An auxiliary channel with a synchronous data rate of 200 bps may also be optionally supported. The primary and auxiliary data rates are determined during phase 4 of modem start-up according to the procedures described in paragraphs 11.4 and 12.4 of the reference document (Ref. 54). The auxiliary channel is used only when the call and answer modems have both declared this capability. The primary channel data rates may be asymmetric.

13.12.4.3.2 Symbol Rates. The symbol rate is $S = (a/c) \times 2400 \pm 0.01\%$ 2D symbols per second, where a and c are integers from the set given in Table 13.15 (in which symbol rates are shown rounded to the nearest integer). The symbol rates 2400, 3000, and 3200 are mandatory; 2743, 2800, and 3429 are optional. The symbol rate is selected during phase 2 of modem start-up according to procedures described in paragraph 11.2 or 12.2 of the reference publication. Asymmetric symbol rates are optionally supported and are used only when the call and answer modems have both declared this capability.

13.12.4.3.3 Carrier Frequencies. The carrier frequency is $(d/e) \times S$ Hz, where d and e are integers. One of two carrier frequencies can be selected at each symbol rate, as given in Table 13.16, which provides the values of d and e and the corresponding frequencies rounded to the nearest integer. The carrier frequency is determined during phase 2 of modem start-up according to the procedures given in paragraph 11.2 or 12.2 of the reference publication. Asymmetric carrier frequencies are supported.

TABLE 13.15 Symbol Rates

Symbol Rate, S	a	c
2400	1	1
2743	8	7
2800	7	6
3000	5	4
3200	4	3
3429	10	7

Source: Reference 54, Table 1/V.34, ITU-T Rec.34.

TABLE 13.16   Carrier Frequencies Versus Symbol Rate

	Low Carrier			High Carrier		
Symbol Rate, S	Frequency	d	e	Frequency	d	e
2400	1600	2	3	1800	3	4
2743	1646	3	5	1829	2	3
2800	1680	3	5	1867	2	3
3000	1800	3	5	2000	2	3
3200	1829	4	7	1920	3	5
3429	1959	4	7	1959	4	7

Source: Reference 54, Table 2/V.34, ITU-T Rec. V.34.

13.12.4.3.4   Preemphasis

TRANSMIT SPECTRUM SPECIFICATIONS. The transmit spectrum specifications use a normalized frequency, which is defined as the ratio f/S, where f is the frequency in hertz and S is the symbol rate. The magnitude of the transmitted spectrum conforms to the templates shown in Figures 13.33 and 13.34 for normalized frequencies in the range from $(d/e = 0.45)$ to $(d/e + 0.45)$. The transmitted spectrum is measured using a 600-Ω pure resistive load.

Figure 13.33 requires parameter α, which is given in Table 13.17; and Figure 13.34 requires parameters β and τ, which are given in Table 13.18.

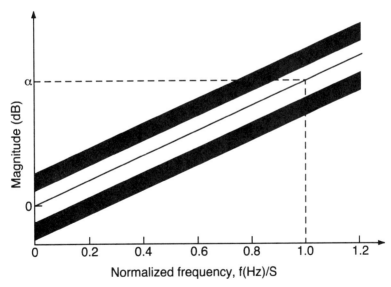

Figure 13.33. Transmit spectra templates for indices 0 to 5. *Note:* Tolerance for transmit spectrum is ± 1 dB. (From Ref. 54, Figure 1/V.34, ITU-T Rec. V.34.)

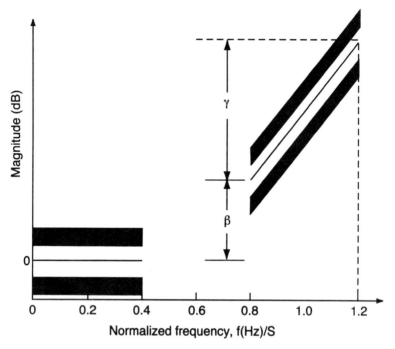

Figure 13.34. Transmit spectra templates for indices 6 to 10. *Note:* Over the range specified, the tolerance for the transmit spectrum magnitude is ±1 dB. (From Ref. 54, Figure 2/V.34, ITU-T Rec. V.34.)

TABLE 13.17   Parameter α for Indices 0 to 5

Index	α
0	0 dB
1	2 dB
2	4 dB
3	6 dB
4	8 dB
5	10 dB

Source: Reference 54, Table 3/V.34, ITU-T Rec. V.34.

SELECTION METHOD. The transmitted spectrum is specified by a numerical index. The index is provided by the remote modem during phase 2 of start-up procedures as defined in paragraph 11.2 or 12.2 of the reference publication.

13.12.4.3.5   Electrical Characteristics of Interchange Circuits. For the primary channel, where an external physical interface is provided, the electrical characteristics will conform to ITU-T Rec. V.10 or V.11. The connector and pole assignments specified in ISO 2110 Amd. 1.0 or ISO/IEC 11569, column "V-Series > 20,000 bps," is used. Alternatively, when the DTE–DCE inter-

TABLE 13.18 Parameters β and τ for Indices 6 to 10

Index	β	γ
6	0.5 dB	1.0 dB
7	1.0 dB	2.0 dB
8	1.5 dB	3.0 dB
9	2.0 dB	4.0 dB
10	2.5 dB	5.0 dB

Source: Reference 54, Table 4/V.34, ITU-T Rec. V.34.

face speed is not designed to exceed 116 kbps, these same connectors may be used with characteristics conforming to ITU-T Rec. V.10 only.

Where an external physical interface is provided for the secondary channel, electrical characteristics are in accordance with ITU-T Rec. V.10.

13.12.4.3.6 Scrambler. A self-synchronizing scrambler is included in the modem for the primary channel. Auxiliary channel data are not scrambled. Each direction of transmission uses a different scrambler. According to the direction of transmission, the generating polynomial is

$$\text{Call mode:} \quad 1 + X^{-18} + X^{-23}$$
$$\text{Answer mode:} \quad 1 + X^{-5} + X^{-23}$$

13.12.4.4 Framing. Figure 13.35 gives an overview of the frame structure. The duration of a superframe is 280 ms and consists of J data frames, where $J = 7$ for symbol rates 2400, 2800, 3000, and 3200, and $J = 8$ for symbol rates 2743 and 3429. A data frame consists of P mapping frames, where P is given in Table 13.19. A mapping frame consists of four four-dimensional (4D) symbol intervals. A 4D symbol interval consists of two 2D symbol intervals. A bit inversion method is used for superframe synchronization.

Mapping frames are indicated by the time index i, where $i = 0$ for the first mapping frame of signal B1 (as defined below) and is incremented by 1 for each mapping frame thereafter.

Sequence B1 consists of one data frame of scrambled 1's at the end of start-up using selected data mode modulation parameters. Bit inversions for superframe synchronization are inserted as if the data frame were the last data frame in a superframe. Prior to transmission of B1, the scrambler, trellis encoder, differential encoder, and precoding filter tap delay line are initialized to zeros.

The 4D symbol intervals are indicated by the time index $m = 4i$, where j $(= 0, 1, 2, 3)$ is a cyclic time index that indicates the position of the 4D symbol interval in a mapping frame. The 2D symbol intervals are indicated by the time index $n = 2m + k$, where k $(= 0, 1)$ is a cyclic time index that indicates the position of the 2D symbol interval in a 4D symbol interval.

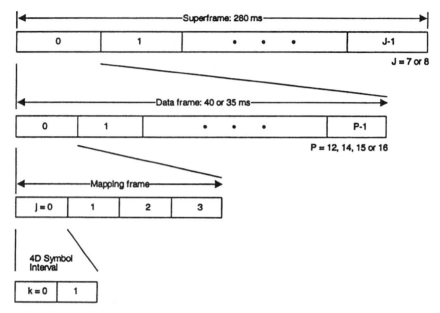

Figure 13.35. Overview of framing and indexing. (From Ref. 54, Figure 3/V.34, ITU-T Rec. V.34.)

13.12.4.4.1 Mapping Frame Switching. An integer number of data bits are transmitted in every data frame. The total number of primary and auxiliary channel data bits transmitted in a data frame is denoted by

$$N = R \times 0.28/J \qquad (13.4)$$

where R is the sum of the primary channel and auxiliary channel data rates.

The total number of (primary and auxiliary) data bits transmitted in a mapping frame varies between $b - 1$ ("low frame") and b ("high frame") bits according to a periodic switching pattern (SWP), of period P, such that the average number of data bits per mapping frame is N/P. The value of b is

TABLE 13.19   Framing Parameters

Symbol Rate, S	J	P
2400	7	12
2743	8	12
2800	7	14
3000	7	15
3200	7	16
3429	8	15

Source: Reference 54, Table 7/V.34, ITU-T Rec. V.34.

defined as the smallest integer not less than N/P. The number of high frames in a period is the remainder:

$$r = N - (b - 1)P \quad \text{for } 1 \le r \le P \tag{13.5}$$

SWP is represented by 12–16-bit binary numbers where 0 and 1 represent low and high frames, respectively. The leftmost bit corresponds to the first mapping frame in a data frame. The rightmost bit is always 1.

SWP may be derived using an algorithm that employs a counter as follows. Prior to each data frame the counter is set to zero. The counter is incremented by r at the beginning of each mapping frame. If the counter is less than P, send a low frame; otherwise, send a high frame and decrement the counter by P.

Table 13.20 gives the values for b and SWP for all combinations of data rate and symbol rate. In the table, SWP is represented by a hexadecimal number. For example, at 19,200 bps and symbol rate 3000, SWP is 0421 (hex) or 0000 0100 0010 0001 (binary).

TABLE 13.20 **[b, Switching Pattern (SWP)] as a Function of Data Rate and Symbol Rate**

Data Rate, R	2400 sym/s P = 12		2743 sym/s P = 12		2800 sym/s P = 14		3000 sym/s P = 15		3200 sym/s P = 16		3429 sym/ P = 15	
	b	SWP	b	SWP	b	SWP	b	SWP	b	SWP	b	SWP
2400	8	FFF	—	—	—	—	—	—	—	—	—	—
2600	9	6DB	—	—	—	—	—	—	—	—	—	—
4800	16	FFF	14	FFF	14	1BB7	13	3DEF	12	FFFF	12	0421
5000	17	6DB	15	56B	15	0489	14	1249	13	5555	12	36DB
7200	24	FFF	21	FFF	21	15AB	20	0421	18	FFFF	17	3DEF
7400	25	6DB	22	56B	22	0081	20	3777	19	5555	18	0889
9600	32	FFF	28	FFF	28	0A95	26	2D6B	24	FFFF	23	14A5
9800	33	6DB	29	56B	28	3FFF	27	0081	25	5555	23	3F7F
12 000	40	FFF	35	FFF	35	0489	32	7FFF	30	FFFF	28	7FFF
12 200	41	6DB	36	56B	35	1FBF	33	2AAB	31	5555	29	1555
14 400	48	FFF	42	FFF	42	0081	39	14A5	36	FFFF	34	2D6B
14 600	49	6DB	43	56B	42	1BB7	39	3FFF	37	5555	35	0001
16 800	56	FFF	49	FFF	48	3FFF	45	3DEF	42	FFFF	40	0421
17 000	57	6DB	50	56B	49	15AB	46	1249	43	5555	40	36DB
19 200	64	FFF	56	FFF	55	1FBF	52	0421	48	FFFF	45	3DEF
19 400	65	6DB	57	56B	56	0A95	52	3777	49	5555	46	0889
21 600	72	FFF	63	FFF	62	1BB7	58	2D6B	54	FFFF	51	14A5
21 800	73	6DB	64	56B	63	0489	59	0081	55	5555	51	3F7F
24 000	—	—	70	FFF	69	15AB	64	7FFF	60	FFFF	56	7FFF
24 200	—	—	71	56B	70	0081	65	2AAB	61	5555	57	1555
26 400	—	—	—	—	—	—	71	14A5	66	FFFF	62	2D6B
26 600	—	—	—	—	—	—	71	3FFF	67	5555	63	0001
28 800	—	—	—	—	—	—	—	—	72	FFFF	68	0421
29 900	—	—	—	—	—	—	—	—	73	5555	68	36DB

Source: Reference 54, Table 8/V.34, ITU-T Rec. V.34.

TABLE 13.21 Auxiliary Channel Multiplexing Parameters

Symbol Rate, S	W	P	AMP
2400	8	12	6DB
2743	7	12	56B
2800	8	14	15AB
3000	8	15	2AAB
3200	8	16	5555
3429	7	15	1555

Source: Reference 54, Table 9/V.34, ITU-T Rec. V.34.

13.12.4.4.2 Multiplexing of Primary and Auxiliary Channel Bits. The auxiliary channel bits are time division multiplexed with the scrambled primary channel bits.

The number of auxiliary channel bits transmitted per data frame is $W = 8$ at symbol rates 2400, 2800, 3000, and 3200, and $W = 7$ at symbol rates 2743 and 3429. In each mapping frame, the bit $I1_{i,0}$ is used to send either an auxiliary channel bit or a primary channel bit according to the auxiliary channel multiplexing pattern, AMP, of period P (see Figure 13.36). AMP can be represented as a P-bit binary number where a 1 indicates that an auxiliary channel bit is sent and a 0 indicates that a primary channel bit is sent. AMP depends only on the symbol rate and is given in Table 13.21 as a hexadecimal number. The leftmost bit corresponds to the first mapping frame in a data frame.

The auxiliary channel multiplexing pattern may be derived using an algorithm similar to the algorithm for SWP, the frame switching pattern. Prior to each data frame, a counter is set to zero. The counter is incremented by W at the beginning of each mapping frame. If the counter is less than P, a primary channel bit is sent; otherwise, an auxiliary channel bit is sent, and the counter is decremented by P.

13.12.4.5 Encoder. Figure 13.36 is a functional block diagram of the encoder.

13.12.4.5.1 Signal Constellation. Signal constellations consist of complex-valued signal points that lie on a two-dimensional rectangular grid.

All signal constellations used with the V.34 modem are subsets of a 960-point superconstellation. Figure 13.37 shows one-quarter of the points in the superconstellation. These points are labeled with decimal integers between 0 and 239. The point with the smallest magnitude is labeled 0, the point with the next larger magnitude is labeled 1, and so on. When two or more points have the same magnitude, the point with the greatest imaginary component is taken first. The full superconstellation is the union of the four quarter-constellations obtained by rotating the constellation in Figure 13.37 by 0°, 90°, 180°, and 270°.

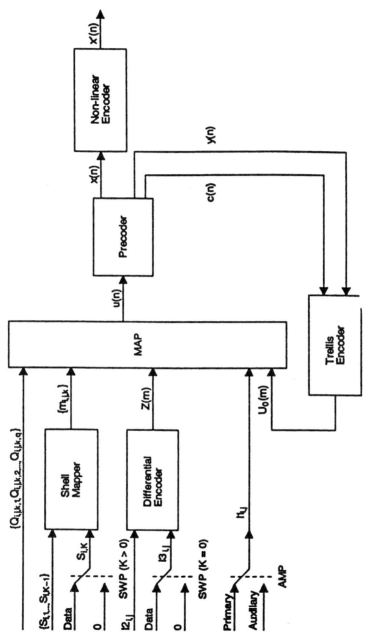

Figure 13.36. Encoder block diagram. (From Ref. 54, Figure 4/V.34, ITU-T Rec. V.34.)

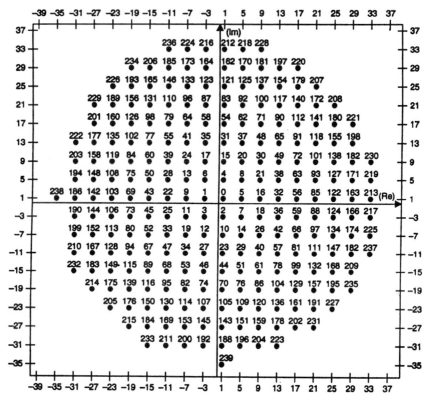

Figure 13.37. One-quarter of the points in the superconstellation. (From Ref. 54, Figure 5/V.34, ITU-T Rec. V.34.)

A signal constellation with L points consists of $L/4$ points from the quarter-constellation in Figure 13.37 with labels 0 through $L/4 - 1$, and the $3L/4$ points that are obtained by 90°, 180°, and 270° rotations of these signal points.

13.13 DATA TRANSMISSION ON THE DIGITAL NETWORK

There is an incompatibility between standard computer data transmission rates and digital network transmission rates.* Computer data rates are based on the relationship 75×2^n bps such as 1200, 2400, 4800, 9600, 14,400, 28,800, and 33,600 bps. These rates have no relationship to 64 kbps, the standard digital voice channel. See CCITT Recs. V. 5 and V.6 (Refs. 55 and 56).

---

*ISDN is an exception, being based on 64-kbps data rates.

One solution to the problem is to use one of the data modems described in Section 13.12 on the analog side of the PCM channel bank. This certainly is one of the least elegant approaches, but it is practical and straightforward. For many users it is the only alternative. Also, some PCM channel banks come equipped with data ports for specific data rates such as 9600 bps. Other approaches are CCITT Rec. V.110 and AT&T's DDS. For an in-depth discussion of the techniques, see Section 3.3.8 (Chapter 3).

13.14 INTEGRATED SERVICES DIGITAL NETWORKS (ISDN)

13.14.1 Introduction

Integrated services digital networks have been developed to ease integration of all telecommunication services, except full motion video, on one basic digital channel, namely, 64 kbps. Whereas 4 kHz is the basic building block of the analog network, 64 kbps is the basic building block of ISDN. The ISDN basic building block is designed to serve, among other services (Ref. 15):

- Digital voice
- High-speed data, both circuit and packet switched
- Telex/teletext
- Telemetry
- Facsimile
- Slow-scan video and compressed video

The goal of ISDN is to provide an integrated facility to incorporate each of the services listed on a common 64-kbps channel and/or a combination of 64- and 16-kbps channels. ISDN assumes that an all-digital network is in place up to and including a subscriber's serving exchange. It also assumes that the network has implemented CCITT Signaling System No. 7.

In the United States, Bellcore modified the current ITU-T Recommendations on ISDN to make ISDN more marketable, to better meet American requirements, and to make it more amenable to the two-wire subscriber plant. There is a National ISDN No. 1, No. 2, and No. 3, which we will define at the end of this section. We also devote several pages on the technical differences for the basic rate between the ITU-T version and the Bellcore version.

13.14.2 ISDN User Channels

Here we are looking from the end user into the network. We consider two user classes: residential and commercial. The following are standard transmission structures for user access links:

B-channel: 64 kbps
D-channel: 16 kbps

The B-channel is the basic user channel and serves any one of the following traffic types:

- PCM-based digital voice channel
- Computer digital data, either circuit or packet switched
- A mix of multiplexed lower data rate traffic, such as vocoded (digital) low data rate voice and lower data rate computer data

However, in this category, the traffic must have the same destination.

The D-channel is a 16-kbps channel. It serves not only as the user signaling channel but also as a lower speed data connectivity to the network.

The H-channels have the following bit rates:

H0:   384 kbps

H1 consisting of the following two subsets:
      H11:   1536 kbps
      H12:   1920 kbps

The H-channel does *not* carry signaling information. Its purpose is to provide service for higher user data rates, such as digitized program channels, compressed video for teleconferencing, fast facsimile, and packet-switched data bit streams.

13.14.3   Basic and Primary Rate User Interfaces

The *basic* interface structure is composed of two B-channels and a D-channel and is commonly referred to as 2B + D. The D-channel at this interface is 16 kbps. The B-channels may be used independently (i.e., two different simultaneous connections).

Appendix I to CCITT Rec. I.412 (Ref. 57) states that the basic access may also be B + D or D.

The *primary* rate B-channel interface structures are composed of N B-channels and one D-channel, where the D-channel in this case is 64-kbps. There are two primary B-channel data rates:

1.544 Mbps = 23B + D
2.048 Mbps = 30B + D

For the user–network access arrangement containing multiple interfaces, it is possible for the D-channel in one structure to carry signaling for

B-channels in another primary rate structure without an activated D-channel. When a D-channel is not activated, the designated time slot may or may not be used to provide an addition B-channel, depending on the situation, such as 24B with 1.544 Mbps.

There are a number of H-channel interface structures covered in CCITT Rec. I.412 (Ref. 57). For the H0 structure, a D-channel may or may not be present and, if present, is always 64 kbps. At the 1.544-Mbps rate interface, the H0 channel structures are 4H0 or 3H0 + D. The H11 and H12 structures use D-channels from other structures to carry signaling, if required.

13.14.4   User Access and Interface

Figure 13.38 shows generic ISDN user connectivity to the network. Here we mean an interface with a link that connects a user to his/her digital serving exchange. We can select either basic or primary service (i.e., 2B + D, 23B + D, or 30B + D) to connect to the ISDN network.

The objectives of any digital interface design, and specifically of ISDN access and interface, are the following:

- Electrical and mechanical specification
- Channel structure and access capabilities
- User–network protocols
- Maintenance and operation
- Performance
- Services

Figure 13.39 shows the ISDN reference model. It delineates interface points for the user. In the figure, NT1, or network termination 1, provides the physical layer interface; it is essentially equivalent to OSI layer 1. The

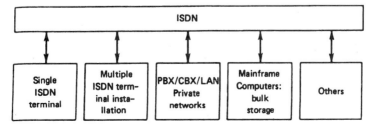

Figure 13.38. Generic ISDN user connectivity to the network.

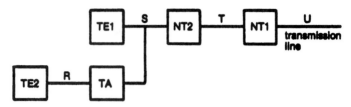

Figure 13.39. ISDN reference model.

functions of the physical layer include the following:

- Transmission facility termination
- Layer-1 maintenance functions and performance monitoring
- Timing
- Power transfer
- Layer-1 multiplexing
- Interface termination, including multidrop termination employing layer-1 contention resolution

Network termination 2 (NT2) can be broadly associated with OSI layers 1, 2, and 3. Among the examples of equipment that provide NT2 functions are user terminal controllers, local area networks (LANs), and private automatic branch exchanges (PABXs). Among the NT2 functions are the following:

- Layer-1, -2, and -3 protocols processing
- Multiplexing (layers 2 and 3)
- Switching
- Concentration
- Interface termination and other layer-1 functions
- Maintenance functions

A distinction must be drawn between North American and European practices. As we are aware, the major telecommunication administrations are national monopolies that are government controlled. In the United States and Canada they are private enterprises, often very competitive. Thus in Europe the customer interface is at the T interface and NT1 is considered part of the PSTN, even though it is physically on customer premises. In North America, the interface with the PSTN is called the U interface in Figure 13.39, which is briefly described near the end of this section. On some North American reference documents, NT2 does not even exist, its interface functions being incorporated in NT1. The basic rate U interface is two-wire.

TE1 in Figure 13.39 is the terminal equipment and has an interface that must comply with the ISDN user–network interface specifications. Terminal equipment (TE) covers functions broadly belonging to OSI layer 1 and higher OSI layers. Among this equipment are digital telephones, computer workstations (data terminal equipment, DTE), and other devices in the user-end equipment category.

TE2 refers to equipment that does *not* meet ISDN terminal–network interface specifications and that requires interface modifications to adapt the equipment to ISDN. A terminal equipment adapter (TA) provides the necessary conversion functions to permit TE2-type terminal equipment to interface with ISDN.

Reference points S, T, and R are used to identify the interface available at those points. S and T are identical electrically and mechanically and from the point of view of protocol. Point R relates to the TA interface or, in essence, is the interface of the nonstandard (i.e., non-ISDN) device.

13.14.5   Layer-1 Interface: Basic Rate

The S and T interface or layer-1 physical interface requires a balanced metallic transmission medium in each direction of transmission supporting 192 kbps. This is the NT interface shown in Figure 13.39. The 192 kbps is made up of 2B + D, which equals 144 kbps; the remaining 48 kbps are overhead bits.

Layer 1 provides the following services to layer-2 ISDN operation:

- The transmission capability by means of appropriately encoded bit streams for both B- and D-channels and also any timing and synchronization functions that may be required
- The signaling capability and the necessary procedures to enable customer terminals and/or network terminating equipment to be deactivated when required, and reactivated when required
- The signaling capability and necessary procedures to allow terminals to gain access to the common resource of the D-channel in an orderly fashion while meeting performance requirements of the D-channel signaling system
- The signaling capability and procedures and necessary functions at layer 1 to enable maintenance functions to be performed
- An indication to the higher layers of the status of layer 1

The frame structure of 2B + D operation is given in Figure 13.40. Note in the figure that in both directions of transmission the bits are grouped into frames of 48 bits each. The frame structure is identical for all configurations, whether point-to-point or point-to-multipoint. However, the frame structures are different for each direction of transmission. Explanatory notes to Figure 13.40 are given in Table 13.22.

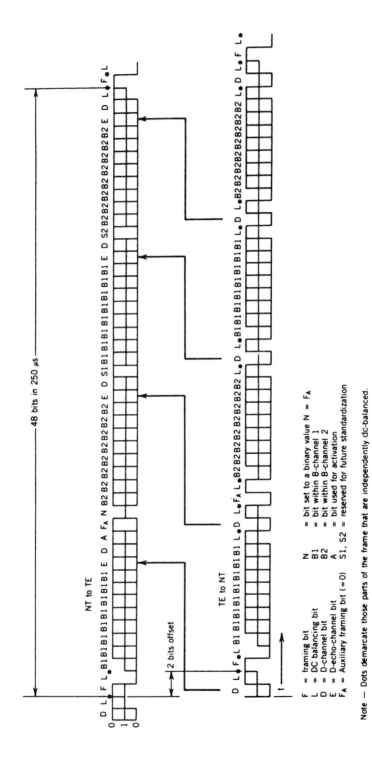

Figure 13.40. Frame structure at reference points S and T. (From Ref. 58, Figure 3/I.430, CCITT Rec. I.430.)

F = framing bit
L = DC balancing bit
D = D-channel bit
E = D-echo-channel bit
F_A = Auxiliary framing bit (=0)

N = bit set to a binary value N = F_A
B1 = bit within B-channel 1
B2 = bit within B-channel 2
A = bit used for activation
S1, S2 = reserved for future standardization

Note — Dots demarcate those parts of the frame that are independently dc-balanced.

TABLE 13.22   Notes on Bit Positions and Groups in Figure 13.40

Bit Position	Group
Terminal to Network: Each frame consists of the following group of bits; each individual group is dc balanced by its last bit (L-bit).	
1 and 2	Framing signal with balance bit
3–11	B1-channel with balance bit (first octet)
12 and 13	D-channel bit with balance bit
14 and 15	Auxiliary framing with balance bit
16–24	B2-channel with balance bit (first octet)
25 and 26	D-channel bit with balance bit
27–35	B1-channel with balance bit (second octet)
36 and 37	D-channel bit with balance bit
38–46	B2-channel with balance bit (second octet)
47 and 48	D-channel with balance bit
Network to Terminal: Frames transmitted by the network (NT) contain an echo channel (E-bits) used to retransmit the D-bits received from the terminals. The D-echo channel is used for D-channel access control. The last bit of the frame (L-bit) is used for balancing each complete frame. The bits are grouped as follows:	
1 and 2	Framing signal with balance bit
3–10	B1-channel (first octet)
11	E-, D-echo-channel bit
12	D-channel bit
13	Bit A used for activation
14	F_A auxiliary framing bit
15	N bit [a]
16–23	B2-channel (first octet)
24	E-, D-echo-channel bit
25	D-channel bit
26	S1, reserved for future standardization[b]
27–34	B1-channel (second octet)
35	E-, D-echo-channel bit
36	D-channel bit
37	S2, reserved for future standardization[b]
38–45	B2-channel (second octet)
46	E-, D-echo-channel bit
47	D-channel bit
48	Frame balance bit

[a]As defined in Section 6.3 of Ref. 58.
[b]S1 and S2 are set to binary 0.

Source: Reference 58, Tables 2 and 3/I.430, CCITT Rec. I.430.

Binary values   0 1 0 0 1 1 0 0 0 1 1

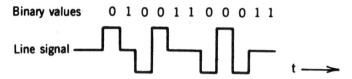

Figure 13.41. ISDN pseudoternary line code.

The line code for both directions of transmission is pseudoternary coding with 100% pulsewidth, as shown in Figure 13.41. Coding is performed such that a binary 1 is represented by a no-line signal, whereas a binary 0 is represented by a positive or negative pulse. The first binary signal following the framing balance bit is the same polarity as the framing balance bit. Subsequent binary 0's alternate in polarity. A balance bit is a 0 if the number of 0's following the previous balance bit is odd. A balance bit is binary 1 if the number of binary 0's following the previous balance bit is even.

The NT derives its timing from the network clock. The TE synchronizes its bit, octet, and frame timing from the received bit stream from the NT and uses the derived timing to synchronize its transmitted signal.

13.14.6   Layer-1 Interface: Primary Rate

This interface is applicable for the 1.544- or 2.048-Mbps data rates.

13.14.6.1   *Interface at 1.544 Mbps*

13.14.6.1.1   Bit Rate and Synchronization

NETWORK CONNECTION CHARACTERISTICS. The network delivers (except as noted below) a signal synchronized from a clock having a minimum accuracy of 1×10^{-11} (stratum 1; see Chapter 4 for stratum definition). When synchronization by a stratum 1 clock has been interrupted, the signal delivered by the network to the interface will have a minimum accuracy of 4.6×10^{-6} (stratum 3).

While in normal operation, the TE1/TA/NT2 transmits a 1.544-Mbps signal having an accuracy equal to that of the received signal by locking the frequency of its transmitter signal to the long-term average of the incoming 1.544-Mbps signal, or by providing equal signal frequency accuracy from another source. ITU-T Rec. I.431 (Ref. 59) advises against this latter alternative.

RECEIVER BIT STREAM SYNCHRONIZED TO A NETWORK CLOCK

Receiver Requirements. Receivers of signals across interface I_a operate with an average transmission rate in the range of 1.544 Mbps ± 4.6 ppm. However, operation with a received signal transmission rate in the range of 1.544 Mbps ± 32 ppm is required in any maintenance state controlled by signals/messages passed over the m-bits and by AIS

TABLE 13.23 Digital Interface at 1.544 Mbps

Bit rate:		1544 kbps
Pair(s) in each direction of transmission:		One symmetrical pair
Code:		B8ZS[a]
Test load impedance:		100-Ω resistive
Nominal pulse shape:		See pulse mask[b]
Signal level[b,c]:	Power at 772 kHz	+12 dBm to +19 dBm
	Power at 1544 kHz	At least 25 dB below the power at 772 kHz

[a]B8ZS is modified AMI code in which eight consecutive binary 0's are replaced with 000+−0−+ if the preceding pulse was positive (+) and with 000−+0+− if the preceding pulse was negative (−).

[b]The pulse mask and power level requirements apply at the end of a pair having a loss at 772 kHz of 0–1.5 dB.

[c]The signal level is the power level measured in a 3-kHz bandwidth at the output port for an all binary 1's pattern transmitted.

Source: Reference 59, Table 4/I.431, ITU-T Rec. I.431.

(alarm indication signal). In normal operation the bit stream is synchronized to stratum 1.

Transmitter Requirements. The average transmission rate of signals transmitted across interface I_a by the associated equipment is the same as the average transmission rate of the received bit stream.

Note: The I_a and I_b interfaces are located at the input/output port of TE or NT.

TE1/TA OPERATING BEHIND AN NT2 THAT IS NOT SYNCHRONIZED TO A NETWORK CLOCK

Receiver Requirements. Receivers of signals across interface I_a operate with a transmission rate in the range of 1.544 Mbps ± 32 ppm.

Transmitter Requirements. The transmitted signal across interface I_a is synchronized to the received bit stream.

SPECIFICATION OF OUTPUT PORTS. The signal specification for output ports is summarized in Table 13.23.

13.14.6.1.2 Frame Structure. The frame structure is shown in Figure 13.42.
Each time slot consists of consecutive bits, numbered 1 through 8. Each frame is 193 bits long and consists of an F-bit (framing bit) followed by 24 consecutive time slots. The frame repetition rate is 8000 frames per second.
Table 3.4 shows the multiframe structure (called *extended superframe* in the United States*), which is 24 frames long. It takes advantage of the more advanced search algorithms for frame alignment. There are 8000 F-bits (frame alignment bits) transmitted per second (i.e., 8000 frames a second, 1

---

*Note that there are some small differences between the CCITT defined 1.544-Mbps superframe and the ANSI/Bellcore specifications.

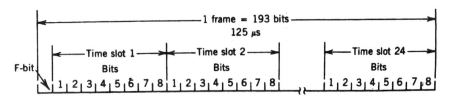

Figure 13.42. Frame structure of 1.544-Mbps interface.

F-bit per frame). In 24 frames, with these new strategies, only 6 bits are required for frame alignment as shown in Table 3.4. The remaining 18 bits are used as follows: There are 12 m-bits used for control and maintenance. The 6 e-bits are used for CRC6 error checking.

13.14.6.1.3 Time-Slot Assignment. Time slot 24 is assigned to the D-channel, when this channel is present.

A channel occupies an integer number of time slots and in the same time-slot positions in every frame. A B-channel may be assigned any time slot in the frame, an H_0-channel may be assigned any six slots in a frame in numerical order (not necessarily consecutive), and an H_{11}-channel may be assigned slots 1 to 24. The assignments may vary on a call-by-call basis.

CODES FOR IDLE CHANNELS, IDLE SLOTS, AND INTERFRAME TIME FILL. A pattern including at least three binary 1's in an octet is transmitted in every time slot that is not assigned a channel (e.g., time slots awaiting channel assignment on a per-call basis, residual slots on an interface that is not fully provisioned) and in every time slot of a channel that is not allocated to a call in both directions. Interframe (layer 2) time fill consists of contiguous HDLC flays transmitted on the D-channel when its layer 2 has no frames to send.

13.14.6.2 Interface at 2.048 Mbps

13.14.6.2.1 Frame Structure. There are 8 bits per time slot and 32 time slots per frame, numbered 0 through 31. The number of bits per frame is 256 (i.e., 32×8) and the frame repetition rate is 8000 frames per second. Time slot 0 provides frame alignment and time slot 16 is assigned to the D-channel when that channel is present. A channel occupies an integer number of time slots and the same time-slot position in every frame. A B-channel may be assigned any time slot in the frame and an H_0-channel may be assigned any six time slots, in numerical order, not necessarily consecutive. The assignment of time slots may vary on a call-by-call basis. An H_{12}-channel is assigned time slots 1 to 15 and 17 to 31 in a frame. Time slots 1 to 31 provide bit-sequence-independent transmission. See Chapter 3 for more complete discussion of DS1 and E-1.

13.14.6.2.2 Timing Considerations. The NT derives its timing from the network clock. The TE synchronizes its timing (bit, octet, framing) from the signal received from the NT and synchronizes accordingly the transmitted signal. In an unsynchronized condition—that is, when the access that normally provides network timing is unavailable—the frequency deviation of the free-running clock shall not exceed ± 50 ppm. A TE shall be able to detect and to interpret the input signal within a frequency range of ± 50 ppm.

Any TE that provides more than one interface is declared to be a multiple access TE and is capable of taking the synchronizing clock frequency from its internal clock generator from one or more than one access (or all access links) and synchronize the transmitted signal at each interface accordingly.

13.14.6.2.3 Codes for Idle Channels and Idle Time Slots, Interframe Fill. A pattern including at least three binary 1's in an octet is transmitted in every time slot that is not assigned to a channel (e.g., time slots awaiting channel assignment on a per-call basis, residual slots on an interface that is not fully provisioned) and in every time slot of a channel that is not allocated to a call in both directions.

Interframe (layer 2) time fill consists of contiguous HDLC flags (01111110), which are transmitted on the D-channel when its layer 2 has no frames to send.

Frame alignment and CRC procedures can be found in ITU-T Rec. G.706, paragraph 4.

13.14.7   BRI Differences in the United States

Bellcore and ANSI prepared ISDN BRI standards fairly well modified from the CCITT I Recommendation counterparts. The various PSTN administrations in the United States are at variance with most other countries. Furthermore, it was Bellcore's intention to produce equipment that was cost effective and marketable and that would easily interface with existing North American telephone plants.

One point, of course, is where the telephone company responsibilities end and customer responsibilities begin. This is called the "U" interface (see Figure 13.39). The tendency toward a two-wire interface rather than a four-wire interface is another. Line waveform is yet another. Rather than a pseudoternary line waveform, the United States uses 2B1Q. The line bit rate is 160 kbps rather than that recommended by CCITT, namely, 192 kbps. More detailed explanation of North American practice follows. The 2B + D + overhead frames differ significantly. Bellcore uses the generic term DSL for digital subscriber line in its specifying document for National ISDN 1.

13.14.7.1 DSL Bandwidth Allocation and Frame Structure. The DSL, or *digital subscriber line*, modulation rate is 80 kilobauds, equivalent to 160 kbps. The effective 160-kbps signal is divided up into 12 kbps for synchronization words, 144 kbps for 2B + D of customer data, and 4 kbps of DSL overhead.

The synchronization technique used is based on transmission of nine (quaternary) symbols every 1.5 ms, followed by 216 bits of 2B + D data and 6 bits of overhead. The synchronization word provides a robust method of conveying line timing and establishes a 1.5-ms DSL "basic frame" for multiplexing subrate signals. Every eighth synchronization word is inverted (i.e., the 1's become 0's and the 0's become 1's) to provide a boundary for a 12-ms superframe composed of eight basic frames. This 12-ms interval defines an appropriate block of customer data for performance monitoring and permits a more efficient suballocation of the overhead bits among various operational functions.

The 48 overhead bits, sometimes referred to as "maintenance" or M-bits, available per superframe are allocated into 24 bits for a 2-kbps eoc, 12 bits for a crc covering the 1728 bits of customer data in a superframe, plus 8 particular overhead bits, and one febe bit for communicating block errors detected by the crc to the far end. In determining relative positions of the various overhead bit functions, febe and crc bits were placed away from the start of the superframe to allow time for the crc calculation associated with the preceding superframe to be completed.

The eoc is the *embedded operations channel* and permits network operations systems to access essential operations functionality in the NT1. There are 12 cyclic redundancy check (crc) bits within each superframe, and they cover the 1728 bits of 2B + D customer data plus the eight "M4" bits in the previous superframe. This provides a means of block error check. Two "crc bits" are assigned to each of basic frames 3 through 8. These BRI frame concepts are illustrated in Figures 13.43a and 13.43b.

The febe bit or, "far end block error" bit, is assigned to each superframe. The febe bit indicates whether a block error was detected in the 2B + D customer data in a preceding superframe in the opposite direction of transmission. An outgoing febe bit with a value of 1 indicates that the crc of the preceding incoming superframe matched the value computed from the transmitted data; a value of 0 indicates a mismatch, which, in turn, indicates one or more bit errors.

13.14.7.2   2B + D Customer Data Bit Pattern.

There are 216 2B + D bits placed in each 1.5-ms basic frame, for a customer data rate of 144 kbps. The bit pattern (before conversion to quaternary form and after reconversion to binary form) for the 2B + D data is

$$B_1 B_1 B_1 B_1 B_1 B_1 B_1 B_1 B_2 B_2 B_2 B_2 B_2 B_2 B_2 B_2 DD$$

where B_1 and B_2 are bits from the B_1- and B_2-channels and D is a bit from the D-channel. This 18-bit pattern is repeated 12 times per DSL basic frame.

[8 × 1.5-ms "Basic Frames" → 12-ms Superframe]

		FRAMING	2B + D	Overhead Bits (M1–M6)					
	Quat Positions	1–9	10–117	118s	118m	119s	119m	120s	120m
	Bit Positions	1–18	19–234	235	236	237	238	239	240
Super-frame #	Basic Frame #	Sync Word	2B + D	M1	M2	M3	M4	M5	M6
A	1	ISW	2B + D	eoc_{a1}	eoc_{a2}	eoc_{a3}	act	1	1
	2	SW	2B + D	eoc_{dm}	eoc_{i1}	eoc_{i2}	dea	1	febe
	3	SW	2B + D	eoc_{i3}	eoc_{i4}	eoc_{i5}	1	crc_1	crc_2
	4	SW	2B + D	eoc_{i6}	eoc_{i7}	eoc_{i8}	1	crc_3	crc_4
	5	SW	2B + D	eoc_{a1}	eoc_{a2}	eoc_{a3}	1	crc_5	crc_6
	6	SW	2B + D	eoc_{dm}	eoc_{i1}	eoc_{i2}	1	crc_7	crc_8
	7	SW	2B + D	eoc_{i3}	eoc_{i4}	eoc_{i5}	1	crc_9	crc_{10}
	8	SW	2B + D	eoc_{i6}	eoc_{i7}	eoc_{i8}	1	crc_{11}	crc_{12}
B, C, ...									

NT1-to-LT superframe delay offset by LT-to-NT1 superframe by 60 ± 2 quats (about 0.75 ms).
All bits other than the Sync Word are scrambled.

"1" = reserved bit for future standard; set = 1
( )$_m$, ( )$_s$ "magnitude" bit and "sign" bit for given quat
act = activation bit
crc = cyclic redundancy check: covers 2B + D & M4
dea = deactivation bit

eoc = embedded operations channel
a = address bit
dm = data/message indicator
i = information (data/message)
febe = far end block error bit

Figure 13.43a. LT to NT1 2B1Q superframe technique and overhead bit assignments. (From Ref. 60, Figure 3.1a.)

13.14.7.3 The DSL Line Code—2B1Q.

The average power of a 2B1Q transmitted signal is between +13 and +14 dBm over a frequency band from 0 Hz to 80 kHz, with the nominal peak of the largest pulse being 2.5 volts. The maximum signal power loss at 40 kHz is about 42 dB. As mentioned earlier, the bit rate is 160 kbps and the modulation rate is 80 kilobauds.

13.14.7.4 The 2B1Q Waveform.

It is convenient to express the 2B1Q waveform as +3, +1, −1, −3 because this indicates symmetry about zero, equal spacing between states, and convenient integer magnitudes. The block synchronization word (SW) contains nine quaternary elements repeated every 1.5 ms:

$$+3, +3, -3, -3, -3, +3, -3, +3, +3$$

[8 × 1.5-ms "Basic Frames" → 12-ms Superframe]

		FRAMING	2B + D	Overhead Bits (M1–M6)					
	Quat Positions	1–9	10–117	118s	118m	119s	119m	120s	120m
	Bit Positions	1–18	19–234	235	236	237	238	239	240
Super-frame #	Basic Frame #	Sync Word	2B + D	M1	M2	M3	M4	M5	M6
1	1	ISW	2B + D	eoc_{a1}	eoc_{a2}	eoc_{a3}	act	1	1/nib
	2	SW	2B + D	eoc_{dm}	eoc_{i1}	eoc_{i2}	ps_1	1	febe
	3	SW	2B + D	eoc_{i3}	eoc_{i4}	eoc_{i5}	ps_2	crc_1	crc_2
	4	SW	2B + D	eoc_{i6}	eoc_{i7}	eoc_{i8}	ntm	crc_3	crc_4
	5	SW	2B + D	eoc_{a1}	eoc_{a2}	eoc_{a3}	cso	crc_5	crc_6
	6	SW	2B + D	eoc_{dm}	eoc_{i1}	eoc_{i2}	1	crc_7	crc_8
	7	SW	2B + D	eoc_{i3}	eoc_{i4}	eoc_{i5}	1	crc_9	crc_{10}
	8	SW	2B + D	eoc_{i6}	eoc_{i7}	eoc_{i8}	1	crc_{11}	crc_{12}
2, 3, ...									

NT1-to-LT superframe delay offset by LT-to-NT1 superframe by 60 ± 2 quats (about 0.75 ms).
All bits other than the Sync Word are scrambled.

"1" = reserved bit for future standard; set = 1
$(\ )_m, (\ )_s$ "magnitude" bit and "sign" bit for given quat
act = activation bit
crc = cyclic redundancy check: covers 2B + D & M4
cso = cold start only bit
dea = deactivation bit
febe = far end block error bit

eoc = embedded operations channel
    a = address bit
    dm = data/message indicator
    i = information (data/message)
    ntm = NT1 in Test Mode bit
    nib = network indicator bit from LULT and LUNT to LT
    ps_1, ps_2 = power status bits
        = 1 from NT1 to LT

Figure 13.43b. NT1 to LT 2B1Q superframe technique and overhead bit assignment. (From Ref. 60, Figure 3-1b.)

The 2B1Q waveform is shown below:

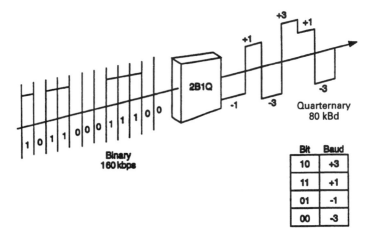

Bit	Baud
10	+3
11	+1
01	-1
00	-3

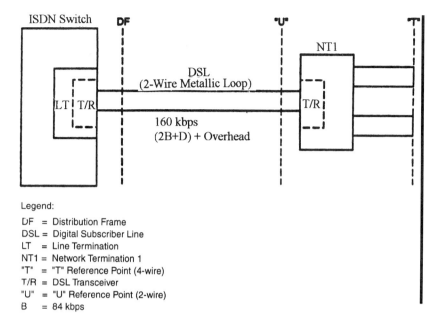

Figure 13.44. ISDN basic access rate configuration. (From Ref. 61, Figure 12-20.)

13.14.7.5 Line Configuration. One basic access is provided by a single DSL interface. This is shown in Figure 13.44. The two-wire interface is shown on the right-hand side of the figure. Operation on this two-wire line is full-duplex. To avoid interference between the transmitted and received signals, an echo canceler with hybrid is used. Echo cancellation involves adaptively forming a replica of the echo signal arriving at a receiver from its local transmitter and subtracting it from the signal at the input of the receiver (Ref. 61).

13.14.8   National ISDN 1, 2, and 3

13.14.8.1 National ISDN 1. As set out in Bellcore document SR-NWT-001937 (Ref. 62), National ISDN 1 identifies an available and maintainable ISDN platform and associated package of features that has been developed in the 1991–1992 time frame. This is essentially the initial offering of the Bellcore client companies (BCC).

13.14.8.2 National ISDN 2. National ISDN 2 represents the second step in the evolution of ISDN in North America and was introduced in the second half of 1993. Among the long list of capabilities in Bellcore SR-NWT-002120 (Ref. 63) is packet-switched data on the B-channel.

13.14.8.3 National ISDN 3. National ISDN 3 continues the evolution of ISDN in the United States. Refer to Bellcore SR-NWT-002457 (Ref. 64). It was introduced in early 1994.

13.15   AN OVERVIEW OF ATM AND ITS TRANSMISSION CONSIDERATIONS

13.15.1   Introduction

ATM (asynchronous transfer mode) is the first true attempt to integrate voice, data, and video on a switched basis and transported on a single transmission medium. It works on a packet concept, and ATM packets are very small, just 53 octets long. Instead of packets, they are called cells. Figure 13.45 illustrates the concept where cells from each medium source are "mixed and matched" for transmission down a common transport medium.

Typically, these ATM cells can be transported on SONET/SDH, on E1/T1, or on a radio/fiber medium just transporting cells without an underlying format. ATM on underlying format structures is discussed at the end of this section.

Philosophically, voice and data are worlds apart regarding time sensitivity. Voice cannot wait for long processing and ARQ delays. Most types of data can. So ATM must distinguish the type of service such as constant bit rate (CBR) and variable bit rate (VBR) services. Voice service is typical of constant bit rate or CBR service.

Signaling is another area of major philosophical difference. In data communications, "signaling" is carried out within the header of a data frame (or packet). As a minimum the signaling will have the destination address, and quite often the source address as well. Also, this signaling information will be repeated over and over again on a long data file that is heavily segmented. On a voice circuit, a connectivity is set up and the destination address, and possibly the source address, is sent just once during call setup. There is also some form of circuit supervision to keep the circuit operational throughout the duration of a telephone call. ATM is a compromise, stealing a little from each of these separate worlds.

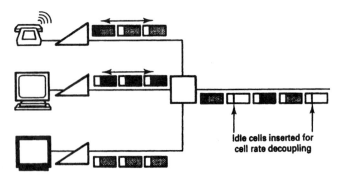

Figure 13.45. An ATM transport link simultaneously carries a mix of voice, data, and image information. (Courtesy of Hewlett-Packard Co.)

Like voice telephony, ATM is fundamentally a connection-oriented telecommunication system. Here we mean that a connection must be established between two stations before data can be transferred between them. An ATM connection specifies the transmission path, allowing ATM cells to self-route through an ATM network. Being connection-oriented also allows ATM to specify a guaranteed quality of service (QoS) for each connection.

By contrast, most LAN protocols are connectionless. This means that LAN nodes simply transmit traffic when they need to, without first establishing a specific connection or route with the destination node.

Because ATM uses a connection-oriented protocol, cells are allocated only when the originating user requests a connection. This allows ATM to efficiently support a network's *aggregate* demand by allocating cells on demand based on immediate user need. Indeed, it is this concept that lies in the heart of the word *asynchronous*. Let us explain. The cell repetition rate is synchronous and constant, and, of course, some or many of the cells are idle. The utilization of the cells (i.e., that they carry traffic or are part of a voice connection) is asynchronous. Here we mean that cells are used on an "as-needed basis" up to some maximum. This is the maximum cell repetition rate.

13.15.2 User–Network Interface (UNI) Configuration and Architecture

ATM is the underlying packet technology of broadband ISDN (B-ISDN). At times in this section we will use the terms ATM and B-ISDN interchangeably. Figures 13.46 and 13.47 interrelate the two. Figure 13.46 associates the B-ISDN access reference configuration with the ATM user–network interface (UNI). Note the similarities with this figure and Figure 13.39. The only difference is that the block nomenclature has a "B" placed in front to indicate *broadband*. Figure 13.47 is the traditional ITU-T Rec. I.121 (Ref. 65) B-ISDN protocol reference model showing the extra layer necessary for the several services.

Returning to Figure 13.46, we see that there is an upper part and a lower part. The lower part shows the UNI boundaries. The upper part is the B-ISDN reference configuration with four interface points. The interfaces at reference points U_B, T_B, and S_B are standardized. These interfaces support all B-ISDN services.

There is only one interface per B-NT1 at the U_B and one at the T_B reference points. The physical medium is point-to-point (in each case) in the sense that there is only one receiver in front of one transmitter.

One or more interfaces per B-NT2 are present at the S_B reference point. The interface at the S_B reference point is point-to-point at the physical layer in the sense that there is only one receiver in front of one transmitter and may be point-to-point with other layers.

Consider now the functional groupings in Figure 13.46. B-NT1 includes functions broadly equivalent to OSI layer 1, the physical layer. These

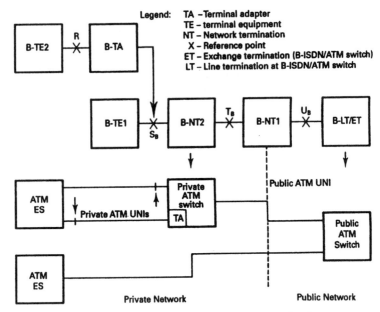

Figure 13.46. ATM reference model and user–network interface (UNI) configuration. (From Refs. 66, 67, and 68.)

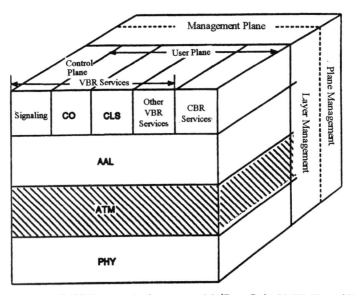

Figure 13.47. B-ISDN protocol reference model. (From Refs. 66, 70, 71, and 75.)

functions include the following:

- Line transmission termination
- Interface handling at T_B and U_B
- Operations, administration, and maintenance (OAM) functions

The B-NT2 functional group includes functions broadly equivalent to OSI layer 1 and higher OSI layers. The B-NT2 may be concentrated or distributed. In a particular access arrangement, the B-NT2 functions may consist of physical connections. Examples of B-NT2 functions are as follows:

- Adaptation functions for different media and topologies
- Cell delineation
- Concentration; buffering
- Multiplexing and demultiplexing
- OAM functions
- Resource allocation
- Signaling protocol handling

The functional group B-TE (TE stands for terminal equipment) also includes functions of OSI layer 1 and higher OSI layers. Some of these functions are as follows:

- User–user and user–machine dialogue and protocol
- Protocol handling for signaling
- Connection handling to other equipment
- Interface termination
- OAM functions

B-TE1 has an interface that complies with the B-ISDN interface. B-TE2, however, has a noncompliant B-ISDN interface. Compliance refers to ITU-T Recs. I.413 and I.432 as well as ANSI T1.624 (Refs. 67, 68, 72).

The terminal adapter (B-TA) converts the B-TE2 interface into a compliant B-ISDN user–network interface.

Four bit rates are specified at the U_B, T_B, and S_B interfaces based on ANSI T1.624 (Ref. 72):

51.840 Mbps	(SONET STS-1)
155.520 Mbps	(SONET STS-3 and SDH STM-1)
622.080 Mbps	(SONET STS-12 and SDH STM-4)
44.736 Mpbs	(DS3)

The following definitions refer to Figure 13.47:

User Plane (in other literature called the U-plane). The user plane provides for the transfer of user application information. It contains the physical layer, ATM layer, and multiple ATM adaptation layers required for different service users such as constant bit rate service (CBR) and variable bit rate service (VBR).

Control Plane (in other literature called the C-plane). The control plane protocols deal with call establishment and call release and other connection control functions necessary for providing switched services. The C-plane structure shares the physical and ATM layers with the user plane as shown in Figure 13.47. It also includes ATM adaptation layer (AAL) procedures and higher-layer signaling protocols.

Management Plane (in other literature called the M-plane). The management plane provides management functions and the capability to exchange information between the user plane and the control plane. The management plane contains two sections: layer management and plane management. The layer management performs layer-specific management functions, while the plane management performs management and coordination functions related to the complete system.

13.15.3 The ATM Cell—Key to Operation

13.15.3.1 ATM Cell Structure. An ATM cell consists of 53 octets, 5 of which make up the header and 48* octets are in the payload or "info" portion of the cell. Figure 13.48 shows an ATM cell stream delineating the 5-octet header and 48-octet information field (payload) of each cell. Figure 13.49 shows the detailed structure of cell headers at the user−network interface (UNI) (Figure 13.49*a*) and at the network−network interface (NNI)[†] (Figure 13.49*b*).

We digress a moment to discuss why a 53-octet cell was standardized. The cell header contains only 5 octets. It was shortened as much as possible, containing the minimum address and control functions for a working system. It is also non-revenue-bearing overhead. It is the information field that contains the revenue-bearing payload. For efficiency, we would like the payload to be as long as possible. Yet the ATM designer team was driven to shorten the payload as much as possible. The issue in this case was what is called *packetization delay*. This is the amount of time required to fill a cell at a rate of 64 kbps—that is, the rate to fill the cell with digitized voice samples. According to Ref. 73, the design team was torn between efficiency and

---

*There are several exceptions to these 5- and 48-cell values. A number of the ATM adaptation layer (AAL) configurations have 6 octets in the header and 47 octets in the payload.

[†]NNI is variously called network−node interface or network−network interface. It is the interface between two network nodes or switches.

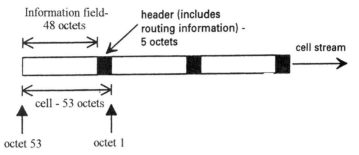

Figure 13.48. An ATM cell stream illustrating the basic makeup of a cell.

packetization delay. One school of thought fought for a 64-octet cell, and another argued for a 32-octet cell size. Thus the ITU-T opted for a fixed-length 53-octet compromise.

Now let us return to the discussion of the ATM cell and its headers. The left-hand side of Figure 13.49 shows the structure of a UNI header, whereas the right-hand side illustrates the NNI header. The only difference is the presence of the GFC field in the UNI header. The following paragraphs define each header field. By removing the GFC field, the NNI has four additional bits for addressing.

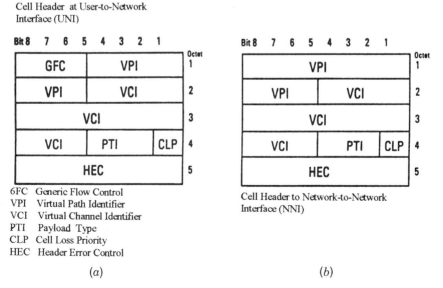

Figure 13.49. Basic ATM header structures: (a) UNI cell header structure and (b) NNI header structure.

GFC—Generic Flow Control. The GFC field contains 4 bits. When the GFC function is not used, the value of this field is 0000. This field has local significance only and can be used to provide standardized local flow control functions on the customer side. In fact the value encoded in the GFC is not carried end-to-end and will be overwritten by ATM switches (i.e., the NNI interface).

Two modes of operation have been defined for operation of the GFC field. These are *uncontrolled access* and *controlled access*. The "uncontrolled access" mode of operation is used in the early ATM environment. This mode has no impact on the traffic that a host generates. Each host transmits the GFC field set to all zeros (0000). In order to avoid unwanted interactions between this mode and the "controlled access" mode, where hosts are expected to modify their transmissions according to the activity of the GFC field, it is required that all CPE (customer premise equipment) and public network equipment monitor the GFC field to ensure that attached equipment is operating in "uncontrolled mode." A count of the number of nonzero GFC fields should be measured for nonoverlapping intervals of 30,000 ± 10,000 cell times. If ten or more nonzero values are received within this interval, an error is indicated to layer management (Ref. 66).

Routing Field (VPI/VCI). Twenty-four bits are available for routing a cell. There are 8 bits for virtual path identifier (VPI) and 16 bits for virtual channel identifier (VCI). Preassigned combinations of VPI and VCI values are given in Table 13.24. Other preassigned values of VPI and VCI are for further study according to the ITU-T Organization. The VCI value of zero is not available for user virtual channel identification. The bits within the VPI and VCI fields used for routing are allocated using the following rules:

- The allocated bits of the VPI field are contiguous.
- The allocated bits of the VPI field are the least significant bits of the VPI field, beginning at bit 5 of octet 2.
- The allocated bits of the VCI field are contiguous.
- The allocated bits of the VCI field are the least significant bits of the VCI field, beginning at bit 5 of octet 4.

Payload Type (PT) Field. Three bits are available for payload type identification. Table 13.25 gives the payload type identifier (PTI) coding. The main purpose of the PTI is to discriminate between user cells (i.e., cells carrying user information) and nonuser cells. The first four code groups (000–011) are used to indicate user cells. Within these four, the second and third (010 and 011) are used to indicate that congestion has been experienced. The fifth and sixth code groups (100 and 101) are used for VCC level management functions.

TABLE 13.24   Combinations of Preassigned VPI, VCI, and CLP Values at the UNI

Use	VPI	VCI	PT	CLP
Meta-signaling	XXXXXXXX	00000000 00000001	0A0	C
(refer to Rec. I.311)	(Note 1)	(Note 5)		
General broadcast signaling	XXXXXXXX	00000000 00000010	0AA	C
(refer to Rec. I.311)	(Note 1)	(Note 5)		
Point-to-point signaling	XXXXXXXX	00000000 00000101	0AA	C
(refer to Rec. I.311)	(Note 1)	(Note 5)		
Segment OAM F4 flow cell	YYYYYYYY	00000000 00000011	0A0	A
(refer to Rec. I.610)	(Note 2)	(Note 4)		
End-to-end OAM F4 flow cell	YYYYYYYY	00000000 00000100	0A0	A
(refer to Rec. I.610)	(Note 2)	(Note 4)		
Segment OAM F5 flow cell	YYYYYYYY	ZZZZZZZZ ZZZZZZZZ	100	A
(refer to Rec. I.610)	(Note 2)	(Note 3)		
End-to-end OAM F5 flow cell	YYYYYYYY	ZZZZZZZZ ZZZZZZZZ	101	A
(refer to Rec. I.610)	(Note 2)	(Note 3)		
Resource management cell	YYYYYYYY	ZZZZZZZZ ZZZZZZZZ	110	A
(refer to Rec. I.371)	(Note 2)	(Note 3)		
Unassigned cell	00000000	00000000 00000000	BBB	0

The GFC field is available for use with all of these combinations.
A Indicates that the bit may be 0 or 1 and is available for use by the appropriate ATM layer function.
B Indicates the bit is a "don't care" bit.
C Indicates the originating signaling entity shall set the CLP bit to 0. The value may be changed by the network.
Note 1. XXXXXXXX: Any VPI value. For VPI value equal to 0, the specified VCI value is reserved for user signaling with the local exchange. For VPI values other than 0, the specified VCI value is reserved for signaling with other signaling entities (e.g., other users or remote networks).
Note 2. YYYYYYYY: Any VPI value.
Note 3. ZZZZZZZZ ZZZZZZZZ: Any VCI value other than 0.
Note 4. Transparency is not guaranteed for the OAM F4 flows in a user-to-user VP.
Note 5. The VCI values are preassigned in every VPC at the UNI. The usage of these values depends on the actual signaling configurations. (See ITU-T Rec. I.311.)
Source: Reference 69, Table 2/I.361, ITU-T Rec. I.361.

TABLE 13.25   PTI Coding

	PTI Coding	Interpretation
Bits	4 3 2	
	0 0 0	User data cell, congestion not experienced. ATM-user-to-ATM-user indication = 0
	0 0 1	User data cell, congestion not experienced. ATM-user-to-ATM-user indication = 1
	0 1 0	User data cell, congestion experienced. ATM-user-to-ATM-user indication = 0
	0 1 1	User data cell, congestion experienced. ATM-user-to-ATM-user indication = 1
	1 0 0	OAM F5 segment associated cell
	1 0 1	OAM F5 end-to-end associated cell
	1 1 0	Resource management cell
	1 1 1	Reserved for future functions

Source: Reference 69, para. 2.2.4, ITU-T Rec. I.361.

Any congested network element, upon receiving a user data cell, may modify the PTI as follows. Cells received with PTI = 000 or PTI = 010 are transmitted with PTI = 010. Cells received with PTI = 001 or PTI = 011 are transmitted with PTI = 011. Noncongested network elements should not change the PTI.

Cell Loss Priority (CLP) Field. Depending on network conditions, cells where the CLP is set (i.e., CLP value is 1) are subject to discard prior to cells where the CLP is not set (i.e., CLP value is 0). The concept here is identical with that of frame relay and the DE (discard eligibility) bit. ATM switches may tag CLP = 0 cells detected by the UPC (usage parameter control) to be in violation of the traffic contract by changing the CLP bit from 0 to 1.

Header Error Control (HEC) Field. The HEC is an 8-bit field and it covers the entire cell header. The code used for this function is capable of either single-bit error correction or multiple-bit error detection. Briefly, the transmitting side computes the HEC field value. The receiver has two modes of operation as shown in Figure 13.50. In the default mode there is the capability of single-bit error correction. Each cell header is examined and, if an error is detected, one of two actions takes place. The action taken depends on the state of the receiver. In the *correction mode*, only single-bit errors can be corrected and the receiver switches to the *detection mode*. In the "detection mode," all cells with detected header errors are discarded. When a header is examined and found not to be in error, the receiver switches to the "correction mode." The term *no action* in Figure 13.50 means no correction is performed and no cell is discarded.

Figure 13.51 is a flowchart showing the consequences of errors in the ATM cell header. The error protection function furnished by the HEC provides for both recovery from single errors and a low probability of delivery of cells with errored headers under bursty error conditions. ITU-T Rec. I.432 (Ref. 67) states that error characteristics of fiber-optic transmission systems

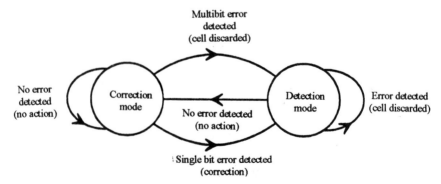

Figure 13.50. HEC: receiver modes of operation. (Based on Ref. 67, ITU-T Rec. I.432.)

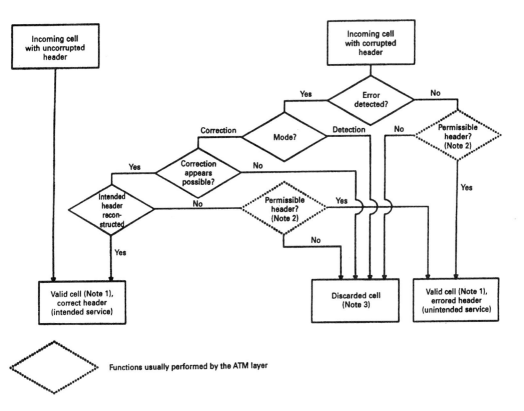

Figure 13.51. Consequences of errors in an ATM cell header. (From Ref. 67, Figure 12/I.432, ITU-T Rec. I.432.)

appear to be a mix of single-bit errors and relatively large burst errors. Thus for certain transmission systems the error correction capability might not be invoked.

13.15.3.2 Idle Cells. Idle cells cause no action at a receiving node except for cell delineation including HEC verification. They are inserted and extracted by the physical layer in order to adapt the cell flow rate to the available payload capacity of the transmission media. This is called *cell rate decoupling.* Idle cells are identified by the standardized pattern for the cell header as shown in Table 13.26.

TABLE 13.26 Header Pattern for Idle Cell Identification

	Octet 1	Octet 2	Octet 3	Octet 4	Octet 5
Header pattern	00000000	00000000	00000000	00000001	HEC = Valid code 01010010

Source: Reference 67, Table 4/I.432, ITU-T Rec. I.432.

The content of the information field is 01101010 repeated 48 times for an idle cell.

There is some variance in this area between the ITU-T Organization documentation and that of the ATM Forum. McDysan and Spohn in Ref. 73 point out the following. ITU-T Rec. I.321 places this function in the TC (transmission convergence) sublayer of the PHY (physical layer) and uses idle cells, whereas the ATM Forum places it in the ATM layer and uses unassigned cells. This presents a potential low-level incompatibility if different systems use different cell types for cell rate decoupling.

13.15.4 Cell Delineation and Scrambling

13.15.4.1 Delineation and Scrambling Objectives. Cell delineation allows identification of the cell boundaries. The cell header error control (HEC) field achieves cell delineation. Keep in mind that the ATM signal must be self-supporting in that it has to be transparently transported on every network interface without any constraints from the transmission systems used. Scrambling is used to improve security and robustness of the HEC cell delineation mechanism discussed below. In addition, it helps the randomizing of data in the information field for possible improvement in transmission performance.

Any scrambler specification must not alter the ATM header structure, header error control, and cell delineation algorithm.

13.15.4.2 Cell Delineation Algorithm. Cell delineation is performed by using the correlation between the header bits to be protected (32 bits or 4 octets) and the HEC octet, which are the relevant control bits (8 bits) introduced in the header using a shortened cyclic code with the generating polynomial $X^8 + X^2 + X + 1$.

Figure 13.52 shows the state diagram of the HEC cell delineation method. A discussion of the figure is given as follows:

1. In the HUNT state, the delineation process is performed by checking bit by bit for the correct HEC (i.e., syndrome equals zero) for the assumed header field. For the cell-based* physical layer, prior to scrambler synchronization, only the last 6 bits of the HEC are used for

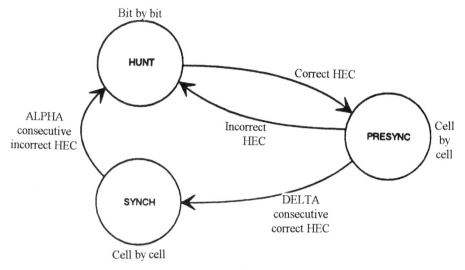

Figure 13.52. Cell delineation state diagram. (From Ref. 67, Figure 13/I.432, ITU-T Rec. I.432.)

cell delineation checking. For the SDH-based* interface, all 8 bits are used for acquiring cell delineation. Once such an agreement is found, it is assumed that one header has been found, and the method enters the PRESYNCH state. When octet boundaries are available within the receiving physical layer prior to cell delineation as with the SDH-based interface, the cell delineation process may be performed octet by octet.

2. In the PRESYNCH state, the delineation process is performed by checking cell by cell for the correct HEC. The process repeats until the correct HEC has been confirmed *Delta* times consecutively. If an incorrect HEC is found, the process returns to the HUNT state.

3. In the SYNCH state the cell delineation will be assumed to be lost if an incorrect HEC is obtained *Alpha* times consecutively.

The parameters *Alpha* and *Delta* are chosen to make the cell delination process as robust and secure as possible while satisfying QoS (quality of service) requirements. Robustness depends on *Alpha* when it is against false misalignments due to bit errors; and robustness depends on *Delta* when it is against false delineation in the resynchronization process.

For the SDH-based physical layer, values of *Alpha* = 7 and *Delta* = 6 are suggested by the ITU-T Organization (Rec. I.432 [Ref. 67]); and for cell-based physical layer, values of *Alpha* = 7 and *Delta* = 8 are suggested. Figures 13.53 and 13.54 give performance information of the cell delineation

---

*Only cell-based and SDH-based interfaces are covered by current ITU-T recommendations. Besides these, we will cover cells riding on other transport means at the end of this chapter.

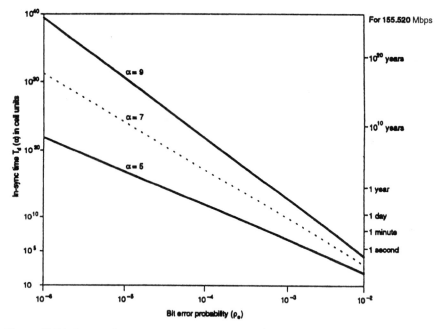

Figure 13.53. In-sync time versus bit error probability. (From Ref. 67, Figure B1/I.432, ITU-T Rec. I.432.)

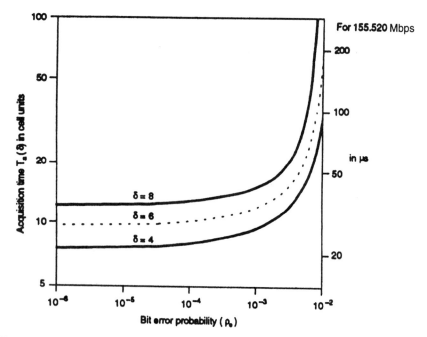

Figure 13.54. Acquisition time versus bit error probability. (From Ref. 67, Figure B.2/I.432, ITU-T Rec. I.432.)

algorithm in the presence of random bit errors, for various values of *Alpha* and *Delta*.

13.15.5 Quality of Service (QoS)

13.15.5.1 ATM Service Quality Review. A basic performance measure for any digital data communication system is bit error rate (BER). Well-designed fiber-optic links will predominate now and into the foreseeable future. We may expect BERs from such links on the order of 1×10^{-10} and with end-to-end performance better than 1×10^{-9} (Ref. 74).* Thus other performance issues may dominate the scene. These may be called ATM unique QoS items, namely:

- Cell transfer delay
- Cell delay variation
- Cell loss ratio
- Mean cell transfer delay
- Cell error ratio
- Severely errored cell block ratio
- Cell misinsertion rate

13.15.5.2 QoS Parameter Descriptions

Cell Transfer Delay. In addition to the normal delay through network elements and lines, extra delay is added to an ATM network at an ATM switch. The cause of the delay at this point is the statistical asynchronous multiplexing. By using this method, two cells can be directed toward the same output of an ATM switch or cross-connect, resulting in output contention.

The result is that one cell (or more) is held in a buffer until the next available opportunity to continue transmission. We can see that the second cell will suffer additional delay. The delay of a cell will depend on the amount of traffic within a switch and thus the probability of contention.

The asynchronous path of each ATM cell also contributes to cell delay. Cells can be delayed one or many cell periods, depending on traffic intensity, switch sizing, and the transmission path taken through the network.

Cell Delay Variation. By definition, ATM traffic is asynchronous, magnifying transmission delay. Delay is also inconsistent across the network. It can be a function of time (i.e., a moment in time), network design/switch design (such as buffer size), and traffic characteristics at that moment of time. The result is cell delay variation (CDV).

---

*These seem like ambitious goals if end-to-end. They should be considered in light of ITU-T Recs. G.821 and G.826. This topic is discussed in Chapter 4.

CDV can have several deleterious effects. The dispersion effect, or spreading out, of cell interarrival times can impact signaling functions or the reassembly of cell user data. Another effect is called *clumping*. This occurs when the interarrival times between transmitted cells shorten. One can imagine how this could affect the instantaneous network capacity and how it can impact other services using the network.

There are two performance parameters associated with cell delay variation: 1-point cell delay variation (1-point CDV) and 2-point cell delay variation (2-point CDV).

The 1-point CDV describes variability in the pattern of cell arrival events observed at a single boundary with reference to the negotiated peak rate $1/T$ as defined in ITU-T Rec. I.371 (Ref. 75). The 2-point CDV describes variability in the pattern of cell arrival events as observed at the output of a connection portion (MP_2) with reference to the pattern of the corresponding events observed at the input to the connection portion (MP_1).

Cell Loss Ratio. Cell loss may not be uncommon in an ATM network. There are two basic causes of cell loss: error in cell header or network congestion.

Cells with header errors are automatically discarded. This prevents misrouting of errored cells, as well as the possibility of privacy and security breaches.

Switch buffer overflow can also cause cell loss. It is in these buffers that cells are held in prioritized queues. If there is congestion, cells in a queue may be discarded selectively in accordance with their level of priority. Here enters the CLP (cell loss priority) bit discussed in Section 13.15.3.1. Cells with this bit set to 1 are discarded in preference to other, more critical cells. In this way, buffer fill can be reduced to prevent overflow (Ref. 74).

Cell loss ratio is defined for an ATM connection as

$$\text{Lost cells}/\text{total transmitted cells}$$

Lost and transmitted cells counted in severely errored cell blocks should be excluded from the cell population in computing cell loss ratio (Ref. 66).

Mean Cell Transfer Delay. Mean cell transfer delay is defined as the arithmetic average of a specified number of cell transfer delays.

Cell Error Ratio. Cell error ratio is defined as for an ATM connection:

$$\text{Errored cells}/(\text{successfully transferred cells} + \text{errored cells})$$

Successfully transferred cells and errored cells contained in cell blocks counted as severely errored cell blocks should be excluded from the population used in calculating cell error ratio.

Severely Errored Cell Block Ratio. The severely errored cell block ratio for an ATM connection is defined as

Total severely errored cell blocks/total transmitted cell blocks

A cell block is a sequence of N cells transmitted consecutively on a given connection. A severely errored cell block outcome occurs when more than M errored cells, lost cells, or misinserted cell outcomes are observed in a received cell block.

For practical measurement purposes, a cell block will normally correspond to the number of user information cells transmitted between successive OAM cells. The size of a cell block is to be specified.

Cell Misinsertion Rate. The cell misinsertion rate for an ATM connection is defined as

Misinserted cells/time interval

This rate may be expressed equivalently as the number of misinserted user information cells per virtual connection.

A misinserted cell is a received cell that has no corresponding transmitted cell on that connection. Cell misinsertion on a particular connection can be caused by an undetected or miscorrected error in the header of a cell originated on a different connection or by an incorrectly programmed translation of VPI or VCI values for cells originated on a different connection [66, 74, 80]

13.15.6   Transporting ATM Cells

13.15.6.1 In the DS3 Frame. One of the most popular higher-speed digital transmission systems in North America is DS3 operating at a nominal transmission rate of 45 Mbps. It is also being widely implemented for the transport of SMDS.* The system used to map ATM cells into the DS3 format is the same as that used for SMDS.

DS3 uses the *physical layer convergence protocol* (PLCP) to map ATM cells into its bit stream. A DS3 PLCP frame is shown in Figure 13.55.

There are 12 cells in a frame. Each cell is preceded by a 2-octet framing pattern (A1, A2), to enable the receiver to synchronize cells. After the framing pattern there is an indicator consisting of one of 12 fixed patterns used to identify the cell location within the frame (POI). This is followed by an octet of overhead information used for path management. The entire frame is then padded with 13 or 14 nibbles (a nibble = 4 bits) of trailer to bring the transmission rate up to the exact DS3 rate. The DS3 frame has a 125-μs duration.

---

*SMDS—switched multimegabit data service, an offering of the Bell related companies, GTE, and so on. Its primary purpose is for LAN interconnect through the PSTN.

I PLCP Framing		I PO	I POH	I PLCP Payload	I
A1	A2	P11	Z6	First ATM Cell	
A1	A2	P10	Z5	ATM Cell	
A1	A2	P9	Z4	ATM Cell	
A1	A2	P8	Z3	ATM Cell	
A1	A2	P7	Z2	ATM Cell	
A1	A2	P6	Z1	ATM Cell	
A1	A2	P5	X	ATM Cell	
A1	A2	P4	B1	ATM Cell	
A1	A2	P3	G1	ATM Cell	
A1	A2	P2	X	ATM Cell	
A1	A2	P1	X	ATM Cell	
A1	A2	P0	C1	Twelfth ATM Cell	Trailer

I 1 Octet I   1 Octet   I 1 Octet   I 1 Octet   I 53 Octets   I   13 or 14   I
                                                   Object of BIP-8 Calculation   Nibbles

POI	Path Overhead Indicator
POH	Path Overhead
BIP-8	Bit Interleaved Parity - 8
X	Unassigned - Receiver required to ignore
A1, A2	Frame Alignment

Figure 13.55. Format of a DS3 PLCP frame. (From Ref. 74. Courtesy of Hewlett-Packard Co.)

DS3 has to contend with network slips (added/dropped frames to accommodate synchronization alignment). Thus PLCP is padded with a variable number of stuff (justification) bits to accommodate possible timing slips. The Cl overhead octet indicates the length of padding. The BIP (bit-interleaved parity) checks the payload and overhead functions for errors and performance degradation. This performance information is transmitted in the overhead.

13.15.6.2   DS1 Mapping. One approach to mapping ATM cells into a DS1 frame is to use a similar procedure as used on DS3 with PLCP. In this case only 10 cells are bundled into a frame, and two of the Z overheads are removed. The padding in the frame is set at 6 octets. The entire frame takes 3 ms to transmit and spans many DS1 ESF (extended superframe) frames. This mapping is shown in Figure 13.56. Note the reference to L2_PDU, taken directly from SMDS (see Ref. 61, Chapter 14). One must also consider the arithmetic. Each DS1 time slot is 8 bits long, 1 octet. There are then 24 octets in a DS1 frame. This, of course, can lead to the second method of carrying ATM cells in DS1, by directly mapping in ATM cells octet for octet (time slot). This is done in groups of 53 octets (1 cell) and would, by necessity, cross DS1 frame boundaries to accommodate an ATM cell.

13.15.6.3   E1 Mapping. E1 PCM has a 2.048-Mbps transmission rate. An E1 frame has 256 bits representing 32 channels or time slots, 30 of which

1	1	1	1	◄────── 53 Octets ──────►
A1	A2	P9	Z4	L2_PDU
A1	A2	P8	Z3	L2_PDU
A1	A2	P7	Z2	L2_PDU
A1	A2	P6	Z1	L2_PDU
A1	A2	P5	F1	L2_PDU
A1	A2	P4	B1	L2_PDU
A1	A2	P3	G1	L2_PDU
A1	A2	P2	M2	L2_PDU
A1	A2	P1	M1	L2_PDU
A1	A2	P0	C1	L2_PDU

OH Byte	Function
A1, A2	Framing bytes
PS-P0	Path overhead Identifier Bytes

PLCP Path Overhead Bytes

Z4-Z1	Growth Bytes
F1	PLCP Path User Channel
B1	BIP-8
G1	PLCP Status
M2-M1	SMDS Control Information
C1	Cycle/Stuff Counter Byte

Trailer = 6 Octets

3 ms

Figure 13.56. DS1 mapping with PLCP. (From Ref. 74. Courtesy of Hewlett-Packard Co.)

carry traffic. Time slots (TS) 0 and 16 are reserved. TS0 is used for synchronization and TS16 for signaling. The E1 frame is shown in Figure 13.57. From bits 9 to 128 and from bits 137 to 256 may be used for ATM cell mapping.

ATM cells can also be mapped directly into special E3 and E4 frames. The first has 530 octets available for cells (i.e., 10 cells) and the second has 2160 octets (not evenly divisible).

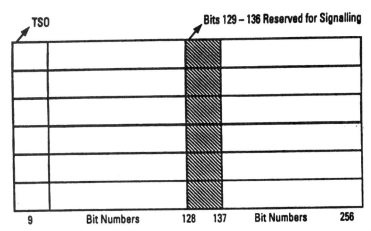

Figure 13.57. Mapping ATM cells directly into E1. (From Ref. 74. Courtesy of Hewlett Packard Co.)

13.15.6.4   *Mapping ATM Cells into SDH*

At the STM-1 Rate (155.520 Mbps). SDH is described in Chapter 4. Figure 13.58 shows the mapping procedure. The ATM cell stream is first mapped into the C-4 and then mapped into the VC-4 container along with the VC-4 path overhead. The ATM cell boundaries are aligned with the STM-1 octet boundaries. Since the C-4 capacity (2340 octets) is not an integer multiple of the cell length (53 octets), a cell may cross a C-4 boundary.

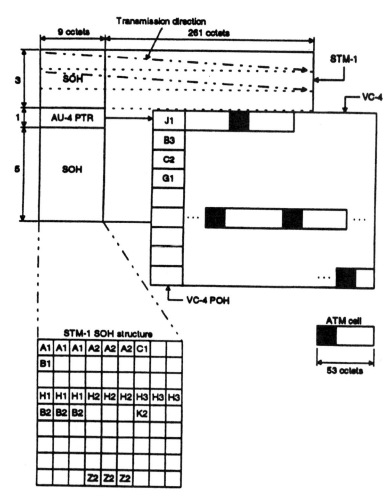

Figure 13.58. Mapping ATM cells in the 155.520-Mbps frame structure (STM-1) for SDH-based UNI. (From Ref. 67, Figure 8/I.432, ITU-T Rec. I.432.)

The AU-4 pointer (octets H1 and H2 in the SOH) is used for finding the first octet of the VC-4.

At the STM-4 Rate (622.080 Mbps). As shown in Figure 13.59, the ATM cell stream is first mapped into C-4-4c and then packed into the VC-4-4c container along with the VC-4-4c path overhead (POH). The ATM cell boundaries are aligned with the STM-4 octet boundaries. Since the C-4-4c capacity (9360 octets) is not an integer multiple of the cell length (53 octets), a cell may cross a C-4-4c boundary.

The AU pointers are used for finding the first octet in the VC-4-4c.

13.15.6.5 *Mapping ATM Cells into SONET.* ATM cells are mapped directly into the SONET payload (49.54 Mbps). As with SDH, the payload in

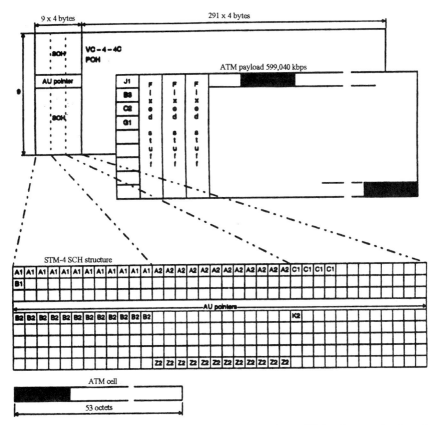

Figure 13.59. Mapping ATM cells into the 622.080-Mbps STM-4 frame structure for SDH-based UNI. (From Ref. 67, Figure 10/I.432, ITU-T Rec. I.432.)

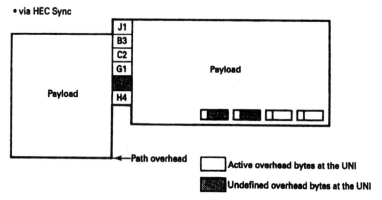

Figure 13.60. Mapping ATM cells into a SONET STS-1 frame. (From Ref. 74. Courtesy of Hewlett-Packard Co.)

octets is not an integer multiple of the cell length, and thus a cell may cross a STS frame boundary. This mapping concept is shown in Figure 13.60. The H4 pointer can indicate where the cells begin inside the STS frame. Another approach is to identify cell headers, and thus the first cell in the frame.

13.16 DATA TRANSMISSION ON TWISTED WIRE PAIR

13.16.1 Introduction

Twisted wire pair is an excellent transmission medium when employed within certain constraints. It is commonly used in building wiring for the transmission of voice and data. Wire pair must be viewed as a capacitor to understand its distance and bit rate constraints. By definition we have two conductors separated by insulator(s). As we know, capacitance varies directly with the area of the conductor and inversely with distance between the conductors (among other parameters). A wire pair also is resistive. Thus we have an RC circuit. The longer the pair, the greater the value of C (the capacitance) and R (the resistance). As a pulse is applied to the pair, it must charge the capacitor. This takes a finite amount of time. We now have the problem of that "time" exceeding the duration of a bit (the pulsewidth). The duration of a bit is inversely proportional to the bit rate. The more capacitance (the longer the wire pair), the longer time it takes to charge the capacitor. We can now see that both length of the pair and bit rate are constraining factors. If we can reduce the capacitance per unit length of a wire pair, we can increase the bit rate and/or extend the pair.

13.16.2 Characteristics of a Wire Pair

A twisted wire pair has the following characteristics:

- Copper conductor diameter between 20 AWG (American wire gauge) and 26 AWG
- A twist rate of 2–12 twists per foot for each pair
- A color-coding scheme
- A characteristic impedance of between 90 and 110 Ω

These wire pairs may be made up into cables of 25, 50, 100, or 300 pairs. Wire pairs are susceptible to electromagnetic interference (EMI), and the twisting imparted to the cable tends to mitigate the effects of EMI (i.e., increases the cable's immunity to EMI). Pairs in a cable are color-coded. The color coding not only identifies the conductors that make up a pair but also identifies what pair in sequence is relative to other pairs within a multipair cable. Color coding also distinguishes polarity inside a pair (which wire lead is positive and which is negative). Two opposing colors are used. A single pair may have one lead blue on a white background and the other white on a blue background. The five primary colors used here are white, red, black, yellow, and violet.

Cables are broken down into groups of five wire pairs. The primary colors represent positive polarity, whereas secondary colors represent negative polarity. The secondary colors are blue, orange, green, brown, and slate. The primary color represents the group, and the secondary color represents the pair in the group. For example, the fifth group with pairs 21 through 25 has a primary color of violet and pair 21 has blue, 22 has orange, and so forth.

If a cable is *plenum-rated*, it is encased with a material certified not to give off toxic fumes when heated or burned. The PVC insulation on wire pairs is notorious for giving off deadly toxic fumes when heated or burned. A plenum is a special air passageway in a building for ventilation.

13.16.3 Categories of Unshielded Twisted Pair (UTP)

The categories of unshielded twisted pair (UTP) as described by EIA/TIA (Ref. 76) are defined below:

Category 1 and 2. These cables are not commonly used for high data transmission rate data applications. They are usually used for voice and low data applications and are not covered by EIA/TIA standards.

Category 3. This designation applies to cables currently specified by EIA/TIA. The characteristics are specified up to 16 MHz. Category 3 is used for voice and data transmission rates up to 10 Mbps and typically finds application for IEEE 802.3 (10BASE-T) and IEEE 802.5 at 4 Mbps.

Category 4. The characteristics of these cables are specified up to 20 MHz. They are intended to be used for voice and data transmission rates up to 16 Mbps, with typical application for IEEE 802.5 at 16 Mbps.

Category 5. The characteristics of these cables are specified up to 100 MHz. They are intended to be used for voice and data transmission rates up to and including 100 Mbps, and typically may be used for FDDI on wire pair (TPDDI).

13.16.4   General Characteristics of Specified UTP

These specified UTP cables consist of four unshielded twisted pairs of 24-AWG thermoplastic insulated conductors enclosed in a thermoplastic jacket. Four-pair, 22-AWG cables that also meet the specified requirements may be used. There are two new categories of UTP, called enhanced unshielded twisted pair (EUTP) cables, which are now available off-the-shelf. These are specifically used for LANs because they provide improved signal-to-crosstalk noise margins necessary for LAN systems operating at 16 Mbps with cable distances of 100 m between hub and stations.

13.16.5   UTP Transmission Characteristics

The mutual capacitance of any pair at 1 kHz at 20°C will not exceed 20 nF per 305 m (1000 ft) for category 3 cable and will not exceed 17 nF per 305 m for categories 4 and 5 cables.

Categories 3, 4, and 5 UTP have a characteristic impedance of 100 ohms ± 15% in the frequency range from 1 MHz up to the highest referenced frequency. In practice, UTP cables exhibit some structural variations in geometry depending on the manufacturing process and how the cable was installed.

The attenuation of UTP is commonly derived from swept frequency signal level measurements at the output of a 305-m (1000-ft) length of cable. The maximum attenuation of any pair will be less than the values given in Table 13.27 in the frequency range from 0.772 MHz to the highest referenced frequency.

Near-end crosstalk (NEXT) loss decreases as frequency increases. The minimum NEXT coupling loss for any pair combination at room temperature will be greater than the value determined using the following formula:

$$\text{NEXT}(F) > \text{NEXT}(0.772) - 15\log(F/0.772)$$

for all frequencies in the range from 0.772 MHz to the highest referenced frequency for a 305-m (1000-ft) length of cable.

TABLE 13.27   Attenuation Characteristics of UTP Cable

| (MHz) | Maximum Attenuation (dB per 305 m [1000 ft] at 20°C) Frequency | | |
	Category 3	Category 4	Category 5
0.064	2.8	2.3	2.2
0.256	4.0	3.4	3.2
0.512	5.6	4.6	4.5
0.772	6.8	5.7	5.5
1.0	7.8	6.5	6.3
4.0	17	13	13
8.0	26	19	18
10.0	30	22	20
16.0	40	27	25
20.0	—	31	28
25	—	—	32
31.25	—	—	36
62.5	—	—	52
100	—	—	67

Note: The attenuation values for frequencies of 0.512 MHz and below are provided for information only. These values are intended for engineering design purposes and are not required for conformance testing.
Source: Reference 77, Table 10-2. Reprinted with permission.

The NEXT value at 0.772 is 43 dB for category 3 cable, 58 dB for category 4 cable, and 64 dB for category 5 cable (Ref. 77). Pair twist lengths are selected by the manufacturer to meet the crosstalk requirements. The twist rate is in the range of 2–12 twists per foot. The dc resistance will not exceed 28.6 Ω per 305 m (1000 ft).

REVIEW EXERCISES

1. Distinguish between analog and digital transmission and extend the discussion to binary digital transmission.

2. Distinguish between a binary 1 and binary 0: voltage, mark/space, paper tape, and frequency shift keying.

3. What important function in the telephone network is carried out by a 1-bit code?

4. A certain binary source code has 6 bits. How many distinct characters/functions can such a code represent? How many for an 8-bit code?

5. Name at least three nonprinting functions that are represented by a binary sequence.

6. In a particular case we are using even parity with ASCII. Give the value of bit number 8 in each case:

$$
\begin{array}{ccccccc}
0 & 1 & 1 & 0 & 1 & 1 & 0 \\
1 & 1 & 0 & 1 & 0 & 1 & 1 \\
0 & 0 & 1 & 0 & 0 & 0 & 0 \\
0 & 0 & 0 & 1 & 0 & 1 & 0
\end{array}
$$

Suppose we use odd parity. What would be the value of bit 8 in that case?

7. Define throughput. List at least five "items" that will reduce throughput.

8. Distinguish between error detection and error correction.

9. Define two generic methods of detecting errors and compare their advantages and disadvantages.

10. Define two methods of correcting errors. Compare the two and cite trade-offs.

11. When using cyclic redundancy check (CRC), what does the block check count/character (BCC) contain?

12. Define and compare the three types of automatic repeat request (ARQ) described in the text. Give advantages and disadvantages of each.

13. What error performance can be expected on a data link riding on a PSTN digital channel (assuming direct mapping)? On a LAN? According to CCITT Rec. G.821?

14. Draw a neutral waveform for the sequence 10101101. Directly underneath draw a polar waveform, using the same binary values.

15. Draw the same sequence for a return-to-zero (RZ) waveform.

16. Distinguish between start–stop and synchronous transmission.

17. The stop element in start–stop transmission has three standard durations in use today. Give the value of "unit intervals" (bits) of each. Define the standard start element.

18. What are the three basic causes of errors in a data transmission system?

19. Describe how transition timing works.

20. Analyze clock stability versus bit rate versus error rate for the following:

Bit Rate	Bit Error Rate (BER)	Clock Stability
75 bps	1×10^{-5}	?
?	1×10^{-6}	1×10^{-11}/mo
2400 bps	?	1×10^{-8}/mo
10 Mbps	?	1×10^{-9}/mo
100 Mbps	1×10^{-10}	?

21. What are the two basic types of distortion we must deal with in data/telegraph transmission?

22. Where, in a pulse, do we assume sampling takes place?

23. Name at least seven topics that deal with data interface.

24. What are the three distinct phases of a telephone call? Relate these to a data connection.

25. What do segmentation and reassembly involve? What is encapsulation?

26. List the seven open system interconnection (OSI) layers.

27. What is a service access point (SAP)?

28. Give the four most important characteristics of the physical layer.

29. Draw a generalized data link layer frame. Relate the "flag" field to the start and stop elements of asynchronous transmission.

30. With EIA RS-232 and CCITT Recs. V.10/V.11, in a certain bit period we find +4 V dc on the data signal lead. Is it a mark or space, a 1 or a 0?

31. CCITT uses the term *double current*. What does the term signify?

32. What is the primary difference between EIA RS-422 and RS-423 (CCITT V.10 and V.11)?

33. How do we transmit a nonreturn-to-zero (NRZ) digital data waveform on a standard analog channel?

34. There are three basic ways of modulating a signal. What are they?

35. Name some advantages of frequency shift keying (FSK) over amplitude shift keying (ASK); over phase shift keying (PSK).

36. Describe quaternary PSK (QPSK) in relation to binary PSK (BPSK). With QPSK, how many bits are transmitted per transition?

37. Give at least three advantages of phase modulation when transmitting a digital waveform compared to amplitude modulation (AM) and frequency modulation (FM).

38. What is the difference between coherent PSK and differential PSK when viewed from the receive side?

39. If the envelope delay distortion (EDD) between 800 and 2600 Hz is 1000 μs, all other impairments disregarded, what is the maximum modulation rate the channel will support?

40. List the four types of noise described in the text. Define each in one sentence.

41. How is impulse noise specified?

42. Why is the CCIT specification on frequency translation error so stringent?

43. Shannon (Ref. 33) stated that the capacity of a channel in bits per second was a function of what two parameters?

44. A modem transmits 2400 bps using QPSK modulation. What is the modulation rate in bauds?

45. What is the cause of intersymbol interference? Could there be still another cause?

46. An equalizer on a voice channel modem tends to nullify the effects of what two basic impairments on a voice channel? How does it accomplish this?

47. Describe how voice frequency carrier telegraph (VFCT) operates.

48. Why would a modem using phase shift keying (PSK) use 1800 Hz as the tone frequency rather than some other tone frequency, such as 1000 Hz?

49. How can we have full-duplex data operation on a two-wire facility?

50. What is the modulation rate in bauds of a CCITT V.29 modem operating at 9600 bps?

51. Define trellis-coded modulation. What are its merits?

52. What is bit packing? What would be the bit packing of a V.34 modem operating at 28,800 bps?

53. Digital data conventionally is transmitted with bit rates related to 75×2^n, where n is an integer including zero. What is (are) the problem(s) of transmitting such data rates on the public switched digital network?

54. Supply at least two approaches on how one might transmit computer data on the digital network when such data operate at a line bit rate related to 75×2^n?

55. How does integrated services digital network (ISDN) address the problem expressed in exercise 53?

56. What are the two underlying service rates (bit rates) offered by ISDN? How are they expressed?

57. Give at least two applications of the D-channel.

58. Identify at least five telecommunication user services ISDN will support.

59. What is the bit rate of a B-channel?

60. What type of digital network signaling is so vital to the operation of ISDN?

61. In the ISDN reference model, what would the interface NT12 signify? Discuss NT1, NT2, and why we would have NT12.

62. How is noncompatible ISDN user equipment interface handled? Name the appropriate element in the ISDN user model.

63. Describe the ISDN baseband line code.

64. What is the gross bit rate of 2B + D at the user interface? Allocate the bits to their functions.

65. How does an ISDN network termination (NT) derive its timing?

66. Name at least four differences between ISDN specified by CCITT and U.S. practice (BRI only).

67. What is the underlying objective of ATM?

68. Differentiate between connection-oriented and connectionless service.

69. If the ATM cell rate is synchronous, why do we call it the asynchronous transfer mode?

70. How many octets are in an ATM cell? How many of these are in the header and how many in the payload? (Careful, there is a catch here!)

71. Give the two distinct functions of the HEC field.

72. Name at least five QoS items (parameters) dealing with ATM performance.

73. Why is cell delay variation so important?

74. Name at least five ways to transport ATM cells.

75. What is the maximum published (TIA) bit rate on Category 5 wire pair?

76. How does loss vary with frequency for UTP? (Linear or exponential?)

Extra credit: With wire pair, why is bit rate so sensitive to length?

REFERENCES

1. *The New IEEE Standard Dictionary of Electrical and Electronic Terms*, 5 ed., IEEE-Std-100-1992, IEEE Press, New York, 1992.

2. "Equivalence Between Binary Notation Symbols and the Significant Conditions of a Two-Condition Code," CCITT Rec. V.1, Vol. VIII.1, IXth Plenary Assembly, Melbourne, 1988.

3. "Electrical Characteristics for Unbalanced Double-Current Interchange Circuits for General Use with Integrated Circuit Equipment in the Field of Data Communications," ITU-T Rec. V.10, ITU, Geneva, Mar. 1993.

4. "Electrical Characteristics for Balanced Double-Current Interchange Circuits for General Use with Integrated Circuit Equipment in the Field of Data Communications," ITU-T Rec. V.11, Vol. VIII.1, ITU, Geneva, Mar. 1993.

5. "Interface Between Data Terminal Equipment and Data Circuit Terminating Equipment Employing Serial Binary Data Interchange," EIA/TIA-232E, TIA, Arlington, VA, 1991.

6. *Reference Data for Radio Engineers*, 6th ed., Howard W. Sams, Indianapolis, IN, 1977.

7. "International Telegraph Alphabet No. 2," CCITT Rec. S.1, Vol. VII.1, IXth Plenary Assembly, Melbourne, 1988.

8. *Coded Character Set—7-bit American National Standard Code for Information Interchange (7-bit ASCII)*, ANSI X.3.4-1986, ANSI, New York, 1986.

9. J. E. McNamara, *Technical Aspects of Data Communications*, 2nd ed., Digital Equipment Corporation, Maynard, MA, 1982.

10. Roger L. Freeman, *Practical Data Communications*, Wiley, New York, 1995.

11. *Standard for Start–Stop Signal Quality Between Data Terminal Equipment and Nonsynchronous Data Communication Equipment*, EIA RS-404, EIA, Washington, DC, Mar. 1973.

12. "Standard Limits for Transmission Quality of Data Transmission," CCITT Rec. V.50, Vol. VIII.1, IXth Plenary Assembly, Melbourne, 1988.

13. "Limits for Maintenance of Telephone-Type Circuits Used for Data Transmission," CCITT Rec. V. 53, Vol. VIII.1, IXth Plenary Assembly, Melbourne, 1988.

14. "Standard for Specifying Signal Quality for Transmitting and Receiving Data Processing Terminal Equipments Using Serial Data Transmission at the Interface of Nonsynchronous Data Communications Equipment," EIA RS-363, EIA, Washington, DC, May 1969.

15. W. Stallings, *Handbook for Computer Communication Standards*, Vol. I, Macmillan, New York, 1987.

16. "Open Systems Interconnection (OSI) Service Definition Conventions," CCITT Rec. X.210, Vol. VIII. 3, IXth Plenary Assembly, Melbourne, 1988.

17. "Information Technology—Open Systems Interconnection—Basic Reference Model—The Basic Model," ITU-T Rec. X.200, ITU, Geneva, July 1994.

18. "Electrical Characteristics for Balanced Double-Current Interchange Circuits for Interfaces with Data Signaling Rates Up to 52 Mbps," ITU-T Rec. V.12, ITU, Geneva, Aug. 1995.

19. *Electrical Characteristics of Balanced Voltage Digital Interface Circuits*, EIA/TIA-422-B, TIA, Washington, DC, April 1994.

20. *Electrical Characteristics of Unbalanced Voltage Digital Interface Circuits*, EIA/TIA-423-B, TIA, Washington, DC, May 1995.

21. "Error Performance of an International Digital Connection Forming Part of an Integrated Services Digital Network," CCITT Rec. G.821, Vol. III.5, IXth Plenary Assembly, Melbourne, 1988.

22. "List of Definitions of Interchange Circuits Between Data Terminal Equipment (DTE) and Data Circuit-Terminating Equipment (DCE)," CCITT Rec. V.24, Vol. VIII.1, IXth Plenary Assembly, Melbourne, 1988.

23. "Electrical Characteristics for Unbalanced Double-Current Interchange Circuits," ITU-T Rec. V.28, ITU, Geneva, Mar. 1993.

24. "General Data Communication Interface Layer 1 Specification," CCITT Rec. V.230, Vol. VIII.1, IXth Plenary Assembly, Melbourne, 1988.

25. *Electrical Characteristics of Digital Interface Circuits*, U.S. Military Standard, MIL-STD-188-114B, U.S. Department of Defense, Washington, DC, July 1991.

26. "Electrical Characteristics for Single-Current Interchange Circuits Controlled by Contact Closure," CCITT Rec. V.31, Vol. VIII.1, IXth Plenary Assembly, Melbourne, 1988.

27. "Interface Between Data Terminal Equipment (DTE) and Data Circuit-Terminating Equipment (DCE) for Start–Stop Transmission Services on Public Data Networks," CCITT Rec. X.20, Vol. VIII.2, IXth Plenary Assembly, Melbourne, 1988.

28. "Interface Between Data Terminal Equipment (DTE) and Data Circuit-Terminating Equipment (DCE) for Synchronous Operation on Public Data Networks," CCITT Rec. X.21, ITU, Geneva, Sept. 1992.

29. "Interface Between Data Terminal Equipment (DTE) and Data Circuit-Terminating Equipment (DCE) for Terminals Operating in the Packet Mode and Accessing a Packet-Switched Telephone Network or an Integrated Services Digital Network or a Circuit-Switched Public Data Network," ITU-T Rec. X.32, ITU, Geneva, Mar. 1993.

30. *High Speed 25-Position Interface for Data Terminal Equipment and Data Circuit-Terminating Equipment*, EIA/TIA-530-A, TIA, Washington, DC, June 1992.

31. *Telecommunication Transmission Engineering*, vol. 2, 2nd ed., AT&T, New York, 1977.

32. Bell System Technical Reference, *Transmission Parameters Affecting Voiceband Data Transmission—Description and Parameters*, Pub. 41008, AT&T, New York, 1974.

33. C. E. Shannon, "A Mathematical Theory of Communication," *Bell Syst. Tech. J.*, vol. 27, pp. 379–423, July 1948; pp. 623–656, Oct. 1948.

34. "Characteristics of Special Quality International Leased Circuits with Special Bandwidth Conditioning," CCITT Rec. M.1020, Vol. IV-2, IXth Plenary Assembly, Melbourne, 1988.

35. "Characteristics of Special Quality Leased Circuits with Basic Bandwidth Conditioning," CCITT Rec. M.1025, Vol. IV-2, IXth Plenary Assembly, Melbourne, 1988.

36. H. Nyquist, "Certain Topics in Telegraph Transmission Theory," *Trans. AIEE*, vol. 47, pp. 617–644, Apr. 1928.

37. "Numbering of International VFT Channels," Table 2/R-70 bis, CCITT Rec. R.70 bis, Vol. VII, IXth Plenary Assembly, Melbourne, 1988.

38. "Voice Frequency Telegraphy on Radio Circuits," CCITT Rec. R.39, Vol. VII, IXth Plenary Assembly, Melbourne, 1988.

39. Roger L. Freeman, *Reference Manual for Telecommunication Engineering*, 2nd ed., Wiley, New York, 1994.

40. "300 Bits per Second Duplex Modem Standardized for Use in the General Switched Telephone Network," CCITT Rec. V.21, Vol. VIII.1, IXth Plenary Assembly, Melbourne, 1988.

41. "1200 Bits per Second Duplex Modem Standardized for Use in the General Switched Telephone Network or 2-Wire Leased Telephone-Type Circuits," CCITT Rec. V.22, Vol. VIII.1, IXth Plenary Assembly, Melbourne, 1988.

42. "2400 Bits per Second Duplex Model Using Frequency Division Technique Standardized for Use on the General Switched Telephone Network and on Point-to-Point 2-Wire Leased Telephone-Type Circuits," CCITT Rec. V.22 bis, Vol. VIII.1, IXth Plenary Assembly, Melbourne, 1988.

43. "600/1200 Baud Modem Standardized for Use on the General Switched Telephone Network," CCITT Rec. V.23, Vol. VIII.1, IXth Plenary Assembly, Melbourne, 1988.

44. "2400 Bits per Second Modem Standardized for Use on 4-Wire Leased Telephone-Type Circuits," CCITT Rec. V.26, Vol. VIII.1, IXth Plenary Assembly, Melbourne, 1988.

45. "2400/1200 Bits per second Modem Standardized for Use in the General Switched Telephone Network," CCITT Rec. V.26 bis, Vol. VIII.1, IXth Plenary Assembly, Melbourne, 1988.

46. "4800 bits per Second Modem with Manual Equalizer Standardized for Use on Leased Telephone-Type Circuits," CCITT Rec. V.27, Vol. VIII.1, IXth Plenary Assembly, Melbourne, 1988.

47. "4800/2400 Bits per Second Modem with Automatic Equalizer Standardized for Use on Leased Telephone-Type Circuits," CCITT Rec. V.27 bis, Vol. VIII.1, IXth Plenary Assembly, Melbourne, 1988.

48. "4800/2400 Bits per Second Modem Standardized for Use in the General Switched Telephone Network," CCITT Rec. V.27 ter, Vol. VIII.1, IXth Plenary Assembly, Melbourne, 1988.

49. "9600 Bits per Second Modem Standardized for Use on Point-to-Point 4-Wire Telephone-Type Circuits," CCITT Rec. V.29, Vol. VIII.1, IXth Plenary Assembly, Melbourne, 1988.

50. "A Family of 2-Wire, Duplex Modems Operating at Data Signaling Rates Up to 9600 bps for Use on the General Switched Telephone Network or Leased Telephone-Type Circuits," CCITT Rec. V.32, Vol. VIII.1, IXth Plenary Assembly, Melbourne, 1988.

51. "14,400 Bits per Second Modem Standardized for Use on Point-to-Point 4-Wire Leased Telephone-Type Circuits," CCITT Rec. V.33, Vol. VIII.1, IXth Plenary Assembly, Melbourne, 1988.

52. Gottfried Ungerboeck, "Trellis-Coded Modulation with Redundant Sets—Parts I and II," *IEEE Commun. Mag.*, vol. 27, no. 2, Feb. 1987.

53. Andrew J. Viterbi et al., "A Pragmatic Approach to Trellis-Coded Modulation," *IEEE Commun. Mag.*, vol. 27, no. 7, Feb. 1987.

54. "A Modem Operating at Data Signalling Rates Up to 28,800 bps for Use on the General Switched Telephone Network and on Leased Point-to-Point 2-Wire Telephone-Type Circuits," ITU-T Rec. V.34, ITU, Geneva, Sept. 1994.

55. "Standardization of Data Signaling Rates for Synchronous Data Transmission in the General Switched Telephone Network," CCITT Rec. V.5, Vol. VIII.1, IXth Plenary Assembly, Melbourne, 1988.

56. "Standardization of Data Signaling Rates for Synchronous Data Transmission on Leased Telephone-Type Circuits," CCITT Rec. V.6, Vol. VIII.1, IXth Plenary Assembly, Melbourne, 1988.

57. "ISDN User–Network Interface Structure and Access Capabilities," CCITT Rec. I.412, Vol. III.8, IXth Plenary Assembly, Melbourne, 1988.

58. "Basic User–Network Interface: Layer 1 Specification," CCITT Rec. I.430, Vol. III.8, IXth Plenary Assembly, Melbourne, 1988.

59. "Primary Rate User–Network Interface—Layer 1 Specification," ITU-T Rec. I.431, ITU, Geneva, Mar. 1993.

60. *ISDN Basic Access Transport System Requirements*, Technical Reference TR-TSY-000397, Issue 1, Bellcore, Piscataway, NJ, Oct. 1988.

61. *BOC Notes on the LEC Networks—1994*, Issue 2, SR-TSV-002275, Bellcore, Piscataway, NJ, Apr. 1994.

62. *National ISDN 1*, Special Report, SR-NWT-001937, Issue 1, Bellcore, Piscataway, NJ, Feb. 1991 (with supplement, Feb. 1993).

63. *National ISDN-2*, Special Report, SR-NWT-002120, Issue 1, Bellcore, Piscataway, NJ, May 1992 (with revision 1, June 1993).

64. *National ISDN 3*, Special Report, SR-NWT-002457, Issue 1, Bellcore, Piscataway, NJ, Dec. 1993.

65. "Broadband Aspects of ISDN," CCITT Rec. I.121, ITU, Geneva, 1991.

66. *ATM User–Network Interface Specification, Version 3.0*, The ATM Forum, PTR Prentice Hall, Englewood Cliffs, NJ, 1993.

67. "B-ISDN User–Network Interface—Physical Layer Specification," ITU-T Rec. I.432, ITU, Geneva, Mar. 1993.

68. "B-ISDN User–Network Interface," CCITT Rec. I.413, ITU, Geneva, 1991.

69. "B-ISDN ATM Layer Specification," ITU-T Rec. I.361, ITU, Geneva, Mar. 1993.

70. "B-ISDN Protocol Reference Model and Its Application," CCITT Rec. I.321, ITU, Geneva, 1991.

71. William Stallings, *ISDN, Broadband ISDN with Frame Relay and ATM*, 3rd ed., Prentice Hall, Englewood Cliffs, NJ, 1995.

72. *Broadband ISDN User–Network Interfaces—Rates and Formats Specifications*, ANSI T1.624-1993, ANSI, New York, 1993.

73. D. E. McDysan and D. L. Spohn, *ATM Theory and Application*, McGraw-Hill, New York, 1995.

74. "Broadband Testing Technologies," Hewlett-Packard Seminar, H-P, Burlington, MA, Oct. 1993.

75. "Traffic Control and Congestion Control in B-ISDN," ITU-T Rec. I.371, ITU, Geneva, 1993.

76. *Commercial Building Telecommunication Wiring Standard*, EIA/TIA-368, TIA, Washington, DC, 1991.

77. *Technical Systems Bulletin Additional Cable Specifications for Unshielded Twisted Pair Cables*, EIA/TIA TSB-36, TIA, Washington, DC, 1991.

78. "Common Long Haul and Tactical Communication System Technical Standards," MIL-STD-188-100, U.S. Department of Defense, Washington, DC, Nov. 15, 1972.

79. W. R. Bennett and J. R. Davey, *Data Transmission*, McGraw-Hill, New York, 1965.

80. *B-ISDN ATM Layer Cell Transfer-Performance Parameters*, ANSI T1-511-1994, American National Standards Institute, New York, 1994.

14

COMMUNITY ANTENNA
TELEVISION (CABLE TELEVISION)

14.1 OBJECTIVE AND SCOPE

The principal thrust of community antenna television (CATV) is entertainment. Lately CATV has taken on some new dimensions. It is indeed a broadband medium, providing up to 1 GHz of bandwidth at customer premises. It was originally a unidirectional system, from the point of origin, which we call the headend, toward customer premises. It does, however, have the capability of being a two-way system by splitting the band, say, from 5 to 50 MHz for upstream traffic (i.e., toward the headend) and the remainder for downstream traffic (i.e., from the headend to customer premises). Our interest will primarily focus on two-way traffic, usually voice and data.

First, however, conventional CATV will be described, including the concept of supertrunks and HFC (hybrid fiber–coaxial cable) systems. We will involve the reader with such topics as wideband amplifiers in tandem, optimum amplifier gain, IM noise, beat noise, and cross-modulation products. System layout, hubs, and last-mile or last 100-ft considerations will also be covered. There will also be a brief discussion of the conversion to a digital system using some of the compression techniques described in Chapter 15. We will show that a CATV system capacity of up to 500 channels can be accommodated on a digital HFC layout.

14.2 THE EVOLUTION OF CATV

14.2.1 The Beginnings

Broadcast television, as we know it, was in its infancy around 1948. Fringe area problems were much more acute in that period. By fringe area, we mean areas with poor or scanty signal coverage. A few TV users in fringe areas found that if they raised their antennas high enough and improved antenna gain characteristics, an excellent picture could be received. These users were the envy of the neighborhood. Several of these people who were familiar with RF signal transmission employed signal splitters so that their neighbors could share the excellent picture service. It was soon found there was a limit on how much signal splitting could be done before signal levels got so low that they were snowy or unusable.

Remember that each time a signal splitter (even power split) is added, neglecting insertion losses, the TV signal drops 3 dB. Then someone got the bright idea of amplifying the signal before splitting. Now some real problems arose. One-channel amplification worked fine, but two channels from two antennas with signal combining became difficult. Now we are dealing with comparatively broadband amplifiers. Among the impairments we can expect from broadband amplifiers and their connected transmission lines (coaxial cable) are the following:

- Poor frequency response. Some part of the received band had notably lower levels than other parts. This is particularly true as the frequency increases. In other words, there was fairly severe amplitude distortion. Thus equalization became necessary.
- The mixing of two or more RF signals in the system caused intermodulation products and "beats" (harmonics), which degraded reception.
- When these TV signals carried modulation, cross-modulation (Xm) products degraded or impaired reception.

Several small companies were formed to sell these "improved" television reception services. Some of the technicians working for these companies undertook ways of curing the ills of broadband amplifiers.

These were coaxial cable systems, where a headend with a high tower received signals from several local television broadcasting stations, amplified the broadband signals, and distributed the results to CATV subscribers. A subscriber's TV set was connected to the distribution system, and the signal received looked just the same as if it was taken off the air with its own antenna. In fringe areas signal quality, however, was much better than own-antenna quality. The key to everything was that no changes were required in the user's TV set. The cable system was just an extension of his/her TV set antenna. This simple concept is shown in Figure 14.1.

Note in Figure 14.1 that home A is in the shadow of a mountain ridge and receives a weakened diffracted signal off the ridge and a reflected signal off a lake. Here is the typical multipath scenario resulting in ghosts in A's TV screen. The picture is also snowy, meaning it is noisy, as a result of a poor carrier-to-noise ratio. Home B extended the height of the antenna to be in line-of-sight of the TV transmitting antenna. Its antenna is of higher gain; thus it is more discriminating against unwanted reflected and diffracted signals. Home B has an excellent picture without ghosts. Home B shares its fine signal with home A by use of a 3-dB power split (P).

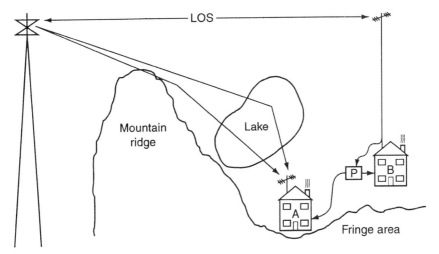

Figure 14.1. CATV initial concept. P = power split.

14.2.2 Early System Layouts

Figure 14.2 illustrates an early CATV distribution system (ca. 1968). Taps and couplers (power splits) are not shown. These systems provided from 5 to 12 channels. A microwave system brought in channels from distant cities (50–150 miles). We had direct experience with the Atlantic City, New Jersey, system where channels were brought in from Philadelphia and New York by microwave (MW). A 12-channel system would occupy the entire assigned VHF band (i.e., channels 2–13).

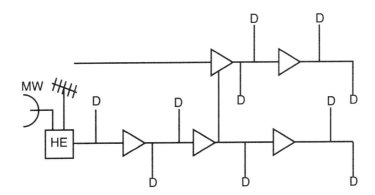

HE = head end
D = drop wire to residence
MW = microwave connectivity

Figure 14.2. An early CATV distribution system.

As UHF TV stations began to appear, a new problem arose for the CATV operator. It was incumbent on that operator to keep the bandwidth as narrow as possible. One approach was to convert UHF channels to vacant VHF channel allocations at the headend.

Satellite reception at the headend doubled or tripled the number of channels that could be available to the CATV subscriber. Each satellite has the potential of adding 24 channels to the system. Note how the usable cable bandwidth is "broadened" as channels are added. We assume contiguous channels across the band, starting at 55 MHz. For 30 channels, we have 55–270 MHz; for 35 channels, 55–300 MHz; for 40 channels, 55–330 MHz; for 62 channels, 55–450 MHz; and for 78 channels, 55–550 MHz. These numbers of channels were beyond the capability of many TV sets of the day. Set-top converters were provided that converted all channels to a common channel, an unoccupied channel, usually channel 2, 3, or 4, to which the home TV set is tuned. This approach is still very prevalent today.

In the next section we discuss impairments and measures of system performance. In Section 14.4, hybrid fiber–coaxial cable systems are addressed. The replacement of coaxial cable trunk by fiber-optic cable made a major stride to improved performance and better reliability/availability.

14.3   SYSTEM IMPAIRMENTS AND PERFORMANCE MEASURES

14.3.1   Overview

A CATV headend places multiple TV and FM (from 30 to 125) carriers on a broadband coaxial cable trunk and distribution system. The objective is to deliver a signal-to-noise ratio (S/N) of from 42 to 45 dB at a subscriber's TV set. From previous chapters we would expect such impairments as the accumulation of thermal and intermodulation noise. We find that CATV technicians use the term *beat* to mean intermodulation (IM) products. For example, there is triple beat distortion, defined by Grant (Ref. 1) as "spurious signals generated when three or more carriers are passed through a non-linear circuit (such as a wideband amplifier)." The spurious signals are sum and difference products of any three carriers, sometimes referred to as "beats." Triple beat distortion is calculated as a voltage addition.

The wider the system bandwidth is and the more RF carriers transported on that system, the more intermodulation distortion, "triple beats," and cross-modulation we can expect. We can also expect combinations of all of the above, such as *composite triple beat* (CTB), which represents the pile up of beats at or near a single frequency.

Grant (Ref. 1) draws a dividing line at 21 TV channels. On a system with 21 channels or less, one must expect cross-modulation (Xm) to predominate. Above 21 channels, CTB will predominate.

14.3.2 dBmV and Its Applications

The dBmV was introduced in Section 1.4.3. We have 0 dBmV defined as 1 millivolt across an impedance of 75 Ω. Note that 75 Ω is the standard impedance used in CATV. From the power law,

$$P_{\mathrm{W}} = E^2/R \qquad\qquad (0.001) \qquad (14.1)$$

$$0\ \text{dBmV} = 0.0133 \times 10^{-6}\ \text{watt or } 0.0133\ \mu\text{W}$$

By definition then, 0.0133 *watt* is +60 dBmV.

If 0 dBmV = 0.0133×10^{-6} watt and 0 dBm = 0.001 watt, and gain in dB = $10\log(P_1/P_2)$, or in this case $10\log[0.001/(0.0133 \times 10^{-6})]$, then 0 dBm = +48.76 dBmV.

Remember that, when working with dB in the voltage domain, we are working with the E^2/R relationship, where $R = 75\ \Omega$. With this in mind (Section 1.4.3), the definition of dBmV is

$$\text{dBmV} = 20\log\left(\frac{\text{voltage in millivolts}}{1\ \text{millivolt}}\right) \qquad (14.2)$$

If a signal level is 1 volt at a certain point in a circuit, what is the level in dBmV?

$$\text{dBmV} = 20\log(1000/1) = +60\ \text{dBmV}$$

If we are given a signal level of +6 dBmV, to what voltage level does this correspond?

$$+6\ \text{dBmV} = 20\log(X_{\mathrm{mV}}/1\ \text{mV})$$

Divide through by 20:

$$6/20 = \log(X_{\mathrm{mV}}/1\ \text{mV})$$

$$\text{antilog}(6/20) = X_{\mathrm{mV}}$$

$$X_{\mathrm{mV}} = 1.995\ \text{mV or 2 millivolts or 0.002 volt}$$

These signal voltages are rms (root mean square) volts. For peak voltage, divide by 0.707. If you are given peak signal voltage and wish the rms value, multiply by 0.707.

14.3.3 Thermal Noise in CATV Systems

The lowest noise levels permissible in a CATV system—at antenna output terminals, at repeater (amplifier) inputs, or at a subscriber's TV set—without producing snowy pictures, are determined by thermal noise.

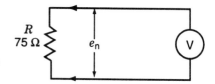

Figure 14.3. Resistor model for thermal noise voltage, e_n.

Consider the following, remembering we are in the voltage domain. Any resistor or source that looks resistive over the band of interest, including antennas, amplifiers, and long runs of coaxial cable, generates thermal noise. In the case of a resistor, the noise level can be calculated based on Figure 14.3.

To calculate the noise voltage, e_n, use the following formula:

$$e_n = (4RBk)^{1/2} \qquad (14.3)$$

where e_n = rms noise voltage
  R = resistance in ohms (Ω)
  B = bandwidth (Hz) of the measuring device (electronic voltmeter, V)
  k = a constant equal to 40×10^{-16} at standard room temperature*

Let the bandwidth, B, of a NTSC TV signal be rounded to 4 MHz. The open circuit noise voltage for a 75-Ω resistor is

$$e_n = (4 \times 75 \times 4 \times 10^{-16})^{1/2}$$
$$= 2.2 \ \mu V \text{ rms}$$

Figure 14.4 shows a 2.2-μV noise-generating source (resistor) connected to a 75-Ω (noiseless) load. Only half of the voltage (1.1 μV) is delivered to the load. Thus the noise input to 75 Ω is 1.1 μV rms or -59 dBmV. This is the basic noise level, the minimum that will exist in any part of a 75-Ω CATV system. The value -59 dBmV will be used repeatedly in the following (Ref. 2).

The noise figure of typical CATV amplifiers ranges between 7 and 9 dB (Ref. 3).

14.3.4 Signal-to-Noise (S/N) Ratio Versus Carrier-to-Noise Ratio (C/N) in CATV Systems

We have been using S/N and C/N many times in previous chapters. In CATV systems S/N has a slightly different definition as follows (Ref. 2):

This relationship is expressed by the "signal-to-noise ratio," which is the difference between the signal level measured in dBmV, and the noise level, also measured in dBmV, both levels being measured at the same point in the system.

---

*This value can be derived from Boltzmann's constant (Chapter 1) at room temperature (68°F or 290 kelvins).

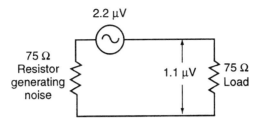

Figure 14.4. Minimum noise model.

S/N can be related to C/N on CATV systems as

$$C/N = S/N + 4.1 \text{ dB} \qquad (14.4)$$

This is based on work by Carson (Ref. 4), where the basis is "noise just perceptible" by a population of TV viewers, NTSC 4.2-MHz bandwidth TV signal. Here the S/N is 39 dB and the C/N is 43 dB. Adding noise weighting improvement (6.8 dB), we find

$$S/N = C/N + 2.7 \text{ dB} \qquad (14.5)$$

It should be noted that S/N is measured where the signal level is peak-to-peak and the noise level is rms. For C/N, both the carrier and noise levels are rms. These values are based on a VSB-AM TV signal with 87.5% modulation index.

For comparison, consider another series of tests conducted by the Television Allocations Study Organization (TASO) and published in their report to the U.S. FCC in 1959. Their ratings, corrected for a 4-MHz bandwidth, instead of the 6-MHz bandwidth that was used previously, are shown below (Ref. 2):

	TASO Picture Rating	**S/N Ratio**
1.	Excellent, no perceptible noise	45 dB
2.	Fine (snow just perceptible)	35 dB
3.	Passable (snow definitely perceptible, but not objectionable)	29 dB
4.	Marginal (snow somewhat objectionable)	25 dB

Once a tolerable noise level is determined, the levels required in a CATV system can be specified. If the desired S/N has been set at 43 dB at a subscriber TV set, the minimum signal level required at the first amplifier would be -59 dBmV $+ 43$ dB or -16 dBmV, considering thermal noise only. Actual levels would be quite a bit higher because of the noise generated by subsequent amplifiers in cascade.

It has been found that the optimum gain of a CATV amplifier is about 22 dB. When the gain is increased, intermodulation/cross-modulation products become excessive. For gains below this value, thermal noise increases, and system length is shortened or the number of amplifiers must be increased —neither of which is desirable.

There is another rule-of-thumb of which we should be cognizant. Every time the gain of an amplifier is increased 1 dB, intermodulation products and "beats" increase their levels by 2 dB. And the converse is true: every time gain is decreased 1 dB, IM products and beat levels are decreased by 2 dB.

With most CATV systems, coaxial cable trunk amplifiers are identical. This, of course, eases noise calculations. We can calculate the noise level at the output of one trunk amplifier. This is

$$N_V = -59 \text{ dBmV} + \text{NF}_{\text{dB}} \tag{14.6}$$

where NF is the noise figure of the amplifier in dB.

In the case of two amplifiers in cascade (tandem), the noise level (voltage) is

$$N_V = -59 \text{ dBmV} + \text{NF}_{\text{dB}} + 3 \text{ dB} \tag{14.7}$$

If we have M identical amplifiers in cascade, the noise level (voltage) at the output of the last amplifier is

$$N_V = -59 \text{ dBmV} + \text{NF}_{\text{dB}} + 10 \log M \tag{14.8}$$

This assumes that all system noise is generated by the amplifiers, and none is generated by the intervening sections of coaxial cable.

Example 1. A CATV system has 30 amplifiers in tandem; each amplifier has a noise figure of 7 dB. Assume that the input of the first amplifier is terminated in 75 Ω resistive. What is the thermal noise level (voltage) at the last amplifier output?

Use equation 14.8:

$$N_V = -59 \text{ dBmV} + 7 \text{ dB} + 10 \log 30$$
$$= -59 \text{ dBmV} + 7 \text{ dB} + 14.77 \text{ dB}$$
$$= -37.23 \text{ dBmV}$$

For carrier-to-noise ratio (C/N) calculations, we can use the following procedures. To calculate the C/N at the output of one amplifier,

$$C/N = 59 \text{ dBmV} - \text{NF}_{\text{dB}} + \text{input level (dBmV)} \tag{14.9}$$

Example 2. If the input level of a CATV amplifier is $+5$ dBmV and its noise figure is 7 dB, what is the C/N at the amplifier output?

Use equation 14.9:

$$C/N = 59 \text{ dBmV} - 7 \text{ dB} + 5 \text{ dBmV}$$
$$= 57 \text{ dB}$$

With N cascaded amplifiers, we can calculate the C/N at the output of the last amplifier, assuming all the amplifiers are identical, by the following equation:

$$C/N_L = C/N(\text{single amplifier}) - 10 \log N \qquad (14.10)$$

Example 3. Determine the C/N at the output of the last amplifier with a cascade (in tandem) of 20 amplifiers, where the C/N of a single amplifier is 62 dB.
    Use equation 14.10:

$$C/N_L = 62 \text{ dB} - 10 \log 20$$
$$= 62 \text{ dB} - 13.0 \text{ dB}$$
$$= 49 \text{ dB}$$

Another variation for calculating C/N is when we have disparate C/N ratios in different parts of a CATV system. CATV systems are usually of a tree topology, where the headend is the base of the tree. There is a trunk branching to limbs and further branching out to leaf stems supporting many leaves. In the case of CATV, there is the trunk network, bridger, and line extenders with different gains, and possibly different noise figures.
    To solve this problem we turn back to Chapter 7, Section 7.4.8.3. Rewriting equation 7.26:

$$\left(\frac{C}{N}\right)_{sys} = \frac{1}{(C/N_1)^{-1} + (C/N_2)^{-1} + \cdots + (C/N_n)^{-1}} \qquad (14.11)$$

Example 4. Out of a cascade of 20 trunk amplifiers the C/N was 49 dB; out of a bridger, $C/N = 63$ dB; and out of two line extenders, $C/N = 61$ dB. Calculate the C/N at the end of the system described.
    Use equation 14.11, but first convert each dB value to a value in decimal units and then take its inverse $(1/X)$.

49 dB: antilog(49/10) = 79,433 $1/X = 12{,}589 \times 10^{-9}$

63 dB: antilog(63/10) = 2,000,000 $1/X = 500 \times 10^{-9}$

61 dB: antilog(61/10) = 1,258,925 $1/X = 794 \times 10^{-9}$

We can now sum the inverses because they all have the same exponent: Sum $= 13{,}883 \times 10^{-9}$. Take the inverse: 72,030.

$$10 \log(72{,}030) = 48.57 \text{ dB}$$

Remember that the "sum" of C/N series values must be something less than the worst value of the series. That is a way of self-checking.

14.3.5   The Problem of Cross-Modulation (Xm)

Many specifications for TV picture quality are based on the judgment of a population of viewers. One example was the TASO ratings for picture quality given earlier. In the case of cross-modulation (cross-mod or Xm) and CTB (composite triple beat), acceptable levels are -51 dB for Xm and -52 dB for CTB. These are good guideline values (Ref. 1).

Xm is a form of third-order distortion so typical of a broadband, multicarrier system. Xm varies with the operating level of an amplifier in question and the number of TV channels being transported. It is derived from the amplifier manufacturer specifications. The manufacturer will specify a value for Xm (in dB) for several numbers of channels and for a particular level. The level in the specification may not be the operating level of a particular system. To calculate Xm for an amplifier to be used in a given system, using manufacturer's specifications, the following formula applies:

$$Xm_a = Xm_{spec} + 2(OL_{oper} - OL_{spec}) \qquad (14.12)$$

where $Xm_a = $ Xm for the amplifier in question
$\qquad Xm_{spec} = $ Xm specified by the manufacturer of the amplifier
$\qquad OL_{oper} = $ desired operating output signal level (dBmV)
$\qquad OL_{spec} = $ manufacturer's specified output signal level

We spot the "2" multiplying factor and relate it to our earlier comments: namely, increase the operating level 1 dB, third-order products increase 2 dB, and the contrary applies for reducing signal level. As we said, Xm is a form of third-order product.

Example 1. Suppose a manufacturer tells us that for an Xm of -57 dB for a 35-channel system, the operating level should be $+50.5$ dBmV. We want a longer system and use an operating level of $+45$ dBmV. What Xm can we expect under these conditions?

Use equation 14.12:

$$Xm_a = -57 \text{ dB} + 2(+45 \text{ dBmV} - 50.5 \text{ dBmV})$$

$$= -68 \text{ dB}$$

CATV trunk systems have numerous identical amplifiers. To calculate Xm for N amplifiers in cascade (tandem), our approach is similar to that of thermal noise: namely,

$$Xm_{sys} = Xm_a + 20 \log N \qquad (14.13)$$

where N = number of identical amplifiers in cascade
  Xm_a = Xm for one amplifier
  Xm_{sys} = Xm value at the end of the cascade

Example 2. A certain CATV trunk system has 23 amplifiers in cascade where Xm_a is -88 dB. What is Xm_{sys}?
  Use equation 14.13:

$$Xm_{sys} = -88 \text{ dB} + 20 \log 23$$
$$= -88 + 27$$
$$= -61 \text{ dB}$$

To combine unequal Xm values, we turn to a technique similar to equation 14.11, but now, because we are in the voltage domain, we must divide through by 20 rather than 10 when converting logarithms to equivalent numerics.

Example 3. At the downstream end of our trunk system the Xm was -58 dB, and taking the bridger/line extender system alone, their Xm combined is -56 dB. Now from the headend through the trunk and bridger/line extender system, what is the Xm_{sys}? Convert -56 dB and -58 dB to their equivalent decimal numerics and invert $(1/X)$:

$$-56 \text{ dB: antilog} - 56/20 = 0.001584$$
$$-58 \text{ dB: antilog} - 58/20 = 0.001259$$
$$\text{Sum} = 0.002843$$

Take 20 log of this value:

$$Xm_{sys} = -50.9 \text{ dB}$$

14.3.6 Gain and Levels for CATV Amplifiers

Setting both gain and level settings for CATV broadband amplifiers is like walking a tightrope. If levels are set too low, thermal noise will limit system length (i.e., number of amplifiers in cascade). If levels are set too high, system length will be limited by excessive CTB and cross-modulation (Xm). On trunk amplifiers available gain is between 22 and 26 dB (Ref. 1). Feeder amplifiers will usually operate at higher gains, trunk systems at lower gains. Feeder amplifiers usually operate in the range of 26–32-dB gain with output levels in the range of $+47$ dBmV. Trunk amplifiers have gains of 21–23 dB, with output levels in the range of $+32$ dBmV. If we wish to extend the length of the trunk plant, we should turn to using lower loss cable. Using fiber optics in the trunk plant is even a better alternative (see Section 14.4).

The gains and levels of feeder systems are purposefully higher. This is the part of the system serving customers through taps. These taps are passive and draw power. Running the feeder system at higher levels improves tap efficiency. Because feeder amplifiers run at higher gain and with higher levels, the number of these amplifiers in cascade must be severely limited to meet CTB and cross-modulation requirements at the end user.

14.3.7 The Underlying Coaxial Cable System

The coaxial cable employed in CATV plant is nominally 75 Ω. A typical response curve for such cable ($\frac{7}{8}$-in., air dielectric) is illustrated in Figure 14.5. This frequency response of coaxial cable is called "tilt" in the CATV industry.

For 0.5-in. cable, the loss per 100 ft at 50 MHz is 0.52 dB; for 550 MHz, 1.85 dB. Such cable systems require equalization. The objective is to have a comparatively "flat" frequency response across the entire system. An equalizer is a network that presents a mirror image of the frequency response curve, introducing more loss at the lower frequencies and less loss at the higher frequencies. These equalizers are often incorporated with an amplifier.

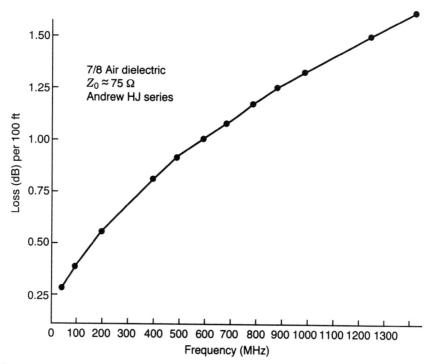

Figure 14.5. Attenuation–frequency response for $\frac{7}{8}$-in. coaxial cable, air dielectric, $Z_0 = 75$ Ω, Andrew HJ series helix (Ref. 5).

Equalizers are usually specified for a certain length of coaxial cable, where length is measured in dB at the highest frequency of interest. Grant (Ref. 1) describes a 13-dB equalizer for a 300-MHz system, which is a corrective unit for a length of coaxial cable having 13-dB loss at 300 MHz. This would be equivalent to approximately 1000 ft of $\frac{1}{2}$-in. coaxial cable. Such a length of cable would have 5.45-dB loss at 54 MHz and 13-dB loss at 300 MHz. The equalizer would probably present a loss of 0.5 dB at 300 MHz and 8.1 dB at 54 MHz.

14.3.8 Taps

A *tap* is similar to a directional coupler. It is a device inserted into a coaxial cable, that diverts a predetermined amount of its input energy to one or more tap outputs for the purpose of feeding a TV signal into subscriber drop cables. The remaining balance of the signal energy is passed on down the distribution system to the next tap or distribution amplifier. The concept of the tap and its related distribution system is shown in Figure 14.6.

Taps are available to feed 2, 4, or 8 service drops from any one unit. Many different types of taps are available to serve different signal levels that appear along a CATV cable system. Commonly, taps are available in 3-dB increments. For two-port taps, the following tap losses may be encountered: 4, 8, 11, 14, 17, 20, and 23 dB. The insertion loss for the lower value tap loss may be on the order of 2.8 dB and once the tap loss exceeds 26 dB, the insertion is 0.4 dB and remains so as tap values increase. Another important tap parameter is isolation. Generally, the higher the tap loss, the better the isolation. With an 8-dB tap loss, the isolation may only be 23 dB, but with 29-dB tap loss (two-port taps), the isolation can be as high as 44 dB. Isolation in this context is the isolation between the two tap ports to minimize undesired interference from a TV set on one tap to the TV set on the other tap.

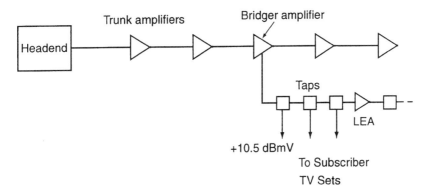

Figure 14.6. A simplified layout of a CATV system showing its basic elements. The objective is to provide a +10.5-dBmV signal level at the drops (tap outputs). LEA = line extender amplifier.

For example, a line voltage signal level is +34.8 dBmV entering a tap. The tap insertion loss is 0.4 dB so the level of the signal leaving the tap to the next tap or extender amplifier is +34.4 dBmV. The tap is two-port. We know we want at least a +10.5 dBmV at the port output. Calculate +34.8 dBmV − X dB = +10.5 dBmV. X then = 24.3 dB, which would be the ideal tap loss value. Taps are not available off-the-shelf at that loss value, the nearest value being 23 dB. Thus the output at each tap port will be +34.8 dBmV − 23 dB = 11.8 dBmV.

14.4 HYBRID FIBER–COAX (HFC) SYSTEMS

The following advantages accrue by replacing the coaxial cable trunk system with optical fiber:

- Reduces the number of amplifiers required per unit distance to reach the furthest subscriber
- Results in improved C/N and reduced CTB and Xm levels
- Also results in improved reliability (i.e., by reducing the number of active components)
- Has the potential to greatly extend a particular CATV serving area

One disadvantage is that a second fiber link has to be installed for the reverse direction, or a form of wave division multiplex (WDM) is required, when two-way operation is required and/or for the CATV management system (used for monitoring the health of the system, amplifier degradation, or failure).

This concept is shown in Figure 14.7. Figure 14.8 illustrates a HFC system where there are no more than three amplifiers to reach any subscriber tap. Note that with this system layout there cannot be a catastrophic failure. For the loss of an amplifier, only one-sixteenth of the system is affected in a worst case scenario; with the loss of a fiber link, the worst case would be one-sixth of the system.

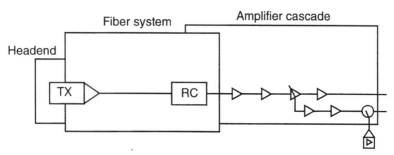

Figure 14.7. The concept of a hybrid fiber–coaxial cable CATV system. TX = fiber-optic transmitter, RC = fiber-optic receiver.

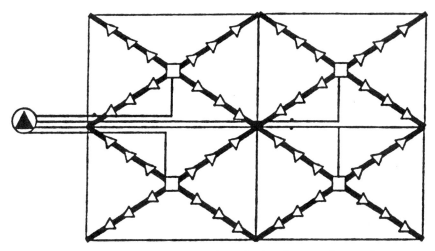

Figure 14.8. HFC system layout for optimal performance (one-way).

14.4.1 Design of the Fiber Optic Portion of a HFC System

Before proceeding with this section, it is recommended that the reader turn back to Chapter 12 for a review of the principles of fiber-optic transmission.

There are two approaches to fiber-optic transmission of analog CATV signals. Both approaches take advantage of the intensity modulation characteristics of the fiber-optic source. Instead of digital modulation of the source, amplitude modulation (analog) is employed. The most common method takes the entire CATV spectrum as it would appear on a coaxial cable and uses that as the modulating signal. The second method also uses analog amplitude modulation, but the modulating signal is a grouping of subcarriers that are each frequency modulated. One off-the-shelf system multiplexes in a broad FDM configuration, eight television channels, each on a separate subcarrier. Thus a 48-channel CATV system would require six fibers, each with eight subcarriers (plus 8 or 16 audio subcarriers).

14.4.1.1 Link Budget for an AM System. Assume a model using a distributed feedback (DFB) laser with an output of +5 dBm coupled to the pigtail. The receiver is a PINFET, where the threshold is −5 dBm. This threshold will derive approximately 52-dB S/N in a video channel. Compared to digital operation, the C/N is around 49.3 dB, assuming the S/N value is noise-weighted (see Section 14.3.4). This is a very large C/N value and leaves only 10 dB to be allocated to fiber, splice loss, and margin. If we allocated 2 dB for the link margin, only 8 dB is left for fiber/splice loss. At 1550-nm operation, assuming a conservative 0.4-dB/km fiber/splice loss, the maximum distance to the coax hub or fiber-optic repeater is only 8/0.4 or

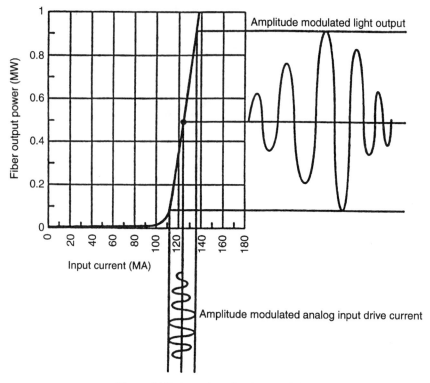

Figure 14.9. Laser transfer characteristics.

20 km. Of course, if we employ a EDFA (erbium-doped fiber amplifier) with, say, only a 20-dB gain, this distance can be extended by 20/0.4 or 50 km. Figure 14.9 illustrates a typical laser transfer characteristic showing the amplitude-modulated input drive.

Typical design goals for the video/TV output of a fiber-optic trunk are:

CNR = 50 dB

Composite second-order products (CSO) = -62 dBc

CTB = -65 dBc

One common technique used on HFC systems is to employ optical couplers where one fiber trunk system feeds several hubs. A hub is a location where the optical signal is converted back to an electrical signal for transmission on coaxial cable. Two applications of optical couplers are illustrated in Figure 14.10. Keep in mind that a signal split not only includes splitting the power

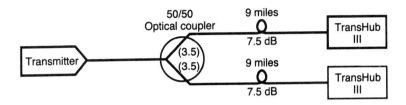

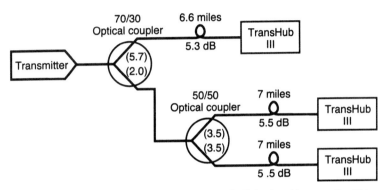

Figure 14.10. Two-way and three-way splits of a light signal transporting TV.

but also the insertion loss of the coupler. The values shown in parentheses give the loss in the split branches (e.g., 5.7 dB and 2.0 dB).

14.4.1.2 FM Systems. FM systems are more expensive than their AM counterparts but provide improved performance. EIA/TIA-250C (Ref. 6), discussed in Chapter 15, specifies a signal-to-noise ratio of 67 dB for short-haul systems. With AM systems it is impossible to achieve this S/N, whereas a well-designed FM system can conform to EIA/TIA-250C. AM systems are degraded by dispersion on the fiber link; FM systems much less so. FM systems can also be extended further. FM systems are available with 8, 16, or 24 channels, depending on the vendor.

Figure 14.11 shows an eight-channel per fiber frequency plan, and Figure 14.12 is a transmit block diagram for the video portion of the system. Figure 14.13 illustrates a typical FM/fiber hub. Figure 14.14 shows the link performance of a FM system and how we can achieve a S/N of 67 dB and better.

As illustrated in Figure 14.12, each video and audio channel must be broken out separately at the head end. Each of these channels must FM modulate its own subcarrier (see Figure 14.11). It should be noted that there is a similar but separate system for the associated aural (audio) channels with 30 MHz spacing, beginning at 70 MHz. These audio channels may be

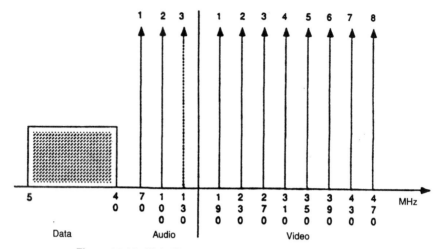

Figure 14.11. Eight-TV-channel frequency plan for FM system.

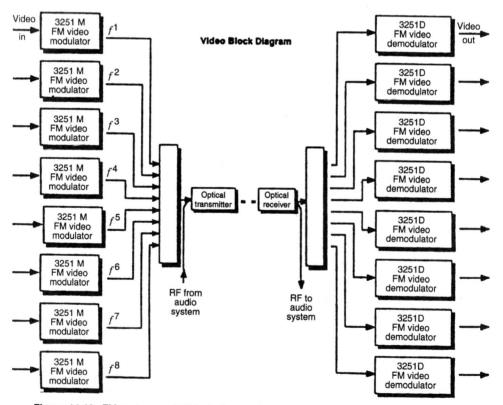

Figure 14.12. FM system model block diagram for video transmission subsystem. (Courtesy of ADC Video Corp.)

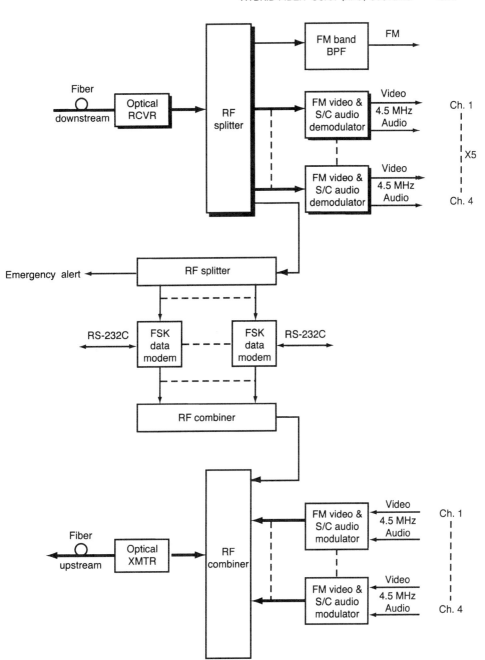

Figure 14.13. A typical FM/fiber hub.

multiplexed before transmission. Each video carrier occupies a 40-MHz slot. These RF carriers, audio and video, are combined in a passive network. The composite RF signal intensity-modulates a laser diode source. Figure 14.13 shows a typical fiber/FM hub.

Calculation of Video S/N for FM System. Given the C/N ratio (CNR) for a particular FM system, the S/N of a TV video channel may be calculated as follows:

$$\text{SNRw} = K + \text{CNR} + 10 \log \frac{B_{\text{IF}}}{B_{\text{F}}} + 20 \log \frac{1.6 \, \Delta F}{B_{\text{F}}} \qquad (14.14)$$

where K = a constant ($\sim$ 23.7 dB) made of weighting network, deemphasis, and rms to p-p conversion factors

$\quad$ CNR = carrier-to-noise ratio in the IF bandwidth

$\quad B_{\text{IF}}$ = IF bandwidth

$\quad B_{\text{F}}$ = baseband filter bandwidth

$\quad \Delta F$ = sync tip-to-peak white (STPW) deviation

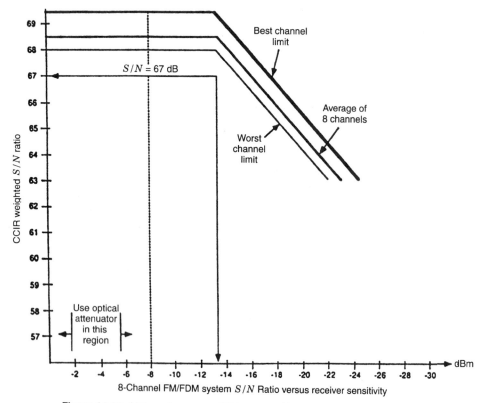

Figure 14.14. Link performance of FM system. (Courtesy of ADC Video Corp.)

TABLE 14.1   Link Budget for AM Fiber Link

.s.s.s.s. Distance (mi)	Distance (km)	Fiber Loss/km	Total Fiber Loss	Splice Loss/2 km	Total Splice Loss	Total Path Loss	Link Budget	Link Margin
		Mileage, Losses, and Margins—1310 nm						
12.40	19.96	0.5 dB	9.98	0.1 dB	1.00	10.98	13.00	2.02
15.15	24.38	0.4 dB	9.75	0.1 dB	1.22	10.97	13.00	2.03
17.00	27.36	0.35 dB	9.58	0.1 dB	1.37	10.94	13.00	2.06
		Mileage, Losses, and Margins—1550 nm						
22.75	36.61	0.25 dB	9.15	0.1 dB	1.83	10.98	13.00	2.02

With $\Delta F = 4$ MHz, $B_{IF} = 30$ MHz, and $B_F = 5$ MHz, the SNRw is improved by approximately 34 dB above CNR.

Example. If the C/N on a FM fiber link is 32 dB, what is the S/N for a TV video channel using the values given above? Use equation 14.14:

$$S/N = 23.7 \text{ dB} + 32 \text{ dB} + 10\log(30/5) + 20\log(1.6 \times 4/5)$$

$$= 23.7 + 32 + 7.78 + 2.14$$

$$= 65.62 \text{ dB}$$

Figure 14.14 illustrates the link performance of a FM fiber-optic system for video channels.

Table 14.1 shows typical link budgets for a HFC AM system.

14.5   DIGITAL TRANSMISSION OF CATV

14.5.1   Approaches

There are two approaches to transmit TV, both audio and video, digitally. The first is to transport raw, uncompressed video. The second method is to transport compressed video. Each method has advantages and disadvantages. Some advantages and disadvantages are application-driven. For example, if the objective is digital to the residence/office, compressed TV may be the most advantageous.

14.5.2   Transmission of Uncompressed Video on CATV Trunks

Video, as we discuss in Chapter 15, is an analog signal. It is converted to a digital format using techniques with some similarity to the 8-bit PCM, that was covered in Chapter 3. A major difference is in the sampling. Broadcast quality TV is generally *oversampled*. Here we mean that the sampling rate is

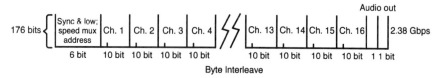

Figure 14.15. Typical frame structure on a single fiber in a CATV trunk. (Courtesy of ADC Video Systems, Meriden, CT.)

greater than the Nyquist rate. The Nyquist rate, as we remember, requires the sampling rate to be twice the highest frequency of interest. In our case this is 4.2 MHz, the bandwidth of a TV signal. Thus the sampling rate is greater than 8.4×10^6 samples per second.

One example is the ADC Video Systems scheme, which uses a sampling rate of 13.524×10^6 samples per second. One option is an 8-bit system, another is a 10-bit system. The resulting equivalent bit rates are 108.192 Mbps and 135.24 Mbps, respectively. The 20-kHz audio channel is sampled at 41,880 samples per second and uses 16-bit PCM. The resulting bit rate is 2.68 Mbps for four audio channels (quadraphonic).

ADC Video Systems multiplexes and frames a 16-TV-channel configuration for transmission over a fiber-optic trunk in a HFC system. The bit rate on each fiber is 2.38 Gbps. The frame structure is illustrated in Figure 14.15 and Figure 14.16 is an equipment block diagram for a 16-channel link.

A major advantage of digital transmission is the regeneration capability just as it is in PSTN 8-bit PCM. As a result, there is no noise accumulation on the digital portion of the network. These digital trunks can be extended hundreds of miles or more. The complexity is only marginally greater than a FM system. The 10-bit system can easily provide a S/N ratio at the conversion hub of 67 dB in a video channel and a S/N value of 63 dB with an 8-bit system. With uncompressed video, BER requirements are not very stringent because video contains highly redundant information.

14.5.3   Compressed Video

Motion picture experts group (MPEG) compression is widely used today. A common line bit rate for MPEG is 1.544 Mbps. Allowing 1 bit per hertz of bandwidth, BPSK modulation, and a cosine roll-off of 1.4, the 1.544-Mbps TV signal can be effectively transported in a 2-MHz bandwidth. Certainly 1000-MHz coaxial cable systems are within the state-of-the-art. With simple division we can see that 500-channel CATV systems are technically viable. If the modulation scheme utilizes 16-QAM (4 bits/Hz theoretical), three 1.544-Mbps compressed channels can be accommodated in a 6-MHz slot. We select 6 MHz because it is the current RF bandwidth assigned for one NTSC TV channel.

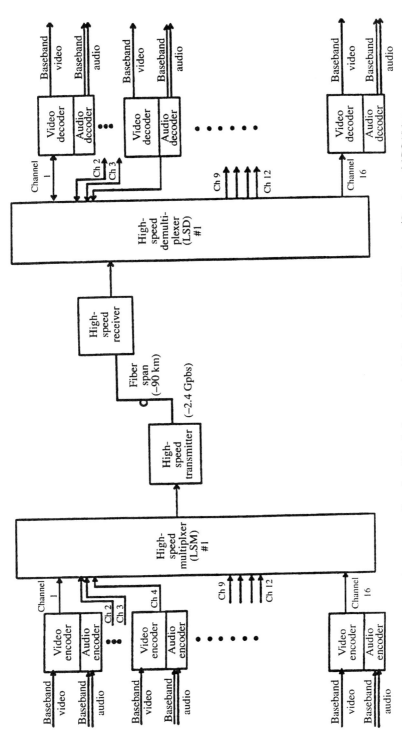

Figure 14.16. Functional block diagram of a 16-channel digital TV system. (Courtesy of ADC Video Systems, Meriden, CT.)

995

14.6   TWO-WAY CATV SYSTEMS

14.6.1   Introduction

Figures 14.17*a* and 14.17*b* are two views of the CATV spectrum as they would appear on coaxial cable. Of course, with conventional CATV systems, each NTSC television channel is assigned 6 MHz as shown in Chapter 15, Figure 15.7.

In Figure 14.17*a* only 25 MHz is assigned for upstream services. Not all of this bandwidth may be used for voice and data. A small portion should be set aside for upstream telemetry from active CATV equipment. On the other hand, downstream has 60 MHz assigned. In this day of the Internet, this would be providential, for the majority of the traffic would be downstream.

Comments on Figures 14.17a and 14.17b. Large guardbands isolate upstream from downstream TV and other services, 24 MHz in Figure 14.17*a*

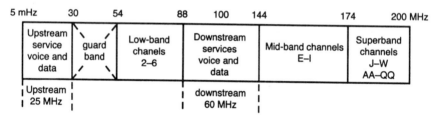

Figure 14.17a. CATV spectrum based on Ref. 1, showing additional upstream and downstream services. Note the bandwidth imbalance between upstream and downstream. (Adapted from Ref. 1.)

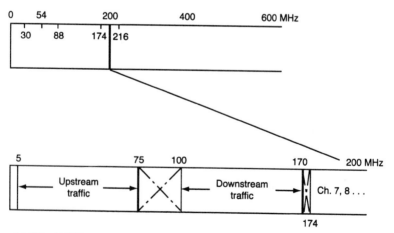

Figure 14.17b. CATV spectrum with equal upstream and downstream bandwidths for other services.

and 25 MHz in Figure 14.17*b*. A small guardband was placed in the slot from 170 to 174 MHz to isolate downstream data and voice signals from conventional CATV television.

We assume the voice service will be POTS (plain old telephone service) and that both the data and voice will be digital.

In another approach, downstream voice, data, and special video are assigned the band 550–750 MHz, which is the highest frequency segment portion of this system (Ref. 9). In this case, we are dealing with a 750-MHz system.

The optical fiber trunk terminates in a node or hub. This is where the conversion from optical to the standard CATV coaxial cable format occurs. Let a node serve four groupings of subscribers, each with a coaxial cable with the necessary amplifiers, line extenders, and taps. Such subscriber groups consist of 200–500 terminations (TV sets). Assume each termination has upstream service using the band 5–30 MHz (Figure 14.17*a*). In our example, the node has four incoming 5–30-MHz bands, one for each coaxial cable termination. It then converts each of these bands to a higher frequency slot 25 MHz wide in a frequency division configuration for backhaul on a return fiber.

In one scheme, at the headend, each 25-MHz slot is demultiplexed and the data and voice traffic are segregated for switching/processing.

Access by data, voice, and special video users of the upstream and downstream assets is another question. There are many ways this can be accomplished. One unique method suggested by a consultant is to steal a page from the AMPS cellular radio specifications. Because we have twice the bandwidth available, and because there are no handovers required, there are no fading and shadowing effects and no multipath; thus a much simpler system can be developed. For data communications, CDPD can be applied directly. Keep in mind that each system only serves 500 users as a maximum. Those 500 users are allocated 25 MHz of bandwidth (one-way).

An interesting exercise is to divide 25 MHz by 500. This tells us we can allot each user 50 kHz full-period. By taking advantage of the statistics of calling (usage), we could achieve a bandwidth multiplier of from 4 to 10 times by using forms of concentration. However, upstream video, depending on the type of compression, might consume a large portion of this spare bandwidth.

There are many other ways a subscriber can gain access. DAMA techniques, where AMPS cellular is one, are favored.

Suppose we were to turn to a digital format using standard 8-bit PCM. Allowing 1 bit/Hz and dividing 25 MHz by 64 kHz, we find only some 390 channels available. Keep in mind that these simple calculations are not accurate if we dig a little further. For instance, how will we distinguish one channel from another unless we somehow keep them in the frequency domain, where each channel is assigned a 64-kHz slot? This could be done by using QPSK modulation, which will leave some spare bandwidth for filter roll-off and guardbands.

One well thought out approach is set forth by the IEEE 802.14 (Ref. 8) committee. This is covered in Section 14.7.

Why not bring fiber directly into the home or to the desk in the office? The most convincing argument is economic. A CATV system interfaces with a home/office TV set by means of the set-top box. As we discussed earlier, the basic function of this box is to convert incoming CATV channels to a common channel on the TV set, usually either channel 2, 3, or 4. Now we will ask much more of this "box." It is to terminate the fiber in the AM system as well as to carry out channel conversion. The cost of a set-top in 1997 dollars should not exceed $300. With AM fiber to the home, the cost target will be exceeded.

The reason that the driving factor is the set-top box is the multiplier effect. In this case we would be working with multipliers of, say, 500 (subscriber) by > $300. For a total CATV network, we could be working with 100,000 or more customers. Given the two-way and digital options, both highly desirable, the set-top box might exceed $1000 (1997 dollars) even with mass production. This amount is excessive.

14.6.2 Impairments Peculiar to Upstream Service

14.6.2.1 More Thermal Noise Upstream than Downstream. Figure 14.18 shows a hypothetical layout of amplifiers in a CATV distribution system for two-way operation. In the downstream direction, broadband amplifiers point

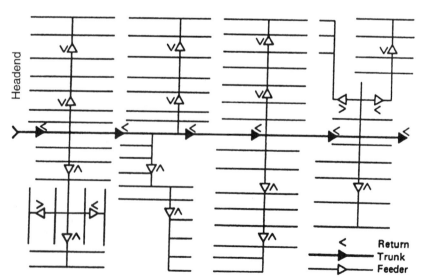

Figure 14.18. Trunk/feeder system layout for two-way operation. (From Ref. 1. Reprinted with permission.)

outward, down trunks and out distribution cables. In the upstream direction, the broadband amplifiers point inward toward the headend and all their thermal noise accumulates and concentrates at the headend. This can account for from 3- to 20-dB additional noise upstream at the headend where the upstream demodulation of voice and data signals takes place. Fortunately, the signal-to-noise ratio requirements for good performance for voice and data are much less stringent than for video, which compensates to a certain extent for this additional noise.

14.6.2.2 Ingress Noise. This noise source is peculiar to CATV system. It basically derives from the residence/office TV sets that terminate the system. Parts 15.31 and 15.35 of the FCC Rules and Regulations govern such unintentional radiators. These rules have not been rigidly enforced.

One problem is that the 75-Ω impedance match between the coaxial cable and the TV set is poor. Thus not only all radiating devices in the TV set, but other radiating devices nearby in residences and office buildings couple back through the TV set into the CATV system in the upstream direction. This type of noise is predominant in the lower frequencies, that band from 5 to 30 MHz that carries the upstream signals. As frequency increases, ingress noise intensity decreases. Fiber-optic links in a HFC configuration provide some isolation.

14.7 TWO-WAY VOICE AND DATA OVER CATV SYSTEMS ACCORDING TO IEEE 802.14 COMMITTEE STANDARD

14.7.1 General

The narrative in this section is based on a draft edition of IEEE Standard 802.14 dated March 11, 1997 (Ref. 8) and subsequent narrative kindly provided by the chairman of the IEEE 802.14 committee (Ref. 9). See also Ref. 7. The model for the standard is a hybrid fiber–coaxial cable (HFC) system with a service area of 500 households. The actual household number may vary from several hundred to a few thousand, depending on penetration rate.

Two issues limit the depth and completeness of our discussion: (1) reasonable emphasis and inclusion of details versus page count and (2) because of the draft nature of the reference document, many parameters have not been quantified. Figure 14.19 is a pictorial overview of the 802.14 system.

The IEEE 802.14 supports voice, video, file transfer, and interactive data services across an international set of networks. These are represented by switched data services such as ATM, variable length data services such as CSMA/CD (Ethernet), near constant bit rate services such as MPEG digital video systems (Ref. 10), and very low latency data services such as virtual circuits or STM (e.g., E1 and T1 families of PCM formats). Instantaneous

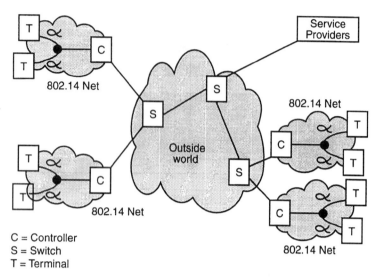

Figure 14.19. A pictorial overview of 802.14 networks and their relationship with the outside world. (From Ref. 8, IEEE 802.14, draft R2, p. 14.)

data rates and actual throughput are no longer limited by the protocol but are, rather, a function of the network traffic engineering and the theoretical limit of the media and the modulation schemes that are employed. This generates a wide range of QoS parameters that must be supported simultaneously in order to create a scaleable, multiservice delivery system.

The 802.14 system can be envisioned as having OSI layers 1 and 2. Layer 1, of course is the physical layer, and layer 2, from a LAN viewpoint, covers the "lower portion" of the data link layer. This portion covers the functions of medium access control (MAC), which is discussed in the next subsection. The "upper portion" of the data link layer is the LLC (logical link control). In our discussion in the next subsection. we will use the acronym PHY for the physical layer.

14.7.2   Overview of the Medium Access Control (MAC)

As overall controller of the actual transmission and reception of information, the MAC must account for the unique physical topology constraints of the network while guaranteeing the required QoS for each type of data to be transported. The network consists of a multicast or broadcast downstream from the headend to individual subscriber groups and multiple allocated and contention upstream channels. The downstream channel consists of a single wideband, high symbol rate channel, composed of six-octet time allocation units. A single unit can be assigned an idle pattern or multiple units can be used to create ATM cells, variable length fragments, or MPEG video streams.

The 802.14 system allows for multiple simultaneous downstream channels as well. The upstream consists of multiple channels divided in time into a series of *minislots*. These represent the smallest orthogonal unit of data allocation. There is enough time in one minislot for the transmission of eight octets of data plus PHY overhead and guard time. Multiple minislots can be concatenated in an upstream channel to create larger packet data units such as ATM cells, variable length fragments, or even MPEG video streams. Because of the varying amount of overhead and guard time required by different physical layers, the number of minislots required for the transmission of any data stream will vary from one upstream PHY to another.

The use of minislots with independent in-channel/in-band control messaging creates a flexible architecture. This flexibility allows one to change traffic flow patterns of the network and to fully integrate multiple channels and time slots.

14.7.3 Overview of the Physical Layer (PHY)

Similar to the MAC architecture, the physical topology of the hybrid fiber–coax (HFC) plant allows for multicast downstream and multiple converging upstream paths. The 802.14 specification does not specify data types and resident topology. Constraints are defined and resolved while allowing the architecture to adapt on a session by session basis. Two distinctly different downstream PHYs are supported. Each type is centered around an existing coding and modulation standard: ITU-T Rec. J.83 (Ref. 12) Annex A/C, which is adopted for European cable systems, and Annex B, which is adopted for North American cable systems. In addition to these standards, 802.14 specifies modulation, coding sequence, scrambling method, symbol rates, synchronization, physical layer timing, message length and formats, transmitter power, and resolution characteristics.

14.7.3.1 Subsplit/Extended Subsplit Frequency Plan. The majority of CATV systems, particularly in North America, are upgraded to subsplit* (5–30 MHz) or extended subsplit (5–42 MHz) operation. This scenario represents the worst-case design in terms of ingress noise and availability of reverse channel (upstream) bandwidth. If the design can be deployed in this configuration, the infrastructure upgrade cost for cable systems will be minimal. In the future, the availability of midsplit or highsplit cable plants will enable a PHY with enhanced performance.

---

*Subsplit—a frequency division scheme that allows bidirectional traffic on a single cable. Reverse path signals come to the headend from 5 to 30 MHz and up to 42 MHz on newer systems. Forward path signals go from the headend to end-users from 54 MHz to the upper frequency limit of the system in question.

14.7.4 Other General Information

14.7.4.1 *Frequency Reuse in the Upstream Direction.* The assumption is made that the coaxial cable traffic for each service area will be able to use the entire 5–30/42-MHz reverse bandwidth. This could be done either by use of a separate return fiber for each service area or by use of a single fiber for several service areas, whose return traffic streams would be combined using block frequency translation at the fiber node.

14.7.4.2 *Up to 160-km Round-Trip Cable Distance.* The distance coverage of the system can be influenced by several factors such as the fiber-optic technology employed and the coaxial cable distribution topology. The limiting factor could be the number of active amplifiers and the resulting noise parameters that must be bounded for an optimal physical layer design.

14.7.5 Medium Access Control

The MAC controls the allocation of upstream and downstream bandwidth.

14.7.5.1 *Logical Topology.* The logical topology of the CATV plant imposed some significant constraints on the 802.14 protocol design. For example, classical collision detection is impossible to do reliably on the cable plant since a station can only hear transmissions by the HC (headend controller) and not by other stations. Even the detection of collisions by the HC is not entirely reliable. The protocol had to take into account the fact that round-trip delay from a station to the HC and back can be as high as 400 μs.

Figure 14.20 shows the elements of MAC (medium access control) topology. Each station has amplifiers in each direction that restrict the data flow unidirectionally. The path from the HC to the stations is referred to as the *downstream path*. All stations on the network receive the same downstream path. It is incumbent on the station to filter out messages that are not addressed to itself.

The path from a set of user stations to the HC is called the *upstream path*. In the upstream direction any station can transmit but only the HC can

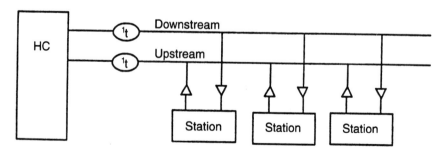

Figure 14.20. Elements of MAC topology.

receive. Diplex amplifiers prevent one station from listening to the transmission of another station. A single network may have several upstream channels to which stations may be assigned. Each station must be capable of changing its upstream channel at the request of the HC. The HC must be able to simultaneously receive all upstream channels within the network.

14.7.5.2 *MAC Framing and Synchronization.*
All stations must be slaved to a master timing source that resides at the HC. To provide a time base to all stations that is synchronized correctly, the HC broadcasts a time-stamped cell to all stations at periodic intervals (~ 2 ms). The HC can adjust each station's timebase through messages so that all stations are synchronized in time. For any two stations on the network, it is important that, if they both decide to transmit at a given network time, both transmissions will arrive at the HC at the same instant.

14.7.5.3 *Channel Hierarchies*

14.7.5.3.1 *Downstream Hierarchy.*
The downstream is composed of six-octet allocation units. A single unit can be assigned the idle pattern or multiple units can be used to create ATM cells or variable length fragments. Some of the ATM cells will carry MAC messages in the form of information elements described further in the following. All MAC messaging is done in ATM cells with certain header values shown in Table M-9 of the reference publication (Ref. 8). This hierarchy is illustrated in Figure 14.21.

14.7.5.3.2 *Upstream Channel Hierarchy.*
The upstream channel is a multiple access medium. For each upstream channel in the network there is a group of stations that share the assigned bandwidth. Each upstream channel is divided in time into a series of mini (time) slots. A minislot has the time capacity for the transmission of eight octets of data plus PHY overhead and guard time. A PDU (protocol data unit) that only occupies a single minislot is termed a *minipdu*. Minipdus are used primarily for contention opportunities to request bandwidth (bit rate capacity).

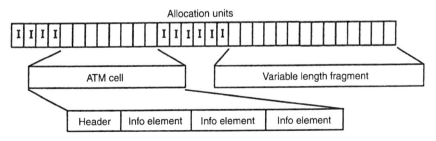

Figure 14.21. Downstream message hierarchy. I shows idle pattern. (From Ref. 8, Figure M-2, IEEE Std. 802.14, draft R2.)

Multiple minislots can be concatenated in an upstream channel to create larger PDUs such as ATM cells and variable length fragments. Due to the varying amounts of overhead and guard time required by different physical layers, the number of minislots required to transmit an ATM cell will vary from one upstream PHY to another. There is a requirement that an integral number of minislots be used to transmit an ATM cell. The allocation of partial minislots by the HC is prohibited.

Similar to the downstream, some ATM cells will be used to carry MAC messages from a station to the HC in the form of information elements.

14.7.5.4   Station Addressing. There are two alternative station addressing mechanisms: MAC address and local ID (LI).

14.7.5.4.1   MAC Address. Each station has an IEEE 802.14 administered 48-bit MAC address. This address uniquely identifies each station. The MAC address follows the rules specified in the document "Organizationally Unique Identifiers."

14.7.5.4.2   Local ID (LI). The MAC address is not normally used in the 802.14 protocol. Instead, one or more 14-bit local identifiers (LI) are assigned to the station by the HC. The first identifier assigned to a station is called the primary local identifier and is bound to the station's MAC address. The primary LI is used for MAC, security, and PHY control as well as data transfer.

Local identifiers are also used to identify broadcast and multicast groups. The syntax of the LI does not differentiate between its use as a single station identifier and its use as a multicast identifier; thus all single station identifiers, multicast identifiers, and broadcast identifiers are assigned by the HC from the same 14-bit space.

Local identifiers are meaningful only within a single 802.14 network and may not be used by entities outside the local 802.14 network to address entities within the local 802.14 network.

14.7.5.4.3   Multicast Local Identifiers. Multicasting within the network is based logically on the 48-bit MAC address and physically on the 14-bit LI. When a multicast group is formed, either by a station or the HC, the HC allocates a local identifier to the multicast group and associates it internally with a 48-bit multicast MAC address. As stations are added, by 48-bit MAC address, to the multicast group, the HC must send the new station a message to add the LI of the multicast group to its list of recognized local identifiers. If the multicast group is encryped, then the HC must also inform the new station of the keys that are pertinent to the multicast group. These messages are sent in unicast cells directly to the primary LI of the new station.

When a station is removed from a multicast group, again by 48-bit MAC address, the HC sends the station a message instructing it to no longer recognize the LI of the multicast group. If the multicast group is to continue operation, then the HC assigns new encryption keys to the group when stations have been removed. The message to remove a local identifier from the recognized group is unicast to the station's primary LI. Any encryption key distribution that occurs after the removal of stations from a multicast group is unicast to each station remaining in the multicast group rather than multicast to the group as a whole.

14.7.5.5 Common Elements of Upstream and Downstream Messaging

14.7.5.5.1 Information Elements. Both upstream and downstream MAC messages contain the concept of an *information element*. Information elements are carried in the payload of ATM cells. The general format for information elements is illustrated in Figure 14.22. This format is constructed so that unrecognized elements can be skipped over in the event that an implementation-dependent element is encountered that is not understood by the receiving station.

Each information element has an element ID (EID) field that identifies the purpose of the element. The meaning of a particular value of EID varies between upstream and downstream messages and between virtual circuits. The element length (ELEN) field contains the length of the payload area in octets. The general fields of information elements are described in Table 14.2.

14.7.5.6 Upstream Bandwidth Control Formats. Minislots are the basic unit of upstream bandwidth allocation. The HC determines the function of each minislot in the upstream and relays this information to all stations using the downstream control messages. The upstream PDUs are constructed by concatenating multiple minislots together to create larger units of data. The upstream can be viewed as a "sea of minislots."

Each minislot consists of eight octets of data plus PHY overhead. The length of the guard time is adjusted within the PHY by the headend such that an integral number of minislots are equal in length to an ATM slot.

Figure 14.22. Information element, general format.

TABLE 14.2 Information Element General Fields

Field	Usage	Size
EID	Type of information element	8 bits
ELEN	Short format length field	8 bits
Payload	Content area of information element	Variable

Source: Reference 8, Table M-2, IEEE 802.14, draft R2.

Each minislot is implicitly assigned a 16-bit identifier by the HC. The signature cell includes the minislot number of the next minislot to be transmitted in HC time. The minislot identifier is incremented for each subsequent minislot. The number will wrap back to zero at a network-dependent value. The identifier of any minislot is unique and understood by the HC and all stations participating in the network.

14.7.5.6.1 Request Minislot PDU (RMPDU). RMPDUs occupy minislots and are used to contend for access to the shared upstream medium. The format of a RMPDU is shown in Figure 14.23.

LENGTH (LEN). The length field is an 8-bit quantity that indicates the size of the allocation requested by the station. This size is added to the currently pending allocation requests for the requesting station. If the request type indicates ATM cells, then size is interpreted as a number of ATM cells. If the request type indicates variable length packet fragments, then the size is interpreted as a number of minislots. For variable length fragment requests, the HC attempts to allocate the largest possible number of contiguous minislots. For network defined minislot PDU requests, the length is interpreted as a number of minislots and the headend may allocate the minislots with no particular constraints.

RTY	QID	LEN	Reserved	CRC

Field	Usage	Size
RTY	Request type (see Table 14.3)	8 bits
QID	Queue identifier	24 bits
LEN	Size of allocation requested	8 bits
Reserved	Reserved (zero fill)	8 bits
CRC	CCITT-CRC16	16 bits

Figure 14.23. Request minislot PDU format. (From Ref. 8 Figure M-4, IEEE 802.14, draft R2.) The eight-bit request type field is shown in Table 14.3.

TABLE 14.3   Request Type Field

RTY	Requested Type
0 × 00	ATM PDU
0 × 01	Variable length fragments
0 × 02	Network defined minipdus

Note: All other types are reserved.

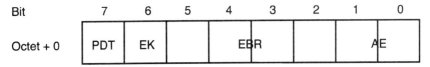

Figure 14.24. PDU-type octet.

REQUEST QUEUE IDENTIFIER (QID). The QID is a 24-bit field that is used to uniquely identify the station making the request. The most significant bits of the request identifier always match the station's locally administered short address. The remaining bits of the request can be used by the station to identify a virtual circuit or queue associated with the request.

14.7.5.7 *Upstream PDU Formats.* PDUs are used to transfer data between a station and the headend controller. Each PDU begins with a prefix octet, which is used to indicate the type of PDU, and continues with the body of the PDU.

The PDU-type octet for upstream burst transmission is shown in Figure 14.24.

14.7.5.7.1 *Upstream ATM Cell PDU (APDU).* The structure of an ATM cell as carried by the 802.14 network is illustrated in Figure 14.25. The NNI cell format is the only type of cell carried by the network, since the GFC field is

0	EK	EBR	SLI	VCI	PT	CLP	HEC	Payload

Field	Usage	Size
PDT	PDU type (0 = ATM)	1 bit
EK	Encryption key even/odd identifier	1 bit
EBR	Extended bandwidth request field	4 bits
SLI	Source station local ID	14 bits
VCI	Virtual circuit identifier	16 bits
PT	Payload type indicator	3 bits
CLP	Cell loss priority	1 bit
HEC	Header error check	8 bits
Payload	ATM payload field	48 octets

Figure 14.25. Upstream ATM cell format. (From Ref. 8, Figure M-6, IEEE 802.14, draft R2.)

not present. In the upstream direction all APDUs are directed to the HC. Section 13.15 of this text provides a description of ATM.

14.7.5.7.2 Upstream Variable Length Packet Fragment PDU (VLPDU). Variable length fragmentation allows a more efficient transport of LLC traffic.* Variable length fragmentation is optional in both user modems and headend controllers and either can choose to operate without this facility. The maximum upstream fragment[†] size is determined by the PHY.

FRAGMENT FORMAT. The variable length fragment format is shown in Figure 14.26.

14.7.5.7.3 Upstream MAC Layer Management Messages. Upstream management messages are used to communicate with the HC to manage the network. Multiple upstream management information elements may be embedded into a single ATM cell but management elements are not allowed to cross cell boundaries. The ATM header values indicate a MAC layer management message. The first byte of an upstream management cell contains the number of message elements.

14.7.6 Downstream Operation

14.7.6.1 Idle Pattern. When there are no data to transmit, the HC continuously sends IDLE segments. An idle segment is a variable length fragment with a destination address of 0×0000 (NULL) and a LEN field of 0×000 (no payload). The SEG field is 0×3 corresponding to a complete frame. C/U is 1, indicating a control frame. All other bits except the HEC are 0. The HEC is calculated in the normal manner. An idle pattern is shown below:

1	0	0000	0×000	0×000	00000	1	11	HEC

14.7.6.2 Synchronization to Downstream Data. When a station first begins listening to a downstream data stream it must first acquire synchronization with the PDU headers. The recommended method according to IEEE 802.14 is to begin looking for HEC values consistent with the previous 4 bytes. When two have been found, the distance between them should be consistent with the length of the first PDU (53 bytes if an ATM cell, LEN + 6 bytes if a variable length fragment). If the distance is not consistent, the first should be thrown away and the search for a second header

---

*LLC, logical link control, is the upper part of the link layer for LANs based on the IEEE 802 series specifications. It interfaces with the MAC layer, which resides just below it in OSI layer 2. The MAC layer interfaces with layer 1, the PHY or physical layer. LLC accepts user traffic from user OSI layer 3.

[†]Think of a fragment as a frame. The term fragment derives from the TCP/IP protocol suite.

1	EK	EBR	SLI	LEN	00000	C/U	SEG	HEC	Payload

Field	Usage	Size
PDT	PDU type = 1, variable length fragment	1 bit
EK	Encryption key even/odd identifier	1 bit
EBR	Extended bandwidth request field	4 bits
SLI	Source station local identifier	14 bits
LEN	Length of fragment	12 bits
Reserved	Zero fill	5 bits
C/U	Control/user indicator	1 bit
SEG	Segmentation control	2 bits
HEC	Header error check	8 bits
Payload	Fragment payload	LEN octets

EK = encryption designator. This field indicates whether the even or odd encryption key is used to encrypt the payload.

EBR = extended bandwidth request. It allows the sending station to request additional upstream bandwidth for variable length fragments. The value of the EBR is multiplied by 16 to determine the number of octets requested. This request is added to the currently outstanding variable length fragment bandwidth for this station.

SLI = source station local identifier. Its field contains the primary local identifier of the sending station.

LEN = payload length in octets.

SEG = segment control field, which indicates the relative position of the fragment within the complete LLC frame as shown in Table 14.4.

C/U = control/user field, which indicates, if nonzero, that this packet contains control information.

HEC = header error check and is calculated as described in Section 13.15 (Chapter 13).

Figure 14.26. Upstream variable length fragment format. (From Ref. 8, Figure M-7, IEEE 802.14, draft R2.)

TABLE 14.4 Segmentation Control Field

SEG	Fragment Type
00	Continuation segment
01	Start of frame
10	End of frame
11	Complete frame (no segmentation)

Source: Reference 8, Table M-5, IEEE 802.14, draft R2.

continues. Once two headers have been detected, separated by the proper distance, the search should continue in the same manner for additional headers. Detection of three consistent headers can be considered a lock on synchronization alignment.

As reception continues, the consistency of the downstream flow is checked. Failure of two consistency checks in a row constitutes a loss of synchronization alignment, and the station should begin to acquire header synchronization.

14.7.6.3 *Downstream Signature Cell.* Signature cells are transmitted periodically by the HC to indicate existence and parameters of the 802.14 network and to provide a timing reference for all stations. The signature cell is addressed to the broadcast local identifier. The header of a signature cell contains specific values indicating a MAC layer management cell as described in Table M-9, "Preassigned Cell Header Values for Use by the MAC and TC Layers," of the reference publication (Ref. 8). The format of the payload portion of the signature cell is illustrated in Table 14.5.

14.7.6.4 *Invitation.* An invitation cell is addressed to the acquisition multicast address. This element invites unregistered stations to attempt ranging and registration at a specific time. The header values of an invitation cell are for a MAC layer management cell. The format of the payload of an invitation cell is shown in Table 14.6.

14.7.6.5 *Downstream Bandwidth Management Cells.* Bandwidth management messages are carried in ATM cells. Each ATM cell can carry an

TABLE 14.5 Signature Cell Fields

Field Size	Mnemonic	Field Meaning
6 Octets	SIG	Network signature = "802.14"
2 Octets	VER	Protocol version number
6 Octets	TS	Time stamp
2 Octets	NMS	Next minislot number
2 Octets	MMS	Maximum minislot number

SIG is the channel signature, which serves as a consistency check that this is an IEEE 802.14 network. The signature consists of the 6 ASCII octets "802.14."

VER is the protocol version number made up of 2 bytes that make up a major and a minor version number, which identifies the network protocol in force uniquely.

TS is the time stamp. It specifies the time the cell began transmission in headend. Time is given in HC time units and is inserted by the PHY/TC layer during transmission. The station uses the time stamps in signature cells to synchronize its local clock with the HC clock.

NMS is the next minislot number. It specifies the number of the next minislot in headend time. At the headend this corresponds to the number of the minislot that will be received by the headend at some arbitrary interval after transmission of the signature cell. To a station in the ranging process it is sufficient to mark the next minislot interval in local time with this number. The HC adjusts the station timebase in the invitation response message.

MMS is the maximum minislot number. It specifies the maximum value of the minislot index. When the minislot index reaches this value, the next minislot is numbered zero.

Source: Reference 8, Table M-6, IEEE 802.14, draft R2.

TABLE 14.6   Invitation Cell Fields

Field Size	Mnemonic	Field Meaning
2 Octets	CTY	Cell type = 0, invitation
2 Octets	FMS	First minislot of ranging window
2 Octets	LMS	Last minislot of ranging window
Variable	PRF	Ranging channel PHY profile

FMS and LMS. First and last minislot numbers defining the boundaries of the ranging window. The HC enlarges the actual window to account for the round-trip delay.

PRF is the PHY profile, which describes a single upstream channel that is logically connected to the current downstream channel. The PHY also contains the following:

Flags. The flag byte of each channel information block contains binary flags denoting properties of the upstream channel.

CBR Allowed (CBRA). The CBR flags, if set, denote that this upstream channel can carry CBR services and guarantee proper QoS. (*Note:* CBR = constant bit rate services, a term used in ATM. A typical constant bit rate service is voice. The other ATM service is VBR, variable bit rate service. Typical in this category are data.)

Variable Length Fragments Allowed (VLFA). The VLFA flag denotes, if set, that this upstream channel can carry variable length fragments. If this flag is not set, then this upstream channel will consist of only ATM traffic.

PHY Parameter Block. The PHY parameter block contains information about the physical aspects of the upstream channel.

Carrier Frequency. Carrier frequency of the upstream channel is in hertz.

Symbol Rate. The symbol rate of the upstream channel is in symbols/second.

Modulation Information. Modulation type of the upstream channel.

Source: Reference 8, Table M-7, IEEE 802.14, draft R2.

---

integral number of information elements. Information elements are not allowed to cross cell boundaries. Layer management cells always use the circuit VCI–BWMGT. Layer management cells may be addressed either to group/global station addresses or to a particular station. The types of information elements that can be contained within bandwidth management cells are allocation grants, request minislot allocations, and request feedback.

14.7.6.5.1   General Format. Each bandwidth management cell begins with an octet that specifies the number of information elements contained in the cell. The octet is followed by the information elements themselves. Each information element begins with an octet that specifies the type of information element. Bandwidth information elements follow the same general format as upstream management elements described in Section 14.7.5.3.2.

Grant information elements indicate the allocation of upstream bandwidth to a particular station and/or circuit.

ATM cell grant has the format shown in Figure 14.27.

QID, the request queue ID field, corresponds to the request ID of the original allocation request. This is the only connection between the request and the allocation message. The starting minislot number (SMS) specifies the first minislot at which transmission can commence. NMS specifies the number of minislots allocated.

	0x01	0x06	QID	SMS	NMS

Field	Usage	Size
EID	Element ID = 0x01, ATM cell grant	8 bits
ELEN	Element length = 6	8 bits
QID	Request queue identifier	24 bits
SMS	Starting minislot number	16 bits
NMS	Number of minislots granted	8 bits

Figure 14.27. ATM cell grant element. (From Ref. 8, Figure M-15, IEEE 802.14, draft R2.)

Variable length fragment grant has the same general format as an ATM cell grant message except that the element type is different. In addition, the number of minislots allocated may not be an even multiple of an ATM cell length. A variable length fragment grant message format is shown in Figure 14.28.

14.7.6.5.2 Request Minislot Allocation Element. Request minislot allocation elements specify the location of minislots available for contention as well as parameters about these minislots, which aid in the collision resolution

	0x02	0x06	QID	SMS	NMS

Field	Usage	Size
EID	Element ID = 0x02, variable length fragment grant	8 bits
ELEN	Element length = 6	8 bits
QID	Request queue identifier	24 bits
SMS	Starting minislot number	16 bits
NMS	Number of minislots granted	8 bits

Figure 14.28. Variable length fragment grant. (From Ref. 8, Figure M-16, IEEE 802.14, draft R2.)

algorithm. The format of a request minislot allocation information element is shown in Figure 14.29. Request minislot allocation elements are variable length structures that designate a number of clusters of minislots as being available for contention mode access.

Feedback information elements provide a closed loop in the request process. All bandwidth requests (piggyback, request PDU, and CBR) are explicitly acknowledged by the HC immediately. A timeout slightly longer than the round-trip delay of the network allows the requesting station to detect the loss of a request in the upstream channel.

14.7.7 Physical Layer Description for CATV HFC Networks

14.7.7.1 Overview of the PHY. The PHY of 802.14 supports asymmetrical bidirectional transmission of signals in a CATV HFC network. The network is a point-to-multipoint, tree branch access network in the downstream direction, and multipoint-to-point, bus access network in the upstream direction. The downstream transmission originates at the headend node and is transmitted to all end-user nodes located at the tips of the branches in the tree and branch network. An upstream transmission originates from an end-user node and reaches the headend node through a multipoint-to-point access network where the access medium is shared by all end-user nodes that are communicating with the same headend.

An example CATV HFC network topology is shown in Figure 14.30. In this case we see a multiple of 5–42-MHz upstream channels that are frequency division multiplexed in the fiber node (FN) and that are then transmitted via a single fiber trunk to the CATV headend. This operation is called frequency stacking. *Frequency stacking* is not a requirement in a CATV HF topology.

14.7.7.2 Downstream Physical Layer Specification. There are two distinctly different PHYs supported by the 802.14 standard. These PHYs are called type A and type B downstream PHYs. The principal difference between the two is the coding method used for FEC. The FEC for type A downstream is based only on RS (Reed–Solomon) block coding. On the other hand, the FEC for type B downstream PHY is based on a concatenated coding method with outer RS coding and inner trellis-coded modulation (TCM). Each downstream PHY type has different modes of operation.

14.7.7.2.1 Downstream Spectrum Allocation. The frequencies from 63 MHz up to the upper frequency limit supported by the CATV cable plant (e.g., 750 MHz) are allocated for downstream transmission. Within this band is a channelized approach (i.e., frequency slots 6 or 8 MHz wide) from the headend to end-user nodes. Standard CATV frequency plans are assumed.

The topology model of the system is shown in Figure 14.30.

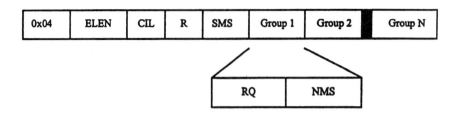

Field	Usage	Size
EID	Element ID = 0x04, minislot allocation	8 bits
ELEN	Element length	8 bits
CIL	Contention interleave identifier	4 bits
R	Contention entry range	12 bits
SMS	Starting minislot number	16 bits
RQ	RQ number for this group	8 bits
NMS	Number of minislots in this group	8 bits

ELEN = element length, which specifies the length of the structure (not including octet zero). The value of ELEN is given by ELEN = 4 + 2*n), where n is the number of repeating descriptors.

CIL = contention interleave, which designates the contention station machine to which the descriptor applies. It is possible for a network to have up to 16 interleaved contention accesses active at a given time, although each station must have a separate contention state machine for each interleave it wishes to contend concurrently. It is anticipated that most networks will have only a few contention interleaves.

R = entry persistence. For stations entering the contention process (RQ = 0), the entry presistence (R) specifies the probability that the station will contend for access. The station calculates a random number between zero and R. If this number falls within the group of minislots designated RQ = 0, then the station contends for access in that minislot. If the random number falls outside the range of RQ = 0 minislots, then the station does not contend for access within this allocation. This allows the HC to lower the probability of excessive collisions during heavy contention periods.

SMS = starting minislot number, which is the first minislot belonging to this contention cluster.

RQ = ternary tree request queue number, which is used to implement the ternary tree collision resolution algorithm. A RQ field of zero indicates the number of minislots that new stations may use to enter the contention process. For each group of minislots designated to a particular RQ value, a separate group descriptor is included. This figure shows a minislot group descriptor that occupies only octets 6 and 7.

NMS = number of minislots, a field that specifies the number of minislots belonging to this minislot group.

Figure 14.29. Request minislot allocation element. (From Ref. 8, Figure M-17, IEEE 802.14, draft R2.)

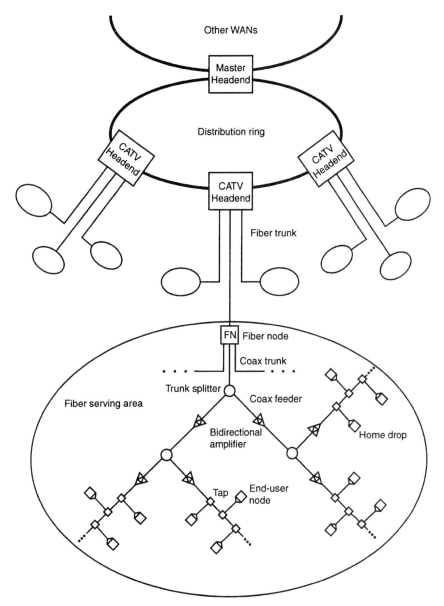

Figure 14.30. A model for HFC CATV serving area topology. (From Ref. 8, Figure P-31, IEEE 802.14, draft R2.)

14.7.7.2.2   Propagation Delay and Delay Variation. The propagation delay for optical fiber is nominally 5 μs/km and for coaxial cable, 4 μs/km. The propagation delay introduced by the downstream transmission medium should be budgeted such that the total round-trip delay between the headend and the end-user station should be a maximum of TBD (to be determined) milliseconds.*

14.7.7.3   *Type A Downstream PHY.* The type A downstream PHY supports two modes of operation: 64- and 256-QAM. Figure 14.31 is a functional block diagram for the type A downstream PHY. A block coding approach based on the shortened RS (Reed–Solomon) coding is used. A convolutional interleaver mitigates the effects of burst noise.

14.7.7.3.1   Transport Framing Structure. A type A downstream PHY supports a framing structure based on transporting MPEG-2 packets. Figure 14.32 illustrates this framing structure. Each MPEG-2 packet consists of 187 bytes of payload preceded with 1 byte of "sync" delimiter. There are a total of eight sync delimiters whose contents are in accordance with the MPEG-2 transport layer definition (Ref. 12).

14.7.7.3.2   Constellations for Type A Downstream PHY. Figure 14.33 illustrates a type A PHY constellation for 64- and 256-QAM waveforms.

14.7.7.4   *Type B Downstream PHY.* Type B downstream PHY supports two modes of operation: 64- and 256-QAM. Figure 14.34 is a functional block diagram for the type B downstream PHY. The coding strategy for type B downstream PHY differs from the type A downstream PHY in that a concatenated coding method with outer RS coding and inner trellis-coded modulation (TCM) is used.

The coding for type B downstream PHY is specified as a combination of four layers: RS coding, interleaving, scrambling, and trellis-coded modulation. The specific FEC synchronization is completely internal to type B downstream PHY and, as such, is specified in the RS encoder.

14.7.7.5   *Downstream Carrier Frequencies.* The downstream carrier frequencies are selected in accordance with the following:

$$f_c = (n \times 250 \text{ kHz}) \pm 8 \text{ kHz}$$

where n is an integer such that 63 MHz $\leq f_c \leq$ 803 MHz.

---

*If the round-trip (loopback) system extension is 160 km (quoted above) and seven-eights of the system is optical fiber, then the optical fiber portion is 140×5 μs $= 600$ μs; $+ 40 \times 4$ μs $= 160$ μs for a total of 0.760 ms.

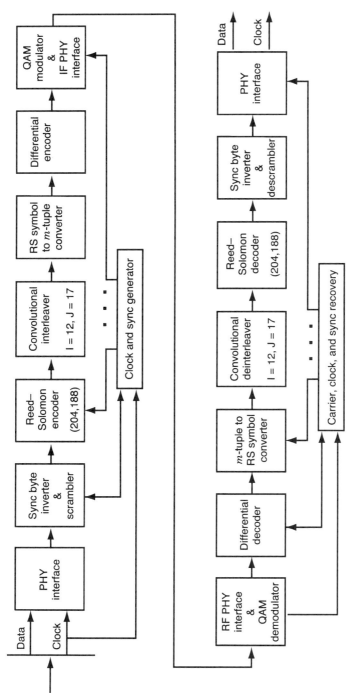

Figure 14.31. A functional block diagram of type A downstream PHY. (From Ref. 8, Figure P-32, IEEE 802.14, draft R2.)

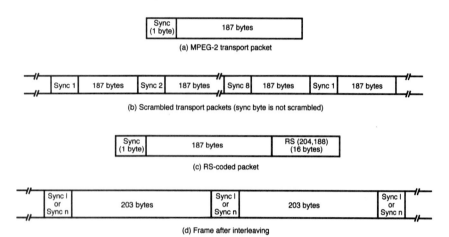

Figure 14.32. Transport framing structure for type A downstream PHY. (From Ref. 8, Figure P-33, IEEE 802.14, draft R2.)

14.7.7.6 Spurious Levels. Spurious levels are less than -57 dBc for in-band ($f_c \pm 3$ MHz), less than -62 dBc for adjacent channel ($f_c + 3$ to $f_c + 9$ and $f_c - 9$ to $f_c - 2$), and less than -67 dBc for other channels (47 MHz to 1 GHz).

14.7.7.7 Transmitted Signal Levels. The type B downstream PHY is capable of transmitting a signal on the cable between the ranges of $+50$ and $+61$ dBmV.

14.7.8 Upstream Physical Layer Specification

14.7.8.1 Upstream Spectrum Allocation. The subsplit band (i.e., frequencies between 5 and 42 MHz) is allocated for upstream transmission. In some cable plants, additional frequency bands for upstream transmission are intended for future use, called "midsplit" and "highsplit" bands. The midsplit extends from 5 to 108 MHz and the highsplit covers the range between 5 and 174 MHz. In some locations, the original subsplit band is modified as 5–50 MHz, 5–65 MHz, and 5–48 MHz in North America, Europe, and Japan, respectively.

14.7.8.2 Upstream Channel Spacing. Channel spacing depends on the modulation rate employed. The minimum channel spacing is

$$(1 + \alpha) \times R_{S(min)}$$

where α and $R_{S(min)}$ denote spectral roll-off factor and minimum symbol rate, respectively.

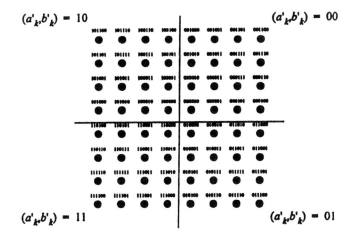

(a) 64-QAM for type A downstream PHY

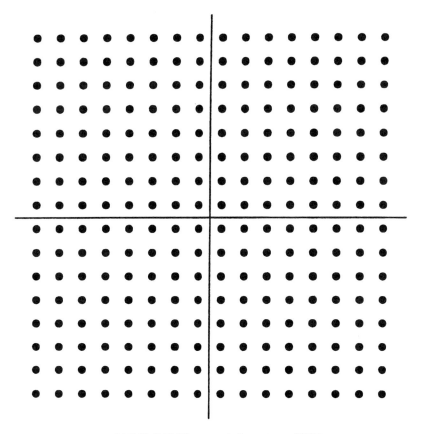

(b) 256-QAM for type A downstream PHY

Figure 14.33. Constellations for type A downstream PHY. (From Ref. 8, Figure P-37, IEEE 802.14, draft R2.)

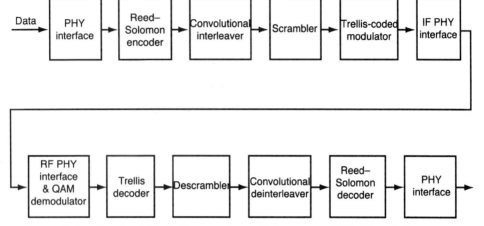

Figure 14.34. Functional block diagram of type B downstream PHY. (From Ref. 8, Figure P-40, IEEE 802.14, draft R2.)

14.7.8.3 *Carrier Frequencies.* Carrier frequencies, f_c, for upstream transmission are selected such that

$$f_c = n \times (32 \text{ kHz})$$

14.7.8.4 *Burst Message Lengths and Formats.* The upstream PHY shall support a channel burst parameter set per upstream transmission burst type. The channel burst parameters are given in Table 14.7.

14.7.8.5 *Resolution of Upstream Start of Transmission.* The upstream PHY layer supports a near continuous mode of transmission in the sense that the "postcursor" of one burst may completely overlap the "precursor" of the

TABLE 14.7 Upstream Channel Burst Parameters

Parameter	Options
Modulation	QPSK, 16 QAM
Differential encoding	On/off
Symbol rate (in megabauds)	0.256, 0.512, 1.024, 2.048, 4.096
Preamble length	see Sec. P-13.7 of Ref. 8
Preamble content	Programmable, 1024 bits
FEC on/off	On/off
FEC codeword length, k (in bytes)	TBD[a]
FEC error correction, t (in bytes)	0 to 10
Burst length, m (in minislots)	TBD
Last codeword length	Fixed or shortened
Guard time	TBD

[a]TBD is "to be determined".

Source: Reference 8, Table P-13, IEEE 802.14, draft R2.

following burst, resulting in a nonzero transmitted signal envelope. The upstream transmissions from various stations are scheduled by the headend such that the center of the last symbol of one burst and the center of the first symbol of the preamble of an immediately following burst are separated by at least five symbols. The guard time is greater than or equal to the duration of TBD symbols plus the maximum timing error that is contributed by both station and headend.

14.7.8.6 Timing and Synchronization. The headend transmits in the downstream time-stamp messages that are used by a station to establish upstream TDMA synchronization. The transmission rate of these messages and their regularity are TBD.

14.7.8.6.1 Inaccuracy Tolerance. In order to properly synchronize the upstream transmissions originating from different stations in the TDMA mode, a ranging offset is applied by the station as a delay correction value to the headend time acquired at the station. This process is called *ranging*. The ranging offset is an advancement equal roughly to the round-trip delay of the station from the headend. Upon successful reception of one or more upstream transmissions from a station, the headend provides the station with a feedback message containing this ranging offset. The accuracy of the ranging offset should be no worse than TBD symbol duration, and resolution thereof is TBD of headend time increment. After the first iteration of ranging, the headend continues to send ranging adjustments when necessary to the station. A negative value for the ranging adjustment indicates that the ranging offset at that station is to be decreased, resulting in later times of transmission at the station. The station implements the ranging adjustment with a resolution of, at most, 1 symbol duration for the symbol rate in use for the given burst. In addition, the accuracy of the station burst transmission timing is TBD $\pm$ TBD symbol relative to the minislot boundaries that are derived at the station based on ideal processing of time-stamp message signals received from the headend.

14.7.8.7 Modulation and Bit Rates. QPSK and 16-QAM are modulation choices for upstream transmission.
Modulation rate refers to the rate at which the modulation symbols are transmitted over the physical medium. The unit of modulation rate is the *baud* (i.e., modulation symbols/second). Data rate refers to the rate at which the data bits are transmitted over the physical medium. The unit of data rate is *bits/second* (bps).
Table 14.8 tabulates the upstream data rates and modulation rates supported by the 802.14 standard.

14.7.8.8 Constellations. QPSK and 16-QAM constellations are to be used for upstream transmission. The 16-QAM capability is an option at the

TABLE 14.8   IEEE 802.14 Standard Data and Modulation Rates

Data Rate (Mbps)	QPSK Modulation Rate (Mbaud)	16-QAM Modulation Rate (Mbaud)
0.512	0.256	N/A
1.024	0.512	0.256
2.048	1.024	0.512
4.096	2.048	1.024
8.192	4.096	2.048
16.384	N/A	4.096

Source: Reference 8, Table P-14, IEEE 802.14, draft R2.

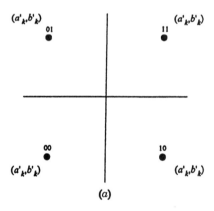

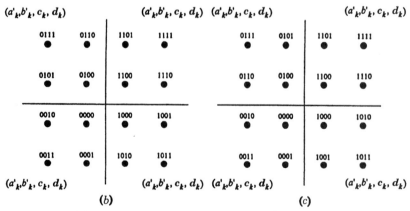

Figure 14.35. (*a*) QPSK, Gray or differentially encoded; (*b*) 16-QAM, differentially encoded; and (*c*) 16-QAM, Gray encoded.

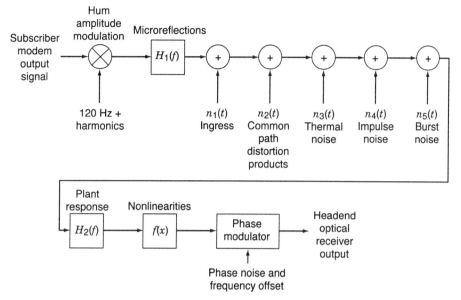

Figure 14.36. IEEE 802.14 upstream channel model. (Ref. 11).

headend cable modem but is mandatory for the station cable modem. For each modulation type, there are two options of encoding bits coming out of the Reed–Solomon block encoder into constellation symbols: (1) differential encoding and (2) Gray encoding.* The constellation diagrams are given in Figure 14.35.

14.7.8.9 IEEE 802.14 Upstream Channel Model. Figure 14.36 illustrates the upstream channel model detailing all the expected impairments.

REVIEW EXERCISES

1. Define a CATV headend. What are its functions?

2. Why must a CATV headend have a "high" tower?

3. List at least three impairments we can expect from a broadband CATV amplifier (downstream).

4. A common signal splitter divides a signal in half, splitting it into two equal power levels. If the input to a 3-dB splitter were − 7 dBm (in the

*Gray coding is discussed in Chapter 5.

power domain), the output would be -10 dBm. Is this really a true statement? Use common sense here.

5. What was the purpose of LOS microwave in earlier CATV systems?

6. What made such microwave facilities obsolete?

7. Why would a CATV system designer wish to keep the total bandwidth as narrow as possible?

8. What is the primary purpose of current set-top converters?

9. What does the term *beat* mean in CATV parlance?

10. Define *composite triple beat*.

11. A signal level is measured at 0.5 Vrms at some point in a CATV system. What is the equivalent value in dBmV?

12. What dBmV level can we expect in the CATV minimum noise model?

13. When calculating S/N, what is the value of the commonly used noise weighting factor in dB?

14. For CATV amplifiers, it has been found that there is an optimum gain value in the vicinity of 22 dB. Why is this? Examine two issues.

15. We have ten CATV identical amplifiers in cascade in a CATV system. The noise figure of an amplifier in this case is 7 dB. All terminations are 75 Ω and all noise derives from amplifiers (i.e., none from the coaxial cable). What would be the noise level in dBmV at the output of the last amplifier?

16. What is the system C/N, when there is a cascade of 18 trunk amplifiers with the following C/N values: 49 dB at the output of the 18th amplifier, 60 dB out a bridger, and 61 dB out of two line extenders?

17. What is an acceptable level (down) for Xm?

18. A certain system with 22 amplifiers in cascade has an Xm of -89 dB. What is the XM_{sys}?

19. Why are levels of a feeder system generally higher than the trunk system?

20. Define a *tap*.

21. Give three advantages of a HFC CATV system over a straight coaxial cable system.

22. Give two ways in which return (upstream) links can be handled.

23. Differentiate and characterize FM and AM optical fiber CATV super trunks.

24. Why does an AM fiber system have so few dB (often around only 10 dB) allocated for trunk/splice losses? (Back in Chapter 11 on fiber optics, we saw 40–50 dB allocated for fiber cable and fusion splice losses for PSTN digital transmission.)

25. Digitized raw video can be expected to have about what bit rate? Turning to Chapter 15, we find that video is sampled at a rate that is related to the color subcarrier, 3.58 MHz, not the highest modulating frequency. Oversample 4×.

26. From a bandwidth viewpoint, why is upstream at a disadvantage over downstream?

27. Considering exercise 26, how do we get around that disadvantage with a HFC system?

28. Explain qualitatively why upstream operation is at a noise disadvantage compared to downstream. Think thermal noise for this answer.

29. Give a general definition of ingress noise and its origin. Explain its cumulative effects upstream.

30. List at least four telecommunication services (media) that the IEEE 802.14 specification supports.

31. With IEEE 802.14, in a one-sentence statement differentiate upstream and downstream operation at the PHY layer.

32. In the IEEE 802.14 system, where does the master timing source reside and how does it disseminate time?

33. What is the purpose of *ranging*?

34. At the MAC layer of an 802.14 system, there are two types of information carrying units. What are they?

35. What is the purpose of the idle pattern for 802.14 downstream?

36. What is the basic unit of upstream bandwidth allocation for an 802.14 system?

37. How does an 802.14 station become synchronized?

38. Differentiate type A and type B downstream PHY for an 802.14 network.

39. What are the two types of modulation that may be used upstream for an 802.14 network?

40. List at least four impairments expected upstream in an 802.14 system.

REFERENCES

1. William O. Grant, *Cable Television*, 3rd ed., GWG Associates, Schoharie, NY, 1994.
2. Ken Simons, *Technical Handbook for CATV Systems*, 3rd ed., Jerrold Electronics Corp., Hatboro, PA, 1968.
3. Eugene R. Bartlett, *Cable Television Technology and Operations*, McGraw-Hill, New York, 1990.
4. D. N. Carson, "CATV Amplifiers: Figure of Merit and the Coefficient System," 1966 *IEEE International Convention Record, Part I, Wire and Data Communications*, pp. 87–97, IEEE, New York, Mar. 1966.
5. *System Planning, Product Specifications and Services*, Catalog #36, Andrew Corp., Orland Park, IL, 1994.
6. *Electrical Performance for Television Transmission Systems*, EIA/TIA-250C, Telecommunication Industry Association, Washington, DC, 1989.
7. *Digital Video Transmission Standard for Cable Television*, SCTE DVS 031, Society of Cable Telecommunications Engineers, Inc., Exton, PA, Oct. 25, 1996.
8. *Multimedia Modem Protocol for Hybrid Fiber–Coax Metropolitan Area Networks*, IEEE Std 802.14, draft R2, IEEE, New York, Mar. 1997.
9. *Lightwave Buyers' Guide Issue*, Pennwell Publishing Co., Tulsa OK, Mar. 15, 1997.
10. "Information Technology: Generic Coding of Moving Pictures and Associated Audio: Systems," ISO/IEC 13818-1, MPEG-2, ITU-T Rec. H.222.0, ISO, Geneva, Nov. 1994.
11. Private communication, Robert Fuller, Chairman, IEEE 802.14 Committee, Apr. 4, 1997.
12. "Digital Multi-Programme Systems for Television, Sound and Data Services for Cable Distribution," ITU-T Rec. J.83, ITU, Geneva, Sept. 1995.

15

VIDEO TRANSMISSION: TELEVISION

15.1 INTRODUCTION

The objective of this chapter is twofold: to provide the reader with an appreciation of how a video transmission system operates and to outline the basic aspects of video transmission, both analog and digital. "Basic aspects" in this context include standard transmission methods, impairments and their mitigation, system testing techniques, and U.S. and international standards.

In the early part of this chapter we concentrate on analog video transmission. Video from a TV camera with its subsequent application to a TV receiver is entirely analog in nature. From the video source to the end-user, video images may be converted to a digital format. So between the source and the end-user receiver display, the analog video signal is more and more often converted to digital, frequently using some sort of PCM technique. In the digital portion of the transmission link, compression can take place, anywhere from 3:1 to 30:1 or better.

Standard, conventional broadcast-quality TV is a bandwidth hog, requiring anywhere from 4 to 8 MHz of bandwidth to accommodate the analog video signal plus a color subcarrier and (an) aural channel(s). HDTV, in theory, will require much more bandwidth. However, HDTV will not be discussed because world standards have not been established.

Our interest in video transmission derives from the fact that the PSTN and private/enterprise networks transport video for broadcasters or other entities. Toward the end of the chapter there is an overview of MPEG-2, probably the leading method of digital television compression. Using an M-ary modulation scheme, the digital signal can be accommodated in a 6-MHz bandwidth. In the enterprise network environment, conference television has become very important. This chapter covers several methods of conference television transmission.

Leaving aside CATV, which is covered in Chapter 14, our interest focuses on point-to-point transmission. Television broadcast, except where it impacts point-to-point transmission, is left to other texts.

15.2   AN APPRECIATION OF VIDEO TRANSMISSION

A video transmission system must deal with four factors when transmitting images of moving objects:

- A perception of the distribution of luminance or simply the distribution of light and shade
- A perception of depth or a three-dimensional perspective
- A perception of motion relating to the first two factors above
- A perception of color (hues and tints)

Monochrome TV deals with the first three factors. Color TV includes all four factors.

A video transmission system must convert these three (or four) factors into electrical equivalents. The first three factors are integrated to an equivalent electric current or voltage whose amplitude is varied with time. Essentially, at any one moment it must integrate luminance from a scene in the three dimensions (i.e., width, height, and depth) as a function of time. And time itself is still another variable, for the scene is changing in time.

The process of integration of visual intelligence is carried out by *scanning*. The horizontal detail of a scene is transmitted continuously and the vertical detail discontinuously. The vertical dimension is assigned discrete values that become the fundamental limiting factor in a video transmission system.

The scanning process consists of taking a horizontal strip across the image on which discrete square elements called pels or pixels (picture elements) are scanned from left to right. When the right-hand end is reached, another, lower, horizontal strip is explored, and so on until the whole image has been scanned. Luminance values are translated on each scanning interval into voltage and current variations and are transmitted over the system. The concept of scanning by this means is shown in Figure 15.1.

The National Television Systems Committee (U.S.) (NTSC) practice divides an image into 525 horizontal scanning lines. It is the number of scanning lines that determines vertical detail or resolution of a picture.

When discussing picture resolution, the aspect ratio is the width-to-height ratio of the video image (see Section 15.2.1). The aspect ratio used almost universally is 4 : 3. In other words, a TV image 12 in. wide would necessarily be 9 in. high. Thus an image divided into 525 (491) vertical elements would then have 700 (652) horizontal elements to maintain an aspect ratio of 4 : 3. The numbers in parentheses represent the practical maximum active lines

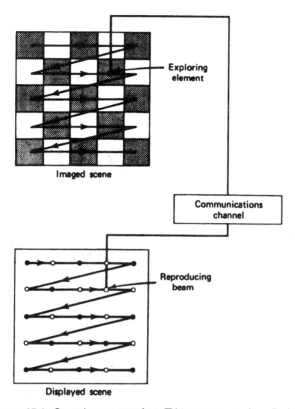

Figure 15.1. Scanning process from TV camera to receiver display.

and elements. Therefore the total number of elements approaches something on the order of 250,000. We reach this number because, in practice, vertical detail reproduced is 64–87% of the active scanning lines. A good halftone engraving may have as many as 14,400 elements per square inch, compared to approximately 3000 elements per square inch for a 9 by 12 in. TV image.

Motion is another variable factor that must be transmitted. The sensation of continuous motion in standard TV video practice is transmitted to the viewer by a successive display of still pictures at a regular rate similar to the method used in motion pictures. The regulate rate of display is called the *frame rate*. A frame rate of 25 frames per second will give the viewer a sense of motion, but on the other hand he/she will be disturbed by luminance flicker (bloom and decay), or the sensation that still pictures are "flicking" on screen one after the other. To avoid any sort of luminance flicker sensation, the image is divided into two closely interwoven (interleaving) parts, and each part is presented in succession at a rate of 60 frames per second, even though *complete* pictures are still built up at a 30 frame-per-second rate. It should be noted that interleaving improves resolution as well as apparent persistence of the cathode ray tube (CRT) by tending to reinforce the scanning spots. It has been found convenient to equate flicker frequency to power line frequency. Hence in North American practice, where power line frequency is 60 Hz, the flicker is 60 frames per second. In Europe it is 50 frames per second to correspond to the 50-Hz line frequency used there.

Following North American practice, some other important parameters derive from the previous paragraphs:

1. A field period is $\frac{1}{60}$ s. This is the time required to scan a full picture on every horizontal line.
2. The second scan covers the lines not scanned on the first period, offset one-half horizontal line.
3. Thus $\frac{1}{30}$ s is required to scan all lines on a complete picture.
4. The transmit time of exploring and reproducing scanning elements or spots along each scanning line is $\frac{1}{15,750}$ s (525 lines in $\frac{1}{30}$ s) = 63.5 μs.
5. Consider that about 16% of the 63.5 μs is consumed in flyback and synchronization. Accordingly, only about 53.3 μs are left per line of picture to transmit information.

What will be the bandwidth necessary to transmit images so described? Consider the worst case, where each scanning line is made up of alternate black and white squares, each the size of the scanning element. There would be 652 such elements. Scan the picture, and a square wave will result with a positive-going square for white and a negative for black. If we let a pair of adjacent square waves be equivalent to a sinusoid (see Figure 15.2), then the baseband required to transmit the image will have an upper cutoff of about 6.36 MHz, permitting no degradation in the intervening transmission system. The lower limit will be a dc or zero frequency.

Figure 15.2. Development of a sinusoid wave from the scan of adjacent squares.

1 Hz

15.2.1 Additional Definitions

Picture Element (pixel or pel). "The smallest area of a television picture capable of being delineated by an electric signal passed through the system or part thereof" (Ref. 26). A picture element has four important properties:

- P_v, the vertical height of the picture element
- P_h, the horizontal length of the picture element
- P_a, the *aspect ratio* of the picture element
- N_p, the total number of picture elements in an entire picture

The value of N_p is often used to compare TV systems.

In digital TV, a picture consists of a series of digital values that represent the points along the scanning path of an image. The digital values represent discrete points and we call these pixels (pels).

The resolution of a digital image is determined by its pixel counts, horizontal and vertical. A typical computer picture display might have 640 × 480 pixels.

Aspect Ratio. The ratio of the frame width to the frame height. This ratio is defined by the active picture. For standard NTSC television and PAL television and computers, the aspect ratio is 4:3 (1.33:1). Widescreen movies have a 16:9 aspect ratio, and HDTV is expected to also use a 16:9 aspect ratio.

15.3 THE COMPOSITE SIGNAL

The word *composite* is confusing in the TV industry. On one hand, composite may mean the combination of the full video signal plus the audio subcarrier; the meaning here is narrower. Composite in this case deals with the transmission of video information as well as the necessary synchronizing information.

Consider Figure 15.3. An image made up of two black squares is scanned. The total time for the line is 63.5 μs, of which 53.3 μs are available for the transmission of actual video information and 10.2 μs are required for synchronization and flyback.

During the retrace time or flyback it is essential that no video information be transmitted. To accomplish this, a blanking pulse is superimposed on the video at the camera. The blanking pulse carries the signal voltage into the reference black region. Beyond this region in amplitude is the "blacker than black" region, which is allocated to the synchronizing pulses. The blanking level (pulse) is shown in Figure 15.3.

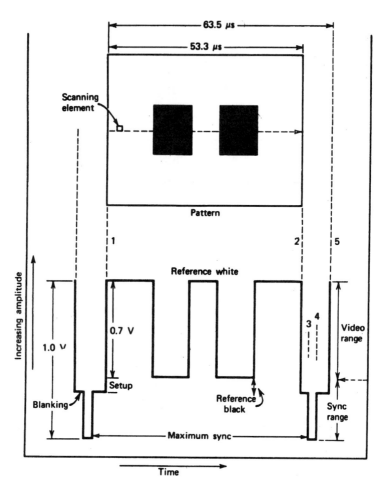

Figure 15.3. Breakdown in time of a scan line.

The maximum signal excursion of a composite video signal is 1.0 V. This 1.0 V is a video/TV reference and is always taken as a peak-to-peak measurement. The 1.0 V may be reached at maximum synchronizing voltage and is measured between synchronizing "tips."

Of the 1.0-V peak, 0.25 V is allotted for the synchronizing pulses and 0.05 V for the setup, leaving 0.7 V to transmit video information. Therefore the video signal varies from 0.7 V for the white-through-gray tonal region to 0 V for black. The best way to describe the actual video portion of a composite signal is to call it a succession of rapid nonrepeated transients.

The synchronizing portion of a composite signal is exact and well defined. A TV/video receiver has two separate scanning generators to control the position of the reproducing spot. These generators are called the horizontal and vertical scanning generators. The horizontal one moves the spot in the X or horizontal direction, and the vertical in the Y direction. Both generators control the position of the spot on the receiver and must in turn be controlled from the camera (transmitter) synchronizing generator to keep the receiver in step (synchronization).

The horizontal scanning generator in the video receiver is synchronized with the camera synchronizing generator at the end of each scanning line by means of horizontal synchronizing pulses. These are the synchronizing pulses shown in Figure 15.3, and they have the same polarity as the blanking pulses.

When discussing synchronization and blanking, we often refer to certain time intervals. These are discussed as follows:

- The time at the horizontal blanking pulse, 2–5 in Figure 15.3, is called the *horizontal synchronizing interval.*
- The interval 2–3 in Figure 15.3 is called the *front porch.*
- The interval 4–5 is the *back porch.*

The intervals are important because they provide isolation for overshoots of video at the end of scanning lines. Figure 15.4 illustrates the horizontal synchronizing pulses and corresponding porches.

The vertical scanning generator in the video/TV receiver is synchronized with the camera (transmitter) synchronizing generator at the end of each field by means of vertical synchronizing pulses. The time interval between successive fields is called the vertical interval. The vertical synchronizing pulse is built up during this interval. The scanning generators are fed by differentiation circuits. Differentiation for the horizontal scan has a relatively short time constant (RC) and that for the vertical a comparatively long time constant. Thus the long-duration vertical synchronization may be separated from the comparatively short-duration horizontal synchronization. This method of separation of synchronization, known as *waveform separation*, is standard in North America.

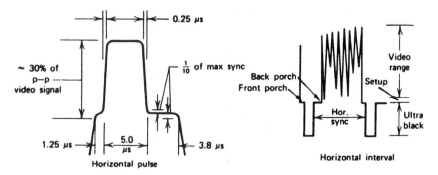

Figure 15.4. Sync pulses and porches.

In the composite video signal (North American standards) the horizontal synchronization has a repetition rate of 15,750 frames per second, and the vertical synchronization has a repetition rate of 60 frames per second (Refs. 1, 2).

15.4   CRITICAL VIDEO TRANSMISSION IMPAIRMENTS

The nominal video baseband is divided into two segments: the high-frequency (HF) segment, above 15,750 Hz, and the low-frequency (LF) segment, below 15,750 Hz. Impairments in the HF segment are operative (in general) along the horizontal axis of the received video image. LF impairments are generally operative along the vertical image axis. Table 15.1 lists critical impairments and their corresponding causes.

The preceding sections presented a short explanation of the mechanics of video transmission for a video camera directly connected to a receiving device for display. The primary concern of this chapter, however, is to describe and discuss the problems of point-to-point video transmission. Often the entity responsible for the point-to-point transport of TV programs is not the same entity that originated the image transmitted, except in the case of that link directly connecting the studio to a local transmitter (STL). A great deal of the "why" has now been covered. The remaining parts of the chapter discuss the video transmission problem (the "how") on a point-to-point basis. The medium employed for this purpose may be radiolink, satellite link, coaxial cable, fiber-optic cable, or specially conditioned wire pairs.

15.5   CRITICAL VIDEO PARAMETERS

15.5.1   General

Raw video baseband transmission requires excellent frequency response, in particular from dc to 15 kHz and extending to 4.2 MHz for North American systems and to 5 MHz for European systems. Equalization is extremely important. Few point-to-point circuits are transmitted at baseband because transformers are used for line coupling, which deteriorate low-frequency response and make phase equalization very difficult.

To avoid low-frequency deterioration, cable circuits transmitting video have resorted to the use of carrier techniques and frequency inversion using vestigial sideband (VSB) modulation. However, if raw video baseband is transmitted, care must be taken in preserving its dc component (Ref. 4).

15.5.2   Transmission Standard—Level

Standard power levels have developed from what is roughly considered to be the input level to an ordinary TV receiver for a noise-free image. This is a

TABLE 15.1   Critical Video Transmission Impairments

Impairment	Cause
HF Segment	
Undistorted echos	Cyclic gain and phase deviation throughout the passband
Distorted echos	Nonlinear gain and phase deviations, especially in the higher end of the band
HF cutoff effects, ringing	1. Limited bandpass distorts and shows picture transitions, causing overshoot and undershoot
	2. This type of distortion (ringing) may also show up on test pattern from lack of even energy distribution and reduced resolution
Porch distortion (poor reproduction of porches)	Poor attenuation and phase distortion
Porch displacement	Zero wander, dc-restored devices such as clamper circuits
Smearing (blurring of the vertical edges of objects)	1. Coarse variations in attenuation and phase
	2. Quadrature distortion
LF Segment	
dc Suppression, zero wander, distorted image	Lack of clamping
LF roll-off (gradual shading from top to bottom)	Poor clamping, deterioration of coupling networks
Streaking	Phase and attenuation distortion, usually from transmission medium
Nonlinear Distortion	
Nonlinearity in the extreme negative region, resulting in horizontal striations or streaking	Compression of synchronizing pulses
Impairments Due to Noise	

Noise, in this case, may be considered an undesirable visual sensation. Noise is considered to consist of three types: single frequency, random, and impulse.

Unwanted pattern in received picture	Single-frequency noise
Pattern in alternate fields	Single-frequency noise as an integral multiple of field frequency (50 or 60 Hz)
Horizontal or vertical bars	Single-frequency noise
It should be noted that the LF region is very sensitive to single-frequency noise	
Picture graininess, "snow"	As random noise increases, graininess of picture increases

TABLE 15.1   (*Continued*)

Impairment	Cause
Impairments Due to Noise	

Note: System design objective of 47-dB signal-to-weighted-noise ratio or 44-dB signal-to-flat-noise ratio, based on 4.2-MHz video bandwidth, meets Television Allocation Study Organization (TASO) rating of excellent picture. Peak noise should be 37 dB below peak video. Signal-to-noise ratios are taken at peak synchronizing tips to rms noise.

Noise "hits," momentary loss of synchronization, momentary rolling, momentary masking of picture	Impulse noise

Note: Bell System limit at receiver input −20-dB reference to 1-V peak-to-peak signal level point, 1 hit/min. Large amplitudes of impulse noise often masked in black.

Weak, extraneous image superimposed on main image	Strong crosstalk
Nonsynchronization of two images causes violent horizontal motion	
Effect is most noticeable in line and field synchronization intervals	

Note: Limiting loss in crosstalk coupling path should be 58 dB or greater for equal signal levels, design objective 61 dB. Crosstalk for video may be defined as the coupling between two TV channels.

Sources: References 2 and 3.

---

1 mV across 75 Ω. With this as a reference, TV levels are given in dBmV. For RF and carrier systems carrying video, the measurement refers to rms voltage. For raw video it is 0.707 of instantaneous peak voltage, usually taken on synchronizing tips.

The signal-to-noise ratio is normally expressed for video transmission as

$$\frac{S}{N} = \frac{\text{peak signal (dBmV)}}{\text{rms noise (dBmV)}} \tag{15.1}$$

TASO picture ratings (4-MHz bandwidth) are related to the signal-to-noise ratio (RF) as follows (Ref. 5):

1. Excellent (no perceptible snow)   45 dB
2. Fine (snow just perceptible)   35 dB
3. Passable (snow definitely perceptible but not objectionable)   29 dB
4. Marginal (snow somewhat objectionable)   25 dB

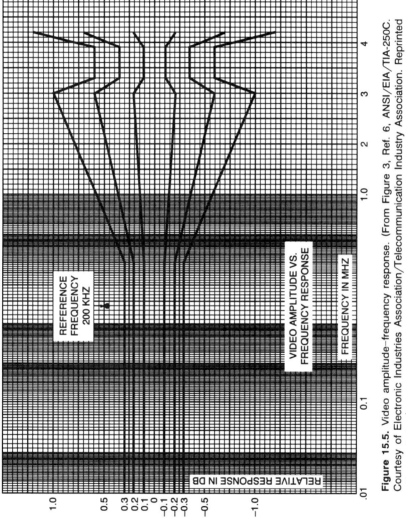

Figure 15.5. Video amplitude–frequency response. (From Figure 3, Ref. 6, ANSI/EIA/TIA-250C. Courtesy of Electronic Industries Association/Telecommunication Industry Association. Reprinted with permission.)

1037

15.5.3   Other Parameters

For black and white video systems there are four critical transmission parameters:

1. Amplitude–frequency response.
2. EDD (group delay).
3. Transient response.
4. Noise (thermal, intermodulation distortion [IM], crosstalk, and impulse).

Color transmission requires consideration of two additional parameters:

5. Differential gain.
6. Differential phase.

A description of amplitude–frequency response (attenuation distortion) may be found in Section 1.9.3. Because video transmission involves such wide bandwidths compared to the voice channel and because of the very nature of video itself, both phase and amplitude requirements are much more stringent.

Transient response is the ability of a system to "follow" sudden, impulsive changes in signal waveform. It usually can be said that if the amplitude–frequency and phase characteristics are kept within design limits, the transient response will be sufficiently good.

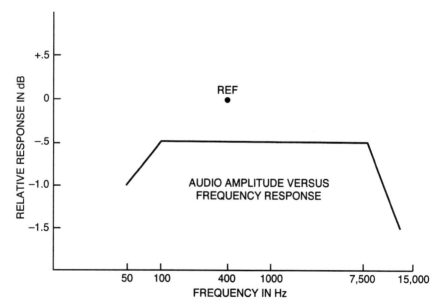

Figure 15.6. Amplitude versus frequency response for the audio or aural channel. (From Ref. 6, Figure 18, ANSI/EIA/TIA-250C. Courtesy of Electronic Industries Association/Telecommunication Industry Association. Reprinted with permission.)

TABLE 15.2   End-to-End Performance Parameter Summary

Characteristic	Short Haul	Medium Haul	Satellite	Long Haul	End-to-End
Video Signal Channel					
Amplitude response as a function of frequency			See Figure 15.5		
Chrominance-to-luminance gain inequality	±2 IRE	±4 IRE	±4 IRE	±7 IRE	±7 IRE
Chrominance-to-luminance delay inequality	±20 ns	±33 ns	±26 ns	±54 ns	±60 ns
Envelope delay response as a function of frequency			Provisional 50 ns between 200 kHz and 3.6 MHz		
Field time waveform distortion			3 IRE maximum from flat		
Line time waveform distortion	0.5 IRE	1 IRE	1 IRE	1.5 IRE	2 IRE
Short time waveform distortion	2% SD	2% SD	2% SD	3% SD	3% SD
Damped low-frequency distortion (bounce)					
Without terminal clamping			35 IRE overshoot with 5-s settling		
With terminal clamping			8 IRE overshoot with 3-s settling		
Line-by-line dc offset (piano keying)			2 IRE		
Insertion gain and variation					
Gain			0 dB		
Variation	±0.15 dB	±0.3 dB	±0.2 dB	±0.45 dB	±0.5 dB
Luminance nonlinearity	2%	4%	6%	8%	10%
Differential gain	2%	5%	4%	8%	10%
Differential phase	0.7°	1.3°	1.5°	2.5°	3°
Chrominance-to-luminance intermodulation	1 IRE	2 IRE	2 IRE	4 IRE	4 IRE
Chrominance nonlinear gain	1 IRE	2 IRE	2 IRE	4 IRE	5 IRE
Chrominance nonlinear phase	1°	2°	2°	4°	5°
Dynamic gain of video signal	2 IRE	3 IRE	4 IRE	5 IRE	6 IRE
Dynamic variation of the synchronizing signal	1.2 IRE	1.6 IRE	2 IRE	2.4 IRE	2.8 IRE
Transient synchronizing signal nonlinearity	1 IRE	2 IRE	3 IRE	4 IRE	5 IRE
Signal-to-noise ratio	67 dB	60 dB	56 dB	54 dB	54 dB
Signal to low-frequency noise ratio	53 dB	48 dB	50 dB	44 dB	43 dB
Signal to periodic-noise ratio	67 dB	62 dB	63 dB	58 dB	57 dB
Availability of the video signal			99.99%		
Audio Signal Channel					
Amplitude response as a function of frequency			See Figure 15.6		
Total harmonic distortion			0.5%		
Signal-to-noise ratio	66 dB	65 dB	58 dB	57 dB	56 dB
Insertion gain			0 dB		
Gain difference between stereo A and B channels					
50–100 Hz			1.0 dB		
100–7500 Hz			0.5 dB		
7500 Hz to 15 kHz			1.0 dB		

TABLE 15.2   *(Continued)*

Characteristic	Short Haul	Medium Haul	Satellite	Long Haul	End-to-End
		Audio Signal Channel			
Phase difference between stereo A and B channels					
50–100 Hz			10°		
100–7500 Hz			3°		
7500 Hz to 15 kHz			10°		
Crosstalk coupling loss between stereo A and B channels					
50 Hz to 15 kHz			56 dB		
Availability of the audio signal			99.99%		
Audio to video transmission time differential			25-ms lead to 40-ms lag		

Notes: ANSI/EIA/TIA-250C is valid only for North American NTSC television. The video signal occupies the spectrum of at least 4.2 MHz.

Definitions

Short-Haul Transmission System. A system for the transmission of television signals between two points, possibly via a number of repeaters. The distance between the two points can range from a few feet to about 20 route miles.

Medium-Haul Transmission System. A system for the transmission of television signals between two points, possibly via a number of repeaters. The distance between the two points can range from about 20 miles to about 150 route miles.

Long-Haul Transmission System. A system for the transmission of television signals between two points via a number of repeaters. The distance between the two points can range from about 150 route miles to about 3000 route miles.

Satellite Transmission System. A system for the transmission of television signals between a transmitting earth station and a receiving earth station via a satellite repeater.

Digital Transmission System. A system for the transmission of television signals that digitally encodes the television signal at or near the point of entry and reconstructs (decodes) the television signal at or near the point of exit.

End-to-End Network. An interconnection of systems for the transmission of television signals between two points. An end-to-end network includes long-haul or satellite systems in tandem with combinations of medium- and short-haul analog systems or digital systems.

IRE. Defined as $\frac{1}{100}$ part of the luminance (blanking to reference white) range. The zero IRE is at the blanking level and 100 IRE at reference white level. IRE below blanking level is referred to in negative values.

Average Picture Level (APL). The average signal level during active scanning time, excluding blanking and synchronizing signals, integrated over a frame period. It is expressed as a percentage of the blanking to reference white range (100 IRE).

Source: Reference 6, Table 1, ANSI/EIA/TIA-250C.

Noise is described in Section 1.9.6. Differential gain is the variant in the gain of the transmission system as the video signal level varies (i.e., as it traverses the extremes from black to white). Differential phase is any variation in phase of the color subcarrier as a result of a changing luminance level. Ideally, variations in the luminance level should produce no changes in either the amplitude or the phase of the color subcarrier. Table 15.2 summarizes end-to-end critical transmission parameters for video.

15.6 VIDEO TRANSMISSION STANDARDS (CRITERIA FOR BROADCASTERS)

The following outlines video transmission standards from the point of view of broadcasters (i.e., as emitted from TV broadcast transmitters). Figure 15.7 illustrates the components of the emitted wave (North American practice).

15.6.1 Basic Standards

Tables 15.3A and 15.3B give a capsule summary of some basic national standards as taken from ITU-R Rec. BT.470-4 (Ref 27).

15.6.2 Variances to National Standards

See Tables 15.3C and 15.3D for variances to national standards given in Section 15.6.1. Table 15.3E serves as a supplement to these tables.

15.6.3 Color Transmission

Three color transmission standards exist:

NTSC   National Television System Committee (North America, Japan)
SECAM   Sequential color and memory (Europe)
PAL   Phase alternation line (Europe)

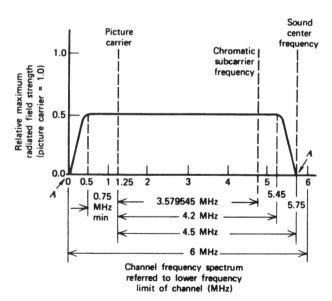

Figure 15.7. RF amplitude characteristics of TV picture transmission. Field strength at points *A* shall not exceed 20 dB below picture carrier. Drawing not to scale.

TABLE 15.3A   United States Standard (Refs. 7, 8)

Channel width	
(Transmission)	6 MHz
Video	4.2 MHz
Aural	$\pm$ 25 kHz
(see Figure 15.7)	
Picture carrier location	1.25 MHz above lower boundary of channel
Modulation	AM composite picture and synchronizing signal on visual carrier together with FM audio signal on audio carrier
Scanning lines	525 per frame, interlaced 2 : 1
Scanning sequence	Horizontally from left to right, vertically from top to bottom
Horizontal scanning frequency	15,750 Hz for monochrome, or 2/455 $\times$ chrominance subcarrier, = 15,734.264 $\pm$ 0.044 Hz for NTSC color transmission
Vertical scanning frequency	60 Hz for monochrome, or 2/525 $\times$ horizontal scanning frequency for color = 59.95 Hz
Blanking level	Transmitted at 75 $\pm$ 25% of peak carrier level
Reference black level	Black level is separated from blanking level by 7.5 $\pm$ 2.5% of video range from blanking level to reference white level
Reference white level	Luminance signal of reference white is 12.5 $\pm$ 2.5% of peak carrier
Peak-to-peak variation	Total permissible peak-to-peak variation in one frame due to all causes is less than 5%
Polarity of transmission	Negative; a decrease in initial light intensity causes an increase in radiated power
Transmitter brightness response	For monochrome TV, RF output varies in an inverse logarithmic relation to brightness of scene
Aural transmitter power	Maximum radiated power is 20% (minimum 10%) of peak visual transmitter power

The systems are similar in that they separate the luminance and chrominance information and transmit the chrominance information in the form of two color difference signals, which modulate a color subcarrier transmitted within the video band of the luminance signal. The systems vary in the processing of chrominance information.

In the NTSC system, the color difference signals I and Q amplitude-modulate subcarriers that are displaced in phase by $\pi/2$, giving a suppressed carrier output. A burst of the subcarrier frequency is transmitted during the horizontal back porch to synchronize the color demodulator.

In the PAL system, the phase of the subcarrier is changed from line to line, which requires the transmission of a switching signal as well as a color burst.

TABLE 15.3B   Basic European Standard (Refs. 7, 8)

Channel width (transmission)	7 or 8 MHz
Video	5, 5.5, and 6 MHz
Aural	FM, ± 50 kHz
Picture carrier location	1.25 MHz above lower boundary of channel

Note: VSB transmission is used, similar to North American practice

Modulation	AM composite picture and synchronizing signal on visual carrier together with FM audio signal on audio carrier
Scanning lines	625 per frame, interlaced 2 : 1
Scanning sequence	Horizontally from left to right, vertically from top to bottom
Horizontal scanning frequency	15,625 Hz $\pm$ 0.1%
Vertical scanning frequency	50 Hz
Blanking level	Transmitted at 75 $\pm$ 2.5% of peak carrier level
Reference black level	Black level is separated from blanking by 3–6.5% of peak carrier
Peak white level as a percentage of peak carrier	10–12.5%
Polarity of transmission	Negative; a decrease in initial light intensity causes an increase in radiated power
Aural transmitter power	Maximum radiated power is 20% of peak visual power

In the SECAM system, the color subcarrier is frequency modulated alternately by the color difference signals. This is accomplished by an electronic line-to-line switch. The switching information is transmitted as a line-switching signal.

15.6.4   Standardized Transmission Parameters* (Point-to-Point TV)

Interconnection at Video Frequencies

Impedance	75 Ω unbalanced or 124 Ω balanced (resistive)
Return loss	No less than 30 dB
Nominal signal amplitude	1 V peak to peak (monochrome)
Nominal signal amplitude	1.25 V peak to peak, maximum (composite color)
Polarity	Black-to-white transitions, positive going

*Based on Refs. 9 and 10.

TABLE 15.3C Basic Characteristics of Video and Synchronizing Signals

Item	Characteristics	System (see Table 15.3E for country code)								ITU-R Rec. BT.472 (Note 2)
		M	N (Note 1)	B, G	H	I	D, K	K1	L	
1	Number of lines per picture (frame)	525	625	625	625	625	625	625	625	625
2	Frequency, nominal value (field/s) (Note 3)	60 (59.94)	50	50	50	50	50	50	50	50
3	Line frequency f_H and tolerance when operated non-synchronously (Hz) (Notes 3 and 4)	15,750 (15,734.264) ±0.0003%	15,625 ±0.15% (±0.00014%)	15,625 (Note 5) ±0.02% (±0.0001%)	15,625 ±0.02% (±0.0001%)	15,625 ±0.00002% (Note 6)	15,625 (Note 5) ±0.02% (±0.0001%)	15,625 ±0.03% (±0.0001%)	15,625 ±0.02% (±0.0001%)	15,625 ±0.02% (±0.0001%)
3(a)	Maximum variation rate of line frequency valid for monochrome transmission (%/s) (Notes 7 and 8)	0.15		0.05	0.05	0.05	0.05	0.05	0.05	
4 (Note 9)	Blanking level (reference level)	0	0	0	0	0	0	0	0	
	Peak white level	100	100	100	100	100	100	100	100	
	Synchronizing level	−40	−40 (−43)	−43	−43	−43	−43	−43	−43	
	Difference between black and blanking level	7.5 ± 2.5 (Note 10)	7.5 ± 2.5 (0)	0	0	0	0−7	0 (color) 0−7(mono.)	0 (color) 0−7(mono.)	0^{+5}_{-0}
	Peak level including chrominance signal	120	133	133 (Note 11)		133	115 (Note 12)	115 (Note 12)	124 (Note 12)	
5	Assumed gamma of display device for which precorrection of monochrome signal is made	2.2	2.2 (2.8)		2.8 (Note 13)				(Note 14)	

6	Nominal video band-width (MHz)	4.2	4.2	5	5.5	6	6	6	5.0 or 5.5 or 6.0
7	Line synchronization					See Table 1-1 of Ref. 7			
8	Field synchronization					See Table 1-2 of Ref. 7			

Note 1. The values in parentheses apply to the combination N/PAL used in Argentina (PAL = phase alternation line).

Note 2. Figures are given for comparison.

Note 3. Figures in parentheses are valid for color transmission.

Note 4. In order to take full advantage of precision offset when the interfering carrier falls in the sideband of the upper video range (greater than 2 MHz) of the wanted signal, a line-frequency stability of at least 2×10^{-7} is necessary.

Note 5. The exact value of the tolerance for line frequency when the reference of synchronism is being changed requires further study.

Note 6. When the reference of synchronism is being changed, this may be relaxed to 15,625 $\pm$ 0.02%.

Note 7. These values are not valid when the reference of synchronism is being changed.

Note 8. Further study is required to define maximum variation rate of line frequency for color transmission. See in this regard CCIR, 1978–82. In the U.K. this is 0.1 Hz/s (CCIR 1982–86b).

Note 9. It is also customary to define certain signal levels in 625-line systems, as follows:

Synchronizing level = 0
Blanking level   = 30
Peak white level   = 100

For this scale, the peak level including chrominance signal for system D, K/SECAM (SECAM = sequential color and memory) = 110.7. (See CCIR, 1982–86a.)

Note 10. The values used in Japan are 0^{+10}_{-0}.

Note 11. Value applies to PAL signals.

Note 12. Values apply to SECAM signals. For program exchange the value is 115.

Note 13. Assumed value for overall gamma is approximately 1.2. The gamma of the picture tube is defined as the slope of the curve giving the logarithm of the luminance reproduced as a function of the logarithm of the video signal voltage when the brightness control of the receiver is set so as to make this curve as straight as possible in a luminance range corresponding to a contrast of at least 1/40.

Note 14. In Rec. ITU-R BT.472, a gamma value for the picture signal is given as approximately 0.4.

Source: Reference 7, Table I, ITU-R Rec. BT.470-4.

Interconnection at Intermediate Frequency (IF)

Impedance	75 Ω unbalanced
Input level	0.3 V rms
Output level	0.5 V rms
IF up to 1 GHz	35 MHz
IF above 1 GHz	70 MHz

Signal-to-weighted noise: 53 dB

15.7 METHODS OF PROGRAM CHANNEL TRANSMISSION FOR TELEVISION

Composite transmission normally is used on broadcast and community antenna television (CATV [cable]) distribution. Video and audio carriers are "combined" before being fed to the radiating antenna for broadcast. These audio subcarriers are described in Section 15.6.

For point-to-point transmission on coaxial cable, radiolink, and earth station systems, the audio program channel is generally transmitted separately from its companion video providing the following advantages:

- Individual channel level control
- Greater control over crosstalk
- Increased guardband between video and audio
- Saves separation at broadcast transmitter
- Leaves TV studio at separate channel
- Permits individual program channel preemphasis

15.8 TRANSMISSION OF VIDEO OVER RADIOLINKS

15.8.1 General

Telephone administrations increasingly are expanding their offerings to include other services. One such service is to provide point-to-point broadcast-quality video relay on a lease basis over radiolinks. As covered earlier in this chapter, video transmission requires special consideration.

The following paragraphs summarize the special considerations a planner must taken into account for video transmission over radiolinks.

Raw video baseband modulates the radiolink transmitter. The aural channel is transmitted on a subcarrier well above the video portion. The overall subcarriers are themselves frequency modulated. Recommended subcarrier frequencies may be found in CCIR Rec. 402-2 (Ref. 11) and Rep. 289-4 (Ref. 12).

TABLE 15.3D Characteristics of the Radiated Signals Monochrome and Color

Item	Characteristics	M	N (Note 1)	B, G	H	I	D, K	K1	L^a
1	Frequency spacing Nominal radio-frequency channel bandwidth (MHz)	6	6	B:7 G:8	8	8	8	8	8
2	Sound carrier relative to vision carrier (MHz)	+4.5 (Note 2)	+4.5	+5.5 ±0.001 (Notes 3–6)	+5.5	+5.9996 ±0.0005 (Note 7)	+6.5 ±0.001	+6.5	+6.5
3	Nearest edge of channel relative to vision carrier (MHz)	−1.25	−1.25	−1.25	−1.25	−1.25	−1.25	−1.25	−1.25
4	Nominal width of main sideband (MHz)	4.2	4.2	5	5	5.5	6	6	6
5	Nominal width of vestigial sideband (MHz)	0.75	0.75	0.75	1.25	1.25	0.75	1.25	1.25
6	Minimum attenuation of vestigial sideband (dB at MHz) (Note 8)	20(−1.25) 42(−3.58)	20(−1.25) 42(−3.5)	20(−1.25) 20(−3.0) 30(−4.43) (Note 9)	20(−1.75) 20(−3.0)	20(−3.0) 30(−4.43)	20(−1.25) 30(−4.33) ±0.1) (Notes 10, 11)	20(−2.7) 30(−4.3)	15(−2.7) 30(−4.3)
7	Type and polarity of vision modulations	C3F neg.	C3F neg.	C3F neg.	C3F neg.	C3F neg.	c3F neg.	ref.: 0(+0.8) C3F neg.	ref.: 0(+0.8) C3F pos.
8	Levels in the radiated signal (% of peak carrier) Synchronizing level	100	100	100	100	100	100	100	< 6
	Blanking level	72.5 to 77.5 (75 ± 2.5)	72.5 to 77.5 (75 ± 2.5)	75 ± 2.5 (Note 12)	72.5 to 77.5	76 ± 2	75 ± 2.5	75 ± 2.5	30 ± 2
	Difference between black level and blanking level	2.88 to 6.75 (Note 13)	2.88 to 6.75	0 to 2 (nominal)	0 to 7	0 (nominal)	0 to 4.5 (Note 14)	0 to 4.5	0 to 4.5
	Peak white-level	10 to 15	10 to 15 (10 to 12.5)	10 to 15 (Notes 12, 15)	10 to 12.5	20 ± 2	10 to 12.5 (Notes 16, 17)	10 to 12.5	100 (≈ 110) (Note 18)
9	Type of sound modulation	F3E	F3E	F3E	F3E	F3E	F3E	F3E	A3E
10	Frequency deviation (kHz)	±25	±25	±50	±50	±50	±50	±50	
11	Preemphasis for modulation (μs)	75	75	50	50	50	50	50	
12	Ratio of effective radiated powers of vision and (primary) sound (Note 19)	10/1 to 5/1 (Note 20)	10/1 to 5/1	20/1 to 10/1 (Notes 3, 6, 21, 22)	5/1 to 10/1	5/1 10/1 (Notes 7, 23) 20/1 (Note 24)	10/1 to 5/1 (Note 25)	10/1	10/1 10/1 to 40/1 (Note 26)

1047

TABLE 15.3D (Continued)

Item	Characteristics	M	N (Note 1)	B, G	H	I	D, K	K1	L[a]
13	Precorrection for receiver group delay characteristics at medium video frequencies (ns)	0	$\left\{\begin{array}{l}1\ \text{MHz}\ 0\pm100\\ 1\ \text{MHz}\ 0\pm100\\ 1\ \text{MHz}\ 0\pm60\end{array}\right\}$ Note 27				Note 28		
14	Precorrection for receiver group-delay characteristics at color subcarrier frequency (ns)	-170 (nominal)	$\left(-170\begin{array}{l}+60\\ -40\end{array}\right)$ $\begin{array}{l}-170\\ (\text{nominal})\\ (\text{Note 27})\end{array}$				Note 29		

[a]See Table 15.3E for country codes.

Note 1. The values in brackets apply to the combination N/PAL used in Argentina.

Note 2. In Japan, the values +4.5 ± 0.001 are used.

Note 3. In the Federal Republic of Germany, Italy, the Netherlands, and Switzerland a system of two sound carriers is used, the frequency of the second carrier being 242.1875 kHz above the frequency of the first sound carrier. The ratio between vision/sound erp for this second carrier is 100/1. For further information on this system see Rec. ITU-BS.707. For stereophonic sound transmissions a similar system is used in Australia with vision/sound power ratios being 20/1 and 100/1 for the first and second sound carriers respectively.

Note 4. New Zealand uses a sound carrier displaced 5.4996 ± 0.0005 MHz from the vision carrier.

Note 5. The sound carrier for single carrier sound transmissions in Australia may be displaced 5.5 ± 0.005 MHz from the vision carrier.

Note 6. In Denmark, Finland, New Zealand, Sweden, and Spain a system of two sound carriers is used. In Iceland and Norway the same system is being introduced. The second carrier is 5.85 MHz above the vision carrier and is DQPSK modulated with 728 kbps sound and data multiplex. The ratios between vision/sound power are 20/1 and 100/1 for the first and second carrier respectively. For further information, see Rec. ITU-R BS.707.

Note 7. In the United Kingdom, a system of two sound carriers is used. The second sound carrier is 6.552 MHz above the vision carrier and is DQPSK modulated with a 728 kbps sound and data multiplex able to carry two sound channels. The ratio between vision and sound erp for the second carrier is 100/1.

Note 8. In some cases, low-power transmitters are operated without vestigial-sideband filter.

Note 9. For B/SECAM and G/SECAM: 30 dB at −4.33 MHz, within the limits of ±0.1 MHz.

Note 10. In some countries, members of the Organization for International Radio and Television (OIRT), additional specifications are in use:
  (a) not less than 40 dB at −4.286 MHz ± 0.5 MHz,
  (b) 0 dB from −0.75 MHz to +6.0 MHz,
  (c) not less than 20 dB at ±6.375 MHz and higher.
     Reference: 0 dB at +1.5 MHz.

Note 11. In the People's Republic of China, the attenuation value at the point (-4.33 ± 0.1) has not yet been determined.

Note 12. Australia uses the nominal modulation levels specified for system I.

Note 13. In Japan, the value of 0 to 6.75 have been adopted.

Note 14. In the People's Republic of China, the values 0 to 5 have been adopted.

Note 15. Italy is considering the possibility of controlling the peak white-level after weighting the video frequency signal by a low-pass filter, so as to take account only of those spectrum components of the signal that are likely to produce intercarrier noise in certain receivers when the nominal level is exceeded. Studies should be continued with a view to optimizing the response of the weighting filter to be used.

Note 16. The former USSR has adopted the value 15 ± 2%.

Note 17. A new parameter "white level with subcarrier" should be specified at a later date. For that parameter, Russia has adopted the value of 7 ± 2%.

Note 18. The peak white-level refers to a transmission without color sub-carrier. The figure in brackets corresponds to the peak value of the transmitted signal, taking into account the color subcarrier of the respective color television system.

Note 19. The values to be considered are:

• the r.m.s. value of the carrier at the peak of the modulation envelope for the vision signal. For system L, only the luminance signal is to be considered. (See Note 15);

• the r.m.s. value of the unmodulated carrier for amplitude-modulated and frequency-modulated sound transmissions.

Note 19. In Japan, a ratio of 1/0.15 to 1/0.35 is used. In the United States, the sound carrier erp is not to exceed 22% of the peak authorized vision erp.

Note 20. It may be that the Austrian Administration will continue to use a 5/1 power ratio in certain cases, when necessary.

Note 22. Recent studies in India confirm the suitability of a 20/1 ratio of effective radiated powers of vision and sound. This ratio still enables the introduction of a second sound carrier.

Note 23. The ratio 10/1 is used in the Republic of South Africa and in the United Kingdom.

Note 24. In the People's Republic of China, the value 10/1 has been adopted.

Note 25. In the United Kingdom it is planned to make a limited use of a ratio of 20/1 for the primary sound carrier on an experimental basis.

Note 26. In France, experimental.

Note 27. In Germany and the Netherlands the correction for receiver group-delay characteristics is made according to curve B in Figure 3a of the reference document. Spain uses curve A. The OIRT countries using the B/SECAM and G/SECAM systems use a nominal precorrection of 90 ns at medium video frequencies. In Sweden, the precorrection is 0 ± 40 ns up to 3.6 MHz. For 4.43 MHz, the correction is −170 ± 20 ns and for 5 MHz it is −350 ± 80 ns. In New Zealand the precorrection increases linearly from 0 ± 20 ns at 0 MHz to 60 ± 50 ns at 2.25 MHz, follows curve A of Fig. 3a from 2.25 MHz to 4.43 MHz and then decreases linearly to −300 ± 75 ns at 5 MHz. In Australia, the nominal precorrection follows curve A up to 2.5 MHz, then decreases to 0 ns at 3.5 MHz, −170 ns at 4.43 MHz and −280 ns at 5 MHz. Based on studies on receivers in India, the receiver group delay pre-equalization proposed to be adopted in India at 1 MHz, 2 MHz, 3 MHz, 4.43 MHz, and 4.8 MHz is +125 ns, +150 ns, +142 ns, −75 ns, and −200 ns respectively. In Denmark, the precorrections at 0, 0.25, 1.0, 2.0, 3.0, 3.8, 4.43, and 4.8 Hz are 0, +5, +53, +75, 0, −170, and 400 ns.

Note 28. Not yet determined. The Czechoslovak Socialist Republic proposes +90 ns (nominal value).

Note 29. Not yet determined. The Czechoslovak Socialist Republic proposes +25 ns (nominal value).

Source: Reference 7, Table III, ITU-R Rec. BT. 470-3.

TABLE 15.3E   Supplement to Tables 15.3C and 15.3D

Country	System Used in Bands[a]	
	VHF	UHF
Germany, Austria, Netherlands	B/PAL	G/PAL
Argentina	N/PAL	N/PAL
Australia	B/PAL	B/PAL
Belgium	B/PAL	G/PAL
Brazil	M/PAL	M/PAL
Canada, Chile	M/NTSC	M/NTSC
Ethiopia	B, G/PAL	G/PAL
China (PRC)	D/PAL	D/PAL
Colombia	M/NTSC	M
Korea, Costa Rica	M/NTSC	M/NTSC
Denmark, Yugoslavia, Algeria	B/PAL	G/PAL
Egypt	B/PAL	G/PAL
Spain	B/PAL	G/PAL
United States, Mexico, Peru	M/NTSC	M/NTSC
Finland	B/PAL	G/PAL
France	L/SECAM	L/SECAM
Greece	B/SECAM	G/SECAM
Hungary, Russian Federation	D/SECAM	K/SECAM
India, Indonesia	B/PAL	
Ireland	I/PAL	I/PAL
Iceland	B/PAL	G
Israel, Italy	B/PAL	G/PAL
Japan	M/NTSC	M/NTSC
Malaysia	B/PAL	G/PAL
Morocco	B/SECAM	G/SECAM
Norway, New Zealand, Pakistan	B/PAL	G/PAL
Gambia	I/PAL	I/PAL
Poland, Portugal, Croatia	B/PAL	G/PAL
United Kingdom	(Does not use VHF)	I/PAL
South Africa	I/PAL	I/PAL
Sweden, Switzerland, Turkey	B/PAL	G/PAL
Czech Republic	D/SECAM	K/SECAM
Estonia	B/PAL, D/SECAM	G/PAL, K/SECAM
Uruguay	N/PAL	
Venezuela, Cuba	M/NTSC	M/NTSC
Iran	B/SECAM	G/SECAM
Uganda	B/PAL	

[a]PAL = Phase line alternation (Europe).
NTSC = National Television Systems Committee (North America and Japan).
SECAM = Sequential color and memory (Europe).
Source: Adapted from Ref. 7, table in Appendix 1 to Annex 1, ITU-R Rec. BT.470-3.

15.8.2   Bandwidth of the Baseband and Baseband Response

One of the most important specifications in any radiolink system transmitting video is frequency response. A system with cascaded hops should have essentially a flat bandpass in each hop. For example, if a single hop is 3 dB down at 6 MHz in the resulting baseband, a system of five such hops would be 15 dB down. A good single hop should be ± 0.25 dB or less out to 8 MHz.

The most critical area in the baseband for video frequency response is in the low-frequency area of 15 kHz and below. Cascaded radiolink systems used in transmitting video must consider response down to 10 Hz.

Modern radiolink equipment used to transport video operates in the 2-GHz band and above. The 525-line video requires a baseband in excess of 4.2 MHz plus available baseband above the video for the aural channel. Desirable characteristics for 525-line video then would be a baseband at least 6 MHz wide. For 625-line TV, 8 MHz would be required, assuming that the aural channel would follow the channelization recommended by CCIR Rec. 402-2 (Ref. 11).

15.8.3 Preemphasis

Preemphasis–deemphasis characteristics are described in CCIR Rec. 405-1 (Ref. 13).

15.8.4 Differential Gain

Differential gain is the difference in gain of the radio relay system as measured by a low-amplitude, high-frequency (chrominance) signal at any two levels of a low-frequency (luminance) signal on which it is superimposed. It is expressed in percentage of maximum gain. Differential gain shall not exceed the amounts indicated below at any value of APL (average picture level) between 10% and 90%:

- Short haul   2%
- Medium haul   5%
- Satellite   4%
- Long haul   8%
- End-to-end   10%

Based on ANSI/EIA/TIA-250C (Ref. 6). Also see CCIR Rec. 567-3 (Ref. 9).

15.8.5 Differential Phase

Differential phase is the difference in phase shift through the radio relay system exhibited by a low-amplitude, high-frequency (chrominance) signal at any two levels of a low-frequency (luminance) signal on which it is superimposed. Differential phase is expressed as the maximum phase change between any two levels. Differential phase, expressed in degrees of the high-frequency sine wave, shall not exceed the amounts indicated below at any value of APL between 10% and 90%:

- Short haul   0.5°
- Medium haul   1.3°

- Satellite   1.5°
- Long haul   2.5°
- End-to-end   3.0°

Based on ANSI/EIA/TIA-250C (Ref. 6).

15.8.6 Signal-to-Noise Ratio (10 kHz to 5.0 MHz)

The video signal-to-noise ratio is the ratio of the total luminance signal level (100 IRE units) to the weighted rms noise level. The noise referred to is predominantly thermal noise in the 10 kHz to 5.0 MHz range. Synchronizing signals are not included in the measurement. The EIA states that there is a difference of less than 1 dB between 525-line systems and 625-line systems.

As stated in the ANSI/EIA/TIA-250C standard, the signal-to-noise ratio shall not be less than the following:

- Short haul   67 dB
- Medium haul   60 dB
- Satellite   56 dB
- Long haul   54 dB
- End-to-end   54 dB

and, for the low-frequency range (0–10 kHz), the signal-to-noise ratio shall not be less than the following:

- Short haul   53 dB
- Medium haul   48 dB
- Satellite   50 dB
- Long haul   44 dB
- End-to-end   43 dB

15.8.7 Radiolink Continuity Pilot

For video transmission the continuity pilot is always above the baseband. CCIR recommends an 8.5-MHz pilot. Refer to CCIR Rec. 401-2 (Ref. 14).

15.9 TV TRANSMISSION BY SATELLITE RELAY

Table 15.4 provides general guidance on the basic performance requirements for the transmission of broadcast-type TV signals via satellite relay based on CCIR recommendations.

TABLE 15.4   Satellite Relay TV Performance

Parameters	Space Segment	Tereestrial Link[a]	End-to- End Values
Nominal impedance	75 Ω		
Return loss	30 dB		
Nonuseful dc component	0.5 V		
Nominal signal amplitude	1V		
Insertion gain	0 ± 0.25 dB	0 ± 0.3 dB	0 ± 0.5dB
Insertion gain variation (1s)	± 0.1 dB	± 0.2 dB	± 0.3 dB
Insertion gain variation (1h)	± 0.25 dB	± 0.3 dB	± 0.5 dB
Signal-to-continuous random noise	53 dB[b]	58 dB	51 dB
Signal-to-periodic noise (0–1 (kHz)	50 dB	45 dB	39 dB
Signal-to-periodic noise (1 kHz–6 MHz)	55 dB	60 dB	53 dB
Signal-to-impulse noise	25 dB	25 dB	25 DB[c]
Crosstalk between channels (undistorted)	58 dB	64 dB	56 dB
Crosstalk between channels (undifferentiated)	50 dB	56 dB	48 dB
Luminance nonlinear distortion	10%	2%	12%
Chrominance nonlinear distortion (amplitude)	3.5%	2%	5%
Chrominance nonlinear distortion (phase)	4°	2°	6°
Differential (gain x or y)	10%	5%	13%
Differential phase (x or y)	3°	2°	6°
Chrominance –luminance intermodulation	±4.5%	±2%	±5%
Steady-state sync pulse nonlinear distortion	+5–10%	±5%	+10–15%
Transient sync pulse nonlinear distortion	20%		
Field-time waveform distortion	6%	2%	10%
Line-time waveform distortion	3%	2%	4%
Short-time waveform distortion (pulse /bar)	100 ± 12%	100 ± 6%[c]	100 ± 18%[c]
Short-time waveform distortion (pulse lobes)	3%[d]	1.5%[d]	4.5%[d]
Chrominance–luminance gain inequality	±10%	±6%	±13%
Chrominance –luminance delay inequality	±50 ns	±60 ns	±90 ns
Gain–frequency characteristic (0.15–6 MHz)	±0.5 dB	±0.5 dB[e]	±1.0 dB[e]
Delay–frequency characteristic (0.15–6 MHz)	±50 ns	±50 ns[e]	±105 ns[e]

[a]Connecting earth station to national technical control center.

[b]In cases where the receive earth station is colocated with the broadcaster's premises, a relaxation of up to 3 dB in video signal-to-weighted-noise ratio may be permissible. In this context, the term *colocated* is intended to represent the situation where the noise contribution of the local connection is negligible.

[c]Law of addition not specified in CCIR. Rec. 567.

[d]The pulse lobes are contained within a mask of the type shown in Figure 29a in CCIR Rec. 567. The figures in the space segment column is the amplitude of the mask in Figure 29a for times ≤ −800 ns and ≥ 800 ns. The figures in the terrestrial link and end-to-end value columns are corresponding amplitudes of scaled versions of the mask.

[e]Highest frequency: 5 MhZ.

Source: Reference 15, CCIR Rep. 965-1.

15.10 BASIC TESTS FOR VIDEO QUALITY

15.10.1 Window Signal

The window signal when viewed on a picture monitor is a large square or rectangular white area with a black background. The signal is actually a sine-squared pulse. As such it has two normal levels, reference black and reference white. The signal usually is adjusted so that the white area covers one-fourth to one-half the total picture width and one-fourth to one-half the total picture height. This is done in order to locate the maximum energy content of the signal in the lower portion of the frequency band.

A number of useful checks derive from the use of a window signal and a picture monitor. These include the following:

1. *Continuity or Level Check.* With a window signal of known white level, the peak-to-peak voltage of the signal may be read on a calibrated oscilloscope using a standard roll-off characteristic (e.g., EIA).
2. *Sync Compression or Expansion Measurements.* Comparison of locally received window signals with that transmitted from the distant end with respect to white level and horizontal synchronization on calibrated oscilloscopes using standard roll-off permits evaluation of linearity characteristics.
3. *Test and Adjustment.* Can be made at clamper amplifiers and low-frequency equalizers to minimize streaking by observing the test signal on scopes using the standard roll-off characteristics at both the vertical and the horizontal rates.
4. *Indication of Ringing.* With a window signal the presence of ringing may be detected by using properly calibrated wideband oscilloscopes and adjusting the horizontal scales to convenient size. Both amplitude and frequency of ringing may be measured by this method.

15.10.2 Sine-Squared Test Signal

The sine-squared test signal is a pulse type of test signal that permits an evaluation of amplitude–frequency response, transient response, envelope delay, and phase. An indication of the HF amplitude characteristic can be determined by the pulse width and height, and the phase characteristic by the relative symmetry about the pulse axis. However, this test signal finds its principal application in checking transient response and phase delay. The sine-squared signal is far more practical than a square wave test signal to detect overshoot and ringing. The pulse used for checking video systems should have a repetition rate equal to the line frequency and a duration, at half amplitude, equal to one-half the period of the nominal upper cutoff frequency of the system.

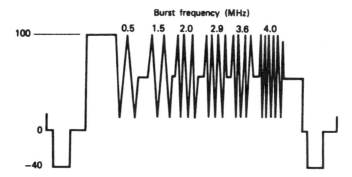

Figure 15.8. Multiburst signal (horizontal frequencies normally used).

15.10.3 Multiburst

This test signal is used for a quick check of gain at a few determined frequencies. A common form of multiburst consists of a burst of peak white (called white flag) that is followed by bursts of six sine wave frequencies from 0.5 to 4.0 MHz (for NTSC systems) plus a horizontal synchronizing pulse. All these signals are transmitted during one-line intervals. The peak white or white flag serves as a reference. For system checks a multiburst signal is applied to the transmit end of a system.

At the receiving point the signal is checked on an oscilloscope. Measurements of peak-to-peak amplitudes of individual bursts are indicative of gain. A multiburst image on an oscilloscope gives a quick check of amplitude–frequency response and changes in setup. Figure 15.8 illustrates a typical NTSC-type multiburst signal. The y axis of the figure is in IRE units.*

15.10.4 Stair Steps

For the measurement of differential phase and gain, a 10-step stair-step signal is often used. Common practice (in the United States) is to superimpose 3.6 MHz on the 10 steps that extend progressively from black to white level. The largest amplitude sine wave block is adjusted on the oscilloscope to 100 standard (i.e., IRE) divisions and is made a reference block. Then the same 3.6-MHz sine waves from the other steps are measured in relation to the reference black. Any difference in amplitudes of the other blocks represents differential gain. By the use of a color analyzer in conjunction with the above, differential phase may also be measured.

The stair-step signal may be used as a linearity check without the sine wave signal added. The relative height between steps is in direct relation to signal compression or nonlinearity.

---

*IRE standard scale is a linear scale for measuring, in arbitrary IRE units, the relative amplitude of the various components of a television signal. The scale varies from −40 (sync peaks, max. carrier) to +120 (zero carrier). Reference white is +100 and blanking is 0.

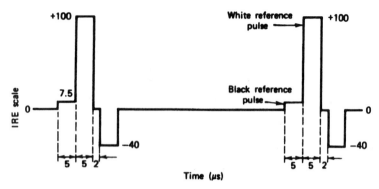

Figure 15.9. Typical vertical-interval reference signal.

15.10.5 Vertical Interval Test Signals (VITS)

VITS makes use of the vertical retrace interval for the transmission of test signals. The FCC specifies the interval for use in the United States as the last 12 μs of lines 17 through 20 of the vertical blanking interval of each field. For whichever interval boundaries are specified, test signals transmitted in the interval may include reference modulation levels, signals designed to check performance of the overall transmission system or its individual components, and cue and control signals related to the operation of TV broadcast stations. These signals are used by broadcasters because, by necessity, they are inserted at the point of origin. Standard test signals are used as described previously or with some slight variation, such as multiburst, window, and stair step. Some broadcasters use vertical-interval reference signals. Figure 15.9 shows one currently in use in the United States.

15.10.6 Test Patterns

Standard test patterns, especially those inserted at a point of program origination, provide a simple means of determining transmission quality. The distant viewer, knowing the exact characteristics of the transmitted image, can readily detect distortion(s). Standard test patterns such as the EIA test pattern used widely in the United States, with a properly adjusted picture monitor, can verify the following:

- Horizontal linearity
- Vertical linearity
- Contrast
- Aspect ratio
- Interlace
- Streaking
- Ringing
- Horizontal and vertical resolution

15.10.7 Color Bars

Color bar test signals are used by broadcasters for the adjustment of their equipment, including color monitors (Ref. 6, 16). Color bars may also be sent over transmission facilities for test purposes. The color bar also may be used to test color transmission using a black and white monitor by examining gray densities of various bars, depending on individual colors. A wideband A-scope horizontal presentation can show whether or not the white reference of the luminance signal and the color information have the proper amplitude relationships. The color bar signal may further be observed on a vector display oscilloscope (chromascope), which allows measurement of absolute amplitude and phase angle values. It also can be used to measure differential phase and gain.

15.11 DIGITAL TELEVISION

15.11.1 Introduction

Our concern in this chapter is television transmission. The PSTN trunk network in developed nations is digital, and we expect the entire network, including the local serving area segments, to be all-digital by the year 2000. Television transmission has been a little slower to turn to digital, but we believe by the year 2000 or just beyond the CATV plant will be digital. All of conference television is digital. Direct broadcast satellite (DBS) is all-digital.

Digital television often uses a PCM format, similar to what was discussed in Chapter 3. This is an 8-bit PCM format derived from sampling, quantization, and coding. The quantization is linear. Without compression we can expect data rates at 108 or 114 Mbps for NTSC video.

In this section we will discuss several digitizing techniques for TV video and provide an overview of some bit rate reduction methods including derivatives of MPEG-2. Several types of conference television (video teleconferencing) will also be covered.

15.11.2 Basic Digital Television

15.11.2.1 Two Coding Schemes. There are two distinct digital coding methods for color television: component and composite coding. For our discussion there are four components that make up a color video signal. These are R for red, G for green, B for blue, and Y for luminance. The output signals of a TV camera are converted by a linear matrix into luminance (Y) and two color-difference signals, R-Y and B-Y.

With the component method these signals are individually digitized by an A/D converter. The resulting digital bit streams are then combined with overhead and timing by means of a multiplexer for transmission over a single medium such as a wire pair or coaxial cable.

Composite coding, as the term implies, directly codes the entire video baseband. The derived bit stream has a notably lower bit rate than that for component coding.

CCIR Rep. 646-4 (Ref. 17) compares the two coding techniques. The advantages of separate-component coding are the following:

- The input to the circuit is provided in separate component form by the signal sources (in the studio).
- The component coding is adopted generally for studios, and the inherent advantages of component signals for studios must be preserved over the transmission link in order to allow downstream processing at a receiving studio.
- The country receiving the signals via an international circuit uses a color system different from that used in the source country.
- The transmission path is entirely digital, which fits in with the trend toward all-digital systems that is expected to continue.

The advantages of transmitting in the composite form are the following:

- The input to the circuit is provided in the composite form by the signal sources (at the studio).
- The color system used by the receiving country, in the case of an international circuit, is the same as that used by the source country.
- The transmission path consists of mixed analog and digital sections.

15.11.2.2   Development of a PCM Signal. As in Chapter 3, there are three stages in the development of a PCM format from an analog signal. These are sampling, quantization, and coding. The same three stages are used in the development of a digital TV signal from its analog counterpart.

The calculation of the sampling rate is based on the color subcarrier frequency called f_{sc}. For NTSC television the color subcarrier is at 3.58 MHz. In some cases the sampling rate is three times this frequency ($3f_{sc}$), in other cases four times the color subcarrier frequency ($4f_{sc}$). For PAL television, the color subcarrier is 4.43 MHz. Based on 8-bit PCM words, the bit rates are $3 \times 3.58 \times 10^6 \times 8 = 85.92$ Mbps and $4 \times 3.58 \times 10^6 \times 8 = 114.56$ Mbps. In the case of PAL transmission systems using $4f_{sc}$, the uncompressed bit rate for the video is $4 \times 4.43 \times 10^6 \times 8 = 141.76$ Mbps. These values are for composite coding.

In the case of component coding there are two separate digital channels. The luminance channel is 108 Mbps and the color-difference channel is 54 Mbps (for both NTSC and PAL systems).

In any case, linear quantization is employed, and a S/D (signal-to-distortion ratio) of better than 48 dB can be achieved with 8-bit coding. The coding is pure binary or twos complement encoding.

15.11.3   Bit Rate Reduction—Compression Techniques

In Section 15.11.2 raw video transmitted digitally required bit rates from 86 to 142 Mbps per video channel (no audio). Leaving aside studio-to-transmitter links and CATV supertrunks, it is incumbent on the transmission engineer to reduce these bit rates without sacrificing picture quality if there is any route-distance requirement involved.

CCIR Rep. 646-4 (Ref. 17) covers three basic bit rate reduction methods:

* Removal or horizontal and vertical blanking intervals
* Reduction of sampling frequency
* Reduction of the number of bits per sample

We will only cover the last method.

15.11.3.1   Reduction of the Number of Bits per Sample. There are three methods that may be employed for bit rate reduction of digital television by reducing the number of bits per sample. These may be used singly or in combination:

* Predictive coding, sometimes called differential PCM
* Entropy coding
* Transform coding

Differential PCM, according to CCIR, has so far emerged as the most popular method. The prediction process required can be classified into two groups. The first one is called *intraframe* or *intrafield* and is based only on the reduction of spatial redundancy. The second group is called *interframe* or *interfield* and is based on the reduction of temporal redundancy as well as spatial redundancy.

15.11.3.2   Specified Bit Rate Reduction Techniques. References 17 and 18 provide the following information on some of the digital video compression techniques:

Intraframe Coding. Intraframe coding techniques provide compression by removing redundant information within each video frame. These techniques rely on the fact that images typically contain a great deal of similar information; for example, a one-color background wall may occupy a large part of each frame. By taking advantage of this redundancy, the amount of data necessary to accurately reproduce each frame may be reduced.

Interframe Coding. Interframe coding is a technique that adds the dimension of time to compression by taking advantage of the similarity between adjacent frames. Only those portions of the picture that have changed since the previous picture frame are communicated. Interframe coding systems do not transmit detailed information if it has not changed from one frame to the next. The result is a significant increase in transmission efficiency.

Intraframe and Interframe Coding Used in Combination. Intraframe and interframe coding used together provide a powerful compression technique. This is achieved by applying intraframe coding techniques to the image changes that occur from frame to frame. That is, by subtracting image elements between adjacent frames, a new image remains that contains only the differences between the frames. Intraframe coding, which removes similar information within a frame, is applied to this image to provide further reduction in redundancy.

Motion Compensation Coding. To improve image quality at low transmission rates, a specific type of interframe coding motion compensation is commonly used. Motion compensation applies the fact that most changes between frames of a video sequence occur because objects move. By focusing on the motion that has occurred between frames, motion compensation coding significantly reduces the amount of data that must be transmitted.

Motion compensation coding compares each frame with the preceding frame to determine the motion that has occurred between the two frames. It compensates for this motion by describing the magnitude and direction of an object's movement (e.g., a head moving right). Rather than completely regenerating any object that moves, motion compensation coding simply commands that the existing object be moved to a new location.

Once the motion compensation techniques estimate and compensate for the motion that takes place from frame-to-frame, the differences between frames are smaller. As a result, there is less image information to be transmitted. Intraframe coding techniques are applied to this remaining image information.

Motion Compensation Transform Coding and Hierarchical Vector Quantization Coding. Motion Compensated Transform (MCT) is proprietary technology of PictureTel Corp. This technology has significantly improved upon the previously described traditional technology. MCT combines the interframe coding technique, motion compensation, with the intraframe technique, transform coding, to greatly reduce the amount of information that must be transmitted to achieve a given level of video quality. PictureTel's Hierarchical Vector Quantization (HVQ) Coding further advanced compression technology. For example, HVQ transmitted at 112 kbps is comparable to the video quality of MCT transmitted at 224 kbps. Like MCT, HVQ compression technology employs motion compensation coding. However, it replaces cosine transform coding with hierarchical vector quantization.

In transform coding, images are divided into blocks prior to image compression. Under certain conditions, transmission blocks can result in image distortion or breaking of the image into a block pattern, particularly in background portions of the picture.

HVQ greatly improves upon transform coding, significantly enhancing image quality by eliminating the artifacts and other distortions inherent in transform coding. With HVQ, the image is coded not in terms of discrete blocks, but on the basis of different levels of image resolution, arranged in a hierarchy from low to high. For each frame, HVQ transmits the image in as much detail as possible. In areas of the image containing motion, any image alterations resulting from compression are much less easily perceived than the more visible effects of transform coding (e.g., blocking) because, with HVQ, they are only the result of reductions in resolution.

15.11.4 MPEG-2* Compression and Related Standards

15.11.4.1 Introduction. This section if based on the ATSC (Advanced Television System Committee) version of MPEG-2, which is used primarily for terrestrial broadcasting and cable TV.

15.11.4.2 Overview. The objective of the ATSC standard (Ref. 19–22) is to describe a system for the transmission of high-quality video, audio, and ancillary services over a single 6-MHz channel. The system delivers about 19 Mbps of throughput on a 6-MHz broadcasting channel and about 38 Mbps on a 6-MHz CATV channel. The video source, which is encoded, can have a resolution as much as five times better than conventional NTSC television resolution. This means that a bit rate reduction factor of 50 or higher is required. The system must be efficient in utilizing the channel capacity by exploiting complex video and audio reduction technology. The objective is to represent the video, audio, and data sources with as few bits as possible while preserving the level of quality required for a given application.

A block diagram of the system is shown in Figure 15.10. We call this the system model, which consists of three subsystems:

- Source coding and compression
- Service multiplex and transport
- RF/transmission subsystem

Of course *source coding and compression* refers to bit rate reduction methods (data compression), that are appropriate for the video, audio, and ancillary digital data bit streams. The ancillary data include control data, conditional access control data, and data associated with the program audio and video services, such as closed captioning. Ancillary data can also refer to independent program services. The digital television system uses MPEG-2 video

---

* MPEG = Motion Picture Experts Group.

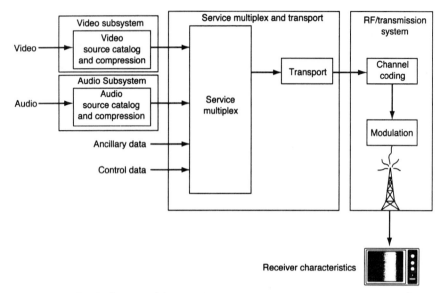

Figure 15.10. Block diagram of the digital terrestrial television broadcasting model. (Based on ITU-R Task Group 11.3 model. From Ref. 22, Figure 4.1. Reprinted with permission.)

stream syntax for coding of the video and Digital Audio Compression Standard, called AC-3, for the coding of the audio.

Service multiplex and transport refers to dividing the bit stream into packets of information, the unique identification of each packet or packet type, and appropriate methods of multiplexing video bit stream packets, audio bit stream packets, and ancillary data bit stream packets into a single data stream. A prime consideration in the system transport design was interoperability among the digital media, such as terrestrial broadcasting, cable distribution, satellite distribution, recording media, and computer interfaces. MPEG-2 transport stream syntax was developed for applications where channel bandwidth is limited and the requirement for efficient channel transport was overriding. Another aspect of the design was interoperability with ATM transport systems.

RF/transmission deals with channel coding and modulation. The input to the channel coder is the multiplexed data stream from the service multiplex unit. The coder adds overhead to be used by the far-end receiver to reconstruct the data from the received signal. At the receiver we can expect that this signal has been corrupted by channel impairments. The resulting bit stream out of the coder modulates the transmitted signal. One of two modes can be used by the modulator: 8-VSB for the terrestrial broadcast mode and 16-VSB for the high data rate mode.

15.11.4.3 Video Compression and Decompression. The ATSC standard is based on a specific subset of MPEG-2 algorithmic elements and is

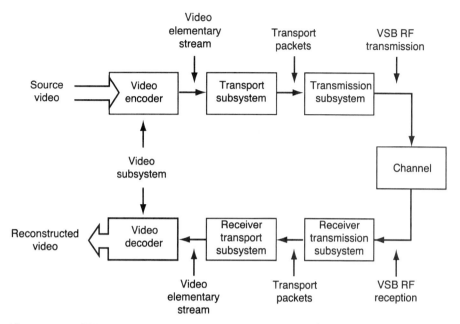

Figure 15.11. Video coding in relation to the ATSC system. (From Ref. 22, Figure 5.1. Reprinted with permission.)

based on the Main Profile. The Main Profile includes three types of frames for prediction (I-frames, P-frames, and B-frames), and an organization of luminance and chrominance samples (designated 4 : 2 : 0) within the frame. The Main Profile is limited to a compressed data rate of no more than 80 Mbps.

Figure 15.11 is a simplified block diagram of signal flow for the ATSC system.

Video preprocessing converts the analog input signals to digital samples in such a form needed for later compression. The analog input signals are red (R), green (G), and blue (B).

Table 15.5 lists the compression formats covered by the ATSC standard. The following explains some of the items in the table. *Vertical lines* refers to the number of active lines in the picture. *Pixels* are the number of pixels

TABLE 15.5   ATSC Compression Formats

Vertical Lines	Pixels	Aspect Ratio	Picture Rate
1080	1920	16:9	60I, 30P, 24P
720	1280	16:9	60P, 30P, 24P
480	704	16:9 and 4:3	60P, 60I, 30P, 24P
480	640	4:3	60P, 60I, 30P, 24P

Source: Reference 22, Table 5.1. Reprinted with permission.

TABLE 15.6   Sampling Rate Summary

Line Format	Total Lines per Frame	Total Samples per Line	Sampling Frequency	Frame Rate
1080 line	1125	2200	74.25 MHz	30.00 fps[a]
720 line	750	1650	74.25 MHz	60.00 fps
480 line	525	858	13.5 MHz	59.94 Hz
(704 pixels)				field rate

[a]fps = frames per second.

during the active line. *Aspect ratio* refers to the picture aspect ratio. *Picture rate* gives the number of frames or fields per second. Regarding picture rate values, P refers to progressive scanning and I refers to interlaced scanning. It should be noted that both 60.00-Hz and 59.95-Hz (i.e., $60 \times 1000/1001$) picture rates are allowed. Dual rates are permitted at 30 Hz and 24 Hz.

Sampling Rates. Three active line formats are considered: 1080, 720, 483. Table 15.6 summarizes the sampling rates.

For the 480-line format, there may be 704 or 640 pixels in an active line. If the input is based on ITU-R Rec. BT.601-4 (Ref. 23), it will have 483 active lines with 720 pixels in each active line. Only 480 of the 483 active lines are used for encoding. Only 704 of the 720 pixels are used for encoding: the first eight and the last eight are dropped. The 480-line, 640-pixel format corresponds only to the IBM VGA graphics format and may be used with ITU-R BT.601-4 (Ref. 23) sources by employing appropriate resampling techniques.

Sampling precision is based on the 8-bit sample.

Colorimetry means the combination of color primaries, transfer characteristics, and matrix coefficients. The standard accepts colorimetry that conforms to SMPTE.* Video inputs corresponding to ITU-R BT.601-4 may have SMPTE 274M or 170M colorimetry.

The input video consists of the RGB components that are matrixed into luminance (Y) and chrominance (Cb and Cr) components using a linear transformation by means of a 3×3 matrix. Of course the luminance carries picture intensity information (black-and-white) and the chrominance components contain the color. There is a high degree of correlation of the original RGB components, whereas the resulting Y, Cb, and Cr have less correlation and can be coded efficiently.

In the coding process, advantage is taken of the differences in the ways humans perceive luminance and chrominance. The human visual system is less sensitive to the high frequencies in the chrominance components than to the high frequencies in the luminance component. To exploit these characteristics the chrominance components are low-pass filtered and subsampled by a factor of 2 along with the horizontal and vertical dimensions, thus producing

* SMPTE = Society of Motion Picture and Television Engineers.

chrominance components that are one-fourth the spatial resolution of the luminance components.

Macroblocks and Slices. Information blocks are organized into macroblocks. A macroblock consists of four blocks of luminance and two chroma blocks (Cb and Cr). A macroblock contains 256 luminance samples, 64 Cb samples, and 64 Cr samples for a total of 384 samples per macroblock. Table 15.7 breaks out macroblocks for the three line formats of video considered. One or more contiguous macroblocks within the same row are grouped together to form slices. Macroblocks are arranged in a slice the same as a TV video raster scan, from left to right.

In tight compression techniques such as the ATSC MPEG-2, errors can propagate, which is highly undesirable. A coded bit stream consists of variable length codeword; thus any uncorrected transmission error will cause the loss of codeword alignment. A start codeword begins each slice. Since the MPEG/ATSC codeword assignment guarantees that no legal combination of codewords can emulate the start code, the slice start code can be used to regain codeword alignment after an error. When an error does occur, the decoder skips to the start of the next slice and resumes correct decoding.

The slice is the minimum unit for resynchronization after an error. There is a trade-off between number of slices and picture quality. The more slices, the more error recovery is improved. But this removes bits that could otherwise be used to improve picture quality.

In the ATSC/MPEG-2 system, the beginning of a slice, with possibly several slices across the row, is the initial macroblock of every horizontal row.

Pictures, Groups of Pictures, and Sequences. The individual frame or picture is the primary coding unit of a video sequence. A video picture consists of the collection of slices that constitute the active picture area.

A video sequence, which consists of a grouping of one or more consecutive pictures, starts with a sequence header and is terminated by an end-of-sequence code in the bit stream. A video sequence can contain additional sequences headers. Any sequence header can serve as an entry point. An *entry point* is a point in the coded video bit stream after which a decoder can become properly initialized and correctly parse the bit stream syntax.

TABLE 15.7 Number of Macroblocks

Line Format	Samples per Line	Rows of Macroblocks	Macroblocks per Row
1080 line	1920	68[a]	120
720 line	1280	45	80
480 line	704	30	44

[a]Including last row that adds eight dummy lines to create the 1088 lines for coding.

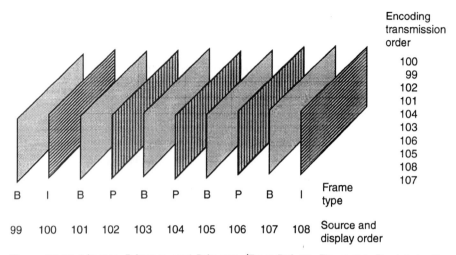

Encoding
transmission
order

100
99
102
101
104
103
106
105
108
107

B   I   B   P   B   P   B   P   B   I   Frame
                                        type

99 100 101 102 103 104 105 106 107 108   Source and
                                                  display order

Figure 15.12. I-frames, P-frames, and B-frames. (From Ref. 22, Figure 5.4. Reprinted with permission.)

A *group of pictures* (GOP) is one or more pictures (frames) that are combined. A GOP provides boundaries for interpicture coding and registration of time code. GOPs are optional within both the MPEG-2 and ATSC systems.

Figure 15.12 shows a time sequence of video frames consisting of intra-coded pictures (I-frames), predictive coded pictures (P-frames), and bidirectionally coded pictures (B-frames).

I-Frames. A TV image is highly redundant both inside a single frame and among frames in a temporal manner. The compression process takes advantage of these redundancies. When only the spatial redundancy within a single frame is exploited, it is called intraframe coding. It does not take advantage of the temporary correlation addressed by temporal prediction, which is referred to as interframe coding. Frames that do not use any interframe coding are called *I-frames* (where I denotes *intraframe* coded). The ATSC/MPEG-2 compression system uses both intraframe and interframe coding.

The use of periodic I-frames facilitates receiver initialization and channel acquisition (e.g., when the receiver is turned on or the channel is changed). The decoder takes advantage of the intraframe coding mode, when noncollectible errors occur. An initial frame must be available at the decoder to start the prediction loop with motion compensated prediction. Thus a mechanism is required so that if the decoder loses synchronization, it can rapidly acquire tracking.

P-Frames. P-frames (where P stands for prediction) are those frames where the prediction is in the forward direction only. This means that predictions for the P-frame are formed only from pixels in the most recently decoded I-frame or P-frame. These forward-predicted frames use interframe coding techniques to improve overall compression efficiency and picture quality. P-frames may include portions that are only intraframe coded. Within a P-frame, each macroblock can be either forward-predicted or intraframe coded.

B-Frames. A B-frame includes prediction from a previous frame and from a future frame. B denotes bidirectionality. The referenced previous or future frames are sometimes called *anchor frames*. In all cases they are either I-frames or P-frames.

The concept behind the B-frame prediction is that there is correlation both with frames that occur in the past and with frames that occur in the future. As a consequence, if a future frame is available to the decoder, a superior prediction can be achieved. This saves bits and improves performance. However, some undesirable consequences can occur by using future frames:

- A B-frame cannot be used for predicting future frames.
- The transmission order of frames is different from the displayed order of frames.
- The coder and decoder must reorder the video frames thereby increasing the total latency.

In Figure 15.12 it should be noted that there is one B-frame between each pair of I/P-frames. Each frame is labeled with both its display order and transmission order. The reason that I- and P-frames are transmitted out of sequence is that the video decoder has both anchor frames decoded and available for prediction.

Motion Estimation. The compression algorithm used with the ATSC/MPEG-2 system depends on creating an estimate of the image being compressed, and subtracting from the image being compressed the pixel values of the estimate or prediction. When the estimates are good, the subtraction will leave a small residue amount to be transmitted. If the prediction were perfect, such as with still pictures, the residue would be zero and no information need be sent.

If the estimate is not close to zero for some or many pixels, such differences represent information that needs to be sent so that the far-end decoder can reconstruct a correct image. When there are large prediction differences, we may assume there has been sharp and rapid motion.

Part of the data compression process is *motion compensated prediction.* Motion compensation refers to the fact that the locations of the macroblock sized regions in the reference frame can be offset to account for local motions. These macroblock offset are known as *motion vectors*.

The estimation of interframe displacement is calculated with half-pel* precision in both horizontal and vertical dimensions. What this means is that the displaced macroblock from the previous frame can be shifted by noninteger displacements. This will require interpolation to compute the values of displaced picture elements at locations not in the original array of samples. Estimates for half-pel locations are computed by averages of adjacent cell samples.

Motion vectors within a slice are differenced such that the first value for the motion vector is transmitted directly, and the following sequence of motion vector differences is transmitted using variable length codes (VLCs).

Encoder Prediction Loop. The encoder *prediction loop* lies at the heart of the ATSC/MPEG-2 compression technique. Analysis of the prediction loop is the best approach to explain how the different algorithmic elements combine to achieve compression.

There is a prediction function in the prediction loop that estimates, or predicts, the picture values of the next picture to be encoded in the sequence of successive pictures that constitute a TV program. Such a prediction is based on previous information that was available in the loop that was derived from earlier pictures.

The core of predictive coding is the subtraction of the predicted picture values from the new picture to be coded. The objective is to do such a good job with the prediction function that the remainder values when subtracting are zero or nearly zero.

The prediction differences are computed separately for the luminance and the chrominance components before further processing. Figure 15.13 is a functional block diagram of the encoder prediction loop.

Spatial Transform Block—DCT. The prediction differences are grouped into 8-by-8 blocks and a spatial transform is applied to the blocks of difference values. In the intraframe case, the spatial transform is applied to the raw, undifferenced picture data. The luminance and two chrominance components are transformed separately. The spatial transform used is the DCT or discrete cosine transform.

Figure 15.13 shows an IDCT (inverse discrete cosine transform) function. Applying the IDCT, in principle, would yield exactly the same array as the original. In that sense, transforming the data does not modify the data but merely represents those data in a different form.

Quantizer. There is a quantizer matrix that contains relative coefficient precision information derived from an 8-by-8 array of DCT coefficients. Two types of quantizer matrix are supported. One is used for macroblocks that are intraframe coded. The other is used for non-intraframe-coded macroblocks.

---

* Pel and pixel are synonymous.

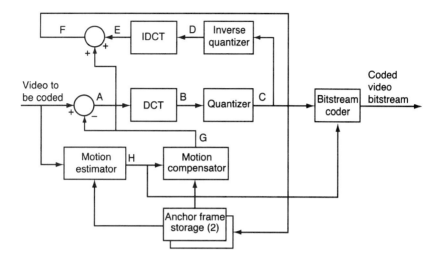

A	Pixel-by-pixel prediction errors
B	Transformed blocks of prediction errors (DCT coefficients)
C	Prediction error DCT coefficients in quantitized form
D	Quantitized prediction error DCT coefficients in standard form
E	Pixel-by-pixel prediction errors, designed by quantization
F	Reconstructed pixel values, degraded by quantization
G	Motion compensated predicted pixel values
H	Motion vectors

Figure 15.13. Functional block diagram of the encoder prediction loop. DCT = discrete cosine transform; IDCT = inverse discrete cosine transform. (From Ref. 22, Figure 5.5. Reprinted with permission.)

The video coding system defines default values for both the intraquantizer and the nonintraquantizer matrices. Either or both of the quantizer matrices can be overridden at the picture level by transmitting the appropriate arrays of 64 values. Any quantizer overrides stay in effect until the following sequence start code.

The transform coefficients represent the bulk of the actual coded video information. They are quantized to various degrees of coarseness. The appearance of some portions of the picture will be more affected than others due to the loss of precision through coefficient quantization. There is a quantizer scale factor that governs the overall level of quantization to vary with each macroblock. In such a way entire macroblocks can be quantized more coarsely, which results in decreasing the number of bits needed to represent the picture.

The quantizer scale factor is multiplied by the corresponding value in the appropriate quantizer matrix to form the quantizer step size for each coefficient other than the DC coefficient of intraframe coded blocks. Quantization of DC coefficients of intracoded blocks is unaffected by the quantizer scale factor and is only governed by the $(0, 0)$ element of the intraquantized matrix, which is always set to 8.

Inverse Quantizer. Coded coefficients are decoded and an 8-by-8 block of quantized coefficients is reconstructed at the decoder. Each of the 64 coefficients is *inverse quantized* in accordance with the prevailing quantizer matrix, quantizer scale, and frame type. A block of 64 DCT coefficients is the result of inverse quantization.

IDCT—Inverse Spatial Transform Block. The decoded and inverse quantized coefficients are organized as 8-by-8 blocks of DCT coefficients and to each block the inverse discrete cosine transform (IDCT) is applied. A new array of pixel values or pixel difference values results. These correspond to the output of the subtraction at the beginning of the prediction loop. The values will be pixel differences if the prediction loop was in the interframe mode. On the other hand, if the mode was in the intraframe mode, the inverse transform will directly produce actual pixel values.

Motion Compensator. For the case where a portion of the picture has not moved, then the subtraction of the old portion from the new portion will produce a zero pixel difference or nearly so. Of course, this is the objective of the prediction.

In the case where there has been movement in that portion of an image under consideration, the direct pixel-by-pixel differences will generally not be zero. It could be statistically very large. The motion in most natural scenes, however, is organized, and in most cases can be approximated locally as a translation. For this reason the coding system allows for *motion compensated* prediction. Here macroblock-size regions in the reference frame may be translated vertically and horizontally with respect to the macroblock being predicted, to compensate for local motion. The pixel-by-pixel differences between the current macroblock and the motion compensated prediction are transformed by the DCT and quantized using the composition of the nonintraquantizer matrix and the quantizer scale factor. These quantized coefficients are then coded.

Anchor Frames. The entire frame is coded in the case of I-frames without reference to any other frames. P-frames, on the other hand, are referenced to the most recently decoded I- or P-frame. B-frames use two prediction reference frames. One of the reference frames occurs earlier than the coded frame in display order. This can be used for forward prediction. The other reference frame occurs later in the display order. This one can be used for backward prediction.

The encoder has four options for a given macroblock within a B-frame: forward prediction, backward prediction, bidirectional prediction, and intraframe coding. In the case of bidirectional prediction, the forward and backward predictors are averaged and then subtracted from the target macroblock to form the prediction error. This prediction error is then transformed, quantized, and transmitted in the usual manner.

Image Refresh. As we are now aware, a given picture may be transmitted by describing the differences between it and one or two previously transmitted pictures. To make this scheme work, decoders need an image refresh of a valid picture upon tuning to a new channel or to be reinitialized when errors have corrupted reception. Some limit must be placed on the consecutive predictions that can be performed at a decoder to control the buildup of errors due to IDCT mismatch.

Purposefully, a video coding system does not completely specify the results of the IDCT operation. This can result in the decoded pictures drifting away from those in the encoder if many successive predictions are used, even in the absence of transmission errors. The amount of drift is controlled by requiring that each macroblock be coded without prediction (i.e., intracoded) at least once in any 132 consecutive frames. We could call this the image refresh rate.

Discrete Cosine Transform. The temporal correlation in the sequence of image frames is exploited by the predictive coding in the ATSC compression algorithm. Motion compensation is a refinement of that temporal condition that allows a coder to account for apparent motions in the image that can be estimated.

Another source of correlation that represents redundancy in the image data is the spatial correlation within the image frame or field. This spatial correlation of images, including parts of images that contain apparent motion, can be accounted for by a spatial transform of the prediction differences. For the case of intraframe coding (I-frames) where there is no attempt at prediction, the spatial transform applies to the actual picture data. The effect of this spatial transform is to concentrate a large portion of the signal energy in a few transform coefficients.

The image prediction residual pixels are represented by their DCT coefficients in order to exploit the spatial correlation in intraframe and predicted portions of the image. For typical images, a large portion of the energy is concentrated in a few of these coefficients. Thus only a few of these coefficients need to be coded without seriously affecting picture quality. The DCT was selected because it has good energy compaction properties and, in addition, results in real coefficients. Also, there are numerous computational algorithms for its implementation.

Entropy Coding of Video Data. As we have seen, quantization creates an efficient discrete representation for the data to be transmitted. Using the same approach as in the PCM discussed in Chapter 3, codeword assignment takes the quantized values and produces a digital bit stream to be transmitted. In Chapter 3, uniform length codewords were employed (i.e., 8-bit words). Under such an approach, every quantized value would be represented by the same number of bits. However, a greater efficiency can be achieved, in terms of bit rate, by using *entropy coding*. Entropy coding exploits the statistical properties of the signal to be encoded. A signal, whether it is a pixel value or transform coefficient, carries a certain amount of information, or entropy, based on the probability of different possible values or events occurring. For example, an event that occurs infrequently conveys much more new information than one that occurs often. By taking advantage of the fact that some events occur more often than others, the average bit rate may be reduced.

One of the most common entropy schemes used in video compression is *Huffman coding*. The basis of Huffman coding is a codebook that is generated, which can approach the minimum average description length (in bits) of events, given the probability of distribution of all the events. Shorter length codewords are assigned to events that happen more frequently and longer length codewords are assigned to events that happen less frequently.

Most of the transform coefficients are frequently quantized to zero in video compression. There may be a few nonzero low-frequency coefficients and a sparse scattering of nonzero high-frequency coefficients, but the great majority of coefficients may have been quantized to zero. To exploit this phenomenon, the two-dimensional array of transform coefficients is reformatted and prioritized into a one-dimensional sequence through either a zigzag or alternate scanning sequence. This results in most of the important nonzero coefficients (in terms of energy and visual perception) being grouped together early in the sequence. They will be followed by long runs of coefficients that are quantized to zero. Such zero-valued coefficients can be represented efficiently through *run-length encoding*. With run-length encoding, the number (run) of consecutive zero coefficients before a nonzero coefficient value is encoded followed by the nonzero coefficient value can be exploited. The run-length and coefficient value can be entropy coded, either separately or jointly. The scanning separates most of the zero and nonzero coefficients into groups, thereby enhancing the efficiency of the run-length encoding process. Also, an end-of-block (EOB) marker is used to signify when all of the remaining coefficients in the sequence are equal to zero. Such an approach can be very efficient, yielding a significant degree of compression.

Decoder Block Diagram. The video decoder is shown in Figure 15.14. From left to right in the figure, the incoming coded bit stream is stored in the channel buffer. These bits are removed from the buffer by a variable length decoder (VLD).

The 8-by-8 arrays of quantized DCT coefficients are reconstructed by decoding run-length amplitude codes and appropriately distributing the coef-

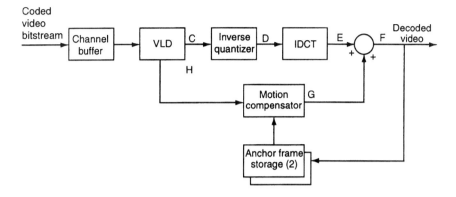

C	Prediction error DCT coefficients in quantized from
D	Quantized prediction error DCT coefficients in standard form
E	Pixel-by-pixel prediction errors, degraded by quantization
F	Reconstructed pixel values, degraded by quantization
G	Motion compensated predicted pixel values
H	Motion vectors

Figure 15.14. Functional block diagram of the video decoder. (From Ref. 22, Figure 5.7. Reprinted with permission.)

ficients in accordance with the scan type used. These coefficients are dequantized and transformed by the inverse discrete cosine transform (IDCT) to obtain pixel values or prediction errors.

In the case of interframe prediction the decoder uses the received motion vectors to perform the same prediction as was done in the encoder. These prediction errors are summed with the results of motion compensated prediction to produce the appropriate pixel values.

When transmission errors occur, the decoder may act to minimize the perceived picture degradation.

15.11.4.4 Audio System Overview. Figure 15.15 is a functional block diagram of the audio subsystem. The audio encoder(s) is (are) responsible for generating the audio elementary stream(s), which contains encoded representations of the baseband audio input signals. The flexibility of the transport system allows multiple audio elementary streams to be delivered to a receiver. At the receiver the audio subsystem decodes the elementary stream(s) back into baseband audio.

An audio program source is encoded by the digital television audio encoder. The encoder output is a stream of bits that represent the audio source. It is referred to as an *audio elementary stream*. The transport subsystem packetizes the audio data into PES (packetized elementary stream) packets, which are then further packetized into transport packets. The transport packets modulate a RF carrier for transmission to the far-end

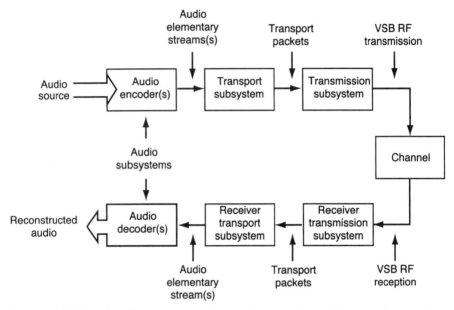

Figure 15.15. Functional block diagram of the audio subsystem within the digital television system. (From Ref. 22, Figure 6.1. Reprinted with permission.)

receiver. At that receiver, the signal is demodulated by the receiver transmission subsystem. The demodulated signal, now consisting of audio packets, is then converted into an audio elementary stream, which is then decoded by the digital television audio decoder.

The audio system accepts up to six audio channels per audio program bit stream. The six audio channels are left, center, right, left surround, right surround, and low-frequency enhancement (LFE). Multiple audio elementary bit streams may be conveyed by the transport system.

The bandwidth of the LFE channel is limited to 120 Hz. The main channel bandwidths are limited to 20 kHz.

The audio sampling frequency (main channel) is 48 kHz, locked to the 27-MHz system clock. In general, input signals are quantized to 16-bit resolution. The audio system can convey audio signals with up to 24-bit resolution.

The audio bit stream consists of a repetition of audio frames, which are referred to as AC-3 sync frames. An AC-3 sync frame is shown in Figure 15.16. Each AC-3 sync frame is a self-contained entity consisting of synchronization information (SI), bit stream information (BSI), 32 ms of encoded audio, and a CRC tail for error detection. Each sync frame is the same size and contains six encoded audio blocks. The sync frame may be called an audio access unit. Within the SI is a 16-bit sync word, an indication of the audio sample rate (48 kHz for the digital television system), and an indication of the size of the audio frame, which tells us the bit rate.

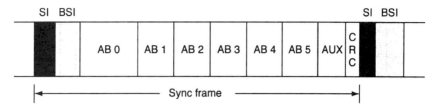

SI BSI

| | AB 0 | AB 1 | AB 2 | AB 3 | AB 4 | AB 5 | AUX | C R C |

SI BSI

|←——————————— Sync frame ———————————→|

Figure 15.16. The AC-3 synchronization frame. (From Ref. 22, Figure 6.3. Reprinted with permission.)

15.11.4.5 Service Multiplex and Transport Systems. The organization of a digital television transmitter–receiver and the location of the transport subsystem are illustrated in Figure 15.17. The subsystem formats the encoded bits and multiplexes the different components of the program for transmission. The far-end receiver side recovers the bit streams for the individual decoders and carries out error detection.

The data transport is based on fixed-length packets that are identified by headers. Each header identifies a particular application bit stream, which is

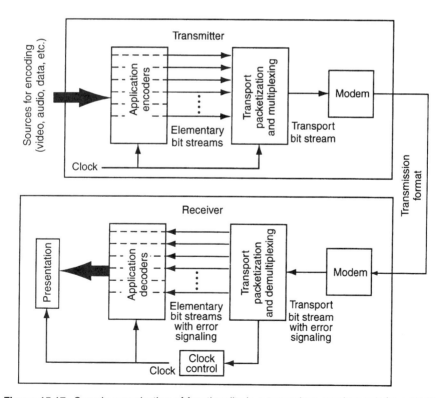

Figure 15.17. Sample organization of functionality in a transmitter–receiver pair for a single digital television program. (From Ref. 22, Figure 8.1. Reprinted with permission.)

also called an *elementary bit stream*. This forms the payload of the packet. These applications include video, audio, data, program, and system control information. The video and audio elementary bit streams are enclosed in a variable-length packet structure, which is called the *packetized elementary stream* (PES).

Moving up one layer in functionality of the general organization of bit streams, elementary bit streams sharing a common time base are multiplexed, along with control data, into *programs*. A program is analogous to a channel in the NTSC system because it contains all of the video and other information required to make up a complete television program. These programs and an overall system control data bit stream are then asynchronously multiplexed for transmission.

Packetization Approach and Functionality. Figure 15.18 shows the transport packet format. Each of these fixed-length packets are 188 bytes long and are based on MPEG-2 transport syntax and semantics. The contents of each packet are identified by packet headers. The header structure is layered and is somewhat similar to OSI link and transport layers.

The *link layer*, which contains 4 bytes, carries out packet synchronization, packet identification, and error handling. Packet synchronization is enabled by the first byte in the packet. Packet identification is contained in a 13-byte field called the PID or packet ID field. This function is the basic mechanism for multiplexing and demultiplexing bit streams. It identifies packets belonging to an elementary or control bit stream. The location of the PID field in the header is fixed. Once packet synchronization is established, extraction of packets corresponding to a particular elementary bit stream is very simple to achieve by filtering packets based on PIDs.

Error detection is carried out at the packet layer in the decoder through the use of the continuity_counter field. The value of this field, at the transmit end, cycles from 0 through 15 for all packets with the same PIDs that carry a data payload. At the far-end receiver, under normal conditions, the reception of packets in a PID stream with a discontinuity in continuity_counter value indicates that data have been lost in transmission. At the decoder, the transport processor can signal the decoder for the particular elementary bit stream about the loss of data.

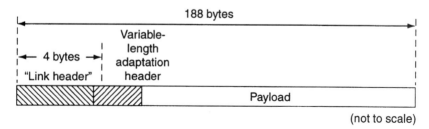

(not to scale)

Figure 15.18. Transport packet format. (From Ref. 22, Figure 8.2. Reprinted with permission.)

One method to improve robustness of very important data is to transmit duplicate packets containing the data. Among the family of very important data that contribute to the smooth and continuous operation of the system we will find adaptation headers, time stamps, and program maps. At the decoder the duplicate packets are either used, if the original packet was in error, or are dropped.

Synchronization and Timing. Received data are expected to be processed at the same rate at which they were generated at the transmitter; otherwise there can be buffer overflow or underflow. It can also cause loss of presentation and/or display information. For conventional analog NTSC television, clock and synchronization derive directly from picture synchronizing pulses. It is done quite differently for compressed digital television. For one thing, the amount of data generated for the compressed digital picture is variable, based on the picture coding approach and complexity. Timing cannot be derived directly from the start of picture data and there is really no natural concept of synchronism between transmission and display in a compressed digital television system.

Timing information is carried in the adaptation (layer) headers of selected packets to serve as reference timing at the decoder. A sample of a 27-MHz clock is carried in the program_clock_reference (PCR) field, which indicates the expected time at the completion of the reading of that field from the bit stream at the transport decoder. The local running clock phase at the decoder is compared to the PCR value in the bit stream at the instant at which it is obtained, to determine whether the decoding process is synchronized. The PCR from the bit stream does not directly change the phase of the local clock but only serves as an input to adjust the local clock rate. The audio and video sample clocks in the decoder system are locked to the system clock derived from the PCR values.

15.11.4.6 RF/Transmission Systems

Overview. There are two vestigial sideband (VSB) transmission modes: 8 VSB for simulcast terrestrial broadcast mode and 16 VSB for a high data rate mode. These two modes share the same pilot, symbol rate, data frame structure, interleaving, Reed–Solomon coding, and synchronization pulses. Both modes can be accommodated in a 6-MHz channel.

To maximize the service area, the terrestrial broadcast mode incorporates both a NTSC rejection filter in the receiver and trellis coding. The transmitter precodes with a trellis code. The trellis decoder is switched to a trellis code corresponding to the encoder trellis code concatenated with the filter when the NTSC rejection filter is activated in the receiver. The decoder only requires one A/D converter and a real (not complex) equalizer at the symbol rate of 10.76 megasamples per second.

Table 15.8 lists the basic parameters for the two VSB transmission modes.

TABLE 15.8   Parameters for VSB Transmission Modes

Parameter	Terrestrial Mode	High Data Rate Mode
Channel bandwidth	6 MHz	6MHz
Excess bandwidth	11.5%	11.5%
Symbol rate	10.76 Msymbols/s	10.76 Msymbols/s
Bits per symbol	3	4
Trellis FEC	2/3 rate	None
Reed–Solomon FEC	T = 10 (207,187)	T = 10 (207,187)
Segment length	832 symbols	832 symbols
Segment sync	4 symbols per segment	4 symbols per segment
Frame sync	1 per 313 segments	1 per 313 segments
Payload data rate	19.28 Mpbs	38.57 Mbps
NTSC cochannel rejections	NTSC rejection filter in receiver	N/A
Pilot power contribution	0.3 dB	0.3 dB
C/N threshold	14.9 dB	28.3 dB

Source: Reference 22, Table 9.1. Reprinted with permission.

Bit Rate Delivered to a Transport Decoder by the Transmission Subsystem.
The exact symbol rate of the transmission subsystem is given by

$$\frac{4.5 \times 684}{286} = 10.76 \text{ MHz} \tag{15.2}$$

The symbol rate is locked in frequency to the transport rate. The transmission subsystem carries 2 information bits per trellis-coded symbol, so the gross modulation payload is

$$10.6 \times 2 = 21.52 \text{ Mbps} \tag{15.3}$$

Equation 15.3 is adjusted for the overhead of the data segment sync, data field sync, and Reed–Solomon FEC. Now the net payload bit rate of the 8 VSB terrestrial transmission system becomes:

$$21.52 \text{ Mbps} \times \frac{312}{313} \times \frac{828}{832} \times \frac{187}{207} = 19.28 \text{ Mbps} \tag{15.4}$$

The factor 312/313 accounts for the data field sync overhead of one data segment per field. The factor 828/832 accounts for the data segment sync overhead of four symbol intervals per data segment, and the factor 187/207 accounts for the Reed–Solomon FEC overhead of 20 bytes per data segment.

To calculate the net payload bit rate of the high data rate mode, the same exercise is carried out as formula 15.4, except that 16 VSB carries four information bits per symbol. Thus the net bit rate is twice that of the 8 VSB terrestrial mode, or

$$19.28 \text{ Mbps} \times 2 = 38.57 \tag{15.5}$$

To calculate the net bit rate of the transport decoder, it is necessary, however, to account for the fact that the MPEG-2 sync bytes are removed from the data stream input to the 8 VSB transmitter. This amounts to the removal of 1 byte per data segment. These MPEG sync bytes are then reconstituted at the output of the 8 VSB receiver. Then the net bit rate as seen by the transport decoder is

$$19.28 \text{ Mbps} \times \frac{188}{187} = 19.39 \text{ Mbps} \tag{15.6}$$

And the net bit rate seen by the transport decoder for the high data rate mode is

$$19.39 \text{ Mbps} \times 2 = 38.78 \text{ Mbps} \tag{15.7}$$

Performance Characteristics of the Terrestrial Broadcast Mode. The terrestrial VSB system can operate with a signal-to-additive-white Gaussian-noise ratio of 14.9 dB. The four-state segment error probability curve for 8 VSB operation is shown in Figure 15.19. This shows a segment error probability of 1.93×10^{-4}, which is equivalent to 2.5 segment errors per second. Measurement has established this value as the threshold of visibility errors.

15.11.4.7 *Receiver Characteristics.* Reference 22 points out that at the time Doc. A/54 was prepared, only one prototype receiver was fabricated and tested. The receiver planning factors are summarized in Table 15.9.

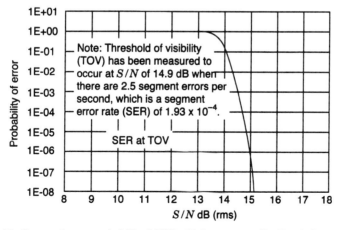

Figure 15.19. Segment error probability, 8 VSB with four-state trellis, Reed–Solomon coding (207, 187). (From Ref. 22, Figure 9.1. Reprinted with permission.)

TABLE 15.9 Receiver Planning Factors

Planning Factors	Low VHF	High VHF	UHF
Antenna impedance (ohms)	75	75	75
Bandwidth (MHz)	6	6	6
Thermal noise (dBm)	−106.2	−106.2	−106.2
Noise figure (dB)	10	10	10
Frequency (MHz)	69	194	615
Antenna factor (dBm/dB μ)	−111.7	−120.7	−130.7
Line loss (dB)	1	2	4
Antenna gain (dB)	4	6	10
Antenna F/ratio (dB)	10	12	14

Source: Table 10.1. Reprinted with permission.

15.12 CONFERENCE TELEVISION*

15.12.1 Introduction

Video conferencing (conference television) systems have seen phenomenal growth since the early 1990's. Many of the world's corporations have branches and subsidiaries that are widely dispersed. Rather than pay travel expenses to send executives to periodic meetings at one central location, video conferencing is used, saving money on the travel budget.

The video and telecommunications technology has matured in the intervening period to make video conferencing cost effective. Among these developments we include:

- Video compression techniques
- Eroding cost of digital processing
- The arrival of the all-digital network

Proprietary video conferencing systems normally use lower line rates than conventional broadcast TV. Whereas conventional broadcast TV systems have line rates at 525/480 line (NTSC countries) or 625/580 for PAL/SECAM countries, proprietary video conferencing systems use either 256/240 or 352/288 lines. For the common applications of conference television (e.g., meetings and demonstrations), the reduced resolution is basically unnoticeable.

One of the compression schemes widely used for video conference systems is based on ITU-T Rec. H.261 (Ref. 24), titled "Video Codec for Audiovisual Services at px64 kbps," which is described in the following.

---

*Section 15.12 is based on Ref. 24, CCITT Rec. H.261.

15.12.2   The px64-kbps Codec

The px64 codec has been designed for use with some of the ISDN data rates, specifically B-channel (64 kbps), H_0 (384 kbps), and H_{11}/H_{12} (1.536/1.984 Mbps) for the equivalent DS1/E1 data rates. A functional block diagram of the codec (coder/decoder) is shown in Figure 15.20. However, the px64 system uses standard line rates (i.e., 525/625 lines) rather than the reduced line rate structure mentioned earlier.

One of the most popular rates is 384 kbps, which is 6 DS0 or E0 channels. However, it is not unusual to find numerous systems operating at 64/56 kbps.

15.12.2.1   pX64 Compression Overview

Sampling Frequency. Pictures are sampled at an integer multiple of the video line rate. The sampling clock and network clock are asynchronous.

Source Coding Algorithm. Compression is based on interpicture prediction to utilize temporal redundancy and transform coding of the remaining signal to reduce spatial redundancy. The decoder has motion compensation capability, allowing optional incorporation of this technique in the coder. There is optional forward error correction available based on the BCH $(511, 493)$ code. The codec can support multipoint operation.

15.12.2.2   Source Coder. The coder operates on noninterlaced pictures occurring 30,000/1001 (approximately 29.97) times per second. The tolerance on the picture frequency is ± 50 ppm.

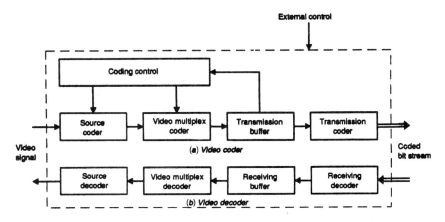

Figure 15.20. Outline functional block diagram of the px64 video codec. (From Ref. 24, Figure 1/H.261, ITU-T Rec. H.261.)

As in Section 15.11.4, pictures are coded as one luminance and two color difference components (Y, Cb, and Cr). Refer to CCIR Rec. 601-4 (Ref. 23) for their components and codes representing their sampled values. For example,

Black = 16
White = 235
Zero color difference = 128
Peak color difference = 16 and 240

The values given are nominal values and the coding algorithm functions with input values of 1 through 254. Two picture scanning formats have been specified.

For the first format (CIF), the luminance structure is 352 pels per line, 288 lines per picture in an orthogonal arrangement. The color-difference components are sampled at 176 pels per line, 144 lines per picture, orthogonal. Figure 15.21 shows the color-difference samples being sited such that the block boundaries coincide with luminance block boundaries. The picture area covered by these numbers of pels and lines has an aspect ratio of 4:3 and corresponds to the active portion of the local standard video input.

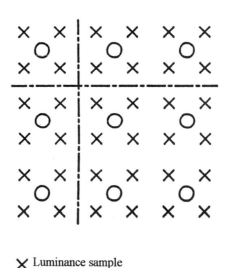

× Luminance sample

○ Chrominance sample

— ·· — Block edge

Figure 15.21. Positioning of luminance and chrominance samples. (From Ref. 24, Figure 2/H.261, ITU-T Rec. H.261.)

It should be noted that the number of pels per line is compatible with sampling the active portions of the luminance and color-difference signals from 525- or 625-line sources at 6.75 or 3.375 MHz, respectively. These frequencies have a simple relationship with those in CCIR Rec. 601-4 (Ref. 23).

The second format, called quarter-CIF or QCIF, has half the number of pels and half the number of lines used in CIF. All codecs must be able to operate using QCIF.

A means is provided to restrict the maximum picture rate of encoders by having at least 0, 1, 2, or 3 nontransmitted pictures between transmitted pictures. Both CIF/QCIF and this minimum number of nontransmitted frames shall be selectable externally.

Video Source Coding Algorithm. A block diagram of the coder is illustrated in Figure 15.22. The principal functions are prediction, block transformation, and quantization. The prediction error (INTER mode) or the input picture (INTRA mode) is subdivided into 8-pel-by-8-pel line blocks, which are segmented as transmitted or nontransmitted, Furthermore, four luminance blocks and two spatially corresponding color-difference blocks are combined to form a macroblock. Transmitted blocks are transformed and the resulting coefficients are quantized and then variable length coded.

Motion compensation is optional. The decoder will accept one vector per macroblock. The components, both horizontal and vertical, of these motion vectors have integer values not exceeding ± 15. The vector is used for all four luminance blocks in the macroblock. The motion vector for both color-difference blocks is derived by halving the component values of the macroblock vector and truncating the magnitude parts toward zero to yield integer components.

A positive value of the horizontal or vertical component of the motion vector signifies that the prediction is formed from pels in the previous picture, which are spatially to the right or below the pels being predicted. Motion vectors are restricted such that all pels referenced by them are within the coded picture area.

Loop Filter. A two-dimensional spatial filter may be used in the prediction process. The filter operates on pels within a predicted 8-by-8 block. It is separable into one-dimensional horizontal and vertical functions. Both are nonrecursive carrying coefficients of $\frac{1}{4}, \frac{1}{2}, \frac{1}{4}$ except at block edges, where one of the taps would fall outside the block. In this case the one-dimensional filter is changed to have coefficients of $0, 1, 0$. There is rounding to 8-bit integer values at the two-dimensional filter output and full arithmetic precision is retained. Rounding upward is used where values whose fractional part is one-half. The filter is switched on/off for all six blocks in a macroblock according to the macroblock type. There are ten types of macroblocks such as INTRA, INTER, INTER + MC (motion compensation), and INTER + MC + FIL (filter).

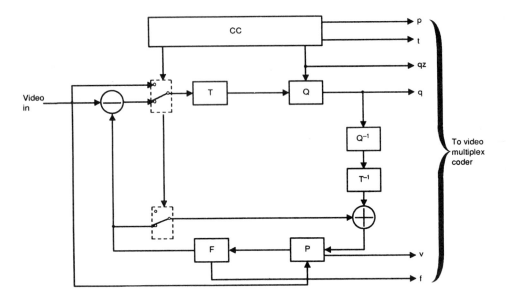

T	Transform
Q	Quantizer
P	Picture memory with motion compensated variable delay
F	Loop filter
CC	Coding control
p	Flag for INTRA/INTER
t	Flag for transmitted or not .
qz	Quantizer indication
q	Quantizing index for transform coefficients
v	Motion vector
f	Switching on/off of the loop filter

Figure 15.22. Functional block diagram of the source coder. (From Ref. 24, Figure 3/H.261, ITU-T Rec. H.261.)

Discrete Cosine Transform. The transmitted blocks are first processed by a separable two-dimensional discrete cosine transform, which is 8 by 8 in size. There is an output range of the inverse transform from -256 to $+255$ after clipping to be represented by 9 bits.

Quantization. There are 31 quantizers for all other coefficients except the INTRA dc coefficient, which has just 1. The decision levels are not defined in CCITT Rec. H.261. The INTRA dc coefficient is nominally the transform value linearly quantized with a step size of 8 and no dead-zone. The other 31 quantizers are nominally linear but with a central dead-zone around zero and with a step size of an even value in the range of 2–62.

Clipping of Reconstructed Picture. Clipping functions are inserted to prevent quantization distortion of transform coefficient amplitudes causing arithmetic overflow in the encoder and decoder loops. The clipping function is applied to the reconstructed picture. This picture is formed by summing the prediction and the prediction error as modified by the coding process. When resulting pel values are less than 0 or greater than 255, the clipper changes them to 0 and 255, respectively.

Coding Control. To control the rate of generation of coded video data, several parameters may be varied. These parameters include processing prior to source coder, the quantizer, block significance criterion, and temporal subsampling. When invoked, temporal subsampling is performed by discarding complete pictures.

Forced Updating. Forced updating is achieved by forcing the use of the INTRA mode of the coding algorithm. Recommendation H.261 does not define the update pattern. For the control of accumulation of inverse transform mismatch error, a macroblock should be updated forcibly at least once per every 132 times it is transmitted.

15.12.2.3 Video Multiplex Coder. The video multiplex is arranged in a hierarchical structure with four layers. From top to bottom, these layers are:

- Picture
- Group of blocks (GOB)
- Macroblock (MB)
- Block

For further description of these layers, consult Ref. 24.

15.13   FRAME STRUCTURE FOR A 64–1920-kbps CHANNEL TRANSPORT FOR VIDEO CONFERENCING

15.13.1   Objective

This section outlines one method that has been standardized for the transport of digital video conferencing signals. It is based on ITU-T Rec. H.221 (Ref. 25).

15.13.2   Basic Principle

An overall transmission channel of 64–1920 kbps is dynamically subdivided into lower rates suitable for the transport of audio, video, and data and for telematic purposes. The transmission channel is derived by synchronizing and

Bit number									
1	2	3	4	5	6	7	8 (SC)		
								1	Octet number
S	S	S	S	S	S	S	FAS	:	
u	u	u	u	u	u	u		8	
b	b	b	b	b	b	b		9	
-	-	-	-	-	-	-	BAS	:	
c	c	c	c	c	c	c		16	
h	h	h	h	h	h	h		17	
a	a	a	a	a	a	a	ECS	:	
n	n	n	n	n	n	n		24	
n	n	n	n	n	n	n		25	
e	e	e	e	e	e	e		.	
l	l	l	l	l	l	l		.	
#	#	#	#	#	#	#	#	.	
1	2	3	4	5	6	7	8	80	

FAS   Frame alignment signal

BAS   Bit-rate allocation signal

ECS   Encryption control signal

Figure 15.23. Frame structure of a single 64-kbps channel, which we may call a B-channel. (From Ref. 25, Figure 1/H.221, ITU-T Rec. H.221.)

ordering transmissions over from one to six B-connections,* from one to five H_0-connections,† or an H_{11} or H_{12} connection. The initial connection is the first connection established and it carries the initial channel in each direction. The additional connections carry additional channels. The total rate of transmitted information is called the *transfer rate*. The transfer rate can be less than the capacity of the overall transmission channel.

A single 64-kbps channel is structured into octets transmitted at an 8-kHz rate. Each bit position of the octets may be regarded as a subchannel of 8 kbps. This is shown in Figure 15.23. The service channel (SC) resides in the eighth subchannel, which consists of several parts that are described below.

We can regard an H_0, H_{11}, H_{12} channel as consisting of a number of 64-kbps time slots (TS). This is illustrated in Figure 15.24. The lowest numbered time slot is structured exactly as described above for a single 64-kbps channel; whereas the other time slots have no such structure. All channels have a frame structure in the case of multiple B or H_0 channels; the

---

*This is the ISDN B-channel, which is the same as an E0 channel of 64 kbps and a DS0 channel cleared of signaling bit slots. The use of ISDN nomenclature has no bearing on whether these channels are ISDN derived or DS1/E1 subrate derived.
†The H_0 connection is the 384-kbps channel or 6 DS0 or 6 E0 64-kbps channel.

125 microseconds

1	2	3	4	5	6	7	. . .	. . .	6n–2	6n–1	6n

H_0	$n = 1$
H_{11}	$n = 4$
H_{12}	$n = 5$

Audio + service channel

1	2	3	4	5	6	7	8		Octet number
S	S	S	S	S	S	S	FAS	1	
u	u	u	u	u	u	u		:	
b	b	b	b	b	b	b		8	
-	-	-	-	-	-	-	BAS	9	
c	c	c	c	c	c	c		:	
h	h	h	h	h	h	h	S	16	
a	a	a	a	a	a	a	u	17	
n	n	n	n	n	n	n	b		
n	n	n	n	n	n	n	-		
e	e	e	e	e	e	e	c		
l	l	l	l	l	l	l	h		
							a		
							n		
							n		
							e		
							l		
#	#	#	#	#	#	#	#	80	
1	2	3	4	5	6	7	8		

Figure 15.24. Frame structure for higher-rate single channels such as H_0, H_{11}, and H_{12}. (From Ref. 25, Figure 2/H.221, ITU-T Rec. H.221.)

initial channel controls most functions across the overall transmission, while the frame structure in the additional channels is used for synchronization, channel numbering, and related controls. The term *I-channel* is applied to the initial or only B-channel, to time slot 1 of initial or only H_0 channels, and to TS1 or H_{11}, H_{12} channels.

15.13.3 Frame Alignment Signal (FAS)

The FAS structures the I-channel and other framed 64-kbps channels into frames 80 octets long and multiframes (MF) of 16 frames each. Each multiframe is divided into eight 2-frame submultiframes (SMF). In the case of this discussion based on ITU-T Rec. H.221, the term *frame alignment signal* (FAS) refers to bits 1–8 of the SC (service channel) in each frame. The concepts of FAS introduced in Chapters 3 and 4 are similar to the ones discussed here. Thus we can expect that in addition to framing and multi-framing information, control and alarm information may be inserted in the FAS, as well as error check information to control end-to-end error performance and to check frame alignment validity. Other time slots (TS) are aligned to the first TS. Bits are transmitted to the line in order, bit 1 first.

The FAS is transmitted and received in the least significant bit of the octet within each 125-μs frame period (e.g., in an ISDN basic rate or primary rate interface), when an 8-kHz network clock is provided. Network timing is essential when interworking between the audiovisual terminal and the telephone is required. On the receive side, FAS should be sought in all bit positions. In the case where the received FAS position conflicts with the network octet timing, the FAS position is given priority. This may happen when the receiver uses network octet timing while the transmitter does not, as in the case where a terminal uses codecs separate from the ISDN terminal adaptor, or when interworking takes place between 64-kbps and 56-kbps terminals.

When not provided by the network, the FAS can be used to derive receive octet timing. However, in the latter case, the terminal cannot transmit FAS with correct alignment into the octet timed part of the network and cannot intercommunicate with terminals that rely only on network timing for octet alignment.

15.13.4   Bit Rate Allocation Signal (BAS)

BAS refers to bits 9–16 of the SC in each frame. This signal allows the transmission of codewords to describe the capability of a terminal to structure the capacity of the channel or synchronized multiple channels in various ways, and to command a receiver to demultiplex and make use of the constituent signals in such structures. The BAS is also used for indications and controls.

15.13.5   Encryption Control Signal (ECS)

A future encryption capability may require a dedicated transmission channel. When allocating bits 17–24 of the service channel, the 800 bps anticipated could be provided. This results in a reduction of variable data and video transmission rates herein by 800 bps. The 800 bps is referred to as the ECS channel.

15.13.5.1   Remaining Capacity. There is surplus capacity (including the rest of the service channel) in the case of a 64-kbps connectivity carried in bits 1–8 of each octet. This remaining capacity may carry a variety of signals within the framework of multimedia service, under control of the BAS. The following are some examples:

- Voice encoded at 56 kbps using a truncated form of PCM covered in CCITT Rec. G.711, either A-law or μ-law.
- Voice encoded at 16 kbps and video at 46.4 kbps.
- Voice encoded at 56 kbps with a bandwidth of 50–7000 Hz (subband ADPCM based on CCITT Rec. G.722). The coding algorithm is also

able to work at 48 kbps; data can then be dynamically inserted up to 14.4 kbps.

- Still pictures coded at 56 kbps.
- Data at 56 kbps inside an audiovisual session (such as file transfer for communicating between personal computers).

15.13.6   Frame Alignment

An 80-octet frame length produces an 80-bit word in the service channel. These 80 bits are numbered 1–80. As shown in Figure 15.25, bits 1–8 of the service channel in every frame constitute the frame alignment signal (FAS). The content of the FAS is as follows:

- Multiframe structure (described below)
- Frame alignment word (FAW)
- A-bit
- E- and C-bits (see in the following)

The four C-bits are used for a CRC-4 computation; the E-bit, the "error" bit, is an indication of a far-end error. The FAW consists of 0011011 in bit positions 2–8 of the FAS in even frames, complemented by a "1" in bit position 2 of the succeeding odd frame. The A-bit of the I-channel is set to 0 whenever the receiver is in multiframe alignment; otherwise it is set to 1.

15.13.7   Multiframe Structure

The multiframe structure is shown in Figure 15.26. Each multiframe contains 16 consecutive frames numbered 0–15 divided into eight submultiframes of two frames each. The multiframe alignment signal is in the form of 001011 and is located in bit 1 of frames 1-3-5-7-9-11. Bit 1 of frame 15 is for future use and its value is fixed at 0.

To number frames in descending order, bit 1 of frames 0-2-4-6 may be used for a modulo-16 counter. The least significant bit is transmitted in frame 0, and the most significant bit in frame 6. At the receive end, the multiframe numbering is used to equalize the differential delay of separate connections and to synchronize the receive signals.

In the cases of initial and additional channels for multiple B or multiple H_0 communications, the multiframe numbering is mandatory. It may or may not be inserted for single B or single H_0 or H_{11}/H_{12}, where synchronization between multiple channels is not required. Bit 1 of frame 8 is set to 1 when multiframes are numbered and is set to 0 when they are not.

Bit 1 of frames 10-12-13 must be used to number each channel in a multiconnection structure so that the distant receiver can place the octets received in each 125-μs period in the correct order. Information bits in the

				Bit number			
	1	2	3	4	5	6	7	8
Successive frames	1							8
Even frames	(Note 1)	0	0	1	1	0	1	1
				Frame alignment word (Note 2)			
Odd frames	(Note 1)	1 (Note 2)	A (Note 3)	E (Note 4)	C1	C2	C3	C4

Figure 15.25. Assignment of bits 1–8 of the service channel in each frame. (From Ref. 25, Figure 3/H.221, ITU-T Rec. H.221.)

NOTES

1 See 2.2 and Figure 4.

2 The first seven bits of the frame alignment word are in the even frames. The eighth bit of the FAW in the odd frame is the complement of the first FAW bit in order to avoid simulation of FAW by a frame-repetitive pattern.

3 A-bit: loss of multiframe alignment indication (0 = alignment; 1 = loss).

4 The use of bits E and C1–C4 is described in 2.6 (0 = no error or cyclic redundancy check (CRC) not in use; 1 = error).

Sub-multiframe (SMF)	Frame	Bits 1 to 8 of the service channel in every frame							
		1	**2**	**3**	**4**	**5**	**6**	**7**	**8**
SMF1	0	N1	0	0	1	1	0	1	1
	1	0	1	A	- E	C1	C2	C3	C4
SMF2	2	N2	0	0	1	1	0	1	1
	3	0	1	A	E	C1	C2	C3	C4
SMF3	4	N3	0	0	1	1	0	1	1
	5	1	1	A	E	C1	C2	C3	C4
SMF4	6	N4	0	0	1	1	0	1	1
	7	0	1	A	E	C1	C2	C3	C4
SMF5	8	N5	0	0	1	1	0	1	1
	9	1	1	A	E	C1	C2	C3	C4
SMF6	10	L1	0	0	1	1	0	1	1
	11	1	1	A	E	C1	C2	C3	C4
SMF7	12	L2	0	0	1	1	0	1	1
	13	L3	1	A	E	C1	C2	C3	C4
SMF8	14	TEA	0	0	1	1	0	1	1
	15	R	1	A	E	C1	C2	C3	C4

(Multiframe spans all SMF rows)

L1-L3   Channel number, least significant bit in L1

Channel	L3 ·	L2	L1
Initial	0	0	1
Second	0	1	0
Third	0	1	1
...	..	..	..
Sixth	1	1	0

R   Reserved for future use set to 0.

A, E, C1-C4   As in Figure 3.

N1-N4   Used for multiframe numbering as described in 2.2; set to 0 while numbering is inactive.

	N4	N3	N2	N1	
Multiframe number	0	0	0	0	0	(or numbering inactive)
	1	0	0	0	1	
	2	0	0	1	0	
	..	..	..	..	..	
	15	1	1	1	1	

N5   Indicates whether multiframe numbering is active (N5 =1) or inactive (N5 = 0).

TEA   The terminal equipment alarm is set to 1 the outgoing signal while an internal terminal fault exists such that it cannot receive and act on the incoming signal. Otherwise it is set to 0.

Figure 15.26. Assignment of bits 1–8 of the service channel of each frame in a multiframe. (From Ref. 25, Figure 4/H.221, ITU-T Rec. H.221.)

multiframe should be validated by, for example, being received consistently for three multiframes.

15.13.8 Loss and Recovery of Frame Alignment

Frame alignment is assumed lost when three consecutive frame alignment words have been received with an error.

Frame alignment is declared recovered when the following sequence is detected:

- For the first time, the presence of the correct first 7 bits of the frame alignment word
- For the eighth bit, the frame alignment word in the following frame is detected verifying that bit 2 is a 1
- for the second time, the presence of the correct first 7 bits of the frame alignment word in the next frame

If frame alignment is achieved but multiframe alignment cannot be achieved, then frame alignment should be sought at another position. When the frame alignment is lost, the A-bit of the next frame is set to 1 in the transmit direction.

Multiframe alignment is defined to have been lost when three consecutive multiframe alignment signals have been received in error. It is defined to have been recovered when the multiframe alignment signal has been received with no error in the next multiframe. When multiframe alignment is lost, even when an unframed mode is received, the A-bit of the next odd frame is set to 1 in the transmit direction. When multiframe alignment is regained, it is set to 0. It is reset in additional channels when multiframe alignment and synchronism with the initial channel are gained.

15.13.9 Search for Frame Alignment Signal (FAS)

There are two methods that may be used to search for frame alignment: sequential and parallel. In the sequential method, eight possible bit positions for FAS are tried. When FAS is lost after being validated, the search must resume starting from the previously validated bit position. In the parallel method, a sliding window, shifting one bit for each bit period, may be used. In that case, when frame alignment is lost, the search must resume starting from the bit position next to the previously validated one.

REVIEW EXERCISES

1. What four factors must be dealt with by a color video transmission system transmitting images of moving objects?

2. Describe scanning: (1) horizontally and (2) vertically.

3. Define a pel or pixel (beyond the meaning of the acronym).

4. What do we mean by an aspect ratio of 4 : 3? Suppose the width of an image is 12 in. What is its height?

5. NTSC divides an image into how many horizontal lines? How many horizontal lines are there in European practice?

6. Given an aspect ratio of 4 : 3, an image with 525 vertical lines (elements) would have how many horizontal elements?

7. How do we achieve the sensation of motion in TV? Relate this to frame rate and flicker.

8. Describe interleaving (regarding TV video) and explain why it is used.

9. Define a field period regarding North American (NTSC) practice.

10. In North American practice the time to scan a line is 63.5 μs. This time interval consists of two segments. What are they?

11. What is the standard maximum voltage excursion of a video signal? Just what are we measuring here? (What TV signal component?)

12. How is frame rate related to power line frequency or "flicker frequency"?

13. Give two definitions of composite signal.

14. At a TV receiver, what signal-to-noise ratio S/N (dB) is required for an excellent picture?

15. How is S/N normally measured and in what units? (We mean here the signal and the noise.)

16. What are the two major additional impairments that must be considered when transmitting a color signal?

17. What type of modulation is used to transmit video? For an audio subcarrier? For a chrominance subcarrier?

18. Give four generalized differences in video transmission between general European practice and North American/Japanese practice.

19. How is a program channel commonly handled when TV is transported by line-of-sight (LOS) microwave, satellite, and other media when specifically to be used by broadcasters (not necessarily cable TV)?

20. For digital TV based on PCM techniques, calculate at least two sampling rates.

21. Why is S/N end-to-end specified as 54 dB when only 45 dB is required at a user TV receiver for an excellent picture?

22. Differentiate composite coding from component coding. Name one major advantage of component coding.

23. What type of coding is nearly universally used to digitize video for transmission?

24. Give the three basic bit rate reduction methods for digitized TV.

25. What are the desirable digital line rates for TV? Why use those rates and not some other rate(s)?

26. What are the two basic types of redundancy encountered in a TV video system?

27. Distinguish intraframe coding from interframe coding. What type of redundancy does each address?

28. To what voltage value does 0 dBmV correspond? When we give a value in dBmV, we must also state another parameter. What is this important parameter?

29. How does DPCM (alone) achieve bit rate reduction?

30. How does motion compensation coding reduce bit rate?

31. What is the general range of bit rates achieved for broadcast-quality TV when effective bit rate reduction techniques are employed?

32. Give the two bit rates at the output of a MPEG-2 coder based on the ATSC specification. Relate each bit rate to an application.

33. How many bits per sample are used with MPEG-2/ATSC PCM?

34. What are the two basic components of the video portion of a modern TV image?

35. Define a *slice* in MPEG-2/ATSC coding.

36. Differentiate among I-frames, P-frames, and B-frames.

37. What is a *video sequence*?

38. What are *anchor frames* and why are they necessary?

39. Regarding compression, explain *motion compensation*.

40. Discuss *image refresh* and necessary minimum image refresh rates.

41. What is the basic idea or concept behind *entropy coding*?

42. Describe *Huffman coding* in two sentences.

43. Describe the two VSB transmission modes used in MPEG-2/ATSC compression.

44. What three factors made video conferencing very popular?

REFERENCES

1. *Fundamentals of Television Transmission*, Bell System Practices, Section AB 96.100, American Telephone & Telegraph Co., New York, Mar. 1954.

2. *Television Systems Descriptive Information—General Television Signal Analysis*, Bell System Practices, Section 318-015-100, no. 3, American Telephone & Telegraph Co., New York, Jan. 1963.

3. K. Blair Benson and Jerry C. Whitaker, *Television Engineering Handbook*, McGraw-Hill, New York, 1992.

4. Andrew F. Inglis and Arch C. Luther, *Video Engineering*, 2nd ed., McGraw-Hill, New York, 1996.

5. K. Simons, *Technical Handbook for CATV Systems*, 3rd ed., General Instrument–Jerrold Electronics Corp., Hatboro, PA, 1980.

6. *Electrical Performance for Television Transmission Systems*, ANSI/EIA/TIA-250C, EIA/TIA, Washington, DC, Jan. 1990.

7. "Television Systems," ITU-R Rec. BT.470-4, 1995 BT Series, ITU, Geneva, 1995.

8. *Reference Data for Engineers*: *Radio, Electronics, Computer & Communications*, 8th ed., Sams–Prentice-Hall, Carmel, IN, 1993.

9. "Transmission Performance of Television Circuits Designed for Use in International Connections," CCIR Rec. 567-3, Vol. XII, ITU, Geneva, 1990.

10. *Code of Federal Regulations* (*47*) *Telecommunication, Part* 73 (FCC Rules and Regulations), FCC, Washington, DC, Oct. 1994.

11. "The Preferred Characteristics of a Single Sound Channel Simultaneously Transmitted with a Television Signal on an Analogue Radio-Relay System," CCIR Rec. 402-2, Vol. IX, Part 1, ITU, Geneva, 1990.

12. "The Preferred Characteristics for the Simultaneous Transmission of Television and a Maximum of Four Sound Channels on Analogue Radio-Relay Systems," CCIR Rep. 289-4, Part 1, Vol. IX, CCIR, Dubrovnik, 1986.

13. "Pre-emphasis Characteristics for Frequency Modulation Radio-Relay Systems for Television," CCIR Rec. 405-1, Vol. IX, Part 1, ITU, Geneva, 1990.

14. "Frequencies and Deviations of Continuity Pilots for Frequency-Modulation Radio-Relay Systems for Television and Telephony," CCIR Rec. 401-2, Vol. IX, Part 1, ITU, Geneva, 1990.

15. "Transmission Performance of Television Circuits over Systems in the Fixed Satellite Service," CCIR Rep. 965-1, Annex to Vol. XII, ITU, Geneva, 1990.

16. L. E. Weaver, *Television Video Transmission Measurements*, Marconi Instruments, St. Albans, Herts, England, 1972.

17. "Digital or Mixed Analogue-and-Digital Transmission of Television Signals," CCIR Rep. 646-4, Annex to Vol. XII, ITU, Geneva, 1990.

18. "Digital Transmission of Component-Coded Television Signals at 30–34 Mbps and 45 Mbps," CCIR Rep. 1235, Annex to Vol. XII, ITU, Geneva, 1990.

19. "Generic Coding of Moving Pictures and Associated Audio Information: Systems," ITU-T Rec. H.222.0 (MPEG-2), ITU, Geneva, 1996.

20. *A Compilation of Advanced Television Systems Committee Standards*, Advanced Television Systems Committee (ATSC), Washington, DC, Oct. 1996.
21. *ATSC Digital Television Standard*, Doc. A/53, ATSC, Washington, DC, Sept. 1995.
22. *Guide to the Use of the ATSC Digital Television Standard*, Doc. A/54, ATSC, Washington, DC, Oct. 1995.
23. "Encoding Parameters of Digital Television for Studios," ITU-R Rec. BT.601-4, 1994 BT Series Volume, ITU, Geneva, 1994.
24. "Video Codec for Audiovisual Services at px64 kbps," CCITT Rec. H.261, ITU, Geneva, 1990.
25. "Frame Structure for a 64 to 1920 kbps Channel in Audiovisual Teleservices," ITU-T Rec. H.221, ITU, Geneva, 1993.
26. *The IEEE Standard Dictionary of Electrical and Electronics Terms*, 6th ed., IEEE Std-100-96, IEEE, New York, 1996.

16

FACSIMILE COMMUNICATION

16.1 INTRODUCTION

Over the past decade facsimile has very nearly become ubiquitous. Facsimile equipment is commonly found in all businesses and in many residences. Its application today is essentially document transmission where all that is required is black and white copy, no tonal grays. Prior to about 1980, facsimile was analog, usually a FM signal that was compatible with the analog voice channel. Today it is universally digital from the near-end scanner output to the far-end recorder input. Conventional data modems or digital service units are used to convert or condition the signal for transport on the voice channel.

This chapter covers facsimile scanning and recording. It then has an overview of CCITT Group 3 and Group 4 facsimile systems and their compression techniques, including entropy coding applications. The definition of facsimile is stretched to include the transmission of still pictures. This includes the Joint Photographic Experts Group (JPEG) standard as embodied in CCITT Rec. T.81 (Ref. 9).

My first experience with facsimile was aboard ship in the 1950s. It was used for the reception of weather maps. Sensitized paper was carefully wrapped around a drum. The drum rotated and an electric stylus "burned" a facsimile image on the paper. The standard transmission rate was 20 minutes per page. Present Group 3 facsimile is around 40 seconds per page, and Group 4 can produce a page a second. Since present facsimile is nearly ubiquitous and considering the transmission speeds, Telex has been pretty much displaced.

16.2 BASIC FACSIMILE OPERATION

16.2.1 General

A facsimile system consists of some method of converting graphic copy on paper to an electrical equivalent signal suitable for transmission on a telephone pair (or other narrowband media), the connection of the pair/tele-phone circuit and transmission to the distant end-user, and the recording/printing of the copy on paper by that user. Eight basic functions or processes are involved (Ref. 10):

1. Scanning.
2. A/D conversion.
3. Digital processing including compression.
4. Modem function as any other data modem.
5. Transmission and switching carried out by the PSTN as any other telephone call.
6. Modem function (demodulate).
7. Digital processing, decompression.
8. Recording/printing.

Figure 16.1 is a simplified block diagram of a facsimile system showing these eight functions. There is a scanning function carried out by a photoelectrical transducer, which is similar in many respects to the scanning used in video (see Chapter 15). The output of the scanner is analog, which is then converted to digital by an A/D converter. The digital signal is processed and compressed. The resulting 1's and 0's are transmitted via a modem or other conditioning device to a standard telephone line. At the receive end, the signal is demodulated by a companion modem, processed (which includes decompression to the original digital signal), and then digital-to-analog converted. The resulting analog signal is fed to a facsimile recorder for printing.

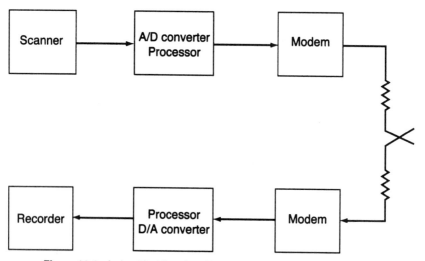

Figure 16.1. A simplified functional block diagram of a facsimile system.

The complete assembly as shown in Figure 16.1 is incorporated directly into the facsimile equipment. The other functions of the processors are compression and operational. The operational functions include setting up a circuit, synchronization and phasing of the transmit side with the receive side, maintaining the circuit, and finally terminating the connection when facsimile exchange is completed. These operational features are based on ITU-T Rec. T.30.

We will first discuss scanning and recording.

16.2.2 Scanning

The scanner is a photoelectric transducer that produces an electrical analog signal representing the graphic copy to be transmitted. With conventional facsimile, scanning is carried out by one of three basic methods, which are essentially mechanical in variation:

- A spot of light scans a fixed graphic copy.
- The copy moves across a fixed spot of light.
- Both the copy and the scan spot may move, usually at right angles to each other and at different speeds.

There are two common approaches of "lighting" the copy. Both employ the technique of bouncing (reflecting) light off the graphic material to be transmitted. One approach to scanning projects a tiny spot of light onto the surface of the printed copy. The reflection of the spot is then picked up directly by a photoelectric cell or other transducer.

The other approach is the flood projection technique whereby the printed copy is illuminated with diffuse light in the area of scan. The reflected light is then optically projected through a very small aperture onto the cathode of a photoelectric transducer.

In modern electromechanical facsimile systems, cylinder scanning has practically disappeared and flatbed scanning is almost universal. Flatbed scanners use a feed mechanism. The copy to be transmitted is fed into a slot where a feed mechanism takes over, slowly advancing the copy through the machine. Because all machines compress, and the compression is based on black density, the paper passing through the machine will slow or even stop over dense black areas. This control is used to avoid buffer overflow. Once sufficient bits are pulled out of the buffer, the paper will start to move again through the scanner until the transmission of the page is completed.

The electronics of facsimile scanners is based on the photoelectric cell, the photomultiplier tube, or the photodiode. They differ primarily in the relative signal strength of their outputs in relation to scan spot intensity.

The charge-coupled-device (CCD) electronic scanning, which previously had been exclusively in the realm of cathode ray tubes (CRTs), now offers promise in the area of facsimile scanning, making it particularly attractive for direct digital transmission of a facsimile signal.

The electrical output of a conventional scanner may also be digital. In fact, the trend is more and more toward digital facsimile systems. In its simplest form of digital transmission, where the signal represents either black or white, a decision circuit may provide a series of marks or 1's for black copy and spaces or 0's for white copy. If, in addition, shades of gray are to be transmitted as well, a level quantizing scheme with subsequent coding is required. This, of course, in most respects is similar to the quantizing (based on fixed signal thresholds) and coding described in Chapter 3.

16.2.3 Recording / Printing

Recording in a facsimile system is the reproduction (e.g., printing) of visual copy of graphic material from an electric signal. Recorders remain electromechanical, just as their scanner counterparts are. And like scanners, recorders are available in cylinder configurations and flatbed. Certain desktop versions are on the market that are semicylindrical, allowing continuous paper feed. There are four basic electromechanical recording methods in use today (Ref. 1):

- Electrolytic
- Electrothermal
- Electropercussive
- Electrostatic

Two types of facsimile recording do not fit into the above categories. One prominent German manufacturer uses an offset process. The other process involves the use of modulation of a fine spray of ink directed to the surface of plain paper. Yet another process, which is well established, is not electromechanical and is used in the facsimile reproduction of newspapers.

Electrolytic Recording. Electrolytic recording is one of the most popular methods of facsimile recording, yet one of the oldest. It requires a special type of recording paper that is actually an electrolyte-saturated material. When an electric current passes through it, the material tends to discolor. The amount of discoloration or darkness is a function of the current passing through the paper.

To record an image, the electrolytic paper is passed between two electrodes. One is a fixed electrode, a backplate or platen on the machine; the other is a moving stylus. Horizontal lines of varying darkness appear on the paper as the stylus sweeps across the sheet. As each recorded horizontal line is displaced one line width per sweep of the stylus, a "printed" pattern begins to take shape. This pattern is a facsimile of the original pattern transmitted from the distant-end scanner.

The more practical helix–blade technique of electrolytic recording is now favored over the stylus–backplate method. The concept is basically the same, except that the two electrodes in this case consist of a special drum containing a helix at the rear and a stationary blade in the front. The drum–helix makes one complete revolution per scanning line. The rotating drum moving the helix carries out the same function as the moving stylus of the more conventional electrolytic recorders.

Electrothermal (Electroresistive) Recording. Electrothermal facsimile recording is misnamed "thermal" because the recording process gives the appearance of being a "heat" process or burning. More properly, it should be called electroresistive. It is similar to the electrolytic process in that the recording paper is interposed between two electrodes (i.e., stylus and backplate). The recorded pattern is made by an electric arc passing through the paper from one electrode to the other. A special recording paper is required, the type varying from equipment to equipment. The paper has a white coating that is decomposed by the current passing through it, the amount of decomposition being a function of the current passing between the electrodes at any moment in time. A major characteristic of electrothermal recording is the high contrast that can be achieved. A true electrothermal process is evolving using specially treated paper and a resistive heat element that responds to rapid temperature changes as a function of signal current.

Electropercussive Recording. Electropercussive recording is a technique similar to that of recording audio on a record. In this case, an amplified facsimile signal is fed to an electromagnetic transducer that actuates a stylus in response to the electrical signal variations of the facsimile signal. When a sheet of carbon paper is interposed between the stylus and a sheet of plain white paper, a carbon copy impression is made on the plain paper by the vibrations of the stylus in accordance with the signal variations. The intensity of the darkness of the copy varies in proportion to the variation in strength of the picture signal. An advantage of this type of recording is that no special recording paper is required. Other names for electropercussive recording are *impact* or *impression* recording. It is also known by the term *pigment transfer*.

Electrostatic Recording. There are essentially two types of electrostatic recording used in facsimile. One kind is based on printing the facsimile image from a CRT by means of xerography techniques requiring only plain paper. The other kind is a direct copy method requiring a specially coated paper. One type of electrostatic recorder is called a *pin* printer. The advantage of direct electrostatic recording is its exceptional capability of reproducing gray tonal qualities.

FACSIMILE COMMUNICATION

16.3 FUNDAMENTAL SYSTEM INTERFACE AND OPERATION

16.3.1 General

With older drum-type facsimile systems, if we connected a scanner and a recorder back-to-back, there were three basic interface requirements that concerned us:

- Phasing compatibility
- Synchronization
- Index of cooperation

The last item, index of cooperation, dealt essentially with drum sizes. Because we no longer deal with rotating drums, the index of cooperation can be neglected.

Phasing and synchronization are still important.

16.3.2 Phasing and Synchronization

Proper phasing and synchronization are vital factors in conventional facsimile transmission. Both are time-domain functions. Phasing and synchronization of the far-end facsimile receiver with the near-end transmitting scanner permit the reassembly of picture elements in the same spatial order as when the picture was scanned by the transmitter:

- Phasing assures that the receiving recorder stylus coincides with the transmitter in time and position on the copy at the start of transmission.
- Synchronization keeps the two this way throughout the transmission of a single graphic copy.

These functions and the entire automatic facsimile operation are based on ITU-T Rec. T.30 (Ref. 2).

16.3.3 Automatic Facsimile Operation Based on ITU-T Rec. T.30*

This section will only deal with fully automatic operation as applied to CCITT Group 3. The five operational phases of a facsimile call based on ITU-T T.30 are shown in Figure 16.2.

16.3.3.1 Description of Phases

Phase A—Call Establishment. A facsimile-equipped telephone is dialed, and that dialed far-end telephone goes off-hook.

*Section 16.3.3 has been extracted from Ref. 2, ITU-T Rec. T.30.

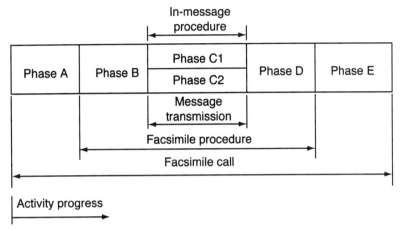

Figure 16.2. Time sequence of a facsimile call.

Phase B—Premessage Procedure. This procedure consists of the identification of capabilities (both stations) and the commanding of the selected conditions as well as confirmation of accepted conditions.

The identification section includes group identification, confirmation for reception, and (optionally) subscriber identification and nonstandard facilities identification.

The command section includes group command, phasing/training, and synchronization. The following are optional commands:

- Nonstandard facilities command
- Subscriber identification command
- Polling (send) command
- Line conditioning
- Echo suppressor disabling

Phase C1—In-Message Procedure. This procedure takes place at the same time as message transmission and controls the complete signaling for in-message procedure, for example, in-message synchronization, error detection, and correction and line supervision.

Phase C2—Message Transmission. This is the transmission of facsimile "data," the binary information governed by CCITT Recs. T.4 and T.6 or possibly proprietary transmission techniques.

Phase D—Postmessage Procedure. This procedure includes the following:

- End-of-message signaling
- Confirmation signaling

- Multipage signaling
- End-of-facsimile procedure

Phase E—Call Release. This is the standard telephone circuit take-down or on-hook procedure.

16.3.3.2 Phases B, C, and D Facsimile Procedures. The operational procedures in these phases may be carried out in either a tonal technique or a binary-coded technique. The simple tonal technique is straightforward but has limited capabilities. The binary-coded technique can cover a wide variety of capabilities and is particularly adapted for comprehensive automatic functions and internal processing (e.g., redundancy reduction) for faster transmission rates and for security features (encryption).

Interaction between near-end and far-end facsimile terminals in phase B should use the following guidelines. When available, binary-coded signaling should be tried first. The interaction steps are as follows:

1. An unattended called station answers a facsimile call with the CED signal (CED = called station identification). At 1.8–2.5 s after the called station is connected to the line, it sends a continuous 2100 Hz ± 15 Hz tone for a duration of not less than 2.6 s and not more than 4 s.
2. The unattended calling station indicates a call with the CNG signal (CNG = calling station tone: 1100 Hz on 0.5 s, off 3 s).
3. Whenever there is a binary-coded signaling capability, the called station starts with binary-coded signaling.
4. Facsimile terminals with tonal capability will start tonally.
5. Facsimile terminals with both binary-coded and tonal signaling capabilities send a sequence of signals, the first being binary coded, and the second and all following signals being a composite of binary-coded and tonal information.
6. If a calling station reacts tonally, then tonal signaling goes on through all procedures.

Figure 16.3 is a flow diagram showing phases B–D. The following abbreviations/acronyms are used in the flowchart:

T = transmitter

R = receiver

T1, T2, T3,... = timers

GI indicates group (Group 1, Group 2,...). GI1 1650-Hz tone, on 1.5 s, off 3 s. GI2 1850-Hz tone, on 1.5 s, off 3 s.

GC = group command. GC1, Group 1 1300-Hz tone at least 1.5-s and less than 10-s duration. GC2, Group 2 2100-Hz tone, same duration.

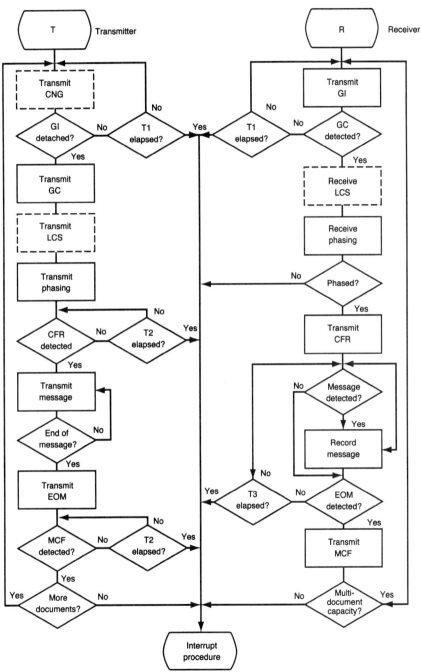

Note 1. T1 = 30–40 s.
Note 2. T2 = 3 s minimum.
Note 3. T3 = 1 s minimum.
Note 4. Broken boxes indicate signal not used in all methods.

Figure 16.3. Flow diagram, tonal control. (From Ref. 2, Figure 9/T.30, ITU-T Rec. T.30.)

LCS = line conditioning signal. Format in CCITT Rec. T.3.

CFR = confirmation to receive. Indicates receiver is phased and ready to receive at least one page. Group 1 1850-Hz tone, 3-s duration; Group 2 1650-Hz tone, 3-s duration.

MCF = message confirmation signal. Same tone frequencies and durations as CFR.

EOM = end of message. 1100-Hz tone, on 3 s immediately following message.

16.3.3.3 *Binary-Coded Signaling.* HDLC frame structure* is used for all binary-coded facsimile control procedures. The basic HDLC structure consists of a frame subdivided into fields. It provides for frame labeling, error checking, and confirmation of correctly received information. A typical HDLC frame structure as applied to facsimile control is illustrated in Figure 16.4.

As shown in Figure 16.4 there is a *preamble* that precedes all binary-coded signaling whenever a new transmission of information begins in any direction. The preamble assures that all elements of the communication channel, such as echo suppressors, are properly conditioned so that subsequent data may be passed unimpaired. There are two types of preamble:

- Binary-coded signaling at 300 bps using a series of flag sequences for 1 s.
- For the optional binary-coded procedure at 2400 bps, the preamble is a long modem training sequence defined in CCITT Rec. T.4 (Ref. 4).

Flag Sequence. The 8-bit HDLC flag sequence (01111110) denotes the beginning and end of the frame. When applied in this facsimile application, the flag sequence is used to establish bit and frame synchronization. The trailing flag of one frame may be the leading flag of the following frame. The continued transmission of the flag sequence may be used to signal the distant terminal that the machine remains on-line but is not presently prepared to proceed with the facsimile procedure.

Address Field. This field would only be used in a multipoint arrangement. For point-to-point operation, the field is set at all 1's.

Control Field. The 8-bit HDLC control field provides the capability of encoding the commands and responses unique to facsimile control procedures. Its format is 1100X000. X = 0 for nonfinal frames within the T.30 procedure; X = 1 for final frames. A final frame is the last frame transmitted prior to an expected response from the distant station.

*HDLC = high-level data link control. There is an excellent description of HDLC in Ref. 3, Section 5.6.

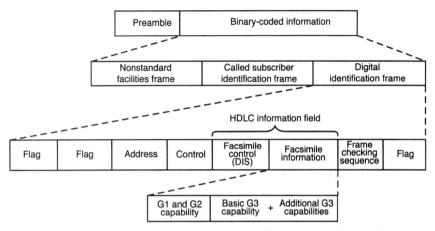

Figure 16.4. HDLC frame structure as applied to facsimile control. (From Ref. 2, Figure 16/T.30, ITU-T Rec. T.30.)

Information Field. This field is variable in length and contains specific information for the control and message interchange between two facsimile stations. There are two parts of this field: the FCF (facsimile control field) and FIF (facsimile information field). Due to space limitation, only sample commands and responses will be provided in this section.

FACSIMILE CONTROL FIELD. This field is the first 8 or 16 bits inside the HDLC information field. The FCF with 16 bits is only used for the optional T.4 error correction mode. This field contains the complete information regarding the type of information being exchanged and the position in the overall sequence. The bit assignments within the FCF are as follows:

Where X appears as the first bit of the FCF, X is defined as follows:

- X is set to 1 by the station that receives a valid DIS (digital identification signal).
- X is set to 0 by the station that receives a valid and appropriate response to a DIS.
- X remains unchanged until the station again enters the beginning of phase B.

Initial Identification. From the called to the calling station.

Format: 0000XXXX (general format)

1. Digital identification signal (DIS). Characterizes the standard CCITT capabilities of the called terminal.

Format 00000001

2. Called subscriber identification (CSI). This is an optional signal that may be used to provide specific identity of the called subscriber by its international telephone number.

Format 00000010

3. Nonstandard facilities (NSF). This is an optional signal that may be used to identify specific user requirements that are not covered by T recommendations.

Format 00000100

Command to Send. From a calling station wishing to be a receiver of a called station that is capable of transmitting.

Format: 1000XXXX   (general format)

1. Digital transmit command (DTC). The digital command response to the standard capabilities identified by the DIS.

Format: 10000001

2. Calling subscriber identification (CIG). This optional signal indicates that the following FIF information is an identification of that calling station. It may be used to provide additional security to the facsimile procedure.

Format: 10000010

3. Nonstandard facilities command (NSC). This optional signal is the digital command response to the information contained in the NSF signal.

Format: 10000100

4. Password (PWD). This optional signal indicates that the following FIF information is a password for the polling mode. It may be used to provide additional security to the facsimile procedure. PWD is only sent if bit 50 in DIS is set.

Format: 10000011

5. Selective polling (SEP). This optional signal indicates that the following FIF information is a subaddress for the polling mode. It may be used to

indicate that a specific document shall be polled at the called side. SEP is only sent if bit 47 in DIS is set.

Format: 10000101

Command to Receive. From the transmitter to the receiver.

Format: X100XXXX   (general format)

1. Digital command signal (DCS). Digital setup command responding to the standard capabilities identified by the DIS.

Format: X1000001

2. Transmitting subscriber identification (TSI). This optional signal indicates that the following FIF information is the identification of the transmitting station. It may be used to provide additional security to the facsimile procedure.

Format: X1000010

3. Nonstandard facilities setup (NSS). This optional signal is the digital command response to the information contained in the NSC or NSF signal.

Format: X1000100

4. Subaddress (SUB). This optional signal indicates that the following FIF information is a subaddress in the called subscriber's domain. It may be used to provide additional routing information in the facsimile procedure. SUB is only sent if bit 50 in DIS is set.

Format: X1000011

5. Password (PWD). This optional signal indicates that the following FIF information is a password for transmission. PWD is only sent if bit 50 in DIS is set.

Format: X1000101

6. Training check (TCF). This digital command is sent through the T.4* modulation system to verify training and to give a first indication of the acceptability of the channel for the data.

*T.4 is ITU-T Rec. T.4, which defines CCITT Group 3 facsimile transmission. It is described in Section 16.4.1.

Format: A series of 0's for 1.5 seconds

No HDLC frame is required for this command.

7. Continue to correct (CTC). This digital command is only used in the optional T.4 error correction mode.

Premessage Response Signals. From the receiver to the transmitter.

Format: X010XXXX   (general format)

1. Confirmation to receive (CFR). A digital response confirming that the entire premessage procedure has been completed and the message transmissions may commence.

Format: X0100001

2. Failure to train (FTT). A digital response rejecting Group 3 training signal and requesting a retraining.

Format: X0100010

3. Response for continue to correct (CTR). This digital response is only used in the optional T.4 error correction mode.

Postmessage Commands. From transmitter to receiver.

Format: X111XXXX   (general format)

1. End-of-message (EOM). Indicates the end of a complete page of facsimile information and to return to the beginning of phase B.

Format: X1110001

2. Multipage signal (MPS). Indicates the end of a complete page of facsimile information and to return to the beginning of phase C upon receipt of confirmation.
Format: X1110010

3. End of procedures (EOP). Indicates the end of a complete page of facsimile information and to further indicate that no further documents are forthcoming and to proceed to phase E, upon receipt of a confirmation.

Format: X1110100

4. Procedure interrupt–end of message (PRI-EOM). Indicates the same as an EOM command with the additional optional capability of requesting operator intervention. If operator intervention is accomplished, further facsimile procedures commence at the beginning of phase B.

Format: X1111010

5. Procedure interrupt–multipage signal (PRI-MPS). Indicates the same as MPS command with the additional optional capability of requesting operator intervention. If operator intervention is accomplished, further facsimile procedures commence at the beginning of phase B.

Format: X1111010

Postmessage Responses. From the receiver to the transmitter.

Format: X011XXXX   (general format)

1. Message confirmation (MCF). Indicates that a complete message has been satisfactorily received and additional messages may follow. This is a positive response to MPS, EOM, EOP, receiver ready (RR), and partial page signal (PPS).

Format: X0010001

2. Retrain positive (RTP). Indicates that a complete message has been received and that additional messages may follow after retransmission of training and/or phasing and CFR.

Format: X0110011

3. Retrain negative (RTN). Indicates that the previous message has not been satisfactorily received. However, further receptions may be possible, provided training and/or phasing are retransmitted.

Format: X0110010

4. Procedure interrupt negative (PIN). Indicates that the previous (or in-process) message has not been satisfactorily received and that further transmissions are not possible without operator intervention.

Format: X00110101

Other Line Control Signals. For the purpose of handling errors and controlling the state of the line.

Format: X101XXXX   (general format)

1. Disconnect (DCN). This command indicates the initiation of phase E, call release. This command requires no response.

Format: X1011111

2. Command repeat (CRP). This optional response indicates that the previous command was received in error and should be repeated in its entirety (i.e., optional frames included).

16.4   DIGITAL TRANSMISSION OF FACSIMILE

16.4.1   Introduction

ITU-T Rec. T.4 (Ref. 4) has almost universal application today. This specification (recommendation) covers what is commonly called *Group 3*. It was designed primarily for the transmission of the European (ISO) A4 page but works equally as well with the North American standard $8\frac{1}{2} \times 11$-in. page. Group 3 equipment can transmit a page under 60 seconds, depending on the black density on the page. A normal typed page takes from 30 to 40 seconds to transmit. This speed is also dependent on the data rate used by the modem (see Figure 16.1). The values given are for 9600-bps transmission.

Section 16.5 covers CCITT Group 4 based on CCITT Rec. T.6 (Ref. 5). Group 4 was designed to operate on the 64-kbps channel with 1-second per page capability.

16.4.2   Equipment Dimensions—DIN A4

In the vertical direction the standard resolution is 3.85 lines/mm $\pm 1\%$. The optional higher resolution is 7.7 lines/mm $\pm 1\%$.

The standard scan line length is 215 mm and is made up of 1728 pixels (black and white elements). There are two optional scan line lengths, 255 and 303 mm, which are made up of 2048 and 2432 pixels, respectively.

16.4.3   Transmission Time per Total Coded Scan Line

The standard minimum time duration to scan a standard line is 20 ms. A line consists of Data bits plus any required Fill bits and End-of-line (EOL) bits. The Data bits give the information on the line's pixel content; Fill bits complete unfinished lines out to 215 mm, the standard line length, and ensure minimum line transmission time is met. The EOL bit sequence carries out the line feed and carriage return function used in teleprinter service or the "soft-return" command on computers.

TABLE 16.1 Terminating Codes

White Run Length	Codeword	Black Run Length	Codeword
0	00110101	0	0000110111
1	000111	1	010
2	0111	2	11
3	1000	3	10
4	1011	4	011
5	1100	5	0011
6	1110	6	0010
7	1111	7	00011
8	10011	8	000101
9	10100	9	000100
10	00111	10	0000100
11	01000	11	0000101
12	001000	12	0000111
13	000011	13	00000100
14	110100	14	00000111
15	110101	15	000011000
16	101010	16	0000010111
17	101011	17	0000011000
18	0100111	18	0000001000
19	0001100	19	00001100111
20	0001000	20	00001101000
21	0010111	21	00001101100
22	0000011	22	00000110111
23	0000100	23	00000101000
24	0101000	24	00000010111
25	0101011	25	0000001100
26	0010011	26	000011001010
27	0100100	27	000011001011
28	0011000	28	000011001100
29	00000010	29	000011001101
30	00000011	30	000001101000
31	00011010	31	000001101001
32	00011011	32	000001101010
33	00010010	33	000001101011
34	00010011	34	000011010010
35	00010100	35	000011010011
36	00010101	36	000011010100
37	00010110	37	000011010101
38	00010111	38	000011010110
39	00101000	39	000011010111
40	00101001	40	000001101100
41	00101010	41	000001101101
42	00101011	42	000011011010
43	00101100	43	000011011011
44	00101101	44	000001010100
45	00000100	45	000001010101

TABLE 16.1 *(Continued)*

White Run Length	Codeword	Black Run Length	Codeword
46	00000101	46	000001010110
47	00001010	47	000001010111
48	00001011	48	000001100100
49	01010010	49	000001100101
50	01010011	50	000001010010
51	01010100	51	000001010011
52	01010101	52	000000100100
53	00100100	53	000000110111
54	00100101	54	000000111000
55	01011000	55	000000100111
56	01011001	56	000000101000
57	01011010	57	000001011000
58	01011011	58	000001011001
59	01001010	59	000000101011
60	01001011	60	000000101100
61	00110010	61	000001011010
62	00110011	62	000001100110
63	00110100	63	000001100111

Source: Reference 4, Table 1/T.4, ITU-T Rec. T.4.

There is an optional two-dimensional coding scheme where the total scan line is defined as the sum of the Data bits plus the Fill bits plus the EOL and a tag bit.

Recommendation T.4 has optional 10- and 5-ms minimum transmission times of the total scan line. It notes that with the 10-ms option, the minimum transmission time of the total coded scan line is the same both for standard resolution and for the optional higher resolution.

16.4.4   Coding Scheme

In CCITT Rec. T.4, two coding schemes are given: one-dimensional and an optional two-dimensional. We will only briefly describe the one-dimensional scheme here. Both schemes use what is called *run length coding*.

A line of data is composed of a series of variable length codewords. Each codeword represents a run length of either all white or all black. White runs and black runs alternate. Again, 1728 picture elements (pixels) represent one horizontal scan line 215 mm long.

To ensure black–white synchronization at the far-end receiver, all Data lines begin with a white-run-length codeword. If the actual line begins with a black run, a white run length of zero is sent. Black or white run lengths up to the maximum length of one scan line (1728 pixels) are defined by the codewords shown in Tables 16.1 and 16.2. There are two types of codewords used: terminating codewords and make-up codewords. Each run length is

TABLE 16.2   Make-up Codes

White Run Lengths	Codeword	Black Run Lengths	Codeword
64	11011	64	0000001111
128	10010	128	000011001000
192	010111	192	000011001001
256	0110111	256	000001011011
320	00110110	320	000000110011
384	00110111	384	000000110100
448	01100100	448	000000110101
512	01100101	512	0000001101100
576	01101000	576	0000001101101
640	01100111	640	0000001001010
704	011001100	704	0000001001011
768	011001101	768	0000001001100
832	011010010	832	0000001001101
896	011010011	896	0000001110010
960	011010100	960	0000001110011
1024	011010101	1024	0000001110100
1088	011010110	1088	0000001110101
1152	011010111	1152	0000001110110
1216	011011000	1216	0000001110111
1280	011011001	1280	0000001010010
1344	011011010	1344	0000001010011
1408	011011011	1408	0000001010100
1472	010011000	1472	0000001010101
1536	010011001	1536	0000001011010
1600	010011010	1600	0000001011011
1664	011000	1664	0000001100100
1728	010011011	1728	0000001100101
EOL	000000000001	EOL	000000000001

Note: It is recognized that machines exist that accommodate larger paper widths maintaining the standard horizontal resolution. This option has been provided for by the addition of the make-up code set defined as follows:

Run Length (Black and White)	Make-up Codes
1792	00000001000
1856	00000001100
1920	00000001101
1984	000000010010
2048	000000010011
2112	000000010100
2176	000000010101
2240	000000010110
2304	000000010111
2368	000000011100
2432	000000011101
2496	000000011110
2560	000000011111

Source: Reference 4, Table 2/I.4, ITU-T Rec. T.4.

represented by either one terminating codeword or one make-up codeword followed by a terminating codeword.

Run lengths in the range of 0–63 pels (pixels) are encoded with their appropriate terminating codeword. Note that there is a different list of codewords for black and white run lengths (see Table 16.1).

Run lengths in the range of 64–1728 pixels are encoded first by the make-up codeword representing the run length that is equal to or shorter than required. This then is followed by the terminating codeword representing the difference between the required run length represented by the make-up codeword (see Table 16.2).

The end-of-line (EOL) sequence is a unique bit sequence that can never be found within a valid line of Data. If it is imitated due to an error burst, resynchronization may be required. This same bit sequence is used prior to the first Data line of a page. The bit sequence format is 000000000001.

When there is a pause in the message flow, Fill is transmitted. Fill may be inserted between a line of data and an EOL, but never within a line of Data. Fill must be added to ensure that the transmission time of Data, Fill, and EOL is not less than the minimum transmission time of the total coded scan line established in the premessage control procedure. The format of Fill is a variable length string of 0's.

When scanning reaches the end of a document, a RTC (return-to-control) signal is sent. The RTC signal consists of six consecutive EOLs. Following the RTC signal, the transmitter will send the postmessage commands in the frame format and the data signaling rate of the control signals, which are defined in CCITT Rec. T.30 (Ref. 2). See Section 16.3.3.

Figures 16.5 and 16.6 illustrate the relationship of the signals described above. Figure 16.5 shows several scan lines of data starting at the beginning of a transmitted page, and Figure 16.6 shows the last coded scan line of a page.

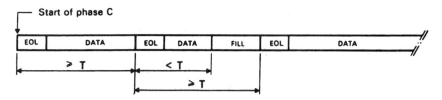

Figure 16.5. Several scan lines of data starting at the beginning of a transmitted page. (From Ref. 4, Figure 1/T.4, ITU-T Rec. T.4.)

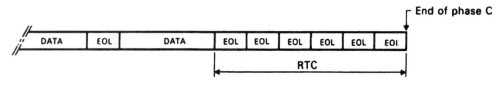

Figure 16.6. The last coded scan line of a page. (From Ref. 4, Figure 2/T.4, ITU-T Rec. T.4.)

16.5   FACSIMILE CODING SCHEMES AND CODING CONTROL FUNCTIONS FOR GROUP 4 FACSIMILE EQUIPMENT

16.5.1   Introduction

Group 4 document transmission involves major reduction of image redundancy by means of a two-dimensional coding scheme. This results in much faster transmission. However, at present, CCITT Rec. T.6 (Ref. 5) only covers black and white transmission; there is no provision for gray scales. The coding schemes utilized assume forward error correction (FEC) coding to maintain a bit error rate (BER) of 1×10^{-5} or better (Ref. 7). CCITT Rec. T.6 appears essentially unchanged in U.S. Standard FIPS 150 (Ref. 6) and in Section 4.2 of ITU-T Rec. T.4 (Ref. 4).

16.5.2   Basic Coding Scheme

16.5.2.1   Principle of the Coding Scheme. The coding scheme uses a two-dimensional line-by-line coding method in which the position of each changing picture element on the current coding line is coded with respect to the position of a corresponding reference element situated on either the coding line or the reference line that is immediately above the coding line. After the coding line has been coded, it becomes the reference line for the next coding line. The reference line for the first coding line on a page is an imaginary white line.

A changing picture element is defined as an element whose *color* (i.e., black or white) is different from that of the previous element along the scan line. An example of this concept is shown in Figure 16.7.

16.5.2.2   Coding Modes. There are three coding modes. The mode selected depends on the coding procedure, as described in Section 16.5.3, which is used to code the position of each changing element along the coding

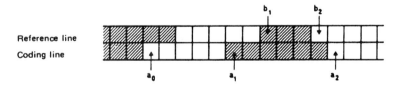

a₀ : The reference or starting changing element on the coding line. At the start of the line a₀ is set on an imaginary white changing element situated just before the first element on the line. During the coding of the coding line, the position of a₀ is defined by the previous coding mode.

a₁ : The next changing element to the right of a₀ on the coding line.

a₂ : The next changing element to the right of a₁ on the coding line.

b₁ : The first changing element on the reference line to the right of a₀ and of opposite colour to a₀.

b₂ : The next changing element to the right of b₁ on the reference line.

Figure 16.7. Changing picture elements. (From Ref. 5, Figure 1/T.6, CCITT Rec. T.6.)

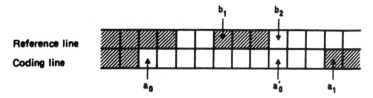

Figure 16.8. Pass mode. (From Ref. 5, Figure 2/T.6, CCITT Rec. T.6.)

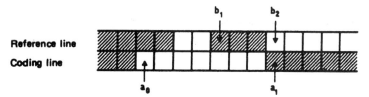

Figure 16.9. An example not corresponding to a pass mode. (From Ref. 5, Figure 3/T.6, CCITT Rec. T.6.)

line. Examples of the three coding modes are given in Figures 16.8, 16.9, and 16.10.

The *pass mode* is shown in Figure 16.8. This mode is identified when the position of b_2 lies to the left of a_1. However, the state where b_2 occurs just above a_1, as shown in Figure 16.9, is not considered a pass mode.

When the *vertical mode* is identified, the position of a_1 is coded relative to the position of b_1. The relative distance a_1b_1 can take on one of seven values: $V(0)$, $V_R(1)$, $V_R(2)$, $V_R(3)$, $V_L(1)$, $V_L(2)$, and $V_L(3)$, each of which is represented by a separate codeword. The subscripts R and L indicate that a_1 is to the right or left, respectively, of b_1; the number in brackets indicates the value of the distance a_1b_1 (see Figure 16.10).

When the *horizontal mode* is identified, both run lengths a_0a_1 and a_1a_2 are coded using codewords $H + M(a_0a_1) + M(a_1a_2)$. H is the flag codeword 001 taken from the two-dimensional code table, Table 16.3. $M(a_0a_1)$ and

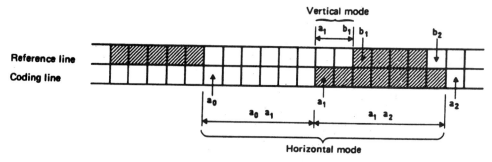

Figure 16.10. Vertical mode and horizontal mode. (From Ref. 5, Figure 4/T.6, CCITT Rec. T.6.)

TABLE 16.3   Code Table

Mode	Elements to Be Coded		Notation	Codeword
Pass	b_1, b_2		P	0001
Horizontal	a_0a_1, a_1a_2		H	$001 + M(a_0a_1) + M(a_1a_2)$[a]
Vertical	a_1 just under b_1	$a_1b_1 = 0$	$V(0)$	1
	a_1 to the right of b_1	$a_1b_1 = 1$	$V_R(1)$	011
		$a_1b_1 = 2$	$V_R(2)$	000011
		$a_1b_1 = 3$	$V_R(3)$	0000011
	a_1 to the left of b_1	$a_1b_1 = 1$	$V_L(1)$	010
		$a_1b_1 = 2$	$V_L(2)$	000010
		$a_1b_1 = 3$	$V_L(3)$	0000010
Extension				0000001xxx

[a]Code $M(\ )$ of the horizontal mode represents the codewords in Table 16.2.
Source: Reference 5, Table 1/T.6, CCITT Rec. T.6.

$M(a_1a_2)$ are codewords that represent the length and "color" of the runs a_0a_1 and a_1a_2, respectively, and are taken from the appropriate white or black run-length code tables (Table 16.1).

16.5.3   Coding Procedure

The coding procedure (shown in the coding flow diagram in Figure 16.11), identifies the coding mode that is to be used to code each changing element along the coding line. When one of the three coding modes has been identified according to step 1 or step 2 as shown in the figure, an appropriate codeword is selected from the code table given in Table 16.3. The steps of the coding procedure are described below:

Step 1

1. If a pass mode is identified, this is coded using the word 0001 (Table 16.3). After this processing, picture element a_0' just under b_2 is regarded as the new starting picture element a_0 for the next coding (see Figure 16.8).
2. If a pass mode is not detected, then proceed to step 2.

Step 2

1. Determine the absolute value of the relative distance a_1b_1.
2. If $[a_1b_1] \leq 3$, as shown in Table 16.3, a_1b_1 is coded by the vertical mode, after which position a_1 is regarded as the new starting picture element a_0 for the next coding.
3. If $[a_1b_1] > 3$, as shown in Table 16.3, following horizontal mode code 001, a_0a_1 and a_1a_2 are respectively coded by one-dimensional run-length coding.

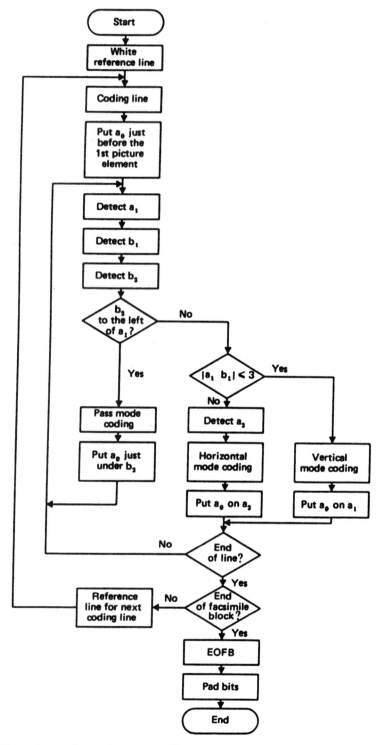

Figure 16.11. Coding flow diagram. (From Ref. 5, Figure 6/T.6, CCITT Rec. T.6.)

Run lengths in the range of 0–63 pixels are encoded with their appropriate terminating codeword from Table 16.1. Note that there is a different list of codewords for black and white run lengths. Run lengths in the range of 64–2623 pixels are encoded first by the make-up codeword representing the run length that is nearest, not longer, to that required. This is then followed by a terminating codeword representing the difference between the required run length and the run length represented by the make-up code. Run lengths longer than or equal to 2624 pixels are coded first by the make-up code of 2560. If the remaining part of the run (after the first make-up code of 2560) is 2560 pixels or greater, additional make-up code(s) of 2560 are issued until the remaining part of the run becomes less than 2560 pixels. Then the remaining part of the run is encoded by terminating code or by make-up code plus terminating code according to the range mentioned previously.

After this processing, position a_2 is regarded as the new starting picture element a_0 for the next coding.

To process the first picture element, the first starting element a_0 on each coding line is imaginarily set at a position just before the first picture element and is regarded as a white picture element.

The first run length on a line $a_0 a_1$ is replaced by $a_0 a_1 - 1$. Therefore, if the first actual run is black and is deemed to be coded by the horizontal mode coding, then the first codeword $M(a_0 a_1)$ corresponds to an imaginary white run of zero length.

To process the last picture element, the coding of the line continues until the position of the imaginary changing element situated just after the last actual element has been coded. This may be coded as a_1 or a_2. Also, if b_1 and/or b_2 are not detected at any time during the coding of the line, they are positioned on the imaginary changing element situated just after the last actual picture element on the reference line.

16.6 MILITARY DIGITAL FACSIMILE SYSTEMS*

16.6.1 Definitions

Type I facsimile equipment provides for the transmission and reception of an image with black and white information only. *Type II facsimile equipment* provides for the transmission and reception of an image with shades of gray, as well as black and white. *CCITT Group 3 facsimile equipment* (Section 16.4) provides for the transmission and reception of an image with black and white information, as defined by ITU-T Recs. T.4 and T.30 (Refs. 4 and 2, respectively) (FED-STD-1062 and FED-STD-1063). ITU-T Rec. T.4 and FED-STD-1062 incorporate means for reducing redundant information in the document signal prior to the modulation process.

Basic mode is where the transmitter for Type I and Type II facsimile does not pause after calling the receiving unit to wait for an acknowledgment

---

*Section 16.6 is excerpted from Ref. 7, MIL-STD-188-161A.

TABLE 16.4   Codewords and Signaling Sequences for Type I and Type II Facsimile Equipment

Name	Make-up
Beginning of intermediate line pair (BILP)	0000000000000011
Beginning of line pair (BOLP)	0000000000000010
End of line (EOL)	000000000001
End of message (EOM)	16 consecutive S_1 codewords
Not end of message ($\overline{EOM}$)	16 consecutive inverted S_1 codewords
Return to control (RTC)	EOL EOL EOL EOL EOL EOL
Start of message (SOM)	$S_1 S_0 X$ clock periods $S_0 S_1$ (where X is the number of clock periods between the pairs of codewords)
S_0	111100010011010
S_1	111101011001000
Fill	Variable length string of 0's
Stuffing	Variable length string of 1's
Preamble	Variable length string of all 1's or all 0's

Source: Reference 7, MIL-STD-188-161A.

before transmitting an image in the simplex and broadcast mode of operation. *Handshake mode* is where the transmitter for Type I, Type II, and CCITT Group 3 facsimiles pauses after calling the receiving unit to wait for an acknowledgment before transmitting an image in the duplex mode of operation.

Synchronization codewords and signaling sequences used in Type I and Type II facsimile are defined in Table 16.4.

16.6.2   Technical Requirements for Type I and Type II Facsimile Equipment

Transmission rates are bit-by-bit synchronous at data rates of 2400, 4800, and 9600 bps, and 16 and 32 kbps, with timing provided by an external clock. Digital interfaces shall comply with the applicable requirements of MIL-STD-188-114 (Chapter 13). The functional interchange circuits are shown in Table 16.5.

Image Parameters

Scan line length: 215 mm, left justified

Resolution: three switch-selectable standards for horizontal and vertical resolutions:

- 3.85 lines/mm (vertical) by 1728 black and white pixels along the horizontal scan line (*Note*: This is a nominal medium resolution of 100×200 lines per inch.)

TABLE 16.5 Functional Interchange Circuits

Circuit	Direction[a]
Request to send	From DTE to DCE
Clear to send	From DCE to DTE
Receive input control	From DTE to DCE
Send data	From DTE to DCE
Receive data	From DCE to DTE
Send timing	From DCE to DTE
Receive timing	From DCE to DTE
Send common	Return
Receive common	Return
Signal ground	Ground

[a]DTE = data terminal equipment; DCE = data communication equipment.

- 3.85 lines/mm (vertical) by 864 black and white pixels along the horizontal scan line (*Note*: This is the nominal low resolution of 100×100 lines per inch.)
- 7.7 lines/mm by 1728 pixels along the horizontal scan line (*Note*: This is a nominal high resolution of 200×200 lines per inch.)

Tolerance of image parameters: $\pm 1\%$

Scanning direction: from left to right and from top to bottom

Scanning line transmission time: the minimum scanned line transmission time is 20 ms

Contrast levels: black and white

Document dimensions: input documents up to a maximum of 215 mm wide and 1000 mm long. Documents up to 230 mm wide may be accepted into scanner, but only 215 mm of the document will be scanned

16.6.3 Image Coding Schemes

The facsimile equipment shall be capable of operating in three modes: uncompressed, compressed, and compressed with forward error correction (FEC).

In the uncompressed mode, facsimile data are transmitted pixel by pixel, with logic 1 representing black. Each line of the output data consists of a synchronization code followed by the number of pixels as previously specified in "Resolution".

In the compressed mode, facsimile data are transmitted after compression by the redundancy algorithm (see Section 16.4 and CCITT Rec. T.4). A line of data is composed of a series of variable length codewords. Each codeword represents a run length of either all white or all black. White runs and black runs alternate. All data lines begin with a white-run-length codeword to ensure that the receiver maintains color synchronization. A white run length of zero is sent if the actual scan line begins with a black run. Black or white run lengths may be up to a maximum of one scanning line (1728 pixels) and are defined in Tables 16.1 and 16.2 in Section 16.4.4. For descriptions of run lengths dealing with EOL (end of line) and Fill, see Section 16.4.

In the compressed mode with FEC, facsimile data are further processed by a channel coder and bit interleaving buffer to provide this FEC. The channel coder uses a Bose–Chaudhuri–Hocquenghen (BCH) code which, in this case, has the capability of correcting 2 errored bits per block.

16.7 DIGITAL COMPRESSION AND CODING OF CONTINUOUS-TONE STILL IMAGES—JPEG*

16.7.1 Introduction

This section is based on CCITT Rec. T.81 (Ref. 9). In its introduction it states: "The requirements which these processes must satisfy to be useful for specific image communications applications such as *facsimile, Videotex* and *audiographic conferencing* are defined in CCITT Rec. T.80 [Ref. 8] .... In addition to the applications addressed by CCITT and ISO/IEC, the JPEG committee has developed a compression standard to meet the needs of other applications as well, including desktop publishing, graphic arts, medical imaging and scientific imaging."

In this fourth edition we deleted discussions on freeze-frame video in the video/television chapter. We believe that equipments meeting CCITT Rec. T.81 will provide a superior replacement for freeze-frame video. We have included it in this chapter on facsimile because of its application for facsimile and because it treats the transmission of *still* pictures.

The discussion that follows is applicable to continuous-tone, grayscale or color, digital still image data. It has a wide range of applications where compressed images are required. It is not applicable to bilevel image data.

The JPEG, in CCITT Rec. T.81,

- Specifies processes for converting source image data to compressed image data
- Specifies processes for converting compressed image data to reconstructed image data
- Gives guidance on how to implement these processes in practice
- Specifies coded representations for compressed data

From here on in this section we will refer to the JPEG specification rather than CCITT Rec. T.81, realizing that Rec. T.81 is the embodiment of the JPEG specification.

*JPEG = Joint Photographic Experts Group.

16.7.1.1 *Selected Definitions*

Abbreviated Format. A representation of compressed image data that is missing some or all of the table specifications required for decoding, or a representation of table-specification data without frame headers, scan headers, and entropy-coded segments.

AC Coefficient. Any DCT coefficient for which the frequency is not zero in at least one dimension.

Component. One of the two-dimensional arrays that comprise an image.

Conditioning Table. The set of parameters that select one of the defined relationships between prior coding decisions and the conditional probability estimates in arithmetic coding.

Continuous-Tone Image. An image whose components have more than one bit per sample.

Data Unit. An 8×8 block of samples of one component in DCT-based processes; a sample in lossless processes.

DC Coefficient. The DCT coefficient for which the frequency is zero in both dimensions.

(DCT) Coefficient. The amplitude of a specific cosine basis function; may refer to an original DCT coefficient, to a quantized coefficient, or to a dequantized coefficient.

Entropy Encoding. A lossless procedure that converts a sequence of input symbols into a sequence of bits such that the average number of bits per symbol approaches the entropy of the input symbols.

Frame. A group of one or more scans (all using the same DCT-based or lossless process) through the data of one or more of the components in an image.

Hierarchical. A mode of operation for coding an image in which the first frame for a given component is followed by frames that code the differences between the source data and the reconstructed data from the previous frame for that component. Resolution changes are allowed between frames.

Huffman Encoding. An entropy encoding procedure that assigns a variable length code to each input symbol.

Huffman Table. The set of variable length codes required in a Huffman encoder and Huffman decoder.

Interleaved. The descriptive term applied to repetitive multiplexing of small groups of data units from each component in a scan in a specific order.

Lossless. A descriptive term for encoding and decoding processes and procedures in which the output of the decoding procedure(s) is identical to the input to the encoding procedure(s).

Marker. A 2-byte code in which the first byte is hexadecimal FF (X'FF') and the second byte is a value between 1 and hexadecimal FE (X'FE').

Minimum Coded Unit (MCU). The smallest group of data units that is coded.

Parameters. Fixed length integers 4, 8, or 16 bits in length, used in compressed data formats.

Quantization Table. The set of 64 quantization values used to quantize DCT coefficients.

Restart Interval. The integer number of MCUs processed as an independent sequence within a scan.

Successive Approximation. A progressive coding process in which the coefficients are coded with reduced precision in the first scan, and precision is increased by 1 bit with each succeeding scan.

Table Specification Data. The coded representation from which the tables used in the encoder and decoder are generated and their destinations specified.

16.7.2 Lossy and Lossless Compression

There are two classes of encoding and decoding processes included in the JPEG specification: *lossy* and *lossless* processes. Those based on the *discrete cosine transform* (DCT) are lossy, thereby allowing substantial compression to be achieved while producing a reconstructed image with high visual fidelity to the encoder's source image.

The *baseline sequential* process is the simplest DCT-based coding process. There are additional DCT-based processes that extend the baseline sequential process to a broader range of application. The baseline decoding process is required to be present in any decoder using *extended DCT-based decoding processes* in order to provide a default decoding capability.

Lossless encoding and decoding processes are the second class of coding. This second class is not based on the DCT and is provided to meet the needs of applications requiring lossless compression. They are also used independently of any of the DCT-based processes. Table 16.6 at the end of this section summarizes the relationship among these lossy and lossless coding processes.

The amount of compression provided by any of the various processes is dependent on the characteristics of the particular image being compressed, as well as on the picture quality desired by the application and the desired speed of compression and decompression.

16.7.3 DCT-Based Coding

Figure 16.12 is highly simplified but shows all the principal procedures for all encoding processes based on DCT. In this case it shows the special case of a

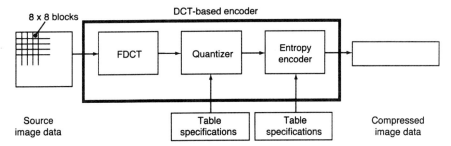

Figure 16.12. A simplified functional block diagram of a DCT-based encoder. (From Ref. 9, Figure 4, CCITT Rec. T.81.)

single-component image. The simplification is valid because all processes specified in this JPEG specification operate on each image component independently.

One will note the similarity between techniques used with JPEG and those used with MPEG-2 (Chapter 15). For example, in the encoding process with JPEG, the input component's *samples* are grouped into *8 × 8 blocks*, and each block is transformed by the *forward DCT* (FDCT) into a set of 64 values referred to as DCT coefficients. One of these values is referred to as the *DC coefficient* and the other 63 as the *AC coefficients*.

Each of the 64 coefficients is then *quantized* using one of 64 corresponding values from a *quantization table* (determined by one of the table specifications illustrated in Figure 16.12).

In the next processing step the DC coefficient and the 63 AC coefficients are prepared for *entropy coding*, as shown in Figure 16.13. The current

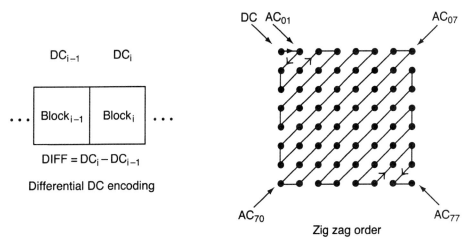

Figure 16.13. Preparation of quantized coefficients for entropy coding. (From Ref. 9, Figure 5, ITU-T Rec. T.81.)

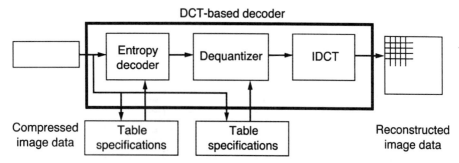

Figure 16.14. Simplified functional block diagram of the DCT-based decoder. (From Ref. 9, Figure 6, ITU-T Rec. T.81.)

quantized DC coefficient is predicted from the previous quantized DC coefficient, and the difference is encoded. The 63 quantized AC coefficients undergo no such differential encoding but are converted into a one-dimensional zigzag sequence, also illustrated in Figure 16.13.

In the third processing stage the data are compressed still further by entropy encoding. One of two entropy coding procedures can be used: *Huffman encoding* and *arithmetic encoding*. Of course, if Huffman encoding is employed, *Huffman table* specifications must be provided to the encoder. If arithmetic encoding is used, arithmetic coding *conditioning table* specifications may be provided, otherwise the default conditioning table specifications are used.

Figure 16.14 illustrates the principal procedures for all DCT-based decoding processes. Each step shown in this figure performs essentially the inverse of its corresponding main procedure within the encoder. The entropy decoder decoded the zigzag sequence of quantized DCT coefficients. After *dequantization* the DCT coefficients are transformed to an 8 × 8 block of samples by the *inverse DCT* (IDCT).

16.7.4   Lossless Coding

Figure 16.15 outlines the principal procedures of the lossless encoding process. Figure 16.16 illustrates how a *predictor* combines the reconstructed values of up to three neighborhood samples at positions *a*, *b*, and *c* to form a prediction of the sample at position *x*. This prediction is then subtracted from the actual value of the sample at position *x*, and the difference is losslessly entropy-coded by either Huffman or arithmetic coding.

Another approach uses a slightly modified procedure, where the precision of the input samples is reduced by one or more bits prior to the lossless coding. Such an approach achieves higher compression than the lossless process (but lower compression than the DCT-based processes for equivalent visual fidelity) and limits the reconstructed image's worst-case sample error to the amount of input precision reduction.

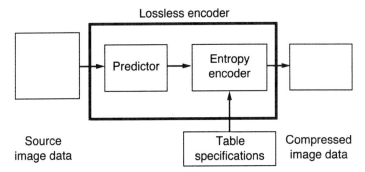

Figure **16.15.** Simplified functional block diagram of a lossless encoder. (From Ref. 9, Figure 7, ITU-T Rec. T.81.)

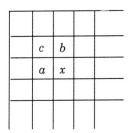

Figure **16.16.** Three-sample prediction neighborhood. (From Ref. 9, Figure 8, ITU-T Rec. T.81.)

16.7.5 Modes of Operation

The various coding processes are defined by four distinct modes of operation. These are *sequential DCT-based*, *progressive DCT-based*, *lossless*, and *hierarchical*. A JPEG implementation is not required to provide all four. The lossless mode was described in Section 16.7.4. The other three modes of operation are compared as follows.

For the sequential DCT-based mode, 8×8 sample blocks are typically input block-by-block from left to right and block-row by block-row from top to bottom. The next processing step is to transform a block by the forward DCT, then quantize and prepare for entropy coding. Now all 64 of its quantized DCT coefficients can be immediately entropy encoded and output as part of the compressed image data (as described in Section 16.7.3), thereby minimizing coefficient storage requirements.

In the case of the progressive DCT-based mode, the 8×8 blocks are typically encoded in the same order (as above), but in multiple *scans* through the image. To do this an image-sized coefficient memory buffer is added between the quantizer and entropy encoder (this buffer is not shown in Figure 16.12). A block's coefficients are stored in that buffer as each block is transformed by the forward DCT and quantized. The DCT coefficients in the buffer are then partially encoded in each of multiple scans. Figure 16.17

Progressive

Sequential

Figure 16.17. Progressive versus sequential presentation. (From Ref. 9, Figure 9, ITU-T Rec. T.81.)

shows the typical sequence of image presentation at the output of the decoder for sequential versus progressive modes of operation.

Within a scan there are two procedures by which the quantized coefficients in the buffer may be partially encoded. First, only a specified band of coefficients need be encoded from the zigzag sequence. This procedure is called *spectral selection*. It is called this because each band typically contains coefficients that occupy a lower or higher part of the frequency spectrum for that 8×8 block. Second, the coefficients within the current band need not be encoded to their full (quantized) accuracy within each scan. A specified number of most significant bits are encoded first upon a coefficient's first encoding. The less significant bits are encoded on subsequent scans. This procedure is called *successive approximation*. Either procedure may be used separately or they may be mixed in flexible combinations.

An image is encoded as a sequence of frames in the hierarchical mode. These frames provide *referenced reconstructed components* that are usually needed for prediction in subsequent frames. *Differential frames* encode the difference between source components and reference reconstructed components except for the first frame for a given component. The coding of differences may be carried out using only lossless processes, only DCT-based processes, or DCT-based processes with a final lossless process for each component. As shown in Figure 16.18, *downsampling* and *upsampling filters* may be used to provide a pyramid of spatial resolutions. Alternatively, the

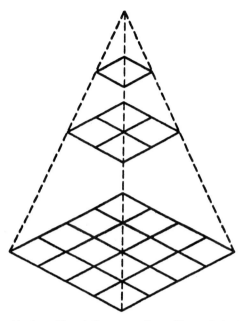

Figure 16.18. Hierarchical multiresolution encoding. (From Ref. 9, Figure 10, ITU-T Rec. T.81.)

hierarchical mode can be used to improve the quality of the reconstructed components at a given spatial resolution.

The hierarchical mode is useful in environments that have multiresolution requirements. It offers a progressive presentation similar to the progressive DCT-based mode. It also offers the capability of progressive coding to a final lossless stage.

16.7.6 Entropy Coding Alternatives

There are two entropy coding alternatives used with JPEG compression techniques: Huffman and arithmetic coding. The Huffman code tables used are determined by table specifications at both the coder and decoder, as one would expect. Arithmetic coding procedures use arithmetic coding conditioning tables, which may also be determined by a table specification. No default values for Huffman tables are specified, so that applications may choose tables appropriate for their own environments. Default tables are defined for the arithmetic coding conditioning.

The baseline sequential process uses Huffman coding, while the extended DCT-based and lossless processes may use either Huffman or arithmetic coding.

16.7.7 Sample Precision

For lossless processes the sample precision is specified from 2 to 16 bits. For DCT-based processes, on the other hand, two alternative sample precisions are specified: either 8 or 12 bits per sample. Applications that use samples with other precisions can use either 8- or 12-bit precision by shifting their source image samples appropriately. The baseline process uses only 8-bit precision. Greater precision, such as 12-bit source image samples, is likely to need greater processor resources than those that handle 8-bit source images. Thus JPEG specifies separate normative requirements for 8-bit and 12-bit DCT-based processes.

16.7.8 Multiple-Component Control

Besides the encoding and decoding processes covered previously—namely, those that operate on the sample values in order to achieve compression—there are other major parts as well. These are the procedures that control the order in which the image data from multiple components are processed to create the compressed data, and which ensure that the proper set of table data is applied to the proper *data units* in the image. A data unit may be defined as a sample for lossless processes and an 8 × 8 block of samples for DCT-based processes.

16.7.8.1 *Interleaving Multiple Components.* An example of how an encoding process selects between multiple source image components as well as multiple sets of table data, when performing its coding procedure, is shown in Figure 16.19. The source image in this example consists of three components, A, B, and C, and there are two sets of table specifications. Note that this simplified view does not distinguish between the quantization tables and entropy encoding tables.

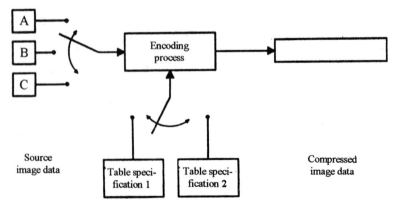

Figure 16.19. Component-interleave and table-switching control. (From Ref. 9, Figure 11, ITU-T Rec. T.81.)

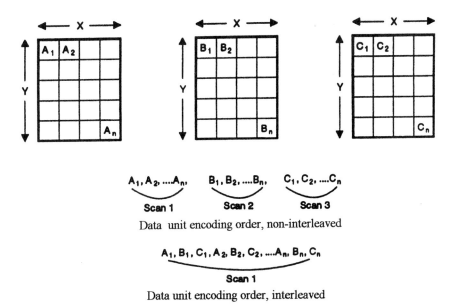

Figure 16.20. Interleaved versus noninterleaved coding order. (From Ref. 9, Figure 12, ITU-T Rec. T.81.)

In the sequential mode, encoding is *noninterleaved*. This is true if the encoder compresses all image data units in component A before beginning component B, and then in turn all of B before C. Encoding is *interleaved* if a coder compresses a data unit from A, a data unit from B, and a data unit from C, then back to A, and so on. Figure 16.20 shows these three alternatives where a case of all three image components have identical dimensions: X columns by Y lines, for a total of n data units each.

Source image components with different horizontal and vertical dimensions can be handled by these control procedures.

16.7.9 Structure of Compressed Data

For both lossy and lossless encoding, the compressed data are described by a uniform structure and a set of *parameters* for all modes of operation (sequential, progressive, lossless, and hierarchical). Special 2-byte *markers* identify the various parts of the compressed data. Some markers are followed by particular sequences of parameters, as in the case of table specifications, *frame header*, or *scan header*. Other functions such as marking the start-of-image and end-of-image are used without parameters. When a marker is associated with a particular sequence of parameters, the marker and its parameter comprise a *marker segment*.

The data created by the entropy encoder are also segmented. Entropy-coded data segments are isolated by one particular marker called the *restart marker*. The outputs of the encoder are the restart markers, intermixed with the entropy-coded data, at regular *restart intervals* of the source image data. Restart markers can be identified without having to decode the compressed data to find them. The advantage of being independently decoded is that they have application-specific uses such as parallel encoding or decoding, isolation of data corruptions, and semirandom access of entropy-coded segments.

There are three compressed data formats:

- The interchange format
- The abbreviated format for compressed image data
- The abbreviated format for table-specification data

These three data formats are described below.

16.7.9.1 *Interchange Format.* The interchange format includes the marker segments for all quantization and entropy-coding table specifications needed for the decoding process, in addition to certain required marker segments and the entropy-coded segments. Regardless of how each environment internally associates tables with compressed image data, this guarantees that a compressed image can cross the boundary between application environments.

16.7.9.2 *Abbreviated Format for Compressed Image Data.* The abbreviated format for compressed image data is identical to the interchange format, except that it does not include all tables required for decoding. However, it may include some of them. This format is intended for use within applications where alternative mechanisms are available for supplying some or all of the table-specification data needed for decoding.

16.7.9.3 *Abbreviated Format for Table-Specification Data.* Only table-specification data are contained in this format. It is a means by which the application may install the tables required in the decoder to subsequently reconstruct one or more images.

16.7.10 Image, Frame, and Scan

Compressed data consist of only one image. In the cases of sequential and progressive coding processes, an image contains only one frame. An image contains multiple frames for the hierarchical mode.

A frame contains one or more scans. A scan contains a complete encoding of one or more image components in the case of sequential processes. In

Figure 16.20 the frame consists of three scans when noninterleaved and one scan if all three components are interleaved together. In another case, a frame could consist of two scans: one with a noninterleaved component and the other with two components interleaved.

In the case of progressive processes, a scan contains a partial encoding of all data units from one or more image components. Components are not to be interleaved in the progressive mode, except for the DC coefficients in the first scan for each component of a progressive frame.

16.7.11 Summary of Coding Processes

Table 16.6 provides a summary of the essential characteristics of the various coding processes specified in ITU-T Rec. T.81. The full specification of these processes is contained in Annexes F, G, H, and J of the reference publication.

TABLE 16.6 Summary of Essential Characteristics of Coding Processes

Baseline Process (Required for all DCT-Based Decoders)

DCT-based process
Source image: 8-bit samples within each component
Sequential
Huffman coding: 2 AC and 2 DC tables
Decoders shall process scans with 1, 2, 3, and 4 components
Interleaved and noninterleaved scans

Extended DCT-Based Processes

DCT-based process
Source image: 8-bit or 12-bit samples
Sequential or progressive
Huffman or arithmetic coding: 4 AC and 4 DC tables
Decoders shall process scans with 1, 2, 3, and 4 components
Interleaved and noninterleaved scans

Lossless Processes

Predictive process (not DCT-based)
Source image: P-bit samples ($2 \leq P \leq 16$)
Sequential
Huffman or arithmetic coding: 4 DC tables
Decoders shall process scans with 1, 2, 3, and 4 components
Interleaved and noninterleaved scans

Hierarchical Processes

Multiple frames (nondifferential and differential)
Uses extended DCT-based or lossless processes
Decoders shall process scans with 1, 2, 3, and 4 components
Interleaved and noninterleaved scans

Source: Reference 9, Table 1, ITU-T Rec. T.81.

REVIEW EXERCISES

1. List and describe the eight functions or processes involved in the transmission and reception of facsimile.

2. Name the five basic *operational* functions during a facsimile communication.

3. Name the three methods of scanning a page for facsimile transmission.

4. Describe the four basic recording methods for facsimile.

5. Differentiate phasing and synchronization and discuss the purpose of each.

6. What are the two basic transmission techniques used with current facsimile systems?

7. What is the purpose of the *preamble* in binary-coded facsimile transmission?

8. What is the purpose of the *flag sequence* in binary facsimile transmission?

9. Group 3 (CCITT Rec. T.4) transmits a page in under 60 seconds. What really governs its transmission rate? Assume the channel bit rate is 9600 bps in all cases.

10. With Chapter 15 as background, what type of coding is *run length encoding*?

11. Define a pixel (not just the acronym).

12. For Group 3 facsimile, what are the maximum number of pixels for a scan line 215 mm long?

13. What are the two types of codewords used with Group 3 (and Group 4) facsimile?

14. When a scan has reached the end of the last scan line in a document, what happens next? How does the receiver know that this is the end of the page?

15. Define a *changing picture element*.

16. What is the basic difference between Group 3 and Group 4 facsimile?

17. What does run length encoding do for us?

18. In military facsimile, what is the difference between Type I and Type II equipment?

19. The JPEG specification is for the transmission of what types of images?

20. Describe the operation of a DCT as used in JPEG type compression.

21. List and briefly describe the two types of entropy coding used with the JPEG compression technique.

22. List the four different modes for JPEG operation.

23. How many bits per sample are used for the two basic types of JPEG compression?

24. What is the function of a *marker* in JPEG compression?

25. For all modes but the hierarchical mode, how many frames make up an image?

REFERENCES

1. D. M. Costigan, *Electronic Delivery of Documents and Graphics*, Van Nostrand Reinhold, New York, 1978.

2. "Procedures for Document Transmission in the General Switched Telephone Network," ITU-T Rec. T.30, ITU, Geneva, 1994.

3. Roger L. Freeman, *Practical Data Communications*, Wiley, New York, 1995.

4. "Standardization of Group 3 Facsimile Apparatus for Document Transmission," ITU-T Rec. T.4, ITU, Geneva, 1994.

5. "Facsimile Coding Schemes and Coding Control Functions for Group 4 Facsimile Apparatus," CCITT Rec. T.6, Fascicle VII.3, IXth Plenary Assembly, Melbourne, 1988.

6. *Facsimile Coding Schemes and Coding Control Functions for Group 4 Facsimile Apparatus*, FIPS Pub. 150, U.S. Department of Commerce, NIST, Washington, DC, Nov. 1988.

7. "Interoperability and Performance Standards for Digital Facsimile Equipment," MIL-STD-188-161A with Notice 1, U.S. Department of Defense, Washington, DC, Mar. 1989.

8. "Common Components for Image Compression and Communication—Basic Principles," CCITT Rec. T.80, ITU, Geneva, Sept. 1992.

9. "Information Technology—Digital Compression and Coding of Continuous-Tone Still Images—Requirements and Guidelines," CCITT Rec. T.81, ITU, Geneva, Sept. 1982.

10. Roger L. Freeman, *Reference Manual for Telecommunication Engineering*, Wiley, New York, 1994, with Updates '95 and '96.

ACRONYMS

AAL	ATM adaptation layer
AC, ac	alternating current
ACK	acknowledge, acknowledgment
A/D	analog-to-digital
ADM	add–drop multiplex
ADPCM	adaptive differential pulse code modulation
ADSL	asymmetric digital subscriber line
AGC	automatic gain control
AIFM	adjacent channel interference fade margin
AIS	alarm indication signal
ALBO	automatic line build-out
ALE	automatic link establishment
AM	amplitude modulation
AMD	automatic message display
AMI	alternate mark inversion
AMP	auxiliary channel multiplexing pattern
AMPS	advanced mobile phone system
AMSC	American Mobile Satellite Corporation
ANSI	American National Standards Institute
APC	automatic power control; automatic phase control
APD	avalanche photodiode
APDU	ATM cell protocol data unit
APL	average picture level
ARPA	Advanced Research Projects Agency
ARQ	automatic repeat request
ASK	amplitude shift keying
ASCII	American Standard Code for Information Interchange
ATB	all trunks busy
ATM	asynchronous transfer mode

ATSC	Advanced Television System Committee
AT&T	American Telephone & Telegraph (Corp.)
ATU	ADSL terminal unit
AU	administrative unit
AUG	administrative unit group
AutoVon	automatic voice (operated) network
AWG	American wire gauge
AWGN	additive white Gaussian noise
2B1Q	2 binary to 1 quaternary
B3ZS, B6ZS, B8ZS	binary 3 zero substitution; binary 6 zero substitution; binary 8 zero substitution
BAS	bit rate allocation signal
BBC	British Broadcasting Corporation
BBE	background block error
BBER	background block error ratio
BCC	block check count (character), Bellcore client companies
BCD	binary-coded decimal
BCH	Bose−Chaudhuri−Hocquenghem (a family of block codes)
BCI	bit count integrity
BER	bit error rate; bit error ratio
BERT	bit error rate test
BFSK	binary frequency shift keying
BH	busy hour
BIP	bit interleaved parity
B-ISDN	broadband ISDN
BITE	built-in test equipment
BITS	building integrated timing supply
BLOS	beyond line-of-sight
BNZS	binary n-zeros substitution
BORSCHT	battery feed, overvoltage protection, ringing, signaling, coding, hybrid, test
BPV	bipolar violation
BPSK	binary phase shift keying
BRA, BRI	basic rate (interface)
BSI	bit stream information
BSTJ	*Bell System Technical Journal*
BT	bridged tap
BTR	bit timing recovery
BWR	bandwidth ratio
CATV	community antenna television
CBR	constant bit rate
CBX	computer-based exchange

CCD	charge-coupled device
CCIR	International Consultive Committee for Radio
CCITT	International Consultive Committee for Telephone and Telegraph
CDMA	code division multiple access
CDPD	cellular digital packet data
CDV	cell delay variation
CED	called station identification (fax)
CGSA	cellular geographic serving area
CELP	codebook-excited linear predictive (coder)
CEPT	Conference European Post and Telegraph (from the French)
CFM	composite fade margin
C/I	carrier-to-interference (ratio)
CLP	cell loss priority
CLR	circuit loudness rating
CMI	coded mark inversion
C/N_0	carrier-to-noise in 1-Hz bandwidth
C/N, CNR	carrier-to-noise ratio
CNG	calling station tone (fax)
CNT	Canadian National Telephone (Company)
CODEC, codec	coder−decoder
compander	compressor−expander
COT	central office terminal
CPU	central processing unit
CRC, crc	cyclic redundancy check
CR/BTR	carrier recovery/bit timing recovery
CRE	corrected reference equivalent
CREG	concentrated range extension with gain
CRPL	Central Radio Propagation Laboratory
CRT	cathode ray tube
CSA	carrier serving area
CSC	common signaling channel
CSMA, CSMA/CD	carrier sense multiple access, carrier sense multiple access with collision detection
CSO	composite second-order (products)
CT	cordless telephone
C/T	carrier (level)-to-noise temperature ratio
CTB	composite triple beat
CU	crosstalk unit
CVSD	continuous variable-slope delta modulation
CW	carrier wave; continuous wave
D/A	digital-to-analog
DAMA	demand assignment multiple access
DA/TDMA	demand assignment TDMA

DAR	digital (distortion, dispersion) adaptive receiver
DARPA, DARPANET	Defense Advanced Research Projects Agency (network)
dB	decibel
dBc	decibel referenced to the carrier level
dBd	decibel referenced to a dipole (antenna)
dBi	decibel over an isotropic (antenna)
dBm	decibel referenced to a milliwatt
DBM	data block mode
dBmP	dBm psophometrically weighted
dBmV	decibel referenced to a millivolt
dBm0	dBm referenced to the zero test level point (0 TLP)
DBPSK	differential binary PSK
dBr	decibels above or below "reference"
dBrnC	decibel reference noise C-message weighted
DBS	direct broadcast satellite
dBμ	decibel referenced to a microvolt
dBW	decibel referenced to 1 watt
dBx	crosstalk coupling in dB above reference coupling, which is a crosstalk coupling loss of 90 dB
DC, dc	direct current
DCE	data communication equipment; data circuit-terminating equipment
DCPBH	double-channel planar buried heterostructure
DCS	digital cross-connect (system)
DCT	discrete cosine transform
DDS	digital data system
DE	discard eligibility
DECT	digital European cordless telephone
DFB	distributed feedback (laser)
DFI	digital facility interface
DFM	dispersive fade margin
DFMR	reference dispersive fade margin
DL	data link
DLC	digital loop carrier
DLP	decode level point
DM	delta modulation; degraded minute
DMT	discrete multitone (modulation)
DMW	digital milliwatt
dNp	decineper
DQPSK	differential QPSK
DRS	digital reference signal
DSCS	Defense Satellite Communication System

DS	direct sequence
DSI	digital speech interpolation
DSL	digital subscriber loop
DS0, DS1, DS1C, DS2, ...	"digital system" 0, 1, 1C, etc.; the North American PCM hierarchy
DSU	digital service unit
DTE	data terminal equipment
DTM	data text mode
EB	errored block
EBCDIC	extended binary-coded decimal interchange code
E_b/I_0	energy per bit to interference density ratio
E_b/N_0	energy per bit to noise spectral density ratio
EC	earth curvature, earth coverage
ECS	encryption control signal
EDC	error detection code
EDD	envelope delay distortion
EDFA	erbium-doped fiber amplifier
EFS	error-free second
EIA	Electronics Industries Association
EID	element identifier
EIFM	external interference fade margin
EIRP	effective (equivalent) isotropically radiated power
ELED	edge-emitting light-emitting diode
ELEN	element length
ELP	encode level point
EMI	electromagnetic interference
EML	expected measured loss
EOB	end of block
eoc	embedded operations channel
EOL	end of line (fax)
EOM	end of message
EPL	equivalent peak level
ERL	echo return loss
ERP	effective radiated power
ES	end section; errored second
ESF	extended superframe
ESR	errored-second ratio
EUTP	enhanced unshielded twisted pair
FAS	frame alignment signal
FAW	frame alignment word
FCC	Federal Communications Commission (U.S.)
FCF	facsimile control field

FCS	frame check sequence
FDCT	forward discrete coded transform
FDD	frequency division duplex
FDDI	fiber distributed data interface
FDM	frequency division multiplex
FDMA	frequency division multiple access
FEAC	far-end alarm and control (signals)
febe, FEBE	far-end block error
FET	field effect transistor
FEXT	far-end crosstalk
FEC	forward error correction
FIF	facsimile information field
FLTSAT	(fleet sat), U.S. Navy satellites
FLTR	filter
FM	frequency modulation
FN	fiber node
FOT	Frequence optimum de travail (French for optimum working frequency)
FP	Fabry-Perot
FPLMTS	future public land mobile telecommunication system
FPS	frames per second; framing pattern sequence
FS	fixed station
FSK	frequency shift keying
FSL	free-space loss
FWM	four-wave mixing
GBLC	Gaussian band-limited channel
Gbps	gigabits per second
GEO	geostationary earth orbit
GFC	generic flow control
GHz	gigahertz ($Hz \times 10^9$)
GMSK	Gaussian minimum shift keying
GMT	Greenwich mean time
GOP	group of pictures
GPS	geographical positioning system
GSM	"group system mobile" (from the French) (the digital European cellular scheme), also called Global System for Mobile Communications
GSTN	general switched telecommunications network
G/T	gain (antenna)-to-noise temperature ratio
HC	headend controller
HCDS	high-capacity digital service

HDB3	high-density binary 3
HDLC	high-level data link control
HDSL	high-speed digital subscriber line
HEC	header error control
HF	high frequency; also the radio frequency band 3–30 MHz
HFC	hybrid fiber–coaxial cable
HPA	high-power amplifier
HPF	highest probable frequency
HRP	hypothetical reference path
HTU	high bit-rate terminal unit
HU	high usage (route[s])
HVQ	hierarchical vector quantization
Hz	hertz
I and Q	in-phase and quadrature
IBM	International Business Machine (Inc.)
IBPD	in-band power difference
ICL	inserted connection loss
IDCT	inverse discrete cosine transform
IDLC	integrated digital loop carrier
IDR	intermediate data rate
IEEE	Institute of Electrical and Electronics Engineers
IF	intermediate frequency
IFRB	International Frequency Registration Board (ITU)
IG	international gateway
ILD	injection laser diode
IM	intermodulation
InGaAsP	Indium gallium arsenide phosphorus
INMARSAT	International Marine Satellite (consortium)
INTELSAT	International Telecommunication Satellite (consortium)
I/O	input/output (device)
IONCAP	ionospheric communications analyses and prediction program
IP	internet protocol
IRE	Institute of Radio Engineers; a unit of measure used with TV video
IRL	isotropic receive level
ISC	international switching center
ISDN	integrated services digital networks
ISI	intersymbol interference
ISL	intersatellite link
ISM	industrial, scientific, and medical (band)

ISO	International Standards Organization
ISU	IRIDIUM subscriber unit
ITA	International Telegraph Alphabet
ITU	International Telecommunication Union
IXEC	interexchange carrier
JPEG	Joint Photographic Experts Group
kbps	kilobits per second
kHz	kilohertz
km	kilometer
LAD	linear amplitude dispersion
LAN	local area network
LATA	local access and transport area
LAP, LAPB, LAPD	link access protocol; link access protocol, B-channel; link access protocol, D-channel
LBO	line build-out
LD	laser diode
LEA	line extender amplifier
LEC	local exchange carrier
LED	light-emitting diode
LEN	length
LEO	low earth orbit
LF	low frequency
LFE	low-frequency enhancement
LHCP	left-hand circular polarized
LI	local ID or local identifier
Lincompex	link compression−expansion
LLC	logical link control
ln	log to the natural base
LNA	low-noise amplifier
LO	local oscillator
LOF	loss of frame
LOH	line overhead
LORAN	Long Range Navigation
LOS	loss of signal; line-of-sight
LP	log periodic (antenna)
LPE	liquid phase epitaxy
LPI	low probability of intercept
LQA	link quality assessment
LR	loudness rating
LRC	longitudinal redundancy check
LRD	long-route design
LSB	lower sideband

LSS	local serving switch
LST	local sidereal time
LSTR	listener sidetone rating
LUF	lowest usable frequency
mA	milliampere
MAC	media access control
MAN	metropolitan area network
MBA	multiple-beam antenna
MBC	meteor burst communication
Mbps	megabits per second
MCT	motion-compensated transform
MCU	minimum coded unit
MEA	multiple exposure allowance
MEO	medium earth orbit
MERCAST	merchant marine broadcast (U.S. Navy)
MF	multiframe
MF/HF	medium-frequency/high-frequency radio
MFSK	multilevel or M-ary FSK
MPSK	multilevel or M-ary PSK
MHz	megahertz
MILSTAR	military strategic-tactical radio (satellite)
MLRD	modified long-route design
MOVCD	metal organic chemical vapor deposition
MPEG	Motion Picture Expert's Group
ms	millisecond
MS	mobile station
MSC	mobile switching center
MSL	mean sea level
MTBF	mean time between failures
MTIE	maximum time interval error
MTSO	mobile telephone switching office
MTTR	mean time to repair
MUF	maximum usable frequency
mW	milliwatt
MW, M/W	microwave
N/A	not applicable
NA	numerical aperture
NACK	negative acknowledgment
N-AMPS	narrowband AMPS
NARS	North Atlantic Radio System
NASA	National Aeronautics and Space Administration
NATO	North Atlantic Treaty Organization
NBS	National Bureau of Standards (now NIST)

NEP	noise equivalent power
NEXT	near-end crosstalk
NF	noise figure
NIST	National Institutes of Standards and Technology
NLR	noise load ratio
nm	nautical mile; nanometer
NNI	network–network or network–node interface
NPR	noise power ratio
NRZ	nonreturn to zero
ns, nsec	nanosecond
NT	network termination
NTSC	National Television Systems Committee
NVI	near-vertical incidence
NWT	Northwest Territories (Canada)
OAM, OA&M	operation and maintenance; operation, administration & maintenance
OC	optical carrier (OC-1, OC-3, . . .)
OIM	operations interface module
OLL	open-loop loss
OLR	overall loudness rating
ORE	overall reference equivalent
OSI	open systems interconnection
OSP	outside plant
OWF	optimum working frequency
PA	power amplifier
PABX	private automatic branch exchange
PACS	personal access communication services
PAL	phase alternation line
PAM	pulse amplitude modulation
PBX	private branch exchange
PC	personal computer
PCM	pulse code modulation
PCR	program clock reference
PCS	personal communications services
PDF	power distribution function
PDH	plesiochronous digital hierarchy
PDU	protocol data unit
PEL	picture element
PEP	path end point, peak envelope power
PES	packetized elementary stream
pfd	picofarad; power flux density
PHEMT	pseudomorphic high electron mobility transistor

PHS	personal handyphone system
PHY	refers to the physical layer (OSI layer 1)
PIN	p-intrinsic-n
PIXEL	picture element (same as pel)
PLCP	physical layer convergence protocol
PLL	phase-lock loop
PM	phase modulation
PMD	polarization mode dispersion
POH	path overhead
POI	path overhead identifier
POTS	plain old telephone service
PRS	primary reference source
PS, ps	personal station; picosecond
PSK	phase shift keying
PSTN	public switched telecommunication network
PTI	payload type indicator (identifier)
PTT	post, telephone and telegraph
pW	picowatt
pWp	picowatt psophometrically weighted
QAM	quadrature amplitude modulation
qdu	quantization distortion unit
QID	queue identifier
QoS	quality of service
QPR	quadrature partial response
QPSK	quadrature phase shift keying
QSY	a "Q" signal meaning to change frequency
quat	quaternary symbol
QVI	quasi-vertical incidence
RBER	residual bit error rate
RC	resistance capacitance (time constant)
RCVR	receiver
RD	resistance design
RDI	remote defect indication
RE	reference equivalent
RELP	residual excited linear predictive (coder)
RF	radio frequency
RFI	radio frequency interference
RGB	red, green, blue
RH	relative humidity
RHCP	right-hand circular polarization
RL	return loss
RLL	radio local loop
RLR	receive loudness rating

RMPDU	request minislot protocol data unit
rms	root mean square
RPC1	radio paging code 1
RRD	revised resistance design
RRE	receive reference equivalent
RS	Reed–Solomon (code)
RSL	receive signal level
RT	remote terminal
RTC	return to control
RZ	return to zero
SAC	Strategic Air Command
SAP	service access point
SAR	search and rescue; segmentation and reassembly
SBC	subband coding
SC	service channel
SCADA	supervisory control and data acquisition
SCPC	single channel per carrier
S/D	signal-to-distortion ratio
SDH	synchronous digital hierarchy
SDU	service data unit
SECAM	sequential color and memory
SER	symbol error rate
SES	severely errored second
SESR	severely errored second ratio
SFF	single-frequency fade
SHF	super high frequency (3000–30,000 MHz)
SI	synchronous information
SIC	station identification code
SID	sudden ionospheric disturbance
SINAD	signal + noise + distortion-to-noise + distortion ratio
SLA	semiconductor laser amplifier
SLIC	subscriber line interface card
SLM	single longitudinal mode
SLR	send loudness rating
SM	service module
SMDS	switched multimegabit data service
SMF	submultiframe
SMPTE	Society of Motion Picture and Television Engineers
S/N, SNR	signal-to-noise ratio
SOH	section overhead
SONET	synchronous optical network
SPDU	session protocol data unit

SPE	synchronous payload envelope
SRDL	subrate digital loop(s)
SREJ	selective reject
SRS	stimulated Raman scattering
SSA	solid state amplifier
SSB, SSBSC	single sideband; single-sideband suppressed carrier
SSN	sunspot number
SS/TDMA	switched satellite TDMA
STL	studio-to-transmitter link
STM	synchronous transport module
STMR	sidetone masking rating
STS	synchronous transport signal
SWP	switching pattern
TA	terminal adapter
TASO	Television Allocation Study Organization
TBD	to be determined
TCP	transmission control protocol
TCM	trellis-coded modulation
TCXO	temperature-compensated crystal oscillator
TDD	time division duplex
TDM	time division multiplex
TDMA	time division multiple access
TE	terminal equipment
TELR	talker echo loudness rating
TFM	thermal fade margin
THz	terahertz (10^{12} Hz)
TIA	Telecommunication Industry Association
TIE	timing interval error
TLP	test level point
TOA	takeoff angle
TOD	time of day
TPDU	transport protocol data unit
T/R	transmit/receive
TRE	transmit reference equivalent
TRI-TAC	Tri (service)-tactical (U.S. armed forces)
TS	time slot
TSG	timing signal generator
TT&C	telemetry, tracking & command
TU	tributary unit
TUG	tributary unit group
TW	traveling wave
TWT	traveling-wave tube

UDLC	universal digital loop carrier
UHF	ultra-high frequency (300–3000 MHz)
UI	unit interval
UNI	user–network interface
UPC	usage parameter control
USB	upper sideband
USAF	United States Air Force
μs	microsecond
UT	universal time
UTC	universal time coordinated (universal coordinated time)
UTP	unshielded twisted pair
μV	microvolt
μW	microwatt
UW	unique word

VASP	virtual analog switching point
VBR	variable bit rate
VC	virtual container
VCI	virtual channel identifier
VDC	voltage direct current
VF	voice frequency
VFCT, VFTG	voice frequency carrier telegraph
VHF	very high frequency (30–300 MHz)
VLC	variable length code
VLD	variable length decoder
VLPDU	variable length protocol data unit
VNL	via net loss
VNLF	via net loss factor
VPI	virtual path identifier
VRC	vertical redundancy checking
VSAT	very small aperture terminal
VSB	vestigial sideband
VSWR	voltage standing wave ratio
VT	virtual tributary
VU	volume unit

WACS	wireless access communication system
WAN	wide area network
WBHF	wideband high frequency
WDM	wave(length) division multiplex
WLAN	wireless local area network
WLL	wireless local loop

WWII	World War II
WWV	call letters of NIST radio station
Xm	cross-modulation
xmtr	transmitter
ZBTSI	zero byte time slot interchange

INDEX

Boldface denotes in-depth coverage of a topic.
Italic denotes a definition of a term